Genetic Polymorphism and Disease

Genetic polymorphisms are important determinants of phenotypic variations and may modulate the risk of or even cause various diseases, including genetic disorders and multifactorial diseases. Genetic polymorphisms also serve as important genetic, population, and evolutionary markers that allow the study of genetic and evolutionary aspects of individuals, populations, and organisms and aid in tracing the evolutionary and parental lineages. Genetic polymorphisms in low-penetrance genes are responsible for the alterations in the gene expression of critical signal transduction proteins and metabolic enzymes. Some of these polymorphisms are linked to increased susceptibility to various diseases, especially cancers, cardiovascular diseases, immune disorders, and neurological pathologies. This book collates the reviews on the roles played by polymorphisms in critical metabolic, signal transduction, cell cycle, or DNA repair genes either directly or indirectly in the disease mechanisms. The focus is on various techniques for identifying the various single nucleotide polymorphisms (SNPs). Polymorphism studies document the effects SNPs and their expressions have on the functionality of the enzymes and proteins.

Key Features

- Describes the genetic polymorphism and its various types
- Discusses the role of genetic polymorphism in modulating the risk of various human diseases
- Explores various molecular techniques used for detecting genetic polymorphism
- Discusses the role of SNPs in modulating the susceptibility to human diseases
- Provides a genetic basis for individual variations in response to therapeutics

Genetic Polymorphism and Disease

Edited by
Syed Sameer Aga, Mujeeb Zafar Banday, and Saniya Nissar

CRC Press
Taylor & Francis Group
Boca Raton London New York

CRC Press is an imprint of the
Taylor & Francis Group, an **informa** business

First edition published 2023
by CRC Press
6000 Broken Sound Parkway NW, Suite 300, Boca Raton, FL 33487-2742

and by CRC Press
4 Park Square, Milton Park, Abingdon, Oxon, OX14 4RN

CRC Press is an imprint of Taylor & Francis Group, LLC

Library of Congress Cataloging-in-Publication Data
Names: Aga, Syed Sameer, editor. | Banday, Mujeeb Zafar, editor. | Nissar, Saniya, editor.
Title: Genetic polymorphism and disease / edited by Syed Sameer Aga, Mujeeb
Zafar Banday, and Saniya Nissar.
Description: First edition. | Boca Raton, FL : CRC Press, 2022. | Includes bibliographical references and index.
Identifiers: LCCN 2022015654 (print) | LCCN 2022015655 (ebook) |
ISBN 9781032153162 (hbk) | ISBN 9781032159058 (pbk) | ISBN 9781003246244 (ebk)
Subjects: LCSH: Genetic polymorphisms. | Genetic disorders.
Classification: LCC RB155.5 .G4583 2022 (print) | LCC RB155.5 (ebook) |
DDC 616/.042—dc23/eng/20220722
LC record available at https://lccn.loc.gov/2022015654
LC ebook record available at https://lccn.loc.gov/2022015655

ISBN: 9781032153162 (hbk)
ISBN: 9781032159058 (pbk)
ISBN: 9781003246244 (ebk)

DOI: 10.1201/9781003246244

Typeset in Times
by codeMantra

Dedicated to

Our Ever-Lasting Friends & Family

without whom this feat was impossible

Contents

Foreword

I feel greatly honored to write a Foreword for the book *Genetic Polymorphism and Disease* edited by Dr Syed Sameer Aga and his team. I know Dr Syed Sameer Aga since long through our scientific collaborations concerning various research projects. I always feel proud of his scientific achievements, including his quality research and publications, and I consider this book another milestone in this regard. I believe and I am hopeful that with their hard work, dedication, and knowledge, Dr Syed Sameer Aga and his team will continue their scientific journey and reach more and more significant scientific milestones. The subject of polymorphism is very controversial and important from disease pathogenesis point of view and is difficult to understand. The importance of the subject can be judged from the fact that numerous papers are being published regularly from different parts of the world.

This book discusses various aspects of genetic polymorphisms and discusses in detail their role in various diseases. The introductory chapter of this book written by editors provides a comprehensive description of various types of genetic polymorphisms, the methods used in their detection, and the function of genetic polymorphisms that will prove to be a valuable resource toward understanding various aspects of genetic polymorphisms. The major portion of this book consists of various chapters written by a galaxy of authors comprehensively discussing the role of genetic polymorphisms in general and single nucleotide polymorphisms in particular in modulating the risk of various diseases, including the complex and multifactorial diseases such as cancer and diabetes. This book also contains chapters that discuss various SNPs databases and provide a detailed description of various available tools for the analysis of SNPs. I hope this book will prove to be a valuable scientific contribution that will add to our understanding of genetic polymorphisms and the role they play in various diseases, including modulating the risk of or susceptibility to diseases that will further aid in the management and treatment of these diseases. I am sure that this book can be very useful for cancer biologists, researchers, academicians working on cancer genetics, pharmacologists, and students of sciences and allied subjects.

To conclude, with immense pleasure, I present this Foreword and introduce the book *Genetic Polymorphism and Disease* edited by a team of dedicated and hardworking researchers, and I wish this book, the editors, and the contributors a great success.

Nisar A. Chowdri
Former Head, Gen & Min Access Surgery/Colorectal Division
Department of General and Min Access Surgery & Colorectal Surgery
Sheri Kashmir Institute of Medical Sciences, Srinagar, J&K, India, 190011
President Elect, Association of Colon & Rectal Surgeons of India (ACRSI) 2022
Founder Chairman: J&K State Chapter of Association of Surgeons of India

Foreword

I feel highly honored to write a Foreword for the book *Genetic Polymorphism and Disease* edited and contributed by one of my former colleagues, Dr Syed Sameer Aga and his team. This book focuses on the role of genetic polymorphisms in modulating the susceptibility to various diseases. This book contains a good number of chapters that discuss the role of genetic polymorphisms in general and single nucleotide polymorphisms (SNPs) in particular in various diseases or disorders, including various types of cancers, diabetes, cardiovascular diseases, chronic inflammatory diseases, malaria, migraine, Parkinson's disease, and autoimmune disorders, among others. The first chapter of this book "Genetic Polymorphisms – Classification, Structure, Detection and Function" authored by the editors themselves provides a broad overview of genetic polymorphisms and discusses in detail various aspects of genetic polymorphisms, their classification, and most importantly, their function that lays a solid foundation for understanding the impact of genetic polymorphisms in modulating the susceptibility to various diseases discussed in subsequent chapters of this book. The second chapter of this book "Clinical Relevance of Pharmacogenomics: A Road to Personalized Medicine" discusses the role and impact of SNPs in the application of the field of pharmacogenomics in general and personalized medicine in particular for the treatment of human diseases. The 12th chapter, "Single Nucleotide Polymorphism-Related Computational Resources Part I: Databases", and the 13th chapter, Single Nucleotide Polymorphism-Related Computational Resources Part II: Algorithms, provide a detailed overview of the SNP databases and describe various tools including predictors and web resources required for the prediction and analysis of SNPs. The rest of the chapters in this book discuss the role of genetic polymorphisms, mainly the SNPs, in modulating the risk of or susceptibility to various diseases mentioned already.

I congratulate the editors for coming up with this important and highly informative compilation of chapters that fully justify the title of the book. I whole heartedly wish the editors and contributing authors of this book a huge success. I wish the editors a successful and fruitful scientific journey ahead, and I hope they will continue their effort in regularly coming up with publications, including books that will serve as a resource of valuable scientific knowledge.

Dr Sheikh A. Saeed
Professor of Physiology
Bahria University Medical & Dental College, Karachi, Pakistan
Email: s.abdulsaeed@gmail.com

Preface

Human genome encompasses all the genetic information necessary to guide numerous life-sustaining cellular processes. The set of instructions or information contained within the genome determines all the phenotypic properties of an individual and is responsible for the differences in the phenotypic characteristics of an individual in comparison with other individuals. The human genome is highly conserved to the extent that all the humans share at least 99.5% sequence similarity and exhibit only a difference of around 0.1%–0.5%. This small difference between different genomes along with epigenetic modifications determines almost all the phenotypic diversity or differences that exist between any two individuals within and/or between populations. The small differences that exist between different genomes constitute the genetic variations that include two main types, the commonly occurring genetic polymorphisms and the less common genetic variations termed as mutations. Genetic polymorphism that represents the most dynamic and most common form of genetic variations present in the human genome includes several major types such as single nucleotide polymorphisms (SNPs), small indels, variable number tandem repeats (VNTRs), short tandem repeats (STRs), copy number variations (CNVs; deletions or duplications), copy neutral variations (inversions and translocations), and transposable element insertions. Genetic polymorphisms are important determinants of phenotypic variations and may modulate the risk of or even cause various diseases, including genetic disorders and multifactorial diseases. Genetic polymorphisms also serve as important genetic, population, and evolutionary markers that allow the study of genetic and evolutionary aspects of individuals, populations, and organisms and aid in tracing the evolutionary and parental lineages. Among the various types of polymorphisms, SNPs constitute one of the most important and most common type of genetic polymorphism that plays a significant role in determining various inter-individual phenotypic differences and exhibits a greater role in predicting, revealing, or determining the risk of development and progression of various human diseases. In addition to their role in modulating the inter-individual differences in the susceptibility toward various diseases, SNPs may also play a role in determining the inter-individual differences in responses toward environmental stresses and toward the treatment of various diseases by drugs. Therefore, a deep understanding of genetic polymorphisms in general and SNPs in particular and the methods or techniques that allow their comprehensive detection, analysis, and characterization is essential to understand the evolutionary mechanisms, population dynamics, and genetics and to understand disease mechanisms even on a personalized or individual level that enables the development of personalized or precision medicine.

Despite the emergence of new and high-throughput techniques or methods that allow rapid detection and analysis of genetic polymorphisms, the field of genomics that deals with the study of structural and functional characteristics of genome including those of genetic polymorphisms such as SNPs and their application to various fields of human significance such as those that study human evolution, human disease, and human population dynamics is still emerging and a work in progress. Despite the ever-increasing number of studies that evaluate the role of genetic polymorphisms including SNPs in various complex and multifactorial diseases such as cancer and diabetes, the knowledge of the actual role played by SNPs in the regulation or determination of inter-individual phenotypic differences including the differences in the susceptibility toward various diseases and toward the treatment of various diseases by drugs is still emerging and a work in progress much like the field of genomics itself. There is a need to bring forth in a clear concise manner the already known facts and findings as much as there is a need for novel and innovative studies that may enable the more clear and comprehensive understanding regarding the structural and functional significance of genetic polymorphisms, including SNPs. The book *Genetic Polymorphism and Disease* is an effort in this direction that aims to present the outcome of numerous studies regarding the role of SNPs in causing or in modulating the susceptibility toward various diseases in a clear and comprehensive manner.

This book provides an overview of the significant role played by genetic polymorphisms in general and SNPs in particular in modulating the risk of development and progression of various human diseases. This book consists of 25 chapters. The introductory chapter of this book (*Genetic Polymorphisms – Classification, Structure, Detection and Function*) gives a detailed description of various types of genetic polymorphisms, their functional effects, and the techniques used to detect two most common types of genetic polymorphisms – SNPs and CNVs. The second chapter (*Clinical Relevance of Pharmacogenomics – a Road to Personalized Medicine*) gives a brief overview of pharmacogenomics, its effect on pharmacokinetics and pharmacodynamics, and its application in therapeutics. The third chapter of this book (*Diabetes and Role of SNPs in Diabetes Susceptibility*) discusses the possible association between SNPs in various genes and the risk of development of diabetes. The fourth chapter of this book (*Genetic Polymorphisms Associated with Innate Immune Response to Malaria*) describes the role played by SNPs in modulating the host immune response to malaria and the outcome of the disease. The fifth chapter (*Xenobiotic Metabolic Enzyme Polymorphisms and Cancer Susceptibility*) discusses the role of SNPs in three major classes of xenobiotic-metabolizing enzymes, GSTs, NATs, and CYPs, in modulating the risk of development of various cancers. The sixth chapter of this book (*Genetic Polymorphisms in Matrix Metalloproteinase (MMP) Genes and Cancer*) discusses the role of some prominent members of the MMP family in the pathogenesis of various cancers and reviews the possible association of SNPs in several MMP genes with the risk of development of various cancers. The seventh chapter of this book (*Vitamin D Receptor Polymorphisms and Diseases*) describes the role of various SNPs in vitamin D receptor (VDR) gene in modulating the risk of development of various diseases. The eighth chapter of this book (*Genetic Polymorphisms in Fatty Acid Metabolism Genes and Cancer Susceptibility*) discusses the possible role of functional polymorphisms in fatty acid metabolism genes in modulating the risk of various cancers. The ninth chapter of this book (*Single Nucleotide Polymorphisms and Parkinson's Disease Susceptibility*) discusses the SNPs in various genes involved in dopamine metabolism, oxidative stress, detoxification metabolism, and inflammatory processes and their impact on the risk of development of Parkinson's disease. The tenth chapter of this book (*Genetic Polymorphisms: Role in Drug Metabolism and Effect on Clinical Pharmacy*) discusses the role of genetic polymorphisms in modulating the expected response to therapeutic medications, as well as their possible impact on pharmaceutical safety and efficacy. The 11th chapter of this book (*Role of Genetic Polymorphisms in Chronic Inflammatory Disease*) discusses the role of several important gene SNPs that have a direct bearing on risk modulation of different chronic inflammatory diseases. The 12th chapter of this book (*Single Nucleotide Polymorphism-Related Computational Resources Part I: Databases*) discusses the available SNP-related resources, along with many supplementary links redirecting to corresponding web servers and their biological relevance or functional aspects of diseases, especially cancers. The 13th chapter of this book (*Single Nucleotide Polymorphism-Related Computational Resources Part II: Algorithms*) sums up and provides links for a range of web servers, predictors, and applications in SNP prediction. The 14th chapter of this book (*Role of Single Nucleotide Polymorphisms (SNPs) in Common Migraine*) discusses the possible role of genetic polymorphisms in various genes in modulating the susceptibility of migraine in light of the available literature and meta-analyses. The 15th chapter of this book (*The Role of Vitamin D Single Nuclear Polymorphisms in Cancer Susceptibility*) gives a brief overview of the role of vitamin D-related metabolites in the cancer development and discusses the possible role of vitamin D-related SNPs in modulating the risk of development of various cancers. The 16th chapter of this book (*Role of Genetic Polymorphisms in Lipid Metabolism Pathways in Modulating Cancer Promotion*) reviews the role of SNPs in several fatty acid metabolism-related genes in modulating the risk of development of cancer and its prognosis. The 17th chapter of this book (*Toll-Like Receptor (TLR) Polymorphisms in Prostate Cancer*) discusses the possible role of TLR SNPs in modulating the risk of development of prostate cancer. The 18th chapter of this book (*Role of Single Nucleotide Polymorphism of Cytochrome P450 Genes in Cancer Susceptibility*) discusses the role of CYP gene SNPs in lung, breast, liver, gastric, colorectal, prostate, and head-and-neck cancers. The 19th chapter of this book (*Role of Single Nucleotide Polymorphisms of Inflammatory Molecules in Susceptibility to Cardiovascular Diseases (CVDs)*) discusses the general role of inflammatory molecules in CVD and the role of SNPs in inflammatory molecules coding gene in the initiation and development of CVD. The 20th chapter of this book (*Role of Genetic Polymorphisms in the Etiopathogenesis of Esophageal Cancer*) discusses mutations caused due to various risk factors related to the progression of EC in various populations and focuses on genome variations, gene–environment interactions, and tumorigenesis. The 21st chapter of this book (*Cytokine Polymorphisms and Their Role in Modulating Cancer Risk*) discusses the functional polymorphisms in various cytokine genes and their possible role in modulating the risk of development of various cancers. The 22nd chapter of this book (*Gastric Cancer and Genetic Polymorphisms in Modulating the Susceptibility*) summarizes the association of the most extensively studied polymorphisms in some of the candidate genes involved in immunogenic, DNA repair, cell signaling, and detoxification pathways, which could possibly lead to progression and development of gastric cancer. The 23rd chapter of this book (*Cytokine Gene Polymorphisms and Their Role in Autoimmune Diseases*) details the possible role of genetic polymorphisms in modulating the risk of development of autoimmune diseases. The 24th chapter of this book (*Genomic Disturbances Associated with Penile Cancer and Novel Therapies*) briefly discusses the most updated knowledge on PeCa regarding the understanding of the current molecular markers and their application in monitoring and predicting the disease. The 25th and last chapter of this book (*Role of Genetic Polymorphisms in Gastric Cancer Susceptibility*) provides a brief insight into the genetic profile of gastric cancer, with an emphasis on how SNPs influence gastric cancer propensity and outcomes.

We hope that the book *Genetic Polymorphism and Disease* will serve as a valuable resource of research-based knowledge that will add to our understanding of genetic polymorphisms in general and SNPs in particular and their role in modulating the susceptibility toward various diseases. We would like to thank each author who contributed to this book without whom this knowledgeable effort would not have been possible. In the end, we would like to thank our publisher for believing in us and giving us the opportunity to work with them that resulted in this book.

Acknowledgments

First, we are thankful to our Creator for blessing us with the ability to think and work.

We would like to express our sincere gratitude to all the authors, from all walks of life, who contributed to this book and had a proactive attitude to share their immense knowledge. Their worthy contribution, patience, and trust have made this book possible.

Many thanks to the researchers whose work/data and transparency helped us to write this book. Their hard work has inspired us to take up this task.

With great sense of gratitude, we want to thank our mentors who helped in shaping our knowledge and research ideas right from our graduate school days to the current life.

We would like to thank Dr Chuck Crumly and Ms Kara Roberts at CRC Press Ltd. for their valuable advice and suggestions through all stages of this book.

We are immensely grateful to our families for always being so encouraging and supportive through every thick and thin.

Finally, we thank our family and friends for being the moral support ever and always.

Dr Syed Sameer Aga
Dr Mujeeb Zafar Banday
Dr Saniya Nissar
Infinity Biochemistry Pvt. Ltd.
infinitybiochem2020@gmail.com
March 2022

Editors

Dr Syed Sameer Aga is currently working as Associate Professor of Biochemistry in the Department of Basic Medical Sciences, College of Medicine, King Saud Bin Abdulaziz University for Health Sciences, Jeddah, Saudi Arabia. He is also serving as the Chairman of the Quality Assurance Unit in College of Medicine, KSAU-HS. Dr Syed is a lifetime member of many societies related to genetics, molecular biology, and medical education like the Royal Society of Biology, Biochemical Society, Association of Clinical Biochemists of India, Society of Indian Academy of Medical Genetics, Middle East Molecular Biology Sources, Asian Council of Science Editors, and Saudi Society of Medical Education, etc. He is also serving as the Associate Editor of five journals: *Frontiers in Oncology – Molecular Biosciences*, *BioMed Research International*, *Education Research International*, *World Journal of Biological Chemistry*, and *Current Drug Metabolism*. He also serves as a full-time reviewer for various journals including *European Journal of Cancer Prevention* (Lippincott Williams & Wilkins), *PLOS One & PLOS Genetics* (Public Library of Science), *World Journal of Gastroenterology* (Baishideng Publishing), *BMC Medical Genetics* (BMC-Springer), *Gene & Meta Gene* (Elsevier Publishing), *Bioscience Reports* (Portland Press), *Journal of Clinical Laboratory Analysis* (John Wiley & Sons), and *Tumor Biology* (Sage Journals). His research interests are focused on cancer biology, and he has published more than 100 research and review articles in highly reputed journals with an h-index of 22 and has authored 10 books.

Dr Mujeeb Zafar Banday is presently working as a Senior Researcher and Lecturer in the Department of Biochemistry, Govt. Medical College, Srinagar. Dr Banday has completed his PhD, MPhil, and MSc in Biochemistry at the University of Kashmir, J&K, India. He is serving as the review editor and associate reviewer for the journal *Frontiers in Oncology – Cancer Epidemiology and Prevention* (Frontiers Group) and as a full-time reviewer for various journals including *Gene* (Elsevier Publishing), *Bioscience Reports* (Portland Press), *BioMed Research International* (Hindawi Publishing), and *Toxicology and Environmental Health Sciences Review* (Springer Nature). He has authored more than 25 research and review articles and 3 books.

Dr Saniya Nissar is presently working as a Senior Researcher and Lecturer in the Department of Biochemistry, Govt. Medical College, Srinagar. She has completed her PhD, MPhil, and MSc in clinical biochemistry from the University of Kashmir, J&K, India. She works as freelance scientific content writer with many scientific magazines. She has published more than 30 research and review articles and has authored 4 books.

Contributors

Amir Abushouk
Department of Basic Medical Sciences
College of Medicine, King Saud Bin Abdulaziz University for Health Sciences (KSAU-HS), King Abdullah International Medical Research Center (KAIMRC), Ministry of National Guard Health Affairs (MNGHA), Jeddah, Kingdom of Saudi Arabia

Syed Sameer Aga
Molecular Diseases & Diagnostics Division
Infinity Biochemistry Pvt. Ltd. Srinagar, Kashmir, India
and
Department of Basic Medical Sciences Department
College of Medicine
King Saud bin Abdulaziz University for Health Sciences (KSAU-HS)
King Abdullah International Medical Research Center (KAIMRC), Ministry of National Guard Health Affairs (NGHA)
Jeddah, Kingdom of Saudi Arabia

Kulsum Akhtar
Department of Clinical Biochemistry, Sher-I-Kashmir Institute of Medical Sciences, Soura, Srinagar, India

Moudi M. Alasmari
Department of Basic Medical Sciences, College of Medicine, King Saud bin Abdulaziz University for Health Sciences (KSAU-HS), King Abdullah International Medical Research Center (KAIMRC), Ministry of National Guard Health Affairs (NGHA), Jeddah, Kingdom of Saudi Arabia

Aarif Ali
Department of Clinical Biochemistry, School of Biological Sciences, University of Kashmir, Hazratbal, Jammu and Kashmir, India

Mansour Al Qurashi
Department of Basic Medical Sciences, College of Medicine, King Saud bin Abdulaziz University for Health Sciences (KSAU-HS), King Abdullah International Medical Research Center (KAIMRC), Ministry of National Guard Health Affairs (NGHA), Jeddah, Kingdom of Saudi Arabia

Bashayr M. Alsuwayni
Department of Clinical Pharmacy, College of Pharmacy, King Saud University, Riyadh, Saudi Arabia

Azher Arafah
Department of Clinical Pharmacy, College of Pharmacy, King Saud University, Riyadh, Saudi Arabia

Abdullah A. Awadh
Department of Basic Medical Sciences
College of Medicine, King Saud bin Abdulaziz University for Health Sciences (KSAU-HS), King Abdullah International Medical Research Center (KAIMRC), Ministry of National Guard Health Affairs (NGHA), Jeddah, Kingdom of Saudi Arabia

Ruqaya Aziz
Department of Biochemistry, SKIMS Medical College & Associated Hospital, Srinagar, India

Mujeeb Zafar Banday
Department of Biochemistry,
Government Medical College
Karan Nagar, Srinagar, India
and
Molecular Diseases & Diagnostics Division
Infinity Biochemistry Pvt. Ltd. Srinagar, Kashmir, India

Payam Behzadi
Department of Microbiology, Shahr-e-Qods Branch, Islamic Azad University, Tehran, Iran

Ishrat Parveiz Bhat
Department of Clinical Biochemistry, Sher-I-Kashmir Institute of Medical Sciences, Soura
Srinagar, India

Nadiem Nazir Bhat
Division of Veterinary Biochemistry,
Sher-e-Kashmir University of Agricultural Sciences & Technology of Kashmir
Shuhama Alustang, Kashmir, Jammu and Kashmir, India

Amjad Faihan Binlebdah
Department of Clinical Pharmacy, College of Pharmacy, King Saud University, Riyadh, Saudi Arabia

María Juliana Chaves-Medina
UROGIV Research Group, School of Medicine, Universidad del Valle, Cali, CO

Sayequa Dandoti
Department of Biology, Deanship of Preparatory Year, Imam Abdulrahman Bin Faisal University, Dammam, Saudi Arabia

Qazi Danish
Department of Biochemistry, Deanship of Preparatory Year, Sri Pratap (SP) College, Cluster University, Srinagar, India

Showkat Ahmad Dar
Bioinformatics Center, CSIR Institute of Microbial Technology, Chandigarh, India

Vishal Kumar Deb
Department of Microbiology, School of Life and Allied Sciences (SLAS), Baba Farid Institute of Technology (BFIT), Hemvati Nandan Bahuguna Garhwal Central University, Uttarakhand, India

Sunia Faiz
Department of Biochemistry, Government Medical College Srinagar & Associated Hospitals, Karan Nagar, Srinagar, Jammu and Kashmir, India

Iqra Farooq
Department of Biochemistry, Government Medical College Srinagar & Associated Hospitals, Karan Nagar, Srinagar, Jammu and Kashmir, India

Razaz Felemban
Department of Basic Medical Sciences, College of Medicine, King Saud bin Abdulaziz University for Health Sciences (KSAU-HS), King Abdullah International Medical Research Center (KAIMRC), Ministry of National Guard Health Affairs (NGHA), Jeddah, Kingdom of Saudi Arabia

Herney Andrés García-Perdomo
Director "UROGIV" Research Group, School of Medicine, Universidad del Valle, Cali, CO
Division of Urology/Urooncology, Department of Surgery, School of Medicine, Universidad del Valle, Cali, CO

Alqassem Hakami
Department of Basic Medical Sciences, College of Medicine, King Saud bin Abdulaziz University for Health Sciences (KSAU-HS), King Abdullah International Medical Research Center (KAIMRC), Ministry of National Guard Health Affairs (NGHA), Jeddah, Kingdom of Saudi Arabia

Ehtishamul Haq
Department of Biotechnology, University of Kashmir, Hazratbal, Srinagar, Jammu and Kashmir, India

Shobana John
Department of Pharmacy Practice, Karpagam College of Pharmacy, Coimbatore, India, Affiliated to the Tamil Nadu Dr. MGR Medical University, India

Manmeet Kaur
Department of Biotechnology, Guru Nanak Dev University, Amritsar, Punjab, India

Racha Kerek
Faculty of Sciences, Department of Biochemistry, Jinan University, Tripoli, Lebanon

Johra Khan
Department of Medical Laboratory Sciences, College of Applied Medical Sciences, Majmaah University
Majmaah, Saudi Arabia

Mosin S. Khan
Department of Biochemistry, Government Medical College Srinagar & Associated Hospitals, Karan Nagar, Srinagar, Jammu and Kashmir, India
and
Government Medical College Baramulla & Associated Hospitals, Kanth Bagh, Baramulla, Jammu and Kashmir, India

Bader H. Khawaji
Basic Medical Sciences, College of Medicine, King Saud bin Abdulaziz University for Health Sciences (KSAU-HS), King Abdullah International Medical Research Center (KAIMRC), Ministry of National Guard Health Affairs (NGHA), Jeddah, Kingdom of Saudi Arabia

Sachin Kumar
Department of Pharmacology, Delhi Pharmaceutical Sciences and Research University, Govt. of NCT of Delhi, New Delhi, India

Vikram Kumar
Faculty of Biotechnology, Amity Institute of Biotechnology, Amity University Rajasthan, Jaipur, India

Pallawi Kumari
Department of Bioscience & Biotechnology
Banasthali Vidyapith University,
Tonk, Rajasthan, India

Suhail S. Lone
Department of Biochemistry, Government Medical College Srinagar & Associated Hospitals, Karan Nagar, Srinagar, Jammu & Kashmir, India

Sabhiya Majid
Department of Biochemistry & Multidisciplinary Research Unit (MRU-ICMR), Government Medical College Srinagar & Associated Hospitals, Karan Nagar, Srinagar, Jammu and Kashmir, India

Israa A. Malli
Department of Basic Medical Sciences, College of Medicine, King Saud bin Abdulaziz University for Health Sciences (KSAU-HS), King Abdullah International Medical Research Center (KAIMRC), Ministry of National Guard Health Affairs (NGHA), Jeddah, Kingdom of Saudi Arabia

Syed Mudassar
Department of Clinical Biochemistry, Sher-I-Kashmir Institute of Medical Sciences, Soura, Srinagar, India

Nisar A. Naikoo
Department of Biochemistry, Government Medical College, Karan Nagar, India

Amre Nasr
Department of Basic Medical Sciences, College of Medicine, King Saud Bin Abdulaziz University for Health Sciences (KSAU-HS), King Abdullah International Medical Research Center (KAIMRC), Ministry of National Guard Health Affairs (MNGHA), Jeddah, Kingdom of Saudi Arabia

Saniya Nissar
Department of Biochemistry
Government Medical College
Srinagar, India
and
Molecular Diseases & Diagnostics Division
Infinity Biochemistry Pvt. Ltd
Srinagar, Kashmir, India
Infinity Biochemistry LLC
Srinagar, India.

Nusrath Yasmeen
Department of Biotechnology, Amity Institute of Biotechnology, Amity University Rajasthan, Jaipur, India

Faheem Hyder Pottoo
Department of Pharmacology, College of Clinical Pharmacy, Imam Abdul Rahman Bin Faisal University
Dammam, Saudi Arabia

Waseem Qureshi
Registrar Academics, Government Medical College Srinagar & Associated Hospitals, Karan Nagar, Srinagar, Jammu and Kashmir, India

Ahmad Rabbani
Department of Microbiology, School of Life and Allied Sciences (SLAS), Baba Farid Institute of Technology (BFIT), Hemvati Nandan Bahuguna Garhwal Central University, Uttarakhand, India

Sanjeev Rana
Human Genomics Research Group, Shri Mata Vaishno Devi University (SMVDU), Jammu and Kashmir, Katra, India

Fouzia Rashid
Department of Clinical Biochemistry, University of Kashmir, Srinagar, Kashmir, India

Roohi Rasool
Department of Immunology and Molecular Medicine, Sher-i-Kashmir Institute of Medical Sciences
Srinagar, India

Muneeb U. Rehman
Department of Clinical Pharmacy, College of Pharmacy, King Saud University, Riyadh, Saudi Arabia

Sheikh A. Saeed
Department of Physiology, Bahria University Medical & Dental College, Karachi, Pakistan

Nouroz Sehar
Centre for Translational and Clinical Research, School of Chemical & Life Sciences, Jamia Hamdard, New Delhi, India

Shagufta Sheikh
Department of Biochemistry, Government Degree College for Women, Anantnag, Jammu and Kashmir, India

Rajiv Kumar Tonk
Department of Pharmacology, Delhi Pharmaceutical Sciences and Research University, Govt. of NCT of Delhi, New DelhI, India

Hilal A. Wani
Department of Biochemistry, Government Degree College, Bandipora, India

Syed Douhath Yousuf
Department of Clinical Biochemistry, Sher-I-Kashmir Institute of Medical Sciences, Srinagar, India

1

Genetic Polymorphisms – Classification, Structure, Detection and Function

Mujeeb Zafar Banday and Saniya Nissar
Department of Biochemistry, Government Medical College, Karan Nagar, Srinagar, Kashmir, India
Infinity Biochemistry Pvt. Ltd., Srinagar, Kashmir, India

Syed Sameer Aga
Basic Medical Sciences, College of Medicine, King Saud bin Abdulaziz University for Health Sciences (KSAU-HS), King Abdullah International Medical Research Center (KAIMRC), Ministry of National Guard Health Affairs (NGHA), Jeddah, Kingdom of Saudi Arabia
Infinity Biochemistry Pvt. Ltd. Srinagar, Kashmir, India

CONTENTS

ABSTRACT Human genome represents the entire DNA present in a viable human cell put together that contains all the information required to guide numerous life-sustaining cellular processes. The Human Genome Project (HGP) was undertaken mainly with the aim of establishing a reference genome deciphering, in detail, the basic structure, organization and function of all the genes present in the human genome, including the instructions required for the growth, development, function and survival of human beings. The HGP also revealed the highly conserved nature of human genome because of which all the humans share at least 99.5% sequence similarity and exhibit only a difference of around 0.1%–0.5%. This small difference between different genomes along with epigenetic modifications determines almost all the phenotypic diversity or differences that exist between any two individuals within and/or between populations. The genetic variations represent the variability that occurs in a human genome in comparison with other genomes and the reference genome established by the HGP. These genetic variations include two main types: the more commonly occurring genetic variations termed as genetic polymorphisms and the rare, less commonly occurring and often disease-causing genetic variations termed as mutations. Genetic polymorphism represents the most dynamic and most common form of genetic variations that occur in the human genome. It includes several major types of annotated genetic variations such as single nucleotide polymorphisms (SNPs),small indels, variable number of tandem repeats (VNTRs),short tandem repeats (STRs),copy number variations (CNVs; deletions or duplications),copy neutral variations (inversions and translocations) and transposable (mobile) element insertions. Genetic polymorphisms can directly lead to or may influence the phenotypic variations, including the difference in traits such as height; color of eyes, hair, and skin; and, and response to pathogens, chemicals, drugs and vaccines, and may modulate the susceptibility to or even cause various diseases including genetic disorders and multifactorial diseases. The methods or techniques used for the detection, analysis and characterization of genetic polymorphisms are diverse and dynamic. These techniques include several traditional techniques and their modifications that allow low- to medium-throughput detection and analysis and several other techniques including the new and emerging techniques such as next-generation sequencing (NGS)-based techniques that enable high-throughput detection and analysis of genetic variations.

KEYWORDS: *Genetic Polymorphism; Human Genome Project; Simple Nucleotide Variation; Structural Variations; Single Nucleotide Polymorphism; Copy Number Variation; Variable Number of Tandem Repeats; Short Tandem Repeats; Transposable Element Insertions.*

DOI: 10.1201/9781003246244-1

1.1 Introduction

Nucleic acids, deoxyribonucleic acid (DNA) and ribonucleic acid (RNA) along with complex carbohydrates, lipids and proteins, are essential biological macromolecules or biomacromolecules that form the basis of almost all forms of life known today. DNA constitutes the genetic or hereditary determinant in almost all the known organisms including plants and animals and some viruses. DNA carries all the essential instructions that guide numerous cellular processes that sustain life. These instructions are stored in the DNA in the form of a sequence or genetic code generated through repeated, but differential arrangement or order of four chemical or nitrogenous bases, also known as nucleobases. These include the purine bases, adenine (A) and guanine (G), and the pyrimidine bases, cytosine (C) and thymine (T). Each of these bases individually bonds to a five-carbon sugar known as 2′-deoxyribose, which in turn attaches to a phosphate group to form the fundamental structural unit of DNA known as nucleotide. Thus, depending on the nitrogenous base present, there are four types of nucleotides, which link through phosphodiester covalent linkage between sugar and phosphate to form polynucleotide chain known as DNA strand. Each DNA molecule structurally consists of two such strands held together by hydrogen bonds between the nitrogenous bases in a specific manner where adenine (A) base pairs with thymine (T) and cytosine (C) base pairs with guanine (G) to form a spiral ladder-shaped structure known as double helix. The backbone or sides of this ladder are composed of alternating sugar phosphate moieties, whereas base pairs constitute the inner portion, steps or rungs of this ladder. The differential arrangement, order or sequence of the nucleotides determines the biological instruction or information contained in the DNA, and both the strands of DNA carry the same instruction.

In the cells of eukaryotic organisms, DNA is mostly present in the form of long complex nucleoprotein structures known as chromosomes and is mostly located in the nucleus, hence termed as nuclear DNA. In addition, some DNA, in simple small circular forms and in multiple numbers, is present in mitochondria and chloroplasts and termed as mitochondrial DNA (mtDNA) and chloroplast DNA (cpDNA), respectively. The process of DNA replication whereby the DNA and hence the chromosomes are duplicated ensures that each daughter cell produced from a typical cell division receives a complete set of chromosomes and hence all the DNA of the parent cell. The nucleotides are not only the structural units of DNA, but they also form the functional unit of DNA, termed as gene. Typically, genes consist of specific or distinct arrangement, order or sequence of the nucleotides. Genes are variable in terms of the nature or type and number of bases, which carry specific biological instructions. These instructions include the information required for the synthesis of messenger RNA (mRNA) molecules and subsequently the polypeptides or proteins they encode or express. These instructions also include the information required for the synthesis of specific types of RNAs, which do not express polypeptides or proteins such as small noncoding RNAs (sncRNAs) and long noncoding RNAs (lncRNAs).Genes thus determine the phenotype or observable trait of a cell or an organism and constitute the physical, fundamental and functional determinant of heredity. The entire DNA present in a viable cell put together, which contains all the instructions required for all the cellular or biological functions, constitutes the genome of that cell. In the cells of eukaryotic organisms, since most of the DNA is present in the form of chromosomes, a complete set of chromosomes in a single cell of these organisms constitute the genome in these organisms. Almost all the cells in multicellular eukaryotic organisms contain a whole copy of the genome. In human beings, almost all the nucleated cells contain two sets of identical chromosomes, termed as homologous chromosomes, with each chromosome in the homologous pair contributed by one of the two parent cells. In humans, the nucleus of somatic cells contains 23 pairs of chromosomes or 46 chromosomes. Out of these, 22 pairs are autosomal chromosomes (non-sex chromosomes), whereas the remaining pair consists of two allosomes (sex chromosomes). The allosome pair consists of either two copies of X chromosome, each inherited from maternal cell (ovum or egg) and paternal cell (sperm) in the case of females, or one copy each of X chromosome and Y chromosome inherited from the maternal cell and the paternal cell, respectively, in the case of males. However, the cell inherits the entire mtDNA from ovum (egg cell) because only the ovum but not the sperm retains the mitochondria during fertilization. Each of the 22 paternal autosomal chromosomes pairs with its corresponding maternal chromosome during meiosis to form the homologous chromosome pair. However, allosomes X and Y are not homologous and therefore a paternal cell typically contains only 22 pairs of homologous chromosomes and a non-homologous chromosome pair consisting of X and Y chromosomes, whereas a maternal cell contains 23 pairs of homologous chromosomes, which include a homologous pair of X allosomes. Since almost all the nucleated cells in humans are diploid, the human genome is also diploid; however, the germ cells, which include sperm and ovum, are haploid and therefore contain haploid genome. The human genome also includes the mtDNA. However, the term human genome referred to in the 'Human Genome Project' (HGP) includes the haploid genome, that is, the total DNA contained in 23 chromosomes or half of the total DNA content present in the nucleus of diploid somatic cells, and also includes the mtDNA (Ruddle 2004, Gartler 2006).

The HGP began on October 1, 1990, and concluded in April 2003 with a rough draft version of the human genome announced in between on June 26, 2000. The HGP was undertaken mainly with the aim of establishing a reference genome that could act as a standard for research and testing in the area of genomics and involved three major steps: first, determining the arrangement, order or sequence of all the bases or nucleotides present in each of the 22 autosomes, X and Y allosomes and the mtDNA to generate a single contiguous sequence in each case; second, mapping the location of all the genes on all the chromosomes; and finally, generating linkage maps for tracking inherited traits including genetic disease traits from one generation to another. The DNA sequence and other data established in the HGP, usually referred to as human reference genome, represent composite information, which in no way corresponds to any actual single human being. The

HGP deciphered, in detail, the basic structure, organization and function of all the genes present in the human genome, and the information obtained represents a fundamental set of heritable instructions required for the growth, development, function and survival of human beings. However, this information does not represent a complete human being. In short, the HGP reference genome represents a haploid contiguous mosaic consensus sequence derived from several individuals. The HGP established that the haploid (somatic) human genome and the mtDNA consist of approximately 3.1 billion (3.1×10^9) base pairs or 3,100 Mbp (megabase pairs) (Lander, Linton et al. 2001, Venter, Adams et al. 2001, IHGSC 2004). However, the real human genome is diploid, and therefore, the human genome present in the nucleus of every somatic cell and the mtDNA actually contains twice the number established by HGP, that is, 2×3.1 billion ($2 \times 3.1 \times 10^9$), which equals 6.2 billion (6.2×10^9) base pairs or 6,200 Mbp. Further, the number of bases present in each human chromosome ranges from approximately 4.6×10^7 to 2.5×10^8 base pairs [46,709,983 base pairs (chromosome 21) to 248,956,422 base pairs (chromosome 1)]. The mtDNA consists of 16,569 base pairs (Taylor and Turnbull 2005).

An important finding from the HGP was that the human genome contains 26,588 protein-coding genes (Venter, Adams et al. 2001) or 31,000 protein-coding genes (Lander, Linton et al. 2001). Both these estimates were significantly less than the initial average estimate of 100,000 (50,000 – 140,000) protein-coding genes (Pennisi 2003, Salzberg 2018). However, this number decreased to between 20,000 and 25,000 (~24,000) protein-coding genes when the final draft of the human genome was released in 2004 (IHGSC 2004). However, with the evolution of modern sequencing techniques and the availability of techniques and resources to identify the protein-coding genes, the number has decreased and the current number ranges from about 19,901 to 21,306 protein-coding genes. The protein-coding genes constitute only about 1% – 1.5% of the human genome (Clamp, Fry et al. 2007, Ezkurdia, Juan et al. 2014, Salzberg 2018). It has also emerged that the human genome is more complex, more dynamic and more diverse in its structure and function than was previously thought (Karki, Pandya et al. 2015).

The human genome also contains genes that code for specific types of RNAs, but not proteins. These genes, also known as noncoding genes, encompass those genes that code for various types of noncoding RNAs (ncRNAs), that is, the RNAs that do not express polypeptides or proteins. These include two major classes: small noncoding RNAs (sncRNAs, size 18 – 200 nucleotides) and long noncoding RNAs (lncRNAs, size greater than 200 nucleotides). These two classes of ncRNAs encompass several types such as ribosomal RNAs (rRNAs), transfer RNAs (tRNAs), small interfering RNAs (siRNAs), Piwi-interacting RNAs (piRNAs), tRNA-derived small RNAs (tsRNAs), telomere-associated RNAs [Telomeric repeat–containing RNA (TERRA) and Telomerase RNA Component (TERC)], small nuclear RNAs (snRNAs), small nucleolar RNAs (snoRNAs) and microRNAs (miRNAs/small temporal RNAs/stRNAs) (Dozmorov, Giles et al. 2013, Ma, Bajic et al. 2013, Ransohoff, Wei et al. 2018, Salzberg 2018). A study has estimated the total number of RNA genes to be around 25,525, which include approximately 18,484 genes that code for lncRNAs, but does not include about 2,300 known miRNA genes and a varied number of genes, which code for various types of small RNAs. This brings the total number of genes to be around 46,831 (21,306 protein-coding genes, 18,484 lncRNA genes, 2,694 antisense RNA genes and 4,347 miscellaneous RNA genes), which represent 323,827 total gene transcripts at an average of 6.9 transcripts per gene (Salzberg 2018, Alles, Fehlmann et al. 2019). Another study has estimated the total number of genes to be around 43,162, which include 21,306 protein-coding genes and 21,856 noncoding genes, which represent 323,827 total gene transcripts at an average of 7.5 transcripts per gene. These numbers include 4,998 new genes (1,178 protein-coding genes and 3,819 noncoding genes) and additional 97,511 novel splice variants of protein-coding genes in comparison with the most recent available catalogs of human genome (Pertea, Shumate et al. 2018, Willyard 2018). Further, vast regions of DNA contain sequence segments that do not code for either proteins or RNA and these include regulatory DNA sequences (such as those present in gene promoter and gene enhancer/silencer regions), introns, several other types of functional elements and sequence regions with no known function (Lander, Linton et al. 2001, IHGSC 2004, Pennisi 2012).

Another important finding from the HGP is that the human genome has remained largely well conserved throughout the evolution and all the humans share at least 99.5% sequence similarity; that is, the genomes of any two human individuals are at least 99.5% identical and differ by only about 0.5%. The sequence dissimilarities or genetic variations, which constitute around 0.1% –0.5% of human genome, along with epigenetic modifications (an important attribute of human genome after nucleotide sequence), determine almost all the phenotypic diversity or differences that exist between any two individuals within and/or between populations. These phenotypic variations include difference in traits such as height; color of eyes, hair and skin; susceptibility to various diseases; and even response to pathogens, chemicals, drugs and vaccines among numerous others (Jorde and Wooding 2004, Abecasis, Auton et al. 2012, Auton, Brooks et al. 2015, Karki, Pandya et al. 2015).

The human genomic variability or genetic variations in human genome include the simple and complex sequence dissimilarities or sequence variations that occur in a human genome in comparison with other genomes in general and the reference genome established by the HGP in particular. These variations, which can occur in nuclear DNA or mtDNA, consist of two main types: simple nucleotide variations (SNVs) and structural variations (SVs). The SNVs, which result in sequence variations, include single nucleotide polymorphisms (SNPs), small indels (insertions and deletions) and tandem repeats or tandem repeat polymorphisms; variable number of tandem repeats (VNTRs) and short tandem repeats (STRs). The SVs involve genomic rearrangements, which result in structural alterations, and include copy number variations or CNVs (deletions or duplications), copy neutral variations (inversions and translocations), transposable (mobile) element insertions and more complex chromosomal rearrangements and aneuploidies (abnormal chromosome numbers) (Weischenfeldt, Symmons et al. 2013).

The 1000 Genomes Project has established an exhaustive resource, or a comprehensive catalog of genetic variations present in the human genome. The 1000 Genomes Project (TGP/1KGP/1000GP), which began in January 2008 and concluded in 2015, was undertaken with the aim of studying and characterizing almost all types of genetic variations present in the human genome including SNPs, SVs and their haplotype contexts across different populations worldwide. The TGP used a combination of three high-throughput technical approaches: low-coverage whole genome sequencing (WGS), high-coverage or deep whole exome sequencing and high-density microarray SNP genotyping to sequence the whole genomes of 2,504 unrelated and related sets of individuals representing 26 populations from world's five continental regions: Africa (AFR), Americas (AMR), Europe (EUR), South Asia (SAS) and East Asia (EAS). The TGP identified and characterized more than 88 million variations, which include 84.7 million SNPs, 3.6 million short or small indels (insertions and deletions) and 60,000 SVs (Auton, Brooks et al. 2015). The TGP established that the difference between a typical genome and the reference genome involves 4.1 – 5.0 million genomic sites. However, the SNPs and short or small indels constitute more than 99.99% of these variations, but affect relatively a smaller number of bases in comparison with the SVs. The SVs, numbering around 2,100 – 2,500 in a typical genome, affects around 20 million bases or nucleotides. The SVs include around 1,000 large deletions, 915 Alu insertions, 160 CNVs, 128 long interspersed nuclear element (LINE/LINE-1/L1) insertions, 51 short interspersed nuclear element (SINE)/VNTR/Alu (SVA) composite retrotransposon insertions, 4 mtDNA-like sequences in the nucleus or nuclear mtDNA sequences/variants (NUMTs) and 10 inversions. Of all the variations present in a typical genome, 149 – 182 rare variant sites are associated with protein truncation and affect around 1% of the protein-coding genes. Further, around 10,000 – 12,000 variant sites affect peptide sequence and 459,000 – 565,000 variant sites are either present in or overlap known gene regulatory regions including promoters, transcription factor binding sites, untranslated regions (UTRs), enhancers/silencers and insulators (Lander 2011, Auton, Brooks et al. 2015, Sudmant, Rausch et al. 2015, Srinivasan, Clements et al. 2016). Further, the TGP, through the merger of individual datasets from 2,504 genomes, created an integrated or unified data release or map. This map includes 42,279 bi-allelic deletions, 16,631 mobile element insertions (MEIs), 6,025 bi-allelic duplications, 2,929 multi-allelic copy number variations (mCNVs), 786 inversions and 168 nuclear mtDNA segments (NUMTs), also known as nuclear mtDNAs or nuclear DNA of mitochondrial origin. The 16,631 MEIs include 12,748 Alu insertions, 3,048 L1 element insertions and 835 SVA composite retrotransposon insertions (Sudmant, Rausch et al. 2015). In all, a single diploid human genome encompasses a median of approximately 18.4 Mbp of SVs with size greater than 50 kilobase pairs (kbp). However, CNVs including duplications with a median length greater than 10 kb are more common (Birney and Soranzo 2015). Overall, in a typical haploid human genome, an estimated one genetic variation occurs every 100 – 300 base pairs. Further, an estimated 150 rare genetic variations are present in the protein-coding regions of a typical haploid human genome and these variations affect less than 1% of the protein-coding genes (Lander 2011, Gonzaga-Jauregui, Lupski et al. 2012, Srinivasan, Clements et al. 2016). All the numbers or estimated numbers of different genetic variations mentioned here are the ones that have emerged from the HGP and the TGP; however, these numbers have underwent a significant change since then due to the findings from numerous subsequent studies. The genetic variations in human genome arise due to various processes such as recombination, mutation, gene immigration and successive and complex processes of evolution (natural selection) (Birney and Soranzo 2015). However, the mutations together with natural selection have emerged as the major source of genetic variations. The genetic variations give rise to an important genomic characteristic or phenomenon known as genetic polymorphism. Genetic polymorphism encompasses most of the genetic variations present across the genomes including the human genome. It constitutes the most dynamic and most prevalent form of variability present across the human genome.

1.2 Genetic Polymorphisms

Genetic polymorphisms include SNVs and simple or complex SVs, which are commonly present across the genomes of individuals, groups of individuals or populations. Genetic polymorphisms describe the existence of a particular genetic locus of the genome in two or more possible alternative forms, termed as alleles with each allele present in the individuals, groups or populations under study at a frequency, which is appreciable. The frequency of the allele, which is less common, termed as minor allele, has been arbitrarily fixed at 1% or more, and consequently, the frequency of the most common allele, termed as major allele, should be 99% or less (Brookes 1999, Schildgen and Schildgen 2013). The arbitrary frequency limits, established based on findings from numerous studies and prior to the advent of next-generation sequencing (NGS) techniques, have been used to distinguish more common genetic polymorphic variations (genetic polymorphisms) from rare, often disease-causing genetic variations termed as mutations (Schildgen and Schildgen 2013). Genetic polymorphisms, therefore, enable the segregation of common genetic variations into distinct alternative forms or alleles within the individuals, groups or populations under study, and the widespread occurrence of these genetic variations clearly points toward the role of evolution in their maintenance and propagation within the human population. Genetic polymorphisms can directly lead to or may influence the phenotypic variations including the difference in traits such as height; color of eyes, hair and skin; susceptibility to various diseases; and even response to pathogens, chemicals, drugs and vaccines, among numerous others. However, they can have a total neutral effect, as is the case with many polymorphisms in general and those present in the noncoding regions of the genome in particular (Jorde and Wooding 2004, Daly 2010, Abecasis, Auton et al. 2012, Auton, Brooks et al. 2015, Karki, Pandya et al. 2015). The HGP in general and the TGP in particular have generated a comprehensive catalog of genetic variations including genetic polymorphisms present in the human genome. The development and use of high-throughput and affordable NGS techniques has greatly accelerated the process of genome annotation, which involves the identification and functional characterization of genomic sequences including those of genetic variations (Sudmant, Rausch et al. 2015, Srinivasan, Clements et al. 2016).

1.2.1 Types of Genetic Polymorphisms

The genetic polymorphism encompasses several major types of annotated genetic variations. These include two main types: SNVs and the SVs. The SNVs include SNPs, small indels (insertions and deletions) and tandem repeats or tandem repeat polymorphisms; VNTRs and STRs. The SVs include CNVs (deletions or duplications), copy neutral variations (inversions and translocations) and transposable (mobile) element insertions (Figure 1.1) (Weischenfeldt, Symmons et al. 2013, Srinivasan, Clements et al. 2016). This chapter describes in detail the two major types of genetic polymorphisms, SNPs and CNVs, and provides a brief description of some of the other types of genetic polymorphisms.

1.2.1.1 Single Nucleotide Polymorphisms (SNPs)

The single nucleotide polymorphism (SNP), pronounced as snip, is a SNV that results from the substitution of one nucleotide by another at a particular location within the genome. SNPs represent the most common type of human genetic polymorphism or genetic variation and constitute more than 90% of all the human genetic variations (Risch 2000, Ding 2007, Orr and Chanock 2008). The human genome on average consists of one SNP after every 300 base pairs (bp), and the usual range of occurrence of SNPs varies from 100 to 300 bp. The haploid (somatic) human genome consists of approximately 3.1 billion (3.1×10^9) base pairs, and the occurrence of one SNP after every 300 bp reveals the presence of an estimated ten million SNPs in the haploid human genome (Risch 2000, Orr and Chanock 2008, Lander 2011, Sudmant, Rausch et al. 2015). The SNPs also represent one of the major sources of inter-chromosomal genetic variations between two identical chromosomes that constitute a homologous chromosome pair within the human genome. These SNPs occur at an average frequency of about one SNP after every 1,000 bp. Thus, almost three million SNPs contribute to single nucleotide variations or differences between the two haploid complements of a diploid human genome (Brookes 1999, Orr and Chanock 2008, Sudmant, Rausch et al. 2015). Every SNP present at a particular chromosomal location within a diploid human genome results from the substitution of one nucleotide by another, and therefore, every SNP occurs in either of the two alternative allelic forms that depend on the presence of one or the other alternative allele at that particular chromosomal location. The two alternative alleles that give each SNP two alternative genotypic forms include the major allele and the minor allele. The major allele, also known as wild or more common allele, is the allelic form of a SNP that is present more abundantly within a specific population. The minor allele, also known as variant or less common allele, is the allelic form of a SNP that is present less abundantly or minimally within a specific population. Due to the diploid nature of human genome, certain percentage

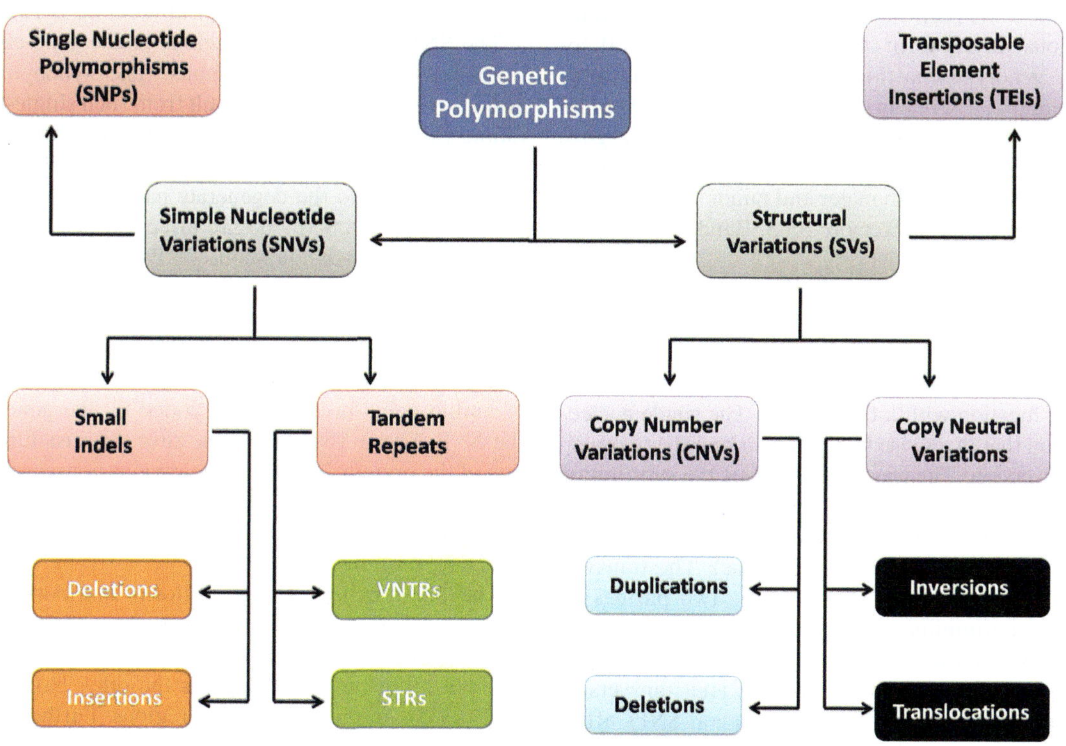

FIGURE 1.1 Types of genetic polymorphisms. Major types of annotated genetic variations include single nucleotide polymorphisms (SNPs), small indels, variable number of tandem repeats (VNTRs), short tandem repeats (STRs), copy number variations (CNVs; deletions or duplications), copy neutral variations (inversions and translocations) and transposable (mobile) element insertions (TEIs).

of a specific population contains both the allelic forms of a SNP within the same individual. The term 'homozygous wild' refers to the genotypic form of SNP that contains the major allele, the term 'homozygous variant' refers to the genotypic form of SNP that contains the minor allele, whereas the term 'heterozygous' refers to the genotypic form of SNP that contains both the major allele and the minor allele. The diploid human genome contains two identical chromosomes paired together as a homologous chromosome pair. Therefore, every SNP has two copies, one each on each of the two identical chromosomes of a homologous chromosome pair. When defined in terms of the diploid nature of human genome, the homozygous wild genotype describes a condition in which both the copies of the SNP, one each on each of the two identical chromosomes of a homologous chromosome pair, contain the major or wild allele. The homozygous variant genotype describes a condition in which both the copies of the SNP, one each on each of the two identical chromosomes of a homologous chromosome pair, contain the minor or variant allele. The heterozygous genotype describes a condition in which the SNP on one of the two identical chromosomes of a homologous chromosome pair contains the major or wild allele, whereas the SNP on the other chromosome of a homologous chromosome pair contains the minor or variant allele. In other words, the heterozygous genotype describes a condition in which the major allele of SNP occurs on one chromosome, whereas the minor allele of SNP occurs on the other chromosome of a homologous chromosome pair. For example, a study by our group evaluated the association of interleukin-1β (IL1β)-511T/C SNP with the risk of colorectal cancer in ethnic Kashmiri population. In this study, IL1β-511TT or TT genotype constitutes the homozygous wild form of the SNP, IL1β-511CC or CC genotype constitutes the homozygous variant form of this SNP, whereas the IL1β-511TC or TC genotype constitutes the heterozygous form of this SNP. Therefore, a SNP can occur within a specific population in three genotypic forms: homozygous wild (homozygous for major allele), homozygous variant (homozygous for minor allele) and heterozygous (both major and minor alleles present together) (Crawford and Nickerson 2005, Banday, Mir et al. 2017).

1.2.1.1.1 Types of Single Nucleotide Polymorphisms (SNPs)

SNPs result from the substitution of one nucleotide by another at a particular location within the genome. The single nucleotide substitution that gives rise to a SNP can be either a transition or a transversion. The transition involves the substitution of one purine by another purine (adenine to guanine or guanine to adenine) or one pyrimidine by another pyrimidine (cytosine to thymine or thymine to cytosine). The transversion involves the substitution of any of the two purines by any of the two pyrimidines or any of the two pyrimidines by any of the two purines. Accordingly, two types of SNPs exist: transition SNPs or simply transitions and transversion SNPs or simply transversions. The transversion SNPs affect the physical and chemical structure of DNA more severely in comparison with the transition SNPs. Further, the transition SNPs are more common in comparison with the transversion SNPs and constitute almost two-thirds of all the SNPs. There are two possible reasons behind the predominance of transitions over transversions. The first reason is that the SNPs within the human genome may be under the strong influence of natural selection processes described further in this chapter. The second reason is that around 70% – 80% of the cytosine nucleotides in CpG sequences present mostly in the regions of genome termed as CpG or CG islands undergo methylation to generate 5-methylcytosines. The 5-methylcytosines frequently undergo spontaneous deamination and mutate to thymine nucleotides. The methylation-induced transition of cytosine to thymine contributes significantly to the predominance of transitions over transversions (Figure 1.2) (Brookes 1999, Shastry 2009, Robert and Pelletier 2018).

The single-base substitution or the substitution of one nucleotide by another that characterizes a SNP can occur either in the coding regions or in the noncoding regions of the gene or in the regions between the genes (intergenic regions). Accordingly, based on the location, two types of SNPs occur in the genes: the coding region SNPs and the noncoding region SNPs. The coding region SNPs, as the name indicates, are located in the coding regions of the gene known as exons, that is, the regions that code for proteins. The noncoding region SNPs, as the name indicates, are located in the noncoding regions of the gene, that is, the regions that do not code for proteins. The noncoding regions of a gene starting from 5′-end to 3′-end include 5′-gene enhancer/silencer region, promoter region, 5′-untranslated region (5′-UTR), introns, 3′-UTR and 3′-gene enhancer/silencer region. Accordingly, the noncoding region SNPs include 5′-enhancer/silencer region SNPs, promoter region SNPs, 5′-UTR SNPs, intronic SNPs, 3′-UTR SNPs and 3′-enhancer/silencer region SNPs (Figure 1.2) (Aerts, Wetzels et al. 2002, Gonzaga-Jauregui, Lupski et al. 2012).

The SNPs in the coding regions of the gene can be either synonymous or non-synonymous. The synonymous SNPs, also known as silent SNPs, result from such single-base substitutions that do not lead to any change in the sequence of amino acids in the protein coded by the gene containing the SNP, and this occurs due to the degenerate nature of the triplet codon. The synonymous SNPs do not affect or alter the structure of protein. The non-synonymous SNPs include missense SNPs and nonsense SNPs. The missense SNPs result from such single-base substitutions that change one codon to another non-degenerate codon that codes for a different amino acid. The non-synonymous missense SNPs change the sequence of amino acids in the protein coded by the gene containing the SNP. These SNPs affect or alter the structure and function of the protein. The non-synonymous nonsense SNPs result from such single-base substitutions that generate a premature stop codon. These SNPs lead to the formation of truncated (inactive) proteins (Figure 1.2).

The SNPs exhibit almost uniform distribution across the coding regions and the noncoding regions of the gene. Almost half (50%) of the SNPs that are present in the genes occur in their coding regions, whereas the other half (50%) of the SNPs occur in their noncoding regions. Further, the synonymous SNPs constitute almost half (50%) of the coding region SNPs, whereas the non-synonymous SNPs constitute the other half (50%) of the coding region SNPs. Therefore, overall only one-fourth (25%) of all the SNPs that are present in the

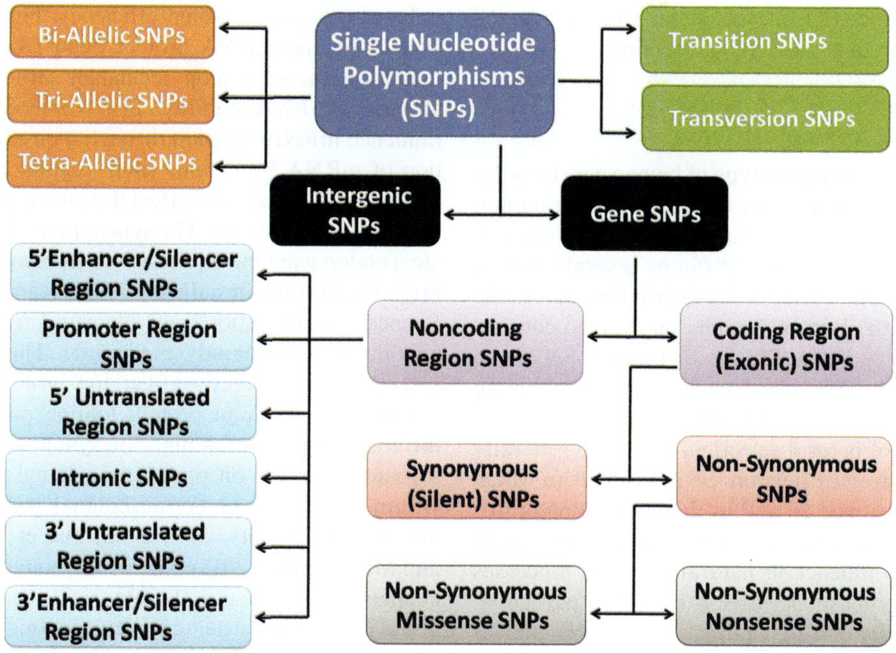

FIGURE 1.2 Types of single nucleotide polymorphisms (SNPs). The various types of SNPs include exonic SNPs, intergenic SNPs, 5′-enhancer/silencer region SNPs, promoter region SNPs, 5′-UTR SNPs, intronic SNPs, 3′-UTR SNPs and 3′-enhancer/silencer region SNPs. The SNPs also occur as transition SNPs and transversion SNPs.

genes are missense SNPs or nonsense SNPs and actually lead to a change in the sequence of amino acids in the protein or lead to the formation of truncated (inactive) proteins, respectively. Furthermore, although most of the SNPs present across the human genomes are bi-allelic, the tri-allelic SNPs and tetra-allelic SNPs also occur (Brookes 1999, Sudmant, Rausch et al. 2015). The most prevalent SNPs, the bi-allelic SNPs, involve interchangeable two-nucleotide-based substitutions at a particular location across the genomes; that is, one nucleotide substitutes another nucleotide at that particular location in an interchangeable manner. The tri-allelic SNPs involve interchangeable three- nucleotide-based substitutions at a particular location across the genomes; that is, two nucleotides alternatively substitute a third nucleotide at that particular location in an interchangeable manner. The Phase III of 1000 Genomes Project (Phase III TGP) identified more than 271,934 tri-allelic SNPs through whole genome sequencing of 2,504 unrelated and related sets of individuals representing 26 populations from world's five continental regions: Africa (AFR), Americas (AMR), Europe (EUR), South Asia (SAS) and East Asia (EAS) (Phillips, Amigo et al. 2020). The tetra-allelic SNPs involve interchangeable four-nucleotide-based substitutions at a particular location across the genomes; that is, three nucleotides alternatively substitute a fourth nucleotide at that particular location in an interchangeable manner. The Phase III TGP identified more than 961 tetra-allelic SNPs through whole genome sequencing of 2,504 unrelated and related sets of individuals representing 26 populations from world's five continental regions: Africa (AFR), Americas (AMR), Europe (EUR), South Asia (SAS) and East Asia (EAS) (Phillips, Amigo et al. 2015). Normally, the bi-allelic SNPs exist or occur at the levels of individuals as well as populations, whereas the tri-allelic SNPs and the tetra-allelic SNPs exist or occur at the level of populations only. This differential occurrence is due to the reason that within a single individual or a genome, only one nucleotide can substitute the other nucleotide at a given location and give rise to bi-allelic SNPs. However, within a given population two or three nucleotides can substitute a specific nucleotide at a given location and give rise to tri-allelic SNPs and tetra-allelic SNPs, respectively, in addition to the bi-allelic SNPs. For example, in one genome or an individual of a specific population, adenine (A) may substitute thymine (T) at a specific location and give rise to bi-allelic thymine (T) to adenine (A) SNP and this nucleotide substitution generates two alleles, one corresponding to thymine and the other corresponding to cytosine, and this occurs at the level of an individual. In another genome or individual of the same population, cytosine (C) may substitute the thymine (T) at this specific location. The substitution of specifically located thymine (T) by adenine (A) in one individual and the substitution of the same thymine (T) by cytosine (C) in another individual in a given population give rise to tri-allelic thymine (T) to adenine (A) and thymine (T) to cytosine (C) SNP. These three-nucleotide-based substitutions charactering this tri-allelic SNP generate three alleles, one corresponding to thymine (T), one to adenine (A) and another to cytosine (C) nucleotide. Further, in another genome or individual of the same population, guanine (G) may substitute the thymine (T) at this specific location. The substitution of this specifically located thymine (T) by adenine (A) in one individual and by cytosine (C) in another individual and the substitution of the same thymine (T) by guanine (G) in another individual in a given population give rise to tetra-allelic thymine (T) to adenine (A), thymine (T) to cytosine (C) and thymine (T) to guanine (G) SNP. These four-nucleotide-based substitutions charactering this tetra-allelic SNP generate four alleles one corresponding to thymine (T), one to adenine (A), one to cytosine (C) and the fourth to guanine (G) nucleotide.

1.2.1.1.2 Single Nucleotide Polymorphisms (SNPs) and Natural Selection

SNPs represent the most common type of human genetic polymorphism and exhibit uneven distribution across the human genome. The frequencies of occurrence of SNPs within two different regions of the genome can differ by several hundred folds. However, SNPs also exhibit wide distribution across the human genome and the human genome on average consists of one SNP after every 100 to 300 bp that makes SNPs one of the most uniformly distributed genetic variation among diverse variation types across the human genome. Due to the strong influence of natural selection, SNPs are generally stable and have no adverse effect on the individuals in which they occur. The occurrence as well as the distribution of different types of SNPs within the human genome appears to be under the strong influence of natural selection processes in general and purifying or negative selection in particular. The transversion SNPs and the coding region SNPs appear to be under the stronger selection processes in comparison with the transition SNPs and the noncoding region SNPs, respectively. Similarly, the non-synonymous missense SNPs and the non-synonymous nonsense SNPs appear to be under the stronger selection processes in comparison with the synonymous (silent) SNPs. The purifying selection is responsible for the removal of deleterious genetic variations including the deleterious SNPs that include transversion SNPs, coding region SNPs and non-synonymous missense SNPs and the non-synonymous nonsense SNPs. As a result, the transversion SNPs and the non-synonymous SNPs occur less frequently in comparison with the transition SNPs and the synonymous SNPs, respectively. Similarly, due to strong natural selection, the frequency of occurrence of SNPs in the coding regions of the gene is much lower in comparison with the frequency of occurrence of SNPs in the noncoding regions of the gene (Brookes 1999, Sudmant, Rausch et al. 2015).

1.2.1.1.3 Effect of Single Nucleotide Polymorphisms (SNPs) on Gene Function

The effect of SNPs on gene function often manifests through its effect on the phenotypic expression of the gene under study and depends largely on the location of the SNPs within different regions of the gene or the genomic regions adjacent to the gene. The SNPs present in the coding regions of the gene or exons can directly modulate the structure and/or function of proteins through their effect on gene transcription and translation. The synonymous SNPs by virtue of their null effect on the sequence of amino acids in the protein do not affect or alter the structure of protein and under normal conditions exhibit no or slightly deleterious effect on protein structure and function. However, these SNPs can influence or alter the structure and stability, function and splicing of messenger RNAs (mRNAs). The synonymous SNPs through their effect on the stability of mRNA secondary structures such as stem-loops and helices can influence the translational efficiency and subsequently the mRNA stability. For example, the stem-loop secondary structure through its sequestering effect on the start codon can influence the translational efficiency. Further, the ribosomes translating the mRNA protect them from degradation, and therefore, the stem-loop secondary structure can indirectly affect the mRNA stability. The synonymous SNPs can also influence mRNA function through their effect on the recognition of mRNA by various small regulatory RNAs. The synonymous SNPs can also affect the rate or pace of translation of mRNAs by ribosomes. The synonymous SNPs can also generate a codon used infrequently, but its recognition by a specific rare tRNA results in stalled translation and, if there is no interference or intervention from downstream sequences, the polypeptide folds differently every time. The synonymous SNPs can, therefore, directly affect the protein function through initiation of different protein folding pathways that lead to differential protein folding. The synonymous SNPs thus have a profound effect on protein translation and protein folding within translation or cotranslational protein folding (Sauna and Kimchi-Sarfaty 2011, Buhr, Jha et al. 2016, Kristofich and Morgenthaler 2018). The non-synonymous SNPs lead to a change in the sequence of amino acids and, therefore, directly affect or alter the structure, conformation and function of the protein. However, usually the protein remains functionally viable, but its activity can differ significantly from its wild type. Alternatively, the non-synonymous SNPs can generate a stop codon from a sense codon that stops mRNA translation and results in premature termination of the polypeptide chain. As a result, the translational process generates a truncated protein that is usually non-functional.

The noncoding regions of the gene or genome, especially the regions upstream or downstream from the transcription initiation site within a gene, play vital roles in regulating the gene function. The SNPs present in these regions with gene regulatory functions can directly modulate gene transcription and expression usually in an allele-specific manner through their modifying effects on the binding affinities and interactions of various nuclear factors including transcription factors and regulatory proteins. These SNPs known as expression SNPs (eSNPs) when present in the promoter region of a specific gene can modulate gene transcription and expression through alteration of a number of promoter-related activities. These include the binding of various transcription factors and repressors to the promoter sequence thereby affecting promoter activity. The SNPs can also modulate other processes such as promoter clearance and the epigenetic modifications including DNA methylation and histone modifications that further influence gene transcription and expression. The SNPs present in the 5′-UTR of the gene can affect its transcription through modulation of binding of various transcription factors and subsequently affect the translation of mRNA to proteins. The SNPs present in the introns can modulate gene transcription and expression through their effect on mRNA splicing. The intronic SNPs lead to the generation of different splice (splicing) variants that directly affect the translation process and protein function. These SNPs can either promote or disrupt the binding of various trans-acting factors to cis-regulatory elements (CREs) and subsequently affect the function of CREs that in turn alters gene transcription and expression. The intronic SNPs also affect the gene regulatory functions of lncRNAs. The 3′-UTRs of the gene contain binding sites for small regulatory RNAs known as miRNAs. The miRNAs negatively regulate gene expression (translation) by directly binding to the 3′-UTRs of target mRNAs. The 3′-UTR of a particular mRNA can have multiple sites for the binding of the same miRNA that enhances the degree of gene regulatory functions of miRNAs. The SNPs present in the 3′-UTR of the gene can affect the binding of miRNAs and regulate their degradation. As a result, the 3′-UTR SNPs directly affect the translation of mRNA to proteins. The SNPs present in the 5′-enhancer/silencer regions and 3′-enhancer/silencer regions can modulate the effect or regulatory role of these regions on/in gene function including gene expression. The SNPs present in the regions of the genome distant from the genes either promote or inhibit gene transcription and expression through long-range cis-regulatory effects (Brookes 1999, Shastry 2009, Karki, Pandya et al. 2015, Robert and Pelletier 2018). Through their effect on gene transcription and expression, the SNPs subsequently influence various physiological and pathophysiological (disease) processes including the modulation of the susceptibility or risk to various diseases. The extent or nature of these effects depends largely on the location of the SNPs within different regions of the gene in particular or other regions of the genome, especially those adjacent to the genes. Therefore, the location of a SNP within the gene or genome is important in determining the effect of that SNP on a particular phenotype. Consequently, the effect of SNP on phenotype changes with the change in the location of SNP within the gene or genome.

1.2.1.1.4 Methods for Detection and Analysis of Single Nucleotide Polymorphisms (SNPs)

The methods or techniques used for the detection, analysis and characterization of SNPs are diverse and dynamic. As with most of the genetic polymorphisms, the detection and analysis of SNPs involves two broad approaches: screening the sample DNA sequences or genomes for detection of new, novel or previously unknown or unreported SNPs and screening the sample DNA sequences or genomes for the detection of previously known or reported SNPs. The screening of the sample DNA sequences or genomes for the detection and analysis of previously unknown or unreported SNPs and previously known or reported SNPs further involves two broad approaches that operate at three different levels. These include the detection and analysis of SNPs at the level of single gene or a group of genes, at the level encompassing all the exons in a genome (exome level) and at the level encompassing the entire genome (global or whole genome level). However, the detection and analysis of SNPs at the exome and whole genome levels, largely, shares a common approach. Therefore, the detection and analysis of whole exome level and whole-genome-level SNPs uses similar methods although some modifications may be required. Thus, there are two major categories of methods or approaches used for SNP detection and analysis: the targeted approach and the global, random or whole approach. The targeted approach enables the detection and analysis of SNPs in a single gene or a group of genes present either adjacent to each other on the same chromosome or spread across the entire genome. The global or whole approach can be either global or whole exome approach, which enables the detection and analysis of SNPs present in all the exons of a particular genome including the human genome, or it can be global or whole genome approach, which allows the detection and analysis of SNPs spread globally or across the entire genome. There are two broad categories of techniques or methods available to genotype SNPs, and these include the low-to-medium throughput techniques and high-throughput techniques. The low-to-medium throughput techniques include several traditional methods and their modifications. The prominent among these techniques include polymerase chain reaction-restriction fragment length polymorphism (PCR-RFLP), polymerase chain reaction-denaturing gradient gel electrophoresis (PCR-DGGE), polymerase chain reaction-temporal temperature gel electrophoresis (PCR-TTGE), polymerase chain reaction-single-strand conformation polymorphism (PCR-SSCP), polymerase chain reaction-chemical cleavage of mismatch (PCR-CCM), allele-specific (AS) polymerase chain reaction (AS-PCR) methods and multiplex single-base primer extension (SBE) methods. The high-throughput methods mostly include the latest and emerging techniques. The prominent among these techniques include pyrosequencing, TaqMan assay (5′-nuclease allelic discrimination assay), SNP microarray (SNP array or SNP chip) technique and sequencing techniques, in particular, NGS techniques. The low-to-medium throughput techniques usually allow the targeted approach that involves the detection and analysis of SNPs in a single gene or a group of genes present either adjacent to each other on the same chromosome or spread across the entire genome. The high-throughput techniques such as SNP array and NGS techniques in addition to targeted approach also allow the whole approach. The whole approach includes the whole exome approach that enables the detection and analysis of SNPs present in all the exons of a particular genome including the human genome and the whole genome approach that allows the detection and analysis of SNPs spread across the entire genome.

1.2.1.1.4.1 Polymerase Chain Reaction-Restriction Fragment Length Polymorphism (PCR-RFLP) Technique The polymerase chain reaction-restriction fragment length polymorphism (PCR-RFLP) is a rapid, accurate and one of the simplest, inexpensive and most commonly used targeted approach techniques to genotype previously known or reported SNPs and point mutations. This method, as the name indicates, involves the endonuclease- or restriction enzyme-mediated digestion of specific PCR-amplified genomic DNA sequences. The detection using this method exploits a usually observed single-nucleotide-level genomic phenomenon whereby the change in a single nucleotide that characterizes a SNP or a point mutation creates a new site for a restriction enzyme, known as restriction site, or alternatively abolishes an existing restriction site. Consequently, the target DNA sequence either acquires the capacity of cleavage, or loses the ability of cleavage by a specific restriction enzyme or endonuclease. The PCR-RFLP method consists of two main steps: the PCR amplification of the target DNA using specific primers and the digestion of PCR-amplified DNA products using specific restriction enzymes. If the change in a single nucleotide that characterizes a SNP abolishes a restriction site or if the change in a single nucleotide that characterizes a restriction site creating SNP is absent, then there is no restriction digestion of the PCR-amplified target DNA sequence with endonuclease.

The digestion reaction therefore generates a single fragment, the original PCR product, and this digestion pattern is termed as homozygous wild. If the change in a single nucleotide that characterizes a SNP is present in both the parental alleles and creates a restriction site, then the restriction digestion of the PCR-amplified target DNA sequence with particular endonuclease generates two or more fragments shorter in length than the original PCR product. This digestion pattern is termed as homozygous variant. If the change in a single nucleotide that characterizes a SNP is present in only one of the two parental alleles and creates a restriction site, then the restriction digestion of the PCR-amplified target DNA sequence with particular endonuclease generates multiple DNA fragments. The multiple DNA fragments generated include the original PCR product similar to that of homozygous wild digestion pattern and two or more fragments shorter in length than the original PCR product similar to the homozygous variant digestion pattern. This digestion pattern is termed as heterozygous (Ota, Fukushima et al. 2007). For example, a study by our group evaluated the association of interleukin-1β (IL1β)-511T/C SNP with the risk of colorectal cancer in ethnic Kashmiri population. In this study, the genotyping of IL1β-511T/C SNP by *Ava*I restriction enzyme yielded a non-cleaved 305 bp fragment in homozygous wild (IL1β-511TT or TT) condition; two fragments in homozygous variant (IL1β-511CC or CC) condition, 190 and 115 bp in size; and three fragments in heterozygous (IL1β-511TC or TC) condition, 305,190 and 115 bp in size (Figure 1.3) (Banday, Mir et al. 2017). The digestion products of PCR-RFLP technique are visualized using gel electrophoresis with ethidium bromide (EtBr) as staining dye. In some scenarios where the presence of SNP does not create a restriction site, a mismatch base incorporated into the site encompassing the SNP under study using specific primers can create a site for specific restriction endonuclease that allows the detection of either the presence or the absence of the SNP. For example, a study by our group evaluated the association of tumor necrosis factor-α (TNF-α)-308G/A SNP with the risk of colorectal cancer in ethnic Kashmiri population. In this study, we used a primer with mismatched base to create a site for *Nco*I restriction enzyme and the digestion by *Nco*I endonuclease indicated the presence of TNF-α-308GG genotype and the absence of TNF-α-308AA genotype or TNF-α-308G/A SNP (Banday, Balkhi et al. 2016). However, the incorporation of mismatched base to create a site for specific restriction endonuclease to allow the detection of the presence or absence of SNP is not always possible. The main advantages of PCR-RFLP technique are its simple, low-cost procedure and less operational time that ranges from few hours to a single day. The major disadvantage of PCR-RFLP technique is that it cannot detect SNPs or point mutations that neither create nor abolish the recognition sites for specific restriction endonucleases.

1.2.1.1.4.2 Polymerase Chain Reaction-Denaturing Gradient Gel Electrophoresis (PCR-DGGE) Technique The PCR-DGGE is another important technique that uses a combination of PCR and gel electrophoresis and predominantly enables the detection of previously known or reported SNPs. It also allows the detection and analysis of previously unknown or unreported SNPs. The basis of this technique is that the process of denaturation of double-stranded DNA (dsDNA) molecule is highly dependent on its nucleotide sequence. The variation in nucleotide sequence that may arise due to the difference of even a single nucleotide gives DNA molecules or DNA fragments unique melting profiles or rates of denaturation that allows the distinction of one DNA molecule or DNA fragment from another (Fischer and Lerman 1983). The process of denaturation involves the partial or complete separation of dsDNA molecules into two single polynucleotide strands and occurs through the disruption of hydrogen bonds holding the two strands together. These hydrogen bonds exist between complementary bases, three bonds between cytosine (C) and guanine (G) and two bonds between adenine (A) and thymine (T). The rate of denaturation depends on the size or length of the DNA molecule and relative proportions of complementary base pairs, cytosine-guanine and adenine-thymine. The rate of denaturation is slower in case of DNA molecule with large size or length and higher proportion, content or percentage of cytosine-guanine complementary base pair. In vitro denaturation involves exposure of DNA molecule to denaturing agents such as heat, increasing pH (through the use of alkali such as NaOH) and specific chemicals such as urea and detergents. The heat-induced denaturation of DNA molecule is termed as melting and occurs abruptly at a specific temperature termed as melting temperature (T_m), defined as the temperature at which half (50%) of the DNA occurs in

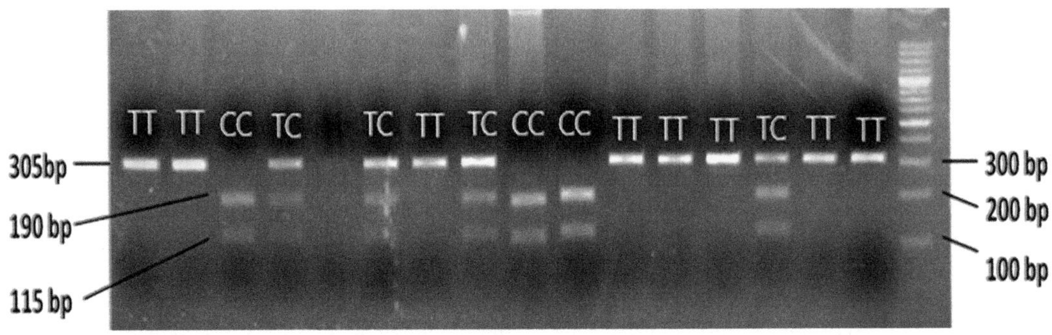

FIGURE 1.3 The pictorial representation of the genotyping of IL1β-511T/C SNP by PCR-RFLP method: IL1β-511TT or TT genotype represents homozygous wild condition, IL1β-511CC or CC genotype represents homozygous variant condition, and IL1β-511TC or TC genotype represents heterozygous condition (Banday, Mir et al. 2017).

melted state. The process of DNA denaturation involves the generation of unzipped terminals and non-terminal denaturing (melting) loops or domains. Therefore, the dsDNA molecules or DNA fragments, when exposed to denaturing agents, undergo varying degrees of denaturation depending on their length and their nucleotide sequence. The differences in the degree or extent of denaturation of DNA molecules or DNA fragments are detectable through various techniques including polyacrylamide gel electrophoresis (PAGE). The denaturation of dsDNA molecules or DNA fragments retards their electrophoretic mobility within polyacrylamide gel consisting of a gradient of denaturants, and consequently, the DNA molecule with a higher degree of denaturation migrates relatively slowly in comparison with the DNA molecule with a lower degree of denaturation. The DNA molecules with varying degrees of denaturation, therefore, retard or immobilize at different positions within the gel and appear as separate bands.

In the PCR-DGGE technique, the PCR-amplified dsDNA fragments undergo electrophoresis in a polyacrylamide gel that consists of an increasing gradient of denaturants such as urea and formamide. Due to the denaturing gradient that exists within the gel, the DNA fragments undergo retardation in their mobility or immobilize as soon as the DNA fragments reach the region within the gel that corresponds to the minimum concentration of denaturing agent required to trigger or initiate the denaturation of DNA fragments under study. The change of even a single nucleotide as in point mutations or SNPs affects the denaturation profile of the PCR-amplified DNA fragment, and consequently, the DNA fragment that contains the point mutation or SNP retards or immobilizes at a position within the gel that is different from DNA fragment that lacks the point mutation or SNP. Furthermore, the denaturation profile and subsequently the electrophoretic mobility of a homoduplex DNA fragment are profoundly different from its heteroduplex DNA copy that contains the base mismatch as a result of a point mutation or a SNP. The heteroduplex DNA due to the presence of a base mismatch denatures earlier and at a much lower concentration of denaturants in comparison with the homoduplex DNA, and it occupies positions within the gel that are more farther apart from each other (Myers, Lumelsky et al. 1985). In order to ensure maximum electrophoretic separation of DNA fragments, it is essential to ensure their partial rather than complete denaturation. This is because the completely denatured DNA fragments have similar electrophoretic mobility, which leads to poor resolution of DNA samples within the gel. The partial rather than complete denaturation of DNA in DGGE is achievable through several methods, but the most common procedure involves the incorporation of 'GC clamp' into the DNA fragments under analysis. The GC clamp that consists of a long stretch of guanine (G) and cytosine (C) nucleotides has higher temperature of melting than normal DNA that contains randomly distributed nucleotides. The GC clamp adds a high melting region to otherwise normal, relatively low melting DNA fragment and ensures partial rather than complete denaturation of DNA. The GC clamp remains annealed even if the rest of the DNA fragment that contains the region of interest undergoes denaturation, and this improves the electrophoretic resolution of DNA fragments in DGGE (Sheffield, Cox et al. 1989). In case of PCR-generated DNA fragments, the incorporation of GC clamp is achievable through the addition of GC-rich stretch to the 3′-end of either the forward primer, or the reverse primer. Currently, several software applications such as Bio-Rad's WinMelt™ (for Windows OS) or MacMelt™ (for Macintosh OS) are available, which allow the researcher to simulate or model the melting pattern of DNA fragments under analysis and aid in optimizing the process of incorporation of GC clamp.

The use of denaturing agents such as urea and formamide in different concentrations enables the generation of polyacrylamide gels with varying denaturing gradients, and this allows the use of DGGE technique over a wide detection range. Further, the DGGE technique, when carried out at a constant temperature of around 60°C, shortens the denaturation time of DNA fragments and the distance they traverse within the gel. The choice to generate denaturing gradients either in parallel direction or in perpendicular direction gives further flexibility and options to optimize the detection process. The resolution of DGGE for the detection and analysis of SNPs usually ranges between 100 and 1,000 bp. Low-percentage (6% – 8%) polyacrylamide gels are suitable for the separation and analysis of larger DNA fragments, whereas high-percentage (10% – 15%) polyacrylamide gels are suitable for the separation and analysis of smaller DNA fragments. For example, a 12% polyacrylamide gel is suitable for the separation and analysis of DNA fragments, 300 – 400 bp in size (Knapp, Cadavid et al. 1997). The DGGE in addition to the detection and analysis of SNPs and point mutations that involve a single nucleotide also allows the detection and analysis of polymorphisms encompassing one or more regions within the gene, such as across multiple exons. The DGGE technique also allows precise determination of nucleotide sequence when used in combination with DNA sequencing. One of the major limitations of this technique is the lack of reproducibility that arises largely due to the difficulties associated with generating the same chemical denaturing gradient repeatedly.

1.2.1.1.4.3 Polymerase Chain Reaction-Temporal Temperature Gel Electrophoresis (PCR-TTGE) Technique

The PCR-TTGE is another important technique similar to PCR-DGGE technique that uses a combination of PCR and gel electrophoresis and predominantly enables the detection of previously known or reported SNPs. It also allows the detection and analysis of previously unknown or unreported SNPs. The basis or principle of this technique is similar to that of DGGE and uses chemical denaturants such as urea to generate denaturing conditions, but instead of chemical denaturing gradient used in DGGE, the TTGE technique uses a temporal temperature gradient in combination with a uniform concentration of denaturants in order to achieve electrophoretic separation of DNA fragments under analysis. The temperature gradient used in TTGE technique is temporal and is generated through a gradual increase in temperature with time rather than spatial, which differentiates this technique from temperature gradient gel electrophoresis (TGGE). The TGGE technique, unlike the TTGE technique, uses a temperature gradient that exists from the top to bottom of the polyacrylamide gel and is generated through specialized equipment to denature and separate DNA molecules. Therefore, in the TTGE technique, a uniform temperature exists throughout the polyacrylamide gel at any given time and changes uniformly with time throughout the gel.

As in the DGGE technique, the dsDNA molecules or DNA fragments in this technique undergo varying degrees of denaturation depending on their length and their nucleotide sequence, when exposed to denaturing agents including increasing temperature. This technique exploits the differences in the degree or extent of denaturation of DNA molecules or DNA fragments at different temperatures. As in the DGGE, the DNA molecules with varying degrees of denaturation, therefore, retard or immobilize at different positions within the gel and appear as separate bands. The TTGE technique is highly sensitive and can detect the change of even a single nucleotide, which makes this technique highly suitable for the detection of point mutations and SNPs in the nuclear DNA as well as mtDNA. This technique is relatively simple and provides information in a straightforward, conclusive manner, and more importantly, it is easily reproducible unlike the DGGE technique (Yoshino, Nishigaki et al. 1991, Henco, Harders et al. 1994).

1.2.1.1.4.4 Polymerase Chain Reaction-Single-Strand Conformation Polymorphism (PCR-SSCP) Technique The PCR-SSCP is another sensitive technique that uses a combination of PCR and gel electrophoresis to enable the detection and analysis of known or previously reported and unknown or previously unreported point mutations, small indels (insertions and deletions) and SNPs (Orita, Suzuki et al. 1989, Kukita, Tahira et al. 1997). The basis of this technique is that the intra-strand secondary and tertiary structures of single-stranded DNA molecules or DNA fragments and subsequently their conformation depend on their nucleotide sequence and microenvironment that includes factors such as temperature, pH, ionic strength and buffer conditions. Under similar microenvironmental conditions, the difference in nucleotide sequence that includes the difference of even a single nucleotide gives single-stranded DNA molecules or DNA fragments a unique conformation. The nucleotide sequence-dependent conformational differences between single-stranded DNA molecules or DNA fragments are detectable through their unique rate of migration within a non-denaturing polyacrylamide gel. The unique rates of migration of conformationally different DNA molecules or DNA fragments within a non-denaturing polyacrylamide gel allow the distinction of one DNA molecule or DNA fragment from another purely based on the differences in their nucleotide sequence, provided other factors including microenvironmental conditions and size of the DNA molecule or DNA fragment are similar, uniform or constant. The DNA molecules with variation in nucleotide sequence and subsequently the conformation, therefore, retard or immobilize at different positions within the polyacrylamide gel and appear as separate bands. The changes in factors that define the microenvironment such as temperature, pH, ionic strength and buffer conditions also affect the conformation of DNA molecules or DNA fragments and the conformational differences between them. The changes in these factors either disrupt or preserve the conformation of DNA molecules or DNA fragments and consequently decrease or increase the sensitivity and resolution of this technique. For example, lower temperature and pH generally preserve the conformation of DNA molecules or DNA fragments and make them more distinguishable, which increases the sensitivity and resolution of this technique (Hayashi 1991).

The PCR-SSCP technique, in brief, involves the PCR amplification of specific DNA regions to generate DNA fragments usually of 300 bp or less in size. The PCR-amplified DNA fragments then undergo heat-induced denaturation, for example, at 96°C for 5 minutes or 90°C for 3 minutes, to generate single-stranded DNA fragments followed by snap cooling, for example, on ice for 10 minutes, in order to prevent the single-stranded DNA fragments from reannealing. The snap-cooled single-stranded DNA fragments loaded onto non-denaturing polyacrylamide gels, usually 8% – 10%, then undergo electrophoresis and separate into distinct bands based on the nucleotide sequence-dependent differences in conformation. Several methods are available that enable the in-gel visualization of SSCP-separated single-stranded DNA fragment bands and include the commonly used silver staining method. Alternatively, the PCR products labeled with radioisotopes such as radio-labeled deoxynucleotide triphosphates (dNTPs) or radio-labeled primers and fluorescent dyes such as rhodamine 110 (R110) or rhodamine 6G (Basic Red 1 or R6G) allow in-gel visualization of SSCP-separated single-stranded DNA fragment bands. Several modifications of the SSCP technique are available, including capillary electrophoresis (CE)-based SSCP analysis and post-labeling automated CE-based SSCP analysis (PLACE-SSCP), and these modified SSCP methods are rapid, allow automation and increase the efficiency, reproducibility, flexibility and sensitivity of this technique (Inazuka, Wenz et al. 1997). The PCR-SSCP technique is a simple, rapid, cost-effective, easily reproducible and highly sensitive method with a detection rate of 70% – 95% for SNPs and point mutations depending on the DNA region under analysis. The major limitation of this technique is the decrease in the sensitivity for the detection of SNPs and point mutations in DNA fragments larger than 300 bp in size. This is because as the size of the DNA fragment increases, the effect of the presence of SNP or point mutation on its conformation decreases (Hayashi 1991, Sheffield, Beck et al. 1993).

1.2.1.1.4.5 Polymerase Chain Reaction-Chemical Cleavage of Mismatch (PCR-CCM) Technique The PCR-CCM method is an important technique that enables the detection of known or previously reported and unknown or previously unreported genetic variations including SNPs, point mutations and small indels (insertions and deletions) in PCR-amplified DNA fragments of size ranging from few 100 bp to greater than 1,000 bp (1 kb). This technique also gives an accurate location of these genetic variations within the DNA fragments under analysis. The basis of detection and analysis of mutations and genetic polymorphisms including SNPs using this technique is the formation of heteroduplex DNA molecule and preferential chemical modification, specifically oxidation, of cytosine and thymine residues by specific chemicals when they are present in single-stranded DNA molecule in comparison with when they are part of dsDNA. The formation of heteroduplex or hybrid DNA occurs when the homozygous wild DNA fragments and homozygous variant DNA fragments anneal to each other. The homozygous wild DNA fragments and homozygous variant DNA fragments that form the heteroduplex DNA may belong to two different individuals each having an alternative form of the homozygous genotype, or they may arise from a single individual or genome with heterozygous genotype for the genetic variation under analysis. The heteroduplex DNA characteristically contains sites with mismatched bases that exist as single nucleotides because the mismatched bases in one strand of the heteroduplex DNA do not hydrogen bond to uncomplimentary bases or nucleotides in the other strand of the heteroduplex DNA. The deoxynucleotides including cytosine and thymine nucleotides present in base mismatch sites in a heteroduplex DNA functionally behave in a manner similar to deoxynucleotides present in single-stranded DNA molecule rather than the deoxynucleotides present at normal sites in the dsDNA molecule.

In this technique, pyrimidines, cytosine and thymine nucleotides present at mismatch base sites in the heteroduplex DNA, undergo chemical modifications on exposure to certain chemicals, which opens up the pyrimidine ring structure and allows the detection of known or unknown genetic variations. The cytosine residues present at mismatch base sites in the heteroduplex DNA undergo chemical modification on exposure to hydroxylamine (NH_2OH), which forms adducts across its carbon 5 and carbon 6 double bonds. The thymine residues present at mismatch base sites in the heteroduplex DNA undergo oxidation across its carbon 5 and carbon 6 double bonds on exposure to osmium tetroxide (OsO_4) or potassium permanganate ($KMnO_4$). After chemical modification of mismatched pyrimidines, the heteroduplex DNA undergoes cleavage at the site of modified mismatched cytosine and thymine bases on exposure to another chemical, piperidine. The chemical piperidine induces cleavage of sugar phosphate backbone of DNA at the site of modified mismatched bases through β-elimination. The cleaved DNA fragments then undergo electrophoretic separation, and the relative position or size of the DNA fragment bands reveals the presence and location of the known or unknown genetic variations including SNPs and point mutations. This technique allows the detection of mismatched purines, adenine and guanine bases using complementary probes that analyze their complementary base partners, thymine and cytosine (Cotton, Rodrigues et al. 1988, Cotton 1999).

The PCR-CCM is a rapid, robust, cost-effective, reliable, easily reproducible, low-to-medium throughput technique that allows the complete solid-phase and liquid-phase detection of known and unknown sequence variations of multiple types along with their position in DNA fragments several kb in size. This is unlike the previously discussed techniques that allow detection of limited genetic variations in DNA fragments up to 1 kb in size. The specificity of the two chemical modification reactions allows direct inference of mismatch sequence that adds to the capability of this technique to detect potentially functional unknown sequence variations in addition to the known sequence variations and thereby reduces the need for sequencing the samples under study for these variations. The use of differential fluorescent labeling on forward and reverse primers not only allows the detection of genetic variations, but also determines their location whether the detected genetic variations are on the leading strand or the lagging strand. This also enables multiplexing that along with the uniform assay conditions that characterize PCR-CCM technique makes it a diagnostically important method that allows the analysis of multitude of genes through screening of a small number of DNA samples. The PCR-CCM technique finds its use in the detection and analysis of known genetic variations, including the mutations with a well-established causative role in the pathogenesis of several diseases such as colorectal cancer, Alzheimer's disease, hemophilia A and B, and phenylketonuria. The major disadvantages of this technique are the use of toxic or potentially toxic chemicals in steps involving chemical modification of pyrimidines and cleavage of sugar phosphate backbone of DNA.

1.2.1.1.4.6 Allele-Specific (AS) Polymerase Chain Reaction (AS-PCR) Technique and Its Modifications The AS-PCR has different names such as amplification refractory mutation system (ARMS), allele-specific amplification (ASA) and PCR allele-specific amplification or PCR amplification of specific alleles (PASA). The AS-PCR is a simple technique that uses a combination of PCR and gel electrophoresis and enables the detection of known or previously reported and unknown or previously unreported genetic variations including SNPs, point mutations and small indels (insertions and deletions). The detection and analysis of mutations and genetic polymorphisms including SNPs using this technique relies on the use of allele-specific (AS) primers that allow amplification of target DNA sequence only if one or the other allelic form of the SNP is present. This technique uses three primers, one common primer and two AS primers. The 5′-end of the common primer is complementary to the 3′-ends of both the allelic forms of the SNP under study. The 3′-ends of two AS primers are complementary to the 5′-end of either of the two allelic forms of the SNP under study. In other words, the 3′-end of one AS primer is complementary to the 5′-end of wild (major) allelic form of the SNP, whereas the 3′-end of the other AS primer is complementary to the 5′-end of variant (minor) allelic form of the SNP under study. The AS-PCR technique, therefore, involves setting up two parallel PCRs using in each setup the common target DNA, the common primer and one of the two AS primers. The outcome of PCR amplifications visualized through agarose gel electrophoresis allows the detection of specific allelic form of the SNP present in the target DNA. The outcome of PCR amplifications enables to determine whether the SNP occurs in the target DNA in its wild allelic form or in its variant allelic form or whether the SNP is present in the target DNA in its heterozygous form, that is, both wild and variant allelic forms present together. If the SNP is present in the target DNA in its wild allelic form, the amplification occurs only in the reaction mixture that contains primer specific to the wild allelic form. The amplification in the other reaction mixture that contains primer specific to variant allelic form does not occur, is greatly retarded or occurs improperly, resulting in the formation of non-specific products including primer dimers. Further, if the SNP is present in the target DNA in its heterozygous form, then proper amplification occurs in both the reaction mixtures, revealing the occurrence of both the wild allelic form and the variant allelic form of the SNP in the target DNA (Newton, Graham et al. 1989, Hassan, Ahmad et al. 2013, Hanafi, Hassan et al. 2014).

The AS-PCR in its basic form is an efficient, reliable, rapid and cost-effective technique for the detection of SNPs, point mutations and small indels in target DNA molecules. However, the allelic specificity of primers that arises from the difference of a single nucleotide between the two primers is not fully reliable and sufficient to discriminate clearly between the two allelic forms of the SNP or point mutation under study because in practice, some degree of amplification may occur even with mismatched primers giving false-positive results. There is possibility of false-negative results as well. Further, this technique requires that two parallel PCRs be set up, each specific to one allelic form of the SNP. All these factors make this technique unsuitable and unreliable for large-scale detection of or screening for polymorphisms and mutations. However, with proper modifications along with the provision for kinetic and real-time monitoring, this technique allows accurate, rapid and cost-effective detection of SNPs and measurement of allelic frequencies on a relatively larger scale in comparison with the basic form of AS-PCR technique (Ahmadian, Gharizadeh et al. 2001). One of the modifications of the AS-PCR technique, known as PCR amplification of multiple specific alleles, allows SNP genotyping in a single tube. This modified AS-PCR technique also involves the use of two AS primers, but the design or sequence of one of these primers includes an additional GC-rich sequence or GC clamp that increases the melting temperature (T_m) of one of the two PCR products in comparison with the other. The technique makes use of a DNA-intercalating dye such as SYBR-Green I to enable the detection of dsDNA and uses fluorescent probes to distinguish between the two PCR products in a single-tube reaction mixture. Further, the technique uses the Stoffel fragment of *Taq* DNA polymerase to increase the discrimination of the 3′-end of the allelic specific primer for the non-specific DNA template that lacks its complementary base or contains the mismatched base that is the binding target of another allelic specific primer. The Stoffel fragment minimizes non-specific PCR amplification as it extends the primer only when its 3′-end fully matches the target sequence and it usually yields the amplified PCR fragments less than 100 bp in size. The use of GC clamp, fluorescent probes and Stoffel fragment besides ensuring highly specific amplification of target DNA sequence also allows an easy distinction of one PCR product from another. The procedure involves the measurement of melting temperatures of the two PCR products upon completion of PCR amplification and uses real-time fluorescence detection to determine and analyze the melting curves of the PCR products to distinguish one PCR product from another. This form of AS-PCR technique is highly accurate and allows a rapid single-closed-tube automated analysis, but this method requires expensive equipment and often expensive sequence-specific fluorescently tagged primers or sequence-specific probes and requires the amplified PCR products to be smaller in size. Further, the GC clamp in some cases depending upon the sequence can result in inhibition of PCR amplification. This technique is ideal for genotyping of individual SNPs or point mutations, but is not highly desirable when there is a need to screen samples for multiple SNPs or point mutations (Dutton and Sommer 1991, Ririe, Rasmussen et al. 1997, Germer and Higuchi 1999).

Another modification of the AS-PCR technique involves the use of specific DNA oligonucleotide sequences that possess catalytic activity. These DNA oligonucleotide sequences, known as DNA enzymes, deoxyribozymes or DNAzymes, possess ribonuclease activity and thus are capable of cleaving RNA molecules. This modified AS-PCR technique also involves the use of two AS primers, but the design or sequence of one of these primers includes a 10- to 23-nucleotide-long DNAzyme antisense sequence. This DNAzyme antisense sequence generates a sense DNAzyme sequence or active DNAzyme sequence only when there is occurrence of PCR amplification in the reaction mixture that reveals the presence of allelic form of the SNP specific to that primer in the target DNA. The technique uses a chimeric RNA/DNA reporter probe that contains a fluorescent dye molecule and a quencher dye molecule attached to the DNA strands and separated from each other by the RNA strand that constitutes the DNAzyme cleavage site. If the PCR amplification occurs, it generates the active DNAzyme that cleaves the RNA strand and separates the DNA strand fragments that hold the fluorescent dye molecule and quencher dye molecule closer to each other. With PCR amplification in progress, there is gradual accumulation of PCR products and the concomitant generation of active DNAzyme sequence, which cleaves the RNA sequence region between the fluorescent dye molecule and quencher dye molecule that holds the fluorescent dye and quencher dye pair together. Consequently, fluorescent dye molecule and quencher dye molecule separate from each other and there is a gradual increase in the fluorescence signal. The real-time monitoring of increase in fluorescence signal reveals the progress of PCR amplification and the accumulation of PCR products in the reaction mixture. The DNAzyme AS-PCR technique is a highly efficient and specific technique for the detection of SNPs. The DNAzyme sequence and chimeric RNA/DNA reporter probe are suitable for use in any SNP assay with specific allele-specific primers. Since this technique allows real-time monitoring of PCR amplification and accumulation of PCR products, the direct quantification of PCR DNA products is possible. This technique, however, does not allow multiplexing, and for multiple SNP genotyping assays, there is need to setup multiple reactions using multiple allele-specific primers (Todd, Fuery et al. 2000).

Another modification of the AS-PCR technique involves the use of five different primers, two AS primers each with a different 5′-end tail, a reverse primer and two universal energy-transfer (ET)-labeled primers. Each allele-specific primer consists of a single allele-specific nucleotide at its 3′-end and a different tail sequence consisting of 21 nucleotides at its 5′-end. The middle regions of the allele-specific primers contain a sequence segment of 16 – 21 nucleotides that is complementary to a specific region within the region or gene of interest. The optimized length of this sequence segment is such that the primers have the melting temperature (Tm) in the range of 69°C – 75°C. The design of reverse primers includes sequences that cannot form inverted repeats and overlaps with the other primers, two AS primers and two universal ET-labeled primers, and this prevents mispriming. The design of each of the two universal ET-labeled primers includes a 3′-priming end with sequence identical to the 5′-end tail of the one or the other allele-specific primer. The 5′-end of each of these ET-labeled primers consists of a hairpin loop that contains a fluorophore such as sulforhodamine (red dye) in one primer and fluorescein (green dye) in the other primer. The hairpin loop in both ET-labeled primers contains a quencher such as 4-(dimethylamino)-azobenzene-4′-carboxylic acid (Dabcyl). Under quiescent conditions, the hairpin loop structure of the ET-labeled primers holds the fluorescent dye molecule and quencher dye molecule closer to each other and, as a result, there is low-level emission of background fluorescence as the fluorophore directly transmits most of its energy to the nearby quencher rather than emitting it as fluorescence. If the PCR amplification occurs, the ET-labeled primer loses its hairpin loop structure as it linearizes and incorporates into the PCR product. With PCR amplification in progress, there is gradual accumulation of PCR products and the concomitant linearization of ET-labeled primer and its incorporation into PCR product that separates the fluorescent dye molecule from the quencher dye molecule. Consequently, there is gradual increase in fluorescence signal. The real-time monitoring of increase in fluorescence signal reveals the progress of PCR amplification and the accumulation of PCR products in the reaction mixture. This technique uses two different fluorophores to enable detection and, if necessary, quantification of each of the homozygous allelic forms of the SNP or genetic variation under study. If there is emission of both the green and the red fluorescence, it reveals the presence of heterozygous form or genotype of the genetic variation under study. This technique enables SNP genotyping in a single-tube setup without the need for parallel reaction and in a relatively cost-effective manner (Myakishev, Khripin et al. 2001).

Another modification of the AS-PCR technique, known as bidirectional PCR amplification of specific alleles (PASA) or bidirectional AS-PCR technique, involves bidirectional amplification in which the amplification of one allelic form of SNP occurs in one direction and the amplification of the other allelic form of SNP occurs in another direction. This technique uses four different primers: two allele-specific inner primers and two outer primers. Each allele-specific inner primer consists of a single destabilizing mismatched nucleotide at or near its 3′-end that increases the specificity of the primer for one allelic form and decreases its specificity for other allelic form of SNP or genetic variation under study. Each allele-specific inner primer also contains a different non-complementary GC-rich tail sequence consisting of 10 nucleotides at its 5′-end. The significance of this 5′-end GC rich tail sequence gets a mention later in this section. The middle regions of the inner allele-specific primers contain a short sequence segment that is complementary to a specific region within the region or gene of interest. The design of two outer primers includes sequences that are complementary to two different sites that are at different distances from the nucleotide that determines the allelic form within the region or gene of interest. This enables the differentiation of PCR products that result from amplifications occurring in opposite direction from one another. This technique generates two different PCR products each under homozygous wild condition or homozygous variant condition. In each genotypic scenario (homozygous wild or homozygous variant condition), one PCR product results from the amplification of one or the other inner AS primer-outer primer pair and the other PCR product results from the amplification of outer primer pair. The PCR product that results from the amplification of outer primer pair is similar for both the allelic forms (homozygous wild or homozygous variant form) of the SNP. This PCR product generated in all genotypic scenarios (homozygous wild, homozygous variant and heterozygous conditions) serves as positive control for the PCR amplification. The PCR product that results from the amplification of one inner AS primer-outer primer pair is different from the PCR product that results from the amplification of the other inner AS primer-outer primer pair, and this enables the differentiation of two allelic forms from each other when visualized through agarose gel electrophoresis. This technique generates all the three PCR products under heterozygous condition. For example, if we designate the primer specific to homozygous wild allele as W and that to homozygous variant allele as V and the corresponding outer primers as A and B, respectively, then the following products result from the PCR amplification depending on the allelic forms present. Under homozygous wild condition, the PCR amplification will generate two products, WA and AB, whereas under homozygous variant condition, the PCR amplification will generate two products, VB and AB. The PCR amplification will generate all the three PCR products, WA,VB and AB, under heterozygous condition. The 5′-end non-complementary GC-rich tail sequence mentioned earlier enables the quick switch of PCR from less efficient amplification of genomic DNA to highly efficient amplification of PCR products or amplicons generated in the previous amplification cycle. The 5′-end tail also prevents the allele-specific PCR products or amplicons generated in the previous amplification cycle (WA and VB from example) from acting as primers for a larger DNA template such as AB (from example) or genomic DNA, a process known as megapriming. The bidirectional PASA technique is a rapid, highly efficient and accurate method for the detection of SNPs. However, this technique requires two parallel polymerase chain reactions to be set up, each specific to one allelic form of the SNP. Further, this technique involves multiple primers, which makes it difficult and sometimes impossible to optimize the PCR for an efficient or at least sufficient amplification of all the target DNA sequence fragments (Liu, Thorland et al. 1997, Waterfall and Cobb 2002).

The other modification of the AS-PCR technique involves the use of allele-specific primers that differ in length from one another and generate products of different size that are easily distinguishable from each other through agarose gel electrophoresis or PAGE. However, this version of AS-PCR technique often requires extensive optimization of PCR conditions, including the optimization of molar concentrations of the AS primers (Dutton and Sommer 1991, Okimoto and Dodgson 1996).

1.2.1.1.4.7 Multiplex Single-Base Primer Extension Methods The single-base primer extension (SBE) method also known as minisequencing method is a highly flexible, highly sensitive, robust, very specific and accurate method based on allelic discrimination mechanism. This method uses SBE and standard capillary electrophoresis (CE) for SNP genotyping, including tetra-allelic SNPs. This method involves the use of a small number of simple-to-design primers and probes. The SBE method enables the detection of known or previously reported genetic variations including SNPs, point mutations and small indels (insertions and deletions) in PCR-amplified DNA fragments. The SBE method uses two types of primers, multiplex PCR primers and SBE primers, and based on these primers, it involves two types of PCR-based amplifications. The design of multiplex PCR primers allows them to anneal very close to, but not include the target SNP site. The target SNPs therefore occur in the regions between the primer-binding sites, included in the amplicon, but not in the primer sequences or the primer-binding sites. The SBE method uses multiplex PCR primers for initial multiplex PCR amplification that amplifies the DNA sequence fragments encompassing the target SNP sites. The amplified DNA fragments or amplicons are usually smaller in size, around 50 bp, but necessarily include the target SNP. The smaller size of amplicons leads to higher PCR amplification efficiencies as they enable efficient amplification even with starting DNA samples that are degraded and low in concentration or copy number. The amplified DNA fragments or amplicons generated by multiplex PCR undergo purification to remove any unincorporated primers and dNTPs. The purified multiplex PCR amplicons then undergo SBE PCR. The design of SBE PCR primers is similar to that of a typical sequencing primer and targets the purified multiplex PCR-amplified products and uses them as template. The SBE PCR primer binds to the purified PCR amplicon template in a 5′- to 3′-orientation. The 3′-end of the SBE primer binds to the template upstream of, but immediately adjacent or next to the target SNP site. The SBE PCR uses dideoxyribonucleotides (ddNTPs) instead of deoxyribonucleotides (dNTPs). The ddNTPs lack the 3′-OH group required for the formation of a phosphodiester bond and include ddATP, ddGTP, ddCTP and ddTTP corresponding to dATP, dGTP, dCTP and dTTP, respectively. When each ddNTP, labeled with a different fluorescent dye (earlier labeled radioactively), usually green (ddATP), black (ddGTP), blue (ddCTP) and red (ddTTP), adds onto the primer annealed to template DNA fragment, it emits a fluorescent signal typical of the fluorescent dye that labels it. During SBE PCR, the SBE PCR primer binds to the purified PCR amplicon template. The SBE PCR uses ddNTPs instead of dNTPs. The DNA polymerase-mediated random, but complementary incorporation of ddNTPs onto the template terminates the extension at that position after the addition of just one complementary base because there is no phosphodiester bond formation beyond the position of the incorporated ddNTP. During the SBE PCR, the SBE PCR primer binds to the purified PCR amplicon template and the DNA polymerase incorporates the fluorescently labeled ddNTP onto the template. The addition of fluorescently labeled ddNTP occurs immediately downstream or onto the base next after the annealed primer and complementary to the base or nucleotide in the SNP site that corresponds to the one or the other allelic form of the SNP. The type of fluorescence emitted allows the identification of incorporated nucleotide and the complementary nucleotide to which it binds on the purified PCR amplicon template, and this reveals the allelic form of the SNP present in the DNA sample under analysis. The SBE PCR products undergo purification to remove any unincorporated primers and dNTPs that could interfere with detection and analysis. The purified SBE PCR products then undergo CE that generates electropherogram and allows the detection and analysis or genotyping of SNPs. The analysis of electropherograms requires commercially available software applications and enables identification of SNP locus based on the length of SBE primer and identification of SNP allele based on the distinct fluorescent dye used to label one of the four ddNTPs (Sokolov 1990, Pastinen, Kurg et al. 1997, Syvänen 1999, Fiorentino, Magli et al. 2003, Vallone, Just et al. 2004, Phillips, Salas et al. 2007).

The multiplex SBE PCR coupled with CE allows simultaneous genotyping of multiple SNPs. The multiplex SBE PCR protocol involves the addition of a nonbinding tail sequence, usually a poly-T or multiple repeats of AGCT sequence to the 5′-end of the SBE (sequencing) primer. The use of tail sequence comprising of poly-T or multiple repeats of AGCT sequence results in altered electrophoretic mobility of the SBE primers and subsequently SBE PCR products and allows distinction of one SBE PCR product from another in a multiplex reaction. A difference of three to four bases between two SBE PCR primers allows a sufficient resolution to distinguish one SBE PCR product from another using capillary gel electrophoresis (CGE). However, the electrophoretic mobility and consequently the electrophoretic separation of small SBE PCR primers and SBE PCR products also depend on the primer sequence and the specific fluorescent dye that labels the incorporated ddNTP. The length of a SBE PCR primer varies from 20 to 80 nucleotides. The capacity of automated oligonucleotide synthesizers largely defines the limits of SBE PCR primer length. The multiplex SBE PCR in comparison with single SBE PCR allows simultaneous genotyping of multiple SNPs that minimizes the cost, effort and time required for this genotyping method. It also reduces the amount of the DNA sample required for this assay. The multiplex SBE PCR may allow simultaneous genotyping of up to 30 SNPs spread across the genome, but this method has inherent limitations in simultaneous genotyping of more than five SNPs. These inherent limitations include increased frequency of primer dimer formation, increased formation of undesired PCR-amplified products, unequal amplification efficiency and requirement of extensive, often difficult-to-achieve optimizations in primer and probe design and reaction condition optimizations (Sokolov 1990, Pastinen, Kurg et al. 1997, Syvänen 1999, Fiorentino, Magli et al. 2003, Vallone, Just et al. 2004, Phillips, Salas et al. 2007).

1.2.1.1.4.8 Pyrosequencing The in vivo or in vitro DNA polymerase-catalyzed elongation of DNA involves the incorporation or addition of deoxynucleoside triphosphates (nucleotides) onto the 3′-end of the extending primer annealed to the DNA template strand, and the incorporation of each nucleotide into the growing DNA strand releases one inorganic pyrophosphate (PPi) moiety as a by-product. Pyrosequencing is a luminescence-based DNA sequencing technique that converts the PPi moiety generated during DNA polymerase-catalyzed elongation of DNA into visible light. This visible light generated whenever a new nucleotide is incorporated into the growing DNA strand is measurable in real-time and at any given time is proportional to and reveals the number and nature of nucleotides incorporated into DNA. This technique detects the nature, order and number of nucleotides incorporated into the growing DNA strand and consequently reveals the sequence of the DNA strand that acts as template for the addition of new nucleotides and elongation of the new strand. Pyrosequencing allows qualitative and quantitative sequence read of nucleotides ranging in numbers from a single nucleotide up to more than 200 nucleotides. The simultaneous and parallel pyrosequencing involving tens of thousands of sequencing events enables assembled sequence read of nucleotides up to 20 Mb in length in a single run. This technique uses a cascade of enzyme-catalyzed reactions that convert each PPi released on incorporation of a nucleotide into the growing DNA strand to ATP. The ATP produced from inorganic pyrophosphate participates in a luciferase reaction that generates visible light with amount or intensity proportion to the amount of ATP channelized into luciferase reaction. In this manner, this technique links the process of incorporation of new nucleotides into the growing or elongating strand to the proportional generation of visible light that is measurable in real-time and gives the sequence of the template strand (Ronaghi, Uhlén et al. 1998, Ronaghi 2001).

Pyrosequencing uses three different types of enzymatic cascades to convert PPi into measurable visible light via the formation of ATP. These include two three-enzyme cascades and one four-enzyme cascade. One of the three-enzyme cascades consists of enzymes, DNA polymerase, sulfurylase or ATP sulfurylase (ATPS) and luciferase, whereas the other three-enzyme cascade consists of enzymes, DNA polymerase, pyruvate phosphate dikinase (PPDK) and luciferase. The four-enzyme cascade consists of enzymes, DNA polymerase, sulfurylase, luciferase and apyrase. The general procedure of pyrosequencing consists of three main steps. The first step involves the PCR amplification of the genomic DNA sequence segment under study. The quality and quantity of PCR products are vital to ensure the reliability, accuracy and reproducibility of this sequencing technique. The second step consists of template preparation that involves separating the two complimentary strands of double-stranded PCR product into two single-stranded DNA sequence fragments, isolating the single-stranded DNA fragments, maintaining the separated or isolated DNA fragments in single-stranded conformation and annealing or hybridization of the sequencing primers to these single-stranded DNA fragments. The technique uses the high immobilization capability of solid-phase matrix consisting of streptavidin-coated Sepharose beads and denaturing conditions to isolate the single-stranded DNA fragments and to immobilize and maintain the isolated DNA fragments in single-stranded conformation. This technique under normal circumstances uses only one of the strands as template. To achieve this immobilization, the PCR amplification carried out initially involves the use of three primers: two sequence-specific amplification primers (reverse and forward primers) and a universal biotinylated primer or biotinylated M13 primer. The use of the third biotinylated primer ensures the subsequent isolation and immobilization of single-stranded DNA fragments through highly specific and one of the strongest known non-covalent interactions, biotin-streptavidin interaction ($K_d = 10^{-14}$ to 10^{-15}M). The third step involves the DNA polymerase-catalyzed incorporation of user-defined nucleotide onto the 3′-end of the extending primer annealed to the DNA template strand. The incorporation of each nucleotide releases an inorganic pyrophosphate (PPi) moiety. The enzyme ATP sulfurylase used in one of the three-enzyme cascades and in the four-enzyme cascade converts the PPi to ATP using adenosine 3′-phosphosulfate (APS) as substrate. In the other three-enzyme cascade, the enzyme PPDK converts inorganic pyrophosphate to ATP using phosphoenolpyruvate (PEP) as substrate. The ATP produced from inorganic pyrophosphate participates in a luciferase reaction that generates visible light with amount or intensity linearly proportion to the amount of ATP channelized into luciferase reaction and the nucleotides incorporated into the growing DNA strand. A photomultiplier tube, a photodiode and the more commonly used charge-coupled device (CCD) camera detects and measures the amount of light generated that in turn reveal the sequence. The two three-enzyme cascades use washing to remove the unreacted extra remaining nucleotides, whereas the four-enzyme cascade uses the enzyme apyrase to remove the excess of nucleotides. The removal of the unreacted extra remaining nucleotides prepares the reaction mixture for the addition of next nucleotide and is an essential step for this cyclic process of sequencing by synthesis (SBS) that characterizes the pyrosequencing technique. The pyrosequencing instrument detects the light generated on incorporation of each nucleotide into the growing strand that enables the sequencing of the template strand. The pyrosequencing instrument generates unprocessed raw sequencing data in the form of Pyrograms, and the processing of this data by pyrosequencing software reveals the sequence of the DNA fragment and the genotype of the SNP under analysis. The automated pyrosequencing instruments mainly use four-enzyme cascades that make this enzyme cascade a commonly used pyrosequencing enzyme system. However, the three-enzyme cascades allow application of pyrosequencing in large-scale procedures and the use of washing ensures no leftover nucleotides remain from the previous sequencing cycle (Langaee and Ronaghi 2005, Royo, Hidalgo et al. 2007).

The pyrosequencing comes with many advantages. The pyrosequencing is a rapid, highly accurate and flexible technique that allows parallel processing using simple automation. This technique has minimal requirement for labeled reagents such as labeled nucleotides and labeled primers and is not dependent on gel electrophoresis-enabled visualization. Pyrosequencing is a highly useful and versatile technique for the detection and analysis of known or previously reported and unknown or previously unreported genetic variations including SNPs, mutations, tri-allelic variants and small indels (insertions and deletions). This technique also allows determination of allele frequencies and relative quantitative determination of the amount of each allelic form (allele quantification assays) of SNPs or other genetic variations, provided the different DNA samples are equimolar. The multiplex pyrosequencing enables the simultaneous detection and analysis or genotyping of SNPs on multiple sites within the same PCR-amplified DNA target sequence fragment or on different PCR-amplified DNA fragments (Guo, Qi et al. 2003, Lavebratt and Sengul 2006). Pyrosequencing is a frequently used technique for the detection and analysis of DNA sequence regions undergoing loss of heterozygosity (LOH). Besides, this technique finds a wide use in sequencing of CpG islands, analysis of DNA methylation within CpG islands, heteroplasmic DNA sequencing, resequencing of specific genomic regions, cDNA sequencing, sequencing of sample co-infected with multiple viruses and determination of secondary DNA structures. Pyrosequencing is a useful technique for molecular haplotyping of SNPs and mutations. Although pyrosequencing is a suitable technique for low-to -medium throughput analysis of SNPs and mutations, it is not a desirable and well-suited technique for large-scale throughput analysis. Moreover, this technique requires high cost input in particular, with procedures that require analysis of a large number of SNPs, and therefore, this technique is not cost-effective when compared to other genotyping techniques. Further, although pyrosequencing allows the real-time follow-up of reaction progress, in essence, the technique generates real analysis at the end of the reaction. This technique also requires well-optimized PCR amplification program, and the technique requires very high quality of the starting DNA as the poor quality adversely affects the accuracy of the results (Loeuillet, Weale et al. 2007, Marsh 2007, Lee, Issa et al. 2008).

1.2.1.1.4.9 TaqMan Assay (5′-Nuclease Allelic Discrimination Assay) The TaqMan SNP genotyping assay, also known as the 5′-nuclease allelic discrimination assay or simply 5′-nuclease reaction, is a closed-tube high-throughput technique that is highly suitable and frequently used for the simultaneous detection and analysis of multiple, but smaller number of SNPs in a large sample population. The basis of this technique is that the cleavage of a fluorescent reporter dye from a short allele-specific oligonucleotide (ASO) probe mediated by intrinsic 5′-nuclease (exonuclease) activity of Taq DNA polymerase generates a fluorescent signal or leads to increased fluorescence that reveals the presence of the allelic form of SNP complementary to the probe. This technique uses two sequence-specific (forward and reverse) PCR primers and two oligonucleotide probes each labeled with a different fluorescent reporter dye and each specific to one or the other allelic form of the SNP. Each TaqMan assay probe consists of a unique fluorescent reporter dye (donor fluorophore) attached to its 5′-end and an acceptor fluorophore (fluorescence quencher) attached to its 3′-end. Each probe contains a different fluorescent reporter dye such that each fluorescent reporter dye emits fluorescence of different wavelengths, but the fluorescence quencher is common for both the probes. The commonly used fluorescent reporter dyes include 6-carboxyfluorescein (6-FAM), 5- and 6-carboxyfluorescein (5, 6-FAM), 2′-chloro-7′-phenyl-1,4-dichloro-6-carboxy fluorescein (VIC), 4,7,2′,7′-tetrachloro-6-carboxy fluorescein (TET) and 6-carboxy-4′,5′-dichloro-2′,7′-dimethoxy fluorescein (JOE). All these reporter dyes emit fluorescence of different wavelengths. The commonly used fluorescence quenchers that emit little or no fluorescence include 6-carboxytetramethylrhodamine (TAMRA) and non-fluorescent quencher (NFQ). The most commonly used combination involves the use of FAM and VIC as two unique fluorescent reporter dyes and TAMRA as fluorescence quencher. The length of each probe is short enough to allow the fluorescent reporter dye and fluorescence quencher to be in close proximity to each other in the intact uncleaved probe whether in solution or when hybridized. The close proximity between the reporter dye and the quencher enables the quencher to absorb the fluorescence emitted by the reporter dye. The suppression of fluorescence emitted by reporter dye by fluorescence quencher primarily occurs through Förster (fluorescence) resonance energy transfer (FRET). Often, the design of TaqMan assay probe consists of a minor groove binder (MGB) attached near its 3′-end. The MGBs are small crescent-shaped molecules or groups that fit into the minor grooves of dsDNA. The MGBs greatly enhance the stability and specificity of probes to enable efficient hybridization. Each TaqMan assay probe consists of a sequence that is highly complementary to one or the other allelic form of the SNP such that each of the two probes uniquely labeled with a fluorescent reporter dye is complementary to one of the two allelic forms of the SNP. The region of the target DNA sequence that is complementary to the sequence of each probe lies between the regions where the primers bind to the target sequence of template DNA. The design of TaqMan assay probes usually is such that each probe has a higher annealing temperature in comparison with each of the PCR primers. This ensures that the probes will bind or hybridize to target sequence in the template DNA only after the primer extension phase of the PCR begins. The use of MGB in probes makes it possible to have shorter sequence probes with higher annealing temperatures.

During the primer extension phase of the PCR, the probe hybridizes to the target sequence in the template DNA between the regions complimentary to each of the two PCR primers. When the target sequence in the template DNA to which one of the two probes hybridizes is perfectly complementary to the sequence of the hybridized probe, the 5′-nuclease (exonuclease) activity of Taq DNA polymerase cleaves the fluorescent reporter dye from the probe and releases it into the solution that separates it from the fluorescence quencher. The release of the fluorescent reporter dye and its separation from the fluorescence quencher results in a significant increase in fluorescence that corresponds to one of the two fluorescent reporter dyes and reveals the presence of the allelic form of the SNP specific to either the labeled probe or the other differently labeled probe. However, when the target sequence in the template DNA to which one of the two probes hybridizes contains a single-base mismatch to the sequence of the hybridized probe, the probe binds or hybridizes to the target region in the template DNA inefficiently and with low specificity and stability. The strand displacement that occurs during the primer extension phase of the PCR destabilizes the inefficiently bound or inefficiently hybridized probe or even releases the intact probe from the template. As a result, there is little or no cleavage of the probe due to the 5′-nuclease (exonuclease) activity

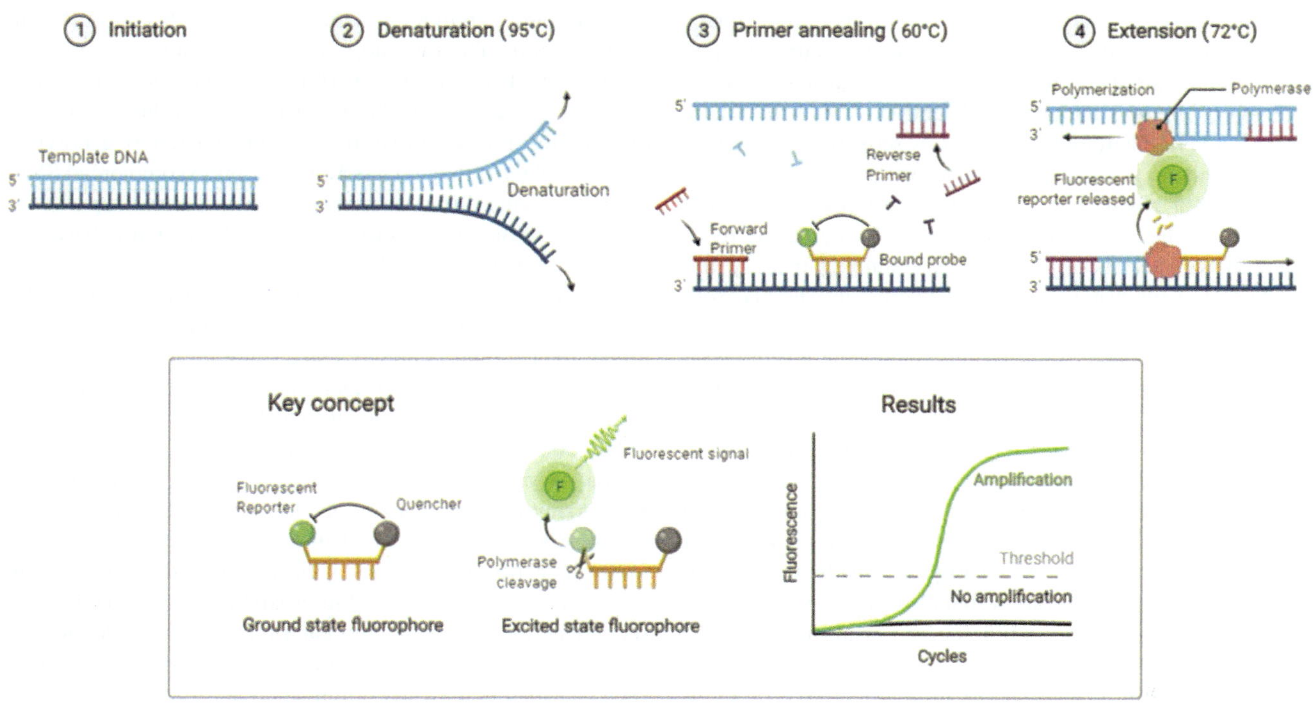

FIGURE 1.4 The pictorial representation of the TaqMan SNP genotyping assay. The TaqMan SNP genotyping assay uses the cleavage of a fluorescent reporter dye from a short allele-specific oligonucleotide (ASO) probe mediated by intrinsic 5′-nuclease (exonuclease) activity of Taq DNA polymerase that generates a fluorescent signal or leads to increased fluorescence that reveals the presence of the allelic form of SNP complementary to the probe.

of Taq DNA polymerase, and consequently, there is little or no increase in fluorescence. Therefore, a significant increase in fluorescence occurs only when the target sequence is perfectly complementary to the sequence of the hybridized probe revealing the presence of the allelic form of the SNP specific to the probe. The nature of the fluorescent signal emitted by the reaction mixture reveals the allelic form or the genotype of the SNP. The increase in fluorescent signal corresponding to both the fluorescent reporter dyes used to label the probes reveals the presence of both the allelic forms of the SNPs in the given DNA sample. For example, consider A and B represent two allelic forms of the SNP and the fluorescent reporter dyes FAM and VIC label, respectively, the probes complementary or corresponding to A and B alleles. The increase in blue fluorescent signal corresponding to FAM fluorescent reporter dye reveals the presence of 'A' allele of the SNP in homozygous AA genotype, whereas the increase in green fluorescent signal corresponding to VIC fluorescent reporter dye reveals the presence of 'B' allele of the SNP in homozygous BB genotype. The increase in both the blue and green fluorescent signals, which indicates the successful hybridization and subsequent efficient cleavage of both the FAM- and VIC-labeled probes, reveals the presence of heterozygous AB genotype of the SNP. The TaqMan SNP genotyping assay is an advanced, high-throughput, widely used and thoroughly validated SNP genotyping technique. This technique is easier to optimize and tolerates the differences in the quality and quantity of the sample DNA. The TaqMan assay allows automation and is highly accurate with very low genotyping errors. The technique involves a closed-tube setup and requires no end-point processing, and the results involve the relatively simple detection and analysis of emitted fluorescence. The only possible disadvantage or deterrent in the use of this technique is the cost of TaqMan probes (Holland, Abramson et al. 1991, Latif, Bauer-Sardina et al. 2001, Livak 2003) (Figure 1.4).

1.2.1.1.4.10 SNP Microarray (SNP Array or SNP Chip) Technique The SNP microarray or SNP array technique is a DNA microarray-based high-throughput genotyping method that allows qualitative, quantitative and simultaneous analysis of a large number of pre-defined and known or previously reported genetic variations including SNPs and CNVs. The SNP microarray involves hybridization of fluorescently labeled fragmented single-stranded DNA sequences to arrays that contain hundreds of thousands of unique allele-specific oligonucleotide (ASO) probe sequences each complimentary or specific to and designed to bind to target DNA sequences that contain regions corresponding to one or the other allelic form of the SNPs under study. The binding of ASO probe to a matched or mismatched target DNA sequence generates a fluorescent hybridization signal whose intensity depends on the amount of target DNA sequence present in the sample and the degree of affinity between the hybridized ASO probe and target DNA sequence that indicates matched or mismatched pairing. A specialized instrumentation system detects and records the intensity of each emitted fluorescent signal that results from each hybridization reaction or the binding of each probe to its target DNA sequence. A software application using computational algorithms simultaneously processes and analyzes extensively the complex and raw fluorescent signal intensity data consisting of fluorescent signals from thousands of hybridization reactions and converts them into inferences that reveal the presence or absence of one or the other allelic forms or the presence of both allelic forms of the SNPs under study. The design of an array includes at least two probes for each SNP to enable the simultaneous detection of both the allelic forms of a SNP. Therefore, a single array setup reveals whether a SNP is homozygous for one allele or homozygous for the other allele or whether both the alleles of a SNP are present together in a heterozygous form. The SNP microarray technique allows analysis of a single DNA sample per array and compares the intensities of emitted fluorescent signals from numerous hybridization reactions with the intensities of a cluster of emitted fluorescent signals obtained from a setup consisting of reference samples such as whole study population. The comparison of intensities of emitted fluorescent signals from sample DNA with the intensities of emitted fluorescent signals from reference samples gives a log ratio that generates the inferences and reveals the presence or absence of one or the other allelic form or the presence of both allelic forms of the SNPs under study. As mentioned already, the intensities of fluorescent hybridization signals depend on the amount of target DNA sequence and the degree of affinity between the hybridized ASO probe and target DNA. Therefore, the SNP array generates two types of inferences based on emitted fluorescent signals: the total fluorescence obtained from the intensities of fluorescent signals from both the alleles put together and the relative fluorescence obtained from the comparison of intensity of fluorescent signal emitted by one allele to the intensity of fluorescent signal emitted by another allele. The latter reveals the allelic ratio. The normalized intensity of fluorescent signal emitted by each allele plotted against one another generates any of the three types of fluorescent intensity clusters, each of which corresponds to one of the three different allelic or genotypic forms in which the particular SNP occurs in the DNA sample under study. For example, if we consider A and B represent the two allelic forms of the SNP under study, then one type of the fluorescent intensity cluster corresponds to homozygous AA genotype, the other type of fluorescent intensity cluster corresponds to homozygous BB genotype, and the third type of fluorescent intensity cluster corresponds to heterozygous AB genotype of the SNP. A single SNP array analysis will generate only one of the three fluorescent intensity clusters depending on whether the SNP occurs in homozygous AA genotypic form, homozygous BB genotypic form or heterozygous AB genotypic form. Another version of SNP microarray based on the use of allele-specific primer (ASP) extension or single-base extension to differentiate one allelic form of the SNP from another uses ASPs instead of ASO probes discussed above (McCarroll, Kuruvilla et al. 2008, LaFramboise 2009).

The emergence of SNP databases such as HapMap and their continuing upgrades have enabled an easy access to the updated information regarding the number of SNPs present in the human genome, their mapped location within the genome and their relevance to or association with various human diseases and traits. All this has revolutionized the SNP microarray technique. The data from SNP databases containing the information regarding large numbers of SNPs have enabled the generation and commercialization of high-throughput human SNP microarray chips, each of which allows simultaneous detection and genotyping of millions of SNPs and CNVs in a single setup. The two most widely used and commercially available SNP microarray platforms include the Affymetrix SNP microarray and the Illumina BeadArray, both of which offer more or less similar throughput and report genotypic accuracy of more than 99.5%. However, these commercial SNP array platforms differ from each other in several aspects, in particular in the methods they use for differentiating between the two allelic forms of a SNP.

The Affymetrix SNP microarray, the first commercially available SNP array, started from an array that contained probes to genotype 1494 SNPs on a single chip. Gradually, the SNP genotyping capacity increased to 10,000 (10k) SNPs on a single chip and from 10,000 (10k) to 100,000 (100K) and then to 500,000 (500K) SNPs on a single chip. The current version of the Affymetrix SNP microarray, the Affymetrix Human SNP Array 6.0, contains probes for genotyping close to one million (906,600) SNPs, and in addition, it contains probes for the detection and analysis of close to one million (946,000) CNVs. The Affymetrix SNP microarray uses a set of four to six probes for each SNP, and each probe is 25 nucleotide or 25-mer long. A single probe set consists of at least two probes, each of which is perfectly complementary to one or the other allelic form of the SNP under study. The probe set also consists of probes that contain a single mismatch and are not perfectly complementary to either of the two allelic forms of the SNP. These probes aid in the assessment and measurement of background noise. The Affymetrix SNP microarray technique uses the process of differential hybridization or differential binding of each of the probes in a 25-mer probe set to one or the other allelic form of the SNP in a perfectly matched condition or a mismatched condition to reveal SNP genotype. The basis of differential hybridization is the difference in melting temperatures between the hybridized matched probes and targets and hybridized mismatched probes and targets. Therefore, the probes used in the Affymetrix SNP microarray need to be highly effective in differentiating between the matched targets and mismatched targets. The sequence of probes used in the design of Affymetrix SNP microarray largely depends on the local sequence of the SNPs or the sequence encompassing the SNPs under study. Therefore, the Affymetrix SNP microarray does not include probes targeting regions with unusual local sequences such as those with a very low or very high GC content and even the tagging SNPs such as the SNPs that occur in the regulatory regions or in the coding regions of the gene. Therefore, the Affymetrix SNP microarray technique uses a hybrid approach for the selection of SNPs: half of the targeted SNPs are tagging SNPs, and half are random SNPs. The reason behind this exclusion is that at a specific temperature of hybridization, strong differential hybridization between matched probes and targets and between mismatched probes and targets is not achievable in case of unusual local SNP sequences. In Affymetrix SNP microarray technique, all the probes (in millions) hybridize to their target sequences under similar conditions of buffer, temperature and time period making this array technique an ideal method for high-throughput SNP and CNV genotyping. If we consider A and B represent the two allelic forms of the SNP under study, PM represents the perfectly matched probe, and MM represents the mismatched probe, then the Affymetrix SNP microarray technique generates four possible types of fluorescence intensities, PM_A, PM_B, MM_A and MM_B. PM_A corresponds to the binding of A-allele of the SNP to its perfectly matched probe, PM_B corresponds to the binding of B-allele of the SNP to its perfectly matched probe, MM_A corresponds to the binding of A-allele of the SNP to its mismatched probe, and MM_B corresponds to the binding of B-allele of the SNP to its mismatched probe. This quartet of intensities of fluorescent signals constitutes the raw data. A software application using computational algorithms processes these raw data and converts it into inferences that reveal the presence or absence of one or the other genotypic forms of the SNP: homozygous AA genotype, homozygous BB genotype or heterozygous AB genotype (LaFramboise 2009).

The Illumina BeadArray is another commercially available SNP array that started from an array containing probes to genotype 100,000 (100k) SNPs on a single chip. Gradually, the SNP genotyping capacity increased to 240,000 (240k) SNPs on a single chip and then to 317,000 (317k) SNPs on a single chip. From 317,000 (317k), the SNP genotyping capacity increased to 550,000 (550K) and then to 650,000 (650K) SNPs on a single chip. The current version of the Illumina BeadArray, the Illumina BeadArray High-Density Human 1 M-Duo or the Illumina BeadArray HumanHap1M contains ASP probes for genotyping one million SNPs. The Illumina BeadArray technique uses the process of ASP extension or single-base extension to differentiate one allelic form of the SNP from another unlike the differential hybridization of the Affymetrix SNP microarray. This technique uses ASPs that undergo primer extension instead of ASO probes that undergo differential hybridization. The ASP undergoes extension and emits fluorescent signal only when the 3′-end of primer pairs to its perfectly matched or fully complementary sequence on the target DNA fragments rather than when the 3′-end of primer pairs to a mismatched sequence on the sample DNA fragment. This technique uses differentially and fluorescently labeled dNTPs for primer extension reaction. The fidelity of this enzyme-catalyzed ASP extension to differentiate one allelic form of the SNP from another is extremely high and independent of the local sequence of the SNPs or the sequence encompassing the SNPs under study. This makes the Illumina BeadArray technique less dependent on the local sequence of the SNPs and increases its capacity of SNP genotyping by making it applicable to increased number of SNPs. However, unlike the Affymetrix SNP microarray, this technique involves additional steps of primer extension and staining to allow the detection of signals that complicates the possibility of full automation of this technique. The Illumina BeadArray technique uses 50 nucleotide or 50-mer long probes for target binding or capture and enzyme-catalyzed primer extension all of which involve hybridization. Further, the design of each Illumina BeadArray consists of a unique layout of beads or probes in each SNP chip that requires an extra step to decode the geometric arrangement of the beads specific to the target SNPs. The Illumina BeadArray technique unlike the Affymetrix SNP microarray gives more importance to the tagging SNPs such as the SNPs that occur in the regulatory regions or in the coding regions of the gene in comparison with the random SNPs (LaFramboise 2009).

The SNP microarray technique is a high-throughput method that allows simultaneous detection of large number of known genetic variations such as SNPs and CNVs in a single DNA sample. This technique allows automation, requires minimal manual steps and is highly accurate with very low genotyping errors. This technique is highly suitable and widely used for whole genome association studies (WGAS) or genome-wide association studies (GWAS) and requires very little amount of starting DNA sample and genotypes very large number of SNPs in a single experiment. The SNP microarray technique is relatively cost-effective with low to moderate cost in comparison with the other more commonly used high-throughput NGS-based techniques. Further, this technique allows relatively easier and less time-consuming analysis of the generated data and does not require extensive knowledge of bioinformatics unlike the NGS-based techniques that involve time-consuming multistep and complex data analysis not possible without extensive knowledge of bioinformatics. In SNP microarray technique, the calling of genotypes, that is, the process of determining the genotype of the SNP, is much easier and the researcher has to refer to genotype calling pipeline of commercial SNP chip manufacturers, Affymetrix or Illumina, following the instructions in the user guide or the commercial SNP chip manufacturers itself facilitate the process of genotype calling. Another advantage of this technique is that the commercial SNP chip manufacturers, Affymetrix or Illumina, provide the option to custom-design the array to allow inclusion of specific genomic regions or genomic regions of interest in the array. This technique also allows a flexible choice of probes for genotyping a specific number of SNPs based on the genotyping requirements of the user. A major disadvantage of this technique is that it allows the analysis of only the known genomic regions to genotype previously reported or known genetic variations including SNPs. Further, the design and production of SNP microarray is complex and expensive that does not allow any change in the set of probes or markers once selected to detect specific number of and specifically located SNPs. In other words, the array once designed and produced does not allow any further change (Clevenger, Chavarro et al. 2015).

1.2.1.1.4.11 Sequencing Techniques The sequencing process involves identifying the four nitrogenous bases, also known as nucleobases or nucleotides, and determining their differential arrangement or order in the nucleotide polymers, known as nucleic acids that include DNA and RNA. The DNA and RNA sequencing involves determining the differential arrangement of purine nitrogenous bases, adenine (A) and guanine (G), and pyrimidine nitrogenous bases, cytosine (C) and thymine (T) in case of DNA and cytosine (C) and uracil (U) in case of RNA.

1.2.1.1.4.11.1 First-Generation Sequencing (1GS) Techniques

The sequencing process started in the year 1977 with the development of two sequencing techniques: the Sanger dideoxy synthesis method and the Maxam-Gilbert chemical cleavage method. The Sanger dideoxy synthesis method, also known as Sanger chain-termination DNA sequencing method, chain-termination method, conventional sequencing method or simply Sanger sequencing method, is sequencing-by-synthesis (SBS) method that uses a DNA polymerase to generate DNA sequence reads from the template obtained from the sample DNA under analysis. This technique uses specific primer extension, chain termination and electrophoretic separation to detect incorporation of a new nucleotide that reveals the sequence of polynucleotides or DNA fragments longer in size than oligonucleotides.

The Sanger sequencing method uses specifically modified dNTPs, known as dideoxyribonucleotides (ddNTPs), for chain-termination step. These ddNTPs lack the 3′-OH group required for the formation of a phosphodiester bond and include ddATP, ddGTP, ddCTP and ddTTP corresponding to dATP, dGTP, dCTP and dTTP, respectively. The DNA polymerase-mediated random, but complementary incorporation of ddNTPs into a growing polynucleotide chain terminates the extension at that position because there is no phosphodiester bond formation beyond the position of the incorporated ddNTP. When each ddNTP, labeled with a different fluorescent dye (earlier labeled radioactively), usually green (ddATP), black (ddGTP), blue (ddCTP) and red (ddTTP), is incorporated into the growing polynucleotide chain, it emits a fluorescent signal typical of the fluorescent dye that labels it. Briefly, the reaction mixture for chain-termination PCR in the Sanger sequencing method consists of double-stranded sample DNA (PCR product), sequencing primer (forward or reverse primer), dNTPs (dATP, dGTP, dCTP and dTTP), fluorescently labeled ddNTPs (ddATP, ddGTP, ddCTP and ddTTP) and thermostable DNA polymerase. The first step in the Sanger sequencing method involves the denaturation of double-stranded sample PCR product DNA into single-stranded DNA molecules. The single-stranded DNA then anneals to oligonucleotide sequencing primer that undergoes DNA polymerase-catalyzed extension or elongation using dNTPs to generate a new dsDNA product. The reaction mixture also contains small quantities of each of the four chain-terminating ddNTPs. Therefore, the DNA polymerase can incorporate either dNTPs or ddNTPs on a random basis.

The DNA polymerase-catalyzed primer extension or elongation continues using dNTPs until a random incorporation of a specific ddNTP into the growing nucleotide strand terminates the extension of the growing polynucleotide chain at the site of ddNTP incorporation. Since the incorporation of ddNTPs and the concomitant chain termination occur randomly, the repeated chain-termination steps in a chain-termination PCR generate numerous copies of target DNA sequence fragments of varying lengths. The gel electrophoresis separates the randomly generated DNA fragments based on size, and the use of differentially and fluorescently labeled ddNTPs enables the identification of dideoxyribonucleotide that cause the chain termination. The Sanger dideoxy synthesis method allows both manual sequencing and automated sequencing and uses radioactively labeled or fluorescently labeled ddNTPs for the detection or visualization in sequencing gels in case of manual sequencing or fluorescently labeled ddNTPs in case of sequencing carried out by automated sequencers. The Sanger sequencing method formed the basis of first-generation of automated DNA sequencers developed by Applied Biosystems (AB) (now part of Thermo Fisher Scientific Corporation, USA). The automated sequencers are fitted with lasers that detect the differential fluorescence emitted by ddNTPs and generate a specific fluorescent peak, for example green, black, blue or red, depending on the incorporated ddNTP to reveal the sequence of the target DNA fragment.

The current generations of the automated DNA sequencers use fluorescent dye-labeled chain terminators or fluorescent terminator dyes and CE and generate accurate sequences ranging in length from 600 to 1,000 base pairs. The software applications enable the analysis and interpretation of the sequencing data generated by these automated sequencers (Sanger, Nicklen et al. 1977). The Maxam-Gilbert method involves two main steps: specific chemical modification of radio-labeled bases or nucleotides followed by chemical cleavage of the sugar phosphate backbone of DNA at sites adjacent to the chemically modified bases. The denaturing PAGE separates the different fragments generated by the chemical cleavage of DNA based on size, and autoradiography allows the visualization by generating the pattern of PAGE-separated DNA fragments on an X-ray film (Maxam and Gilbert 1977). The Sanger sequencing method emerged as the most prevalent and highly adopted sequencing technique in comparison with the Maxam-Gilbert method because of several advantages. These include high reliability, relative technical simplicity, low use of toxic or hazardous chemicals, low or no use of radioactivity and relative ease in scale-up and automation. Despite several modifications to the original Maxam-Gilbert method including the elimination of toxic or hazardous chemicals from the protocol and the emergence of new latest sequencing techniques, the Sanger sequencing method remains a standard in sequencing.

The Sanger sequencing technique is a highly accurate technique for precise low- to-medium-throughput DNA sequencing of multiple genes, specific genes, gene fragments and other regions of the genome. This technique considered as a gold standard (benchmark) method for the detection and analysis of known and unknown genomic alterations including point mutations and small indels (insertions or deletions) is a robust and widely used technique for SNP genotyping. The Sanger sequencing technique serves as a frequently used orthogonal method to corroborate the findings including the SNP genotyping data from the latest sequencing techniques. This is because the Sanger sequencing technique remains nearly unmatched in its accuracy and its ability to generate longer sequence reads. This technique allows sequencing including SNP genotyping of any region of the genome including the regions of the genome that remain weakly accessed or poorly analyzed by the latest sequencing techniques such as the genomic regions rich in GC content. The Sanger sequencing technique serves as an important and frequently used sequencing technique in diagnostics, and for years, its use has revealed the mutational profile of tumors and enabled the identification of causative genetic variations. Therefore, this technique remains an integral part of clinical genomics. However, the Sanger sequencing technique is highly costly and labor-intensive in comparison with other multiplex sequencing or genotyping techniques and the latest sequencing methods that make this technique unsuitable for high-throughput sequencing procedures like WGS. This technique has low sensitivity and does not allow precise quantification.

The original Sanger sequencing method, the original Maxam-Gilbert method, and some of their subsequent modifications excluding the development of automated sequencers constitute the **first-generation sequencing (FGS/1GS) techniques.** The invention and development of automated sequencers based on Sanger sequencing technique paved the way for the subsequent generations of sequencing techniques, known as **next-generation sequencing (NGS) techniques**. Therefore, the development of the original Sanger sequencing method marked the beginning of FGS, whereas its full automation that led to the development of automated sequencers laid the foundation of **NGS**.

1.2.1.1.4.11.2 Next-Generation Sequencing (NGS) or Massively Parallel Sequencing (MPS)

The NGS techniques include the multiple generations of sequencing techniques that emerged after the FGS or Sanger sequencing technique. The NGS techniques include second-generation sequencing (SGS/2GS) techniques, third-generation sequencing (TGS) techniques and the new emerging fourth-generation sequencing (4GS) techniques.

1.2.1.1.4.11.2.1 Second-Generation Sequencing (SGS/2GS)

The second-generation (next-generation) sequencing (SGS/2GS/NGS) also known as massive (massively) parallel sequencing (MPS) represents sequencing methods that encompass a number of sequencing techniques besides the inceptive sequencing technique of the generation, the automated Sanger sequencing. The second-generation next-generation sequencing techniques allow rapid, cost-effective, large-scale or high-throughput sequencing, for example, sequencing of whole genomes including the whole human genome. The NGS techniques enable rapid pace and optimal costs of sequencing using chemical reactions, optical detections and in silico analyses that run in a parallel manner as massively parallel processes, hence the name massively parallel sequencing. The SGS technique uses either of the two methods: sequencing by hybridization (SBH) and sequencing by synthesis (SBS). Besides these two methods, several other methods are also available and in use in various NGS platforms.

The SBH is an indirect sequencing method that uses specific oligonucleotide sequences (probes) that hybridize to the labeled complementary sequence regions of the target or to-be-sequenced DNA. The process involves repeated hybridization followed by washing to remove non-hybridized DNA that enables identification of the labeled DNA fragments with sequence complementary to the oligonucleotide probes that indirectly reveals the sequence of the hybridized DNA fragments. The identification of DNA fragments with sequence complimentary to the probes through repeated hybridization and washing steps allows generation of a large contiguous sequence from the sequence overlapping pattern revealed through the analysis of target DNA-probe hybridization spots. The SBH enables extensively parallel sequence determination that allows cost-effective, high-throughput sequencing procedures using miniaturized arrays of oligonucleotide probes that target specific DNA sequences. The modified SBH methods based on DNA ligation allow relatively small number of short probes to effectively probe and identify hundreds of thousands of longer sequence fragments in the target DNA. The sequencing techniques based on SBH method and its modifications use arrays of probes to analyze target DNA sequences and reveal sequence of DNA fragments several kilobase pairs in length. The SBH method-based sequencing techniques allow the sequencing of novel genomic DNA sequences for which no reference sequence is available, a process known as de novo sequencing, and also the sequencing of genomic DNA sequences for which reference sequence is available, a process known as resequencing. The SBH method-based sequencing techniques allow the identification of known and unknown SNPs, mutations and chromosome abnormalities including structural genetic variations such as CNVs, deletions, duplications and chromosomal rearrangements (Drmanac, Drmanac et al. 2002, Qin, Schneider et al. 2012).

The SBS method-based NGS techniques use modified Sanger sequencing method that differs from the original Sanger sequencing method in three major aspects. First, the modified Sanger sequencing method does not use dideoxyribonucleotides (ddNTPs) for chain termination; although some of the SBS-based sequencing techniques use reversible chain-terminating nucleotides (chain terminators). The SBS method involves repeated cycles of synthesis or incorporation of labeled nucleotides into the growing polynucleotide chain followed by the detection or identification of the incorporated labeled nucleotide using direct or camera-based imaging techniques.

The use of reversible chain terminators allows transient blocking of synthesis or incorporation of new nucleotides into the growing polynucleotide chain to enable the detection or identification of the previously incorporated nucleotide. Once the incorporation of nucleotide and its detection is complete, the removal of reversible chain terminator or temporary synthesis blocker allows the process of synthesis or incorporation of the next nucleotide into the growing polynucleotide sequence to begin again. The use of fluorescently labeled reversible chain terminators allows rapid detection or identification of the incorporated nucleotide through rapid direct imaging. Second, the modified Sanger sequencing method generates shorter DNA sequences, called reads, ranging in length from 25 to 500 base pairs in comparison with the original Sanger sequencing method that generates reads 600 – 1,000 bases (nucleotides) in length. Most of the NGS platforms based on SBS method use short-read DNA sequencers that generate DNA sequences or reads, ranging in length from 25 to 300 base pairs. These reads are very short in comparison with the millions of base pairs in a typical human chromosome that range from approximately 40 million base pairs (Chromosome 22) to 240 million base pairs (Chromosome 1). Therefore, the SBS method-based NGS techniques use MPS of hundreds of millions of short-read DNA sequences that enable extensive sequence coverage and generate high-throughput accurate sequence based on the detection or identification of the consensus sequence. Further, in order to overcome the problem of short-read DNA sequences and increase the sequence read lengths, many SGS platforms use paired (mate) end mapping method that involves generation of two sequence reads from opposite ends of a long DNA fragment or DNA insert to be sequenced. Although with this approach and using different methods, the SGS platforms can generate fragments ranging from few hundred base pairs to few (2 – 3) kilobase pairs, the most effective and frequently used approaches generate fragments few hundred base pairs in length. In contrast, the Sanger sequencing method uses cloning to generate DNA fragments or DNA inserts. The cloning using bacterial plasmids and bacterial artificial chromosomes (BACs) allows generation of DNA fragments or DNA inserts ranging in length from 2 to 150 kbp. Therefore, the SBS method-based NGS techniques allow limited length or size of both the sequence reads and inserts in comparison with the Sanger sequencing method. Third, the modified Sanger sequencing method is relatively less accurate in comparison with the original Sanger sequencing method that remains a highly accurate sequencing method (Mardis 2008, Fuller, Middendorf et al. 2009, Metzker 2010).

The various SGS techniques include whole genome sequencing (WGS), whole exome sequencing (WES), target sequencing, RNA sequencing (RNA-seq), chromatin immunoprecipitation sequencing (ChIP-Seq), RNA immunoprecipitation sequencing (RIP-seq), methylome (methylation) sequencing (methyl-seq or bisulfite-seq) and metagenomic sequencing. The various computational approaches or methods used in these SGS techniques include split read methods, read depth (depth of coverage) methods, read-pair or paired (mate) end mapping methods, whole exome assembly and whole genome assembly methods (Medvedev, Stanciu et al. 2009). The basic procedure used by various SBS-based SGS techniques involves distribution of target DNA molecules or DNA molecules to be sequenced into millions of wells or chambers in case of emulsion phase reaction or tethering or immobilization to specific spots on a solid substrate in case of solid-phase reaction. The DNA molecules undergo amplification either by emulsion or solid-phase PCR or by isothermal rolling circle amplification (RCA) methods. The amplified DNA molecules then undergo synthesis that involves incorporation of nucleotides into the growing polynucleotide chain. The use of fluorescently labeled nucleotides that emit fluorescence when incorporated into the growing polynucleotide chain or specific chemical reactions associated with the incorporation of nucleotides allows the detection or identification of the nucleotide through direct or camera-based imaging techniques. The SGS/2GS/NGS technique uses various commercial sequencing platforms based on different sequencing methods. These include 454 pyrosequencing (now discontinued) and GS Junior platforms (454/Roche) based on pyrosequencing method. The HiSeq, NovaSeq, NextSeq, MiSeq and MiniSeq platforms (Illumina/Solexa) based on SBS method that use removable or reversible fluorescently labeled chain terminators. The Sequencing by Oligonucleotide Ligation and Detection (SOLiD) platform (Life Technologies/Applied Biosystems now part of Thermo Fisher Scientific Inc., USA) based on modified SBH method that use DNA ligation. The Retrovolocity platform (BGI) based on modified SBH method known as the DNA nanoball sequencing (DNBSEQ™), a PCR-free rolling circle replication method that uses patterned nanoarrays and a combination of hybridization and ligation. The Ion Torrent System includes Ion Proton, Ion Personal Genome Machine (PGM), Ion S5 and ION S5 XL platforms (Life Technologies/Thermo Fisher Scientific Inc., USA) based on detection of hydrogen ions (protons) as changes in voltage or voltage spikes using semiconductor chips. These SGS platforms use either solid-phase PCR (SP-PCR) as in case of HiSeq, NovaSeq, NextSeq, MiSeq and MiniSeq platforms (Illumina) or emulsion PCR (Em-PCR) as in case of 454 pyrosequencing (Roche), SOLiD (Thermo Fisher Scientific Inc.,USA) and Ion Torrent platforms (Thermo Fisher Scientific Inc.,USA). Each of these SGS platforms differs in various sequencing and other characteristics besides the basic method used. These include maximum output per run, read lengths per run, number of reads per run, time taken for each run, raw error rate (accuracy of base calling), cost per base pair and cost of the platform itself. Further, each of these SGS platforms has its own inherent issues that affect the overall accuracy of the sequencing process and choice of the platform. The combined use of several of these sequencing platforms enables to overcome the limitations associated with each of these sequencing platforms (Buermans and den Dunnen 2014).

The SGS techniques use short-read sequencing approaches and enable cost-effective and highly accurate sequencing that makes these techniques highly suitable for clinical and diagnostic studies that involve detection of known or unknown genetic variations and for population studies. The SGS techniques generate or use short-read DNA sequences that require generation of hundreds of millions of short-read DNA sequences to enable extensive sequence coverage. Even with hundreds of millions of these short-read DNA sequences, the cost per nucleotide is lower by several orders of magnitude in comparison with the Sanger sequencing method. However, the NGS techniques are prone to context-dependent sequencing errors, some of which are not rectifiable even through the increase in the number of sequence reads or through the application of second best call approach. Further, the NGS techniques have limited accuracy in mapping repetitive elements, in differentiating between paralogous sequences, in determining the correct number of consecutive nucleotides in homopolymer nucleotide sequences and in phasing allelic variants. These NGS techniques also possess inherent bias toward GC sequences (GC bias) that adversely affects the genome assembly process especially in case of low sequence coverage in GC-rich or GC-poor genomic regions. Further, the SGS techniques are prone to PCR-induced amplification biases and the varied template extension leads to dephasing. Due to these reasons, the Sanger sequencing method is in routine use in combination with SGS and acts as an orthogonal method to corroborate the findings of various NGS technique-based platforms including the findings related to genomic alterations such as SNP genotyping data (Mardis 2008, Fuller, Middendorf et al. 2009, Metzker 2010).

1.2.1.1.4.11.2.2 Third-Generation Sequencing (TGS/3GS) or Large Fragment Single Molecule Sequencing

The TGS/3GS represents sequencing methods that enable sequencing of a single DNA molecule or DNA from a single cell using the physical properties of DNA. The TGS techniques allow single DNA molecule sequencing that eliminates the requirement for DNA amplification prior to sequencing or synthesis process. The TGS techniques like the SGS techniques allow MPS, but generate long-read DNA sequence lengths (10 – 50 kb or longer) in comparison with SGS techniques (25 – 500 base pairs). The TGS technique uses various commercial sequencing platforms based on different sequencing methods. The most prominent among these include PacBio sequencing also known as singe-molecule real-time (SMRT) sequencing (Pacific Biosciences, PacBio) based on SBS method. The SMRT sequencing-based PacBio RS II and PacBio Sequel sequencers use SMRT cells that contain 150,000 to one million ultra-microwells. Each ultra-microwell known as zero-mode waveguide (ZMW) consists of an engineered DNA polymerase immobilized at its bottom. The single strand of the target DNA molecule binds to the immobilized DNA polymerase. The technique uses phospho-linked fluorescent dyes to label four nucleotides uniquely or differentially. The incorporation of each of the fluorescently labeled nucleotides into the growing polynucleotide chain emits a different fluorescence depending on the nucleotide incorporated, and an imaging system captures the emitted fluorescence to detect or identify the incorporated nucleotide. The imaging system captures the fluorescence and detects or identifies the nucleotide in real time during its incorporation into the growing polynucleotide chain. The PacBio or SMRT sequencing allows sequencing of very large DNA fragments, 30 – 50 kb or longer in size, and thus generates long sequence reads for genome assemblies including whole genome assemblies. The PacBio sequencing inherently is prone to errors that limit its usefulness in the genotyping of genetic variations including point mutations and SNPs. However, the technique allows repeated sequencing and repeated alignment of individual sequence reads that generates highly accurate consensus sequence reads. Further, the PacBio SMRT sequencing allows rapid detection of different modifications including epigenetic modifications, for example, through identification of methylation sites (Flusberg, Webster et al. 2010, Pareek, Smoczynski et al. 2011, Fang, Munera et al. 2012, Murray, Clark et al. 2012, Rhoads and Au 2015, Zhang, Sun et al. 2015).

The other major TGS platforms include Helicos sequencing and HeliScope true single-molecule sequencing (tSMS) (Helicos Biosciences) and nanopore sequencing (Oxford Nanopore Technologies, ONT). The Helicos sequencing and HeliScope true single-molecule sequencing (tSMS) (Helicos Biosciences) based on combination of SBH and SBS methods uses a DNA polymerase and involve shearing of DNA followed by addition of a 3′-poly-A tail to the DNA fragments obtained from shearing process. The poly-A-tailed sheared DNA fragments then hybridize or attach to surfaces of disposable glass flow cells through 5′-dT (50) (poly (oligo)-dT) anchors. The technique uses differentially and fluorescently labeled nucleotides along with reversible chain-terminating nucleotides. The incorporation of one of the four fluorescently labeled nucleotides followed by a pause in extension using reversible chain-terminating nucleotides allows detection or imaging of the incorporated nucleotide by the HeliScope sequencer. The cyclic extension process resumes after the detection or identification of the incorporated nucleotide and the removal of the reversible chain terminator and continues until the completion of sequencing process. The HeliScope sequencer allows the process of SBS of hundreds of thousands of molecules to run in parallel. The HeliScope sequencers generate sequence read lengths with an average length of 35 bases and ranging in size from 25 to 60 bases. The HeliScope sequencing allows rapid, efficient and accurate RNA-seq to generate quantitative cell and tissue transcriptome, genome sequencing including human genome sequencing and detection or identification of point mutations and SNPs (Harris, Buzby et al. 2008, Pushkarev, Neff et al. 2009, Pareek, Smoczynski et al. 2011).

The nanopore sequencing uses either of the two types of nanopore systems: the biological nanopore system or the solid-state sensor-based nanopore system, each of which consists of electrically resistant nanopores that allow the passage of nucleic acid, DNA or RNA molecules through them. The passage of DNA or RNA molecules through nanopores changes or impedes the conductivity or flow of ion currents through the pores. The changes in the flow of current or the electrical impedance across the pores depends on the type or the shape of the molecules translocating or passing through the nanopores. Each of the four nucleotides that constitute nucleic acids, DNA and RNA due to their distinct shapes, produce or generate characteristic changes in the flow of current or electrical impedance when they pass through these nanopores. The detectors integrated into the sequencers detect these characteristic changes in the flow of current through the nanopores in real time to identify the individual nucleotides passing through the nanopore and ultimately sequence the target DNA template. The biological nanopore system, based on biological membrane architecture, consists of pore-forming transmembrane beta-barrel proteins embedded in synthetic lipid bilayer membranes that generate or form the nanopores. The proteins used to generate pores include modified alpha-hemolysin (α-hemolysin), a channel protein normally released by Gram-positive bacterium *Staphylococcus aureus*, and Mycobacterium smegmatis porin A (MspA) normally produced by another Gram-positive bacterium *Mycolicibacterium smegmatis*. The modifications include covalent attachment of an exonuclease (exonuclease I) to the extracellular end of the alpha-hemolysin protein and covalent attachment of cyclodextrin sensor to the inside surface of the nanopore formed by the protein. The exonuclease sequentially cleaves the nucleotides in the DNA loaded onto the membrane, and the cleaved nucleotides pass through the nanopore enabling their detection and identification. The biological nanopore system makes use of highly processive DNA polymerase phi29 as motor protein, and other proteins such as DNA helicase and oligonucleotides that bind DNA strands and differential concentration of ions across the membrane to enable unwinding of DNA and facilitate or drive the passage or translocation of nucleotides through the nanopores. The passage of single-stranded nucleic acids, DNA or RNA, through the nanopores occurs at a constant rate and continues for tens of thousands of nucleotides. The passage of a nucleotide through the pore impedes the flow of the current applied through the nanopore, and each of the four nucleotides generates a characteristic disruption pattern. The characteristic disruption in the flow of current recorded in real time detects the distinct nucleotide responsible for the disruption and subsequently reveals the sequence of the DNA or RNA strand. The solid-state sensor-based nanopore system consists of solid substrate made up of various metals or alloys. The pores of nanometer size or nanopores in the solid substrate allow the single-stranded nucleic acids, DNA or RNA, to pass through them. The sequencers based on the nanopore sequencing that use biological (protein) nanopores include portable MinION, benchtop GridION and PromethION (Oxford Nanopore Technologies, ONT). The Nanopore sequencers generate longer sequence read lengths with an average length of 10 kb although theoretically this technique allows very large DNA fragments with size hundreds of kilobase pairs to pass through the nanopore revealing their sequence. The nanopore sequencing allows detection or identification of different modifications including epigenetic modifications such as methylation. This sequencing also allows haplotyping, SNP genotyping, identification of mutations, PCR-free cDNA sequencing, direct RNA sequencing and high degree of scalability at a low cost. This sequencing has proven applicability in the field of metagenomics and environmental surveillance. The portable MinION in the International Space Station (ISS) enables sequencing of bacterial and viral genomes to reveal or identify their strains. Further, the nanopore sequencing requires minimal sample preparation and generates long sequence reads. There is no requirement for amplification and ligation prior to sequencing that avoids largely the bias associated with determining the size and content of the GC sequence regions. However, currently the nanopore sequencing technique has a very high error rate. Much of the errors originate from the problems associated with optimizing the speed or rate of DNA or RNA passage or translocation through the nanopore to allow reliable and accurate measurement of change in the conductivity of the ion current through the pore (Ammar, Paton et al. 2015, Greninger, Naccache et al. 2015, Laver, Harrison et al. 2015, Cao, Ganesamoorthy et al. 2016, Schmidt, Vogel et al. 2017). Besides the above-discussed techniques, the other emerging sequencing techniques include direct imaging of DNA nucleotide sequence using transmission electron microscopy (TEM) (Halcyon Molecular and ZS genetics), direct imaging of DNA nucleotide sequence using scanning tunneling microscopy (STM) (Revo) and transistor-mediated DNA sequencing (IBM). Discussing these sequencing techniques in detail is beyond the scope of this chapter.

The emerging or TGS techniques largely based on single-molecule sequencing offer several advantages over the current SGS techniques and the previous FGS techniques. The TGS techniques are cost-effective, require minimal amount of starting sample material (theoretically, a single DNA or RNA molecule) and usually do not require amplification and in most cases not even ligation that simplifies and minimizes sample preparation and largely avoids the bias associated with determining the size and content of the GC sequence regions. The TGS techniques allow higher-throughput sequencing with faster turnaround time. The TGS techniques generate longer sequence read lengths with average lengths in kilobase pairs (kbp), which enhances de novo sequencing and assembly and allows direct haplotyping and phasing of the whole chromosome. The TGS techniques allow repeated sequencing and repeated alignment of individual sequence reads, which generates highly accurate consensus sequence reads that enable a rapid and efficient detection or identification of rare genetic variations, point mutations, SNPs and other genetic variations including CNVs. Further, the TGS techniques allow rapid and accurate detection or identification of different modifications including epigenetic modifications such as methylation. However, the accuracy of most of the TGS techniques is still lower in comparison with the FGS and SGS techniques although the accuracy is increasing with improvements and refinements in the basic methods.

The NGS techniques including the SGS and TGS allow rapid and accurate detection and analysis of SNPs. In addition to the detection of known or reported SNPs, these techniques enable discovery of new SNPs. The SNP detection and discovery using NGS techniques finds application in various processes involved in genetic analysis, including analysis of genetic diversity, construction of linkage maps, disease association studies and disease association mapping and marker-assisted selection. However, several inherent characteristics associated with genome and factors associated with the sequencing techniques itself decrease the efficiency and accuracy of SNP detection and discovery using these techniques. The genomic characteristics that affect the efficiency and accuracy of SNP detection and discovery include the presence of repeat sequence elements and paralogous sequences. The factors associated with the sequencing techniques that affect the SNP genotyping and discovery include sequencing errors that lead to errors in base calling and misaligned reads and incomplete reference genomic sequences that affect the accuracy of base calling (Altshuler, Pollara et al. 2000, Treangen and Salzberg 2011).

1.2.1.1.5 Applications of Single Nucleotide Polymorphisms (SNPs)

SNPs represent the most common type of and widely distributed sequence variation across the human genome. The SNPs because of their common occurrence and wide distribution serve as crucial physical or genetic markers that allow mapping of diverse human diseases on the human genomic map. Numerous studies clearly establish the important role of various types of sequence variations including SNPs in predicting various diseases or revealing the predisposition of carrier individuals to such diseases. Around half of the SNPs present in the genes are located within their noncoding regions, and almost half of those present in coding regions are synonymous (silent) SNPs. These SNPs possess very low penetrance and do not cause any disease by themselves. Although the non-synonymous missense SNPs and the non-synonymous nonsense SNPs may play a direct role in the pathogenesis of a disease, even these SNPs are generally stable. Nevertheless, numerous studies including meta-analyses clearly reveal various possible affects that the SNPs exert over the health and development of an individual. Numerous studies clearly establish the direct or indirect role that the SNPs play in modulating the risk or susceptibility of an individual to various multifactorial diseases including cancer, obesity, diabetes mellitus, atherosclerosis, multiple sclerosis, rheumatoid arthritis and asthma (Kruglyak 1999, Goldstein and Cavalleri 2005, Shastry 2009, Lander 2011, Karki, Pandya et al. 2015, Sudmant, Rausch et al. 2015, Ramírez-Bello 2019).

The SNPs provide a valuable tool for comparative genomic studies, including the population and evolutionary genomic studies. The SNPs are crucial for understanding the genetic behavior or genetics of populations including human populations and serve as important genetic tools that help in understanding and uncovering the mechanisms of evolution. Numerous studies reveal the important role played by SNPs in the emergence and establishment of the divergent evolutionary pathway undertaken by humans in comparison with other primates including chimpanzees. Studies reveal that most of the present-age SNPs found in the human genome trace their origin to the times after the evolutionary divergence of humans from other primates, but before the divergence of humans into various distinct populations (Mountain, Lin et al. 1992). This is clear from the observation that most of the present-age SNPs found in the human genome do not occur in other primates, and the majority of SNPs found in the human genome, around 85%, occur commonly across all the distinct human populations (Barbujani, Magagni et al. 1997). However, within humans, there occurs a significant heterogeneity in SNPs across populations of different races or ethnicities and this heterogeneity helps in revealing the genealogy and ethnicity of different individuals or human populations (Goldstein and Weale 2001, Goldstein and Cavalleri 2005, Weir, Cardon et al. 2005).

The SNPs serve as valuable tools in the field of pharmacogenomics. This field studies the effect of genes in general and genes that encode drug targets and drug-metabolizing enzymes in particular and the effect of SNPs within these genes on drug metabolism, drug response and drug efficacy (Alwi 2005). The pharmacogenomic studies that evaluate the association of various variations in genes involved in drug metabolism and drug response frequently use SNPs as molecular markers. Studies reveal that various controllable and uncontrollable factors such as individual response to drugs, optimal drug dosage and adverse drug reactions influence drug metabolism and drug efficacy. A direct consequence of these observed effects is that the same treatment regimens or protocols do not work effectively or even optimally in every case. This brought to fore the need for a new approach that takes into consideration the effect of relevant genes and variations within these genes such as SNPs on drug metabolism, drug response and drug efficacy. The need for a new approach gave birth to the field of precision medicine or personalized medicine that studies or analyzes the effect of genome and genetic variations such as SNPs on various crucial drug-related processes to reveal or determine the most suitable drug for the treatment of an individual suffering from a particular disease. Studies reveal the important and indispensable role that SNPs serve as tools or molecular markers in this new and emerging field of personalized or precision medicine. Numerous studies reveal the significant role that the SNPs play in modulating the response to drugs. In this regard, the SNPs in genes involved in the metabolism of xenobiotics including drugs, in particular, the SNPs in the genes that belong to cytochrome p450 (CYP) superfamily, serve as indispensable tools that enable the prediction of drug response and help in the prevention of adverse drug reactions (Katara 2014, Chaudhary, Singh et al. 2015). The application of SNP databases, in particular, the HapMap project database to the field of personalized medicine, allows the study and analysis of various aspects of a disease in an individual-specific manner such as susceptibility to a disease, disease screening, diagnosis and prognosis. It further allows the stratification of patients into groups, drug dosage estimation, drug response monitoring and evaluation and prediction of resistance to specific drugs (Laing, Hess et al. 2011). The use of SNP-based genomic maps in the field of oncology allows the design and development of personalized therapies for targeted treatment of benign and aggressive cancers based on the SNPs that occur in genes involved in the disease (Rodríguez-Antona and Taron 2015). Therefore, a thorough analysis of genetic polymorphisms in general and the SNPs in particular is crucial for understanding the evolutionary mechanisms and population dynamics including population genetic behavior. This analysis is also indispensable in understanding the disease mechanisms and disease dynamics that are crucial for the design and development of personalized precision medicine.

1.2.1.2 Copy Number Variations (CNVs)

Copy number variations (CNVs) are structural genetic variations, which involve the duplications (gains) or deletions (losses) of chromosomal segments or genomic sequence stretches ranging in size from one kilobase (1 Kb) to several megabases (Mb). In CNVs, a different copy number of a particular chromosomal or genomic sequence segment is present in a specific genome in comparison with the standard or reference genome. A CNV means the presence of a particular chromosomal or genomic sequence segment in copy numbers, which differ among the genomes of different individuals and in comparison with the reference genome (Feuk, Carson et al. 2006, Freeman, Perry et al. 2006, Henrichsen, Chaignat et al. 2009, Schrider and Hahn 2010, Auton, Brooks et al. 2015). CNVs are commonly found in humans (Sebat, Lakshmi et al. 2004, Conrad, Andrews et al. 2006, McCarroll, Hadnott et al. 2006, Redon, Ishikawa et al. 2006). It also occurs commonly in the genomes of other mammalian species such as rhesus macaques (Lee and Jeon 2008), chimpanzees (Perry, Dominy et al. 2007, Perry, Ben-Dor et al. 2008), cows (Liu, Hou et al. 2010), dogs (Chen, Swartz et al. 2009, Nicholas, Cheng et al. 2009) and mice (Graubert, Cahan et al. 2007, She, Cheng et al. 2008). CNVs are also present in the genomes of other eukaryotic organisms such as chickens (Griffin, Robertson et al. 2008), Drosophila melanogaster (Dopman and Hartl 2007, Emerson, Cardoso-Moreira et al. 2008), maize (Springer, Ying et al. 2009), *Arabidopsis thaliana* (Ossowski, Schneeberger et al. 2008), *Caenorhabditis elegans* (Maydan, Lorch et al. 2010) and *Saccharomyces cerevisiae* (Carreto, Eiriz et al. 2008). Although the SNPs and short indels constitute more than 99.99% of all the human genetic variations, the CNVs contribute more significantly to the total sequence variation and hence the genetic diversity between two typical human individuals largely because CNVs affect more number of bases or nucleotides in comparison with SNPs and short indels. A HapMap Project study, which included 270 different multiethnic genomic samples, identified almost 1,500 (1,447) distinct CNVs which combined together span or affect around 360 Mb which amount to almost 12% (11.6%) of the entire human haploid genome (Redon, Ishikawa et al. 2006). CNVs arise more often or frequently than SNPs or short indels, and the assessment of genome-wide CNV maps reveals their heterogeneous,

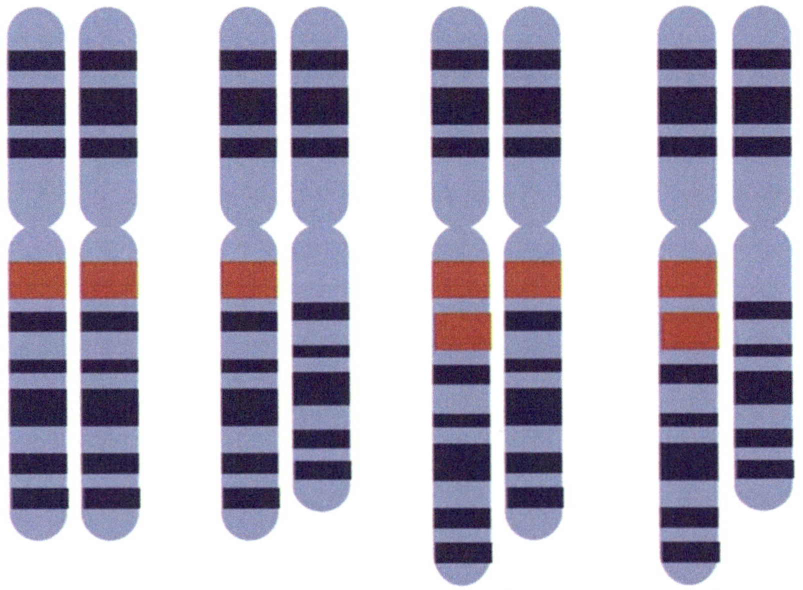

FIGURE 1.5 The pictorial representation of copy number variations (CNVs) in a pair of homologous chromosomes. 'A' represents normal copy number or no CNV; 'B' represents deletion CNV; 'C' represents duplication CNV; and 'D' represents same locus duplication and deletion CNV.

widespread occurrence throughout the human genome. Almost 35% of all the genes in the human genome show partial or full presence of CNVs, and many of these CNVs are functionally significant (Gonzaga-Jauregui, Lupski et al. 2012). Further, the estimates reveal that the proportion of any of the given individual human chromosome, potentially affected by, or which may contain CNVs, range from 6% to 19% or even more (Redon, Ishikawa et al. 2006). A population-specific study revealed that the variation in copy number between the genomes of any two individuals encompasses almost 0.2% of the haploid human genome, which equals approximately 6.2Mb (McCarroll, Kuruvilla et al. 2008). Furthermore, CNVs contribute to the variations in the protein-coding genes between the genomes of any two individuals through full overlap of almost 105 genes and partial overlap of approximately 367 genes (Conrad, Pinto et al. 2010). The recently updated list of CNVs consists of almost 2.4 million (2,391,408) different CNVs mapped to or distributed across more than 200,000 identified genomic sites or regions across different individuals and populations (MacDonald, Ziman et al. 2014). The sequence and structure of CNVs may be simple or complex. The simple CNVs involve simple tandem duplications, whereas the complex CNVs are segmental and embody complex multi-site deletions or duplications of homologous genomic sequences (Freeman, Perry et al. 2006, Redon, Ishikawa et al. 2006) (Figure 1.5).

1.2.1.2.1 Types of Copy Number Variations (CNVs)

Based on their sequence, structure and the mechanisms involved in their formation, that is, cellular origin, the CNVs consist of two major categories: recurrent CNVs and non-recurrent CNVs. The recurrent CNVs are of similar size and share similar or fixed breakpoints, endpoints or junctions; that is, they show clustering of endpoints with endpoints confined to few positions in the genome. As a result, these CNVs follow the same interval within the genomes of different individuals and thus exhibit a recurrent character. The breakpoints or endpoints of recurrent CNVs connect to or occur in segmental duplications or low copy repeats, which are the extensively homologous regions of the genome. The recurrent CNVs are mostly present in the regions of genome, which contain low copy repeats (LCRs) in general and large LCRs in particular. In other words, LCRs serve as the hotspots of recurrent CNVs. The recurrent CNVs occur as reciprocal deletions and duplications, multi-allelic CNVs (mCNVs), same locus duplication and deletion CNVs and complex CNVs and constitute approximately 20% – 40% of all CNVs (Gu, Zhang et al. 2008, Hastings, Lupski et al. 2009, Conrad, Pinto et al. 2010).

The non-recurrent CNVs are of dissimilar size, have unique endpoints, distinct from each other and independent of segmental duplications or LCRs, and thus exhibit a non-recurrent character. The breakpoints in non-recurrent CNVs are frequently micro-homologous and involve as low as 2 – 15 base pairs, whereas other breakpoints are blunt-ended or may encompass short insertions (Redon, Ishikawa et al. 2006, Gu, Zhang et al. 2008, Hastings, Lupski et al. 2009, Conrad, Pinto et al. 2010). The occurrence of micro-homology associated with breakpoints, which characterize the non-recurrent CNVs, though appear to be a chance event, may actually be non-random as studies reveal higher degree of micro-homology than could be expected from random events (Vissers, Bhatt et al. 2009, Conrad, Pinto et al. 2010, Newman, Hermetz et al. 2015). The non-recurrent CNVs of different sizes may occur in the same genomic region to varying extent and may thus overlap and share a sequence segment from the same region of the genome termed as common region of overlap or the smallest region of overlap (SRO). The occurrence of these CNVs in the same genomic region results in a wide array of clustered CNVs each with a unique breakpoint. The non-recurrent CNVs constitute the major proportion of CNVs present across the human genome. Most of the non-recurrent CNVs are simple and occur as deletions or sequential (tandem) duplications. Other non-recurrent CNVs are complex and occur as duplication and deletion events within the same locus or interval with interruption by inversions or normal sequences (Redon, Ishikawa et al. 2006, Gu, Zhang et al. 2008, Hastings, Lupski et al. 2009, Conrad, Pinto et al. 2010, Weckselblatt and Rudd 2015). Some non-recurrent CNVs are more complex and may consist of breakpoints, which encompass short insertions from a different region of the genome or may consist of a combination of deletions, inversions, duplications and triplications interspersed with stretches of normal unmutated sequence (Chen, Chuzhanova et al. 2005, Sheen, Jewell et al. 2007, Vissers, Stankiewicz et al. 2007, Gajecka, Gentles et al. 2008). Further, the non-recurrent complex CNVs may exist as a wide array of SVs clustered in the same region of the genome (Conrad, Pinto et al. 2010, Liu, Erez et al. 2011).

Besides the recurrent and non-recurrent CNVs, there is another CNV category that consists of simple CNVs that are generated through rearrangement of VNTRs and retrotransposition mechanisms. All these categories of CNVs together encompass a wide variety of structural genetic variations earlier known as copy number polymorphisms (CNPs), intermediate-sized variants (ISVs) and large-scale copy number variants (LCVs). However, the SVs which arise due to Class I transposonal (retroposonal) insertions such as those resulting from 6 Kb KpnI element insertions are not classified as CNVs (Redon, Ishikawa et al. 2006).

1.2.1.2.2 Copy Number Variations (CNVs) in Human Genome

The distribution of CNVs within the human genome follows a widespread, but heterogeneous and non-random pattern. CNVs are commonly present in the regions of genome, which contain segmental duplications (SDs), also known as low copy repeats (LCRs) (Figure 1.6). Segmental duplications are multiple copies (≥2) of genomic sequences larger than 1 Kb (usually 10 – 300 Kb) in size, which share more than 90% (usually 95% – 97%) similarity or identity in sequence and are present on the same chromosome or separate non-homologous chromosomes. Segmental duplications constitute around 5.2% of the entire human genome (Bailey, Yavor et al. 2001, She, Jiang et al. 2004). The frequent presence of CNVs in the regions of genome, which contain segmental duplications (SDs), indicates that significant proportion of CNVs, as already mentioned, involve SDs. However, the mere presence of segmental duplications in any region of the genome does not necessarily indicate the occurrence of CNVs because only those segmental duplications, which vary in copy number in comparison with the reference genome, can be categorized as CNVs (Feuk, Carson et al. 2006). Further, the CNVs, which occur in genomic regions that contain segmental duplications, can be recurrent or non-recurrent depending on their mechanism of formation. An enrichment of large multi-allelic CNVs, same locus duplication and deletion CNVs and complex CNVs occurs in genomic regions that contain SDs. Furthermore, CNVs are more likely to be associated with intra-chromosomal segmental duplications in comparison with inter-chromosomal segmental duplications. As already mentioned, the breakpoints in non-recurrent CNVs are independent of segmental duplications or LCRs, but still an enrichment of non-recurrent CNVs occurs in the LCR-rich regions of genome or in adjacent genomic regions.

These LCRs may be present in direct orientation or opposite (inverted) orientation, and this adds further to the structural complexity of the genomic regions involved (Redon, Ishikawa et al. 2006). The deletion CNVs are present in the coding as well as the noncoding regions of the genome. These deletion CNVs may involve loss or deletion of both the copies of the specific genomic region and thus occur as homozygous deletion CNVs. Alternatively, they may involve loss or deletion of only one copy of the specific genomic region and thus occur as heterozygous deletion CNVs. Within the coding regions of the

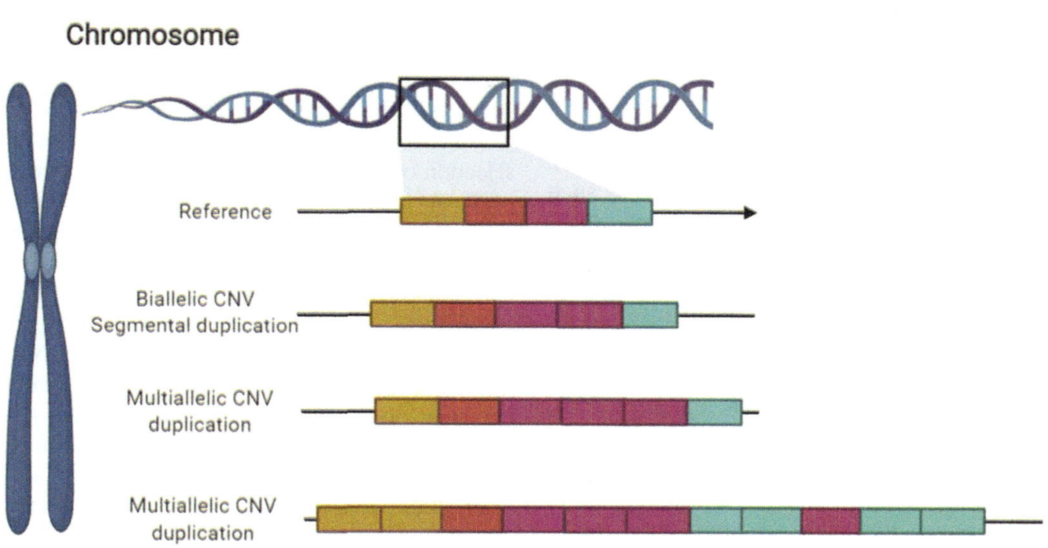

FIGURE 1.6 The pictorial representation of duplication copy number variations (CNVs) with respect to the reference genome. The picture shows bi-allelic CNV duplication and multi-allelic CNV duplications.

genome or genes, these deletions occur in both the exons and introns. Some of the deletion CNVs involves the loss of single or multiple exons, and such CNVs occur in genes involved in olfaction and metabolism of drugs (xenobiotics) and steroids. Some CNVs that involve the loss of coding regions of the gene or even the deletion of whole gene also exist. However, such CNVs are of rare occurrence (Conrad, Andrews et al. 2006, McCarroll, Hadnott et al. 2006). Therefore, the genetic variability due to CNVs involves duplications in some genomic regions and deletions in others or both at the same locus, and these regions of the genome involve both the coding and noncoding sequences. The presence of duplication and deletion CNVs within the coding sequences or genes of normal genome thus indicates that even the duplications and deletions can be associated with normal phenotypic expression rather than high penetrance early-onset genetic diseases or disorders (Buckland 2003). However, many large duplication as well as deletion CNVs when associated with particular genes such as the dosage-sensitive genes may increase the susceptibility to or even cause various genetic diseases or disorders (Schuster-Böckler, Conrad et al. 2010).

A large number of CNVs that occur in large gene clusters or large families of multiple genes as well as in individual genes stand annotated. The identification and characterization of CNVs in the form of deletions as well as duplications in gene clusters/gene families occurred even before the emergence of NGS. These gene clusters/gene families include major histocompatibility complex *(MHC)* class III gene cluster (Ghanem, Uring-Lambert et al. 1988), amylase *(AMY)* gene cluster (Groot, Bleeker et al. 1989), antimicrobial β-defensin *(DEFB)* gene cluster *(DEFB-CN)* (Hollox, Armour et al. 2003), olfactory receptor (OR) gene family (Trask, Friedman et al. 1998), opsin gene family (Nathans, Thomas et al. 1986) and human leukocyte antigen *(HLA)-DR* gene loci (Zhang, Degli-Esposti et al. 1990). The human amylase *(AMY)* gene cluster, in particular, salivary amylase *(AMY1)* gene, is among the most polymorphic genomic regions with regard to the presence of CNVs. The size and the copy number of the *AMY* gene polymorphic region vary from 150 to 425 Kb and from 6 to 15, respectively, among the DNA samples from different individuals or populations. The *AMY* gene cluster CNVs may play a role in adaptation to varied diets and may be associated with obesity as well (Freeman, Perry et al. 2006, Perry, Dominy et al. 2007). CNVs were also characterized much earlier in genes such as cytochrome P450 2D6 *(CYP2D6)* gene (Heim and Meyer 1992), rhesus blood group D antigen *(RHD)* gene (Colin, Chérif-Zahar et al. 1991), cholinergic receptor nicotinic alpha 7 subunit *(CHRNA7)* gene (Riley, Williamson et al. 2002), complement component 4 *(C4)* gene (Schneider, Carroll et al. 1986) and golgin-related gene linked to promyelocytic leukemia *(PML)* gene [golgin linked to PML (GLP) gene] (Gilles, Goy et al. 2000). Many of these mentioned genes or gene clusters are known to be involved in the sensory perception of smell (olfaction) and adaptation to or dynamic interactions with varying environmental conditions such as changes in the dietary conditions or exposure to chemicals and pathogens (Conrad, Pinto et al. 2010).

1.2.1.2.3 Copy Number Variations (CNVs) and Natural Selection

It has emerged from various studies that the occurrence, size as well as the distribution of different types of CNVs within the human genome may be under the strong influence of natural selection processes in general and purifying (negative) selection in particular (Conrad, Andrews et al. 2006, Redon, Ishikawa et al. 2006, Korbel, Urban et al. 2007, Conrad, Pinto et al. 2010, Mills, Walter et al. 2011). The studies reveal that the deletion CNVs may be subject to stronger selection processes in comparison with the duplication CNVs of the same or even larger size. The deletion CNVs are under increased purifying selection and are mainly enriched in the noncoding regions of the genome away from the genes, in comparison with the duplication CNVs (Korbel, Urban et al. 2007, Mills, Walter et al. 2011). As a result, the deletion CNVs are smaller in size (on average, three times shorter) and less frequent in occurrence across the human genome in comparison with duplication CNVs. The studies reveal a marked deficiency of deletion CNVs in the coding sequences in comparison with the noncoding sequences and in exons in comparison with introns. In contrast, even the large multicopy duplication CNVs are present in good numbers in the coding as well as noncoding regions, which indicates that duplication CNVs are under a weaker selection pressure (Locke, Sharp et al. 2006).

As the purifying selection is responsible for the removal of deleterious genetic variations including mutations, the stronger negative selection pressure indicates that coding region deletion CNVs are more detrimental in comparison with the duplication CNVs. However, the duplication CNVs encompassing the whole genes such as the dosage-sensitive genes involved in developmental processes are detrimental and may be under strong purifying selection (Schuster-Böckler, Conrad et al. 2010). Further, the CNVs including the large segmental duplications present in the intergenic and noncoding regions of the genome can influence the gene function and may be under strong purifying selection if the functional changes associated with such CNVs are deleterious (Stranger, Forrest et al. 2007, Itsara, Cooper et al. 2009). However, various studies report that many CNVs in individual genes or large gene clusters/large gene families, associated with neutral, advantageous or even marginally or moderately deleterious changes in gene function may be under positive (directional) natural selection or balancing natural selection rather than the usual purifying natural selection. For example, the higher copy number of human salivary amylase *(AMY1)* gene in populations that consume starch-rich diets in comparison with those that consume starch-deficient diets contributes to adaptive positive natural selection (Freeman, Perry et al. 2006, Perry, Dominy et al. 2007). Another example is of deletion CNV in UGT2B17 gene, which is under positive natural selection in East Asians and balancing natural selection in Europeans (Xue, Sun et al. 2008). This UGT2B17 gene deletion CNV may increase the risk of or susceptibility to prostate cancer (Karypidis, Olsson et al. 2008) and osteoporosis (Yang, Chen et al. 2008) and may therefore be under purifying selection in other populations (Xue, Sun et al. 2008). The UGT2B17 gene encodes UDP glucuronosyltransferase family 2 member B17 enzyme that catalyzes glucuronidation, an intermediate reaction in steroid metabolism. Glucuronidation involves glucuronic acid transfer from uridine diphosphoglucuronic acid (UDP-glucuronic acid) to various substrates including steroid hormones (such as androgens) and lipophilic drugs (Karypidis, Olsson et al. 2008). Various studies including Gene Ontology (GO) analysis reveal that CNVs are generally biased toward genomic regions or genes associated with biological pathways that belong to the peripheral layer of cellular signaling network (Redon, Ishikawa et al. 2006, Kim, Korbel et al. 2007, Conrad, Pinto et al. 2010). In contrast, the CNVs are generally biased away from genomic regions or genes associated with biological pathways that belong to the central layer of cellular signaling network (Futreal, Coin et al. 2004, Hurles 2004, Redon, Ishikawa et al. 2006). Accordingly, CNVs exhibit enrichment in genomic regions or genes involved in various cellular processes such as sensory perception of smell (olfaction), cellular adhesion, cellular recognition, cellular transport, cellular communication, cellular response to chemical stimuli and various neurophysiological and immunological processes (Redon, Ishikawa et al. 2006, Kim, Korbel et al. 2007, Conrad, Pinto et al. 2010). Similarly, CNVs are less frequent in occurrence and smaller in size in genomic regions or genes involved in cellular processes such as regulation and processing of nuclear information, cellular proliferation (cell cycle), metabolism, kinase-catalyzed cellular reactions (phosphorylation) and gene dosage-sensitive cellular processes including development, oncogenesis and tumor suppression (Futreal, Coin et al. 2004, Hurles 2004, Redon, Ishikawa et al. 2006).

1.2.1.2.4 Potential Mechanisms Involved in the Formation of Copy Number Variations (CNVs)

The CNVs within the human genome arise through the processes of recombination, mutation and repair that occur through two main mechanisms: homologous recombination (HR) and non-homologous recombination/repair (NHR). The homologous recombination (HR) mechanisms responsible for the generation of CNVs include non-allelic (unequal or ectopic) homologous recombination (NAHR) and single-strand annealing (SSA). The non-homologous recombination (NHR) mechanisms involved in the generation of CNVs include two main types: non-homologous non-replicative recombination mechanisms and non-homologous replicative recombination mechanisms. The non-homologous non-replicative recombination occurs through various mechanisms including non-homologous end joining (NHEJ), micro-homology-mediated end joining (MMEJ) and breakage-fusion-bridge cycle, whereas the non-homologous replicative recombination occurs through mechanisms which include replication slippage (template switching), fork stalling and template switching (FoSTeS) and micro-homology-mediated break-induced replication (MMBIR). Besides these mechanisms, mobile element insertion (MEI), mainly retrotransposition, and VNTR rearrangements are other mechanisms involved in the formation of CNVs (Stankiewicz and Lupski 2002, Lee, Carvalho et al. 2007, Gu, Zhang et al. 2008, Kidd, Cooper et al. 2008, Hastings, Ira et al. 2009, Conrad, Pinto et al. 2010). Of all the mechanisms involved in the formation of CNVs in the human genome, NAHR has emerged as the main or primary mechanism followed by NHEJ. The other mechanisms, which contribute significantly to the generation of CNVs, include MMEJ, FoSTeS and MMBIR. The replication slippage, SSA, MEI and VNTR rearrangement appear to play only a minor role in the generation of CNVs in the human genome. However, in stricter sense, the relative contribution of each of these mechanisms toward the formation of CNVs is not fully clear yet (Gu, Zhang et al. 2008, Kidd, Cooper et al. 2008, Conrad, Pinto et al. 2010). In addition, it appears that there is variation even in the dominance of one mechanism over the other across the genome (Cardoso-Moreira and Long 2010).

1.2.1.2.4.1 Non-allelic Homologous Recombination (NAHR)

The non-allelic homologous recombination (NAHR), which involves crossover and subsequent recombination between two non-allelic highly homologous genomic regions known as SDs or LCRs, is responsible for the generation of most of the CNVs including the major proportion of recurrent CNVs. This process often occurs during meiosis, but there is mitotic NAHR as well. The process involves abnormal alignment, often termed as misalignment, of two copies of non-allelic LCRs instead of allelic versions during mitotic or meiotic cell division, which facilitates crossover and subsequent genomic rearrangement and recombination that results in the generation of a novel variant such as CNV in the subsequent cellular generation. The non-allelic LCRs therefore act as hotspots of NAHR and are directly involved in the generation of various forms of genomic rearrangements including CNVs (Lupski 1998, Stankiewicz and Lupski 2002, Shaw and Lupski 2004, Gu, Zhang et al. 2008). The efficient or optimal occurrence of NAHR event is highly dependent on the presence of genomic segments of a minimum length, which share extremely high degree of similarity or identity in sequence between the two participant LCRs. These highly homologous genomic segments, known as minimal efficient processing segments (MEPS), are extremely important for intra-chromosomal as well as inter-chromosomal NAHR events, which may occur during either mitosis or meiosis. The importance of MEPS is evident from the studies, which have shown that even a single to few base pair change results in decreased efficiency and frequency of both intra-chromosomal and inter-chromosomal NAHR (Rubnitz and Subramani 1984, Waldman and Liskay 1988). Further, the MEPS involved in mitotic NAHR may be different from the MEPS involved in meiotic NAHR (Steinmann, Cooper et al. 2007). The recombination event between two non-allelic LCRs, which involve sister chromatids or homologous chromosomes, results in a duplication as well as a deletion. If the recombination event involves two non-allelic LCRs located on the same or single chromatid, only a deletion results (Turner, Miretti et al. 2008) and this deletion involves circular intermediates (van Loon, Miller et al. 1994). However, NAHR is predominantly involved in the generation of duplications and deletions. The frequent association of CNVs with or their presence in the regions of genome, which contain SDs, indicates that the occurrence of NAHR in these regions of genome is an important mechanism for the generation of CNVs (Kim, Lam et al. 2008).

1.2.1.2.4.2 Non-homologous End Joining (NHEJ) The non-homologous end joining (NHEJ), which is activated frequently in response to DNA double-strand breaks (DSBs), is an important recombination mechanism responsible for the generation of a major proportion of non-recurrent CNVs, mostly deletions (van Gent and van der Burg 2007). The NHEJ-mediated DNA repair can occur in any region of the genome where DSBs may arise, and it does not require the presence of LCRs, MEPS, repetitive sequences or any sequence homology (Gu, Zhang et al. 2008, Lieber 2008). The NHEJ DNA repair mechanism is a multistep process and starts with the detection or identification of DSBs followed by molecular bridging in which the two broken ends of DNA are brought together. The bridging is followed by the modification step in which the broken ends of DNA are altered to make them compatible and fit for ligation, which is the final step in the NHEJ recombination (Weterings and van Gent 2004). The modification step in NHEJ often involves cleavage at the repair site, which results in the deletion of a small number of base pairs (usually 1 – 4 bp) from the two broken ends of DNA (Inoue, Osaka et al. 2002, Shaw and Lupski 2004, Hastings, Ira et al. 2009, Hastings, Lupski et al. 2009). Alternatively, this step may involve insertion or addition of free nucleotides (sequence fragments) of chromosomal, mitochondrial or retrotransposonal origin to the broken ends of the DNA (Hastings, Ira et al. 2009, Hastings, Lupski et al. 2009, Lieber, Gu et al. 2010). This deletion or insertion event produces a 'sequence scar' also known as 'information or DNA scar' at the repair site, and this scar is an important characteristic of NHEJ repair process (Gu, Zhang et al. 2008, Lieber 2008, Stankiewicz and Lupski 2010). NHEJ recombination more often results in deletions; insertions and chromosomal translocations are also common; however, NHEJ may result in duplications as well and this involves the formation of complex DNA intermediates (Lee, Inoue et al. 2006, Lee and Lupski 2006). The MMEJ is another recombination mechanism activated in response to DNA DSBs, but unlike NHEJ, it is dependent on the presence of micro-homologous genomic sequence, usually 5 – 25 bp long, within the two broken ends of DNA to align them together for subsequent joining or annealing process. MMEJ like NHEJ involves cleavage at the repair site, which results in sequence deletions at the two broken ends of DNA. MMEJ is also an important mechanism involved in the generation of non-recurrent CNVs and almost invariably results in deletion. However, MMEJ recombination may be involved in inversions, translocations and other complex chromosomal rearrangements as well (McVey and Lee 2008).

1.2.1.2.4.3 Replication Slippage or Template Switching Replication slippage or template switching, a non-homologous replicative recombination mechanism, is one of the simplest mechanisms involved in the generation of CNVs. During the process of DNA replication, the synthesis of lagging strand in the form of Okazaki fragments renders a segment of lagging-strand template single stranded. This single-stranded lagging-strand template may contain short stretches of identical or homologous sequence and may encompass secondary structures. Further, the lagging-strand synthesis occurs in a discontinuous manner in the form of Okazaki fragments. During the synthesis of lagging strand, the 3′-end of the primer or the lagging strand can shift or slip to another identical sequence segment on the template strand and continue replication leaving the bypassed or original sequence segment uncopied. This slippage in replication often results in deletion (Albertini, Hofer et al. 1982, Hastings, Ira et al. 2009, Hastings, Lupski et al. 2009, Conrad, Pinto et al. 2010). However, through modifications of the basic process, the replication slippage can also lead to duplications and this may or may not involve exchange and recombination between sister chromatids (Lovett 2004). Further, the replication slippage process or the variations resulting from it remain confined to a very small region of genome (1 – 2 Kb), which constitutes a single replication fork. Due to this size limitation, the contribution of replication slippage in the generation of CNVs is minimal. However, replication slippage is one of the primary mechanisms involved in the generation of smaller variations such as the ones that present as VNTRs (Hastings, Lupski et al. 2009).

1.2.1.2.4.4 Fork Stalling and Template Switching (FoSTeS) Fork stalling and template switching (FoSTeS) is an important non-homologous replicative recombination mechanism and is considered the third important mechanism, after NAHR and NHEJ, for the generation of CNVs. FoSTeS is considered the primary mechanism for the generation of those complex non-recurrent CNVs, which cannot be explained by other two major recombination mechanisms, NAHR and NHEJ. FoSTeS involves interfork template switching, which occurs between multiple replication forks, rather than intrafork template switching, which remains restricted to a single replication fork as is seen in replication slippage (Lee, Carvalho et al. 2007, Hastings, Lupski et al. 2009). During lagging-strand synthesis, the presence of LCRs, repetitive DNA sequences or secondary structures on the single-stranded template or DNA abrasions, DNA chemical modifications such as alkylation or scarcity of deoxynucleotide triphosphates (dNTPs) may stall or block the progression of replication forks, which may lead to a pause in replication (Viguera, Canceill et al. 2001, Minca and Kowalski 2011). Under such conditions, the lagging strand detaches from the replication fork on the template strand and shifts or switches to another accessible active replication fork in close physical proximity, but not necessarily in the primary structure or sequence. The new replication fork or site on the template strand to which the lagging strand attaches and anneals contains short sequence stretches identical or homologous to the original site, which means the old and the new replication fork sites should be micro-homologous to each other. Once annealed to the new replication fork on template, the template-driven synthesis of lagging strand is started again (Lee, Carvalho et al. 2007, Hastings, Lupski et al. 2009). The annealing that follows the switching by lagging strand from one replication fork to other results in juxtaposition or collocation of sequence segments and creates a join point instead of a breakpoint in the genomic region involved. Depending on the position of the new replication fork site to which switching occurs relative to original replication fork site, FoSTeS may result in duplication, deletion, inversion or translocation. FoSTeS results in duplication if the new replication fork site to which switching occurs is located upstream of the original site, and leads to deletion if the new replication fork site is located downstream of the original replication fork site. Further, the switching strand may anneal to either the leading or the lagging strand in the new replication fork and the replication fork can progress in either direction relative to the direction in the original replication fork. FoSTeS may thus result in the generation of new variant genomic segment with same (direct) orientation or inverted (opposite) orientation relative to the orientation in the original replication fork.

The inverted orientation results in inversions. The template switching to a micro-homologous site on a different chromosome results in translocations. The complex genomic rearrangements including complex non-recurrent CNVs arise when the major events involved in FoSTeS, that is, lagging-strand detachment, template switching, annealing to new site followed by replicative synthesis, happen multiple times in a series-wise manner in a particular region of genome (Lee, Carvalho et al. 2007, Gu, Zhang et al. 2008, Hastings, Lupski et al. 2009). FoSTeS is a replicative recombination mechanism and occurs during the synthesis or S phase of the cell cycle. The errors and repair mechanisms in replication have been mostly associated with small genetic mutations confined to one or few genes. FoSTeS brings forth a new paradigm in this regard that the errors in replication can result in large and complex genetic variations or rearrangements encompassing genomic regions spanning thousands or even millions of base pairs in size (Shaw and Lupski 2004, Gu, Zhang et al. 2008).

1.2.1.2.4.5 Micro-homology-Mediated Break-Induced Replication (MMBIR) The micro-homology-mediated break-induced replication (MMBIR), another template switching mechanism, is based on break-induced replication (BIR) repair process and is involved in the repair of one-ended or single-sided DSBs. During replication, under cellular stress conditions such as hypoxia or starvation, the advancing replication forks may encounter DNA abrasions such as single-stranded break or nick in the template strand, which may stall or block their progression and lead to their collapse, a process in which an arm splits from the fork. The collapse of replication fork can also result from endonuclease activity. The replication fork collapse manifests as or generates a single-sided or one-ended DNA DSB. Under such conditions, the 5′-terminal of the dsDNA from the collapsed replication fork is resected to reveal the single-stranded 3′-terminal. This single-stranded 3′-terminal shifts or switches to another accessible template, a single-stranded DNA segment in close physical proximity to the 3′-terminal of DNA. The new template site to which the single-stranded 3′-terminal from the collapsed replication fork attaches and anneals contains very short sequence stretches (1 – 3 bp or more) identical or homologous to the original site. In other words, the old and the new sites should share micro-homology.

The single-stranded new DNA template may be the provided by the lagging-strand template in other replication forks, secondary structures, excision repair tracts and transcription sites. The annealing of single-stranded 3′-terminal from the collapsed replication fork to the new micro-homologous template creates a micro-homologous junction. At this junction, a new replication fork emerges and the replication process that involves the synthesis of lagging and leading strands starts again (Gu, Zhang et al. 2008, Hastings, Ira et al. 2009, Hastings, Lupski et al. 2009). However, the replication fork and the replication process itself have low (short) processivity and replication fork after extension of the broken terminal by some length may collapse again. As a result, the same processes starting from the resection of 5′-terminal to the generation of new low (short) processivity replication fork occur again. However, any of these low processivity replication forks can develop into a highly (fully) processive replication fork and, in such cases, the replication can proceed to the end of replicon or chromosome. The repeated occurrence of these processes in a single event leads to the generation of complex CNVs, including complex non-recurrent CNVs. Depending on the position of new replication fork site at micro-homologous junction relative to the original site where fork collapse occurs, MMBIR may result in duplication, deletion, inversion or translocation as in case of FoSTeS. MMBIR results in duplication if the new replication fork site is located upstream of the original site, and leads to deletion if the new replication fork site is located downstream of the original replication fork site. Relative to the original replication fork, the orientation of the displaced single-stranded 3′-terminal may be similar (direct) or inverted (opposite) in the progressing replication fork. The inverted orientation results in inversions. Again, as is the case with FoSTeS, the template switching to a micro-homologous site on a different chromosome results in translocations. Downstream loss of heterozygosity (LOH) may arise if the MMBIR event involves a homologous chromosome instead of sister chromatid. MMBIR is one of the main mechanisms responsible for the generation of non-recurrent duplication CNVs, non-recurrent triplication CNVs and non-recurrent complex CNVs (Malkova, Ivanov et al. 1996, Smith, Llorente et al. 2007, Gu, Zhang et al. 2008, Hastings, Ira et al. 2009, Hastings, Lupski et al. 2009, Zhang, Khajavi et al. 2009).

The mobile element insertion (MEI), mainly in the form of retrotransposition, and VNTRs are the other two mechanisms involved in the generation of CNVs, mostly the duplication CNVs. However, their overall contribution toward the generation of CNVs in human genome is small (Hastings, Ira et al. 2009, Hastings, Lupski et al. 2009, Conrad, Pinto et al. 2010, Treangen and Salzberg 2011). These two mechanisms have been discussed elsewhere in this chapter.

1.2.1.2.5 Methods for Detection and Analysis of Copy Number Variations (CNVs)

The methods or techniques used for the detection, analysis and characterization of CNVs, overlapping CNVs and overlapping copy number variant regions (CNVRs) are diverse and dynamic. These methods range from conventional cytogenetic techniques such as Giemsa (G)-stained or G-banded karyotyping (approximate resolution 5 – 10 Mb) to high-throughput sequencing techniques (approximate resolution 1 – 700 bp) and several other methods capable of CNV detection and analysis in the intermediate resolution range (Carter 2007). Most of the currently used CNV detection methods are based on second-generation DNA sequencing approaches, also known as NGS or MPS, which encompass highly accurate and sensitive techniques for the detection and analysis of genetic variations including SVs such as CNVs. The first-generation DNA sequencing approaches, which predominantly include Sanger sequencing method and its modifications and to a lesser extent Maxam-Gilbert sequencing methods, that have proven their usefulness in varied genomic fields are, however, not suitable for the detection and analysis of CNVs. The detection and analysis of CNVs using techniques based on second-generation DNA sequencing can be carried out at three different levels: at the level of single gene or a group of genes, at the level encompassing all the exons in a genome (exome level) and at the level encompassing the entire genome (whole genome level). However, the detection and analysis of CNVs at the exome and whole genome levels, largely, share a common approach. Therefore, the detection and analysis of whole exome level and whole-genome-level CNVs uses similar methods; however, some modifications may be required. Thus, there are two major categories of methods or approaches used for CNV detection and analysis: the targeted approach and the whole approach. The targeted approach enables the detection and analysis of CNVs in a single gene or a group of genes either present adjacent to each other on the same chromosome or spread across the entire genome. The whole approach can be either whole exome approach, which enables the detection and analysis of CNVs present in all the exons of a particular genome including the human genome, or it can be whole genome approach, which allows the detection and analysis of CNVs spread across the entire genome. The various methods used in the targeted approach CNV detection and analysis include hybridization-based methods, molecular cytogenetic techniques, Southern blot hybridization technique, PCR-based methods, quantitative real-time PCR (qPCR or RTPCR), multiplex amplifiable probe hybridization (MAPH), multiplex ligation-dependent probe amplification (MLPA) and multiplex amplicon quantification (MAQ). The various methods used in the whole exome and whole genome approaches include microarray techniques and NGS techniques. Each of the methods that allow the detection and characterization of CNVs has their own advantages and limitations. The selection of a particular method takes into account various factors including cost, level of analysis or throughput, the nature of genome sequences under analysis, resolution required and the available laboratory facilities.

1.2.1.2.5.1 Targeted Approach for CNV Detection (Gene - or Locus-Specific CNV Detection)

1.2.1.2.5.1.1 Molecular Cytogenetic Techniques

The various molecular cytogenetic techniques, which emerged through the combination of conventional cytogenetic methods with molecular procedures, in particular, in situ hybridization, are valuable tools for the detection and analysis of small and large chromosomal alterations or rearrangements including chromosomal aberrations and non-random SVs such as CNVs. These techniques include fluorescent in situ hybridization (FISH) and its various modifications, multicolor-FISH (mFISH), which further consists of two distinct methods: spectral karyotyping (SKY) and multiplex-FISH (M-FISH). More importantly, it includes comparative genomic hybridization (CGH), which is one of the main techniques used for the detection and analysis of CNVs. Each of the molecular cytogenetic techniques – FISH, SKY, M-FISH and CGH – has certain limitations, but together, these techniques provide an efficient means for identifying and studying the CNVs.

1.2.1.2.5.1.2 Fluorescent In Situ Hybridization (FISH)

FISH, originally developed as a technique for physical mapping of chromosomes to outline genes, involves the use of fluorescent- or non-fluorescent [haptens, biotin (vitamin H) or digoxigenin] - labeled probes, which bind or hybridize, with a high level of complementarity, to the target genomic sequences on the chromosomal DNA. The general procedure involves the denaturation of probe and target chromosomal DNA and subsequent binding or hybridization of the denatured probe to its complimentary site or genomic sequence on the target chromosomal DNA. The hybridization of the probe to its complementary genomic sequence generates a fluorescent signal, detectable by means of a fluorescent (fluorescence) microscope or a non-fluorescent signal, detectable through immunological or enzymatic methods (Langer-Safer, Levine et al. 1982, Lichter, Difilippantonio et al. 1993, Carter 2007). The generation of probes for FISH involves various methods such as random primer extension labeling (random-primed labeling), nick translation and PCR. FISH often involves the use of fluorescent DNA probes, which usually are cloned human genomic sequences or human DNA fragments, 80 – 200 Kb in size obtained from bacterial artificial chromosome (BAC) vectors. The labeling or tagging of DNA probes involves the incorporation of deoxynucleotides conjugated with fluorophores, which include traditional dyes such as cyanines, coumarins, fluorescein and rhodamine and new ones such as Eterneon and Atto dyes. The various fluorophores differ in several characteristics such as size, incorporation or conjugation efficiency, hydrophilicity and photostability. The probes are generally of three types: centromeric probes, locus-specific probes and painting probes. The chromosome-specific centromeric probes, obtained from cloned alphoid tandemly repeated DNA sequences, present vicinal to centromere of each chromosome, target and bind to specific alpha satellite repeat (ASR) or beta satellite repeat (BSR) sequences present within the centromeric region of each chromosome. The locus-specific probes target and bind specific sequence segments or genes and enable the detection and analysis of SVs or chromosomal aberrations including deletion, duplication and fusion. The locus-specific multicolor probes enable the simultaneous detection of these various SVs in several genomic regions in a single FISH assay. The painting probes target and bind specific chromosomal bands, a whole chromosome or an entire genome and find use in modified FISH techniques discussed later in this section. The detection of SVs such as CNVs (chromosomal segment duplications or deletions), chromosomal translocations and inversions by FISH involves the evaluation of binding or hybridization of fluorescent probes (specific fluorophore-tagged DNA sequences) with complimentary target genomic sequences present in test samples, which usually consist of interphase nuclei or metaphase chromosomes fixed onto the microscopic slides. The analysis of generated fluorescent signals, detected through a fluorescent microscopy-based imaging system, gives the nature, position (site) and magnitude of the hybridization process, which enables the identification of deletions, duplications or chromosomal translocations. The enhanced hybridization at a particular sequence or site in the test genomic sample in comparison with the reference genomic sample, detected through an increase or spike in the intensity of the fluorescent signal, indicates the presence of one or more highly similar copies of a particular sequence or site in the test sample in comparison with the reference sample. This reveals chromosomal segment duplication relative to the reference sample.

Similarly, the diminished hybridization at a particular sequence or site in the test genomic sample in comparison with the reference genomic sample, detected through a decrease or dip in the intensity of the fluorescent signal, indicates the absence of a particular sequence or site in the test sample in comparison with the reference sample. This reveals chromosomal segment deletion relative to the reference sample. Further, the detection of a shift in the relative position of hybridization spot or signal in test genomic sample in comparison with the reference genomic sample indicates the presence of translocations or inversions, the two important chromosomal structural rearrangements. In order to minimize the interference by repetitive interspersed elements in target chromosomal DNA and probes and to increase hybridization signal-to-noise ratio (SNR) in order to obtain low-noise FISH, the probes competitively hybridized with or pre-annealed to Cot 1 DNA are used. Cot 1 DNA, usually 50 – 300 bp in size, consists of DNA enriched for recurrent repetitive sequences such as short interspersed elements (SINEs), long interspersed elements (LINEs) and sequence homology segments and provides an effective means to decrease noise by blocking non-specific hybridization. Cot1 DNA blocking reagent blocks repetitive sequences such as SINEs, LINEs and sequence homology among members of the same gene family when added to the hybridization solution. FISH is a useful, highly specific, sensitive and extensively used technique, which allows the direct detection and analysis of deletion and duplication CNVs and their genotyping (whether homozygous or heterozygous) in genomic samples consisting of both interphase nuclear DNA and metaphase chromosomal DNA. The direct use of interphase nuclear DNA eliminates the need for time- and resource-consuming cell culture. Further, FISH allows the detection and analysis of hybridization signals at the level of a single cell (Klinger, Landes et al. 1992, Ried, Landes et al. 1992, McCarroll, Kuruvilla et al. 2008, Xu, Li et al. 2014). More than its usefulness in the detection and analysis of CNVs, FISH is a highly valuable technique for rapid screening and diagnosis of microdeletion and microduplication syndromes, syndromes resulting in subtelomeric chromosomal rearrangements (subtelomeric deletions and duplications) and common abnormalities in chromosome number (aneuploidies) (Xu, Li et al. 2014). The resolution of FISH for the detection and analysis of SVs including CNVs depends on the probe size, which usually ranges between 80 and 200 Kb, the size of BAC-cloned fragments. This resolution is almost 50 times the resolution achievable with high-resolution conventional cytogenetic G-banded karyotyping (Cui, Shu et al. 2016). However, the FISH technique cannot detect smaller chromosomal variations less than 50 kb in size including smaller sequence rearrangements, deletions and duplications. Further, for karyotype and other chromosomal analyses, prior knowledge regarding the chromosome structure and sequence is an absolute requirement for designing an appropriate probe for analysis. These limitations led to the development of modified FISH techniques including mFISH and CGH.

1.2.1.2.5.1.3 Multicolor-FISH (mFISH)

The multicolor-FISH (mFISH) involves the use of chromosome-specific painting probes or chromosome paints, which target and bind specific chromosomal bands or regions, a whole chromosome or an entire genome (complete karyotype), for example, an entire set of 24 human chromosomes. The chromosome painting probes are a collection or an assortment of fluorescently labeled DNA sequences derived from specific chromosomal regions, a whole chromosome or an entire genome, consisting of a complete haploid set of chromosomes. Depending on the origin and degree of compositeness of the collection, these labeled DNA sequences target and hybridize to different specific sequences and label specific chromosomal regions, a whole chromosome end to end or an entire genome (complete karyotype). Once hybridized to target chromosomal regions or chromosomes, these probes label or paint the entire length of targets in an even uniformly distributed manner generating a clear distinct detection signal to reveal a painted chromosomal region, a painted whole chromosome or a painted entire karyotype, a phenomenon called chromosome painting. The chromosome painting probes which consist of chromosomal DNA sequences are usually prepared from flow-sorted chromosomes for most of the applications or from micro-dissected chromosomes (or chromosomal regions) for labeling specific chromosomal regions. The degenerate oligonucleotide-primed (random-primed) polymerase chain reaction (DOP-PCR) or linker adaptor (A-T linker adapter)-mediated polymerase chain reaction (LA-PCR) amplifies the generated chromosomal DNA sequences (Vooijs, Yu et al. 1993, Carter 2007, Ferguson-Smith and Trifonov 2007). The preparation of painting probes from flow-sorted chromosomes involves the use of dual-laser fluorescence-activated flow cytometer system, which sorts freshly prepared suspensions of 300 – 500 chromosomes. This technique generates probes with high purity, highest possible resolution or better coverage of target chromosomes. DOP-PCR and LA-PCR procedures can achieve more than 90% amplification of the generated chromosomal DNA sequences (Lichter, Cremer et al. 1988, Pinkel, Landegent et al. 1988). The preparation of painting probes from micro-dissected chromosomes involves the use of inverted microscope along with suitable chromosome micromanipulators to generate high-resolution probes. These probes are useful in labeling chromosomal regions, which contain repetitive sequences, such as centromeres and telomeres (Meltzer, Guan et al. 1992).

The mFISH includes two distinct methods: spectral karyotyping (SKY) and multiplex-FISH (M-FISH). Both these methods use different combinations or varying amounts of finite number (usually five) of fluorophores each with distinct emission spectra to produce 24 distinct color combinations or patterns to label all the chromosomes in the human genome or karyotype, which gives each of the 24 human chromosomes or each homologous pair of chromosomes a unique color combination. The mFISH, through unique color combination labeling, therefore, enables the visualization of all the chromosomes simultaneously. SKY and M-FISH differ in the methods used to detect and discriminate between each of the distinct color combinations or patterns, which label each chromosome pair uniquely. The SKY technique, based on Fourier transform spectral imaging (FTSI), uses charge-coupled device (CCD)-based imaging system to capture the images. An interferometer connected to an epifluorescence microscope analyzes the captured images. The image processing software then imputes a pseudo-color to each of the spectrally unique color combination, which enables the visualization of all the uniquely colored chromosomes simultaneously. The M-FISH uses a high sensitivity monochrome camera to capture the images involving the emission spectra, which pass through a series of specific or defined band-pass (excitation and emission) filters or filter sets. The image processing software superimposes and categorizes the captured images based on combinations of fluorophores used for labeling to generate a picture of unique color combination-labeled chromosomes. The mFISH is a highly specific and sensitive technique for the detection, assessment or analysis of small and large or complex non-random structural chromosomal rearrangements such as deletions and amplifications, inter-chromosomal rearrangements (translocations), numerical chromosomal aberrations and marker chromosomes such as Philadelphia chromosome, the marker for chronic myeloid leukemia (CML). This technique also allows the characterization of genomic alterations in tumor cell lines. The mFISH is a useful technique for the detection and analysis of CNVs including complex and deletion CNVs. However, the mFISH technique cannot detect intra-chromosomal SVs or aberrations including duplications, inversions and insertions (Ried, Landes et al. 1992, Schröck, du Manoir et al. 1996, Speicher, Gwyn Ballard et al. 1996). Another important modified FISH technique is comparative genomic hybridization (CGH). The CGH overcomes most of the limitations associated with FISH or mFISH and allows genome-wide detection and analysis of chromosomal SVs including CNVs such as those that involve simple or complex losses or gains of specific chromosomal regions. This technique uses a single hybridization step even without direct visualization of the chromosomes (Kallioniemi, Kallioniemi et al. 1992, Bejjani, Saleki et al. 2005).

1.2.1.2.5.1.4 Comparative Genomic Hybridization (CGH)

The comparative genomic hybridization (CGH) is a semi-quantitative modified FISH technique, which uses two genomes, test and reference (control), and two distinct fluorescent dyes usually with red and green chromophores. These two distinct fluorescent dyes label differentially the DNA sequence fragments (probes) generated from the test and reference (control) genomic samples. The CGH allows the test genomic sequence fragments (probes) labeled with one fluorescent dye, for example, green, and the reference genomic sequence fragments (probes) labeled with another fluorescent dye, for example, red, to hybridize simultaneously, but competitively to the complementary genomic sequences on the target human chromosomes, which consist of normal metaphase chromosomes. CGH involves comparing the intensities of hybridization of the test and reference genomic sequence fragments with the complimentary sequences along the length of the target metaphase chromosome. The difference in the hybridization intensities measured as the ratio of test sample fluorescence to reference sample fluorescence reflect the variations in the copy number between the test sample and the reference sample at particular chromosomal loci and subsequently the whole chromosome or the entire genome depending on the probes used. The difference in the fluorescence intensities between the test and the reference genomic samples, plotted linearly for hybridization across the length of each chromosome, reveals the copy number changes. The similarities in hybridization intensities indicate equal representation of a particular chromosomal region in both the test and the reference genomic samples reflected by orange fluorescence in case of green and red fluorescent dyes for test and reference samples, respectively. The increase in hybridization intensity of the test sample probes reflected by an increase in fluorescence of the test sample labeling fluorochrome (more green) and a decrease in fluorescence of the reference sample labeling fluorochrome (less red), that is, increased test sample-to-reference sample fluorescence ratio, indicates the amplification or gain of particular chromosomal region in test genomic samples. The decrease in hybridization intensity of the test sample probes reflected by a decrease in fluorescence of the test sample labeling fluorochrome (less green) and an increase in fluorescence of the reference sample labeling fluorochrome (more red), that is, decreased test sample-to-reference sample fluorescence ratio, indicates the loss or deletion of particular chromosomal regions in test genomic samples. A digital imaging and analysis system captures the fluorescent images of the test sample and reference sample probes hybridized to metaphase chromosomes and calculates the test sample-to-reference sample fluorescence ratio to reveal the copy number of the sequences under evaluation. The data from several experimental setups, combined and analyzed together, increase the accuracy and sensitivity of the CGH evaluation and generate comparative copy number profiles, which enables the generation of map depicting the changes in copy number of specific chromosomal DNA sequences. The CGH is a highly sensitive and efficient technique to detect and analyze unbalanced structural chromosomal rearrangements and provides an effective method to screen whole genome for CNV detection. The structural chromosomal rearrangements detected by CGH involve changes in copy number of particular DNA sequences and include loss or gain of particular chromosomal regions. However, the CGH cannot detect balanced structural chromosomal rearrangements including some CNVs. These structural chromosomal rearrangements involve no change in copy number of particular DNA sequences and include inversions and translocations. The CGH is an important technique in the field of cancer cytogenetics as it allows the detection and characterization of loss or gain of specific chromosomal regions associated with the pathogenesis of various cancers. This technique works well even with the DNA samples extracted from formalin-fixed paraffin-embedded (FFPE) samples (Kallioniemi, Kallioniemi et al. 1992, Pinkel, Segraves et al. 1998, Bejjani, Saleki et al. 2005). The CGH can detect the decrease or increase in copy numbers (unbalanced structural changes) of particular chromosomal sequences at a resolution of 5 – 10 Mb, which is largely dependent on the non-linear, complex DNA packaging along the length of the target metaphase chromosomes (Kirchhoff, Gerdes et al. 1998, Lichter, Joos et al. 2000). However, a modification of standard (traditional) CGH, known as array-CGH (aCGH), which involves the use of microarrays instead of normal human metaphase chromosomes, has greatly improved the resolution of this technique with detection size limit down to the scale of few nucleotides (bases pairs). This technique uses human genome-specific bacterial artificial chromosome (BAC) library, complementary DNA (cDNA) or oligonucleotides to generate DNA segments or probes for microarrays (Schena, Shalon et al. 1995, Lucito, Healy et al. 2003).

1.2.1.2.5.1.5 Southern Blot Hybridization Technique

Southern blot hybridization or Southern blotting technique, developed by Edward M. Southern in 1975, is a highly useful technique for the detection of certain structural chromosomal rearrangements including various types of CNVs (Southern 1975, Mellars and Gomez 2011). The basic Southern blot hybridization procedure involves the fragmentation or digestion of cloned DNA, PCR products or genomic DNA with restriction endonuclease enzymes followed by separation or fractionation of digested DNA sample, based on size, by agarose gel electrophoresis. The separated or fractionated DNA then undergoes in-gel denaturation or separation into individual strands, for example, by alkaline (sodium hydroxide) treatment. The next step involves the transfer (blotting) and immobilization of the separated and denatured DNA samples onto a nitrocellulose or a nylon membrane. The blotting, which occurs via capillary action or electroelution, preserves the relative location of the DNA samples by creating an arrangement of the separated DNA samples on the membrane similar to what was on the gel before transfer. The size of the DNA fragments greatly affects the efficiency of the transfer (blotting) (Mellars and Gomez 2011). The DNA fragments 1 Kb or less in size transfer onto the membrane almost completely, but for DNA fragments greater than 15 Kb, the efficiency drops to around 20%. The treatment of the agarose gel encompassing the separated DNA fragments with dilute hydrochloric acid (HCl, 0.25N) prior to blotting breaks the large DNA into smaller fragments, which greatly increases the efficiency of the transfer. The treatment of the blotted membrane with a blocking solution containing blocking agents such as salmon sperm DNA or dry milk blocks the empty regions on the membrane and prevents non-specific binding. After blocking, the labeled specific DNA sequence fragments or probes added onto the membrane bind or hybridize to membrane-bound complementary target DNA to produce hybridization signals, which allow visualization and analysis.

The labeling or tagging of DNA probes involves mainly three methods: radioactive labeling, chromogenic labeling and chemiluminescent labeling. The radioactive labeling involves the incorporation of ^{32}P- or ^{35}S- labeled deoxynucleotides for radioactive labeling, and the radioactive signal detectable through autoradiography and visualized on X-ray film gives the pattern of hybridization. The chromogenic labeling involves labeling the DNA probes with an enzyme such as alkaline phosphatase (AP), which acts on chromogenic substrates, 5-bromo-4-chloro-3-indolyl phosphate (BCIP) and nitro blue tetrazolium (NBT), to produce blue-purple insoluble product or precipitate (5, 5′-dibromo-4, 4′-dichloro-indigo), which reveals the intensity and the pattern of hybridization. The chemiluminescent labeling involves labeling the DNA probes with digoxigenin (DIG), fluorescein isothiocyanate (FITC) or biotin. Horseradish peroxidase (HRP)- or alkaline phosphatase (AP)-conjugated anti-DIG antibody for DIG-labeled probe, anti-FITC antibody for FITC-labeled probe and streptavidin for biotin-labeled probe act on chemiluminogenic or chemiluminescent substrate to produce chemiluminescence signal (light). X-ray (photographic) film or an imaging system captures the chemiluminescent emission as image to reveal the intensity and the pattern of hybridization. Southern blotting technique enables the detection and analysis of structural chromosomal rearrangements including simple and complex CNVs and is particularly useful for the detection of amplifications and deletions including deletion CNVs (Aldred, Hollox et al. 2005, Hollox, Detering et al. 2009). This technique detects decrease or increase in hybridization at a particular DNA sequence (site) or altered mobility of a particular DNA sequence fragment, which results from the generation of different-sized DNA fragments from restriction digestion, in the test genomic sample in comparison with the reference genomic sample. These differences in the test and genomic samples with regard to a particular sequence (site) detected through a decrease (dip) or an increase (spike) in the intensity of detection signals reveal the structural chromosomal variations such as deletions, duplications or amplifications at that particular sequence (site). Southern blotting technique can detect structural chromosomal rearrangements including CNVs at a resolution of 5 – 500 kb, which is higher than molecular cytogenetic techniques such as FISH (Lee and Jeon 2008). Southern blotting is a useful highly specific and sensitive technique. However, the uneven or incomplete transfer of DNA from the gel to the membrane or incomplete washing of membrane to remove unbound (excess of) probes or DNA can affect the sensitivity and specificity of hybridization, which affects the intensities of detection signals and may lead to misinterpretation of these hybridization spot intensities. Careful probe design and labeling, selection of appropriate detection methods, optimization of washing steps and use of restriction enzymes, which can generate different-sized DNA fragments, can minimize the aforementioned undesirable affects.

However, the time-consuming, labor-intensive procedure (workflow) and the need for large quantity of high-quality or highly pure genomic DNA remain the major limitations of Southern blotting technique (Hoebeeck, Speleman et al. 2007). Southern blotting alone is not an accurate technique for the detection of complex structural chromosomal rearrangements including complex CNVs; however, in combination with pulsed-field gel electrophoresis (PFGE) instead of simple gel electrophoresis, it can accurately resolve DNA sequence fragments greater than 12 Mb in size and detect complex copy number changes. In PFGE technique, the electric field applied to the gel alters periodically, which leads to continuous changes in the orientation of DNA sequence fragments. This enables the size-dependent resolution or separation of large DNA sequence fragments unlike the conventional gel electrophoresis in which the large random coiled DNA sequence fragments unravel or uncoil in order to migrate through small pores in the gel matrix in an almost size-independent manner, which leads to incomplete separation or decreased resolution (Herschleb, Ananiev et al. 2007). The PFGE technique in combination with Southern blotting procedure allows the detection and characterization of large or complex genomic rearrangements including deletions, duplications, amplifications, inversions, insertions and translocations, many of which constitute complex CNVs (Aitman, Dong et al. 2006).

1.2.1.2.5.1.6 Quantitative Real-Time Polymerase Chain Reaction (qPCR)

The quantitative real-time polymerase chain reaction (qPCR) is a highly sensitive technique used extensively for the detection and analysis of gene or chromosomal locus-specific CNVs. It allows accurate, qualitative and quantitative evaluation of target genomic regions to detect and genotype copy numbers and change in copy numbers including deletions and/or duplications within a single gene or chromosomal locus (Chen, Chuzhanova et al. 2005, Redon, Ishikawa et al. 2006, Fanciulli, Norsworthy et al. 2007, Nakajima, Ohtani et al. 2007, Wu, Savelli et al. 2007).

The quantitative real-time polymerase chain reaction (qPCR), also known as quantitative PCR or real-time PCR (RTPCR), is a high-throughput PCR technique based on real-time measurement of the accumulation of PCR product or amplicon. This technique enables the real-time monitoring or assessment of the kinetics or progress of the PCR in terms of degree of amplification of the target DNA sequence by allowing simultaneous detection, quantification and analysis of the product formed after each reaction cycle. In contrast, qualitative real-time PCR enables only the real-time detection (presence or absence), but not the quantification of the target DNA sequence, whereas the conventional PCR allows only the endpoint detection and semi-quantitative estimation of the PCR product or amplicon. The RTPCR uses qPCR systems, which in general consist of thermal cyclers or PCR systems linked to optical instruments for the detection of fluorescent reaction signals and appropriate software suite to allow real-time detection, quantification and analysis. The RTPCR uses two types of fluorescence-based methods to detect, quantify and analyze the accumulation of PCR products or amplicons in a qPCR mixture. These include dsDNA-intercalating and fluorophore (small fluorescent molecule)-labeled oligonucleotides (Heid, Stevens et al. 1996, Wong and Medrano 2005, Hoebeeck, Speleman et al. 2007). The dsDNA-intercalating fluorescent dyes include SYBR-Green I (SG), EvaGreen (EG), ResoLight (RL), LCGreen, SYTO, BEBO, BOXTO and EtBr.

These fluorescent dyes bind strongly to the dsDNA in comparison with the weak or no binding in case of single-stranded DNA (ssDNA). On binding to dsDNA, these dyes exhibit a tremendous increase in fluorescence. For example, when bound to dsDNA, the fluorescence of SYBR-Green I increases by more than 1,000 times, that of EvaGreen by more than 70 times and that of EtBr by almost 20 times. Further, these fluorescent dyes can be saturating or non-saturating. The saturating dyes such as SYBR-Green I inhibit the PCR amplification when they reach a saturation level, which corresponds to the generation of a certain quantity of PCR product. The non-saturating dyes such as EvaGreen do not reach a saturation level practically and the PCR amplification continues until the dye binds all the available sites on DNA and the process continues until any of the reagents in the PCR mix depletes. During the extension step of PCR-based amplification, the dyes bind to each of the newly generated copies of target dsDNA sequence, which generates fluorescent signals for detection and analysis. The major limitation of dsDNA-intercalating dyes is that they bind in a non-specific manner not only to specific or desired dsDNA PCR products, but also to non-specific dsDNA PCR products and primer dimers, which generate non-specific amplification fluorescent signal that appears with the product-specific fluorescent signal and reduce the overall accuracy of the detection and quantification process. However, the melt or melting curve analysis such as high-resolution melting (HRM) after the completion of PCR enables the analysis of amplicon and amplification signal to evaluate the specificity in order to rule out non-specific amplification. The melt curve assays analyze the discrete changes in fluorescent signals resulting from DNA denaturation to generate a high-resolution detection and quantification data. Further, since these dyes bind along the length of the amplicon, the longer amplicons produce stronger or more intense signals, which can lead to saturation of detection signals. In addition, it is not possible to differentiate between fluorescent signals generated from the PCR amplification of different target DNA sequences in the same PCR mixture. Because of these limitations, the DNA-intercalating dyes are unsuitable for use in multiplex qPCRs and the amplification of different sequence targets requires parallel or multiple reactions setups, which increases the probability of errors in detection and quantification. The major advantages of DNA-intercalating dyes are that they allow a low-cost simple qPCR setup, which requires only simple primers, and there is no need to design probes. Further, these dyes allow analysis of multiple target sequences in a single experiment, but in a parallel setup using microarray technique. The melt curve analysis, which involves studying the interaction of DNA-intercalating dyes with PCR products or amplicons, allows the evaluation of reaction specificity (Wong and Medrano 2005, Gudnason, Dufva et al. 2007, Navarro, Serrano-Heras et al. 2015).

The fluorophore (small fluorescent molecule)-labeled oligonucleotides include three major categories of labeled probes: primer-probes, probes and nucleic acid analogues. The primer-probes include Hairpin primer-probes [Scorpions, Light Upon Extension (LUX) and Amplifluor (Sunrise)], Angler primer-probes and Cyclicon primer-probes. The probes include hydrolysis probes (TaqMan, TaqMan-Minor Groove Binders (MGBs) and Snake assay) and hybridization probes (ResonSense probes, hairpin molecular beacon (MB) probes, Yin-Yang (displacing) probes, MGB-conjugated probes (MGB-Eclipse and MGB-Pleiades), hybridization beacon (HyBeacon) probes and HybProbes or fluorescence resonance energy transfer (FRET) probes. The nucleic acid analogues include zip nucleic acids (ZNAs), locked nucleic acids (LNAs), peptide nucleic acids (PNAs), tiny-molecular beacon probes and non-natural bases (Plexor primers). The fluorophores used to label probes and primers are of two main types: reporter (donor) fluorophores and quencher (acceptor) fluorophores. The design of most of the labeled probes and primers involves covalent linking of reporter and quencher fluorophores. This ensures that in the absence of target DNA sequence segments, the quencher fluorophores accept or quench fluorescence from reporter fluorophores, which minimizes or limits the fluorescence signal from labeled probes or primers. When the labeled probe or primer binds to their specific target DNA sequence, the reporter and quencher fluorophores separate, which prevents the quenching of fluorescence from the reporter and generates fluorescent signal for detection and analysis. The major limitation of fluorescent DNA probes and primers is that they are complex in design and are more expensive in comparison with the DNA-intercalating dyes. The major advantages of fluorescent DNA probes and primers are that they bind only to their specific target DNA sequences and generate only the target amplification-specific fluorescent signal that enables accurate detection and quantification of the target amplicon. Further, due to the specific binding and generation of target amplification-specific fluorescent signal, it is easy to differentiate between fluorescent signals generated from the PCR amplification of different target DNA sequences in the same PCR mixture. This capability makes the fluorescent DNA probes and primers highly suitable for use in single-tube multi-target multiplex qPCRs (Wong and Medrano 2005, Navarro, Serrano-Heras et al. 2015).

The basic principle of real-time PCR is that the fluorescent signal from DNA-intercalating dyes or fluorescent DNA probes and primers gives the proportional amount of target dsDNA sequence fragment present in a PCR mixture at a given time. The fluorescent signal shows an increase as PCR amplification progresses and generates more quantity of the target DNA. Until any of the reagents in the PCR mix depletes, the PCR amplification proceeds and generates new copies of target dsDNA exponentially with a corresponding exponential increase in fluorescent signal as well. The quantification of the original template concentration and dsDNA subsequently formed at any given time in a real-time PCR depends on this proportional relationship between the quantity of dsDNA and the fluorescent signal generated. This relationship between the quantity of target DNA sequence and the generated fluorescent signal, expressed in terms of cycle threshold (C_t) value, gives an accurate estimation of the quantity of target DNA present in the starting test sample. The cycle threshold (C_t) value is the number of PCR cycles required to obtain a fluorescent signal with intensity greater than the background threshold fluorescence. Therefore, the qPCR, which starts with a higher quantity of the starting target DNA in the test sample, generates more quantity of target DNA in lesser time in comparison with qPCR, which starts with a lower quantity of the starting target DNA. Consequently, the intensity of the fluorescent signal increases much faster and crosses the threshold fluorescence intensity in lesser time to yield a low C_t value for the qPCR with a higher starting quantity of the target DNA. In comparison, for the qPCR with a lower quantity of the starting target DNA, the intensity of the fluorescent signal increases slowly and takes more time to cross the threshold fluorescence intensity and therefore yields a high C_t value. The target DNA may be a gene of interest (GOI) or region (sequence) of interest (ROI) (Wong and Medrano 2005, Hoebeeck, Speleman et al. 2007). The qPCR usually uses two methods to determine the absolute or relative quantity of the target DNA from C_t values. One of these methods, the comparative C_t ($\Delta\Delta C_t$) method, uses the equation $2^{-\Delta\Delta Ct}$ to compare the C_t values of target DNA segment and reference (control) gene between the control sample and the test sample. However, this method requires the PCR efficiencies of both the target DNA segment and reference gene to be similar or at least comparable to each other (Wong and Medrano 2005, Schmittgen and Livak 2008). Another method, known as standard curve method, involves amplifying different but known quantities of target DNA or standard DNA sample (usually obtained through serial dilution) to obtain corresponding C_t values. These C_t values plotted against the log of target DNA concentrations give the standard curve, which enables the conversion of C_t values into relative quantities to reveal the concentrations of unknown quantities of target DNA. The standard curve method gives the linear range, efficiency and reproducibility of qPCR. However, this method allows the detection and quantification of the unknown samples only if their concentration lies within the range of the curve, which in turn depends on the different quantities used for generation of the standard curve (Liu and Saint 2002, Wong and Medrano 2005).

The major limitation of qPCR technique is its inability to amplify large number of target sequences simultaneously, which makes it unsuitable for multiple quantifications including multiplex or simultaneous CNV analysis of more than one gene or chromosomal locus. Further, various factors including sample preparation, sample storage and integrity of the sample DNA greatly affect the outcome of qPCR analysis. The probability of underestimation or overestimation of copy numbers is also very high. However, the use of replicates minimizes these errors and increases both the accuracy and precision of overall analysis. The qPCR is a highly sensitive and specific technique and allows accurate measurement of the copy numbers of genes and DNA segments over a wide range of size and magnitude. The measurement of gene copy numbers enables evaluation of gene dosage, which is important to study amplifications. This technique allows accurate allelic discrimination, which makes it a valuable analytical tool for the detection and analysis of SNPs and mutations, and it enables simultaneous analysis of multiple samples and provides results faster and in real time (Heid, Stevens et al. 1996, Wong and Medrano 2005, Hoebeeck, Speleman et al. 2007).

1.2.1.2.5.1.7 Multiplex Amplifiable Probe Hybridization (MAPH)

The multiplex amplifiable probe hybridization (MAPH) method is a highly sensitive PCR-based targeted approach technique used extensively for the detection and analysis of gene or chromosomal locus-specific CNVs. This technique uses fluorescently labeled oligonucleotide probes designed to bind or hybridize to specific regions of the genome or the sample DNA. The basis of this technique is the quantitative estimation of the fluorescently labeled oligonucleotide probes after they bind or hybridize to their target sequence(s) in the immobilized target or sample DNA. The technique uses PCR amplification and the separation of amplified PCR products by denaturing CE coupled with fluorescence measurement to detect or identify and quantify the hybridized probes that bind or hybridize to their target sequence(s) in the sample DNA. The original method used radioactive labeling instead of fluorescent labeling and used slab agarose or PAGE in place of CE. The amount or number of copies of each PCR-amplified probe is proportional to and reveals the copy number of the target DNA sequence specific to that probe in the sample DNA. The technique allows simultaneous analysis of multiple regions of the genome through the use of multiple probes that target specific regions of the sample DNA (Armour, Sismani et al. 2000, Sellner and Taylor 2004, Patsalis, Kousoulidou et al. 2005).

The probes used in this technique are of different size, usually 100 – 600 bp, to allow their distinction by PCR amplification and electrophoresis and consist of two distinct regions, target-specific sequence region and universal primer sequence or 5′-sequence region identical in all the probes. The generation of these probes involves the PCR amplification of target sequences from the genomic or sample DNA using the primer pairs that contain identical 5′-universal primer sequence or 5′-identical sequence regions known as MAPH sequences. This method generates probes that are different in size and sequence barring the 5′-universal primer sequence region or 5′-identical sequence region that allows the subsequent PCR amplification of all the probes simultaneously through a multiplex single-tube setup using a single pair of universal primers. The different probes generated from multiple PCRs using this approach when mixed in appropriate quantities create a probe mix for use in hybridization reaction. The basic procedure of this technique involves the immobilization of the target DNA onto a membrane such as nylon membrane followed by overnight incubation with multiple target sequence-specific probes with 5′-identical sequence regions to allow the probes to bind or hybridize to the target regions in the sample DNA. The washing step that follows the overnight incubation and hybridization process removes the unbound probes. The subsequent heat-mediated denaturation step disrupts the hydrogen bonds that hold the target sequences and probes together and releases the hybridized probes into the solution. A specific volume or aliquot from the solution containing the released and previously hybridized probes then undergoes PCR amplification in a single-tube setup using a single pair of universal primers. The CE separates the PCR-amplified probes based on size. The measurement and analysis of intensities of fluorescent signals from the PCR-amplified and CE-separated probes allows their detection and quantification that enables the genotyping of CNVs. An increase in the intensity of the fluorescent signal corresponding to a particular probe and consequently a particular DNA sequence and represented by an increase in the height or area of the fluorescence peak in comparison with a reference or control DNA sample reveals the presence of duplications. Similarly, a decrease in the intensity of the fluorescent signal corresponding to a particular probe and consequently a particular DNA sequence and represented by a decrease in the height or area of the fluorescence peak in comparison with a reference or control DNA sample reveals the presence of deletions. The MAPH technique is a cost-effective and highly accurate multiplex CNV typing method that allows simultaneous analysis or genotyping of up to 40 multiple regions of the genome or sample DNA through the use of multiple probes. The simultaneous analysis of even up to 100 multiple regions is also possible with careful optimization. The technique enables the detection of even small CNVs with a size of around 150 bp. The array technique based on MAPH method offers an alternative method to array-CGH method discussed later on in this chapter. Further, the presence of unpredicted sequence polymorphisms in the genomic or sample DNA regions containing the probe binding sites does not affect the process of probe hybridization and consequently the operation and outcomes of this technique. However, this technique is labor-intensive and the specificity of the probes used in this technique is highly dependent on the sequence, which makes this technique unsuitable for analyzing genomic regions with unusual structural properties (Armour, Sismani et al. 2000, Schouten, McElgunn et al. 2002, Sellner and Taylor 2004, Patsalis, Kousoulidou et al. 2005).

1.2.1.2.5.1.8 Multiplex Ligation-Dependent Probe Amplification (MLPA)

The multiplex ligation-dependent probe amplification (MLPA) method is another highly sensitive PCR-based targeted approach technique that allows an accurate detection and analysis of gene or chromosomal locus-specific CNVs in a cost-effective manner. This technique like MAPH technique also uses fluorescently labeled oligonucleotide probes designed to bind or hybridize to specific regions of the genome or the sample DNA. However, unlike MAPH technique, the MLPA technique uses oligonucleotide probes, each of which consists of two half probes: the 5′-end half probe (5′-MLPA probe) and 3′-end half probe (3′-MLPA probe). Each half probe in turn consists of two distinct regions: an inner target-specific sequence region and an outer universal primer sequence region. Each half probe or only one of the two half probes of an MLPA probe also contains a stuffer sequence of distinct length that gives each full-length or complete probe and their PCR-amplified product a unique length that allows their differentiation based on size through denaturing CE. The inner target-specific sequence regions of the two half probes of an MLPA probe allows them to bind or hybridize to their respective contiguous target sequences in the sample DNA. Once hybridized to their respective perfectly complimentary target sequences in the sample DNA, the half probes undergo ligation to generate full-length or complete MLPA probe. The hybridized full-length probe consisting of two ligated half probes then undergoes PCR amplification, and the subsequent separation of PCR-amplified probes by CE allows their quantitative estimation to determine their copy number and consequently the copy number of their target DNA sequences. The ligation and subsequent PCR amplification of the half probes occurs only when the half probes hybridize to their perfectly matched sequences on the target or sample DNA, and even a single-base mismatch disrupts this whole process. As in MAPH assay, each of the probes used in the MLPA assay is different in size and sequence barring the 5′-universal primer sequence or 5′-identical sequence region. The 5′-identical sequence region of the probes allows the subsequent PCR amplification of all the probes simultaneously through a multiplex single-tube setup using a single pair of universal primers that consists of one fluorescently labeled primer and one unlabeled primer, and both possess a universal primer sequence region. The multiplex PCR amplification of MLPA probes using universal primer pair generates amplified probes each with a distinct size that allows their separation based on size using CE. The MLPA technique allows the PCR amplification of only the probes that undergo ligation, and therefore, the amount or number of copies of each ligated probe and subsequently the amount or number of copies of each PCR-amplified probe are proportional to and reveal the copy number of the target DNA sequence specific to that probe in the sample DNA.

The basic procedure of this technique involves the mixing of certain volumes of probe mix solution containing multiple target sequence-specific oligonucleotide half probes and the genomic or sample DNA solution. The subsequent heat-mediated denaturation step followed by incubation allows the two half probes of each full-length oligonucleotide probe to bind or hybridize to the contiguous target sequence regions in the sample DNA. The half probes hybridized to complementary contiguous target sequence then undergo ligation to generate full-length or complete MLPA probe. The subsequent heat-mediated denaturation step disrupts the hydrogen bonds that hold the target sequences and ligated or full-length probes together and releases the hybridized and ligated probes into the solution. As in MAPH technique, a specific volume or aliquot from the solution containing the released and previously hybridized probes then undergoes PCR amplification in a single-tube setup using a single pair of universal primers. The CE separates the PCR-amplified probes based on size. The measurement and analysis of intensities of fluorescent signals from the PCR-amplified and CE-separated probes allows their detection and quantification that enables the genotyping of CNVs. As in MAPH technique, an increase in the intensity of the fluorescent signal corresponding to a particular probe and consequently a particular DNA sequence and represented by an increase in the height or area of the fluorescence peak in comparison with a reference or control DNA sample reveals the presence of duplications. Similarly, a decrease in the intensity of the fluorescent signal corresponding to a particular probe and consequently a particular DNA sequence and represented by a decrease in the height or area of the fluorescence peak in comparison with a reference or control DNA sample reveals the presence of deletions (Schouten, McElgunn et al. 2002, Sellner and Taylor 2004).

The MLPA technique is a cost-effective and highly accurate and sensitive multiplex CNV typing method that allows simultaneous analysis or genotyping of up to 60 multiple regions of the genomic or sample DNA through the use of multiple probes. This technique uses very little quantity of genomic or sample DNA. Further, the use of ligation step makes this technique highly suitable for the detection and analysis of genetic variations including CNVs within genomic regions with high sequence identity, including the SDs. The high sensitivity of ligation reaction allows the MLPA probes to differentiate between two sequence regions that differ from each other by only a single nucleotide in the sequence region encompassing the ligation site. This technique is relatively easier and less labor-intensive in comparison with the MAPH technique and enables the detection of even small CNVs with a size of around 60 bp, which makes it a highly suitable method for the detection of small duplications as well as small deletions such as those confined to a single exon. A major limitation of this technique is that the presence of unpredicted sequence polymorphisms in the genomic or sample DNA regions in and around the probe ligation sites affects the process of probe hybridization and consequently the operation and outcomes of this technique, and this effect may manifest as false-positive detection of deletions (Schouten, McElgunn et al. 2002, Sellner and Taylor 2004).

1.2.1.2.5.1.9 Pyrosequencing

Pyrosequencing, as described already in Section 1.2.1.1.4.8, is a luminescence-based DNA sequencing technique that converts the PPi moiety generated during DNA polymerase-catalyzed elongation of DNA into visible light. This visible light generated whenever a new nucleotide is incorporated into the growing DNA strand is measurable in real-time and at any given time is directly proportional to and reveals the number and nature of nucleotides incorporated into the DNA. This technique usually uses a four-enzyme cascade consisting of enzymes, DNA polymerase, sulfurylase, luciferase and apyrase and involves the conversion of each PPi released on incorporation of a nucleotide into the growing DNA strand to ATP. The ATP produced from PPi participates in a luciferase reaction that generates visible light with amount or intensity proportion to the amount of ATP channelized into luciferase reaction. In this manner, this technique links the process of incorporation of new nucleotides into the growing or elongating strand to the proportional generation of visible light that is measurable in real-time and gives the sequence of the template strand (Ronaghi, Uhlén et al. 1998, Ronaghi 2001). Pyrosequencing is a highly useful technique for the detection and analysis of SNPs, mutations, tri-allelic variations and small indels (insertions or deletions) and is a highly sensitive technique used extensively for the detection and analysis of gene or chromosomal locus-specific CNVs. Many studies report the detection and analysis of CNVs using this technique such as the detection of CNVs in cytochrome P450 2D6 (CYP2D6) in humans and CNVs of dominant white/KIT locus in pigs. However, the pyrosequencing technique often underestimates the copy numbers if the copy numbers are particularly high, but it is still a highly useful technique (Pielberg, Day et al. 2003, Zackrisson and Lindblom 2003).

1.2.1.2.5.2 Whole Approach for CNV Detection (Genome-Wide and Exome-Wide CNV Detection)

1.2.1.2.5.2.1 Microarray Techniques (Array-CGH and SNP-Array)

The microarray techniques for the detection of structural chromosomal rearrangements including CNVs use DNA microarrays, which consist of specific collection of small DNA segments, known as probes, arrayed or immobilized in an ordered manner as microscopic spots, each containing picomolar amounts of DNA, onto a solid surface such as silicon, plastic or glass slides to serve as evaluation targets. The microarray techniques used extensively for the study of CNVs include array-CGH and SNP-array.

1.2.1.2.5.2.1.1 Array-CGH (aCGH)

The array-CGH (aCGH), a modification of the standard CGH, involves the use of DNA microarrays instead of normal human metaphase chromosomes. This technique uses DNA segments or probes of different sizes depending on the resolution desired for the detection and analysis of specific structural chromosomal rearrangements including CNVs. The DNA segment probes, which determine the resolution of aCGH, are variable in size and depend on the method and source used for their generation. The length, number and distribution of the probes determine the resolution of the array, which represents a particular region of the genome or the whole genome. Besides resolution, the two other important factors that affect the potential of aCGH to detect CNVs and other structural chromosomal rearrangements include hybridization SNR and the characteristics of probe response. The probes used for aCGH include BAC clone probes, fosmid and cosmid clone probes, cDNA clone probes, genomic PCR product probes and oligonucleotide probes (Schena, Shalon et al. 1995, Lucito, Healy et al. 2003). The genomic clones provide a resolution range, which varies between 80 and 200 Kb for probes generated from BACs and 20 – 50 kb for probes generated from fosmid and cosmid clones. The probes generated from BACs are not suitable for the detection of copy number changes smaller than 50 kb in size. Similarly, probes generated from fosmid and cosmid clones do not work efficiently and accurately beyond the resolution of 15 kb, so they are not suitable for the detection of copy number changes including CNVs smaller than 15 kb in size. An array consisting of over 30,000 overlapping BAC clones, which samples or tiles the human genome, is commercially available (Ishkanian, Malloff et al. 2004, Fiegler, Redon et al. 2006). The probes generated from cDNA clones provide a better resolution range of 0.5 – 2 kb, which is the level of a single gene or a portion of the gene. However, for the genome-wide detection of CNVs, these probes produce a variable or uneven resolution due to the non-uniform distribution of genes across the genome. Further, the mismatched hybridization between the cDNA clone probes and the genomic DNA decreases the SNR and reduces probe response to changes in fluorescence ratio, which subsequently affects the accuracy of the interpretations (Pollack, Perou et al. 1999, Porkka, Saramäki et al. 2002, Squire, Pei et al. 2003, Ylstra, van den Ijssel et al. 2006). The probes based on genomic PCR products enable a more extensive genomic coverage and genomic analysis at a higher resolution range of 100 bp to 1.5 kb. However, the major limitations of genomic PCR product-based probes are the poor SNR and higher cost of probe generation, in particular, for genome-wide arrays (Mantripragada, Buckley et al. 2003, Dhami, Coffey et al. 2005, Ylstra, van den Ijssel et al. 2006). The clone and genomic PCR product-based arrays can have more than 60,000 microscopic DNA-probe spots immobilized onto a glass slide, which represents or samples the entire genome once every 50 Kb. In other words, the immobilized probes represent the genomic regions that are 50 Kb apart. However, it is not practically viable to visualize all the 60,000 DNA spots, which limits the resolution of this array. The oligonucleotide probes deliver the highest possible resolution with the limit to detect structural chromosomal rearrangements including CNVs down to the scale of few nucleotides. The arrays based on oligonucleotide probes synthesized as overlapping nucleotides, usually 25 – 85 bp in size, with a single nucleotide shift across the genomic segment under analysis can achieve almost a single-nucleotide-level resolution. The very high resolution makes oligonucleotide array-CGH a very useful method for the analysis of specific genomic regions for smaller chromosomal rearrangements including small translocations and CNVs. However, oligonucleotide aCGH is not suitable for genome-wide detection of CNVs as this entails the generation of a very large number of probes, which is prohibitively a difficult and expensive process. Further, compared to BAC clone probe array, it offers a much better resolution, but the hybridization SNR is poor. An array consisting of 2.1million oligonucleotide probes for analysis of human genome is commercially available (Lucito, Healy et al. 2003, Carvalho, Ouwerkerk et al. 2004, Carter 2007, Gribble, Kalaitzopoulos et al. 2007). A single aCGH assay is equivalent to conducting thousands of FISH procedures at once. The array-CGH technique allows simultaneous detection of diverse structural chromosomal rearrangements and aberrations including deletions, duplications, translocations, amplifications and aneuploidies (in embryos). This technique can analyze any target sequence or region that constitutes the array. The target regions include specific regions of the genome, whole genome or specific regions on the chromosome. The chromosomal regions include pericentromeric and subtelomeric regions and chromosomal regions containing loci of microdeletion and microduplication syndromes. Although the aCGH is the method of choice for CNV detection, it is not a reliable technique for accurate detection of CNVs that encompass SDs, or LCRs because the same fluorescence signal results from hybridization at multiple locations, which share the LCR-associated sequence homology (Coughlin, Scharer et al. 2012).

1.2.1.2.5.2.1.2 SNP Microarray (SNP Array or SNP Chip) Technique

The SNP microarray or SNP array technique, as described already in Section 1.2.1.1.4.10, is a DNA microarray-based high-throughput genotyping method that allows qualitative, quantitative and simultaneous analysis of a large number of pre-defined and known reported genetic variations including SNPs and CNVs. The SNP microarray involves hybridization of fluorescently labeled fragmented single-stranded DNA sequences to arrays that contain hundreds of thousands of unique ASO probe sequences each complimentary or specific to and designed to bind to target DNA sequences that contain regions corresponding to one or the other allelic form of the SNPs under study. If we consider A and B represent the two allelic forms of the SNP under study, this technique generates any of the three types of fluorescent intensity clusters. Each of the three types of fluorescent intensity clusters corresponds to one of the three different allelic or genotypic forms: homozygous AA genotype, homozygous BB genotype or heterozygous AB genotype, in which the particular SNP occurs in the DNA sample under study. As mentioned already, the binding of ASO probe to a matched or mismatched target DNA sequence generates a fluorescent hybridization signal whose intensity depends on the amount of target DNA sequence present in the sample in addition to the degree of affinity between the hybridized ASO probe and target DNA sequence that indicates matched or mismatched pairing. This means that the intensity of fluorescent hybridization signal generated from the hybridization of a specific probe to its target sequence in a specific region increases with an increase in the quantity of the target DNA sequence or the DNA sequence fragment that harbors the target sequence. This feature of the SNP microarray makes it a highly suitable technique for the detection and analysis of CNVs at the sequence regions that contain the SNP targeted in the array. Further, the currently available commercial SNP chips contain hundreds of thousands of singleton non-polymorphic copy number probes that are not dependent on the sequence properties or local sequence of the SNPs targeted in the array. For example, the Affymetrix SNP microarray contains probes for the detection and analysis of close to one million (946,000) CNVs (Cooper, Zerr et al. 2008, McCarroll, Kuruvilla et al. 2008).

The SNP microarray allows the detection of various types of deletion events. Going back to the example where we consider A and B as the two allelic forms, the detection of a CNV deletion generates three more types of fluorescent intensity clusters in addition to the three types of fluorescent intensity clusters mentioned above. The simultaneous generation of reduced intensity clusters corresponding to allele A and allele B indicates heterozygous loss or deletion. The generation of reduced intensity cluster corresponding to allele A, but no signal corresponding to allele B indicates hemizygous deletion of allele A (A_). The generation of reduced intensity cluster corresponding to allele B, but no signal corresponding to allele A indicates hemizygous deletion of allele B (B_). The total absence of any signal or no generation of fluorescent intensity cluster indicates homozygous deletion or loss of either the AA genotype (_ _) or BB genotype (_ _) depending on the genotype of the SNP under study. The SNP microarray also allows detection of CNVs involving duplication or gain events of different types, homozygous gains AAAA, AAAB, AABB, ABBB and BBBB and hemizygous gains AAA, AAB, ABB and BBB. In case of homozygous gain events, there is generation of four fluorescent intensity clusters in each case with intensity cluster type dependent on the alleles involved. In case of hemizygous gain events, there is generation of three fluorescent intensity clusters in each case with intensity cluster type dependent on the alleles involved (LaFramboise 2009).

1.2.1.2.5.2.2 Sequencing Techniques

1.2.1.2.5.2.2.1 First-Generation Sequencing Techniques

The sequencing techniques, as described in Section 1.2.1.1.4.11.1, include the FGS, SGS and TGS techniques. The DNA sequencing enables precise determination of the nucleotide sequence in the DNA, and the analysis of the DNA sequence allows the detection of genetic variations including CNVs. The FGS techniques, in particular, the Sanger sequencing technique, considered a gold standard in sequencing and for the detection of genetic variations including CNVs, allow rapid and accurate detection of CNVs. However, the detection of CNVs using FGS techniques highly depends on the types of CNVs and their length. The detection of CNVs also depends on their position with reference to the annealing sites of the sequencing primers used for the amplification. For example, the allelic dropout results if the deletion CNVs present in the DNA sequence region under analysis affect one of the primer-binding sequences. Further, if the deletion CNVs encompass the DNA sequence region that includes the entire DNA region amplified by the sequencing primers, then the sequencing technique falsely reveals the presence of hemizygous alleles instead of homozygous alleles. Another drawback of the sequencing techniques including the FGS and SGS techniques is that the presence of unknown or uncharacterized homologous copies of the genomic regions under analysis such as the pseudoexons and pseudogenes reveals the false presence of genetic variations and contributes to inaccurate false-positive results (Kubiritova, Gyuraszova et al. 2019). However, it is possible to some extent to minimize the effect of various factors that adversely affect the accuracy of the sequencing and consequently the process of accurate detection of genetic variations including CNVs. These include analysis of specific regions of the genome for CNV detection and analysis by another alternative method in addition to sequencing or by two or more methods based on different principles and susceptible to different types of errors, limitations and inherent biases.

1.2.1.2.5.2.2.2 Second-Generation Sequencing Techniques

The SGS techniques allow the detection and analysis of CNVs at three different levels: at the level of a single gene or a group of genes (targeted sequencing), at the level encompassing all the exons in a genome (exome level or whole exome sequencing) and at the level encompassing the entire genome (genome level or whole genome sequencing) (Wang, Nettleton et al. 2014). The detection and analysis of CNVs using SGS/NGS techniques involves two major approaches: the resequencing approach and the de novo sequence assembly approach (Figure 1.7).

The resequencing approach involves the precise determination of the nucleotide sequence in a DNA sample or DNA of an individual and comparison of the generated sequence with that of the known sequence of reference genome through the alignment of the generated sequence reads with that of the reference genomic sequence. The alignment of sequence reads from sample DNA with that of the reference genomic sequence reveals the difference between the two sequences and through prediction analysis; using powerful computational techniques based on complex algorithms reveals the presence of genetic variations including SVs such as CNVs. The NGS-based resequencing approach reveals clearly and accurately the structure and specific location of genetic variations including CNVs. The resequencing approach using NGS techniques allows detection and analysis of a large number of genetic variations including CNVs in a single sequencing event (Alkan, Coe et al. 2011). However, the accurate detection and characterization of SVs including CNVs using the resequencing approach is a difficult process due to the inherent limitations of the sequencing techniques and certain biological

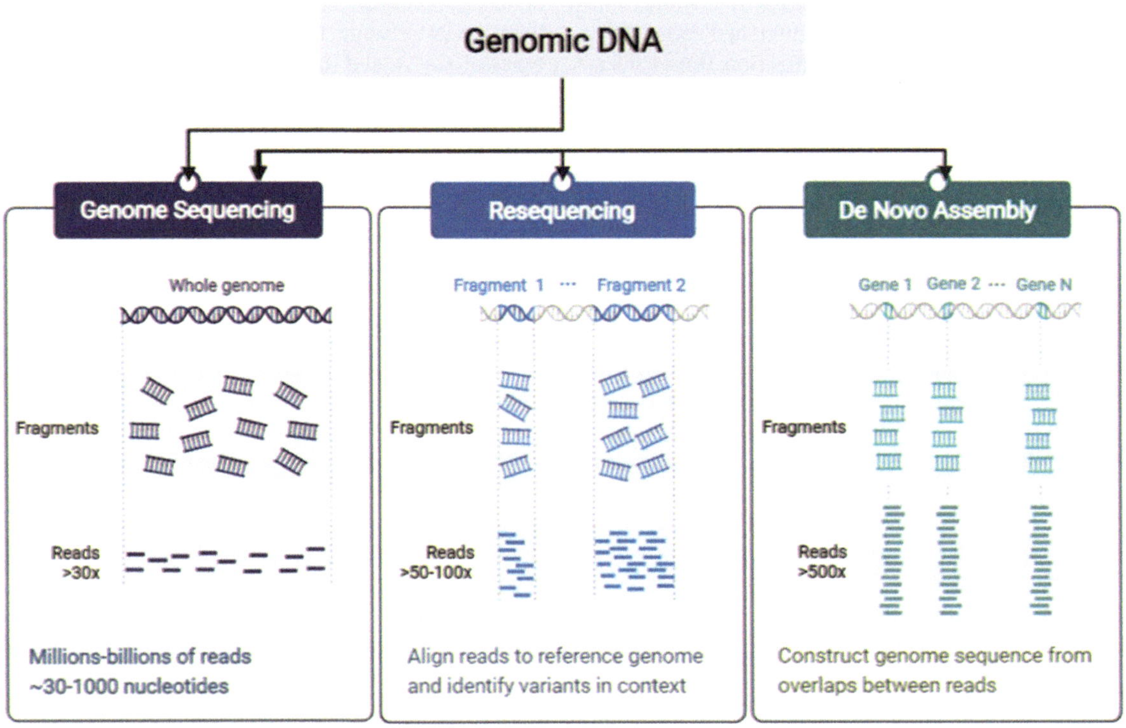

FIGURE 1.7 Detection of CNVs using NGS techniques. The detection of CNVs using SGS/NGS techniques involves two major approaches: the resequencing approach and the de novo sequence assembly approach.

characteristics of genomic sequences including the sequence regions that consist of SVs. The limitations of NGS techniques include the sequencing errors, short read lengths, short insert sizes and GC bias, and they adversely affect the accuracy and applicability of NGS techniques in the detection and analysis of SVs such as CNVs. Further, the presence of CNVs, in particular recurrent CNVs and complex CNVs, and certain biological characteristics commonly associated with CNVs such as enrichment for repetitive sequences near the breakpoints contribute to the inherent difficulty associated with the prediction, detection and analysis of CNVs through the alignment of the generated sequence reads with that of the reference genomic sequence in the resequencing approach. However, the emergence of complex algorithm-based computational approaches to a large extent aids in overcoming these limitations (Kidd, Cooper et al. 2008, Medvedev, Stanciu et al. 2009, Alkan, Coe et al. 2011).

The de novo sequence assembly approach applies powerful computational techniques based on complex algorithms that allow highly advanced statistical and bioinformatic analysis to assemble or construct the genomic sequences from the overlaps between the sequence reads without the use of reference genome or genomic sequence. The comparison of assembled or constructed genomic sequences with either the reference genome or the other individual assembled or constructed genomic sequences allows the identification of genetic variations including CNVs. The successful and accurate de novo assembly approach enables an efficient and accurate identification and characterization of CNVs (Kosugi, Momozawa et al. 2019). However, the NGS techniques generate short sequence reads, and the current human genome sequence assembly consists of hundreds of thousands of intermediate-sized DNA sequence assemblies based on a series of overlapping DNA sequences known as contigs. These factors along with the inherent difficulty to achieve accurate assembly of genomic sequences in the absence of reference genome or genomic sequence affect the accuracy of detection and analysis of genetic variations including CNVs using the de novo assembly approach (Schatz, Delcher et al. 2010).

The various NGS-based computational approaches used for the detection or identification and characterization of SVs such as simple and complex CNVs include read depth methods, split read methods, read-pair or paired end mapping (PEM) methods and whole genome sequence assembly. Apart from whole genome assembly the most accurate method available currently under NGS platform for the detection or identification of SVs including CNVs is the read depth method or depth of coverage method. PEM method represents the other sensitive and accurate approach for CNV detection or identification. A method known as Genome STRucture In Populations (Genome STRiP) that combines both these methods allows the most sensitive and most accurate detection or identification of the SVs including CNVs (Mills, Walter et al. 2011). The read depth method or depth of coverage method is a useful and highly accurate approach for the determination of absolute copy number of a specific DNA sequence segment. This method reveals the copy number in terms of the number of sequence reads that map to a single specific region in the genome. The number or amount of sequence reads that align or map to a specific region in the genome reveals the number of copies of that specific sequence read in the genome under study or, in other words, the number of times that specific sequence read exists in the genome under study. This method detects the CNVs by analyzing the differences between the sample genome and the reference genome in the number or amount of sequence reads that align or map to a specific region. The increased number of or gain in specific sequence reads or the increase in specific read depth coverage in the sample genome in comparison with the reference genome reveals the presence of duplications. On the other hand, the decreased number of or loss or absence of specific sequence reads or reduced or null specific read depth coverage in the sample genome in comparison with the reference genome reveals the presence of deletions. The increase in the number of sequence reads or increase in read depth coverage that reveals the amplification or duplication of specific sequence reads is proportional to the number of copies of the duplicated sequence segment in the sample genome. The decrease in the number of sequence reads or reduction in read depth coverage by half reveals the presence of heterozygous deletions, whereas the total absence of specific sequence reads reveals the presence of homozygous deletions. The read depth method enables the accurate detection of CNVs present in the complex regions of the genome, and it allows an accurate and rapid detection of complex and multi-allelic CNVs. Further, to enable the detection of smaller CNVs and those with high copy number in the sample genome, a deeper sequence read depth (SRD) analysis provides a rapid and highly accurate approach. The read depth method analyzes the sequence reads at random that gives a variable coverage and presents data as distribution, and therefore, this method uses distributions, mostly the Poisson distribution of sequencing reads to detect and characterize the CNVs (Medvedev, Stanciu et al. 2009, Sudmant, Kitzman et al. 2010).

The split read method allows direct detection or identification of structural variations including CNVs through identification of split reads or split sequence reads. The split reads are sequence reads with split alignment to the reference genome or with alignment to the reference genome in two parts. The identification of split reads involves detection of reads with structural variant or CNV breakpoints. This method identifies multiple split reads that share or contain the same SV or CNV breakpoint in order to minimize false-positive results and increase the accuracy of the CNV prediction or detection. This method involves independent alignment of two split sequence reads to the reference genome. Therefore, the accurate detection or prediction of sequence reads is possible only in case of sequence reads with length enough to allow unique and stringent alignment to the reference genome. In order to increase the accuracy of prediction or detection of CNVs, the split read methods use paired reads and require one of the sequence reads in the pair to align fully or for full-length to the reference genome. The method uses the end of the split sequence read with full-length alignment to the reference genome to search for alignments corresponding to the other split sequence read of the pair. This method allows accurate detection or prediction only when the length of split reads is more than 50 bp (Maher, Kumar-Sinha et al. 2009, Ye, Schulz et al. 2009).

The read-pair or paired end mapping (PEM) methods involve generation of paired sequence reads from the opposite ends of a large-sized DNA fragment or clone obtained from an individual genome and alignment of these sequence reads to the reference genome. The alignment of paired sequence reads to the reference genome mostly generates concordant sequence read pairs, and it also generates discordant sequence read pairs. The concordant sequence read pair consists of sequence reads that align to the reference genome at a distance from each other equal to the length of the DNA fragment under analysis. The discordant sequence read pair consists of sequence reads that align to the reference genome at a distance from each other shorter than or longer than the length of the DNA fragment or align to the reference genomic sequences located on different chromosomes. The generation of discordant sequence read pairs reveals the presence of structural variants including CNVs, but may also arise due to sequencing errors. The generation of discordant sequence read pairs that align to the reference genome at a distance from each other shorter than the length of the DNA fragment under analysis reveals the presence of duplication CNVs. The generation of discordant sequence read pairs that align to the reference genome at a distance from each other longer than the length of the DNA fragment under analysis reveals the presence of deletion CNVs. The generation of discordant sequence read pairs that align to the reference genomic sequences located on different chromosomes reveals the presence of translocations. Similarly, the generation of other types of discordant sequence read pairs enables the detection of inversions and transpositions. The presence of clusters of discordant sequence read pairs that reveal the same SV or CNV allows the distinction of real structural variations including CNVs from sequencing errors. The read-pair or paired end mapping (PEM) methods use various complex algorithms to detect the clusters of discordant sequence read pairs that allow the distinction between the real SV or CNV and sequencing errors. The read-pair or paired end mapping (PEM) method is the most common NGS approach in use to detect or identify the presence of SVs and CNVs using resequencing (Hormozdiari, Alkan et al. 2009, Maher, Kumar-Sinha et al. 2009, Sindi, Helman et al. 2009, Ye, Schulz et al. 2009, Quinlan, Clark et al. 2010).

The computational approaches discussed earlier, the read depth methods, split read methods and read-pair or paired end mapping (PEM) methods involve the alignment of sequence reads under analysis to the reference genome. These methods enable accurate detection or identification of complex and multi-allelic CNVs not possible with FGS and allow the comprehensive mapping of CNVs in the human genome through characterization of CNV breakpoints. However, these computational methods nevertheless useful and accurate for the detection or identification of CNVs have their own limitations. The read depth methods cannot accurately characterize the breakpoints of CNVs, whereas split read methods and read-pair or paired end mapping methods possess relatively limited sensitivity and accuracy for the detection and analysis of duplication CNVs. The genome sequence assembly-based methods, in particular the whole genome sequence assembly, offer a more comprehensive and most accurate approach for the detection, identification and characterization of SVs including CNVs. The genome sequence assembly-based methods analyze and compare a series of longer overlapping DNA sequence segments of an individual genome, known as contigs. The whole genome sequence assembly through the application of sophisticated or complex algorithms and highly intensive computing approaches allows the most accurate detection and characterization of complex and multi-allelic CNVs. This method allows the most accurate characterization of CNV breakpoints at almost a single-nucleotide-level resolution. The other approach in the genome sequence assembly is the whole exome sequencing assembly method that allows the detection, identification and characterization of SVs including CNVs in only the portions of genome that code for proteins. The approach analyzes only the exons that encompass most of the pathogenic genetic variations present in the genome. This method allows the accurate and relatively more cost-effective detection and characterization of those SVs or CNVs that are biologically more relevant (Bartha and Győrffy 2019, Magi, Bolognini et al. 2019).

The TGS techniques are fast emerging as rapid, cost-effective and accurate methods for the detection and characterization of CNVs. The TGS techniques allow generation of longer sequence reads and use new genome sequence assembly algorithms that enable more accurate and stringent alignment of sequence reads even for complex regions of the genome and for genomic regions encompassing LCRs (segmental duplications) and other repetitive sequences relevant to the formation and occurrence of structural variants including CNVs (Eid, Fehr et al. 2009, Tham, Tirado-Magallanes et al. 2020).

1.2.1.2.6 Functional Significance and Clinical Impact of Copy Number Variations (CNVs)

The effect of CNVs on gene function like that of SNPs often manifests through their effect on the phenotypic expression of the gene under study and depends largely on the location of the CNVs within the gene or the genomic regions adjacent to the gene. The CNVs occur in both the coding and the noncoding regions of the genome. The duplication CNVs as well as deletion CNVs in the coding regions of the genome can modulate gene expression and function through direct disruption of the coding sequences of the gene, through alteration of gene dosage and through perturbation of long-range regulation of gene expression. The deletion CNVs in the coding regions of the genome (regions of genome that harbor genes) under homozygous condition lead to partial or total gene knockout that manifests as complete loss or disruption of gene expression and function. The heterozygous deletion CNVs as well as duplication CNVs or amplifications in the coding regions of the genome also affect gene expression and function. The CNVs in the noncoding regions of the genome may also affect gene expression and function and therefore functionally modulate the genome. The CNVs in the coding regions of the genome and in noncoding regions, in particular, in regions that flank the genes, may affect the tissue-level gene expression or the tissue transcriptome even on a global scale and impact posttranscriptional gene regulation, and this effect can extend over whole of the chromosome that contains the affected regions. The CNVs mediate these effects through the modulation of the degree and time course of gene expression and possibly involve the role of lncRNAs and sncRNAs such as miRNAs, tRNAs, piRNAs and snoRNAs that occur within these CNVs. In other words, these effects on gene expression and function result from the CNVs of genomic regions that encode sncRNAs and lncRNAs. The CNVs depending on their location within the genome may also disrupt the spatial organization of the genome through alteration of chromatin interaction domains or topologically associating domains (TADs). These disruptions affect the gene enhancer interactions and ultimately disrupt normal gene expression and function. In addition, the CNVs affect gene expression and function through many unknown or poorly understood mechanisms. Therefore, the CNVs in the coding regions of the genome and in the noncoding regions, especially the regions flanking the genes, alter the phenotypic expression of genes through modulation of gene transcription and subsequently gene translation. The modulation of gene transcription and gene translation may either affect the susceptibility to or even cause various diseases or disorders (Merla, Howald et al. 2006, Henrichsen, Chaignat et al. 2009, Henrichsen, Vinckenbosch et al. 2009, Chaignat, Yahya-Graison et al. 2011, Lupiáñez, Kraft et al. 2015, Nowakowska 2017, Kumaran, Krishnan et al. 2018, Liu, Gu et al. 2019).

The CNVs represent an important and predominant source of genetic variations across the human genome. The CNVs include both non-pathogenic (benign or normal) genetic variations and pathogenic genetic variations and profoundly affect human health and disease states. Most of the CNVs, in particular, the recurrent CNVs like that of other genetic variations, are categorizable into various types depending on their functional effect including their clinical impact on the function of human genome. The various types of CNV categories depending on their functional effect and clinical impact include benign CNVs, likely benign CNVs, variant of uncertain significance (VUS) CNVs, likely pathogenic CNVs and pathogenic CNVs. However, this categorization is often complex and difficult, but very useful in analyzing the functional impact of CNVs on human genome and in understanding the potential role of CNVs in the pathogenesis of various diseases. The non-pathogenic or benign CNVs encompass genes that exhibit less conservation and more variability in copy numbers. The pathogenic CNVs are often large, extend across multiple genes and gene regulatory regions and frequently encompass dosage-sensitive genes such as those involved in developmental processes and genes that exhibit higher degree of evolutionary conservation in copy numbers. The dosage sensitivity of genes plays a vital role in determining the non-pathogenic or pathogenic nature of CNVs; the CNVs that encompass dosage-sensitive genes are frequently pathogenic, whereas the CNVs that encompass genes less sensitive to dosage are often non-pathogenic (Richards, Aziz et al. 2015, Nowakowska 2017, Rice and McLysaght 2017).

The pathogenic CNVs usually occur in the form of duplications (amplifications) and deletions. Depending on various characteristics such as size and type of genomic regions involved, location within the genome, the number of genes they harbor and the presence of known or potential risk factors, the pathogenic CNVs modulate the susceptibility to or even cause a number of diseases or disorders. These diseases or disorders include genetic diseases with Mendelian inheritance (Mendelian diseases), contiguous gene disorders, chromosomal aberrations or disorders, and sporadic and complex multifactorial diseases (Shaffer and Lupski 2000, Stankiewicz and Lupski 2002).

The CNVs through modulation of single gene dosage can affect the susceptibility to various diseases or disorders such as psoriasis (β-defensin gene; chromosomal location,8p23.1), systemic lupus erythematosus (SLE) [C4A and C4B genes; chromosomal location, 6p21.33], glomerulonephritis [Fc gamma receptor IIIB (FCGR3B) gene; chromosomal location,1q23.3] and acquired immunodeficiency syndrome (AIDS) [C-C motif chemokine ligand 3 like 1 (CCL3L1) gene; chromosomal location,17q12] (Gonzalez, Kulkarni et al. 2005, Aitman, Dong et al. 2006, Hollox, Huffmeier et al. 2008, Pereira, Perazzio et al. 2019). The CNVs in the form of small duplications and deletions that affect a single gene modulate the susceptibility to or even cause a number of monogenic diseases. These monogenic diseases include various X-chromosome-linked disorders such as X-linked Alport syndrome [collagen type IV alpha 5 chain (COL4A5) gene; chromosomal location, Xq22.3], X-linked juvenile retinoschisis [retinoschisin 1(RS1) gene; chromosomal location, Xp22.13], mucopolysaccharidosis type II or Hunter syndrome [iduronate 2-sulfatase (IDS) gene; chromosomal location,Xq28], Duchenne muscular dystrophy [dystrophin (DMD) gene; chromosomal location, Xp21.2-p21.1], androgen-insensitivity syndrome [androgen receptor (AR) gene; chromosomal location, Xq12], Menkes disease (MD) [ATPase copper transporting alpha (ATP7A) gene; chromosomal location, Xq21.1, and X-linked recessive hypoparathyroidism (HYPX or XLHPT), interstitial insertion/deletion, chromosomal location, Xq27.1 near SRY-box transcription factor 3 (SoX3) gene] (Cooper, Ball et al. 1998, Tümer, Birk Møller et al. 2003, Bowl, Nesbit et al. 2005, Vissers, Veltman et al. 2005). The large duplications and deletions often affect the chromosomal regions encompassing a contiguous set of genes, often dosage-sensitive developmental genes, and modulate the susceptibility to or even cause a number of rare genomic disorders, referred to as microduplication and microdeletion disorders. These rare genomic disorders include Smith-Magenis syndrome (deletion, chromosomal location,17p11.2), DiGeorge (velocardiofacial) syndrome (deletion, chromosomal location,22q11.2),Williams-Beuren (Williams) syndrome (deletion, chromosomal location,7q11.23), Potocki-Lupski syndrome (duplication, chromosomal location,17p11.2), Prader-Willi syndrome (PWS) and Angelman syndrome (AS) (deletion, imprinting center defect or uniparental disomy, chromosomal location,15q11.2-q13), CHARGE syndrome (deletion, chromosomal location,8q12.2), Cri du chat syndrome (deletion, chromosomal location, 5p15), Charcot-Marie-tooth disease type 4B2 (deletion, chromosomal location,11p15.4),Wolf-Hirschhorn syndrome (deletion, chromosomal location, 4p16.2 and 4p16.3), spinal muscular atrophy (SMA) (deletion, chromosomal location, 5q13.2), juvenile nephronophthisis (deletion, chromosomal location, 2q13),[deletion, chromosomal location, AZFc (azoospermia factor c) region, Y chromosome], infantile spasms (deletion, chromosomal location,7q11.23-q21.11), Charcot-Marie-Tooth disease type 1A (CMT1A) (duplication, chromosomal location,17p12), Cockayne-Pelizaeus-Merzbacher disease or Pelizaeus-Merzbacher disease (duplication, chromosomal location, Xq22.2), Alzheimer's disease (duplication, chromosomal location, 21q21.3) and autism (duplication, chromosomal location,7q11.23) (Konrad, Saunier et al. 1996, Potocki, Chen et al. 2000, Kuroda-Kawaguchi, Skaletsky et al. 2001, Bi, Yan et al. 2002, Bayés, Magano et al. 2003, Baldini 2004, Iafrate, Feuk et al. 2004, Tuzun, Sharp et al. 2005, McCarroll, Hadnott et al. 2006, Marshall, Young et al. 2008, Sanders, Ercan-Sencicek et al. 2011, Rovelet-Lecrux, Legallic et al. 2012, Wirth, Karakaya et al. 2020).

The sporadic and complex multifactorial diseases result from a combinatorial effect of various genetic and environmental factors. The CNVs can modulate the susceptibility to or even play an important role in the pathogenesis of various complex multifactorial diseases. The CNVs are emerging as potential missing links that could explain, at least partially, the missing heritability of complex diseases (Manolio, Collins et al. 2009). The studies reveal proven or potential role of CNVs in modulating the susceptibility to or even cause various complex multifactorial diseases. These complex diseases include cancers such as acute lymphoblastic leukemia (ALL), nasopharyngeal carcinoma (NPC), squamous cell lung cancer, colorectal cancer (CRC), breast cancer, ovarian cancer and prostate cancer (Paris, Andaya et al. 2004, Weir, Zhao et al. 2004, Mullighan, Goorha et al. 2007, Leary, Lin et al. 2008, Tse, Su et al. 2011, Yoshihara, Tajima et al. 2011, Cancer Genome Atlas Research Network 2012). These complex diseases also include neuropsychiatric disorders, neurodevelopmental disorders and neurodegenerative disorders such as autism spectrum disorders (ASDs), schizophrenia, Alzheimer's disease and Parkinson's disease (Lee and Lupski 2006, Sebat, Lakshmi et al. 2007, Szatmari, Paterson et al. 2007, ISC 2008, Stefansson, Rujescu et al. 2008, Fanciulli, Petretto et al. 2010, Stankiewicz and Lupski 2010, Sekar, Bialas et al. 2016, Martin, Tammimies et al. 2019, Martin, Wain et al. 2020). The studies also report proven or potential role of CNVs in cardiovascular diseases such as congenital heart disease (CHD) (Fahed, Gelb et al. 2013, Sailani, Makrythanasis et al. 2013, Glessner, Li et al. 2020), bone-related diseases such as osteoporosis (OP) (Rivadeneira, Styrkársdottir et al. 2009, Deng, Zhao et al. 2010), obesity (Rivadeneira, Styrkársdottir et al. 2009, Deng, Zhao et al. 2010, Speliotes, Willer et al. 2010, Wheeler, Huang et al. 2013), developmental disorders (Park, Nam et al. 2019), autoimmune disorders, inflammatory bowel disease (IBD) including Crohn's disease (Fellermann, Stange et al. 2006, Fanciulli, Petretto et al. 2010, Stankiewicz and Lupski 2010, Ellinghaus, Bethune et al. 2015, Frenkel, Bernstein et al. 2019) and human immunodeficiency virus (HIV) infection (Fanciulli, Petretto et al. 2010, Stankiewicz and Lupski 2010).

The CNVs like SNPs serve as valuable tools in the field of pharmacogenomics. The CNVs have a profound effect on genes in general and genes that encode drug targets and drug-metabolizing enzymes in particular. Studies reveal the role of CNV-mediated inter-individual gene expression variability in modulating the inter-individual differences in the response to drugs (pharmacological or drug response), and therefore, CNVs serve as important tools or molecular markers in the field of personalized or precision medicine. For example, cytochrome P450 2D6 (CYP2D6) gene encodes the enzyme cytochrome P450 monooxygenase that plays a vital role in phase I reactions of metabolism and elimination of a large number of clinically used drugs. The CYP2D6-catalyzed phase I reactions of drug metabolism involves chemical modification processes such as hydroxylation and dealkylation and occurs through oxidation, reduction and hydrolysis. The CNVs encompassing this gene vary in number from 0 to 13, which significantly affects the plasma concentrations of various drugs including antidepressants, and consequently modulate the response to these drugs. The CNVs may result in non-responsiveness of individuals to antidepressant therapy that potentially leads to increased incidences of suicide. The CNVs may also result in higher toxicity following the treatment with opioid analgesic, codeine (Johansson, Lundqvist et al. 1993, Wang, Yang et al. 2009, Johansson and Ingelman-Sundberg 2011). Similarly, the gene sulfotransferase 1A1 (SULT1A1) gene encodes the enzyme sulfotransferase that catalyzes the sulfate conjugation of a wide array of hydroxyl or amine group containing acceptor molecules using 3′-phospho-5′-adenylyl sulfate (PAPS) as sulfonate donor. The acceptor molecules that undergo sulfotransferase-catalyzed sulfate conjugation include xenobiotics including the drugs minoxidil and acetaminophen. The sulfate conjugation constitutes the phase II reactions of drug metabolism. The CNVs encompassing SULT1A1gene may significantly increase the risk of adverse drug reactions and chemical carcinogenesis (Daniels and Kadlubar 2013, Palli, Rizzolo et al. 2013).

The human microbiome plays a vital role in various physiological processes and significantly affects the health including physical and mental well-being of an individual. The genetic variations including CNVs also affect the diversity and composition of human microbiome. For example, studies reveal the significant association of duplication CNVs in human salivary amylase (AMY1) gene with increased population of starch-degrading microorganisms that produce higher amounts of short-chain fatty acids (SFAs). The studies also reveal the association of duplication CNVs in human AMY1 gene with increased abundance of oral *Porphyromonas gingivalis,* anaerobic Gram-negative bacteria involved in the pathogenesis of inflammatory tooth disease, periodontitis. The duplication CNVs in human AMY1 gene demonstrates the physiological and pathophysiological significance of non-pathogenic CNVs (Mohajeri, Brummer et al. 2018, Poole, Goodrich et al. 2019).

1.2.1.3 Copy Neutral Variations

Copy number neutral variations are structural genetic variations that involve no net loss or gain of chromosomal segments or genomic sequence stretches unlike the CNVs. The copy number neutral variations include inversions and balanced translocations. The inversion genetic variations or simply inversions result from double breaks in a chromosomal segment followed by end-to-end chromosomal ligation of the broken chromosomal segment, but in an inverted manner. In other words, the inversions involve breaking and subsequent flipping of a DNA sequence stretch or a chromosomal segment between the two ends or break points. Studies report that somatic inversions arise in human genome through recombination events that occur during mitosis. The inversions usually result from NAHR that involves crossover and subsequent recombination between two non-allelic highly homologous or identical genomic repeat sequence regions, but the recombination occurs in inverted orientation. The most important factors that affect the rate of recombination and generation of inversion genetic variations include the size of the DNA sequence stretch or a chromosomal segment involved in recombination and the distance between these recombining regions. The inversions occur either in the coding regions or in the noncoding regions of the gene or in the regions between the genes (intergenic regions). The inversions do not disrupt the original gene balance, but they change the gene order that may disrupt normal gene function. The inversions in the coding regions of the gene or exons may disrupt the normal gene function by changing the amino acid sequence, generation of nonsense or missense mutations or alteration of splicing process through change in donor and acceptor splicing sites. The inversion polymorphisms refer to the variability regarding the presence of inverted sequences in a particular genomic region that occurs in a human genome in comparison with other genomes and the reference genome established by the HGP. As with other types of polymorphisms including SNPs and CNVs, the occurrence as well as the distribution of different types of inversions within the human genome appears to be under the strong influence of natural selection processes in general and purifying or negative selection in particular. As already mentioned, the inversions in the coding regions of the gene or exons may disrupt the normal gene function. The disruption of normal gene function may modulate the susceptibility to or cause many genetic diseases with Mendelian inheritance and complex multifactorial diseases. The most well-known example of pathogenic inversion occurs in the distal region of F8 gene (chromosomal location, Xq28). The F8 gene encodes coagulation factor VIIIC that plays a vital role in normal blood clotting. The inversion in distal region of F8 gene results in loss of function mutation in the gene that manifests as severe deficiency of coagulation factor VIIIC that leads to the most common chromosome X-linked coagulation disease known as hemophilia A. The F8 gene inversion is the causative event responsible for more than 20% of all hemophilia A cases and accounts for about half of highly severe cases of the disease. One of the most common polymorphic inversion events are reported in a 900 kb sequence segment on chromosome 17 (chromosomal location,17q21.31) that increases the predisposition to deletions within the affected chromosomal segment and are responsible for 17q21.31 microdeletion syndrome, also known as Koolen-de Vries syndrome (KdVS) (Lakich, Kazazian et al. 1993, Koolen, Sharp et al. 2008).

The translocations involve transfer of chromosomal segment(s) to a non-homologous (heterologous) chromosome or to new regions on the same chromosome. The translocations occur either in a non-reciprocal manner that involves one-way transfer of chromosomal segments to a non-homologous chromosome, or in a reciprocal manner that involves two-way transfer of chromosomal segments between two non-homologous chromosomes. Further, the reciprocal translocations may occur in a balanced manner that involves no net loss or gain of chromosomal segments or genomic sequence stretches, or may occur in an unbalanced manner that results in loss of chromosomal segments or genomic sequence stretches. The balanced translocations represent the copy neutral variations. The translocation polymorphism refers to the variability regarding the presence or absence of exchange of specific chromosomal segments that occurs in a human genome in comparison with other genomes and the reference genome established by the HGP. The translocations occur either in the coding regions or in the noncoding regions of the gene or in the regions between the genes (intergenic regions). The translocations generate new linkage relationships between the genes, and the translocations that occur in coding regions of the genome are generally deleterious and rarely advantageous for the carrier individual. As with other types of polymorphisms including SNPs and CNVs, the occurrence as well as the distribution of different types of translocations within the human genome appears to be under the strong influence of natural selection processes in general and purifying or negative selection in particular. The effect of inversions and translocations on gene function like other polymorphisms depends on their location within the genome. The inversions and translocations located in coding regions and noncoding regulatory regions of the gene directly affect gene expression through creation of transcript breakpoints, rearrangement of regulatory sequences and generation of position effects that can affect the distant genomic regions even up to 1.5 Mb farther from the affected gene or genomic region. The SVs including inversions and translocations may also affect gene expression through disruption of interactions between cis- and trans-regulatory elements of the gene or genomic region (Harewood, Schütz et al. 2010, Harewood, Chaignat et al. 2012).

1.2.1.4 Small Indel (Insertion and Deletion) Polymorphisms

The insertion and deletion polymorphism, abbreviated as indel or I/D, is a genetic variation that results from insertion or deletion of a specific nucleotide sequence involving a specific region of the gene or genome. In humans, the length of inserted or deleted sequence varies from 1 to 10,000 base pairs. However, small indels generally include the inserted or deleted sequences that vary in length from 1 to 100 base pairs. The insertions and deletions (indels) are not as frequent as SNPs, but they affect equal number of bases or nucleotides as that of SNPs. The indels represent the second most frequent type of human genetic polymorphism or genetic variation after SNPs and constitute around 15% – 21% of all the human genetic variations. The indels are present in high abundance with widespread occurrence across the entire human genome, including all the 24 chromosomes (Mullaney, Mills et al. 2010). The TGP identified and characterized more than 3.6 million short or small indels (insertions and deletions) (Auton, Brooks et al. 2015).

The indels occur in many functionally important regions of the human genome that includes many genes and therefore directly affect many human traits and modulate the susceptibility to or even cause many diseases that include Mendelian diseases and complex multifactorial diseases. For example, an indel gene polymorphism involving a deletion of three base pairs within the cystic fibrosis transmembrane conductance regulator (CFTR) gene causes cystic fibrosis (Kosorok, Wei et al. 1996). The indels can modulate the susceptibility to or even play an important role in the pathogenesis of various complex multifactorial diseases. The studies reveal proven or potential role of indels in modulating the susceptibility to or even cause various complex multifactorial diseases including cancers such as acute myeloid leukemia (AML) besides other cancers (Falini, Mecucci et al. 2005, Ye, Wang et al. 2016). The indels in the noncoding regions of the gene affect gene regulation, expression and function through modulation of gene transcription, gene silencing and gene splicing. The noncoding indels may also modulate chromosome fragility. The indels in the coding regions of the genes result in two types of genetic variations or mutations: frameshift (FS) mutations (variations) and non-frameshift (NFS) mutations (variations). The indel gene polymorphism in the coding regions of the genes involving insertion or deletion of sequence segments containing such number of base pairs that are not a multiple of three results in frameshift (FS) mutation. In contrast, the indel gene polymorphism in the coding regions of the genes involving insertion or deletion of sequence segments containing such number of base pairs that are a multiple of three results in non-frameshift (FS) mutation. The frameshift mutation changes or shifts the sequence-reading frame starting from the nucleotide position where insertion or deletion occurs. Consequently, the affected gene or coding DNA segment codes for an entirely different sequence of amino acids and the shift in sequence-reading frame may generate a stop codon that leads to premature chain termination. The generation of a stop codon that leads to premature chain termination may activate nonsense-mediated mRNA decay (NMD) surveillance pathway. The non-frameshift mutation does not change or shift the sequence-reading frame, and therefore, the affected gene or coding DNA segment does not code for different amino acids except for insertion or deletion of one or more amino acids depending on the indel and length of inserted or deleted sequence; the other amino acids encoded by the gene remain unaffected (Scharplatz, Puhan et al. 2004, Brogna and Wen 2009).

The most cited example that reveals the functional significance of indel gene polymorphism is the insertion and/or deletion of 287 bp (250 or 289 bp according to other studies) *Alu* repeat sequence element in the intron 16 of ACE gene that encodes angiotensin I-converting enzyme (ACE). The angiotensin I-converting enzyme (ACE) gene (size 21 kb; chromosomal location,17q23.3) consists of 26 exons and 25 introns. The indel polymorphism of ACE gene (rs4646994) based on the presence or absence of *Alu* repeat sequence possibly affects the expression and function of ACE gene and consequently the physiological level and activity of angiotensin I-converting enzyme. The ACE gene rs4646994 indel polymorphism consists of two alleles, insertion (I) allele and deletion (D) allele, and generates three genotypic forms: insertion/insertion (II) homozygous genotype, insertion/deletion (ID) heterozygous genotype and deletion/deletion (DD) homozygous genotype. The presence of deletion (D) allele that corresponds to the absence of *Alu* repeat sequence element results in higher expression of ACE gene, higher physiological level and higher activity of ACE enzyme and increased production of angiotensin II in comparison with insertion (I) allele that corresponds to the presence of *Alu* repeat sequence element. The presence of deletion/deletion (DD) homozygous genotype results in two times higher expression of ACE gene, higher physiological level and higher activity of ACE enzyme in comparison with insertion/insertion (II) homozygous genotype (Rigat, Hubert et al. 1990, Scharplatz, Puhan et al. 2004, Castellon and Hamdi 2007, Sameer, Syeed et al. 2010).

1.2.1.5 Genomic or DNA Repetitive Sequences and Repetitive Sequence Polymorphisms

DNA repeats or DNA repetitive elements consist of nucleotide sequences that occur multiple times or exist as multiple copies within the genome. DNA repeat sequences are present in high abundance in the human genome and constitute more than two-thirds of the entire human genome (de Koning, Gu et al. 2011). The DNA repetitive elements or sequences consist of two major classes of repetitive sequences: tandem repeats and interspersed repeats. The tandem repeat polymorphisms and interspersed repeat polymorphisms refer to the polymorphisms resulting from the variations or difference in these repeats that occurs in a human genome in comparison with other genomes or the reference genome.

1.2.1.5.1 Tandem Repeat Polymorphisms (TRPs)

The tandem repetitive sequences or tandem repeats (TRs) consist of tandem or end-to-end repeats or repeat units of two or more nucleotides, for example, repeats of dinucleotides, trinucleotides, tetranucleotides, pentanucleotides or hexanucleotides arranged adjacent to each other in a direct or inverted manner. The number of copies of nucleotide repeats (nucleotide repeat motifs) within a tandem repetitive sequence ranges from two to several thousand. The tandem repetitive sequences, also known as satellite DNAs because of their discovery as small distinct satellite band during density gradient centrifugation, are highly repetitive and constitute almost 30% of the entire human genome. In eukaryotes including humans, the TRs, although dispersed throughout the genome, mostly occur in the regions of genome that constitute the heterochromatin. The heterochromatin describes the cytologically dense and relatively gene poor regions of the chromosome that constitute centromeres and telomeres and play a vital role in silencing the expression of genes that occur in euchromatin. Therefore, the centromeres and telomeres consist of DNA highly enriched in tandem repetitive sequences. In the context of genes, the TRs occur either in the coding regions or in the noncoding regions of the gene or in the regions between the genes (intergenic regions). The number of nucleotide repeats that characterize a particular TR sequence may vary between any two individuals. The tandem repeat polymorphism (TRP) refers to the variability in the number of nucleotide repeats within a particular tandem repeat that occurs in a human genome in comparison with other genomes and the reference genome established by the HGP. The term length polymorphism also describes TRP since this polymorphism arises because of differences in the length of the repeat sequence that in turn results from difference in the number of repeats. The TRPs result from various DNA-level processes including replication slippage, unequal crossover and DNA double-strand breaks repair (DSBR) (Gabriel, Schaffner et al. 2002, Iafrate, Feuk et al. 2004, O'Dushlaine, Edwards et al. 2005, Richard, Kerrest et al. 2008).

The TRs occur in the coding and noncoding regions of many genes, and therefore, the polymorphisms in these TRs like other genetic polymorphisms directly affect the gene structure and function resulting in the modulation of gene expression and affect the concentration and function of the encoded proteins. The changes in gene expression affect many human traits and modulate the susceptibility to or even cause many diseases that include genetic diseases with Mendelian inheritance (Mendelian diseases) and complex multifactorial diseases although the frequency of occurrence and the effect on gene function are less pronounced in comparison with other genetic polymorphisms. The TRs in the coding regions of the genes result in two types of genetic variations or mutations: frameshift (FS) mutations (variations) and non-frameshift (NFS) mutations (variations). The TRs that consist of end-to-end repeats of dinucleotides, tetranucleotides, pentanucleotides, and so on, are not a multiple of three and result in frameshift (FS) mutation. In contrast, the tandem repeats that consist of end-to-end repeats of trinucleotides, hexanucleotides and so on, are a multiple of three and result in non-frameshift (FS) mutation. The translation of tandem repeats of nucleotides in the coding regions may generate long runs of amino acids, provided the nucleotide repeats do not generate a stop codon (Iafrate, Feuk et al. 2004, O'Dushlaine, Edwards et al. 2005, Richard, Kerrest et al. 2008).

As with other types of polymorphisms including SNPs and CNVs, the occurrence as well as the distribution of different types of tandem repeats within the human genome appears to be under the strong influence of natural selection processes in general and purifying or negative selection in particular. The TRs are present in the coding regions of the gene or exons, but mostly, they occur in introns and in the regions between the genes. Most of the coding region and intergenic region TRs involve genes that do not play a direct role in the etiology and pathogenesis of genetic diseases with Mendelian inheritance or Mendelian diseases. Further, the tandem repeats that consist of end-to-end repeats of trinucleotides, hexanucleotides, and so on, occur more frequently in the coding regions of the gene. In particular, these tandem repeats occur in the regions that code for long runs of amino acids, for example, polyhistidine tracts, where the increase or decrease in the number of their nucleotide repeats alters the length of amino acid runs, which is a non-lethal phenomenon. In contrast, the TRs that consist of end-to-end repeats of dinucleotides, tetranucleotides, pentanucleotides, and so on, occur less frequently in the coding regions of gene because they result in frameshift (FS) mutations, which is a lethal phenomenon (Subramanian, Mishra et al. 2003, Salichs, Ledda et al. 2009).

Three major subtypes of the TRs occur within the genomes, and these include macrosatellites, minisatellites and microsatellites that differ in the length of repeat units, sequence repeats or repeats. The macrosatellites and minisatellites together constitute VNTRs, whereas the microsatellites represent STRs. The variable number of tandem repeat polymorphisms

(VNTRPs) and short tandem repeat polymorphisms (STRPs) represent the polymorphisms resulting from the variations or difference in VNTRs and STRs, respectively, that occurs in a human genome in comparison with other genomes or the reference genome established by the HGP.

1.2.1.5.1.1 Variable Number of Tandem Repeats (VNTRs) As mentioned already, the macrosatellites and minisatellites together constitute VNTRs. The macrosatellites or macrosatellite VNTRs represent the tandem repetitive sequences that consist of multiple sequence repeats with each sequence repeat unit longer than 100 bp in size. The macrosatellites mostly occur near or in specific regions of the genome that constitute centromeres and telomeres and show enrichment in Y chromosome that contains multiple sequence repeat units each with a length of 171bp. The total length of macrosatellites varies from 100kb to more than 10 Mb. Therefore, the centromeres, telomeres and to some extent chromosome Y consist of DNA highly enriched in macrosatellite VNTRs. The alphoid family of TRs that occur in the heterochromatin DNA of centromeres in human chromosomes is an example of macrosatellite VNTRs. Each sequence repeat unit in alphoid DNA is around 170 bp in size and shows consensus sequence although the sequence may vary by almost 40% across different genomes. The total length of the alphoid DNA ranges from 0.5 to 10 Mb. The alphoid DNA occurs within the human genome in a chromosome-specific manner and seems to play an important structural role in the cellular chromosome cycle. The minisatellites or minisatellite VNTRs represent the tandem repetitive sequences that consist of multiple sequence repeats with each sequence repeat unit ranging in length from 10 to 100 bp. The total length of minisatellites varies from 1 to about 30kb. The minisatellites like macrosatellites are mostly present near or in specific regions of the genome that constitute centromeres and telomeres (pericentric and subtelomeric regions) and show enrichment in Y chromosome, but they occur, although to a lesser extent, in other regions of the genome. The numbers of sequence repeat units that constitute a minisatellite typically varies from 5 to 50 repeats, but a minisatellite may contain up to 1,500 sequence repeat units. The macrosatellites and minisatellites exhibit a high rate of mutation mostly in the form of random deletions and expansions, but they also encompass highly consensus sequences that exhibit no mutation or variation. For example, the number of sequence repeats in a Y chromosome minisatellite marker, MSY1, shows a variation of 2% – 11% between generations, whereas another Y chromosome minisatellite marker, MSY2, occurs in only three variable forms. Therefore, the number and to some extent the sequence of repeat units that constitute macrosatellites and minisatellites vary across different genomes and each genome has a distinct number of sequence repeats except for identical twins. The VNTRs exhibit diversity across individuals, populations and organisms. Due to these variations, the VNTRs exhibit a high degree of RFLPs across genomes of different individuals, populations and organisms and restriction digestion by restriction enzymes that target sequences flanking the VNTRs generates restriction fragments of variable lengths unique or specific to each genome. Therefore, the analysis of restriction digestion fragments, for example, by Southern blotting, generates a characteristic pattern of bands specific for an individual, population or an organism. Since the macrosatellites and minisatellites exhibit Mendelian inheritance, the analysis of characteristic banding pattern allows paternal and maternal identification. The analysis of polymorphisms in macrosatellites and minisatellites allows identification of individuals and paternal and maternal lineages. The analysis allows characterization and identification of populations and organisms. The macrosatellites and minisatellites at a specific locus may occur in two or more allelic forms depending on the variation in the number of sequence repeat units within and across the genomes. The VNTRs occur in the coding as well as noncoding regions of the genome and like other genetic polymorphisms may directly affect the gene structure and function and modulate the susceptibility to or even cause many diseases. However, most of the VNTRs are non-pathogenic and only few actually cause a disease. The examples of VNTRs include 2 – 10 repeats of a 48 bp sequence segment in exon 3 of dopamine D4 receptor; 9, 10 or 12 repeats of a 17 bp sequence segment in intron 2 of serotonin transporter; and 3 – 11 (usually 9 and 10) repeats of a 40 bp sequence segment in 3′-UTR of dopamine transporter (Iafrate, Feuk et al. 2004, O'Dushlaine, Edwards et al. 2005, Richard, Kerrest et al. 2008, Hua-Van, Le Rouzic et al. 2011).

1.2.1.5.1.2 Short Tandem Repeats (STRs) The short tandem repeats (STRs), simple sequence repeats (SSRs) or microsatellites represent the tandem repetitive sequences that consist of multiple sequence repeats with each sequence repeat unit ranging in length usually from 1 to 5 bp, but may consist up to 9 bp (less than 10). The total length of microsatellites is usually less than 100 bp, but this length can extend up to 150 bp or more. The numbers of sequence repeat units that constitute a microsatellite typically varies from 5 to 60 repeats, but a microsatellite may contain up to 200 sequence repeat units. The most commonly occurring sequence repeat units in STRs include A, CA, AG, AAN and AAAN (N represents any nucleotide). The CA sequence repeat units represent the most commonly occurring sequence repeat units within the human genome. The CA repeats occur within the human genome after every 30 – 60kb of sequence interval, and an estimated 50,000 – 100,000 genomic sites contain CA microsatellites. The number of identified CA microsatellites within the human genome stands at more than 8,000, and the number of sequence repeat units that constitute a human genome CA microsatellite varies from 10 to 60 repeats. The phase I and subsequent phases of TGP identified almost 700,000 – 1,000,000 microsatellites through whole genome sequencing of 2,504 unrelated and related sets of individuals representing 26 populations from world's five continental regions: Africa (AFR), Americas (AMR), Europe (EUR), South Asia (SAS) and East Asia (EAS) (Ellegren 2004, Willems, Gymrek et al. 2014). The dinucleotide and tetranucleotide sequence repeat units constitute around 75% of total simple sequence repeats or microsatellites present in the human genome, whereas the rest of the microsatellites consist of trinucleotide, pentanucleotide and hexanucleotide repeats. The microsatellites exhibit non-random occurrence in the coding as well as noncoding regions of many genes. The microsatellite TRs that consist of end-to-end repeats of trinucleotides, hexanucleotides, and so on, occur more frequently in the coding regions of the gene because they may cause less deleterious NFS mutations in comparison with other repeats. The microsatellite TRs that consist of end-to-end repeats of dinucleotides mostly occur in noncoding regulatory regions of the gene including UTRs and promoters (Bolton, Ross et al. 2013, Willems, Gymrek et al. 2014). The microsatellite like macrosatellites and minisatellites exhibit a high rate of variation or mutation mostly in the form of random deletions and expansions and the rate of mutation or variation, but they also encompass highly consensus sequences that exhibit no mutation or variation. The microsatellites show a mutation rate that ranges from 10^{-3} to 10^{-6} per cell generation, and this rate is 10 – 10,000 times higher than the rate of mutation in other regions of the genome excluding the regions that contain repeat sequences. However, some microsatellites show a mutation rate greater than 10^{-2}, whereas other microsatellites are resistant to variation and show no mutation (Ellegren 2004, Verstrepen, Jansen et al. 2005). The mutations in microsatellites or STRs as well as VNTRs discussed previously arise mainly due to homology-dependent or homology-directed DSBR mechanisms such as break-induced replication (BIR) and synthesis-dependent strand annealing (SDSA). In addition to these mechanisms, the mutations may arise due to replication slippage or slipped-strand mispairing particularly in case of STRs and VNTRs that have a tendency to form loops (Pâques and Haber 1999, Viguera, Canceill et al. 2001, Lopes, Ribeyre et al. 2006). The microsatellites due to their high diversity and higher tendency to undergo mutations serve as better genetic or DNA markers in comparison with RFLPs for mapping diseases, in particular, monogenic diseases, and for the identification of individuals and paternal and maternal lineages and even characterization and identification of populations and organisms. Many characteristics make microsatellites better DNA markers in comparison with RFLPs. The microsatellites occur more frequently in comparison with RFLPs, and they are present in multi-allelic form in comparison with the bi-allelic nature of RFLPs. The bi-allelic genetic polymorphisms including RFLPs allow a maximum heterozygosity of 50%, whereas the heterozygosity of microsatellites ranges from 70% to 90%, which means that at maximum 50% of a given population may be heterozygous for a specific RFLP, whereas a maximum of 90% of a given population may be heterozygous for a specific microsatellite. The degree of heterozygosity is a direct measure of the genetic information including inheritance that is extractable using a specific genetic marker; the more the heterozygosity of a DNA marker such as genetic polymorphism, the more the information that the genetic analysis using the given marker will reveal. The STRs may play a major role in the etiology and pathogenesis of diseases such as fragile X syndrome, Friedreich's ataxia and Huntington's disease. The examples of STRs include the 6 – 50 repeats of CGG sequence in the fragile X mental retardation protein translational regulator 1 (FMR1) gene that plays a role in the etiology and pathogenesis of fragile X syndrome. Another example of microsatellite is the 10 – 35 (usually 11 – 14) repeats of a 3 bp (trinucleotide) repeats of CAG sequence segment in HTT gene that codes for Huntington protein. The mutations in this microsatellite in the form of CAG repeat expansion or increase in the number of CAG repeats causes Huntington's disease. An increase in the number of CAG repeats to at least 38 results in this disease. The individuals with Huntington's disease contain 36 – 120 repeats of CAG sequence. The individuals with 36 – 39 repeats of CAG sequence may or may not develop Huntington's disease, whereas the individuals with 40 or more repeats of CAG sequence usually develop this disorder. The CAG repeat expansion or the increase in the number of CAG repeats generates abnormally long Huntington protein containing extra glutamine residues. This abnormally long Huntington protein undergoes cleavage and generates toxic fragments that accumulate within the neurons and cause their dysfunction. The neuronal dysfunction consequently leads to neuronal death in specific regions of the brain that results in signs and symptoms characteristic of this disease. The CTG repeat expansion that occurs within the 3′-UTR of dystrophia myotonica protein kinase (DMPK) gene represents another example of pathogenic repeat expansion (Pearson, Nichol Edamura et al. 2005, Usdin 2008).

1.2.1.5.2 Interspersed Repeat Polymorphisms

The interspersed repeats, interspersed DNA repeats or transposable elements (TEs) consist of DNA repeat sequences that occur in a scattered and non-adjacent manner throughout the human genome unlike the tandem repeats that show enrichment in specific regions of the genome. The TEs occur abundantly in human genome and are present in almost all the regions of the genome including the regions that contain individual genes or families of genes and represent a major source of intermediately repetitive DNA sequences. The interspersed repeats or TEs are also highly repetitive and constitute more than 50% of the entire human genome, and almost 50% –70% of the human genome owes its evolutionary origin to the TEs. The TEs by themselves and through their capacity to mobilize contribute in a highly significant manner to the variability of the human genome. Consequently, the TEs affect the structure and function of the human genome including the gene function, which in turn may affect many human traits and modulate the susceptibility to or even cause many diseases that include genetic diseases with Mendelian inheritance (Mendelian diseases) and complex multifactorial diseases (Lander, Linton et al. 2001, de Koning, Gu et al. 2011, Ray and Batzer 2011, Sudmant, Rausch et al. 2015, Hancks and Kazazian 2016).

The TE sequences represent the remnants of past insertion events of mobile DNA elements, most of whom are now incapable of mobilization or transposition. However, few families of transposable elements occur within the human genomes that are still capable of mobilization or transposition. These families of active TEs include LINEs, SINEs and SVAs. The LINEs consist of three main classes, but LINE1 (L1) elements represent the most commonly occurring LINE type. The LINEs are about 500 bp to 10 kb (usually 6 – 8 kb) in length and constitute almost 17% – 21% of the entire human genome. The SINEs are around 100 – 500 bp in length and mostly represented by Alu elements (named after restriction enzyme *AluI*) that are around 280 –300 bp in size and arise from 7SL RNA, an abundantly occurring ncRNA that functions as a component of signal recognition particle (SRP). The SINEs constitute almost 13% of the entire human genome.

The SVA elements represent a hominid-specific family of composite transposons that derives its name from its constituent component sequences, endogenous retroviral sequence – SINE (S), VNTR (V) and Alu (A) (Batzer and Deininger 1991, Brouha, Schustak et al. 2003, Salem, Kilroy et al. 2003, Wang, Xing et al. 2005). The LINEs, SINEs and SVAs all represent retrotransposons, that is, the type or family of TEs that propagate, mobilize or transpose through the involvement of an RNA intermediate and occur via a copy-paste amplification mechanism that involves reverse transcription of an RNA sequence intermediate. The LINEs including L1s contain all the genetic information required to encode the enzymes including reverse transcriptase and transposase that enable their own reverse transcription and subsequent transposition and therefore represent autonomous retrotransposons. The retrotransposition of SINEs (Alu elements) and SVAs occurs via trans-mobilization mechanism and uses the enzymatic machinery of LINEs (Moran, Holmes et al. 1996, Dewannieux, Esnault et al. 2003, Salem, Kilroy et al. 2003).

The transposable (mobile) element insertion (TEI) polymorphisms refer to the variability in the number of copies or insertions of a particular TE within a human genome in comparison with other genomes and the reference genome established by the HGP. The number of copies of a transposable element may vary from few to millions depending on the type, but generally, each transposable element occurs in abundance within the human genome. For example, the human genome may contain more than a million copies of Alu elements and one Alu element occurs every 5 – 10 kb of genomic segment. The Alu elements may occur within the coding as well as noncoding regions of the genome including introns. The human reference genome consists of around 500,000 copies of inactive, truncated L1 elements and around 100 – 150 active L1 loci that include the highly abundant L1 loci known as hot-L1s. The L1 elements and Alu elements represent the most common type of TEI in the human genome. The studies report that somatic and germline L1 insertions occur within the human genome at a frequency much higher than expected. The L1 insertion elements occur in the human genome in dimorphic form, and studies reveal that two individuals or genomes on average show a TEI polymorphic difference at 285 L1 insertion sites. The total number of dimorphic L1 elements in a given population ranges from 3,000 to 10,000 (Iskow, McCabe et al. 2010, Ewing and Kazazian 2011). As with other types of polymorphisms including SNPs, CNVs, VNTRs and STRs, the occurrence as well as the distribution of different types of tandem repeats within the human genome appears to be under the strong influence of natural selection processes in general and purifying or negative selection in particular. Depending on the location of transposable element insertions (TEIs) within the genome, the TEI polymorphisms may affect the genome structure and function differently. The TEIs contribute to several physiological and pathophysiological processes. The studies report involvement of TEIs in physiological processes such as innate immunity and placental development (Dupressoir, Lavialle et al. 2012, Chuong, Elde et al. 2016, Rodriguez-Martin, Alvarez et al. 2020). The presence of TEIs in coding regions as well as in noncoding regulatory regions of the gene may affect gene expression and function. The TEIs may generate recombination hotspots, may change the gene copy numbers or may change or disrupt the gene structure and modulate the susceptibility to or even cause many genetic diseases with Mendelian inheritance and complex multifactorial diseases such as cancer.

The studies report the association or linkage of more than a hundred TEIs with Mendelian genetic diseases and hereditary cancers. The studies reveal the role of somatic L1 insertions in modulating the immune response to various cancers and consequently their role in the pathogenesis of various cancers including epithelial cancers. The somatic L1 element insertions occur in almost 50% of human cancers, but play a little or no role in the development of these cancers although they may still affect the gene function. The somatic L1 element insertions also induce and/or promote the generation of other types of germline and somatic SVs in the genome over and above the usual L1 retrotranspositional insertions; however, the actual role in or the contribution of these SVs to the pathogenesis of various cancers is still unknown (Lee, Iskow et al. 2012, Helman, Lawrence et al. 2014). A very recent Pan-Cancer Analysis of Whole Genomes (PCAWG) project that analyzed 2,954 cancer genomes has identified 19,166 somatic TEIs that affected 35% of the total samples. The study revealed the somatic L1 insertions as the primary or most abundant SV in esophageal adenocarcinoma and the second most abundant SV in colorectal cancer and head-and-neck cancer. The study also revealed the somatic L1 insertion-induced deletions of Mb-sized chromosomal regions. The deleted chromosomal regions may contain tumor-suppressor genes. The somatic L1 insertions and other TEIs may also induce complex chromosomal segment translocations and large-scale genomic duplications. The somatic L1 insertions may induce highly enhanced amplification of oncogenes through breakage-fusion-bridge (BFB) cycle, also known as breakage-rejoining-bridge (BRB) cycle. The somatic TEIs may affect the gene structure and function through various mechanisms including both DNA- and RNA-level (transcription and translation level) mechanisms such as insertional mutagenesis, modulation of RNA expression and induction of functional premature polyadenylation that results in aberrant mRNA splicing (Cordaux and Batzer 2009, Hancks and Kazazian 2016). The TEIs like that of SNPs and CNVs serve as important tools or molecular markers in the field of personalized or precision medicine and may serve as potential targets for the development of personalized drugs. The use of a personalized antisense oligonucleotide based drug to treat or manage successfully the Batten disease, which is a rare and fatal neurodegenerative disease that results from insertion of SVA element around 2 kb in size and leads to exon trapping and subsequent aberrant mRNA splicing, provides an example (Kim, Hu et al. 2019). The germline TEIs may generate novel individual-specific insertion patterns in the form of polymorphic TE (polyTE) insertions that are inheritable and may serve as important tools for studying human ancestry and evolution. The studies report on the analysis of human ancestry and evolution mostly based on Alu element retrotransposon insertions and few studies report such analysis based on L1 element insertions, but no study reports such analysis using SVA elements. The phase III of TGP (November 2014) identified 16,192 polyTE loci through whole genome sequencing of 2,504 unrelated and related sets of individuals representing 26 populations from world's five continental regions: Africa (AFR), Americas (AMR), Europe (EUR), South Asia (SAS) and East Asia (EAS) (Ray and Batzer 2011, Rishishwar, Tellez Villa et al. 2015).

1.3 Conclusions

The dynamic, complex and interdisciplinary field of modern genomics is revolutionizing our understanding of human genome structure, function and evolution through the application of new and emerging techniques that allow high-throughput and accurate analysis of various structural and functional characteristics of the human genome. The field of genomics includes various distinct areas of study including DNA and RNA sequencing, global RNA expression analysis, cell and tissue transcriptomics, identification and genotyping of genetic variations including polymorphisms and mutations and computational and bioinformatic analysis of genetic and systemic data. However, the structural and functional characterization of genetic polymorphisms lies at the core of human genome analysis using genomics. The genetic polymorphisms occur in the human genome in various forms and constitute the most predominant source of genetic variability that occurs in a human genome in comparison with other genomes and the reference genome established by the HGP. Because of their presence in varied forms and widespread occurrence within the human genome that encompasses many functionally important regions of the human genome including many genes and gene regulatory regions, the genetic polymorphisms profoundly affect the gene structure and function. The effect of genetic polymorphisms on gene structure and function can directly lead to or may influence the phenotypic expression and modulate the susceptibility to or even cause various diseases including genetic disorders and multifactorial diseases. The study of effects of genetic polymorphisms on drug metabolism, drug response and drug efficacy gave birth to the fields of precision medicine or personalized medicine and pharmacogenomics that allow the development of tailored personalized medicines. The use of tailored personalized medicines in clinical practice has minimized the exposure of patients to less or non-effective drugs and minimizes or prevents the adverse effects of drugs. To conclude, the techniques used for the detection, analysis and characterization of genetic polymorphisms are diverse and dynamic. These techniques differ in various characteristics besides the basic method used, and each of these techniques has its own inherent issues that affect the overall accuracy and influence the choice of the method. The careful and combined use of several of these techniques taking into consideration their advantages and limitations enables optimized use of each of these methods and increases the overall efficiency and accuracy of the detection of genetic variations including genetic polymorphisms.

ABBREVIATIONS

aCGH:	Array-CGH
AS-PCR:	Allele-specific polymerase chain reaction
BACs:	Bacterial artificial chromosomes
BFB cycle:	Breakage-fusion-bridge cycle
BIR:	Break-induced replication
BRB cycle:	Breakage-rejoining-bridge cycle
ChIP-Seq:	Chromatin immunoprecipitation sequencing
CNV:	Copy number variation
DNA:	Deoxyribonucleic acid
Em-PCR:	Emulsion PCR
FGS:	First-generation sequencing
FoSTeS:	Fork stalling and template switching
GWAS:	Genome-wide association studies
HGP:	Human Genome Project
HR:	Homologous recombination
HRM:	High-resolution melting
IHGSC:	International Human Genome SequencingConsortium
IL1β:	Interleukin-1β
ISVs:	Intermediate-sized variants
LCRs:	Low copy repeats
LCVs:	Large-scale copy number variants
LINE/LINE-1/L1:	Long interspersed nuclear element
lncRNA:	Long noncoding RNA
MAPH:	Multiplex amplifiable probe hybridization
MAQ:	Multiplex amplicon quantification
MEI:	Mobile element insertion
mFISH:	Multicolor-FISH
M-FISH:	Multiplex-FISH
miRNA:	microRNA
MLPA:.	Multiplex ligation-dependent probe amplification
MMBIR:	Micro-homology-mediated break-induced replication
MMEJ:	Micro-homology-mediated end joining
MPS:	Massively parallel sequencing
mRNA:	Messenger RNA
NAHR:	Non-allelic homologous recombination
NGS:	Next-generation sequencing
NHEJ:	Non-homologous end joining
NHR:	Non-homologous recombination/repair
NUMTs:	Nuclear mitochondrial DNA sequences/variants
PCR-CCM:	Polymerase chain reaction-chemical cleavage of mismatch
PCR-DGGE:	Polymerase chain reaction-denaturing gradient gel electrophoresis
PCR-RFLP:	Polymerase chain reaction-restriction fragment length polymorphism
PCR-SSCP:	Polymerase chain reaction-single-strand conformation polymorphism
PCR-TTGE:	Polymerase chain reaction-temporal temperature gel electrophoresis
qPCR or RTPCR:	Quantitative real-time PCR
RFLPs:	Restriction fragment length polymorphisms
RIP-seq:	RNA immunoprecipitation sequencing
RNA:	Ribonucleic acid
RNA-seq:	RNA sequencing

SBE:	Single-base primer extension
SBH:	Sequencing by hybridization
SBS:	Sequencing by synthesis
SDs:	Segmental duplications
SDSA:	Synthesis-dependent strand annealing
SGS:	Second-generation sequencing
SINE:	Short interspersed nuclear element
SKY:	Spectral karyotyping
sncRNA:	Small noncoding RNA
SNP:	Single nucleotide polymorphism
SNR:	Signal-to-noise ratio
SNV:	Simple nucleotide variation
SP-PCR:	Solid-phase PCR
SSA:	Single-strand annealing
SSR:	Simple sequence repeat
STR:	Short tandem repeat
SVA:	Composite retrotransposon insertion consisting of sequences SINE (S), VNTR (V) and Alu (A)
TEs:	Transposable elements
TGP:	1000 Genomes Project
TGS/3GS:	Third-generation sequencing
TNF:	Tumor necrosis factor
UTR:	Untranslated region
VNTR:	Variable number of tandem repeat
WES:	Whole exome sequencing
WGAS:	Whole genome association studies

REFERENCES

Abecasis, G. R., A. Auton, L. D. Brooks, M. A. DePristo, R. M. Durbin, R. E. Handsaker, H. M. Kang, G. T. Marth and G. A. McVean (2012). An integrated map of genetic variation from 1,092 human genomes. *Nature* **491**(7422): 56–65.

Aerts, J., Y. Wetzels, N. Cohen and J. Aerssens (2002). Data mining of public SNP databases for the selection of intragenic SNPs. *Hum Mutat* **20**(3): 162–173.

Ahmadian, A., B. Gharizadeh, D. O'Meara, J. Odeberg and J. Lundeberg (2001). Genotyping by apyrase-mediated allele-specific extension. *Nucleic Acids Res* **29**(24): E121.

Aitman, T. J., R. Dong, T. J. Vyse, P. J. Norsworthy, M. D. Johnson, J. Smith, J. Mangion, C. Roberton-Lowe, A. J. Marshall, E. Petretto, M. D. Hodges, G. Bhangal, S. G. Patel, K. Sheehan-Rooney, M. Duda, P. R. Cook, D. J. Evans, J. Domin, J. Flint, J. J. Boyle, C. D. Pusey and H. T. Cook (2006). Copy number polymorphism in Fcgr3 predisposes to glomerulonephritis in rats and humans. *Nature* **439**(7078): 851–855.

Albertini, A. M., M. Hofer, M. P. Calos and J. H. Miller (1982). On the formation of spontaneous deletions: the importance of short sequence homologies in the generation of large deletions. *Cell* **29**(2): 319–328.

Aldred, P. M., E. J. Hollox and J. A. Armour (2005). Copy number polymorphism and expression level variation of the human alpha-defensin genes DEFA1 and DEFA3. *Hum Mol Genet* **14**(14): 2045–2052.

Alkan, C., B. P. Coe and E. E. Eichler (2011). Genome structural variation discovery and genotyping. *Nat Rev Genet* **12**(5): 363–376.

Alles, J., T. Fehlmann, U. Fischer, C. Backes, V. Galata, M. Minet, M. Hart, M. Abu-Halima, F. A. Grässer, H. P. Lenhof, A. Keller and E. Meese (2019). An estimate of the total number of true human miRNAs. *Nucleic Acids Res* **47**(7): 3353–3364.

Altshuler, D., V. J. Pollara, C. R. Cowles, W. J. Van Etten, J. Baldwin, L. Linton and E. S. Lander (2000). An SNP map of the human genome generated by reduced representation shotgun sequencing. *Nature* **407**(6803): 513–516.

Alwi, Z. B. (2005). The use of SNPs in pharmacogenomics studies. *Malays J Med Sci* **12**(2): 4–12.

Ammar, R., T. A. Paton, D. Torti, A. Shlien and G. D. Bader (2015). Long read nanopore sequencing for detection of HLA and CYP2D6 variants and haplotypes. *F1000Res* **4**: 17.

Armour, J. A., C. Sismani, P. C. Patsalis and G. Cross (2000). Measurement of locus copy number by hybridisation with amplifiable probes. *Nucleic Acids Res* **28**(2): 605–609.

Auton, A., L. D. Brooks, R. M. Durbin, E. P. Garrison, H. M. Kang, J. O. Korbel, J. L. Marchini, S. McCarthy, G. A. McVean and G. R. Abecasis (2015). A global reference for human genetic variation. *Nature* **526**(7571): 68–74.

Bailey, J. A., A. M. Yavor, H. F. Massa, B. J. Trask and E. E. Eichler (2001). Segmental duplications: organization and impact within the current human genome project assembly. *Genome Res* **11**(6): 1005–1017.

Baldini, A. (2004). DiGeorge syndrome: an update. *Curr Opin Cardiol* **19**(3): 201–204.

Banday, M. Z., H. M. Balkhi, Z. Hamid, A. S. Sameer, N. A. Chowdri and E. Haq (2016). Tumor necrosis factor-α (TNF-α)-308G/A promoter polymorphism in colorectal cancer in ethnic Kashmiri population – a case control study in a detailed perspective. *Meta Gene* **9**: 128–136.

Banday, M. Z., A. H. Mir, A. S. Sameer, N. A. Chowdri and E. Haq (2017). Interleukin-1β (IL-1β) -31C/T and -511T/C promoter single nucleotide polymorphism in colorectal cancer in ethnic Kashmiri population – a case control study. *Meta Gene* **12**: 94–103.

Barbujani, G., A. Magagni, E. Minch and L. L. Cavalli-Sforza (1997). An apportionment of human DNA diversity. *Proc Natl Acad Sci U S A* **94**(9): 4516–4519.

Bartha, Á. and B. Győrffy (2019). Comprehensive outline of whole exome sequencing data analysis tools available in clinical oncology. *Cancers (Basel)* **11**(11):1725.

Batzer, M. A. and P. L. Deininger (1991). A human-specific subfamily of Alu sequences. *Genomics* **9**(3): 481–487.

Bayés, M., L. F. Magano, N. Rivera, R. Flores and L. A. Pérez Jurado (2003). Mutational mechanisms of Williams-Beuren syndrome deletions. *Am J Hum Genet* **73**(1): 131–151.

Bejjani, B. A., R. Saleki, B. C. Ballif, E. A. Rorem, K. Sundin, A. Theisen, C. D. Kashork and L. G. Shaffer (2005). Use of targeted array-based CGH for the clinical diagnosis of chromosomal imbalance: is less more?*Am J Med Genet A* **134**(3): 259–267.

Bi, W., J. Yan, P. Stankiewicz, S. S. Park, K. Walz, C. F. Boerkoel, L. Potocki, L. G. Shaffer, K. Devriendt, M. J. Nowaczyk, K. Inoue and J. R. Lupski (2002). Genes in a refined Smith-Magenis syndrome critical deletion interval on chromosome 17p11.2 and the syntenic region of the mouse. *Genome Res* **12**(5): 713–728.

Birney, E. and N. Soranzo (2015). Human genomics: the end of the start for population sequencing. *Nature* **526**(7571): 52–53.

Bolton, K. A., J. P. Ross, D. M. Grice, N. A. Bowden, E. G. Holliday, K. A. Avery-Kiejda and R. J. Scott (2013). STaRRRT: a table of short tandem repeats in regulatory regions of the human genome. *BMC Genomics* **14**: 795.

Bowl, M. R., M. A. Nesbit, B. Harding, E. Levy, A. Jefferson, E. Volpi, K. Rizzoti, R. Lovell-Badge, D. Schlessinger, M. P. Whyte and R. V. Thakker (2005). An interstitial deletion-insertion involving chromosomes 2p25.3 and Xq27.1, near SOX3, causes X-linked recessive hypoparathyroidism.*J Clin Invest* **115**(10): 2822–2831.

Brogna, S. and J. Wen (2009). Nonsense-mediated mRNA decay (NMD) mechanisms.*Nat Struct Mol Biol* **16**(2): 107–113.

Brookes, A. J. (1999). The essence of SNPs.*Gene* **234**(2): 177–186.

Brouha, B., J. Schustak, R. M. Badge, S. Lutz-Prigge, A. H. Farley, J. V. Moran and H. H. Kazazian, Jr. (2003). Hot L1s account for the bulk of retrotransposition in the human population. *Proc Natl Acad Sci U S A* **100**(9): 5280–5285.

Buckland, P. R. (2003). Polymorphically duplicated genes: their relevance to phenotypic variation in humans.*Ann Med* **35** (5): 308–315.

Buermans, H. P. and J. T. den Dunnen (2014). Next generation sequencing technology: advances and applications.*Biochim Biophys Acta* **1842**(10): 1932–1941.

Buhr, F., S. Jha, M. Thommen, J. Mittelstaet, F. Kutz, H. Schwalbe, M. V. Rodnina and A. A. Komar (2016). Synonymous codons direct cotranslational folding toward different protein conformations.*PLoS Genet* **61**(3): 341–351.

Cancer Genome Atlas Research Network, T. (2012). Comprehensive genomic characterization of squamous cell lung cancers. *Nature* **489**(7417): 519–525.

Cao, M. D., D. Ganesamoorthy, A. G. Elliott, H. Zhang, M. A. Cooper and L. J. Coin (2016). Streaming algorithms for identification of pathogens and antibiotic resistance potential from real-time MinION(TM) sequencing. *Gigascience* **5**(1): 32.

Cardoso-Moreira, M. M. and M. Long (2010). Mutational bias shaping fly copy number variation: implications for genome evolution.*Trends Genet* **26**(6): 243–247.

Carreto, L., M. F. Eiriz, A. C. Gomes, P. M. Pereira, D. Schuller and M. A. Santos (2008). Comparative genomics of wild type yeast strains unveils important genome diversity. *BMC Genomics* **9**: 524.

Carter, N. P. (2007). Methods and strategies for analyzing copy number variation using DNA microarrays.*Nat Genet* **39**(7 Suppl): S16–21.

Carvalho, B., E. Ouwerkerk, G. A. Meijer and B. Ylstra (2004). High resolution microarray comparative genomic hybridisation analysis using spotted oligonucleotides.*J Clin Pathol* **57**(6): 644–646.

Castellon, R. and H. K. Hamdi (2007). Demystifying the ACE polymorphism: from genetics to biology.*Curr Pharm Des* **13**(12): 1191–1198.

Chaignat, E., E. A. Yahya-Graison, C. N. Henrichsen, J. Chrast, F. Schütz, S. Pradervand and A. Reymond (2011). Copy number variation modifies expression time courses.*Genome Res* **21**(1): 106–113.

Chaudhary, R., B. Singh, M. Kumar, S. K. Gakhar, A. K. Saini, V. S. Parmar and A. K. Chhillar (2015). Role of single nucleotide polymorphisms in pharmacogenomics and their association with human diseases.*Drug Metab Rev* **47**(3): 281–290.

Chen, J. M., N. Chuzhanova, P. D. Stenson, C. Férec and D. N. Cooper (2005). Intrachromosomal serial replication slippage in trans gives rise to diverse genomic rearrangements involving inversions.*Hum Mutat* **26**(4): 362–373.

Chen, W. K., J. D. Swartz, L. J. Rush and C. E. Alvarez (2009). Mapping DNA structural variation in dogs.*Genome Res* **19**(3): 500–509.

Chuong, E. B., N. C. Elde and C. Feschotte (2016). Regulatory evolution of innate immunity through co-option of endogenous retroviruses.*Science* **351**(6277): 1083–1087.

Clamp, M., B. Fry, M. Kamal, X. Xie, J. Cuff, M. F. Lin, M. Kellis, K. Lindblad-Toh and E. S. Lander (2007). Distinguishing protein-coding and noncoding genes in the human genome. *Proc Natl Acad Sci U S A* **104**(49): 19428–19433.

Clevenger, J., C. Chavarro, S. A. Pearl, P. Ozias-Akins and S. A. Jackson (2015). Single nucleotide polymorphism identification in polyploids: a review, example, and recommendations.*Mol Plant* **8**(6): 831–846.

Colin, Y., B. Chérif-Zahar, C. Le Van Kim, V. Raynal, V. Van Huffel and J. P. Cartron (1991). Genetic basis of the RhD-positive and RhD-negative blood group polymorphism as determined by Southern analysis. *Blood* **78**(10): 2747–2752.

Conrad, D. F., T. D. Andrews, N. P. Carter, M. E. Hurles and J. K. Pritchard (2006). A high-resolution survey of deletion polymorphism in the human genome. *Nat Genet* **38**(1): 75–81.

Conrad, D. F., D. Pinto, R. Redon, L. Feuk, O. Gokcumen, Y. Zhang, J. Aerts, T. D. Andrews, C. Barnes, P. Campbell, T. Fitzgerald, M. Hu, C. H. Ihm, K. Kristiansson, D. G. Macarthur, J. R. Macdonald, I. Onyiah, A. W. Pang, S. Robson, K. Stirrups, A. Valsesia, K. Walter, J. Wei, C. Tyler-Smith, N. P. Carter, C. Lee, S. W. Scherer and M. E. Hurles (2010). Origins and functional impact of copy number variation in the human genome. *Nature* **464**(7289): 704–712.

Cooper, D. N., E. V. Ball and M. Krawczak (1998). The human gene mutation database. *Nucleic Acids Res* **26**(1): 285–287.

Cooper, G. M., T. Zerr, J. M. Kidd, E. E. Eichler and D. A. Nickerson (2008). Systematic assessment of copy number variant detection via genome-wide SNP genotyping. *Nat Genet* **40**(10): 1199–1203.

Cordaux, R. and M. A. Batzer (2009). The impact of retrotransposons on human genome evolution. *Nat Rev Genet* **10**(10): 691–703.

Cotton, R. G. (1999). Mutation detection by chemical cleavage. *Genet Anal* **14**(5–6): 165–168.

Cotton, R. G., N. R. Rodrigues and R. D. Campbell (1988). Reactivity of cytosine and thymine in single-base-pair mismatches with hydroxylamine and osmium tetroxide and its application to the study of mutations. *Proc Natl Acad Sci U S A* **85**(12): 4397–4401.

Coughlin, C. R., 2nd, G. H. Scharer and T. H. Shaikh (2012). Clinical impact of copy number variation analysis using high-resolution microarray technologies: advantages, limitations and concerns. *Genome Med* **4**(10): 80.

Crawford, D. C. and D. A. Nickerson (2005). Definition and clinical importance of haplotypes. *Annu Rev Med* **56**: 303–320.

Cui, C., W. Shu and P. Li (2016). Fluorescence in situ hybridization: cell-based genetic diagnostic and research applications. *Front Cell Dev Biol* **4**: 89.

Daly, A. K. (2010). Pharmacogenetics and human genetic polymorphisms. *Biochem J* **429**(3): 435–449.

Daniels, J. and S. Kadlubar (2013). Sulfotransferase genetic variation: from cancer risk to treatment response. *Drug Metab Rev* **45**(4): 415–422.

de Koning, A. P., W. Gu, T. A. Castoe, M. A. Batzer and D. D. Pollock (2011). Repetitive elements may comprise over two-thirds of the human genome. *PLoS Genet* **7**(12): e1002384.

Deng, F. Y., L. J. Zhao, Y. F. Pei, B. Y. Sha, X. G. Liu, H. Yan, L. Wang, T. L. Yang, R. R. Recker, C. J. Papasian and H. W. Deng (2010). Genome-wide copy number variation association study suggested VPS13B gene for osteoporosis in Caucasians. *Osteoporos Int* **21**(4): 579–587.

Dewannieux, M., C. Esnault and T. Heidmann (2003). LINE-mediated retrotransposition of marked Alu sequences. *Nat Genet* **35**(1): 41–48.

Dhami, P., A. J. Coffey, S. Abbs, J. R. Vermeesch, J. P. Dumanski, K. J. Woodward, R. M. Andrews, C. Langford and D. Vetrie (2005). Exon array CGH: detection of copy-number changes at the resolution of individual exons in the human genome. *Am J Hum Genet* **76**(5): 750–762.

Ding, C. (2007). 'Other' applications of single nucleotide polymorphisms. *Trends Biotechnol* **25**(7): 279–283.

Dopman, E. B. and D. L. Hartl (2007). A portrait of copy-number polymorphism in Drosophila melanogaster. *Proc Natl Acad Sci U S A* **104**(50): 19920–19925.

Dozmorov, M. G., C. B. Giles, K. A. Koelsch and J. D. Wren (2013). Systematic classification of non-coding RNAs by epigenomic similarity. *BMC Bioinformatics* **14** (Suppl 14): S2.

Drmanac, R., S. Drmanac, G. Chui, R. Diaz, A. Hou, H. Jin, P. Jin, S. Kwon, S. Lacy, B. Moeur, J. Shafto, D. Swanson, T. Ukrainczyk, C. Xu and D. Little (2002). Sequencing by hybridization (SBH): advantages, achievements, and opportunities. *Adv Biochem Eng Biotechnol* **77**: 75–101.

Dupressoir, A., C. Lavialle and T. Heidmann (2012). From ancestral infectious retroviruses to bona fide cellular genes: role of the captured syncytins in placentation. *Placenta* **33**(9): 663–671.

Dutton, C. and S. S. Sommer (1991). Simultaneous detection of multiple single-base alleles at a polymorphic site. *Biotechniques* **11**(6): 700–702.

Eid, J., A. Fehr, J. Gray, K. Luong, J. Lyle, G. Otto, P. Peluso, D. Rank, P. Baybayan, B. Bettman, A. Bibillo, K. Bjornson, B. Chaudhuri, F. Christians, R. Cicero, S. Clark, R. Dalal, A. Dewinter, J. Dixon, M. Foquet, A. Gaertner, P. Hardenbol, C. Heiner, K. Hester, D. Holden, G. Kearns, X. Kong, R. Kuse, Y. Lacroix, S. Lin, P. Lundquist, C. Ma, P. Marks, M. Maxham, D. Murphy, I. Park, T. Pham, M. Phillips, J. Roy, R. Sebra, G. Shen, J. Sorenson, A. Tomaney, K. Travers, M. Trulson, J. Vieceli, J. Wegener, D. Wu, A. Yang, D. Zaccarin, P. Zhao, F. Zhong, J. Korlach and S. Turner (2009). Real-time DNA sequencing from single polymerase molecules. *Genome Biol* **323**(5910): 133–138.

Ellegren, H. (2004). Microsatellites: simple sequences with complex evolution. *Nat Rev Genet* **5**(6): 435–445.

Ellinghaus, D., J. Bethune, B. S. Petersen and A. Franke (2015). The genetics of Crohn's disease and ulcerative colitis–status quo and beyond. *Scand J Gastroenterol* **50**(1): 13–23.

Emerson, J. J., M. Cardoso-Moreira, J. O. Borevitz and M. Long (2008). Natural selection shapes genome-wide patterns of copy-number polymorphism in Drosophila melanogaster. *Science* **320**(5883): 1629–1631.

Ewing, A. D. and H. H. Kazazian, Jr. (2011). Whole-genome resequencing allows detection of many rare LINE-1 insertion alleles in humans. *Genome Res* **21**(6): 985–990.

Ezkurdia, I., D. Juan, J. M. Rodriguez, A. Frankish, M. Diekhans, J. Harrow, J. Vazquez, A. Valencia and M. L. Tress (2014). Multiple evidence strands suggest that there may be as few as 19,000 human protein-coding genes. *Hum Mol Genet* **23**(22): 5866–5878.

Fahed, A. C., B. D. Gelb, J. G. Seidman and C. E. Seidman (2013). Genetics of congenital heart disease: the glass half empty. *Circ Res* **112**(4): 707–720.

Falini, B., C. Mecucci, E. Tiacci, M. Alcalay, R. Rosati, L. Pasqualucci, R. La Starza, D. Diverio, E. Colombo, A. Santucci, B. Bigerna, R. Pacini, A. Pucciarini, A. Liso, M. Vignetti, P. Fazi, N. Meani, V. Pettirossi, G. Saglio, F. Mandelli, F. Lo-Coco, P. G. Pelicci and M. F. Martelli (2005). Cytoplasmic nucleophosmin in acute myelogenous leukemia with a normal karyotype. *N Engl J Med* **352**(3): 254–266.

Fanciulli, M., P. J. Norsworthy, E. Petretto, R. Dong, L. Harper, L. Kamesh, J. M. Heward, S. C. Gough, A. de Smith, A. I. Blakemore, P. Froguel, C. J. Owen, S. H. Pearce, L. Teixeira, L. Guillevin, D. S. Graham, C. D. Pusey, H. T. Cook, T. J. Vyse and T. J. Aitman (2007). FCGR3B copy number variation is associated with susceptibility to systemic, but not organ-specific, autoimmunity. *Nat Genet* **39**(6): 721–723.

Fanciulli, M., E. Petretto and T. J. Aitman (2010). Gene copy number variation and common human disease. *Clin Genet* **77**(3): 201–213.

Fang, G., D. Munera, D. I. Friedman, A. Mandlik, M. C. Chao, O. Banerjee, Z. Feng, B. Losic, M. C. Mahajan, O. J. Jabado, G. Deikus, T. A. Clark, K. Luong, I. A. Murray, B. M. Davis, A. Keren-Paz, A. Chess, R. J. Roberts, J. Korlach, S. W. Turner, V. Kumar, M. K. Waldor and E. E. Schadt (2012). Genome-wide mapping of methylated adenine residues in pathogenic Escherichia coli using single-molecule real-time sequencing. *Nat Biotechnol* **30**(12): 1232–1239.

Fellermann, K., D. E. Stange, E. Schaeffeler, H. Schmalzl, J. Wehkamp, C. L. Bevins, W. Reinisch, A. Teml, M. Schwab, P. Lichter, B. Radlwimmer and E. F. Stange (2006). A chromosome 8 gene-cluster polymorphism with low human beta-defensin 2 gene copy number predisposes to Crohn disease of the colon. *Am J Hum Genet* **79**(3): 439–448.

Ferguson-Smith, M. A. and V. Trifonov (2007). Mammalian karyotype evolution. *Nat Rev Genet* **8**(12): 950–962.

Feuk, L., A. R. Carson and S. W. Scherer (2006). Structural variation in the human genome. *Nat Rev Genet* **7**(2): 85–97.

Fiegler, H., R. Redon, D. Andrews, C. Scott, R. Andrews, C. Carder, R. Clark, O. Dovey, P. Ellis, L. Feuk, L. French, P. Hunt, D. Kalaitzopoulos, J. Larkin, L. Montgomery, G. H. Perry, B. W. Plumb, K. Porter, R. E. Rigby, D. Rigler, A. Valsesia, C. Langford, S. J. Humphray, S. W. Scherer, C. Lee, M. E. Hurles and N. P. Carter (2006). Accurate and reliable high-throughput detection of copy number variation in the human genome. *Genome Res* **16**(12): 1566–1574.

Fiorentino, F., M. C. Magli, D. Podini, A. P. Ferraretti, A. Nuccitelli, N. Vitale, M. Baldi and L. Gianaroli (2003). The minisequencing method: an alternative strategy for preimplantation genetic diagnosis of single gene disorders. *Mol Hum Reprod* **9**(7): 399–410.

Fischer, S. G. and L. S. Lerman (1983). DNA fragments differing by single base-pair substitutions are separated in denaturing gradient gels: correspondence with melting theory. *Proc Natl Acad Sci U S A* **80**(6): 1579–1583.

Flusberg, B. A., D. R. Webster, J. H. Lee, K. J. Travers, E. C. Olivares, T. A. Clark, J. Korlach and S. W. Turner (2010). Direct detection of DNA methylation during single-molecule, real-time sequencing. *Nat Methods* **7**(6): 461–465.

Freeman, J. L., G. H. Perry, L. Feuk, R. Redon, S. A. McCarroll, D. M. Altshuler, H. Aburatani, K. W. Jones, C. Tyler-Smith, M. E. Hurles, N. P. Carter, S. W. Scherer and C. Lee (2006). Copy number variation: new insights in genome diversity. *Genome Res* **16**(8): 949–961.

Frenkel, S., C. N. Bernstein, M. Sargent, Q. Kuang, W. Jiang, J. Wei, B. Thiruvahindrapuram, E. Spriggs, S. W. Scherer and P. Hu (2019). Genome-wide analysis identifies rare copy number variations associated with inflammatory bowel disease. *PLoS One* **14**(6): e0217846.

Fuller, C. W., L. R. Middendorf, S. A. Benner, G. M. Church, T. Harris, X. Huang, S. B. Jovanovich, J. R. Nelson, J. A. Schloss, D. C. Schwartz and D. V. Vezenov (2009). The challenges of sequencing by synthesis. *Nat Biotechnol* **27**(-11): 1013–1023.

Futreal, P. A., L. Coin, M. Marshall, T. Down, T. Hubbard, R. Wooster, N. Rahman and M. R. Stratton (2004). A census of human cancer genes. *Nat Rev Cancer* **4**(3): 177–183.

Gabriel, S. B., S. F. Schaffner, H. Nguyen, J. M. Moore, J. Roy, B. Blumenstiel, J. Higgins, M. DeFelice, A. Lochner, M. Faggart, S. N. Liu-Cordero, C. Rotimi, A. Adeyemo, R. Cooper, R. Ward, E. S. Lander, M. J. Daly and D. Altshuler (2002). The structure of haplotype blocks in the human genome. *Science* **296**(5576): 2225–2229.

Gajecka, M., A. J. Gentles, A. Tsai, D. Chitayat, K. L. Mackay, C. D. Glotzbach, M. R. Lieber and L. G. Shaffer (2008). Unexpected complexity at breakpoint junctions in phenotypically normal individuals and mechanisms involved in generating balanced translocations t(1;22)(p36;q13). *Genome Res* **18**(11): 1733–1742.

Gartler, S. M. (2006). The chromosome number in humans: a brief history. *Nat Rev Genet* **7**(8): 655–660.

Germer, S. and R. Higuchi (1999). Single-tube genotyping without oligonucleotide probes. *Genome Res* **9**(1): 72–78.

Ghanem, N., B. Uring-Lambert, M. Abbal, G. Hauptmann, M. P. Lefranc and G. Lefranc (1988). Polymorphism of MHC class III genes: definition of restriction fragment linkage groups and evidence for frequent deletions and duplications. *Hum Genet* **79**(3): 209–218.

Gilles, F., A. Goy, Y. Remache, K. Manova and A. D. Zelenetz (2000). Cloning and characterization of a Golgin-related gene from the large-scale polymorphism linked to the PML gene. *Genomics* **70**(3): 364–374.

Glessner, J. T., J. Li, A. Desai, M. Palmer, D. Kim, A. M. Lucas, X. Chang, J. J. Connolly, B. Almoguera, J. B. Harley, G. P. Jarvik, M. D. Ritchie, P. M. A. Sleiman, D. M. Roden, D. Crosslin and H. Hakonarson (2020). CNV Association of Diverse Clinical Phenotypes from eMERGE reveals novel disease biology underlying cardiovascular disease. *Int J Cardiol* **298**: 107–113.

Goldstein, D. B. and G. L. Cavalleri (2005). Genomics: understanding human diversity. *Nature* **437**(7063): 1241–1242.

Goldstein, D. B. and M. E. Weale (2001). Population genomics: linkage disequilibrium holds the key. *Curr Biol* **11**(14): R576–579.

Gonzaga-Jauregui, C., J. R. Lupski and R. A. Gibbs (2012). Human genome sequencing in health and disease. *Annu Rev Med* **63**: 35–61.

Gonzalez, E., H. Kulkarni, H. Bolivar, A. Mangano, R. Sanchez, G. Catano, R. J. Nibbs, B. I. Freedman, M. P. Quinones, M. J. Bamshad, K. K. Murthy, B. H. Rovin, W. Bradley, R. A. Clark, S. A. Anderson, J. O'Connell R, B. K. Agan, S. S. Ahuja, R. Bologna, L. Sen, M. J. Dolan and S. K. Ahuja (2005). The influence of CCL3L1 gene-containing segmental duplications on HIV-1/AIDS susceptibility. *Science* **307**(5714): 1434–1440.

Graubert, T. A., P. Cahan, D. Edwin, R. R. Selzer, T. A. Richmond, P. S. Eis, W. D. Shannon, X. Li, H. L. McLeod, J. M. Cheverud and T. J. Ley (2007). A high-resolution map of segmental DNA copy number variation in the mouse genome. *PLoS Genet* **3**(1): e3.

Greninger, A. L., S. N. Naccache, S. Federman, G. Yu, P. Mbala, V. Bres, D. Stryke, J. Bouquet, S. Somasekar, J. M. Linnen, R. Dodd, P. Mulembakani, B. S. Schneider, J. J. Muyembe-Tamfum, S. L. Stramer and C. Y. Chiu (2015). Rapid metagenomic identification of viral pathogens in clinical samples by real-time nanopore sequencing analysis. *Genome Med* **7**: 99.

Gribble, S. M., D. Kalaitzopoulos, D. C. Burford, E. Prigmore, R. R. Selzer, B. L. Ng, N. S. Matthews, K. M. Porter, R. Curley, S. J. Lindsay, J. Baptista, T. A. Richmond and N. P. Carter (2007). Ultra-high resolution array painting facilitates breakpoint sequencing. *J Med Genet* **44**(1): 51–58.

Griffin, D. K., L. B. Robertson, H. G. Tempest, A. Vignal, V. Fillon, R. P. Crooijmans, M. A. Groenen, S. Deryusheva, E. Gaginskaya, W. Carré, D. Waddington, R. Talbot, M. Völker, J. S. Masabanda and D. W. Burt (2008). Whole genome comparative studies between chicken and turkey and their implications for avian genome evolution. *BMC Genomics* **9**: 168.

Groot, P. C., M. J. Bleeker, J. C. Pronk, F. Arwert, W. H. Mager, R. J. Planta, A. W. Eriksson and R. R. Frants (1989). The human alpha-amylase multigene family consists of haplotypes with variable numbers of genes. *Genomics* **5**(1): 29–42.

Gu, W., F. Zhang and J. R. Lupski (2008). Mechanisms for human genomic rearrangements. *Pathogenetics* **1**(1): 4.

Gudnason, H., M. Dufva, D. D. Bang and A. Wolff (2007). Comparison of multiple DNA dyes for real-time PCR: effects of dye concentration and sequence composition on DNA amplification and melting temperature. *Nucleic Acids Res* **35**(19): e127.

Guo, D. C., Y. Qi, R. He, P. Gupta and D. M. Milewicz (2003). High throughput detection of small genomic insertions or deletions by pyrosequencing. *Biotechnol Lett* **25**(20): 1703–1707.

Hanafi, S., R. Hassan, R. Bahar, W. Z. Abdullah, M. F. Johan, N. D. Rashid, N. F. Azman, A. Nasir, S. Hassan, R. Ahmad, A. Othman, M. I. Ibrahim, S. Sukeri, S. Sulong, S. Yusoff, N. S. Mohamad, A. Hussein, R. Hassan, N. Yusoff, B. H. Yahaya, E. Ismail, N. K. Yussof, S. Salleh and B. A. Zilfalil (2014). Multiplex amplification refractory mutation system

(MARMS) for the detection of β-globin gene mutations among the transfusion-dependent β-thalassemia Malay patients in Kelantan, Northeast of Peninsular Malaysia. *Am J Blood Res* **4**(1): 33–40.

Hancks, D. C. and H. H. Kazazian, Jr. (2016). Roles for retrotransposon insertions in human disease. *Mob DNA* **7**: 9.

Harewood, L., E. Chaignat and A. Reymond (2012). Structural variation and its effect on expression. *Methods Mol Biol* **838**: 173–186.

Harewood, L., F. Schütz, S. Boyle, P. Perry, M. Delorenzi, W. A. Bickmore and A. Reymond (2010). The effect of translocation-induced nuclear reorganization on gene expression. *Genome Res* **20**(5): 554–564.

Harris, T. D., P. R. Buzby, H. Babcock, E. Beer, J. Bowers, I. Braslavsky, M. Causey, J. Colonell, J. Dimeo, J. W. Efcavitch, E. Giladi, J. Gill, J. Healy, M. Jarosz, D. Lapen, K. Moulton, S. R. Quake, K. Steinmann, E. Thayer, A. Tyurina, R. Ward, H. Weiss and Z. Xie (2008). Single-molecule DNA sequencing of a viral genome. *Science* **320**(-5872): 106–109.

Hassan, S., R. Ahmad, Z. Zakaria, Z. Zulkafli and W. Z. Abdullah (2013). Detection of β-globin Gene Mutations Among β-thalassaemia Carriers and Patients in Malaysia: application of multiplex amplification refractory mutation system-polymerase chain reaction. *Malays J Med Sci* **20**(-1): 13–20.

Hastings, P. J., G. Ira and J. R. Lupski (2009). A microhomology-mediated break-induced replication model for the origin of human copy number variation. *PLoS Genet* **5**(1): e1000327.

Hastings, P. J., J. R. Lupski, S. M. Rosenberg and G. Ira (2009). Mechanisms of change in gene copy number. *Nat Rev Genet* **10**(8): 551–564.

Hayashi, K. (1991). PCR-SSCP: a simple and sensitive method for detection of mutations in the genomic DNA. *PCR Methods Appl* **1**(1): 34–38.

Heid, C. A., J. Stevens, K. J. Livak and P. M. Williams (1996). Real time quantitative PCR. *Genome Res* **6**(10): 986–994.

Heim, M. H. and U. A. Meyer (1992). Evolution of a highly polymorphic human cytochrome P450 gene cluster: CYP2D6. *Genomics* **14**(1): 49–58.

Helman, E., M. S. Lawrence, C. Stewart, C. Sougnez, G. Getz and M. Meyerson (2014). Somatic retrotransposition in human cancer revealed by whole-genome and exome sequencing. *Genome Res* **24**(7): 1053–1063.

Henco, K., J. Harders, U. Wiese and D. Riesner (1994). Temperature gradient gel electrophoresis (TGGE) for the detection of polymorphic DNA and RNA. *Methods Mol Biol* **31**: 211–228.

Henrichsen, C. N., E. Chaignat and A. Reymond (2009). Copy number variants, diseases and gene expression. *Hum Mol Genet* **18**(R1): R1–8.

Henrichsen, C. N., N. Vinckenbosch, S. Zöllner, E. Chaignat, S. Pradervand, F. Schütz, M. Ruedi, H. Kaessmann and A. Reymond (2009). Segmental copy number variation shapes tissue transcriptomes. *Nat Genet* **41**(4): 424–429.

Herschleb, J., G. Ananiev and D. C. Schwartz (2007). Pulsed-field gel electrophoresis. *Nat Protoc* **2**(3): 677–684.

Hoebeeck, J., F. Speleman and J. Vandesompele (2007). Real-time quantitative PCR as an alternative to Southern blot or fluorescence in situ hybridization for detection of gene copy number changes. *Methods Mol Biol* **353**: 205–226.

Holland, P. M., R. D. Abramson, R. Watson and D. H. Gelfand (1991). Detection of specific polymerase chain reaction product by utilizing the 5′----3′ exonuclease activity of Thermus aquaticus DNA polymerase. *Proc Natl Acad Sci U S A* **88**(16): 7276–7280.

Hollox, E. J., J. A. Armour and J. C. Barber (2003). Extensive normal copy number variation of a beta-defensin antimicrobial-gene cluster. *Am J Hum Genet* **73**(3): 591–600.

Hollox, E. J., J. C. Detering and T. Dehnugara (2009). An integrated approach for measuring copy number variation at the FCGR3 (CD16) locus. *Hum Mutat* **30**(3): 477–484.

Hollox, E. J., U. Huffmeier, P. L. Zeeuwen, R. Palla, J. Lascorz, D. Rodijk-Olthuis, P. C. van de Kerkhof, H. Traupe, G. de Jongh, M. den Heijer, A. Reis, J. A. Armour and J. Schalkwijk (2008). Psoriasis is associated with increased beta-defensin genomic copy number. *Nat Genet* **40**(1): 23–25.

Hormozdiari, F., C. Alkan, E. E. Eichler and S. C. Sahinalp (2009). Combinatorial algorithms for structural variation detection in high-throughput sequenced genomes. *Genome Res* **19**(7): 1270–1278.

Hua-Van, A., A. Le Rouzic, T. S. Boutin, J. Filée and P. Capy (2011). The struggle for life of the genome's selfish architects. *Biol Direct* **6**: 19.

Hurles, M. (2004). Gene duplication: the genomic trade in spare parts. *PLoS Biol* **2**(7): E206.

Iafrate, A. J., L. Feuk, M. N. Rivera, M. L. Listewnik, P. K. Donahoe, Y. Qi, S. W. Scherer and C. Lee (2004). Detection of large-scale variation in the human genome. *Nat Genet* **36**(9): 949–951.

IHGSC (2004). International Human Genome Sequencing Consortium I finishing the euchromatic sequence of the human genome. *Nature* **431**(7011): 931–945.

Inazuka, M., H. M. Wenz, M. Sakabe, T. Tahira and K. Hayashi (1997). A streamlined mutation detection system: multicolor post-PCR fluorescence labeling and single-strand conformational polymorphism analysis by capillary electrophoresis. *Genome Res* **7**(11): 1094–1103.

Inoue, K., H. Osaka, V. C. Thurston, J. T. Clarke, A. Yoneyama, L. Rosenbarker, T. D. Bird, M. E. Hodes, L. G. Shaffer and J. R. Lupski (2002). Genomic rearrangements resulting in PLP1 deletion occur by nonhomologous end joining and cause different dysmyelinating phenotypes in males and females. *Am J Hum Genet* **71**(4): 838–853.

ISC (2008). International Schizophrenia Consortium. Rare chromosomal deletions and duplications increase risk of schizophrenia. *Nature* **455**(7210): 237–241.

Ishkanian, A. S., C. A. Malloff, S. K. Watson, R. J. DeLeeuw, B. Chi, B. P. Coe, A. Snijders, D. G. Albertson, D. Pinkel, M. A. Marra, V. Ling, C. MacAulay and W. L. Lam (2004). A tiling resolution DNA microarray with complete coverage of the human genome. *Nat Genet* **36**(3): 299–303.

Iskow, R. C., M. T. McCabe, R. E. Mills, S. Torene, W. S. Pittard, A. F. Neuwald, E. G. Van Meir, P. M. Vertino and S. E. Devine (2010). Natural mutagenesis of human genomes by endogenous retrotransposons. *Cell* **141**(7): 1253–1261.

Itsara, A., G. M. Cooper, C. Baker, S. Girirajan, J. Li, D. Absher, R. M. Krauss, R. M. Myers, P. M. Ridker, D. I. Chasman, H. Mefford, P. Ying, D. A. Nickerson and E. E. Eichler (2009). Population analysis of large copy number variants and hotspots of human genetic disease. *Am J Hum Genet* **84**(2): 148–161.

Johansson, I. and M. Ingelman-Sundberg (2011). Genetic polymorphism and toxicology–with emphasis on cytochrome p450. *Toxicol Sci* **120**(1): 1–13.

Johansson, I., E. Lundqvist, L. Bertilsson, M. L. Dahl, F. Sjöqvist and M. Ingelman-Sundberg (1993). Inherited amplification of an active gene in the cytochrome P450 CYP2D locus as a cause of ultrarapid metabolism of debrisoquine. *Proc Natl Acad Sci U S A* **90**(24): 11825–11829.

Jorde, L. B. and S. P. Wooding (2004). Genetic variation, classification and 'race'. *Nat Genet* **36**(11 Suppl): S28–33.

Kallioniemi, A., O. P. Kallioniemi, D. Sudar, D. Rutovitz, J. W. Gray, F. Waldman and D. Pinkel (1992). Comparative genomic hybridization for molecular cytogenetic analysis of solid tumors. *Science* **258**(5083): 818–821.

Karki, R., D. Pandya, R. C. Elston and C. Ferlini (2015). Defining "mutation" and "polymorphism" in the era of personal genomics. *BMC Med Genomics* **8**: 37.

Karypidis, A. H., M. Olsson, S. O. Andersson, A. Rane and L. Ekström (2008). Deletion polymorphism of the UGT2B17 gene is associated with increased risk for prostate cancer and correlated to gene expression in the prostate. *Pharmacogenomics J* **8**(2): 147–151.

Katara, P. (2014). Single nucleotide polymorphism and its dynamics for pharmacogenomics. *Interdiscip Sci* **6**(2): 85–92.

Kidd, J. M., G. M. Cooper, W. F. Donahue, H. S. Hayden, N. Sampas, T. Graves, N. Hansen, B. Teague, C. Alkan, F. Antonacci, E. Haugen, T. Zerr, N. A. Yamada, P. Tsang, T. L. Newman, E. Tüzün, Z. Cheng, H. M. Ebling, N. Tusneem, R. David, W. Gillett, K. A. Phelps, M. Weaver, D. Saranga, A. Brand, W. Tao, E. Gustafson, K. McKernan, L. Chen, M. Malig, J. D. Smith, J. M. Korn, S. A. McCarroll, D. A. Altshuler, D. A. Peiffer, M. Dorschner, J. Stamatoyannopoulos, D. Schwartz, D. A. Nickerson, J. C. Mullikin, R. K. Wilson, L. Bruhn, M. V. Olson, R. Kaul, D. R. Smith and E. E. Eichler (2008). Mapping and sequencing of structural variation from eight human genomes. *Nature* **453**(7191): 56–64.

Kim, J., C. Hu, C. Moufawad El Achkar, L. E. Black, J. Douville, A. Larson, M. K. Pendergast, S. F. Goldkind, E. A. Lee, A. Kuniholm, A. Soucy, J. Vaze, N. R. Belur, K. Fredriksen, I. Stojkovska, A. Tsytsykova, M. Armant, R. L. DiDonato and J. Choi (2019). Patient-customized oligonucleotide therapy for a rare genetic disease. **381**(17): 1644–1652.

Kim, P. M., J. O. Korbel and M. B. Gerstein (2007). Positive selection at the protein network periphery: evaluation in terms of structural constraints and cellular context. *Proc Natl Acad Sci U S A* **104**(51): 20274–20279.

Kim, P. M., H. Y. Lam, A. E. Urban, J. O. Korbel, J. Affourtit, F. Grubert, X. Chen, S. Weissman, M. Snyder and M. B. Gerstein (2008). Analysis of copy number variants and segmental duplications in the human genome: evidence for a change in the process of formation in recent evolutionary history. *Genome Res* **18**(12): 1865–1874.

Kirchhoff, M., T. Gerdes, H. Rose, J. Maahr, A. M. Ottesen and C. Lundsteen (1998). Detection of chromosomal gains and losses in comparative genomic hybridization analysis based on standard reference intervals. *Cytometry* **31**(3): 163–173.

Klinger, K., G. Landes, D. Shook, R. Harvey, L. Lopez, P. Locke, T. Lerner, R. Osathanondh, B. Leverone, T. Houseal, et al. (1992). Rapid detection of chromosome aneuploidies in uncultured amniocytes by using fluorescence in situ hybridization (FISH). *Am J Hum Genet* **51**(1): 55–65.

Knapp, L. A., L. F. Cadavid, M. E. Eberle, S. J. Knechtle, R. E. Bontrop and D. I. Watkins (1997). Identification of new mamu-DRB alleles using DGGE and direct sequencing. *Immunogenetics* **45**(3): 171–179.

Konrad, M., S. Saunier, L. Heidet, F. Silbermann, F. Benessy, J. Calado, D. Le Paslier, M. Broyer, M. C. Gubler and C. Antignac (1996). Large homozygous deletions of the 2q13 region are a major cause of juvenile nephronophthisis. *Hum Mol Genet* **5**(3): 367–371.

Koolen, D. A., A. J. Sharp, J. A. Hurst, H. V. Firth, S. J. Knight, A. Goldenberg, P. Saugier-Veber, R. Pfundt, L. E. Vissers, A. Destrée, B. Grisart, L. Rooms, N. Van der Aa, M. Field, A. Hackett, K. Bell, M. J. Nowaczyk, G. M. Mancini, P. J. Poddighe, C. E. Schwartz, E. Rossi, M. De Gregori, L. L. Antonacci-Fulton, M. D. McLellan, 2nd, J. M. Garrett, M. A. Wiechert, T. L. Miner, S. Crosby, R. Ciccone, L. Willatt, A. Rauch, M. Zenker, S. Aradhya, M. A. Manning, T. M. Strom, J. Wagenstaller, A. C. Krepischi-Santos, A. M. Vianna-Morgante, C. Rosenberg, S. M. Price, H. Stewart, C. Shaw-Smith, H. G. Brunner, A. O. Wilkie, J. A. Veltman, O. Zuffardi, E. E. Eichler and B. B. de Vries (2008). Clinical and molecular delineation of the 17q21.31 microdeletion syndrome. *J Med Genet* **45**(11): 710–720.

Korbel, J. O., A. E. Urban, J. P. Affourtit, B. Godwin, F. Grubert, J. F. Simons, P. M. Kim, D. Palejev, N. J. Carriero, L. Du, B. E. Taillon, Z. Chen, A. Tanzer, A. C. Saunders, J. Chi, F. Yang, N. P. Carter, M. E. Hurles, S. M. Weissman, T. T. Harkins, M. B. Gerstein, M. Egholm and M. Snyder (2007). Paired-end mapping reveals extensive structural variation in the human genome. *Science* **318**(5849): 420–426.

Kosorok, M. R., W. H. Wei and P. M. Farrell (1996). The incidence of cystic fibrosis. *Stat Med* **15**(5): 449–462.

Kosugi, S., Y. Momozawa, X. Liu, C. Terao, M. Kubo and Y. Kamatani (2019). Comprehensive evaluation of structural variation detection algorithms for whole genome sequencing. *Genome Biol* **20**(1): 117.

Kristofich, J. and A. B. Morgenthaler (2018). Synonymous mutations make dramatic contributions to fitness when growth is limited by a weak-link enzyme. PLoS Genet **14**(8): e1007615.

Kruglyak, L. (1999). Prospects for whole-genome linkage disequilibrium mapping of common disease genes. *Nat Genet* **22**(2): 139–144.

Kubiritova, Z., M. Gyuraszova, E. Nagyova, M. Hyblova, M. Harsanyova, J. Budis, R. Hekel, J. Gazdarica, F. Duris, L. Kadasi, T. Szemes and J. Radvanszky (2019). On the critical evaluation and confirmation of germline sequence variants identified using massively parallel sequencing. *J Biotechnol* **298**: 64–75.

Kukita, Y., T. Tahira, S. S. Sommer and K. Hayashi (1997). SSCP analysis of long DNA fragments in low pH gel. *Hum Mutat* **10**(5): 400–407.

Kumaran, M., P. Krishnan, C. E. Cass and R. Hubaux (2018). Breast cancer associated germline structural variants harboring small noncoding RNAs impact post-transcriptional gene regulation. *Sci Rep* **8**(1): 7529.

Kuroda-Kawaguchi, T., H. Skaletsky, L. G. Brown, P. J. Minx, H. S. Cordum, R. H. Waterston, R. K. Wilson, S. Silber, R. Oates, S. Rozen and D. C. Page (2001). The AZFc region of the Y chromosome features massive palindromes and uniform recurrent deletions in infertile men. *Nat Genet* **29**(3): 279–286.

LaFramboise, T. (2009). Single nucleotide polymorphism arrays: a decade of biological, computational and technological advances. *Nucleic Acids Res* **37**(13): 4181–4193.

Laing, R. E., P. Hess, Y. Shen, J. Wang and S. X. Hu (2011). The role and impact of SNPs in pharmacogenomics and personalized medicine. *Curr Drug Metab* **12**(5): 460–486.

Lakich, D., H. H. Kazazian, Jr., S. E. Antonarakis and J. Gitschier (1993). Inversions disrupting the factor VIII gene are a common cause of severe haemophilia A. *Nat Genet* **5**(3): 236–241.

Lander, E. S. (2011). Initial impact of the sequencing of the human genome. *Nature* **470**(7333): 187–197.

Lander, E. S., L. M. Linton, B. Birren, C. Nusbaum, M. C. Zody, J. Baldwin, K. Devon, K. Dewar, M. Doyle, W. FitzHugh, R. Funke, D. Gage, K. Harris, A. Heaford, J. Howland, L. Kann, J. Lehoczky, R. LeVine, P. McEwan, K. McKernan, J. Meldrim, J. P. Mesirov, C. Miranda, W. Morris, J. Naylor, C. Raymond, M. Rosetti, R. Santos, A. Sheridan, C. Sougnez, N. Stange-Thomann, N. Stojanovic, A. Subramanian, D. Wyman, J. Rogers, J. Sulston, R. Ainscough, S. Beck, D. Bentley, J. Burton, C. Clee, N. Carter, A. Coulson, R. Deadman, P. Deloukas, A. Dunham, I. Dunham, R. Durbin, L. French, D. Grafham, S. Gregory, T. Hubbard, S. Humphray, A. Hunt, M. Jones, C. Lloyd, A. McMurray, L. Matthews, S. Mercer, S. Milne, J. C. Mullikin, A. Mungall, R. Plumb, M. Ross, R. Shownkeen, S. Sims, R. H. Waterston, R. K. Wilson, L. W. Hillier, J. D. McPherson, M. A. Marra, E. R. Mardis, L. A. Fulton, A. T. Chinwalla, K. H. Pepin, W. R. Gish, S. L. Chissoe, M. C. Wendl, K. D. Delehaunty, T. L. Miner, A. Delehaunty, J. B. Kramer, L. L. Cook, R. S. Fulton, D. L. Johnson, P. J. Minx, S. W. Clifton, T. Hawkins, E. Branscomb, P. Predki, P. Richardson, S. Wenning, T. Slezak, N. Doggett, J.-F. Cheng, A. Olsen, S. Lucas, C. Elkin, E. Uberbacher, M. Frazier, R. A. Gibbs, D. M. Muzny, S. E. Scherer, J. B. Bouck, E. J. Sodergren, K. C. Worley, C. M. Rives, J. H. Gorrell, M. L. Metzker, S. L. Naylor, R. S. Kucherlapati, D. L. Nelson, G. M. Weinstock, Y. Sakaki, A. Fujiyama, M. Hattori, T. Yada, A. Toyoda, T. Itoh, C. Kawagoe, H. Watanabe, Y. Totoki, T. Taylor, J. Weissenbach, R. Heilig, W. Saurin, F. Artiguenave, P. Brottier, T. Bruls, E. Pelletier, C. Robert, P. Wincker, A. Rosenthal, M. Platzer, G. Nyakatura, S. Taudien, A. Rump, D. R. Smith, L. Doucette-Stamm, M. Rubenfield, K. Weinstock, H. M. Lee, J. Dubois, H. Yang, J. Yu, J. Wang, G. Huang, J. Gu, L. Hood, L. Rowen, A. Madan, S. Qin, R. W. Davis, N. A. Federspiel, A. P. Abola, M. J. Proctor, B. A. Roe, F. Chen, H. Pan, J. Ramser, H. Lehrach, R. Reinhardt, W. R. McCombie, M. de la Bastide, N. Dedhia, H. Blöcker, K. Hornischer, G. Nordsiek, R. Agarwala, L. Aravind, J. A. Bailey, A. Bateman, S. Batzoglou, E. Birney, P. Bork, D. G. Brown, C. B. Burge, L. Cerutti, H.-C. Chen, D. Church, M. Clamp, R. R. Copley, T. Doerks, S. R. Eddy, E. E. Eichler, T. S. Furey, J. Galagan, J. G. R. Gilbert, C. Harmon, Y. Hayashizaki, D. Haussler, H. Hermjakob, K. Hokamp, W. Jang, L. S. Johnson, T. A. Jones, S. Kasif, A. Kaspryzk, S. Kennedy, W. J. Kent, P. Kitts, E. V. Koonin, I. Korf, D. Kulp, D. Lancet, T. M. Lowe, A. McLysaght, T. Mikkelsen, J. V. Moran, N. Mulder, V. J. Pollara, C. P. Ponting, G. Schuler, J. Schultz, G. Slater, A. F. A. Smit, E. Stupka, J. Szustakowki, D. Thierry-Mieg, J. Thierry-Mieg, L. Wagner, J. Wallis, R. Wheeler, A. Williams, Y. I. Wolf, K. H. Wolfe, S.-P. Yang, R.-F. Yeh, F. Collins, M. S. Guyer, J. Peterson, A. Felsenfeld, K. A. Wetterstrand, R. M. Myers, J. Schmutz, M. Dickson, J. Grimwood, D. R. Cox, M. V. Olson, R. Kaul, C. Raymond, N. Shimizu, K. Kawasaki, S. Minoshima, G. A. Evans, M. Athanasiou, R. Schultz, A. Patrinos, M. J. Morgan, C. International Human Genome Sequencing, C. f. G. R. Whitehead Institute for Biomedical Research, C. The Sanger, C. Washington University Genome Sequencing, U. D. J. G. Institute, C. Baylor College of Medicine Human Genome Sequencing, R. G. S. Center, Genoscope, U. M. R. Cnrs, I. o. M. B. Department of Genome Analysis, G. T. C. S. Center, C. Beijing Genomics Institute/Human Genome, T. I. f. S. B. Multimegabase Sequencing Center, C. Stanford Genome Technology, T. University of Oklahoma's Advanced Center for Genome, G. Max Planck Institute for Molecular, L. A. H. G. C. Cold Spring Harbor Laboratory, G. B. G. R. C. f. Biotechnology, G. *Genome Analysis, U. S. N. I. o. H. Scientific management: National Human Genome Research Institute, C. Stanford Human Genome, C. University of Washington Genome, K. U. S. o. M. Department of Molecular Biology, D. University of Texas Southwestern Medical Center at, U. S. D. o. E. Office of Science and T. The Wellcome (2001). Initial sequencing and analysis of the human genome. *Nature* **409**(6822): 860–921.

Langaee, T. and M. Ronaghi (2005). Genetic variation analyses by pyrosequencing.*Mutat Res* **573**(1–2): 96–102.

Langer-Safer, P. R., M. Levine and D. C. Ward (1982). Immunological method for mapping genes on Drosophila polytene chromosomes. *Proc Natl Acad Sci U S A* **79**(14): 4381–4385.

Latif, S., I. Bauer-Sardina, K. Ranade, K. J. Livak and P. Y. Kwok (2001). Fluorescence polarization in homogeneous nucleic acid analysis II: 5′-nuclease assay. *Genome Res* **11**(3): 436–440.

Lavebratt, C. and S. Sengul (2006). Single nucleotide polymorphism (SNP) allele frequency estimation in DNA pools using pyrosequencing. *Nat Protoc* **1**(6): 2573–2582.

Laver, T., J. Harrison, P. A. O'Neill, K. Moore, A. Farbos, K. Paszkiewicz and D. J. Studholme (2015). Assessing the performance of the Oxford Nanopore Technologies MinION. *Biomol Detect Quantif* **3**: 1–8.

Leary, R. J., J. C. Lin, J. Cummins, S. Boca, L. D. Wood, D. W. Parsons, S. Jones, T. Sjöblom, B. H. Park, R. Parsons, J. Willis, D. Dawson, J. K. Willson, T. Nikolskaya, Y. Nikolsky, L. Kopelovich, N. Papadopoulos, L. A. Pennacchio, T. L. Wang, S. D. Markowitz, G. Parmigiani, K. W. Kinzler, B. Vogelstein and V. E. Velculescu (2008). Integrated analysis of homozygous deletions, focal amplifications, and sequence alterations in breast and colorectal cancers. *Proc Natl Acad Sci U S A* **105**(42): 16224–16229.

Lee, E., R. Iskow, L. Yang, O. Gokcumen, P. Haseley, L. J. Luquette, 3rd, J. G. Lohr, C. C. Harris, L. Ding, R. K. Wilson, D. A. Wheeler, R. A. Gibbs, R. Kucherlapati, C. Lee, P. V. Kharchenko and P. J. Park (2012). Landscape of somatic retrotransposition in human cancers. *Science* **337**(6097): 967–971.

Lee, E. S., J. P. Issa, D. B. Roberts, M. D. Williams, R. S. Weber, M. S. Kies and A. K. El-Naggar (2008). Quantitative promoter hypermethylation analysis of cancer-related genes in salivary gland carcinomas: comparison with methylation-specific PCR technique and clinical significance. *Clin Cancer Res* **14**(9): 2664–2672.

Lee, J. A., C. M. Carvalho and J. R. Lupski (2007). A DNA replication mechanism for generating nonrecurrent rearrangements associated with genomic disorders. *Cell* **131**(7): 1235–1247.

Lee, J. A., K. Inoue, S. W. Cheung, C. A. Shaw, P. Stankiewicz and J. R. Lupski (2006). Role of genomic architecture in PLP1 duplication causing Pelizaeus-Merzbacher disease. *Hum Mol Genet* **15**(14): 2250–2265.

Lee, J. A. and J. R. Lupski (2006). Genomic rearrangements and gene copy-number alterations as a cause of nervous system disorders. *Neuron* **52**(1): 103–121.

Lee, J. H. and J. T. Jeon (2008). Methods to detect and analyze copy number variations at the genome-wide and locus-specific levels. *Cytogenet Genome Res* **123**(1–4): 333–342.

Lichter, J. B., M. J. Difilippantonio, A. J. Pakstis, P. J. Goodfellow, D. C. Ward and K. K. Kidd (1993). Physical and genetic maps for chromosome 10. *Genomics* **16**(2): 320–324.

Lichter, P., T. Cremer, C. J. Tang, P. C. Watkins, L. Manuelidis and D. C. Ward (1988). Rapid detection of human chromosome 21 aberrations by in situ hybridization. *Proc Natl Acad Sci U S A* **85**(24): 9664–9668.

Lichter, P., S. Joos, M. Bentz and S. Lampel (2000). Comparative genomic hybridization: uses and limitations. *Semin Hematol* **37**(4): 348–357.

Lieber, M. R. (2008). The mechanism of human nonhomologous DNA end joining. *J Biol Chem* **283**(1): 1–5.

Lieber, M. R., J. Gu, H. Lu, N. Shimazaki and A. G. Tsai (2010). Nonhomologous DNA end joining (NHEJ) and chromosomal translocations in humans. *Subcell Biochem* **50**: 279–296.

Liu, G. E., Y. Hou, B. Zhu, M. F. Cardone, L. Jiang, A. Cellamare, A. Mitra, L. J. Alexander, L. L. Coutinho, M. E. Dell'Aquila, L. C. Gasbarre, G. Lacalandra, R. W. Li, L. K. Matukumalli, D. Nonneman, L. C. Regitano, T. P. Smith, J. Song, T. S. Sonstegard, C. P. Van Tassell, M. Ventura, E. E. Eichler, T. G. McDaneld and J. W. Keele (2010). Analysis of copy number variations among diverse cattle breeds. *Genome Res* **20**(5): 693–703.

Liu, H., X. Gu, G. Wang, Y. Huang, S. Ju, J. Huang and X. Wang (2019). Copy number variations primed lncRNAs deregulation contribute to poor prognosis in colorectal cancer. *Sci Rep* **11**(16): 6089–6108.

Liu, P., A. Erez, S. C. Nagamani, S. U. Dhar, K. E. Kołodziejska, A. V. Dharmadhikari, M. L. Cooper, J. Wiszniewska, F. Zhang, M. A. Withers, C. A. Bacino, L. D. Campos-Acevedo, M. R. Delgado, D. Freedenberg, A. Garnica, T. A. Grebe, D. Hernández-Almaguer, L. Immken, S. R. Lalani, S. D. McLean, H. Northrup, F. Scaglia, L. Strathearn, P. Trapane, S. H. Kang, A. Patel, S. W. Cheung, P. J. Hastings, P. Stankiewicz, J. R. Lupski and W. Bi (2011). Chromosome catastrophes involve replication mechanisms generating complex genomic rearrangements. *Cell* **146**(6): 889–903.

Liu, Q., E. C. Thorland, J. A. Heit and S. S. Sommer (1997). Overlapping PCR for bidirectional PCR amplification of specific alleles: a rapid one-tube method for simultaneously differentiating homozygotes and heterozygotes. *Genome Res* **7**(4): 389–398.

Liu, W. and D. A. Saint (2002). A new quantitative method of real time reverse transcription polymerase chain reaction assay based on simulation of polymerase chain reaction kinetics. *Anal Biochem* **302**(1): 52–59.

Livak, K. J. (2003). SNP genotyping by the 5′-nuclease reaction. *Methods Mol Biol* **212**: 129–147.

Locke, D. P., A. J. Sharp, S. A. McCarroll, S. D. McGrath, T. L. Newman, Z. Cheng, S. Schwartz, D. G. Albertson, D. Pinkel, D. M. Altshuler and E. E. Eichler (2006). Linkage disequilibrium and heritability of copy-number polymorphisms within duplicated regions of the human genome. *Am J Hum Genet* **79**(2): 275–290.

Loeuillet, C., M. Weale, S. Deutsch, M. Rotger, N. Soranzo, J. Wyniger, G. Lettre, Y. Dupré, D. Thuillard, J. S. Beckmann, S. E. Antonarakis, D. B. Goldstein and A. Telenti (2007). Promoter polymorphisms and allelic imbalance in ABCB1 expression. *Pharmacogenet Genom* **17**(11): 951–959.

Lopes, J., C. Ribeyre and A. Nicolas (2006). Complex minisatellite rearrangements generated in the total or partial absence of Rad27/hFEN1 activity occur in a single generation and are Rad51 and Rad52 dependent. *Mol Cell Biol* **26**(17): 6675–6689.

Lovett, S. T. (2004). Encoded errors: mutations and rearrangements mediated by misalignment at repetitive DNA sequences. *Mol Microbiol* **52**(5): 1243–1253.

Lucito, R., J. Healy, J. Alexander, A. Reiner, D. Esposito, M. Chi, L. Rodgers, A. Brady, J. Sebat, J. Troge, J. A. West, S. Rostan, K. C. Nguyen, S. Powers, K. Q. Ye, A. Olshen, E. Venkatraman, L. Norton and M. Wigler (2003). Representational oligonucleotide microarray analysis: a high-resolution method to detect genome copy number variation. *Genome Res* **13**(10): 2291–2305.

Lupiáñez, D. G., K. Kraft, V. Heinrich, P. Krawitz, F. Brancati, E. Klopocki, D. Horn, H. Kayserili, J. M. Opitz, R. Laxova, F. Santos-Simarro, B. Gilbert-Dussardier, L. Wittler, M. Borschiwer, S. A. Haas, M. Osterwalder, M. Franke, B. Timmermann, J. Hecht, M. Spielmann, A. Visel and S. Mundlos (2015). Disruptions of topological chromatin domains cause pathogenic rewiring of gene-enhancer interactions. *Cell* **161**(5): 1012–1025.

Lupski, J. R. (1998). Genomic disorders: structural features of the genome can lead to DNA rearrangements and human disease traits. *Trends Genet* **14**(10): 417–422.

Ma, L., V. B. Bajic and Z. Zhang (2013). On the classification of long non-coding RNAs. *RNA Biol* **10**(6): 925–933.

MacDonald, J. R., R. Ziman, R. K. Yuen, L. Feuk and S. W. Scherer (2014). The Database of Genomic Variants: a curated collection of structural variation in the human genome. *Nucleic Acids Res* **42**(Database issue): D986–992.

Magi, A., D. Bolognini, N. Bartalucci, A. Mingrino, R. Semeraro, L. Giovannini, S. Bonifacio, D. Parrini, E. Pelo, F. Mannelli, P. Guglielmelli and A. Maria Vannucchi (2019). Nano-GLADIATOR: real-time detection of copy number alterations from nanopore sequencing data. *Bioinformatics* **35**(21): 4213–4221.

Maher, C. A., C. Kumar-Sinha, X. Cao, S. Kalyana-Sundaram, B. Han, X. Jing, L. Sam, T. Barrette, N. Palanisamy and A. M. Chinnaiyan (2009). Transcriptome sequencing to detect gene fusions in cancer. *Nature* **458**(7234): 97–101.

Malkova, A., E. L. Ivanov and J. E. Haber (1996). Double-strand break repair in the absence of RAD51 in yeast: a possible role for break-induced DNA replication. *Proc Natl Acad Sci U S A* **93**(14): 7131–7136.

Manolio, T. A., F. S. Collins, N. J. Cox, D. B. Goldstein, L. A. Hindorff, D. J. Hunter, M. I. McCarthy, E. M. Ramos, L. R. Cardon, A. Chakravarti, J. H. Cho, A. E. Guttmacher, A. Kong, L. Kruglyak, E. Mardis, C. N. Rotimi, M. Slatkin, D. Valle, A. S. Whittemore, M. Boehnke, A. G. Clark, E. E. Eichler, G. Gibson, J. L. Haines, T. F. Mackay, S. A. McCarroll and P. M. Visscher (2009). Finding the missing heritability of complex diseases. *Nature* **461**(7265): 747–753.

Mantripragada, K. K., P. G. Buckley, C. Jarbo, U. Menzel and J. P. Dumanski (2003). Development of NF2 gene specific, strictly sequence defined diagnostic microarray for deletion detection. *J Mol Med (Berl)* **81**(7): 443–451.

Mardis, E. R. (2008). Next-generation DNA sequencing methods. *Annu Rev Genomics Hum Genet* **9**: 387–402.

Marsh, S. (2007). Pyrosequencing applications. *Methods Mol Biol* **373**: 15–24.

Marshall, C. R., E. J. Young, A. M. Pani, M. L. Freckmann, Y. Lacassie, C. Howald, K. K. Fitzgerald, M. Peippo, C. A. Morris, K. Shane, M. Priolo, M. Morimoto, I. Kondo, E. Manguoglu, S. Berker-Karauzum, P. Edery, H. H. Hobart, C. B. Mervis, O. Zuffardi, A. Reymond, P. Kaplan, M. Tassabehji, R. G. Gregg, S. W. Scherer and L. R. Osborne (2008). Infantile spasms is associated with deletion of the MAGI2 gene on chromosome 7q11.23-q21.11. *Am J Hum Genet* **83**(1): 106–111.

Martin, C. L., K. E. Wain, M. T. Oetjens, K. Tolwinski, E. Palen, A. Hare-Harris, L. Habegger, E. K. Maxwell, J. G. Reid, L. K. Walsh, S. M. Myers and D. H. Ledbetter (2020). Identification of neuropsychiatric copy number variants in a health care system population. *JAMA Psychiatry* **77**(12): 1276–1285.

Martin, J., K. Tammimies, R. Karlsson, Y. Lu, H. Larsson, P. Lichtenstein and P. K. E. Magnusson (2019). Copy number variation and neuropsychiatric problems in females and males in the general population. *Am J Med Genet B Neuropsychiatr Genet* **180**(6): 341–350.

Maxam, A. M. and W. Gilbert (1977). A new method for sequencing DNA. *Proc Natl Acad Sci U S A* **74**(2): 560–564.

Maydan, J. S., A. Lorch, M. L. Edgley, S. Flibotte and D. G. Moerman (2010). Copy number variation in the genomes of twelve natural isolates of Caenorhabditis elegans. *BMC Genomics* **11**: 62.

McCarroll, S. A., T. N. Hadnott, G. H. Perry, P. C. Sabeti, M. C. Zody, J. C. Barrett, S. Dallaire, S. B. Gabriel, C. Lee, M. J. Daly and D. M. Altshuler (2006). Common deletion polymorphisms in the human genome. *Nat Genet* **38**(1): 86–92.

McCarroll, S. A., F. G. Kuruvilla, J. M. Korn, S. Cawley, J. Nemesh, A. Wysoker, M. H. Shapero, P. I. de Bakker, J. B. Maller, A. Kirby, A. L. Elliott, M. Parkin, E. Hubbell, T. Webster, R. Mei, J. Veitch, P. J. Collins, R. Handsaker, S. Lincoln, M. Nizzari, J. Blume, K. W. Jones, R. Rava, M. J. Daly, S. B. Gabriel and D. Altshuler (2008). Integrated detection and population-genetic analysis of SNPs and copy number variation. *Nat Genet* **40**(10): 1166–1174.

McVey, M. and S. E. Lee (2008). MMEJ repair of double-strand breaks (director's cut): deleted sequences and alternative endings. *Trends Genet* **24**(11): 529–538.

Medvedev, P., M. Stanciu and M. Brudno (2009). Computational methods for discovering structural variation with next-generation sequencing.*Nat Methods***6**(11 Suppl): S13–20.

Mellars, G. and K. Gomez (2011). Mutation detection by Southern blotting. *Methods Mol Biol* **688**: 281–291.

Meltzer, P. S., X. Y. Guan, A. Burgess and J. M. Trent (1992). Rapid generation of region specific probes by chromosome microdissection and their application.*Nat Genet***1**(1): 24–28.

Merla, G., C. Howald, C. N. Henrichsen, R. Lyle, C. Wyss, M. T. Zabot, S. E. Antonarakis and A. Reymond (2006). Submicroscopic deletion in patients with Williams-Beuren syndrome influences expression levels of the nonhemizygous flanking genes.*Am J Hum Genet***79**(2): 332–341.

Metzker, M. L. (2010). Sequencing technologies – the next generation.*Nat Rev Genet***11**(1): 31–46.

Mills, R. E., K. Walter, C. Stewart, R. E. Handsaker, K. Chen, C. Alkan, A. Abyzov, S. C. Yoon, K. Ye, R. K. Cheetham, A. Chinwalla, D. F. Conrad, Y. Fu, F. Grubert, I. Hajirasouliha, F. Hormozdiari, L. M. Iakoucheva, Z. Iqbal, S. Kang, J. M. Kidd, M. K. Konkel, J. Korn, E. Khurana, D. Kural, H. Y. Lam, J. Leng, R. Li, Y. Li, C. Y. Lin, R. Luo, X. J. Mu, J. Nemesh, H. E. Peckham, T. Rausch, A. Scally, X. Shi, M. P. Stromberg, A. M. Stütz, A. E. Urban, J. A. Walker, J. Wu, Y. Zhang, Z. D. Zhang, M. A. Batzer, L. Ding, G. T. Marth, G. McVean, J. Sebat, M. Snyder, J. Wang, K. Ye, E. E. Eichler, M. B. Gerstein, M. E. Hurles, C. Lee, S. A. McCarroll and J. O. Korbel (2011). Mapping copy number variation by population-scale genome sequencing.*Nature***470**(7332): 59–65.

Minca, E. C. and D. Kowalski (2011). Replication fork stalling by bulky DNA damage: localization at active origins and checkpoint modulation.*Nucleic Acids Res***39**(7): 2610–2623.

Mohajeri, M. H., R. J. M. Brummer, R. A. Rastall, R. K. Weersma, H. J. M. Harmsen, M. Faas and M. Eggersdorfer (2018). The role of the microbiome for human health: from basic science to clinical applications.*Eur J Nutr***57**(Suppl 1): 1–14.

Moran, J. V., S. E. Holmes, T. P. Naas, R. J. DeBerardinis, J. D. Boeke and H. H. Kazazian, Jr. (1996). High frequency retrotransposition in cultured mammalian cells.*Cell***87**(5): 917–927.

Mountain, J. L., A. A. Lin, A. M. Bowcock and L. L. Cavalli-Sforza (1992). Evolution of modern humans: evidence from nuclear DNA polymorphisms.*Philos Trans R Soc Lond B Biol Sci***337**(1280): 159–165.

Mullaney, J. M., R. E. Mills, W. S. Pittard and S. E. Devine (2010). Small insertions and deletions (INDELs) in human genomes.*Hum Mol Genet***19**(R2): R131–136.

Mullighan, C. G., S. Goorha, I. Radtke, C. B. Miller, E. Coustan-Smith, J. D. Dalton, K. Girtman, S. Mathew, J. Ma, S. B. Pounds, X. Su, C. H. Pui, M. V. Relling, W. E. Evans, S. A. Shurtleff and J. R. Downing (2007). Genome-wide analysis of genetic alterations in acute lymphoblastic leukaemia.*Nature***446**(7137): 758–764.

Murray, I. A., T. A. Clark, R. D. Morgan, M. Boitano, B. P. Anton, K. Luong, A. Fomenkov, S. W. Turner, J. Korlach and R. J. Roberts (2012). The methylomes of six bacteria.*Nucleic Acids Res***40**(22): 11450–11462.

Myakishev, M. V., Y. Khripin, S. Hu and D. H. Hamer (2001). High-throughput SNP genotyping by allele-specific PCR with universal energy-transfer-labeled primers.*Genome Res***11**(1): 163–169.

Myers, R. M., N. Lumelsky, L. S. Lerman and T. Maniatis (1985). Detection of single base substitutions in total genomic DNA.*Nature*313(6002): 495–498.

Nakajima, T., H. Ohtani, T. Naruse, H. Shibata, J. I. Mimaya, H. Terunuma and A. Kimura (2007). Copy number variations of CCL3L1 and long-term prognosis of HIV-1 infection in asymptomatic HIV-infected Japanese with hemophilia. *Immunogenetics***59**(10): 793–798.

Nathans, J., D. Thomas and D. S. Hogness (1986). Molecular genetics of human color vision: the genes encoding blue, green, and red pigments.*Science***232**(4747): 193–202.

Navarro, E., G. Serrano-Heras, M. J. Castaño and J. Solera (2015). Real-time PCR detection chemistry. *Clin Chim Acta* **439**: 231–250.

Newman, S., K. E. Hermetz, B. Weckselblatt and M. K. Rudd (2015). Next-generation sequencing of duplication CNVs reveals that most are tandem and some create fusion genes at breakpoints.*Am J Hum Genet***96**(2): 208–220.

Newton, C. R., A. Graham, L. E. Heptinstall, S. J. Powell, C. Summers, N. Kalsheker, J. C. Smith and A. F. Markham (1989). Analysis of any point mutation in DNA. The amplification refractory mutation system (ARMS).*Nucleic Acids Res***17**(7): 2503–2516.

Nicholas, T. J., Z. Cheng, M. Ventura, K. Mealey, E. E. Eichler and J. M. Akey (2009). The genomic architecture of segmental duplications and associated copy number variants in dogs.*Genome Res***19**(3): 491–499.

Nowakowska, B. (2017). Clinical interpretation of copy number variants in the human genome. *J Appl Genet* **58**(4): 449–457.

O'Dushlaine, C. T., R. J. Edwards, S. D. Park and D. C. Shields (2005). Tandem repeat copy-number variation in protein-coding regions of human genes.*Genome Biol***6**(8): R69.

Okimoto, R. and J. B. Dodgson (1996). Improved PCR amplification of multiple specific alleles (PAMSA) using internally mismatched primers.*Biotechniques***21**(1): 20–22, 24, 26.

Orita, M., Y. Suzuki, T. Sekiya and K. Hayashi (1989). Rapid and sensitive detection of point mutations and DNA polymorphisms using the polymerase chain reaction.*Genomics***5**(4): 874–879.

Orr, N. and S. Chanock (2008). Common genetic variation and human disease. *Adv Genet* **62**: 1–32.

Ossowski, S., K. Schneeberger, R. M. Clark, C. Lanz, N. Warthmann and D. Weigel (2008). Sequencing of natural strains of Arabidopsis thaliana with short reads.*Genome Res***18**(12): 2024–2033.

Ota, M., H. Fukushima, J. K. Kulski and H. Inoko (2007). Single nucleotide polymorphism detection by polymerase chain reaction-restriction fragment length polymorphism.*Nat Protoc***2**(11): 2857–2864.

Palli, D., P. Rizzolo, I. Zanna, V. Silvestri, C. Saieva, M. Falchetti, A. S. Navazio, V. Graziano, G. Masala, S. Bianchi, A. Russo, S. Tommasi and L. Ottini (2013). SULT1A1 gene deletion in BRCA2-associated male breast cancer: a link between genes and environmental exposures?*J Cell Mol Med***17**(5): 605–607.

Pâques, F. and J. E. Haber (1999). Multiple pathways of recombination induced by double-strand breaks in Saccharomyces cerevisiae.*Microbiol Mol Biol Rev***63**(2): 349–404.

Pareek, C. S., R. Smoczynski and A. Tretyn (2011). Sequencing technologies and genome sequencing.*J Appl Genet***52**(4): 413–435.

Paris, P. L., A. Andaya, J. Fridlyand, A. N. Jain, V. Weinberg, D. Kowbel, J. H. Brebner, J. Simko, J. E. Watson, S. Volik, D. G. Albertson, D. Pinkel, J. C. Alers, T. H. van der Kwast, K. J. Vissers, F. H. Schroder, M. F. Wildhagen, P. G. Febbo, A. M. Chinnaiyan, K. J. Pienta, P. R. Carroll, M. A. Rubin, C. Collins and H. van Dekken (2004). Whole genome scanning identifies genotypes associated with recurrence and metastasis in prostate tumors.*Hum Mol Genet***13**(13): 1303–1313.

Park, K. B., K. E. Nam, A. R. Cho, W. Jang, M. Kim and J. H. Park (2019). Effects of copy number variations on developmental aspects of children with delayed development.*Ann Rehabil Med***43**(2): 215–223.

Pastinen, T., A. Kurg, A. Metspalu, L. Peltonen and A. C. Syvänen (1997). Minisequencing: a specific tool for DNA analysis and diagnostics on oligonucleotide arrays.*Genome Res***7**(6): 606–614.

Patsalis, P. C., L. Kousoulidou, C. Sismani, K. Männik and A. Kurg (2005). MAPH: from gels to microarrays.*Eur J Med Genet***48**(3): 241–249.

Pearson, C. E., K. Nichol Edamura and J. D. Cleary (2005). Repeat instability: mechanisms of dynamic mutations.*Nat Rev Genet***6**(10): 729–742.

Pennisi, E. (2003). Human genome. A low number wins the GeneSweep Pool. *Science* **300**(5625): 1484.

Pennisi, E. (2012). Genomics. ENCODE project writes eulogy for junk DNA. *Genome Biol* **337**(6099): 1159, 1161.

Pereira, K. M. C., S. Perazzio, A. G. A. Faria, E. S. Moreira, V. C. Santos, M. Grecco, N. P. da Silva and L. E. C. Andrade (2019). Impact of C4, C4A and C4B gene copy number variation in the susceptibility, phenotype and progression of systemic lupus erythematosus. **59**(1): 36.

Perry, G. H., A. Ben-Dor, A. Tsalenko, N. Sampas, L. Rodriguez-Revenga, C. W. Tran, A. Scheffer, I. Steinfeld, P. Tsang, N. A. Yamada, H. S. Park, J. I. Kim, J. S. Seo, Z. Yakhini, S. Laderman, L. Bruhn and C. Lee (2008). The fine-scale and complex architecture of human copy-number variation. *Am J Hum Genet* **82**(3): 685–695.

Perry, G. H., N. J. Dominy, K. G. Claw, A. S. Lee, H. Fiegler, R. Redon, J. Werner, F. A. Villanea, J. L. Mountain, R. Misra, N. P. Carter, C. Lee and A. C. Stone (2007). Diet and the evolution of human amylase gene copy number variation. *Nat Genet* **39**(10): 1256–1260.

Pertea, M., A. Shumate, G. Pertea, A. Varabyou, F. P. Breitwieser, Y. C. Chang, A. K. Madugundu, A. Pandey and S. L. Salzberg (2018). CHESS: a new human gene catalog curated from thousands of large-scale RNA sequencing experiments reveals extensive transcriptional noise. *Genome Biol* **19**(1): 208.

Phillips, C., J. Amigo, Á. Carracedo and M. V. Lareu (2015). Tetra-allelic SNPs: informative forensic markers compiled from public whole-genome sequence data. *Forensic Sci Int Genet* **19**: 100–106.

Phillips, C., J. Amigo, A. O. Tillmar, M. A. Peck, M. de la Puente, J. Ruiz-Ramírez, F. Bittner, Š. Idrizbegović, Y. Wang, T. J. Parsons and M. V. Lareu (2020). A compilation of tri-allelic SNPs from 1000 Genomes and use of the most polymorphic loci for a large-scale human identification panel. *Forensic Sci Int Genet* **46**: 102232.

Phillips, C., A. Salas, J. J. Sánchez, M. Fondevila, A. Gómez-Tato, J. Alvarez-Dios, M. Calaza, M. C. de Cal, D. Ballard, M.

V. Lareu and A. Carracedo (2007). Inferring ancestral origin using a single multiplex assay of ancestry-informative marker SNPs. *Forensic Sci Int Genet* **1**(3–4): 273–280.

Pielberg, G., A. E. Day, G. S. Plastow and L. Andersson (2003). A sensitive method for detecting variation in copy numbers of duplicated genes. *Genome Res* **13**(9): 2171–2177.

Pinkel, D., J. Landegent, C. Collins, J. Fuscoe, R. Segraves, J. Lucas and J. Gray (1988). Fluorescence in situ hybridization with human chromosome-specific libraries: detection of trisomy 21 and translocations of chromosome 4. *Proc Natl Acad Sci U S A* **85**(23): 9138–9142.

Pinkel, D., R. Segraves, D. Sudar, S. Clark, I. Poole, D. Kowbel, C. Collins, W. L. Kuo, C. Chen, Y. Zhai, S. H. Dairkee, B. M. Ljung, J. W. Gray and D. G. Albertson (1998). High resolution analysis of DNA copy number variation using comparative genomic hybridization to microarrays. *Nat Genet* **20**(2): 207–211.

Pollack, J. R., C. M. Perou, A. A. Alizadeh, M. B. Eisen, A. Pergamenschikov, C. F. Williams, S. S. Jeffrey, D. Botstein and P. O. Brown (1999). Genome-wide analysis of DNA copy-number changes using cDNA microarrays. *Nat Genet* **23**(1): 41–46.

Poole, A. C., J. K. Goodrich, N. D. Youngblut, G. G. Luque, A. Ruaud, J. L. Sutter, J. L. Waters, Q. Shi, M. El-Hadidi, L. M. Johnson, H. Y. Bar, D. H. Huson, J. G. Booth and R. E. Ley (2019). Human salivary amylase gene copy number impacts oral and gut microbiomes. *Cell Host Microbe* **25**(-4): 553–564.e557.

Porkka, K., O. Saramäki, M. Tanner and T. Visakorpi (2002). Amplification and overexpression of Elongin C gene discovered in prostate cancer by cDNA microarrays. *Lab Invest* **82**(5): 629–637.

Potocki, L., K. S. Chen, S. S. Park, D. E. Osterholm, M. A. Withers, V. Kimonis, A. M. Summers, W. S. Meschino, K. Anyane-Yeboa, C. D. Kashork, L. G. Shaffer and J. R. Lupski (2000). Molecular mechanism for duplication 17p11.2 – the homologous recombination reciprocal of the Smith-Magenis microdeletion. *Nat Genet* **24**(1): 84–87.

Pushkarev, D., N. F. Neff and S. R. Quake (2009). Single-molecule sequencing of an individual human genome. *Nat Biotechnol* **27**(9): 847–850.

Qin, Y., T. M. Schneider and M. P. Brenner (2012). Sequencing by hybridization of long targets. *PLoS One* **7**(5): e35819.

Quinlan, A. R., R. A. Clark, S. Sokolova, M. L. Leibowitz, Y. Zhang, M. E. Hurles, J. C. Mell and I. M. Hall (2010). Genome-wide mapping and assembly of structural variant breakpoints in the mouse genome. *Genome Res* **20**(5): 623–635.

Ramírez-Bello, J. (2019). Role of genetic variability in Mendelian and multifactorial diseases. *Gac Med Mex* **155**(5): 463–470.

Ransohoff, J. D., Y. Wei and P. A. Khavari (2018). The functions and unique features of long intergenic non-coding RNA. *Nat Rev Mol Cell Biol* **19**(3): 143–157.

Ray, D. A. and M. A. Batzer (2011). Reading TE leaves: new approaches to the identification of transposable element insertions. *Genome Res* **21**(6): 813–820.

Redon, R., S. Ishikawa, K. R. Fitch, L. Feuk, G. H. Perry, T. D. Andrews, H. Fiegler, M. H. Shapero, A. R. Carson, W. Chen, E. K. Cho, S. Dallaire, J. L. Freeman, J. R. González, M. Gratacòs, J. Huang, D. Kalaitzopoulos, D. Komura, J. R. MacDonald, C. R. Marshall, R. Mei, L. Montgomery, K. Nishimura, K. Okamura, F. Shen, M. J. Somerville, J. Tchinda, A. Valsesia, C. Woodwark, F. Yang, J. Zhang, T. Zerjal, J. Zhang, L. Armengol, D. F. Conrad, X. Estivill, C. Tyler-Smith, N. P. Carter, H. Aburatani, C. Lee, K. W. Jones, S. W. Scherer and M. E. Hurles (2006). Global variation in copy number in the human genome. *Nature* **444**(7118): 444–454.

Rhoads, A. and K. F. Au (2015). PacBio sequencing and its applications. *Genom Proteom Bioinform* **13**(5): 278–289.

Rice, A. M. and A. McLysaght (2017). Dosage sensitivity is a major determinant of human copy number variant pathogenicity. *Nat Commun* **8**: 14366.

Richard, G. F., A. Kerrest and B. Dujon (2008). Comparative genomics and molecular dynamics of DNA repeats in eukaryotes. *Microbiol Mol Biol Rev* **72**(4): 686–727.

Richards, S., N. Aziz, S. Bale, D. Bick, S. Das, J. Gastier-Foster, W. W. Grody, M. Hegde, E. Lyon, E. Spector, K. Voelkerding and H. L. Rehm (2015). Standards and guidelines for the interpretation of sequence variants: a joint consensus recommendation of the American College of Medical Genetics and Genomics and the Association for Molecular Pathology. *Nat Commun* **17**(5): 405–424.

Ried, T., G. Landes, W. Dackowski, K. Klinger and D. C. Ward (1992). Multicolor fluorescence in situ hybridization for the simultaneous detection of probe sets for chromosomes 13, 18, 21, X and Y in uncultured amniotic fluid cells. *Hum Mol Genet* **1**(5): 307–313.

Rigat, B., C. Hubert, F. Alhenc-Gelas, F. Cambien, P. Corvol and F. Soubrier (1990). An insertion/deletion polymorphism in the angiotensin I-converting enzyme gene accounting for half the variance of serum enzyme levels. *J Clin Invest* **86**(-4): 1343–1346.

Riley, B., M. Williamson, D. Collier, H. Wilkie and A. Makoff (2002). A 3-Mb map of a large segmental duplication overlapping the alpha7-nicotinic acetylcholine receptor gene (CHRNA7) at human 15q13-q14. *Genomics* **79**(2): 197–209.

Ririe, K. M., R. P. Rasmussen and C. T. Wittwer (1997). Product differentiation by analysis of DNA melting curves during the polymerase chain reaction. *Anal Biochem* **245**(2): 154–160.

Risch, N. J. (2000). Searching for genetic determinants in the new millennium. *Nature* **405**(6788): 847–856.

Rishishwar, L., C. E. Tellez Villa and I. K. Jordan (2015). Transposable element polymorphisms recapitulate human evolution. *N Engl J Med* **6**: 21.

Rivadeneira, F., U. Styrkársdottir, K. Estrada, B. V. Halldórsson, Y. H. Hsu, J. B. Richards, M. C. Zillikens, F. K. Kavvoura, N. Amin, Y. S. Aulchenko, L. A. Cupples, P. Deloukas, S. Demissie, E. Grundberg, A. Hofman, A. Kong, D. Karasik, J. B. van Meurs, B. Oostra, T. Pastinen, H. A. Pols, G. Sigurdsson, N. Soranzo, G. Thorleifsson, U. Thorsteinsdottir, F. M. Williams, S. G. Wilson, Y. Zhou, S. H. Ralston, C. M. van Duijn, T. Spector, D. P. Kiel, K. Stefansson, J. P. Ioannidis and A. G. Uitterlinden (2009). Twenty bone-mineral-density loci identified by large-scale meta-analysis of genome-wide association studies. *Nat Genet* **41**(11): 1199–1206.

Robert, F. and J. Pelletier (2018). Exploring the impact of single-nucleotide polymorphisms on translation. *Front Genet* **9**: 507.

Rodríguez-Antona, C. and M. Taron (2015). Pharmacogenomic biomarkers for personalized cancer treatment. *J Intern Med* **277**(2): 201–217.

Rodriguez-Martin, B., E. G. Alvarez, A. Baez-Ortega, J. Zamora and F. Supek (2020). Pan-cancer analysis of whole genomes identifies driver rearrangements promoted by LINE-1 retrotransposition. **52**(3): 306–319.

Ronaghi, M. (2001). Pyrosequencing sheds light on DNA sequencing. *Genome Res* **11**(1): 3–11.

Ronaghi, M., M. Uhlén and P. Nyrén (1998). A sequencing method based on real-time pyrophosphate. *Science* **281**(5375): 363, 365.

Rovelet-Lecrux, A., S. Legallic, D. Wallon, J. M. Flaman, O. Martinaud, S. Bombois, A. Rollin-Sillaire, A. Michon, I. Le Ber, J. Pariente, M. Puel, C. Paquet, B. Croisile, C. Thomas-Antérion, M. Vercelletto, R. Lévy, T. Frébourg, D. Hannequin and D. Campion (2012). A genome-wide study reveals rare CNVs exclusive to extreme phenotypes of Alzheimer disease. *Eur J Hum Genet* **20**(6): 613–617.

Royo, J. L., M. Hidalgo and A. Ruiz (2007). Pyrosequencing protocol using a universal biotinylated primer for mutation detection and SNP genotyping. *Nat Proto* **2**(7): 1734–1739.

Rubnitz, J. and S. Subramani (1984). The minimum amount of homology required for homologous recombination in mammalian cells. *Mol Cell Biol* **4**(11): 2253–2258.

Ruddle, F. H. (2004). Theophilus painter: first steps toward an understanding of the human genome. *J Exp Zool A Comp Exp Biol* **301**(5): 375–377.

Sailani, M. R., P. Makrythanasis, A. Valsesia, F. A. Santoni, S. Deutsch, K. Popadin, C. Borel, E. Migliavacca, A. J. Sharp, G. Duriaux Sail, E. Falconnet, K. Rabionet, C. Serra-Juhé, S. Vicari, D. Laux, Y. Grattau, G. Dembour, A. Megarbane, R. Touraine, S. Stora, S. Kitsiou, H. Fryssira, C. Chatzisevastou-Loukidou, E. Kanavakis, G. Merla, D. Bonnet, L. A. Pérez-Jurado, X. Estivill, J. M. Delabar and S. E. Antonarakis (2013). The complex SNP and CNV genetic architecture of the increased risk of congenital heart defects in Down syndrome. *Genome Res* **23**(9): 1410–1421.

Salem, A. H., G. E. Kilroy, W. S. Watkins, L. B. Jorde and M. A. Batzer (2003). Recently integrated Alu elements and human genomic diversity. *Mol Biol Evol* **20**(8): 1349–1361.

Salichs, E., A. Ledda, L. Mularoni, M. M. Albà and S. de la Luna (2009). Genome-wide analysis of histidine repeats reveals their role in the localization of human proteins to the nuclear speckles compartment. *PLoS Genet* **5**(3): e1000397.

Salzberg, S. L. (2018). Open questions: how many genes do we have? *BMC Biol* **16**(1): 94.

Sameer, A. S., N. Syeed, S. A. Tak, S. Bashir, S. Nissar and M. A. Siddiqi (2010). ACE I/D polymorphism in hypertensive patients of Kashmiri population. *Cardiol Res* **1**(1): 1–7.

Sanders, S. J., A. G. Ercan-Sencicek, V. Hus, R. Luo, M. T. Murtha, D. Moreno-De-Luca, S. H. Chu, M. P. Moreau, A. R. Gupta, S. A. Thomson, C. E. Mason, K. Bilguvar, P. B. Celestino-Soper, M. Choi, E. L. Crawford, L. Davis, N. R. Wright, R. M. Dhodapkar, M. DiCola, N. M. DiLullo, T. V. Fernandez, V. Fielding-Singh, D. O. Fishman, S. Frahm, R. Garagaloyan, G. S. Goh, S. Kammela, L. Klei, J. K. Lowe, S. C. Lund, A. D. McGrew, K. A. Meyer, W. J. Moffat, J. D. Murdoch, B. J. O'Roak, G. T. Ober, R. S. Pottenger, M. J. Raubeson, Y. Song, Q. Wang, B. L. Yaspan, T. W. Yu, I. R. Yurkiewicz, A. L. Beaudet, R. M. Cantor, M. Curland, D. E. Grice, M. Günel, R. P. Lifton, S. M. Mane, D. M. Martin, C. A. Shaw, M. Sheldon, J. A. Tischfield, C. A. Walsh, E. M. Morrow, D. H. Ledbetter, E. Fombonne, C. Lord, C. L. Martin, A. I. Brooks, J. S. Sutcliffe, E. H. Cook, Jr., D. Geschwind, K. Roeder, B. Devlin and M. W. State (2011). Multiple recurrent de novo CNVs, including duplications of the 7q11.23 Williams syndrome region, are strongly associated with autism. *Neuron* **70**(5): 863–885.

Sanger, F., S. Nicklen and A. R. Coulson (1977). DNA sequencing with chain-terminating inhibitors. *Proc Natl Acad Sci U S A* **74**(12): 5463–5467.

Sauna, Z. E. and C. Kimchi-Sarfaty (2011). Understanding the contribution of synonymous mutations to human disease. *Nat Rev Genet* **12**(10): 683–691.

Scharplatz, M., M. A. Puhan, J. Steurer and L. M. Bachmann (2004). What is the impact of the ACE gene insertion/deletion (I/D) polymorphism on the clinical effectiveness and adverse events of ACE inhibitors?–Protocol of a systematic review. *BMC Med Genet* **5**: 23.

Schatz, M. C., A. L. Delcher and S. L. Salzberg (2010). Assembly of large genomes using second-generation sequencing. *Genome Biol* **20**(9): 1165–1173.

Schena, M., D. Shalon, R. W. Davis and P. O. Brown (1995). Quantitative monitoring of gene expression patterns with a complementary DNA microarray. *Science* **270**(5235): 467–470.

Schildgen, V. and O. Schildgen (2013). How is a molecular polymorphism defined? *Cancer* **119**(9): 1608.

Schmidt, M. H., A. Vogel, A. K. Denton and B. Istace (2017). De novo assembly of a new Solanum pennellii accession using nanopore sequencing. *Gigascience* **29**(10): 2336–2348.

Schmittgen, T. D. and K. J. Livak (2008). Analyzing real-time PCR data by the comparative C(T) method. *Nat Protoc* **3**(6): 1101–1108.

Schneider, P. M., M. C. Carroll, C. A. Alper, C. Rittner, A. S. Whitehead, E. J. Yunis and H. R. Colten (1986). Polymorphism of the human complement C4 and steroid 21-hydroxylase genes. Restriction fragment length polymorphisms revealing structural deletions, homoduplications, and size variants. *J Clin Invest* **78**(3): 650–657.

Schouten, J. P., C. J. McElgunn, R. Waaijer, D. Zwijnenburg, F. Diepvens and G. Pals (2002). Relative quantification of 40 nucleic acid sequences by multiplex ligation-dependent probe amplification. *Nucleic Acids Res* **30**(12): e57.

Schrider, D. R. and M. W. Hahn (2010). Gene copy-number polymorphism in nature. *Proc Biol Sci* **277**(1698): 3213–3221.

Schröck, E., S. du Manoir, T. Veldman, B. Schoell, J. Wienberg, M. A. Ferguson-Smith, Y. Ning, D. H. Ledbetter, I. Bar-Am, D. Soenksen, Y. Garini and T. Ried (1996). Multicolor spectral karyotyping of human chromosomes. *Science* **273**(-5274): 494–497.

Schuster-Böckler, B., D. Conrad and A. Bateman (2010). Dosage sensitivity shapes the evolution of copy-number varied regions. *PLoS One* **5**(3): e9474.

Sebat, J., B. Lakshmi, D. Malhotra, J. Troge, C. Lese-Martin, T. Walsh, B. Yamrom, S. Yoon, A. Krasnitz, J. Kendall, A. Leotta, D. Pai, R. Zhang, Y. H. Lee, J. Hicks, S. J. Spence, A. T. Lee, K. Puura, T. Lehtimäki, D. Ledbetter, P. K. Gregersen, J. Bregman, J. S. Sutcliffe, V. Jobanputra, W. Chung, D. Warburton, M. C. King, D. Skuse, D. H.

Geschwind, T. C. Gilliam, K. Ye and M. Wigler (2007). Strong association of de novo copy number mutations with autism. *Science* **316**(5823): 445–449.

Sebat, J., B. Lakshmi, J. Troge, J. Alexander, J. Young, P. Lundin, S. Månér, H. Massa, M. Walker, M. Chi, N. Navin, R. Lucito, J. Healy, J. Hicks, K. Ye, A. Reiner, T. C. Gilliam, B. Trask, N. Patterson, A. Zetterberg and M. Wigler (2004). Large-scale copy number polymorphism in the human genome. *Science* **305**(5683): 525–528.

Sekar, A., A. R. Bialas, H. de Rivera, A. Davis, T. R. Hammond, N. Kamitaki, K. Tooley, J. Presumey, M. Baum, V. Van Doren, G. Genovese, S. A. Rose, R. E. Handsaker, M. J. Daly, M. C. Carroll, B. Stevens and S. A. McCarroll (2016). Schizophrenia risk from complex variation of complement component 4. *Nature* **530**(7589): 177–183.

Sellner, L. N. and G. R. Taylor (2004). MLPA and MAPH: new techniques for detection of gene deletions. *Hum Mutat* **23**(-5): 413–419.

Shaffer, L. G. and J. R. Lupski (2000). Molecular mechanisms for constitutional chromosomal rearrangements in humans. *Genet Med* **34**: 297–329.

Shastry, B. S. (2009). SNPs: impact on gene function and phenotype. *Methods Mol Biol* **578**: 3–22.

Shaw, C. J. and J. R. Lupski (2004). Implications of human genome architecture for rearrangement-based disorders: the genomic basis of disease. *Hum Mol Genet* **13 Spec No 1**: R57–64.

She, X., Z. Cheng, S. Zöllner, D. M. Church and E. E. Eichler (2008). Mouse segmental duplication and copy number variation. *Nat Genet* **40**(7): 909–914.

She, X., Z. Jiang, R. A. Clark, G. Liu, Z. Cheng, E. Tuzun, D. M. Church, G. Sutton, A. L. Halpern and E. E. Eichler (2004). Shotgun sequence assembly and recent segmental duplications within the human genome. *Nature* **431**(7011): 927–930.

Sheen, C. R., U. R. Jewell, C. M. Morris, S. O. Brennan, C. Férec, P. M. George, M. P. Smith and J. M. Chen (2007). Double complex mutations involving F8 and FUNDC2 caused by distinct break-induced replication. *Hum Mutat* **28**(12): 1198–1206.

Sheffield, V. C., J. S. Beck, A. E. Kwitek, D. W. Sandstrom and E. M. Stone (1993). The sensitivity of single-strand conformation polymorphism analysis for the detection of single base substitutions. *Genomics* **16**(2): 325–332.

Sheffield, V. C., D. R. Cox, L. S. Lerman and R. M. Myers (1989). Attachment of a 40-base-pair G+C-rich sequence (GC-clamp) to genomic DNA fragments by the polymerase chain reaction results in improved detection of single-base changes. *Proc Natl Acad Sci U S A* **86**(1): 232–236.

Sindi, S., E. Helman, A. Bashir and B. J. Raphael (2009). A geometric approach for classification and comparison of structural variants. *Bioinformatics* **25**(12): i222–230.

Smith, C. E., B. Llorente and L. S. Symington (2007). Template switching during break-induced replication. *Nature* **447**(-7140): 102–105.

Sokolov, B. P. (1990). Primer extension technique for the detection of single nucleotide in genomic DNA. *Nucleic Acids Res* **18**(12): 3671.

Southern, E. M. (1975). Detection of specific sequences among DNA fragments separated by gel electrophoresis. *J Mol Biol* **98**(3): 503–517.

Speicher, M. R., S. Gwyn Ballard and D. C. Ward (1996). Karyotyping human chromosomes by combinatorial multi-fluor FISH. *Nat Genet* **12**(4): 368–375.

Speliotes, E. K., C. J. Willer, S. I. Berndt, K. L. Monda, G. Thorleifsson, A. U. Jackson, H. Lango Allen, C. M. Lindgren, J. Luan, R. Mägi, J. C. Randall, S. Vedantam, T. W. Winkler, L. Qi, T. Workalemahu, I. M. Heid, V. Steinthorsdottir, H. M. Stringham, M. N. Weedon, E. Wheeler, A. R. Wood, T. Ferreira, R. J. Weyant, A. V. Segrè, K. Estrada, L. Liang, J. Nemesh, J. H. Park, S. Gustafsson, T. O. Kilpeläinen, J. Yang, N. Bouatia-Naji, T. Esko, M. F. Feitosa, Z. Kutalik, M. Mangino, S. Raychaudhuri, A. Scherag, A. V. Smith, R. Welch, J. H. Zhao, K. K. Aben, D. M. Absher, N. Amin, A. L. Dixon, E. Fisher, N. L. Glazer, M. E. Goddard, N. L. Heard-Costa, V. Hoesel, J. J. Hottenga, A. Johansson, T. Johnson, S. Ketkar, C. Lamina, S. Li, M. F. Moffatt, R. H. Myers, N. Narisu, J. R. Perry, M. J. Peters, M. Preuss, S. Ripatti, F. Rivadeneira, C. Sandholt, L. J. Scott, N. J. Timpson, J. P. Tyrer, S. van Wingerden, R. M. Watanabe, C. C. White, F. Wiklund, C. Barlassina, D. I. Chasman, M. N. Cooper, J. O. Jansson, R. W. Lawrence, N. Pellikka, I. Prokopenko, J. Shi, E. Thiering, H. Alavere, M. T. Alibrandi, P. Almgren, A. M. Arnold, T. Aspelund, L. D. Atwood, B. Balkau, A. J. Balmforth, A. J. Bennett, Y. Ben-Shlomo, R. N. Bergman, S. Bergmann, H. Biebermann, A. I. Blakemore, T. Boes, L. L. Bonnycastle, S. R. Bornstein, M. J. Brown, T. A. Buchanan, F. Busonero, H. Campbell, F. P. Cappuccio, C. Cavalcanti-Proença, Y. D. Chen, C. M. Chen, P. S. Chines, R. Clarke, L. Coin, J. Connell, I. N. Day, M. den Heijer, J. Duan, S. Ebrahim, P. Elliott, R. Elosua, G. Eiriksdottir, M. R. Erdos, J. G. Eriksson, M. F. Facheris, S. B. Felix, P. Fischer-Posovszky, A. R. Folsom, N. Friedrich, N. B. Freimer, M. Fu, S. Gaget, P. V. Gejman, E. J. Geus, C. Gieger, A. P. Gjesing, A. Goel, P. Goyette, H. Grallert, J. Grässler, D. M. Greenawalt, C. J. Groves, V. Gudnason, C. Guiducci, A. L. Hartikainen, N. Hassanali, A. S. Hall, A. S. Havulinna, C. Hayward, A. C. Heath, C. Hengstenberg, A. A. Hicks, A. Hinney, A. Hofman, G. Homuth, J. Hui, W. Igl, C. Iribarren, B. Isomaa, K. B. Jacobs, I. Jarick, E. Jewell, U. John, T. Jørgensen, P. Jousilahti, A. Jula, M. Kaakinen, E. Kajantie, L. M. Kaplan, S. Kathiresan, J. Kettunen, L. Kinnunen, J. W. Knowles, I. Kolcic, I. R. König, S. Koskinen, P. Kovacs, J. Kuusisto, P. Kraft, K. Kvaløy, J. Laitinen, O. Lantieri, C. Lanzani, L. J. Launer, C. Lecoeur, T. Lehtimäki, G. Lettre, J. Liu, M. L. Lokki, M. Lorentzon, R. N. Luben, B. Ludwig, P. Manunta, D. Marek, M. Marre, N. G. Martin, W. L. McArdle, A. McCarthy, B. McKnight, T. Meitinger, O. Melander, D. Meyre, K. Midthjell, G. W. Montgomery, M. A. Morken, A. P. Morris, R. Mulic, J. S. Ngwa, M. Nelis, M. J. Neville, D. R. Nyholt, C. J. O'Donnell, S. O'Rahilly, K. K. Ong, B. Oostra, G. Paré, A. N. Parker, M. Perola, I. Pichler, K. H. Pietiläinen, C. G. Platou, O. Polasek, A. Pouta, S. Rafelt, O. Raitakari, N. W. Rayner, M. Ridderstråle, W. Rief, A. Ruokonen, N. R. Robertson, P. Rzehak, V. Salomaa, A. R. Sanders, M. S. Sandhu, S. Sanna, J. Saramies, M. J. Savolainen, S. Scherag, S. Schipf, S. Schreiber, H. Schunkert, K. Silander, J. Sinisalo, D. S. Siscovick, J. H. Smit, N. Soranzo, U. Sovio, J. Stephens, I. Surakka, A. J. Swift, M. L. Tammesoo, J. C. Tardif, M. Teder-Laving, T. M. Teslovich, J. R. Thompson, B. Thomson, A. Tönjes, T. Tuomi, J. B. van Meurs, G. J. van Ommen, V. Vatin, J. Viikari, S. Visvikis-Siest, V. Vitart, C. I. Vogel, B. F. Voight, L. L. Waite, H. Wallaschofski, G. B. Walters, E. Widen, S. Wiegand, S. H. Wild, G. Willemsen, D. R. Witte, J. C. Witteman, J. Xu, Q. Zhang, L. Zgaga, A.

Ziegler, P. Zitting, J. P. Beilby, I. S. Farooqi, J. Hebebrand, H. V. Huikuri, A. L. James, M. Kähönen, D. F. Levinson, F. Macciardi, M. S. Nieminen, C. Ohlsson, L. J. Palmer, P. M. Ridker, M. Stumvoll, J. S. Beckmann, H. Boeing, E. Boerwinkle, D. I. Boomsma, M. J. Caulfield, S. J. Chanock, F. S. Collins, L. A. Cupples, G. D. Smith, J. Erdmann, P. Froguel, H. Grönberg, U. Gyllensten, P. Hall, T. Hansen, T. B. Harris, A. T. Hattersley, R. B. Hayes, J. Heinrich, F. B. Hu, K. Hveem, T. Illig, M. R. Jarvelin, J. Kaprio, F. Karpe, K. T. Khaw, L. A. Kiemeney, H. Krude, M. Laakso, D. A. Lawlor, A. Metspalu, P. B. Munroe, W. H. Ouwehand, O. Pedersen, B. W. Penninx, A. Peters, P. P. Pramstaller, T. Quertermous, T. Reinehr, A. Rissanen, I. Rudan, N. J. Samani, P. E. Schwarz, A. R. Shuldiner, T. D. Spector, J. Tuomilehto, M. Uda, A. Uitterlinden, T. T. Valle, M. Wabitsch, G. Waeber, N. J. Wareham, H. Watkins, J. F. Wilson, A. F. Wright, M. C. Zillikens, N. Chatterjee, S. A. McCarroll, S. Purcell, E. E. Schadt, P. M. Visscher, T. L. Assimes, I. B. Borecki, P. Deloukas, C. S. Fox, L. C. Groop, T. Haritunians, D. J. Hunter, R. C. Kaplan, K. L. Mohlke, J. R. O'Connell, L. Peltonen, D. Schlessinger, D. P. Strachan, C. M. van Duijn, H. E. Wichmann, T. M. Frayling, U. Thorsteinsdottir, G. R. Abecasis, I. Barroso, M. Boehnke, K. Stefansson, K. E. North, M. I. McCarthy, J. N. Hirschhorn, E. Ingelsson and R. J. Loos (2010). Association analyses of 249,796 individuals reveal 18 new loci associated with body mass index. *Nat Genet* **42**(11): 937–948.

Springer, N. M., K. Ying, Y. Fu, T. Ji, C. T. Yeh, Y. Jia, W. Wu, T. Richmond, J. Kitzman, H. Rosenbaum, A. L. Iniguez, W. B. Barbazuk, J. A. Jeddeloh, D. Nettleton and P. S. Schnable (2009). Maize inbreds exhibit high levels of copy number variation (CNV) and presence/absence variation (PAV) in genome content. *PLoS Genet* **5**(11): e1000734.

Squire, J. A., J. Pei, P. Marrano, B. Beheshti, J. Bayani, G. Lim, L. Moldovan and M. Zielenska (2003). High-resolution mapping of amplifications and deletions in pediatric osteosarcoma by use of CGH analysis of cDNA microarrays. *Genes Chromosomes Cancer* **38**(3): 215–225.

Srinivasan, S., J. A. Clements and J. Batra (2016). Single nucleotide polymorphisms in clinics: fantasy or reality for cancer? *Crit Rev Clin Lab Sci* **53**(1): 29–39.

Stankiewicz, P. and J. R. Lupski (2002). Genome architecture, rearrangements and genomic disorders. *Trends Genet* **18**(-2): 74–82.

Stankiewicz, P. and J. R. Lupski (2010). Structural variation in the human genome and its role in disease. *Annu Rev Med* **61**: 437–455.

Stefansson, H., D. Rujescu, S. Cichon, O. P. Pietiläinen, A. Ingason, S. Steinberg, R. Fossdal, E. Sigurdsson, T. Sigmundsson, J. E. Buizer-Voskamp, T. Hansen, K. D. Jakobsen, P. Muglia, C. Francks, P. M. Matthews, A. Gylfason, B. V. Halldorsson, D. Gudbjartsson, T. E. Thorgeirsson, A. Sigurdsson, A. Jonasdottir, A. Jonasdottir, A. Bjornsson, S. Mattiasdottir, T. Blondal, M. Haraldsson, B. B. Magnusdottir, I. Giegling, H. J. Möller, A. Hartmann, K. V. Shianna, D. Ge, A. C. Need, C. Crombie, G. Fraser, N. Walker, J. Lonnqvist, J. Suvisaari, A. Tuulio-Henriksson, T. Paunio, T. Toulopoulou, E. Bramon, M. Di Forti, R. Murray, M. Ruggeri, E. Vassos, S. Tosato, M. Walshe, T. Li, C. Vasilescu, T. W. Mühleisen, A. G. Wang, H. Ullum, S. Djurovic, I. Melle, J. Olesen, L. A. Kiemeney, B. Franke, C. Sabatti, N. B. Freimer, J. R. Gulcher, U. Thorsteinsdottir, A. Kong, O. A. Andreassen, R. A. Ophoff, A. Georgi, M. Rietschel, T. Werge, H. Petursson, D. B. Goldstein, M. M. Nöthen, L. Peltonen, D. A. Collier, D. St Clair and K. Stefansson (2008). Large recurrent microdeletions associated with schizophrenia. *Nature* **455**(7210): 232–236.

Steinmann, K., D. N. Cooper, L. Kluwe, N. A. Chuzhanova, C. Senger, E. Serra, C. Lazaro, M. Gilaberte, K. Wimmer, V. F. Mautner and H. Kehrer-Sawatzki (2007). Type 2 NF1 deletions are highly unusual by virtue of the absence of nonallelic homologous recombination hotspots and an apparent preference for female mitotic recombination. *Am J Hum Genet* **81**(6): 1201–1220.

Stranger, B. E., M. S. Forrest, M. Dunning, C. E. Ingle, C. Beazley, N. Thorne, R. Redon, C. P. Bird, A. de Grassi, C. Lee, C. Tyler-Smith, N. Carter, S. W. Scherer, S. Tavaré, P. Deloukas, M. E. Hurles and E. T. Dermitzakis (2007). Relative impact of nucleotide and copy number variation on gene expression phenotypes. *Science* **315**(5813): 848–853.

Subramanian, S., R. K. Mishra and L. Singh (2003). Genome-wide analysis of microsatellite repeats in humans: their abundance and density in specific genomic regions. *Genome Biol* **4**(2): R13.

Sudmant, P. H., J. O. Kitzman, F. Antonacci, C. Alkan, M. Malig, A. Tsalenko, N. Sampas, L. Bruhn, J. Shendure and E. E. Eichler (2010). Diversity of human copy number variation and multicopy genes. *Science* **330**(6004): 641–646.

Sudmant, P. H., T. Rausch, E. J. Gardner, R. E. Handsaker, A. Abyzov, J. Huddleston, Y. Zhang, K. Ye, G. Jun, M. H. Fritz, M. K. Konkel, A. Malhotra, A. M. Stütz, X. Shi, F. P. Casale, J. Chen, F. Hormozdiari, G. Dayama, K. Chen, M. Malig, M. J. P. Chaisson, K. Walter, S. Meiers, S. Kashin, E. Garrison, A. Auton, H. Y. K. Lam, X. J. Mu, C. Alkan, D. Antaki, T. Bae, E. Cerveira, P. Chines, Z. Chong, L. Clarke, E. Dal, L. Ding, S. Emery, X. Fan, M. Gujral, F. Kahveci, J. M. Kidd, Y. Kong, E. W. Lameijer, S. McCarthy, P. Flicek, R. A. Gibbs, G. Marth, C. E. Mason, A. Menelaou, D. M. Muzny, B. J. Nelson, A. Noor, N. F. Parrish, M. Pendleton, A. Quitadamo, B. Raeder, E. E. Schadt, M. Romanovitch, A. Schlattl, R. Sebra, A. A. Shabalin, A. Untergasser, J. A. Walker, M. Wang, F. Yu, C. Zhang, J. Zhang, X. Zheng-Bradley, W. Zhou, T. Zichner, J. Sebat, M. A. Batzer, S. A. McCarroll, R. E. Mills, M. B. Gerstein, A. Bashir, O. Stegle, S. E. Devine, C. Lee, E. E. Eichler and J. O. Korbel (2015). An integrated map of structural variation in 2,504 human genomes. *Nature* **526**(7571): 75–81.

Syvänen, A. C. (1999). From gels to chips: "minisequencing" primer extension for analysis of point mutations and single nucleotide polymorphisms. *Hum Mutat* **13**(1): 1–10.

Szatmari, P., A. D. Paterson, L. Zwaigenbaum, W. Roberts, J. Brian, X. Q. Liu, J. B. Vincent, J. L. Skaug, A. P. Thompson, L. Senman, L. Feuk, C. Qian, S. E. Bryson, M. B. Jones, C. R. Marshall, S. W. Scherer, V. J. Vieland, C. Bartlett, L. V. Mangin, R. Goedken, A. Segre, M. A. Pericak-Vance, M. L. Cuccaro, J. R. Gilbert, H. H. Wright, R. K. Abramson, C. Betancur, T. Bourgeron, C. Gillberg, M. Leboyer, J. D. Buxbaum, K. L. Davis, E. Hollander, J. M. Silverman, J. Hallmayer, L. Lotspeich, J. S. Sutcliffe, J. L. Haines, S. E. Folstein, J. Piven, T. H. Wassink, V. Sheffield, D. H. Geschwind, M. Bucan, W. T. Brown, R. M. Cantor, J. N. Constantino, T. C. Gilliam, M. Herbert, C. Lajonchere,

D. H. Ledbetter, C. Lese-Martin, J. Miller, S. Nelson, C. A. Samango-Sprouse, S. Spence, M. State, R. E. Tanzi, H. Coon, G. Dawson, B. Devlin, A. Estes, P. Flodman, L. Klei, W. M. McMahon, N. Minshew, J. Munson, E. Korvatska, P. M. Rodier, G. D. Schellenberg, M. Smith, M. A. Spence, C. Stodgell, P. G. Tepper, E. M. Wijsman, C. E. Yu, B. Rogé, C. Mantoulan, K. Wittemeyer, A. Poustka, B. Felder, S. M. Klauck, C. Schuster, F. Poustka, S. Bölte, S. Feineis-Matthews, E. Herbrecht, G. Schmötzer, J. Tsiantis, K. Papanikolaou, E. Maestrini, E. Bacchelli, F. Blasi, S. Carone, C. Toma, H. Van Engeland, M. de Jonge, C. Kemner, F. Koop, M. Langemeijer, C. Hijmans, W. G. Staal, G. Baird, P. F. Bolton, M. L. Rutter, E. Weisblatt, J. Green, C. Aldred, J. A. Wilkinson, A. Pickles, A. Le Couteur, T. Berney, H. McConachie, A. J. Bailey, K. Francis, G. Honeyman, A. Hutchinson, J. R. Parr, S. Wallace, A. P. Monaco, G. Barnby, K. Kobayashi, J. A. Lamb, I. Sousa, N. Sykes, E. H. Cook, S. J. Guter, B. L. Leventhal, J. Salt, C. Lord, C. Corsello, V. Hus, D. E. Weeks, F. Volkmar, M. Tauber, E. Fombonne, A. Shih and K. J. Meyer (2007). Mapping autism risk loci using genetic linkage and chromosomal rearrangements. *Nat Genet* **39**(3): 319–328.

Taylor, R. W. and D. M. Turnbull (2005). Mitochondrial DNA mutations in human disease. *Nat Rev Genet* **6**(5): 389–402.

Tham, C. Y., R. Tirado-Magallanes, Y. Goh, M. J. Fullwood, B. T. H. Koh, W. Wang, C. H. Ng, W. J. Chng, A. Thiery, D. G. Tenen and T. Benoukraf (2020). NanoVar: accurate characterization of patients' genomic structural variants using low-depth nanopore sequencing. *Genome Biol* **21** (1): 56.

Todd, A. V., C. J. Fuery, H. L. Impey, T. L. Applegate and M. A. Haughton (2000). DzyNA-PCR: use of DNAzymes to detect and quantify nucleic acid sequences in a real-time fluorescent format. *Clin Chem* **46**(5): 625–630.

Trask, B. J., C. Friedman, A. Martin-Gallardo, L. Rowen, C. Akinbami, J. Blankenship, C. Collins, D. Giorgi, S. Iadonato, F. Johnson, W. L. Kuo, H. Massa, T. Morrish, S. Naylor, O. T. Nguyen, S. Rouquier, T. Smith, D. J. Wong, J. Youngblom and G. van den Engh (1998). Members of the olfactory receptor gene family are contained in large blocks of DNA duplicated polymorphically near the ends of human chromosomes. *Hum Mol Genet* **7**(1): 13–26.

Treangen, T. J. and S. L. Salzberg (2011). Repetitive DNA and next-generation sequencing: computational challenges and solutions. *Nat Rev Genet* **13**(1): 36–46.

Tse, K. P., W. H. Su, M. L. Yang, H. Y. Cheng, N. M. Tsang, K. P. Chang, S. P. Hao, Y. Yao Shugart and Y. S. Chang (2011). A gender-specific association of CNV at 6p21.3 with NPC susceptibility. *Hum Mol Genet* **20**(14): 2889–2896.

Tümer, Z., L. Birk Møller and N. Horn (2003). Screening of 383 unrelated patients affected with Menkes disease and finding of 57 gross deletions in ATP7A. *Hum Mutat* **22**(6): 457–464.

Turner, D. J., M. Miretti, D. Rajan, H. Fiegler, N. P. Carter, M. L. Blayney, S. Beck and M. E. Hurles (2008). Germline rates of de novo meiotic deletions and duplications causing several genomic disorders. *Nat Genet* **40**(1): 90–95.

Tuzun, E., A. J. Sharp, J. A. Bailey, R. Kaul, V. A. Morrison, L. M. Pertz, E. Haugen, H. Hayden, D. Albertson, D. Pinkel, M. V. Olson and E. E. Eichler (2005). Fine-scale structural variation of the human genome. *Nat Genet* **37**(7): 727–732.

Usdin, K. (2008). The biological effects of simple tandem repeats: lessons from the repeat expansion diseases. *Genome Res* **18**(7): 1011–1019.

Vallone, P. M., R. S. Just, M. D. Coble, J. M. Butler and T. J. Parsons (2004). A multiplex allele-specific primer extension assay for forensically informative SNPs distributed throughout the mitochondrial genome. *Int J Legal Med* **118**(3): 147–157.

van Gent, D. C. and M. van der Burg (2007). Non-homologous end-joining, a sticky affair. *Oncogene* **26**(56): 7731–7740.

van Loon, N., D. Miller and J. P. Murnane (1994). Formation of extrachromosomal circular DNA in HeLa cells by nonhomologous recombination. *Nucleic Acids Res* **22**(13): 2447–2452.

Venter, J. C., M. D. Adams, E. W. Myers, P. W. Li, R. J. Mural, G. G. Sutton, H. O. Smith, M. Yandell, C. A. Evans, R. A. Holt, J. D. Gocayne, P. Amanatides, R. M. Ballew, D. H. Huson, J. R. Wortman, Q. Zhang, C. D. Kodira, X. H. Zheng, L. Chen, M. Skupski, G. Subramanian, P. D. Thomas, J. Zhang, G. L. Gabor Miklos, C. Nelson, S. Broder, A. G. Clark, J. Nadeau, V. A. McKusick, N. Zinder, A. J. Levine, R. J. Roberts, M. Simon, C. Slayman, M. Hunkapiller, R. Bolanos, A. Delcher, I. Dew, D. Fasulo, M. Flanigan, L. Florea, A. Halpern, S. Hannenhalli, S. Kravitz, S. Levy, C. Mobarry, K. Reinert, K. Remington, J. Abu-Threideh, E. Beasley, K. Biddick, V. Bonazzi, R. Brandon, M. Cargill, I. Chandramouliswaran, R. Charlab, K. Chaturvedi, Z. Deng, V. Di Francesco, P. Dunn, K. Eilbeck, C. Evangelista, A. E. Gabrielian, W. Gan, W. Ge, F. Gong, Z. Gu, P. Guan, T. J. Heiman, M. E. Higgins, R. R. Ji, Z. Ke, K. A. Ketchum, Z. Lai, Y. Lei, Z. Li, J. Li, Y. Liang, X. Lin, F. Lu, G. V. Merkulov, N. Milshina, H. M. Moore, A. K. Naik, V. A. Narayan, B. Neelam, D. Nusskern, D. B. Rusch, S. Salzberg, W. Shao, B. Shue, J. Sun, Z. Wang, A. Wang, X. Wang, J. Wang, M. Wei, R. Wides, C. Xiao, C. Yan, A. Yao, J. Ye, M. Zhan, W. Zhang, H. Zhang, Q. Zhao, L. Zheng, F. Zhong, W. Zhong, S. Zhu, S. Zhao, D. Gilbert, S. Baumhueter, G. Spier, C. Carter, A. Cravchik, T. Woodage, F. Ali, H. An, A. Awe, D. Baldwin, H. Baden, M. Barnstead, I. Barrow, K. Beeson, D. Busam, A. Carver, A. Center, M. L. Cheng, L. Curry, S. Danaher, L. Davenport, R. Desilets, S. Dietz, K. Dodson, L. Doup, S. Ferriera, N. Garg, A. Gluecksmann, B. Hart, J. Haynes, C. Haynes, C. Heiner, S. Hladun, D. Hostin, J. Houck, T. Howland, C. Ibegwam, J. Johnson, F. Kalush, L. Kline, S. Koduru, A. Love, F. Mann, D. May, S. McCawley, T. McIntosh, I. McMullen, M. Moy, L. Moy, B. Murphy, K. Nelson, C. Pfannkoch, E. Pratts, V. Puri, H. Qureshi, M. Reardon, R. Rodriguez, Y. H. Rogers, D. Romblad, B. Ruhfel, R. Scott, C. Sitter, M. Smallwood, E. Stewart, R. Strong, E. Suh, R. Thomas, N. N. Tint, S. Tse, C. Vech, G. Wang, J. Wetter, S. Williams, M. Williams, S. Windsor, E. Winn-Deen, K. Wolfe, J. Zaveri, K. Zaveri, J. F. Abril, R. Guigó, M. J. Campbell, K. V. Sjolander, B. Karlak, A. Kejariwal, H. Mi, B. Lazareva, T. Hatton, A. Narechania, K. Diemer, A. Muruganujan, N. Guo, S. Sato, V. Bafna, S. Istrail, R. Lippert, R. Schwartz, B. Walenz, S. Yooseph, D. Allen, A. Basu, J. Baxendale, L. Blick, M. Caminha, J. Carnes-Stine, P. Caulk, Y. H. Chiang, M. Coyne, C. Dahlke, A. Mays, M. Dombroski, M. Donnelly, D. Ely, S. Esparham, C. Fosler, H. Gire, S. Glanowski, K. Glasser, A. Glodek, M. Gorokhov, K. Graham, B. Gropman, M. Harris,

J. Heil, S. Henderson, J. Hoover, D. Jennings, C. Jordan, J. Jordan, J. Kasha, L. Kagan, C. Kraft, A. Levitsky, M. Lewis, X. Liu, J. Lopez, D. Ma, W. Majoros, J. McDaniel, S. Murphy, M. Newman, T. Nguyen, N. Nguyen, M. Nodell, S. Pan, J. Peck, M. Peterson, W. Rowe, R. Sanders, J. Scott, M. Simpson, T. Smith, A. Sprague, T. Stockwell, R. Turner, E. Venter, M. Wang, M. Wen, D. Wu, M. Wu, A. Xia, A. Zandieh and X. Zhu (2001). The sequence of the human genome. *Science* **291**(5507): 1304–1351.

Verstrepen, K. J., A. Jansen, F. Lewitter and G. R. Fink (2005). Intragenic tandem repeats generate functional variability. *Nat Genet* **37**(9): 986–990.

Viguera, E., D. Canceill and S. D. Ehrlich (2001). Replication slippage involves DNA polymerase pausing and dissociation. *Embo J* **20**(10): 2587–2595.

Vissers, L. E., S. S. Bhatt, I. M. Janssen, Z. Xia, S. R. Lalani, R. Pfundt, K. Derwinska, B. B. de Vries, C. Gilissen, A. Hoischen, M. Nesteruk, B. Wisniowiecka-Kowalnik, M. Smyk, H. G. Brunner, S. W. Cheung, A. G. van Kessel, J. A. Veltman and P. Stankiewicz (2009). Rare pathogenic microdeletions and tandem duplications are microhomology-mediated and stimulated by local genomic architecture. *Hum Mol Genet* **18**(19): 3579–3593.

Vissers, L. E., P. Stankiewicz, S. A. Yatsenko, E. Crawford, H. Creswick, V. K. Proud, B. B. de Vries, R. Pfundt, C. L. Marcelis, J. Zackowski, W. Bi, A. G. van Kessel, J. R. Lupski and J. A. Veltman (2007). Complex chromosome 17p rearrangements associated with low-copy repeats in two patients with congenital anomalies. *Hum Genet* **121**(6): 697–709.

Vissers, L. E., J. A. Veltman, A. G. van Kessel and H. G. Brunner (2005). Identification of disease genes by whole genome CGH arrays. *Adv Rheumatol* **14 Spec No. 2**: R215–223.

Vooijs, M., L. C. Yu, D. Tkachuk, D. Pinkel, D. Johnson and J. W. Gray (1993). Libraries for each human chromosome, constructed from sorter-enriched chromosomes by using linker-adaptor PCR. *Am J Hum Genet* **52**(3): 586–597.

Waldman, A. S. and R. M. Liskay (1988). Dependence of intrachromosomal recombination in mammalian cells on uninterrupted homology. *Mol Cell Biol* **8**(12): 5350–5357.

Wang, B., L. P. Yang, X. Z. Zhang, S. Q. Huang, M. Bartlam and S. F. Zhou (2009). New insights into the structural characteristics and functional relevance of the human cytochrome P450 2D6 enzyme. *Drug Metab Rev* **41**(4): 573–643.

Wang, H., D. Nettleton and K. Ying (2014). Copy number variation detection using next generation sequencing read counts. *BMC Bioinformatics* **15**: 109.

Wang, H., J. Xing, D. Grover, D. J. Hedges, K. Han, J. A. Walker and M. A. Batzer (2005). SVA elements: a hominid-specific retroposon family. *J Mol Biol* **354**(4): 994–1007.

Waterfall, C. M. and B. D. Cobb (2002). SNP genotyping using single-tube fluorescent bidirectional PCR. *Biotechniques* **33**(1): 80, 82–84, 86 passim.

Weckselblatt, B. and M. K. Rudd (2015). Human structural variation: mechanisms of chromosome rearrangements. *Trends Genet* **31**(10): 587–599.

Weir, B., X. Zhao and M. Meyerson (2004). Somatic alterations in the human cancer genome. *Cancer Cell* **6**(5): 433–438.

Weir, B. S., L. R. Cardon, A. D. Anderson, D. M. Nielsen and W. G. Hill (2005). Measures of human population structure show heterogeneity among genomic regions. *Genome Res* **15**(11): 1468–1476.

Weischenfeldt, J., O. Symmons, F. Spitz and J. O. Korbel (2013). Phenotypic impact of genomic structural variation: insights from and for human disease. *Nat Rev Genet* **14**(2): 125–138.

Weterings, E. and D. C. van Gent (2004). The mechanism of non-homologous end-joining: a synopsis of synapsis. *DNA Repair (Amst)* **3**(11): 1425–1435.

Wheeler, E., N. Huang, E. G. Bochukova, J. M. Keogh, S. Lindsay, S. Garg, E. Henning, H. Blackburn, R. J. Loos, N. J. Wareham, S. O'Rahilly, M. E. Hurles, I. Barroso and I. S. Farooqi (2013). Genome-wide SNP and CNV analysis identifies common and low-frequency variants associated with severe early-onset obesity. *Nat Genet* **45**(5): 513–517.

Willems, T., M. Gymrek, G. Highnam, D. Mittelman and Y. Erlich (2014). The landscape of human STR variation. *Genome Res***2 4**(11): 1894–1904.

Willyard, C. (2018). New human gene tally reignites debate. *Nature* **558**(7710): 354–355.

Wirth, B., M. Karakaya, M. J. Kye and N. Mendoza-Ferreira (2020). Twenty-five years of spinal muscular atrophy research: from phenotype to genotype to therapy, and what comes next. *Annu Rev Genomics Hum Genet* **21**: 231–261.

Wong, M. L. and J. F. Medrano (2005). Real-time PCR for mRNA quantitation. *Biotechniques* **39**(1): 75–85.

Wu, Y. L., S. L. Savelli, Y. Yang, B. Zhou, B. H. Rovin, D. J. Birmingham, H. N. Nagaraja, L. A. Hebert and C. Y. Yu (2007). Sensitive and specific real-time polymerase chain reaction assays to accurately determine copy number variations (CNVs) of human complement C4A, C4B, C4-long, C4-short, and RCCX modules: elucidation of C4 CNVs in 50 consanguineous subjects with defined HLA genotypes. *J Immunol* **179**(5): 3012–3025.

Xu, F., L. Li, V. P. Schulz, P. G. Gallagher, B. Xiang, H. Zhao and P. Li (2014). Cytogenomic mapping and bioinformatic mining reveal interacting brain expressed genes for intellectual disability. *Mol Cytogenet* **7**(1): 4.

Xue, Y., D. Sun, A. Daly, F. Yang, X. Zhou, M. Zhao, N. Huang, T. Zerjal, C. Lee, N. P. Carter, M. E. Hurles and C. Tyler-Smith (2008). Adaptive evolution of UGT2B17 copy-number variation. *Am J Hum Genet* **83**(3): 337–346.

Yang, T. L., X. D. Chen, Y. Guo, S. F. Lei, J. T. Wang, Q. Zhou, F. Pan, Y. Chen, Z. X. Zhang, S. S. Dong, X. H. Xu, H. Yan, X. Liu, C. Qiu, X. Z. Zhu, T. Chen, M. Li, H. Zhang, L. Zhang, B. M. Drees, J. J. Hamilton, C. J. Papasian, R. R. Recker, X. P. Song, J. Cheng and H. W. Deng (2008). Genome-wide copy-number-variation study identified a susceptibility gene, UGT2B17, for osteoporosis. *Am J Hum Genet* **83**(6): 663–674.

Ye, K., M. H. Schulz, Q. Long, R. Apweiler and Z. Ning (2009). Pindel: a pattern growth approach to detect break points of large deletions and medium sized insertions from paired-end short reads. *Bioinformatics* **25**(21): 2865–2871.

Ye, K., J. Wang, R. Jayasinghe, E. W. Lameijer, J. F. McMichael, J. Ning, M. D. McLellan, M. Xie, S. Cao, V. Yellapantula, K. L. Huang, A. Scott, S. Foltz, B. Niu, K. J. Johnson, M. Moed, P. E. Slagboom, F. Chen, M. C. Wendl and L. Ding (2016). Systematic discovery of complex insertions and deletions in human cancers. *Nat Med* **22**(1): 97–104.

Ylstra, B., P. van den Ijssel, B. Carvalho, R. H. Brakenhoff and G. A. Meijer (2006). BAC to the future! or oligonucleotides: a perspective for micro array comparative genomic hybridization (array CGH). *Nucleic Acids Res* **34**(2): 445–450.

Yoshihara, K., A. Tajima, S. Adachi, J. Quan, M. Sekine, H. Kase, T. Yahata, I. Inoue and K. Tanaka (2011). Germline copy number variations in BRCA1-associated ovarian cancer patients. *Genes Chromosomes Cancer* **50**(3): 167–177.

Yoshino, K., K. Nishigaki and Y. Husimi (1991). Temperature sweep gel electrophoresis: a simple method to detect point mutations. *Nucleic Acids Res* **19**(11): 3153.

Zackrisson, A. L. and B. Lindblom (2003). Identification of CYP2D6 alleles by single nucleotide polymorphism analysis using pyrosequencing. *Eur J Clin Pharmacol* **59**(7): 521–526.

Zhang, F., M. Khajavi, A. M. Connolly, C. F. Towne, S. D. Batish and J. R. Lupski (2009). The DNA replication FoSTeS/MMBIR mechanism can generate genomic, genic and exonic complex rearrangements in humans. *Nat Genet* **41**(7): 849–853.

Zhang, W., Z. Sun, B. Menghe and H. Zhang (2015). Short communication: single molecule, real-time sequencing technology revealed species- and strain-specific methylation patterns of 2 Lactobacillus strains. *J Dairy Sci* **98**(5): 3020–3024.

Zhang, W. J., M. A. Degli-Esposti, T. J. Cobain, P. U. Cameron, F. T. Christiansen and R. L. Dawkins (1990). Differences in gene copy number carried by different MHC ancestral haplotypes. Quantitation after physical separation of haplotypes by pulsed field gel electrophoresis. *J Exp Med* **171**(6): 2101–2114.

2

Clinical Relevance of Pharmacogenomics – a Road to Personalized Medicine

Nusrath Yasmeen
Department of Biotechnology, Amity Institute of Biotechnology, Amity University Rajasthan, Jaipur, India

Shobana John
Department of Pharmacy Practice, Karpagam College of Pharmacy, Coimbatore, India, Affiliated to the Tamil Nadu Dr. MGR Medical University, India

Vikram Kumar
Faculty of Biotechnology, Amity Institute of Biotechnology, Amity University Rajasthan, Jaipur, India

Moudi M. Alasmari
Basic Medical Sciences, College of Medicine, King Saud bin Abdulaziz University for Health Sciences (KSAU-HS), King Abdullah International Medical Research Center (KAIMRC), Ministryof National Guard Health Affairs (NGHA), Jeddah, Kingdom of Saudi Arabia

CONTENTS

ABSTRACT Pharmacogenomics (PGx) is a scientific discipline that has evolved over a period of 50 years parallel to DNA sequencing and molecular technology, unraveling the truth behind the influence of genes on drug outcomes. It uses the understanding and in-depth knowledge of pharmacology clubbed with genomic data to individualize/personalize drug treatment and successfully implement the paradigm "*right drug to right patient at right dose*" by limiting adverse drug events and enhancing drug efficacy. Consequently, the application of PGx-based therapies will overcome the burden of the trial-and-error method of traditional drug prescription, thereby reducing the time spent in the treatment, and helps in improving the prognosis of the disease. Furthermore, validation of PGx helps minimize the time spent and the financial burden implicated during the clinical trials of the new drug development process. Currently, several validated examples exist that can substantiate the significance of PGx application in clinical practice. Despite such significant advancements in the field of PGx, several challenges prevail, and the ethical elements and clinical guidelines essential for PGx utilization and implementation in clinical practice are still in the phase of dormancy. This chapter provides a comprehensive overview pertaining to the current existing literature on pharmacogenomics that will aid in augmenting the existing knowledge, and decision making of healthcare professionals toward individualized drug therapy.

KEYWORDS: *Pharmacogenomics; Drug-Metabolizing Enzymes; Genetic Polymorphism; Drug Transporters; Drug Outcomes; Personalized Medicine.*

DOI: 10.1201/9781003246244-2

2.1 Introduction

2.1.1 Definition and Historical Perspective of Pharmacogenomics

The management of diseases using drugs described as "pharmacotherapy" is known to humans since the dawn of civilizations. The guidelines on which modern pharmacotherapy is based are formulated in accordance with the evidence gathered from the results of several clinical trials, which involve highly selected individuals/participants in the experimental procedures. However, in clinical practice the actual patient population comprises the elderly, people with kidney and liver disorders, those with varied comorbidities and those who are highly heterogeneous (Bank, Swen & Guchelaar, 2014). The inter-individual variability in patients presents several challenges to modern pharmacotherapy, such as no response to drug therapy and adverse drug reactions. Indeed, such a variability in drug response observed by several scientists is high (Tomalik-Scharte et al., 2008). Drug efficacy of nearly 50%–75% has been reported for many drugs. Earlier, the difference in drug response among patients was addressed by the trial-and-error method by the physicians, and the concept "one size fits all" was very prevalent, which might be due to insufficient screening methods and knowledge pertaining to them.

Pharmacogenetics/pharmacogenomics (PGx) is a burgeoning branch of science with enormous potential toward developing individualized/tailored medicine with the prime aim to provide personalized health care to humans and successfully implement the motto "maximize efficacy and minimize toxicity" (Elewa & Awaisu, 2019; Ma, Lee & Kuo, 2012). It is a well-known fact that drug response varies among individuals; while some individuals may respond well, others may show toxicity at the same prescribed dose. The heterogeneity in drug outcomes compelled physicians to brainstorm and evolve suitable solutions with constant research being conducted in this direction. It was in 510 B.C. when Pythagoras first observed inter-individual variability, where he noticed that some individuals who consumed fava beans developed toxic effects, while others didn't; later, the adverse effect was recognized as hemolytic anemia. This paved the way for modern-day pharmacogenetics.

However, significant contributions to the field of genetics and its influence on drug outcomes were made by Motulsky, who for the first time in 1957 identified and reported two genotype-dependent drug-induced adverse effects: One is hemolytic anemia, which is due to G6PD deficiency, and the other is prolonged apnea, which is due to pseudocholinesterase deficiency following anesthesia (Motulsky, 1957). The term "pharmacogenetics" was later coined by Friedrich Vogel in 1959 (Motulsky & Qi, 2006). Later, the works of several scientists led to the development of pharmacogenetics/pharmacogenomics (PGx) as a modern scientific discipline. The Human Genome Project was launched in the year 1990 and completed in 2004. This project aimed to identify all human DNA base pairs and genes. Several complicated theories such as human evolution, etiology, pathophysiology of diseases and gene-environment interactions were well understood with the sequencing of human genome (Lopes & Sela, 2017). Figure 2.1 describes the timeline that garnered several milestones in the development of PGx as a scientific discipline. The scientific fraternity has proposed several definitions for pharmacogenetics/pharmacogenomics (PGx), and the most widely accepted definition is the following: Pharmacogenetics focuses on how genetic variations in a single gene, especially those causing polymorphisms in genes related to drug-metabolizing enzymes (DMEs), receptors and transporters, cause variability in drug response. PGx is defined as "the study of genetic causes of individual variations in drug response," whereas the term pharmacogenomics is a combination of two scientific disciplines pharmacology (science of

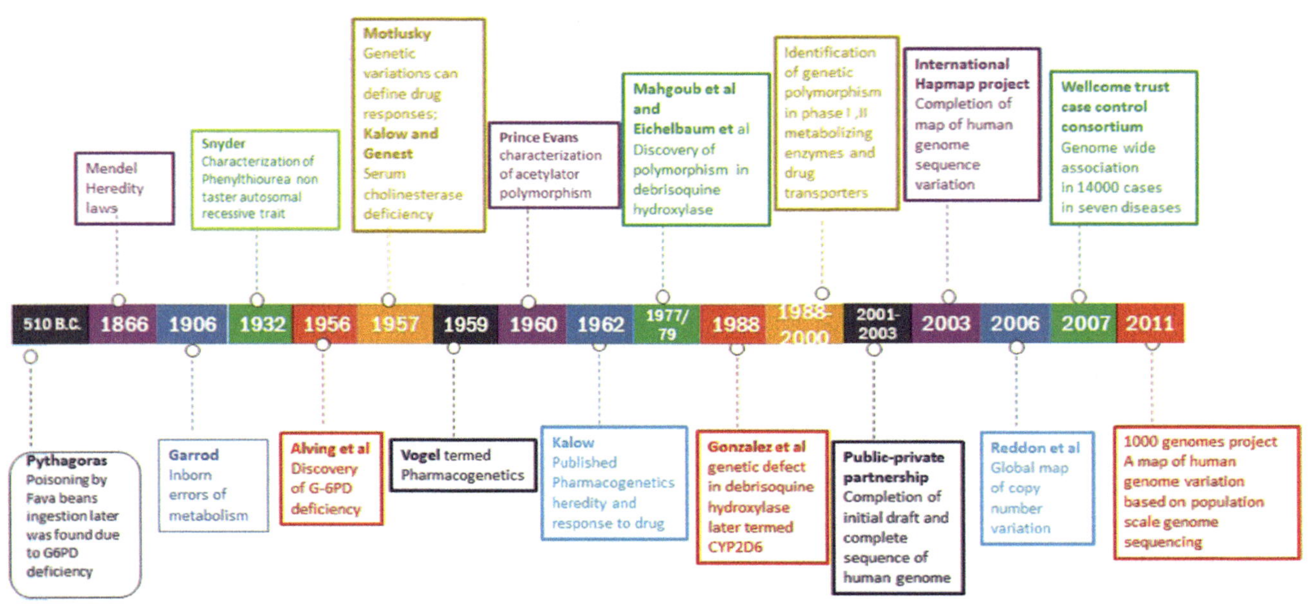

FIGURE 2.1 The milestones achieved in the field of pharmacogenetics over time.

drugs) and genomics (genes and their functions) (Algahtani, 2020). Pharmacogenomics primarily deals with genetic variation in multiple genes, variation in genes encoding enzymes involved in relevant pathways and eventually the entire genome and its influence on drug response. Furthermore, it is defined as "the genome wide analysis of genetic determinants of drug efficacy and toxicity" and deals with the genes present in all chromosomes (Raj, 2019).

Pharmacogenomics includes DNA variation (inherited and acquired) and RNA variations, as well as variations in gene expression and epigenetic regulators (Marsh & McLeod, 2004). The terms pharmacogenetics and pharmacogenomics are so interrelated that their differences are deemed to be arbitrary, and hence, both the terms are used interchangeably (Aroke & Hicks, 2019). Breakthrough progress has been made in the field of pharmacogenetics/pharmacogenomics with the implementation of the Human Genome Project and advanced screening and diagnostic genetic testing. The revolutionary achievements in PGx have led to the archetypal shift from the concept of "one size fits all" to a more precise, individualized or personalized regime of pharmacotherapy. The motto behind the development of PGx is to provide "the right drug to the right patient at the right dose." The personalization of drug therapy is currently deemed the sole responsibility/ability of the prescriber/physician, to order, obtain and interpret the pharmacogenomics information of the patients. However, PGx utilization from bench to bedside is associated with certain hurdles such as heterogeneity of the disease, multifaceted genetic makeup of the individuals, dearth of pharmacogenomic biomarkers and inadequate knowledge and availability of pharmacogenetic tests for screening (Malsagova et al., 2020). Consequently, considerable progress has been made during the previous years to overcome the above-mentioned hurdles and make PGx readily available to minimize adverse effects and costs related to drugs.

This chapter presents the key principles of PGx with a focus on genetic polymorphism in drug-metabolizing enzymes, receptors and transporters along with its application in varied specialties such as oncology, pain management, cardiology and diabetology in a very comprehensive manner.

2.2 Influence of Pharmacogenomics on Pharmacokinetics and Pharmacodynamics

The pharmacokinetic (PK) and pharmacodynamic (PD) properties of a drug are mainly responsible for variable drug outcomes in individual patients; subsequently, these properties are influenced by genetic variance of the genes related to drug-metabolizing enzymes and transporters. Single nucleotide polymorphisms (SNPs), that is, single base pair variation or variations in haplotypes encoding the genes for PK-PD pathways, influence the drug response of an individual (Lin, 2007). Currently, PGx testing is used to comprehend the various PK and PD properties that mainly focus on the metabolism of different medicines. Pharmacogenomic results on metabolism indicate whether a patient is prone to a lack of response due to a lower drug concentration, a normal response, or toxicities from a higher concentration. Other than metabolic enzymes, genetic polymorphisms of drug transporters and drug targets affect the drug response.

2.2.1 Role of Pharmacogenomics in Phase I Drug-Metabolizing Enzymes

The action/response of the parent drug and/or its metabolites is dependent on drug disposition (ADME – absorption, distribution, metabolism, and excretion of pharmacokinetics) that is in turn influenced by the genetic variability in encoding genes responsible for ADME. Drugs are metabolized primarily in two sets of reactions: phase I and phase II. Significantly, the phase I metabolic transformations are catalyzed by cytochrome P450 (CYP450) enzymes, which are harbored in the liver. There are nearly 18 CYPs families, which are further divided into 44 subfamilies with 57 total genes, of which CYP1, CYP2 and CYP3 only are of functional importance in phase I reactions. It is reported that approximately 75% of the prescribed drugs are partly or completely metabolized by CYP3A, CYP2D6 and CYP2C. Genetic polymorphisms of functional significance have been reported in the enzymes of CYP450 family, such as CYP2D6, CYP2C9, CYP2C19, CYP3A4 and CYP3A5, and also they are reported in UDP glucuronosyltransferases (UGTs), TPMT and NAT2 (Ahmed et al., 2016).

Genetic polymorphisms greatly influence the activity of the CYP family of enzymes, which are of great interest to scientists as they might lead to several adverse drug reactions (Preissner et al., 2013). If an allelic variant is found in at least 1% in a population, it is often termed as a pharmacogenetic polymorphism. Although CYP3A4 catalyzes majority of the drugs, it is quite surprising to note that it is unaffected by genetic polymorphisms; it might have been related to insignificant genetic mutations in the genes encoding this enzyme. On the other hand, CYP2D6 is a highly polymorphic CYP enzyme that metabolizes nearly one-quarter of the drugs (Ingelman-Sundberg et al., 2007).

Multiallelic genetic polymorphisms, which are influenced by ethnicity, are critically important for the functioning of CYPs, especially 2D6, 2C19, 2C9, 2B6, 3A5 and 2A6 types, creating variable pharmacogenetic phenotypes. These phenotypes alter drug-metabolizing capacity and are often referred to as poor (PM), intermediate (IM), extensive and ultra-rapid metabolizers (UM). The PM are at enhanced risk of toxicity, when the drug has narrow therapeutic index or is metabolized by allelic variant enzyme. Similarly, the UM achieve subtherapeutic concentrations of the drug at the prescribed dose and might report toxicity (Zanger & Schwab, 2013). CYP1A2 also exhibits genetic polymorphisms, and two allelic variants of the several are extensively studied and include CYP1A2*1C (rs2069514) and CYP1A2*1K (rs2069526, rs12720461 and rs762551). In a meta-analysis done by Wang et al., it was mentioned that CYP1A2 rs762551 polymorphism could serve as a diagnostic marker and help unravel the association between the allelic variant and cancer susceptibility across various ethnic groups (Wang et al., 2012). Human CYP 2C family enzymes are highly polymorphic, and its isoform includes CYP2C8, CYP2C9 and CYP2C19, each contributing differently to drug metabolism.

Clinically relevant polymorphisms include CYP2C8*2 (rs11572103, 11054A >T), CYP2C8*3 (rs11572080, 2130G >A and rs10509681, 30411A >G) and CYP2C8*4 (rs1058930: c.792C.G, p. Ile264Met). The allelic variant CYP2C8*2 is more commonly detected in Africans, while CYP2C8*3 and CYP2C8*4 are reported to be found in Caucasians. The presence of CYP2C8*3 allele had a profound effect on the disposition of (R)-ibuprofen in Caucasians. In a study done by García-Martín et al., it was reported that individuals with CYP2C8*1/*3 and CYP2C9*1/*2 allelic variants had decreased ibuprofen clearance (García-Martín et al., 2004). CYP2C9 is an essentially important isoform of CYP enzyme that is critically involved in several drug–drug interactions in this subfamily. The variant alleles CYP2C9*2 and CYP2C9*3 represent the metabolic variations of PM phenotype. Hence, drugs with narrow therapeutic index such as warfarin should be prescribed at lower doses in homozygous variant individuals (Daly et al., 2018). The evidence of existence of alleles of types CYP2C9*2, CYP2C9*3 and VKORC1 (–1639 G>A) and their influence on variations in dose response in varied ethnic population to the drug warfarin helped in updating the drug label by US FDA in 2007 and then again in 2010 (Finkelman et al., 2011). Altered enzyme activity was a result of genetic polymorphism of CYP2A6 gene, which had SNPs in the forms CYP2A6*2, CYP2A6*4, CYP2A6*5 and CYP2A6*20. Subsequently, the identified polymorphic variants of CYP2A6*6, CYP2A6*7, CYP2A6*9, CYP2A6*10, CYP2A6*11, CYP2A6*12, CYP2A6*17, CYP2A6*18 and CYP2A6*19 exhibited diminished enzyme activity (http://www.cypalleles.ki.se). The effect of these allelic variants was best explained observing the metabolism of coumarin and nicotine, which exhibit a diminished enzyme activity and altered metabolism (Xia et al., 2018). Drug therapy may be difficult to personalize if the drug metabolism involves CYP450 3A4 isoenzyme as this might exhibit both drug-related induction and inhibition (dependent on drug's half-life), further complicating the drug therapy. Close drug monitoring is required when dealing with such drugs.

2.2.2 Role of Pharmacogenomics in Phase II Drug-Metabolizing Enzymes

The knowledge about the existence of the genotype–phenotype relationship/allelic variations along with their functional confirmations is essential in generating personalized medicine utilizing PGx. Consequently, SNP studies related to drug receptors, enzymes and transporters evolved, which enhanced the microdetails related to genetic makeup and inter-individual variability in drug response.

Phase II drug metabolism reactions are mainly conjugation reactions, which are catalyzed by prominent enzymes such as N-acetyltransferases 1 and 2 (NAT1 and NAT2), sulfotransferases (SULTs), uridine diphosphate glucuronosyltransferases (UGTs), glutathione S-transferases (GSTs), thiopurine S-methyltransferase (TPMT) and catechol-O-methyltransferase (COMT) (Arbitrio et al. 2016). N-acetyltransferase 2 (NAT2) has a significant role in the metabolism of several drugs such as paracetamol, isoniazid and sulfonamides, and it also contributes to the detoxification of environmental toxicants. Depending on the presence of allelic variants of NAT2, individuals have varied metabolic effects on drugs and are grouped as slow, intermediate and fast acetylators. Several research studies reported an association between NAT2 and AL risk. A meta-analysis done by Zhu et al. reported that the allelic variant NAT2*5 (rs1801280; 341T>C) was found to increase the risk of AL, whereas the polymorphic forms NAT2*11 (rs1799929; 481C>T) and NAT2*7 (rs1799931; 587G>A) were reported to significantly decrease the AL risk (Zhu et al., 2019). In the research study done by Zahra et al., it is reported that Arabs are slow acetylators and prone to drug-induced liver injuries (DILI) with different metabolic rates of isoniazid (INH-anti-tubercular drug); hence, the dosage regimen of INH, which is mainly metabolized by NAT2, should be adjusted in tuberculosis patients. They also reported a significant correlation between NAT2*5 and anemia in cisplatin-treated ovarian cancer patients (Zahra et al., 2020).

In 2005, FDA advocated genotyping for UGT1A1*28 as a strong correlation between UGT1A1*28/*28 and risk of neutropenia was observed in patients taking irinotecan dose for colorectal cancer (Sim et al., 2013). TPMT is an enzyme that catalyzes thiopurine drugs via S-methylation, e.g., 6-mercaptopurine (6-MP), azathioprine (AZA) and 6-thioguanine (6-TG) (Schaeffeler et al. 2004). Nearly 24 polymorphisms have been identified for TPMT till date, of which four are very crucial: TPMT*2 (rs1800662; 238G>C), TPMT*3A (rs1800460; 460G>A and rs1142345; 719A>G), TPMT*3B (rs1800460; 460G>A) and TPMT*3C (rs1142345; 719A>G). Patients with allelic variants of TPMT taking 6-MP for the treatment of inflammatory bowel disease were reported to exhibit a higher risk of developing leukopenia (Evans 2004) Similarly, in a study done by Ansari et al., it was reported that myelotoxicity and gastric intolerance were prominent adverse drug effects seen in heterozygous TPMT patients taking AZA for inflammatory bowel disease than in TPMT wild-type patients (Ansari et al., 2008). TPMT variant-positive genotypes are reported to be good predictors of risk of bone marrow toxicity and overall adverse drug responses (ADRs), depending on the sensitivity and specificity of variants (Abaji & Krajinovic, 2017). Table 2.1 depicts the impact of pharmacogenomics on phase II drug-metabolizing enzymes.

2.2.3 Role of Pharmacogenomics on Drug Transporters and Drug Targets

Drug transporters are essential for the movement of drugs and their metabolites in and out of cells. Genetic variants of drug transporters have been identified to play a crucial role in the pharmacokinetic and pharmacodynamic parameters of drug therapy and ultimately might influence the safety and efficacy of the drugs. The most widely studied drug transporter and their variants include those of ATP-binding cassette (ABC) transporters, such as ABCB1 (P-glycoprotein), ABCC2 (MRP2) and ABCG2 (BCRP). They also include solute carriers (SLC) such as SLCO1B1 (OATP1B1) and SLC22A1 (OCT1) involved in the uptake of the drugs. Variations in genes encoding drug transporters alter protein expression influencing their role toward drug exposure and hence altering toxicity (Franke et al., 2010).

TABLE 2.1

Impact of Pharmacogenomics on Phase II Drug-Metabolizing Enzymes

Phase II Drug-Metabolizing Enzyme	Genetic Variants	Example of Drug	Impact of Polymorphisms on Drug Outcome	References
N-acetyltransferases (NATs)	NAT2*5 (rs1801280;341T>C), NAT2*6 (rs1799930;590G>A), NAT2*7 (rs1799931;587G>A), NAT2*14 (rs1801279;191G>A), NAT2*11 (rs1799929;481C>T), NAT2*12 (rs1208;803A>G) and NAT2*13 (rs1041983;282C>T)	Isoniazid, dapsone, procainamide, sulfamethazine, sulfapyridine and hydralazine	Enhanced drug effect	Hein and Millner (2021), Mortensen et al. (2011)
Sulfotransferases (SLUTs)	SULT1A1*1 (wild type), SULT1A1*2 (rs9282861), and SULT1A1*3 (rs1801030)	Apixaban	Remains unclear	Shnayder et al. (2021)
Uridine diphosphate glucuronosyltransferases (UGTs)	UGT1A1*28(rs8175347;(TA)6>7 UGT1A7*2(387T>G; 392G>A) UGT1A7*3(387T>G; 392G>A; 622T>C) UGT1A7*4(rs1126802;622T>C) UGT1A7*12(rs7586110;57T>G)	Morphine	Increased effect	MacKenzie and Hall (2017)
Glutathione S-transferases (GSTs)	GSTP1*B (rs1695;313A>G), GSTM1, GSTT1	Cisplatin	Increased effect	Roco et al. (2014)
Thiopurine S-methyltransferase (TPMT)	TPMT*2 (rs1800662;238G>C), TPMT*3A (rs1800460;460G>A and rs1142345;719A>G), TPMT*3B (rs1800460;460G>A) and TPMT*3C (rs1142345;719A>G)	Azathioprine; 6-mercaptopurine	Increased risk of drug toxicity	Katara and Kuntal (2016)
Catechol-O-methyl transferase (COMT)	rs4680; 472G>A rs3788319; 5289A>G rs737865; 4239T>C	Epinephrine calcium-channel blockers	Decreased effect	Xu et al. (2017)

Among the several drug transporters identified, MRPs/ABCCs, P-glycoprotein (P-gp, MDR1, ABCB1), breast cancer resistance protein (BCRP/ABCG2) and bile salt export pump (BSEP/ABCB11) show a significant contribution to drug disposition (Li et al., 2019).

ABCB1 is highly polymorphic, with ethnicity-dependent allelic variants. This can be well explained through the following example. SNP 3435C >T is seen more prominently, i.e., nearly 60%–72%, in Asians, but less frequently observed in Caucasians, nearly 34%–42%. The effects of genetic variants of ABCB1 were studied, but had controversial results. However, it was reported that in the A989T variant nearly 50% reduction in transport activity was observed. It was also studied that African population might experience decreased incidence of renal carcinoma due to less frequent occurrence of the MDR1 3435T allele. Li et al. reported that genetic variation in rs6498588 (ABCC1) and rs12720066 (ABCB1) alter metabolite disposition and lead to irinotecan-induced neutropenia (Li et al., 2018). ABCC1 and ABCC2 are other prominent ABC transporters that might need the help of co-transporter glutathione for effective functioning. G671V variant and variant V417I of ABCC2 are known to cause drug resistance in cancer and HIV patients (Robey et al., 2018). ABCG2/BCRP is an ABC transporter that is generally termed as "half-type" transporter. The functional relevance of the polymorphism in ABCG2/BCRP was widely studied and was reported to affect drug efflux. SNP 421C>A (Q141K) of ABCG2/BCRP altered the expression levels and minimized its function when compared to wild-type variant. Furthermore, two variants 376C>T and 421T/C>A when combined together drastically reduced the ABCG2/BCRP activity (Noguchi, Katayama & Sugimoto, 2014). There are several families of SLC transporters, which include OATPs or organic anion transporting polypeptides encoded by SLCO genes, OCTs or organic cation transporters encoded by SLC22A genes, OCTN or organic cation transporter novel type encoded by SLC22Agenes, CNTs encoded by SLC28A genes, ENTs encoded by SLC29Agenes, MATEs or multidrug and toxin extruders encoded by SLC47Agenes and PEPTs or peptide transporters encoded SLC15A genes (Hagos et al., 2019). The OATPC*5 and OATPC*9 allelic variants are known to be associated with a decreased OATPC substrate uptake.

The effect of genetic polymorphism on drug response is very complicated; cell-based studies can measure this effect in terms of overexpression/underexpression (Hauser et al., 2018). Genetic polymorphisms in drug receptors, for instance in beta-adrenergic receptor, have been reported to modify the sensitivity of patients prescribed beta-agonists therapy, hence altering the pharmacodynamics of the drug response. Genetic variants/polymorphisms identified in drug receptor can modify their expression, function and response turnover when binding to drugs. Several genetic variations and their consequences in drug targets/receptors have been investigated, and some of them are illustrated in Table 2.2.

TABLE 2.2
Impact of Pharmacogenomics on Drug Targets/Receptors

Drug Receptor	Genetic Variants	Example of Drug	Impact of Polymorphisms on Drug Outcome	References
β2-Adrenergic receptor	Multiple polymorphisms	Albuterol Ventolin	Poor asthma control	Zhao et al. (2019)
5-HT2A serotonergic receptor	102T/C and His452/His452	Clozapine	Increased drug response	Arranz et al. (2000), Yoshida and Müller (2019)
HER2 receptor	Ile655Val	Herceptin/trastuzumab	Trastuzumab-induced cardiotoxicity	Udagawa and Zembutsu (2020)
α1A-adrenergic receptor	1475C→T SNP	Tamsulosin	Inconsistent data related to drug response	Mochtar et al. (2006)
AT1 angiotensin II receptor	AGTR1SNP is rs5186 (also termed the A1166C variant)	Losartan	Inconsistent data related to drug response	Insel et al. (2007), Mottl et al. (2008)
β1-Adrenergic receptor	Ser49Gly, Arg389Gly	Propranolol, atenolol, metoprolol, carvedilol	Increased drug effect	Katsarou et al. (2018), Ahles and Engelhardt (2014)
Glycoprotein IIIa	GPIIIa (PIA1/A2)	Antiplatelet drugs	Decreased drug effect	Siest et al. (2004)

2.3 Applications of Pharmacogenomics in Therapeutics

2.3.1 Pharmacogenomics in Cancer Therapy

The pharmacogenomics of cancer therapy is one of the important frontiers in personalized medicine. Substantial evidence for the validity and utility of pharmacogenomic testing is available for many anti-cancer drugs, and pre-emptive testing is recommended in most situations and mandated for some (Aminkeng et al. 2016). Some genes have strong evidence, such as the loss of functional alleles TPMT*2, TPMT*3A and TPMT*3C, associated with thiopurine-induced myelosuppression. Similarly, polymorphisms in nucleoside diphosphate-linked moiety X-type motif 15 (NUDT15) exhibit significant clinical relevance. NUDT15 dephosphorylates TGTP and TdGTP active metabolites into less toxic products and protects cell from DNA damage of their toxic effects. The allelic variants of NUDT15 alleles *2 and*3 are of clinical significance. Based on the polymorphisms in TPMT and NUDT15, the Royal Dutch Pharmacists Association – Pharmacogenomics Working Group (DPWG) and Clinical Pharmacogenetics Implementation Consortium (CPIC) published guidelines for dose optimization of thiopurine-related drugs. Accordingly, the dosing guidelines recommended for the 6-MP and AZA by DPWG are 90% reduction for PM and 50% for IM, and those given by CPIC are to begin with an initial 10% of the standard dose thrice in a week rather than daily for the PM and between 30% and 80% of the standard dose daily for IM (Relling et al., 2013; Nagy et al., 2020).

Similarly, DPYD variations such as DPYD*2A and DPYD*13 have been linked to 5-FU-related hematological and gastrointestinal toxicity (Dean & Kane, 2021; Meulendijks et al. 2015). Irinotecan-induced neutropenia and diarrhea are linked to UGT1A1*28 alleles, which impede glucuronidation and raise SN-38, the active form of irinotecan (Innocenti et al., 2004; Matsuoka & Ando, 2015). In Caucasians, the UGT1A1*28 allele is found in nearly 29%–45% (Liu et al. 2014),It is found in 42%–51% of Africans, and it is also found in a much lesser proportion of approximately 16% among Asians (Ozdemir et al. 2007; Fujii et al., 2019). CYP2D6 is required for the conversion of tamoxifen, a prodrug, to the active metabolite endoxifen, and a non-functional CYP2D6*10 results in poor efficacy and prognosis (Goetz et al., 2018; de Vries Schultink et al., 2018).

The ABCB1 gene, which encodes P-gp, has been associated with anthracycline-induced cardiotoxicity, paclitaxel-induced peripheral neuropathy, neutropenia (Abdelfattah et al., 2021) and also vincristine-induced neurotoxicity (Ceppi et al. 2014). Aside from these, an association between CYP2B6 and cyclophosphamide and SLCO1B1 and methotrexate (MTX) has been demonstrated in a few studies (Liu et al., 2017). Cyclophosphamide is an alkylating prodrug. It has been confirmed that patients carrying CYP2B6*6 alleles can exhibit lower pharmacological action and fewer side effects with cyclophosphamide (Johnson et al. 2013; Veal et al., 2016). Likewise, the SLCO1B1 affects the clearance of MTX (Ramsey et al. 2013). There have also been a few unusual varieties discovered; for example, the variants of the CYP3A4 gene have also been related to a greater prevalence and severity of paclitaxel-induced peripheral neuropathy (Sissung et al. 2006). Although the implementation of PGx is having a lot of challenges, 50% of US FDA-approved biomarkers are oncology related, and they are considered the best in precision medicine.

2.3.2 Pharmacogenomics in Neurological Disorders

Among all the US FDA-approved PGx testing, 15% of the testing is for centrally acting drugs. CYP450 enzymes such as CYP2C9, CYP2C19, CYP2D6, ABCB1 and UGT2B7 are found to be correlated to many CNS drugs-induced toxicities and changes in the therapeutic response. For instance, CYP2C19 is involved in the metabolism of amitriptyline, imipramine and clomipramine (TCAs); SSRIs such as citalopram, escitalopram and sertraline; and some benzodiazepines such as diazepam and clobazam. In accordance with the CPIC guidelines, patients carrying allelic variants with receded functional activities, i.e., CYP2C19*2 and CYP2C19*3, should avoid TCA or might need a dose adjustment to avoid severe adverse events (Hicks et al.,2013, 2017).

Another significant advancement in ADR-pharmacogenetics is the link between HLA B* 15:02 and carbamazepine or aromatic antiepileptic drugs (AEDs)-induced severe cutaneous adverse drug reactions (SCARs) such as Stevens-Johnson syndrome (SJS), toxic epidermal necrolysis (TEN), drug reaction with eosinophilia and systemic syndrome (DRESS) and exfoliative dermatitis (ED) (Chang et al. 2011). The genetic biomarkers, or pharmacogenetic (PGx) markers, may vary based on ethnicity, phenotypes and drugs. For instance, many countries/ethnicities in Asia have confirmed the strong association between HLA B*15:02 and CBZ-induced SJS/TEN, such as Han Chinese in Taiwan (Wang, 2011; Chung et al., 2004), Thailand (Yampayon et al., 2017; Tassaneeyakul et al. 2010), Malaysia (Chang et al., 2017), Vietnamese (Van Nguyen et al., 2015), Indonesia (Yuliwulandari et al., 2017) and India (Mehta et al., 2009).

However, in European and Japanese populations, the involvement of HLA A*31:01 for the same CBZ-induced SJS/TEN has been proved (McCormack et al., 2011; Kaniwa et al., 2013). Likewise, CBZ-induced DRESS is associated with HLA B*51:01, and no association was found with HLA B*15:02; thus, HLA B*51:01 becomes phenotypic specific (DRESS) (Manuyakorn et al., 2020). The association strength between the causative allele and the drug-induced reactions may also differ based on the drug. For example, the association between HLA B*15:02 and PHT-induced SJS/TEN is weaker than it is reported for the CBZ (Chang et al., 2017; Tempark et al., 2022). FDA has recommended to screen patients for genotypic variant HLA B*15:02 before initiating CBZ therapy.

Another known example of atypical anti-psychotic pharmacogenetics is risperidone. It is a prodrug and is converted to its active metabolite, 9-hydroxy risperidone, by the enzyme CYP2D6(Wang, et al., 2007; Janssen et al., 1988) The CYP2D6 gene, which is largely expressed in human liver cells but is also highly expressed in the brain, encodes this enzyme. Genetic mutations or loss of functional alleles in this gene may affect the risperidone metabolism; for example, IM and PM may have a higher plasma exposure than UM (Cui et al., 2020; Puangpetch et al., 2016).

2.3.3 Pharmacogenomics in Respiratory Disorders

Drugs acting on the respiratory system are not studied much in terms of their efficacy and ADR with a genetic predisposition. However, some studies found an association between beta (β)-2 adrenergic receptor (ADRB2) genes in asthma patients. TUBG1 (tubulin gamma 1 gene), PDGFD (platelet-derived growth gene), RAPGEF5 (Rap guanine nucleotide exchange factor 5 gene) and TBCD (tubulin folding cofactor D) are all being investigated for their links to various ADRs.

According to studies, polymorphisms in the glucocorticoid receptor gene, NR3C1, increase glucocorticoid sensitivity and corticosteroid dependency (Krupoves et al., 2011; Baker et al., 2012). Other than this, genetic variants of genes such as CRHR1 (corticotrophin-releasing hormone receptor), STIP1 (adaptor protein within the intracellular glucocorticoid receptor heterocomplex) and TBX21 (a transcription factor that influences T-lymphocyte production) are linked with sensitivity of inhaled corticosteroid in asthmatic patients (Tantisira, Lake, et al., 2004; Hawkins et al., 2009; Tantisira, Hwang, et al., 2004).

Polymorphisms of some metabolic genes are studied in this category. For instance, genetic variants of UGT1A9 have been associated with a significant increase in UGT activity, thereby reducing the amount of circulating bioactive medroxyprogesterone acetate (MPA) (Kuypers et al., 2005). Polymorphisms within CYP genes CYP3A5 and CYP3A5 have been shown to influence the bioavailability of tacrolimus (Tavira et al., 2011; Birdwell et al., 2015). The existence of SNPs in the MDR1 gene, which encodes a cellular efflux pump protein, may also affect tacrolimus absorption (Soria et al., 2010). A medicine's overall efficacy is determined by a balance of expected benefits and potential hazards. Adverse drug responses (ADRs) in asthmatics must also be taken into account.

2.3.4 Pharmacogenomics in Diabetes Mellitus

Inter-individual variation in therapy response exists in type 2 diabetes mellitus (T2D), whether measured in terms of efficacy or adverse medication effects. There have already been numerous reports of T2D gene-drug interactions. Sulfonylureas (SUs) such as tolbutamide, glimepiride and glipizide are mainly metabolized by CYP2C9. The CYP2C9*2/*3 non-functional alleles slow down the metabolism of these drugs (Ragia et al., 2009; Huang & Florez, 2011).

Another pharmacogenetic example of oral hypoglycemic drugs (OHDs) is the SLC22A1 gene and metformin. Metformin is often regarded as the first-line treatment in type 2 diabetes. This gene encodes the organic cation transporter 1 (OCT1), which is involved in metformin absorption. SLC22A1 and SLC22A2 mutations have been associated with changes in the amino acids Arg61Cys, Gly401Ser and Gly465Arg, which have been connected to decreased metformin transportation and hence therapeutic efficacy (Distefano & Watanabe, 2010). In a study done by Umamasheswaran et al., it was reported that SNPs rs6622342 AA and rs594709 in the SLC22A1 gene (OCT1) enhance the response to metformin treatment in Chinese and South Indian T2D patients (Umamaheswaran et al., 2015). Also, evidence from a study done on Chinese T2D patients shows that the allelic variant SLC22A2 808G>T (rs316019) was associated significantly with an increased reduction in HbA1c levels and decreased renal clearance. Allelic variants c.596C>T (rs201919874), c.602C>T and c.808G>T (rs316019) have been found to be associated with decreased renal clearance of metformin (Xie et al., 2018).

Phase II metabolism of canagliflozin requires the presence of UGT1A9. UGT1A9*3 allele carriers had a 1.45 higher canagliflozin AUC than wild-type allele carriers (Francke et al., 2015). After therapy with DPP IV inhibitors, GLP 1R patients with at least one minor A allele exhibited lower HbA1c levels than GG carriers (Han et al., 2016). DPP4 TT genotype carriers demonstrated higher DPP IV activity with sitagliptin treatment than the CC and CT genotype groups (Wilson et al., 2017). PNPLA3 CG genotype is found to be more substantially connected to HBA1C decline and AST and ALT levels after alogliptin treatment than CC genotype (Kan et al., 2016).

2.3.5 Pharmacogenomics in Cardiovascular Disorders

Warfarin clearance is associated with the CYP2C9 gene, while warfarin sensitivity is linked to the VKORC1 gene, which regulates vitamin K oxidation status. Pharmacogenetic variations in these genes can affect drug responsiveness and metabolism (Johnson et al., 2011). Statins are commonly used drugs in cardiovascular diseases. Although many ADRs are reported with statins, hepatotoxicity, GI upset, cataracts and statin-induced myopathy (SIM) are common with prevalence rates ranging from 7% to 30% (Björnsson, 2017; du Souich et al., 2017). SIM is found to be the major reason for its discontinuation (Ward et al., 2019). Data suggest that genetics influences the intolerance and development of SIM, at least for some statins.

The SLCO1B1 gene encodes the organic anion polypeptide transporter 1B1, which delivers most statins to the liver. SIM has been linked to the polymorphism 521T>C (p. Val174Ala). Simvastatin has the maximum data on its association with genetic testing findings among the statin categories listed in the CPIC guidelines. Patients with the 521CT or CC genotypes, for example, were prescribed lower doses of simvastatin (usually 20 mg per day) (Ramsey et al., 2014). SIM can also be caused by genetic variations of CYP450 enzymes and the ABC transporter protein.

Clopidogrel is another example of CVS pharmacogenetics. Clopidogrel is a platelet aggregation inhibitor that acts as a prodrug. PON1 and CYP450 enzymes convert clopidogrel to its active form. Gene mutations in the genes that encode these enzymes affect the availability of the active form of clopidogrel. CYP450 mutations, on the other hand, have stronger evidence connecting them to clopidogrel metabolism than PON1. Among the several enzymes involved, CYP2C19 variations are crucial. Variants CYP2C19*2 or *3 in CYP2C19, the primary enzyme involved in clopidogrel metabolism significantly reduced its AUC_{0-24}. Where the frequency of these alleles is higher, the impact of this effect is more severe. For example, CYP2C19*2 is more common in South Asians (32.5%) than in East Asians (31%), although CYP2C19*3 is more common in East Asians (6%) than in South Asians (0.4%) (Pereira et al., 2019). Individuals who inherit the defective CYP2C19 variant form have an increased risk of cardiac, especially if they have a history of atrial fibrillation (AF) or percutaneous coronary intervention (PCI) (Scott et al., 2013). The loss of functional alleles of CYP2D6 genetic variations decreases the metabolism of beta-blockers such as propranolol, timolol and metoprolol (Weeke & Roden, 2013, Petersen et al., 2011). Beta-blockers are linked to the ADBR1 and ADBR2 genes as well (Bijl et al., 2009).

2.3.6 Pharmacogenomics in Pain

Codeine (prodrug) is converted into morphine by O-demethylation, which is mediated by CYP2D6. In UM, genetic polymorphisms such as the CYP2D6*1/*1 and *1/*2 alleles will result in subtherapeutic concentrations. Furthermore, because the gene UGT2B7 is involved in the glucuronidation of morphine, carriers of the UGT2B7*2/*2 genotype will have lower enzyme activity and a higher risk of toxicity. ABCB1 (drug transporter) and OPRM1 are two important potential genetic factors to opioid efficacy and side effects (receptor interaction). For example, the GA genotype of the OPRM1 gene is wild type; thus, if a patient has the GG genotype in this gene, the effectiveness of morphine to cure pain will be reduced, so the daily dose should be raised for such people (Klepstad et al., 2004).

The polymorphism rs1799971 of OPRM1 is responsible for persistent chronic low back pain (CLBP) more prominent in women than in men (Hasvik et al., 2014). Another study reported that CYP2D6 pharmacogenetics helped in improving pain management, wherein patients with CYP2D6 phenotypes were classified as PM, IM or UM. The CYP2D6 gene, which is largely expressed in human liver cells and in the brain, encodes this enzyme. These patients were recommended to use drugs that are non-CYP2D6 substrates, as PM have lower levels of morphine and O-desmethyltramadol (metabolites of codeine and tramadol) and UM exhibited enhanced toxicity. This example substantiates that PGx-guided treatment is beneficial as it reduced the pain intensity well in comparison with routine care patients (Yamamoto et al., 2019). Similarly, variations in the development of respiratory depression have been linked to ABCB1 genetic polymorphism (Sadhasivam et al., 2015; Kirchheiner et al., 2007). Traditional NSAIDs such as flurbiprofen and celecoxib may have CYP2C9 polymorphisms that affect their analgesic efficacy and toxicity. To avoid severe cardiovascular and gastrointestinal complications, PM with the CYP2C9*3/*3 genotype should be prescribed half the low dose (Carbonell et al., 2010).

2.4 Conclusions and Future Perspectives

The current research with regard to whole genome SNP profiling, gene expression studies, complete genetic profile and haplotyping has contributed significantly toward our understanding about the influence of gene association/variants on drug efficacy and helped us to predict the adverse drug events even more promptly than ever before. Consequently, these studies helped in updating the drug labels by authorized regulatory bodies. This ultimately led to the evolution of pharmacogenomics, which is clubbed with genomics and epigenetic studies along with pharmacology. Pharmacogenomics paved the way for scrutinized study about variability in drug response among patients of varied ethnicities and helped us resolve and overcome the therapeutic paradigm "one size fits all". It also led to several significant advancements in the disease diagnosis, helped in enhancing prognosis of disease in patients and improved their quality of life through the development of personalized medicine, overcoming several ramifications in drug discovery and drug development. Subsequently, it helped to cater the individualized needs of patients by modified dosage regimens. The influence of SNPs on the drug metabolism and the solutions offered by PGx are worth taking a note of and have helped several patients. However, the optimization of dosage regimen still is far-fetched to a common man, which might be attributed to insufficient pharmacogenomics testing, lack of awareness among patients and healthcare professionals, lack of guidelines for pharmacogenomic testing, etc. The data gathered form the Human Genome Project and genome-wide association studies must be appropriately interpreted and made available to the public and concerned authorities via a database and enable the development of clinical tools for patient evaluation. The success of precision medicine development depends on the proper in-depth understanding and utilization of PGX and genetic code and the involvement of health authorities to develop guidelines and policies so that it can be easily understandable and available to all, making it beneficial both at individual level and at governmental level.

ABBREVIATIONS

5-FU: 5-Fluorouracil
ABC: ATP-binding cassette
ADBR2: Beta-2 adrenergic receptor genes
ADRs: Adverse drug reactions
AEDs: Antiepileptic drugs
AF: Atrial fibrillation
AGEP: Acute generalized exanthematous pustulosis
ALT: Alanine transaminase
AST: Aspartate aminotransferase
AUC: Area under the curve
BCRP: Breast cancer resistance protein
BSEP: Bile salt export pump
CBZ: Carbamazepine
COMT: Catechol-O-methyl transferase
CPIC: Clinical Pharmacogenetics Implementation Consortium
CRHR 1: Corticotrophin-releasing hormone receptor
CYP450: Cytochrome P450
DMEs: Drug-metabolizing enzymes
DRESS: Drug reaction with eosinophilia and systemic symptoms
DYPD: Dihydropyrimidine dehydrogenase
G6PD: Glucose-6-phosphate dehydrogenase
GPCR: G protein-coupled receptor
GSTs: Glutathione S-transferases
HLA: Human leukocyte antigen
MATE: Multidrug and toxin extruder
MDR: Multidrug resistance
NAT: N-Acetyltransferase
NSAIDs: Non-steroidal anti-inflammatory drugs
OATPs: Organic anion transporting polypeptides
OCTs: Organic cation transporters
OCTN: Organic cation transporter novel type
OPRM: Opioid receptor mu 1
PCI: Percutaneous coronary intervention
PD: Pharmacodynamics
PDGFD: Platelet-derived growth gene
PEPTs: Peptide transporters
PGx: Pharmacogenomics
PK: Pharmacokinetics
PM: Poor metabolizers
RAPGEF: Rap guanine nucleotide exchange factor 5 gene
SCARs: Severe cutaneous adverse drug reactions
SJS: Stevens-Johnson syndrome
SLCO1B1: Solute carrier organic anion transporter family member 1B1
SNP: Single nucleotide polymorphism
SSRIs: Selective serotonin reuptake inhibitors
STIP1: Stress-induced phosphoprotein 1
SULT: Sulfotransferase
TBCD: Tubulin folding cofactor D
TBX21: Transcription factor that influences T-lymphocyte production
TCAs: Tricyclic antidepressants
TEN: Toxic epidermal necrolysis
TPMT: Thiopurine S-methyltransferase
TUBG: Tubulin gamma 1 gene
UGT: UDP glucuronosyltransferase
UM: Ultra-rapid metabolizers
US FDA: US Food and Drug Administration

REFERENCES

Abaji R, Krajinovic M. Thiopurine S-methyltransferase polymorphisms in acute lymphoblastic leukemia, inflammatory bowel disease and autoimmune disorders: influence on treatment response. *Pharmacogenomics and Personalized Medicine*. 2017;10:143–156. https://doi.org/10.2147/PGPM.S108123

Abdelfattah NM, Solayman MH, Elnahass Y, Sabri NA. ABCB1 single nucleotide polymorphism genotypes as predictors of paclitaxel-induced peripheral neuropathy in breast cancer. *Genetic Testing and Molecular Biomarkers*. 2021;25(7):-471–477. https://doi.org/10.1089/gtmb.2021.0014

Ahles A, Engelhardt S. Polymorphic variants of adrenoceptors: pharmacology, physiology, and role in disease. *Pharmacological Reviews*. 2014;66(3):598–637. https://doi.org/10.1124/pr.113.008219

Ahmed S, Zhou Z, Zhou J, Chen SQ. Pharmacogenomics of drug metabolizing enzymes and transporters: relevance to precision medicine. *Genomics, Proteomics & Bioinformatics*. 2016;14(5):298–313. http://dx.doi.org/10.1016/j.gpb.2016.03.008

Algahtani M. Knowledge, perception, and application of pharmacogenomics among hospital pharmacists in Saudi Arabia. *Risk Management and Healthcare Policy*. 2020;13:1279–1291. https://doi.org/10.2147/RMHP.S267492

Aminkeng F, Ross CJ, Rassekh SR, Hwang S, Rieder MJ, et al. Recommendations for genetic testing to reduce the incidence of anthracycline-induced cardiotoxicity. *British Journal of Clinical Pharmacology*. 2016;82(3):683–695. https://doi.org/10.1111/bcp.13008

Ansari A, Arenas M, Greenfield SM, Morris D, Lindsay J et al. Prospective evaluation of the pharmacogenetics of azathioprine in the treatment of inflammatory bowel disease. *Alimentary Pharmacology & Therapeutics*. 2008;28(8):-973–983. https://doi.org/10.1111/j.1365-2036.2008.03788.x

Arbitrio M, Di Martino MT, Scionti F, Agapito G, Guzzi PH, et al. DMET™ (Drug Metabolism Enzymes and Transporters): a pharmacogenomic platform for precision medicine. *Oncotarget*. 2016;7(33):54028–54050. https://doi.org/10.18632/oncotarget.9927

Aroke EN, Hicks TL. Pharmacogenetics of postoperative nausea and vomiting. *Journal of PeriAnesthesia Nursing*. 2019;34(-6):1088–1105. https://doi.org/10.1016/j.jopan.2019.03.007.

Arranz MJ, Munro J, Birkett J, Bolonna A, Mancama D, et al. Pharmacogenetic prediction of clozapine response. *The Lancet.* 2000;355(9215):1615–1616. https://doi.org/10.1016/S0140-6736(00)02221-2

Baker AC, Green TL, Chew VW, Tung K, Amini A, et al. Enhanced steroid response of a human glucocorticoid receptor splice variant. *Shock.* 2012;38(1):11–17. https://doi.org/10.1097/SHK.0b013e318257c0c0

Bank PC, Swen JJ, Guchelaar HJ. Pharmacogenetic biomarkers for predicting drug response. *Expert Review of Molecular Diagnostics.* 2014;14(6):723–735. https://doi.org/10.1586/14737159.2014.923759

Bijl MJ, Visser LE, Van Schaik RH, Kors JA, Witteman JC, et al. Genetic variation in the CYP2D6 gene is associated with a lower heart rate and blood pressure in β-blocker users. *Clinical Pharmacology & Therapeutics.* 2009;85(1):45–50. https://doi.org/10.1038/clpt.2008.172

Birdwell KA, Decker B, Barbarino JM, Peterson JF, Stein CM, et al. Clinical Pharmacogenetics Implementation Consortium (CPIC) guidelines for CYP3A5 genotype and tacrolimus dosing. *Clinical Pharmacology & Therapeutics.* 2015;98(1):19–24. https://doi.org/10.1002/cpt.113

Björnsson ES. Hepatotoxicity of statins and other lipid-lowering agents. *Liver International.* 2017;37(2):173–178. https://doi.org/10.1111/liv.13308

Carbonell N, Verstuyft C, Massard J, Letierce A, Cellier C, et al. CYP2C9* 3 Loss-of-function allele is associated with acute upper gastrointestinal bleeding related to the use of NSAIDs other than aspirin. *Clinical Pharmacology & Therapeutics.* 2010;87(6):693–698. https://doi.org/10.1038/clpt.2010.33

Ceppi F, Langlois-Pelletier C, Gagné V, Rousseau J, Ciolino C, et al. Polymorphisms of the vincristine pathway and response to treatment in children with childhood acute lymphoblastic leukemia. *Pharmacogenomics.* 2014;15(8):-1105–1116. https://doi.org/10.2217/pgs.14.68

Chang CC, Ng CC, Too CL, Choon SE, Lee CK, et al. Association of HLA-B* 15: 13 and HLA-B* 15: 02 with phenytoin-induced severe cutaneous adverse reactions in a Malay population. *The Pharmacogenomics Journal.* 2017;17(2):-170–173. https://doi.org/10.1038/tpj.2016.10

Chang CC, Too CL, Murad S, Hussein SH. Association of HLA-B* 1502 allele with carbamazepine-induced toxic epidermal necrolysis and Stevens–Johnson syndrome in the multi-ethnic Malaysian population. *International Journal of Dermatology.* 2011;50(2):221–224. https://doi.org/10.1111/j.1365-4632.2010.04745.x

Chung WH, Hung SI, Hong HS, Hsih MS, Yang LC, et al. A marker for Stevens–Johnson syndrome. *Nature.* 2004;428(-6982):486. https://doi.org/10.1038/428486a

Cui Y, Yan H, Su Y, Wang L, Lu T, Zhang D, Yue W. CYP2D6 genotype-based dose recommendations for risperidone in Asian people. *Frontiers in Pharmacology.* 2020;11:936. https://doi.org/10.3389/fphar.2020.00936

Daly AK, Rettie AE, Fowler DM, Miners JO. Pharmacogenomics of CYP2C9: functional and clinical considerations. *Journal of Personalized Medicine.* 2018;8(1):1. https://doi.org/10.3390/jpm8010001

de Vries Schultink AH, Huitema AD, Beijnen JH. Therapeutic DRUG MONITORING of endoxifen as an alternative for CYP2D6 genotyping in individualizing tamoxifen therapy. *The Breast.* 2018;42:38–40 https://doi.org/10.1016/j.breast.2018.08.100

Dean L, Kane M. *Fluorouracil therapy and DPYD genotype In: Medical Genetics Summaries.* National Center for Biotechnology Information (US), Bethesda (MD); 2021. PMID: 28520376

DiStefano JK, Watanabe RM. Pharmacogenetics of anti-diabetes drugs. *Pharmaceuticals.* 2010;3(8):2610–2646. https://doi.org/10.3390/ph3082610

du Souich P, Roederer G, Dufour R. Myotoxicity of statins: mechanism of action. *Pharmacology & Therapeutics.* 2017;175:1–6 https://doi.org/10.1016/j.pharmthera.2017.02.029

Elewa H, Awaisu A. Pharmacogenomics in pharmacy practice: current perspectives. *Integrated Pharmacy Research & Practice.* 2019;8:97–104. https://doi.org/10.2147/IPRP.S180154

Evans WE. Pharmacogenetics of thiopurine S-methyltransferase and thiopurine therapy. *Therapeutic Drug Monitoring.* 2004;26(2):186–191.

Finkelman, BS, Gage, BF, Johnson, JA, Brensinger, CM, Kimmel, SE. Genetic warfarin dosing: tables versus algorithms. *Journal of the American College of Cardiology.* 2011; 57(-5):612–618. https://doi.org/ 10.1016/j.jacc.2010.08.643

Francke, S, Mamidi, RN, Solanki, B, Scheers, E, Jadwin, A, Favis, R., Devineni, D. In vitro metabolism of canagliflozin in human liver, kidney, intestine microsomes, and recombinant uridine diphosphate glucuronosyltransferases (UGT) and the effect of genetic variability of UGT enzymes on the pharmacokinetics of canagliflozin in humans. *The Journal of Clinical Pharmacology.* 2015;55(9):1061–1072. https://doi.org/10.1002/jcph.506

Franke RM, Gardner ER, Sparreboom A. Pharmacogenetics of drug transporters. *Current Pharmaceutical Design.* 2010;16(2):-220–230. https://doi.org/10.2174/138161210790112683

Fujii H, Yamada Y, Watanabe D, Matsuhashi N, Takahashi T, Yoshida K, Suzuki A. Dose adjustment of irinotecan based on UGT1A1 polymorphisms in patients with colorectal cancer. *Cancer Chemotherapy and Pharmacology.* 2019;83(-1):123–129. https://doi.org/10.1007/s00280-018-3711-8

García-Martín E, Martínez C, Tabarés B, Frías J, Agúndez JA. Interindividual variability in ibuprofen pharmacokinetics is related to interaction of cytochrome P450 2C8 and 2C9 amino acid polymorphisms. *Clinical Pharmacology & Therapeutics.* 2004;76(2):119–127. https://doi.org/10.1016/j.clpt.2004.04.006.

Goetz MP, Sangkuhl K, Guchelaar HJ, Schwab M, Province M, et al. Clinical Pharmacogenetics Implementation Consortium (CPIC) guideline for CYP2D6 and tamoxifen therapy. *Clinical Pharmacology & Therapeutics.* 2018;103(5):770–777. https://doi.org/10.1002/cpt.1007

Hagos FT, Adams SM, Poloyac SM, Kochanek PM, Horvat CM, et al. Membrane transporters in traumatic brain injury: pathological, pharmacotherapeutic, and developmental implications. *Experimental Neurology.* 2019;317:10–21. https://doi.org/10.1016/j.expneurol.2019.02.011

Han E, Park HS, Kwon O, Choe EY, Wang HJ, et al. A genetic variant in GLP1R is associated with response to DPP-4 inhibitors in patients with type 2 diabetes. *Medicine.* 2016;95(44):e5155. https://doi.org/10.1097/MD.0000000000005155

Hasvik E, Iordanova Schistad E, Grøvle L, Julsrud Haugen A, Røe C, et al. Subjective health complaints in patients with lumbar radicular pain and disc herniation are associated with a sex-OPRM1 A118G polymorphism interaction: a prospective 1-year observational study. *BMC Musculoskeletal Disorders.* 2014;15(1):1–6. https://doi.org/10.1186/1471-2474-15-161

Hauser AS, Chavali S, Masuho I, Jahn LJ, Martemyanov KA, et al. Pharmacogenomics of GPCR drug targets. *Cell.* 2018;172(1–2):41–54. https://doi.org/10.1016/j.cell.2017.11.033

Hawkins GA, Lazarus R, Smith RS, Tantisira KG, Meyers DA, et al. The glucocorticoid receptor heterocomplex gene STIP1 is associated with improved lung function in asthmatic subjects treated with inhaled corticosteroids. *Journal of Allergy and Clinical Immunology.* 2009;123(6):1376–1383. https://doi.org/10.1016/j.jaci.2009.01.049

Hein DW, Millner LM. Arylamine N-acetyltransferase acetylation polymorphisms: paradigm for pharmacogenomic-guided therapy-a focused review. *Expert Opinion on Drug Metabolism & Toxicology.* 2021;17(1):9–21. https://doi.org/10.1080/17425255.2021.1840551

Hicks JK, Sangkuhl K, Swen JJ, Ellingrod L, Müller DJ, et al. Clinical pharmacogenetics implementation consortium guideline (CPIC) for CYP2D6 and CYP2C19 genotypes and dosing of tricyclic antidepressants: 2016 update. *Clinical Pharmacology Therapeutics.* 2017;102(1):37–44. https://doi.org/10.1002/cpt.597

Hicks JK, Swen JJ, Thorn CF, Sangkuhl K, Kharasch ED, Clinical Pharmacogenetics Implementation Consortium, et al. Clinical Pharmacogenetics Implementation Consortium guideline for CYP2D6 and CYP2C19 genotypes and dosing of tricyclic antidepressants. *Clinical Pharmacology Therapeutics.* 2013;93(5):402–408. https://doi.org/10.1038/clpt.2013.2

Huang C, Florez JC. Pharmacogenetics in type 2 diabetes: potential implications for clinical practice. *Genome Medicine.* 2011;3(11):1–9. https://doi.org/10.1186/gm292

Ingelman-Sundberg M, Sim SC, Gomez A, Rodriguez-Antona C. Influence of cytochrome P450 polymorphisms on drug therapies: pharmacogenetic, pharmacoepigenetic and clinical aspects. *Pharmacology & Therapeutics.* 2007;116(3):496–526. https://doi.org/10.1016/j.pharmthera.2007.09.004

Innocenti F, Undevia SD, Iyer L, Xian Chen P, Das S, et al. Genetic variants in the UDP-glucuronosyltransferase 1A1 gene predict the risk of severe neutropenia of irinotecan. *Journal of Clinical Oncology.* 2004;22(8):1382–1388. https://doi.org/10.1200/JCO.2004.07.173

Insel, PA, Tang, CM, Hahntow, I, Michel, MC. Impact of GPCRs in clinical medicine: monogenic diseases, genetic variants and drug targets. *Biochimica et Biophysica Acta (BBA)-Biomembranes,* 2007;1768(4):994–1005. https://doi.org/10.1200/JCO.2004.07.173

Janssen PA, Niemegeers CJ, Awouters F, Schellekens KH, Megens AA, Meert TF. Pharmacology of risperidone (R 64 766), a new antipsychotic with serotonin-S2 and dopamine-D2 antagonistic properties. *Journal of Pharmacology and Experimental Therapeutics.* 1988;244(2):685–693.

Johnson JA, Gong L, Whirl-Carrillo M, Gage BF, Scott SA, et al. Clinical Pharmacogenetics Implementation Consortium: clinical pharmacogenetics implementation consortium guidelines for CYP2C9 and VKORC1 genotypes and warfarin dosing. *Clinical Pharmacology Therapeutics.* 2011;90(4):625–629. https://doi.org/10.1038/clpt.2011.185

Johnson GG, Lin K, Cox TF, Oates M, Sibson DR, et al. CYP2B6* 6 is an independent determinant of inferior response to fludarabine plus cyclophosphamide in chronic lymphocytic leukemia. *Blood, The Journal of the American Society of Hematology.* 2013;122(26):4253–4258. https://doi.org/10.1182/blood-2013-07-516666

Kan, H, Hyogo, H, Ochi, H, Hotta, K, Fukuhara T, et al. (2016). Influence of the rs738409 polymorphism in patatin-like phospholipase 3 on the treatment efficacy of non-alcoholic fatty liver disease with type 2 diabetes mellitus. *Hepatology Research.* 2016;46(3):146–153. https://doi.org/10.1111/hepr.12552

Kaniwa N, Sugiyama E, Saito Y, Kurose K, Maekawa K, et al. Specific HLA types are associated with antiepileptic drug-induced Stevens–Johnson syndrome and toxic epidermal necrolysis in Japanese subjects. *Pharmacogenomics.* 2013;14(15):1821–1831. https://doi.org/10.2217/pgs.13.180

Katara P, Kuntal H. TPMT polymorphism: when shield becomes weakness. Interdisciplinary Sciences. *Computational Life Sciences.* 2016;8(2):150–155. https://doi.org/10.1007/s12539-015-0111-1

Katsarou MS, Karathanasopoulou A, Andrianopoulou A, Desiniotis V, Tzinis E, et al. Beta 1, beta 2 and beta 3 adrenergic receptor gene polymorphisms in a southeastern European population. *Frontiers in Genetics.* 2018;9:560. https://doi.org/10.3389/fgene.2018.00560

Kirchheiner J, Schmidt H, Tzvetkov M, Keulen JT, Lötsch J, Roots I, Brockmöller J. Pharmacokinetics of codeine and its metabolite morphine in ultra-rapid metabolizers due to CYP2D6 duplication. *The pharmacogenomics Journal.* 2007;7(4):257–265. https://doi.org/10.1038/sj.tpj.6500406

Klepstad P, Rakvåg TT, Kaasa S, Holthe M, Dale O, et al. The 118 A> G polymorphism in the human μ-opioid receptor gene may increase morphine requirements in patients with pain caused by malignant disease. *Acta Anaesthesiologica Scandinavica.* 2004;48(10):1232–1239. https://doi.org/10.1111/j.1399-6576.2004.00517.x

Krupoves A, Mack D, Deslandres C, Seidman E, Amre DK. Variation in the glucocorticoid receptor gene (NR3C1) may be associated with corticosteroid dependency and resistance in children with Crohn's disease. *Pharmacogenetics and Genomics.* 2011;21(8):454–460. https://doi.org/10.1097/FPC.0b013e3283476a01

Kuypers DR, Naesens M, Vermeire S, Vanrenterghem Y. The impact of uridine diphosphate–glucuronosyltransferase 1A9 (UGT1A9) gene promoter region single-nucleotide polymorphisms T—275A and C—2152T on early mycophenolic acid dose-interval exposure in de novo renal allograft recipients. *Clinical Pharmacology & Therapeutics.* 2005;78(4):351–361. https://doi.org/10.1016/j.clpt.2005.06.007

Li M, Seiser EL, Baldwin RM, Ramirez J, Ratain MJ, et al. ABC transporter polymorphisms are associated with irinotecan pharmacokinetics and neutropenia. *The Pharmacogenomics Journal.* 2018;18(1):35–42. https://doi.org/10.1038/tpj.2016.75

Li Y, et al. Current trends in drug metabolism and pharmacokinetics. *Acta Pharmaceutica Sinica B.* 2019;9(6):1113–1144. https://doi.org/10.1016/j.apsb.2019.10.001

Lin JH. Pharmacokinetic and pharmacodynamic variability: a daunting challenge in drug therapy. *Current*

Drug Metabolism. 2007;8(2):109–136. https://doi.org/10.2174/138920007779816002

Liu X, Cheng D, Kuang Q, Liu G, Xu W. Association of UGT1A1* 28 polymorphisms with irinotecan-induced toxicities in colorectal cancer: a meta-analysis in Caucasians. *The Pharmacogenomics Journal*. 2014;14(2):120–129. https://doi.org/10.1038/tpj.2013.10

Liu SG, Gao C, Zhang RD, Zhao XX, Cui L, et al. Polymorphisms in methotrexate transporters and their relationship to plasma methotrexate levels, toxicity of high-dose methotrexate, and outcome of pediatric acute lymphoblastic leukemia. *Oncotarget*. 2017;8(23):37761. https://doi.org/10.18632/oncotarget.17781

Lopes GS, Sela Y. Human Genome Project. In: Zeigler-Hill V., Shackelford T. (eds) *Encyclopedia of Personality and Individual Differences*. Springer, Cham; 2017. https://doi.org/10.1007/978-3-319-28099-8_724-1

Ma JD, Lee KC, Kuo GM. Clinical application of pharmacogenomics. *Journal of Pharmacy Practice*. 2012;25(4):417–427. https://doi.org/10.1177/0897190012448309

MacKenzie M, Hall R. Pharmacogenomics and pharmacogenetics for the intensive care unit: a narrative review. Canadian *Journal of Anesthesia/Journal canadien d'anesthésie*. 2017;64(1):45–64. https://doi.org/10.1007/s12630-016-0748-1

Malsagova KA, Butkova TV, Kopylov AT, Izotov AA, Potoldykova NV, et al. Pharmacogenetic testing: a tool for personalized drug therapy optimization. *Pharmaceutics*. 2020; 12(12):1240. https://doi.org/10.3390/pharmaceutics12121240

Manuyakorn, W., et al., Association of HLA genotypes with phenytoin induced severe cutaneous adverse drug reactions in Thai children. *Epilepsy Research*. 2020;162:106321. https://doi.org/10.1016/j.eplepsyres.2020.106321

Marsh S, McLeod HL. Cancer pharmacogenetics. *British Journal Cancer*. 2004;90(1):8–11. https://doi.org/10.1038/sj.bjc.6601487

Matsuoka A, Ando Y. UGT1A1 genotyping for proper use of irinotecan. Rinsho byori. *The Japanese Journal of Clinical Pathology*. 2015;63(7):876–882.

McCormack, M, Alfirevic, A, Bourgeois, S, Farrell, JJ., Kasperavičiūtė, D, et al. HLA-A*3101 and carbamazepine-induced hypersensitivity reactions in Europeans. *New England Journal of Medicine*. 2011;364(12):1134–1143. https://doi.org/10.1056/nejmoa1013297

Mehta TY, Prajapati LM, Mittal B, Joshi CG, Sheth JJ, et al. Association of HLA-B'1502 allele and carbamazepine-induced Stevens. Johnson syndrome among Indians. *Indian Journal of Dermatology, Venereology & Leprology*. 2009;75(6). https://doi.org/10.4103/0378-6323.57718

Meulendijks D, Henricks LM, Sonke GS, Deenen MJ, Froehlich TK, et al. Clinical relevance of DPYD variants c. 1679T> G, c. 1236G> A/HapB3, and c. 1601G> A as predictors of severe fluoropyrimidine-associated toxicity: a systematic review and meta-analysis of individual patient data. *The Lancet Oncology*. 2015;16(16):1639–1650. https://doi.org/10.1016/S1470-2045(15)00286-7

Mochtar CA, Laan W, Van Houwelingen KP, Franke B, De La Rosette JJ, et al. Polymorphisms in the α1A-adrenoceptor gene do not modify the short-and long-term efficacy of α1-adrenoceptor antagonists in the treatment of benign prostatic hyperplasia. *BJU International*. 2006;97(4):852–855. https://doi.org/10.1111/j.1464-410X.2006.05998.x

Mortensen HM, Froment A, Lema G, Bodo JM, Ibrahim M, et al. Characterization of genetic variation and natural selection at the arylamine N-acetyltransferase genes in global human populations. *Pharmacogenomics*. 2011;12(11):1545–1558. https://doi.org/10.2217/pgs.11.88

Mottl AK, Shoham DA, North KE. Angiotensin II type 1 receptor polymorphisms and susceptibility to hypertension: a HuGE review. *Genetics in Medicine*. 2008;10(8):560–574. https://doi.org/10.1097/GIM.0b013e3181809613

Motulsky AG. Drug reactions, enzymes, and biochemical genetics. *Journal of the American Medical Association*. 1957;165(7):835–837.

Motulsky AG, Qi M. Pharmacogenetics, pharmacogenomics and ecogenetics. *Journal of Zhejiang University. Science. B*. 2006;7(2):169–170. https://doi.org/10.1631/jzus.2006.B0169

Nagy M, Attya M, Patrinos GP. Unraveling heterogeneity of the clinical pharmacogenomic guidelines in oncology practice among major regulatory bodies. *Pharmacogenomics*. 2020;21(17):1247–1264. https://doi.org/10.2217/pgs-2020-0056

Noguchi, K, Katayama, K, Sugimoto, Y. Human ABC transporter ABCG2/BCRP expression in chemo resistance: basic and clinical perspectives for molecular cancer therapeutics. *Pharmacogenomics and Personalized Medicine*. 2014;7:-53–64. https://doi.org/10.2147/PGPM.S38295

Ozdemir V, Bertilsson L, Miura J, Carpenter E, Reist C, et al. CYP2D6 genotype in relation to perphenazine concentration and pituitary pharmacodynamic tissue sensitivity in Asians: CYP2D6–serotonin–dopamine crosstalk revisited. *Pharmacogenetics and Genomics*. 2007;17(5):339–347. https://doi.org/10.1097/FPC.0b013e32801a3c10

Pereira NL, Rihal CS, So DY, Rosenberg Y, Lennon RJ, et al. Clopidogrel pharmacogenetics: state-of-the-art review and the TAILOR-PCI Study. *Circulation: Cardiovascular Interventions*. 2019;12(4):007811. https://doi.org/10.1161/CIRCINTERVENTIONS.119.007811

Petersen, M, et al.. Association of beta-adrenergic receptor polymorphisms and mortality in carvedilol-treated chronic heart-failure patients. *British Journal of Clinical Pharmacology*. 2011;71(4): 556–565. https://doi.org/10.1111/j.1365-2125.2010.03868.x

Preissner SC, Hoffmann MF, Preissner R, Dunkel M, Gewiess A, Preissner S. Polymorphic cytochrome P450 enzymes (CYPs) and their role in personalized therapy. *PLoS One*. 2013;8(-12):e82562. https://doi.org/10.1371/journal.pone.0082562

Puangpetch A, Vanwong N, Nuntamool N, Hongkaew Y, Chamnanphon M, Sukasem C. CYP2D6 polymorphisms and their influence on risperidone treatment. *Pharmacogenomics and Personalized Medicine*. 2016;9:131–147. https://doi.org/10.2147/PGPM.S107772

Ragia G, Petridis I, Tavridou A, Christakidis D, Manolopoulos VG. Presence of CYP2C9* 3 allele increases risk for hypoglycemia in Type 2 diabetic patients treated with sulfonylureas. *Pharmacogenomics*. 2009;10(11):1781–1787. https://doi.org/10.2217/pgs.09.96

Raj GM. Pharmacogenetics, pharmacogenomics, and personalized medicine. In: *Introduction to Basics of Pharmacology and Toxicology*. Springer, Singapore; 2019:235–259. https://doi.org/10.1007/978-981-32-9779-1_15

Ramsey LB, Johnson SG, Caudle KE, Haidar CE, Voora D, et al. The Clinical Pharmacogenetics Implementation Consortium guideline for SLCO1B1 and simvastatin-induced myopathy: 2014 update. *Clinical Pharmacology & Therapeutics.* 2014;96(4):423–428. https://doi.org/10.1038/clpt.2014.125

Ramsey LB, Panetta JC, Smith C, Yang W, Fan Y, et al. Genome-wide study of methotrexate clearance replicates SLCO1B1. *Blood, The Journal of the American Society of Hematology.* 2013;121(6):898–904. https://doi.org/10.1182/blood-2012-08-452839

Relling, MV, Gardner, EE, Sandborn, WJ, Schmiegelow, K, Pui, CH, et al. Clinical Pharmacogenetics Implementation Consortium guidelines for thiopurine methyltransferase genotype and thiopurine dosing: 2013 update. *Clinical Pharmacology & Therapeutics.* 2013;93(4):324–325. https://doi.org/10.1038/clpt.2013.4

Robey RW, Pluchino KM, Hall MD, Fojo AT, Bates SE, Gottesman MM. Revisiting the role of ABC transporters in multidrug-resistant cancer. *Nature Reviews Cancer.* 2018;18(7):452–464. https://doi.org/10.1038/s41568-018-0005-8

Roco Á, Cayún J, Contreras S, Stojanova J, Quiñones L. Can pharmacogenetics explain efficacy and safety of cisplatin pharmacotherapy? *Frontiers in Genetics.* 2014;5:391. https://doi.org/10.3389/fgene.2014.00391

Sadhasivam S, Chidambaran V, Zhang X, Meller J, Esslinger H, Zhang K, Martin LJ, McAuliffe J. Opioid-induced respiratory depression: ABCB1 transporter pharmacogenetics. *The Pharmacogenomics Journal.* 2015;15(2):119–126. https://doi.org/10.1038/tpj.2014.56

Scott SA et al. Clinical Pharmacogenetics Implementation Consortium guidelines for CYP2C19 genotype and clopidogrel therapy: 2013 update. *Clinical Pharmacology & Therapeutics.* 2013;94(3): 317–323. https://doi.org/10.1038/clpt.2013.105

Schaeffeler E, Fischer C, Brockmeier D, Wernet D, Moerike K, et al. Comprehensive analysis of thiopurine S-methyltransferase phenotype–genotype correlation in a large population of German-Caucasians and identification of novel TPMT variants. *Pharmacogenetics and Genomics.* 2004;14(7):407–417. https://doi.org/10.1097/01.fpc.0000114745.08559.db

Shnayder NA, Petrova MM, Shesternya PA, Savinova AV, Bochanova EN, et al. Using pharmacogenetics of direct oral anticoagulants to predict changes in their pharmacokinetics and the risk of adverse drug reactions. *Biomedicines.* 2021;9(-5):451. https://doi.org/10.3390/biomedicines9050451

Siest G, Jeannesson E, Berrahmoune H, Maumus S, Marteau JB, et al. Pharmacogenomics and drug response in cardiovascular disorders. *Pharmacogenomics.* 2004;5(7):779–802. https://doi.org/10.1517/14622416.5.7.779

Sim SC, Kacevska M, Ingelman-Sundberg M. Pharmacogenomics of drug-metabolizing enzymes: a recent update on clinical implications and endogenous effects. *The Pharmacogenomics Journal.* 2013;13(1):1–1. https://doi.org/10.1038/tpj.2012.45

Sissung TM, Mross K, Steinberg SM, Behringer D, Figg WD, et al. Association of ABCB1 genotypes with paclitaxel-mediated peripheral neuropathy and neutropenia. *European Journal of Cancer.* 2006;42(17):2893–2896. https://doi.org/10.1016/j.ejca.2006.06.017

Soria ML, Berga JK, Catalán SB, Payá JM, Mateu LP, Torres NJ. Genetic polymorphisms and individualized tacrolimus dosing. *Transplantation Proceedings.* 2010;42(8):3031–3033. https://doi.org/10.1016/j.transproceed.2010.08.001

Tantisira KG, Hwang ES, Raby BA, Silverman ES, Lake SL et al. TBX21: a functional variant predicts improvement in asthma with the use of inhaled corticosteroids. *Proceedings of the National Academy of Sciences.* 2004;101(52):18099–18104. https://doi.org/10.1073/pnas.0408532102

Tantisira KG, Lake S, Silverman ES, Palmer LJ, Lazarus R, et al. Corticosteroid pharmacogenetics: association of sequence variants in CRHR1 with improved lung function in asthmatics treated with inhaled corticosteroids. *Human Molecular Genetics.* 2004;13(13):1353–1359. https://doi.org/10.1093/hmg/ddh149

Tassaneeyakul W, Tiamkao S, Jantararoungtong T, Chen P, Lin SY, et al. Association between HLA-B* 1502 and carbamazepine-induced severe cutaneous adverse drug reactions in a Thai population. *Epilepsia.* 2010;51(5):926–930. https://doi.org/10.1111/j.1528-1167.2010.02533.x

Tavira B, Garciá EC, Díaz-Corte C, Ortega F, Arias M, et al. Pharmacogenetics of tacrolimus after renal transplantation: analysis of polymorphisms in genes encoding 16 drug metabolizing enzymes. *Clinical Chemistry and Laboratory Medicine.* 2011;49(5):825–833. https://doi.org/10.1515/CCLM.2011.143

Tempark, T., John, S., Rerknimitr, P., Satapornpong, P., & Sukasem, C. Drug-Induced Severe Cutaneous Adverse Reactions: Insights Into Clinical Presentation, Immunopathogenesis, Diagnostic Methods, Treatment, and Pharmacogenomics. *Frontiers in Pharmacology.* 2022; 13:832048. https://doi.org/10.3389/fphar.2022.832048

Tomalik-Scharte D, Lazar A, Fuhr U, Kirchheiner J. The clinical role of genetic polymorphisms in drug-metabolizing enzymes. *The Pharmacogenomics Journal.* 2008;8(1): 4–15. https://doi.org/10.1038/sj.tpj.6500462

Udagawa C, Zembutsu H. Pharmacogenetics for severe adverse drug reactions induced by molecular-targeted therapy. Cancer Science. 2020;111(10):3445–3457. https://doi.org/10.1111/cas.14609

Umamaheswaran G, Praveen RG, Damodaran SE, Das AK, Adithan C. Influence of SLC22A1 rs622342 genetic polymorphism on metformin response in South Indian type 2 diabetes mellitus patients. *Clinical and Experimental Medicine.* 2015;15(4):511–517. https://doi.org/10.1007/s10238-014-0322-5

Van Nguyen D, Chu HC, Van Nguyen D, Phan MH, Craig T, et al. HLA-B* 1502 and carbamazepine-induced severe cutaneous adverse drug reactions in Vietnamese. *Asia Pacific Allergy.* 2015;5(2):68–77. https://doi.org/10.5415/apallergy.2015.5.2.68

Veal GJ, Cole M, Chinnaswamy G, Sludden J, Jamieson D, et al. Cyclophosphamide pharmacokinetics and pharmacogenetics in children with B-cell non-Hodgkin's lymphoma. *European Journal of Cancer.* 2016;55:56–64. https://doi.org/10.1016/j.ejca.2015.12.007

Wang L, Yu L, Zhang AP, Fang C, Du J, et al. Serum prolactin levels, plasma risperidone levels, polymorphism of cytochrome P450 2D6 and clinical response in patients with schizophrenia. *Journal of Psychopharmacology.* 2007;21(8):837–842. https://doi.org/10.1177/0269881107077357

Wang H, Zhang Z, Han S, Lu Y, Feng F, Yuan J. CYP1A2 rs762551 polymorphism contributes to cancer susceptibility: a meta-analysis from 19 case-control studies. *BMC Cancer.* 2012;12:528. https://doi.org/10.1186/1471-2407-12-528

Wang Q, Zhou JQ, Zhou LM, Chen ZY, Fang ZY, et al. Association between HLA-B* 1502 allele and carbamazepine-induced severe cutaneous adverse reactions in Han people of southern China mainland. *Seizure.* 2011;20(6):446–448. https://doi.org/10.1016/j.seizure.2011.02.003.

Ward, NC, Watts, GF, Eckel, RH. Statin toxicity: mechanistic insights and clinical implications. *Circulation Research.* 2019;124(2): 328–350. https://doi.org/10.1161/CIRCRESAHA.118.312782

Weeke P, Roden DM. Pharmacogenomics and cardiovascular disease. *Current Cardiology Reports.* 2013;15(7):376. https://doi.org/10.1007/s11886-013-0376-0

Wilson JR, Shuey MM, Brown NJ, Devin JK. Hypertension and type 2 diabetes are associated with decreased inhibition of dipeptidyl peptidase-4 by sitagliptin. *Journal of the Endocrine Society.* 2017;1(9):1168–1178. https://doi.org/10.1210/js.2017-00312

Xia R, Zhu X, Chen Z, He Y, Kang L, Yuan D, Jin T, Wang L. Genetic polymorphisms of the drug-metabolizing enzyme cytochrome P450 2A6 in a Tibetan Chinese population. *International Journal of Clinical and Experimental Pathology.* 2018;11(10):5024–5033.

Xie F, Chan JC, Ma RC. Precision medicine in diabetes prevention, classification and management. *Journal of Diabetes Investigation.* 2018;9(5):998–1015. https://doi.org/10.1111/jdi.12830

Xu J, Boström AE, Saeed M, Dubey RK, Waeber G, et al. A genetic variant in the catechol-O-methyl transferase (COMT) gene is related to age-dependent differences in the therapeutic effect of calcium-channel blockers. *Medicine.* 2017;96(30). https://doi.org/10.1097/MD.0000000000007029

Yamamoto PA, Conchon Costa AC, Lauretti GR, De Moraes NV. Pharmacogenomics in chronic pain therapy: from disease to treatment and challenges for clinical practice. *Pharmacogenomics.* 2019;20(13):971–982. https://doi.org/10.2217/pgs-2019-0066

Yampayon K, Sukasem C, Limwongse C, Chinvarun Y, Tempark T, et al. Influence of genetic and non-genetic factors on phenytoin-induced severe cutaneous adverse drug reactions. *European Journal of Clinical Pharmacology.* 2017;73(7):855–865. https://doi.org/10.1007/s00228-017-2250-2

Yoshida K, Müller DJ. Pharmacogenetics of antipsychotic drug treatment: update and clinical implications. *Molecular Neuropsychiatry.* 2019;5(1):1–26. https://doi.org/10.1159/000492332

Yuliwulandari R, Kristin E, Prayuni K, Sachrowardi Q, Suyatna FD, et al. Association of the HLA-B alleles with carbamazepine-induced Stevens–Johnson syndrome/toxic epidermal necrolysis in the Javanese and Sundanese population of Indonesia: the important role of the HLA-B75 serotype. *Pharmacogenomics.* 2017;18(18):1643–1648. https://doi.org/10.2217/pgs-2017-0103

Zahra MA, Kandeel M, Aldossary SA, Al-Taher A. Study on genotyping polymorphism and sequencing of N-acetyltransferase 2 (NAT2) among Al-Ahsa population. *BioMed Research International.* 2020:1–9. https://doi.org/10.1155/2020/8765347

Zanger UM, Schwab M. Cytochrome P450 enzymes in drug metabolism: regulation of gene expression, enzyme activities, and impact of genetic variation. *Pharmacology & Therapeutics.* 2013;138(1):103–141. https://doi.org/10.1016/j.pharmthera.2012.12.007

Zhao S, Zhang W, Nie X. Association of β2-adrenergic receptor gene polymorphisms (rs1042713, rs1042714, rs1042711) with asthma risk: a systematic review and updated meta-analysis. *BMC Pulmonary Medicine.* 2019;19(1):1–7. https://doi.org/10.1186/s12890-019-0962-z

Zhu X, Liu Y, Chen G, Guo Q, Zhang Z, et al. Association between NAT2 polymorphisms and acute leukemia risk: a meta-analysis. *Medicine (Baltimore).* 2019;98(12):e14942. https://doi.org/10.1097/MD.0000000000014942

3

Diabetes and Role of SNPs in Diabetes Susceptibility

Johra Khan
Department of Laboratory Sciences, College of Applied Medical Sciences, Majmaah University, Majmaah, Saudi Arabia

Sayequa Dandoti
Department of Biology, Deanship of Preparatory Year, Imam Abdulrahman Bin Faisal University, Dammam, Saudi Arabia

CONTENTS

DOI: 10.1201/9781003246244-3

ABSTRACT When pancreas of a person does not produce enough insulin or the tissue or cells show resistance to insulin, this chronic disease condition is known as diabetes. It's a complex chronic disease that is responsible for 1.6 million deaths around the globe each year. There are many factors responsible for the development of this disease, such as genetic, environmental and mental stresses. All of them are correlated to each other at molecular level. In this chapter, we try to understand the types and role of SNPs in diabetes predisposition.

KEYWORDS: *Diabetes Mellitus; Polymorphism; β-Cell Dysfunction; HLA; Risk Factors; RNA Damage; SNPs.*

3.1 Introduction

Diabetes is a group of metabolic diseases that occur due to hyperglycemia which results from either defect in insulin secretion, insulin resistance by target tissues, or both. The three common forms of diabetes are type 1 diabetes, type 2 diabetes and gestational diabetes (Association, 2020). The most common of these is type 2 diabetes, which will become the seventh leading cause of death worldwide by 2030 (Choubey et al., 2020). Around the world, about 422 million people have diabetes and 1.6 million die every year due to this disease (Maniruzzaman et al., 2020).

Single nucleotide polymorphism (SNP) is a powerful tool in search of disease susceptibility loci in whole genome. SNPs are normally found throughout human DNA, and they occur at a frequency of 1 in 1,000 nucleotides (Dayeh et al., 2013). Some SNPs are unique from individual to individual where many are commonly found in a disease condition. Till now, around 100 million SNPs have been discovered by different scientists around the world, and not all of them affect the susceptibility to disease (Ban et al., 2010). Most SNPs have no effect on our health, but some of them when occur within a gene or in a regulatory region can play an important role in disease occurrence by affecting gene function. SNPs can also be used in predicting drug responses, risk of disease occurrence and finding susceptibility to environmental factors (Fontaine-Bisson et al., 2010).

Numerous studies established an association of genes with all types of diabetes. There are more than 1,000 different SNPs recorded, and several molecular biology methods have been developed, such as PCR, PCR-RFLP and AS-PCR for the analysis of SNPs associated with diabetes (de Miguel-Yanes et al., 2011). In this chapter, we discuss different SNPs associated with type 1, type 2 and gestational diabetes with different methods developed for prediction and genotyping of diabetes-associated SNPs.

3.2 Types

Diabetes in global scenario is expected to increase up to 693 million by 2045, and it is estimated that around 49.7%, which is almost half of the total population, are living with undiagnosed diabetes (Robert & Al Dawish, 2020). The basic classification of diabetes was first proposed by the American Diabetes Association (ADA) in the year 1997 into three basic types: 1) type 1, 2) type 2, and 3) gestational diabetes mellitus and some authors consider fourth as specific type of diabetes due to other causes (Barhate & Kulkarni, 2018). An acceleratory hypothesis, proposed by Wilkin, argues that type 1 and type 2 diabetes are both the same disorder occurring due to insulin defect, which differ only by tempo. A faster tempo expresses a more susceptible genotype and an early predisposition. This hypothesis related insulin resistance to increased incidences of overweight and obesity, which is the core reason for all increasing cases of diabetes (DiMenna & Arad, 2018).

Ninety percent of diabetic populations in different countries are having type 2 diabetes, which is a highly heterogeneous category (Buryk, 2014; Ahlqvist et al. 2018). A recent classification of diabetes divides it into five clusters based on risk factors:

1. **Autoimmune Diabetes (Type 1 Diabetes)**: People with insulin deficiency with the presence of glutamate decarboxylase antibodies. Six percent of the total diabetic population falls under this cluster.
2. **Insulin-Deficient Diabetes**: Eighteen percent people with defect in insulin production and having high risk of diabetic retinopathy.
3. **Insulin-Resistant Diabetes**: People who produce insulin, but not responding to it correctly. Mostly found in patients with high BMI or overweight issues. The frequency of its occurrence is around 15% of the total diabetic patients. They have a high risk of developing diabetic kidney disease.
4. **Obesity-Related Diabetes**: Patients with obesity, but no insulin resistance. Approximately 22% patients come under this cluster.
5. **Age-Related Diabetes**: Due to older age issues with glucose control. The highest percentage (39%) of diabetic patients has age-related diabetes.

3.2.1 Risk Factors

The knowledge of risk factors associated with diabetes becomes more important when 1 out of 4 does not know that they are living with diabetes. Risk factors for diabetes depend on the type of diabetes. In T1D, body produces no insulin and 5% of all diabetic patients suffer from this condition. T1D risk increases if someone in family, a parent or a sibling has T1D (Oram et al., 2014). Exposures to viral diseases or the presence of damaged immune cells (autoantibody) is associated with the risk factors for T1D. Geographical locations are also found to play an important role in the increased risk of T1D; for example, people of Finland and Sweden are found to have high percentage of T1D in comparison with other types of diabetes (Borchers et al., 2010).

The risk of developing T2D depends upon two basic factors: genetics and lifestyle. A person of age 45 becomes a candidate of T2D, especially people with ethnic background from Africa, Alaska Native, Indian, Asian American, Hispanic/Latino, Native Hawaiian or Pacific Islander (Patterson et al., 2019). Females who had gestational diabetes during pregnancy or gave birth to a baby with weight more than 9 pounds are at a high risk of developing T2D. More than 80% of overweight people have T2D (Fujioka, 2015). A sedate or inactive lifestyle increases manifold the risk of developing T2D, but if you have a sedate lifestyle with obesity, then the person becomes prime candidate to develop T2D (Strasser, 2013). The risk of developing gestational diabetes increases with advanced age ≥ 35 years, family history of diabetes, overweight or obesity with short stature <1.50 m, excess fat deposition in central body area (around waist and hip), excessive weight gain during pregnancy, hypertension, etc., including the most common factor of all type of diabetes, i.e., sedentary lifestyle (Pons et al., 2015).

3.3 Classification and Challenges

To classify an individual with diabetes depends on the characteristics present at the time of diagnosis; many times, an individual doesn't fit into a single class of diabetes. The broad classification of diabetes can be done as diabetes mellitus and diabetes insipidus. The classification of diabetes mellitus is clearly explained in Figure 3.1. Diabetes insipidus is caused due to the deficiency of the antidiuretic hormone arginine vasopressin (AVP). Diabetes mellitus is the most common form of diabetes. The general classification of diabetes divides diabetes into four types based on the factors responsible for the development of the disease condition (Association, 2015).

1. Type 1 diabetes mellitus (includes idiopathic type).
2. Type 2 diabetes mellitus.
3. Gestational diabetes mellitus (GDM).
4. Other specific types of diabetes caused or associated with certain specific conditions, pathologies and/or disorders.

3.3.1 Type 1 Diabetes Mellitus

3.3.1.1 Autoimmune Type 1 Diabetes

A condition that develops due to destruction of pancreatic β-cells as an effect of T cell-mediated inflammatory response (insulitis) or a humoral (B cell) response. Of young individuals in the age group of 1–14 years, 80%–90% are diagnosed with this type (C. Wang et al., 2010). The autoimmune background is considered as the hallmark of T1D. It includes islet cell autoantibodies, autoantibodies to insulin (IAA), glutamic acid decarboxylase (GAD, GAD65), protein tyrosine phosphatase (IA2 and IA2β) and zinc transporter protein (ZnT-8A). Their presence can be detected in the serum of the patient's blood (Donath et al., 2010). Research shows T1D has strong HLA relations, with linkage to DR and DQ genes. HLA-DR/DQ alleles can be either predisposing to or protective of T1D condition (Donath et al., 2010; Halban et al., 2010).

3.3.1.2 Idiopathic Type 1 Diabetes

As the name suggests, it's a rare type of T1D with unknown origin. It is mostly found in Asian and African populations suffering from erratic degrees of insulin deficiency and episodic ketoacidosis (Schenck et al., 2018). Fulminant type 1 diabetes is a non-immune-mediated T1D with the development of ketoacidosis with hyperglycemia (≥288 mg/dL) followed by untraceable serum C-peptide level, indicating endogenous insulin production (Okamoto et al., 2016). It is mainly found in East Asian countries with 20% acute onset of T1D. The known factors for the onset include both genetics and viral response: The anti-viral immune response of body may initiate beta-cell destruction via hastened immune reaction and no detectable autoantibodies in serum against pancreatic β-cells (Imagawa et al., 2012).

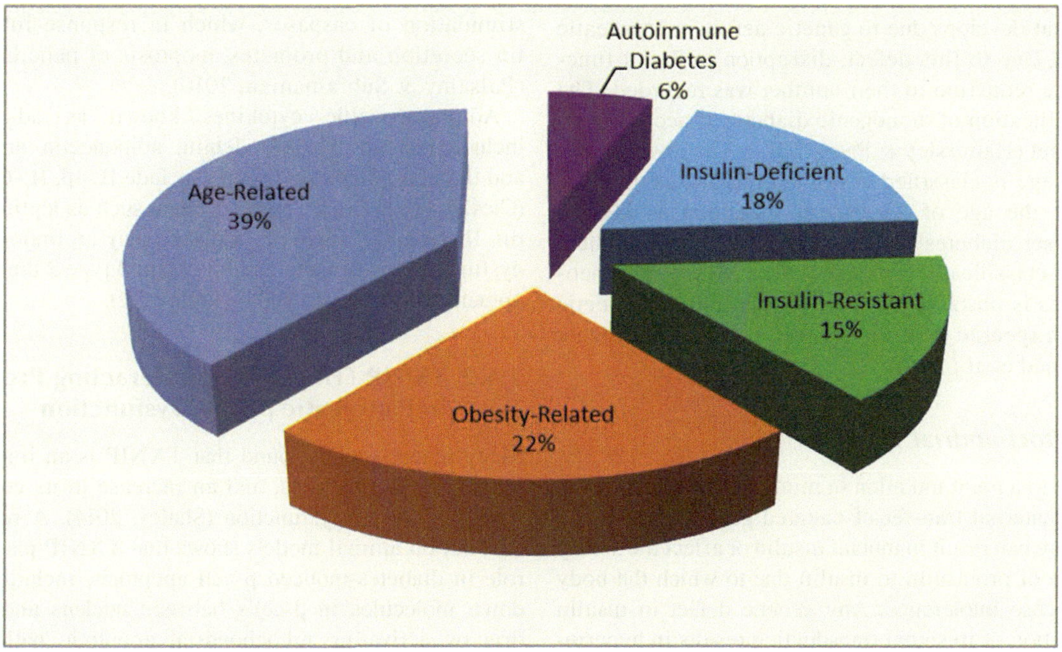

FIGURE 3.1 Frequency of five different clusters of diabetes with percentage in total diabetic patients.

3.3.2 Type 2 Diabetes Mellitus

This is the common type of diabetes found in the age group of 20–80 years. According to IDF 2019 report, 14 million more men than women in the age group of 40–59 will increase in number, making a total of 592 million diabetes cases by the year 2035, with an annual increase of 10.1% prevalence rate (Saeedi et al., 2019). T2D has much manifestation, which includes obesity, increased triglyceride level and decreased HDL cholesterol, nephropathy, essential hypertension, non-alcoholic fatty liver, systemic inflammation, postprandial lipemia and remnant lipoprotein accumulation (Thomas et al., 2019).

3.3.3 Gestational Diabetes Mellitus (GDM)

GDM is defined as the recognition of glucose intolerance (any degree) with onset or during pregnancy. Depending on the population studied, GDM prevalence can range between 1% and 14% of all pregnancy and around 200,000 cases of GDM annually. A female with marked obesity, strong family history of diabetes and previous history of GDM is at a high risk of developing GDM. They should undergo glucose testing at their first prenatal visit to hospital. In high-risk patients, glucose testing should be redone between 24 and 28 weeks of gestation, if not detected in the early stage. A fasting glucose level >126 mg/dL or a random glucose level >200 mg/dL indicates the threshold to be diagnosed as diabetes. Further testing for confirmation can be OGTT (oral glucose tolerance test) or GCT (glucose challenge test), and a result of a glucose threshold level >140 mg/dL (7.8 mmol/L) identifies GDM (Association 2004).

3.3.4 Other Specific Types

3.3.4.1 Monogenic Diabetes

It's a type that develops due to genetic defect in pancreatic β-cells gene. Due to this defect, disruption in β-cell function or a huge reduction in their number was recorded. The further classification of monogenic diabetes is according to age of its onset (Hattersley & Patel, 2017). The onset before 6 months of age is classified as neonatal diabetes, and the onset before the age of 25 years is classified as MODY (maturity-onset diabetes of the young). Many researchers consider the classification of diabetes as MODY and neonatal diabetes is obsolete and it is used in relating genetic etiology with specific treatment implications of monogenic diabetes (Amed et al., 2010).

3.3.4.2 Mitochondrial Diabetes

It occurs due to a point mutation in mitochondrial DNA associated with maternal transfer of mutated gene and deafness. Mutations here can result in mutant insulin or affect the ability of conversion of proinsulin to insulin due to which the body develops glucose intolerance. Any genetic defect in insulin receptor function or in signal transduction results in hyperinsulinemia, later causing hyperglycemia and serving as diabetes (Mazzaccara et al., 2012).

3.3.4.3 Diabetes of Exocrine Pancreas

Diffused injury due to damage of β-cells of the pancreas also causes diabetes; this damage can be due to trauma, pancreatic carcinoma, pancreatic infections and pancreatectomy. Accumulation of fat, atrophy of exocrine pancreas or pancreatic steatosis and cystic fibrosis can also lead to diabetes. Table 3.1 shows the classification and diagnostic categories of diabetes (Rodriguez-Calvo et al., 2014).

3.4 Genetic Background

Like several other human diseases, diabetes mellitus has complex patterns of heredity traits, which get superimposed due to the effect of diet, level of exercise and other environmental factors. The two most common forms of diabetes T1D and T2D occur by a combination of both genetic and environmental factors, whereas diabetes due to mutation in mitochondrial DNA and MODY are directly inherited forms (Ozougwu et al., 2013).

In diabetes, the two most studied pathways are (1) β-cell dysfunction pathway and (2) insulin signaling pathway.

3.4.1 β-Cell Dysfunction Pathway

The role of β-cell dysfunction in T1D and T2D progression is different. In the case of T1D, the β-cells get distorted due to autoimmune-mediated apoptosis, causing loss of insulin production. The important role in this process is played by inflammatory cytokines (Eguchi et al., 2012). An increase in circulating cytokines, free fatty acids (FFAs) and continuous hyperglycemia play a crucial role in the β-cell dysfunction and β-cell mass decline, which results in the progression of T2D (Giacca et al., 2011). A prolonged exposure of β-cell to the risk factors encourages unnecessary production of reactive oxygen species (ROS) and stimulation of caspases, which in response inhibits insulin secretion and promotes apoptosis of pancreatic β-cells (Palsamy & Subramanian, 2010).

Adipose-specific cytokines known as adipocytokines include resistin, visfatin, leptin, adiponectin and omentin, and the non-adipose cytokines include IL-1β, IL-6 and TNFα (Cieślak et al., 2015). Some of them such as leptin, adiponectin, IL-1β and resistin are found to play an important role in dysfunctioning of pancreatic β-cell and type 2 diabetes development (Bruun et al., 2004) (Figure 3.2).

3.4.2 TXNIP (Thioredoxin-Interacting Protein) in Pancreatic β-Cell Dysfunction

Many recent studies found that TXNIP is an important factor for β-cell functions, and an increase in its concentration results in β-cell dysfunction (Shalev, 2014). A research carried out on animal models shows that TXNIP plays a critical role in diabetes-induced β-cell apoptosis, including shutting down molecules in β-cells between nucleus and mitochondria by activating mitochondrial apoptotic pathway (Zhao et al., 2015). TXNIP is also responsible for glucotoxicity, and higher expressions suppress the gene function responsible for

TABLE 3.1

Classification of Diabetes with Diagnostic Categories

Type of Diabetes	Factors Responsible for Onset	Diagnostic Categories
Type 1 diabetes	Autoimmune, idiopathic	Two methods for diagnosis are commonly used: 1. One-step method: OGTT measurement without prior screening of serum and plasma glucose. 2. Two-step method: Initial screening of serum and plasma glucose after 2 hours of 70 g oral glucose load (OGL) followed by confirmatory test OGTT (oral glucose tolerance test).
Type 2 diabetes	Insulin resistance, obesity	
Gestational diabetes	Family history, obesity	
Other specific types		
a. Genetic defects of cell function	Mutation in chromosome no. 12, mutation in glucokinase gene on chromosome 7p, mutation in HNF-4α (chromosome 20q), gene IPF-1 (chromosome 12), HNF-1α (MODY1-14) (chromosome no. 2,7,13), neonatal diabetes mutation in HYMAI (hydatidiform mole associated and imprinted) gene, zinc finger protection (Mazzaccara et al.) gene, PACAP1 (pituitary adenylate cyclase-activating polypeptide 1) gene, ATP-binding cassette subfamily C member 8 (ABCC8) gene (sulfonylurea receptor 1 (SUR1) gene) and potassium voltage-gated channel subfamily J member 11 (KCNJ11 or Kir6)	
b. Genetic defects in insulin action	Type A insulin resistance, leprechaunism, Rabson-Mendenhall syndrome, lipoatrophic diabetes	Reference range for confirmation of diabetes: 1. Level of plasma glucose concentration 200 mg/dL (11.1 mmol/L) with symptoms of diabetes including polyuria, polydipsia and unexplained weight loss. 2. FPG 126 mg/dL (7.0 mmol/L) with fasting for at least 8 hours. 3. Two-hour post-load glucose 200 mg/dL (11.1 mmol/L) during an OGTT using 75 g anhydrous glucose dissolved in water.
c. Diseases of the exocrine pancreas	Pancreatitis, trauma/pancreatectomy, neoplasia, cystic fibrosis, hemochromatosis, fibrocalculous pancreatopathy	
d. Endocrinopathies	Acromegaly, Cushing's syndrome, glucagonoma, pheochromocytoma, hyperthyroidism, somatostatinoma, aldosteronoma	
e. Drug-induced diabetes	Vacor, pentamidine, nicotinic acid, glucocorticoids, thyroid hormone, diazoxide, adrenergic agonists, thiazides, Dilantin, interferon	
f. Infections	Congenital rubella, cytomegalovirus	
g. Uncommon forms of immune-mediated diabetes	"Stiff-man" syndrome, anti-insulin receptor antibodies	
h. Other genetic syndromes sometimes associated with diabetes	Down's syndrome, Klinefelter's syndrome, Turner's syndrome, Wolfram's syndrome, Friedreich's ataxia, Huntington's chorea, Laurence-Moon-Biedl syndrome, Myotonic dystrophy, porphyria, Prader-Willi syndrome	
i. Gestational diabetes mellitus (GDM)	Glucose intolerance	

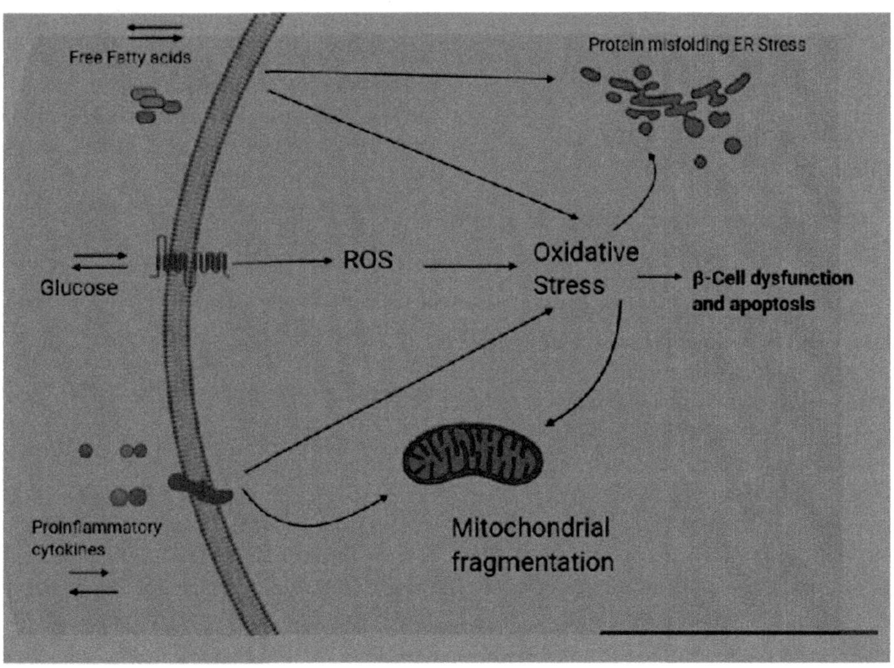

FIGURE 3.2 Mechanism of β-cell dysfunction promoted by cytokines, high glucose and free fatty acids.

the regulation of insulin production in β-cells. TXNIP is also found to induce the expression of miR-204 that inhibits secretion of insulin by targeting the expression of MAFA, which is a transcription factor for Pro-inculin (Shalev, 2008). The sequence of TXNIP-miR204-MAFA creates a sequence of events in the pathogenesis of β-cell dysfunction (Chen et al., 2008).

3.4.3 Insulin Signaling Pathway in Diabetes

The insulin signaling pathway begins with binding of insulin to insulin receptors, which in response trigger phosphorylation of insulin receptor substrate (IRS) creating a docking site at the membrane for PI3K (Mima et al., 2011) which helps in phosphorylation of lipid phosphatidylinositol 4,5-bisphosphate (PIP2) (Kubota et al., 2016) that activates phosphoinositide which is a protein-dependent kinase 1 (PDK) and recruits protein kinase B or Akt. Akt-mediated phosphorylation of AS160 (Rab GTPase-activating protein) helps in the absorption of glucose in muscles and adipocyte cells, which allow translocation of GLUT4 to plasma membrane (Kubota et al., 2016).

mTOR is a main regulator of Akt and p70s6k that help in the growth and survival of cell. mTORC1 is recognized by a raptor, which identifies TOR signaling motifs (Zeng et al., 2007). mTORC1 is found to phosphorylate p70s6k at threonine 389 (Thr389) in response to which PI3K signaling is inhibited. mTORC1 is also responsible for the phosphorylation of 4E that prevents the connection of 4E-BP to eukaryotic initiation factor eIF, which in turn controls many mRNAs responsible for cell survival and growth-controlling mechanisms (Demeulder et al., 2013). An increase in the production of any such mRNAs encoding CCAAT with enhancer-binding proteins (C/EBPs: C/EPBα, C/EPBβ and C/EPBδ) and adipocyte differentiation transcription factors results in adipogenesis, the outcome of which is seen as obesity and insulin resistance (Dal Col et al., 2008).

3.5 Polymorphisms and Susceptibility to Diabetes Disease

As discussed above, diabetes is a complex metabolic syndrome affecting millions of people around the world. It is a rapidly growing public health concern. Different factors such as genetic, epigenetic, developmental and environmental factors are involved; among them, genetic components are found to be crucial, playing a role in the initiation and development of diabetes. The interactivity between substantial genetic background and abruptly changing environment has led to an increased prevalence of diabetes in recent decades (Qi et al., 2006).

Human genome has gigantic amount of genetic variations, and as mentioned in previous chapters, SNPs are the most common type of genetic variations (Buraczynska et al., 2011; Unoki et al., 2008). These SNPs may be unique or may occur in many individuals. The function of a gene is affected when SNPs occur in that gene or in a regulatory region of that gene, thus playing a direct role in disease susceptibility (Buraczynska et al., 2011). Therefore, these SNPs have proven to be important in human health studies. The role played by SNPs in modulating the risk of diabetes is very distinctive. In this chapter, we would be discussing the association of SNPs with T1D, T2D and gestational diabetes (Törn et al., 2015).

Several genome-wide association studies (GWAS) have established the association of polymorphisms of various genes in modulating the risk of diabetes in almost all the populations around the world. The identification of SNPs and detailed knowledge about the relationship between these variant genes' activity and the risk of disease is important to translate them into novel therapeutic targets (Nair et al., 2010). The following SNPs have been studied and are still undergoing research to further elucidate their association with diabetes.

3.5.1 T1D

3.5.1.1 SNPs in Diabetes Type 1

Type 1 diabetes (T1D) is an immune-associated or immune-mediated destruction of pancreatic β-cells. T1D was largely found in children and adolescents, but it's found that symptomatic onset is not restricted to any particular age group (Plagnol et al., 2011). The classic trio (polydipsia, polyuria and polyphagia) symptoms for the onset of diabetes along with overt hyperglycemia with immediate exogenous insulin need are diagnostic symptoms of T1D in children and adolescents. As the occurrence of T1D is polygenic in nature, till now around 60 genes and 40 loci have been known to affect its susceptibility. The HLA region loci are found to affect susceptibility the most, and HLA class II is the most associated one of HLA type (P. Wang et al., 2010).

The rate of T1D diversity is linked to different patterns of HLA genotype in a population. Many studies showed an association between HLA class II DR, and DP and DQ genotypes are related to many factors like environment and ethnicities. DRB103:-01-DQB102:01 and DRB04:01-DQB103:02 are high-risk HLA haplotypes, which are found high in African American population, but very low in Japan and Southeast Asian ethnic groups. Haplotypes DRB104:05-DQB104:01 and DRB109:01-DQB103:03 were found high in Japanese and Korean, whereas DRB104:05-DQB104:01 and DRB109:01-DQB103:03 were reported in Arab ethnic group (McKinney et al., 2013). TPA1 gene (transporter associated with antigen processing), which is located in the class II region between DQB1 and DPB1, is associated with the process and preservation of endogenous antigen. Deficiency and mutation of TAP1 gene create a barrier in the processing and transpiration of endogenous antigen, resulting in an autoimmune disease such as T1D.

3.5.2 Type 2 Diabetes

Caused due to the defects in insulin function or insulin secretion or both, T2D is found to be the most common among various types of diabetes. T2D is developed as a result of interaction between environmental factors and genetic component. Environmental factors such as obesity, lifestyle, weight, nutritional factors, toxins and stress are known to influence the development of T2D. However, even with the same exposure to same environmental factors, everyone is not affected in the same way. Some people are at more risk than others due to the variation in their genes. Multiple genes are involved in its onset and development. Insulin resistance has been proposed to be a major driver of progression to T2D. However, most of the validated genetic variants are involved in β-cell function.

Genetic predisposition was found to play a key role in T2D, and GWAS have identified multiple SNPs linked with T2D in various ethnic groups (Yamauchi et al., 2010). Further investigation is needed to find the contribution of non-genetic factors such as age, gender, BMI, smoking, drinking and ethnicity to the risk of T2D (Basile et al., 2014). The following are the genes found most mutated in T2D and thus assumed to have a role in higher risk of and susceptibility to T2D. The association of SNPs in these genes with T2D had consistently been demonstrated in multiple populations (Imamura et al., 2016).

It is reported that if a person has a genotype for all these SNPs, then he/she is estimated to be at double risk of acquiring T2D.

3.5.2.1 *TCF7L2*

The chromosome location of TCF7L2 (transcription factor 7-like 2) is 10q25.2-q25.3. It produces a high-mobility group (HMG) box-containing transcription factor composed of 217,432 bases and 619 amino acids, having 67919Da. TCF7L2 is involved in Wnt signaling pathway (Tsai, 2011).

TCF7L2 has the largest substantial effect on T2D susceptibility compared to various T2D susceptibility genes. Variation in TCF7L2 is involved in insulin secretion, synthesis and processing. Alteration in TCF7L2 expression affects pancreatic islet function (possibly by dysregulating proglucagon gene expression), causing reduced insulin secretion and increased risk of type 2 diabetes. TCF7L2 polymorphism is associated with impaired proinsulin conversion. However, Noberga (2013) tried to drive attention away from pancreas and stated that the physiological role of TCF7L2 in other tissues (such as brain, liver, skeletal muscle, fat and bones) might be the reason underlying the association of genetic variants with T2D risk (Palmer et al., 2011).

The SNP TCF7L2 (rs7903146: C>T or C>G in intron region) is extensively studied and associated with T2D risk. It is reported in the GWAS of Wellcome Trust Case Control Consortium (2012) that two SNPs rs7903146 and rs4506565 have 92% correlation and have approximately equivalent power to estimate T2D risk. TCF7L2 SNPs have been shown to be associated with type 2 diabetes mellitus (T2MD) in multiple ethnic groups (Huang et al., 2012). A study on T2D pathogenesis and TCF7L2 variations found association of six SNPs of TCF7L2 (rs7903146, rs4506565, rs12255372; G>A/G>T) and TCF7L2 (rs11196205, rs7901695; T>C, rs7895340 (G>A)) with T2D; however, it couldn't establish this relationship with other two SNPs rs290487 (C>T) and rs11196218. TCF7L2 rs12255372 [OR=2.082, 95% CI=1.112–3.898, p=0.022] and rs7903146 [OR=1.589, 95% CI=0.978–2.582, p=0.061] were associated with T2D (Tabara et al., 2009). Another study on T2D pathogenesis and TCF7L2 variations found an association of six SNPs of TCF7L2 (rs7903146, rs12255372, rs11196205, rs7901695, rs7895340 and rs4506565) with T2D; however, it couldn't establish this relationship with other two SNPs (rs290487 and rs11196218). TCF7L2 (rs12255372; G>T) polymorphism showed the highest OR for type 2 diabetes [OR=1.714, 95% CI=1.298–2.263] (Isakova et al., 2019b).

3.5.2.2 *IGF2BP2*

IGF2BP2 (insulin-like growth factor 2 mRNA-binding protein 2) is located at chromosome 3q27.2, containing 181,926 bases. It codes for a protein that regulates IGF2 (insulin-like growth factor 2) mRNA translation by binding to its 5′-UTR. The protein it translates into is termed as insulin-like growth factor 2 mRNA-binding protein 2, and it contains 599 amino acids. This gene plays a crucial role in metabolism, and thus, variation in this gene is associated with susceptibility to diabetes (diabetes mellitus; non-insulin-dependent and rare diabetes mellitus type 2). rs4402960 polymorphism is with a G>T substitution in intron 2, rs6777038 has C>T, and rs16860234 has A>C substitutions (Nemr et al., 2012).

The SNP of IGF2BP2 (rs4402960) has been associated with an increased risk of T2D. With IGF2BP2 rs4402960 variant, a decrease in insulin secretion was also evident, which points toward susceptibility to T2D (Kommoju et al., 2013). Many studies have been conducted to establish solidified association between SNP (rs4402960) and T2D, but the results of different studies are found to be inconsistent. GWAS found that IGF2BP2 rs4402960 and rs1470579 polymorphisms were associated with the risk of type 2 diabetes mellitus (T2DM); thus, IGF2BP2 is considered as an important gene involved in T2DM. Contrarily, other studies found no association between IGF2BP2 and T2DM (Rodriguez et al., 2010).

Varying results suggest that the association might depend on external factors such as ethnicity. Jia et al. (2011) reported a significant association of rs4402960 with the risk of T2DM among Europeans [odds ratio (OR)=1.12, 95% confidence interval (CI)=1.10–1.15, p=0.000], East Asians [OR=1.15, 95% CI=1.12–1.19, p=0.000] and South Asians [OR=1.14, 95% CI=1.08–1.20, p=4.88×10–6] (Jia et al., 2011); however, they couldn't establish the association among African population. Along with rs4402960, Duesing et al. (2016) obtained novel evidence of association between SNP rs9826022 (in the three downstream regions) and T2D [OR=1.53, 95% CI=1.22–1.91, p=0.0002] (Duesing et al., 2016). A similar study on Asian Indian Sikhs by Sanghera et al. (2012) concluded this association [OR=1.37, 95% CI=1.04–1.82, p=0.027]. Other studies also strongly confirmed the association of IGF2BP2 SNPs (rs4402960 and rs1470579) with T2DM susceptibility in Asian, Han Chinese and Danish populations (Prakash et al., 2012).

3.5.2.3 *CDKAL1*

CDKAL1 (CDK5 regulatory subunit-associated protein 1-like 1) gene belongs to methylthiotransferase family, with 697,948 bases, and is mapped on chromosome 6p22.3 (Brambillasca et al., 2012). It encodes a cyclin-dependent kinase 5 regulatory subunit-associated protein 1 (CDK5RAP1)-like 1, which has its effect on the glucose-dependent regulation of the insulin secretion. It is also involved in gene expression and tRNA processing (Brambillasca et al., 2012). The protein it codes for catalyzes the methylthiolation of N6-threonylcarbamoyladenosine (t(6)A), leading to the formation of 2-methylthio-N6-threonylcarbamoyladenosine (ms(2)t(6)A) at position 37 in tRNAs that read codons beginning with adenine. Palmer et al. (2017) reported the novel function of CDKAL1 in mitochondrial activity (Hu et al., 2010).

CDKAL1 variants are found to be associated with impaired insulin secretion and impaired β-cell function (Groenewoud et al., 2008). The SNP of CDKAL1 (rs7754840; G>A/G>C/G>T) is widely studied and associated with T2D (Saxena et al., 2007; Scott et al., 2007). A study on Japanese population associated rs7754840 [OR=1.28, 95% CI=1.17–1.41, $p=4.5\times10^{-7}$] and rs7756992 (A>G/A>T) in intron 5 [OR=1.27, 95% CI=1.15–1.40, $p=9.8\times10^{-7}$] with T2D susceptibility (Horikawa et al., 2008). The SNP of CDKAL1 (rs7756992) is associated with T2D in Icelandic population [OR=1.23, p=0.00021], Philadelphia population [OR=1.18, p=0.073], European population [OR=1.2, $p=7.7\times10^{-9}$] and Hong Kong population [OR=1.25, p=0.00018] (Steinthorsdottir et al., 2007). A new SNP of CDKAL1 (rs35612982; T>A/T>C) was detected to be associated with T2D in Han Chinese population [OR=1.34, p=0.0010] (Tian et al., 2019). However, another study reported no significant association of SNPs of CDKAL1 (rs7754840, rs7756992, rs4712523 and rs10946398; A>C) with T2D susceptibility in Pima Indians (Rong et al., 2009).

3.5.2.4 *FTO*

FTO (fat mass obesity-associated) gene is located on chromosome 16q12.2. It is 456,821 bases long containing 9 exons. It codes for enzyme alpha-ketoglutarate-dependent dioxygenase, which plays a role in glucose/energy metabolism and DNA damage reversal (Ahmad et al., 2011). FTO is mainly expressed in brain, but also detected in several other tissues, for example adipose tissue and pancreatic beta-cells. Studies in humans have indicated its strong association with body mass index, obesity risk and type 2 diabetes. Amid other SNPs analyzed in the study, FTO is found to have the strongest association with metabolic traits. The exact physiological function of this gene is not known (Brunkwall et al., 2013).

Several SNPs are found in FTO. The four most common SNPs (namely FTO rs9939609, rs8050136, rs1421085 and rs17817499) are located in intron 1 of the FTO (rs8050136) gene; the polymorphism is with a C>A substitution in intron. Ying Yang et al. (2017) showed a region-related association between FTO (rs8050136 and rs9939609) and T2DM. According to Kamura et al. (2016), rs1558902 is associated with T2D in Japanese men and T2D susceptibility is mediated through its effect on the BMI (Gao et al., 2010; Kamura et al., 2016). FTO (rs8050136) is well studied in different populations. A strong association of rs9939609 with T2D [OR=1.46, 95% CI=1.11–1.93, p=0.007] was reported in Asian Indians, after adjusting for age, sex and BMI. A similar study on rs8050136 in Pima Indians gave similar results [OR=1.20, 95% CI=1.02–1.41, p=0.03]. A recent study on young Arab population demonstrated that rs3751812 in the FTO gene is the key SNP associated with the risk of overweight/obesity. The association between SNPs in the FTO gene and type 2 diabetes was shown to be due to obesity. Other studies reported a lack of association between FTO SNPs and T2D (Ahmad et al., 2011).

3.5.2.5 *SLC30A8*

SLC30A8 (solute carrier family 30 (zinc transporter), member 8) gene is 226,535 base pairs long and located on chromosome 8q24.11. This gene is found to be highly expressed in pancreas, particularly in islets of Langerhans, and is also found in kidney. It codes for zinc transporter 8 (ZnT-8) having 369 amino acids and 40755 Da molecular mass. ZnT-8 is involved in the accumulation of zinc from cytoplasm into the intracellular vesicles. ZnT-8 is localized in insulin-containing secretory granules in pancreatic beta-cells (Salem et al., 2014).

Allelic variants of this gene exist that confer susceptibility to diabetes mellitus and non-insulin-dependent diabetes mellitus (NIDDM). SLC30A8 is a key factor for insulin secretion and processing. Studies suggest that SLC30A8 SNPs contribute to T2D susceptibility by causing defects in insulin secretion. However, a study on SLC30A8 (rs13266634; C>A/C>T) reported that it causes impaired proinsulin conversion, but has no effect on insulin secretion (Kirchhoff et al., 2008). Several studies have been conducted to report the association of variants of this gene with diabetes susceptibility. It is found that three SNPs are playing a key role in diabetes susceptibility, i.e., rs13266634, rs7002176 (T>A) and rs1995222 (G>A). A study has found a strong association of rs13266634 with T2D in Icelandic [OR=1.19, 95% CI=1.08–1.31, p=0.0006], Philadelphia [OR=1.51, 95% CI=1.25–1.81, $p=1.5\times10^{-5}$], European [OR=1.15, 95% CI=1.08–1.22, $p=3.3\times10^{-6}$] and Hong Kong [OR=1.19, 95% CI=1.06–1.33, p=0.0035] populations (Steinthorsdottir et al., 2007). However, Rong and colleagues reported no significant association between rs7002176 [OR=0.93, 95% CI=0.81–1.08, p=0.34], rs13266634 [OR=1.04, 95% CI=0.84–1.29, p=0.71] and rs1995222 [OR=1.07, 95% CI=0.90–1.27, p=0.42] and T2D among Pima Indians (Rong et al., 2009).

These three SNPs have been associated with T2D among other populations such as European, Asian, Arabian and American. However, Kulkarni et al. (2016) found no association between SLC30A8 variants and T2D in Mexican American families. A study on rs13266634 among Chinese population confirmed its association with T2D and stated that "C allele" carriers are at high risk. Another detailed study on SLC30A8 (rs13266634) reported that it causes impaired proinsulin conversion and has no effect on insulin secretion (Kulkarni et al. 2016).

3.5.2.6 *KCNJ11*

KCNJ11 (potassium inwardly-rectifying channel, subfamily J, member 11) is located at chromosome 11p15.1, having 4,101 bases with no intron. It codes for an inward-rectifier potassium ion channel (Kir6.2) protein of 390 amino acids. The Kir6.2 protein with the sulfonylurea receptor 1 (SUR1) forms the KATP channel located at the plasma membrane of pancreatic beta-cells. The KATP channel mediates insulin secretion and has a wide range of physiological responses. Mutations in KCNJ11 are associated with insulin resistance, defects in insulin secretion and β-cell dysfunction. The genetic variants can cause DM, as it affects KATP channel activity leading to defective insulin secretion, ultimately causing DM. Studies have found a total of 219 SNPs in KCNJ11, out of which six (rs5219, rs5215, rs5210, rs5218, rs886288 and rs2285676.) are associated with diabetes. Among these, the association of two SNPs of KCNJ11 (rs5219; T>A/T>C/T>G) and KCNJ11 (rs5215- C>T) with T2D has extensively been investigated, and both are located in exon 1 of KCNJ11. A study by Makhzoom et al. (2019) reported the association of T2D risk with rs5219 [OR=3.81, 95% CI=1.24–14.17, p=0.035] in Syrian population. Studies on Japanese and Iranian populations established the association of rs5219 [OR=1.32, 95% CI=1.09–1.60, p=0.004 and OR=2.5, 95% CI=1.01–6.14, p=0.048, in KK homozygotes, respectively] with the prevalence of T2D.

Similarly, many studies supported association of rs5219 and rs5215 with T2D in different groups; however, there are few studies reporting no association of these SNPs with T2D, in certain groups. rs5210 has also caught attention. Few studies concluded its association with T2D, whereas few others don't.

3.5.2.7 PPARG

PPARG (peroxisome proliferator-activated receptor gamma) gene is located on chromosome 3p25.2 and is 146,989 bases long containing 9 exons. It has two transcripts: PPARG1 and PPARG2. PPARG1 is expressed in various tissues such as adipose, skeletal muscle, heart, liver and large intestine, and PPARG2 is only expressed in adipose tissues (Ahmad et al., 2016).

It is predominantly expressed in adipose tissue and codes for 505-amino acid-containing protein peroxisome proliferator-activated receptor gamma, which mainly regulates adipocyte differentiation and hence is termed as "key regulator" of adipogenesis. The PPARG2 gene also plays roles in glucose homeostasis, lipid metabolism, obesity, insulin sensitivity, T2DM and various adipocyte-specific genes (Cai et al., 2011).

Many genome-wide linkage studies have associated it with T2D (Burton et al., 2007; Saxena et al., 2007; Scott et al., 2007; Sladek et al., 2007b). Mutations within PPARG gene cause dysregulated adipogenesis and decreased capacity to seize lipid in CLD (cytoplasmic lipid droplet), leading to lipodystrophy, lipid deposition and metabolic dysfunction, ultimately increasing the risk of diabetes (Wafer et al., 2017; Ahamed et al., 2021). Three SNPs are more commonly associated with T2D susceptibility: PPARG (rs11715073; C>G) and PPARG (rs3892175; G>A/G>T) from PPARG1, and PPARG2 (rs1801282; C>G) (Sanghera et al., 2010). A prospective study identified a significant association of two PPARG SNPs with T2D in FDPS (Finnish Diabetes Prevention Study): PPARG (rs17036314; G>C in intron 2) [OR=1.48, 95% CI=1.02–2.14, p=0.038] and PPARG (rs1801282 in exon B) [OR=1.49, 95% CI=1.03–2.18, p=0.037], but did not report a significant association of SNPs of PPARG (rs1152003; G>C in 3′-flanking region) [OR=0.93, 95% CI=0.65–1.34, p=0.696], PPARG (rs3856806; C>T in exon 6) [OR=1.36, 95% CI=0.93–1.97, p=0.109] and other three in intron 5 PPARG (rs4135263; T>C) [OR=1.24, 95% CI=0.83–1.86, p=0.296], PPARG (rs2972162; T>A/T>C) [OR=1.04, 95% CI=0.67–1.60, p=0.918] and PPARG (rs2938395; G>A/G>C) [OR=1.15, 95% CI=0.78–1.70, p=0.355] (Kilpeläinen et al., 2008). Another study on Asian Indian Sikhs and Han Chinese confirmed the association of PPARG (rs1801282) [OR=0.13, 95% CI=0.03–0.56, p=0.0007 and OR=1.30, 95% CI=1.00–1.68, p=0.050, respectively] with T2D (Sanghera et al., 2010; Wen et al., 2010).

3.5.2.8 CDKN2A

CDKN2A (cyclin-dependent kinase inhibitor 2A) gene is located at chromosome 9p21.3 having 27,550 bases. It codes for cyclin-dependent kinase inhibitors of 156 amino acids, involved in the metabolism of proteins. CDKN2A is expressed in many tissues (Omori et al., 2008).

This gene has different transcript variants with difference in their first exons. SNP rs10811661 with T>C substitution is found to be the most common. A study on CDKN2A [rs10811661; OR=1.26, 95% CI=1.12–1.43, $p=1\times10^{-4}$] on Han Chinese cohort reported a significant association with T2D susceptibility. Another study on middle-aged and young healthy group suggested the role of rs10811661 in impaired pancreatic beta-cell function and validated its association with T2D [OR=1.30, 95% CI=1.16–1.47, $p=1\times10^{-5}$]. One more study confirmed the association of CDKN2 A [rs10811661; OR=1.227, 95% CI=1.016–1.482, p=0.034] with T2D in Danish subjects. Its exact mechanism in T2D susceptibility needs further investigation as studies found CDKN2A (rs10811661 and rs564398) has no effect on proinsulin conversion and insulin secretion. However, Min Xu et al. (2010) concluded that CHKN2 A (rs10811661 and rs564398) has an association with T2D through beta-cell mass reduction and insulin release impairment (Ng et al., 2008). A study on another SNP of CDKN2A (rs10965250) on Saudi population associated it with T2D. A study on Asian ancestors of Hong Kong and Korea also reported the association of SNPs with T2D: CDKN2A (rs564398) [OR=1.06, 95% CI=0.95–1.18, p=0.284], CDKN2A (rs1333040) [OR=1.03, 95% CI=0.96–1.11, p=0.402] and CDKN2A (rs10757278) [OR=1.09, 95% CI=1.02–1.17, p=0.015].

3.5.2.9 HHEX

HHEX (hematopoietically expressed homeobox) gene is a protein-coding gene located at chromosome 10q23.33, contains 4 exons and spans 5,728 bases of genomic sequence. It codes for a 270-amino acid-long member of homeobox family of transcription factors, involved in Wnt signaling for cell growth and development. It is highly expressed in hematopoietic tissues and also in liver, gall bladder and appendix (Grarup et al., 2007).

Many studies have concluded that HHEX gene has a strong biological claim for role in T2D pathogenesis. Sladek et al. (2007) identified four novel risk loci for type 2 diabetes in a GWAS of the French population and reported an association of HHEX with T2D (Hinney et al., 2007; Sladek et al., 2007a). Since then, HHEX (rs1111875; T>C), rs5015480; T>C) and rs7923837 A>G) are being extensively evaluated in regard to their association with diabetes. SNPs of HHEX (rs1111875 and rs7923837) are present on the 3′-flanking region of the gene (Hu et al., 2009).

Variants of the HHEX gene might influence the susceptibility to T2DM by altering insulin degradation or insulin sensitivity or insulin secretion. Many studies and pooled meta-analysis associated the variants of this gene with T2D in various ethnic groups, especially in Japanese population. SNPs of HHEX (rs7923837, rs5015480 and rs1111875) were also significantly associated with T2D in Japanese population. SNP rs7923837 has been associated with T2D in several studies. A similar study on Japanese population reported the association of HHEX (rs1111875; G allele) [OR=1.42, 95% CI=1.13–1.78, p=0.0024] and (rs7923837; G allele) [OR=1.66, 95% CI=1.28–2.15, p=0.00014] with T2D susceptibility. Another study on Japanese population reported the association of HHEX (rs1111875) [OR=1.27, 95% CI=1.14–1.40, $p=1.4\times10^{-4}$] and HHEX (rs7923837) [OR=1.27, 95% CI=1.13–1.43, $p=1\times10^{-4}$] with T2D susceptibility. Another study on HHEX in Dutch population reported the association of HHEK (rs1111875; G allele) [OR=1.68, 95% CI=1.19–2.35, p=0.003] and HHEK (rs7923837; G allele) [OR=1.57, 95% CI=1.08–2.27, p=0.017] with 36% and 33% risk of T2D, respectively (van Vliet-Ostaptchouk et al., 2008). Pivovarova et al. (2009) validated the association of rs1111875 [OR=1.526, 95% CI=0.88–2.62, p=0.13] and rs7923837 [OR=1.93, 95% CI=1.08–3.42, p=0.03] with T2D in German population. HHEX (rs5015480) has also been significantly associated with T2D in various groups (Ragvin et al., 2010). The SNP of HHEX (rs5015480) has also been significantly associated with T2D in various groups (Han et al., 2010; Zhou et al., 2010; Ryoo et al., 2011; Votsi et al., 2017; Sull et al., 2018).

3.5.2.10 SREBF1

SREBF1 (sterol regulatory element-binding transcription factor 1) is a protein-coding gene on chromosome 17p11.2, spanning 26,619 bases with 22 exons and 20 introns. It codes for 1147-amino acid-long SREBP1 (sterol regulatory element-binding protein 1), a transcriptional factor required for carbohydrate and lipid metabolism. Two isoforms produce from this gene, i.e., SREBP1a and SREBP1c (also called ADD-1), due to alternative promoters and splicing (not to confuse with SREBP2, which is produced from a separate gene SREBF2 on 22q13) (Kim and Spiegelman 1996). SREBP1a activates genes of fatty acid and cholesterol synthesis pathways and is expressed in the cell line, intestine and spleen, whereas SREBP1c is mainly expressed in liver, muscle and adipose tissue where it facilitates storage of fatty acids (excess nutrients) as triglycerides and mediates transcriptional effects of insulin on genes encoding enzymes involved in glycolysis, lipogenesis and gluconeogenesis. SREBP1c has the ability to activate lipogenic genes, glucokinase and hexokinase and thereby induces both glucose utilization and lipid metabolism, indicating that a low level of SREBP1c contributes to the pathogenesis of insulin resistance and type 2 diabetes (Liu et al., 2002). Considering functions, it is speculated that genetic variations of the SREBF1 gene could be associated with obesity and/or obesity-related metabolic dysfunctions such as dyslipidemia, insulin resistance and type 2 diabetes. A study revealed 19 SNPs in SREBF1 gene; among those, the SNP (G>C) located on exon 18c is associated with type 2 diabetes (p=0.03). Another study reported the association of SREBF1 (rs2297508) [OR=1.17, 95% CI=1.05–1.30, p=0.003], SREBF1 (rs1889018) [OR=1.12, 95% CI=1.01–1.24, p=0.04], SREBF1 (rs11868035) [OR=1.19, 95% CI=1.07–1.33, p=0.002] and SREBF1 (rs4925118) [OR=1.12, 95% CI=0.95–1.33, p=0.2] with T2D in Danish subjects. Harding et al. (2006) validated the association of rs11868035 [OR=1.20, 95% CI=1.04–1.38, p=0.015] and SNPs at 5′-region (i.e., rs2236513, rs6502618 and rs1889018) [OR≥1.21, p≤0.006] with T2D susceptibility (Harding et al., 2006b) (Table 3.2).

3.5.3 Gestational Diabetes Mellitus (GDM)

It is one of the most common perinatal complications where women's body faces resistance to insulin or lack of insulin causing carbohydrate intolerance and hyperglycemia during pregnancy and returns to normal after the delivery. GDM seems to be caused by the same physiological and genetic factors that cause other types of diabetes. GDM gives opportunity for early diagnosis of diabetes as women with GDM are at high risk of picking up diabetes outside pregnancy. There is insulin desensitizing by placental hormones; therefore, beta-cells increase insulin secretion during pregnancy to counteract this insulin resistance. Hence, beta-cell dysfunction and placental hormones are key causes of GDM.

Several studies reported that numerous SNPs known to be associated with increased risk of type 2 diabetes are also associated with increased risk of gestational diabetes. A study in a cohort of women with GDM history and women with glucose tolerance showed the association of CDKAL1 (rs7756992) [OR=1.22, 95% CI=1.01–1.49, p=0.04]; CDKN2A (rs10811661) [OR=1.12, 95% CI=0.87–1.45, p=0.39]; HHEX (rs1111875) [OR=1.16, 95% CI=0.96–1.4, p=0.12]; IGF2BP2 (rs4402960) [OR=1.18, 95% CI=1.01–1.38, p=0.03]; TCF7L2 (rs7903146) [OR=1.45, 95% CI=1.19–1.75, p=0.00013]; PPARG (rs1801282) [OR=1.06, 95% CI=0.81–1.41, p=0.74]; and KCNJ11 (rs5219) [OR=1.17, 95% CI=0.97–-1.41, p=0.10] with GDM (Lauenborg et al., 2009). Similarly, a study in Korean women reported the association of CDKAL1 (rs7754840) [OR=1.55, 95% CI=1.34–1.79, $p=4.17\times10^{-9}$]; CDKN2A (rs10811661) [OR=1.49, 95% CI=1.29–1.72, $p=1.05\times10^{-7}$]; HHEX (rs7923837) [OR=1.27, 95% CI=1.09–1.49, p=0.003]; IGF2BP2 (rs4402960) [OR=1.18, 95% CI=1.01–1.38, p=0.03]; SLC30A8 (rs13266634) [OR=1.24, 95% CI=1.07–1.43, p=0.005]; TCF7L2 (rs7903146) [OR=1.58, 95% CI=1.03–2.43, p=0.038]; and FTO (rs9939609) [OR=1.26, 95% CI=1.05–1.51, p=0.012] with GDM (Cho et al., 2009). Ye et al. (2016) reported that TCF7L2 SNPs are likely to be helpful in GDM prediction as they found their strong association with GDM: TCF7L2 (rs290487) [OR=2.686, p=0.002] and TCF7L2 (rs7094463; A>G/A>T) [OR=0.429, p=0.005] (Ye et al., 2016). A GWAS by Kwak et al. (2012) concluded the strong association of CDKAL1 (rs7754840) [OR=1.518, $p=6.65\times10^{-16}$] with GDM (Kwak et al., 2012). Another study suggested that the SNPs of CDKAL1 (rs9295478 and rs6935599; A>G/A>T) have a significant association with GDM (Wang et al., 2019).

TABLE 3.2

Primarily Found Genes Significantly Associated with T2D Susceptibility Based on GWAS, GWAS Meta-Analysis and Linkage Studies

Gene	Cytogenetic Band	Probable Mechanism	SNPs	Base pair Substitution	References
ABCC8	11p15.1	Impaired regulation of insulin secretion	rs985136	C>A/C>G	Chavali et al. (2011)
			rs1799854	G>A/G>C	
			rs757110	C>A/C>T	
			rs7124355	A>G	
			rs722341	G>A	
ACAD10	12q24.12	Insulin resistance	rs659964	G>A/G>C	Hanson et al. (2014)
ADIPOQ	3q27.3	Insulin sensitivity in muscle and liver	rs1501299	G>C/G>T	Isakova et al. (2019a, 2019b)
ADAMTS9	3p14.1	Skeletal muscle insulin sensitivity	rs4607103	C>G/C>T	Hotta et al. (2012)
ADRB3	8p11.23	Obesity	rs4994	A>G	Cruz et al. (2010)
APOM	6p21.33	Cellular cholesterol accumulation causing beta-cell dysfunction	rs707921	C>A	Zhou et al. (2011)
			rs707922	-	
			rs805264	G>A	
			rs805297	C>A/C>G	
			rs9404941	T>C	
ARHGEF11	1q23.1	Affects insulin secretion and insulin action	rs12136088	-	Fu et al. (2007)
CDC123/CAMK1D	10p14-p13	Insulin secretion, insulin sensitivity, beta-cell function	rs12779790	A>G	Chen et al. (2013)
CDKAL1	6p22.3	Impaired insulin secretion, impaired beta-cell function	rs7754840	G>A/G>C/G>T	Tian et al. (2019)
			rs4712523	A>G/A>T	
			rs4712524	A>G	
			rs10946398	A>C	
			rs35612982	T>A/T>C	
			rs10440833	T>A/T>G	
			rs7756992	A>G/A>T	Wu et al. (2008)
			rs9465871	T>C	
CDKN2A	9p21.3	Impaired beta-cell function, protein metabolism	rs564398	T>C	Hotta et al. (2012)
			rs10811661	T>A/T>C	
C2CD4A	15q22.2	Reduced function of beta-cells	rs1370176	-	Cui et al. (2011)
			rs1436953	C>T	
			rs7172432	A>G	
DNER	2q36.3	Impaired insulin response	rs1861612	G>A/G>C	Hanson et al. (2014)
EXT2	11p11.2	Insulin resistance	rs3740878	T>C/T>G	Sladek et al. (2007, (2007a, 2007b) and Hotta et al. (2012)
			rs11037909	T>C	
			rs1113132	G>A/G>C	
FTO	16q12.2	BMI, obesity risk	rs9939609	T>A	Yang et al. (2017)
			rs8050136	C>A	
FABP2	4q28	Insulin resistance, dyslipidemia and obesity	rs1799883	T>A/T>C /T>G	Sikhayeva et al. (2017)
HHEX	10q23.33	altering insulin degradation, insulin sensitivity, insulin secretion	rs1111875	C>G/C>T	Hotta et al. (2012)
			rs7923837	G>A/G>T	Xu et al. (2010)
			rs5015480	C>T	
HNF1A	12q24.31	Dysregulated beta-cell development	rs7953249	G>A/G>C	Chavali et al. (2011)

(*Continued*)

TABLE 3.2 (*Continued*)

Primarily Found Genes Significantly Associated with T2D Susceptibility Based on GWAS, GWAS Meta-Analysis and Linkage Studies

Gene	Cytogenetic Band	Probable Mechanism	SNPs	Base pair Substitution	References
HNF1B/TCF2	17q12	Dysregulated beta-cell development Impaired embryonic pancreas development	rs2229295 rs1800929 rs12450628	G>A/G>T T>A/T>C/T>G C>T	Goda et al. (2015) Chavali et al. (2011)
HNF4A	20q13.12	Beta-cell function	rs1884613 rs2144908 rs6093978	C>A/C>G/C>T G>A -	Chavali et al. (2011)
IGF2BP2	3q27.2	Beta-cell function, reduced insulin secretion	rs4402960 rs1470579	G>T A>C	Rao et al. (2016)
ISL1	5q11.1	Impairment of insulin expression	rs991216 rs1017	C/G A/T	Chavali et al. (2011)
INSM1	20p11.23	Reduced insulin secretion, glucose intolerance	rs10470054	-	
JAZF1	7p15.2-p15.1	Glucose regulation, lipid homeostasis	rs864745	T>C	Chen et al. (2013)
KCNQ1	11p15.5-p15.4	Beta-cell function	rs163182 rs8181588 rs2237892	G>A/G>C T>A/T>C C>T	Cui et al. (2011) Hanson et al. (2014) Hotta et al. (2012)
KCNJ11	11p15.1	Impaired insulin secretion, insulin resistance, beta-cell function	rs5219 rs5215	T>A/T>C/T>G C>T	Chavali et al. (2011)
KIF6	6p21.2	Modulating lipid metabolism	rs6901022	C>A	Hanson et al. (2014)
MTNR1B	11q14.3	Increased blood glucose level, impairment in early insulin release	rs10830962 rs10830963 rs1387153	C>A/C>G/C>T C>G/C>T C>G/C>T	Kwak et al. (2012) Xu et al. (2010)
MAP3K8	10p11.23	Obesity, insulin resistance	rs590855	C>T	Hanson et al. (2014)
NOTCH2	1p12	Insulin secretion, insulin sensitivity, growth and development of pancreas	rs10923931	G>T	Hotta et al. (2012)
PPARG	3p25.2	Insulin sensitivity, glucose homeostasis, lipid metabolism, obesity, impaired adipogenesis	rs1801282	C>G	
PTPN22	1p13.2	Insulin resistance	1858c	-	Bottini et al. (2004)
PTPRD	9p24.1-p23	Decreased insulin receptor signaling	rs832234	T>A	Hanson et al. (2014)
PAX6	11p13	Beta-cell dysfunction	rs628224	-	Chavali et al. (2011)
PLCE1	10q23.33	Acute insulin secretion	rs11187815	-	Hanson et al. (2014)
SREBF1	17p11.2	Insulin resistance	rs6502618 rs11868035 rs4925118 rs2236513 rs1889018	A>G G>A - - G>A	Harding et al. (2006, 2006a, 2006b)
SLC30A8	8q24.11	Impaired insulin secretion, impaired beta-cell function	rs13266634 rs2466295 rs6469675 rs2466293	C>A/C>T - - -	Wenzlau et al. (2008) Xu (2010)

(*Continued*)

TABLE 3.2 (*Continued*)

Primarily Found Genes Significantly Associated with T2D Susceptibility Based on GWAS, GWAS Meta-Analysis and Linkage Studies

Gene	Cytogenetic Band	Probable Mechanism	SNPs	Base pair Substitution	References
TCF7L2	10q25.2-q25.3	Beta-cell function, reduced insulin secretion	rs7903146	C>T/C>G	Tong et al. (2009)
			rs7901695	T>C	
			rs12255372	G>A/G>T	
			rs11196205	G>A/G>C/G>T	
THADA	2p21	Lower beta-cell mass	rs7578597	T>C	Hotta et al. (2012)
TSPAN8/LGR5	12q21.1	Decreased insulin release	rs7961581	C>T	Sikhayeva et al. (2017)
VEGFA	6p21.1	Insulin secretion	rs9472138	C>T	Hotta et al. (2012)
WFS1	4p16.1	Glucose/energy metabolism	rs10010131	A>G	Chen et al. (2013)

The contents of the table are also derived from PubMed, SNPedia, RefSeq and dbSNP by querying for the rs number of the respective polymorphism and diabetes.

A significant association of GDM was found with ADIPOQ (rs266729; C>A/C>G/C>T) [OR=1.56, 95% CI=1.10–2.23, p=0.013], SSR1-RREB1 (rs9505118; A>G) [OR=1.41, 95% CI=1.01–1.97, p=0.046] and CDKN2A (rs10811661) [OR=1.46, 95% CI=1.03–2.08, p=0.035] (Kasuga et al., 2017).

Other than these SNPs, the following gene SNPs are reported to be associated with GDM.

HNF1B (hepatocyte nuclear factor 1 homeobox B) also known as TCF2 (transcription factor 2) is 58,675 bases long and is located at chromosome 17q12. The gene codes for hepatocyte nuclear factor 1-beta, which is 557 amino acids long with a molecular mass of 61,324 Da. It acts as a transcription factor by binding to specific regions of DNA and regulating activity of other genes. It is found in many tissues and organs such as pancreas, liver, kidney and lungs and is believed to have a role in their development. Mutations in this gene result in renal cysts and diabetes syndrome and non-insulin-dependent diabetes mellitus.

A study on Danish women reported the association of HNF1B rs7501939 (C>G/C>T) [OR=1.22, 95% CI=1.02–1.47, p=0.029] with GDM (Lauenborg et al., 2009). Tarnowski et al. (2018) reported the association of HNF1B rs4430796 (A>C/A>G/A>T) [OR=1.39, 95% CI=1.06–1.83, p=0.018] with a higher risk of GDM (Tarnowski et al., 2018).

3.5.3.1 *WFS1*

WFS1 (wolframin ER transmembrane glycoprotein) gene is 33,417 bases long with 10 exons, located on chromosome 4p16.1. It codes for wolframin, an 890-amino acid-long transmembrane protein having a molecular mass of 100,292 Da. It is present ubiquitously in the endoplasmic reticulum, ovary, brain, pancreas, heart and few other tissues and is in vivo expressed in insulinoma beta-cell lines. Glucose/energy metabolism is among its related pathways, suggesting its association with diabetes. A study done on the SNP of WFS1 (rs10010131; A>G) [OR=0.87, 95% CI=0.73–1.05, p=0.14] couldn't report a significant association with GDM (Lauenborg et al., 2009). Several studies have been done on WFS1 with regard to Wolfram syndrome and on its SNPs in relation to T2D, but there is very limited knowledge about WFS1 risk loci associated with GDM.

3.5.3.2 *MTNR1B*

MTNR1B (melatonin receptor 1B) is located on chromosome 11q14.3 containing 16,591 bases. It codes for melatonin receptor MT2 or 1B having 362 amino acids and 40,188 Da molecular mass. It is primarily expressed in the retina and brain. The variants of this gene have shown association with T2D and GDM. MTNR1B, rs10830962 (C>A/C>G/C>T), is significantly associated with the risk of GDM [OR=1.454; $p=2.49\times10^{-13}$] (Kwak et al., 2012). The SNPs of MTNR1B rs1387153 (C>G/C>T) [OR=1.38, 95%CI=1.14–1.66, $p=7.6\times10^{-4}$] and rs10830963 (C>G/C>T) [OR=1.62, 95%CI=1.34–1.96, $p=4.5\times10^{-7}$] are also found to be strongly associated with GDM (Huopio et al., 2013). rs1387153 (C>G/C>T) [OR=1.68, 95% CI=1.09–2.60, p=0.019] and rs10830963 [OR=1.81, 95% CI=1.17–2.78, p=0.007] are found to be associated with GDM in Saudi women (Alharbi et al., 2019). A case-control study on Caucasians found a significant association of rs10830963 [OR=1.64, 95% CI=1.38–1.97, p=0.006] with GDM (Rosta et al., 2017). However, Chao Li et al. (2013) reported no association between rs10830962 and GDM, but found GDM association with rs10830963 [OR=1.22, 95% CI=1.04–1.48, p=0.037] in Han Chinese population.

3.6 Conclusions

In conclusion, this study is a review of previously reported association between genetic variants and diabetes susceptibility. It is compelling to know that diabetes-associated genetic variants play a role in distinct and overlapping regulatory pathways. Interconnection of T1D-associated variants directs that they are able to influence immune response gene expression in a particular tissue and specific cell type. The SNPs associated with T2D are either missense coding SNPs or regulatory noncoding SNPs, leading to a direct change in amino acids within a protein.

T2D-associated SNPs contribute to impaired proinsulin conversion, insulin secretion, insulin action, beta-cell function and obesity and disrupt homeostasis. Studies on GDM risk loci are limited, and there is a scope for rigorous research in this area.

In spite of striking progress and study and outcomes, the results of the association studies remain indecisive. This might be due to differences in allele frequency, genetic background, lifestyle, environmental exposures and different ethnic groups and relatively small sample size. Untangling these effects requires precise empirical studies, first to illustrate the genetic basis of individual immunological response and second to determine the correlations between gene expression and the mechanisms they are involved in. Studies based on different ethnic groups need to be conducted. Re-evaluation and confirmation of the currently identified genes is also necessary due to inconsistent results. Moreover, most of the genes are investigated because they are presumed to have a role in diabetes pathogenesis based on their function. Thus, the "candidate gene approach" must be focused to detect novel disease susceptibility genes. Intense research must be done to develop gene therapy for diabetes, and it should be specific to ensure that the treatment of a particular gene will not have adverse effects on other genes.

These studies are important for the management of major diseases such as diabetes, and their translation into clinical practices will enable to reduce incidence and provide an understanding of treatment response, pre-diabetic prognosis and disease development.

ABBREVIATIONS

ADA:	American Diabetes Association
AVP:	Arginine vasopressin
BMI:	Body mass index
CDK:	Cyclin-dependent kinase
CLD:	Cytoplasmic lipid droplet
FFAs:	Free fatty acids
FTO:	Fat mass obesity-associated
GAD:	Glutamic acid decarboxylase
GWAS:	Genome-wide association studies
HLA:	Human leukocyte antigen
IAA:	Insulin autoantibodies
IRS:	Insulin receptor substrate
MODY:	Maturity-onset diabetes of the young
NIDDM:	Non-insulin-dependent diabetes mellitus
PACAP:	Pituitary adenylate cyclase-activating polypeptide
PCR:	Polymerase chain reaction
PDK:	Protein-dependent kinase
PPARG:	Peroxisome proliferator-activated receptor gamma
RFLP:	Restriction fragment length polymorphism
ROS:	Reactive oxygen species
SNP:	Single nucleotide polymorphism
SREBF1:	Sterol regulatory element-binding transcription factor 1
SUR:	Sulfonylurea receptor
T1D:	Type 1 diabetes
T2D:	Type 2 diabetes
TCF7L2:	Transcription factor 7-like 2
TXNIP:	Thioredoxin-interacting protein
ZFP:	Zinc finger protection

REFERENCES

Ahamed, S. N., Khan, J., Rahamathulla, S., Venkateswarulu, T., & Krupanidhu, S. (2021). Metabolic and molecular effects of edible oils on PPAR modulators in rabbit liver. *Advancements in Life Sciences*, 8(2), 154–159.

Ahlqvist, E., Hjort, R., Carlsson, P. O., Grill, V., Groop, L., Martinell, M., Carlsson, S. (2018). Overweight, obesity and the risk of LADA: results from a Swedish case–control study and the Norwegian HUNT Study. *Diabetologia*, 61(6), 1333–1343.

Ahmad, Tariq, I-Min Lee, Guillaume Paré, Daniel I. Chasman, Lynda Rose, Paul M. Ridker, and Samia Mora. (2011) Lifestyle interaction with fat mass and obesity-associated (FTO) genotype and risk of obesity in apparently healthy US women. *Diabetes Care* 34(3), 675–680.

Amed, S., Dean, H. J., Panagiotopoulos, C., Sellers, E. A., Hadjiyannakis, S., Laubscher, T. A., Dannenbaum, D., Shah, B. R., Booth, G. L., & Hamilton, J. K. (2010). Type 2 diabetes, medication-induced diabetes, and monogenic diabetes in Canadian children: a prospective national surveillance study. *Diabetes Care*, 33(4), 786–791.

American Diabetes Association. (2004). Dyslipidemia management in adults with diabetes. *Diabetes Care, 27*(suppl_1), s68–s71.

Association, A. D. (2015). 2. Classification and diagnosis of diabetes. *Diabetes Care*, 38(Supplement 1), S8–S16.

Association, A. D. (2020). 2. Classification and diagnosis of diabetes: Standards of Medical Care in Diabetes-2020. *Diabetes Care*, 43(Suppl 1), S14.

Ban, H.-J., Heo, J. Y., Oh, K.-S., & Park, K.-J. (2010). Identification of type 2 diabetes-associated combination of SNPs using support vector machine. *BMC Genetics*, 11(1), 26.

Barhate, R., & Kulkarni, P. (2018). Analysis of classifiers for prediction of type ii diabetes mellitus. *2018 Fourth International Conference on Computing Communication Control and Automation (ICCUBEA)*, pp. 1–6. IEEE.

Basile, K. J., Johnson, M. E., Xia, Q., & Grant, S. F. (2014). Genetic susceptibility to type 2 diabetes and obesity: follow-up of findings from genome-wide association studies. *International Journal of Endocrinology*, Article ID 769671, 13 pages. http://dx.doi.org/10.1155/2014/769671.

Borchers, A. T., Uibo, R., & Gershwin, M. E. (2010). The geo-epidemiology of type 1 diabetes. *Autoimmunity Reviews*, 9(5), A355–A365.

Bruun, J. M., Lihn, A. S., Madan, A. K., Pedersen, S. B., Schiøtt, K. M., Fain, J. N., & Richelsen, B. (2004). Higher production of IL-8 in visceral vs. subcutaneous adipose tissue. Implication of nonadipose cells in adipose tissue. *American Journal of Physiology-Endocrinology and Metabolism*, 286(1), E8–E13.

Buraczynska, M., Zukowski, P., Buraczynska, K., Mozul, S., & Ksiazek, A. (2011). Renalase gene polymorphisms in patients with type 2 diabetes, hypertension and stroke. *Neuromolecular Medicine*, 13(4), 321–327.

Buryk, M. (2014). Type 1 or type 2 diabetes? What's in a name. The controversy and conundrum of diabetes type in youth and the public health impact University of Pittsburgh].

Chen, J., Saxena, G., Mungrue, I. N., Lusis, A. J., & Shalev, A. (2008). Thioredoxin-interacting protein: a critical link between glucose toxicity and β-cell apoptosis. *Diabetes*, 57(4), 938–944.

Choubey, D. K., Paul, S., Shandilya, S., & Dhandhania, V. K. (2020). Implementation and analysis of classification algorithms for diabetes. *Current Medical Imaging*, 16(4), 340–354.

Cieślak, M., Wojtczak, A., & Cieślak, M. (2015). Role of pro-inflammatory cytokines of pancreatic islets and prospects of elaboration of new methods for the diabetes treatment. *Acta Biochimica Polonica*, 62(1), 15–21.

Dal Col, J., Zancai, P., Terrin, L., Guidoboni, M., Ponzoni, M., Pavan, A., Spina, M., Bergamin, S., Rizzo, S., & Tirelli, U. (2008). Distinct functional significance of Akt and mTOR constitutive activation in mantle cell lymphoma. *Blood, The Journal of the American Society of Hematology*, 111(10), 5142–5151.

Dayeh, T. A., Olsson, A. H., Volkov, P., Almgren, P., Rönn, T., & Ling, C. (2013). Identification of CpG-SNPs associated with type 2 diabetes and differential DNA methylation in human pancreatic islets. *Diabetologia*, 56(5), 1036–1046.

de Miguel-Yanes, J. M., Shrader, P., Pencina, M. J., Fox, C. S., Manning, A. K., Grant, R. W., Dupuis, J., Florez, J. C., D'Agostino, R. B., & Cupples, L. A. (2011). Genetic risk reclassification for type 2 diabetes by age below or above 50 years using 40 type 2 diabetes risk single nucleotide polymorphisms. *Diabetes Care*, 34(1), 121–125.

Demeulder, B., Zarrinpashneh, E., Ginion, A., Viollet, B., Hue, L., Rider, M. H., Vanoverschelde, J.-L., Beauloye, C., Horman, S., & Bertrand, L. (2013). Differential regulation of eEF2 and p70S6K by AMPKalpha2 in heart. *Biochimica et Biophysica Acta (BBA)-Molecular Basis of Disease*, 1832(6), 780–790.

DiMenna, F. J., & Arad, A. D. (2018). Exercise as 'precision medicine' for insulin resistance and its progression to type 2 diabetes: a research review. *BMC Sports Science, Medicine and Rehabilitation*, 10(1), 1–23.

Donath, M. Y., Böni-Schnetzler, M., Ellingsgaard, H., Halban, P. A., & Ehses, J. A. (2010). Cytokine production by islets in health and diabetes: cellular origin, regulation and function. *Trends in Endocrinology & Metabolism*, 21(5), 261–267.

Eguchi, K., Manabe, I., Oishi-Tanaka, Y., Ohsugi, M., Kono, N., Ogata, F., Yagi, N., Ohto, U., Kimoto, M., & Miyake, K. (2012). Saturated fatty acid and TLR signaling link β cell dysfunction and islet inflammation. *Cell Metabolism*, 15(4), 518–533.

Fontaine-Bisson, B., Renström, F., Rolandsson, O., Payne, F., Hallmans, G., Barroso, I., Franks, P. W., & The MAGIC investigators. (2010). Evaluating the discriminative power of multi-trait genetic risk scores for type 2 diabetes in a northern Swedish population. *Diabetologia*, 53(10), 2155–2162.

Fujioka, K. (2015). Current and emerging medications for overweight or obesity in people with comorbidities. *Diabetes, Obesity and Metabolism*, 17(11), 1021–1032.

Giacca, A., Xiao, C., Oprescu, A. I., Carpentier, A. C., & Lewis, G. F. (2011). Lipid-induced pancreatic β-cell dysfunction: focus on in vivo studies. *American Journal of Physiology-Endocrinology and Metabolism*, 300(2), E255–E262.

Groenewoud, M.J., Dekker, J.M., Fritsche, A., Reiling, E., Nijpels, G., Heine, R.J., Maassen, J.A., Machicao, F., Schäfer, S.A., Häring, H.U. and t Hart, L.M. (2008). Variants of CDKAL1 and IGF2BP2 affect first-phase insulin secretion during hyperglycaemic clamps. *Diabetologia*, 51(9), 1659–1663.

Halban, P. A., Ehses, J. A., Krycer, J. R., Sharpe, L. J., Luu, W., Brown, A. J., Nader, N., Chrousos, G. P., Kino, T., & Onaka, T. (2010). Cytokine production by islets in health and diabetes: cellular origin, regulation and function. *Trends in Endocrinology & Metabolism*, 21(5), 261–267.

Hattersley, A. T., & Patel, K. A. (2017). Precision diabetes: learning from monogenic diabetes. *Diabetologia*, 60(5), 769–777.

Hu, C., Zhang, R., Wang, C., Wang, J., Ma, X., et al. (2009) *PPARG, KCNJ11, CDKAL1, CDKN2A-CDKN2B, IDE-KIF11-HHEX, IGF2BP2* and *SLC30A8* Are Associated with Type 2 Diabetes in a Chinese Population. *PLoS ONE*, 4, e7643.

Imagawa, A., Hanafusa, T., Awata, T., Ikegami, H., Uchigata, Y., Osawa, H., Kawasaki, E., Kawabata, Y., Kobayashi, T., & Shimada, A. (2012). Report of the Committee of the Japan Diabetes Society on the Research of Fulminant and Acute-onset Type 1 Diabetes Mellitus: new diagnostic criteria of fulminant type 1 diabetes mellitus (2012). *Diabetology International*, 3(4), 179–183.

Imamura, M., Takahashi, A., Yamauchi, T., Hara, K., Yasuda, K., Grarup, N., Zhao, W., Wang, X., Huerta-Chagoya, A., & Hu, C. (2016). Genome-wide association studies in the Japanese population identify seven novel loci for type 2 diabetes. *Nature Communications*, 7(1), 1–12.

Kilpelainen, T.O., Bingham, S.A., Khaw, K.T. et al. (2008) Association of variants in the PCSK1 gene with obesity in the EPIC-Norfolk study. *Human Molecular Genetics* 18:3496–3501.

Kubota, N., Kubota, T., Kajiwara, E., Iwamura, T., Kumagai, H., Watanabe, T., Inoue, M., Takamoto, I., Sasako, T., & Kumagai, K. (2016). Differential hepatic distribution of insulin receptor substrates causes selective insulin resistance in diabetes and obesity. *Nature Communications*, 7(1), 1–16.

Li, H., T. O. Kilpeläinen, C. Liu, J. Zhu, Y. Liu, C. Hu, Z. Yang et al. (2012) Association of genetic variation in FTO with risk of obesity and type 2 diabetes with data from 96,551 East and South Asians. *Diabetologia* 55(4), 981–995.

Makhzoom, Osama, Younes Kabalan, and Faizeh Al-Quobaili. (2019) Association of KCNJ11 rs5219 gene polymorphism with type 2 diabetes mellitus in a population of Syria: a case-control study. *BMC Medical Genetics* 20(1): 1–6.

Maniruzzaman, M., Rahman, M. J., Ahammed, B., & Abedin, M. M. (2020). Classification and prediction of diabetes disease using machine learning paradigm. *Health Information Science and Systems*, 8(1), 7.

Mazzaccara, C., Iafusco, D., Liguori, R., Ferrigno, M., Galderisi, A., Vitale, D., Simonelli, F., Landolfo, P., Prisco, F., & Masullo, M. (2012). Mitochondrial diabetes in children: seek and you will find it. *PLoS One*, 7(4), e34956.

McKinney, D. M., Southwood, S., Hinz, D., Oseroff, C., Arlehamn, C. S. L., Schulten, V., Taplitz, R., Broide, D., Hanekom, W. A., & Scriba, T. J. (2013). A strategy to determine HLA class II restriction broadly covering the DR, DP, and DQ allelic variants most commonly expressed in the general population. *Immunogenetics*, 65(5), 357–370.

Mima, A., Ohshiro, Y., Kitada, M., Matsumoto, M., Geraldes, P., Li, C., Li, Q., White, G. S., Cahill, C., & Rask-Madsen, C. (2011). Glomerular-specific protein kinase C-β-induced insulin receptor substrate-1 dysfunction and insulin resistance in rat models of diabetes and obesity. *Kidney International*, 79(8), 883–896.

Nair, A. K., Sugunan, D., Kumar, H., & Anilkumar, G. (2010). Case-control analysis of SNPs in GLUT4, RBP4 and STRA6: association of SNPs in STRA6 with type 2 diabetes in a South Indian population. *PloS One*, 5(7), e11444.

Nemr R., Echtay, A., Dashti, E.A., Almawi, A.W., Al-Busaidi, A.S., et al. (2012). Strong association of common variants in the IGF2BP2 gene with type 2 diabetes in Lebanese Arabs.

Diabetes Research and Clinical Practice, DOI: 10.1016/j.diabres.2011.1012.1026.

Nobrega, M. A. (2013). TCF7L2 and glucose metabolism: time to look beyond the pancreas. *Diabetes*, 62(3), 706–708.

Okamoto, M., Okamoto, M., Gotoh, K., Masaki, T., Ozeki, Y., Ando, H., Anai, M., Sato, A., Yoshida, Y., & Ueda, S. (2016). Fulminant type 1 diabetes mellitus with anti-programmed cell death-1 therapy. *Journal of Diabetes Investigation*, 7(6), 915–918.

Oram, R. A., Jones, A. G., Besser, R. E., Knight, B. A., Shields, B. M., Brown, R. J., Hattersley, A. T., & McDonald, T. J. (2014). The majority of patients with long-duration type 1 diabetes are insulin microsecretors and have functioning beta cells. *Diabetologia*, 57(1), 187–191.

Ozougwu, J., Obimba, K., Belonwu, C., & Unakalamba, C. (2013). The pathogenesis and pathophysiology of type 1 and type 2 diabetes mellitus. *Journal of Physiology and Pathophysiology*, 4(4), 46–57.

Palmer, C.J., Bruckner, R.J., Paulo, J.A., Kazak, L., Long, J.Z., Mina, A.I., Deng, Z., LeClair, K.B., Hall, J.A., Hong, S. and Zushin, P.J.H. (2017). Cdkal1, a type 2 diabetes susceptibility gene, regulates mitochondrial function in adipose tissue. *Molecular Metabolism*, 6(10), 1212–1225.

Palsamy, P., & Subramanian, S. (2010). Ameliorative potential of resveratrol on proinflammatory cytokines, hyperglycemia mediated oxidative stress, and pancreatic β-cell dysfunction in streptozotocin-nicotinamide-induced diabetic rats. *Journal of Cellular Physiology*, 224(2), 423–432.

Patterson, C. C., Harjutsalo, V., Rosenbauer, J., Neu, A., Cinek, O., Skrivarhaug, T., Rami-Merhar, B., Soltesz, G., Svensson, J., & Parslow, R. C. (2019). Trends and cyclical variation in the incidence of childhood type 1 diabetes in 26 European centres in the 25 year period 1989–2013: a multicentre prospective registration study. *Diabetologia*, 62(3), 408–417.

Plagnol, V., Howson, J. M., Smyth, D. J., Walker, N., Hafler, J. P., Wallace, C., Stevens, H., Jackson, L., Simmonds, M. J., & Bingley, P. J. (2011). Genome-wide association analysis of autoantibody positivity in type 1 diabetes cases. *PLoS Genetics*, 7(8), e1002216.

Qi, L., van Dam, R. M., Meigs, J. B., Manson, J. E., Hunter, D., & Hu, F. B. (2006). Genetic variation in IL6 gene and type 2 diabetes: tagging-SNP haplotype analysis in large-scale case-control study and meta-analysis. *Human Molecular Genetics*, 15(11), 1914–1920.

Ren, Jing, Alina Isakova, Drew Friedmann, Jiawei Zeng, Sophie M. Grutzner, Albert Pun, Grace Q. Zhao et al. (2019) Single-cell transcriptomes and whole-brain projections of serotonin neurons in the mouse dorsal and median raphe nuclei. *Elife* 8, e49424.

Reus-Pons, M., Mulder, C. H., Kibele, E. U., & Janssen, F. (2018). Differences in the health transition patterns of migrants and non-migrants aged 50 and older in southern and western Europe (2004–2015). *BMC Medicine*, 16(1), 1–15.

Robert, A. A., & Al Dawish, M. A. (2020). The worrying trend of diabetes mellitus in Saudi Arabia: an urgent call to action. *Current Diabetes Reviews*, 16(3), 204–210.

Rodriguez-Calvo, T., Ekwall, O., Amirian, N., Zapardiel-Gonzalo, J., & von Herrath, M. G. (2014). Increased immune cell infiltration of the exocrine pancreas: a possible contribution to the pathogenesis of type 1 diabetes. *Diabetes*, 63(11), 3880–3890.

Saeedi, P., Petersohn, I., Salpea, P., Malanda, B., Karuranga, S., Unwin, N., Colagiuri, S., Guariguata, L., Motala, A. A., & Ogurtsova, K. (2019). Global and regional diabetes prevalence estimates for 2019 and projections for 2030 and 2045: results from the International Diabetes Federation Diabetes Atlas. *Diabetes Research and Clinical Practice*, 157, 107843.

Saxena, R., Voight, B.F., Lyssenko, V., Burtt, N.P., de Bakker, P.I., Chen, H., Roix, J.J., Kathiresan, S., Hirschhorn, J.N., Daly, M.J. and Hughes, T.E. (2007). Genome-wide association analysis identifies loci for type 2 diabetes and triglyceride levels. *Science*, 316(5829), 1331–1336.

Schenck, S., Rosenbauer, J., Niewerth, M., Klotsche, J., Minden, K., Schwarz, T., Foeldvari, I., Horneff, G., Weller-Heinemann, F., & Holl, R. W. (2018). Comorbidity of type 1 diabetes mellitus in patients with juvenile idiopathic arthritis. *The Journal of Pediatrics*, 192, 196–203.

[Record #9 is using a reference type undefined in this output style.]

Shalev, A. (2014). Minireview: Thioredoxin-interacting protein: regulation and function in the pancreatic β-cell. *Molecular Endocrinology*, 28(8), 1211–1220.

Steinthorsdottir, V., Thorleifsson, G., Reynisdottir, I., Benediktsson, R., Jonsdottir, T., Walters, G.B., Styrkarsdottir, U., Gretarsdottir, S., Emilsson, V., Ghosh, S. and Baker, A. (2007). A variant in CDKAL1 influences insulin response and risk of type 2 diabetes. *Nature Genetics*, 39(6), 770–775.

Strasser, B. (2013). Physical activity in obesity and metabolic syndrome. *Annals of the New York Academy of Sciences*, 1281(1), 141.

Tabara, Y., Osawa, H., Kawamoto, R., Onuma, H., Shimizu, I., Miki, T. & Makino, H. (2009). Replication study of candidate genes associated with type 2 diabetes based on genome-wide screening. Diabetes, 58(2), 493–498.

Thomas, R., Halim, S., Gurudas, S., Sivaprasad, S., & Owens, D. (2019). IDF Diabetes Atlas: a review of studies utilising retinal photography on the global prevalence of diabetes related retinopathy between 2015 and 2018. *Diabetes Research and Clinical Practice*, 157, 107840.

Törn, C., Hadley, D., Lee, H.-S., Hagopian, W., Lernmark, Å., Simell, O., Rewers, M., Ziegler, A., Schatz, D., & Akolkar, B. (2015). Role of type 1 diabetes-associated SNPs on risk of autoantibody positivity in the TEDDY Study. *Diabetes*, 64(5), 1818–1829.

Tsai, P.-J. (2011). *The Development of Type 2 Diabetes Mellitus (Type 2 DM) in California Twins*. University of Southern California, ProQuestLLC.

Unoki, H., Takahashi, A., Kawaguchi, T., Hara, K., Horikoshi, M., Andersen, G., Ng, D. P., Holmkvist, J., Borch-Johnsen, K., & Jørgensen, T. (2008). SNPs in KCNQ1 are associated with susceptibility to type 2 diabetes in East Asian and European populations. *Nature Genetics*, 40(9), 1098–1102.

Wang, C., Guan, Y., & Yang, J. (2010). Cytokines in the progression of pancreatic β-cell dysfunction. *International Journal of Endocrinology*, Article ID 515136, 10 pages, doi:10.1155/2010/515136.

Wang, P., Sidney, J., Kim, Y., Sette, A., Lund, O., Nielsen, M., & Peters, B. (2010). Peptide binding predictions for HLA DR, DP and DQ molecules. *BMC Bioinformatics*, 11(1), 568.

Xu, Xiaohua, Zhekang Ying, Ming Cai, Zhaobin Xu, Yuanjing Li, Silis Y. Jiang, Kevin Tzan et al. "Exercise ameliorates high-fat diet-induced metabolic and vascular dysfunction, and increases adipocyte progenitor cell population in brown adipose tissue. (2011). *American Journal of Physiology-Regulatory, Integrative and Comparative Physiology*, 300(5): R1115–R1125.

Xu, Min, Yufang Bi, Yu Xu, Bing Yu, Yun Huang, Lina Gu, Yaohua Wu et al. (2010) Combined effects of 19 common variations on type 2 diabetes in Chinese: results from two community-based studies. PLoS one 5(11), e14022.

Yamauchi, T., Hara, K., Maeda, S., Yasuda, K., Takahashi, A., Horikoshi, M., Nakamura, M., Fujita, H., Grarup, N., & Cauchi, S. (2010). A genome-wide association study in the Japanese population identifies susceptibility loci for type 2 diabetes at UBE2E2 and C2CD4A-C2CD4B. *Nature Genetics*, 42(10), 864–868.

Yang, Y., Liu, B., Xia, W., Yan, J., Liu, H.Y., Hu, L. and Liu, S.M. (2017). FTO genotype and type 2 diabetes mellitus: spatial analysis and meta-analysis of 62 case-control studies from different regions. *Genes*, 8(2), p.70.

Zeggini, E., Scott, L.J., Saxena, R., Voight, B.F., Marchini, J.L., Hu, T., de Bakker, P.I., Abecasis, G.R., Almgren, P., Andersen, G. and Ardlie, K. (2008). Meta-analysis of genome-wide association data and large-scale replication identifies additional susceptibility loci for type 2 diabetes. *Nature Genetics*, 40(5), 638–645.

Zeng, Z., Sarbassov, D. D., Samudio, I. J., Yee, K. W., Munsell, M. F., Ellen Jackson, C., Giles, F. J., Sabatini, D. M., Andreeff, M., & Konopleva, M. (2007). Rapamycin derivatives reduce mTORC2 signaling and inhibit AKT activation in AML. *Blood*, 109(8), 3509–3512.

Zhao, Y.-C., Zhu, J., Song, G.-Y., & Li, X.-S. (2015). Relationship between thioredoxin-interacting protein (TXNIP) and islet β-cell dysfunction in patients with impaired glucose tolerance and hypertriglyceridemia. *International Journal of Clinical and Experimental Medicine*, 8(3), 4363.

4

Genetic Polymorphisms Associated with Innate Immune Response to Malaria

Amre Nasr and Amir Abushouk
Department of Basic Medical Sciences, College of Medicine, King Saud Bin Abdulaziz University for Health Sciences (KSAU-HS), King Abdullah International Medical Research Center (KAIMRC), Ministry of National Guard Health Affairs (MNGHA), Jeddah, Kingdom of Saudi Arabia

CONTENTS

ABSTRACT Malaria is a parasitic infectious disease caused by five species of *Plasmodium* (*P.*) transmitted to humans through mosquito bites. Among these species, *P. falciparum*, the most prevalent in Africa, and *P. vivax* constitute the greatest threat to health. The genotypes of the Fc gamma receptors (FcγRs), Toll-like receptors (TLRs), tumor necrosis factor (TNF), nitric oxide (NO) and heme oxygenase (HO) have been reported to play a key role in innate immune response against malaria parasites through increase or decrease in the susceptibility to malaria infections. This chapter emphasizes the association between FcγRs, TLRs, TNF, NO and HO genes' polymorphisms and malaria infection.

KEYWORDS: *Malaria; IgG Subclass; FcγR; Susceptibility.*

4.1 Introduction to the Biology of Gene Polymorphism

Deoxyribonucleic acid (DNA) sequences significantly vary with population. Most DNA variations have a slight apparent functional impact, known as DNA polymorphisms. By agreement, polymorphism is a variation in the DNA sequence that takes place in ≥1% of the population genetics. Sometimes, these changes in single nucleotide polymorphisms (SNPs) by substitution of a nucleotide at an exact location in the gene sequences increase or decrease the susceptibility to specific diseases, which is used for several purposes in molecular medicine.

This chapter will highlight the association of Fc gamma receptor (FcγR) genetics with malaria infection.

4.1.1 Malaria

Malaria is a serious health issue in sub-Saharan Africa and different regions in Asia and South America. The World Health Organization (WHO) in 2018 estimated that the number of malaria cases was about 228 million cases globally (lower 95% confidence interval (CI) 206–258 million) (WHO, 2019), with 405,000 deaths. Most of them (67%) were children aged under 5 years (WHO, 2019) and pregnant women.

Malaria is a disease caused by the obligatory intraerythrocytic protozoan parasites of the genus *Plasmodium*. Individuals may be affected by a single or multiple *Plasmodium* (*P.*) types that have a different prevalence in the world depending on the region. These species include *P. falciparum*, *P. vivax*, *P. ovale*, *P. malariae* and *P. knowlesi*. The *Plasmodium* parasites are

DOI: 10.1201/9781003246244-4

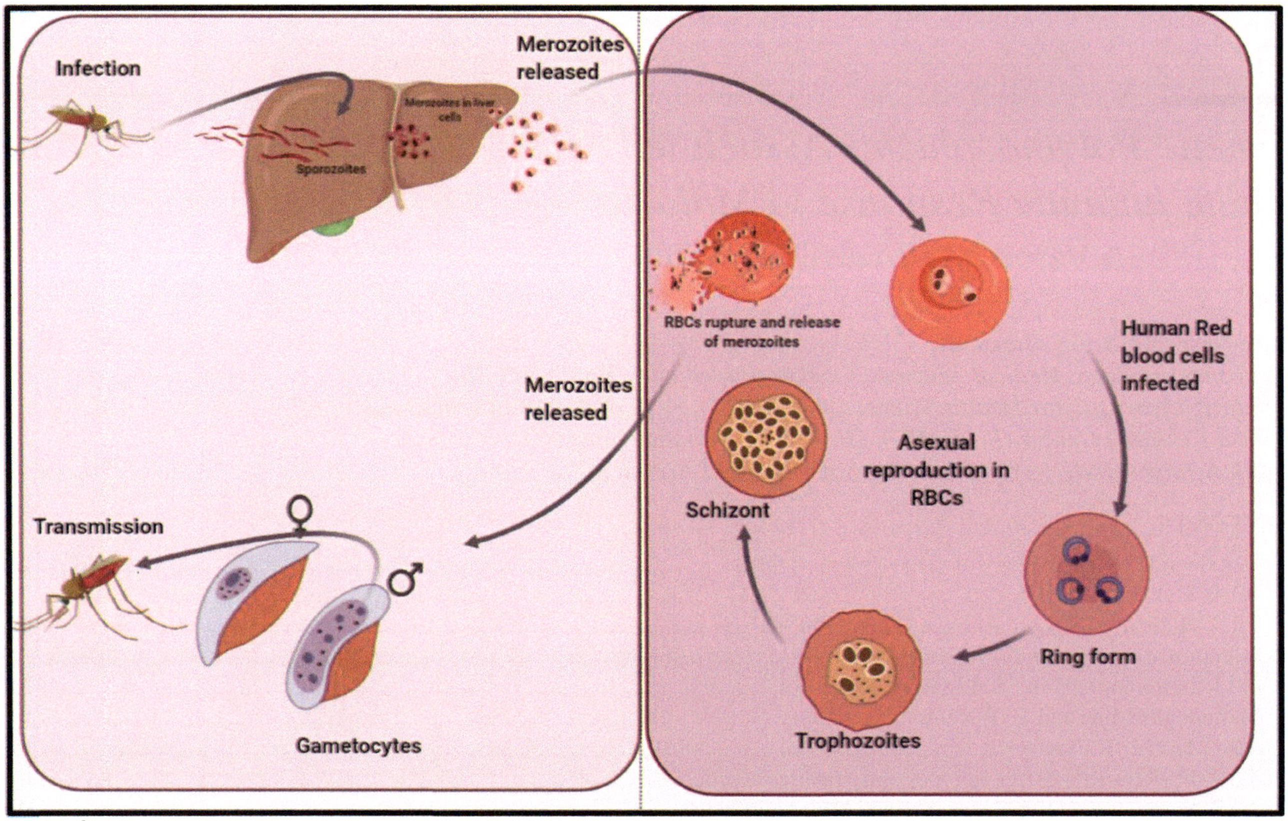

FIGURE 4.1 The life cycle of *P. falciparum* in the human (host) and mosquito (vector).

usually transmitted from an infected person by the female Anopheles mosquitoes; also, the infection may result from direct contact with contaminated blood products, i.e., transfusion malaria. *P. falciparum* is the predominant of the five *Plasmodium* species and is accountable for the deaths and most of the clinical presentations related to malaria. In 2019, most of the global mortality is associated with *P. falciparum* (61%) and 31% with *P. vivax* (WHO, 2019) (Figure 4.1).

Many factors can contribute to malaria susceptibilities, such as environmental factors, naturally acquired immunity and host genetic factors. To understand the genetic susceptibility and/or resistance of the host to malaria, it is necessary to recognize the nature of the human immune response and its interaction with parasitic infection. This chapter focuses on the analysis of some of the genetic variables that contribute to the host immune response to malaria and the outcome of the disease.

4.1.2 Human Leukocyte Antigen (HLA) and Malaria

HLA genes encode different sets of 22 proteins accountable for interceding host-parasite interactions. HLA is a cell surface glycoprotein encoded by six classical genes (HLA class I: A, B, C and HLA class II: DR, DQ and DP) expressed over the cells to present the processed antigenic peptide to T cells, which play an important role in immune response (Abbas et al., 2014; Mota and Rodriguez, 2017). HLA class I is expressed on all nucleated cells to present the processed antigenic peptide to be recognized by T cell receptor (TCR), including CD8+ cytotoxic T cells (CTL), NKG2D receptor of several immune effector cells and natural killer (NK) cells (Molinero et al., 2002; Vivier et al., 2002), which play a role in cellular immunity (Abbas et al., 2014; Mota and Rodriguez, 2017). HLA class II is expressed over antigen-presenting cells (APCs), which include dendritic cells, macrophage and B cells to be recognized by CD4+ T-helper (Th), which are responsible for humoral immunity by secreting specific antibodies (Abbas et al., 2014; Mota and Rodriguez, 2017).

A previous case-control study by Adrian Hill and associates suggested that the HLA class I (HLA-Bw53) allele and HLA class II haplotype (DRB1*1302-DQB1*0501) were independently associated with protection (Hill et al., 1991). Another similar study in West Africa suggested that the HLA-A2.1 allele, which is an epitope from the thrombospondin-related anonymous protein (TRAP), is a naturally processed circumsporozoite (CS) antigenic protein to be presented over MHC class I to be recognized by CTL from naturally exposed humans. The second alleles HLA-B7 and HLA-B35 restricted the epitope for the pre-erythrocytic (liver) stage for *P. falciparum* (CS antigenic protein), which will cognate with altered peptide ligand (APL) in CTL that may occur in natural CTL responses to *P. falciparum*. The two variant alleles of HLA may explain the role of CTL against malaria parasites in severe liver-stage malaria in West African children (Doolan and Good, 1992; Lalvani et al., 1996; Udhayakumar et al., 1997).

A study including Thai malaria patients revealed that the HLA-B46 allele and DRB1*1001 haplotypes were statistically significantly associated with the severity of malaria (Hananantachai et al., 2005). A recent review by Tukwasibwe and associates (Tukwasibwe et al., 2020) summarized the human killer cell immunoglobulin-like receptors (KIRs) expressed over NK cells usually inhibit the functional activity of NK when they can be recognized by HLA class I (Yindom et al., 2012). They were strongly associated with higher *P. falciparum* parasite prevalence among Ugandan patients (Digitale et al., 2020).

Based on a recent review article published by Alicia Sanchez-Mazas (Sanchez-Mazas, 2020), the HLA alleles frequency in several study populations suggested that the HLA alleles A*30:01, A*33:01, DPB1*17:01 and DRB1*04:01 play an important role in susceptibility to malaria infection, and the HLA alleles B*53:01, DQB1*05:01, DRB1*01:01 and DRB1*13:02 are related to resistance to malaria infection (Garcia et al., 2013; Sabbagh et al., 2013). A recent genetics modeling study suggested that the HLA-B*53:01 shows a protective effect against *P. falciparum* infection in Africa population (Sanchez-Mazas et al., 2017).

These factors can be summarized into the following:

I. Malaria endemicity indicates disease prevalence and can be classified into four different areas depending on age, spleen rate and parasite rate.
   a. **Holoendemic Areas**: Malaria transmission occurs the whole year.
   b. **Hyperendemic Areas**: Malaria transmission is intense, but with periods of no transmission during the dry season.
   c. **Mesoendemic Areas**: Malaria transmission is a seasonal transmission.
   d. **Hypendemic Areas**: Malaria transmission is unstable.

II. Changes in antibodies with infection.

III. Host genetic factors.

4.2 Structure and Function of Immunoglobulins

Immunoglobulins (Ig) or antibodies are heterodimeric proteins expressed over B cell as receptors or can be circulated in the serum. Structurally, Ig is composed of 4 chains, 2 identical heavy chains (each chain composed of 440–550 amino acids) and 2 identical light chains (each chain composed of 220 amino acids). Each light chain has one "variable" and one "constant" region. The heavy chains on human Ig also consist of variable and constant regions. Depending on these regions, the Ig can be further classified into five isotypes, which are IgA, IgD, IgG, IgE and IgM (Warmerdam et al., 1991). In general, the functions of antibodies are (1) antigen recognition molecules (proteins) of B cells. (2) Neutralization of the antigens: Most neutralizing antibodies in the blood are of IgG isotype, and in the mucosal organs, they are of IgA isotype. (3) Complement activation: The antibodies activate mainly the classical complement pathway. IgM and IgG3 are the most effective complement activators, followed by IgG1 and IgG2, while IgG4 cannot activate complement. (4) Opsonization: IgG1 and IgG3 bind with high affinity to Fc receptors expressed on phagocytic cells such as macrophage and neutrophils and mediate opsonization. (5) Activation of other immune cells, such as NK cells and eosinophils, to play an important role in antibody-dependent cellular cytotoxicity (ADCC) or antibody-dependent cellular inhibitions (ADCI) in malaria patients.

4.2.1 Immunoglobulins and Malaria Protection

Immunoglobulins or antimalarial antibodies can protect the human body from the infection, and this has previously been demonstrated by serum transfer studies in humans (Cohen et al., 1961) and animal models (Orjih et al., 1981).

There are many factors that contribute to controlling malaria infection, through specific antimalarial antibodies that can mediate protection in clinical malaria. Antibodies play a crucial part in defense against malarial parasites within the asexual blood stage (Nasr et al., 2007, 2011, 2013; Sabchareon et al., 1991). The antibody level of response to specific malaria antigen targets is usually decreased in the absence of malaria infection or re-infection, which is believed to be related to immune memory against malaria (Achtman et al., 2005; Alonso et al., 2005). Data highlight the important role of IgG in the defense against malaria infection and susceptibility to malaria (Iriemenam et al., 2009; Jerne, 1955; Kinyanjui et al., 2007; Marsh et al., 1989; Munde et al., 2017). Many studies suggested that the IgG1 and IgG3 were considered as cytophilic antibodies playing a key role in protection against *P. falciparum*, while IgG2 and IgG4 were thought to block protective mechanisms (Adamou et al., 2019; Aucan et al., 2000; Kwenti et al., 2019; Nasr et al., 2013, 2007). It is demonstrated that in some Sudanese tribe's higher levels of specific IgG2 and lower levels of IgG4 antibodies play an important role in malarial protective immunity properties (Nasr et al., 2007). Similar results have been reported in children in Burkina Faso and Gabon (Aucan et al., 2000; Ntoumi et al., 2005).

4.3 Immunoglobulins and Fc Receptors

Fc immunoglobulin (Ig) domain receptors have been identified on the cell surface of most immune cells. Fc receptors (FcRs) serve a vital role in immune regulation as they are required to align cell immunity with humoral (antibody-mediated) immune responses. This link will activate the antibodies-dependent cellular immunity (Koenderman, 2019; Kubagawa et al., 2019). Specific Fc receptors (FcRs) are expressed on the cell membrane for most of the immune cells to bind specific class and subclass of immunoglobulin, and each antibody has a specific receptor binding with a different affinity, including IgA

(FcαR), IgD (FcδR), IgE (FcεR), IgG (FcγR) and IgM (FcμR) (Koenderman, 2019; Kubagawa et al., 2019).

4.3.1 The FcγR Family

In this chapter, we will deal with IgG receptors' gene polymorphisms, specifically FcγRIIa, FcγRIIIa and FcγRIIIb, with more detailed information on each of them in the subsequent section.

The FcγR is a cell member receptor for IgG antibody expressed in most of the effector cells of the immune system. The FcγR consists of three subfamilies of receptors, labeled in Figure 4.2. They are classified into FcγR (I, II and III) (Ravetch and Bolland, 2001). FcγRI (CD64), which includes subclass FcγR (Ia, Ib and Ic), results from SNPs in humans (Ravetch and Bolland, 2001).

4.3.2 FcγRIIa-R/H131 Genotypes and Malaria Infection

The FcγR is an activating receptor that prompts the activation of phagocytic cells to induce the opsonization mechanism within ADCI (Adu et al., 2012; van Sorge et al., 2003). Several studies suggested that the FcγRIIa-R/R131 genotypes were significantly associated with susceptibility to severe malaria (Nasr et al., 2007; Ouma et al., 2006). Previous studies in the high malaria transmission area suggested that the FcγRIIa-R/R131 genotypes and R131 allele were involved in parasite clearance and also play a protective role against severe malaria (Cooke et al., 2003; Omi et al., 2002; Shi et al., 2001). In addition, a study in Kenya also showed the same finding (Bruhns et al., 2009). FcγRIIa-H/H131 genotypes and H131 allele were statistically associated with the protection against severe malaria (Nasr et al., 2007; Sinha et al., 2008). This protective role is due to the ability of FcγRIIa-H/H131 genotypes and H131 allele to bind the cytophilic IgG1 and IgG3 subclasses (Nasr et al., 2007, 2009). In addition, malaria patients carrying FcγRIIa-H131 genotypes significantly associated with increased levels of IgG2 antibodies demonstrated protection against malaria (Nasr et al., 2007, 2009). FcγRIIa-H/H131 genotypes are associated with an elevated level of antimalarial IgG2, which is actively engaged in parasite clearance and also reduces the risk of severe malaria (Aucan et al., 2000; Nasr et al., 2007, 2009; Ntoumi et al., 2005). IgG2 is a weak stimulator for the classical complement pathway. In an earlier finding from Sudan, IgG3 and IgG2 have been shown to play a protective role against malaria infection, which may activate ADCI through FcγR in malarial infection.

4.3.3 FcγRIIIa-F/V176 Genotypes and Malaria Infection

The FcγRIIIa prompts the activation of phagocytic cells with two alleles, the 176V and the 176F (Ravetch and Kinet, 1991). FcγRIIIa-F/V176 genotypes have an influence on the binding of the IgG and IgG subclasses. The FcγRIIIa-V/V176 allele has a higher binding affinity for IgG1 and IgG3, as compared

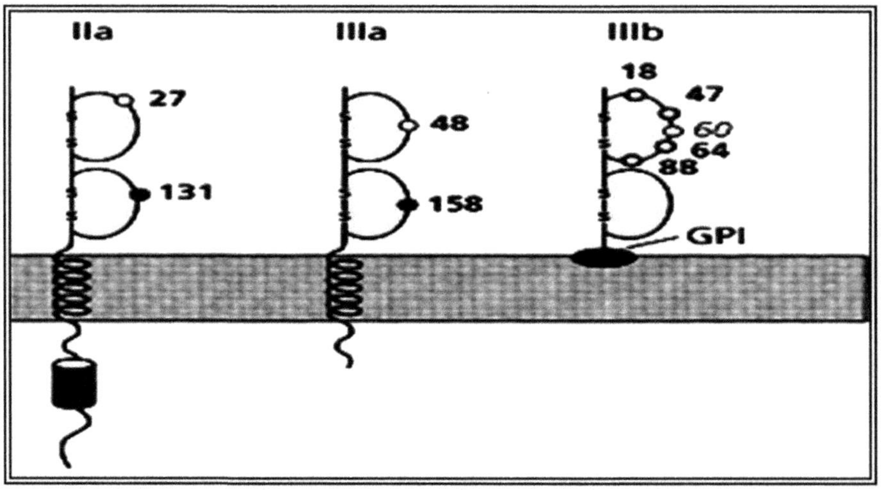

FIGURE 4.2 This figure is adopted from (Ravetch and Bolland, 2001) and shows gene polymorphisms. The polymorphisms in FcγRIIa in position 131 – the amino acid arginine (R) is replaced with amino acid histidine (H). The FcγRIIIa with the point mutation at position 158, where the amino acid valine (V) is substituted by the amino acid phenylalanine (F). The FcγRIIIb with mutation in four different amino acid substations in positions 18, 47, 64 and 88, which results in glycosylation differences, affecting receptor affinity (Hulett and Hogarth, 1994; Hulett et al., 1995) (Table 4.1).

TABLE 4.1
The Binding Affinity for IgG Subclasses and FcγR Family

FcγR	IgG1	IgG2	IgG3	IgG4
FcγRI (CD64) (Nasr, 2008)	K_d of 10^{-8} M	$K_d >> 10^{-9}$ M	K_d of 10^{-9} M	$K_d >> 10^{-9}$ M
FcγRII (CD32) (Nasr, 2008)	$K_d > 10^{-7}$ M	Non-existent	$K_d > 10^{-7}$ M	Non-existent
FcγRIII (CD16) (Galon et al., 1997)	$K_d > 10^{-3}$ M	$K_d >>> 10^{-3}$ M	$K_d > 10^{-5}$ M	$K_d >> 10^{-3}$ M

to FcγRIIIa-F/F176 (Koene et al., 1997), which is important in the immune response against malaria infection. A study in western Kenya suggested that FcγRIIIa-F/V176 genotypes were not associated with susceptibility to severe malarial anemia in children patients infected with malaria infection (Munde et al., 2017). A previous study in Thai revealed that FcγRIIIa-F/V176 genotypes were statistically associated with the severity of malaria disease (Omi et al., 2002). FcγRIIIa-F/F genotypes were significantly less binding to IgG subclasses (1, 3 and 4) compared to FcγRIIIa-V/V genotypes, which bind to cytophilic IgG subclasses with high affinity [34]. FcγRIIIa-F/F176 genotypes led to the opsonization of malaria parasites clearance [34]. In general, FcγRIIIa is expressed over NK cells and can bind to IgG, which results in the release of pro-inflammatory cytokines interluken-1 beta (IL-1β), interferon-gamma (IFN-γ) and tumor necrosis factor-alpha (TNF-α) (Lyke et al., 2004).

4.3.4 FcγRIIIb-NA1/NA2 Genotypes and Malaria Infection

FcγRIIIb-NA1/NA2 genotypes are expressed over neutrophils with phenotypes (NA1 and NA2) in four different positions in the genome, resulting in different possibilities to bind the phagocytic cells. FcγRIIIb-NA2/NA2 genotypes were associated with reduced phagocytosis relative to FcγRIIIb-NA1/NA1 (Ouma et al., 2012; Salmon et al., 1990). Another finding in Ghanaian children suggested that the FcγRIIIb-NA2/NA2 genotype was statistically significantly associated with clinical malaria (Adu et al., 2012). A published research thesis in 2010 demonstrated that the FcγRIIIb-NA2/NA2 was associated with susceptibility to severe and uncomplicated malaria among Ghanaian children (Adu, 2010).

4.4 Toll-Like Receptors' Polymorphisms, Malaria Susceptibility and Pathogenesis

Toll-like receptors (TLRs) are a group of pathogen recognition receptors that detect pathogen-associated molecular patterns (PAMPs). TLRs play an important role in innate immunity, especially in pathogen recognition within the early phase of activation the inflammation. The signaling with TLR in the progression or protection from malaria has been described by many of the previous studies investigating the TLR gene polymorphisms and association of disease outcome. Previous studies in African children suggested that polymorphisms in TLR-4 (Mockenhaupt et al., 2006) and TLR-9 (Munde et al., 2012; Omar et al., 2012) were correlated with severe malaria, symptomatic malaria, and severe malarial anemia. In addition to that TLR-9 gene polymorphism was associated with high parasite density (Leoratti et al., 2008). Another study in Burundian children suggested that the TLR-9 gene polymorphism has been associated with susceptibility to malaria. However, it is not associated with the severity of the disease (Leoratti et al., 2008). Another case-control study in individuals from many countries with aggressive pneumococcal disease, malaria, tuberculosis and bacteremia suggested that no differences in associations were detected with TLR-4 and Toll-interleukin 1 receptor domain-containing adaptor protein (TIRAP) polymorphisms between mild and complicated malaria cases, a protective association with TIRAP Ser180L heterozygosity and malaria disease (Khor et al., 2007). Taking together, much of the contradictory data suggest that changes in the expression of TLR-2, TLR-4 and TLR-9 could potentially affect the inflammatory immune response and disease outcome. Several studies have investigated the potential selective pressure of malaria exposure on the presence of particular polymorphisms.

4.5 Tumor Necrosis Factor (TNF)

TNF is a potent pro-inflammatory mediator shown to contribute to anti-*Plasmodium* action resulting in intra-erythrocyte death of parasites and therefore a marked decrease in parasitemia (Clark et al., 1981). Increased serum TNF is a frequent finding in malaria and is correlated in several studies with malaria severity and parasitemia, reflecting its role in malaria pathogenesis (McGuire et al., 1998; Mordmüller et al., 1997; Shaffer et al., 1991). Mutations and allelic variants of proteins expressed in red blood cells have repeatedly been associated with increased genetic effects. Many of the SNP promotors involved in the regulation of TNF expression are correlated with the degree of parasitemia and with elevated levels of anti-*P. falciparum* IgG (Flori et al., 2005; Nguyen et al., 2017; Wilson et al., 1997).

Several reports revealed the role of TNF promoter variants in the clinical outcome of malaria, suggesting a wide range of genetic effects of TNF on the disease at different stages, as summarized in Table 4.2. The precise mechanism of TNF in malaria pathogenesis remains uncertain, and further investigations are necessary.

TABLE 4.2

Genetic Effects of Tumor Necrosis Factor (TNF) Variant SNPs on Malaria

TNF Variants	Genetic Effects
TNF (rs1799964, C; -1031)	Vulnerability to severe malaria (Dunstan et al., 2012).
TNF (rs1800630, C; -863)	Severe malaria in pregnancy (Nasr et al., 2014).
TNFP-D (rs1799724, T; -857)	Vulnerability to cerebral malaria (Ubalee et al., 2001).
TNF (rs673, A; -244)	Increased parasitemia (Nguyen et al., 2017).
TNF (rs3093664, A; 1034)	Progression to clinical malaria (Flori et al., 2005).

4.6 Nitric Oxide (NO)

Most cells in the body produce nitric oxide (NO), which is a free radical involved in various physiological processes and inflammatory reactions, but its vascular endothelium production is particularly important in regulating blood flow. Reduced NO bioavailability is documented to contribute to the pathogenesis of severe malaria. However, increased NO showed a protective effect on cerebral malaria through inhibition of systemic inflammation and enhancement of endothelial cell function (Anstey et al., 1999; Gramaglia et al., 2006; Sobolewski et al., 2005; Weinberg et al., 2008).

Nitric oxide synthase-2 is a protein-coding gene for the inducible nitric oxide synthase (iNOS), which is responsible for production of nitric oxide in different tissues. Various studies have shown an association between NOS2 polymorphism variants and the course of malaria (Gudo et al., 2013). Variants that increase the bioavailability of NO have shown positive impacts on clinical presentation of malaria, such as reduced incidence of uncomplicated malaria, increased frequency of asymptomatic malaria (rs3794767 and rs3729508) and, more generally, lower incidence of severe malaria involving obvious protection against cerebral malaria (rs6505469) (de Jesus Trovoada et al., 2014; Hobbs et al., 2002). Strong evidence indicates that NO works through anti-inflammatory pathways to interrupt anti-parasitic reactions, raise susceptibility to infection, and also protect against stronger inflammatory reactions correlated with the progression in clinical manifestations of malaria.

4.7 Heme Oxygenase

Heme oxygenase (HO) is an enzyme that breaks down the heme to produce biliverdin, iron and carbon monoxide and comprises two main isoforms, HO-1 and HO-2 (Maines, 1997). The blood phase of malaria infection may be associated with massive destruction of red cells and, as a result, increased circulating heme, which provokes inflammation and produces extreme toxic effects on tissues. Heme oxygenase-1 (HO-1) encoded with the HMOX1 gene showed anti-inflammatory activity and protection from the adverse effects of malaria hemolysis and protection against severe and cerebral malaria in various laboratory models (Parmley et al., 1984; Seixas et al., 2009). Inversely, HO-1 has been documented to augment the murine hepatic stage of malaria infection, suggesting that an optimum concentration of HO-1 is needed to balance disease progression and protection from infection (Epiphanio et al., 2008).

Induction of HO-1 in several diseases, including malaria, has been shown to be influenced by the length of GT-repeat alleles. Short repeats related to severe malaria and gene polymorphisms showed increased mortality and predisposition to severe and cerebral malaria (Sambo et al., 2010; Shibahara, 2003; Takeda et al., 2005). In contrast, the long GT-repeat polymorphisms demonstrated protection from mild malaria in some cases (Mendonça et al., 2012).

4.8 Conclusions

Significant efforts have been made to detect the host genetic factors related to innate immune response that may play a key role in the protection and/or pathogenesis of malaria. Many factors can contribute to genetics-associated findings such as the different malaria transmission settings and malaria epidemic. Furthermore, different ethnic populations associated with variations in the genetic background may significantly contribute to the FcγR, TLR, TNF, NO and HO gene polymorphisms and susceptibility/protection to severe malaria.

ABBREVIATIONS

ADCC:	Antibody-dependent cellular cytotoxicity
ADCI:	Antibody-dependent cellular inhibitions
DNA:	Deoxyribonucleic acid
FcγR:	Fragment crystallizable gamma receptor
HLA:	Human leukocyte antigen
HO:	Heme oxygenase
IFN-γ:	Interferon-gamma
Ig:	Immunoglobulins
IL-1β:	Interluken-1 beta
NK cells:	Natural killer cells
NO:	Nitric oxide
SNPs:	Single nucleotide polymorphisms
TLRs:	Toll-like receptors
TNF:	Tumor necrosis factor

REFERENCES

Abbas, A.K., Lichtman, A.H., Pillai, S., 2021. *Cellular and Molecular Immunology E-Book*. Elsevier Health Sciences, Philadelphia.

Achtman, A., Bull, P., Stephens, R., Langhorne, J., 2005. Longevity of the immune response and memory to blood-stage malaria infection. *Immunology and Immunopathogenesis of Malaria*. Springer-verlag berlin heidelberg, Simon Rallison, Heidelberg, pp. 71–102.

Adamou, R., Dechavanne, C., Sadissou, I., d'Almeida, T., Bouraima, A., Sonon, P., Amoussa, R., Cottrell, G., Le Port, A., Theisen, M., 2019. *Plasmodium falciparum* merozoite surface antigen-specific cytophilic IgG and control of malaria infection in a Beninese birth cohort. *Malaria Journal* 18, 194.

Adu, B., 2010. Immunological and Genetic Correlates of Immunity to *Plasmodium falciparum* Malaria. A (M.Phil.) Thesis Submitted to the Department of Biochemistry and Biotechnology, Kwame Nkrumah University of Science and Technology, Kumasi http://ir.knust.edu.gh/handle/123456789/481.

Adu, B., Dodoo, D., Adukpo, S., Hedley, P.L., Arthur, F.K., Gerds, T.A., Larsen, S.O., Christiansen, M., Theisen, M., 2012. Fc gamma receptor IIIB (FcgammaRIIIB) polymorphisms are associated with clinical malaria in Ghanaian children. *PLoS One* 7, e46197.

Alonso, P.L., Sacarlal, J., Aponte, J.J., Leach, A., Macete, E., Aide, P., Sigauque, B., Milman, J., Mandomando, I., Bassat, Q.,

2005. Duration of protection with RTS, S/AS02A malaria vaccine in prevention of *Plasmodium falciparum* disease in Mozambican children: single-blind extended follow-up of a randomised controlled trial. *The Lancet* 366, 2012–2018.

Anstey, N.M., Granger, D.L., Hassanali, M.Y., Mwaikambo, E.D., Duffy, P.E., Weinberg, J.B., 1999. Nitric oxide, malaria, and anemia: inverse relationship between nitric oxide production and hemoglobin concentration in asymptomatic, malaria-exposed children. *The American Journal of Tropical Medicine and Hygiene* 61, 249–252.

Aucan, C., Traore, Y., Tall, F., Nacro, B., Traore-Leroux, T., Fumoux, F., Rihet, P., 2000. High immunoglobulin G2 (IgG2) and low IgG4 levels are associated with human resistance to *Plasmodium falciparum* malaria. *Infect and Immunity* 68, 1252–1258.

Bruhns, P., Iannascoli, B., England, P., Mancardi, D.A., Fernandez, N., Jorieux, S., Daëron, M., 2009. Specificity and affinity of human Fcγ receptors and their polymorphic variants for human IgG subclasses. *Blood* 113, 3716–3725.

Clark, I., Virelizier, J., Carswell, E., Wood, P., 1981. Possible importance of macrophage-derived mediators in acute malaria. *Infection and Immunity* 32, 1058–1066.

Cohen, S., McGregor, I., Carrington, S., 1961. Gamma-globulin and acquired immunity to human malaria. *Nature* 192, 733–737.

Cooke, G.C., Aucan, C., Walley, A.J., Segal, S., Greenwood, B.M., Kwiatkowski, D.P., Hill, A.V.S., 2003. Association of Fcgamma receptor IIa (CD32) polymorphism with severe malaria in West Africa. *The American Journal of Tropical Medicine and Hygiene* 69, 565–568.

de Jesus Trovoada, M., Martins, M., Mansour, R.B., do Rosário Sambo, M., Fernandes, A.B., Gonçalves, L.A., Borja, A., Moya, R., Almeida, P., Costa, J., 2014. NOS2 variants reveal a dual genetic control of nitric oxide levels, susceptibility to Plasmodium infection, and cerebral malaria. *Infection and Immunity* 82, 1287–1295.

Digitale, J.C., Callaway, P.C., Martin, M., Nelson, G., Viard, M., Rek, J., Arinaitwe, E., Dorsey, G., Kamya, M., Carrington, M., 2020. Inhibitory KIR ligands are associated with higher P. falciparum parasite prevalence. medRxiv.

Doolan, D.L., Good, M.F., 1992. *Plasmodium falciparum* CS protein-prime malaria vaccine candidate: definition of the human CTL domain and analysis of its variation. *Memórias do Instituto Oswaldo Cruz* 87, 241–247.

Dunstan, S.J., Rockett, K.A., Quyen, N.T.N., Teo, Y.Y., Thai, C.Q., Hang, N.T., Jeffreys, A., Clark, T.G., Small, K.S., Simmons, C.P., 2012. Variation in human genes encoding adhesion and proinflammatory molecules are associated with severe malaria in the Vietnamese. *Genes & Immunity* 13, 503–508.

Epiphanio, S., Mikolajczak, S.A., Gonçalves, L.A., Pamplona, A., Portugal, S., Albuquerque, S., Goldberg, M., Rebelo, S., Anderson, D.G., Akinc, A., 2008. Heme oxygenase-1 is an anti-inflammatory host factor that promotes murine plasmodium liver infection. *Cell Host & Microbe* 3, 331–338.

Flori, L., Delahaye, N., Iraqi, F., Hernandez-Valladares, M., Fumoux, F., Rihet, P., 2005. TNF as a malaria candidate gene: polymorphism-screening and family-based association analysis of mild malaria attack and parasitemia in Burkina Faso. *Genes & Immunity* 6, 472–480.

Galon, J., Robertson, M.W., Galinha, A., Maziéres, N., Spagnoli, R., Fridman, W.H., Sautès, C., 1997. Affinity of the interaction between Fcgamma receptor type III (FcγRIII) and monomeric human IgG subclasses. Role of FcγRIII glycosylation. *European Journal of Immunology* 27, 1928–1932.

Garcia, A., Milet, J., Courtin, D., Sabbagh, A., Massaro, J.D., Castelli, E.C., Migot-Nabias, F., Favier, B., Rouas-Freiss, N., Donadi, E.A., 2013. Association of HLA-G 3′ UTR polymorphisms with response to malaria infection: a first insight. *Infection, Genetics and Evolution* 16, 263–269.

Gramaglia, I., Sobolewski, P., Meays, D., Contreras, R., Nolan, J., Frangos, J., Intaglietta, M., van der Heyde, H.C., 2006. Low nitric oxide bioavailability contributes to the genesis of experimental cerebral malaria. *Nature Medicine* 12, 1417–1422.

Gudo, E.S., Prista, A., Jani, I.V., 2013. Impact of asymptomatic *Plasmodium falciparum* parasitemia on the imunohematological indices among school children and adolescents in a rural area highly endemic for Malaria in southern Mozambique. *BMC Infectious Diseases* 13, 244.

Hananantachai, H., Patarapotikul, J., Ohashi, J., Naka, I., Looareesuwan, S., Tokunaga, K., 2005. Polymorphisms of the HLA-B and HLA-DRB1 genes in Thai malaria patients. *Japanese Journal of Infectious Diseases* 58, 25–28.

Hill, A.V., Allsopp, C.E., Kwiatkowski, D., Anstey, N.M., Twumasi, P., Rowe, P.A., Bennett, S., Brewster, D., McMichael, A.J., Greenwood, B.M., 1991. Common west African HLA antigens are associated with protection from severe malaria. *Nature* 352, 595–600.

Hobbs, M.R., Udhayakumar, V., Levesque, M.C., Booth, J., Roberts, J.M., Tkachuk, A.N., Pole, A., Coon, H., Kariuki, S., Nahlen, B.L., 2002. A new NOS2 promoter polymorphism associated with increased nitric oxide production and protection from severe malaria in Tanzanian and Kenyan children. *The Lancet* 360, 1468–1475.

Hulett, M.D., Hogarth, P.M., 1994. Molecular basis of Fc receptor function. *Advances in Immunology* 57, 1–127.

Hulett, M.D., Witort, E., Brinkworth, R.I., McKenzie, I.F., Hogarth, P.M., 1995. Multiple regions of human Fc gamma RII (CD32) contribute to the binding of IgG. *Journal of Biological Chemistry* 270, 21188–21194.

Iriemenam, N.C., Khirelsied, A.H., Nasr, A., ElGhazali, G., Giha, H.A., Elhassan, A.E.T.M., Agab-Aldour, A.A., Montgomery, S.M., Anders, R.F., Theisen, M., Troye-Blomberg, M., Elbashir, M.I., Berzins, K., 2009. Antibody responses to a panel of *Plasmodium falciparum* malaria blood-stage antigens in relation to clinical disease outcome in Sudan. *Vaccine* 27, 62–71.

Jerne, N.K., 1955. The natural-selection theory of antibody formation. *Proceedings of the National Academy Sciences of United States of America* 41, 849–857.

Khor, C.C., Chapman, S.J., Vannberg, F.O., Dunne, A., Murphy, C., Ling, E.Y., Frodsham, A.J., Walley, A.J., Kyrieleis, O., Khan, A., 2007. A Mal functional variant is associated with protection against invasive pneumococcal disease, bacteremia, malaria and tuberculosis. *Nature Genetics* 39, 523–528.

Kinyanjui, S.M., Conway, D.J., Lanar, D.E., Marsh, K., 2007. IgG antibody responses to *Plasmodium falciparum* merozoite antigens in Kenyan children have a short half-life. *Malaria Journal* 6, 82.

Koenderman, L., 2019. Inside-out control of Fc-receptors. *Frontiers in Immunology* 10, 544.

Koene, H.R., Kleijer, M., Algra, J., Roos, D., EG Kr. von dem Borne, A., de Haas, M., 1997. FcγRIIIa-158V/F polymorphism influences the binding of IgG by natural killer cell FcγRIIIa, independently of the FcγRIIIa-48L/R/H phenotype. *Blood, The Journal of the American Society of Hematology* 90, 1109–1114.

Kubagawa, H., Honjo, K., Ohkura, N., Sakaguchi, S., Radbruch, A., Melchers, F., Jani, P.K., 2019. Functional roles of the IgM Fc receptor in the immune system. *Frontiers in Immunology* 10, 945.

Kwenti, T.E., Kukwah, T.A., Kwenti, T.D.B., Nyassa, B.R., Dilonga, M.H., Enow-Orock, G., Tendongfor, N., Anong, N.D., Wanji, S., Njunda, L.A., 2019. Comparative analysis of IgG and IgG subclasses against *Plasmodium falciparum* MSP-1 19 in children from five contrasting bioecological zones of Cameroon. *Malaria Journal* 18, 16.

Lalvani, A., Hurt, N., Aidoo, M., Kibatala, P., Tanner, M., Hill, A.V., 1996. Cytotoxic T lymphocytes to *Plasmodium falciparum* epitopes in an area of intense and perennial transmission in Tanzania. *European Journal of Immunology* 26, 773–779.

Leoratti, F.M., Farias, L., Alves, F.P., Suarez-Mútis, M.C., Coura, J.R., Kalil, J., Camargo, E.P., Moraes, S.L., Ramasawmy, R., 2008. Variants in the toll-like receptor signaling pathway and clinical outcomes of malaria. *The Journal of Infectious Diseases* 198, 772–780.

Lyke, K., Burges, R., Cissoko, Y., Sangare, L., Dao, M., Diarra, I., Kone, A., Harley, R., Plowe, C., Doumbo, O., 2004. Serum levels of the proinflammatory cytokines interleukin-1 beta (IL-1β), IL-6, IL-8, IL-10, tumor necrosis factor alpha, and IL-12 (p70) in Malian children with severe *Plasmodium falciparum* malaria and matched uncomplicated malaria or healthy controls. *Infection and Immunity* 72, 5630–5637.

Maines, M.D., 1997. The heme oxygenase system: a regulator of second messenger gases. *Annual Review of Pharmacology and Toxicology* 37, 517–554.

Marsh, K., Otoo, L., Hayes, R.J., Carson, D.C., Greenwood, B.M., 1989. Antibodies to blood stage antigens of *Plasmodium falciparum* in rural Gambians and their relation to protection against infection. *Transactions of the Royal Society of Tropical Medicine and Hygiene* 83, 293–303.

McGuire, W., D'Alessandro, U., Stephens, S., Olaleye, B., Langerock, P., Greenwood, B., Kwiatkowski, D., 1998. Levels of tumour necrosis factor and soluble TNF receptors during malaria fever episodes in the community. *Transactions of the Royal Society of Tropical Medicine and Hygiene* 92, 50–53.

Mendonça, V.R., Luz, N.F., Santos, N.J., Borges, V.M., Gonçalves, M.S., Andrade, B.B., Barral-Netto, M., 2012. Association between the haptoglobin and heme oxygenase 1 genetic profiles and soluble CD163 in susceptibility to and severity of human malaria. *Infection and Immunity* 80, 1445–1454.

Mockenhaupt, F.P., Cramer, J.P., Hamann, L., Stegemann, M.S., Eckert, J., Oh, N.-R., Otchwemah, R.N., Dietz, E., Ehrhardt, S., Schröder, N.W., 2006. Toll-like receptor (TLR) polymorphisms in African children: common TLR-4 variants predispose to severe malaria. *Proceedings of the National Academy of Sciences* 103, 177–182.

Molinero, L., Marcos, C., Mirbaha, F., Fainboim, L., Stastny, P., Zwirner, N., 2002. Codominant expression of the polymorphic MICA alloantigens encoded by genes in the HLA region. *European Journal of Immunogenetics* 29, 315–319.

Mordmüller, B., Metzger, W., Juillard, P., Brinkman, B., Verweij, C., Grau, G., Kremsner, P., 1997. Tumor necrosis factor in *Plasmodium falciparum* malaria: high plasma level is associated with fever, but high production capacity is associated with rapid fever clearance. *European Cytokine Network* 8, 29–35.

Mota, M.M., Rodriguez, A., 2017. *Malaria: Immune Response to Infection and Vaccination*. Springer International Publishing.

Munde, E.O., Okeyo, W.A., Anyona, S.B., Raballah, E., Konah, S., Okumu, W., Ogonda, L., Vulule, J., Ouma, C., 2012. Polymorphisms in the Fc gamma receptor IIIA and toll-like receptor 9 are associated with protection against severe malarial anemia and changes in circulating gamma interferon levels. *Infection and Immunity* 80, 4435–4443.

Munde, E.O., Okeyo, W.A., Raballah, E., Anyona, S.B., Were, T., Ong'echa, J.M., Perkins, D.J., Ouma, C., 2017. Association between Fcγ receptor IIA, IIIA and IIIB genetic polymorphisms and susceptibility to severe malaria anemia in children in Western Kenya. *BMC Infectious Diseases* 17, 289.

Nasr, A., 2008. Single nucleotide polymorphisms related to immune responses in *Plasmodium falciparum* malaria. Wenner-Grens institut för experimentell biologi.

Nasr, A., Allam, G., Hamid, O., Al-Ghamdi, A., 2014. IFN-gamma and TNF associated with severe falciparum malaria infection in Saudi pregnant women. *Malaria Journal* 13, 314.

Nasr, A., Hamid, O., Al-Ghamdi, A., Allam, G., 2013. Anti-malarial IgG subclasses pattern and FcgammaRIIa (CD32) polymorphism among pregnancy-associated malaria in semi-immune Saudi women. *Malaria Journal* 12, 110.

Nasr, A., Iriemenam, N.C., Giha, H.A., Balogun, H.A., Anders, R.F., Troye-Blomberg, M., ElGhazali, G., Berzins, K., 2009. FcgammaRIIa (CD32) polymorphism and anti-malarial IgG subclass pattern among Fulani and sympatric ethnic groups living in eastern Sudan. *Malaria Journal* 8, 43.

Nasr, A., Iriemenam, N.C., Troye-Blomberg, M., Arnot, D., Theander, T.G., Berzins, K., Giha, H.A., Elghazali, G., 2011. Pattern of pre-existing IgG subclass responses to a panel of asexual stage malaria antigens reported during the lengthy dry season in Daraweesh, Sudan. *Scandinavian Journal of Immunology* 74, 390–396.

Nasr, A., Iriemenam, N.C., Troye-Blomberg, M., Giha, H.A., Balogun, H.A., Osman, O.F., Montgomery, S.M., ElGhazali, G., Berzins, K., 2007. Fc gamma receptor IIa (CD32) polymorphism and antibody responses to asexual blood-stage antigens of *Plasmodium falciparum* malaria in Sudanese patients. *Scandinavian Journal of Immunology* 66, 87–96.

Nguyen, T., Baaklini, S., Koukouikila-Koussounda, F., Ndounga, M., Torres, M., Pradel, L., Ntoumi, F., Rihet, P., 2017. Association of a functional TNF variant with *Plasmodium falciparum* parasitaemia in a congolese population. *Genes & Immunity* 18, 152–157.

Ntoumi, F., Flori, L., Mayengue, P.I., Matondo Maya, D.W., Issifou, S., Deloron, P., Lell, B., Kremsner, P.G., Rihet, P., 2005. Influence of carriage of hemoglobin AS and the Fc gamma receptor IIa-R131 allele on levels of immunoglobulin G2 antibodies to *Plasmodium falciparum* merozoite antigens in Gabonese children. *The Journal of Infectious Diseases* 192, 1975–1980.

Omar, A.H., Yasunami, M., Yamazaki, A., Shibata, H., Ofori, M.F., Akanmori, B.D., Shuaibu, M.N., Kikuchi, M., Hirayama, K., 2012. Toll-like receptor 9 (TLR9) polymorphism associated with symptomatic malaria: a cohort study. *Malaria Journal* 11, 168.

Omi, K., Ohashi, J., Patarapotikul, J., Hananantachai, H., Naka, I., Looareesuwan, S., Tokunaga, K., 2002. Fc gamma receptor IIA and IIIB polymorphisms are associated with susceptibility to cerebral malaria. *Parasitology International Journal* 51, 361–366.

Orjih, A.U., Cochrane, A.H., Nussenzweig, R.S., 1981. Active immunization and passive transfer of resistance against sporozoite-induced malaria in infant mice. *Nature* 291, 331–332.

Ouma, C., Davenport, G.C., Garcia, S., Kempaiah, P., Chaudhary, A., Were, T., Anyona, S.B., Raballah, E., Konah, S.N., Hittner, J.B., 2012. Functional Haplotypes of Fc gamma (Fcγ) receptor (FcγRIIA and FcγRIIIB) predict risk to repeated episodes of severe malarial anemia and mortality in Kenyan children. *Human Genetics* 131, 289–299.

Ouma, C., Keller, C.C., Opondo, D.A., Were, T., Otieno, R.O., Otieno, M.F., Orago, A.S., Ong'Echa, J.M., Vulule, J.M., Ferrell, R.E., Perkins, D.J., 2006. Association of FCgamma receptor IIA (CD32) polymorphism with malarial anemia and high-density parasitemia in infants and young children. *The American Journal of Tropical Medicine and Hygiene* 74, 573–577.

Parmley, R.T., Takagi, M., Denys, F.R., 1984. Ultrastructural localization of glycosaminoglycans in human term placenta. *The Anatomial Record* 210, 477–484.

Ravetch, J.V., Bolland, S., 2001. IgG Fc receptors. *Annual Review of Immunology* 19, 275–290.

Ravetch, J.V., Kinet, J.-P., 1991. Fc receptors. *Annual Review of Immunology* 9, 457–492.

Sabbagh, A., Courtin, D., Milet, J., Massaro, J., Castelli, E., Migot-Nabias, F., Favier, B., Rouas-Freiss, N., Moreau, P., Garcia, A., 2013. Association of HLA-G 3′ untranslated region polymorphisms with antibody response against *Plasmodium falciparum* antigens: preliminary results. *Tissue Antigens* 82, 53–58.

Sabchareon, A., Burnouf, T., Ouattara, D., Attanath, P., Bouharoun-Tayoun, H., Chantavanich, P., Foucault, C., Chongsuphajaisiddhi, T., Druilhe, P., 1991. Parasitologic and clinical human response to immunoglobulin administration in falciparum malaria. *The American Journal of Tropical Medicine and Hygiene* 45, 297–308.

Salmon, J., Edberg, J., Kimberly, R., 1990. Fc gamma receptor III on human neutrophils. Allelic variants have functionally distinct capacities. *The Journal of Clinical Investigation* 85, 1287–1295.

Sambo, M.R., Trovoada, M.J., Benchimol, C., Quinhentos, V., Gonçalves, L., Velosa, R., Marques, M.I., Sepulveda, N., Clark, T.G., Mustafa, S., 2010. Transforming growth factor beta 2 and heme oxygenase 1 genes are risk factors for the cerebral malaria syndrome in Angolan children. *PloS One* 5, e11141.

Sanchez-Mazas, A., 2020. A review of HLA allele and SNP associations with highly prevalent infectious diseases in human populations. *Swiss Medical Weekly* 150. https://smw.ch/article/doi/smw.2020.20214.

Sanchez-Mazas, A., Černý, V., Di, D., Buhler, S., Podgorná, E., Chevallier, E., Brunet, L., Weber, S., Kervaire, B., Testi, M., 2017. The HLA-B landscape of Africa: signatures of pathogen-driven selection and molecular identification of candidate alleles to malaria protection. *Molecular Ecology* 26, 6238–6252.

Seixas, E., Gozzelino, R., Chora, Â., Ferreira, A., Silva, G., Larsen, R., Rebelo, S., Penido, C., Smith, N.R., Coutinho, A., 2009. Heme oxygenase-1 affords protection against noncerebral forms of severe malaria. *Proceedings of the National Academy of Sciences* 106, 15837–15842.

Shaffer, N., Grau, G.E., Hedberg, K., Davachi, F., Lyamba, B., Hightower, A.W., Breman, J.G., Nguyen-Dinh, P., 1991. Tumor necrosis factor and severe malaria. *Journal of Infectious Diseases* 163, 96–101.

Shi, Y.P., Nahlen, B.L., Kariuki, S., Urdahl, K.B., McElroy, P.D., Roberts, J.M., Lal, A.A., 2001. Fcg Receptor IIa (CD32) Polymorphism is associated with protection of infants against high-density *Plasmodium falciparum* infection. VII. Asembo Bay cohort project. *Journal of Infectious Diseases* 184, 107–111.

Shibahara, S., 2003. The heme oxygenase dilemma in cellular homeostasis: new insights for the feedback regulation of heme catabolism. *The Tohoku Journal of Experimental Medicine* 200, 167–186.

Sinha, S., Mishra, S., Sharma, S., Patibandla, P., Mallick, P., Sharma, S., Mohanty, S., Pati, S., Mishra, S., Ramteke, B., Bhatt, R.M., Joshi, H., Dash, A., Ahuja, R., Awasthi, S., Indian Genome Variation Consortium, Venkatesh, V., Habib, S., 2008. Polymorphisms of TNF-enhancer and gene for FcgammaRIIa correlate with the severity of falciparum malaria in the ethnically diverse Indian population. *Malaria Journal* 7, 13.

Sobolewski, P., Gramaglia, I., Frangos, J., Intaglietta, M., van der Heyde, H.C., 2005. Nitric oxide bioavailability in malaria. *Trends in Parasitology* 21, 415–422.

Takeda, M., Kikuchi, M., Ubalee, R., Na-Bangchang, K., Ruangweerayut, R., Shibahara, S., Imaiand, S.-i., Hirayama, K., 2005. Microsatellite polymorphism in the heme oxygenase-1 gene promoter is associated with susceptibility to cerebral malaria in Myanmar. *Japanese Journal of Infectious Diseases* 58, 268–271.

Tukwasibwe, S., Nakimuli, A., Traherne, J., Chazara, O., Jayaraman, J., Trowsdale, J., Moffett, A., Jagannathan, P., Rosenthal, P.J., Cose, S., 2020. Variations in killer-cell immunoglobulin-like receptor and human leukocyte antigen genes and immunity to malaria. *Cellular & Molecular Immunology*, 1–8.

Ubalee, R., Suzuki, F., Kikuchi, M., Tasanor, O., Wattanagoon, Y., Ruangweerayut, R., Na-Bangchang, K., Karbwang, J., Kimura, A., Itoh, K., 2001. Strong association of a tumor necrosis factor-α promoter allele with cerebral malaria in Myanmar. *Tissue Antigens* 58, 407–410.

Udhayakumar, V., Ongecha, J.M., Shi, Y.P., Aidoo, M., Orago, A., Oloo, A.J., Hawley, W.A., Nahlen, B.L., Hoffman, S.L., Weiss, W.R., 1997. Cytotoxic T cell reactivity and HLA-B35 binding of the variant *Plasmodium falciparum* circumsporozoite protein CD8+ CTL epitope in naturally exposed Kenyan adults. *European Journal of Immunology* 27, 1952–1957.

van Sorge, N.M., van der Pol, W.L., van de Winkel, J.G., 2003. FcgammaR polymorphisms: implications for function, disease susceptibility and immunotherapy. *Tissue Antigens* 61, 189–202.

Vivier, E., Tomasello, E., Paul, P., 2002. Lymphocyte activation via NKG2D: towards a new paradigm in immune recognition? *Current Opinion in Immunology* 14, 306–311.

Warmerdam, P.A., van de Winkel, J.G., Vlug, A., Westerdaal, N.A., Capel, P.J., 1991. A single amino acid in the second Ig-like domain of the human Fc gamma receptor II is critical for human IgG2 binding. *Journal of Immunology* 147, 1338–1343.

Weinberg, J.B., Lopansri, B.K., Mwaikambo, E., Granger, D.L., 2008. Arginine, nitric oxide, carbon monoxide, and endothelial function in severe malaria. *Current Opinion in Infectious Diseases* 21, 468.

WHO, 2019. *World Malaria Report 2019*. Geneva: World Health Organization.

Wilson, A.G., Symons, J.A., McDowell, T.L., McDevitt, H.O., Duff, G.W., 1997. Effects of a polymorphism in the human tumor necrosis factor α promoter on transcriptional activation. *Proceedings of the National Academy of Sciences* 94, 3195–3199.

Yindom, L.M., Forbes, R., Aka, P., Janha, O., Jeffries, D., Jallow, M., Conway, D., Walther, M., 2012. Killer-cell immunoglobulin-like receptors and malaria caused by *Plasmodium falciparum* in The Gambia. *Tissue Antigens* 79, 104–113.

5

Xenobiotic Metabolic Enzyme Polymorphisms and Cancer Susceptibility

Saniya Nissar
Department of Biochemistry, Government Medical College, Karan Nagar, Srinagar, Kashmir, India.
Infinity Biochemistry Pvt. Ltd. Srinagar, Kashmir, India

Syed Sameer Aga
Basic Medical Sciences, College of Medicine, King Saud bin Abdulaziz University for Health Sciences (KSAU-HS), King Abdullah International Medical Research Center (KAIMRC), Ministry of National Guard Health Affairs (NGHA), Jeddah, Kingdom of Saudi Arabia
Infinity Biochemistry Pvt. Ltd. Srinagar, Kashmir, India

Mujeeb Zafar Banday
Department of Biochemistry, Government Medical College, Karan Nagar, Srinagar, Kashmir, India.
Infinity Biochemistry Pvt. Ltd. Srinagar, Kashmir, India

Fouzia Rashid
Department of Biochemistry, University of Kashmir, Srinagar, Kashmir, India

CONTENTS

DOI: 10.1201/9781003246244-5

ABSTRACT Xenobiotic-metabolizing enzymes (XMEs) – glutathione S-transferases (GSTs), N-acetyltransferases (NATs), and cytochrome P450 (CYP) – are pivotal enzymes required for the biotransformation reactions essential for cellular detoxification and protection of numerous macromolecules from the attack by reactive oxygen species, environmental carcinogens, reactive electrophiles, and chemotherapeutic agents. Genetic polymorphisms in these key enzymes have been found to be associated with the increased genetic instability leading to defective enzyme production and thereby carcinogenesis. It is a known fact that both the processes initiation and progression of the carcinogenesis are influenced by genes involved in the detoxification process. There are a number of published studies that report the relevance of SNPs in XME genes to different types of cancers, especially cancers of breast, lung, gastric, prostate, colorectal, and hepatic tissues. Considering the wide variation in the allele frequency of XME genes in different ethnic populations and their interaction with different carcinogenic pathways, the influence of the genetic polymorphisms of GSTs, NATs, and CYPs in conferring susceptibility to different cancers is deeply reviewed in this chapter.

KEYWORDS: *Cancer; Polymorphism; Genetics; Xenobiotic Metabolic Enzymes; SNP.*

5.1 Introduction

Xenobiotic-metabolizing enzyme (XME) superfamilies play an important role in the maintenance of cellular activity and correspond to protection against cellular damage due to endogenous as well as exogenous toxins. As already known, dietary and environmental carcinogens play a pivotal role in carcinogenesis mechanisms. Toxins and carcinogens that enter our body are metabolized via the xenobiotic pathway, which is an important defense mechanism against a wide variety of carcinogenesis mechanisms. DNA damage caused by different carcinogenesis mechanisms may be prohibited by caretaker genes, and among these, XMEs such as cytochrome P450 (CYP), glutathione S-transferases (GSTs), and N-acetyltransferases (NATs) are thought to have evolved as an adapting response to environmental insult. Therefore, numerous studies have demonstrated that these detoxifying enzymes may play a key role in modulating the risk of cancers.

Almost around 75% of the phase I detoxifying enzymes belong to the CYP family, which are known to catalyze various oxidative reactions and act as monooxygenases, oxidases, and peroxidases (Ingelman-Sundberg et al., 2007), while the phase II enzymes mostly catalyze transfer reaction and hence include NATs, GSTs, sulfotransferases (SULTs), and UDP-glucuronosyltransferases (UGTs) (Singh and Michael, 2009).

The phase I and phase II enzymes are mostly active in the liver, where they catalyze a major part of its detoxifying reactions. However, various CYP and GST isozymes do also exist in other organs and tissues such as esophageal mucosa, thereby suggesting that metabolic activation and detoxification of xenobiotics could also occur locally within the human body (Nakajima et al., 1996). The expression level of the XMEs and consequently their activity which is directly responsible for the metabolism of various xenobiotics vary between individuals as well in various organs of the body. This variability plays a critical role in modulating the carcinogenic effects of various xenometabolites and eventually in individual's susceptibility to cancer. Numerous factors do also affect and add to the variability, such as the environmental induction or inhibition of enzymes by agents such as tobacco smoke and alcohol as well as combination of genetic factors.

The best studied of XMEs are the CYP enzymes, which confer a modest increase in the risk of diverse types of cancers (Agúndez, 2004). Also, members of the GST family are important candidates for involvement in cancer susceptibility, and numerous epidemiological studies have explored their role in a wide variety of common cancers, especially digestive tract cancers (McIlwain et al., 2006; Martínez et al., 2006). In recent decades, NAT enzymes have come to the forefront of being potential risk factors directly involved in the development of environment-related cancers.

The progressive growth in the knowledge of the polymorphisms of the XMEs has led to multiple studies showing association of XMEs and development of cancer risk. Therefore, in this chapter, we have reviewed role of these XMEs in the development and progression of different cancers.

5.2 Glutathione S-Transferases (GSTs)

GSTs are the key phase II biotransformation enzymes that play a pivotal role in cellular detoxification and essentially protecting macromolecules from attack by reactive electrophiles, environmental carcinogens, reactive oxygen species, and chemotherapeutic agents (Strange et al., 2000). GSTs are widely distributed in nature and are present in both prokaryotes and eukaryotes as the principal phase II detoxifying enzymes (Hayes et al., 2005). GSTs are a superfamily of ubiquitous, multifunctional enzymes (GSTs; EC 2.5.1.18), all of which catalyze the addition of the tripeptide glutathione (GSH; g-Glu-Cys-Gly) nucleophile to several hazardous exogenous xenobiotics, including phase I electrophilic and carcinogenic metabolites (Senhaji et al., 2013; Nebert and Vasiliou, 2004; Hayes and Strange, 2000). The reaction results in the neutralization of the electrophilic sites and renders the toxins more water-soluble, thereby facilitating easy elimination from the cells by phase III enzymes (Coleman et al., 1997). Additionally, GSTs can serve as peroxidases, isomerases, and thiol transferases as well (Board et al., 2004). They are also important in the non-catalytic functions such as modulation of signaling processes and non-substrate ligand binding (Axarli et al., 2004).

5.2.1 Polymorphism of GSTs and Cancers

In humans, GST gene polymorphism has been demonstrated to be associated with susceptibility to various diseases. Thus, GST genotypes alone and in combination have been linked with clinical outcome. Several polymorphisms have been identified within each class of GSTs.

The GSTM and GSTT classes both are known to possess a null phenotype (GSTM*0 and GSTT*0) as well, in which individuals lack the catalytically active protein in the cells. The GSTM1*0 allele is observed in ~40%–60% of the Caucasian population (Kassab et al., 2014) and is associated with an increased risk of lung, prostate, gastric, and bladder cancer and is a risk factor for head-and-neck cancer (HNC) as well (Di Pietro et al., 2010; Hashibe et al., 2003). Likewise, the GSTT*0 phenotype also varies between ethnic groups and is found to be highest in Chinese (65%) and lowest in Mexican American (9%) populations (Nelson et al., 1995). The GSTT*0 phenotype is associated with an increased risk of tumors of the head and neck, oral cavity, pharynx, and larynx (Strange and Fryer, 1999; Chenevix-Trench et al., 1995).

The GSTM1 gene possesses four alleles and has been the most widely studied among all GST enzymes. GSTM1 SNP M1*A 0.2 has been reported to be associated with the decreased risk of bladder and breast cancer in Caucasians, while GSTM1*B 0.2 is associated with the decreased risk of pituitary adenomas and GSTM1*0 0.59 has been shown to increase the susceptibility to various cancers such as lung, colon, bladder, and breast cancer. Furthermore, GSTM1*A with an allele frequency of 20% is reported to decrease the risk of bladder cancer (Smith et al., 1995).

GSTP1 gene polymorphism is most often a point mutation SNP (single nucleotide polymorphism) within exon 5 Ile105 Val. Thus, the results of SNP are GSTP1 genotypes Ile/Ile, Ile/Val, and Val/Val. The exchange results in decreased enzymatic activity of the variant protein (Wei et al., 2012; Hezova et al., 2012). About 10%–20% of the Caucasian population are known to be the carriers of GSTP1 null genotype (Hezova et al., 2012). In addition, GSTP1 locus also possesses other SNPs that result in four distinct GSTP1 alleles -GSTP1*A to -GSTP1*D, all of which differ both structurally and functionally. The GSTP1 promoter region contains a TATA box, two SP1 sites, an insulin response element, and an antioxidant response element within an AP1 site (Lo and Ali-Osman, 1998).

GSTP1*A plays a role in the acquisition of resistance to cisplatin (CDDP) by enhancing the capacity of the cell to form platinum–GSH conjugates/CDDP-GSH adducts (Goekkurt et al., 2006). GSTP1*B is an allele in which a single nucleotide (A→G) substitution at position 313 substantially diminishes catalytic activity (Sharma et al., 2014). Homozygosity for GSTP1*B is favorable in the treatment of cancer patients because they have a diminished capacity to detoxify platinum-based anticancer agents (Stoehlmacher et al., 2002). GSTP1*C is an allelic variant that is more predominant in malignant glioma cells and differs from other GSTP1 variants by two transitions resulting in Ile104Val and Ala113Val (Lo and Ali-Osman, 1998). No major functional property has yet been assigned to this polymorphism.

GSTZ1 gene is known to possess four SNPs, namely, GSTZ1*A (Lys32; Arg42; Thr82); GSTZ1*B (Lys32; Gly42; Thr82); GSTZ1*C (Glu32; Gly42; Thr82); and GSTZ1*D (Glu32; Gly42; Met82). The isozyme GSTZ1*A is known to have the highest catalytic activity in contrast to GSTZ1*D, which has been shown to be associated with inborn errors in tyrosine metabolism, along with mutations in other enzymes (McIlwain et al., 2006).

5.2.1.1 GSTs and Head-and-Neck Carcinoma

Head-and-neck squamous cell carcinoma (HNSCC) are among the highly studied cancers having association with GST polymorphisms, because of the fact that HNSCC has a strong link with the carcinogens and pro-carcinogens such as PAHs present in tobacco, which are preferentially deactivated by the GSTs. The habit of consuming tobacco and its products is the key risk factor for HNSCC. For example, death from laryngeal squamous cell carcinoma is more likely to occur in heavy smokers than in non-smokers (Jethwa and Khariwala, 2017). Variations in the expression of GSTs due to heritable genetic polymorphisms probably modulate the process of carcinogenesis by altering the exposure levels of tobacco-derived carcinogenesis. However, many studies that evaluate the association between GST polymorphisms and HNSCC are quite divergent, reporting weak to moderate associations in the risk of the main effect of the gene (Geisler and Olshan, 2001). A recent meta-analysis including 31 studies with 4,635 cases and 5,770 controls and one pooled analysis of nine studies with 2,334 cases and 2,766 controls in HNSCC patients found a high probability that GSTM1 and GSTT1 null genotypes and GSTP1*B may result in a synergetic risk of carcinogenesis in HNSCC (Rebbeck, 1997). Similarly, for GSTM1 null, a study including 63 African-American and 101 Caucasian patients with primary oral cancer, as well as 133 African-American and 213 Caucasian matched control subjects showed that oral cancer risk was higher in heavy smokers (Park et al., 2000). On the other hand, some studies did not find a positive association between GST polymorphisms and HNSCC.

In a review of 11 case-control studies on the relation of GSTM1 null polymorphism with HNSCC, five studies did not show a positive association and for the other remaining six studies, four of the analyzed ones did not find any association (Ruiz et al., 2006). Additionally, another study also suggested that GSTT1 null was not significantly associated with HNSCC (Minard et al., 2006).

Another important study showed that GST polymorphisms modify the susceptibility to HNSCC by demonstrating the importance of gene–environment interaction modulation of the risk. Their results showed an increase in risk of HNSCC in patients with null genotype of GSTM1 (OR: 2.02; 95% CI: 1.32–3.10; $p=0.001$) or GSTT1 (OR: 1.66; 95% CI: 1.02–2.69; $p=0.04$). Additionally, there were 3.5-fold and 4-fold increases in the risk interaction between the null genotypes of GSTM1 or GSTT1 and tobacco and alcohol use, respectively, as compared to non-tobacco chewers and non-drinkers (Singh et al., 2008). Since HNSCC represents an ideal cancer for studying the gene–environment interaction, the divergences of the results may be because of the difficulty in controlling the individual's environmental exposure, especially among those culturally influenced.

5.2.1.2 GSTs and Breast Cancer

GST polymorphisms are well documented in breast cancer (BC) disease affecting thousands of women over the world and are associated with a greater risk. One such study that involved 278 cases and 271 African-American controls and 410 cases and 392 white controls showed some positive genetic associations, and women with GSTM1 null and GSTT1 null genotypes showed an increased risk (Millikan et al., 2000). Mitrunen et al. (2001) associated GSTM1, GSTM3, GSTP1, and GSTT1 polymorphisms with the risk of BC development in 483 patients and observed a positive association with GSTM1 null and GSTM3*B allele. For the GSTP1 Ile105/Ile105 genotype, the observed risk was significantly higher in women with GSTT1 null, where the risk was further strongly increased.

Furthermore, upon the evaluation of new gene combinations, it was found that in the combination GSTM1 null-GSTT1 null-GSTP1 Ile105/Ile105, the risk of developing cancer was increasingly high (Mitrunen et al., 2001). Also, Gudmundsdottir et al. (2001) investigated 388 BC patients and 395 controls, and observed a positive association in patients carrying GSTT1*A combined with p53 mutations between cases and controls (24.6 versus 12.4%; p=0.019) (Gudmundsdottir et al., 2001).

Vogl et al. (2004) revised seven studies that associate GSTM1, GSTT1 and GSTP1*B polymorphisms at the International Collaborative Study on Genetic Susceptibility to Environmental Carcinogens, with 2048 BC patients and 1969 controls. The authors showed a slight increase in risk with GSTM1 null and a moderate increase with GSTT1 null and GSTP1*B heterozygous and homozygous mutants, respectively (Vogl et al., 2004).

A meta-analysis including 59 studies (20,993 cases and 25,288 controls) showed a significantly elevated BC risk with GSTM1 null genotype (OR=1.10, 95% CI=1.04–1.16). In the subgroup analysis by ethnicity, significantly increased risks were found for Caucasians (OR=1.05, 95% CI=1.00–1.10) and Asians (OR=1.21, 95% CI=1.08–1.35) (Qiu et al., 2010). Another meta-analysis including 17 studies (5,323 BC cases and 7196 controls) showed the similar results (OR=1.28, 95% CI: 1.09–1.51), suggesting that the GSTM1 null genotype is a low-penetrant risk factor for developing BC in the Chinese population (Xue et al., 2016).

Like other types of cancer, most of these studies concentrated on genes GSTM1, GSTT1, and GSTP1. For the gene GSTA1, most of the evidence for GSTA1*B polymorphism has a negative association with BC as well as GSTO2*B (McCarty et al., 2007). On the other hand, GSTO1*BB genotype had higher risks of BC when compared to carriers of the GSTO1*AA genotype, with incidence rate ratio=1.62 (95% CI=1.01–2.61). This association was strongest regarding estrogen receptor-positive BC, incidence rate ratio=2.16 (95% CI=1.21–3.84) (Olsen et al., 2008). GSTO1 A140D and GSTO2 N142D gene polymorphisms with BC risk in an Iranian population (OR=1.75, 95% CI=1.07–2.86, p=0.026) (Sharif et al., 2017).

A meta-analysis of 20 studies (4,770 cases and 5,701 controls) revealed an overall significantly increased risk of susceptibility to BC with GSTO2 polymorphism (GG vs. AA: OR=1.20, 95% CI: 1.02–1.41, p=0.116), but GSTO1 polymorphism was found to be insignificant. Also, the subgroup analysis showed that GSTO2 polymorphism significantly increased the cancer risk in Caucasian population (GG vs. AA: OR=1.32, 95% CI 1.06–1.64, p=0.616) and GSTO2 polymorphism was significantly associated with an elevated risk of BC (GG vs. AA OR=1.37, 95% CI: 1.06–1.77; p=0.281) (Xu et al., 2014 a).

5.2.1.3 GSTs and Lung Cancer

GSTP1 is commonly expressed in the pulmonary tissue in order to mediate and handle main metabolic agents such as PAHs that are present in cigarette smoke (Sato et al., 1999). Cote et al. (2009) published a meta-analysis involving 27 studies (8,322 cases and 8,844 controls), which did not show any significant association with GSTP1*B, and a pooled analysis of 15 studies (4,282 cases and 5,032 controls), stratified by ethnicity, revealed a slightly high risk of LC in members who carried the Val105 as compared to Ile105. After stratification by ethnicity and adjustment for age, sex, and smoking status, an increased risk was associated with GSTP1*B presence in Asians, but not in white population.

GSTP1 GG genotype was found to be associated with increased lung cancer susceptibility among younger study participants, but was not associated with an overall increased risk of lung cancer (OR=1.02; 95% CI=0.78–1.34) compared with the GSTP1 AA genotype. Therefore, GSTP1*B may be significantly associated with susceptibility to early-onset LC in older individuals (Miller et al., 2006).

A meta-analysis of six studies with 912 cases and 1,063 controls in Asian populations observed a slight increase in LC risk with GSTM1 and GSTT1 null genotypes, but not with GSTP1*B genotype, suggesting that the LC risk is higher in populations with coal exposure (Hosgood et al., 2007). A study showed the significant association of GSTM1 and GSTT1 with the increase in LC risk. The carriers of GSTM1 null showed a significant risk of LC even among smokers (Lee et al., 2006). A meta-analysis including 34 studies (7,629 cases LC and 10,087 controls) suggested a positive association of GSTT1 null in Asians, but not in Caucasian subjects (Raimondi et al., 2006).

Several meta-analyses showed that the GSTP1 341C/T polymorphism is associated with the risk of lung cancer development. For instance, one meta-analysis reported a significant association (TT vs CC: OR=3.33, 95% CI=1.49–7.44; CT vs CC: OR=1.35, 95% CI=1.10–1.65; dominant model: OR=1.43, 95% CI=1.05–1.96; recessive model: OR=0.31, 95% CI=0.14–0.70) (Yan et al., 2016). Another meta-analysis including 13 case-control studies (2,026 lung cancer cases and 2,451 controls) revealed a significantly increased lung cancer risk with the variant genotypes of GSTP1 polymorphism in the Chinese population (GG vs. AA: OR=1.36, 95% CI=1.01–1.84). A gene-gene interaction analysis showed that there was an interaction between the combination of GSTM1 (or GSTT1) null genotype and GSTP1 (AG+GG) mutant genotype and lung cancer risk in Chinese individuals (Li et al., 2015).

Also a meta-analysis revealed that the GSTM1 null genotype is associated with an increased LC risk in Japanese and lung adenocarcinoma risk in Asians; the GSTT1 null genotype is associated with an increased LC risk in Chinese, and the GSTP1 Ile105Val polymorphism is associated with an increased LC risk in Asians (Zhang et al., 2021).

5.2.1.4 GSTs and Esophageal Cancer

Many studies have investigated the association of GST gene polymorphism and the risk of esophageal cancer (EC), but the findings are controversial (Dura et al., 2013; Du et al., 2018; Liu, 2017). A meta-analysis including 2,236 EC cases and 3,243 controls reported that the GSTM1 null genotype was a significant risk factor for EC in most populations (p heterogeneity < 0.000001, and I (2)=77.0%), particularly in the Asians, but not in the Caucasian population (Lu et al., 2016). Similar results were shown by another study, which concluded that the GSTM1 and GSTT1 null genotypes are associated with increased genetic susceptibility to EC in the overall human population (p heterogeneity < 0.001), particularly among Asians (Ma et al., 2018).

5.2.1.5 GSTS and Gastric Cancer

There are mixed studies that mainly show the association between the respective deletions and the increase in the risk of gastric cancer (GC) in relation to GSTM1 and GSTT1 genes, and most of them are insignificant (Al-Moundhri et al., 2009). However, a study with 304 GC patients and 427 control subjects showed a 1.48-fold increased risk in patients with GSTT1 null genotype, but not with GSTM1, GSTM3, or GSTP1 genotypes. Furthermore, upon stratification by age, it was observed that younger patients with GSTT1 null genotype (<50 years old) showed an increased GC risk (Lan et al., 2001). However, some authors showed that GSTM1 and GSTT1 null genotypes (in addition to the presence of GSTP1*B and GSTO2*B alleles) confer a reduction in susceptibility to GC (Martínez et al., 2006).

A meta-analysis (2,508 GC cases and 4,634 controls) suggested that a synergic relation between GSTM1 and CYP2E1 may account for a proportion of GC cases with GSTT1 polymorphism and that the CYP2E1 PstI/RsaI polymorphism may be a risk factor for GC in Asians (Boccia et al., 2007).

A meta-analysis including 54 individual studies (9,322 cases and 15,118 controls) suggested that the null genotype of GSTM1 may be an important genetic risk factor for GC development (p < 0.001). Also, the dual null genotypes of GSTM1 and GSTT1 might elevate the risk of GC (p=002) (Lao et al., 2014).

5.2.1.6 GSTs and Colorectal Cancer

Many studies are available concerning the polymorphisms of GST genes as risk-modulating factors in different cancers, including gastrointestinal cancer. Gene polymorphisms affect the functioning of GST enzymes at the gene level as well as the protein level, thereby influencing the detoxification process of carcinogens, and consequently, the level of DNA damage; thus, it may have an indirect effect on the risk of development of cancer (Gong et al., 2012). It is believed that GST polymorphisms play a role in cancer development because many carcinogens are inactivated via GST systems. Multiple studies have been carried out from time to time to assess the impact of GST gene polymorphisms in colorectal cancer (CRC) development and progression. The results of these studies are inconsistent and inconclusive. Some research studies suggest there exists no association, some show low risk, whereas others show an increased risk and thus are summarized in Table 5.1.

This inconsistency in results reflects the complexity in the role of GSTs and refers to the fact that the metabolic pathways involved in the carcinogen metabolism are complex and facilitated by the actions of multiple genes.

5.2.1.7 GSTS and Hepatocellular Carcinoma

The hepatocellular carcinoma or HCC accounts for 80%–90% of all liver cancers (El-Serag and Rudolph, 2007). Liver and kidney are two organs that express the highest level of GSTT in the human body. The GSTT1 genotype seems to modulate the hepatic cancer risk, considering the source of the xenobiotic exposure. For example, elevated levels of GSTT1 protein are found in GSTT1*A carriers, which show an elevated risk of development of HCC upon exposure to halogenated solvents (Simic et al., 2009).

A meta-analysis involving 15 studies showed different associations of GSTA1, GSTA4, GSTM1, GSTM2, GSTM3, GSTT1, GSTP1, GSTO1, and GSTO2 polymorphisms in Asian, African, and European populations with the HCC risk. One study observed that only GSTT1 and GSTM1 null carriers show a positive association, however, with weak risk, OR=1.19 and 1.16 (95% CI=0.99–1.44 and 0.89–1.53), respectively (White et al., 2008). A study carried out in India found that GSTT1 null individuals have an increased risk of development of HCC in comparison with the control group, while the GSTM1 null genotype presented a protective effect (Kiran et al., 2008).

A meta-analysis including 41 articles (6,124 cases and 9,781 controls) showed that both the GSTM1 null genotypes (OR=1.37, 95% CI=1.18–1.59) and GSTT1 null genotypes (OR=1.43, 95% CI=1.23–1.66) increased the risk of HCC. Also, the subgroup analysis obtained significant results for GSTM1 null (OR=1.44, 95% CI=1.22–1.71), GSTT1 null (OR=1.48, 95% CI=1.25–1.77), GSTM1-GSTT1 null (OR=1.58, 95% CI=1.19–2.09), among Asians but not in Caucasians or Africans (Li et al., 2019). Similarly, another study reported both the GSTM1 and GSTT1 null genotypes are statistically associated with the development of hepatic cancer (Xiao and Ma, 2012). Also, the GSTM1 and GSTT1 null genotypes significantly increased the risk of HCC in the Chinese population (p<0.001) and Asians (Liu et al., 2013a; Chen et al., 2012).

TABLE 5.1
GST polymorphisms and Risk of Colorectal Cancer

Study	Sample Population	Risk	Concomitance
Loktionov et al. (2001)	206 cases; 355 controls	*GSTM1* and *GSTM3* polymorphisms are associated with a high CRC risk	
Hlavata et al. (2010)	Czech population (495)	Moderate increase in risk by *GSTM1* deletion No risk by *GSTP1* gene polymorphism	Simultaneous deletion of the *GSTM1* and *GSTT1* genes causes a significantly higher risk.
Economopoulos and Sergentanis (2010)	Caucasian population	*GSTM1* as well as *GSTT1* null carriers exhibit an increased CRC risk	
Aghajany-Nasab et al. (2011)	Iran (140 cases; 90 controls)	*GSTM1* null predisposes to the development of CRC in individuals aged over 60	
Wang et al. (2011a)	Indian Hindus (300)	*GSTM1* null increased the risk of rectal cancer *GSTT1* null increased the risk of colon cancer	Predisposing to the development of CRC.
Darazy et al. (2011)	Lebanese population	*GSTM1* null genotype increased the risk of CRC	
Sameer et al. (2011)	Kashmiri population (86 cases and 160 controls)	No significant risk of *GSTP1* Ile105Val polymorphism	
Hezova et al. (2012)	Czech population (197 cases; 218 controls)	No significant risk of development of CRC by *GSTM1* and *GSTT1*	
Khabaz (2012)	Jordan	*GSTP1* Ile105Val polymorphism does not exert any risk of CRC	
Zhao et al. (2012)(	China	*GSTP1,GSTT1*, and *GSTM1 gene* polymorphisms are not colorectal adenoma risk factors	No risk.
Kassab et al. (2013)	Tunisian population (150 cases; 128 controls)	No significant risk with *GSTM1* and *GSTT1* null genotypes, but significant risk of CRC with *GSTP1*	
Nissar et al. (2017)	Kashmiri population (160 cases and 200 controls)	No significant risk with *GSTM1* and *GSTT1* null genotypes	
Klusek et al. (2018)	Polish population (197 CRC patients and 104 healthy controls)	*GSTT1 null* polymorphism correlated with higher tumor grade, and the *GSTM1 null* genotype was associated with more frequent metastasis to lymph nodes	*GST* gene polymorphisms may influence CRC tumor grade and stage.

5.2.1.8 GSTs and Prostate Cancer

GSTM1, GSTT1, and GSTP1 polymorphisms are associated with the carcinogenic development in the prostate. Recently, a meta-analysis has evaluated the relationship of polymorphisms and hypermethylation in GSTs and biochemical recurrence (BCR). The results showed GSTP1 polymorphism and CpG hypermethylation are associated with BCR (HR = 1.94, 95% CI = 1.13–3.34). Also, a possible association was found between GSTM1 null polymorphism and prostate cancer (PC) with borderline significance (HR = 1.29, 95% CI = 0.97–1.71) (Chen et al., 2013).

Another meta-analysis including 29 studies (4,564 cases and 5,464 controls) reported that GSTM1 null increased the PC risk. Upon stratification based on ethnicity, similar results were obtained for Asians and Caucasians, whereas the Africans and African Americans who shared risk similarities (OR = 0.66 and 0.47, respectively) were far from the general population (OR = 1.3). However, no significant association of GSTT1 null (22 studies with 3,837 cases and 4,552 controls) as well as GSTP1*B (24 studies with 5,301 cases and 5,621 controls) polymorphisms with PC risk was observed even after ethnic stratification (Mo et al., 2009).

Similarly, another meta-analysis including 11 studies on GSTM1 null (2,063 cases and 2,625 controls) and GSTT1 null (1,965 cases and 2,554 controls) and 12 studies on GSTP1*B polymorphisms (2,528 cases and 3,076 controls) did not obtain any significant frequency differences between PC patients and healthy controls (Ntais et al., 2005). In another study that included 135 PC patients and 169 controls, a 2.5-fold increased risk was found in GSTM3*AB and BB carriers and cases who were smokers and/or alcoholics demonstrated an association with GSTM3*B genetic variant (Kesarwani et al., 2009).

Another study conducted in Japan suggested that patients with GSTA1*B variants were at high risk of developing PC. However, the genetic polymorphism distribution was not statistically different for genotypes of GSTM1 and GSTP1 when compared to healthy controls. Interestingly, this work also demonstrated an increase in PC risk in carriers of the following combined genotypes: GSTA1*AB or BB and GSTT1 non-deletion (Komiya et al., 2005). Another study showed the GSTP1 Ile/Val polymorphism is involved in the risk of PC development in Kashmiri population (Qadri et al., 2011). Another study observed that both the GSTT1 null and GSTM1 null genotypes are significantly associated with PC risk ($p<0.01$) and the GSTM1/GSTT1 dual null genotype was associated with a higher risk of PC in Asians ($p=0.007$) (Liu et al., 2013b).

Further, a haplotype analysis for cases carrying *GSTP1*C* haplotype, represented by both variant alleles (*GSTP1*Val rs1695+GSTP1*Val rs1138272*), showed a 5.46 times higher risk of PC development compared to individuals with the most frequent haplotype (95% CI=2.56–11.65, $p<0.001$), suggesting a potential role of those variants in PC susceptibility. And the regression analysis (*GSTM1*active*, *GSTT1*null*, *GSTP1*Val rs1695*, and *GSTP1*Val rs1138272*) showed a significant increase in the risk of developing PC, from 3.65-fold in carriers of two risk alleles (95% CI=1.55–8.61, $p=0.003$) to an approximately 12-fold increase in carriers of all four risk alleles (95% CI=3.05–44.93, $p<0.001$) (Santric et al., 2020).

5.2.2 Conclusions

Different studies have provided comprehensive data regarding the diverse association of the GST gene polymorphism and the risk of varied number of cancers, with some of them showing a positive influence on susceptibility to disease risk and others reporting no significant association at all. Overall, there is huge evidence of the role of genetic polymorphism in XMEs, especially GSTs, in the risk of cancer development as well as progression. For example, GSTP1 341C>T polymorphism may contribute to genetic susceptibility to lung cancer among Asian population and GSTM1 and GSTT1 null genotypes confer additional risk of colorectal cancer in Caucasian populations. Moreover, the multiple polymorphic GST genes such as GSTM1/T1 null and, in particular, GSTP1 are thought to influence PC, thereby highlighting the role of gene-gene interactions in human susceptibility to cancer.

5.3 Arylamine N-Acetyltransferases (NATs)

Human arylamine N-acetyltransferases (NATs) (EC 2.3.1.5) are cytosolic conjugating enzymes, which transfer an acetyl group from acetyl coenzyme A to a xenobiotic acceptor substrate (Westwood and Sim, 2007). There are three NAT loci: NAT1, NAT2, and a pseudogene – NATP. NAT1 and NAT2 are products of single, intron-less protein-coding exons of 870-bp open reading frames encoding 290 amino acids and are located on human chromosome 8p22 (Alberg et al., 2004; Morabia et al., 2000), a region frequently displaying loss of heterozygosity in human tumors (*Chang-Claude et al., 2002*). NAT1 and NAT2 share 87% nucleotide homology in the coding region, yielding 55 amino acid differences. NAT1 derives its entire transcript from a single exon, and NAT2 mRNA is derived from both the protein-coding exon and a second non-coding exon of 100 bp located about 8 kb upstream of the translation start site (Butcher et al., 2005).

NATs are phase II metabolic enzymes. They catalyze the acetylation of aromatic and heterocyclic amine carcinogens (Xu et al., 2017). The endogenous substrates for human NAT1 and NAT2 are still to be verified; however, the catabolite of folic acid, p-aminobenzoyl-glutamate, is found to be N-acetylated by NAT1 (Marques et al., 2006), defining its role in folate metabolism. Generally, aromatic amines and hydrazines are the examples of acceptor substrates that are deactivated (N-acetylation) and N-hydroxy-aromatic and heterocyclic amines are activated (O-acetylation) by NAT1 and/or NAT2 (*Smith et al., 1995*). Each of the enzymes has a preferred substrate; NAT1 has increased catalytic activity with 2-aminofluorene, p-aminobenzoic acid, p-aminosalicylic acid, sulfamethoxazole, and sulfanilamide, whereas the substrates such as aminoglutethimide, 2-aminofluorene, hydralazine, and procainamide are preferred by NAT2.

NAT1 and NAT2 also help in catalyzing the intramolecular N, O-acetoacetyl transfer of N-hydroxy-N-acetyl-aromatic amines and catalyze the metabolic activation (via O-acetylation) of aromatic and heterocyclic amine carcinogens (Smith et al., 1995; Majumder et al., 2007). It is often suggested that the human NAT2 activity is the highest in the liver and gastrointestinal tract, whereas the human NAT1 activity is expressed in many extrahepatic tissues (Salazar-González et al., 2018).

The NAT2 gene has been reported to exhibit a polymorphism. Subsequently, the presence of the different alleles in each individual genome produces a broad range of metabolic phenotypes that vary from fully active rapid metabolizers to less active alleles of slow metabolizers (El-Sharkawi et al., 2014). There are remarkable genetic polymorphisms of NAT2 activity and are found to be associated with different levels of susceptibility to develop many kinds of cancers, including bladder cancer (Zhao et al., 2016). The activation or deactivation of the substrate depends on the nature and its kinetic properties; however, the tissue expression of NAT is also an important factor in determining cancer risk at specific sites (Walker et al., 2009). The N- or O-acetylation activity of NAT determines which phenotype is at the highest risk of developing certain types of cancer. For example, the NAT2 slow acetylator phenotype is at a higher risk when N-acetylation is a detoxification step, as in aromatic amine-associated urinary bladder cancer, whereas the NAT2 rapid acetylator phenotype is at a highest risk of cancers in which O-acetylation is an activation step, such as heterocyclic amine-related colon cancer (Hein, 2002).

NATs play an important role in pharmacogenetics because of their key role in cellular homeostasis as well as in gene–environment interactions (Stanley and Sim, 2008). Unexpectedly, isoniazid and other drugs such as sulfamethazine were earlier termed as "polymorphic" substrates for N-acetylation by NATs and divided human populations into rapid, intermediate, and slow acetylator phenotypes, whereas drugs such as p-aminosalicylic acid were termed "monomorphic" (Trizna et al., 1998). However, the biochemical basis for this observation was attributed to substrate specificity and molecular genetics of NAT1 and NAT2 (Morita et al., 1999).

Also, the polymorphism in acetylation and inactivation of the anti-tubercular agent isoniazid residues in human NAT2 makes it one of two polymorphic human NATs. The presence of classical isoniazid slow acetylator shows striking ethnic differences with a frequency that is approximately 50%–60% in Caucasian populations and is much more frequent in Egyptians, but much less frequent in Asians (Hickman et al., 1994).

5.3.1 Polymorphism of NATs and Cancers

The SNPs affect the acetylation rate of the NAT enzymes. The activity of NAT2 is predicted from the combination of the NAT2 alleles. The presence of at least one wild-type allele is considered a fast acetylator, whereas the absence of any wild-type allele is considered as slow acetylator. The two NAT2 phenotypes have been described: fast and slow. Slow acetylators are less efficient in metabolizing aromatic amines, increased toxin burden, and ultimately carcinogenic hallmarks such as DNA adduct formation and oxidative stress and hence may increase the risk of cancer (Kamel et al., 2015). Also, the genetic polymorphism in NAT1 and/or NAT2 may modify the risk of cancer related to exposures to these carcinogens. The metabolic activation of many N-hydroxy heterocyclic amine carcinogens is catalyzed to a greater extent by human NAT2 than NAT1, but tissue-specific expression is also important for the effect of the NAT1 and NAT2 acetylation polymorphisms on cancer risk (Hein, 2006).

Almost seven missense (G191A, T341C, A434C, G590A, A803G, A845C, and G857A) and four silent (T111C, C282T, C481T, and C759T) substitutions have been identified in the NAT2 coding exon. NAT2 alleles containing the G191A, T341C, A434C, G590A, and/or G857A missense substitutions are associated with slow acetylator phenotypes (Murali et al., 2012). NAT2*4 is considered the wild-type allele because of its absence of any of these substitutions and is not the most common allele in many ethnic groups, including Caucasians and Africans. However, NAT2*14 allele including G191A nucleotide substitution was found more frequently in African than in Caucasians (Deloménie et al., 1996). Also, NAT2*7 allele including G857 nucleotide substitution is more frequent in South Indians and Koreans than in other populations, whereas NAT2*5 allele containing T341C nucleotide substitution is less frequent in Korea than in Europe, North America, India, and Africa (Hein, 2006). NAT2 alleles containing the G191A, T341C, A434C, G590A, and/or G857A missense substitutions are associated with slow acetylator phenotype(s), and the G191A substitution is present in African Americans and native Africans, but it is virtually absent in Caucasian populations (Hein et al., 2000).

Similarly, NAT1 allelic frequency has recently been reviewed (Bolt and Thier, 2006). NAT1*4 is the most common wild-type allele in all populations studied to date. It is known that NAT1*10 is more frequent in some ethnic populations such as Mexican and African Americans (Wu et al., 2002) than in Caucasians (Egan et al., 2003) and Asians (Kocabaş et al., 2004). Phenotyping analyses have revealed an association between NAT enzyme activities and the risk of developing several forms of cancer. As genotyping procedures have become available for NAT1 and NAT2 gene variations, hundreds of association studies on NAT polymorphisms and cancer risk have been documented. The presence of different alleles of NAT1 and NAT2 because of specific nucleotide change is shown in Tables 5.2 and 5.3.

TABLE 5.2

Different Alleles of NAt1 and Effect on Enzymes/Proteins

Allele	Nucleotide Change	Amino Acid/Protein Affected
NAT1*3	C1095A	None
NAT1*4	None	None
NAT1*5	G350,351C, G497–499C, A884G, D976, D1105	Arg1173Thr, Arg1663Thr, Glu1673Gln
NAT1*10	T1088A, C1095A	None
NAT1*11	C2344T, A240T, G445A, G459A, T640G, D9 between 1065–1090, C1095A	Val1493Ile, Ser2143Ala
NAT1*14A	G560A, T1088A, C1095A	Arg1873Gln 51
NAT1*14B	G560A	Arg1873Gln
NAT1*15	C559T	Arg1873Stop
NAT1*16	[AAA] immediately after 1091, C1095A	None
NAT1*17	C190T	Arg643Trp
NAT1*18A	D3 between 1064–1087, T1088A, C1095A	None
NAT1*18B	D3 between 1064–1091	None
NAT1*19	C97T	Arg333Stop
NAT1*20	T402C	None
NAT1*21	A613G	Met2053Val
NAT1*22	A752T	Asp2513Val
NAT1*23	T777C	None
NAT1*24	G781A	Glu2613Lys
NAT1*25	A787G	Ile2633Val
NAT1*26A	[TAA] insertion between 1066 and 1091, C1095A	None
NAT1*26B	[TAA] insertion between 1066 and 1091	None
NAT1*27	T21G, T777C	None
NAT1*28	[TAATAA] deletion between 1085 and 1090	None

TABLE 5.3
Different Alleles of NAt2 and Effect on Enzymes/Proteins

Allele (Haplotype)	Nucleotide Change(s)	Amino Acid Change; Catalytic Activity
NAT2*4	None	None; high
NAT2*5A	341T>C; 481C>T	Ile114Thr; low
NAT2*5B	341T>C; 481C>T; 803A>G	Ile114Thr; Lys268Arg; low
NAT2*5C	341T>C; 803A>G	Ile114Thr; Lys268Arg; low
NAT2*6A	282C>T; 590G>A	Arg197Gln; low
NAT2*6B	590G>A	Arg197Gln; low
NAT2*7A	857G>A	Gly286Glu; low
NAT2*7B	282C>T; 857G>A	Gly286Glu; low
NAT2*12A	803A>G	Lys268Arg; high
NAT2*12B	282C>T; 803A>G	Lys268Arg; high
NAT2*12C	481C>T; 803A>G	Lys268Arg; high
NAT2*13	282C>T	None; high
NAT2*14A	191G>A	Arg64Gln; low
NAT2*14B	191G>A; 282C>T	Arg64Gln; low
NAT2*14C	G191A, T341C, C481T	Ile1143Thr, Lys2683Arg Arg643Gln
NAT2*14D	G191A, C282T, G590A	Arg643Gln, Arg1973Gln
NAT2*14E	G191A, A803G	Arg643Gln, Lys2683Arg
NAT2*14F	G191A, T341C, A803G	Arg643Gln, Ile1143Thr, Lys2683Arg
NAT2*14G	G191A, C282T, A803G	Arg643Gln, Lys2683Arg
NAT2*17	A434C	Gln1453Pro
NAT2*18	A845C	Lys2823Thr

5.3.1.1 NATs and Head-and-Neck Cancer

Earlier, the polymorphic studies reported a weak association of the NAT2 slow acetylation genotype with an increased risk of HNC (González et al., 1998; Lei et al., 2002; Bendjemana et al., 2006). However, two studies involving 289 Polish patients and 231 Brazilian patients reported an increased frequency for NAT2 rapid acetylator genes among cancer patients (Gajecka et al., 2005; Marques et al., 2006). The negative association results were later attributed to small sample sizes, or methodological limitations. Thus, diverse studies are conducted in patients with oral and larynx cancer. Henning et al. reported an association of head-and-neck and larynx cancer with the NAT1 genotype, but a weak association with rapid NAT2 acetylation genotype. The risk of cancer development is said to be modified by NAT genotypes due to the reason that the chemical compounds present in tobacco, which is the major risk factor for oral cancer, are inactivated by phase II enzymes such as NATs. A meta-analysis (4,393 cases and 5,321 controls) reported that individuals with slow acetylators of NAT2 polymorphism are at a higher risk of HNC ($p=0.03$) (Mohammadi et al., 2021). Similar results are reported in a meta-analysis, which concludes that slow acetylators might increase HNC susceptibility among Asians (OR =1.78, 95% CI 1.27–2.49) (Zhang et al., 2014).

Unal et al. (2005) reported an increased risk for carriers of some defective alleles such as NAT2*5 and NAT2*6 and decreased risk for carriers of NAT2*7. Other studies did not identify any significant associations (Chen et al., 2001; Majumder et al., 2007).

5.3.1.2 NATs and Breast Cancer

Several studies have addressed the role of NAT gene polymorphism and BC, and the interaction of polymorphisms with environmental risk factors such as smoking in BC risk. Huang et al. identified a marginal association of the NAT2 slow acetylation genotype with cancer risk in postmenopausal women. No association of BC risk with NAT2 polymorphism alone was seen, but an interaction of NAT2 genotype and smoking with an increased risk in postmenopausal women who were slow acetylators and smokers was observed (Ambrosone et al., 1996; Alberg et al., 2004). In contrast, elevated relative risks with both passive and active smoking in fast acetylators, especially among postmenopausal women, have been identified in an independent study (Morabia et al., 2000). Another study reported a marginally increased risk for the combination of slow acetylation plus smoking and, surprisingly, rapid acetylation plus passive smoking, thus suggesting a differential effect of NAT2 in the association between smoking and BC risk (Chang-Claude et al., 2002).

Millikan et al. found no association of NAT genotypes with cancer risk. However, authors identified an association of cancer risk with smoking and the NAT2 rapid genotype in postmenopausal women. Interestingly, this study also reports an association of the NAT1*10 variant allele and smoking with increased risk in postmenopausal women. Another study involving 229 BC patients found no association of NAT1 genotypes and cancer risk (van der Hel et al., 2003).

A meta-analysis involving 26 studies (9,215 cases and 10,443 controls) suggested that there is an overall lack of association between NAT2 genotypes and BC risk (OR=1.026,

95% CI=0.968–1.087); however, NAT2 polymorphisms when combined with heavy smoking history may contribute to BC susceptibility (OR=1.400, 95% CI=1.099–1.784). In the subgroup analysis by ethnicity, increased risks were not found for either Caucasians (OR=1.001, 95% CI=0.938–1.068) or Asians (OR=1.155, 95% CI=0.886–1.506), nor for premenopausal (OR=1.053, 95% CI=0.886–1.252) or postmenopausal women (OR=0.965, 95% CI=0.844–1.104) (Zhang et al., 2010). Other findings obtained in dozens of studied involving several thousands of BC patients, as well as meta-analyses are consistent and conclusive in ruling out a relevant association of NAT polymorphisms and BC risk (Dunning et al., 1999; Ochs-Balcom et al., 2007).

5.3.1.3 NATs and Lung Cancer

It is hypothesized that slow acetylation is a risk factor for LC. Martinez et al. in a study carried out in 108 patients observed a significantly increased frequency for the defect alleles NAT2*5 and NAT2*6 in lung cancer patients, as compared to healthy subjects, although in this study a lack of significant differences in the frequency for slow NAT2 genotypes was observed in lung cancer patients, as compared to healthy subjects. Similar findings were observed in an independent study (Habalová et al., 2005). Oyama et al. (2007) identified an association between slow acetylation genotypes and lung cancer risk, the highest significance being observed among patients diagnosed under 65 years. Also, similar findings were obtained from other studies showing the presence of higher risk among individuals carrying slow acetylator genotypes (Seow et al., 2001; Sørensen et al., 2005; Gemignani et al., 2007; Osawa et al., 2007). However, the opposite result, i.e., rapid acetylators are at an increased risk of development of LC, was also found (Zhou et al., 2002; Chiou et al., 2005).

Further, there was no association between acetylation genotypes, isolated or combined with other genotypes for drug-metabolizing enzymes, and lung cancer risk (Belogubova et al., 2005). A meta-analysis showed a significant association of the NAT1 rapid acetylation genotype, whereas NAT2 genotype was found to be insignificant (Zienoldd iny et al., 2008). Despite negative associations between NAT1 polymorphism and lung cancer, it is significantly related to lung cancer risk (McKay et al., 2008).

In summary, it is thought that the NAT2 slow acetylation genotype causes a marginally increased risk of developing lung cancer irrespective of indefinite conclusive studies (Sørensen et al., 2005, Wikman et al., 2001). Moreover, based on the present evidence, it is also suggested that the NAT polymorphism alone does not constitute a relevant risk factor for lung cancer, but may reinforce the effect of other genetic and/or environmental factors.

5.3.1.4 NATs and Esophageal Cancer

NAT polymorphisms in esophageal cancer is very less studied, and the data available are also very limited. A significant effect of the NAT2 slow acetylation genotype was reported in esophageal squamous cell cancer (OR=2.55; 95% CI=1.15–5.65, p=0.023) (Morita et al., 1999; Shibuta et al., 2001). However, other studies reported an insignificance of NAT1 polymorphism in esophageal cancer (Lee et al., 2000; Jain et al., 2007; Wideroff et al., 2007). Many studies need to be carried out to establish the significance of NAT gene polymorphism and to associate the role of NAT genotypes with the development of EC risk.

5.3.1.5 NATs and Gastric Cancer

Like esophageal cancer, there are limited data on NAT genotypes and GC risk. However, a positive association of the NAT1*10 variant allele with GC risk was reported, but NAT2 genotype was found to be insignificant in susceptibility to GC risk (Katoh et al., 2000). On the contrary, no association between NAT1 polymorphism and GC risk was observed (Wideroff et al., 2007; Agudo et al., 2006). Other studies reported a weak association of the NAT2 rapid acetylation genotypes with increased risk (Ladero et al., 2002), which is consistent with that reported for patients with family history of GC (Hong et al., 2006). Also, many studies did not show any association between the NAT2 genotype and cancer risk (Agudo et al., 2006; Suzuki et al., 2004; Al-Moundhri et al., 2007), and some studies observed the opposite association, indicating that slow NAT2 genotypes combined with the null GSTT1 (Boccia et al., 2007) or GSTM1 (Oda et al., 1999) genotypes increase the risk of GC. In summary, overall findings are consistent and indicate that neither the NAT1 nor the NAT2 genotypes constitute a relevant factor for GC.

5.3.1.6 NATs and Colorectal Cancer

Following the hypothesis that the acetylator status of NAT1 gene may predispose to a determined cancer risk on the basis of a differential effect of N-acetylation as a potential detoxification step and O-acetylation as a potential carcinogen-activation step, it was hypothesized that O-acetylation is more relevant than N-acetylation in colorectal cancer, and therefore, the rapid acetylation genotype is the putative risk factor for colorectal cancer development (Borlak and Reamon-Buettner, 2006; Hein et al., 2002; Hein , 2002).

A number of studies have explained the supposed role of NAT polymorphisms in colorectal cancer risk. A study involving a limited number of patients described a significant association of the rapid NAT1 (Landi et al., 2005). Another study found a significant association of NAT2 acetylator genotype with cancer risk (Huang et al., 2007). Also, an inverse association (i.e., slow acetylation genotypes as the risk factor) has been pointed in a number of studies. Surprisingly, an unexpected association with both a rapid and slow NAT2 variant allele in the same study or increased frequency for a slow allele and decreased frequency for another slow allele among colorectal cancer patients has been reported. Nearly 30 studies failed to detect a statistically significant association (with 95% CI for odds ratios excluding the value 1) of NAT1 or NAT2 genotypes with both colorectal cancer and adenomas (Agúndez, 2008).

Meta-analyses are consistent with a lack of a relevant association of either NAT1 or NAT2 rapid acetylator genotypes with colorectal cancer risk (Chan et al., 2005; Ye and Parry, 2002; Houlston and Tomlinson, 2001). In a meta-analysis involving 20 studies, the odds ratio for rapid NAT2 acetylation genotypes was 1.06 (95% CI 0.97–1.15) (Ye and Parry 2002). Another meta-analysis obtained values equal to 1.05 (95% CI=0.94–1.14) for 10 NAT2 genotyping studies and 1.25 (95% CI=0.96–1.63) for the NAT1*10 allele (Chan et al., 2005). A meta-analysis involving 13,875 cases and 18,879 controls showed no significant association between the NAT2 acetylator status and CRC susceptibility. Also, no increased risks were found in Asians, Europeans, Americans, or Australasians (Liu et al., 2012).

5.3.1.7 NATs and Prostate Cancer

Several studies observed that NAT genes are in the area where the most frequent allelic loss occurs in PC tissue, and thus, it was hypothesized that NAT activities could play a role in the PC risk. However, no differences were identified, and results confirmed a lack of significant association between the NAT2 acetylation genotype and PC risk (Agúndez et al., 1998; Wadelius et al., 1999; Gao et al., 2003; Srivastava and Mittal, 2005). But one study identified a slightly significantly increased risk with slow NAT2 acetylator genotypes (Hamasaki et al., 2003) and another found an association of the combination of the rapid NAT1 plus slow NAT2 genotypes as a risk factor (Hein et al., 2002). Also, the increased risk of PC was observed among Japanese men with the NAT2 slow acetylator phenotype (OR, 1.65; 95% CI, 1.04–2.61) (Koda et al., 2017). The meta-analysis including 10 studies (1,253 cases and 1,722 controls) showed a significant association between NAT2 polymorphism and PC risk in Asians (OR=1.10 95% CI: 0.87–1.39; I(2)=49.1%, p value = 0.039), but not in Caucasians (Gong et al., 2011).

Some studies show a significant association of rapid NAT2 genotypes with PC risk, but not NAT1 (Rovito et al., 2005). Overall, the findings suggest that NAT enzyme activity in the human prostate is independent of the NAT1 and NAT2 genotypes, thus suggesting that the target polymorphism for association studies could be one other than NAT1 or NAT2.

5.3.2 Conclusions

Many epidemiological studies have been carried out to evaluate the association of NAT gene polymorphisms with the risk of developing cancers such as lung cancer, breast cancer, prostate cancer, colorectal cancer, and gastric cancer, but the results remain contradictory whereby some studies show a positive association of NAT gene SNPs with different types of cancers, whereas other studies show a negative relation. Such discrepancies in the results may be attributed either to the ethnic differences or to the gene-to-gene interactions. Usually, the impact of genetic variants might be masked by other unknown casual genes which may be involved in the process of carcinogenesis and are yet to be identified. Thus, the gene-to-gene and gene-to-environment interactions have become very important and must be considered to evaluate the exact role of genetic polymorphisms in cancer development.

5.4 Cytochrome P450 (CYP)

The cytochrome P450 is a superfamily of enzymes that catalyze the metabolism of drugs and other substances. The CYP superfamily consists of 57 CYP genes and 58 pseudogenes, which are arranged in 18 families and 43 subfamilies in humans (Nelson et al., 2011). CYPs are grouped into families and subfamilies according to the similarity of the amino acid sequence. If they share more than 40% identity in amino acid sequence, they will be grouped in the same family, for example CYP2, and if there is more than 55% similarity, the CYP enzymes belong to the same subfamily, for example CYP2A (Sim and Ingelman-Sundberg, 2013). They are located primarily in the liver, small intestine, and kidneys and are expressed in the endoplasmic reticulum membrane, mitochondria, and cell surface (Renaud et al., 2011; Neve and Ingelman-Sundberg, 2010).

CYPs are heme–thiolate proteins (iron containing) of the cytochrome b type and derive the name P450 from the wavelength (450 nm) at which the carbon monoxide derivative of the reduced cytochrome has a maximum absorption. They catalyze different oxidation and reduction reactions and are known as phase I detoxification enzymes. They belong to the enzymes class of monooxygenase that incorporates only one atom of the molecular oxygen into their substrates (Ortiz de Montellano, 2010). The mechanism of action involves binding of the substrate to the catalytic pocket of CYPs containing heme iron, leading to its reduction from ferric to the ferrous state (Chang and Kam, 1999). The reducing equivalents are transferred from NADPH to P450 by a flavoprotein enzyme called the NADPH–cytochrome P450 reductase (P450 reductase). Then the heme containing the active site is bound by the molecular oxygen (Zanger and Schwab, 2013). Thereafter, one oxygen atom is inserted in the substrate molecule and the other atom is reduced to form H_2O. CYPs mediate the conversion and metabolism of the lipophilic exogenous (xenobiotics) and endogenous compounds (cholesterol, testosterone, progesterone, prostaglandin H2, corticosterone, retinoic acid, vitamin D_3, and arachidonic acid), transforming them into hydrophilic or polar compounds which are easily excreted from the body (Guengerich, 2017; Chang and Kam, 1999).

The CYP1 family consists of three genes and two subfamilies: CYP1A1, CYP1A2, and CYP1B1, and the expression of downstream genes are controlled by the AhRs – a ligand-activated transcriptional factor that dimerizes with aryl hydrocarbon receptor nuclear translocator (ARNT), resulting in a complex that binds to xenobiotic response elements (XREs), which is activated most notably by components of incineration products and cigarette smoke (Moorthy et al., 2015). CYP1A2 preferentially metabolizes arylamines and N-heterocyclics, while CYP1A1 and CYP1B1 are most efficient in metabolizing PAHs, which are known to induce tumorigenesis. For example, the benzo[a]pyrene, dibenzo[a]pyrene, 7,12-dimethylbenz[a]anthracene, and 5-methylchrysene are metabolically activated by the CYP1B1 and CYP1A1 (Shimada et al., 2001). The activation of PAHs results in the formation of the reactive and redox-active o-quinones, diolepoxides, or radical cations, which react with the DNA forming the DNA adducts.

The DNA adducts eventually lead to the mutation and initiation of the carcinogenesis (Moorthy et al., 2015). For instance, it has been reported that CYP1B1 that is found in pulmonary and other tissues (except for liver) has a pivotal role in the metabolic activation of the PAHs (Chun and Kim, 2016).

CYPs play an important role in drug metabolism. The duration of action of many drugs depends on their rate of metabolism by the CYPs. These enzymes catalyze the hydroxylation reactions of the drugs such as barbiturate and phenobarbital to increase their hydrophilicity and thus the easy elimination by the kidneys. Other drugs such as ibuprofen and caffeine are metabolized through oxidation reactions carried out by the CYPs (Zanger and Schwab, 2013). About 15 isoforms belonging to CYP families 1, 2, and 3 are accountable for 70%–80% of all phase I metabolism of clinically used drugs (Zanger and Schwab, 2013) and are involved in the biotransformation of a vast diversity of environmental chemicals (approx. 90%), including 66% metabolism reactions of chemical carcinogens (Rendic and Guengerich, 2012).

CYPs are also involved in many biological processes and mediate cholesterol metabolism and bile acid biosynthesis, steroid, and Vitamin D3 synthesis and metabolism. Human CYP P450s that metabolize xenobiotic compounds are almost exclusively in the CYP1–CYP4 and to some degree CYP (5, 8, 19, 21, and 26) families, and the possible forms of CYP 450s reported in humans are shown in Table 5.4 as compiled by Karolinska Institute.

The CYP1 family consists of three genes and two subfamilies: CYP1A1, CYP1A2, and CYP1B1 are controlled by the Ah receptor, which is activated most notably by components of incineration products and cigarette smoke (Moorthy et al., 2015). CYP1A1 and CYP1B1 are most efficient in metabolizing PAHs, while CYP1A2 preferentially metabolizes arylamines and N-heterocyclics. Human CYP2 family consists of isoenzymes which are mostly involved in the drug metabolism: CYP2C8, CYP2C9, CYP2C18, and CYP2C19 together metabolize to varying amounts greater than half of all frequently prescribed drugs. CYP2D6 is known to metabolize more than 75 drugs. CYP2E1 metabolizes acetaminophen, benzene, chloroform, carbon tetrachloride, butadiene, vinyl chloride, etc. The CYP3 family has four members and has the most abundantly expressed P450s in the liver and is regulated via the pregnane X receptor (PXR), a nuclear receptor that can be induced by pregnenolone-related compounds.

TABLE 5.4

Human Cytochrome P450s Families (CYPs) and Their Respective Isoenzymes

CYP Family	Isoenzyme Name
POR	Cytochrome P450 oxidoreductase
CYP1	CYP1A1, CYP1A2, CYP1B1
CYP2	CYP2A6, CYP2A13, CYP2B6, CYP2C8, CYP2C9, CYP2C19, CYP2D6, CYP2E1, CYP2F, CYP2J2, CYP2R1, CYP2S1, CYP2W1
CYP3	CYP3A4, CYP3A5, CYP3A7, CYP3A43
CYP4	CYP4A11, CYP4A22, CYP4B1, CYP4F2
CYP>4	CYP5A1, CYP8A1, CYP19A1, CYP21A2, CYP26A1

CYP3A4 and CYP3A5 are known to metabolize more than 120 frequently prescribed drugs (Shankar and Mehendale, 2014).

5.4.1 Polymorphism of CYPs and Cancers

The genetic polymorphisms in genes encoding P450 enzymes alter the specific actions of affected proteins and thus have a considerable impact on the fate of xenobiotics, since these subtle DNA changes due to SNPs tend to represent the most frequent cause of variations in oxidative metabolism of drugs or bioactivation of toxicants (Lee and Kim, 2013). SNPs of CYPs have been reported to confer risk of cancer susceptibility (Table 5.5).

5.4.1.1 CYPs and Head-and-Neck Cancer

A number of carcinogens present in tobacco smoke and betel quid, such as PAHs, heterocyclic amines, tannins, alkaloids, and polyphenols, are related to HNCs. CYP1A1 is involved in the activation of such carcinogens (Wu et al., 2004; Sharan et al., 2012), and the modulation of CYP1A1 activity may be important in the etiology of HNC. CYP1A1 activity is also modulated by specific polymorphisms in the CYP1A1 gene. One of its common polymorphisms CYP1A1 rs4646903 has been associated with higher induction of CYP1A1

TABLE 5.5

CYP Gene Polymorphisms and Association with Different Types of Cancer Susceptibility

Gene	SNP	Nucleotide Change	Positive Risk
CYP1A1	rs4646903	3801T>C	Esophageal cancer Oral cancer risk Laryngeal cancer risk Prostate cancer Cervical cancer Bladder cancer Lung cancer
CYP1A1	rs1048943	2455A>G	Oral cancer risk Laryngeal cancer risk Esophageal cancer risk Colorectal cancer Lung cancer Prostate cancer Breast cancer
CYP1A1	CYP1A1*4	2453A>C	Endometrial cancer
CYP1A1	rs4646422	4889A>G 6235T>C 11599G>C 113T>C 264T>C	Sporadic breast cancer Sporadic breast cancer Bladder cancer Bladder cancer Breast cancer
CYP1A2	rs762551 rs3569413 rs2470890 rs2472299	−163C>A	Cholangiocarcinoma Breast cancer Lung cancer Lung cancer
CYP2D6	rs1058172	−941 G>A	Hepatocellular cancer
CYP2E1	rs2031920 rs6413432	−1055C>T 179V>I	Hepatocellular cancer Breast cancer Lung cancer Colorectal cancer

enzyme activity and has been well studied. For example, a meta-analysis of CYP1A1 SNP rs4646903 including 4,168 patients and 4,638 controls indicated that carriers of TC and CC genotypes had a 34% increased risk of developing HNCs compared to TT carriers (95% CI=1.15–1.57, $p<0.001$) (Liu et al., 2013c). Also, CYP1A1 CC genotype has been found to be significantly associated with HNC risk ($p=0.045$) (Singh et al., 2015). Furthermore, the meta-analysis study on CYP2E1 (4,951 patients and 6,071 controls) suggests that individuals with the homozygote genotypes of SNP rs2031920 (PstI/RsaI) (OR=1.96; 95% CI: 1.33–2.90) and rs6413432 (DraI) polymorphism (OR=1.56; 95% CI: 1.06–2.27) might be associated with an increased risk of HNC, especially in Asians (Tang et al., 2010). Another meta-analysis including 43 studies reported that CYP2E1 rs2031920 SNP may modify the susceptibility to HNC, particularly among Asians (Zhuo et al., 2016). Also, a significant association between HNC and CYP1A1 rs4646421 (TT) and CYP2D6 rs1058172 (AA) variant carriers (OR 1.75, $p=0.008$ and OR 1.66, $p=0.016$, respectively) is reported in Tunisian population (Khlifi et al., 2014).

Moreover, multiple studies have shown that numerous XME gene polymorphisms influence the susceptibility to developing HNC; for instance, the results of one study indicated that CYP1A1 TC/CC+GSTM1 null and CYP1A1 TC/CC+GSTT1 null genotypes are significantly associated with HNC risk ($p<0.001$ and $p=0.005$, respectively). Similarly, several fold increased risk of HNC ($p<0.001$) was observed among tobacco-betel quid chewers, carrying CYP1A1 TC/CC+GSTM1 null genotypes (Choudhury et al., 2015). Likewise, another study reported a significant association ($p=0.04$) between CYP1A1 A4889G and T6235C plus GSTM1/T1 null genotypes with HNC. The results revealed that the carriers of combined genotypes CYP1A1 6235TC+CC and null GSTT1/M1 had a 16-fold increased risk of HNSCC (Lourenço et al., 2011).

5.4.1.2 CYPs and Breast Cancer

Certain CYP genes, particularly CYP19A1 and CYP1A2, have been associated with BC risk and development in the different populations. A study on Jordanian BC patients reported that the CYP19A1 SNP rs7176005 and the CYP1A2 SNP rs762551 were significantly associated with BC risk (p value$\leq$0.004) (Al-Eitan et al., 2019). CYP1A1 SNPs (rs11636419 and rs17861162) and CYP1A2 SNP rs2470890 were found to be significantly associated with age and menstruation status of BC patients. Furthermore, the study of CYP1A1 SNPs rs1048943 (A4889G) and rs4646903 (T6235C) reported an increased risk of developing grade I and II tumors in BC patients of Brazil. The results showed that the G allele (OR=2.44-fold; 95% CI: 1.48–4.02) and C allele carriers (OR=1.67-fold; 95% CI: 1.03–2.69) are significantly associated with BC risk (Oliveira et al., 2015). Also, SNPs rs11636419 and rs17861162 were associated with the P53 status, whereas SNP rs2470890 was correlated with CYP1A2 protein expression under the co-dominant and dominant models ($p=0.017$ and $p=0.006$, respectively) (Bai et al., 2017). A study on SNP rs1048943 of CYP1A1 showed that the G allele (A2455G) functions as a risk factor for BC among Caucasian subjects (Sergentanis and Economopoulos, 2010), but others showed negative results (Khalili-Tanha et al., 2019). Several studies including meta-analyses strongly indicate that there is no significant association between the CYP1A1 T3801C (rs4646903) and A2455G (rs1048943) polymorphisms and BC risk (Yang and He, 2021; Wu et al., 2017; Kiruthiga et al., 2011).

5.4.1.3 CYPs and Lung Cancer

A meta-analysis (5,074 cases and 6,828 controls) indicates that the CYP2E1 rs2031920 (RsaI) polymorphism is associated with lung cancer risk among Asians and rs6413432 (DraI) decreased the lung cancer risk (Ye et al., 2015). These results are in concordance with another meta-analysis that found a protective effect of the CYP2E1 rs6413432 CC and CD+CC polymorphism for LC (Wang et al., 2009; Xu et al., 2014b). CYP2E1 c1/ c1, DD, and NAT2 slow acetylation, especially when combined with smoking, contributes to the development of lung cancer ($p=0.003$) (Zhang et al., 2014).

Studies have shown that CYP1A1 rs1048943 SNP might play a modest role in susceptibility to lung cancer, especially in Asians, but not in Africans, Caucasians, or other populations (Wang et al., 2011b). Also, significantly increased risks were found for the CYP1A1 rs4646903 (MspI) and rs1048943 SNPs among East Asians, but only marginally significant associations were observed for the rs4646903 SNP, while no significant associations were observed for the rs1048943 SNP in Caucasians and other populations (Chen et al., 2011). A meta-analysis suggests that the Ile462Val polymorphism (rs1048943) of CYP1A1 correlates with increased lung cancer susceptibility in Asian and Caucasian populations and there is an interaction with smoking status (Ji et al., 2012).

5.4.1.4 CYPs and Esophageal Cancer

Numerous studies have shown that the CYP1 Rsa SNP is associated with EC and the frequency of the variant c1 allele was observed to be much higher in patients with esophageal diseases than in healthy individuals (Lin et al., 1998).

A meta-analysis including 1,519 EC cases and 1,962 controls showed a significant association between CYP1A1 rs4646903 (3801T>C) and EC in Chinese population ($p<0.05$). Also the two variants of CYP1A1 (CC and CT combined vs TT) were found to significantly elevate the risk of EC (Zheng et al., 2015). Also various studies including meta-analyses on CYP1A1 have detected a significance of SNP rs1048943 with EC risk in Chinese population (Chen et al., 2016; Liu et al., 2014; Yun et al., 2014). Similarly, a meta-analysis observed a difference in allele frequencies between Caucasian and Asian populations for rs1048943SNP and demonstrated distinct evidence of association of SNP with the risk of EC ($p<0.001$) (Shen et al., 2013). Based on the meta-analysis on cumulative epidemiological evidence for association between genetic polymorphisms and EC susceptibility, the variant genotype of CYP1A1 rs1048943 is rated as a strong candidate having a significant association with EC and ESCC risk development (Tian et al., 2019).

Further, a meta-analysis involving 1,663 cases and 2,603 controls suggests that the CYP2E1 rs2031920 (RsaI/PstI) polymorphism is a risk factor for EC ($p<0.05$) and the c2 allele is a factor that lowers the possibility of EC in the mainland Chinese population (Leng et al., 2012).

5.4.1.5 CYPs and CRC

There are a number of studies that have shown that SNPs in CYP genes are significantly associated with the development of CRC; for example, CYP2E1*5B (rs3813867) and CYP2E1*6 (rs6413432) are reported to be associated with sporadic CRC ($p<0.01$) (Fernandes et al., 2016). A meta-analysis including 6,654 cases and 7,859 controls revealed that the CYP1A1 rs104894 has a significant association with CRC risk among both Asians and Caucasians (Zheng et al., 2012). Similar results were reported for CYP1A1 rs104894 SNP from a meta-analysis (5,336 cases and 6,226 controls) among Europeans ($p<0.05$) (Jin et al., 2011). Further, a meta-analysis including 8,665 cases and 9,953 controls suggested that the CYP1A1 rs104894 may increase the susceptibility to CRC and might contribute to the pathogenesis of disease (Zhu et al., 2016). A study on Polish CRC patients suggested that CYP1A1 Ile462Val (rs1048943) was statistically significant in CRC patients over 50 years old (OR =2.05; 95 CI=1.29–3.28) and more pronounced in the female cases who were <50 years old (Gil et al., 2014).

CYP2E1 gene polymorphisms such as CYP2E1 SNP rs2031920 and 9-bp insertion polymorphism are a risk factor for developing CRC (Peng et al., 2013; Qian et al., 2013). A meta-analysis including 4,592 cases and 5,918 controls determines a significant association between the CYP2E1 96-bp insertion polymorphism and colorectal cancer risk under the allele contrast model and the dominant contrast model ($p=0.05$) (Qian et al., 2013). CYP2E1 rs3813867 (RsaI) c2/c2 and CYP2E1 98-bp heterozygous i/I genotype are found to be significantly associated with an increased risk of CRC in ethnic Kashmiri population ($p=0.01$) (Sameer et al., 2011). Similar results were shown by another study with higher risk among individuals with allele c2. And RsaI c2/c2 genotype was associated with an increased risk of rectal cancer (Silva et al., 2012).

Moreover, the association of CYP1A2*C (rs2069514) and CYP1A2*F (rs762551) polymorphisms with the risk of CRC has been studied. A meta-analysis that included 7,088 cases and 7,568 controls suggested that the CYP1A2 *F and *C polymorphisms are a protective factor against CRC among Asians. The subgroup analysis showed a significant association of CYP1A2*F ($p=0.04$) and CYP1A2*C ($p=0.53$) among Asians, but not among Caucasians (Zhao et al., 2013).

5.4.1.6 CYPs and Prostate Cancer

A meta-analysis study suggested a positive relation between the CYP1A1 MspI polymorphism and the risk of PC among Asians ($p<0.001$), but not in Caucasians or mixed populations (Ding et al., 2013). The same results are shown by other researchers in different populations (Hoidy et al., 2019). Also, the rs4646903 polymorphism is reported to be associated with a significant increase in the risk of PC in both the Asian and Caucasian populations ($p=0.003$), but the CYP1A1 rs1048943 polymorphism had a significant association with PC risk in Asians only ($p=0.001$); however, no statistical significance of CYP1B1 SNPs (rs10012, rs162549, rs1800440, and rs2551188) and the risk of PC was reported (Zhu et al., 2019). However, there are some studies that are contradictory; for example, a study on the Brazilian population showed CYP1A1 genotypes are not associated with the susceptibility to PC or its outcome (Lima et al., 2008). Similarly, no association between CYP1A1 genotypes and Turkish PC cases is reported (Silig et al., 2006). No significant differences in the frequency distributions of CYP1A1 polymorphisms were observed among PC patients of China (Li et al., 2008).

5.4.2 Conclusions

Results from different studies showed that CYP gene polymorphisms play an important role in cancer development as well as progression. Overall, it is observed that CYP1A1 influences breast, lung, colorectal, HCC, and prostate cancers, whereas CYP1E2 is involved in lung and digestive tract cancers. But there are also some contradictory studies that negate their involvement in cancer development; for example, CYP1A1 Msp polymorphism is common among Asian prostate cancer cases, but not in Caucasians and CYP1A1 rs1048943 SNP is susceptible to lung cancer, especially in Asians, but not in Africans, Caucasians, or other populations. Also, the variation in CYP enzymes may affect the process of chemotherapy as they are involved in the metabolism of various anticancer drugs, resulting in undesirable effects, toxicity, and therapeutic failure.

ABBREVIATIONS

ARNT:	Aryl hydrocarbon receptor nuclear translocator
BC:	Breast cancer
BCR:	Biochemical recurrence
CRC:	Colorectal cancer
CYP:	Cytochrome P450
ESCC:	Esophageal squamous cell carcinoma
GSH:	Glutathione
GST:	Glutathione S-transferases
GWAS:	Genome-wide association studies
HCC:	Hepatocellular carcinoma
HNSCC:	Head-and-neck squamous cell carcinoma
NAT:	N-Acetyltransferases
OR:	Odds ratio
p:	p value
PAH:	Polycyclic aromatic hydrocarbon
PC:	Prostate cancer
PXR:	Pregnane X receptor
rs:	Reference SNP cluster ID
SNP:	Single nucleotide polymorphism
XME:	Xenobiotic-metabolizing enzyme
XRE:	Xenobiotic response element

REFERENCES

Aghajany-Nasab M, Panjehpour M, Samiee SM, Rahimi F, Movahedian A. Glutathione S-transferase mu gene variants and colorectal cancer development – use of sequence-specific probes for an Iranian population. *Asian Pac J Cancer Prev*. 2011;12:1511–5.

Agudo A, Sala N, Pera G, Capellá G, Berenguer A, García N, Palli D, Boeing H, Del Giudice G, Saieva C, Carneiro F, Berrino F, Sacerdote C, Tumino R, Panico S, Berglund G, Simán H,

Stenling R, Hallmans G, Martínez C, Amiano P, Barricarte A, Navarro C, Quirós JR, Allen N, Key T, Bingham S, Khaw KT, Linseisen J, Nagel G, Overvad K, Tjonneland A, Olsen A, Bueno-de-Mesquita HB, Boshuizen HC, Peeters PH, Numans ME, Clavel-Chapelon F, Boutron-Ruault MC, Trichopoulou A, Lund E, Bläker H, Jenab M, Ferrari P, Norat T, Riboli E, González CA. No association between polymorphisms in CYP2E1, GSTM1, NAT1, NAT2 and the risk of gastric adenocarcinoma in the European prospective investigation into cancer and nutrition. *Cancer Epidemiol Biomarkers Prev.* 2006 May;15(5):1043–5. doi: 10.1158/1055-9965.EPI-06-0073.

Agúndez JA. Cytochrome P450 gene polymorphism and cancer. *Curr Drug Metab.* 2004 Jun;5(3):211–24. doi: 10.2174/1389200043335621.

Agúndez JA. Polymorphisms of human N-acetyltransferases and cancer risk. *Curr Drug Metab.* 2008 Jul;9(6):520–31. doi: 10.2174/138920008784892083.

Agúndez JA, Martínez C, Olivera M, Gallardo L, Ladero JM, Rosado C, Prados J, Rodriguez-Molina J, Resel L, Benítez J. Expression in human prostate of drug- and carcinogen-metabolizing enzymes: association with prostate cancer risk. *Br J Cancer.* 1998 Nov;78(10):1361–7. doi: 10.1038/bjc.1998.685.

Alberg AJ, Daudt A, Huang HY, Hoffman SC, Comstock GW, Helzlsouer KJ, Strickland PT, Bell DA. N-acetyltransferase 2 (NAT2) genotypes, cigarette smoking, and the risk of breast cancer. *Cancer Detect Prev.* 2004;28(3):187–93. doi: 10.1016/j.cdp.2004.04.001.

Al-Eitan LN, Rababa'h DM, Alghamdi MA, Khasawneh RH. Association of CYP gene polymorphisms with breast cancer risk and prognostic factors in the Jordanian population. *BMC Med Genet.* 2019 Sep 2;20(1):148. doi: 10.1186/s12881-019-0884-x.

Al-Moundhri MS, Al-Kindi M, Al-Nabhani M, Al-Bahrani B, Burney IA, Al-Madhani A, Ganguly SS, Tanira M. NAT2 polymorphism in Omani gastric cancer patients-risk predisposition and clinicopathological associations. *World J Gastroenterol.* 2007 May 21;13(19):2697–702. doi: 10.3748/wjg.v13.i19.2697.

Al-Moundhri MS, Alkindy M, Al-Nabhani M, Al-Bahrani B, Burney IA, Al-Habsi H, Ganguly SS, Tanira M. Combined polymorphism analysis of glutathione S-transferase M1/G1 and interleukin-1B (IL-1B)/interleukin 1-receptor antagonist (IL-1RN) and gastric cancer risk in an Omani Arab Population. *J Clin Gastroenterol.* 2009 Feb;43(2):152–6. doi: 10.1097/MCG.0b013e31815853fa.

Ambrosone CB, Freudenheim JL, Graham S, Marshall JR, Vena JE, Brasure JR, Michalek AM, Laughlin R, Nemoto T, Gillenwater KA, Shields PG. Cigarette smoking, N-acetyltransferase 2 genetic polymorphisms, and breast cancer risk. *JAMA.* 1996 Nov 13;276(18):1494–501.

Axarli IA, Rigden DJ, Labrou NE. Characterization of the ligandin site of maize glutathione S-transferase I. *Biochem J.* 2004 Sep 15;382(Pt 3):885–93. doi: 10.1042/BJ20040298.

Bai X, Xie J, Sun S, Zhang X, Jiang Y, Pang D. The associations of genetic polymorphisms in CYP1A2 and CYP3A4 with clinical outcomes of breast cancer patients in northern China. *Oncotarget.* 2017 Jun 13;8(24):38367–77. doi: 10.18632/oncotarget.16359.

Belogubova EV, Kuligina ESh, Togo AV, Karpova MB, Ulibina JM, Shutkin VA, Hanson KP, Popowski K, Mosyagin I, Cascorbi I, Hirvonen A, Imyanitov EN. 'Comparison of extremes' approach provides evidence against the modifying role of NAT2 polymorphism in lung cancer susceptibility. *Cancer Lett.* 2005 Apr 28;221(2):177–83. doi: 10.1016/j.canlet.2004.11.008.

Bendjemana K, Abdennebi M, Gara S, Jmal A, Ghanem A, Touati S, Boussen H, Ladgham A, Guemira F. Polymorphisme génétique des enzymes gluthation-S transférases et N-acétyltransférases 2 dans le cancer du nasopharynx en Tunisie [Genetic polymorphism of gluthation-S transferases and N-acetyl transferases 2 and nasopharyngeal carcinoma: the Tunisia experience]. *Bull Cancer.* 2006 Mar 1;93(3):297–302.

Board PG, Coggan M, Chelvanayagam G, Easteal S, Jermiin LS, Schulte GK, Danley DE, Hoth LR, Griffor MC, Kamath AV, Rosner MH, Chrunyk BA, Perregaux DE, Gabel CA, Geoghegan KF, Pandit J. Identification, characterization, and crystal structure of the Omega class glutathione transferases. *J Biol Chem.* 2000 Aug 11;275(32):24798–806. doi: 10.1074/jbc.M001706200

Boccia S, Sayed-Tabatabaei FA, Persiani R, Gianfagna F, Rausei S, Arzani D, La Greca A, D'Ugo D, La Torre G, van Duijn CM, Ricciardi G. Polymorphisms in metabolic genes, their combination and interaction with tobacco smoke and alcohol consumption and risk of gastric cancer: a case-control study in an Italian population. *BMC Cancer.* 2007 Nov 8;7:206. doi: 10.1186/1471-2407-7-206.

Bolt HM, Thier R. Relevance of the deletion polymorphisms of the glutathione S-transferases GSTT1 and GSTM1 in pharmacology and toxicology. *Curr Drug Metab.* 2006 Aug;7(6):613–28. doi: 10.2174/138920006778017786

Borlak J, Reamon-Buettner SM. N-acetyltransferase 2 gene polymorphisms in colon and lung cancer patients. *BMC Med NAT2Genet.* 2006 Jul 9;7:58. doi: 10.1186/1471-2350-7-58.

Butcher NJ, Arulpragasam A, Goh HL, Davey T, Minchin RF. Genomic organization of human arylamine N-acetyltransferase Type I reveals alternative promoters that generate different 5'-UTR splice variants with altered translational activities. *Biochem J.* 2005;387(Pt 1):119–127. doi:10.1042/BJ20040903.

Chan AT, Tranah GJ, Giovannucci EL, Willett WC, Hunter DJ, Fuchs CS. Prospective study of N-acetyltransferase-2 genotypes, meat intake, smoking and risk of colorectal cancer. *Int J Cancer.* 2005 Jul 1;115(4):648–52. doi: 10.1002/ijc.20890.

Chang GW, Kam PC. The physiological and pharmacological roles of cytochrome P450 isoenzymes. *Anaesthesia.* 1999 Jan;54(1):42–50. doi: 10.1046/j.1365-2044.1999.00602.x.

Chang-Claude J, Kropp S, Jäger B, Bartsch H, Risch A. Differential effect of NAT2 on the association between active and passive smoke exposure and breast cancer risk. *Cancer Epidemiol Biomarkers Prev.* 2002 Aug;11(8):698–704.

Chen C, Ricks S, Doody DR, Fitzgibbons ED, Porter PL, Schwartz SM. N-Acetyltransferase 2 polymorphisms, cigarette smoking and alcohol consumption, and oral squamous cell cancer risk. *Carcinogenesis.* 2001 Dec;22(12):1993–9. doi: 10.1093/carcin/22.12.1993.

Chen J, Ma L, Peng NF, Wang SJ, Li LQ. A meta-analysis of the relationship between glutathione S-transferases gene polymorphism and hepatocellular carcinoma in Asian population. *Mol Biol Rep.* 2012 Dec;39(12):10383–93. doi: 10.1007/s11033-012-1917-0. Epub 2012 Oct 10.

Chen R, Ren S, Meng T, Aguilar J, Sun Y. Impact of glutathione-S-transferases (GST) polymorphisms and hypermethylation of relevant genes on risk of prostate cancer biochemical recurrence: a meta-analysis. *PLoS One.* 2013 Sep 23;8(9):e74775. doi: 10.1371/journal.pone.0074775.

Chen XP, Xu DF, Xu WH, Ma ZC, Yao J, Fu SM. Association studies of CYP1A1 Exon7 polymorphism and -GSTM1 interaction with esophageal cancer risk: a meta-analysis in the chinese population. *Clin Lab.* 2016 Sep 1;62(9):1795–802. doi: 10.7754/Clin.Lab.2016.160132.

Chen Z, Li Z, Niu X, Ye X, Yu Y, Lu S, Chen Z. The effect of CYP1A1 polymorphisms on the risk of lung cancer: a global meta-analysis based on 71 case-control studies. *Mutagenesis.* 2011 May;26(3):437–46. doi: 10.1093/mutage/ger002. Epub 2011 Mar 14. PMID: 21402622.

Chenevix-Trench G, Young J, Coggan M, Board P. Glutathione S-transferase M1 and T1 polymorphisms: susceptibility to colon cancer and age of onset. *Carcinogenesis.* 1995;16(7):-1655–7. doi: 10.1093/carcin/16.7.1655.

Chiou HL, Wu MF, Chien WP, Cheng YW, Wong RH, Chen CY, Lin TS, Lee H. NAT2 fast acetylator genotype is associated with an increased risk of lung cancer among never-smoking women in Taiwan. *Cancer Lett.* 2005 Jun 1;223(1):93–101. doi: 10.1016/j.canlet.2004.12.015. Epub 2005 Jan 18.

Choudhury JH, Singh SA, Kundu S, Choudhury B, Talukdar FR, Srivasta S, Laskar RS, Dhar B, Das R, Laskar S, Kumar M, Kapfo W, Mondal R, Ghosh SK. Tobacco carcinogen-metabolizing genes CYP1A1, GSTM1, and GSTT1 polymorphisms and their interaction with tobacco exposure influence the risk of head and neck cancer in Northeast Indian population. *Tumour Biol.* 2015 Aug;36(8):5773–83. doi: 10.1007/s13277-015-3246-0. Epub 2015 Feb 28.

Chun YJ, Kim D. Cancer activation and polymorphisms of human cytochrome P450 1B1. *Toxicol Res.* 2016 Apr;32(2):89–93. doi: 10.5487/TR.2016.32.2.089. Epub 2016 Apr 30.

Coleman JOD. Detoxification of xenobiotics by plants: chemical modification and vacuolar compartmentation. *Trends Plant Sci.* 1997;2:144–51.

Cote ML, Chen W, Smith DW, Benhamou S, Bouchardy C, Butkiewicz D, Fong KM, Gené M, Hirvonen A, Kiyohara C, Larsen JE, Lin P, Raaschou-Nielsen O, Povey AC, Reszka E, Risch A, Schneider J, Schwartz AG, Sorensen M, To-Figueras J, Tokudome S, Pu Y, Yang P, Wenzlaff AS, Wikman H, Taioli E. Meta- and pooled analysis of GSTP1 polymorphism and lung cancer: a HuGE-GSEC review. *Am J Epidemiol.* 2009 Apr 1;169(7):802–14. doi: 10.1093/aje/kwn417. Epub 2009 Feb 24

Darazy M, Balbaa M, Mugharbil A, Saeed H, Sidani H, Abdel-Razzak Z. CYP1A1, CYP2E1, and GSTM1 gene polymorphisms and susceptibility to colorectal and gastrin cancer among Lebanese. *GenetTest Mol Biomarkers.* 2011;15:423–9.

Deloménie C, Sica L, Grant DM, Krishnamoorthy R, Dupret JM. Genotyping of the polymorphic N-acetyltransferase (NAT2*) gene locus in two native African populations. *Pharmacogenetics.* 1996 Apr;6(2):177–85. doi: 10.1097/00008571-199604000-00004.

Di Pietro G, Magno LA, Rios-Santos F. Glutathione S-transferases: an overview in cancer research. *Expert Opin Drug Metab Toxicol.* 2010 Feb;6(2):153–70. doi: 10.1517/17425250903427980.

Ding G, Xu W, Liu H, Zhang M, Huang Q, Liao Z. CYP1A1 MspI polymorphism is associated with prostate cancer susceptibility: evidence from a meta-analysis. *Mol Biol Rep.* 2013 May;40(5):3483–91. doi: 10.1007/s11033-012-2423-0. Epub 2013 Mar 9. PMID: 23475304.

Du J, Fang C, Mao Y, Zhao J, Tie Y, Xiang Z. Association between glutathione S-transferase M1 polymorphism and esophageal cancer: a pooled analysis based on Chinese individuals. *Int J Clin Exp Pathol.* 2018 Jan 1;11(1):375–81.

Dunning AM, Healey CS, Pharoah PD, Teare MD, Ponder BA, Easton DF. A systematic review of genetic polymorphisms and breast cancer risk. *Cancer Epidemiol Biomarkers Prev.* 1999 Oct;8(10):843–54.

Dura P, Salomon J, Te Morsche RH, Roelofs HM, Kristinsson JO, Wobbes T, Witteman BJ, Tan AC, Drenth JP, Peters WH. No role for glutathione S-transferase genotypes in Caucasian esophageal squamous cell or adenocarcinoma etiology: an European case-control study. BMC gastroenterol. 2013 Jun 3;13:97. doi: 10.1186/1471-230X-13-97.

Economopoulos KP, Sergentanis TN. STM1, GSTT1, GSTP1, GSTA1 and colorectal cancer risk: a comprehensive meta-analysis. *Eur J Cancer.* 2010;46:1617–31.

Egan KM, Newcomb PA, Titus-Ernstoff L, Trentham-Dietz A, Mignone LI, Farin F, Hunter DJ. Association of NAT2 and smoking in relation to breast cancer incidence in a population-based case-control study (United States). *Cancer Causes Control.* 2003 Feb;14(1):43–51. doi: 10.1023/a:1022517506689.

El-Serag HB, Rudolph KL. Hepatocellular carcinoma: epidemiology and molecular carcinogenesis. *Gastroenterology.* 2007 Jun;132(7):2557–76. doi: 10.1053/j.gastro.2007.04.061.

El-Sharkawi F, El Sabah M, Hassan Z, Khaled H. The biochemical value of urinary metalloproteinases 3 and 9 in diagnosis and prognosis of bladder cancer in Egypt. *J Biomed Sci.* 2014 Aug 19;21(1):72. doi: 10.1186/s12929-014-0072-4.

Fernandes GM, Russo A, Proença MA, Gazola NF, Rodrigues GH, Biselli-Chicote PM, Silva AE, Netinho JG, Pavarino ÉC, Goloni-Bertollo EM. CYP1A1, CYP2E1 and EPHX1 polymorphisms in sporadic colorectal neoplasms. *World J Gastroenterol.* 2016 Dec 7;22(45):9974–83. doi: 10.3748/wjg.v22.i45.9974.

Gajecka M, Rydzanicz M, Jaskula-Sztul R, Kujawski M, Szyfter W, Szyfter K. CYP1A1, CYP2D6, CYP2E1, NAT2, GSTM1 and GSTT1 polymorphisms or their combinations are associated with the increased risk of the laryngeal squamous cell carcinoma. *Mutat Res.* 2005 Jul 1;574(1–2):112–23. doi: 10.1016/j.mrfmmm.2005.01.027. Epub 2005 Apr 15.

Gao JP, Huang YD, Yang GZ, Yang YQ. [Relationship between genetic polymorphisms of metabolizing enzymes and prostate cancer]. *Zhonghua Nan Ke Xue.* 2003 Feb;9(1):32–5. Chinese.

Geisler SA, Olshan AF. GSTM1, GSTT1, and the risk of squamous cell carcinoma of the head and neck: a mini-HuGE review. *Am J Epidemiol.* 2001;154:95–105.

Gemignani F, Landi S, Szeszenia-Dabrowska N, Zaridze D, Lissowska J, Rudnai P, Fabianova E, Mates D, Foretova L, Janout V, Bencko V, Gaborieau V, Gioia-Patricola L, Bellini I, Barale R, Canzian F, Hall J, Boffetta P, Hung RJ, Brennan P. Development of lung cancer before the age of 50: the role of xenobiotic metabolizing genes. *Carcinogenesis.* 2007 Jun;28(6):1287–93. doi: 10.1093/carcin/bgm021. Epub 2007 Jan 27.

Gil J, Gaj P, Misiak B, Ostrowski J, Karpinski P, Jarczyńska A, Kielan W, Sasiadek MM. CYP1A1 Ile462Val polymorphism and colorectal cancer risk in Polish patients. *Med Oncol.* 2014 Jul;31(7):72. doi: 10.1007/s12032-014-0072-y. Epub 2014 Jun 18.

Goekkurt E, Hoehn S, Wolschke C, Wittmer C, Stueber C, Hossfeld DK, Stoehlmacher J. Polymorphisms of glutathione S-transferases (GST) and thymidylate synthase (TS)--novel predictors for response and survival in gastric cancer patients. *Br J Cancer.* 2006 Jan 30;94(2):281–6. doi: 10.1038/sj.bjc.6602891.

Gong C, Hu X, Gao Y, Cao Y, Gao F, Mo Z. A meta-analysis of the NAT1 and NAT2 polymorphisms and prostate cancer: a huge review. *Med Oncol.* 2011 Mar;28(1):365–76. doi: 10.1007/s12032-010-9423-5. Epub 2010 Jan 23.

Gong M, Dong W, Shi Z, Xu Y, Ni W, An R. Genetic polymorphism of GSTM1, GSTT1 and GSTP1 with prostate cancer risk: a meta-analysis of 57 studies. *PLoS One.* 2012;7:e50587.

González MV, Alvarez V, Pello MF, Menéndez MJ, Suárez C, Coto E. Genetic polymorphism of N-acetyltransferase-2, glutathione S-transferase-M1, and cytochromes P450IIE1 and P450IID6 in the susceptibility to head and neck cancer. *J Clin Pathol.* 1998 Apr;51(4):294–8. doi: 10.1136/jcp.51.4.294.

Gudmundsdottir K, Tryggvadottir L, Eyfjord JE. GSTM1, GSTT1, and GSTP1 genotypes in relation to breast cancer risk and frequency of mutations in the p53 gene. *Cancer Epidemiol Biomarkers Prev.* 2001 Nov;10(11):1169–73.

Guengerich FP. Intersection of the roles of cytochrome P450 enzymes with xenobiotic and endogenous substrates: relevance to toxicity and drug interactions. *Chem Res Toxicol.* 2017 Jan 17;30(1):2–12. doi: 10.1021/acs.chemrestox.6b00226. Epub 2016 Aug 11.

Habalová V, Salagovic J, Kalina I, Stubna J. A pilot study testing the genetic polymorphism of N-acetyltransferase 2 as a risk factor in lung cancer. *Neoplasma.* 2005;52(5):364–8.

Hamasaki T, Inatomi H, Katoh T, Aono H, Ikuyama T, Muratani T, Matsumoto T. *Int J Urol.* 2003;10(3):167–73.

Hamasaki T, Inatomi H, Katoh T, Aono H, Ikuyama T, Muratani T, Matsumoto T. NAT2 gene polymorphism as a possible biomarker for susceptibility to prostate cancer in Japanese. *Int J Urol.* 2003;10:167–173. doi: 10.1046/j.1442-2042.2003.00586.x.

Hashibe M, Brennan P, Strange RC, Bhisey R, Cascorbi I, Lazarus P, Oude Ophuis MB, Benhamou S, Foulkes WD, Katoh T, Coutelle C, Romkes M, Gaspari L, Taioli E, Boffetta P. Meta- and pooled analyses of GSTM1, GSTT1, GSTP1, and CYP1A1 genotypes and risk of head and neck cancer. *Cancer Epidemiol Biomarkers Prev.* 2003 Dec;12(12):1509–17.

Hayes JD, Flanagan JU, Jowsey IR. Glutathione transferases. *Annu Rev Pharmacol Toxicol.* 2005;45:51–88. doi: 10.1146/annurev.pharmtox.45.120403.095857.

Hayes JD, Strange RC. Glutathione S-transferase polymorphisms and their biological consequences. *Pharmacology.* 2000 Sep;61(3):154–66. doi: 10.1159/000028396.

Hein DW. Molecular genetics and function of NAT1 and NAT2: role in aromatic amine metabolism and carcinogenesis. *Mutat Res.* 2002 Sep 30;506–507:65–77. doi: 10.1016/s0027-5107(02)00153-7.

Hein DW, Doll MA, Fretland AJ, Leff MA, Webb SJ, Xiao GH, Devanaboyina US, Nangju NA, Feng Y. Molecular genetics and epidemiology of the NAT1 and NAT2 acetylation polymorphisms. *Cancer Epidemiol Biomarkers Prev.* 2000 Jan;9(1):29–42.

Hein DW. N-acetyltransferase 2 genetic polymorphism: effects of carcinogen and haplotype on urinary bladder cancer risk. *Oncogene.* 2006 Mar 13;25(11):1649–58. doi: 10.1038/sj.onc.1209374.

Hein DW, Doll MA, Fretland AJ, Leff MA, Webb SJ, Xiao GH, Devanaboyina US, Nangju NA, Feng Y. Molecular genetics and epidemiology of the NAT1 and NAT2 acetylation polymorphisms. *Cancer Epidemiol Biomarkers Prev.* 2000 Jan;9(1):29–42.

Hezova R, Bienertova-Vasku J, Sachlova M, et al. Common polymorphisms in GSTM1, GSTT1, GSTP1, GSTA1 and susceptibility of colorectal cancer in the Central European population. *Eur J Med Res.* 2012;17:17–22

Hickman D, Risch A, Buckle V, Spurr NK, Jeremiah SJ, McCarthy A, Sim E. Chromosomal localization of human genes for arylamine N-acetyltransferase. Biochem J. 1994 Feb 1;297 (Pt 3)(Pt 3):441–5. doi: 10.1042/bj2970441.

Hlavata I, Vrana D, Smerhovsky Z, et al. Association between exposure-relevant polymorphisms in CYP1B1, EPHX1, NQO1, GSTM1,GSTP1 and GSTT1 and risk of colorectal cancer in a Czech population. *Oncol Rep.*2010;24:1347–53.

Hoidy WH, Jaber FA, Al-Askari MA. Association of CYP1A1 rs1048943 polymorphism with prostate cancer in Iraqi men patients. *Asian Pac J Cancer Prev.* 2019 Dec 1;20 (12):3839–42. doi: 10.31557/APJCP.2019.20.12.3839.

Hong SH, Kim JW, Kim HG, Park IK, Ryoo JW, Lee CH, Sohn YK, Lee JY. [Glutathione S-transferases (GSTM1, GSTT1 and GSTP1) and N-acetyltransferase 2 polymorphisms and the risk of gastric cancer]. *J Prev Med Public Health.* 2006 Mar;39(2):135–40. Korean.

Hosgood HD 3rd, Berndt SI, Lan Q. GST genotypes and lung cancer susceptibility in Asian populations with indoor air pollution exposures: a meta-analysis. *Mutat Res.* 2007 Nov–Dec;636(1–3):134–43. doi: 10.1016/j.mrrev.2007.02.002. Epub 2007 Mar 2.

Houlston RS, Tomlinson IP. Polymorphisms and colorectal tumor risk. *Gastroenterology.* 2001 Aug;121(2):282–301. doi: 10.1053/gast.2001.26265.

Huang CC, Chien WP, Wong RH, Cheng YW, Chen MC, Chou MC, Lee H. NAT2 fast acetylator genotype is associated with an increased risk of colorectal cancer in Taiwan. *Dis Colon Rectum.* 2007 Jul;50(7):981–9. doi: 10.1007/s10350-007-0230-9.

Ingelman-Sundberg M, Sim SC, Gomez A, Rodriguez-Antona C. Influence of cytochrome P450 polymorphisms on drug therapies: pharmacogenetic, pharmacoepigenetic and clinical aspects. *Pharmacol Ther.* 2007 Dec;116(3):496–526. doi: 10.1016/j.pharmthera.2007.09.004. Epub 2007 Oct 9.

Jain M, Kumar S, Lal P, Tiwari A, Ghoshal UC, Mittal B. Association of genetic polymorphisms of N-acetyltransferase 2 and susceptibility to esophageal cancer in north Indian population. *Cancer Invest.* 2007 Aug;25(5):340–6. doi: 10.1080/07357900701358074.

Jethwa AR, Khariwala SS. Tobacco-related carcinogenesis in head and neck cancer. *Cancer Metastasis Rev.* 2017;36 (3):411–23. https://doi.org/10.1007/s10555-017-9689-6.

Ji YN, Wang Q, Suo LJ. CYP1A1 Ile462Val polymorphism contributes to lung cancer susceptibility among lung squamous carcinoma and smokers: a meta-analysis. *PLoS One.*

2012;7(8):e43397. doi: 10.1371/journal.pone.0043397. Epub 2012 Aug 28.

Jin JQ, Hu YY, Niu YM, Yang GL, Wu YY, Leng WD, Xia LY. CYP1A1 Ile462Val polymorphism contributes to colorectal cancer risk: a meta-analysis. *World J Gastroenterol.* 2011 Jan 14;17(2):260–6. doi: 10.3748/wjg.v17.i2.260.

Kamel AM, Ebid GT, Moussa HS. N-Acetyltransferase 2 (NAT2) polymorphism as a risk modifier of susceptibility to pediatric acute lymphoblastic leukemia. *Tumour Biol.* 2015 Aug;36(8):6341–8. doi: 10.1007/s13277-015-3320-7. Epub 2015 Mar 25.

Kassab A, Msolly A, Lakhdar R, Gharbi O, Miled A. Polymorphisms of glutathione-S-transferases M1, T1, P1 and susceptibility to colorectal cancer in a sample of the Tunisian population. *Med Oncol.* 2014;31:760. doi: 10.1007/s12032-013-0760-z.

Katoh T, Boissy R, Nagata N, Kitagawa K, Kuroda Y, Itoh H, Kawamoto T, Bell DA. Inherited polymorphism in the N-acetyltransferase 1 (NAT1) and 2 (NAT2) genes and susceptibility to gastric and colorectal adenocarcinoma. *Int J Cancer.* 2000 Jan 1;85(1):46–9. doi: 10.1002/(sici)1097–0215(20000101)85:1<46::aid-ijc8>3.0.co;2-0.

Kesarwani P, Singh R, Mittal RD. Association of GSTM3 intron 6 variant with cigarette smoking, tobacco chewing and alcohol as modifier factors for prostate cancer risk. *Arch Toxicol.* 2009 Apr;83(4):351–6. doi: 10.1007/s00204-008-0343-5. Epub 2008 Jul 31.

Khabaz MN. The GSTP1 Ile105Val polymorphism is not associated with susceptibility to colorectal cancer. *Asian Pac J Cancer Prev.* 2012;13:2949–53.

Khalili-Tanha G, Barzegar A, Nikbakhsh N, Ansari-Pirsaraei Z. Association of CYP1A1 M2 (A2455G) polymorphism with susceptibility to breast cancer in Mazandaran Province, Northern Iran: a case-control study. *Int J Prev Med.* 2019 Jun 7;10:92. doi: 10.4103/ijpvm.IJPVM_57_18.

Khlifi R, Chakroun A, Hamza-Chaffai A, Rebai A. Association of CYP1A1 and CYP2D6 gene polymorphisms with head and neck cancer in Tunisian patients. *Mol Biol Rep.* 2014;41(4):2591–600. doi: 10.1007/s11033-014-3117-6. Epub 2014 Jan 22. PMID: 24449363.

Kiran M, Chawla YK, Kaur J. Glutathione-S-transferase and microsomal epoxide hydrolase polymorphism and viral-related hepatocellular carcinoma risk in India. *DNA Cell Biol.* 2008 Dec;27(12):687–94. doi: 10.1089/dna.2008.0805.

Kiruthiga PV, Kannan MR, Saraswathi C, Pandian SK, Devi KP. CYP1A1 gene polymorphisms: lack of association with breast cancer susceptibility in the southern region (Madurai) of India. *Asian Pac J Cancer Prev.* 2011;12(8):2133–8.

Klusek J, Nasierowska-Guttmejer A, Kowalik A, Wawrzycka I, Lewitowicz P, Chrapek M, Głuszek S. GSTM1, GSTT1, and GSTP1 polymorphisms and colorectal cancer risk in Polish nonsmokers. *Oncotarget.* 2018 Apr 20;9(30):21224–30. doi: 10.18632/oncotarget.25031.

Kocabaş NA, Sardaş S, Cholerton S, Daly AK, Karakaya AE. N-acetyltransferase (NAT2) polymorphism and breast cancer susceptibility: a lack of association in a case-control study of Turkish population. *Int J Toxicol.* 2004 Jan–Feb;23(1):25–31. doi: 10.1080/10915810490275053.

Koda M, Iwasaki M, Yamano Y, Lu X, Katoh T. Association between NAT2, CYP1A1, and CYP1A2 genotypes, heterocyclic aromatic amines, and prostate cancer risk: a case control study in Japan. *Environ Health Prev Med.* 2017 Oct 24;22(1):72. doi: 10.1186/s12199-017-0681-0. Erratum in: *Environ Health Prev Med.* 2018 Jul 4;23(1):30.

Komiya Y, Tsukino H, Nakao H, Kuroda Y, Imai H, Katoh T. Human glutathione S-transferase A1, T1, M1, and P1 polymorphisms and susceptibility to prostate cancer in the Japanese population. *J Cancer Res Clin Oncol.* 2005 Apr;131(4):238–42. doi: 10.1007/s00432-004-0634-z. Epub 2004 Dec 23.

Ladero JM, Agúndez JA, Olivera M, Lozano L, Rodríguez-Lescure A, Diaz-Rubio M, Benítez J. N-acetyltransferase 2 single-nucleotide polymorphisms and risk of gastric carcinoma. *Eur J Clin Pharmacol.* 2002 May;58(2):115–8. doi: 10.1007/s00228-002-0460-7. Epub 2002 Apr 17.

Lan Q, Chow WH, Lissowska J, Hein DW, Buetow K, Engel LS, Ji B, Zatonski W, Rothman N. Glutathione S-transferase genotypes and stomach cancer in a population-based case-control study in Warsaw, Poland. *Pharmacogenetics.* 2001 Nov;11(8):655–61. doi: 10.1097/00008571-200111000-00003.

Landi S, Gemignani F, Moreno V, Gioia-Patricola L, Chabrier A, Guino E, Navarro M, de Oca J, Capellà G, Canzian F; Bellvitge Colorectal Cancer Study Group. A comprehensive analysis of phase I and phase II metabolism gene polymorphisms and risk of colorectal cancer. *Pharmacogenet Genomics.* 2005 Aug;15(8):535–46. doi: 10.1097/01.fpc.0000165904.48994.3d.

Lao X, Peng Q, Lu Y, et al. Glutathione S-transferase gene GSTM1, gene-gene interaction, and gastric cancer susceptibility: evidence from an updated meta-analysis. *Cancer Cell Int.* 2014;14:127. https://doi.org/10.1186/s12935-014-0127-3

Lee KS, Kim SK. Direct and metabolism-dependent cytochrome P450 inhibition assays for evaluating drug-drug interactions. *J Appl Toxicol.* 2013 Feb;33(2):100–8. doi: 10.1002/jat.1720. Epub 2011 Sep 14.

Lee JM, Lee YC, Yang SY, Shi WL, Lee CJ, Luh SP, Chen CJ, Hsieh CY, Wu MT. Genetic polymorphisms of p53 and GSTP1,but not NAT2, are associated with susceptibility to squamous-cell carcinoma of the esophagus. *Int J Cancer.* 2000 Sep 20;89(5):458–64. doi: 10.1002/1097–0215(20000920)89:5<458::aid-ijc10>3.0.co;2-r

Lee KM, Kang D, Lee SJ, Park SK, Lee KH, Choi JY, Kim SU, Choi H, Choi SH, Kim YW, Hong YC, Cho SH. Interactive effect of genetic polymorphism of glutathione S-transferase M1 and smoking on squamous cell lung cancer risk in Korea. *Oncol Rep.* 2006 Nov;16(5):1035–9

Lei D, Pan X, Guo C, Xu F, Zhang L, Liu D, Luan X. [Relationship between polymorphism of N-acetyltransferase 2 and genetic susceptibility to laryngeal carcinoma]. *Zhonghua Zhong Liu Za Zhi.* 2002 Mar;24(2):154–6. Chinese.

Leng WD, Zeng XT, Chen YJ, Duan XL, Niu YM, Long RP, Luo ZX. Cytochrome P450 2E1 RsaI/PstI polymorphism and risk of esophageal cancer: a meta-analysis of 17 case-control studies. *Exp Ther Med.* 2012 Nov;4(5):938–48. doi: 10.3892/etm.2012.687. Epub 2012 Aug 28.

Li M, Guan TY, Li Y, Na YQ. Polymorphisms of GSTM1 and CYP1A1 genes and their genetic susceptibility to prostate cancer in Chinese men. *Chin Med J (Engl).* 2008 Feb 20;121(4):305–8. PMID: 18304461.

Li S, Xue F, Zheng Y, Yang P, Lin S, Deng Y, Xu P, Zhou L, Hao Q, Zhai Z, Wu Y, Dai Z, Chen S. GSTM1 and GSTT1 null genotype increase the risk of hepatocellular carcinoma: evidence based on 46 studies. *Cancer Cell Int.* 2019 Mar 29;19:76. doi: 10.1186/s12935-019-0792-3.

Li XM, Yu XW, Yuan Y, Pu MZ, Zhang HX, Wang KJ, Han XD. Glutathione S-transferase P1, gene-gene interaction, and lung cancer susceptibility in the Chinese population: an updated meta-analysis and review. *J Cancer Res Ther.* 2015 Jul–Sep;11(3):565–70. doi: 10.4103/0973-1482.163788.

Lima MM Jr, Oliveira MN, Granja F, Trindade AC, De Castro Santos LE, Ward LS. Lack of association of GSTT1, GSTM1, GSTO1, GSTP1 and CYP1A1 polymorphisms for susceptibility and outcome in Brazilian prostate cancer patients. *Folia Biol (Praha).* 2008;54(3):102–8. PMID: 18647550.

Lin DX, Tang YM, Peng Q, Lu SX, Ambrosone CB, Kadlubar FF. Susceptibility to esophageal cancer and genetic polymorphisms in glutathione S-transferases T1, P1, and M1 and cytochrome P450 2E1. *Cancer Epidemiol Biomarkers Prev.* 1998 Nov;7(11):1013–8.

Liu C, Jiang Z, Deng QX, Zhao YN. Meta-analysis of association studies of CYP1A1 genetic polymorphisms with digestive tract cancer susceptibility in Chinese. *Asian Pac J Cancer Prev.* 2014;15(11):4689–95. doi: 10.7314/apjcp.2014.15.11.4689.

Liu D, Liu Y, Ran L, Shang H, Li D. GSTT1 and GSTM1 polymorphisms and prostate cancer risk in Asians: a systematic review and meta-analysis. *Tumour Biol.* 2013b Oct;34(5):2539–44. doi: 10.1007/s13277-013-0778-z. Epub 2013 Jun 28.

Liu H, Fu ZX, Wang CY, Qian J, Xing L, Liu YW. A meta-analysis of the relationship between NAT2 polymorphism and colorectal cancer susceptibility. *Medicina (Kaunas).* 2012;48(3):117–31.

Liu K, Zhang L, Lin X, Chen L, Shi H, Magaye R, Zou B, Zhao J. Association of GST genetic polymorphisms with the susceptibility to hepatocellular carcinoma (HCC) in Chinese population evaluated by an updated systematic meta-analysis. *PLoS One.* 2013a;8(2):e57043. doi: 10.1371/journal.pone.0057043. Epub 2013 Feb 20. Erratum in: *PLoS One.* 2014;9(1). doi:10.1371/annotation/13ee8778-996e-4649-af2e-4390378c2493.

Liu L, Wu G, Xue F, Li Y, Shi J, Han J, Zhang X, Na Y, Zhang H, Tang X, Pu H, Yuan Q, Zhang L, Yang M. Functional CYP1A1 genetic variants, alone and in combination with smoking, contribute to development of head and neck cancers. *Eur J Cancer.* 2013c Jun;49(9):2143–51. doi: 10.1016/j.ejca.2013.01.028. Epub 2013 Feb 22.

Liu YP, Li L. Null genotype of GSTT1 may be associated with esophageal cancer risk: a meta-analysis in Chinese individuals. *Cell Mol Biol (Noisy-le-grand).* 2017 Sep 30;63(-9):59–63. doi: 10.14715/cmb/2017.63.9.11.

Lo HW, Ali-Osman F. Structure of the human allelic glutathione S-transferase-pi gene variant, hGSTP1 C, cloned from a glioblastoma multiforme cell line. *Chem Biol Interact.* 1998;111–112:91–102. doi: 10.1016/s0009-2797(97)00153-1.

Loktionov A, Watson MA, Gunter M, Stebbings WS, Speakman CT, Bingham SA. Glutathione-S-transferase gene polymorphisms in colorectal cancer patients: interaction between GSTM1 and GSTM3 allele variants as a risk-modulating factor. *Carcinogenesis.* 2001;22:1053–60.

Lourenço GJ, Silva EF, Rinck-Junior JA, Chone CT, Lima CS. CYP1A1, GSTM1 and GSTT1 polymorphisms, tobacco and alcohol status and risk of head and neck squamous cell carcinoma. *Tumour Biol.* 2011 Dec;32(6):1209–15. doi: 10.1007/s13277-011-0224-z. Epub 2011 Aug 26.

Lu QJ, Bo YC, Zhao Y, Zhao EJ, Sapa WB, Yao MJ, Duan DD, Zhu YW, Lu WQ, Yuan L. Glutathione S-transferase M1 polymorphism and esophageal cancer risk: an updated meta-analysis based on 37 studies. *World J Gastroenterol.* 2016 Feb 7;22(5):1911–8. doi: 10.3748/wjg.v22.i5.1911.

Ma L, Lan B, Guo L, Nong S, Huang C, Wu Q, Huang Z. GSTM1 and GSTT1 Gene polymorphisms, gene-gene interaction, and esophageal carcinoma risk: evidence from an updated meta-analysis. *Genet Test Mol Biomarkers.* 2018 Jan;22(1):11–19. doi: 10.1089/gtmb.2017.0137. Epub 2017 Dec 7.

Majumder M, Sikdar N, Ghosh S, Roy B. Polymorphisms at XPD and XRCC1 DNA repair loci and increased risk of oral leukoplakia and cancer among NAT2 slow acetylators. *Int J Cancer.* 2007 May 15;120(10):2148–56. doi: 10.1002/ijc.22547

Marques CF, Koifman S, Koifman RJ, Boffetta P, Brennan P, Hatagima A. Influence of CYP1A1, CYP2E1, GSTM3 and NAT2 genetic polymorphisms in oral cancer susceptibility: results from a case-control study in Rio de Janeiro. *Oral Oncol.* 2006 Jul;42(6):632–7. doi: 10.1016/j.oraloncology.2005.11.003. Epub 2006 Feb 20.

Martínez C, Martín F, Fernández JM, García-Martín E, Sastre J, Díaz-Rubio M, Agúndez JA, Ladero JM. Glutathione S-transferases mu 1, theta 1, pi 1, alpha 1 and mu 3 genetic polymorphisms and the risk of colorectal and gastric cancers in humans. *Pharmacogenomics.* 2006 Jul;7(5):711–8. doi: 10.2217/14622416.7.5.711.

McCarty KM, Ryan L, Houseman EA, Williams PL, Miller DP, Quamruzzaman Q, Rahman M, Mahiuddin G, Smith T, Gonzalez E, Su L, Christiani DC. A case-control study of GST polymorphisms and arsenic related skin lesions. *Environ Health.* 2007 Feb 6;6:5. doi: 10.1186/1476-069X-6-5.

McIlwain CC, Townsend DM, Tew KD. Glutathione S-transferase polymorphisms: cancer incidence and therapy. *Oncogene.* 2006 Mar 13;25(11):1639–48. doi: 10.1038/sj.onc.1209373.

McKay JD, Hashibe M, Hung RJ, Wakefield J, Gaborieau V, Szeszenia-Dabrowska N, Zaridze D, Lissowska J, Rudnai P, Fabianova E, Mates D, Foretova L, Janout V, Bencko V, Chabrier A, Hall J, Boffetta P, Canzian F, Brennan P. Sequence variants of NAT1 and NAT2 and other xenometabolic genes and risk of lung and aerodigestive tract cancers in Central Europe. *Cancer Epidemiol Biomarkers Prev.* 2008 Jan;17(1):141–7. doi: 10.1158/1055-9965.EPI-07-0553.

Miller DP, Asomaning K, Liu G, Wain JC, Lynch TJ, Neuberg D, Su L, Christiani DC. An association between glutathione S-transferase P1 gene polymorphism and younger age at onset of lung carcinoma. *Cancer.* 2006 Oct 1;107(7):1570–7. doi: 10.1002/cncr.22124.

Millikan R, Pittman G, Tse CK, Savitz DA, Newman B, Bell D. Glutathione S-transferases M1, T1, and P1 and breast cancer. *Cancer Epidemiol Biomarkers Prev.* 2000 Jun;9(6):567–73.

Millikan RC, Pittman GS, Newman B, Tse CK, Selmin O, Rockhill B, Savitz D, Moorman PG, Bell DA. Cigarette smoking, N-acetyltransferases 1 and 2, and breast cancer risk. *Cancer Epidemiol Biomarkers Prev.* 1998 May;7(5):371–8.

Minard CG, Margaret R, Spitz MR, et al. Evaluation of Glutathione S-transferase polymorphisms and mutagen sensitivity as risk factors for the development of second primary tumors in patients previously diagnosed with early-stage head and neck cancer. *Cancer.* 2006;106:2636–44

Mitrunen K, Jourenkova N, Kataja V, Eskelinen M, Kosma VM, Benhamou S, Vainio H, Uusitupa M, Hirvonen A. Glutathione S-transferase M1, M3, P1, and T1 genetic polymorphisms and susceptibility to breast cancer. *Cancer Epidemiol Biomarkers Prev.* 2001 Mar;10(3):229–36.

Mo Z, Gao Y, Cao Y, Gao F, Jian L. An updating meta-analysis of the GSTM1, GSTT1, and GSTP1 polymorphisms and prostate cancer: a HuGE review. *Prostate.* 2009 May 1;69 (6):662–88. doi: 10.1002/pros.20907.

Mohammadi H, Roochi MM, Sadeghi M, Garajei A, Heidar H, Ghaderi B, Tadakamadla J, Meybodi AA, Dallband M, Mostafavi S, Mostafavi M, Salehi M, Sadeghi-Bahmani D, Brand S. Association of N-*acetyltransferases* 1 and 2 polymorphisms with susceptibility to head and neck cancers-a meta-analysis, meta-regression, and trial sequential analysis. *Medicina (Kaunas).* 2021 Oct 13;57(10):1095. doi: 10.3390/medicina57101095.

Moorthy B, Chu C, Carlin DJ. Polycyclic aromatic hydrocarbons: from metabolism to lung cancer. *Toxicol Sci.* 2015 May;145(1):5–15. doi: 10.1093/toxsci/kfv040.

Morabia A, Bernstein MS, Bouchardy I, Kurtz J, Morris MA. Breast cancer and active and passive smoking: the role of the N-acetyltransferase 2 genotype. *Am J Epidemiol.* 2000 Aug 1;152(3):226–32. doi: 10.1093/aje/152.3.226.

Morita S, Yano M, Tsujinaka T, Akiyama Y, Taniguchi M, Kaneko K, Miki H, Fujii T, Yoshino K, Kusuoka H, Monden M. Genetic polymorphisms of drug-metabolizing enzymes and susceptibility to head-and-neck squamous-cell carcinoma. *Int J Cancer.* 1999 Mar 1;80(5):685–8. doi: 10.1002/(sici)1097-0215(19990301)80:5<685::aid-ijc9>3.0.co;2-w.

Murali M, Manjari T, Madhuri B, Raghavan S, Jain DC, Vivekanandhan S. Genetic polymorphism of NAT2 metabolizing enzymes on phenytoin pharmacokinetics in Indian epileptic patients developing toxicity. *CNS Neurosci Ther.* 2012;18(4):350–8. doi:10.1111/j.1755-5949.2011.00291.x.

Nakajima T, Wang RS, Nimura Y, Pin YM, He M, Vainio H, Murayama N, Aoyama T, Iida F. Expression of cytochrome P450s and glutathione S-transferases in human esophagus with squamous-cell carcinomas. *Carcinogenesis.* 1996 Jul;17(7):1477–81. doi: 10.1093/carcin/17.7.1477.

Nebert DW, Vasiliou V. Analysis of the glutathione S-transferase (GST) gene family. *Hum Genomics.* 2004 Nov;1(6):460–4. doi: 10.1186/1479-7364-1-6-460.

Nelson HH, Wiencke JK, Christiani DC, Cheng TJ, Zuo ZF, Schwartz BS, Lee BK, Spitz MR, Wang M, Xu X, et al. Ethnic differences in the prevalence of the homozygous deleted genotype of glutathione S-transferase theta. *Carcinogenesis.* 1995 May;16(5):1243–5. doi: 10.1093/carcin/16.5.1243.

Nelson DR, Nebert DW. Cytochrome P450 (CYP) Gene Superfamily. In book: Encyclopedia of Life Sciences. January 2011. DOI: 10.1002/9780470015902.a0005667.pub2

Neve EP, Ingelman-Sundberg M. Cytochrome P450 proteins: retention and distribution from the endoplasmic reticulum. *Curr Opin Drug Discov Devel.* 2010 Jan;13(1):78–85.

Ntais C, Polycarpou A, Ioannidis JP. Association of GSTM1, GSTT1, and GSTP1 gene polymorphisms with the risk of prostate cancer: a meta-analysis. *Cancer Epidemiol Biomarkers Prev.* 2005 Jan;14(1):176–81.

Ochs-Balcom HM, Wiesner G, Elston RC. A meta-analysis of the association of N-acetyltransferase 2 gene (NAT2) variants with breast cancer. *Am J Epidemiol.* 2007 Aug 1;166 (3):246–54. doi: 10.1093/aje/kwm066. Epub 2007 May 29.

Oda Y, Kobayashi M, Ooi A, Muroishi Y, Nakanishi I. Genotypes of glutathione S-transferase M1 and N-acetyltransferase 2 in Japanese patients with gastric cancer. *Gastric Cancer.* 1999 Nov;2(3):158–64. doi: 10.1007/s101200050040.

Oliveira CB, Cardoso-Filho C, Bossi LS, Lourenço GJ, Costa-Gurgel MS, Lima CS. Association of CYP1A1 A4889G and T6235C polymorphisms with the risk of sporadic breast cancer in Brazilian women. *Clinics (Sao Paulo).* 2015 Oct;70(10):680–5. doi: 10.6061/clinics/2015(10)04.

Olsen A, Autrup H, Sørensen M, Overvad K, Tjønneland A. Polymorphisms of glutathione S-transferase A1 and O1 and breast cancer among postmenopausal Danish women. *Eur J Cancer Prev.* 2008 Jun;17(3):225–9. doi: 10.1097/CEJ.0b013e3282b6fe1e.

Ortiz de Montellano PR. Hydrocarbon hydroxylation by cytochrome P450 enzymes. *Chem Rev.* 2010 Feb 10;110(2):932–48. doi: 10.1021/cr9002193.

Osawa Y, Osawa KK, Miyaishi A, Higuchi M, Tsutou A, Matsumura S, Tabuchi Y, Tsubota N, Takahashi J. NAT2 and CYP1A2 polymorphisms and lung cancer risk in relation to smoking status. *Asian Pac J Cancer Prev.* 2007 Jan–Mar;8(1): 103–8.

Oyama T, Kawamoto T, Mizoue T, Yasumoto K, Kodama Y, Mitsudomi T. N-acetylation polymorphism in patients with lung cancer and its association with p53 gene mutation. *Anticancer Res.* 1997 Jan–Feb;17(1B):577–81.

Park JL, Muscat JE, Kaur T, et al. Comparison of GSTM polymorphisms and risk for oral cancer between African-Americans and Caucasians. *Pharmacogenetics.* 2000;10:123–31.

Peng H, Xie SK, Huang MJ, Ren DL. Associations of CYP2E1 rs2031920 and rs3813867 polymorphisms with colorectal cancer risk: a systemic review and meta-analysis. *Tumour Biol.* 2013 Aug;34(4):2389–95. doi: 10.1007/s13277-013-0788-x. Epub 2013 Apr 18.

Qadri Q, Sameer AS, Shah ZA, Hamid A, Alam S, Manzoor S, Siddiqi MA. Genetic polymorphism of the glutathione-S-transferase P1 gene (GSTP1) and susceptibility to prostate cancer in the Kashmiri population. *Genet Mol Res.* 2011 Dec 6;10(4):3038–45. doi: 10.4238/2011.December.6.4.

Qian J, Song Z, Lv Y, Huang X. CYP2E1 T7632A and 9-bp insertion polymorphisms and colorectal cancer risk: a meta-analysis based on 4,592 cases and 5,918 controls. *Tumour Biol.* 2013 Aug;34(4):2225–31. doi: 10.1007/s13277-013-0762-7. Epub 2013 May 1.

Qiu LX, Yuan H, Yu KD, Mao C, Chen B, Zhan P, Xue K, Zhang J, Hu XC. Glutathione S-transferase M1 polymorphism and breast cancer susceptibility: a meta-analysis involving 46,281 subjects. *Breast Cancer Res Treat.* 2010 Jun;121(3):703–8. doi: 10.1007/s10549-009-0636-2. Epub 2009 Nov 21.

Raimondi S, Paracchini V, Autrup H, Barros-Dios JM, Benhamou S, Boffetta P, Cote ML, Dialyna IA, Dolzan V, Filiberti R, Garte S, Hirvonen A, Husgafvel-Pursiainen K, Imyanitov EN, Kalina I, Kang D, Kiyohara C, Kohno T, Kremers P, Lan Q, London S, Povey AC, Rannug A, Reszka E, Risch A, Romkes M, Schneider J, Seow A, Shields PG, Sobti RC, Sørensen M, Spinola M, Spitz MR, Strange RC, Stücker I, Sugimura H, To-Figueras J, Tokudome S, Yang P, Yuan JM, Warholm M, Taioli E. Meta- and pooled analysis of GSTT1 and lung cancer: a HuGE-GSEC review. *Am J Epidemiol.* 2006 Dec 1;164 (11):1027–42. doi: 10.1093/aje/kwj321. Epub 2006 Sep 25.

Rebbeck TR. Molecular epidemiology of the human glutathione S-transferase genotypes GSTM1 and GSTT1 in cancer susceptibility. *Cancer Epidemiol Biomarkers Prev.* 1997;6:733–43.

Renaud HJ, Cui JY, Khan M, Klaassen CD. Tissue distribution and gender-divergent expression of 78 cytochrome P450 mRNAs in mice. *Toxicol Sci.* 2011 Dec;124(2):261–77. doi: 10.1093/toxsci/kfr240. Epub 2011 Sep 13. PMID: 21920951.

Rendic S, Guengerich FP. Contributions of human enzymes in carcinogen metabolism. *Chem Res Toxicol.* 2012 Jul 16;25 (7):1316–83. doi: 10.1021/tx300132k. Epub 2012 May 10.

Rovito PM Jr, Morse PD, Spinek K, Newman N, Jones RF, Wang CY, Haas GP. Heterocyclic amines and genotype of N-acetyltransferases as risk factors for prostate cancer. *Prostate Cancer Prostatic Dis.* 2005;8(1):69–74. doi: 10.1038/sj.pcan.4500780.

Ruiz MT, Pavarino-Bertelli E, Maniglia JV, et al. Epidemiologia ebiomarcadores em ca^ncer de cabec¸a epescoc¸o. Head and neck cancer epidemiology and biomarkers. *Arq Cie^nc Sau´de.* 2006;13:34–8.

Salazar-González RA, Turiján-Espinoza E, Hein DW, et al. Arylamine *N*-acetyltransferase 1 in situ *N*-acetylation on CD3+ peripheral blood mononuclear cells correlate with NATb mRNA and NAT1 haplotype. *Arch Toxicol.* 2018;92:661–8. https://doi.org/10.1007/s00204-017-2082-y.

Sameer AS, Nissar S, Qadri Q, Alam S, Baba SM, Siddiqi MA. Role of CYP2E1 genotypes in susceptibility to colorectal cancer in the Kashmiri population. *Hum Genomics.* 2011 Oct;5(6):530–7. doi: 10.1186/1479-7364-5-6-530.

Santric V, Djokic M, Suvakov S, Pljesa-Ercegovac M, Nikitovic M, Radic T, Acimovic M, Stankovic V, Bumbasirevic U, Milojevic B, Babic U, Dzamic Z, Simic T, Dragicevic D, Savic-Radojevic A. *GSTP1* rs1138272 polymorphism affects prostate cancer risk. *Medicina (Kaunas).* 2020 Mar 13;56(3):128. doi: 10.3390/medicina5603012.

Sato M, Sato T, Izumo T, Amagasa T. Genetic polymorphism of drug-metabolizing enzymes and susceptibility to oral cancer. *Carcinogenesis.* 1999 Oct;20(10):1927–31. doi: 10.1093/carcin/20.10.1927.

Senhaji N, Kassogue Y, Fahimi M, Serbati N, Badre W, Nadifi S. Genetic polymorphisms of multidrug resistance gene-1 (MDR1/ABCB1) and glutathione S-transferase gene and the risk of inflammatory bowel disease among Moroccan patients. *Mediators Inflamm.* 2015;2015:248060. doi: 10.1155/2015/248060.

Seow A, Zhao B, Lee EJ, Poh WT, Teh M, Eng P, Wang YT, Tan WC, Lee HP. Cytochrome P4501A2 (CYP1A2) activity and lung cancer risk: a preliminary study among Chinese women in Singapore. *Carcinogenesis.* 2001 Apr;22(4):673–7. doi: 10.1093/carcin/22.4.673.

Sergentanis TN, Economopoulos KP. Four polymorphisms in cytochrome P450 1A1 (CYP1A1) gene and breast cancer risk: a meta-analysis. *Breast Cancer Res Treat.* 2010 Jul;122(2):459–69. doi: 10.1007/s10549-009-0694-5. Epub 2009 Dec 25. Erratum in: *Breast Cancer Res Treat.* 2012 Feb;131(3):1083.

Shankar K, Mehendale HM. Cytochrome P450. *Encyclopedia of Toxicology*: Third Edition. 2014;1125–7. Academic Press, USA. An Elsevier Imprint.

Sharan RN, Mehrotra R, Choudhury Y, Asotra K. Association of betel nut with carcinogenesis: revisit with a clinical perspective. *PLoS One.* 2012;7(8):e42759. doi: 10.1371/journal.pone.0042759. Epub 2012 Aug 13.

Sharif MR, Sharif A, Kheirkhah D, Taghavi Ardakan M, Soltani N. Association of GSTO1 A140D and GSTO2 N142D gene variations with breast cancer risk. *Asian Pac J Cancer Prev.* 2017 Jun 25;18(6):1723–7. doi: 10.22034/APJCP.2017.18.6.1723.

Sharma A, Pandey A, Sharma S, Chatterjee I, Mehrotra R, Sehgal A, Sharma JK. Genetic polymorphism of glutathione S-transferase P1 (GSTP1) in Delhi population and comparison with other global populations. *Meta Gene.* 2014 Jan 20;2:134–42. doi: 10.1016/j.mgene.2013.12.003.

Shen FF, Zhou FY, Xue QS, Pan Y, Zheng L, Zhang H, Wang LD, Zheng HF. Association between CYP1A1 polymorphisms and esophageal cancer: a meta-analysis. *Mol Biol Rep.* 2013 Oct;40(10):6035–42. doi: 10.1007/s11033-013-2713-1. Epub 2013 Sep 25.

Shibuta J, Eto T, Kataoka A, Inoue H, Ueo H, Suzuki T, Barnard GF, Mori M. Genetic polymorphism of N-acetyltransferase 2 in patients with esophageal cancer. *Am J Gastroenterol.* 2001 Dec;96(12):3419–24. doi: 10.1111/j.1572-0241.2001.05276.x.

Shimada T, Oda Y, Gillam EM, Guengerich FP, Inoue K. Metabolic activation of polycyclic aromatic hydrocarbons and other procarcinogens by cytochromes P450 1A1 and P450 1B1 allelic variants and other human cytochromes P450 in Salmonella typhimurium NM2009. *Drug Metab Dispos.* 2001 Sep;29(9):1176–82.

Silig Y, Pinarbasi H, Günes S, Ayan S, Bagci H, Cetinkaya O. Polymorphisms of CYP1A1, GSTM1, GSTT1, and prostate cancer risk in Turkish population. *Cancer Invest.* 2006 Feb;24(1):41–5. doi: 10.1080/07357900500449579.

Silva TD, Felipe AV, Pimenta CA, Barão K, Forones NM. CYP2E1 RsaI and 96-bp insertion genetic polymorphisms associated with risk for colorectal cancer. *Genet Mol Res.* 2012 Sep 3;11(3):3138–45. doi: 10.4238/2012.September.3.2.

Sim SC, Ingelman-Sundberg M. Update on allele nomenclature for human cytochromes P450 and the human cytochrome P450 Allele (CYP-allele) nomenclature database. *Methods Mol Biol.* 2013; 987:251–9. doi: 10.1007/978-1-62703-321-3_21.

Simic T, Savic-Radojevic A, Pljesa-Ercegovac M, Matic M, Mimic-Oka J. Glutathione S-transferases in kidney and urinary bladder tumors. *Nat Rev Urol.* 2009 May;6(5):281–9. doi: 10.1038/nrurol.2009.49.

Singh M, Shah PP, Singh AP, Ruwali M, Mathur N, Pant MC, Parmar D. Association of genetic polymorphisms in glutathione S-transferases and susceptibility to head and neck cancer. *Mutat Res.* 2008 Feb 1;638(1–2):184–94. doi: 10.1016/j.mrfmmm.2007.10.003. Epub 2007 Oct 16.

Singh MS, Michael M. Role of xenobiotic metabolic enzymes in cancer epidemiology. *Methods Mol Biol.* 2009;472:243–64. doi: 10.1007/978-1-60327-492-0_10.

Singh SA, Choudhury JH, Kapfo W, Kundu S, Dhar B, Laskar S, Das R, Kumar M, Ghosh SK. Influence of the CYP1A1 T3801C polymorphism on tobacco and alcohol-associated head and neck cancer susceptibility in Northeast India. *Asian Pac J Cancer Prev.* 2015;16(16):6953–61. doi: 10.7314/apjcp.2015.16.16. 6953.

Smith G, Stanley LA, Sim E, Strange RC, Wolf CR. Metabolic polymorphisms and cancer susceptibility. *Cancer Surv.* 1995;25:27–65.

Sørensen M, Autrup H, Tjønneland A, Overvad K, Raaschou-Nielsen O. Genetic polymorphisms in CYP1B1, GSTA1, NQO1 and NAT2 and the risk of lung cancer. *Cancer Lett.* 2005 Apr 28;221(2):185–90. doi: 10.1016/j.canlet.2004.11.012.

Srivastava DS, Mittal RD. Genetic polymorphism of the N-acetyltransferase 2 gene, and susceptibility to prostate cancer: a pilot study in north Indian population. *BMC Urol.* 2005 Aug 6;5:12. doi: 10.1186/1471-2490-5-12.

Stanley LA, Sim E. Update on the pharmacogenetics of NATs: structural considerations. *Pharmacogenomics.* 2008 Nov;9(11):1673–93. doi: 10.2217/14622416.9.11.1673.

Stoehlmacher J, Park DJ, Zhang W, Groshen S, Tsao-Wei DD, Yu MC, Lenz HJ. Association between glutathione S-transferase P1, T1, and M1 genetic polymorphism and survival of patients with metastatic colorectal cancer. *J Natl Cancer Inst.* 2002 Jun 19;94(12):936–42. doi: 10.1093/jnci/94.12.936.

Strange RC, Fryer AA. The glutathione S-transferases: influence of polymorphism on cancer susceptibility. *IARC Sci Publ.* 1999;148:231–49.

Strange RC, Jones PW, Fryer AA. Glutathione S-transferase: genetics and role in toxicology. *Toxicol Lett.* 2000 Mar 15;112–113:357–63. doi: 10.1016/s0378-4274(99)00230-1.

Suzuki S, Muroishi Y, Nakanishi I, Oda Y. Relationship between genetic polymorphisms of drug-metabolizing enzymes (CYP1A1, CYP2E1, GSTM1, and NAT2), drinking habits, histological subtypes, and p53 gene point mutations in Japanese patients with gastric cancer. *J Gastroenterol.* 2004;39(3):220–30. doi: 10.1007/s00535-003-1281-x.

Tang K, Li Y, Zhang Z, Gu Y, Xiong Y, Feng G, He L, Qin S. The PstI/RsaI and DraI polymorphisms of CYP2E1 and head and neck cancer risk: a meta-analysis based on 21 case-control studies. *BMC Cancer.* 2010 Oct 22; 10:575. doi: 10.1186/1471-2407-10-575.

Tian J, Liu C, Liu G, Zuo C, Chen H. Cumulative evidence for association between genetic polymorphisms and esophageal cancer susceptibility: a review with evidence from meta-analysis and genome-wide association studies. *Cancer Med.* 2019 Mar;8(3):1289–305. doi: 10.1002/cam4.1972. Epub 2019 Feb 21.

Trizna Z, de Andrade M, Kyritsis AP, Briggs K, Levin VA, Bruner JM, Wei Q, Bondy ML. Genetic polymorphisms in glutathione S-transferase mu and theta, N-acetyltransferase, and CYP1A1 and risk of gliomas. *Cancer Epidemiol Biomarkers Prev.* 1998 Jun;7(6):553–5.

Unal M, Tamer L, Akbaş Y, Pata YS, Vayisoglu Y, Degirmenci U, Camdeviren H. Genetic polymorphism of N-acetyltransferase 2 in the susceptibility to laryngeal squamous cell carcinoma. *Head Neck.* 2005 Dec;27(12):1056–60. doi: 10.1002/hed.20284.

van der Hel OL, Peeters PH, Hein DW, Doll MA, Grobbee DE, Kromhout D, Bueno de Mesquita HB. NAT2 slow acetylation and GSTM1 null genotypes may increase postmenopausal breast cancer risk in long-term smoking women. *Pharmacogenetics.* 2003 Jul;13(7):399–407. doi: 10.1097/00008571-200307000-00005.

Vogl FD, Taioli E, Maugard C, Zheng W, Pinto LF, Ambrosone C, Parl FF, Nedelcheva-Kristensen V, Rebbeck TR, Brennan P, Boffetta P. Glutathione S-transferases M1, T1, and P1 and breast cancer: a pooled analysis. *Cancer Epidemiol Biomarkers Prev.* 2004 Sep;13(9):1473–9.

Wadelius M, Autrup JL, Stubbins MJ, Andersson SO, Johansson JE, Wadelius C, Wolf CR, Autrup H, Rane A. Polymorphisms in NAT2, CYP2D6, CYP2C19 and GSTP1 and their association with prostate cancer. *Pharmacogenetics.* 1999 Jun;9(3):333–40. doi: 10.1097/00008571-199906000-00008

Walker K, Ginsberg G, Hattis D, Johns DO, Guyton KZ, Sonawane B. Genetic polymorphism in N-acetyltransferase (NAT): population distribution of NAT1 and NAT2 activity. *J Toxicol Environ Health B Crit Rev.* 2009;12(5–6):440–72. doi: 10.1080/10937400903158383.

Wang J, Jiang J, Zhao Y, et al. Genetic polymorphisms of glutathione S-transferase genes and susceptibility to colorectal cancer: a case-control study in an Indian population. *Cancer Epidemiol.* 2011a;35:66–72.

Wang JJ, Zheng Y, Sun L, Wang L, Yu PB, Li HL, Tian XP, Dong JH, Zhang L, Xu J, Shi W, Ma TY. CYP1A1 Ile462Val polymorphism and susceptibility to lung cancer: a meta-analysis based on 32 studies. *Eur J Cancer Prev.* 2011b Nov;20(6):445–52. doi: 10.1097/CEJ.0b013e328345f937.

Wei B, Xu Z, Zhou Y, Ruan J, Cheng H, Xi B, Zhu M, Jin K, Zhou D, Hu Q, Wang Q, Wang Z, Yan Z, Xuan F, Huang X, Zhang J, Zhou H. Association of GSTM1 null allele with prostate cancer risk: evidence from 36 case-control studies. *PLoS One.* 2012;7(10):e46982. doi: 10.1371/journal.pone.0046982. Epub 2012 Oct 10.

Westwood IM, Sim E. Kinetic characterisation of arylamine N-acetyltransferase from Pseudomonas aeruginosa. BMC Biochem. 2007 Mar 20;8:3. doi: 10.1186/1471-2091-8-3.

White DL, Li D, Nurgalieva Z, El-Serag HB. Genetic variants of glutathione S-transferase as possible risk factors for hepatocellular carcinoma: a HuGE systematic review and meta-analysis. *Am J Epidemiol.* 2008 Feb 15;167(4):377–89. doi: 10.1093/aje/kwm315. Epub 2007 Dec 7.

Wideroff L, Vaughan TL, Farin FM, Gammon MD, Risch H, Stanford JL, Chow WH. GST, NAT1, CYP1A1 polymorphisms and risk of esophageal and gastric adenocarcinomas. *Cancer Detect Prev.* 2007;31(3):233–6. doi: 10.1016/j.cdp.2007.03.004. Epub 2007 Jul 23.

Wikman H, Thiel S, Jäger B, Schmezer P, Spiegelhalder B, Edler L, Dienemann H, Kayser K, Schulz V, Drings P, Bartsch H, Risch A. Relevance of N-acetyltransferase 1 and 2 (NAT1, NAT2) genetic polymorphisms in non-small cell lung cancer susceptibility. *Pharmacogenetics.* 2001 Mar;11(2):157–68. doi: 10.1097/00008571-200103000-00006.

Wu FY, Lee YJ, Chen DR, Kuo HW. Association of DNA-protein crosslinks and breast cancer. *Mutat Res.* 2002 Apr 25;501(-1–2):69–78. doi: 10.1016/s0027-5107(02)00006-4.

Wu H, Ouyang Q, Tian C, Xie N, Cao M, Shui ZR. Cytochrome P450 1A1 (CYP1A1) gene polymorphisms and susceptibility to breast cancer: a meta-analysis in the Chinese population. *Clin Lab.* 2017 Jan 1;63(1):67–72. doi: 10.7754/Clin.Lab.2016.160535.

Wu X, Zhao H, Suk R, Christiani DC. Genetic susceptibility to tobacco-related cancer. *Oncogene.* 2004 Aug 23;23(38):-6500–23. doi: 10.1038/sj.onc.1207811.

Xiao Y, Ma JZ. [Study of the relationship between glutathione S-transferase genetic polymorphisms M1 and T1 and susceptibility to primary liver cancer in Chinese: a meta-analysis]. *Zhonghua Gan Zang Bing Za Zhi.* 2012 Oct;20(10):774–9. Chinese. doi: 10.3760/cma.j.issn.1007-3418.2012.10.014.

Xu YT, Wang J, Yin R, Qiu MT, Xu L, Wang J, Xu L. Genetic polymorphisms in Glutathione S-transferase Omega (GSTO) and cancer risk: a meta-analysis of 20 studies. *Sci Rep*. 2014 Oct 10;4:6578. doi: 10.1038/srep06578.a

Xu L, Yang M, Zhao T, Jin H, Xu Z, Li M, Chen H. The polymorphism of CYP2E1 Rsa I/Pst I gene and susceptibility to respiratory system cancer: a systematic review and meta-analysis of 34 studies. *Medicine (Baltimore)*. 2014 Dec;93(27):e178. doi: 10.1097/MD.0000000000000178.b

Xu X, Mathieu C, Berthelet J, et al. Human arylamine n-acetyltransferase 1 is inhibited by the dithiocarbamate pesticide thiram. *Mol Pharmacol*. 2017;92:358–65. doi:10.1124/mol.117.108662.

Xue CX, He XM, Zou DH. Glutathione S-transferase M1 polymorphism and breast cancer risk: a meta-analysis in the Chinese population. *Clin Lab*. 2016 Nov 1;62(11):2277–84. doi: 10.7754/Clin.Lab.2016.160333.

Yan F, Wang R, Geng L. The 341C/T polymorphism in the GSTP1 gene and lung cancer risk: a meta-analysis. *Genet Mol Res*. 2016 Jul 29;15(3). doi: 10.4238/gmr.15038172.

Yang C, He XF. Evaluation of association studies and a systematic review and meta-analysis of CYP1A1 T3801C and A2455G polymorphisms in breast cancer risk. *PLoS One*. 2021 Apr 28;16(4):e0249632. doi: 10.1371/journal.pone.0249632.

Ye XH, Song L, Peng L, Bu Z, Yan SX, Feng J, Zhu XL, Liao XB, Yu XL, Yan D. Association between the CYP2E1 polymorphisms and lung cancer risk: a meta-analysis. *Mol Genet Genomics*. 2015 Apr;290(2):545–58. doi: 10.1007/s00438-014-0941-2. Epub 2014 Oct 22.

Ye Z, Parry JM. Meta-analysis of 20 case-control studies on the N-acetyltransferase 2 acetylation status and colorectal cancer risk. *Med Sci Monit*. 2002 Aug;8(8):CR558–65.

Yun YX, Wang YP, Wang P, Cui LH, Wang KJ, Zhang JY, Dai LP. CYP1A1 genetic polymorphisms and risk for esophageal cancer: a case-control study in central China. *Asian Pac J Cancer Prev*. 2014 Jan;14(11):6507–12. doi: 10.7314/apjcp.2013.14.11.6507.

Zanger UM, Schwab M. Cytochrome P450 enzymes in drug metabolism: regulation of gene expression, enzyme activities, and impact of genetic variation. *Pharmacol Ther*. 2013 Apr;138(1):103–41. doi: 10.1016/j.pharmthera.2012.12.007. Epub 2013 Jan 16.

Zhang J, Qiu LX, Wang ZH, et al. NAT2 polymorphisms combining with smoking associated with breast cancer susceptibility: a meta-analysis. *Breast Cancer Res Treat*. 2010;123:877–83. doi: 10.1007/s10549-010-0807-1.

Zhang JW, Yu WJ, Sheng XM, Chang FH, Bai TY, Lv XL, Wang G, Liu SZ. Association of CYP2E1 and NAT2 polymorphisms with lung cancer susceptibility among Mongolian and Han populations in the Inner Mongolian region. *Asian Pac J Cancer Prev*. 2014;15(21):9203–10. doi: 10.7314/apjcp.2014.15.21.9203.

Zhang L, Xiang Z, Hao R, Li R, Zhu Y. N-acetyltransferase 2 genetic variants confer the susceptibility to head and neck carcinoma: evidence from 23 case-control studies. *Tumour Biol*. 2014 Apr;35(4):3585–95. doi: 10.1007/s13277-013-1473-9. Epub 2013 Dec 13.

Zhang WP, Yang C, Xu LJ, Wang W, Song L, He XF. Individual and combined effects of GSTM1, GSTT1, and GSTP1 polymorphisms on lung cancer risk: a meta-analysis and re-analysis of systematic meta-analyses. *Medicine (Baltimore)*. 2021 Jul 2;100(26):e26104. doi: 10.1097/MD.0000000000026104.

Zhao M, He XL, Teng XD. Understanding the molecular pathogenesis and prognostics of bladder cancer. *Chin J Cancer Res*. 2016;28:92–8. doi: 10.3978/j.issn.1000-9604.2016.02.05.

Zhao Y, Chen ZX, Rewuti A, Ma YS, Wang XF, Xia Q, Fu D, Han YS. Quantitative assessment of the influence of cytochrome P450 1A2 gene polymorphism and colorectal cancer risk. *PLoS One*. 2013 Aug 12;8(8):e71481. doi: 10.1371/journal.pone.0071481.

Zhao ZQ, Guan QK, Yang FY, Zhao P, Zhou B, Chen ZJ. System review and meta-analysis of the relationships between five metabolic gene polymorphisms and colorectal adenoma risk. *Tumour Biol*. 2012;33:523–35.

Zheng H, Zhao Y. Association of CYP1A1 MspI polymorphism in the esophageal cancer risk: a meta-analysis in the Chinese population. *Eur J Med Res*. 2015 Mar 30;20(1):46. doi: 10.1186/s40001-015-0135-3.

Zheng Y, Wang JJ, Sun L, Li HL. Association between CYP1A1 polymorphism and colorectal cancer risk: a meta-analysis. *Mol Biol Rep*. 2012 Apr;39(4):3533–40. doi: 10.1007/s11033-011-1126-2. Epub 2011 Jun 28.

Zhou W, Liu G, Thurston SW, Xu LL, Miller DP, Wain JC, Lynch TJ, Su L, Christiani DC. Genetic polymorphisms in N-acetyltransferase-2 and microsomal epoxide hydrolase, cumulative cigarette smoking, and lung cancer. *Cancer Epidemiol Biomarkers Prev*. 2002 Jan;11(1):15–21.

Zhu W, Liu H, Wang X, Lu J, Zhang H, Wang S, Yang W. Associations of CYP1 polymorphisms with risk of prostate cancer: an updated meta-analysis. *Biosci Rep*. 2019 Mar 1;39(3):BSR20181876. doi: 10.1042/BSR20181876.

Zhu X, Wang Z, He J, et al. Associations between CYP1A1 rs1048943 A>G and rs4646903 T>C genetic variations and colorectal cancer risk: proof from 26 case-control studies. *Oncotarget*. 2016;7(32):51365–74. doi:10.18632/oncotarget.10331.

Zhuo X, Song J, Liao J, Zhou W, Ye H, Li Q, Xiang Z, Zhang X. Does CYP2E1 RsaI/PstI polymorphism confer head and neck carcinoma susceptibility? A meta-analysis based on 43 studies. *Medicine (Baltimore)*. 2016 Oct;95(43):e5156. doi: 10.1097/MD.0000000000005156.

Zienolddiny S, Campa D, Lind H, Ryberg D, Skaug V, Stangeland LB, Canzian F, Haugen A. A comprehensive analysis of phase I and phase II metabolism gene polymorphisms and risk of non-small cell lung cancer in smokers. *Carcinogenesis*. 2008 Jun;29(6):1164–9. doi: 10.1093/carcin/bgn020. Epub 2008 Feb 6.

6

Genetic Polymorphisms in Matrix Metalloproteinase (MMP) Genes and Cancer

Mujeeb Zafar Banday and Saniya Nissar
Department of Biochemistry, Government Medical College, Karan Nagar, Srinagar, Kashmir, India
Infinity Biochemistry Pvt. Ltd. Srinagar, Kashmir, India

Syed Sameer Aga
Basic Medical Sciences, College of Medicine, King Saud bin Abdulaziz University for Health Sciences (KSAU-HS), King Abdullah International Medical Research Center (KAIMRC), Ministry of National Guard Health Affairs (NGHA), Jeddah, Kingdom of Saudi Arabia
Infinity Biochemistry Pvt. Ltd. Srinagar, Kashmir, India

Ehtishamul Haq
Department of Biotechnology, University of Kashmir, Hazratbal, Srinagar, Jammu and Kashmir, India

Sabhiya Majid
Department of Biochemistry & Multidisciplinary Research Unit (MRU-ICMR), Government Medical College Srinagar & Associated Hospitals, Karan Nagar, Srinagar, Jammu and Kashmir, India

CONTENTS

ABSTRACT Matrix metalloproteinases (MMPs) constitute an important family of endopeptidases in humans and other primates. MMPs together process or cleave almost all the components of the basement membrane (BM) and the extracellular matrix (ECM) and various non-matrix bioactive substrates. Most of the MMP-processed ECM components and soluble protein mediators play crucial roles in numerous cellular signaling pathways and, therefore, regulate a multitude of physiological processes that depend on these cellular signaling pathways. Consequently, dysfunction or dysregulation of MMPs promote various pathophysiological conditions that often manifest as varied diseases or disorders including cancers of almost all the types. MMPs, predominantly through the proteolysis of diverse substrates, play a critical role in almost all the important pro-tumorigenic processes including tumor growth and progression, apoptosis, angiogenesis, activation and promotion of tissue invasion and metastasis and escape of tumor cells from different immunesurveillance mechanisms. Various single nucleotide polymorphisms (SNPs) in MMP genes have the potential to modulate gene transcription,

DOI: 10.1201/9781003246244-6

expression and subsequently serum levels and enzyme activity of MMPs in an allele-specific manner. The allele-specific differences that result in increased or decreased physiological levels and functions of MMPs may possibly modulate the risk of various cancers. In this chapter, we briefly discuss the MMP family, its structure, classification and regulation. We also discuss the role of some prominent members of MMP family in the pathogenesis of various cancers and review the outcome of various association studies that evaluated the role of SNPs in the genes encoding these MMPs in modulating the risk for the development of various cancers.

KEYWORDS: *Matrix Metalloproteinases; Membrane-Type MMPs; Extracellular Matrix; Basement Membrane; Tissue Inhibitors of Matrix Metalloproteinases; Association Studies; Meta-Analysis.*

6.1 Introduction

Human metalloproteases are a large and diverse class of metal-dependent proteolytic enzymes (peptide bond hydrolases) that degrade or cleave peptides and proteins through hydrolytic cleavage of peptide bonds. Metalloproteases include two main categories of enzymes: endopeptidases and exopeptidases. The endopeptidases, also known as proteinases, are proteolytic enzymes that cleave non-terminal peptide bonds, whereas the exopeptidases are proteolytic enzymes that cleave terminal (and penultimate) peptide bonds. Human metalloproteases play critical roles in various physiological process that include development, morphogenesis, cell cycle regulation, cell proliferation, cell migration, cell adhesion, metabolism of hormones, biologically active (bioactive) peptides and drugs such as antibiotics. Due to their involvement in critical physiological processes, the deficient or increased activities of these enzymes often form the basis of various pathophysiological processes such as inflammation, arthritis and neurodegeneration or diseases such as cancer and cardiovascular diseases (Nagase 2001).This large and diverse class of human metalloproteases includes a family of zinc-dependent endopeptidases or proteinases, known as **m**atrix **m**etallo**p**roteinases or matrixins, abbreviated as MMPs.

The matrix metalloproteinase (MMP) endopeptidase family is a member of metzincin superfamily of zinc-dependent metalloproteases that consist of at least six other major members besides MMPs. These include serralysins, snapalysins, pappalysins, leishmanolysins, astacins and adamalysins/reprolysins/ADAMs (a disintegrin and metalloproteinase). The catalyticdomain (active site) of metzincin superfamily members including MMPs consists of characteristic zinc-binding consensus sequence, His-Glu-X-X-His-X-X-Gly-X-X-His (HEXXHXXGXXH), and the pro-domain region consists of cysteine-switch consensus sequence, Pro-Arg-Cys-Gly-X-Pro-Asp (PRCGXPD). The active site Zn(II) ion is coordinately bound to three histidine imidazole nitrogen atoms of the zinc-binding consensus sequence and the sulfhydryl group of the cysteine residue of the pro-domain region. The coordination of active site Zn with cysteine residue prevents Zn from binding to a water molecule, and this is essential to maintain these endopeptidases including MMPs in inactive (latent) pro-enzyme (zymogen) form. The catalytic domain (active site) also consists of a highly conserved methionine-containing sequence, X-X-Met-X-Pro (XXMXP), eight residues downstream or after and directly beneath the histidine-containing zinc-binding consensus sequence. This conserved methionine-containing sequence forms a 1, 4-β-turn. This tight Met-turn creates a shallow hydrophobic cleft, which harbors the active site Zn(II), provides the binding site for substrates and inhibitors and thus constitutes an integral structural component of the catalytic domain (Gomis-Rüth 2003, 2009).

MMPs collectively process or cleave almost all the components of the basement membrane (BM) and the extracellular matrix (ECM), and these include collagens,-non-collagenous glycoproteins (fibronectin, vitronectinand laminins), elastin and various non-matrix bioactive substrates. MMPs through proteolytic cleavage release or shed several membrane-anchored and ECM-sequestered bioactive substratesincluding cytokines, growth factors, proteases and several other factors into the circulation in their soluble and often functionally active forms. MMP-catalyzed proteolysismodulates (activates or inactivates) several soluble protein mediators such as pro-proteases, proteases, protease inhibitors, antimicrobial peptides, clotting factors, chemotactic factors, chemokines, cytokines, chemokine and cytokine receptors, growth factors, hormones and their receptors and cell surface proteins, for example, cell adhesion molecules (CAMs). The various MMP-processed ECM components and soluble protein mediators play functionally important roles in numerous cellular signaling pathways. MMPs, therefore, regulate a multitude of physiological processes that depend on these cellular signaling pathways (Murphy and Nagase 2008, Fanjul-Fernández, Folgueras et al. 2010, Kessenbrock, Plaks et al. 2010, Klein and Bischoff 2011). Due to their direct or indirect involvement in numerous physiological or cellular processes, the in vivo functionsor activities of MMPs including their proteolytic activities are under tight multimechanistic regulationmainly mediated by several types of physiological endogenous inhibitors. These physiological endogenous inhibitors include tissue inhibitors of matrix metalloproteinases (TIMPs), small TIMP-like molecules, α_2-macroglobulin, thrombospondin-1 (THBS1 or TSP1), thrombospondin-2 (THBS2 or TSP2) and reversion-inducing cysteine-rich protein with Kazal motifs (RECK) (Murphy 2011).

MMPs and their physiological endogenous inhibitors play functionally important roles in numerous cellular signaling pathways that are involved in a multitude of physiological processes such as differentiation of stem cells, cellular proliferation, apoptosis, cellular motility and migration, cellular invasion, cell-to-cell adhesion and inflammation. Besides, MMPs play vital roles in various physiological processes that involvetissue remodelingsuch asembryo implantation and development, bone growth and development, nerve outgrowth and development, angiogenesis and vascular remodeling, wound healing, menstruation, ovulation, mammary,uterine andprostate involution (Hadler-Olsen, Fadnes et al. 2011).MMPs and their endogenous inhibitors particularly when dysfunctional or dysregulated promotevarious pathophysiological conditions that often manifest as varied diseases or disorders. These include inflammation, osteoarthritis and rheumatoid arthritis (Troeberg and Nagase 2012, Siebuhr, Bay-Jensen et al. 2013, Nissinen and Kähäri 2014), aneurysms, fibrosis, nephritis and tissue ulceration (Hadler-Olsen, Fadnes et al. 2011). The other examples include cardiovascular disorders such as hypertension and atherosclerosis (Vacek, Rehman et al. 2015), endometriosis and preeclampsia (Becker, Louis et al. 2010, Plaks, Rinkenberger et al. 2013), periodontal diseases (Sapna, Gokul et al. 2014) and nervous system pathologies such as Alzheimer's disease, multiple sclerosis and stroke (Overall and López-Otín 2002, Könnecke and Bechmann 2013). MMPs and their endogenous inhibitors mainly through their BM and ECM proteolysis and remodeling function play a crucial role in the pathogenesis of almost all the types of cancers through modulation of important tumorigenic processes such as tumor invasion and progression, neovascularization (angiogenesis), metastasis and apoptosis (Overall and López-Otín 2002, Kessenbrock, Plaks et al. 2010, Hadler-Olsen, Winberg et al. 2013). Generally, MMPs are present outside the cells in extracellular space and mediate extracellular proteolytic processes, but evidence is emerging regarding the occurrence of MMPs such as MMP1, MMP2 and MMP11 within the cell where they possibly perform intracellular functions including proteolysis (Luo, Mari et al. 2002, Kwan, Schulze et al. 2004, Limb, Matter et al. 2005).

Human MMPs, a large family of zinc-dependent endopeptidases, encoded by 24 different genes, consist of at least 24 multifunctional and structurally related members, numbered sequentially with few intermittent breaks in numbering as MMP1 to MMP3, MMP7 to MMP17, MMP19 to MMP21, and MMP23 to MMP28 largely based on the chronological order of their identification. The list excludes MMPs originally designated as MMP4, MMP5, MMP6 and MMP22 either becausethey were identical to other established members or they were simply experimental artifacts. Humans lack MMP18 that occurs naturally in Xenopus. Further, human MMP family consists of five protein isoforms, one each of MMP16, MMP19, MMP21, MMP23 and MMP28, and includes catalytically inactive MMP1-like protein (Puente, Sánchez et al. 2003, Quesada, Ordóñez et al. 2009, Fanjul-Fernández, Folgueras et al. 2010, Zitka, Kukacka et al. 2010).

6.2 Structural Aspects of Matrix Metalloproteinases

The matrix metalloproteinase (MMP) family consists of structurally related members that share a high sequence homology, common structural features and a characteristic modular design with specific arrangement of domains and inserts. Most but not all MMPs consist of four structurally distinct regions or domains, which from amino (N)-terminus to carboxy (C)-terminus consist of the pre-peptide domain,the pro-peptide domain, the catalytic domain and the hemopexin (HPX)-like domain (Gomis-Rüth 2003, 2009, Klein and Bischoff 2011). The **pre-peptide domain**, also known as pre-domain or signal sequence, consists of 16–52 predominantly hydrophobic aminoacids. The pre-peptide domain is involved in secretion process and directs the pro-enzyme (zymogen) into rough endoplasmic reticulum (RER) during synthesis and regulates its secretion or release from cells into extracellular space. The enzyme signal peptidase cleavesthe pre-peptide domain from the rest of the enzyme after zymogen secretion (Maskos 2005, Klein and Bischoff 2011). The **pro-peptide domain**, also known as pro-domain,is a stretch of 74–91 amino acids and contains a highly conserved cysteine-switch consensus sequence, Pro-Arg-Cys-Gly-X-Pro-Asp (PRCGXPD)that confers auto-inhibitory function to this domain. The pro-domain uses the sulfhydryl (thiol,-SH) group of the conserved cysteine residue to mask the active site that maintains the proteinase in its inactive (latent) pro-enzyme (zymogen) form. The conserved cysteine sulfhydryl group forms a coordination bond with active site Zn(II) ionand prevents its bonding to water molecule. The blocking of this catalytically vital access of water molecule to active site Zn(II) ion maintains the enzyme in its inactive form and prevents catalysis (Gomis-Rüth 2003, 2009, Klein and Bischoff 2011). The **catalytic (CAT) domain**, a stretch of 162–173 amino acids, contains the highly conserved characteristic zinc-binding consensus sequence, His-Glu-X-X-His-X-X-Gly-X-X-His (HEXXHXXGXXH). Three histidine residues from this consensus sequence form a coordination bond with active site Zn(II) ion through three imidazole nitrogen atoms. A large, shallow hydrophobic spherical cleft harbors the active site Zn(II) and provides the binding site for substrates and inhibitors (Skiles, Gonnella et al. 2001, Maskos 2005, Bertini, Calderone et al. 2006). In addition to the active site Zn(II), the catalytic domain contains a structural Zn(II) that forms a coordination bond with one aspartate residue and three histidine residues. The catalytic domain also contains a minimum of two Ca(II) atoms that perform structural function and help in stabilizing the whole structure (Borkakoti 2000).The structure based on the aforementioned specific arrangement of pre-peptide domain, pro-peptide domain and catalytic domain, often designated as **minimal domain structure**, is characteristically present in all the MMPs. In addition to the minimal domain structure, many MMPs contain C-terminal **hemopexin (HPX)-like domain** composed of around 200 amino acids. Structurally, this domain has a β-propeller architecture, which consistsof four symmetrical beta sheets arranged around a central axis in a toroid manner. The HPX-like domain remains connected to the catalytic (CAT) domain through a flexible segment known as linker or hinge segment. This domain mediates pro-enzyme activation, dimerization, substrate recognition, positioning, binding and other protein-protein interactions (Overall 2002, Maskos 2005, Overall and Butler 2007, Piccard, Van den Steen et al. 2007). The **linker (hinge) region** is a proline-rich highly flexible segment, variable in composition and length consisting of 2–72 amino acids. The linker region not only links the HPX-like domain with the catalytic domain, but also regulates the interactions between the linked domains. The linker region plays a vital role in many enzymatic processes including substrate recognition and cleavage, reorientation of catalytic domain during enzyme catalysis, enzyme translocation, enzyme stability and TIMP recruitment (Overall 2002, Overall and Butler 2007, Rosenblum, Merouch et al. 2007).

The structure of most MMPs follows the modular design described already. However, there are several exceptions to this generalized structural organization. For example, MMP2 and MMP9 possess an additional cysteine-rich structural feature known as fibronectin type II domain. Three fibronectin type II domain repeats termed as Col-1, Col-2 and Col-3, present in a head-to-tail manner within the catalytic domain, adjacent to the active site, constitute the collagen-binding domain or CBD. The collagen-binding domain mediates the recognition of and binding to gelatins and collagens (Skiles, Gonnella et al. 2001, Nagase and Fushimi 2008). MMP23, instead of hemopexin (HPX)-like domain, contains a functional cysteine-rich toxin-like domain (TxD) with K^+ channel-blocking property. This domain, known as $MMP23_{TxD}$ domain, possesses a unique cysteine-rich sequence array, Cys-X-Cys-X-Cys-X-Cys-X-Cys-X-Cys, known as cysteine array or CA. In addition to $MMP23_{TxD}$ domain, MMP23 containsa proline-rich immunoglobulin-like domain that shares considerable similarity with immunoglobulin-like C2-type domain or immunoglobulin-like cell adhesion molecule (IgCAM), hence known as IgCAM domain (Skiles, Gonnella et al. 2001, Rosenblum, Meroueh et al. 2007). MMP23 also possesses a carboxy terminal consensus sequence Arg-X-(Lys/Arg)-Arg (RX(K/R)R), which generally functions as a signal sequence for cleavage of precursors, and in case of MMP23, this signal motif is involved in processing of pro-enzyme (Murphy and Nagase 2008). MMP21 pro-domain region contains a furin-like serine proteinases (Fu) motif and the adjacent vitronectin-like insertion (Vn) domain that lies amino terminal to the Fu motif (Skiles, Gonnella et al. 2001, Klein and Bischoff 2011). The cysteine-switch consensus sequence present in MMP26 has a histidine residue instead of an arginine residue. Therefore, the cysteine-switch sequence in MMP26 is Pro-***His***-Cys-Gly-X-Pro-Asp instead of Pro-***Arg***-Cys-Gly-X-Pro-Asp present in other members of the MMP family. Further, the two members of the MMP family, MMP7 and MMP26 known as minimal domain MMPs, possess only minimal domain structureand lack the linker segment and HPX-like domain (Skiles, Gonnella et al. 2001).

Another exception to the general structural organization of MMPs is the small subfamily of MMPs, known as membrane-type MMPs or MT-MMPs, which consists of around seven members that remain bound (attached) to the cell membrane (cell surface) in contrast to the rest of the MMPs which constitute the secreted or soluble type. The MT-MMPs include two main groups or categories: glycosylphosphatidylinositol (GPI)-linked-type MMPs and transmembrane (TM)-type MMPs, which differ in the manner in which they attach to the cell membrane (cell surface). The glycosylphosphatidylinositol (GPI)-linked-type MMPs, also known as glypiated MT-MMPs, which include MMP17 (MT4-MMP) and MMP25 (MT6-MMP), attach to the cell surface through a carboxy terminal GPI moiety often termed as GPI anchor. The transmembrane (TM)-type MMPs contain an additional transmembrane (TM) domain that makes a single pass or span across the cell membrane. The transmembrane domain mediates several important processes such as internalization of enzyme, enzyme processingand shedding of ectodomain and is involved in the interaction of these MMPs with TIMPs. The transmembrane (TM)-type MMPs include two main groups:type I transmembrane (TM) MMPs and type II transmembrane (TM) MMPs. The type I transmembrane (TM) MMPs attach to the cell surface through a TM domain, which consists of a very short carboxy terminal cytoplasmic tail, often termed as cytoplasmic domain or CY domain. The type I transmembrane (TM) MMPs include MMP14 (MT1-MMP), MMP15 (MT2-MMP), MMP16 (MT3-MMP) and MMP24 (MT5-MMP). The type II transmembrane (TM) MMPs attach to the cell surface through an amino terminal region known as type II signal anchor (SA) and include MMP23A and MMP23B. The various domains of MT-MMPs involved in targeting orlocalization of important proteolytic processes to different and specific regions on the cell surface (cell membrane) allow the MT-MMPs to regulate the extracellular micro environment through pericellular proteolysis (Skiles, Gonnella et al. 2001, Klein and Bischoff 2011).

6.3 Classification of Matrix Metalloproteinases

The classification of MMPs involves a multipronged approach. MMPs, depending on their mode of expression, consist of two distinct groups: the secreted- or soluble-type MMPs and membrane-type MMPs or MT-MMPs. The secreted-type MMPs, secreted into the extracellular space, include most of the members of MMP family, whereas the MT-MMPs remain bound (attached) to the cell membrane (cell surface) and include around seven members already discussed in the previous section. The original MMP classification based on the preference for specific substrates and various structural features including cellular localization of these proteinases and from which MMPs derive their commonly used descriptive namesdivides MMPs into five distinct groups designated as collagenases, gelatinases, stromelysins, matrilysins and MT-MMPs (Skiles, Gonnella et al. 2001, Fanjul-Fernández, Folgueras et al. 2010, Klein and Bischoff 2011).

The most widely used classification of MMPs uses the shared and unique structural features of MMPs and divides MMPs into eight distinct groups, which includes five secreted- or soluble-type MMP groups and three MT-MMP groups. The secreted- or soluble-type MMPs consist of minimal domain MMPs, which include MMP7 and MMP26, and simple hemopexin (HPX)-like domain MMPs, which include MMP1, MMP3, MMP8, MMP10, MMP12, MMP13, MMP19, MMP20, MMP27 and MMP18 (in Xenopus not in humans). The secreted-type MMPs also consist of gelatin-binding (collagen-binding domain) MMPs, which include MMP2 and MMP9, furin-activated (furin-like serine proteinases (Fu) motif) MMPs, which include MMP11 and MMP28, and vitronectin-like insertion (Vn) domain MMPs, which include MMP21. The MT-MMPs consist of type I transmembrane (TM) MMPs, which include MMP14 (MT1-MMP), MMP15 (MT2-MMP), MMP16 (MT3-MMP) and MMP24 (MT5-MMP), and glycosylphosphatidylinositol (GPI)-linked-type MMPs, which include MMP17

(MT4-MMP) and MMP25 (MT6-MMP). The MT-MMPs also consist of type II transmembrane (TM) MMPs, which include MMP23A and MMP23B (Skiles, Gonnella et al. 2001, Fanjul-Fernández, Folgueras et al. 2010, Klein and Bischoff 2011).

Another classification approach uses the characteristic domain organization in MMPs and divides MMPs into four distinct groups, (1) gelatinases, (2) matrilysins, (3) furin and furin-like proprotein convertase activatable MMPsand (4) archetypal MMPs (Figure 6.1). The gelatinases include two MMPs: MMP2 or gelatinase A and MMP9 or gelatinase B. The matrilysins also include two MMPs: MMP7 or matrilysin1 (PUMP1) and MMP26 or matrilysin2 (endometase). The furin and furin-like proprotein convertase activatable MMPs consist of three distinct subgroups: (1) secreted-type MMPs, (2) MT-MMPs and (3) type II transmembrane MMPs. The secreted-type furin and furin-like proprotein convertase activatable MMPs include MMP11, MMP21 and MMP28. The MT-MMPs further consist of two subgroups: the type I transmembrane (TM) MMPs and the glycosylphosphatidylinositol (GPI)-linked-type MMPs. The type I transmembrane MMPs include MMP14 (MT1-MMP), MMP15 (MT2-MMP), MMP16 (MT3-MMP) and MMP24 (MT5-MMP). The glycosylphosphatidylinositol-linked-type MMPs include MMP17 (MT4-MMP) and MMP25 (MT6-MMP). The type II transmembrane (TM) MMPs include MMP23A and MMP23B.

The archetypal MMPs consist of three distinct subgroups,which differin their substrate specificities, and include (1) collagenases, (2) stromelysins and (3) other archetypal MMPs. The collagenases preferentially catalyze the single-site cleave of triple helix of interstitial fibrillar type I, type II and type III collagens at a site three-fourths the length from the amino terminus into two collagen segments, three-fourths and one-fourth the size of original molecule. The collagenases include MMP1 or collagenase 1 (interstitial collagenase), MMP8 or collagenase 2 (neutrophil collagenase) and MMP13 (collagenase 3). The stromelysins possess broad specificity catalytic activity. However, unlike collagenases, the stromelysins do not act on interstitial fibrillar type I and type II collagens, but may catalyze, though weakly, the cleavage of interstitial fibrillar type III collagens and also

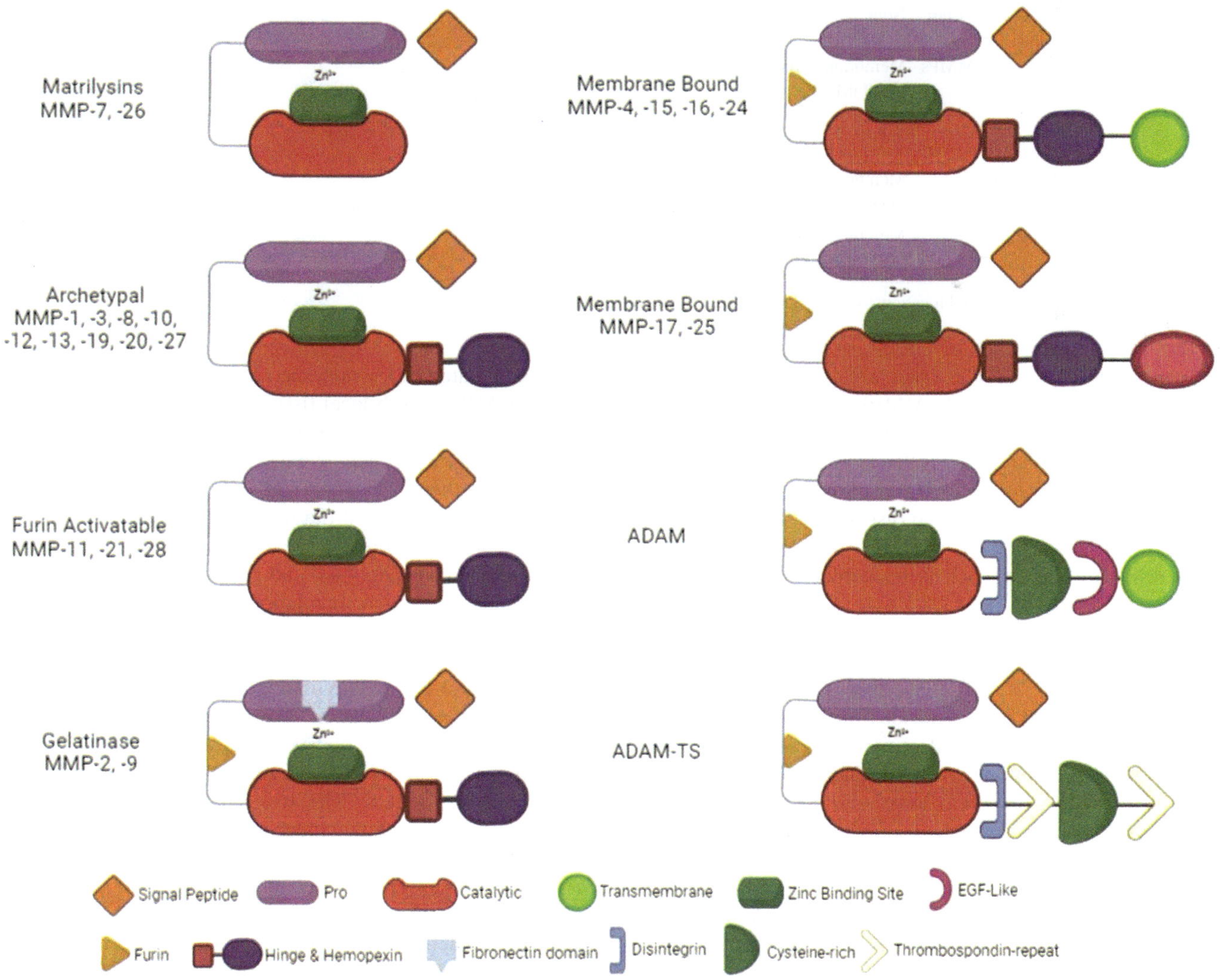

FIGURE 6.1 Diagrammatic representation of characteristic modular design with specific arrangement of domains and inserts in matrix metalloproteinases (MMPs). The diagram also depicts one of the approaches used to classify MMPs.

TABLE 6.1
Tabular Representation of Different Approaches Used in the Classification of Matrix Metalloproteinases (MMPs)

Group Name of MMPs Based on Structure	SubgroupName of MMPs	Individual Designationof MMPs	Alternative Names of MMPs	ChromosomalLocation ofHuman MMPs[a]
MMPs With Minimal Domain Structure	Matrilysins	MMP7	Matrilysin1 or PUMP1	11q22.2
		MMP26	Matrilysin2 or Endometase	11p15.4
MMPs Containing Hemopexin (HPX)-Like Domain	Collagenases	MMP1	Collagenase 1 or Interstitial Collagenase	11q22.2
		MMP8	Collagenase 2 or Neutrophil Collagenase	11q22.2
		MMP13	Collagenase 3	11q22.2
	Stromelysins	MMP3	Stromelysin-1 or Transin-1	11q22.2
		MMP10	Stromelysin-2	11q22.2
		MMP12	Macrophage Metalloelastase or Macrophage Elastase	11q22.2
	Other MMPs	MMP19	RASI-1	12q13.2
		MMP20	Enamelysin	11q22.2
		MMP27		11q22.2
MMPs That Bind Gelatins	Gelatinases	MMP2	GelatinaseA	16q12.2
		MMP9	GelatinaseB	20q13.12
MMPs Activated By Furin or MMPs Containing Furin-Like Serine Proteinases (Fu) Motif	None	MMP11	Stromelysin-3	22q11.23
		MMP28	Epilysin	17q12
MMPs Containing Vitronectin-Like Insertion (Vn) Domain	None	MMP21	None	10q26.2
Type I Transmembrane (TM) MMPs	Membrane-Type (MT)MMPs	MMP14	MT1-MMP	14q11.2
		MMP15	MT2-MMP	16q21
		MMP16	MT3-MMP	8q21.3
		MMP24	MT5-MMP	20q11.22
Glycosylphosphatidylinositol (GPI)-linked-type MMPs	Membrane-Type (MT)MMPs	MMP17	MT4-MMP	12q24.33
		MMP25	MT6-MMP or Leucolysin	16p13.3
Type II Transmembrane (TM) MMPs	Membrane-Type (MT)MMPs	MMP23A	Femalysin	1p36.33
		MMP23B	None	1p36.33

[a] Source of MMP chromosomal locations: www.genecards.org.

act on proteoglycans. The other archetypal MMPs include four loosely grouped MMPs; MMP12, MMP19, MMP20 and MMP27, each of which possesses unique sequence, structural features and substrate preferences due to which they do not fit into any of the other MMP groups (Fanjul-Fernández, Folgueras et al. 2010, Klein and Bischoff 2011). Table 6.1 summarizes the classification of MMPs, and Table 6.2 lists the different substrates for each of the members of MMP family.

6.4 Regulation of Matrix Metalloproteinases

MMPs play a vital role in various physiological and pathophysiological processes predominantly through their role in the proteolysis of various ECM and BM components. In addition to ECM and BM breakdown and remodeling functions, MMPs also mediate several non-ECM and non-BM proteolytic and non-proteolytic functions. MMPs through their diverse roles in ECM- and BM-related processes and non-ECM and non-BM functions serve as crucial regulators ormediators of tissue homeostasis and cellular signaling. The expression, synthesis, secretion and activation of MMPs vary considerably depending on the tissue microenvironment and change dynamically according to local tissue environmental requirements. The dysregulated expression, secretion and uncontrolled activities of MMPs disturb local tissue homeostasis, which manifests as diverse pathophysiological conditions, diseases or disorders. The expression, synthesis, secretion and activation of MMPs are therefore under tight regulation, and this regulationoccurs through four different complex mechanisms, each of which operate at a distinct level. These regulatory mechanisms include transcriptional and post-transcriptional regulation of gene expression, regulation of compartmentalization or extracellular (pericellular) accumulation (localization) of MMPs, regulation through posttranslational pro-enzyme or zymogen activation and regulation by means of inhibition of MMP enzyme activity that involves specific inhibitors, known as tissue inhibitors of matrix metalloproteinases (TIMPs) and non-specific inhibitors such as α_2-macroglobulin, THBS-1, THBS-2 and RECK. The stringent regulation of MMPs through these four different mechanisms ensuresnormal physiological conditions through maintenance of tissue homeostasis. Each of these four regulatory mechanisms contributes differentially to the regulation of different MMPs, and the operation of each of these mechanisms varies according to the local tissue environmental requirements and local physiological conditions (Overall and López-Otín 2002, Ra and Parks 2007).

TABLE 6.2

Substrates for Different Matrix Metalloproteinases (MMPs)

Individual Designation of MMPs	Substrates For MMPs
MMP7	Type III, Type IV, Type V, Type IX, Type X and Type XI collagens, gelatins, elastin, laminin, fibronectin, vitronectin, entactin, aggrecan, decorin, tenascin, pro-α-defensin, E-cadherin, β4 integrin, FasL, pro-tumor necrosis factor-α (pro-TNF-α), proMMP1, pro MMP2, pro MMP9, heparin-binding EGF-like growth factor (HB-EGF), connective tissue growth factor (CTGF), receptor activator of nuclear factor-kappa B ligand (RANKL), casein, plasminogen, myelin basic protein (MBP), transferring, link protein.
MMP26	Type IV collagen, gelatin, fibronectin, vitronectin, fibrinogen, proMMP9, α_1-proteinase inhibitor (α_1-PI), α_2-macroglobulin (α_2-M).
MMP1	Type I, Type II, Type III, Type VII, Type VIII and Type X collagens, gelatins, entactin, perlecan, ovostatin, aggrecan, L-selectin, tenascin C, proteoglycans, pro-TNF-α, pro-interleukin1β (proIL1β), proMMP2, proMMP9, α_2-macroglobulin (α_2-M), α_1-antichymotrypsin, α_1-proteinase inhibitor (α_1-PI), insulin-like growth factor-binding protein (IGFBP)-2,3,5; connective tissue growth factor (CTGF), link protein.
MMP8	Type I, Type II, Type III, Type V, Type VII, Type VIIIand Type X collagens, gelatins, entactin, fibronectin, aggrecan, tenascin, pro-tumor necrosis factor-α (pro-TNF-α), α_1-proteinase inhibitor(α_1-PI), link protein.
MMP13	Type I, Type II, Type III, Type IV, Type VII, Type IX, Type Xand Type XIV collagens, gelatins, laminin, fibronectin, tenascin, osteonectin, perlecan, aggrecan, connective tissue growth factor (CTGF), proMMP9, plasminogen, monocyte chemotactic protein 3 (MCP3), pro-transforming growth factor beta (pro-TGF-β), α_1-antichymotrypsin.
MMP3	Type III, Type IV, Type V, Type IX, Type Xand Type XI collagens, gelatins, laminin, entactin, tenascin, osteonectin, perlecan, decorin, insulin-like growth factor-binding protein 3 (IGFBP3), urokinase-type plasminogen activator (uPA), E-cadherin, heparin-binding EGF-like growth factor (HB-EGF), connective tissue growth factor (CTGF), pro-interleukin1β (proIL1β), α_2-macroglobulin (α_2-M), α_1-antichymotrypsin, α_1-proteinase inhibitor (α_1-PI), proMMP1, proMMP7, proMMP8, proMMP9, proMMP13, plasminogen, link protein.
MMP10	Type III, Type IV, Type V, Type IX, Type Xand Type XI collagens, gelatins, laminin, fibronectin, aggrecan, tenascin, casein, proMMP1, proMMP8, proMMP10, link protein.
MMP12	Type IV collagen, gelatins, laminin, fibronectin, vitronectin, entactin, aggrecan, elastin, nidogen, osteonectin, fibrinogen, fibrin, myelin basic protein (MBP), casein, plasminogen, apolipoprotein A.
MMP19	Type IV collagen, gelatins, laminin, fibronectin, vitronectin, entactin, aggrecan, tenascin, heparin-binding EGF-like growth factor (HB-EGF), connective tissue growth factor (CTGF), urokinase-type plasminogen activator (uPA), E-cadherin, pro-interleukin1β (proIL1β), α_2-macroglobulin (α_2-M), α_1-antichymotrypsin, α_1-proteinase inhibitor (α_1-PI), nidogen, plasminogen, insulin-like growth factor-binding protein 3 (IGFBP3), proMMP1, proMMP7, proMMP8, proMMP9, proMMP13, cartilage oligomeric matrix protein(COMP).
MMP20	Gelatins, aggrecan, amelogenin, cartilage oligomeric matrix protein (COMP).
MMP27	Not specified.
MMP2	Type I, Type IV, Type V, Type VII, Type X, Type XI and Type XIV collagens, gelatins, elastin, laminin, fibronectin, vitronectin, entactin, aggrecan, osteonectin, plasminogen, pro-transforming growth factor beta (pro-TGF-β), fibroblast growth factor receptor 1(FGFR1), monocyte chemotactic protein 3 (MCP3), pro-interleukin1β (proIL1β), insulin-like growth factor-binding protein 5 (IGFBP5), myelin basic protein (MBP), galectin3, link protein.
MMP9	Type III, Type IV, Type V, Type VII, Type Xand Type XII collagens, gelatins, elastin, fibronectin, vitronectin, entactin, aggrecan, pro-transforming growth factor beta (pro-TGF-β), pro-interleukin1β (proIL1β), interleukin2 receptor (IL2R), α_1-proteinase inhibitor (α_1-PI), insulin-like growth factor-binding protein 3 (IGFBP3), intercellular adhesion molecule *1(ICAM1),* plasminogen, N-telopeptide of type I collagen (NTx), ligand for receptor-type protein-tyrosine kinase KIT (KIT ligand or KITLG), myelin basic protein (MBP), galectin3, link protein.
MMP11	Gelatins, laminins, fibronectin, aggrecan, α_2-macroglobulin (α2-M), α_1-proteinase inhibitor (α_1-PI), insulin-like growth factor-binding protein 1 (IGFBP1).
MMP28	Casein.
MMP21	Not specified.
MMP14	Type I, Type II and Type III collagens, gelatins, laminins, fibronectin, vitronectin, entactin, proteoglycans, fibrin, aggrecan, proMMP2, proMMP13, homing cell adhesion molecule (HCAM or CD44), monocyte chemotactic protein 3 (MCP3), tissue transglutaminase (tTG).
MMP15	Collagens, gelatins, laminins, fibronectin, aggrecan, entactin, vitronectin, tenascin, perlecan, nidogen, proMMP2, tissue transglutaminase (tTG).
MMP16	Type III collagen, gelatins, fibronectin, vitronectin, aggrecan, proMMP2, tissue transglutaminase (tTG).
MMP24	Collagens, gelatins, fibronectin, aggrecan, vitronectin, proMMP2, proteoglycans.
MMP17	Gelatins, fibrinogen.
MMP25	Type III collagen, gelatins, laminins, fibronectin, proMMP2, fibrin.
MMP23A	Not specified.
MMP23B	Gelatin.

6.4.1 Transcriptional and Post-Transcriptional Regulation of Gene Expression

The transcriptional and post-transcriptional regulation of gene expression constitutes one of the important mechanisms in the control of MMP expression and function. Under normal quiescent tissue physiological conditions, almost all MMPs except MMP2 exhibit low basal expression and activity due to stringent regulation of each of the MMPs individually at transcriptional and post-transcriptional levels. MMP2 gene exhibits constitutive expression and lacks various regulatory cis-acting elements or transcription factor-binding sites in contrast to other MMPs, which show induced expression in response to various physiological or pathological stimuli and mediate accordingly ECM and BM turnover, remodeling and other processes. The transcriptional regulation of MMP gene expression involvesthe participation of various mediators that act by modulating (stimulating or suppressing) the activity of or flux through various signaling pathways and includediverse growth factors, cytokines and chemokines. The transcriptional regulation of MMP gene expression isalso under regulation by oncogenes, ECM proteins and other mediators such as plant-derived pro-tumorigenic tetracyclic diterpenoid compounds known as phorbol esters. The other types of cellular stimuli such as changes in cell shape, ECM contact and cell stress also modulate the expression of MMP genes at the level of transcription (Yan and Boyd 2007). MMP2 and MMP14 (MT1-MMP) genesalong with TIMP2 gene exhibit wide and constitutive expression in contrast to most other MMPs. The regulation of MMP2 enzyme activity under normal quiescent cellular or tissue physiological conditions primarily occurs at the level of pro-enzyme activation and to some extent through modulation of MMP2 mRNA stability at post-transcriptional level. MMP14 (MT1-MMP) and TIMP2 proteins regulate the activation of MMP2 enzyme that possibly links their expression to that of the MMP2 gene expression (Lohi,Lehti et al. 2000). Further, most MMP genes contain various cis-acting elements in their promoter regions that act as targets or binding sites for diverse trans-acting elements such as transcriptional activators and transcriptional repressors that act to stimulate or repress, respectively, the gene expression and subsequently the enzyme function of MMPs.

The MMP family members share most of these cis-acting elements, and this allows the stringentspatial and temporal regulation of MMP gene expression in a combined manner. Due to the conserved nature of various cis-acting elementsamong the members of the MMP family, the various MMP regulatory mediators such as growth factors, cytokines, chemokines, retinoids, ECM proteins, glucocorticoidsand oncogenes oftenmodulate the expression or repression of MMPs in a combined manner. The most prominent cis-acting elements present in the MMP gene promoter region include binding sites or response elements for specificity protein 1 (Sp1), polyomavirus enhancer-A binding protein-3 (PEA3), activator protein 1(AP1) and nuclear factor-kappaB (NF-κB). The presence of NF-κB-binding site in the MMP gene promoter region crucially brings MMP gene expression under the regulatory control of various immune mediators and links it to another important cellular mediator, the signal transducer and activator of transcription or STAT (Overall and López-Otín 2002).

Besides the conserved or commonly occurring cis-acting elements, several MMPs additionally contain binding sites for various important trans-acting elements that mediate the activation or repression of genes involved in the pathogenesis of various diseases. These cis-acting elements also provide a crucial link between the MMP gene expression and their regulation and dysregulation during the pathogenesis of various diseases including cancer. For example, the transcription factors,transcription factor 4 also known as T-cell factor 4 or immunoglobulin transcription factor 2 (TCF4/ITF2) and lymphoid enhancer-binding factor 1 (LEF1), and several other members of the TCF/LEF family of transcription factors play a crucial role in the activation of Wnt signaling pathway. TCF4 and LEF1mediate the transcriptional activation of various canonical Wnt signaling pathway genes by promoting the recruitment of β-catenin,the main nuclear effector or activator of canonical Wnt signaling, to the enhancer elements of target Wnt signaling pathway genes (Cadigan and Waterman 2012).

The gene promoter sites of several members of the MMP family such as MMP7, MMP12, MMP14 and MMP26 encompass the binding sites for various members of the TCF/LEF family of transcription factors including TCF4 and LEF1. The TCF/LEF transcription factors including TCF4 and LEF1 through β-catenin-mediated activation of Wnt signaling pathway act as main promoters of upregulation of MMP7 gene expression and subsequently increased MMP7 enzyme activity in various cancers, in particular, colorectal cancer. The TCF/LEF transcription factors such as LEF1 through interaction with other MMP regulatory mediators including transcription factors such as PEA3 and AP1 increase the expression of target MMP genes such as MMP7 gene further and promote the interplay between various signaling pathways specific to these transcription factors (Ilyas 2005, Cadigan and Waterman 2012). Several mediators of the Wnt signaling pathwaysuch as Wnt2 induce theupregulation of MMP7 gene expression and increased MMP7 enzyme activity by promoting the stabilization of Wnt signaling nuclear effector, β-catenin, which further increases the flux through the Wnt signaling pathway and leads to a further increase in MMP7 gene expression and MMP7 enzyme activity. The studies report that the increased MMP7 gene expression and MMP7 enzyme activity promote proliferation and invasivenessin various cancers such as colorectal cancer (Le Floch, Rivat et al. 2005). The Wnt signaling nuclear effector β-catenin also promotes increased MMP14 gene expression and MMP14 enzyme activity in lung cancer cells through the involvement of soluble E-cadherin. The β-catenin also induces or promotes the expression of MMP genes independent of the TCF/LEF cis-acting elements and activities of TCF/LEF family of transcription factors, and one of these mechanisms involves the transcriptional activation of activator protein 1(AP1) family of transcription factors such as c-Jun(Ilyas 2005). Another transcription factor family known as Smad (SMAD) family of transcription factors that includes members such as Smad1, Smad2,

Smad3, Smad5 and Smad9 also regulates the expression of MMP genes at transcription level. Smads mediate the transcription of MMP genes through the regulation of transforming growth factor-β (TGF-β) signaling-mediated gene expression. TGF-β has a dual role in regulating the expression of MMP genes at transcriptional level and can act as an activator as well as a repressor of MMP gene transcription. However, TGF-β generally functions in a manner antagonistic to that of interleukin1β (IL1β), and IL1β, in turn, is a potent activator of MMP gene transcription (Leivonen, Chantry et al. 2002).

The post-transcriptional regulation of MMP gene expression involves modulation of mRNA stability in the cytoplasm. The modulation of mRNA stability involves the role of trans-acting RNA-binding factors or proteins that mediate the process through interactions with multiple adenylate-uridylate-rich elements, also known as AU-rich elements or AREs located in the 3′-untranslated regions (3′-UTRs). These proteins, known as ARE-binding proteins (AUBPs), include HU or Hu family of RNA-binding proteins such as HuR that bind to AREs and increase the stability of cytosolic mRNA. In contrast, various destabilizing proteinsor factors such as the AU-rich element RNA-binding factor 1 (AUF1) bind to AREs and decrease the stability of cytosolic mRNA by mediating the recruitment of RNA decay machines, also known as exosomes. The regulation of gene expression of several members of MMP family such as MMP2, MMP9 and MMP13 involves the modulation of the stability of cytosolic MMP mRNAs (Yan and Boyd 2007, Clark, Swingler et al. 2008). The transcriptional and post-transcriptional regulation of MMP gene expression allows a stringent regulation of each member of the MMP family in a tissue-specific and time-specific manner. The transcriptional regulation involves complex interactions between various cis-acting elements in the promoter regions of MMP genes and a wide array of trans-acting elements including transcription factors. These interactions manifest as a complex interplay between various signaling pathways and regulate MMP gene expression according to local physiological requirements of the target tissue.

6.4.2 Inhibition of Matrix Metalloproteinases by Specific and Non-Specific Inhibitors

The post-pro-enzyme activation functions of MMPs including their enzyme activities are under stringent control process that involves their inhibition by physiological endogenous MMP inhibitors. There are two major types of endogenous MMP inhibitors: non-specific MMP inhibitors and highly specific MMP inhibitors. The non-specific MMP inhibitors prominently include general protease inhibitor α2-macroglobulin, a 725 kDa glycoprotein. Due to its abundance in plasma, α2-macroglobulin acts as a major protease inhibitor and inhibits almost all the endopeptidases including MMPs. The α_2-macroglobulin binds endopeptidases including MMPs in an irreversible manner and forms an endopeptidase (MMP)-α_2-macroglobulin complex. The MMP-α_2-macroglobulin complex then binds to α_2-macroglobulin receptor (A2MR) also known as low-density lipoprotein (LDL) receptor-related protein1 (LRP1) or CD91. The α2-macroglobulin receptor acts as a scavenger receptor and mediates irreversible scavenging of MMP-α_2-macroglobulin complex by endocytosis (Baker, Edwards et al. 2002, Murphy 2011). The other major non-specific physiological endogenous MMP inhibitors include reversion-inducing cysteine-rich protein with Kazal motifs (RECK), thrombospondin-1 (THBS1 or TSP1)and thrombospondin-2 (THBS2 or TSP2). RECK is a glycosylphosphatidylinositol (GPI)-anchored glycoprotein, which acts a suppressor of angiogenesis, tumor growth, tumor proliferation, tumor invasion, tumor metastasis and tumorrecurrenceand negatively regulates MMPs. RECK is the only known membrane-bound MMP inhibitor that inhibits several MMPs such as MMP2, MMP9 and MMP14 (Oh, Takahashi et al. 2001, Murphy 2011). Thrombospondin-1 (THBS1 or TSP1) directly binds pro-enzyme forms of MMP7 and MMP9 and inhibits the activation of these MMPs. However, THBS1 has an opposite effect on activated MMP2 and MMP9 and increases their activation (Bein and Simons 2000, Taraboletti, Morbidelli et al. 2000, Murphy 2011). Thrombospondin-2 (THBS2 or TSP2) much like α_2-macroglobulin binds MMP2 to form an MMP2-THBS2 complex. The LRP acts as a scavenger receptor and mediates irreversible scavenging of MMP2-THBS2 complex by endocytosis. Besides these, the other non-specific physiological endogenous MMP inhibitors includeprocollagen C-terminal proteinase enhancer (PCPE), tissue factor path way inhibitor2 (TFPI2), laminin-binding domain of agrin and non-collagenous NC1 domain of type IV collagen. The proteolytic cleavage of PCPE releases C-terminal fragment that is structurally similar to N-terminal MMP inhibitory domain of TIMPs. TFPI2 is a Kunitz-type serine protease with MMP inhibitory activity. The laminin-binding domain of agrin and non-collagenous NC1 domain of type IV collagen are structurally similar to TIMPs and exhibit MMP inhibitory activities (Baker, Edwards et al. 2002, Rodríguez, Morrison et al. 2010, Murphy 2011). The reactive oxygen species (ROS) produced abundantly during inflammatory processes by various cell types of immune systeminhibit immune-specific activities of MMPs (Ra and Parks 2007).

6.4.3 Tissue Inhibitors of Metalloproteinases (TIMPs)

The tissue inhibitors of metalloproteinases (TIMPs) constitute the highly specific physiological endogenous MMP inhibitors. As is the case with MMPs, TIMP family consists of structurally related members that share a high sequence homology, common structural features and a characteristic modular design. In humans, the TIMP family includes four 21–28 kDa proteins encoded by four paralogous genes and designated as TIMP1, TIMP2, TIMP3 and TIMP4 (Brew and Nagase 2010). The sequence homology among TIMPs ranges from 37% to 51%. The least homologous TIMP family member, TIMP1, has a sequence homology of 37% to 41% with other members of TIMP family. The highest homologous pair of TIMP2 and TIMP4 has a sequence homology of 51%. The TIMPs containtwo structurally distinct regions or domains: an amino (N)-terminal domain that consists of 125 amino acids and a C-terminal domain that consists of 65 amino acids. Each of the TIMP domains consists of six similarly spaced and conserved cysteine residues that stabilize the structural conformation of each TIMP domain and subsequently the overall TIMP structure through the formation of three intra-chain or intra-domain disulfide bonds. In case of N-terminal domain, the three intra-chaindisulfide bonds occur between Cys^{1}:Cys^{70}, Cys^{3}:Cys^{100} and Cys^{13}:Cys^{12}. The N-terminal domain of TIMPs binds to the catalytic domain of MMPs and inhibits their function including their enzyme activity. The inhibitory mechanism of MMPs involves the bidentate coordination linkage of N-terminal α-amino group and Cys^{1} carbonyl group of TIMPs to the catalytic zinc ion of MMPs. The formation of this bidentate coordination linkage omits the water molecule from the catalytic center of MMPs and prevents the hydrolysis-mediated activation of MMPs. The other TIMP N-terminal domain amino acids including other cysteine residues also participate in the inhibition of MMPs. The functional significance of C-terminal domain of TIMPs in the inhibition of MMPs is not fully clear yet. However, the C-terminal domain of TIMPs binds tightly to hemopexin (HPX)-like domain of pro-enzyme forms of various MMPs; for example, TIMP2, TIMP3 and TIMP4 bind to proMMP2, whereas TIMP1 and TIMP3 bind to proMMP9. The N-terminal domain of each member of the TIMP family shares a very high sequence homology with N-terminal domain of other members of the TIMP family, and this provides credible evidence regarding the importance of this domain in MMP inhibition. However, the C-terminal domain of TIMPs shows a high degree of variability among different members of TIMP family and this variability gives each TIMP its uniqueness in physiological properties and function.

The TIMP-mediated inhibition of MMPs involves the formation of a temporary non-covalent linkage between TIMP and MMPs. The TIMP: MMP association occurs with a dissociation constant K_d of 10^{-9} to 10^{-10} and in a stoichiometric or molar ratio of 1:1 (Brew and Nagase 2010, Murphy 2011). All the members of the TIMP family, TIMP1, TIMP2, TIMP3 and TIMP4, although capable of inhibiting all the MMPs, exhibit differential ability in inhibiting different members of the MMP family. The TIMPs in general exhibit a markedly low inhibitory activity toward MMP19 and all the MT-MMPs: MMP14 (MT1-MMP), MMP15 (MT2-MMP), MMP16 (MT3-MMP) and MMP24 (MT5-MMP) (Murphy 2011). In contrast to other TIMPs and all MMPs except MMP2, which exhibit inducible expression, the expression of TIMP2 gene occurs in a constitutive manner. TIMP2 is also a highly capable MMP inhibitor and exhibits ability to inhibit all the members of MMP family. TIMP3 enzyme differs from other members of the TIMP family in two major respects: first, TIMP3 enzyme remains attached to the ECM through its N-terminal and C-terminal domains and is not secreted, and second, TIMP3 enzyme in addition to MMP inhibitory activity is capable of inhibiting several members of ADAM family and ADAMTS family together known as adamalysins. The members of ADAM family and ADAMTS family include ADAM17, also known as tumor necrosis factor-α (TNF-α)-converting enzyme or TACE, ADAMTS4, ADAMTS5 and aggrecanases (Brew and Nagase 2010, Murphy 2011). The tissue-specific expression of different members of TIMPs also varies considerably. TIMPs in general exhibit a wide expression across different human tissues. However, the TIMP4 expression occurs mostly in brain, heart, kidney, pancreas, colon, ovary, testes and adipose tissue, whereas eye and kidney are two major sites of TIMP3 expression and certain regions of the brain constitute high expression sites of TIMP1 gene (Murphy 2011). As with MMPs, various soluble mediators such as growth factors and cytokines mediate the regulation of TIMP gene expression thatoccurs at the level of transcription.

The TIMPs play a vital role in regulating the turnover and remodeling ECM predominantly through their role in the inhibition of MMPs and thus mediate various tissue physiological and pathological processes. However, TIMPs perform diverse functions in addition to the inhibition of MMPs. For example, TIMPs through their direct interaction with various cell membrane or cell surface receptors stimulate or promote the activities or flux through various cellsignaling pathways that mediate critical processes such as cellular proliferation, cellular migration, cellular invasionand synaptic plasticity besides inhibition of angiogenesis and activation or inhibition of apoptosis. The TIMPs mediate these cellular processes in MMP inhibition-dependent and MMP inhibition-independent manners. Table 6.3 lists various important characteristics of TIMPs.

6.5 Role of Matrix Metalloproteinases and Polymorphisms in Matrix Metalloproteinase Genes in Cancer

The process of tumorigenesis is highly complex and involves intricate interplay between several altered fundamental cellular processes including self-sustained promotion of cell growth, self-sustained resistance to inhibition of cell growth, inhibition of apoptosis, infinite or sustained replication, induction of sustained neovascularization (angiogenesis) and induction or activation of tissue (Hanahan and Weinberg 2011). MMP enzymes, predominantly through the proteolysis of diverse substrates including various ECM and BM components, play a critical role in almost all the important pro-tumorigenic processes. MMPs play a differential role in tumor growth and progression. MMPs promote tumor growth through direct proteolytic cleavage of ECM and vascular basement membrane proteins and proteolytic activation that generate active forms of a wide array of growth factors such as transforming growth factor alpha (TGF-α) and epidermal growth factor (EGF). MMPs also promotetumor growth through proteolytic cleavage and subsequent inactivation of growth factor inhibitory proteins such as insulin-like growth factor-binding proteins (IGF-BPs) that generate active forms of growth factors such as insulin-like growth factor (IGF).

However, MMPs also inhibit tumor growth through activation of growth factors such as TGF-β, pro-apoptotic mediators such as Fasligand (FasL) and other mediators such as TNF-α. MMPs also mediate the process of apoptosis and play a differential role in its regulation. MMPs promote as well as inhibit apoptosis depending on different cellular conditions. MMPs play a vital, but differential role in mediating the process of angiogenesis. MMPs through direct proteolytic cleavage of

TABLE 6.3
Tabular Representation of General Characteristics of Human Tissue Inhibitors of Metalloproteinases (TIMPs)

Description of General Characteristic	Tissue Inhibitor of Metalloproteinase 1 (TIMP1)	Tissue Inhibitor of Metalloproteinase 2 (TIMP2)	Tissue Inhibitor of Metalloproteinase 3 (TIMP3)	Tissue Inhibitors of Metalloproteinase 4 (TIMP4)
Chromosomal location ofhuman TIMPs[a]	Xp11.3	17q25.3	22q12.3	3p25.2
Number of amino acids[a]	207	220	211	224
Molecular weight (Da/kDa)[a]	23,171 Da (23.2 kDa)	24,399 Da (24.4 kDa)	24,145 Da (24.1 kDa)	25,503 Da (25.5 kDa)
Pattern of expression	Inducible	Constitutive	Inducible	Not specified
Tissues of predominant expression	Bone, ovary	Placenta	Kidney, brain	Heart
Localization	Soluble	Soluble	ECM associated	Soluble
Isoelectric point [pI, IEP or pH(I)]	8.47	6.48	9.14	7.21
Glycosylation profile	Yes	No	Partial	No
MMPs inhibited	Weak inhibition of MMP14, MMP16, MMP19, MMP24	All MMPs	All MMPs	Most MMPs
Other metalloproteinases (MPs) inhibited	ADAM10	ADAM12	ADAM10, ADAM12 ADAM17, ADAM28 ADAM33; ADAMTS1, ADAMTS4 and ADAMTS5; weak inhibition of ADAMTS2	ADAM17d and ADAM28; weak inhibition of ADAM33
Interactions with ProMMPs	ProMMP9	ProMMP2	ProMMP2 and ProMMP9	ProMMP2
Partner mediators	CD63, LRP1 or A2MR, MMP9	$\alpha_3\beta_1$ integrin, LRP1	Angiotensin II receptor EFEMP1 and VEGFR2	Not specified
Effect on apoptosis	Anti-apoptotic	Anti- and pro-apoptotic	Pro-apoptotic	Not specified
Effect on angiogenesis	Inhibitor of apoptosis			

[a] Source: www.genecards.org; ADAM (adisintegrin and metalloproteinase); ADAMTS (adisintegrin and metalloproteinase with thrombospondin motifs), LRP1 (low-density lipoprotein receptor-related protein 1); A2MR (α_2-macroglobulin receptor), EFEMP1 (EGF-containing fibulin-like extracellular matrix protein1), VEGFR2 (vascular endothelial growth factor receptor 2) and angiotensin II receptor.

ECM and vascular basement membrane proteins and proteolytic activation generate active forms of various pro-angiogenic mediators including growth factors such as vascular endothelial growth factor (VEGF) that stimulates the activation of endothelial cells and recruitment of pericytes to induce the formation of new vasculature. In contrast, MMPs also inhibit angiogenesis through various mechanisms such as the activation of angiogenic inhibitors. MMPs play a vital role in the activation and promotion of tissue invasion and metastasis. MMPs also mediate and predominantly promote the escape of tumor cells from different immunesurveillance mechanisms (Egeblad and Werb 2002). Various functional single nucleotide polymorphisms (SNPs) in MMP geneshave the potential to modulate (increase or decrease) MMP gene transcription, expression and subsequently serum levels and enzyme activity of MMPs in an allele-specific manner. However, the SNP-induced allele-specific differences in gene transcription and expression, serum levels and enzyme activitiesof MMPs are very mild or subtle in comparison with the overexpression of oncogenes induced by gain of function mutations or underexpression of tumor suppressor genes induced by loss of function mutations. Nevertheless, the SNP-induced allele-specific differences that result in increased or decreased physiological levels and functions of MMPs persist over a lifetime and possibly modulate the risk for the development of various cancers. In the subsequent sections, we briefly discuss the role of some prominent members of MMP family in the pathogenesis of various cancers and review the outcome of several association studies that evaluated the role of SNPs in MMP genes in modulating the risk for the development of various cancers.

6.5.1 Polymorphisms in Matrix Metalloproteinase7 (MMP7) Gene

Matrix metalloproteinase7 (MMP7) gene (chromosomal location, 11q22.2) encodes a 267 amino acid 29677 Da (29 kDa) MMP7 pro-enzyme, also known as matrilysin1 or putative uterine metalloprotease 1 (PUMP1). The active form of MMP7 enzyme is a 19 or 21 kDa protein that contains only the minimal domain structure. MMP7 exhibits a wide and constitutive expression in the epithelial tissue of prostate, endometrium, airways, pancreas and liver and in the normal, non-inflammed, non-injured glandular (ductal) epithelium of parotid, peribronchial and mammary glands (Harrell, McCawley et al. 2005). The MMP7 gene and its promoter region encompass binding sites for various regulatory mediators including inflammation-related transcription factors, in particular, activator protein1 (AP1), which enable various inflammatory mediators such as growth factors and cytokines to regulate MMP7 gene expression. The substrates of MMP7 predominantly include various ECM and BM components and in addition non-ECM components (Table6.2). These include fibronectin, laminins, gelatins, casein, nidogen-1 (entactin), elastin, cartilage proteoglycans and type IV and type X collagens. MMP7 can also act on and cleave the pro-domain of gelatinases, which results in the activation of proMMP2 and proMMP9 (Wilson and Matrisian 1996). MMP7 is also involved in the proteolytic activation and release of several cell membrane- or cell surface-bound cellular mediators through shedding of their ectodomains. These mediators include epithelial cadherin (uvomorulin, E-cadherin or E-CAD), Fas ligand (FasL), integrin β4 (ITGB4), heparin-binding epidermal growth factor-like growth factor (HB-EGF) and TNF-α (Wilson and Matrisian 1996, Ii, Yamamoto et al. 2006). MMP7, predominantly through the proteolysis of various ECM and BM components, mediate various physiological processes including but not limited to post-pregnancy endometrial involution, wound healing and angiogenesis (Li, Park et al. 2002). MMP7 is a vital component of innate immune homeostasis or response to bacterial infection in lungs and gastrointestinal tract (GIT). The intestinal and lung epithelia constitutively express MMP7, which catalyzes the proteolytic cleavage of pro-α-defensins to generate antimicrobial α-defensin peptides and induces TNF-α secretion. The macrophages and neutrophils in lungs and Paneth cells in small intestine express pro-α-defensin peptides in abundance. The pro-α-defensin peptide-mediated antibacterial innate immune response in lungs and GIT specifically involves epithelial cells and is exclusively dependent on MMP7. Besides the presence of bacterial and other microbial components, hypoxia also promotes MMP7 expression (Wilson, Ouellette et al. 1999, Parks and Shapiro 2001, Burke 2004). MMP7 also mediates various inflammatory immune processes, in particular, the recruitment of inflammatory immune cells such as neutrophils. MMP7 catalyzes the cleavage of syndecan1, a heparan sulfate proteoglycan, and promotes the secretion of its complex with keratinocyte chemoattractant (KC), a Cys-X-Cys motif chemokine. MMP7 also mediates the proteolytic activation and release of E-cadherin. The release of syndecan1-KC complex and E-cadherin facilitates the migration of inflammatory immune cells across the epithelium toward the site of inflammation or injury (Li, Park et al. 2002, McGuire, Li et al. 2003).

MMP7, predominantly through its role in the proteolysis of various ECM and BM components, plays a vital role in various pathophysiological processes such as neoplastic diseases (cancers), muscular, skeletal and bone pathologies, cardiovascular and respiratory disorders and nervous system diseases. MMP7 mediates several important processes involved in the pathogenesis of various cancers. MMP7 is the only reported MMP, which cancer cells express and secrete by themselves unlike other MMPs that arise from stromal cells (Ii, Yamamoto et al. 2006). The overexpression of MMP7 occurs in various epithelial and mesenchymal cancers and in invasive GIT cancers. The overexpression and increase in MMP7 activity promotes neoplastic invasiveness in various cancers including esophageal cancer, pancreatic cancer, liver cancer, gastric cancer, colorectal cancer and ovarian cancer and correlates directly with aggressiveness in these cancers. The overexpression and increase in MMP7 activity correlates significantly with decreased survival, increased tumor growth and invasiveness including involvement of regional lymph nodes and metastasis in these cancers. MMP7 expression is a potential prognostic and predictive marker in undifferentiated early gastric cancer (Yamamoto, Adachi et al. 1999, Adachi, Yamamoto et al. 2001, Jones, Humphreys et al. 2004, Aihara, Mochiki et al. 2005, Tan, Egami et al. 2005, Wang, So et al. 2005). Enhanced neoplastic invasiveness and aggressiveness associated with overexpression and elevated MMP7 activity mainly occurs due to increased ECM and BM proteolysis and remodeling and activation of other MMPs, in particular, proMMP2 and proMMP9 (Ii,Yamamoto et al. 2006, Wang, Reierstad et al. 2006). MMP7 catalyzes the shedding of E-cadherin ectodomain and subsequent release of E-cadherin in its soluble form that leads to paracrine decrease or inhibition of E-cadherin cell adhesion function. The decreased or impaired cell adhesiveness promotes cancer cell motility that leads to enhanced neoplastic invasiveness and metastasis (Noë, Fingleton et al. 2001). The MMP7-catalyzed proteolytic ectodomain shedding activity modulates the bioavailability of soluble and active forms of various mediators involved in tumorigenesis. For example, MMP7 catalyzes proteolytic ectodomain shedding of pro-heparin-binding epidermal growth factor (pro-HB-EGF), which generates the active form HB-EGF. The active HB-EGF binds to erb-b2 receptor tyrosine kinase4 (ErbB4) receptor, activates EGFR signaling that inhibits apoptosis, and promotes proliferation of tumor cells (Yu, Woessner et al. 2002, Ii, Yamamoto et al. 2006). MMP7 catalyzes proteolytic cleavage and subsequent inactivation of insulin-like growth factor-binding proteins or IGF-BPs (IGF-BP1 to IGF-BP6) and catalyzes the activation of ADAM28 that also cleaves IGF-BPs, in particular, IGF-BP3. MMP7-mediated or MMP7-promoted inactivation of IGF-BPs increases the bioavailability of insulin-like growth factor, which promotes growth and survival of cancer cells (Nakamura, Miyamoto et al. 2005).

The functional SNPs in MMP7 gene reported to modulate the MMP7 gene transcription, expression and subsequent MMP7 activity in an allele-specific manner include MMP7 gene rs11568818 (–181 A/G) SNP, rs11568819 (–153 C/T) SNP and rs17886546 (–115 C/T) SNP. Of these SNPs, the two MMP7 gene promoter SNPs, rs11568818 (–181 A/G) SNP and rs11568819 (–153 C/T) SNP, modulate the transcription and expression of MMP7 gene in an allele-specific manner through their modifying effect on the binding affinities and interactions of various nuclear proteins including transcription factors with the MMP7 gene promoter. The occurrence of variant alleles,-181G allele and –153T allele individually or as –181G/-153T haplotype, exhibits an enhanced MMP7 gene expression and MMP7 activity. However, numerous studies report MMP7 gene rs11568818 (–181 A/G) SNP as the most important SNP with the potential to modulate the risk for the development of various cancers (Jormsjö, Whatling et al. 2001, Lugli, Pinti et al. 2005, Vira, Pradhan et al. 2017). The MMP7 gene rs11568818 (-181 A/G) SNP involves substitution of adenine(A) by guanine(G) at -181 bp position within the promoter region and affects the binding affinities and interactions of various nuclear proteins or factors with the promoter sequence. The nuclear proteins or factors interact more favorably and bind with higher affinities to the variant allele 'G', which results in a two- to threefold increase in the expression of MMP7 gene and substantial increase in the activity of MMP7 enzyme in comparison with the wild allele 'A' (Jormsjö, Whatling et al. 2001, Lugli, Pinti et al. 2005, Vira, Pradhan et al. 2017).

Numerous association studies report on the possible role of MMP7 gene rs11568818 (-181 A/G) SNP in modulating the risk for the development of various cancers. However, many studies report no such association. Here, we mention several of these studiesthat report the association of this SNP with the risk of various cancers and the studies that report no significant association. A meta-analysis based on sixteen association studies on European and Asian populations, which included 4,379 cases and 5,369 controls, reported a significant association of MMP7 gene rs11568818 (-181 A/G) SNP with an increased risk of gastric cancer (3studies: 699 cases; 1,164 controls) in dominant genetic model [AG+GG vs. AA: OR, 1.90 (95% Cl,1.43–2.51); p value$_{heterogeneity}$, 0.992; I^2= 0.0; p value, ≤ 0.001] and in allelic genetic model [G vs. A: OR,1.79 (95% Cl,1.37–2.34); p value$_{heterogeneity}$, 0.995; I^2= 0.0; p value, ≤ 0.001], but not in recessive genetic model [GG vs. AG+AA: OR,1.22 (95% Cl,0.26–5.77); p value$_{heterogeneity}$, 0.781; I^2= 0.0]. The same study reported a significant association of MMP7 gene rs11568818 (-181 A/G) SNP with an increased risk of hepatocellular carcinoma, head and neck cancer, breast cancer, lung cancer, esophageal cancer, prostate cancer, cervical and ovarian cancer (8 studies: 2,614 cases; 3,045 controls) in dominant genetic model [AG+GG vs. AA: OR, 1.50 (95% Cl,1.12–2.01); p value$_{heterogeneity}$, 0.002; I^2 = 68.9; p value, ≤ 0.001] and in allelic genetic model [G vs. A: OR, 1.48 (95% Cl, 1.11–1.99); p value$_{heterogeneity}$, 0.001; I^2 = 77.5; p value, ≤ 0.05], but not in recessive genetic model [GG vs. AG+AA: OR,1.60 (95% Cl,0.79–3.22); p value$_{heterogeneity}$, 0.034; I^2 = 56.1]. The study reported no significant association of this SNP with colorectal cancer (5 studies: 1,066 cases; 1,160 controls) in recessive genetic model [GG vs. AG+AA: OR,1.15 (95% Cl, 0.90–1.48); p value$_{\mathbf{heterogeneity}}$,0.168; I^2= 44.0], in dominant genetic model [AG+GG vs. AA: OR, 1.02 (95% Cl,0.84–1.24); p value$_{\mathbf{heterogeneity}}$, 0.236; I^2= 27.8] and in allelic genetic model [G vs. A: OR,1.05 (95% Cl, 0.92–1.20); p value$_{\mathbf{heterogeneity}}$, 0.129; I^2= 43.9]. The same study reported a significant difference in the distribution or frequency of MMP7–181G allele among the populations under study [European: 42.0% (95% Cl, 16.6–67.4%) vs. Asian: 8.80% (95% Cl, 1.6–15.9); p value, 0.0017] (Peng, Cao et al. 2010). Another meta-analysis based on three association studies on Asian population, which included 635 cases and 649 controls reported a significant association of MMP7 gene rs11568818 (-181 A/G) SNP with an increased risk of lung cancer in dominant genetic model [AG+GG vs. AA: OR,1.98 (95% Cl,1.44–2.70); I^2=2.0; p value,<0.001], in heterozygous genetic model [AG vs. AA: OR,1.92 (95% Cl,1.40–2.63); I^2=0.0; p value,<0.001] and in allelic genetic model [G vs. A: OR,1.89 (95% Cl,1.41–2.54); I^2= 29.7; p value,<0.001], but not in recessive genetic model [GG vs. AG+AA: OR,3.81 (95% Cl,0.61–23.86); I^2=0.0; p value,> 0.05] and homozygous genetic model [GG vs. AA: OR,4.11 (95% Cl, 0.67–25.80); I^2=0.0; p value,> 0.05] (Li, Liu et al. 2020).

A hospital-based case-control study on Chinese population, which included 702 cancer cases [258 esophageal squamous cell carcinoma (ESCC) cases, 201 gastric cardiac adenocarcinoma (GCA) cases and 243 non-small cell lung cancer (NSCLC) cases] and 350 healthy matched controls, reported a significant association of MMP7 gene rs11568818 (-181 A/G) SNP with an increased risk of ESCC [AG+GG vs. AA: OR,1.83 (95% Cl,1.12–2.99); p value,0.019],GCA [AG+GG vs. AA: OR,1.96 (95% Cl,1.17–3.29); p value,0.023] and NSCLC [AG+GG vs. AA: OR,2.00 (95% Cl,1.23–3.24); p value, 0.004] (Zhang, Jin et al. 2005). Another case-control study on ethnic Kashmiri population, which included 142 colorectal cancer cases and 184 healthy matched controls, reported a significant association of the variant genotype (GG) of MMP7 gene rs11568818 (-181 A/G) SNP with decreased risk of colorectal cancer [GG vs. AA: OR, 0.43 (95% Cl,0.20–0.90); p value, 0.02; AG vs. AA: OR, z1.26 (95% Cl,0.77–2.06); p value, 0.34] (Banday, Sameer et al. 2016).

A hospital-based case-control study on Polish population,which included 197 prostate cancer cases and 197 healthy matched controls, reported a significant association of the variant genotype (GG) of MMP7 gene rs11568818 (-181 A/G) SNP with a twofold increased risk of prostate cancer in homozygous genetic model [GG vs. AA: OR,2.03 (95% Cl,1.07–3.86); p value,0.03] (Białkowska, Marciniak et al. 2018). Another Polish hospital-based case-control study, which included 184 colorectal cancer cases and 205 healthy matched controls with no family history of malignancy, reported a significant association of MMP7 gene rs11568818 (-181 A/G) SNP with an increased risk of colorectal cancer [GG vs. AA: OR,2.12 (95% Cl, 1.34–3.38); p value, 0.018; AG vs. AA: OR, 1.67 (95% Cl, 1.03–2.72); p value, 0.038]. Further, the study reported a significant association of the variant genotype (GG) of this SNP with more than twofold increased risk of involvement of lymph node [GG vs. AA: OR,2.83 (95% Cl, 1.18–6.79); p value, 0.017] (Dziki, Przybyłowska et al. 2011). However, another hospital-based case-control study on Polish population which studied MMP7 gene rs11568818 (-181 A/G) SNP along with serum zinc levels reported no significant association of this SNP with the risk of breast cancer, lung cancer and colon cancer. This study included 648 unselected cancer cases (299 breast cancer cases, 199 lung cancer cases and 150 colon cancer cases) and 648 healthy matched controls (299 controls for breast cancer, 199 controls for lung cancer and 150 controls for colon cancer). The study reported no significant association of this SNP with the risk of breast cancer [GG vs. AA: OR, 0.68 (95% Cl, 0.42–1.10); p value, 0.11; AG vs. AA: OR, 0.87 (95% Cl, 0.61–1.24); p value, 0.44], lung cancer [GG vs. AA: OR, 1.24 (95% Cl, 0.63–2.44); p value, 0.54; AG vs. AA: OR, 1.03 (95% Cl, 0.65–1.61); p value, 0.91] and colon cancer [GG vs. AA: OR, 1.41 (95% Cl, 0.66–2.99); p value, 0.38; AG vs. AA: OR, 0.96 (95% Cl, 0.56–1.65); p value, 0.90] (Białkowska, Marciniak et al. 2020). In contrast, a study on Mexican population, which included 102 colorectal cancer cases and 125 healthy matched controls, reported a significant association of AA and AG genotypes of MMP7 gene rs11568818 (-181 A/G) SNP with an increased risk of colorectal cancer [AA vs. GG: OR, 3.38 (95% Cl, 1.16–9.84); p value, 0.02;

AG vs. GG: OR, 3.4 (95% Cl, 1.17–9.83); *p* value, 0.01; AG+AA vs. GG: OR, 3.4 (95% Cl,1.2–9.5); *p* value, 0.01] (Moreno-Ortiz, Gutiérrez-Angulo et al. 2014). Another hospital-based case-control study on Chinese population, which included 246 gastric cancer cases and 252 healthy matched controls, reported a significant association of the heterozygous genotype (AG) of MMP7 gene rs11568818 (-181 A/G) SNP with an increased risk of gastric cancer in heterozygous genetic model [AG vs. AA: OR, 3.189 (95% Cl, 1.52–6.67); *p* value, 0.001] and in allelic genetic model [G vs. A: OR, 3.051 (95% Cl, 1.47–6.31); *p* value, 0.002] (Fang, Liang et al. 2013).

6.5.2 Polymorphisms in Matrix Metalloproteinase1 (MMP1) Gene

Matrix metalloproteinase 1 (MMP1) gene (chromosomal location, 11q22.2) encodes a 469 amino acid MMP1 pro-enzyme, also known as collagenase1, interstitial collagenase or fibroblast collagenase. MMP1 synthesized as pro-enzyme exists in two forms: 57 kDa unglycosylated major pro-enzyme form and 61 kDa glycosylated minor pro-enzyme form. There are two corresponding active forms of MMP1 enzyme as well: around 42 kDa unglycosylated major active enzyme form and around 47 kDa glycosylated active enzyme minor form. The generation of active form of MMP1 enzyme occurs through a two-step proteolysis that involves the serine proteases, urokinase (urokinase-type plasminogen activator), or mast cell tryptase and matrix metalloproteinase 3 (MMP3). MMP1 enzyme is inhibitable by TIMPs such as TIMP1 and α-macroglobulins (Pardo and Selman 2005). The human adult tissues under normal non-pathological conditions show little or no MMP1 expression. MMP1 expression mainly occurs during ECM degradation and remodeling associated with various physiological or pathophysiological processes such as tissue regeneration and wound healing, which involve tissue repairand remodeling. Further, as revealed by in vitro experiments, various types of cells including epithelial cells, endothelial cells, macrophages and fibroblasts can secrete MMP1. A wide variety of cellular mediators, which include cytokines, growth factors and hormones, play an important role in regulating the expression of MMP13 gene (Brinckerhoff and Matrisian 2002). The substrates of MMP1 predominantly include various ECM and BM components and in addition non-ECM components (Table 6.2). These include type I, type II, type III, type VII, type VIIIand type X collagens, gelatins, versican, aggrecan, casein, tenascin C (TNC), nidogen, serpins such asantitrypsin and antichymotrypsin, ovostatin, TNF-α and stromal cell-derived factor 1 (SDF1). MMP1-catalyzed proteolysis results in the activation of various growth factors. For example, the degradation of perlecan releases active form of fibroblast growth factor (FGF) and the degradation of insulin-like growth factor-binding proteins (IGF-BPs) such as IGFBP2, IGFBP3 and IGFBP5 releases active form of insulin-like growth factor (IGF). MMP1-catalyzed proteolysis also releases the membrane-bound forms of interleukin 1β (IL1β) and L-selectin and cleaves the pro-domain of gelatinases, which activate proMMP2 and proMMP9 (Whitelock, Murdoch et al. 1996, McCawley and Matrisian 2001). MMP1-mediated specific proteolysis of type I collagen plays a vital role in the movement or migration of epithelial cells such as keratinocytes during the morphogenesis of epithelium (Pilcher, Dumin et al. 1997).

MMP1, predominantly through its role in the proteolysis of various ECM and BM components, plays a vital role in various pathophysiological processes, diseases or disorders. These include nervous system diseases such as Alzheimer's disease (AD) and vascular dementia (VAD) (Flex, Gaetani et al. 2006), muscular, skeletal and bone pathologies such as rheumatoid arthritis (RA) and osteoarthritis (OA), cardiovascular disorders such as aneurysm and atherosclerosis (Galis and Khatri 2002, Reeps, Pelisek et al. 2009), respiratory system disorders such as interstitial pulmonary fibrosis, tuberculosis, emphysema or chronic obstructive pulmonary disease and bronchial asthma (Lomas and Mahadeva 2002, Rosas, Richards et al. 2008, Todorova, Bjermer et al. 2010, Elkington, Shiomi et al. 2011), wound healing and disorders (Muller, Trocme et al. 2008) and Helicobacter pylori infection of gastric epithelium (Krueger, Hundertmark et al. 2006). MMP1 mediates several important processes involved in the pathogenesis of neoplastic diseases or cancers such as lung cancer, colon cancer and breast cancer. The upregulated MMP1 expression reported in advanced stages of almost all cancer types correlates negatively with cancer survival (Brinckerhoff, Rutter et al. 2000). Various studies report on the importance of MMP1 enzyme in the pathogenesis of breast cancer. The upregulation of MMP1 expression in breast cancer resulting from atypical ductal hyperplasia (ADH), known as atypical ductal hyperplasia cancer (ADHC), and altered MMP1 expression in non-ADHC breast cancer possibly suggest an important role of MMP1 enzyme in the pathogenesis of these cancers. The expression pattern of MMP1in breast cancer suggests a possible use of this MMP as a potential diagnostic and prognostic marker (Poola, DeWitty et al. 2005, Köhrmann, Kammerer et al. 2009). The upregulation of MMP1 expression in animal studies suggests the importance of MMP1 in the initiation of mammary gland cancers possibly through stromal ECM remodeling and release of activated growth factors for tumorigenic proliferation (Duffy, Maguire et al. 2000, Hojilla, Mohammed et al. 2003).

The functional SNPs in MMP1 gene reported to modulate MMP1 gene transcription, expression and subsequently MMP1 enzyme activity in an allele-specific manner most prominently includes MMP1 gene rs1799750 (-1607 1G/2G) SNP. The MMP1 gene rs1799750 (-1607 1G/2G or -1607ins/delG) SNP is an insertion/deletion polymorphism, located within the promoter region of MMP1 gene that involves insertion or deletion of guanosine (G) nucleotide at base pair position -1607. The MMP1 gene rs1799750 1G/2G (-1607 1G/2G) SNP creates a 5′-GGA-3′ sequence, which is the core binding site for erythroblast transformation-specific or E26 transformation-specific (Ets) family of transcription factors. The binding of Ets family of transcription factors increases MMP1 gene promoter activity, MMP1 gene expression and MMP1 enzyme activity. The MMP1 gene rs1799750 2G allele results in a more than twofold increase in the activity of MMP1 gene promoter in comparison with MMP1 gene rs1799750 1G allele that subsequently increases the transcription, protein expression and activity of MMP1 enzyme. Various studies report the association of MMP1 gene rs1799750 1G/2G genotype and MMP1 gene rs1799750

2G/2G genotype with an increased expression and activity of MMP1 enzyme in comparison with MMP1 gene rs1799750 1G/1G genotype. The increased expression and activity of MMP1 enzyme leads to enhanced proteolysis of ECM and BM components and non-ECM substrates, which possibly modulates the susceptibility to various diseases including cancer (Rutter, Mitchell et al. 1998, Fujimoto, Parry et al. 2002).

Numerous association studies report on the possible role of MMP1 gene rs1799750 (-1607 1G/2G) SNP in modulating the risk for the development of various cancers. However, many studies report no such association. Here, we mention several of these studiesthat report the association this SNP with the risk of various cancersand the studies that report no significant association. A meta-analysis based on six association studies on Asian and Caucasian populations, which included 3,967 cases and 3,343 controls, reported a significant association of MMP1 gene rs1799750 (-1607 1G/2G) SNP with an increased risk of lung cancer among Asian population in dominant genetic model [1G/2G+2G/2G vs. 1G/1G: OR,1.45 (95% Cl,1.11–1.89); I = 0.0; p value,0.006], homozygous genetic model [2G/2G vs. 1G/1G: OR,1.73 (95% Cl,1.29–2.31); I^2=0.0; p value,<0.001], recessive genetic model [2G/2Gvs. 1G/2G+1G/1G: OR,1.45 (95% Cl,1.21–1.74); I^2=0.0; p value,<0.001] and allelic genetic model [2G vs. 1G: OR,1.34 (95% Cl,1.18–1.53); I^2= 0.0; p value,<0.001], but not in heterozygous genetic model [1G/2G vs. 1G/1G: OR,1.23 (95% Cl,0.93–1.63); I^2=0.0; p value,>0.05]. The study reported no significant association of this SNP among Caucasian population in all genetic models: dominant genetic model [1G/2G+2G/2G vs. 1G/1G: OR,1.04 (95% Cl,0.92–1.18); I^2=0.0; p value,>0.05], recessive genetic model [2G/2Gvs. 1G/2G+1G/1G: OR,0.97 (95% Cl,0.85–1.10); I^2=0.0; p value,>0.05], allelic genetic model [2G vs. 1G: OR,1.00 (95% Cl,0.93–1.09); I^2= 0.0; p value,>0.05], homozygous genetic model [2G/2G vs. 1G/1G: OR,1.01 (95% Cl,0.87–1.18); I^2=0.0; p value,>0.05] and heterozygous genetic model [1G/2G vs. 1G/1G: OR,1.06 (95% Cl,0.93–1.21); I^2=0.0; p value,>0.05]. Further, the study reported no significant association of this SNP in total population in all genetic models: dominant genetic model [1G/2G+2G/2G vs. 1G/1G: OR,1.11 (95% Cl, 0.99–1.24); I^2=31.8; p value,>0.05], recessive genetic model [2G/2Gvs. 1G/2G+1G/1G: OR, 1.13 (95% Cl,0.92–1.39); I^2=67.2; p value, >0.05], allelic genetic model [2G vs. 1G: OR,1.12 (95% Cl,0.97–1.29); I^2= 70.0; p value,>0.05], homozygous genetic model [2G/2G vs. 1G/1G: OR,1.22 (95% Cl,0.95–1.58); I^2=61.6; p value,>0.05] and heterozygous genetic model [1G/2G vs. 1G/1G: OR,1.09 (95% Cl,0.97–1.23); I^2=0.0; p value,>0.05](Li, Liu et al. 2020). Another meta-analysis based on 47 association studies on European and Asian populations, which included 12,399 cases and 12,433 controls, reported a significant overall association of MMP1 gene rs1799750 (-1607 1G/2G) SNP with an increased risk for the development of cancer in dominant genetic model [1G/2G+2G/2G vs. 1G/1G: OR,1.16 (95% Cl,1.10–1.29); p value$_{heterogeneity}$, 0.004; I^2=39.4; p value,< 0.05], in recessive genetic model [2G/2Gvs. 1G/2G+1G/1G: OR,1.29 (95% Cl,1.17–1.44); p value$_{heterogeneity}$, <0.001; I^2= 67.6; p value,< 0.001] and in allelic genetic model [2G vs. 1G: OR,1.21 (95% Cl,1.13–1.31); p value$_{heterogeneity}$, < 0.001; I^2= 67.1; p value,< 0.001]. The study reported a significant association of this SNP with an increased risk of colorectal cancer (7 studies: 1,343 cases; 1,590 controls) in dominant genetic model [1G/2G+2G/2G vs. 1G/1G: OR,1.23 (95% Cl,1.01–1.49); p value heterogeneity, 0.418; I^2= 0.7; p value,< 0.05], in recessive genetic model [2G/2Gvs. 1G/2G+1G/1G: OR,1.48 (95% Cl,1.26–1.74); p value$_{heterogeneity}$, 0.066; I^2= 49.3; p value,< 0.001] and in allelic genetic model [2G vs. 1G: OR,1.36 (95% Cl,1.14–1.63); p value$_{heterogeneity}$, 0.040; I^2= 54.5; p value,< 0.05]. The study reported a significant association of this SNP with an increased risk of head and neck cancer (11 studies: 1,570 cases; 1,925 controls) in dominant genetic model [1G/2G+2G/2G vs. 1G/1G: OR,1.49 (95% Cl,1.09–2.03); p value$_{heterogeneity}$, 0.009; I^2= 57.4; p value,< 0.05], in recessive genetic model [2G/2Gvs. 1G/2G+1G/1G: OR,1.61 (95% Cl,1.26–2.07); p value$_{heterogeneity}$, 0.002; I^2=64.7; p value, < 0.001] and in allelic genetic model [2G vs. 1G: OR,1.43 (95% Cl,1.20–1.69); p value$_{heterogeneity}$, 0.004; I^2= 61.2; p value,< 0.001]. The study reported no significant association of this SNP with the risk of lung cancer (8 studies: 5,880 cases; 4,910 controls) in all genetic models: dominant genetic model [1G/2G+2G/2G vs. 1G/1G: OR,1.09 (95% Cl,0.99–1.19); p value$_{heterogeneity}$, 0.311; I^2= 15.6; p value,> 0.05], recessive genetic model [2G/2Gvs. 1G/2G+1G/1G: OR,1.20(95% Cl,0.96–1.51); p value$_{heterogeneity}$, 0.000; I^2= 81.3; p value,> 0.05] and allelic genetic model [2G vs. 1G: OR,1.11 (95% Cl,0.98–1.26); p value$_{heterogeneity}$, 0.003; I^2= 70.3; p value,> 0.05]. The study reported no significant association of this SNP with the risk of cervical cancer (three studies: 538 cases; 679 controls) in all genetic models: dominant genetic model [1G/2G+2G/2G vs. 1G/1G: OR,1.04 (95% Cl,0.74–1.46); p value$_{heterogeneity}$, 0.178; I^2= 42.1; p value,> 0.05],recessive genetic model [2G/2G vs. 1G/2G+1G/1G: OR,1.14 (95% Cl,0.91–1.44); p value$_{heterogeneity}$, 0.858; I^2= 0.0; p value,> 0.05] and allelic genetic model [2G vs. 1G: OR,1.09 (95% Cl,0.92–1.29); p value$_{heterogeneity}$, 0.456; I^2= 0.0; p value,> 0.05]. The study reported no significant association of this SNP with the risk of ovarian cancer (4 studies: 443 cases; 797 controls) in all genetic models: dominant genetic model [1G/2G+2G/2G vs. 1G/1G: OR,1.21 (95% Cl,0.86–1.71); p value$_{heterogeneity}$, 0.279; I^2 = 22.0; p value,> 0.05], recessive genetic model [2G/2G vs. 1G/2G+1G/1G: OR,0.87 (95% Cl,0.68–1.12); p value$_{heterogeneity}$, 0.420; I^2= 0.0; p value,> 0.05] and allelic genetic model [2G vs. 1G: OR,0.98 (95% Cl,0.82–1.17); p value$_{heterogeneity}$, 0.721; I^2= 0.0; p value,> 0.05]. Further, the study reported no significant association of this SNP with the risk of breast cancer (two studies: 129 cases; 274 controls) in all genetic models: dominant genetic model [1G/2G+2G/2G vs. 1G/1G: OR,1.46 (95% Cl,0.85–2.49); p value$_{heterogeneity}$, 0.919; I^2= 0.0; p value,> 0.05], recessive genetic model [2G/2G vs. 1G/2G+1G/1G: OR,0.85 (95% Cl,0.52–1.41); p value$_{heterogeneity}$, 0.763; I^2= 0.0; p value,> 0.05] and allelic genetic model [2G vs. 1G: OR, 1.07 (95% Cl,0.79–1.45); p value$_{heterogeneity}$, 0.778; I^2= 0.0; p value,> 0.05]. The study reported no significant association of this SNP with the risk of endometrial cancer (two studies: 207 cases; 363 controls) in all genetic models: dominant genetic model [1G/2G+2G/2G vs. 1G/1G: OR,1.11 (95% Cl,0.22–5.53); p value$_{heterogeneity}$, 0.003; I^2= 88.8; p value,> 0.05],recessive genetic model [2G/2G vs. 1G/2G+1G/1G: OR,1.03 (95% Cl,0.73–1.45); p value$_{heterogeneity}$, 0.162; I^2= 48.9; p value,> 0.05] and allelic genetic model [2G vs. 1G: OR,1.07 (95% Cl,0.55–2.08); p value$_{heterogeneity}$, 0.010; I^2= 84.9; p value,> 0.05]. The study reported no significant association of this SNP with the risk of bladder cancer (two studies: 658 cases; 649 controls) in all genetic models: dominant genetic model [1G/2G+2G/2G vs. 1G/1G: OR,1.39 (95% Cl,0.48–4.02);

p value$_{heterogeneity}$, 0.005; I^2= 87.5; *p* value,> 0.05], recessive genetic model [2G/2G vs. 1G/2G+1G/1G: OR,1.70 (95% Cl,0.59–4.89); *p* value$_{heterogeneity}$, 0.001; I^2= 90.1; *p* value,> 0.05] and allelic genetic model [2G vs. 1G: OR,1.53 (95% Cl,0.58–4.02); *p* value$_{heterogeneity}$< 0.001; I^2= 94.9; *p* value,> 0.05]. The study reported a significant association of this SNP with an increased risk of renal cancer (three studies: 374 cases; 558 controls) in recessive genetic model [2G/2G vs. 1G/2G+1G/1G: OR,1.82 (95% Cl,1. 38–2.39); *p* value$_{heterogeneity}$, 0.589; I^2= 0.0; *p* value,< 0.001] and in allelic genetic model [2G vs. 1G: OR,1.43 (95% Cl,1.17–1.75); *p* value$_{heterogeneity,}$0.265; I^2= 24.6; *p* value,< 0.001], but not in dominant genetic model [1G/2G+2G/2G vs. 1G/1G: OR,1.16 (95% Cl,0.79–1.70); *p* value$_{heterogeneity}$,0.650; I^2= 0.0; *p* value,> 0.05]. Further, the study reported a significant association of this SNP with an increased cancer risk in Asian population (26 studies: 6,437 cases; 6,951 controls) indominant genetic model [1G/2G+2G/2G vs. 1G/1G: OR, 1.28 (95% Cl, 1.09–1.49); *p* value$_{heterogeneity}$,0.004; I^2= 47.6; *p* value,< 0.001],in recessive genetic model [2G/2Gvs. 1G/2G+1G/1G: OR,1.28 (95% Cl,1.13–1.47); *p* value$_{heterogeneity}$,< 0.001; I^2= 66.4; *p* value,< 0.001] and in allelic genetic model [2G vs. 1G: OR,1.23 (95% Cl,1.12–1.36); *p* value$_{heterogeneity}$,< 0.001; I^2= 66.1; *p* value,< 0.001]. The study reported no significant association of this SNP with the risk of cancer in European population (15 studies: 4,848 cases; 4,476 controls) in all genetic models: dominant genetic model [1G/2G+2G/2G vs. 1G/1G: OR,1.07 (95% Cl, 0.98–1.18); *p* value$_{heterogeneity}$, 0.225; I^2= 20.5; *p* value,>0.05], recessive genetic model [2G/2G vs. 1G/2G+1G/1G: OR,1.18 (95% Cl,0.99–1.41); *p* value$_{heterogeneity}$< 0.001; I^2= 64.0; *p* value,> 0.05] and allelic genetic model [2G vs. 1G: OR,1.12 (95% Cl,1.01–1.25); *p* value$_{heterogeneity}$,0.001; I^2= 61.9; *p* value,> 0.05]. The study reported no significant association among European population. The study also reported a significant difference between the Asian and European populations with regard to the difference in the distribution or frequency of MMP1 gene rs1799750 2G (-1607 2G) allele [European population, 52% (95% Cl,48–57); Asian population,64% (95% Cl, 61–68); *p* value,0.0001] (Peng, Cao et al. 2010).

A meta-analysis based on four association studies, which included 4,465 cases and 3,487 controls, reported no significant association of MMP1 gene rs1799750 (-1607 1G/2G) SNP with the risk of developing lung cancer in both the dominant genetic model [1G/2G+2G/2G vs. 1G/1G: OR,1.09 (95% Cl,0.99–1.21); *p* value$_{heterogeneity}$,0.4; *p* value,0.09] and the recessive genetic model [2G/2Gvs. 1G/2G+1G/1G: OR, 1.12 (95% Cl,0.85–1.47); *p* value$_{heterogeneity}$,0.0003; *p* value,0.42]. The same meta-analysis based on four association studies in case of dominant genetic model, which included 539 cases and 765 controls, and five association studies in case of recessive genetic model,which included 599 cases and 929 controls, reported a significant association of this SNP with an increased risk of colorectal cancer in dominant genetic model [1G/2G+2G/2G vs. 1G/1G: OR,1.66 (95% Cl,1.14–2.42); *p* value$_{heterogeneity}$,0.96; *p* value,0.008] and recessive genetic model [2G/2Gvs. 1G/2G+1G/1G: OR,1.59 (95% Cl,1.15–2.20); *p* value$_{heterogeneity}$,0.07; *p* value,0.005]. The same meta-analysis based on two association studies,which included 231 cases and 260 controls, reported no significant association of MMP1 gene rs1799750 (-1607 1G/2G) SNP with the risk of developing breast cancer in both the dominant genetic model [1G/2G+2G/2G vs. 1G/1G: OR, 1.16 (95% Cl,0. 76–1.77); *p* value$_{heterogeneity}$, 0.47; *p* value,0.49] and the recessive genetic model [2G/2Gvs. 1G/2G+1G/1G: OR,1.08 (95% Cl,0.66–1.76); *p* value$_{heterogeneity}$, 0.21; *p* value,0.75] (McColgan and Sharma 2009). Another meta-analysis based on nine association studies on Asian and Caucasian populations, which included 4,695 cases and 4,308 controls, reported a significant association of this SNP with an increased risk of lung cancer in dominant genetic model [1G/2G+2G/2G vs. 1G/1G: OR, 1.14 (95% Cl,1.03–1.26); *p* value$_{heterogeneity}$,0.31; *p* value,0.01], but not in recessive genetic model [2G/2G vs. 1G/2G+1G/1G: OR,1.18 (95% Cl,0.97–1.44); *p* value$_{heterogeneity}$,0.0006; *p* value,0.09]. The study reported a significant association of this SNP with an increased risk of lung cancer in Asian population in dominant genetic model [1G/2G+2G/2G vs. 1G/1G: OR,1.47 (95% Cl,1.16–1.87); *p* value$_{heterogeneity}$,0.90; *p* value,0.001] and recessive genetic model [2G/2Gvs. 1G/2G+1G/1G: OR,1.33 (95% Cl,1.14–1.56); *p* value$_{heterogeneity}$,0.0003; *p* value,0.11]. The study reported no significant association of this SNP with the risk of lung cancer among Caucasian population in both the dominant genetic model [1G/2G+2G/2G vs. 1G/1G: OR,1.07 (95% Cl,0. 95–1.20); *p* value$_{heterogeneity}$,0.62; *p*value,0.25] and the recessive genetic model [2G/2Gvs. 1G/2G+1G/1G: OR,1.10 (95% Cl,0. 84–1.44); *p* value$_{heterogeneity}$,0.002; *p* value,0.48] (Hu, Wang et al. 2013).

A hospital-based case-control study on Polish population, which included 648 unselected cancer cases (299 breast cancer cases, 199 lung cancer cases and 150 colon cancer cases) and 648 matched controls (299 controls for breast cancer, 199 controls for lung cancer and 150 controls for colon cancer), studied the association of MMP1 gene rs1799750 (-1607 1G/2G) SNP along with serum zinc levels with the risk of breast cancer, lung cancer and colon cancer. The study reported no significant association of this SNP with the risk of breast cancer [2G/2G vs. 1G/1G: OR,1.08 (95% Cl,0.70–1.67); *p* value,0.73; 1G/2G vs. 1G/1G: OR,1.01 (95% Cl,0.68–1.46); *p* value,0.99], lung cancer [2G/2G vs. 1G/1G: OR,0.77 (95% Cl,0.42–1.39); *p* value,0.38; 1G/2G vs. 1G/1G: OR,0.98 (95% Cl,0.61–1.57); *p* value,0.92] and colon cancer [2G/2G vs. 1G/1G: OR,1.34 (95% Cl,0.70–2.56); *p* value,0.37;1G/2Gvs. 1G/1G: OR,1.49 (95% Cl,0.87–2.55); *p* value,0.15] (Białkowska, Marciniak et al. 2020). Another hospital-based case-control study on Polish population, which included 197 prostate cancer cases and 197 healthy matched controls, reported no significant association of this SNP with prostate cancer [2G/2G vs. 1G/1G: OR,0.73 (95% Cl,0.41–1.28); *p* value,0.27; 1G/2G vs. 1G/1G: OR,1.24 (95% Cl,0.76–2.02); *p* value,0.39] (Białkowska, Marciniak et al. 2018). Another hospital-based case-control study on Spanish population, which included 501 lung cancer cases and 510 healthy matched controls, reported no significant association of MMP1 gene rs1799750 (-1607 1G/2G) SNP with lung cancer in dominant genetic model [1G/2G+2G/2G vs. 1G/1G: OR,0.99 (95% Cl,0.71–1.40); *p* value,0.975], in homozygous genetic model [2G/2G vs. 1G/1G: OR,1.04 (95% Cl, 0.68–1.58); *p* value,0.859], and in heterozygous genetic model [1G/2G vs. 1G/1G: OR,0.97 (95% Cl,0.68–1.40); *p* value,0.884] (González-Arriaga, López-Cima et al. 2008).

6.5.3 Polymorphisms in Matrix Metalloproteinase8 (MMP8) Gene

Matrix metalloproteinase 8 (MMP8) gene (chromosomal location, 11q22.2) encodes a 467 amino acid MMP8 pro-enzyme, also known as neutrophil collagenase or collagenase 2. The relative molecular mass (M_r) of latent form of MMP8 or MMP8 pro-enzyme varies between 50 and 80 kDa depending on the extent of glycosylation. The relative molecular mass (M_r) of active form of MMP8 or MMP8 enzyme correspondingly varies between 40 and 75 kDa depending on the extent of glycosylation. The activation of MMP8 pro-enzyme involves the proteolytic cleavage of pro-domain, catalyzed by other MMPs such as MMP3, MMP7 and MMP10. However, autoproteolysis can also occur that generates a 40 kDa MMP8 enzyme that is catalytically active, but devoid of interstitial fibrillar collagenase activity. MMP8 enzyme is inhibitable by TIMPs such as TIMP1 and TIMP2 (Knäuper, Osthues et al. 1993, Dozier, Escobar et al. 2006). MMP8 expression mainly occurs in mature human polymorphonuclear leukocytes or polymorphonuclear neutrophils (PMNs), the innate immune cells that circulate freely and abundantly in peripheral blood. These cells store MMP8 pro-enzyme within specific granules and release active MMP8 enzyme upon stimulation. The release of active MMP8 from these cells corresponds directly to the degranulation process that plays a vital role in regulating the activity of this MMP (Mainardi, Pourmotabbed et al. 1991, Schettler, Thorn et al. 1991). In addition to PMNs, various other cell types express MMP8 particularly during physiological or pathophysiological processes that lead to inflammation. The cells that express MMP8 include macrophages, T-cells, plasma cells, fibroblasts, endothelial cells, chondrocytes, keratinocytes, smooth muscle cells, colon myofibroblasts, mucosal cells of colon and epithelial cells of bronchi, cornea and gingival sulcus (Hanemaaijer, Sorsa et al. 1997, Tervahartiala, Pirilä et al. 2000, Prikk, Maisi et al. 2001, 2002). PMNs secrete MMP8 that is highly glycosylated, whereas the other cell types secrete the lowly glycosylated form of MMP8. The higher MMP8 glycosylation in PMNs possibly relates to the storage of this proteinase in granules within these cells. The other cell types secrete MMP8 immediately after synthesis without storing or retaining the enzyme within the cells, and this relates to the low glycosylation of this enzyme originating from these non-PMN cells (Smith, Brandt et al. 1994, Kiili, Cox et al. 2002). The polymorphonuclear neutrophils also express a more stable form of MMP8, a membrane-bound form, which in comparison with the secreted form of MMP8 is less sensitive to inhibition by TIMPs (Owen, Hu et al. 2004). Further, the proteolytic cleavage of MMP8 pro-enzyme occurs at two different sites that generate two forms of active MMP8 enzyme differing in amino (N)-terminal residue (phenylalanine, Phe-79, or methionine, Met-80) and catalytic activity. MMP8 enzyme with Phe-79 as N-terminal residue is three times more catalytically active than MMP8 with Met-80 as N-terminal amino acid. The increased catalytic activity also includes an increase in the collagen-specific activity of the enzyme (Knäuper, Osthues et al. 1993, Gioia, Fasciglione et al. 2002).

The substrates of active MMP8 enzyme predominantly include various ECM and BM components. In addition, the substrates of this enzyme also include a wide variety of non-ECM components (Table 6.2). MMP8 enzyme has an efficient collagenase (collagen-degrading) activity and cleaves various types of collagens, in particular, type I, type II and type III fibrillar collagens. MMP8 also cleaves type V, type VII, type VIIIand type X fibrillar collagens, gelatins and non-collagenous substrates such as entactin, fibronectin, tenascin C (TNC), chemokines, serine protease inhibitors or serpins such as α_1-proteinase inhibitor (α_1-PI) and pro-TNF-α. The proteolytic action of MMP8 enzyme activates or deactivates the substrates and results in an increase or decrease in their activities. A wide variety of cellular mediators, which include cytokines such as TNF-α, CD40 ligand (CD40L) or CD154 and IL1β, growth factors and hormones, play an important role in regulating the expression of MMP8 gene (Van den Steen, Husson et al. 2003, Van Den Steen, Wuyts et al. 2003).

MMP8, predominantly through its role in the proteolysis of various ECM and BM components and non-ECM substrates, plays a vital role in various physiological and pathophysiological processes. The animal studies report the involvement of MMP8 in tissue turnover and remodeling during embryogenesis besides postpartum tissue remodeling of uterus (Balbín, Fueyo et al. 1998). MMP8 enzyme through its proteolytic activity plays a critical role in the pathogenesis of various pathophysiological processes, especially inflammation-related conditions, diseases or disorders. These include nervous system diseases such as autoimmune encephalomyelitis (EAE) and bacterial meningitis (BM) (Leppert, Leib et al. 2000, Folgueras, Fueyo et al. 2008), cardiovascular disorders such as aortic aneurysms (AAA) and atherosclerosis (Eliason, Hannawa et al. 2005, Turu, Krupinski et al. 2006), respiratory system disorders such as asthma, pulmonary emphysema, bronchiectasis, cystic fibrosis (CF) andhypersensitivity pneumonitis (Power, O'Connor et al. 1994, Sepper, Konttinen et al. 1995, Betsuyaku, Nishimura et al. 1999, Pardo, Barrios et al. 2000, Prikk, Maisi et al. 2002, Ratjen, Hartog et al. 2002), muscular, skeletal and bone disorders such as rheumatoid arthritis (RA) and osteoarthritis (OA) (Balbín, Fueyo et al. 2003, García-Prieto, González-López et al. 2010) and other disorders such as chronic adult and localized juvenile periodontitis, ulcerative keratolysis, wound healing and disorders such as non-healing ulcers (Nwomeh, Liang et al. 1999, Tervahartiala, Pirilä et al. 2000, O'Brien, Li et al. 2001, Kiili, Cox et al. 2002).

MMP8 enzyme plays a critical role in several important processes involved in the pathogenesis of various cancers. However, the role of this enzyme in cancers is complex and not fully understood. On the one hand,various studies report the involvement of MMP8 enzyme in promoting tumor growth, progression and metastasis in various cancers such as ovarian cancerand head and neck squamous cell carcinoma. The expression and serum levels of MMP8 enzyme correlate positively with tumor grade and advanced tumor stages in ovarian cancerand head and neck cancer and with poor prognosis in ovarian cancer (Moilanen, Pirilä et al. 2002, Stadlmann, Pollheimer et al. 2003, Stenman, Paju et al. 2003, Kuropkat, Duenne et al. 2004). On the other hand, the studies reveal the protective role of MMP8 enzyme in various cancers.

These studies report the involvement of MMP8 enzyme in suppressing tumor growth, progression and metastasis in various cancers including tongue cancer. Further, MMP8 enzyme also increases the susceptibility or sensitivity of cancers to oncolytic viruses, a group of viruses that specifically infect cancer cells and cause their lysis to kill or destroy them. The expression and serum levels of MMP8 enzyme correlate positively with survival in tongue cancer (Mok, Boucher et al. 2007, Gutiérrez-Fernández, Fueyo et al. 2008, Korpi, Kervinen et al. 2008). The animal studies reveal that mice deficient in MMP8 enzyme are more susceptible to skin cancers and develop more aggressive fibrosarcomas in comparison with wild-type mice (Balbín, Fueyo et al. 2003). Further, various studies on cancer cell lines report the association of increased MMP8 expression with decreased metastatic potential or metastasis in breast tumor-derived cells and specific breast cancer cell lines, whereas MMP8 gene knockout increases the metastatic potential or metastasis in specific breast cancer cell lines (Agarwal, Goodison et al. 2003, Montel, Kleeman et al. 2004, Decock, Hendrickx et al. 2008). Thus, the role played by MMP8 enzyme in the pathogenesis of various cancers is contradictory and still not fully clear.

Several SNPs identified in MMP8 gene include 5′-flanking region and promoter SNPs such as MMP8 gene rs11225395 (-799 C/T) SNP, MMP8 geners 1320632 (-381 A/G) SNP and MMP8 geners 17099462 C/ASNP; intronic SNPs such as MMP8 geners1892886 SNP; exonic SNPs such as MMP8 geners1940475 (+259A/G; Lys87Glu) SNP, MMP8 geners35866072 (A/C; Lys460Thr) SNP, MMP8 gene rs34009635 (T/C; Val436Ala) SNP and MMP8 gene rs2155052 (+17C/G) SNP; and 3′-flanking region (3′-FGR) SNPs such as MMP8 gene rs1276284 SNP, besides tagging SNPs such as MMP8 geners 10895353A/G SNP, MMP8 geners 7943404 C/T SNP, MMP8 geners 17099436A/T SNP and MMP8 geners 2508383C/T SNP. The two MMP8 gene promoter SNPs, rs 11225395 (-799 C/T)SNP and rs1320632 (-381 A/G) SNP, modulate the transcription and expression of MMP8 gene in an allele-specific manner through their modifying effect on the binding affinities and interactions of various nuclear proteins including transcription factors with the MMP8 gene promoter. MMP8 gene promoter SNP, rs11225395 (-799 C/T)SNP, involves substitution of cytosine (C) by thymine (T) at -799 bp position with respect to the transcriptional start sitewithin the promoter region and possibly affects the binding affinities and interactions of various nuclear proteins or factors with the promoter sequence. The yet uncharacterized nuclear protein or factor, probably enhancer, interacts more favorably and binds with higher affinity to the variant T allele, but not to the wild C allele, which results in an almost 1.8-fold higherpromoter activity in case of MMP8 gene -799TT genotype in comparison with MMP8 gene -799CC genotype (Decock, Long et al. 2007). MMP8 gene rs1320632 (-381 A/G) SNP and MMP8 gene rs2155052 (+17C/G) SNP also modulate the transcription and expression of MMP8 gene in an allele-specific manner. The minor allele -381G of MMP8 gene rs1320632 SNP and the minor allele +17G of MMP8 gene rs2155052 SNP are in complete linkage disequilibrium. A study reports almost two- to three-fold higherpromoter activity in case of promoter construct consisting of three minor alleles, -799T/-381G/+17G, in comparison with promoter constructconsisting of three major alleles, -799C/-381A/+17C, and other promoter constructs consisting of one or two minoralleles(Wang, Parry et al. 2004). The MMP8 geners 1940475 (+259A/G; Lys87Glu) SNP located in the pro-domain region modifies the structural stability of the pro-domain, which impairs the proteolytic activation of MMP8 enzyme (McLaughlin, Rella et al. 2011). In general, the extensive studies on SNPs in MMP8 gene are not available, and therefore, the functional significance of most of the MMP8 gene SNPs needs elucidation. Further, not many studies are available that report on the possible association of MMP8 gene SNPs with the risk of various cancers. However, various association studies report on the possible role of MMP8 gene SNPs in modulating the risk for the development of various cancers. Here, we mention several of these studies that report the association this SNP with the risk of various cancersand the studies that report no significant association.

A meta-analysis based on fifteen association studies on Asian and non-Asian populations, which included 7375 cases and 8117 controls, reported no significant overall association of MMP8 gene rs11225395 (-799 C/T)SNP with the risk of various cancers in all genetic models: dominant genetic model [CT+TT vs. CC: OR,1.04 (95% Cl,0.93–1.16); p value$_{heterogeneity}$,0.03; p value,0.49], recessive genetic model [TTvs. CT+CC: OR,0.97 (95% Cl,0.83–1.13); I^2=0.0; p value$_{heterogeneity}$,0.05; p value,0.70], allelic genetic model [T vs. C: OR,1.03 (95% Cl,0.95–1.11); p value$_{heterogeneity}$,0.02; p value,0.46], homozygous genetic model [TT vs. CC: OR,1.01 (95% Cl,0.85–1.20); p value$_{heterogeneity}$,0.03; p value,0.90] and heterozygous genetic model [CT vs. CC: OR,1.06 (95% Cl,0.98–1.14); p value$_{heterogeneity}$,0.13; p value,0.14]. The same meta-analysis reported no significant association of MMP8 gene rs11225395 (-799 C/T) SNP with the risk of bladder cancer (threestudies: 1,632 cases; 1,548controls) in all genetic models: dominant genetic model [CT+TT vs. CC: OR,1.02 (95% Cl,0.84–1.25); p value$_{heterogeneity}$,0.58; p value,0.81], recessive genetic model [TTvs. CT+CC: OR,0.83 (95% Cl,0.61–1.12; p value$_{heterogeneity}$,0.15; p value, 0.22], allelic genetic model [T vs. C: OR,0.97 (95% Cl,0.83–1.12); p value$_{heterogeneity}$,0.29; p value,0.67], homozygous genetic model [TT vs. CC: OR,0.85 (95% Cl,0.61–1.17); p value$_{heterogeneity}$,0.14; p value,0.32] and heterozygous genetic model [CT vs. CC: OR,1.08 (95% Cl,0.87–1.33); p value$_{heterogeneity}$,0.84; p value,0. 49]. The same meta-analysis reported no significant association of MMP8 gene rs11225395 (-799 C/T) SNP with the risk of breast cancer (four studies: 8,544 cases; 8,762controls) in all genetic models: dominant genetic model [CT+TT vs. CC: OR,1.07 (95% Cl,0.86–1.31); p value$_{heterogeneity}$,0.04; p value,0.55],recessive genetic model [TTvs. CT+CC: OR,1.02 (95% Cl,0.89–1.16); p value$_{heterogeneity}$,0.25; p value,0.79],allelic genetic model [T vs. C: OR,1.06 (95% Cl,1.00–1.13); p value$_{heterogeneity}$,0.20; p value,0.07], homozygous genetic model [TT vs. CC: OR,1.04 (95% Cl,0.78–1.37); p value$_{heterogeneity}$,0.08; p value,0.80] and heterozygous genetic model [CT vs. CC: OR,1.08 (95% Cl,0.89–1.30); p value$_{heterogeneity}$,0.08; p value,0. 43]. The same meta-analysis also reported no significant association of MMP8 gene rs11225395 (-799 C/T) SNP with the risk of digestive system cancer (three studies, 2,568 cases; 3,210controls) in all genetic models:dominant genetic model [CT+TT vs. CC: OR,1.09 (95% Cl,0.79–1.51);

p value$_{heterogeneity}$,0.04; *p* value,0.59], recessive genetic model [TTvs. CT+CC: OR,0.97 (95% Cl,0.79–1.20); *p* value$_{heterogeneity}$,0.30; *p* value,0.80], allelic genetic model [T vs. C: OR,1.03 (95% Cl,0.83–1.28); *p* value$_{heterogeneity}$,0.05; *p* value,0.78], homozygous genetic model [TT vs. CC: OR,0.99 (95% Cl,0.79–1.24); *p* value$_{heterogeneity}$,0.11; *p* value,0.95] and heterozygous genetic model [CT vs. CC: OR,1.10 (95% Cl,0.80–1.49); *p* value$_{heterogeneity}$,0.07; *p* value,0.57]. The same meta-analysis reported no significant association of MMP8 gene rs11225395 (-799 C/T) SNP with the risk of malignant melanoma, childhood leukemia, ovarian cancer and lung cancer (five studies: 1,910 cases; 2,618 controls) in all genetic models: dominant genetic model [CT+TT vs. CC: OR,1.03 (95% Cl,0.75–1.44); *p* value$_{heterogeneity}$,0.04; *p* value,0.84], recessive genetic model [TTvs. CT+CC: OR,1.08 (95% Cl,0.65–1.79); *p* value$_{heterogeneity}$,0.02; *p* value, 0.78], allelic genetic model [T vs. C: OR,1.07 (95% Cl,0.83–1.39); *p* value$_{heterogeneity}$,0.02; *p* value,0.60], homozygous genetic model [TT vs. CC: OR,1.14 (95% Cl,0.71–1.86); *p* value$_{heterogeneity}$,0.05; *p* value,0.58] and heterozygous genetic model [CT vs. CC: OR,1.01 (95% Cl,0.74–1.37); *p* value$_{heterogeneity}$,0.09; *p* value,0. 96]. The same meta-analysis also reported no significant association of MMP8 gene rs11225395 (-799 C/T)-SNP with the risk of cancers among Asians (nine studies: 8,414 cases; 9,606 controls) in all genetic models: dominant genetic model [CT+TT vs. CC: OR,0.95 (95% Cl,0.87–1.04); *p* value$_{heterogeneity}$,0.34; *p* value,0.26], recessive genetic model [TTvs. CT+CC: OR,0.89 (95% Cl,0.77–1.02); *p* value$_{heterogeneity}$,0.14; *p* value,0.09], allelic genetic model [T vs. C: OR,0.97 (95% Cl,0. 89–1.07); *p* value$_{heterogeneity}$,0.07; *p* value,0.57], homozygous genetic model [TT vs. CC: OR,0.89 (95% Cl,0.77–1.03); *p* value$_{heterogeneity}$,0.18; *p* value,0.12] and heterozygous genetic model [CT vs. CC: OR,0.97 (95% Cl,0.88–1.07); *p* value$_{heterogeneity}$,0.57; *p* value,0.55]. However, the same meta-analysis reporteda significant association of MMP8 gene rs11225395 (-799 C/T)SNP with an increased risk of various cancers among non-Asians (six studies: 6,240 cases; 6,532controls) in dominant genetic model [CT+TT vs. CC: OR,1.21 (95% Cl,1.08–1.35); *p* value$_{heterogeneity}$,0.45; *p* value,<0.01], allelic genetic model [T vs. C: OR,1.11 (95% Cl,1.04–1.19); *p* value$_{heterogeneity}$,0.43; *p* value,<0.01], homozygous genetic model [TT vs. CC: OR,1.22 (95% Cl,1. 05–1.41); *p* value$_{heterogeneity}$,0.39; *p* value,0.01] and heterozygous genetic model [CT vs. CC: OR,1.21 (95% Cl,1.07–1.36); *p* value$_{heterogeneity}$,0.43; *p* value,<0.01], but not in recessive genetic model [TTvs. CT+CC: OR,1.09 (95% Cl,0.96–1.24); *p*value$_{heterogeneity}$,0.17; *p* value,0.19] (Feng, Chen et al. 2019).

Another meta-analysis based on nineteen association studies on Asian, Caucasianand Latin populations, which included 8,140 cases and 10,529 controls, evaluated the role of three MMP8 gene SNPs: MMP8 gene rs11225395 (-799 C/T)SNP, MMP8 geners1940475 (+259A/G; Lys87Glu) SNP and MMP8 geners 35866072 (A/C; Lys460Thr) SNP, in modulating the risk of various cancers. The meta-analysis reported no significant association of MMP8 gene rs11225395 (-799 C/T)SNP (11 studies:4,372 cases; 5,066 controls) with an overall risk of cancer in all genetic models:dominant genetic model [CT+TT vs. CC: OR,0.98 (95% Cl,0.90–1.07); *p* value$_{heterogeneity}$,0.086; *p* value,0.666], recessive genetic model [TTvs. CT+CC: OR,0.94 (95% Cl,0.83–1.07); *p* value$_{heterogeneity}$,0.249; *p* value,0.348], allelic genetic model [T vs. C: OR,0.98 (95% Cl,0.92–1.04); *p* value$_{heterogeneity}$,0.068; *p* value,0.429], homozygous genetic model [TT vs. CC: OR,0.94 (95% Cl,0.82–1.07); *p* value$_{heterogeneity}$,0.097; *p* value,0.362] andheterozygous genetic model [CT vs. CC: OR,1.00 (95% Cl,0.92–1.09); *p* value$_{heterogeneity}$,0.193; *p* value,0.992]. The same meta-analysis reported no significant association of MMP8 gene rs11225395 (-799 C/T)SNP with the risk of bladder cancer (three studies: 816 cases; 774 controls) in all genetic models: dominant genetic model [CT+TT vs. CC: OR,1.02 (95% Cl,0.84–1.25); *p* value$_{heterogeneity}$,0.582; *p* value,0.813], recessive genetic model [TTvs. CT+CC: OR,0.83 (95% Cl,0.61–1.12); *p* value$_{heterogeneity}$,0.151; *p* value,0.219], allelic genetic model [T vs. C: OR,0.97 (95% Cl,0.83–1.12); *p* value$_{heterogeneity}$,0.286; *p* value,0.671], homozygous genetic model [TT vs. CC: OR,0.85 (95% Cl,0.61–1.17); *p* value$_{heterogeneity}$,0.142; *p* value,0.316] andheterozygous genetic model [CT vs. CC: OR,1.08 (95% Cl,0.87–1.33); *p* value$_{heterogeneity}$,0.837; *p* value,0.492]. The same meta-analysis reported no significant association of MMP8 gene rs11225395 (-799 C/T) SNP with the risk of breast cancer, leukemia, lung cancer, oral cancer, ovarian cancer, malignant melanoma, hepatocellular carcinoma and gastric cancer (eight studies: 3,556 cases; 4,292 controls) in all genetic models: dominant genetic model [CT+TT vs. CC: OR,1.00 (95% Cl,0.86–1.16); *p* value$_{heterogeneity}$,0.033; *p* value,0.965], recessive genetic model [TTvs. CT+CC: OR,0.97 (95% Cl,0.84–1.11); *p* value$_{heterogeneity}$,0.329; *p* value,0.631], allelic genetic model [T vs. C: OR,0.99 (95% Cl,0.89–1.11); *p* value$_{heterogeneity}$,0.039; *p* value,0.898], homozygous genetic model [TT vs. CC: OR,0.96 (95% Cl,0.83–1.11); *p* value$_{heterogeneity}$,0.108; *p* value,0.578] and heterozygous genetic model [CT vs. CC: OR,0.99 (95% Cl,0.89–1.09); *p* value$_{heterogeneity}$,0.080; *p* value,0.763]. The same meta-analysis reported no significant association of MMP8 gene rs11225395 (-799 C/T) SNP with the risk of various cancers among Asian population (eight studies: 3,784 cases; 4,540 controls) in all genetic models: dominant genetic model [CT+TT vs. CC: OR,0.96 (95% Cl,0.88–1.04); *p* value$_{heterogeneity}$,0.396; *p* value,0.322], recessive genetic model [TTvs. CT+CC: OR,0.90 (95% Cl,0.78–1.03); *p* value$_{heterogeneity}$,0.401; *p* value,0.111], allelic genetic model [T vs. C: OR,0.95 (95% Cl,0.89–1.01); *p* value$_{heterogeneity}$,0.248; *p* value,0.125], homozygous genetic model [TT vs. CC: OR,0.89 (95% Cl,0.77–1.02); *p* value$_{heterogeneity}$,0.258; *p* value,0.102] and heterozygous genetic model [CT vs. CC: OR,0.98 (95% Cl,0.89–1.08); *p* value$_{heterogeneity}$,0.585; *p* value,0.682]. The same meta-analysis reported no significant association of MMP8 gene rs11225395 (-799 C/T) SNP with the risk of various cancers among Caucasian population (two studies: 537 cases; 489 controls) in all genetic models: dominant genetic model [CT+TT vs. CC: OR,1.30 (95% Cl,1.00–1.70); *p* value$_{heterogeneity}$,0.115; *p* value,0.047], recessive genetic model [TTvs. CT+CC: OR,1.18 (95% Cl,0.85–1.62); *p* value$_{heterogeneity}$,0.236; *p* value,0.322], allelic genetic model [T vs. C: OR,1.18 (95% Cl,0.99–1.41); *p* value$_{heterogeneity}$,0.082; *p* value,0.062], homozygous genetic model [TT vs. CC: OR,1.38 (95% Cl,0.95–1.96); *p* value$_{heterogeneity}$,0.105; *p* value,0.095] and heterozygous genetic model [CT vs. CC: OR,1.28 (95% Cl,0.97–1.69); *p* value$_{heterogeneity}$,0.201; *p* value,0.082]. The same meta-analysis reported no significant association of MMP8 gene rs11225395 (-799 C/T) SNP with the risk of various cancers among Latin population (one study: 51 cases; 37 controls) in all genetic models: dominant genetic model [CT+TT vs. CC: OR,1.30 (95% Cl,1.00–1.70); p value$_{heterogeneity}$,0.115; p value,0.047], recessive genetic model [TTvs. CT+CC: OR,1.18 (95% Cl,0.85–1.62); p value$_{heterogeneity}$,0.236; p value,0.322], allelic genetic model [T vs. C: OR,1.18 (95% Cl,0.99–1.41); p value$_{heterogeneity}$,0.082; p value,0.062], homozygous genetic model [TT vs. CC: OR,1.38 (95% Cl,0.95–1.96); p value$_{heterogeneity}$,0.105; p value,0.095] and heterozygous genetic model [CT vs. CC: OR,1.28 (95% Cl,0.97–1.69); p value$_{heterogeneity}$,0.201; p value,0.082] (Zhang,Zhu et al. 2019).

The same meta-analysis reported no significant association of MMP8 gene rs35866072 (A/C; Lys460Thr) SNP (five studies: 3,019 cases; 3,544 controls) with an overall risk of cancer among Asian population in all genetic models: dominant genetic model [AC+CC vs. AA: OR,0.96 (95% Cl,0.67–1.40); p value$_{heterogeneity}$,0.886; p value,0.848], recessive genetic model [CCvs. AC+AA: OR,0.83 (95% Cl,0.36–1.93); p value$_{heterogeneity}$,0.866; p value,0.669], allelic genetic model [A vs. C: OR,0.94 (95% Cl,0.67–1.32); p value$_{heterogeneity}$,0.905; p value,0.729], homozygous genetic model [CC vs. AA: OR,0.83 (95% Cl,0.36–1.93); p value$_{heterogeneity}$,0.859; p value,0.669] and heterozygous genetic model [AC vs. AA: OR,1.00 (95% Cl,0.66–1.50); p value$_{heterogeneity}$,0.904; p value,0. 994]. The same meta-analysis reported no significant association of MMP8 gene rs35866072 (A/C; Lys460Thr) SNP with the risk of bladder cancer (one study: 375 cases; 375 controls) in all genetic models: dominant genetic model [AC+CC vs. AA: OR,0.76 (95% Cl,0.33–1.76); p value$_{heterogeneity}$,–0.526], recessive genetic model [CCvs. AC+AA: OR,0.75 (95% Cl,0.17–3.37); p value$_{heterogeneity}$,–0.705], allelic genetic model [A vs. C: OR,0.76 (95% Cl,0.37–1.58); p value$_{heterogeneity}$,–0.462], homozygous genetic model [CC vs. AA: OR,0.74 (95% Cl,0.17–3.35); p value$_{heterogeneity}$,–0.700] and heterozygous genetic model [AC vs. AA: OR,0.77 (95% Cl,0.28–2.09); p value$_{heterogeneity}$,–0.610]. The same meta-analysis reported no significant association of MMP8 gene rs35866072 (A/C; Lys460Thr) SNP with the risk of breast cancer, leukemia, lung cancer and oral cancer (four studies: 2,644 cases; 3,169 controls) in all genetic models: dominant genetic model [AC+CC vs. AA: OR,1.02 (95% Cl,0.68–1.54); p value$_{heterogeneity}$,0.886; p value,0.919], recessive genetic model [CCvs. AC+AA: OR,0.87 (95% Cl,0.32–2.42); p value$_{heterogeneity}$, –0.796], allelic genetic model [A vs. C: OR,1.00 (95% Cl,0.68–1.46); p value$_{heterogeneity}$,0.896; p value,0.462], homozygous genetic model [CC vs. AA: OR,0.88 (95% Cl,0.32–2.42); p value$_{heterogeneity}$,–0.799] and heterozygous genetic model [AC vs. AA: OR,1.05 (95% Cl,0.67–1.65); p value$_{heterogeneity}$,0.868; p value,0.823] (Zhang, Zhu et al. 2019).

The meta-analysis reported no significant association of MMP8 gene rs1940475 (+259 A/G; Lys87Glu) SNP (three studies: 749 cases; 1,919 controls) with an overall risk of cancer among Caucasian population in all genetic models: dominant genetic model [AG+GG vs. AA: OR,1.01 (95% Cl,0.83–1.22); p value$_{heterogeneity}$,0.863; p value,0.928], recessive genetic model [GGvs. AG+AA: OR,1.13 (95% Cl,0.93–1.38); p value$_{heterogeneity}$,0.604; p value,0.222], allelic genetic model [A vs. G: OR,1.05 (95% Cl,0.93–1.18); p value$_{heterogeneity}$,0.968; p value,0.430], homozygous genetic model [GG vs. AA: OR,1.11 (95% Cl,0.88–1.42); p value$_{heterogeneity}$,0.953; p value,0.377] and heterozygous genetic model [AG vs. AA: OR,0.96 (95% Cl,0.78-1.18); p value$_{heterogeneity}$,0.647; p value,0.722]. The same meta-analysis reported no significant association of MMP8 gene rs35866072 (A/C; Lys460Thr) SNP with the risk of bladder cancer (two studies: 543 cases; 1,092 controls) in all genetic models: dominant genetic model [AG+GG vs. AA: OR,0.98 (95% Cl,0.78–1.23); p value$_{heterogeneity}$,0.745; p value,0.875], recessive genetic model [GGvs. AG+AA: OR,1.20 (95% Cl,0.94–1.53); p value$_{heterogeneity}$,0.508; p value,0.152], allelic genetic model [A vs. G: OR,1.06 (95% Cl,0.91–1.22); p value$_{heterogeneity}$,0.841; p value,0.458], homozygous genetic model [GG vs. AA: OR,1.13 (95% Cl,0.85–1.51); p value$_{heterogeneity}$,0.810; p value,0.398] and heterozygous genetic model [AG vs. AA: OR,0.92 (95% Cl,0.72–1.17); p value$_{heterogeneity}$,0.544; p value,0.496]. The same meta-analysis reported no significant association of MMP8 gene rs35866072 (A/C; Lys460Thr) SNP with the risk of skin cancer (one study: 206 cases; 827 controls) in all genetic models: dominant genetic model [AG+GG vs. AA: OR,1.08 (95% Cl,0.75–1.56); p value$_{heterogeneity}$,–0.678], recessive genetic model [GGvs. AG+AA: OR,1.02 (95% Cl,0.72–1.43); p value$_{heterogeneity}$,–0.921], allelic genetic model [A vs. G: OR,1.03 (95% Cl,0.83–1.28); p value$_{heterogeneity}$,–0.756], homozygous genetic model [GG vs. AA: OR,1.07 (95% Cl,0.70–1.66); p value$_{heterogeneity}$,–0.745] and heterozygous genetic model [AG vs. AA: OR,1.08 (95% Cl,0.74–1.59); p value$_{heterogeneity}$,–0.684] (Zhang, Zhu et al. 2019).

A hospital-based case-control study on Spanish population, which included 501lung cancer cases and 501 healthy matched controls, studied the association of MMP8 gene rs2155052 (+17C/G) SNP with the risk of lung cancer. The study reported a significant association of this SNP with a decreased risk of lung cancer in dominant genetic model [CG+GG vs. CC: OR,0.65 (95% Cl,0.45–0.93); p value$_{trend}$,0.019; p value,0.019]. On subgroup analysis, the study reported a decreased risk of lung cancer in dominant genetic model among males [CG+GG vs. CC: OR,0.63 (95% Cl,0.43–0.93); p value$_{trend}$,0.021], individuals with positivefamily lung cancer history [CG+GG vs. CC: OR,0.14 (95% Cl,0.03–0.64); p value$_{trend}$,0.011] and ever-smokers [CG+GG vs. CC: OR,0.69 (95% Cl,0.48–0.97); p value$_{trend}$,0.034]. The study reported a significant association of this SNP with a decreased risk of squamous cell carcinoma [CG+GG vs. CC: OR,0.51 (95% Cl,0.31–0.84); p value$_{trend}$,0.008] and small cell carcinoma [CG+GG vs. CC: OR,0.43 (95% Cl,0.20–0.89); p value$_{trend}$,0.023] (González-Arriaga, López-Cima et al. 2008).

Another case-control study on Taiwanese population,which included 358 lung cancer cases and 716 healthy matched controls, studied the association of MMP8 gene rs11225395 (-799 C/T) SNP, MMP8 gene rs35866072 (A/C; Lys460Thr) SNP and MMP8 gene rs34009635 (T/C; Val436Ala) SNP. The study reported no significant association of MMP8 gene rs11225395 (-799 C/T) SNP with the risk of lung cancer in dominant genetic model [CT+TT vs. CC: OR,0.83 (95% Cl,0.59–1.17); p value$_{trend}$,0.5130; p value,0.2807], allelic genetic model [T vs. C: OR,0.84 (95% Cl,0.65–1.38); p value$_{trend}$,0.5130; p value,0.2225], homozygous genetic model [TT vs. CC: OR,0.79(95% Cl,0.51–1.24); p value$_{trend}$,0.5130;

p value,0.3198] and heterozygous genetic model [CT vs. CC: OR,0.87 (95% Cl,0.68–1.23); p $value_{trend}$,0.5130; p value,0.3999]. The study also reported no significant association of MMP8 gene rs35866072 (A/C; Lys460Thr) SNP with the risk of lung cancer in heterozygous genetic model [AC vs. AA: OR,1.53 (95% Cl,0.38–6.81); p value,0.5931] and allelic genetic model [A vs. C: OR,1.54 (95% Cl,0. 49–7.45); p value,0.5924]. The study reported null presence of variant genotype of MMP8 gene rs34009635 (T/C; Val436Ala) SNP in the study population (Shen, Hsia et al. 2017). A case-control study on Taiwanese population, which included 375 bladder cancer cases and 375 healthy matched controls, studied the association of MMP8 gene rs11225395 (-799 C/T) SNP, MMP8 gene rs35866072 (A/C; Lys460Thr) SNP and MMP8 gene rs34009635 (T/C; Val436Ala) SNP. The study reported no significant association of MMP8 gene rs11225395 (-799 C/T) SNP with the risk of bladder cancer in dominant genetic model [CT+TT vs. CC: OR,1.11 (95% Cl,0.88–1.38); p $value_{trend}$,0.6629; p value,0.4217], allelic genetic model [T vs. C: OR,1.03 (95% Cl,0.82–1.17); p $value_{trend}$,0.6629; p value,0.5712], homozygous genetic model [TT vs. CC: OR,1.10 (95% Cl,0.87–1.52); p $value_{trend}$,0.6629; p value,0.9030] and heterozygous genetic model [CT vs. CC: OR,1.13 (95% Cl,0.89–1.44); p $value_{trend}$,0.6629; p value,0.3688]. The study also reported no significant association of MMP8 gene rs35866072 (A/C; Lys460Thr) SNP with the risk of lung cancer in dominant genetic model [AC+CC vs. AA: OR,0.78 (95% Cl,0.36-2.09); p $value_{trend}$,0.8166; p value,0.5252], allelic genetic model [A vs. C: OR,0.77 (95% Cl,0.38–1.46); p $value_{trend}$,0.8166; p value,0.4607], homozygous genetic model [CC vs. AA: OR,1.00 (95% Cl,0.21–4.73); p $value_{trend}$,0.8166; p value,0. 7247] and heterozygous genetic model [AC vs. AA: OR,0.71 (95% Cl,0.31–2.28); p $value_{trend}$,0.8166; p value,0.6094]. The study reported null presence of variant genotype of MMP8 gene rs34009635 (T/C; Val436Ala) SNP in the study population (Tsai, Wang et al. 2018).

A case-control study on Han Chinese population, which included 551 colorectal cancer (CRC) cases and 623 healthy matched controls, studied the association of MMP8 gene rs11225395 (-799 C/T) SNP. The study reported a significant association of MMP8 gene rs11225395 (-799 C/T) SNP with an increased risk of colorectal cancer in dominant genetic model [CT+TT vs. CC: OR,1.31 (95% Cl,1.04–1.66); p value,0.021], recessive genetic model [TT vs. CT+CC: OR,1.61 (95% Cl,1. 00–2.58); p value,0.048], allelic genetic model [T vs. C: OR,1.29 (95% Cl,1.07–1.56); p value,0.007], homozygous genetic model [TT vs. CC: OR,1.76 (95% Cl,1.09-2.85); p value,0.022], but not inheterozygous genetic model [CT vs. CC: OR,1.25 (95% Cl,0.98–1.59); p value,0.073]. On subgroup analysis, the study reported a significant association of this SNP with an increased risk of CRC among males in dominant genetic model [CT+TT vs. CC: OR,1.35 (95% Cl,1.01–1.82); p value,0.046], recessive genetic model [TT vs. CT+CC: OR,1.98 (95% Cl,1.01–3.87); p value,0.047], homozygous genetic model [TT vs. CC: OR,2.19 (95% Cl,1.10–4.34); p value,0.025], but not in heterozygous genetic model [CT vs. CC: OR,1.27 (95% Cl,0.93–1.73); p value,0.129]. The study also reported a significant association of this SNP with an increased risk of CRC among smokers in dominant genetic model [CT+TT vs. CC: OR,1.37 (95% Cl,1.00–1.87); p value,0.050], recessive genetic model [TT vs. CT+CC: OR,2.09 (95% Cl,1.02–4.27); p value,0.040], homozygous genetic model [TT vs. CC: OR,2.31(95% Cl,1.12–4.79); p value,0.024], but not in heterozygous genetic model [CT vs. CC: OR,1.14 (95% Cl,0.83–1.58); p value,0.426]. Further, the study reported a significant association of this SNP with an increased risk of CRC among individuals with age ≥60 years in dominant genetic model [CT+TT vs. CC: OR,1.44 (95% Cl,1.05–1.98); p value,0.025], recessive genetic model [TT vs. CT+CC: OR,2.46 (95% Cl,1.22–4.97); p value,0.012], homozygous genetic model [TT vs. CC: OR,2.72(95% Cl,1.33–5.55); p value,0.006], but not in heterozygous genetic model [CT vs. CC: OR,1.30 (95% Cl,0.93–1.81); p value,0.122] (Tai, Sun et al. 2020).

6.5.4 Polymorphisms in Matrix Metalloproteinase13 (MMP13) Gene

Matrix metalloproteinase 13 (MMP13) gene (chromosomal location, 11q22.2) encodes a 471 amino acid 53,820 Da (54 kDa) MMP13 pro-enzyme, also known as collagenase3. The active form of MMP13 enzyme generated through autocatalysis (autoproteolysis) or by removal of pro-peptide domain is a 48 kDa protein, inhibitable by TIMP1, TIMP2, TIMP3 and through internalization and subsequent proteolysis by fibroblasts and osteoblasts, and this involves MMP13 binding to specific surface receptors present in these cells (Knäuper, López-Otin et al. 1996, Barmina, Walling et al. 1999, Leeman, Curran et al. 2002). Under normal non-pathological conditions, MMP13 expression mainly occurs during the development of bones in human fetus in hypertrophic cartilage chondrocytes, osteoblasts and periosteal cells, but not in osteoclasts and other fetal tissues such as neural tissue, muscle, skin, liver and lungs. The adult tissues under normal non-pathological conditions show little or no expression of MMP13 (Johansson, Saarialho-Kere et al. 1997). A wide variety of cellular mediators play an important role in regulating the expression of MMP13 gene. These include cytokines such as TNF-α, interleukin 1(IL1) and interleukin 6 (IL6); growth factors such as transforming growth factor-β1 (TGF-β1), platelet-derived growth factor (PDGF), insulin-like growth factors (IGF I and IGF II) and basic fibroblast growth factor (bFGF or FGF-β) and hormones, in particular, parathyroid hormone (PTH) (Leeman, Curran et al. 2002).

MMP13 predominantly and more efficiently acts on and cleaves interstitial fibrillar type II collagen,the main collagen constituent of cartilage that makes it suitable for its role in the development and remodeling of bones. The other collagenases, MMP1 and MMP8, in contrast preferentially cleave interstitial fibrillar type III and type I collagens, respectively. MMP13, in contrast to MMP1 and MMP8, also acts with high specificity and efficiency on gelatins,which extends the proteolytic role of MMP13 beyond the first-step single-site cleave of triple helix of interstitial fibrillar collagens. The other substrates of MMP13 include type I, type III, type IV, type IX, type X and type XIV collagens, fibronectin, aggrecan, perlecanand biglycan. It also includes connective tissue growth factor (CTGF), TGF-β, fibrinogen, tenascin C (TNC), fibrillin-1, fibrillin-2 and serpins such as plasminogen activator inhibitor-2 (PAI2 or SerpinB2) and α1-antichymotrypsin (α_1AT) (Table 6.2) (Knäuper, López-Otin et al. 1996, Billinghurst, Dahlberg et al. 1997, Leeman, Curran et al. 2002). MMP13 catalyzes cleavage of interstitial fibrillar type II collagens during intramembranous and endochondral ossification, the two important processes involved in the development and remodeling of bones in human fetus (Johansson, Saarialho-Kere et al. 1997). MMP13, through proteolytic cleavage of fibrinogen and factor XII (Hageman factor), plays a role in the regulation of blood coagulation. MMP13 can trigger a MMP-mediated degradation cascade through activation of several members of MMP family (Billinghurst, Dahlberg et al. 1997, Barmina, Walling et al. 1999, Hiller, Lichte et al. 2000).

MMP13, predominantly through its role in the proteolysis of various ECM and BM components, plays a vital role in various pathophysiological processes such as neoplastic diseases (cancers), cardiovascular disorders such as aneurysm and atherosclerosis, nervous system diseases, muscular, skeletal and bone pathologies such as arthritis and digestive system diseases. MMP13 mediates several important processes involved in the pathogenesis of various cancers such as breast cancer, gastric cancer, chondrosarcomas, fibrosarcoma, squamous cell cancers and basal cell cancers of head and neck (glioblastoma, larynx and vulval cancers) and skin epithelium (melanoma). MMP13, mainly through its role in the proteolysis of various ECM and BM components, promotes tumor growth and invasiveness. Overexpression and increase in MMP13 activity correlates significantly with increased invasiveness, aggressiveness and metastasis in these cancers (Airola, Johansson et al. 1997, Johansson, Airola et al. 1997, Cazorla, Hernández et al. 1998, Uría, Balbín et al. 1998, Johansson, Vaalamo et al. 1999, Ala-Aho, Johansson et al. 2002, Elnemr, Yonemura et al. 2003, Hattori, Nerusu et al. 2003, Zigrino, Kuhn et al. 2009). The inhibition or promotion of tumorigenesis by cellular mediators including factors such as endostatin (a 20 kDa fragment of type XVIII collagen), hypoxia-induced factor-1α (HIF-1α) and relaxin involves the role of MMP13. A study reports that endostatin mediates its inhibitory role in tongue carcinoma by inhibiting the activation and proteolytic function of MMPs such as MMP2, MMP9 and MMP13. Another study shows that hypoxia-induced factor-1α (HIF-1α) promotes tumor growth, malignant progression and invasion by inducing the expression of various genes including MMPs such as MMP13. The relaxin mediated promotion of breast cancer invasiveness involvesupregulation of MMP13 expression (Koong, Denko et al. 2000, Binder, Hagemann et al. 2002, Nyberg, Heikkilä et al. 2003). The expression and activity of MMP13 correlates significantly with the invasive and metastatic potential of cancer stem cells in glioblastoma. MMP13, one of the secreted and highly expressed proteins in breast cancer, also plays a vital role in metastasis of bone in breast cancer. Further, MMP13 expression and activity is a potential new diagnostic, prognostic and predictive marker in breast cancer (Chang, Yang et al. 2009, Inoue, Takahashi et al. 2010).

The functional SNPs in MMP13 gene reported to modulate the MMP13 gene transcription, expression and subsequent MMP13 activity in an allele-specific manner include MMP13 gene rs11225490 T/C SNP, MMP13 gene rs2252070 G/A SNP, MMP13 gene rs17099788 A/G SNP and MMP13 gene rs3758854 G/A SNP. Of these SNPs, MMP13 gene rs2252070 (-77 A/G) SNP located in MMP13 gene promoter region is the most extensively studied SNP. MMP13 gene rs2252070 (-77 A/G) SNP possibly modulates the transcription and expression of MMP13 gene in an allele-specific manner through its modifying effect on the binding affinities and interactions of various nuclear proteins including transcription factors, in particular, specificity protein 1 (Sp1) with the MMP13 gene promoter. MMP13 gene rs2252070 (-77 A/G) SNP, located in the promoter region of MMP13 gene within the potential binding site of transcription factor, specificity protein 1 (Sp1), possibly modulates the affinity of Sp1 to bind to its specific binding sequence in the MMP13 promoter region. The electrophoretic mobility shift assay (EMSA)-based study using biotin-labeled oligonucleotide 'G' allele and 'A' allele probes reports the binding ofspecificity protein 1 to the G allele of MMP13 gene rs2252070 (-77 A/G) SNP, but not to the A allele of this SNP. The luciferase reporter assay using two different allelic constructs based on MMP13 gene rs2252070G allele and MMP13 gene rs2252070A allele reportsa significantly higher activity of luciferase in case of rs2252070G allele in comparison with rs2252070A allele (*p* value, <0.01). These findings indicate that the Sp1 transcription factor binds specifically or more strongly to the MMP13 gene rs2252070G allele in comparison with the MMP13 gene rs2252070A allele that enhances transcription, leads to increase in MMP13 gene expression and subsequently an increase in MMP13 enzyme activity in case of G allele in comparison with A allele (Shi, Xia et al. 2016).

Various association studies report on the possible role of MMP13 gene rs2252070 (-77 A/G) SNP in modulating the risk for the development of various cancers. However, many studies report no such association. Here, we mention several of these studiesthat report the association of this SNP with the risk of various cancersand the studies that report no significant association. A meta-analysis based on three association studies on Asian and Caucasian populations, which included 1,221 cases and 1,225 controls, reported a significant association of MMP13 gene rs2252070 (-77 A/G) SNP with a decreased risk of lung cancer among Asians in heterozygous genetic model [AG vs. AA: OR,0.67 (95% Cl,0.47–0.96); I^2=45.3; *p* value,0.029], but not in other genetic models:dominant genetic model [AG+GG vs. AA: OR,0.70 (95% Cl,0.48–1.01); I^2=55.3; *p* value,>0.05], recessive genetic model [GG vs. AG+AA: OR,1.01 (95% Cl,0.80–1.27); I^2=0.0; *p* value,> 0.05], homozygous genetic model [GG vs. AA: OR,0.76 (95% Cl,0.50–1.16); I^2=49.5; *p* value,>0.05] and allelic genetic model [G vs. A: OR,0.88 (95% Cl,0.72–1.08); I^2= 46.7; *p* value,> 0.05]. The study reported no significant association of this SNP among Caucasian population in all genetic models: dominant genetic model [AG+GG vs. AA: OR,1.14 (95% Cl,0.89–1.46); *p* value,>0.05], recessive genetic model [GG vs. AG+AA: OR,1.09 (95% Cl,0.70–1.69); *p* value,> 0.05], allelic genetic model [G vs. A: OR,1.10 (95% Cl,0.91–1.34); I^2= 46.7; *p* value,> 0.05], homozygous genetic model [GG vs. AA: OR,1.15 (95% Cl,0.73–1.82); *p* value,> 0.05] and heterozygous genetic model [AG vs. AA: OR,1.14 (95% Cl,0.88–1.47); *p*value,0.029]. The study reported no significant association of this SNP in total population in all genetic models: dominant genetic model [AG+GG vs. AA: OR,0.83 (95% Cl,0.56–1.23); I^2=78.3; *p* value,>0.05], recessive genetic model [GG vs. AG+AA: OR,1.02 (95% Cl,0.83–1.26); I^2=0.0; *p* value,> 0.05], allelic genetic model [G vs. A: OR,0.95 (95% Cl,0.79–1.14); I^2= 59.4; *p* value,> 0.05], homozygous genetic model [GG vs. AA: OR,0.87 (95% Cl,0.61–1.24); *p* value,> 0.05] andheterozygous genetic model [AG vs. AA: OR,0.81 (95% Cl,0. 54–1.22); I^2=78.3; *p* value,> 0.05] (Li, Liu et al. 2020).

A hospital-based case-control study on Polish population, which included 648 unselected cancer cases (299 breast cancer cases, 199 lung cancer cases and 150 colon cancer cases) and 648 matched controls (299 controls for breast cancer, 199 controls for lung cancer and 150 controls for colon cancer) and 648 controls, studied the association of MMP13 gene rs2252070 (-77 A/G) SNP along serum zinc levels with the risk of breast cancer, lung cancer and colon cancer.The study reported no significant association of this SNP with the risk of breast cancer [GG vs. AA: OR,0.84 (95% Cl,0. 44–1.57); *p* value,0.58; AG vs. AA: OR,1.12 (95% Cl,0.79–1.59); *p* value,0.53], lung cancer [GG vs. AA: OR,1.26 (95% Cl,0.61–2.62); *p* value,0.53; AG vs. AA: OR,1.38 (95% Cl,0. 89–2.14); *p* value,0.15] and colon cancer [GG vs. AA: OR,0.95 (95% Cl,0.34–2.64); *p* value,0.92;AG vs. AA: OR,0.78 (95% Cl,0.50–1.24); *p* value,0.30] (Białkowska, Marciniak et al. 2020). Another hospital-based case-control study on Polish population, which included 197 prostate cancer cases and 197 healthy matched controls, reported no significant association of this SNP with prostate cancer [GG vs. AA: OR,1.39 (95% Cl,0.64–3.02); *p* value,0.40; AG vs. AA: OR,1.16 (95% Cl,0. 74–1.81); *p* value,0.51] (Białkowska, Marciniak et al. 2018). In contrast, a study on Mexican population, which included 102 colorectal cancer cases and 125 healthy matched controls, reported a significant association of AA and AG genotypes of MMP13 gene rs2252070 (-77 A/G) SNP with an increased risk of colorectal cancer[AA vs. GG: OR,3.2 (95% Cl,1.004–10.2); *p* value,0.04; AG vs. GG: OR,4.08 (95% Cl,1.3–13.07); *p* value,0.01; AG+AA vs. GG: OR,3.6 (95% Cl,1.16–11.12); *p* value,0.02;] (Moreno-Ortiz, Gutiérrez-Angulo et al. 2014).

A study on Han Chinese populations, Jiangsu and Shandong, which included 1588 esophageal squamous cell carcinoma (ESCC) cases and 1,600 controls and adjusted the ORs for age, sex, smoking and alcohol drinking, studied four haplotype tagging single nucleotide polymorphisms (htSNPs), rs11225490 T/C SNP, rs2252070 G/A SNP, rs17099788 A/G SNP and rs3758854 G/A SNP, in MMP13 gene. The study reported an overall significant association of MMP13 gene rs2252070 (-77 A/G) SNP with a decreased risk of ESCC in homozygous genetic model [AA vs. GG: OR,0.73 (95% Cl,0.66–0.81); *p* value,1.8×10^{-6}], heterozygous genetic model [GA vs. GG: OR,0.63 (95% Cl,0. 54–0.74); *p* value, 1.7×10^{-6}] and allelic genetic model [A vs. G: OR,0.70 (95% Cl,0.60–0.83); *p* value,2.1×10^{-5}]. In subgroup analysis, the study reported a significant association of this SNP with a decreased risk of ESCC among Jiangsu subpopulation in homozygous genetic model [AA vs. GG: OR,0.79 (95% Cl,0.65–0.95); *p* value,0.011] and heterozygous genetic model [GA vs. GG: OR,0.65 (95% Cl,0.49–0.88); *p* value,0.004]. The study reported a significant association of this SNP with a decreased risk of ESCC among Shandong subpopulation in homozygous genetic model [AA vs. GG: OR,0.75 (95% Cl,0.65–0.86); *p* value,4.9×10^{-5}] and heterozygous genetic model [GA vs. GG: OR,0.65 (95% Cl,0.53–0.81); *p* value,8.8×10^{-5}]. However, the study reported no significant association of MMP13 gene rs11225490 T/C SNP (allelic genetic model [A vs. G: OR,0.89 (95% Cl,0.74–1.08); *p* value,0.251]), MMP13 gene rs17099788 A/G SNP (allelic genetic model [A vs. G: OR,1.11 (95% Cl,0.87–1.41); *p* value,0.393]) and MMP13 gene rs3758854 G/A SNP (allelic genetic model [A vs. G: OR,1.09 (95% Cl,0.82–1.44); *p* value,0.569]) with a risk of esophageal squamous cell carcinoma (Shi, Xia et al. 2016). Another hospital-based case-control study on Spanish population, which included 501 lung cancer cases and 510 healthy matched controls, reported no significant association of MMP13 gene rs2252070 (-77 A/G) SNP with lung cancer in homozygous genetic model [GG vs. AA: OR,1.23 (95% Cl,0.72–2.11); *p* value,0.452], heterozygous genetic model [AG vs. AA: OR,1.24 (95% Cl,0.91–1.69); *p* value,0.174] and dominant genetic model [AG+GG vs. AA: OR,1.24 (95% Cl,0.92–1.66); *p* value,0.155]. However,on subgroup analysis based on histological cancer type, the study reported a significant association of this SNP with an increased risk of small cell lung carcinoma in homozygous genetic model [GG vs. AA: OR,2.38 (95% Cl,1. 01–5.65); *p* value,0.048] and heterozygous genetic model [AG vs. AA: OR,1.94 (95% Cl,1.09–3.45); *p* value,0.024] (González-Arriaga, López-Cima et al. 2008). Another hospital-based case-control study on ethnic Han Chinese population, which included 245 non-small cell lung carcinoma (NSCLC) cases and 258 healthy matched controls, reported a significant association of this SNP with a decreased risk of non-small cell lung carcinoma (NSCLC) in homozygous genetic model [AA vs. GG: OR,0.56 (95% Cl,0.33–0.94); *p* value,0.03] and in allelic genetic model [A vs. G: OR,0.70 (95% Cl,0.54–0.91); *p* value,0.00] (Li, Jia et al. 2019).

6.5.5 Polymorphisms in Matrix Metalloproteinase3 (MMP3) Gene

Matrix metalloproteinase3 (MMP3) gene (chromosomal location, 11q22.2) encodes a 477 amino acid 53,977Da (54kDa) MMP3 pro-enzyme, also known as stromelysin-1, transin-1, proteoglycanase or progelatinase. The active form of MMP3 enzyme generated through proteolytic cleavage of amino (N)-terminal pro-domain by serine proteases such as plasmin, cathepsin G, neutrophil elastase, matriptase, chymotrypsin and trypsin2 is a 45kDa protein, inhibitable by TIMPs (Moilanen, Sorsa et al. 2003, Visse and Nagase 2003, Jin, Yagi et al. 2006). Various types of cells that express MMP3 include fibroblasts, keratinocytes and chondrocytes. MMP3 enzyme secretion by epithelial cells and fibroblasts often occurs in combination with MMP10, another member of stromelysins subfamily (Sternlicht, Lochter et al. 1999, Visse and Nagase 2003). The substrates of MMP3 enzyme include type III, type IV, type V, type IX, type X and type XIcollagens, gelatins, laminins, elastin, vitronectin, fibronectin, fibrillins, fibrin, fibulin, proteoglycans, osteonectin, tenascin C (TNC) and myelin basic protein (MBP) (Table 6.2). MMP3 cleaves and inactivates α_1-proteinase inhibitor (α_1-PI) and interleukin1β (IL1β). MMP3 enzyme also acts on various membrane-bound or cell surface molecules and releases their soluble and often activated or mature forms through ectodomain shedding. These include TNF-α, TGF-α, TGF-β, heparin-binding EGF-like growth factor (HB-EGF), L-selectin and E-cadherin. MMP3 differs catalytically from collagenases in its ability to cleave type I collagen. MMP3 enzyme-catalyzed proteolysis is crucial for the activation of other MMPs including proMMP1, proMMP7, proMMP8, proMMP9 and proMMP13 (Shapiro, Fliszar et al. 1995, Sternlicht, Lochter et al. 1999, Visse and Nagase 2003, Steenport, Khan et al. 2009). Interleukin1β (IL1β) activity causes MMP3 gene upregulation, whereas dexamethasone and retinoic acid cause MMP3 gene downregulation. The activity of steroid sex hormone progesterone causes MMP3 gene downregulation and inhibits MMP3 enzyme secretion (Osteen, Igarashi et al. 2003).

MMP3, predominantly through its role in the proteolysis of various ECM and BM components and non-ECM substrates, plays a vital role in various physiological and pathophysiological processes. MMP3 enzyme is involved in tissue turnover and remodeling during mammary gland involution, and studies report mammary tissue upregulation of MMP3 gene and increased MMP3 enzyme activity during post-lactation mammary gland involution (Witty, Wright et al. 1995). MMP3 enzyme plays a vital role in various processes involved in wound healing including acute inflammation, parenchyma cell regeneration, cellular proliferation, cellular migration, angiogenesis, wound contraction, tissue turnover and remodeling and, therefore, has a direct role in wound healing and its impairment (Fray, Dickinson et al. 2003, Gill and Parks 2008). MMP3 is also involved in the pathogenesis of various diseases or disorders. These include nervous system diseases such as Parkinson's disease (PD) and Alzheimer's disease; cardiovascular disorders such as aortic aneurysms (AAA) and atherosclerosis; respiratory system disorders such as acute respiratory distress syndrome (ARDS) and asthma; and muscular, skeletal and bone disorders such as rheumatoid arthritis (RA) and osteoarthritis (OA) (Henney, Wakeley et al. 1991, Dahlen, Shute et al. 1999, Yoshihara, Nakamura et al. 2000, Yoshiyama, Asahina et al. 2000, Warner, Beltran et al. 2001, Choke, Cockerill et al. 2005, Choi, Kim et al. 2011).

MMP3 enzyme seems to play a critical role in several important processes involved in the pathogenesis of various cancers, particularly tumor initiation mainly through the activation of other proMMPs, in particular, proMMP9. The increased expression and activity of MMP3 enzyme correlates positively with the activity and distribution of other MMPs, in particular MMP9, in solid tumors. A study shows that MMP3 enzyme promotes the transformation of mammary epithelial cells into invasive mesenchymal cells characteristic of tumors (Thomasset, Lochter et al. 1998). Another study also reports the pro-tumorigenic role of MMP3 enzyme in the initiation of mammary gland tumors, but reports no correlation between MMP3 gene expression and MMP3 enzyme activity in other stages of tumor progression (Sternlicht, Bissell et al. 2000).

MMP3 enzyme also modulates the bioavailability of cytokines and growth factors, which directly affects tumor growth and progression. For example, MMP3 through activation of TGF-β promotes epithelialtomesenchymal transition (EMT), an important process involved in tumor cell migration and tumor invasion (Blavier, Lazaryev et al. 2010). These studies along with several others indicate a pro-tumorigenic role for MMP3 enzyme (Wiesen and Werb 1996, Mercapide, Lopez De Cicco et al. 2003). However, studies also report anti-tumorigenic role of MMP3 enzyme in various cancers including skin cancer (McCawley, Crawford et al. 2004). Further, MMP3 enzyme possibly through its action on apoptosis-related substrates such as Fas ligand (FasL) and TNF-α promotes apoptosis of cancer cells (Matsuno, Yudoh et al. 2001, Wielockx, Lannoy et al. 2001). Therefore, the direct role of MMP3 enzyme in the pathogenesis of various cancers is complex and not fully understood.

The functional SNPs in MMP3 gene reported to modulate MMP3 gene transcription, expression and subsequently MMP3 enzyme activity in an allele-specific manner most prominently includes MMP3 gene rs35068180/rs3025058 (-1171 5A/6A,-1171 ins/delA, formerly -1612 5A/6A) SNP. The MMP3 gene rs35068180/rs3025058 (-1171 5A/6A) SNP is an insertion/deletion polymorphism, located within the promoter region of MMP3 gene that involves insertion or deletion of adenosine (A) nucleotide at base pair position -1171. One of the alleles of this SNP designated as MMP3 gene rs35068180 6A allele consists of a stretch of six adenosine nucleotides (6A), whereas the other allele of this SNP designated as MMP3 gene rs35068180 5A allele consists of a stretch of five adenosine nucleotides (5A). The MMP3 gene rs35068180/rs3025058 (-1171 5A/6A) SNP modulates the binding of a putative repressor protein in such a way that this repressor protein binds preferentially to -1171 6A allele in comparison with-1171 5A allele. The binding of this repressor protein decreases MMP3 gene promoter activity and MMP3 gene expression, which leads to an almost 50% decrease in MMP3 enzyme activity. Therefore, the occurrence of MMP3 gene rs35068180 5A allele results in 50% higher MMP3 gene transcription, protein expression and activity of MMP3 enzyme in comparison with MMP3 gene rs35068180 6A allele. The increased expression and activity of MMP3 enzyme leads to enhanced proteolysis of ECM and BM components and non-ECM substrates, which possibly modulates the susceptibility to various diseases including cancer (Ye, Eriksson et al. 1996, Humphries, Luong et al. 1998, Medley, Kingwell et al. 2003).

A number of association studies report on the possible role of MMP3 gene rs35068180/rs3025058 (-1171 5A/6A) SNP in modulating the risk for the development of various cancers. However, many studies report no such association. Here, we mention several of these studiesthat report the association of this SNP with the risk of various cancersand the studies that report no significant association. A meta-analysis based on three association studies, which included 2,430 cases and 2,023 controls, reported no significant association of MMP3 gene rs35068180 (-1171 5A/6A)SNP with the risk of lung cancer in both the dominant genetic model [5A/6A+6A/6A vs. 5A/5A: OR,1.03 (95% Cl,0.89–1.2); p value$_{heterogeneity,}$0.81; p value 0.66] and the recessive genetic model [6A/6A vs. 5A/6A+5A/5A: OR,0.97 (95% Cl,0.84–1.12); p value$_{heterogeneity,}$ 0.5; p value,0.71]. The same meta-analysis based on four association studies, which included 473 cases and 803 controls indominant genetic model, and three studies, which included 413 cases and 639 controls, reported no significant association of MMP3 gene rs35068180 (-1171 5A/6A)SNP with the risk of colorectal cancer in both the dominant genetic model [5A/6A+6A/6A vs. 5A/5A: OR,0.93 (95% Cl,0.50–1.71); p value$_{heterogeneity}$,0.15; p value,0.81] and the recessive genetic model [6A/6A vs. 5A/6A+5A/5A: OR,1.29 (95% Cl,0.66–2.51); p value$_{heterogeneity}$,0.005; p value,0.45]. The same meta-analysis based on two association studies, which included 586 cases and 603 controls, reported no significant association of MMP3 gene rs35068180 (-1171 5A/6A) SNP with the risk of breast cancer in both the dominant genetic model [5A/6A+6A/6A vs. 5A/5A: OR,0.94 (95% Cl,0.53–1.65); p value$_{heterogeneity}$,0.11; p value,0.83] andthe recessive genetic model [6A/6A vs. 5A/6A+5A/5A: OR,0.71 (95% Cl,0.38–1.33); p value$_{heterogeneity}$,0.08; p value,0.29] (McColgan and Sharma 2009). Another meta-analysis based on three association studies on Asian and Caucasian populations, which included 2,973 cases and 2,207 controls, reported no significant association of MMP3 gene rs35068180 (-1171 5A/6A) SNP with the risk of lung cancer in all genetic models among Asian population: dominant genetic model [5A/6A+6A/6A vs. 5A/5A: OR,0.79 (95% Cl,0.28–2.20)], recessive genetic model [6A/6A vs. 5A/6A+5A/5A: OR,0.97 (95% Cl,0.69–1.38)], allelic genetic model [6A vs. 5A: OR,0.96 (95% Cl,0.71–1.30)], homozygous genetic model [6A/6A vs. 5A/5A: OR,0.79 (95% Cl,0.28–2.21)] and heterozygous genetic model [5A/6A vs. 5A/5A: OR,0.79 (95% Cl,0.28–2.29)]. The study reported no significant association of MMP3 gene rs35068180 (-1171 5A/6A) SNP with the risk of lung cancer in all genetic models among Caucasian population in dominant genetic model [5A/6A+6A/6A vs. 5A/5A: OR,1.01 (95% Cl,0.88–1.16); I^2=0.0], recessive genetic model [6A/6A vs. 5A/6A+5A/5A: OR,0.97 (95% Cl,0.85–1.11); I^2=0.0], allelic genetic model [6A vs. 5A: OR,0.99 (95% Cl,0.91–1.08; I^2=0.0)], homozygous genetic model [6A/6A vs. 5A/5A: OR,0.98 (95% Cl,0.83–1.16); I^2=0.0] and heterozygous genetic model [5A/6A vs. 5A/5A: OR,1.02 (95% Cl,0.88–1.19); I^2=0.0]. The study reported no significant association of MMP3 gene rs35068180 (-1171 5A/6A)SNP with the risk of lung cancer in all genetic models in total population in dominant genetic model [5A/6A+6A/6A vs. 5A/5A: OR,1.01 (95% Cl,0.88–1.15); I^2=0.0], recessive genetic model [6A/6A vs. 5A/6A+5A/5A: OR,0.97 (95% Cl,0.86–1.10); I^2=0.0], allelic genetic model [6A vs. 5A: OR,0.99 (95% Cl,0.91–1.07; I^2=0.0)], homozygous genetic model [6A/6A vs. 5A/5A: OR,0.98 (95% Cl,0.83–1.15); I^2=0.0] and heterozygous genetic model [5A/6A vs. 5A/5A: OR,1.02 (95% Cl,0.88–1.18); I^2=0.0] (Li, Liu et al. 2020). Another meta-analysis based on four association studies reported no significant association of MMP3 gene rs35068180 (-1171 5A/6A)SNP with the risk of lung cancer among Asian population in both the genetic models: dominant genetic model [5A/6A+6A/6A vs. 5A/5A: OR,0.79 (95% Cl,0.28–2.20); p value,0.65] and recessive genetic model [6A/6A vs. 5A/6A+5A/5A: OR,0.97 (95% Cl,0.69–1.38); p value,0.16]. The study reported nosignificant association of MMP3 gene rs35068180 (-1171 5A/6A)SNP with the risk of lung cancer among Caucasian population in both the dominant genetic model [5A/6A+6A/6A vs. 5A/5A: OR,1.00 (95% Cl,0.87–1.15); p value$_{heterogeneity}$,0.76; p value,0.96] and the recessive genetic model [6A/6A vs. 5A/6A+5A/5A: OR,0.96 (95% Cl,0.84–1.10); p value$_{heterogeneity}$,0.67; p value,0.54]. The study reported no significant association of MMP3 gene rs35068180 (-1171 5A/6A) SNP with the risk of lung cancer in total population in both the dominant genetic model [5A/6A+6A/6A vs. 5A/5A: OR,1.00 (95% Cl,0.87–1.14); p value$_{heterogeneity}$,0.876; p value,0.99] and the recessive genetic model [6A/6A vs. 5A/6A+5A/5A: OR,0.96 (95% Cl,0.85–1.09); p value$_{heterogeneity}$,0.53; p value,0.53] (Hu, Wang et al. 2013).

A meta-analysis based on 26 association studies on European and Asian populations, which included 7,098 cases and 7,035 controls, reported no significant overall association of MMP3 gene rs35068180 (-1171 5A/6A) SNP with the risk of cancer in all three genetic models: dominant genetic model [5A/6A+6A/6A vs. 5A/5A: OR,0.98 (95% Cl,0.89–1.08); p value$_{heterogeneity}$,0.671; I^2= 0.0; p value,>0.05], recessive genetic model [6A/6A vs. 5A/6A+5A/5A: OR,1.00 (95% Cl,0.89–1.12); p value$_{heterogeneity}$,0.005; I^2= 46.4; p value,>0.05] and allelic genetic model [6A vs. 5A: OR,1.00 (95% Cl,0.95–1.05); p value$_{heterogeneity}$,0.124; I^2= 24.9; $p$value,>0.05]. The study reported no significant association of this SNP with the risk of colorectal cancer (seven studies: 1,296 cases; 1,543 controls) in all three genetic models: dominant genetic model [5A/6A+6A/6A vs. 5A/5A: OR,0.94 (95% Cl,0.76–1.18); p value$_{heterogeneity}$,0.540; I^2= 0.0; p value,>0.05], recessive genetic model [6A/6A vs. 5A/6A+5A/5A: OR,1.19 (95% Cl,0.91–1.57); p value$_{heterogeneity}$,0.049; I^2= 52.5; p value,>0.05] and allelic genetic model [6A vs. 5A: OR,1.09 (95% Cl,0.96–1.22); p value$_{heterogeneity}$,0.310; I^2= 15.7; p value,>0.05]. The study also reported no significant association of this SNP with the risk of lung cancer (three studies: 2,286 cases; 1,837 controls) in all three genetic models: dominant genetic model [5A/6A+6A/6A vs. 5A/5A: OR,1.01 (95% Cl,0.86–1.18); p value$_{heterogeneity}$,0.718; I^2= 0.0; p value,>0.05], recessive genetic model [6A/6A vs. 5A/6A+5A/5A: OR,0.95 (95% Cl,0.83–1.10); p value$_{heterogeneity}$,0.693; I^2= 0.0; p value,>0.05] and allelic genetic model [6A vs. 5A: OR,0.98 (95% Cl,0.90–1.08); p value$_{heterogeneity}$,0.630; I^2= 0.0; p value,>0.05]. Further, the study reported no significant association of this SNP with the risk of head and neck cancer (five studies: 816 cases; 1,158 controls) in all three genetic models: dominant genetic model [5A/6A+6A/6A vs. 5A/5A: OR,0.90 (95% Cl,0.65–1.24); p value$_{heterogeneity}$,0.508; I^2= 0.0; p value,>0.05], recessive genetic model [6A/6A vs. 5A/6A+5A/5A: OR,0.91 (95% Cl,0.63–1.30); p value$_{heterogeneity}$,0.014; I^2= 67.8; p value,>0.05] and allelic genetic model [6A vs. 5A: OR,0.96 (95% Cl,0.75–1.23); p value$_{heterogeneity}$,0.036; I^2= 61.0; p value,>0.05]. The study reported no significant association of this SNP with the risk of breast cancer (four studies: 875 cases; 949 controls) in all three genetic models: dominant genetic model [5A/6A+6A/6A vs. 5A/5A: OR,1.02 (95% Cl,0.82–1.28); p value$_{heterogeneity}$,0.218; I^2= 32.3; p value,>0.05], recessive genetic model [6A/6A vs. 5A/6A+5A/5A: OR,0.84 (95% Cl,0.68–1.04); p value$_{heterogeneity}$,0.080; I^2= 55.6; p value,>0.05] and allelic genetic model [6A vs. 5A: OR,0.87 (95% Cl,0.67–1.11); p value$_{heterogeneity}$,0.046; I^2= 62.6; p value,>0.05]. The study reported no significant association of this SNP with the risk of ovarian cancer (three studies: 265 cases; 425 controls) in all three genetic models: dominant genetic model [5A/6A+6A/6A vs.

5A/5A: OR,0.90 (95% Cl,0.56–1.44); p value$_{heterogeneity}$,0.174; I^2= 42.9; p value,>0.05], recessive genetic model [6A/6A vs. 5A/6A+5A/5A: OR,1.03 (95% Cl,0.73–1.47); p value$_{heterogeneity}$,0.121; I^2= 52.6; p value,>0.05] and allelic genetic model [6A vs. 5A: OR,0.99 (95% Cl,077–1.27); p value$_{heterogeneity}$,0.494; I^2= 0.0; p value,>0.05]. The study reported no significant association of this SNP with the risk of cancer in Asian population (twelve studies: 2,432 cases; 2,937 controls) in all three genetic models: dominant genetic model [5A/6A+6A/6A vs. 5A/5A: OR,0.89 (95% Cl,0.61–1.31); p value$_{heterogeneity}$,0.728; I^2= 0.0; p value,>0.05], recessive genetic model [6A/6A vs. 5A/6A+5A/5A: OR,1.10 (95% Cl,0.98–1.25); p value$_{heterogeneity}$,0.227; I^2= 22.1; p value,>0.05] and allelic genetic model [6A vs. 5A: OR,1.08 (95% Cl,0.97–1.20); p value$_{heterogeneity}$,0.327; I^2= 12.0; p value,>0.05]. The study further reported no significant association of this SNP with the risk of cancer in European population (fourteen studies: 4666 cases; 4098 controls) in all three genetic models: dominant genetic model [5A/6A+6A/6A vs. 5A/5A: OR,0.99 (95% Cl,0.90–1.09); p value$_{heterogeneity}$,0.421; I^2= 2.6; p value,>0.05], recessive genetic model [6A/6A vs. 5A/6A+5A/5A: OR,0.89 (95% Cl,0.75–1.05); p value$_{heterogeneity}$,0.006; I^2= 55.5; p value,>0.05] and allelic genetic model [6A vs. 5A: OR,0.97 (95% Cl,0.91–1.03); p value$_{heterogeneity}$,0.187; I^2= 29.5; p value,>0.05]. The study also reported a significant difference between the Asian and European populations with regard to the difference in the distribution or frequency of MMP3 gene rs35068180 6A (-1171 6A)allele [European population,52% (95% Cl,51–54); Asian population, 84% (95% Cl,82–86); p value,0.0001] (Peng, Cao et al. 2010).

Another meta-analysis based on 41 association studies on Asian and Caucasian populations, which included 11, 112 cases and 11,091 controls, reported no significant overall association of MMP3 gene rs35068180 (-1171 5A/6A)SNP with the risk of cancer in all genetic models: homozygous genetic model [6A/6A vs. 5A/5A: OR,0.92 (95% Cl,0.84–1.01); I^2=23.7], heterozygous genetic model [5A/6A vs. 5A/5A: OR,0.95 (95% Cl,0.87–1.03); I^2=14.4], dominant genetic model [5A/6A+6A/6A vs. 5A/5A: OR,0.94 (95% Cl,0.87–1.01); I^2= 14.2] and recessive genetic model [6A/6A vs. 5A/6A+5A/5A: OR,0.94 (95% Cl,0.85–1.04); I^2= 56.2]. The study reported a significant overall association of MMP3 gene rs35068180 (-1171 5A/6A) SNP with a decreased risk of gastrointestinal cancer (nine studies) inheterozygous genetic model [5A/6A vs. 5A/5A: OR,0.74 (95% Cl,0.60–0.91); I^2=1.9] and dominant genetic model [5A/6A+6A/6A vs. 5A/5A: OR,0.77 (95% Cl,0.64–0.94); I^2= 29.0], but not in homozygous genetic model [6A/6A vs. 5A/5A: OR,0.86 (95% Cl,0.68–1.09); I^2=57.3]andrecessive genetic model [6A/6A vs. 5A/6A+5A/5A: OR,0.99 (95% Cl,0.70–1.38); I^2= 77.9]. The study reported no significant overall association of MMP3 gene rs35068180 (-1171 5A/6A) SNP with the risk of gynecological cancer (six studies) in all genetic models: homozygous genetic model [6A/6A vs. 5A/5A: OR,0.86 (95% Cl,0.53–1.39); I^2=0.0], heterozygous genetic model [5A/6A vs. 5A/5A: OR,1.00 (95% Cl,0.64–1.55); I^2=26.2], dominant genetic model [5A/6A+6A/6A vs. 5A/5A: OR,0.93 (95% Cl,0.61–1.42); I^2= 0.0] and recessive genetic model [6A/6A vs. 5A/6A+5A/5A: OR,0.92 (95% Cl,0.66–1.29); I^2= 46.4]. The study reported no significant overall association of MMP3 gene rs35068180 (-1171 5A/6A) SNP with the risk of breast cancer (five studies) in all genetic models: homozygous genetic model [6A/6A vs. 5A/5A: OR,0.98 (95% Cl,0.81–1.18); I^2=55.3], heterozygous genetic model [5A/6A vs. 5A/5A: OR,1.06 (95% Cl,0.90–1.25); I^2=0.0], dominant genetic model [5A/6A+6A/6A vs. 5A/5A: OR,1.03 (95% Cl,0.88–1.20); I^2= 9.8] and recessive genetic model [6A/6A vs. 5A/6A+5A/5A: OR,0.87 (95% Cl,0.66–1.13); I^2= 54.1]. The study also reported significant overall association of MMP3 gene rs35068180 (-1171 5A/6A)SNP with the risk of head and neck cancer (three studies) in homozygous genetic model [6A/6A vs. 5A/5A: OR,0.51 (95% Cl,0.29–0.88); I^2=0.0], but not in heterozygous genetic model [5A/6A vs. 5A/5A: OR,0.91 (95% Cl,0.57–1.44); I^2=0.0], dominant genetic model [5A/6A+6A/6A vs. 5A/5A: OR,0.75 (95% Cl,0.49–1.16); I^2= 0.0] and recessive genetic model [6A/6A vs. 5A/6A+5A/5A: OR,0.73 (95% Cl,0.35–1.52); I^2= 80.3]. The study reported no significant overall association of MMP3 gene rs35068180 (-1171 5A/6A) SNP with the risk of hepatocellular carcinoma (three studies) in all genetic models: homozygous genetic model [6A/6A vs. 5A/5A: OR,0.61 (95% Cl,0.24–1.60); I^2=49.2], heterozygous genetic model [5A/6A vs. 5A/5A: OR,0.58 (95% Cl,0.22–1.55);I^2=51.0], dominant genetic model [5A/6A+6A/6A vs. 5A/5A: OR,0.61 (95% Cl,0.24–1.58); I^2= 50.3] and recessive genetic model [6A/6A vs. 5A/6A+5A/5A: OR,0.80 (95% Cl,0.48–1.32); I^2= 66.8]. The study reported no significant overall association of MMP3 gene rs35068180 (-1171 5A/6A) SNP with the risk of lung cancer (six studies) in all genetic models: homozygous genetic model [6A/6A vs. 5A/5A: OR,0.95 (95% Cl,0.81–1.11); I^2=0.0], heterozygous genetic model [5A/6A vs. 5A/5A: OR,0.97 (95% Cl,0.84–1.12); I^2=44.1], dominant genetic model [5A/6A+6A/6A vs. 5A/5A: OR,0.96 (95% Cl,0.84–1.09); I^2= 39.4] andrecessive genetic model [6A/6A vs. 5A/6A+5A/5A: OR,0.99 (95% Cl,0.80–1.22); I^2= 39.8]. The studyalso reported no significant overall association of MMP3 gene rs35068180 (-1171 5A/6A) SNP with the risk of oral cancer (four studies) in all genetic models: homozygous genetic model [6A/6A vs. 5A/5A: OR,0.94 (95% Cl,0.63–1.40); I^2=40.1], heterozygous genetic model [5A/6A vs. 5A/5A: OR,0.93 (95% Cl,0.62–1.38);I^2=35.8], dominant genetic model [5A/6A+6A/6A vs. 5A/5A: OR,0.94 (95% Cl,0.65–1.35); I^2= 25.9] and recessive genetic model [6A/6A vs. 5A/6A+5A/5A: OR,0.94 (95% Cl,0.64–1.38); I^2= 55.1]. The study reported no significant overall association of MMP3 gene rs35068180 (-1171 5A/6A) SNP with the risk of cancer in Caucasian population (18 studies) in all genetic models: homozygous genetic model [6A/6A vs. 5A/5A: OR,0.96 (95% Cl,0.87–1.06); I^2=4.3], heterozygous genetic model [5A/6A vs. 5A/5A: OR,0.97 (95% Cl,0.89–1.06); I^2=33.7], dominant genetic model [5A/6A+6A/6A vs. 5A/5A: OR,0.97 (95% Cl,0.89–1.05); I^2= 1.8] and recessive genetic model [6A/6A vs. 5A/6A+5A/5A: OR,0.95 (95% Cl,0.83–1.09); I^2=49.3]. Further, the study reported a significant overall association of MMP3 gene rs35068180 (-1171 5A/6A) SNP with a decreased risk of cancerin Asian population (23 studies) in homozygous genetic model [6A/6A vs. 5A/5A: OR,0.68 (95% Cl,0.52–0.90); I^2=26.7], heterozygous genetic model [5A/6A vs. 5A/5A: OR,0.75 (95% Cl,0.58–0.98); I^2=0.0] and dominant genetic model [5A/6A+6A/6A vs. 5A/5A: OR,0.69 (95% Cl,0.54–0.88); I^2= 0.5], but not in recessive genetic model [6A/6A vs. 5A/6A+5A/5A: OR,0.93 (95% Cl,0.79–1.09); I^2= 61.8] (Yang,Hu et al. 2014).

6.5.6 Polymorphisms in Matrix Metalloproteinase2 (MMP2) Gene

Matrix metalloproteinase 2 (MMP2) gene (chromosomal location,16q12.2) encodes a 660 amino acid 73,882 Da (73 kDa) MMP2 pro-enzyme, also known as gelatinase A,72 kDa type IV collagenase or 72 kDa gelatinase. The active form of MMP2 enzyme is a 62–67 kDa protein that contains the collagen-binding domain or CBD. The collagen-binding domain, which mediates the recognition of and binding to gelatins and collagens,consists of three cysteine-rich fibronectin type II domain repeats termed as Col-1, Col-2 and Col-3, present in a head-to-tail manner within the catalytic domain, adjacent to the active site (Templeton and Stetler-Stevenson 1991, Skiles, Gonnella et al. 2001, Nagase and Fushimi 2008). MMP2 enzyme, an extensively glycosylated proteinand mainly involved in the maintenance of tissue homeostasis, exhibits a wide and constitutive expression in almost all the tissues.However, MMP2 gene expression predominantly occurs in epithelial cells, endothelial cells and fibroblasts and a significant concentration of this enzyme is detectable in serum even under normal physiological conditions (Templeton and Stetler-Stevenson 1991, Van den Steen, Dubois et al. 2002). MMP2 gene exhibits constitutive expression, and unlike most other MMPs, its promoter lacks binding sites for various regulatory mediators including inflammation-related transcription factors such as activator protein1 (AP1) and, therefore, the regulatory mediators affect or induce MMP2 gene expression to a smaller or negligible extent in comparison with other MMPs (Templeton and Stetler-Stevenson 1991, Xu, Lai et al. 2004). The proMMP2 protein activation possibly occurs through two distinct mechanisms, which involve the participation of either MT1-MMP and TIMP2, or MT2-MMP, urokinase plasminogen activator surface receptor (uPAR or CD87), activated protein C (aPC) and thrombin (Morrison, Butler et al. 2001, Visse and Nagase 2003, Nishida, Miyamori et al. 2008, Koo, Han et al. 2010). MMP2 enzyme is inhibitable by TIMPs such as TIMP3, TIMP4and mainly TIMP2 (Crocker, Pagenstecher et al. 2004). Further, MMP2 enzyme contains around 29 potential phosphorylation sites and a study reports MMP2 enzyme activity modulation by protein kinase C (PKC)-mediated phosphorylation (Sariahmetoglu, Crawford et al. 2007). The substrates of MMP2 predominantly include various ECM and BM components and in addition non-ECM components (Table 6.2). These include fibronectin, laminins, gelatins, casein, nidogen-1 (entactin), elastin, casein, galectin3, aggrecan, plasminogen, myelin basic protein (MBP) and type I, type IV, type V, type VII, type X, type XI and type XIV collagens. MMP2 enzyme is also involved in the proteolytic cleavage and release of several cell membrane- or cell surface-bound cellular mediators through shedding of their ectodomains or release of inhibitory binding factors. These mediators include monocyte chemotactic protein 3(MCP3), proIL1β, pro-transforming growth factor beta (pro-TGF-β), insulin-like growth factor-binding protein 5 (IGFBP5), β-dystroglycan and fibroblast growth factor receptor 1 (FGFR1) (Chakrabarti and Patel 2005, Bozzi, Inzitari et al. 2009).

MMP2, predominantly through its role in the proteolysis of various ECM and BM components, plays a vital role in various pathophysiological processes such as inflammation, neoplastic diseases (cancers), muscular, skeletal and bone pathologies, cardiovascular and respiratory disorders and nervous system diseases. MMP2 enzyme through proteolysis of diverse substrates mediates several processes that may either promote inflammation, or cause its inhibition. MMP2 enzyme activates pro-inflammatory mediators including cytokines such as interleukin1β (IL1β), chemoattractants and growth factors through proteolytic cleavage of pro-forms of these mediators and promotes recruitment of immune cells such as lymphocytes and neutrophils (Schönbeck, Mach et al. 1998). MMP2 enzyme-catalyzed proteolytic cleavage of chemoattractants including chemokines such as monocyte chemoattractant protein3 (MCP3) generates truncated or inactive forms with antagonistic effect on chemokine receptors and inhibitory effect on inflammation. MMP2 enzyme through generation of chemotactic gradient mediates inflammatory cell clearance from tissues (Corry, Rishi et al. 2002, McQuibban, Gong et al. 2002). MMP2 enzyme plays a vital role in important cellular processes such as apoptosis and angiogenesis under normal physiological conditions and in diseased pathophysiological state. However, the role of MMP2 enzyme in these processes isdifferential and depends on the activation state of this enzyme. MMP2 enzymepromotes neovascularization (angiogenesis) through proteolytic cleavage of vascular basement membrane collagens. The partially activated form of MMP2 enzyme that exists as MMP2-MT1-MMP (MMP2-MMP14) complex promotes neovascularization by enhancing the survival and proliferation of vascular endothelial cells through generation of activated growth factors such as vascular endothelial growth factor (VEGF) and by inducing their migration through proteolytic cleavage of cell adhesion molecule, laminin-5. In contrast, the fully active mature MMP2 enzyme inhibits neovascularization through promotion of apoptosis in vascular endothelial cells (Risau 1997, Nguyen, Arkell et al. 2001, Shapiro, Khodalev et al. 2010). MMP2 enzyme is also involved in the pathogenesis of various diseases or disorders. These include nervous system diseases such as multiple sclerosis (Fainardi, Castellazzi et al. 2009), cardiovascular disorders such as atherosclerosis, stroke and myocardial infarction (Volcik, Campbell et al. 2010, Heo, Cho et al. 2011) and muscular, skeletal and bone disorders such as rheumatoid arthritis (RA) (Kim, Choi et al. 2011).

MMP2 enzyme, predominantly through proteolysis of diverse substrates including various ECM and BM components and its role in important cellular processes such as apoptosis and angiogenesis, seems to be critically involved in the pathogenesis of various cancers and mediate several important tumorigenic processes, in particular, invasiveness and metastasis (Björklund and Koivunen 2005, Fingleton 2006). Various studies report MMP2 gene upregulation, increased MMP2 enzyme serum concentrations and activity in almost all the cancers including prostate cancer, thyroid cancer, ovarian cancer, colorectal cancer, melanoma, bone cancer, gastric cancer and breast cancer. The upregulation and increased serum concentrations and activity of MMP2 enzyme in various cancers and its association with poor prognosis in these cancers support the role of MMP2 enzyme as a potential diagnostic and prognostic cancer marker (Gerstein, Sini et al. 2009, Murnane, Cai et al. 2009, Kushlinsky, Solovyov et al. 2010, Littlepage, Sternlicht et al. 2010, Kwon, Cukierman et al. 2011, Rajoria, Suriano et al. 2011).

MMP2 enzyme is involved in the pathogenesis of initial, intermediate as well as advanced phases of cancer. MMP2 enzyme promotes tumor growth, progression and invasion (Björklund and Koivunen 2005). MMP2 enzymethrough proteolytic ectodomain shedding of various cellular mediators including adhesion molecules such as integrins and cadherins mediates epithelial-mesenchymal transition (EMT). MMP2 enzyme through proteolytic cleavage of CAMs such as P-cadherin induces neovascularization that in turn promotestumor growth and progression (Ribeiro, Albergaria et al. 2010, Aresu, Benali et al. 2011). MMP2 enzyme also catalyzes proteolytic activation of TGF-β through degradation of decorin, a natural TGF-β inhibitor or antagonist, and through proteolysis of inactive precursors to generate active TGF-β. The activated TGF-β under normal physiological conditions generally mediates homeostasis and induces apoptosis, however, under pathophysiological conditions; TGF-β promotes tumor growth, progression and invasion (Yu and Stamenkovic 2000). TGF-β also induces expression of MMP2 enzyme in an activating transcription factor 2 (ATF-2)-dependent manner, which further promotes the growth and progression of tumors (Kim, Sohn et al. 2007, Binker, Binker-Cosen et al. 2011). MMP2 enzyme also promotes tumor growth, progression and invasion through modulation of various processes in a pro-tumorigenic manner such as immune response to tumors,tumor cell metabolism, turnover of varied membrane receptors, tumor cell resistance to apoptosis and activation of a wide range of growth factors and cytokines (Chakrabarti and Patel 2005, Bozzi, Inzitari et al. 2009, Gondi, Talluri et al. 2009). The tumor-specific MMP2 expression occurs in stromal cells, malignant cells and immune cells of the tumorand the surrounding tissue (Coussens, Tinkle et al. 2000, Egeblad and Werb 2002). In addition, tumor cells secrete various mediators that act on non-malignant neighboring cells and increase MMP2 gene expression and MMP2 enzyme secretion by these cells, which further promotes tumor growth, progression and invasion (Du, Petritsch et al. 2008). For example, the secretion of a transmembrane glycoprotein, extracellular matrix metalloproteinase inducer (EMMPRIN), also known as basigin (Bsg) or CD147, by tumor cells promotes MMP2 expression in fibroblasts surrounding the tumor (Biswas, Du et al. 2010). Several mediators upregulated in tumor cells such as heat shock protein 90 (Hsp90), a chaperone protein, increase MMP2 enzyme secretion by promoting the cell membrane activation of proMMP2 (Walsh, Larkin et al. 2011).

The various SNPs identified in MMP2 gene include MMP2 gene (-168G/T) SNP, MMP2 gene rs2285053 (-735 C/T, also known as -1015C/T), MMP2 gene rs243864 (-790G/T) SNP, MMP2 gene (-955 C/A) SNP, MMP2 gene rs243865 (-1306 C/T) SNP and MMP2 rs243866 (-1575G/A) SNP located in the promoter region of MMP2 gene and MMP2 gene +220 G/C SNP located in the 5′-untranslated region (5′-UTR). Of these, the functional SNPs in MMP2 gene reported to modulate the MMP2 gene transcription, expressionand MMP2 enzyme activity in an allele-specific manner include MMP2 gene rs243865 (-1306 C/T) SNP and MMP2 gene rs2285053 (-735 C/T) SNP. MMP2 gene rs243865 (-1306 C/T) SNP occurs within the region that encompasses the core recognition and binding site (CCACC box sequence) for the transcription factor, specificity protein1 (Sp1), a zinc finger motif-containing protein. MMP2 gene -1306C to T transition abolishes the Sp1 recognition and binding site. Sp1 transcription factor specifically binds MMP2 gene -1306C allele, but not MMP2 gene -1306T allele. Consequently, MMP2 gene -1306T allele exhibits a marked decrease in MMP2 gene promoter activity, MMP2 gene expression and MMP2 enzyme activity. The promoter activity associated with MMP2 gene -1306C allele is twofold higher in comparison with MMP2 gene -1306T allele (Price, Greaves et al. 2001, Chaudhary, Singh et al. 2010).

Various studies report that MMP2 gene-1306CC genotype has a higher transcriptional activity, which in turn leads to increased MMP2 gene expression and enhanced secretion and activity of MMP2 enzyme in comparison with MMP2 gene-1306CT genotype and/or MMP2 gene-1306TT genotype (Qin, Sun et al. 1999, Yu, Pan et al. 2002). MMP2 gene rs2285053 (-735 C/T) SNP also occurs within the region that encompasses the core recognition and binding site for the transcription factor, specificity protein1 (Sp1), and the substitution of -735C nucleotide by -735T nucleotide abolishes the Sp1 recognition and binding site (C**C**CTCC to C**T**CTCC). As with MMP2 gene rs243865 (-1306 C/T) SNP, Sp1 transcription factor specifically binds MMP2 gene -735C allele, but not MMP2 gene -735T allele. Consequently, MMP2 gene -735T allele exhibits a marked decrease in MMP2 gene promoter activity, MMP2 gene expression and MMP2 enzyme activity. The promoter activity associated with MMP2 gene -735C allele is almost threefold higher in comparison with MMP2 gene -735T allele (Price, Greaves et al. 2001). Further, MMP2 gene rs2285053 (-735 C/T) SNP and MMP2 gene rs243865 (-1306 C/T) SNP are possibly in linkage disequilibrium (LD) and interact with each other to modulate MMP2 gene transcription, expression and MMP2 enzyme activity. Studies report almost sevenfold higher promoter activity and fourfold higher MMP2 mRNA levels associated with MMP2–735C-MMP2–1306C haplotype in comparison with MMP2–735T-MMP2–1306T haplotype (Price, Greaves et al. 2001, Yu, Zhou et al. 2004).

MMP2 enzyme promotes growth, progression and invasion of various types of tumors. Various studies report the association of MMP2 gene -735T allele- and MMP2–1306T allele-related decrease in promoter activity, MMP2 gene expression and MMP2 enzyme activity with reduced risk of various tumors. In contrast, the studies report the association of MMP2 gene -735C allele- and MMP2–1306C allele-related increase in promoter activity, MMP2 gene expression and MMP2 enzyme activity with an increased risk of various tumors. Here, we mention several of these studiesthat report the association of MMP2 gene rs2285053 (-735 C/T) SNP and MMP2 gene rs243865 (-1306 C/T) SNP with the risk of various cancersand the studies that report no significant association. A meta-analysis based on nine association studies on European and Asian populations, which included 3,277 cases and 4,379 controls, reported a significant association of MMP2 gene rs2285053 (-735 C/T) SNP with a decreased risk of lung cancer (two studies: 859 cases; 867 controls) in dominant genetic model [CT+TT vs. CC: OR,0.65 (95% Cl,0.53–0.79); p value$_{heterogeneity}$,0.423; I^2= 0.0; p value,≤ 0.001] and in allelic genetic model [T vs. C: OR,0.72 (95% Cl,0.61–0.85); p value$_{heterogeneity}$,0.533; I^2= 0.0; p value,≤ 0.001], but not in recessive genetic model [TT vs. CT+CC: OR,0.88 (95% Cl,0.56–1.39); p value$_{heterogeneity}$,0.888; I^2= 0.0]. The same study reported a significant association of MMP2 gene rs2285053 (-735 C/T) SNP with a decreased risk of esophageal cancer (two studies: 862 cases; 1,401 controls) in dominant genetic model [CT+TT vs. CC: OR,0.84 (95% Cl,0.70–0.99); p value$_{heterogeneity}$,0.206; I^2= 37.4; p value,≤ 0.05], but not in allelic genetic model [T vs. C: OR,0.87 (95% Cl,0.75–1.01); p value$_{heterogeneity}$,0.370; I^2= 0.0] and recessive genetic model [TT vs. CT+CC: OR,0.90 (95% Cl,0.60–1.35); p value$_{heterogeneity}$,0.764; I^2= 0.0]. The study reported no significant association of MMP2 gene rs2285053 (-735 C/T) SNP with head and neck cancer (two studies: 622 cases; 635 controls) in dominant genetic model [CT+TT vs. CC: OR,0.90 (95% Cl,0.43–1.87); p value$_{heterogeneity}$,0.042; I^2= 75.9]. The other genetic models are not available. The study also reported no significant association of MMP2 gene rs2285053 (-735 C/T) SNP with ovarian cancer,gastric cancer and hepatocellular cancer (three studies: 934 cases; 1,426 controls) in all the three genetic models: recessive genetic model [TT vs. CT+CC: OR,0.96 (95% Cl,0.67–1.38); p value$_{heterogeneity}$,0.749; I^2= 0.0], dominant genetic model [CT+TT vs. CC: OR,0.85 (95% Cl,0. 55–1.30); p value$_{heterogeneity}$,0.003; I^2= 83.1] and allelic genetic model [T vs. C: OR,0.89 (95% Cl,0.63–1.26); p value$_{heterogeneity}$,0.005; I^2= 81.3].The same study reported no significant difference in the distribution or frequency of MMP2–735T allele among the populations under study [European: 17.5% (95% Cl,27.0–62.0) vs. Asian: 23.2% (95% Cl,20.6–25.8); p value,0.0623] (Peng, Cao et al. 2010). Another meta-analysis based on three association studies on Asian and Caucasian populations, which included 1,229 cases and 1,303 controls, reported a significant association of MMP2 gene rs2285053 (-735 C/T) SNPwith a decreased risk of lung cancer in total population in dominant genetic model [CT+TT vs. CC: OR,0.73 (95% Cl,0.58–0.92); I^2= 34.9; p value,0.007], heterozygous genetic model [CT vs. CC: OR,0.73 (95% Cl,0.56–0.95); I^2= 43.1; p value,0.019] and allelic genetic model [T vs. C: OR,0.76 (95% Cl,0.66–0.88); I^2= 0.0; p value,<0.001], but not in homozygous genetic model [TT vs. CC: OR,0.75 (95% Cl,0.51–1.10); I^2=0.0; p value,>0.05] and recessive genetic model [TTvs. CT+CC: OR,0.85 (95% Cl,0.58–1.24); I^2=0.0; p value,>0.05]. The study reported a significant association of this SNP with a decreased risk of lung cancer among Asians in dominant genetic model [CT+TT vs. CC: OR,0.72 (95% Cl,0.53–0.97); I^2= 64.1] and in allelic genetic model [T vs. C: OR,0.76 (95% Cl,0.66–0.87); I^2= 31.1], but not inhomozygous genetic model [TT vs. CC: OR,0.74 (95% Cl,0.50–1.09); I^2=0.0; p value,>0.05],heterozygous genetic model [CT vs. CC: OR,0.72 (95% Cl,0.51–1.01); I^2= 69.4; p value,>0.05] andrecessive genetic model [TTvs. CT+CC: OR,0.84 (95% Cl,0.57–1.24); I^2=0.0; p value,>0.05].The study reported no significant association among Caucasian population in all genetic models: dominant genetic model [CT+TT vs. CC: OR,0.84 (95% Cl,0.42–1.68); p value,>0.05], recessive genetic model [TT vs. CT+CC: OR,1.01 (95% Cl,0.14–7.34); p value,>0.05], allelic genetic model [T vs. C: OR,0.87 (95% Cl,0.47–1.62); p value,>0.05], homozygous genetic model [TT vs. CC: OR,0.97 (95% Cl,0.13–7.09); I^2=0.0; p value,>0.05] and heterozygous genetic model [CT vs. CC: OR,0.83 (95% Cl,0.41–1.70); p value,>0.05](Li,Liu et al. 2020).

A meta-analysis based on three association studies on Asian and Caucasian populations reported a significant association of MMP2 gene rs2285053 (-735 C/T) SNP with a decreased risk of lung cancer among Asian population in dominant genetic model [CT+TT vs. CC: OR,0.63 (95% Cl,0.51–0.77); p value<0.0001], but not in recessive genetic model [TTvs. CT+CC: OR,0.87 (95% Cl,0.55–1.40); p value,0.58]. The study reported no significant association of MMP2 gene rs2285053 (-735 C/T) SNP with the risk of lung cancer among Caucasian population in both the dominant genetic model [CT+TT vs. CC: OR,1.15 (95% Cl,0.91–1.44); p value$_{heterogeneity}$,0.43; p value,0.24] and the recessive genetic model [TT vs. CT+CC: OR,0.58 (95% Cl,0.30–1.11); p value$_{heterogeneity}$,0.29; p value,0.10]. The study reported no significant association of MMP2 gene rs2285053 (-735 C/T) SNP with the risk of lung cancer in total population in both the genetic models: dominant genetic model [CT+TT vs. CC: OR,0.87 (95% Cl,0.53–1.42); p value$_{heterogeneity}$,0.0005; p value,0.56] and recessive genetic model [TT vs. CT+CC: OR,0.76 (95% Cl,0.52–1.11); p value$_{heterogeneity}$,0.34; p value,0.16]. The same meta-analysis based on five association studies reported a significant association of MMP2 gene rs243865 (-1306 C/T) SNP with a decreased risk of lung cancer among Asian population in dominant genetic model [CT+TT vs. CC: OR,0.58 (95% Cl,0.42–0.81); p value$_{heterogeneity}$,0.01; p value,0.001] and recessive genetic model [TTvs. CT+CC: OR,0.58 (95% Cl,0.36–0.95); p value$_{heterogeneity}$,0.99; p value,0.03]. The study reported no significant association of MMP2 gene rs243865 (-1306 C/T) SNP with the risk of lung cancer among Caucasian population in both the dominant genetic model [CT+TT vs. CC: OR,1.00 (95% Cl,0.54–1.86); p value,1.00] and the recessive genetic model [TT vs. CT+CC: OR,2.02 (95% Cl,0.18–22.71); p value,0.57]. The study reported a significant association of MMP2 gene rs243865 (-1306 C/T) SNP with a decreased risk of lung cancer in total population in both the genetic models: dominant genetic model [CT+TT vs. CC: OR,0.63 (95% Cl,0.46–0.88); p value$_{heterogeneity}$,0.004; p value,0.006] and recessive genetic model [TT vs. CT+CC: OR,0.61 (95% Cl,0.38–0.99); p value$_{heterogeneity}$,0.89; p value,0.04] (Hu, Wang et al. 2013). Another meta-analysis based on 26 association studies on European and Asian populations, which included 8,297 cases and 10,566 controls, reported a significant association of MMP2 gene rs243865 (-1306 C/T) SNP with a decreased risk of lung cancer (four studies: 1,804 cases; 1,867 controls) in dominant genetic model [CT+TT vs. CC: OR,0.50 (95% Cl,0.43–0.59); p value$_{heterogeneity}$,0.147; I^2= 44.1; p value,≤ 0.001] and in allelic genetic model [T vs. C: OR,0.55 (95% Cl,0.48–0.63); p value$_{heterogeneity}$,0.123; I^2= 48.1; p value,≤ 0.001], but not in recessive genetic model [TT vs. CT+CC: OR,0.61 (95% Cl,0.37–1.01); p value$_{heterogeneity}$,0.777; I^2= 0.0]. The same study reported a significant association of MMP2 gene rs243865 (-1306 C/T) SNP with a decreased risk of head and neck cancer (three studies: 930 cases; 870 controls) in dominant genetic model [CT+TT vs. CC: OR,0.53 (95% Cl,0.41–0.69); *p* value$_{heterogeneity}$,0.974; I^2= 0.0; *p* value,≤ 0.001], recessive genetic model [TT vs. CT+CC: OR,0.10 (95% Cl,0.01–0.78); *p* value$_{heterogeneity}$,0.907; I^2= 0.0; *p* value,≤ 0.05] and allelic genetic model [T vs. C: OR,0.50 (95% Cl,0.35–0.70); *p* value$_{heterogeneity}$,0.913; I^2= 0.0; *p* value,≤ 0.001].Further, the study reported a significant association of MMP2 gene rs243865 (-1306 C/T) SNP with a decreased risk of esophageal cancer (three studies: 1,050 cases; 1,731 controls) in dominant genetic model [CT+TT vs. CC: OR,0.67 (95% Cl,0.55–0.80); *p* value$_{heterogeneity}$,0.593; I^2= 0.0; *p* value,≤ 0.001] and in allelic genetic model [T vs. C: OR,0.70 (95% Cl,0.59–0.82); *p* value$_{heterogeneity}$,0.606; I^2= 0.0; *p* value,≤ 0.001], but not in recessive genetic model [TT vs. CT+CC: OR,0.64 (95% Cl,0.33–1.25); *p* value$_{heterogeneity}$,0.670; I^2= 0.0]. The same study reported a significant association of MMP2 gene rs243865 (-1306 C/T) SNP with a decreased risk of gastric cancer (four studies: 1,081 cases; 2,476 controls) in dominant genetic model [CT+TT vs. CC: OR,0.61 (95% Cl,0.37–0.99); *p* value$_{heterogeneity}$,0.001; I^2= 85.4; *p* value,≤ 0.05], but not in recessive genetic model [TT vs. CT+CC: OR,0.95 (95% Cl,0.22–4.03); *p* value$_{heterogeneity}$,0.018; I^2= 70.3] and allelic genetic model [T vs. C: OR,0.66 (95% Cl,0.40–1.09); *p* value$_{heterogeneity}$,0.001; I^2= 88.5]. The same study reported no significant association of MMP2 gene rs243865 (-1306 C/T) SNP with colorectal cancer (three studies: 372 cases; 401 controls) in all the three genetic models: recessive genetic model [TT vs. CT+CC: OR,1.32 (95% Cl,0.58–3.01); *p* value$_{heterogeneity}$,0.691; I^2= 0.0], dominant genetic model [CT+TT vs. CC: OR,0.87 (95% Cl,0.46–1.66); *p* value$_{heterogeneity}$,0.077; I^2= 61.1] and allelic genetic model [T vs. C: OR,0.93 (95% Cl,0.50–1.75); *p* value$_{heterogeneity}$,0.053; I^2= 66.1]. The study also reported no significant association of MMP2 gene rs243865 (-1306 C/T) SNP with breast cancer (four studies: 1,590 cases; 1,653 controls) in all the three genetic models:recessive genetic model [TT vs. CT+CC: OR,0.99 (95% Cl,0.72–1.36); *p* value$_{heterogeneity}$,0.301; I^2= 17.9], dominant genetic model [CT+TT vs. CC: OR,0.66 (95% Cl,0.40–1.07); *p* value$_{heterogeneity}$,0.001; I^2= 86.0] and allelic genetic model [T vs. C: OR,0.72 (95% Cl,0.47–1.11); *p* value$_{heterogeneity}$,0.001; I^2= 85.9]. The same study reported no significant association of MMP2 gene rs243865 (-1306 C/T) SNP with hepatocellular cancer, bladder cancer, prostate cancer, cervical cancer and ovarian cancer (five studies: 1,470 cases; 1,568 controls) in all the three genetic models: recessive genetic model [TT vs. CT+CC: OR,0.79 (95% Cl,0.54–1.17); *p* value$_{heterogeneity}$,0.333; I^2= 12.7], dominant genetic model [CT+TT vs. CC: OR,0.99 (95% Cl,0.77–1.27); *p* value$_{heterogeneity}$,0.074; I^2= 53.1] and allelic genetic model [T vs. C: OR,0.99 (95% Cl,0.86–1.13); *p* value$_{heterogeneity}$,0.293; I^2= 19.0]. The study reported no significant difference in the distribution or frequency of MMP2–1306T allele among the populations under study [European: 23.3% (95% Cl,16.5–30.0%) vs. Asian: 13.6% (95% Cl,11.5–15.6%); *p* value,0. 0004] (Peng, Cao et al. 2010).

A meta-analysis based on three association studies on Asian and Caucasian populations, which included 1,034 cases and 1,090controls, reported a significant association of MMP2 gene rs243865 (-1306 C/T) SNPwith a decreased risk of lung cancer in total population in dominant genetic model [CT+TT vs. CC: OR,0.59 (95% Cl,0.39–0.89); I^2= 61.2; *p* value,0.011], heterozygous genetic model [CT vs. CC: OR,0.58 (95% Cl,0.39–0.87); I^2= 56.6; *p* value,0.007] and allelic genetic model [T vs. C: OR,0.64 (95% Cl,0.44–0.93); I^2= 64.1; *p* value,0.020], but not in homozygous genetic model [TT vs. CC: OR,0.52 (95% Cl,0.27–1.02); I^2=0.0; *p* value,>0.05] and in recessive genetic model [TTvs. CT+CC: OR,0.61 (95% Cl,0.31–1.18); I^2=0.0; *p* value,>0.05]. The study reported a significant association of this SNP with a decreased risk of lung cancer among Asians in dominant genetic model [CT+TT vs. CC: OR,0.48 (95% Cl,0.39–0.59); I^2= 0.0; *p* value,<0.001], homozygous genetic model [TT vs. CC: OR,0.48 (95% Cl,0.39–0.60); I^2=0.0; *p* value,<0.001], heterozygous genetic model [CT vs. CC: OR,0.46 (95% Cl,0.23–0.93); I^2= 0.0; *p* value,0.032] and allelic genetic model [T vs. C: OR,0.53 (95% Cl,0.43–0.64); I^2= 0.0; *p* value,<0.001], but not in recessive genetic model [TTvs. CT+CC: OR,0.55 (95% Cl,0.27–1.10); I^2=0.0; *p* value,>0.05]. The study reported no significant association among Caucasian population in all genetic models: dominant genetic model [CT+TT vs. CC: OR,1.00 (95% Cl,0.54–1.86); *p* value,>0.05], recessive genetic model [TTvs. CT+CC: OR,2.02 (95% Cl,0.18–22.71); *p* value,>0.05], allelic genetic model [T vs. C: OR,1.04 (95% Cl,0.60–1.79); *p* value,>0.05], homozygous genetic model [TT vs. CC: OR,2.00 (95% Cl,0.18–22.65); *p* value,>0.05] and heterozygous genetic model [CT vs. CC: OR,0.97 (95% Cl,0.51–1.81); *p* value,>0.05] (Li,Liu et al. 2020). Another meta-analysis based on nine association studies, which included 9,858 cases and 10,871 controls, reported no significant association of MMP2 gene rs243865 (-1306 C/T) SNP with the risk of breast cancer in homozygous genetic model [TT vs. CC: OR,1.03 (95% Cl,0.90–1.19); *p* value,0.66], recessive genetic model [TT vs. CT+CC: OR,1.02 (95% Cl,0.89–1.17); *p* value,0.77], dominant genetic model [CT+TT vs. CC: OR,0.93 (95% Cl,0.78–1.10); *p* value,0.38] and allelic genetic model [T vs. C: OR,0.95 (95% Cl,0.82–1.10); *p* value,0.49]. On subgroup analysis based on ethnicity, the study reported a significant association of MMP2 gene rs243865 (-1306 C/T) SNP with a decreased risk of breast cancer in dominant genetic model [CT+TT vs. CC: OR,0.61 (95% Cl,0.40–0.93); *p* value,0.02], but not in other models among Latin American population. The study reported no significant association among East Asian, European and other populations (Yang, Li et al. 2014). Another meta-analysis based on six association studies, which included 1,920 cases and 1,986 controls, reported a significant association of MMP2 gene rs243865 (-1306 C/T) SNPwith an overall increasedrisk of prostate cancer in dominant genetic model [CT+TT vs. CC: OR,1.16 (95% Cl,1.02–1.32); *p* value,0.026] and allelic genetic model [T vs. C: OR,1.12 (95% Cl,1.00–1.24); *p* value,0.040]. On subgroup analysis based on ethnicity, the study reported a significant association of MMP2 gene rs243865 (-1306 C/T) SNP with an increased risk of prostate cancer among Asian population in dominant genetic model [CT+TT vs. CC: OR,1.66 (95% Cl,1.21–2.28); *p* value,0.002] and allelic genetic model [T vs. C: OR,1.48 (95% Cl,1.13–1.94); *p* value,0.004]. Based on the genotyping method used, the study reported a significant association of MMP2 gene rs243865 (-1306 C/T) SNP with an increased risk of prostate cancer in PCR-RFLP genotyping method in dominant genetic model [CT+TT vs. CC: OR,1.71 (95% Cl,1.23–2.38); *p* value,0.001] and allelic genetic model [T vs. C: OR,1.58 (95% Cl,1.19-2.10); *p* value,0.001]. The study reveals the significant association of MMP2 gene rs243865 (-1306 C/T) SNP with an increased risk of prostate cancer among Asian population (Zhang, Chen et al. 2017).

A meta-analysis based on 29 association studies, which included 8,590 cases and 9,601 controls, reported a significant association of MMP2 gene rs243865 (-1306 C/T) SNP with a decreased risk of various cancers in dominant genetic model [CT+TT vs. CC: OR,0.816 (95% Cl,0.678–0.982); *p* value,0.032] and heterozygous genetic model [CT vs. CC: OR,0.758 (95% Cl,0.637–0.902); *p* value,0.002], but not in homozygous genetic model [TT vs. CC: OR,1.185 (95% Cl,0. 825–1.700); *p* value,0.358],recessive genetic model [TTvs. CT+CC: OR,1.268 (95% Cl,0.897–1.793); *p* value,0.179] and allelic genetic model [T vs. C: OR,0.882(95% Cl,0.738–1.055); *p* value,0.169]. The study reveals the significant association of MMP2 gene rs243865 (-1306 C/T) SNP with a decreased risk of various cancers among different world populations (Haque, Akhter et al. 2015). Another meta-analysis based on two association studies, which included 859 cases and 867 controls, reported a significant association of MMP2 gene rs2285053 (-735 C/T) SNP with a decreased risk of lung cancer in dominant genetic model [CT+TT vs. CC: OR,0.65 (95% Cl,0.53–0.79); p $value_{heterogeneity,}$0.42; *p* value<0.0001], but not in recessive genetic model [TTvs. CT+CC: OR,0.88 (95% Cl,0.56–1.39); p $value_{heterogeneity,}$0.89; *p* value,0.59]. The same meta-analysis based on three association studies,which included 1,641 cases and 1,719 controls, reported a significant association ofMMP2 gene rs243865 (-1306 C/T) SNP with a decreased risk of lung cancer in dominant genetic model [CT+TT vs. CC: OR,0.53 (95% Cl,0.40–0.72); p $value_{heterogeneity,}$0.07; *p* value<0.0001], but not in recessive genetic model [TTvs. CT+CC: OR,0.63 (95% Cl,0.37–1.06); p $value_{heterogeneity,}$0.62; *p* value,0.08]. The same meta-analysis based on three association studies,which included 379 cases and 460 controls, reported no significant association of MMP2 gene rs243865 (-1306 C/T) SNP with the risk of colorectal cancer in both the dominant genetic model [CT+TT vs. CC: OR,0.97 (95% Cl,0.49–1.93); p $value_{heterogeneity,}$0.01; *p* value,0.81] andthe recessive genetic model [TTvs. CT+CC: OR,1.37 (95% Cl,0.61–3.11); p $value_{heterogeneity,}$0.66; *p* value,0.44].The same meta-analysis based on four association studies, which included 1,590 cases and 1,653 controls, reported no significant association of MMP2 gene rs243865 (-1306 C/T) SNP with the risk of breast cancer in both the dominant genetic model [CT+TT vs. CC: OR,0.66 (95% Cl,0. 40–1.07); p $value_{heterogeneity,}$<0.0001; *p* value,0.09] and the recessive genetic model [TTvs. CT+CC: OR,0.95 (95% Cl,0. 54–1.66); p $value_{heterogeneity,}$0.86; *p* value,0.3] (McColgan and Sharma 2009).

A case-controlstudy on ethnic Kashmiri population,which included 142 colorectal cancer cases and 184 healthy matched controls, reported a significant association of the variant genotype (TT) of MMP2 gene rs243865 (–1306 C/T) SNP with a decreased risk of colorectal cancer [CT vs. CC: OR,0.61 (95% Cl,0.37–1.01); *p* value,0.05] (Banday, Sameer et al. 2016). Another but small-scale case-controlstudy on Iranian population, which included 50 prostate cancer cases and 54 matched controls, observed no significant association between MMP2 gene rs243865 (–1306 C/T) SNP and risk of prostate cancer (Shajarehpoor Salavati, Tafvizi et al. 2017).A hospital-based case-controlstudy on Polish population, which included 648 unselected cancer cases (299 breast cancer cases, 199 lung cancer cases and 150 colon cancer cases) and 648 matched controls (299 controls for breast cancer, 199 controls for lung cancer and 150 controls for colon cancer), studied MMP2 gene rs243865 (–1306 C/T) SNP along with serum zinc levels. The study reported no significant association of this SNP with the risk of breast cancer [TT vs. CC: OR,0.56 (95% Cl,0.28–1.11); *p* value,0.10; CT vs. CC: OR,1.08 (95% Cl,0. 77–1.51); *p* value,0.66], lung cancer [TT vs. CC: OR,1.64 (95% Cl,0.56–4.83); *p* value, 0.37;CT vs. CC: OR,1.01 (95% Cl,0. 65–1.55); *p* value,0.98] and colon cancer [TT vs. CC: OR,1.27 (95% Cl,0.46–3.52); *p* value,0.65; CT vs. CC: OR,1.01 (95% Cl,0.59–1.67); *p* value,0.99] (Białkowska, Marciniak et al. 2020). Another hospital-based case-control study on Polish population, which included 197 prostate cancer cases and 197 healthy matched controls, reported no significant association of this SNP with prostate cancer [TT vs. CC: OR,0.76 (95% Cl,0.32–1.82); *p* value,0.54; CT vs. CC: OR,0.95 (95% Cl,0. 61–1.48); *p* value,0.82] (Białkowska, Marciniak et al. 2018).

6.5.7 Polymorphisms in Matrix Metalloproteinase 9 (MMP9) Gene

Matrix metalloproteinase 9 (MMP9) gene (chromosomal location, 20q13.12) encodes a 707 amino acid MMP9 pro-enzyme, also known as gelatinase B, 92 kDa type IV collagenase or 92 kDa gelatinase. The active form of MMP9 enzyme is an 82–85 kDa variably glycosylated protein. MMP9 enzyme, like MMP2 enzyme, contains the collagen-binding domain or CBD, but unlike MMP2 enzyme, it additionally contains a serine- and threonine-rich, abundantly and variably O-glycosylatedand variably N-glycosylated type V collagen-like 54 amino acid linker domain, known as OG-domain, that connects the collagen-binding domain to the hemopexin (HPX)-like domain and gives this enzyme a very high flexibility. The hemopexin (HPX)-like domain and collagen-binding domain (CBD) are involved in substrate binding (Elkins, Ho et al. 2002, Van den Steen, Dubois et al. 2002, Overall and Butler 2007). MMP9 gene expression unlike MMP2 gene expression is not constitutive and ubiquitous, and under normal physiological conditions, only a few cell types, predominantly immune cells and in particular-neutrophils, express MMP9 enzyme. In neutrophils, MMP9 enzyme is stored in granules and covalently linked to neutrophil gelatinase B-associated lipocalin (NGAL) for rapid release upon proper stimulation. However, in several other cell types including tumor cells, the expression of MMP9 gene, unlike MMP2, is highly inducible by various mediators, predominantly inflammatory mediators (Van den Steen, Dubois et al. 2002). The activation of proMMP9 enzyme,a regular event during inflammatory processes, occurs through the proteolytic cleavage of pro-domain mediated by several proteolytic enzymes including serine proteases such as urokinase (urokinase-type plasminogen activator, uPA), trypsin and kallikrein and by other MMPs such as MMP2, MMP3, MMP7, MMP13 and directly through oxidative stress-induced disruption of the cysteine switch (Maeda, Okamoto et al. 1998, Van den Steen, Dubois et al. 2002). MMP9 enzyme is inhibitable by TIMPs such as TIMP3, TIMP4 and mainly TIMP1 (Crocker, Pagenstecher et al. 2004).

The substrates of MMP9 enzyme predominantly include various ECM and BM components and in addition non-ECM components (Table 6.2) and overlap considerably with that of MMP2 enzyme with the notable exception of type I collagen, which MMP9 enzyme is unable to cleave directly. These include fibronectin, gelatins, casein,nidogen-1 (entactin), elastin, casein, galectin3, aggrecan, plasminogen, myelin basic protein (MBP)and type III, type IV, type V, type VII, type X and type XII collagens. MMP9 enzyme is also involved in the proteolytic cleavage and release of several cell membrane- or cell surface-bound cellular mediators through shedding of their ectodomains or release of inhibitory binding factors. These mediators include intercellular adhesion molecule1 (ICAM1), nexin1, monocyte chemotactic protein 3(MCP3), interleukin8 (IL8), proIL1β, IL2 receptor, pro-transforming growth factor beta (pro-TGF-β), insulin-like growth factor-binding protein 3 (IGFBP3), dystroglycanβ-subunit and β2 integrin (Björklund and Koivunen 2005, Chakrabarti and Patel 2005, Vaisar, Kassim et al. 2009). MMP9 gene expression in various types of cells, unlike MMP2 gene expression, is under strong regulation by various mediators, predominantly inflammatory mediators including growth factorssuch as epidermal growth factor (EGF); cytokines such as TNF-αand TGF-β; and adhesion molecules, predominantly activated integrins, because MMP9 gene promoter, unlike MMP2 gene promoter, contains binding sites forvarious regulatory mediators including inflammation-related transcription factors such as SNAI2 (SLUG) and nuclear factor-κB (NF-κB). The cellular signaling cascades, Raf/MEK/ERK and p38 MAPK/MAPK2, directly regulate and maintain MMP9 gene expression (Chakrabarti and Patel 2005, Moon, Cho et al. 2007, Sun, Diamond et al. 2008, Zhao and Benveniste 2008, Beliveau, Mott et al. 2010).

MMP9, predominantly through its role in the proteolysis of various ECM and BM components, plays a vital role in various physiological and pathophysiological processes such as inflammation, neoplastic diseases (cancers), muscular, skeletal and bone pathologies, cardiovascular and respiratory disorders and nervous system diseases. MMP9 enzyme plays a vital role in mediating several processes that regulate inflammatory immune response through proteolysis of diverse substrates that promote inflammation. MMP9 enzyme activates pro-inflammatory mediators including cytokines such as interleukin1β (IL1β), IL8, chemoattractants and growth factors such as TGF-β and promotes recruitment of various types of immune cells such as lymphocytes, neutrophils and dendritic cells (Lemjabbar, Gosset et al. 1999, Chakrabarti and Patel 2005). MMP9 enzyme, predominantly through proteolysis of diverse substrates including various ECM and BM components, plays a vital role in important cellular processes such as angiogenesis under normalphysiological conditions and in diseased pathophysiological state. MMP9 enzyme plays a vital, but differential role in mediating the process of angiogenesis. MMP9 enzyme is one of the main enzymes that through direct proteolytic cleavage of ECM and vascular basement membrane proteins and proteolytic activation generate active forms of various pro-angiogenic mediators including growth factors and other mediatorssuch asvascular endothelial growth factor (VEGF) and ligand for receptor-type protein-tyrosine kinase KIT (KIT ligand or KITLG).

These mediators stimulate the activation of endothelial cells and recruitment of pericytes to induce the formation of new vasculature and promote recruitment of various types of inflammatory immune cells that maintain the correct orientation or positioning of pericytes on the new vasculature (Bergers, Brekken et al. 2000, Ferrara, Gerber et al. 2003, Chantrain, Shimada et al. 2004, Zijlstra, Seandel et al. 2006). MMP9 enzyme also mediates the process of ossification or formation of new bones. Various studies report the association of diminished or null MMP9 gene expression and enzyme activity with impeded formation of new bones due to reduced recruitment of bone-forming cells predominantly osteoclasts and decreased angiogenesis in bone growth plates (Engsig, Chen et al. 2000). In contrast to the previously discussed pro-angiogenic role, MMP9 enzyme through its hemopexin (HPX)-like domain inhibits angiogenesis as revealed through the decrease in invasiveness of glioblastoma cellsin aglioblastomaxenograftmodel. However, the physiological relevance of this finding with regard to humans is not clear yet (Ezhilarasan, Jadhav et al. 2009). MMP9 enzyme also plays a vital role in mediating several processes thatregulate various processes related to human female reproduction including menstrual remodeling of endometrial tissue and embryoimplantation. Various studies report the association of diminished or null MMP9 gene expression and enzyme activity with impaired reproduction (Dubois, Arnold et al. 2000, Van den Steen, Dubois et al. 2002).

MMP9 enzyme is also involved in the pathogenesis of various diseases or disorders. These include nervous system diseases such as multiple sclerosis and Alzheimer's diseases (Van den Steen, Dubois et al. 2002, Lorenzl, Albers et al. 2003, Fainardi, Castellazzi et al. 2009); cardiovascular disorders such as atherosclerosis, stroke and myocardial infarction (Ramos-Fernandez, Bellolio et al. 2011); and muscular, skeletal and bone disorders such as rheumatoid arthritis (RA) (Van den Steen, Grillet et al. 2005, Burrage, Mix et al. 2006). MMP9 enzyme is also involved in the pathogenesis of a wide spectrum of autoimmune disorders including systemic lupus erythematosus (SLE), lupus nephritis, polymyositis, myocarditis, rheumatoid arthritis (RA), chronic obstructive pulmonary disease (COPD), asthma, multiple sclerosis, Sjögren's syndrome (SjS), bullous pemphigoid and autoimmune inner ear disease (AIED) (Noseworthy, Lucchinetti et al. 2000, Hulkkonen, Pertovaara et al. 2004, Matsumoto, Niimi et al. 2005, Fenton, Fismen et al. 2009, Ma, Jiang et al. 2010, Pathak, Goldofsky et al. 2011).

MMP9 enzyme, predominantly through proteolysis of diverse substrates including various ECM and BM components and its role in important cellular processes such as angiogenesis, plays a critical role in the pathogenesis of various cancers and is involved in almost all the important tumorigenic processes including tumor invasion and metastasis (Björklund and Koivunen 2005, Fingleton 2006). MMP9 enzyme induces and sustains angiogenesis through activation of various pro-angiogenic mediators-including growth factors and other mediators such as VEGF and KIT ligand. The sustained angiogenesis or neovascularization in turn is an essential process for the growth and proliferation of tumors (Bergers, Brekken et al. 2000, Ferrara, Gerber et al. 2003, Chantrain, Shimada et al. 2004, Zijlstra, Seandel et al. 2006). MMP9 enzyme impairs cell-cell adhesion through cleavage of protease nexin1 (PN1) serpin and adhesion molecules such as dystroglycanβ-subunit (Bozzi, Inzitari et al. 2009, Xu, McKee et al. 2010) and modulates, often increasing the bioavailability of various growth factors such as TGF-β and EGF. These processes promote tumor growth and contribute to tumor invasion by playing a crucial role in epithelial-mesenchymal transition (EMT) (Ongusaha, Kwak et al. 2004, Aresu, Benali et al. 2011). MMP9 enzyme mediates inhibition of apoptosis through cleavage of KIT ligand and promotes escape of tumor cells from cytotoxic T-cell (CTL) - and natural killer (NK) cell-mediated immunesurveillance through ICAM-1 shedding. Both these processes promote tumor growth, proliferation and invasion (Fiore, Fusco et al. 2002, Meyer, Vollmer et al. 2005). Various studies report MMP9 gene overexpression and increase in MMP9 enzyme serum levels and MMP9 enzyme activity in almost all the human cancers including brain cancer, breast cancer, bladder cancer, colorectal cancer, gastric cancer, melanoma, ovarian cancer, pancreatic cancer and prostate cancer (Egeblad and Werb 2002). Several studies report the significance of MMP9 enzyme as an independent potential prognostic marker in various cancers including breast cancer, colorectal cancer and urinary bladdercancer. These studies also report on the association of MMP9 gene overexpression and the increase in MMP9 enzyme serum levels and MMP9 enzyme activity with poor or worse prognosis and reduced overall survival in all these cancers (Egeblad and Werb 2002, Turpeenniemi-Hujanen 2005, Buhmeida, Bendardaf et al. 2009, Offersen, Knap et al. 2010, Ranogajec, Jakić-Razumović et al. 2012). However, the role of MMP9 enzyme in cancer like many other MMPs is complex and not fully clear yet.

The functional SNP in MMP9 gene reported to modulate MMP9 gene transcription, expression and subsequently MMP9 enzyme serum levels and MMP9 enzyme activity in an allele-specific manner most prominently includes MMP9 gene rs3918242 (-1562C/T) SNP. The MMP9 gene rs3918242 (-1562C/T) SNP modulates the transcription and expression of MMP9 gene in an allele-specific manner through its modifying effect on the binding affinities and interactions of a putative transcriptional repressor protein with its binding site on MMP9 gene promoter. The transcriptional repressor protein has a higher affinity for MMP9 gene-1562CC genotype in comparison with MMP9 gene -1562TT genotype, and the MMP9 gene -1562C to -1562 Ttransition leads to a decrease in the affinity of this transcriptional repressor protein to bind to its specific site on MMP9 gene promoter. The MMP9 gene -1562TT genotype and MMP9 gene-1562CT genotype therefore havea higher promoter activity in comparison with MMP9 gene-1562CC genotype, which manifests as MMP9 gene overexpression and increase in MMP9 enzyme serum levels and MMP9 enzyme activity (Peng, Cao et al. 2010).

Numerous association studies report on the possible role of MMP9 gene rs3918242 (-1562C/T) SNP in modulating the risk for the development of various cancers (Rollin, Régina et al. 2007, Woo, Park et al. 2007, Wu, Wu et al. 2007, Park, Kim et al. 2011), and studies report on this association in different ethnic populations (Woo, Park et al. 2007, Li, Qu et al. 2013). However, many studies report no such association. Here, we mention several of these studies that report the association of this SNP with the risk of various cancers and the studies that report no significant association. A meta-analysis based on three association studies on Asian and Caucasian populations, which included 1,052 cases and 839 controls, reported a significant association of MMP9 gene rs3918242 (-1562C/T) SNP with a decreased risk of lung cancer in total population in homozygous genetic model [TT vs. CC: OR,0.28 (95% Cl,0.13–0.59); I^2=0.0; *p* value,0.001], recessive genetic model [TTvs. CT+CC: OR,0.28 (95% Cl,0.13–0.59); I^2=0.0; *p* value,0.001] and allelic genetic model [T vs. C: OR,0.80 (95% Cl,0.67–0.97); I^2= 0.0; *p* value,0.021], but not in dominant genetic model [CT+TT vs. CC: OR,0.86 (95% Cl,0.69–1.05); I^2= 0.0; *p* value,>0.05] and heterozygous genetic model [CT vs. CC: OR,0.93 (95% Cl,0.75–1.16); I^2= 0.0; *p* value,>0.05]. The study reported a significant association of this SNP with a decreased risk of lung cancer among Caucasian population in homozygous genetic model [TT vs. CC: OR,0.27 (95% Cl,0.12–0.62); I^2=0.0; *p* value,0.002] and recessive genetic model [TT vs. CT+CC: OR, 0.27 (95% Cl,0.12–0.62); I^2=0.0; *p* value,0.002], but not in dominant genetic model [CT+TT vs. CC: OR,0.89 (95% Cl,0.71–1.12); I^2= 0.0; *p* value,>0.05], heterozygous genetic model [CT vs. CC: OR,0.98 (95% Cl,0.77–1.24); I^2= 0.0; *p* value,>0.05] and allelic genetic model [T vs. C: OR,0.83 (95% Cl,0.67–1.02); I^2= 0.0; *p* value,>0.05]. The study reported no significant association of this SNP among Asian population in all genetic models: dominant genetic model [CT+TT vs. CC: OR,0.68 (95% Cl,0.40–1.15); *p* value,>0.05], recessive genetic model [TT vs. CT+CC: OR,0.33 (95% Cl,0.05–1.99); *p* value,>0.05], allelic genetic model [T vs. C: OR,0.68 (95% Cl,0.43–1.08); *p* value,>0.05], homozygous genetic model [TT vs. CC: OR,0.30 (95% Cl,0.05–1.82); *p* value,>0.05] and heterozygous genetic model [CT vs. CC: OR,0.72 (95% Cl,0.42–1.23); *p* value,>0.05] (Li, Liu et al. 2020). Another meta-analysis based on two association studies, which included 333 cases and 440 controls, reported no significant association of MMP9 gene rs3918242 (-1562C/T) SNP with the risk of lung cancer in dominant genetic model [CT+TT vs. CC: OR,1.03 (95% Cl,0.73–1.45); p $value_{heterogeneity}$,0.38; *p* value,0.86] and recessive genetic model [TT vs. CT+CC: OR,0.77 (95% Cl,0.01–60.82); p $value_{heterogeneity}$,0.04; *p* value,0.91]. The same meta-analysis based on five association studies, which included 700 cases and 963 controls in dominant genetic model, and four association studies, which included 575 cases and 836 controls, reported no significant association of MMP9 gene rs3918242 (-1562C/T) SNP with the risk of colorectal cancer in both the dominant genetic model [CT+TT vs. CC: OR,0.90 (95% Cl,0. 68–1.17); *p* value heterogeneity,0.27; *p* value,0.43] and the recessive genetic model [TTvs. CT+CC: OR,0.60 (95% Cl,0.15–2.41); p $value_{heterogeneity}$,0.35; *p* value,0.47]. The same meta-analysis based on two association studies, which included 1,042 cases and 1,046 controls, reported no significant association of MMP9 gene rs3918242 (-1562C/T) SNP with the risk of breast cancer in both the dominant genetic model [CT+TT vs. CC: OR,1.07

(95% CI,0.88–1.30); p value$_{heterogeneity}$,0.6; p value,0.50] and the recessive genetic model [TTvs. CT+CC: OR,1.03 (95% CI,0.16–6.75); p value$_{heterogeneity}$,0.17; p value,0.98] (McColgan and Sharma 2009).

A meta-analysis based on 23 association studies on European and Asian populations, which included 5,098 cases and 6,442 controls, reported no significant association of MMP9 gene rs3918242 (-1562C/T) SNP with the risk of lung cancer (two studies: 333 cases; 440 controls) in all the three genetic models: dominant genetic model [CT+TT vs. CC: OR,1.03 (95% CI,0.73–1.45); p value$_{heterogeneity}$,0.375; I^2= 0.0], recessive genetic model [TT vs. CT+CC: OR,0.77 (95% CI,0.10–60.82); p value$_{heterogeneity}$,0.038; I^2= 76.8] and allelic genetic model [T vs. C: OR,0.93 (95% CI,0.55–1.57); p value$_{heterogeneity}$,0.122; I^2= 58.2]. The study reported no significant association of MMP9 gene rs3918242 (-1562C/T) SNP with the risk of colorectal cancer (five studies: 693 cases; 904 controls) in all the three genetic models:dominant genetic model [CT+TT vs. CC: OR,0.85 (95% CI,0.67–1.08); p value$_{heterogeneity}$,0.426; I^2= 0.0], recessive genetic model [TT vs. CT+CC: OR,0.61 (95% CI,0.20–1.85); p value$_{heterogeneity}$,0.331; I^2= 12.3] and allelic genetic model [T vs. C: OR,0.85 (95% CI,0.67–1.08); p value$_{heterogeneity}$,0.478; I^2= 0.0]. The same study reported no significant association of MMP9 gene rs3918242 (-1562C/T) SNP with the risk of head and neck cancer (five studies: 781 cases; 1,050 controls) in all the three genetic models: dominant genetic model [CT+TT vs. CC: OR,1.13 (95% CI,0.91–1.41); p value$_{heterogeneity}$,0.125; I^2= 44.6], recessive genetic model [TT vs. CT+CC: OR,1.70 (95% CI,0.63–4.60); p value$_{heterogeneity}$,0.393; I^2= 0.0] and allelic genetic model [T vs. C: OR,1.13 (95% CI,0.93–1.37); p value$_{heterogeneity}$,0.212; I^2= 31.5]. The study also reported no significant association of MMP9 gene rs3918242 (-1562C/T) SNP with the risk of breast cancer (two studies: 1,042 cases; 1,046 controls) in all the three genetic models: dominant genetic model [CT+TT vs. CC: OR,1.07 (95% CI,0.88–1.30); p value$_{heterogeneity}$,0.604; I^2= 0.0], recessive genetic model [TT vs. CT+CC: OR,1.56 (95% CI,0.84–2.92); p value$_{heterogeneity}$,0.168; I^2= 47.3] and allelic genetic model [T vs. C: OR,1.09 (95% CI,0.92–1.30); p value$_{heterogeneity}$,0.968; I^2= 0.0]. The same study reported no significant association of MMP9 gene rs3918242 (-1562C/T) SNP with the risk of gastric cancer (three studies: 575 cases; 1,198 controls) in all the three genetic models: dominant genetic model [CT+TT vs. CC: OR,0.99 (95% CI,0.78–1.25); p value$_{heterogeneity}$,0.390; I^2= 0.0], recessive genetic model [TT vs. CT+CC: OR,1.68 (95% CI,0.79–3.57); p value$_{heterogeneity}$,0.958; I^2= 0.0] and allelic genetic model [T vs. C: OR,1.03 (95% CI,0.83–1.27); p value$_{heterogeneity}$,0.496; I^2= 0.0]. The same study reported no significant association of MMP9 gene rs3918242 (-1562C/T) SNP with the risk of bladder cancer, renal cancer, hepatocellular cancer, uterine cancer, ovarian cancer and endometrial cancer (six studies: 1,674 cases; 1,804 controls) in all the three genetic models: dominant genetic model [CT+TT vs. CC: OR,0.93 (95% CI,0.79–1.08); p value$_{heterogeneity}$,0.496; I^2= 0.0], recessive genetic model [TT vs. CT+CC: OR,0.82 (95% CI,0.50–1.34); p value$_{heterogeneity}$,0.324; I^2= 14.2] and allelic genetic model [T vs. C: OR,0.93 (95% CI,0.81–1.06); p value$_{heterogeneity}$,0.366; I^2= 7.9]. The study reported no significant difference in the distribution or frequency of MMP2–1306T allele among the populations under study [European: 15.4% (95% CI,10.2–20.6%) vs. Asian: 13.5% (95% CI,11.8–15.2%); p value,0.3101] (Peng, Cao et al. 2010). Another meta-analysis based on four association studies reported no significant association of MMP9 gene rs3918242 (-1562C/T) SNP with the risk of lung cancer among Asian population in both-the genetic models: dominant genetic model [CT+TT vs. CC: OR,0.94 (95% CI,0.68–1.29); p value$_{heterogeneity}$,0.13; p value,0.70] and recessive genetic model [TTvs. CT+CC: OR,1.21 (95% CI,0.06–25.94); p value$_{heterogeneity}$,0.08; p value,0.90]. The study reported a significant association of MMP9 gene rs3918242 (-1562C/T) SNPwith a decreased risk of lung cancer among Caucasian population in recessive genetic model [TT vs. CT+CC: OR,0.27 (95% CI,0.12–0.62); p value$_{heterogeneity}$,0.38; p value,0.002], but not in the dominant genetic model [CT+TT vs. CC: OR,0.89 (95% CI,0.71–1.12); p value$_{heterogeneity}$,0.72; p value,0.33]. The study reported a significant association of MMP9 gene rs3918242 (-1562C/T) SNP with a decreased risk of lung cancer in total population in recessive genetic model [TT vs. CT+CC: OR,0.38 (95% CI,0.19–0.75); p value$_{heterogeneity}$,0.19; p value,0.005], but not in dominant genetic model [CT+TT vs. CC: OR,0.91 (95% CI,0.75–1.09); p value$_{heterogeneity}$,0.47; p value,0.31](Hu, Wang et al. 2013).

A hospital-based case-control study on ethnic Han Chinese population, which included 245 non-small cell lung carcinoma (NSCLC) cases and 258 healthy matched controls, reported a significant association of MMP9 gene rs3918242 (-1562C/T) SNP with a decreased risk of non-small cell lung carcinoma in heterozygous genetic model [CT vs. CC: OR,0.45 (95% CI,0.29-0.68); p value,0.00]. The study reported a significant association of MMP9 gene rs3918242C (-1562C) allele with an increased risk of non-small cell lung carcinoma (NSCLC) [C vs. T: OR,2.0 (95% CI,1.36–2.95); p value,0.00] (Li, Jia et al. 2019). Another case-control study on ethnic Kashmiri population, which included 142 colorectal cancer cases and 184 healthy matched controls, reported a significant association of the heterozygous genotype (CT) of MMP9 gene rs3918242 (-1562C/T) SNP with an increased risk of colorectal cancer [CT vs. CC: OR,1.88 (95% CI,1.11–3.18); p value,0.02]. Further, the study reported a significant association of the variant allele MMP9 gene-1562T allele with an increased risk of colorectal cancer [T vs. C: OR,1.74 (95% CI,1.15–2.62); p value,0.007] (Banday, Sameer et al. 2016).

6.6 Conclusions

From the review and discussion presented in this chapter,it is amply clear that the MMPs including MMP7, MMP1, MMP8, MMP13, MMP3, MMP2 and MMP9 predominantly through the proteolysis of diverse substrates play a critical rolein almost all the important pro-tumorigenic processes. It is also evident that the various functional SNPs in the genes encoding these MMPs have the potential to modulategene transcription, expression and subsequently serum levels and enzyme activity of MMPs in an allele-specific manner. However, as supported by numerous association studies presented in this chapter, the role of SNPs in MMP genes in modulating the risk for the development of various cancers is evident, but not fully clear. A single possible cause of this outcome is not possible. However, many possible factors may contribute toward this outcome. First, although predominantly they mediate or promote pro-tumorigenic processes, MMPs play a differential role in tumorigenesis and it is notpossible to designate MMPs strictly either as pro-cancerous or as anti-cancerous. Second, MMPs mediate important cellular processes associated with various signaling pathways, but they are not the central or major regulatory mediators that alone can modulate the activity of signaling pathways toward a pro-cancerous or anti-cancerous function. In other words, MMPs can play a role in the processes that modulate the pathogenesis of various cancers, but MMPs have almost no major role in deciding the final fate of any tumorigenic event. Third, the SNPs in general, even in the genes with major regulatory functions critical to tumorigenesis processes, as risk modulatory factors do not act alone. The outcome possibly results from the SNPs acting in conjunction with other SNPs and even with other risk modulatory factors including the presence or absence of mutations or even various cellular mediators. Nevertheless, the role of MMPs and their SNPs in the pathogenesis of various cancers and in modulating the risk of various cancers is a significant and relevant research question in the field of cancer biology.

ABBREVIATIONS

95% Cl:	95% confidence interval
A2MR:	α_2-macroglobulin receptor
AP1:	activator protein 1
BM:	basement membrane
CAMs:	cell adhesion molecules
CTGF:	connective tissue growth factor
ECM:	extracellular matrix
GPI:	glycosylphosphatidylinositol
HB-EGF:	heparin-binding EGF-like growth factor
HPX:	hemopexin
IGFBP:	insulin-like growth factor-binding protein
IL:	interleukin
MBP:	myelin basic protein
MCP:	monocyte chemotactic protein
MMPs:	matrix metalloproteinases
MT-MMPs:	membrane-type MMPs
OR:	odds ratio
RANKL:	receptor activator of nuclear factor-kappaB ligand
RECK:	reversion-inducing cysteine-rich protein with Kazal motifs
Sp1:	specificity protein 1
TGF:	transforming growth factor
THBS:	thrombospondin
TIMPs:	tissue inhibitors of matrix metalloproteinases
TNF:	tumor necrosis factor
uPA:	urokinase-type plasminogen activator
Vn:	vitronectin-like insertion

REFERENCES

Adachi, Y., H.Yamamoto, F.Itoh, Y.Arimura, M.Nishi, T.Endo and K.Imai (2001). Clinicopathologic and prognostic significance of matrilysin expression at the invasive front in human colorectal cancers.*Int J Cancer* **95**(5): 290–294.

Agarwal, D., S.Goodison, B.Nicholson, D.Tarin and V.Urquidi (2003). Expression of matrix metalloproteinase 8 (MMP-8) and tyrosinase-related protein-1 (TYRP-1) correlates with the absence of metastasis in an isogenic human breast cancer model.*Differentiation* **71**(2): 114–125.

Aihara, R., E.Mochiki, T.Nakabayashi, K.Akazawa, T.Asao and H.Kuwano (2005). Clinical significance of mucin phenotype, beta-catenin and matrix metalloproteinase 7 in early undifferentiated gastric carcinoma.*Br J Surg* **92**(4): 454–462.

Airola, K., N.Johansson, A. L.Kariniemi, V. M.Kähäri and U. K.Saarialho-Kere (1997). Human collagenase-3 is expressed in malignant squamous epithelium of the skin.*J Invest Dermatol* **109**(2): 225–231.

Ala-Aho, R., N.Johansson, A. H.Baker and V. M.Kähäri (2002). Expression of collagenase-3 (MMP-13) enhances invasion of human fibrosarcoma HT-1080 cells.*Int J Cancer* **97**(3): 283–289.

Aresu, L., S.Benali, S.Garbisa, E.Gallo and M.Castagnaro (2011). Matrix metalloproteinases and their role in the renal epithelial mesenchymal transition.*Histol Histopathol* **26**(3): 307–313.

Baker, A. H., D. R.Edwards and G.Murphy (2002). Metalloproteinase inhibitors: biological actions and therapeutic opportunities.*J Cell Sci* **115**(Pt 19): 3719–3727.

Balbín, M., A.Fueyo, V.Knäuper, A. M.Pendás, J. M.López, M. G.Jiménez, G.Murphy and C.López-Otín (1998). Collagenase 2 (MMP-8) expression in murine tissue-remodeling processes. Analysis of its potential role in postpartum involution of the uterus.*J Biol Chem* **273**(37): 23959–23968.

Balbín, M., A.Fueyo, A. M.Tester, A. M.Pendás, A. S.Pitiot, A.Astudillo, C. M.Overall, S. D.Shapiro and C.López-Otín (2003). Loss of collagenase-2 confers increased skin tumor susceptibility to male mice.*Nat Genet* **35**(3): 252–257.

Banday, M. Z., A. S.Sameer, A. H.Mir, T. A.Mokhdomi, N. A.Chowdri and E.Haq (2016). Matrix metalloproteinase (MMP) -2, -7 and -9 promoter polymorphisms in colorectal cancer in ethnic Kashmiri population - A case-control study and a mini review.*Gene* **589**(1): 81–89.

Barmina, O. Y., H. W.Walling, G. J.Fiacco, J. M.Freije, C.López-Otín, J. J.Jeffrey and N. C.Partridge (1999). Collagenase-3 binds to a specific receptor and requires the low density lipoprotein receptor-related protein for internalization.*J Biol Chem* **274**(42): 30087–30093.

Becker, C. M., G.Louis, A.Exarhopoulos, S.Mechsner, A. D.Ebert, D.Zurakowski and M. A.Moses (2010). Matrix metalloproteinases are elevated in the urine of patients with endometriosis.*Fertil Steril* **94**(6): 2343–2346.

Bein, K. and M.Simons (2000). Thrombospondin type 1 repeats interact with matrix metalloproteinase 2. Regulation of metalloproteinase activity. *J Biol Chem* **275**(41): 32167–32173.

Beliveau, A., J. D.Mott, A.Lo, E. I.Chen, A. A.Koller, P.Yaswen, J.Muschler and M. J.Bissell (2010). Raf-induced MMP9 disrupts tissue architecture of human breast cells in three-dimensional culture and is necessary for tumor growth in vivo.*Genes Dev* **24**(24): 2800–2811.

Bergers, G., R.Brekken, G.McMahon, T. H.Vu, T.Itoh, K.Tamaki, K.Tanzawa, P.Thorpe, S.Itohara, Z.Werb and D.Hanahan (2000). Matrix metalloproteinase-9 triggers the angiogenic switch during carcinogenesis.*Nat Cell Biol* **2**(10): 737–744.

Bertini, I., V.Calderone, M.Fragai, C.Luchinat, M.Maletta and K. J.Yeo (2006). Snapshots of the reaction mechanism of matrix metalloproteinases.*Angew Chem Int Ed Engl* **45**(47): 7952–7955.

Betsuyaku, T., M.Nishimura, K.Takeyabu, M.Tanino, P.Venge, S.Xu and Y.Kawakami (1999). Neutrophil granule proteins in bronchoalveolar lavage fluid from subjects with subclinical emphysema.*Am J Respir Crit Care Med* **159**(6): 1985–1991.

Białkowska, K., W.Marciniak, M.Muszyńska, P.Baszuk, S.Gupta, K.Jaworska-Bieniek, G.Sukiennicki, K.Durda, T.Gromowski, M.Lener, K.Prajzendanc, A. Łukomska, C. Cybulski, T. Huzarski, J. Gronwald, T. Dębniak, J. Lubiński and A. Jakubowska (2020). Polymorphisms in MMP-1, MMP-2, MMP-7, MMP-13 and MT2A do not contribute to breast, lung and colon cancer risk in polish population. *Hered Cancer Clin Pract* **18**: 16.

Białkowska, K., W.Marciniak, M.Muszyńska, P.Baszuk, S.Gupta, K.Jaworska-Bieniek, G.Sukiennicki, K.Durda, T.Gromowski, K.Prajzendanc, C.Cybulski, T.Huzarski, J.Gronwald, T.Dębniak, R. J.Scott, J.Lubiński and A.Jakubowska (2018). Association of zinc level and polymorphism in MMP-7 gene with prostate cancer in Polish population.*PLoS One* **13**(7): e0201065.

Billinghurst, R. C., L.Dahlberg, M.Ionescu, A.Reiner, R.Bourne, C.Rorabeck, P.Mitchell, J.Hambor, O.Diekmann, H.Tschesche, J.Chen, H.Van Wart and A. R.Poole (1997). Enhanced cleavage of type II collagen by collagenases in osteoarthritic articular cartilage.*J Clin Invest* **99**(7): 1534–1545.

Binder, C., T.Hagemann, B.Husen, M.Schulz and A.Einspanier (2002). Relaxin enhances in-vitro invasiveness of breast cancer cell lines by up-regulation of matrix metalloproteases.*Mol Hum Reprod* **8**(9): 789–796.

Binker, M. G., A. A.Binker-Cosen, H. Y.Gaisano, R. H.de Cosen and L. I.Cosen-Binker (2011). TGF-β1 increases invasiveness of SW1990 cells through Rac1/ROS/NF-κB/IL-6/MMP-2.*Biochem Biophys Res Commun* **405**(1): 140–145.

Biswas, M. H., C.Du, C.Zhang, J.Straubhaar, L. R.Languino and K. C.Balaji (2010). Protein kinase D1 inhibits cell proliferation through matrix metalloproteinase-2 and matrix metalloproteinase-9 secretion in prostate cancer.*Cancer Res* **70**(5): 2095–2104.

Björklund, M. and E.Koivunen (2005). Gelatinase-mediated migration and invasion of cancer cells.*Biochim Biophys Acta* **1755**(1): 37–69.

Blavier, L., A.Lazaryev, X. H.Shi, F. J.Dorey, G. M.Shackleford and Y. A.DeClerck (2010). Stromelysin-1 (MMP-3) is a target and a regulator of Wnt1-induced epithelial-mesenchymal transition (EMT).*Cancer Biol Ther* **10**(2): 198–208.

Borkakoti, N. (2000). Structural studies of matrix metalloproteinases.*J Mol Med (Berl)* **78**(5): 261–268.

Bozzi, M., R.Inzitari, D.Sbardell, S.Monaco, E.Pavoni, M.Gioia, S.Marini, S.Morlacchi, F.Sciandra, M.Castagnola, B.Giardina, A.Brancaccio and M.Coletta (2009). Enzymatic processing of beta-dystroglycan recombinant ectodomain by MMP-9: identification of the main cleavage site.*IUBMB Life* **61**(12): 1143–1152.

Brew, K. and H.Nagase (2010). The tissue inhibitors of metalloproteinases (TIMPs): an ancient family with structural and functional diversity.*Biochim Biophys Acta* **1803**(1): 55–71.

Brinckerhoff, C. E. and L. M.Matrisian (2002). Matrix metalloproteinases: a tail of a frog that became a prince.*Nat Rev Mol Cell Biol* **3**(3): 207–214.

Brinckerhoff, C. E., J. L.Rutter and U.Benbow (2000). Interstitial collagenases as markers of tumor progression.*Clin Cancer Res* **6**(12): 4823–4830.

Buhmeida, A., R.Bendardaf, M.Hilska, Y.Collan, M.Laato, S.Syrjänen, K.Syrjänen and S.Pyrhönen (2009). Prognostic significance of matrix metalloproteinase-9 (MMP-9) in stage II colorectal carcinoma.*J Gastrointest Cancer* **40**(3–4): 91–97.

Burke, B. (2004). The role of matrix metalloproteinase 7 in innate immunity.*Immunobiology* **209**(1–2): 51–56.

Burrage, P. S., K. S.Mix and C. E.Brinckerhoff (2006). Matrix metalloproteinases: role in arthritis.*Front Biosci* **11**: 529–543.

Cadigan, K. M. and M. L.Waterman (2012). TCF/LEFs and Wnt signaling in the nucleus.*Cold Spring Harb Perspect Biol* **4**(11).

Cazorla, M., L.Hernández, A.Nadal, M.Balbín, J. M.López, F.Vizoso, P. L.Fernández, K.Iwata, A.Cardesa, C.López-Otín and E.Campo (1998). Collagenase-3 expression is associated with advanced local invasion in human squamous cell carcinomas of the larynx.*J Pathol* **186**(2): 144–150.

Chakrabarti, S. and K. D.Patel (2005). Matrix metalloproteinase-2 (MMP-2) and MMP-9 in pulmonary pathology.*Exp Lung Res* **31**(6): 599–621.

Chang, H. J., M. J.Yang, Y. H.Yang, M. F.Hou, E. J.Hsueh and S. R.Lin (2009). MMP13 is potentially a new tumor marker for breast cancer diagnosis.*Oncol Rep* **22**(5): 1119–1127.

Chantrain, C. F., H.Shimada, S.Jodele, S.Groshen, W.Ye, D. R.Shalinsky, Z.Werb, L. M.Coussens and Y. A.DeClerck (2004). Stromal matrix metalloproteinase-9 regulates the vascular architecture in neuroblastoma by promoting pericyte recruitment.*Cancer Res* **64**(5): 1675–1686.

Chaudhary, A. K., M.Singh, A. C.Bharti, K.Asotra, S.Sundaram and R.Mehrotra (2010). Genetic polymorphisms of matrix metalloproteinases and their inhibitors in potentially malignant and malignant lesions of the head and neck.*J Biomed Sci* **17**(1): 10.

Choi, D. H., Y. J.Kim, Y. G.Kim, T. H.Joh, M. F.Beal and Y. S.Kim (2011). Role of matrix metalloproteinase 3-mediated alpha-synuclein cleavage in dopaminergic cell death.*J Biol Chem* **286**(16): 14168–14177.

Choke, E., G.Cockerill, W. R.Wilson, S.Sayed, J.Dawson, I.Loftus and M. M.Thompson (2005). A review of biological factors implicated in abdominal aortic aneurysm rupture.*Eur J Vasc Endovasc Surg* **30**(3): 227–244.

Clark, I. M., T. E.Swingler, C. L.Sampieri and D. R.Edwards (2008). The regulation of matrix metalloproteinases and their inhibitors.*Int J Biochem Cell Biol* **40**(6–7): 1362–1378.

Corry, D. B., K.Rishi, J.Kanellis, A.Kiss, L. Z.SongLz, J.Xu, L.Feng, Z.Werb and F.Kheradmand (2002). Decreased allergic lung inflammatory cell egression and increased susceptibility to asphyxiation in MMP2-deficiency.*Nat Immunol* **3**(4): 347–353.

Coussens, L. M., C. L.Tinkle, D.Hanahan and Z.Werb (2000). MMP-9 supplied by bone marrow-derived cells contributes to skin carcinogenesis.*Cell* **103**(3): 481–490.

Crocker, S. J., A.Pagenstecher and I. L.Campbell (2004). The TIMPs tango with MMPs and more in the central nervous system.*J Neurosci Res* **75**(1): 1–11.

Dahlen, B., J.Shute and P.Howarth (1999). Immunohistochemical localisation of the matrix metalloproteinases MMP-3 and MMP-9 within the airways in asthma.*Thorax* **54**(7): 590–596.

Decock, J., W.Hendrickx, U.Vanleeuw, V.Van Belle, S.Van Huffel, M. R.Christiaens, S.Ye and R.Paridaens (2008). Plasma MMP1 and MMP8 expression in breast cancer: protective role of MMP8 against lymph node metastasis.*BMC Cancer* **8**: 77.

Decock, J., J. R.Long, R. C.Laxton, X. O.Shu, C.Hodgkinson, W.Hendrickx, E. G.Pearce, Y. T.Gao, A. C.Pereira, R.Paridaens, W.Zheng and S.Ye (2007). Association of matrix metalloproteinase-8 gene variation with breast cancer prognosis.*Cancer Res* **67**(21): 10214–10221.

Dozier, S., G. P.Escobar and M. L.Lindsey (2006). Matrix metalloproteinase (MMP)-7 activates MMP-8 but not MMP-13. *Med Chem* **2**(5): 523–526.

Du, R., C.Petritsch, K.Lu, P.Liu, A.Haller, R.Ganss, H.Song, S.Vandenberg and G.Bergers (2008). Matrix metalloproteinase-2 regulates vascular patterning and growth affecting tumor cell survival and invasion in GBM.*Neuro Oncol* **10**(3): 254–264.

Dubois, B., B.Arnold and G.Opdenakker (2000). Gelatinase B deficiency impairs reproduction.*J Clin Invest* **106**(5): 627–628.

Duffy, M. J., T. M.Maguire, A.Hill, E.McDermott and N.O'-Higgins (2000). Metalloproteinases: role in breast carcinogenesis, invasion and metastasis.*Breast Cancer Res* **2**(4): 252–257.

Dziki, L., K.Przybyłowska, I.Majsterek, R.Trzciński, M.Mik and A.Sygut (2011). A/G polymorphism of the MMP-7 gene promoter region in colorectal cancer.*Pol Przegl Chir* **83**(11): 622–626.

Egeblad, M. and Z.Werb (2002). New functions for the matrix metalloproteinases in cancer progression.*Nat Rev Cancer* **2**(3): 161–174.

Eliason, J. L., K. K.Hannawa, G.Ailawadi, I.Sinha, J. W.Ford, M. P.Deogracias, K. J.Roelofs, D. T.Woodrum, T. L.Ennis, P. K.Henke, J. C.Stanley, R. W.Thompson and G. R.Upchurch, Jr. (2005). Neutrophil depletion inhibits experimental abdominal aortic aneurysm formation.*Circulation* **112**(2): 232–240.

Elkington, P., T.Shiomi, R.Breen, R. K.Nuttall, C. A.Ugarte-Gil, N. F.Walker, L.Saraiva, B.Pedersen, F.Mauri, M.Lipman, D. R.Edwards, B. D.Robertson, J.D'Armiento and J. S.Friedland (2011). MMP-1 drives immunopathology in human tuberculosis and transgenic mice.*J Clin Invest* **121**(-5): 1827–1833.

Elkins, P. A., Y. S.Ho, W. W.Smith, C. A.Janson, K. J.D'Alessio, M. S.McQueney, M. D.Cummings and A. M.Romanic (2002). Structure of the C-terminally truncated human ProMMP9, a gelatin-binding matrix metalloproteinase.*Acta Crystallogr D Biol Crystallogr* **58**(Pt 7): 1182–1192.

Elnemr, A., Y.Yonemura, E.Bandou, K.Kinoshita, T.Kawamura, S.Takahashi, S.Tochiori, Y.Endou and T.Sasaki (2003). Expression of collagenase-3 (matrix metalloproteinase-13) in human gastric cancer.*Gastric Cancer* **6**(1): 30–38.

Engsig, M. T., Q. J.Chen, T. H.Vu, A. C.Pedersen, B.Therkidsen, L. R.Lund, K.Henriksen, T.Lenhard, N. T.Foged, Z.Werb and J. M.Delaissé (2000). Matrix metalloproteinase 9 and vascular endothelial growth factor are essential for osteoclast recruitment into developing long bones.*J Cell Biol* **151**(4): 879–889.

Ezhilarasan, R., U.Jadhav, I.Mohanam, J. S.Rao, M.Gujrati and S.Mohanam (2009). The hemopexin domain of MMP-9 inhibits angiogenesis and retards the growth of intracranial glioblastoma xenograft in nude mice.*Int J Cancer* **124**(2): 306–315.

Fainardi, E., M.Castellazzi, C.Tamborino, A.Trentini, M. C.Manfrinato, E.Baldi, M. R.Tola, F.Dallocchio, E.Granieri and T.Bellini (2009). Potential relevance of cerebrospinal fluid and serum levels and intrathecal synthesis of active matrix metalloproteinase-2 (MMP-2) as markers of disease remission in patients with multiple sclerosis.*Mult Scler* **15**(5): 547–554.

Fang, W. L., W. B.Liang, L. B.Gao, B.Zhou, F. L.Xiao and L.Zhang (2013). Genetic polymorphisms in Matrix Metalloproteinases -1 and -7 and susceptibility to gastric cancer: an association study and meta-analysis.*Iran J Allergy Asthma Immunol* **12**(3): 203–210.

Fanjul-Fernández, M., A. R.Folgueras, S.Cabrera and C.López-Otín (2010). Matrix metalloproteinases: evolution, gene regulation and functional analysis in mouse models.*Biochim Biophys Acta* **1803**(1): 3–19.

Feng, J., Y.Chen, W.Hua, X.Sun, Y.Chen, Y.Liu, J.Fan, Y.Zhao, L.Zhao, X.Xu and X.Yang (2019). The MMP-8 rs11225395 promoter polymorphism increases cancer risk of non-asian populations: evidence from a meta-analysis.*Biomolecules* **9**(10):570.

Fenton, K., S.Fismen, A.Hedberg, N.Seredkina, C.Fenton, E. S.Mortensen and O. P.Rekvig (2009). Anti-dsDNA antibodies promote initiation, and acquired loss of renal Dnase1 promotes progression of lupus nephritis in autoimmune (NZBxNZW)F1 mice.*PLoS One* **4**(12): e8474.

Ferrara, N., H. P.Gerber and J.LeCouter (2003). The biology of VEGF and its receptors.*Nat Med* **9**(6): 669–676.

Fingleton, B. (2006). Matrix metalloproteinases: roles in cancer and metastasis.*Front Biosci* **11**: 479–491.

Fiore, E., C.Fusco, P.Romero and I.Stamenkovic (2002). Matrix metalloproteinase 9 (MMP-9/gelatinase B) proteolytically cleaves ICAM-1 and participates in tumor cell resistance to natural killer cell-mediated cytotoxicity.*Oncogene* **21**(34): 5213–5223.

Flex, A., E.Gaetani, A. S.Proia, G.Pecorini, G.Straface, F.Biscetti, G.Fioroni, A.Sabusco, R.Flore, P.Tondi, P.Pola and R.Pola (2006). Analysis of functional polymorphisms of metalloproteinase genes in persons with vascular dementia and Alzheimer's disease.*J Gerontol A Biol Sci Med Sci* **61**(10): 1065–1069.

Folgueras, A. R., A.Fueyo, O.García-Suárez, J.Cox, A.Astudillo, P.Tortorella, C.Campestre, A.Gutiérrez-Fernández, M.Fanjul-Fernández, C. J.Pennington, D. R.Edwards, C. M.Overall and C.López-Otín (2008). Collagenase-2 deficiency or inhibition impairs experimental autoimmune encephalomyelitis in mice.*J Biol Chem* **283**(14): 9465–9474.

Fray, M. J., R. P.Dickinson, J. P.Huggins and N. L.Occleston (2003). A potent, selective inhibitor of matrix metalloproteinase-3 for the topical treatment of chronic dermal ulcers.*J Med Chem* **46**(16): 3514–3525.

Fujimoto, T., S.Parry, M.Urbanek, M.Sammel, G.Macones, H.Kuivaniemi, R.Romero and J. F.Strauss, 3rd (2002). A single nucleotide polymorphism in the matrix metalloproteinase-1 (MMP-1) promoter influences amnion cell MMP-1 expression and risk for preterm premature rupture of the fetal membranes.*J Biol Chem* **277**(8): 6296–6302.

Galis, Z. S. and J. J.Khatri (2002). Matrix metalloproteinases in vascular remodeling and atherogenesis: the good, the bad, and the ugly.*Circ Res* **90**(3): 251–262.

García-Prieto, E., A.González-López, S.Cabrera, A.Astudillo, A.Gutiérrez-Fernández, M.Fanjul-Fernandez, E.Batalla-Solís, X. S.Puente, A.Fueyo, C.López-Otín and G. M.Albaiceta (2010). Resistance to bleomycin-induced lung fibrosis in MMP-8 deficient mice is mediated by interleukin-10.*PLoS One* **5**(10): e13242.

Gerstein, E. S., L.Sini, A. B.Ryabov, E. K.Dvorova, A. A.Yurchenko, I. S.Stilidi, N. E.Kushlinskii and M. I.Davydov (2009). Comparative enzyme immunoassay of matrix metalloproteinases-2, -7, -9 and their tissue inhibitor-2 in tumors and plasma of patients with gastric cancer. *Bull Exp Biol Med* **148**(6): 899–902.

Gill, S. E. and W. C.Parks (2008). Metalloproteinases and their inhibitors: regulators of wound healing.*Int J Biochem Cell Biol* **40**(6–7): 1334–1347.

Gioia, M., G. F.Fasciglione, S.Marini, S.D'Alessio, G.De Sanctis, O.Diekmann, M.Pieper, V.Politi, H.Tschesche and M.Coletta (2002). Modulation of the catalytic activity of neutrophil collagenase MMP-8 on bovine collagen I. Role of the activation cleavage and of the hemopexin-like domain.*J Biol Chem* **277**(26): 23123–23130.

Gomis-Rüth, F. X. (2003). Structural aspects of the metzincin clan of metalloendopeptidases.*Mol Biotechnol* **24**(2): 157–202.

Gomis-Rüth, F. X. (2009). Catalytic domain architecture of metzincin metalloproteases.*J Biol Chem* **284**(23): 15353–15357.

Gondi, C. S., L.Talluri, D. H.Dinh, M.Gujrati and J. S.Rao (2009). RNAi-mediated downregulation of MMP-2 activates the extrinsic apoptotic pathway in human glioma xenograft cells.*Int J Oncol* **35**(4): 851–859.

González-Arriaga, P., M. F.López-Cima, A.Fernández-Somoano, T.Pascual, M. G.Marrón, X. S.Puente and A.Tardón (2008). Polymorphism +17 C/G in matrix metalloprotease MMP8 decreases lung cancer risk.*BMC Cancer* **8**: 378.

Gutiérrez-Fernández, A., A.Fueyo, A. R.Folgueras, C.Garabaya, C. J.Pennington, S.Pilgrim, D. R.Edwards, D. L.Holliday, J. L.Jones, P. N.Span, F. C.Sweep, X. S.Puente and C.López-Otín (2008). Matrix metalloproteinase-8 functions as a metastasis suppressor through modulation of tumor cell adhesion and invasion.*Cancer Res* **68**(8): 2755–2763.

Hadler-Olsen, E., B.Fadnes, I.Sylte, L.Uhlin-Hansen and J. O.Winberg (2011). Regulation of matrix metalloproteinase activity in health and disease.*Febs J* **278**(1): 28–45.

Hadler-Olsen, E., J. O.Winberg and L.Uhlin-Hansen (2013). Matrix metalloproteinases in cancer: their value as diagnostic and prognostic markers and therapeutic targets.*Tumour Biol* **34**(4): 2041–2051.

Hanahan, D. and R. A.Weinberg (2011). Hallmarks of cancer: the next generation.*Cell* **144**(5): 646–674.

Hanemaaijer, R., T.Sorsa, Y. T.Konttinen, Y.Ding, M.Sutinen, H.Visser, V. W.van Hinsbergh, T.Helaakoski, T.Kainulainen, H.Rönkä, H.Tschesche and T.Salo (1997). Matrix metalloproteinase-8 is expressed in rheumatoid synovial fibroblasts and endothelial cells. Regulation by tumor necrosis factor-alpha and doxycycline.*J Biol Chem* **272**(50): 31504–31509.

Haque, S., N.Akhter, M.Lohani, A.Ali and R. K.Mandal (2015). Matrix metalloproteinase-2–1306 C>T gene polymorphism is associated with reduced risk of cancer: a meta-analysis. *Asian Pac J Cancer Prev* **16**(3): 889–896.

Harrell, P. C., L. J.McCawley, B.Fingleton, J. O.McIntyre and L. M.Matrisian (2005). Proliferative effects of apical, but not basal, matrix metalloproteinase-7 activity in polarized MDCK cells.*Exp Cell Res* **303**(2): 308–320.

Hattori, Y., K. C.Nerusu, N.Bhagavathula, M.Brennan, N.Hattori, H. S.Murphy, L. D.Su, T. S.Wang, T. M.Johnson and J.Varani (2003). Vascular expression of matrix metalloproteinase-13 (collagenase-3) in basal cell carcinoma.*Exp Mol Pathol* **74**(3): 230–237.

Henney, A. M., P. R.Wakeley, M. J.Davies, K.Foster, R.Hembry, G.Murphy and S.Humphries (1991). Localization of stromelysin gene expression in atherosclerotic plaques by in situ hybridization.*Proc Natl Acad Sci U S A* **88**(18): 8154–8158.

Heo, S. H., C. H.Cho, H. O.Kim, Y. H.Jo, K. S.Yoon, J. H.Lee, J. C.Park, K. C.Park, T. B.Ahn, K. C.Chung, S. S.Yoon and D. I.Chang (2011). Plaque rupture is a determinant of vascular events in carotid artery atherosclerotic disease: involvement of matrix metalloproteinases 2 and 9.*J Clin Neurol* **7**(2): 69–76.

Hiller, O., A.Lichte, A.Oberpichler, A.Kocourek and H.Tschesche (2000). Matrix metalloproteinases collagenase-2, macrophage elastase, collagenase-3, and membrane type 1-matrix metalloproteinase impair clotting by degradation of fibrinogen and factor XII.*J Biol Chem* **275**(42): 33008–33013.

Hojilla, C. V., F. F.Mohammed and R.Khokha (2003). Matrix metalloproteinases and their tissue inhibitors direct cell fate during cancer development.*Br J Cancer* **89**(10): 1817–1821.

Hu, C., J.Wang, Y.Xu, X.Li, H.Chen, H.Bunjhoo, W.Xiong, Y.Xu and J.Zhao (2013). Current evidence on the relationship between five polymorphisms in the matrix metalloproteinases (MMP) gene and lung cancer risk: a meta-analysis. *Gene* **517**(1): 65–71.

Hulkkonen, J., M.Pertovaara, J.Antonen, A.Pasternack, M.Hurme, P.Pöllänen and T.Lehtimäki (2004). Matrix metalloproteinase 9 (MMP-9) gene polymorphism and MMP-9 plasma levels in primary Sjogren's syndrome.*Rheumatology (Oxford)* **43**(12): 1476–1479.

Humphries, S. E., L. A.Luong, P. J.Talmud, M. H.Frick, Y. A.Kesäniemi, A.Pasternack, M. R.Taskinen and M.Syvänne (1998). Thc 5A/6A polymorphism in the promoter of the stromelysin-1 (MMP-3) gene predicts progression of angiographically determined coronary artery disease in men in the LOCAT gemfibrozil study. Lopid Coronary Angiography Trial.*Atherosclerosis* **139**(1): 49–56.

Ii, M., H.Yamamoto, Y.Adachi, Y.Maruyama and Y.Shinomura (2006). Role of matrix metalloproteinase-7 (matrilysin) in human cancer invasion, apoptosis, growth, and angiogenesis.*Exp Biol Med (Maywood)* **231**(1): 20–27.

Ilyas, M. (2005). Wnt signalling and the mechanistic basis of tumour development.*J Pathol* **205**(2): 130–144.

Inoue, A., H.Takahashi, H.Harada, S.Kohno, S.Ohue, K.Kobayashi, H.Yano, J.Tanaka and T.Ohnishi (2010). Cancer stem-like cells of glioblastoma characteristically express MMP-13 and display highly invasive activity.*Int J Oncol* **37**(5): 1121–1131.

Jin, X., M.Yagi, N.Akiyama, T.Hirosaki, S.Higashi, C. Y.Lin, R. B.Dickson, H.Kitamura and K.Miyazaki (2006). Matriptase activates stromelysin (MMP-3) and promotes tumor growth and angiogenesis.*Cancer Sci* **97**(12): 1327–1334.

Johansson, N., K.Airola, R.Grénman, A. L.Kariniemi, U.Saarialho-Kere and V. M.Kähäri (1997). Expression of collagenase-3 (matrix metalloproteinase-13) in squamous cell carcinomas of the head and neck.*Am J Pathol* **151**(2): 499–508.

Johansson, N., U.Saarialho-Kere, K.Airola, R.Herva, L.Nissinen, J.Westermarck, E.Vuorio, J.Heino and V. M.Kähäri (1997). Collagenase-3 (MMP-13) is expressed by hypertrophic chondrocytes, periosteal cells, and osteoblasts during human fetal bone development.*Dev Dyn* **208**(3): 387–397.

Johansson, N., M.Vaalamo, S.Grénman, S.Hietanen, P.Klemi, U.Saarialho-Kere and V. M.Kähäri (1999). Collagenase-3 (MMP-13) is expressed by tumor cells in invasive vulvar squamous cell carcinomas.*Am J Pathol* **154**(2): 469–480.

Jones, L. E., M. J.Humphreys, F.Campbell, J. P.Neoptolemos and M. T.Boyd (2004). Comprehensive analysis of matrix metalloproteinase and tissue inhibitor expression in pancreatic cancer: increased expression of matrix metalloproteinase-7 predicts poor survival.*Clin Cancer Res* **10**(8): 2832–2845.

Jormsjö, S., C.Whatling, D. H.Walter, A. M.Zeiher, A.Hamsten and P.Eriksson (2001). Allele-specific regulation of matrix metalloproteinase-7 promoter activity is associated with coronary artery luminal dimensions among hypercholesterolemic patients.*Arterioscler Thromb Vasc Biol* **21**(11): 1834–1839.

Kessenbrock, K., V.Plaks and Z.Werb (2010). Matrix metalloproteinases: regulators of the tumor microenvironment.*Cell* **141**(1): 52–67.

Kiili, M., S. W.Cox, H. Y.Chen, J.Wahlgren, P.Maisi, B. M.Eley, T.Salo and T.Sorsa (2002). Collagenase-2 (MMP-8) and collagenase-3 (MMP-13) in adult periodontitis: molecular forms and levels in gingival crevicular fluid and immunolocalisation in gingival tissue.*J Clin Periodontol* **29**(3): 224–232.

Kim, E. S., Y. W.Sohn and A.Moon (2007). TGF-beta-induced transcriptional activation of MMP-2 is mediated by activating transcription factor (ATF)2 in human breast epithelial cells.*Cancer Lett* **252**(1): 147–156.

Kim, K. S., H. M.Choi, Y. A.Lee, I. A.Choi, S. H.Lee, S. J.Hong, H. I.Yang and M. C.Yoo (2011). Expression levels and association of gelatinases MMP-2 and MMP-9 and collagenases MMP-1 and MMP-13 with VEGF in synovial fluid of patients with arthritis.*Rheumatol Int* **31**(4): 543–547.

Klein, T. and R.Bischoff (2011). Physiology and pathophysiology of matrix metalloproteases.*Amino Acids* **41**(2): 271–290.

Knäuper, V., C.López-Otin, B.Smith, G.Knight and G.Murphy (1996). Biochemical characterization of human collagenase-3.*J Biol Chem* **271**(3): 1544–1550.

Knäuper, V., A.Osthues, Y. A.DeClerck, K. E.Langley, J.Bläser and H.Tschesche (1993). Fragmentation of human polymorphonuclear-leucocyte collagenase.*Biochem J* **291** (Pt 3): 847–854.

Köhrmann, A., U.Kammerer, M.Kapp, J.Dietl and J.Anacker (2009). Expression of matrix metalloproteinases (MMPs) in primary human breast cancer and breast cancer cell lines: New findings and review of the literature.*BMC Cancer* **9**: 188.

Könnecke, H. and I.Bechmann (2013). The role of microglia and matrix metalloproteinases involvement in neuroinflammation and gliomas.*Clin Dev Immunol* **2013**: 914104.

Koo, B. H., J. H.Han, Y. I.Yeom and D. S.Kim (2010). Thrombin-dependent MMP-2 activity is regulated by heparan sulfate.*J Biol Chem* **285**(53): 41270–41279.

Koong, A. C., N. C.Denko, K. M.Hudson, C.Schindler, L.Swiersz, C.Koch, S.Evans, H.Ibrahim, Q. T.Le, D. J.Terris and A. J.Giaccia (2000). Candidate genes for the hypoxic tumor phenotype.*Cancer Res* **60**(4): 883–887.

Korpi, J. T., V.Kervinen, H.Mäklin, A.Väänänen, M.Lahtinen, E.Läärä, A.Ristimäki, G.Thomas, M.Ylipalosaari, P.Aström, C.Lopez-Otin, T.Sorsa, S.Kantola, E.Pirilä and T.Salo (2008). Collagenase-2 (matrix metalloproteinase-8) plays a protective role in tongue cancer.*Br J Cancer* **98**(4): 766–775.

Krueger, S., T.Hundertmark, T.Kalinski, U.Peitz, T.Wex, P.Malfertheiner, M.Naumann and A.Roessner (2006). Helicobacter pylori encoding the pathogenicity island activates matrix metalloproteinase 1 in gastric epithelial cells via JNK and ERK.*J Biol Chem* **281**(5): 2868–2875.

Kuropkat, C., A. A.Duenne, U.Herz, H.Renz and J. A.Werner (2004). Significant correlation of matrix metalloproteinase and macrophage colony-stimulating factor serum concentrations in patients with head and neck cancer.*Neoplasma* **51**(5): 375–378.

Kushlinsky, N. E., Y. N.Solovyov, I. V.Babkina, E. S.Gershtein and I. V.Bulicheva (2010). Matrix metalloproteinases 2, 7, 9 and tissue inhibitor of matrix metalloproteinase-1 in the sera of patients with bone tumors.*Bull Exp Biol Med* **149**(2): 233–235.

Kwan, J. A., C. J.Schulze, W.Wang, H.Leon, M.Sariahmetoglu, M.Sung, J.Sawicka, D. E.Sims, G.Sawicki and R.Schulz (2004). Matrix metalloproteinase-2 (MMP-2) is present in the nucleus of cardiac myocytes and is capable of cleaving poly (ADP-ribose) polymerase (PARP) in vitro.*FASEB J* **18**(6): 690–692.

Kwon, Y., E.Cukierman and A. K.Godwin (2011). Differential expressions of adhesive molecules and proteases define mechanisms of ovarian tumor cell matrix penetration/invasion.*PLoS One* **6**(4): e18872.

Le Floch, N., C.Rivat, O.De Wever, E.Bruyneel, M.Mareel, T.Dale and C.Gespach (2005). The proinvasive activity of Wnt-2 is mediated through a noncanonical Wnt pathway coupled to GSK-3beta and c-Jun/AP-1 signaling.*FASEB J* **19**(1): 144–146.

Leeman, M. F., S.Curran and G. I.Murray (2002). The structure, regulation, and function of human matrix metalloproteinase-13.*Crit Rev Biochem Mol Biol* **37**(3): 149–166.

Leivonen, S. K., A.Chantry, L.Hakkinen, J.Han and V. M.Kahari (2002). Smad3 mediates transforming growth factor-beta-induced collagenase-3 (matrix metalloproteinase-13) expression in human gingival fibroblasts. Evidence for cross-talk between Smad3 and p38 signaling pathways.*J Biol Chem* **277**(48): 46338–46346.

Lemjabbar, H., P.Gosset, C.Lamblin, I.Tillie, D.Hartmann, B.Wallaert, A. B.Tonnel and C.Lafuma (1999). Contribution of 92 kDa gelatinase/type IV collagenase in bronchial inflammation during status asthmaticus.*Am J Respir Crit Care Med* **159**(4 Pt 1): 1298–1307.

Leppert, D., S. L.Leib, C.Grygar, K. M.Miller, U. B.Schaad and G. A.Holländer (2000). Matrix metalloproteinase (MMP)-8 and MMP-9 in cerebrospinal fluid during bacterial meningitis: association with blood-brain barrier damage and neurological sequelae.*Clin Infect Dis* **31**(1): 80–84.

Li, Q., P. W.Park, C. L.Wilson and W. C.Parks (2002). Matrilysin shedding of syndecan-1 regulates chemokine mobilization and transepithelial efflux of neutrophils in acute lung injury. *Cell* **111**(5): 635–646.

Li, W., M. X.Jia, J. H.Wang, J. L.Lu, J.Deng, J. X.Tang and C.Liu (2019). Association of MMP9-1562C/T and MMP13-77A/G polymorphisms with non-small cell lung cancer in southern Chinese population.*Biomolecules***9**(3):107.

Li, X., C.Liu, R.Ran, G.Liu, Y.Yang, W.Zhao, X.Xie and J.Li (2020). Matrix metalloproteinase family gene polymorphisms and lung cancer susceptibility: an updated meta-analysis.*J Thorac Dis* **12**(3): 349–362.

Li, X., L.Qu, Y.Zhong, Y.Zhao, H.Chen and L.Daru (2013). Association between promoters polymorphisms of matrix metalloproteinases and risk of digestive cancers: a meta-analysis.*J Cancer Res Clin Oncol* **139**(9): 1433–1447.

Limb, G. A., K.Matter, G.Murphy, A. D.Cambrey, P. N.Bishop, G. E.Morris and P. T.Khaw (2005). Matrix metalloproteinase-1 associates with intracellular organelles and confers resistance to lamin A/C degradation during apoptosis.*Am J Pathol* **166**(5): 1555–1563.

Littlepage, L. E., M. D.Sternlicht, N.Rougier, J.Phillips, E.Gallo, Y.Yu, K.Williams, A.Brenot, J. I.Gordon and Z.Werb (2010). Matrix metalloproteinases contribute distinct roles in neuroendocrine prostate carcinogenesis, metastasis, and angiogenesis progression.*Cancer Res* **70**(6): 2224–2234.

Lohi, J., K.Lehti, H.Valtanen, W. C.Parks and J.Keski-Oja (2000). Structural analysis and promoter characterization of the human membrane-type matrix metalloproteinase-1 (MT1-MMP) gene.*Gene* **242**(1–2): 75–86.

Lomas, D. A. and R.Mahadeva (2002). Alpha1-antitrypsin polymerization and the serpinopathies: pathobiology and prospects for therapy.*J Clin Invest* **110**(11): 1585–1590.

Lorenzl, S., D. S.Albers, N.Relkin, T.Ngyuen, S. L.Hilgenberg, J.Chirichigno, M. E.Cudkowicz and M. F.Beal (2003). Increased plasma levels of matrix metalloproteinase-9 in patients with Alzheimer's disease.*Neurochem Int* **43**(3): 191–196.

Lugli, E., M.Pinti, M.Nasi, L.Troiano, N.Prada, C.Mussini, V.Borghi, R.Esposito and A.Cossarizza (2005). MMP-7 promoter polymorphisms do not influence CD4+ recovery and changes in plasma viral load during antiretroviral therapy for HIV-1 infection.*Int J Immunogenet* **32**(5): 269–271.

Luo, D., B.Mari, I.Stoll and P.Anglard (2002). Alternative splicing and promoter usage generates an intracellular stromelysin 3 isoform directly translated as an active matrix metalloproteinase.*J Biol Chem* **277**(28): 25527–25536.

Ma, X., Y.Jiang, A.Wu, X.Chen, R.Pi, M.Liu and Y.Liu (2010). Berberine attenuates experimental autoimmune encephalomyelitis in C57 BL/6 mice.*PLoS One* **5**(10): e13489.

Maeda, H., T.Okamoto and T.Akaike (1998). Human matrix metalloprotease activation by insults of bacterial infection involving proteases and free radicals.*Biol Chem* **379**(2): 193–200.

Mainardi, C. L., T. F.Pourmotabbed and K. A.Hasty (1991). Inflammatory phagocytes and connective tissue degrading metalloproteinases.*Am J Med Sci* **302**(3): 171–175.

Maskos, K. (2005). Crystal structures of MMPs in complex with physiological and pharmacological inhibitors.*Biochimie* **87**(3–4): 249–263.

Matsumoto, H., A.Niimi, M.Takemura, T.Ueda, M.Minakuchi, R.Tabuena, K.Chin, T.Mio, Y.Ito, S.Muro, T.Hirai, S.Morita, S.Fukuhara and M.Mishima (2005). Relationship of airway wall thickening to an imbalance between matrix metalloproteinase-9 and its inhibitor in asthma.*Thorax* **60**(4): 277–281.

Matsuno, H., K.Yudoh, Y.Watanabe, F.Nakazawa, H.Aono and T.Kimura (2001). Stromelysin-1 (MMP-3) in synovial fluid of patients with rheumatoid arthritis has potential to cleave membrane bound Fas ligand.*J Rheumatol* **28**(1): 22–28.

McCawley, L. J., H. C.Crawford, L. E.King, Jr., J.Mudgett and L. M.Matrisian (2004). A protective role for matrix metalloproteinase-3 in squamous cell carcinoma.*Cancer Res* **64**(-19): 6965–6972.

McCawley, L. J. and L. M.Matrisian (2001). Matrix metalloproteinases: they're not just for matrix anymore!*Curr Opin Cell Biol* **13**(5): 534–540.

McColgan, P. and P.Sharma (2009). Polymorphisms of matrix metalloproteinases 1, 2, 3 and 9 and susceptibility to lung, breast and colorectal cancer in over 30,000 subjects.*Int J Cancer* **125**(6): 1473–1478.

McGuire, J. K., Q.Li and W. C.Parks (2003). Matrilysin (matrix metalloproteinase-7) mediates E-cadherin ectodomain shedding in injured lung epithelium.*Am J Pathol* **162**(6): 1831–1843.

McLaughlin, J., J.Rella, A.Bakan, L.Kong, L.Zhu, D.Frederick, S.Yende, R.Ferrell, I.Bahar, S.Shapiro, D.Angus and A.Kaynar (2011). Impact of pro-domain stability of matrix metalloproteinase-8 on the outcome of sepsis. *Crit Care* **15**(Suppl 1):P278. doi: 10.1186/cc9698. Epub 2011 Mar 11.

McQuibban, G. A., J. H.Gong, J. P.Wong, J. L.Wallace, I.Clark-Lewis and C. M.Overall (2002). Matrix metalloproteinase processing of monocyte chemoattractant proteins generates CC chemokine receptor antagonists with anti-inflammatory properties in vivo.*Blood* **100**(4): 1160–1167.

Medley, T. L., B. A.Kingwell, C. D.Gatzka, P.Pillay and T. J.Cole (2003). Matrix metalloproteinase-3 genotype contributes to age-related aortic stiffening through modulation of gene and protein expression.*Circ Res* **92**(11): 1254–1261.

Mercapide, J., R.LopezDe Cicco, J. S.Castresana and A. J.Klein-Szanto (2003). Stromelysin-1/matrix metalloproteinase-3 (MMP-3) expression accounts for invasive properties of human astrocytoma cell lines.*Int J Cancer* **106**(5): 676–682.

Meyer, E., J. Y.Vollmer, R.Bovey and I.Stamenkovic (2005). Matrix metalloproteinases 9 and 10 inhibit protein kinase C-potentiated, p53-mediated apoptosis.*Cancer Res* **65**(10): 4261–4272.

Moilanen, M., E.Pirilä, R.Grénman, T.Sorsa and T.Salo (2002). Expression and regulation of collagenase-2 (MMP-8) in head and neck squamous cell carcinomas.*J Pathol* **197**(1): 72–81.

Moilanen, M., T.Sorsa, M.Stenman, P.Nyberg, O.Lindy, J.Vesterinen, A.Paju, Y. T.Konttinen, U. H.Stenman and T.Salo (2003). Tumor-associated trypsinogen-2 (trypsinogen-2) activates procollagenases (MMP-1, -8, -13) and stromelysin-1 (MMP-3) and degrades type I collagen. *Biochemistry* **42**(18): 5414–5420.

Mok, W., Y.Boucher and R. K.Jain (2007). Matrix metalloproteinases-1 and -8 improve the distribution and efficacy of an oncolytic virus.*Cancer Res* **67**(22): 10664–10668.

Montel, V., J.Kleeman, D.Agarwal, D.Spinella, K.Kawai and D.Tarin (2004). Altered metastatic behavior of human breast cancer cells after experimental manipulation of matrix metalloproteinase 8 gene expression.*Cancer Res* **64**(5): 1687–1694.

Moon, S. K., S. H.Cho, K. W.Kim, J. H.Jeon, J. H.Ko, B. Y.Kim and C. H.Kim (2007). Overexpression of membrane sialic acid-specific sialidase Neu3 inhibits matrix metalloproteinase-9 expression in vascular smooth muscle cells.*Biochem Biophys Res Commun* **356**(3): 542–547.

Moreno-Ortiz, J. M., M.Gutiérrez-Angulo, M.Partida-Pérez, J.Peregrina-Sandoval, R.Ramírez-Ramírez, R.Muñiz-Mendoza, S.Suárez-Villanueva, M.Centeno-Flores, V.Maciel-Gutiérrez, J. E.Cabrales-Vazquez and M. L.Ayala-Madrigal (2014). Association of MMP7–181A/G and MMP13-77A/G polymorphisms with colorectal cancer in a Mexican population.*Genet Mol Res* **13**(2): 3537–3544.

Morrison, C. J., G. S.Butler, H. F.Bigg, C. R.Roberts, P. D.Soloway and C. M.Overall (2001). Cellular activation of MMP-2 (gelatinase A) by MT2-MMP occurs via a TIMP-2-independent pathway.*J Biol Chem* **276**(50): 47402–47410.

Muller, M., C.Trocme, B.Lardy, F.Morel, S.Halimi and P. Y.Benhamou (2008). Matrix metalloproteinases and diabetic foot ulcers: the ratio of MMP-1 to TIMP-1 is a predictor of wound healing.*Diabet Med* **25**(4): 419–426.

Murnane, M. J., J.Cai, S.Shuja, D.McAneny, V.Klepeis and J. B.Willett (2009). Active MMP-2 effectively identifies the presence of colorectal cancer.*Int J Cancer* **125**(12): 2893–2902.

Murphy, G. (2011). Tissue inhibitors of metalloproteinases. *Genome Biol* **12**(11): 233.

Murphy, G. and H.Nagase (2008). Progress in matrix metalloproteinase research.*Mol Aspects Med* **29**(5): 290–308.

Nagase, H. (2001). Metalloproteases.*Curr Protoc Protein Sci* Chapter 21: Unit 21.24.

Nagase, H. and K.Fushimi (2008). Elucidating the function of non catalytic domains of collagenases and aggrecanases. *Connect Tissue Res* **49**(3): 169–174.

Nakamura, M., S.Miyamoto, H.Maeda, G.Ishii, T.Hasebe, T.Chiba, M.Asaka and A.Ochiai (2005). Matrix metalloproteinase-7 degrades all insulin-like growth factor binding proteins and facilitates insulin-like growth factor bioavailability.*Biochem Biophys Res Commun* **333**(3): 1011–1016.

Nguyen, M., J.Arkell and C. J.Jackson (2001). Human endothelial gelatinases and angiogenesis.*Int J Biochem Cell Biol* **33**(-10): 960–970.

Nishida, Y., H.Miyamori, E. W.Thompson, T.Takino, Y.Endo and H.Sato (2008). Activation of matrix metalloproteinase-2 (MMP-2) by membrane type 1 matrix metalloproteinase through an artificial receptor for proMMP-2 generates active MMP-2.*Cancer Res* **68**(21): 9096–9104.

Nissinen, L. and V. M.Kähäri (2014). Matrix metalloproteinases in inflammation.*Biochim Biophys Acta* **1840**(8): 2571–2580.

Noë, V., B.Fingleton, K.Jacobs, H. C.Crawford, S.Vermeulen, W.Steelant, E.Bruyneel, L. M.Matrisian and M.Mareel (2001). Release of an invasion promoter E-cadherin fragment by matrilysin and stromelysin-1.*J Cell Sci* **114**(Pt 1): 111–118.

Noseworthy, J. H., C.Lucchinetti, M.Rodriguez and B. G.Weinshenker (2000). Multiple sclerosis.*N Engl J Med* **343**(13): 938–952.

Nwomeh, B. C., H. X.Liang, I. K.Cohen and D. R.Yager (1999). MMP-8 is the predominant collagenase in healing wounds and nonhealing ulcers.*J Surg Res* **81**(2): 189–195.

Nyberg, P., P.Heikkilä, T.Sorsa, J.Luostarinen, R.Heljasvaara, U. H.Stenman, T.Pihlajaniemi and T.Salo (2003). Endostatin inhibits human tongue carcinoma cell invasion and intravasation and blocks the activation of matrix metalloprotease-2, -9, and -13.*J Biol Chem* **278**(25): 22404–22411.

O'Brien, T. P., Q. J.Li, F.Sauerburger, V. E.Reviglio, T.Rana and M. F.Ashraf (2001). The role of matrix metalloproteinases in ulcerative keratolysis associated with perioperative diclofenac use.*Ophthalmology* **108**(4): 656–659.

Offersen, B. V., M. M.Knap, M. R.Horsman, J.Verheijen, R.Hanemaaijer and J.Overgaard (2010). Matrix metalloproteinase-9 measured in urine from bladder cancer patients is an independent prognostic marker of poor survival.*Acta Oncol* **49**(8): 1283–1287.

Oh, J., R.Takahashi, S.Kondo, A.Mizoguchi, E.Adachi, R. M.Sasahara, S.Nishimura, Y.Imamura, H.Kitayama, D. B.Alexander, C.Ide, T. P.Horan, T.Arakawa, H.Yoshida, S.Nishikawa, Y.Itoh, M.Seiki, S.Itohara, C.Takahashi and M.Noda (2001). The membrane-anchored MMP inhibitor RECK is a key regulator of extracellular matrix integrity and angiogenesis.*Cell* **107**(6): 789–800.

Ongusaha, P. P., J. C.Kwak, A. J.Zwible, S.Macip, S.Higashiyama, N.Taniguchi, L.Fang and S. W.Lee (2004). HB-EGF is a potent inducer of tumor growth and angiogenesis.*Cancer Res* **64**(15): 5283–5290.

Osteen, K. G., T. M.Igarashi and K. L.Bruner-Tran (2003). Progesterone action in the human endometrium: induction of a unique tissue environment which limits matrix metalloproteinase (MMP) expression.*Front Biosci* **8**: d78–86.

Overall, C. M. (2002). Molecular determinants of metalloproteinase substrate specificity: matrix metalloproteinase substrate binding domains, modules, and exosites.*Mol Biotechnol* **22**(1): 51–86.

Overall, C. M. and G. S.Butler (2007). Protease yoga: extreme flexibility of a matrix metalloproteinase.*Structure* **15**(10): 1159–1161.

Overall, C. M. and C.López-Otín (2002). Strategies for MMP inhibition in cancer: innovations for the post-trial era.*Nat Rev Cancer* **2**(9): 657–672.

Owen, C. A., Z.Hu, C.Lopez-Otin and S. D.Shapiro (2004). Membrane-bound matrix metalloproteinase-8 on activated polymorphonuclear cells is a potent, tissue inhibitor of metalloproteinase-resistant collagenase and serpinase.*J Immunol* **172**(12): 7791–7803.

Pardo, A., R.Barrios, M.Gaxiola, L.Segura-Valdez, G.Carrillo, A.Estrada, M.Mejía and M.Selman (2000). Increase of lung neutrophils in hypersensitivity pneumonitis is associated with lung fibrosis.*Am J Respir Crit Care Med* **161**(5): 1698–1704.

Pardo, A. and M.Selman (2005). MMP-1: the elder of the family. *Int J Biochem Cell Biol* **37**(2): 283–288.

Park, K. S., S. J.Kim, K. H.Kim and J. C.Kim (2011). Clinical characteristics of TIMP2, MMP2, and MMP9 gene polymorphisms in colorectal cancer.*J Gastroenterol Hepatol* **26**(2): 391–397.

Parks, W. C. and S. D.Shapiro (2001). Matrix metalloproteinases in lung biology.*Respir Res* **2**(1): 10–19.

Pathak, S., E.Goldofsky, E. X.Vivas, V. R.Bonagura and A.Vambutas (2011). IL-1β is overexpressed and aberrantly regulated in corticosteroid nonresponders with autoimmune inner ear disease.*J Immunol* **186**(3): 1870–1879.

Peng, B., L.Cao, X.Ma, W.Wang, D.Wang and L.Yu (2010). Meta-analysis of association between matrix metalloproteinases 2, 7 and 9 promoter polymorphisms and cancer risk. *Mutagenesis* **25**(4): 371–379.

Peng, B., L.Cao, W.Wang, L.Xian, D.Jiang, J.Zhao, Z.Zhang, X.Wang and L.Yu (2010). Polymorphisms in the promoter regions of matrix metalloproteinases 1 and 3 and cancer risk: a meta-analysis of 50 case-control studies.*Mutagenesis* **25**(1): 41–48.

Piccard, H., P. E.Van den Steen and G.Opdenakker (2007). Hemopexin domains as multifunctional liganding modules in matrix metalloproteinases and other proteins.*J Leukoc Biol* **81**(4): 870–892.

Pilcher, B. K., J. A.Dumin, B. D.Sudbeck, S. M.Krane, H. G.Welgus and W. C.Parks (1997). The activity of collagenase-1 is required for keratinocyte migration on a type I collagen matrix.*J Cell Biol* **137**(6): 1445–1457.

Plaks, V., J.Rinkenberger, J.Dai, M.Flannery, M.Sund, K.Kanasaki, W.Ni, R.Kalluri and Z.Werb (2013). Matrix metalloproteinase-9 deficiency phenocopies features of preeclampsia and intrauterine growth restriction.*Proc Natl Acad Sci U S A* **110**(27): 11109–11114.

Poola, I., R. L.DeWitty, J. J.Marshalleck, R.Bhatnagar, J.Abraham and L. D.Leffall (2005). Identification of MMP-1 as a putative breast cancer predictive marker by global gene expression analysis.*Nat Med* **11**(5): 481–483.

Power, C., C. M.O'Connor, D.MacFarlane, S.O'Mahoney, K.Gaffney, J.Hayes and M. X.FitzGerald (1994). Neutrophil collagenase in sputum from patients with cystic fibrosis.*Am J Respir Crit Care Med* **150**(3): 818–822.

Price, S. J., D. R.Greaves and H.Watkins (2001). Identification of novel, functional genetic variants in the human matrix metalloproteinase-2 gene: role of Sp1 in allele-specific transcriptional regulation.*J Biol Chem* **276**(10): 7549–7558.

Prikk, K., P.Maisi, E.Pirilä, M. A.Reintam, T.Salo, T.Sorsa and R.Sepper (2002). Airway obstruction correlates with collagenase-2 (MMP-8) expression and activation in bronchial asthma.*Lab Invest* **82**(11): 1535–1545.

Prikk, K., P.Maisi, E.Pirilä, R.Sepper, T.Salo, J.Wahlgren and T.Sorsa (2001). In vivo collagenase-2 (MMP-8) expression by human bronchial epithelial cells and monocytes/macrophages in bronchiectasis.*J Pathol* **194**(2): 232–238.

Puente, X. S., L. M.Sánchez, C. M.Overall and C.López-Otín (2003). Human and mouse proteases: a comparative genomic approach.*Nat Rev Genet* **4**(7): 544–558.

Qin, H., Y.Sun and E. N.Benveniste (1999). The transcription factors Sp1, Sp3, and AP-2 are required for constitutive matrix metalloproteinase-2 gene expression in astroglioma cells.*J Biol Chem* **274**(41): 29130–29137.

Quesada, V., G. R.Ordóñez, L. M.Sánchez, X. S.Puente and C.López-Otín (2009). The Degradome database: mammalian proteases and diseases of proteolysis.*Nucleic Acids Res* **37**(Database issue): D239–243.

Ra, H. J. and W. C.Parks (2007). Control of matrix metalloproteinase catalytic activity.*Matrix Biol* **26**(8): 587–596.

Rajoria, S., R.Suriano, A.George, A.Shanmugam, S. P.Schantz, J.Geliebter and R. K.Tiwari (2011). Estrogen induced metastatic modulators MMP-2 and MMP-9 are targets of 3,3'-diindolylmethane in thyroid cancer.*PLoS One* **6**(1): e15879.

Ramos-Fernandez, M., M. F.Bellolio and L. G.Stead (2011). Matrix metalloproteinase-9 as a marker for acute ischemic stroke: a systematic review.*J Stroke Cerebrovasc Dis* **20**(1): 47–54.

Ranogajec, I., J.Jakić-Razumović, V.Puzović and J.Gabrilovac (2012). Prognostic value of matrix metalloproteinase-2 (MMP-2), matrix metalloproteinase-9 (MMP-9) and aminopeptidase N/CD13 in breast cancer patients.*Med Oncol* **29**(2): 561–569.

Ratjen, F., C. M.Hartog, K.Paul, J.Wermelt and J.Braun (2002). Matrix metalloproteases in BAL fluid of patients with cystic fibrosis and their modulation by treatment with dornase alpha.*Thorax* **57**(11): 930–934.

Reeps, C., J.Pelisek, S.Seidl, T.Schuster, A.Zimmermann, A.Kuehnl and H. H.Eckstein (2009). Inflammatory infiltrates and neovessels are relevant sources of MMPs in abdominal aortic aneurysm wall.*Pathobiology* **76**(5): 243–252.

Ribeiro, A. S., A.Albergaria, B.Sousa, A. L.Correia, M.Bracke, R.Seruca, F. C.Schmitt and J.Paredes (2010). Extracellular cleavage and shedding of P-cadherin: a mechanism underlying the invasive behaviour of breast cancer cells.*Oncogene* **29**(3): 392–402.

Risau, W. (1997). Mechanisms of angiogenesis.*Nature* **386**(6626): 671–674.

Rodríguez, D., C. J.Morrison and C. M.Overall (2010). Matrix metalloproteinases: what do they not do? New substrates and biological roles identified by murine models and proteomics.*Biochim Biophys Acta* **1803**(1): 39–54.

Rollin, J., S.Régina, P.Vourc'h, S. Iochmann, C. Bléchet, P. Reverdiau and Y. Gruel (2007). Influence of MMP-2 and MMP-9 promoter polymorphisms on gene expression and clinical outcome of non-small cell lung cancer.*Lung Cancer* **56**(2): 273–280.

Rosas, I. O., T. J.Richards, K.Konishi, Y.Zhang, K.Gibson, A. E.Lokshin, K. O.Lindell, J.Cisneros, S. D.Macdonald, A.Pardo, F.Sciurba, J.Dauber, M.Selman, B. R.Gochuico and N.Kaminski (2008). MMP1 and MMP7 as potential peripheral blood biomarkers in idiopathic pulmonary fibrosis.*PLoS Med* **5**(4): e93.

Rosenblum, G., S.Meroueh, M.Toth, J. F.Fisher, R.Fridman, S.Mobashery and I.Sagi (2007). Molecular structures and dynamics of the stepwise activation mechanism of a matrix metalloproteinase zymogen: challenging the cysteine switch dogma.*J Am Chem Soc* **129**(44): 13566–13574.

Rutter, J. L., T. I.Mitchell, G.Butticè, J.Meyers, J. F.Gusella, L. J.Ozelius and C. E.Brinckerhoff (1998). A single nucleotide polymorphism in the matrix metalloproteinase-1 promoter creates an Ets binding site and augments transcription. *Cancer Res* **58**(23): 5321–5325.

Sapna, G., S.Gokul and K.Bagri-Manjrekar (2014). Matrix metalloproteinases and periodontal diseases.*Oral Dis* **20**(6): 538–550.

Sariahmetoglu, M., B. D.Crawford, H.Leon, J.Sawicka, L.Li, B. J.Ballermann, C.Holmes, L. G.Berthiaume, A.Holt, G.Sawicki and R.Schulz (2007). Regulation of matrix metalloproteinase-2 (MMP-2) activity by phosphorylation. *FASEB J* **21**(10): 2486–2495.

Schettler, A., H.Thorn, B. M.Jockusch and H.Tschesche (1991). Release of proteinases from stimulated polymorphonuclear leukocytes. Evidence for subclasses of the main granule types and their association with cytoskeletal components. *Eur J Biochem* **197**(1): 197–202.

Schönbeck, U., F.Mach and P.Libby (1998). Generation of biologically active IL-1 beta by matrix metalloproteinases: a novel caspase-1-independent pathway of IL-1 beta processing. *J Immunol* **161**(7): 3340–3346.

Sepper, R., Y. T.Konttinen, Y.Ding, M.Takagi and T.Sorsa (1995). Human neutrophil collagenase (MMP-8), identified in bronchiectasis BAL fluid, correlates with severity of disease. *Chest* **107**(6): 1641–1647.

Shajarehpoor Salavati, L., F.Tafvizi and H. K.Manjili (2017). The association between MMP2–1306 C>T (rs243865) polymorphism and risk of prostate cancer. *Ir J Med Sci* **186**(1): 103–111.

Shapiro, S., O.Khodalev, H.Bitterman, R.Auslender and N.Lahat (2010). Different activation forms of MMP-2 oppositely affect the fate of endothelial cells. *Am J Physiol Cell Physiol* **298**(4): C942–951.

Shapiro, S. D., C. J.Fliszar, T. J.Broekelmann, R. P.Mecham, R. M.Senior and H. G.Welgus (1995). Activation of the 92-kDa gelatinase by stromelysin and 4-aminophenylmercuric acetate. Differential processing and stabilization of the carboxyl-terminal domain by tissue inhibitor of metalloproteinases (TIMP). *J Biol Chem* **270**(11): 6351–6356.

Shen, T. C., T. C.Hsia, C. Y.Chao, W. C.Chen, C. Y.Chen, W. C.Chen, Y. T.Lin, C. L.Hsiao, W. S.Chang, C. W.Tsai and D. T.Bau (2017). The Contribution of MMP-8 Promoter Polymorphisms in Lung Cancer.*Anticancer Res* **37**(7): 3563–3567.

Shi, M., J.Xia, H.Xing, W.Yang, X.Xiong, W.Pan, S.Han, J.Shang, C.Zhou, L.Zhou and M.Yang (2016). The Sp1-mediaded allelic regulation of MMP13 expression by an ESCC susceptibility SNP rs2252070. *Sci Rep* **6**: 27013.

Siebuhr, A. S., A. C.Bay-Jensen, D. J.Leeming, A.Plat, I.Byrjalsen, C.Christiansen, D.van de Heijde and M. A.Karsdal (2013). Serological identification of fast progressors of structural damage with rheumatoid arthritis. *Arthritis Res Ther* **15**(4): R86.

Skiles, J. W., N. C.Gonnella and A. Y.Jeng (2001). The design, structure, and therapeutic application of matrix metalloproteinase inhibitors. *Curr Med Chem* **8**(4): 425–474.

Smith, G. N., Jr., K. D.Brandt and K. A.Hasty (1994). Procollagenase is reduced to inactive fragments upon activation in the presence of doxycycline. *Ann N Y Acad Sci* **732**: 436–438.

Stadlmann, S., J.Pollheimer, P. L.Moser, A.Raggi, A.Amberger, R.Margreiter, F. A.Offner, G.Mikuz, S.Dirnhofer and H.Moch (2003). Cytokine-regulated expression of collagenase-2 (MMP-8) is involved in the progression of ovarian cancer. *Eur J Cancer* **39**(17): 2499–2505.

Steenport, M., K. M.Khan, B.Du, S. E.Barnhard, A. J.Dannenberg and D. J.Falcone (2009). Matrix metalloproteinase (MMP)-1 and MMP-3 induce macrophage MMP-9: evidence for the role of TNF-alpha and cyclooxygenase-2. *J Immunol* **183**(12): 8119–8127.

Stenman, M., A.Paju, R.Hanemaaijer, T.Tervahartiala, A.Leminen, U. H.Stenman, Y. T.Konttinen and T.Sorsa (2003). Collagenases (MMP-1, -8 and -13) and trypsinogen-2 in fluid from benign and malignant ovarian cysts.*Tumour Biol* **24**(1): 9–12.

Sternlicht, M. D., M. J.Bissell and Z.Werb (2000). The matrix metalloproteinase stromelysin-1 acts as a natural mammary tumor promoter.*Oncogene* **19**(8): 1102–1113.

Sternlicht, M. D., A.Lochter, C. J.Sympson, B.Huey, J. P.Rougier, J. W.Gray, D.Pinkel, M. J.Bissell and Z.Werb (1999). The stromal proteinase MMP3/stromelysin-1 promotes mammary carcinogenesis.*Cell* **98**(2): 137–146.

Sun, L., M. E.Diamond, A. J.Ottaviano, M. J.Joseph, V.Ananthanarayan and H. G.Munshi (2008). Transforming growth factor-beta 1 promotes matrix metalloproteinase-9-mediated oral cancer invasion through snail expression. *Mol Cancer Res* **6**(1): 10–20.

Tai, J., D.Sun, X.Wang and Z.Kang (2020). Matrix metalloproteinase-8 rs11225395 polymorphism correlates with colorectal cancer risk and survival in a Chinese Han population: a case-control study.*Aging (Albany NY)* **12**(19): 19618–19627.

Tan, X., H.Egami, S.Ishikawa, H.Sugita, H.Kamohara, M.Nakagawa, F.Nozawa, M.Abe and M.Ogawa (2005). Involvement of matrix metalloproteinase-7 in invasion-metastasis through induction of cell dissociation in pancreatic cancer.*Int J Oncol* **26**(5): 1283–1289.

Taraboletti, G., L.Morbidelli, S.Donnini, A.Parenti, H. J.Granger, R.Giavazzi and M.Ziche (2000). The heparin binding 25 kDa fragment of thrombospondin-1 promotes angiogenesis and modulates gelatinase and TIMP-2 production in endothelial cells.*FASEB J* **14**(12): 1674–1676.

Templeton, N. S. and W. G.Stetler-Stevenson (1991). Identification of a basal promoter for the human Mr 72,000 type IV collagenase gene and enhanced expression in a highly metastatic cell line.*Cancer Res* **51**(22): 6190–6193.

Tervahartiala, T., E.Pirilä, A.Ceponis, P.Maisi, T.Salo, G.Tuter, P.Kallio, J.Törnwall, R.Srinivas, Y. T.Konttinen and T.Sorsa (2000). The in vivo expression of the collagenolytic matrix metalloproteinases (MMP-2, -8, -13, and -14) and matrilysin (MMP-7) in adult and localized juvenile periodontitis.*J Dent Res* **79**(12): 1969–1977.

Thomasset, N., A.Lochter, C. J.Sympson, L. R.Lund, D. R.Williams, O.Behrendtsen, Z.Werb and M. J.Bissell (1998). Expression of autoactivated stromelysin-1 in mammary glands of transgenic mice leads to a reactive stroma during early development.*Am J Pathol* **153**(2): 457–467.

Todorova, L., L.Bjermer, A.Miller-Larsson and G.Westergren-Thorsson (2010). Relationship between matrix production by bronchial fibroblasts and lung function and AHR in asthma.*Respir Med* **104**(12): 1799–1808.

Troeberg, L. and H.Nagase (2012). Proteases involved in cartilage matrix degradation in osteoarthritis.*Biochim Biophys Acta* **1824**(1): 133–145.

Tsai, T. H., Y. M.Wang, W. S.Chang, C. W.Tsai, H. C.Wu, H. M.Hsu, Y. C.Wang, H. T.Li, C. L.Gong, D. T.Bau and C. Y.Li (2018). Association of matrix metalloproteinase-8 genotypes with the risk of bladder cancer. *Anticancer Res* **38**(9): 5159–5164.

Turpeenniemi-Hujanen, T. (2005). Gelatinases (MMP-2 and -9) and their natural inhibitors as prognostic indicators in solid cancers. *Biochimie* **87**(3–4): 287–297.

Turu, M. M., J.Krupinski, E.Catena, A.Rosell, J.Montaner, F.Rubio, J.Alvarez-Sabin, M.Cairols and L.Badimon (2006). Intraplaque MMP-8 levels are increased in asymptomatic patients with carotid plaque progression on ultrasound.*Atherosclerosis* **187**(1): 161–169.

Uría, J. A., M.Balbín, J. M.López, J.Alvarez, F.Vizoso, M.Takigawa and C.López-Otín (1998). Collagenase-3 (MMP-13) expression in chondrosarcoma cells and its regulation by basic fibroblast growth factor.*Am J Pathol* **153**(1): 91–101.

Vacek, T. P., S.Rehman, D.Neamtu, S.Yu, S.Givimani and S. C.Tyagi (2015). Matrix metalloproteinases in atherosclerosis: role of nitric oxide, hydrogen sulfide, homocysteine, and polymorphisms.*Vasc Health Risk Manag* **11**: 173–183.

Vaisar, T., S. Y.Kassim, I. G.Gomez, P. S.Green, S.Hargarten, P. J.Gough, W. C.Parks, C. L.Wilson, E. W.Raines and J. W.Heinecke (2009). MMP-9 sheds the beta2 integrin subunit (CD18) from macrophages.*Mol Cell Proteomics* **8**(5): 1044–1060.

Van den Steen, P. E., B.Dubois, I.Nelissen, P. M.Rudd, R. A.Dwek and G.Opdenakker (2002). Biochemistry and molecular biology of gelatinase B or matrix metalloproteinase-9 (MMP-9).*Crit Rev Biochem Mol Biol* **37**(6): 375–536.

Van den Steen, P. E., B.Grillet and G.Opdenakker (2005). Gelatinase B participates in collagen II degradation and releases glycosylated remnant epitopes in rheumatoid arthritis.*Adv Exp Med Biol* **564**: 45–55.

Van den Steen, P. E., S. J.Husson, P.Proost, J.Van Damme and G.Opdenakker (2003). Carboxyterminal cleavage of the chemokines MIG and IP-10 by gelatinase B and neutrophil collagenase.*Biochem Biophys Res Commun* **310**(3): 889–896.

Van Den Steen, P. E., A.Wuyts, S. J.Husson, P.Proost, J.Van Damme and G.Opdenakker (2003). Gelatinase B/MMP-9 and neutrophil collagenase/MMP-8 process the chemokines human GCP-2/CXCL6, ENA-78/CXCL5 and mouse GCP-2/LIX and modulate their physiological activities.*Eur J Biochem* **270**(18): 3739–3749.

Vira, H., V.Pradhan, V.Umare, A.Chaudhary, A.Rajadhyksha, M.Nadkar, K.Ghosh and A.Nadkarni (2017). Role of MMP-7 in the pathogenesis of systemic lupus erythematosus (SLE).*Lupus* **26**(9): 937–943.

Visse, R. and H.Nagase (2003). Matrix metalloproteinases and tissue inhibitors of metalloproteinases: structure, function, and biochemistry.*Circ Res* **92**(8): 827–839.

Volcik, K. A., S.Campbell, L. E.Chambless, J.Coresh, A. R.Folsom, T. H.Mosley, H.Ni, L. E.Wagenknecht, B. A.Wasserman and E.Boerwinkle (2010). MMP2 genetic variation is associated with measures of fibrous cap thickness: The Atherosclerosis Risk in Communities Carotid MRI Study.*Atherosclerosis* **210**(1): 188–193.

Walsh, N., A.Larkin, N.Swan, K.Conlon, P.Dowling, R.McDermott and M.Clynes (2011). RNAi knockdown of Hop (Hsp70/Hsp90 organising protein) decreases invasion via MMP-2 down regulation.*Cancer Lett* **306**(2): 180–189.

Wang, F., S.Reierstad and D. A.Fishman (2006). Matrilysin over-expression in MCF-7 cells enhances cellular invasiveness and pro-gelatinase activation.*Cancer Lett* **236**(2): 292–301.

Wang, F. Q., J.So, S.Reierstad and D. A.Fishman (2005). Matrilysin (MMP-7) promotes invasion of ovarian cancer cells by activation of progelatinase.*Int J Cancer* **114**(1): 19–31.

Wang, H., S.Parry, G.Macones, M. D.Sammel, P. E.Ferrand, H.Kuivaniemi, G.Tromp, I.Halder, M. D.Shriver, R.Romero and J. F.Strauss, 3rd (2004). Functionally significant SNP MMP8 promoter haplotypes and preterm premature rupture of membranes (PPROM).*Hum Mol Genet* **13**(21): 2659–2669.

Warner, R. L., L.Beltran, E. M.Younkin, C. S.Lewis, S. J.Weiss, J.Varani and K. J.Johnson (2001). Role of stromelysin 1 and gelatinase B in experimental acute lung injury.*Am J Respir Cell Mol Biol* **24**(5): 537–544.

Whitelock, J. M., A. D.Murdoch, R. V.Iozzo and P. A.Underwood (1996). The degradation of human endothelial cell-derived perlecan and release of bound basic fibroblast growth factor by stromelysin, collagenase, plasmin, and heparanases.*J Biol Chem* **271**(17): 10079–10086.

Wielockx, B., K.Lannoy, S. D.Shapiro, T.Itoh, S.Itohara, J.Vandekerckhove and C.Libert (2001). Inhibition of matrix metalloproteinases blocks lethal hepatitis and apoptosis induced by tumor necrosis factor and allows safe antitumor therapy.*Nat Med* **7**(11): 1202–1208.

Wiesen, J. F. and Z.Werb (1996). The role of stromelysin-1 in stromal-epithelial interactions and cancer.*Enzyme Protein* **49**(1–3): 174–181.

Wilson, C. L. and L. M.Matrisian (1996). Matrilysin: an epithelial matrix metalloproteinase with potentially novel functions. *Int J Biochem Cell Biol* **28**(2): 123–136.

Wilson, C. L., A. J.Ouellette, D. P.Satchell, T.Ayabe, Y. S.López-Boado, J. L.Stratman, S. J.Hultgren, L. M.Matrisian and W. C.Parks (1999). Regulation of intestinal alpha-defensin activation by the metalloproteinase matrilysin in innate host defense.*Science* **286**(5437): 113–117.

Witty, J. P., J. H.Wright and L. M.Matrisian (1995). Matrix metalloproteinases are expressed during ductal and alveolar mammary morphogenesis, and misregulation of stromelysin-1 in transgenic mice induces unscheduled alveolar development. *Mol Biol Cell* **6**(10): 1287–1303.

Woo, M., K.Park, J.Nam and J. C.Kim (2007). Clinical implications of matrix metalloproteinase-1, -3, -7, -9, -12, and plasminogen activator inhibitor-1 gene polymorphisms in colorectal cancer.*J Gastroenterol Hepatol* **22**(7): 1064–1070.

Wu, C. Y., M. S.Wu, E. P.Chiang, Y. J.Chen, C. J.Chen, N. H.Chi, Y. T.Shih, G. H.Chen and J. T.Lin (2007). Plasma matrix metalloproteinase-9 level is better than serum matrix metalloproteinase-9 level to predict gastric cancer evolution.*Clin Cancer Res* **13**(7): 2054–2060.

Xu, D., C. M.McKee, Y.Cao, Y.Ding, B. M.Kessler and R. J.Muschel (2010). Matrix metalloproteinase-9 regulates tumor cell invasion through cleavage of protease nexin-1. *Cancer Res* **70**(17): 6988–6998.

Xu, E., M.Lai, B.Lv, X.Xing, Q.Huang and X.Xia (2004). A single nucleotide polymorphism in the matrix metalloproteinase-2 promoter is associated with colorectal cancer.*Biochem Biophys Res Commun* **324**(3): 999–1003.

Yamamoto, H., Y.Adachi, F.Itoh, S.Iku, K.Matsuno, M.Kusano, Y.Arimura, T.Endo, Y.Hinoda, M.Hosokawa and K.Imai (1999). Association of matrilysin expression with recurrence and poor prognosis in human esophageal squamous cell carcinoma.*Cancer Res* **59**(14): 3313–3316.

Yan, C. and D. D.Boyd (2007). Regulation of matrix metalloproteinase gene expression.*J Cell Physiol* **211**(1): 19–26.

Yang, L., N.Li, S.Wang, Y.Kong, H.Tang, X.Xie and X.Xie (2014). Lack of association between the matrix metalloproteinase-2-1306C>T polymorphism and breast cancer susceptibility: a meta-analysis. *Asian Pac J Cancer Prev* **15**(12): 4823–4827.

Yang, X., J. W.Hu, M. T.Qiu, M.Li, R.Yin, J.Wang, L.Xu and Q.Zhang (2014). Association of matrix metalloproteinase-3–1171(5A>6A) polymorphism with cancer risk: a meta-analysis of 41 studies. *PLoS One* **9**(1): e87562.

Ye, S., P.Eriksson, A.Hamsten, M.Kurkinen, S. E.Humphries and A. M.Henney (1996). Progression of coronary atherosclerosis is associated with a common genetic variant of the human stromelysin-1 promoter which results in reduced gene expression. *J Biol Chem* **271**(22): 13055–13060.

Yoshihara, Y., H.Nakamura, K.Obata, H.Yamada, T.Hayakawa, K.Fujikawa and Y.Okada (2000). Matrix metalloproteinases and tissue inhibitors of metalloproteinases in synovial fluids from patients with rheumatoid arthritis or osteoarthritis. *Ann Rheum Dis* **59**(6): 455–461.

Yoshiyama, Y., M.Asahina and T.Hattori (2000). Selective distribution of matrix metalloproteinase-3 (MMP-3) in Alzheimer's disease brain. *Acta Neuropathol* **99**(2): 91–95.

Yu, C., K.Pan, D.Xing, G.Liang, W.Tan, L.Zhang and D.Lin (2002). Correlation between a single nucleotide polymorphism in the matrix metalloproteinase-2 promoter and risk of lung cancer. *Cancer Res* **62**(22): 6430–6433.

Yu, C., Y.Zhou, X.Miao, P.Xiong, W.Tan and D.Lin (2004). Functional haplotypes in the promoter of matrix metalloproteinase-2 predict risk of the occurrence and metastasis of esophageal cancer. *Cancer Res* **64**(20): 7622–7628.

Yu, Q. and I.Stamenkovic (2000). Cell surface-localized matrix metalloproteinase-9 proteolytically activates TGF-beta and promotes tumor invasion and angiogenesis. *Genes Dev* **14**(2): 163–176.

Yu, W. H., J. F.Woessner, Jr., J. D.McNeish and I.Stamenkovic (2002). CD44 anchors the assembly of matrilysin/MMP-7 with heparin-binding epidermal growth factor precursor and ErbB4 and regulates female reproductive organ remodeling. *Genes Dev* **16**(3): 307–323.

Zhang, J., X.Jin, S.Fang, R.Wang, Y.Li, N.Wang, W.Guo, Y.Wang, D.Wen, L.Wei, Z.Dong and G.Kuang (2005). The functional polymorphism in the matrix metalloproteinase-7 promoter increases susceptibility to esophageal squamous cell carcinoma, gastric cardiac adenocarcinoma and non-small cell lung carcinoma. *Carcinogenesis* **26**(10): 1748–1753.

Zhang, K., X.Chen, J.Zhou, C.Yang, M.Zhang, M.Chao, L.Zhang and C.Liang (2017). Association between MMP2–1306 C/T polymorphism and prostate cancer susceptibility: a meta-analysis based on 3906 subjects. *Oncotarget* **8**(27): 45020–45029.

Zhang, L. F., L. J.Zhu, W.Zhang, W.Yuan and N. H.Song (2019). MMP-8 C-799T, Lys460Thr, and Lys87Glu variants are not related to risk of cancer. *BMC Med Genet.* **20**(1): 162.

Zhao, X. and E. N.Benveniste (2008). Transcriptional activation of human matrix metalloproteinase-9 gene expression by multiple co-activators. *J Mol Biol* **383**(5): 945–956.

Zigrino, P., I.Kuhn, T.Bäuerle, J.Zamek, J. W.Fox, S.Neumann, A.Licht, M.Schorpp-Kistner, P.Angel and C.Mauch (2009). Stromal expression of MMP-13 is required for melanoma invasion and metastasis. *J Invest Dermatol* **129**(11): 2686–2693.

Zijlstra, A., M.Seandel, T. A.Kupriyanova, J. J.Partridge, M. A.Madsen, E. A.Hahn-Dantona, J. P.Quigley and E. I.Deryugina (2006). Proangiogenic role of neutrophil-like inflammatory heterophils during neovascularization induced by growth factors and human tumor cells. *Blood* **107**(1): 317–327.

Zitka, O., J.Kukacka, S.Krizkova, D.Huska, V.Adam, M.Masarik, R.Prusa and R.Kizek (2010). Matrix metalloproteinases. *Curr Med Chem* **17**(31): 3751–3768.

7

Vitamin D Receptor Polymorphisms and Diseases

Syed Sameer Aga
Department of Basic Medical Sciences, College of Medicine, King Saud bin Abdulaziz University for Health Sciences (KSAU-HS), King Abdullah International Medical Research Center (KAIMRC), Ministry of National Guard Health Affairs (NGHA), Jeddah, Kingdom of Saudi Arabia
Infinity Biochemistry Pvt. Ltd. Srinagar, Kashmir, India.

Mujeeb Zafar Banday and Saniya Nissar
Department of Biochemistry, Government Medical College, Karan Nagar, Srinagar, Kashmir, India.
Infinity Biochemistry Pvt. Ltd. Srinagar, Kashmir, India

Mansour Al Qurashi, Abdullah A. Awadh, Alqassem Y. Hakami, and Israa A. Malli
Department of Basic Medical Sciences, College of Medicine, King Saud bin Abdulaziz University for Health Sciences (KSAU-HS), King Abdullah International Medical Research Center (KAIMRC), Ministry of National Guard Health Affairs (NGHA), Jeddah, Kingdom of Saudi Arabia.

CONTENTS

ABSTRACT Vitamin D is one of the important molecules whose influence on the physiological systems extend far beyond its role as a supplemental molecule for regulation of calcium homeostasis. It exerts its influence via its workhorse vitamin D receptor (VDR), which is known to regulate numerous genes. Genetic polymorphisms in the VDR gene affect the structure of the VDR protein, which has a bearing on its function, leading to various disorders. Numerous single nucleotide polymorphisms (SNPs) of VDR have been reported to increase the susceptibility to various diseases affecting almost all systems of the human body. In this chapter, we focus on the role of the most common SNPs reported in the VDR gene and how these modulate the risk of various human diseases.

KEYWORDS: *Polymorphism; Mutations; Genome-Wide Association Studies; Single Nucleotide Polymorphism; Vitamin D; VDR; Risk Factors.*

DOI: 10.1201/9781003246244-7

7.1 Introduction

Vitamin D is one of the key fat-soluble vitamins, which belongs to the secosteroid family. It exists in two biological forms, ergocalciferol (D2) or cholecalciferol (D3), in the living world (Vahdat, 2020; LoPiccolo and Lim, 2010). D2 form is usually present in the plant- and fish-based diet, while D3 is present in the human body (Vahdat, 2020; Wang, 2009). Both D2 and D3 serve as the precursors for the active form of vitamin D, 1,25-dihydroxyvitamin D ($1,25(OH)_2D$), which functions like a steroid hormone (Lockau and Atkinson, 2018; Pike and Christakos, 2017).

The conversion to the active form ($1,25(OH)_2D$) occurs chiefly in kidneys, but many other organs such as placenta, pancreas, prostate, and breast as well as immune cells such as macrophages do also convert the vitamin to its active form (Lockau and Atkinson, 2018; Berridge, 2015a, b). Vitamin D plays a central role in glucose homeostasis, calcium homeostasis, bone remodeling, fertility, inflammation inhibition, cellular proliferation, and detoxification (Vahdat, 2020; Berridge, 2015a, b; Pike and Christakos, 2017; Wang, 2009). Vitamin D deficiency has been reported to be associated with a variety of diseases such as cancers, gastrointestinal, cardiovascular, neurological, immunological, and metabolic diseases (Zittermann et al., 2021; Nikolac Gabaj et al., 2020; Lockau and Atkinson, 2018; Pfotenhauer et al., 2017; Shang and Sun, 2017; LoPiccolo and Lim, 2010).

The wide variety of biological functions played by vitamin D are via its two specific receptors – (1) a membrane associated rapid-response steroid-binding protein (1,25D3-MARRS) and (2) an intracellular vitamin D receptor (VDR) (Berridge, 2015a, b; Lockau and Atkinson, 2018; Pike and Christakos, 2017). 1,25D3-MARRS mediates the direct signaling pathways via many secondary messengers, e.g., Ca2+, cyclin GMP, MAP kinase, and phosphoinositides, while VDR functions as a DNA-binding transcriptional factor (Christakos et al., 2016; Berridge, 2015a, b). VDR is a member of nuclear receptor family, which, after getting activated by vitamin D, binds to the specific regions in the promoter regions of the genes to initiate their expression and hence regulate the metabolism, proliferation, antioxidant activity, etc. in the living cells (Berridge, 2015a, b; Lockau and Atkinson, 2018).

7.1.1 Vitamin D Production and Metabolism

Vitamin D holds a unique position among all vitamins in its structure, functions, and metabolism. It can be orally ingested via consumption of plant- or fish-based diet or can be synthesized by skin from its precursor on exposure to solar ultraviolet radiation (Christakos et al., 2010, 2012, 2016). There should be a constant supply of vitamin D, as its half-life in the body is just 3 weeks. The dietary supply of vitamin D via non-fortified food is low, so production by the skin becomes the main pathway for its constant replenishment in the body (Pike and Christakos, 2017; Christakos et al., 2016; Berridge, 2015b). The main precursor for the synthesis of vitamin D in the body is 7-dehydrocholesterol, which gets converted in a series of reactions to produce an active molecule referred to as 1a, 25-dihydroxyvitamin D3 [1a, $25(OH)_2D_3$]. These reactions involve three main organs, including skin, kidneys, and liver, and the specialized oxidoreductive enzymes containing cytochrome P450 (CYP) as the active center (Christakos et al., 2016; Berridge, 2015b).

The first reaction of vitamin D biosynthesis involves the photolysis of cutaneous 7-dehydrocholesterol to previtamin D by ultraviolet B radiations after an exposure to sunlight, which undergoes a spontaneous isomerization to form cholecalciferol (D3) (Pike and Christakos, 2017) (**Figure 7.1**). Vitamin D3 is then taken up by the liver, where second conversion reaction is mediated by hepatic 25-hydroxylase (CYP2R1) which hydroxylates it at 25C position to produce to $25(OH)D_3$. This hydroxylase has a very high affinity for vitamin D, and any defect in this enzyme gets reflected in vitamin D dysfunction (Cheng et al., 2004). The third conversion occurs within the kidneys and involves the second and pivotal hydroxylation of the vitamin D, catalyzed by the renal mitochondrial $25(OH)D_3$-1α-hydroxylase (CYP27B1) enzyme. An addition of another hydroxyl group to the C1 position results in the synthesis of an active form of hormone referred to as $1,25(OH)_2D_3$. The reaction catalyzed by renal CYP27B1 is crucial to the production and maintenance of physiological levels of circulating $1,25(OH)_2D_3$, which is why its activity is tightly regulated by endocrine system via calcium and phosphate levels and parathyroid hormone (PTH) (**Figure 7.2**) (Pike and Christakos, 2017; Christakos et al., 2016; Berridge, 2015b).

In the circulation, $25(OH)D_3$ and $1,25(OH)_2D_3$ remain primarily bound to the vitamin D-binding protein (DBP) and albumin, albeit to a lesser extent. A small fraction of the circulating unbound vitamin D is biologically active. The half-life of $25(OH)D_3$ in the circulation is about 10–20 days, which is much higher than that of $1,25(OH)_2D_3$, which has a half-life of just about 10–20 h. This is because of the higher affinity of $25(OH)D_3$ for DBP. The serum levels of $1,25(OH)_2D_3$ are directly dependent upon the availability of $25(OH)D_3$ and is reduced at low level of $25(OH)D_3$ (<25 nmol/L) (Zittermann et al., 2021; Pike and Christakos, 2017; Christakos et al., 2016; Berridge, 2015b). The serum concentration of $1,25(OH)_2D_3$ is regulated within specific limits by the endocrine system in conjunction with the levels of calcium and phosphate. Parathyroid hormone (PTH) and fibroblast growth factor 23 (FGF-23) are two main hormones known to regulate the circulating $1,25(OH)_2D_3$ within relatively narrow physiological ranges. PTH activates 1α-hydroxylase (CYP27B1) enzyme in response to low plasma calcium levels, while FGF-23 is known to inhibit 1α-hydroxylase, but activate 24-hydroxylase (CYP24A1) enzyme (Bacchetta et al., 2013; Zittermann and Ernst, 2016) (**Figure 7.1 and 7.2**).

7.2 Vitamin D Receptor (VDR)

Vitamin D receptor (VDR) is a specialized ligand-dependent transcriptional regulator protein, which belongs to the nuclear receptor superfamily. It functions in a similar way as other steroid receptor family members such as receptors for retinoic acid, thyroid hormones, sex hormones, and adrenal steroids (Christakos et al., 2015).

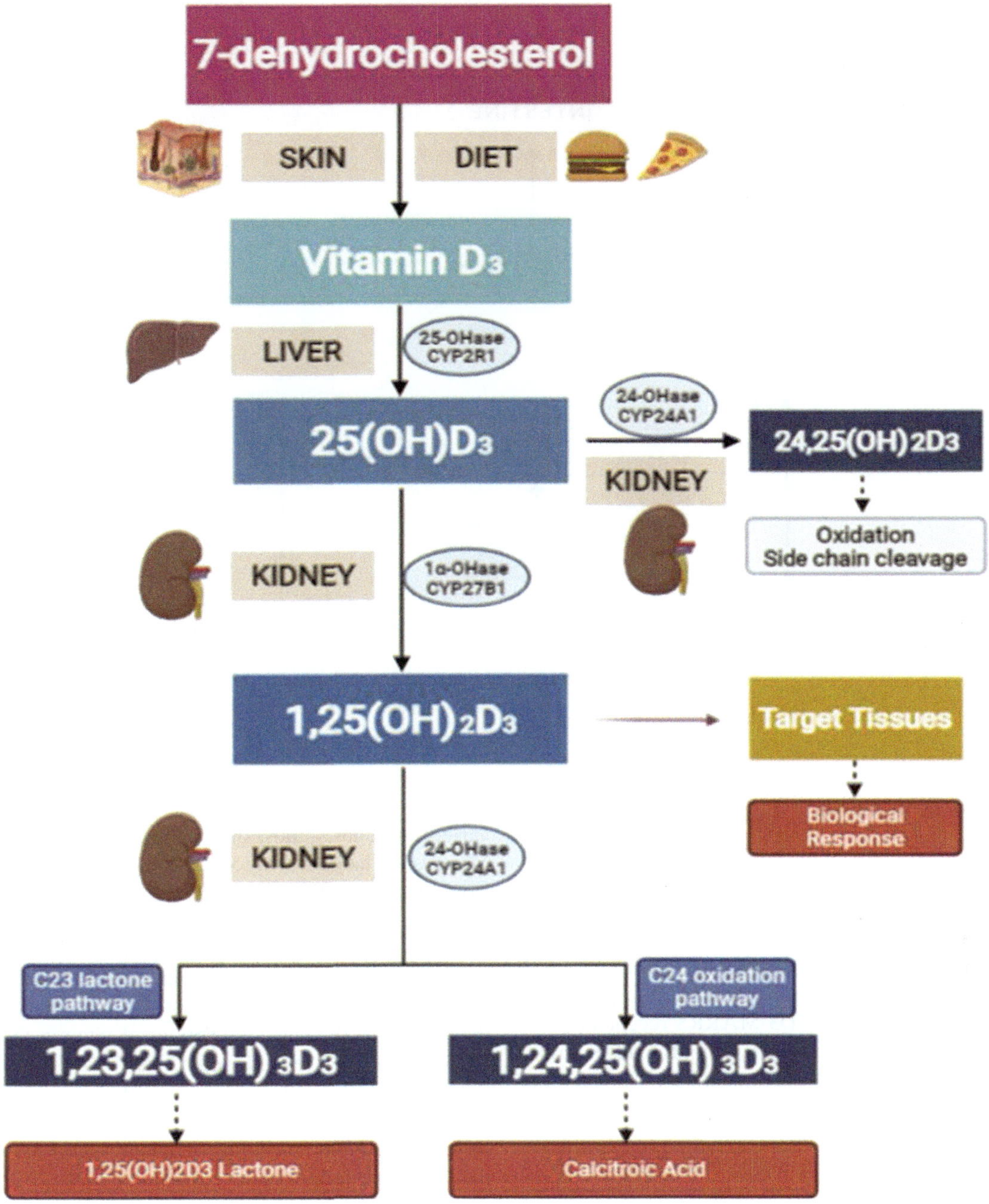

FIGURE 7.1 The metabolic pathway of vitamin D.

The gene which codes for this protein is located on chromosome 12q12-q14 (Pike JW and Christakos, 2017). In humans, the *VDR* gene is comprised of eight coding exons (2–9) and at least six noncoding regions with two regulatory untranslated regions flanking the gene on 5′- and 3′-end (Zella et al., 2006). Human VDR protein is a long chain of ~424–427 amino acids having molecular mass approximately 48 kDa, which functions as an obligate heterodimer with retinoic X receptor (RXR) for activation of vitamin D target genes (Zella et al., 2006; Christakos et al., 2015). In its three-dimensional structure, VDR has two highly conserved domains – an N-terminal cysteine-rich zinc finger DNA-binding domain (DBD) and a C-terminal α-helical ligand-binding domain (LBD). LBD is known to consist of twelve α-helices – H1–H12, of which H12 is critical for interaction and binding of ligands and coactivators (Bikle, 2014).

VDR is activated on binding with $1,25(OH)_2D_3$, initiating a conformational change in the receptor, thereby facilitating an increased interaction with RXR leading to the heterodimerization (Rochel et al., 2000). The VDR-RXR complex together with other coactivators binds to the specific sites (response elements) in the promoter region of the target genes to regulate their transcription and hence expression (Zella et al., 2006; Christakos et al., 2015).

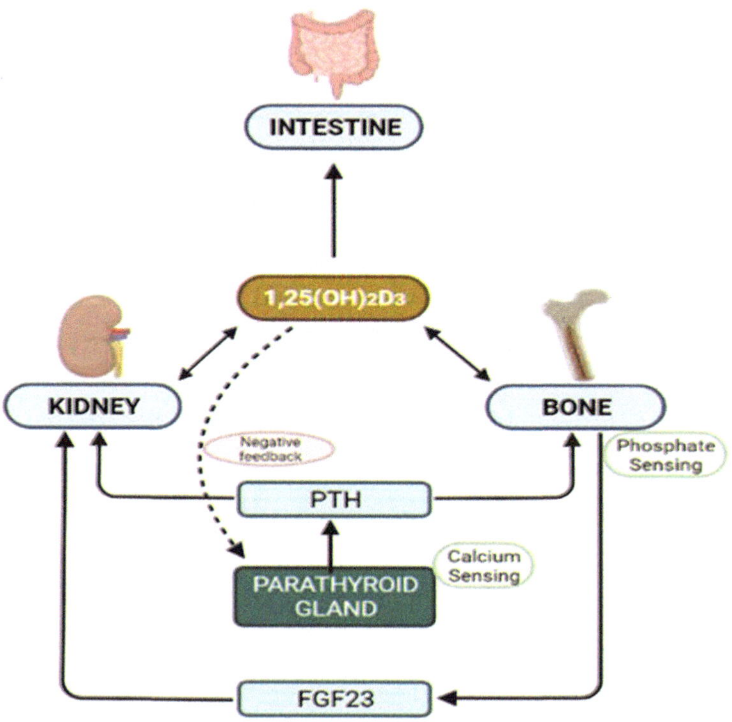

FIGURE 7.2 The regulation of mineral homeostasis.

7.2.1 Vitamin D Signaling

As mentioned above, vitamin D mediates its functions within the human body via two receptors – rapid-response steroid-binding protein (1,25D3-MARRS) and vitamin D receptor (VDR). Hence, there are two distinct signaling pathways for vitamin D. The primary non-genomic physiological role of vitamin D in regulating the mineral homeostasis and calcium-phosphate metabolism is mediated via 1,25D3-MARRS receptors (Pike and Christakos, 2017; Christakos et al., 2016; Berridge, 2015a, b), because of its rapidity and short time effects. However, the long-term genomic effects of vitamin D are mediated by intracellular VDRs (Nagpal et al., 2005).

Soon after its synthesis in the active form $1,25(OH)_2D_3$ in kidneys, vitamin D mediates its non-genomic function by binding to the intracellular VDR. Binding of $1,25(OH)_2D_3$ induces a repositioning of the H12 α-helix of LBD and opens the area for recruitment of the coactivators. H12 is a seat for the ligand-dependent activation function (AF2), which leads to the cooperative and allosteric effects between the DBD and the LBD (Bikle DD, 2014). These interactions allow activated VDR to bind with the primary coactivator RXR to form a heterodimer. VDR-RXR dimer translocates into the nucleus where it binds to the specific upstream promoter regions of the target genes called vitamin D response elements (VDREs). VDR-RXR dimer also interacts with other molecules such as CBP and SRC-1, which facilitate its binding (Berridge, 2015a, b). VDRE is an upstream 15 bp region composed of two directly repeated hexanucleotide sequences AGGTCA separated by a short 3 bp spacer, located in all of the target genes of vitamin D (Ozono et al., 1990) (**Figure 7.3**).

Vitamin D is known to affect the transcription of a wide variety of genes – most of which are activated and some are repressed. The genes which are regulated by vitamin D range from DNA demethylases (such as JMJD3, LSD1, and LSD2), signaling molecules (insulin, cytokines, and tumor necrosis factor) to transport protein components (TRPV5, TRPV6, and PMCA1b) (Pike and Christakos, 2017; Christakos et al., 2016; Berridge, 2015a). The most important genes, which are the bearers of the custodial actions of vitamin D, are nuclear factor-erythroid 2 related factor 2 (Nrf2) and Klotho. Vitamin D together with Nrf2 and Klotho makes up an extensive regulatory network, which ensures the normal homeostasis of cellular functions including calcium signaling, antioxidant defenses, and inflammatory inhibition (Kato, 2000; Haussler et al., 2013).

Nrf2 is a specialized transcription factor with redox-sensitive functionality that activates many genes encoding enzymes for antioxidation mechanisms. Vitamin D regulates the expression of Nrf2 gene and, hence in turn, numerous genes for which Nrf2 functions as a transcription activator (Hayes et al., 2014). Within the nucleus, Nrf2 binds to the specialized antioxidant response element (ARE) of a large number of genes to enhance their expression. The most important of them are the antioxidant and detoxifying enzymes such as catalase, glutamate cysteine ligase (GCL), heme oxygenase-1 (HO-1), NAD(P)H quinone oxidoreductase 1 (NQO1), superoxide dismutase 1 and 2 (SOD1 and SOD2), and thioredoxin (TRX) (Berridge, 2015a, b). Nrf2 is also demonstrated to enhance the transcription of Fos and Jun, which in turn cause increased expression of both VDR and RXRα, thereby enhancing the cellular response to low levels of vitamin D (Bobilev et al., 2011).

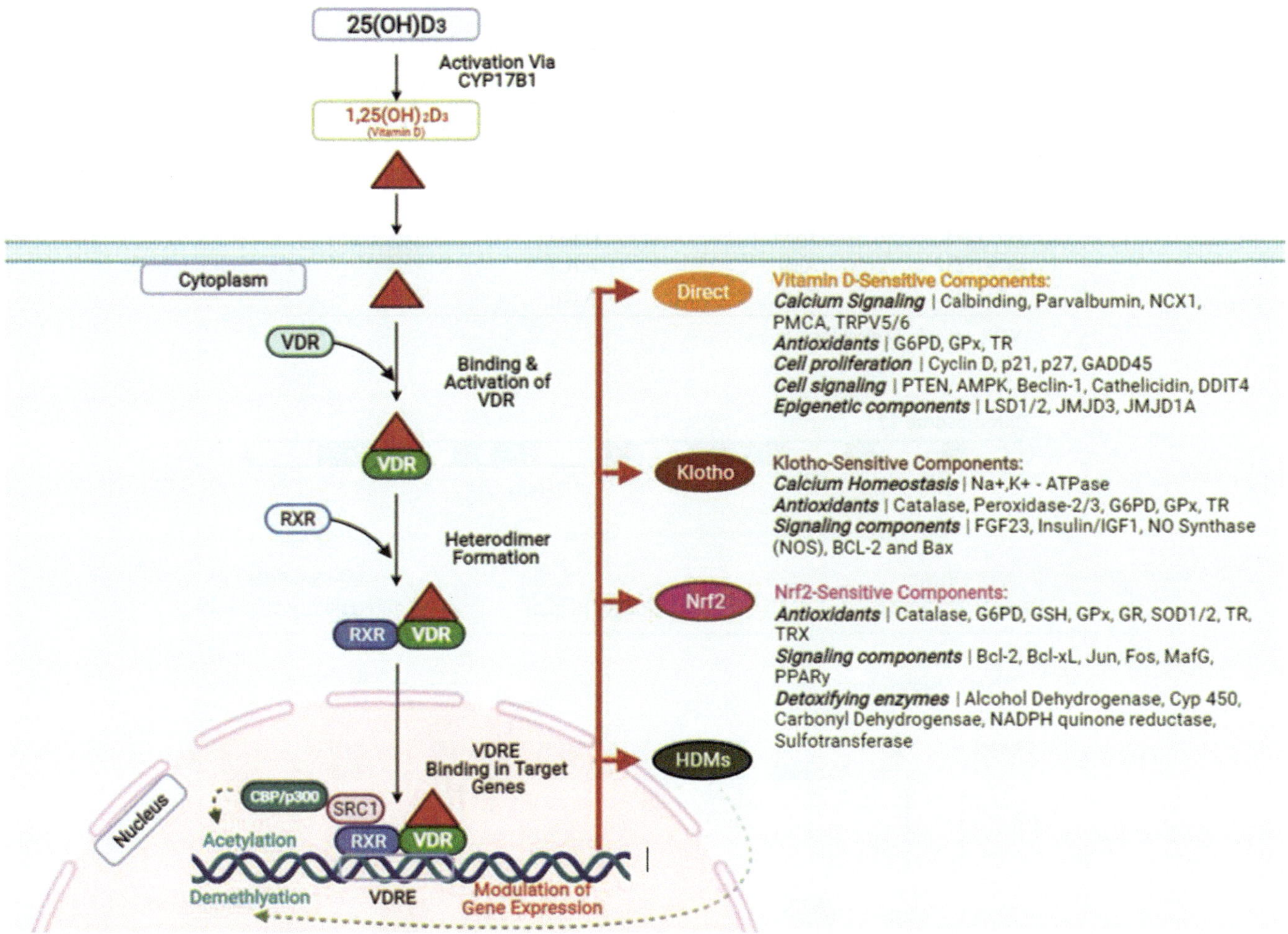

FIGURE 7.3 Vitamin D signaling via vitamin D receptor (VDR).

Klotho is a unique transmembrane protein whose expression from klotho gene is regulated by vitamin D (Forster et al., 2011). Klotho gene is an anti-aging one, disruption of which is known to be linked with multiple disorders such as osteoporosis, osteopenia, cognitive defects, growth retardation, skin atrophy, parkinsonian gait, and impaired hearing (Berridge, 2015a, b; Shiozaki et al., 2008; Yamamoto et al., 2005). Klotho protein is diverse in its functions and influences numerous intracellular signaling pathways such as p53/p21, cyclin AMP, protein kinase C, NO synthase, and FGF23. It is also known to enhance the expression of antioxidant enzymes such as SOD2 and peroxiredoxins (Prx-2 and Prx-3) (Wang and Sun, 2009; Shiozaki et al., 2008).

7.3 VDR Polymorphisms

The human VDR gene is found on the long arm of chromosome 12 (12q13.11). A series of restriction fragment length polymorphisms (RFLPs) have been reported in the human VDR gene. Generally, majority of the polymorphisms found in the VDR gene are found in the regulatory untranslated regions, i.e., 5′-UTR and 3′-UTR, of the gene. The most studied VDR polymorphisms are the seven single nucleotide polymorphisms (SNPs), which are Cdx2 (rs11568820), EcoRV (rs4516035), FokI (rs2228570), Tru9I (rs757343), BsmI (rs1544410), ApaI (rs7975232), and TaqI (rs731236) restriction sites (Makoui et al., 2020; Orlow et al., 2012; Köstner et al., 2009) (**see Table 7.1** and **Figure 7.4**).

7.3.1 rs11568820 (CDX2)

Cdx2 (rs11568820) SNP is in the 5′-untranslated region (UTR) at the 1e locus of the promoter of the VDR gene. The SNP is a change of guanine (G) to adenine (A) within the binding site for an intestinal-specific transcription factor known as caudal-related homeodomain (Cdx2) (Yamamoto et al., 1999). The Cdx2-binding site in VDR promoter influences the vitamin D regulation of calcium absorption in the intestines. The Cdx2 site with the A allele has been demonstrated to have more affinity for binding to the Cdx transcription factor than the one with the G allele (Arai et al., 2001). Thus, the VDR gene with Cdx2 AA phenotype is associated with an increased expression of VDR in the intestines and hence facilitates the

TABLE 7.1

Summary of Six Important VDR Polymorphisms and Their Effects

SNP ID	Alternative Name	Major (Minor) Alleles	Reported Effect
rs11568820	CDX2	C (T)	
rs4516035	EcoRV	T (C)	
rs10735810	FokI	C (T)	
rs1544410	BsmI	C (T)	
rs757343	Tru9I	G (A)	
rs7975232	ApaI	A (C)	
rs731236	TaqI	A (G)	

VDR, vitamin D receptor.

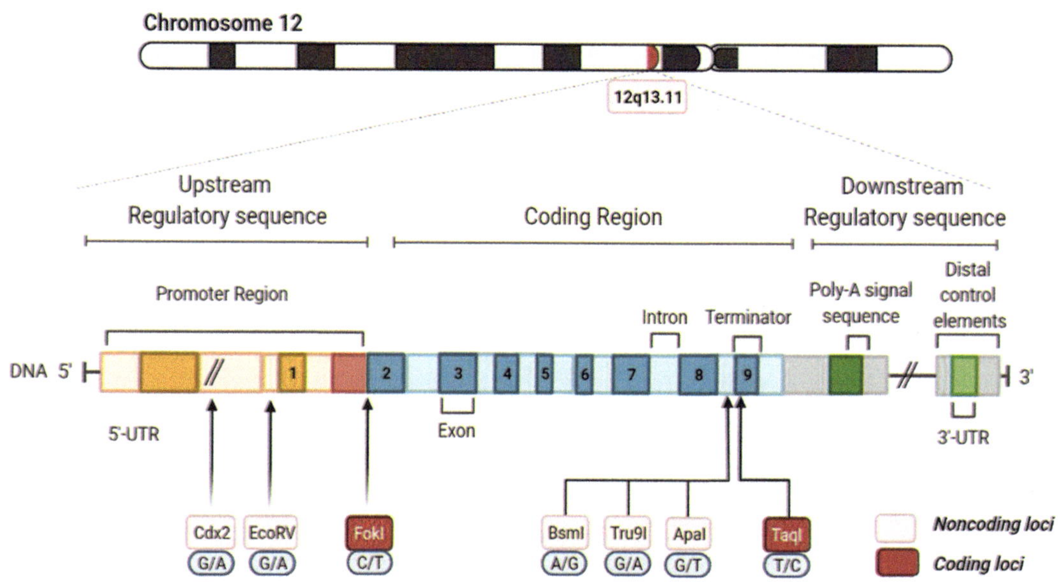

FIGURE 7.4 VDR gene structure and position of known important single nucleotide polymorphisms.

higher expression of calcium transport proteins such as calbindin 9K and 28K and TRPV5 and TRPV6, which results in an enhanced intestinal absorption of calcium manifesting in higher bone mineral density (BMD) (Arai et al., 2001).

7.3.2 rs4516035 (EcoRV)

EcoRV (rs4516035) VDR SNP is also located in the 5′-UTR of the VDR gene, resulting in the change of guanine (G) to adenine (A) (Orlow et al., 2012). This SNP has been demonstrated to affect the docking of transcription factors to the promoter region as it is located within the core GATA-binding site of the promoter. It is the site for binding of glutamyl-transfer RNA amidotransferase subunit A 3 (GATA3) (Mocellin S and Nitti, 2008). This SNP results in the elimination of the binding site; hence, only VDR EcoRV A allele is known to avidly bind with the factor while the binding is absent for the G allele (Schug and Overton, 1997). GATA3 protein has a key regulatory role in the polarization of naive T cells into T-helper 2-type lymphocytes, thereby effecting the anticancer response of immune system (Rengarajan et al., 2000).

7.3.3 rs2228570 (FokI)

FokI (rs2228570) SNP is located at the first potential start codon of the VDR gene, about 10 bp upstream of the actual start codon, and is often referred to as the start codon polymorphism (SCP) (van Etten et al., 2007). This SNP involves the change of cytosine (C) to thymine (T) causing the change of wild sequence of ACG to ATG, thereby creating a new additional start codon upstream from the actual one. Thus, a wild VDR gene containing C allele (also referred to as F allele) generates a shorter VDR protein of 424 amino acid (aa) residues and the variant (f allele) generates a longer one of 427 amino acids (Jurutka et al., 2001).

It has been demonstrated that shorter VDR protein (VDR F allele) is about 1.7 times more active, interacts efficiently with TFIIB and possesses higher transcriptional activity compared with the full-length VDR protein (VDR f/allele) (Arai et al., 1997; Jurutka et al., 2000). FokI SNP is the only polymorphism in VDR that results in the production of the altered protein.

7.3.4 rs757343 (Tru9I)

Tru9I (rs757343) SNP is a novel and less penetrative one, located in between the last two exons (8 and 9) in intron 8,

+443 bp from the end of exon 8 of the VDR gene (Ye et al., 2000). It is one of the important 3′-UTR SNPs, which involve the change of guanine (G) to adenine (A), and has been suggested to regulate the expression of gene via its effect on the enhancer affinity and target area (He and Wang, 2015).

7.3.5 rs1544410 (BsmI), rs7975232 (ApaI) and rs731236 (TaqI)

Three adjacent RFLP SNPs for BsmI (rs1544410), ApaI (rs7975232), and TaqI (rs731236), respectively, are confined near the 3′-UTR of the VDR gene. Located well within the intronic region between exons 8 and 9, these SNPs have been demonstrated to be in strong linkage disequilibrium (LD) with each other (Ručević et al., 2009). The 3′-UTR of the VDR gene serves to regulate the gene expression by primarily affecting the mRNA stability (Saccone et al., 2015). BsmI (rs1544410) and ApaI (rs7975232) are in the intron 8 at the 3'-UTR end of the VDR gene and are both silent SNPs as they do not lead to any change in the amino acid sequence of the encoded protein (Orlow et al., 2012; Uitterlinden et al., 2004a). Irrespective of that, it has been proposed that VDR gene expression gets affected because of differential regulation of mRNA stability (Whitfield et al., 2001).

Taql (rs731236) is another coding region SNP present in the exon 9 toward the 3'-UTR terminus of the VDR gene, resulting in the change of thymine (T) to cytosine (C) (Orlow et al., 2012; Uitterlinden et al., 2004b). It is a silent change in the codon 352 (ATT>ATC; isoleucine) of gene and hence does not affect the structure of the receptor (Vasilovici et al., 2019). However, it has been demonstrated that TT genotype is associated with lower levels of circulating active vitamin D (Valdivielso and Fernandez, 2006; Uitterlinden et al., 2004b).

7.4 VDR SNPs and Diseases

Vitamin D receptor polymorphisms have been reported to play a key role in the predisposition to numerous diseases, some of which are enlisted below (**see Figure 7.5**).

7.4.1 Cardiovascular Diseases

Cardiovascular diseases (CVDs) include a host of disorders involving heart and blood vessels, of which hypertension (HT) and diabetes mellitus (DM) are regarded as two of the most important ones.

FIGURE 7.5 VDR gene polymorphisms and relation with various diseases.

7.4.1.1 Hypertension

Hypertension (HT) is one of the chief cardiovascular ailments affecting about 25%–43% of the general population aged over 18 years globally and is defined as the blood pressure equal to or greater than 140/90 mmHg (Zhou et al., 2008; Danik and Manson, 2012; Nitsa et al., 2018; Gouni-Berthold and Berthold, 2021). Several epidemiological studies have reported that individuals with low blood levels of vitamin D have increased risks of heart disease, stroke, hypertension, and diabetes (Nitsa et al., 2018). Vitamin D has been demonstrated to affect the renin-angiotensin system (RAS) and has favorable effects on vascular smooth muscle cells of the endothelium and cardiomyocytes (Li et al., 2004; Artaza et al., 2011; Nasri et al., 2014).

A recent meta-analysis by Zhu et al. (2019) involving seven reports on 4,011 subjects reported that VDR rs1544410 (BsmI) SNP is associated with susceptibility to hypertension. The meta-analysis was carried out to find a possible association between various VDR SNPs and hypertension. It was found that VDR BsmI AA genotype had lower susceptibility to hypertension in comparison with those carrying GA or GG genotype ($p=0.005$). However, for VDR FokI and VDR ApaI SNPs, no significant association was found with susceptibility to hypertension. Previously, Wang et al. (2013) in their study on 1211 US men had reported that two VDR SNPs – BsmI and FokI – were associated with hypertension in the US population, while Swapna et al. (2011) had reported that VDR FokI SNP was a risk factor for essential hypertension (EH) in the Indian population.

Another pan-world meta-analysis by Nunes et al. (2020) based on six inclusive case-control studies demonstrated that VDR FokI SNP has a modulatory effect on HT. They concluded that the allele f of FokI SNP was associated with significantly lower odds of having HT in global analysis under the additive/homozygote and common/recessive genetic models, particularly in the Asian population. However, the results were not significant for European population. In addition, in a recent case-control study by Caccamo et al. (2020) involving 116 pregnant women with gestational hypertension (GH) and 69 normotensive pregnant women (CTR), the authors concluded that VDR FokI/BsmI – FF/bB haplotype was the most frequent one (29.3%, 34/116) in GH cohort and was responsible for increasing the risk of GH twofold; additionally, 92% of VDR FF/bB containing GH pregnant women were found to suffer from hypovitaminosis D, 27% of which had deficient vitamin D levels in comparison with only 11% of their normotensive counterparts.

7.4.1.2 Diabetes Mellitus

Diabetes mellitus (DM), in any of its forms, is a complex metabolic cardiovascular disorder that is characterized by constant hyperglycemia as a result of defects in either insulin secretion or insulin action, or both. The disease reflects itself as a chronic and heterogeneous dysregulation of macronutrient metabolism, including that of carbohydrates, lipids, and proteins (Banday et al., 2020; Sameer et al., 2019). Numerous biomarkers are currently in use for the effective detection of hyperglycemia and hence for diagnosis of DM (Sameer et al., 2019). The role played by vitamin D via VDRs in pancreas has already been demonstrated in animal models and linked to the inhibition of insulin secretion concomitant with the vitamin D deficiency (Norman AW, 1998). Several population-based studies have also provided evidence for the association of DM with the VDR polymorphisms (De Azevêdo Silva et al., 2013; Zhu et al., 2014; Yu et al., 2016; Hussain et al., 2019; Shaat et al., 2020; Zhai et al., 2020).

In one of the sentinel meta-analysis by Qin et al. (2014) on elucidating the association of VDR BsmI SNP with the risk of type 1 diabetes mellitus (T1DM) by analyzing 23 published articles, it was reported that VDR BsmI bb genotype was associated with T1DM. Furthermore, ethnicity-specific analysis revealed that B allele and bb genotype were associated with T1DM in Asian and Latino populations, but not in Caucasians and in Africans, and B allele and BB genotype were associated with T1DM risk. Another meta-analysis by Yu et al. (2016) involving 30 case-control studies observed that two VDR SNPs – FokI and BsmI – were associated with T2DM, while the association between BsmI SNP and T2DM was weak in two genetic models (Bb vs bb and BB+Bb vs bb) and the association between FokI and T2DM was strong ($p<0.001$). However, novel results were reported by a recent meta-analysis of 39 case-control studies by Zhai et al. (2020). They reported a significant association between VDR SNPs and T1DM risk albeit in an ethnicity-specific manner, with American populations demonstrating a markedly significant negative association between VDR BsmI SNP and the risk of T1DM susceptibility across all genotype models, while European, Asian, and African populations not displaying any such association. Furthermore, VDR FokI was reported to demonstrate an increased risk of T1DM in African population, but decreased risk in European population.

7.4.2 Renal Diseases

Chronic kidney disease (CKD) and end-stage renal disease (ESRD) are important nephrological complications of dysregulated hyperglycemia, which has been reported to be linked with vitamin D or VDR deficiency (Patel et al., 2009; Cardoso and Pereira, 2019; Vahdat S, 2020). Vitamin D activation to its final active form takes place within kidneys, and hence, it has a direct and beneficial effect on kidneys via klotho (predominantly expressed in the distal tubules) in cooperation with fibroblast growth factors (FGFs) (**see Figure 7.3**) (Vahdat, 2020). The essential role played by vitamin D via VDRs and their SNPs in the manifestation of various kidney diseases including nephrolithiasis (NL) has been provided by various population-based studies (Yang et al., 2014; Zhou et al., 2015a, b; Hussain et al., 2019; González-Castro et al., 2019).

Zhou et al. (2015a) in their meta-analysis to explore the relation between various VDR SNPs and the risk of CKD demonstrated that the VDR FokI SNP was associated with the risk of CKD in Asians, but not in overall population nor in Caucasians. Furthermore, in another study, the same authors reported that VDR BsmI SNP was not associated with CKD susceptibility in Asians and in Caucasians (Zhou et al., 2015b).

Similar results were reported in another meta-analysis by Yang et al. (2014), who reported that none of the four VDR SNPs – BsmI, FokI, TaqI, and ApaI – were associated with the risk of ESRD in the overall population, Asians, and Caucasians.

In a recent meta-analysis on 2,777 renal disease cases including NL, diabetic nephropathy (DN), and ESRD, Hussain et al. (2019) reported that VDR TaqI and ApaI SNPs were the risk factors for the renal diseases. However, the results of another meta-analysis reported that VDR BsmI SNP had a protective effect against NL, while two other VDR SNPs – FokI and TaqI – were associated with a decreased risk of NL in the heterozygous model and ApaI had no overall association. Imani et al. (2020) in their meta-analysis of 33 studies to explore the association of VDR SNPs with urolithiasis concluded that none of the studied four VDR SNPs were associated with the disease.

7.4.3 Allergic Diseases

Allergic diseases affecting humans are diverse and include asthma, atopic dermatitis (AD), and allergic rhinitis (Zhang et al., 2020a,b). Both arms of the immune systems – innate and adaptive – play a pivotal role in the pathogenesis of the allergic diseases, and numerous genes have been found to be associated with their susceptibility (Morales et al., 2015). Furthermore, the role of vitamin D in maintaining a healthy immune system becomes more evident with the fact that VDR is expressed by the majority of immune cells including monocytes, macrophages, B and T lymphocytes, and dendritic cells. Additionally, all these immune cells are also found to have an active vitamin D metabolism to locally convert $25(OH)D_3$ into its active form – $1,25(OH)_2D_3$ (Trochoutsou et al., 2015; Sassi et al., 2018; Hufnagl et al., 2018). Vitamin D and its effector VDR have been found be an important modulator of the immune system responses, including macrophage activation, Th1 cytokine production, and Th2 cytokine inhibition, thereby having a bearing on the development of allergic diseases (Sassi et al., 2018).

Evidence on the role of VDR SNPs in allergic diseases came with the meta-analysis of 21 studies by Zhang et al. (2020a,b), who reported that three VDR SNPs – BsmI, TaqI, and ApaI – are associated with conferring higher susceptibility to allergic disease in certain populations, and VDR BsmI and TaqI are associated with elevated susceptibility to asthma in specific populations. VDR FokI SNP, however, was not reported to confer any risk of allergic diseases in any population. Heine et al. (2013) in their case-control study involving 265 AD patients reported the association between three VDR SNPs – BsmI (G allele), ApaI (C allele), and TaqI (T allele) – and the risk of AD, but not FokI. Additionally, VDR GCT haplotype was correlated with severe AD and complementary AAC with protection from AD.

7.4.4 Autoimmune Diseases

Owing to the role played by vitamin D in modulating the immune response of both innate and adaptive arms of immunity, its importance in initiation and progression of several autoimmune diseases such as rheumatoid arthritis (RA), systemic lupus erythematosus (SLE), and inflammatory bowel disease (IBD) is well established in the literature (Christakos et al., 2012; Jeffery et al., 2016; Bizzaro et al., 2017; Nikiphorou et al., 2018; Anaya et al., 2018; Sassi et al., 2018; Szymczak-Tomczak et al., 2019; Harrison et al., 2020; Ruiz-Ballesteros, 2020). In innate immune response, vitamin D participates in the maintenance of intestinal barrier, regulation of tight junctions, and intestinal cell apoptosis, while in adaptive immune responses, it acts as a regulator of T and B cell proliferation via VDR (Hewison, 2012).

Song et al. (2016) in their meta-analysis of seven sentinel studies involving 923 RA cases, for exploring the association of VDR SNPs with susceptibility of RA, reported an association of VDR FokI F allele with RA and FokI SNP with RA in dominant model among Europeans, while no association was found in VDR BsmI and TaqI SNPs. Zhang et al. (2020a,b) in their meta-analysis of 37 studies reported a significant association of several VDR SNPs with RA and SLE, especially in Asians. Mukhtar et al. (2019) in their study on RA patients reported several VDR SNPs to be a risk factor for RA onset.

Bae et al. (2018) in their meta-analysis of 12 studies involving SLE cases reported that VDR SNPs were not associated with SLE susceptibility in overall, European, and Asian populations, but in Arab population, the association of VDR FokI F allele with SLE susceptibility was found to be significant ($p<0.05$). A previous meta-analysis including 11 studies including 1,621 SLE cases reported that VDR BsmI B allele was associated with the onset of SLE in overall population and Asians. Additionally, the reported that FokI FF genotype was correlated with the susceptibility to SLE in Asians, but FokI and TaqI SNPs were not associated with SLE risk among Caucasians. And no significant association was found between VDR ApaI SNP and SLE risk in overall population, Asians, and Caucasians (Mao and Huang, 2014).

7.4.5 Neurological Diseases

Multiple sclerosis (MS) is one of the important chronic, inflammatory, autoimmune-mediated neurodegenerative disorders that affect the central nervous system (CNS) characterized by demyelination of axonal transection (McGinley et al., 2021). Numerous studies and meta-analyses have provided an evidence to the role of vitamin D and VDR SNPs in the susceptibility to MS in different populations across the world (Lu et al., 2018, 2019; Elkama and Karahalil, 2018; Ruiz-Ballesteros, 2020; Scazzone et al., 2021).

Zhang et al. (2019) in their meta-analysis of 27 case-control studies involving 4,879 MS patients reported VCR TaqI SNP to be associated with MS in recessive, homozygous, and codominant model in overall population and Caucasians, and FokI SNP showed association with MS in recessive, dominant, and codominant models in Asians, but no association was found for VDR BsmI and ApaI SNPs. Imani et al. (2019) in their meta-analysis of 30 case-control studies involving MS cases reported a significant association between TaqI polymorphism and MS risk under heterozygote genetic model. In addition, BsmI polymorphisms were found to be associated with a higher risk of MS under recessive model in Asian populations. But, VDR ApaI SNP was found to be linked with a lower risk of MS under recessive and aa vs. AA model in Asian populations.

In another meta-analysis to resolve the conflicting results about the role of VDR SNPs in susceptibility to MS, Mohammadi et al. (2020) reported that VDR ApaI and TaqI are the only SNPs that exhibited a significant association with MS in Iranian population. Tizaoui et al. (2015) in their meta-analysis involving 3300 MS cases of 13 case-control studies reported VDR ApaI SNP to be significantly associated with MS risk in homozygous and codominant models, while BsmI and TaqI SNPs did not display any association.

7.4.6 Hepatic Diseases

One of the important hepatic diseases that have a direct link with vitamin D is the nonalcoholic fatty liver disease (NAFLD). Vitamin D deficiency has been envisaged to affect the disease progression via its extensive anti-fibrotic, anti-inflammatory, and insulin-sensitizing properties (Keane et al., 2019; Barchetta et al., 2020). However, the role played by various VDR SNPs in the susceptibility to liver diseases still remains unexplored.

Apart from these six categories of diseases, which we have discussed in this chapter, there are many other human diseases that have been found to be associated with the VDR SNPs (Uitterlinden et al., 2004; Valdivielso and Fernandez, 2006). However, we have restricted ourselves in discussing those as that would be beyond the scope of this chapter. The association of VDR SNPs with cancers is discussed exhaustively by Dr. Awadh in Chapter 15 of this book.

7.5 Conclusions

The role of vitamin D in the maintenance of human health is of key importance, especially via its mediator vitamin D receptor (VDR), which modulates the functioning of numerous pathways. The genetic polymorphisms of the VDR have been linked to the initiation, development, and progression of many human diseases as discussed in this chapter. Large-scale genome-wide association studies are the need of the hour to fully elucidate the penetrance of the VDR SNPs among different populations of the world and to devise a risk map of the diseases based on the frequency of the VDR SNPs, considering various ethnicities and races.

ABBREVIATIONS

AD: acute dermatitis
Cdx2: caudal-related homeodomain
CNPs: copy number polymorphisms
CNS: central nervous system
CNVs: copy number variants
CVDs: cardiovascular diseases
DBD: DNA-binding domain
DM: diabetes mellitus
DN: diabetic nephropathy
DNA: deoxyribose nucleic acid
ESRD: end-stage renal disease
GCL: glutamate cysteine ligase
GPs: genetic polymorphisms
GVs: genetic variations
GWAS: genome-wide association studies
HO-1: heme oxygenase-1
HT: hypertension
IBD: inflammatory bowel disease
LBD: ligand-binding domain
NAFLD: nonalcoholic fatty liver disease
NL: nephrolithiasis
NQO1: NAD(P)H quinone oxidoreductase 1
RA: rheumatoid arthritis
RFLP: restriction fragment length polymorphism
RXR: retinoid X receptor
SLE: systemic lupus erythematosus
SNP: single nucleotide polymorphism
SOD: superoxide dismutase
TRX: thioredoxin
UTR: untranslated region
VDR: vitamin D receptor
VDRE: vitamin D response element

REFERENCES

Anaya JM, Restrepo-Jiménez P, Ramírez-Santana C. The autoimmune ecology: an update. *Curr Opin Rheumatol.* 2018, 30 (4):350–360. doi: 10.1097/BOR.0000000000000498.

Arai H, Miyamoto K, Taketani Y, Yamamoto H, Iemori Y, Morita K, Tonai T, Nishisho T, Mori S, Takeda E. A vitamin D receptor gene polymorphism in the translation initiation codon: effect on protein activity and relation to bone mineral density in Japanese women. *J Bone Miner Res.* 1997, 12(6):915–921. doi: 10.1359/jbmr.1997.12.6.915.

Arai H, Miyamoto KI, Yoshida M, Yamamoto H, Taketani Y, Morita K, Kubota M, Yoshida S, Ikeda M, Watabe F, Kanemasa Y, Takeda E. The polymorphism in the caudal-related homeodomain protein Cdx-2 binding element in the human vitamin D receptor gene. *J Bone Miner Res.* 2001, 16(7):-1256–1264. doi: 10.1359/jbmr.2001.16.7.1256.

Artaza JN, Contreras S, Garcia LA, Mehrotra R, Gibbons G, Shohet R, Martins D, Norris KC. Vitamin D and cardiovascular disease: potential role in health disparities. *J Health Care Poor Underserved.* 2011;22(4 Suppl):23–38. doi: 10.1353/hpu.2011.0161.

Bacchetta J, Salusky IB, Hewison M. Beyond mineral metabolism, is there an interplay between FGF23 and vitamin D in innate immunity? *Pediatr Nephrol.* 201, 28(4):577–582. doi: 10.1007/s00467-012-2336-0.

Bae SC, Lee YH. Vitamin D receptor FokI, TaqI, and ApaI polymorphisms and susceptibility to systemic lupus erythematosus: an updated meta-analysis. *Clin Rheumatol.* 2018, 37(6):1529–1537. doi: 10.1007/s10067-018-4036-z.

Banday MZ, Sameer AS, Nissar S. Pathophysiology of diabetes: an overview. *Avicenna J Med.* 2020, 10(4):174–188. doi: 10.4103/ajm.ajm_53_20.

Barchetta I, Cimini FA, Cavallo MG. Vitamin D and metabolic dysfunction-associated fatty liver disease (MAFLD): an update. *Nutrients.* 2020 , 12(11):3302. doi: 10.3390/nu12113302.

Berridge MJ. Vitamin D cell signalling in health and disease. *Biochem Biophys Res Commun*. 2015a, 460(1):53–71. doi: 10.1016/j.bbrc.2015.01.008.

Berridge MJ. Vitamin D: a custodian of cell signalling stability in health and disease. *Biochem Soc Trans*. 2015b, 43(3), 349–358. doi: 10.1042/BST20140279.

Bikle DD. Vitamin D metabolism, mechanism of action, and clinical applications. *Chem Biol*. 2014, 21(3):319–329. doi: 10.1016/j.chembiol.2013.12.016.

Bizzaro G, Antico A, Fortunato A, Bizzaro N. Vitamin D and Autoimmune diseases: is vitamin D receptor (VDR) polymorphism the culprit? *Isr Med Assoc J*. 2017, 19(7):438–443.

Bobilev I, Novik V, Levi I, Shpilberg O, Levy J, Sharoni Y, Studzinski GP, Danilenko M. The Nrf2 transcription factor is a positive regulator of myeloid differentiation of acute myeloid leukemia cells. *Cancer Biol Ther*. 2011, 11(3): 317–29. doi: 10.4161/cbt.11.3.14098.

Caccamo D, Cannata A, Ricca S, Catalano LM, Montalto AF, Alibrandi A, Ercoli A, Granese R. Role of Vitamin-D Receptor (VDR) single nucleotide polymorphisms in gestational hypertension development: A case-control study. *PLoS One*. 2020, 15(11):e0239407. doi: 10.1371/journal.pone.0239407.

Cardoso MP, Pereira LAL. Native vitamin D in pre-dialysis chronic kidney disease. *Nefrologia (Engl Ed)*. 2019, 39(1):18–28. English, Spanish. doi: 10.1016/j.nefro.2018.07.004.

Cheng JB, Levine MA, Bell NH, et al. Genetic evidence that the human CYP2R1 enzyme is a key vitamin D 25-hydroxylase. *Proc Natl Acad Sci U S A* 2004, 101(20):7711–7715.

Christakos S, Ajibade DV, Dhawan P, Fechner AJ, Mady LJ. Vitamin D: metabolism. *Endocrinol Metab Clin North Am*. 2010, 39(2):243–353, table of contents. doi: 10.1016/j.ecl.2010.02.002.

Christakos S, Ajibade DV, Dhawan P, Fechner AJ, Mady LJ. Vitamin D: metabolism. *Rheum Dis Clin North Am*. 2012, 38(1):1–11, vii. doi: 10.1016/j.rdc.2012.03.003.

Christakos S, Dhawan P, Verstuyf A, Verlinden L, Carmeliet G. Vitamin D: metabolism, molecular mechanism of action, and pleiotropic effects. *Physiol Rev*. 2016, 96(1):365–408. doi: 10.1152/physrev.00014.2015.

Danik JS and Manson JE. Vitamin D and cardiovascular disease. *Curr Treat Options Cardiovasc Med*. 2012, 14(4):414–424. doi: 10.1007/s11936-012-0183-8.

De Azevêdo Silva J, Guimarães RL, Brandão LA, Araujo J, Segat L, Crovella S, Sandrin-Garcia P. Vitamin D receptor (VDR) gene polymorphisms and age onset in type 1 diabetes mellitus. *Autoimmunity*. 2013, 46(6):382–387. doi: 10.3109/08916934.2013.795952.

Elkama A, Karahalil B. Role of gene polymorphisms in vitamin D metabolism and in multiple sclerosis. *Arh Hig Rada Toksikol*. 2018, 69(1):25–31. doi: 10.2478/aiht-2018-69-3065. PMID: 29604195.

van Etten E, Verlinden L, Giulietti A, Ramos-Lopez E, Branisteanu DD, Ferreira GB, Overbergh L, et al. The vitamin D receptor gene FokI polymorphism: functional impact on the immune system. *Eur J Immunol*. 2007, 37(2):395–405.

Forster RE, Jurutka PW, Hsieh JC, Haussler CA, Lowmiller CL, Kaneko I, Haussler MR, Kerr Whitfield G. Vitamin D receptor controls expression of the anti-aging klotho gene in mouse and human renal cells. *Biochem Biophys Res Commun*. 2011, 414(3):557–562. doi: 10.1016/j.bbrc.2011.09.117.

González-Castro TB, Blachman-Braun R, Hernández-Díaz Y, Tovilla-Zárate CA, Pérez-Hernández N, Moscardi PRM, Alam A, Borgonio-Cuadra VM, Reyes-López PA, Juárez-Rojop IE, López-Narváez ML, Posadas-Sánchez R, Vargas-Alarcón G, Rodríguez-Pérez JM. Association of vitamin D receptor polymorphisms and nephrolithiasis: a meta-analysis. *Gene*. 2019, 711:143936. doi: 10.1016/j.gene.2019.06.026.

Gouni-Berthold I, and Berthold HK. Vitamin D and Vascular Disease. *Curr Vasc Pharmacol*. 2021;19(3):250–268. doi: 10.2174/1570161118666200317151955.

Harrison SR, Li D, Jeffery LE, Raza K, Hewison M. Vitamin D, Autoimmune Disease and Rheumatoid Arthritis. *Calcif Tissue Int*. 2020, 106(1):58–75. doi: 10.1007/s00223-019-00577-2.

Haussler MR, Whitfield GK, Kaneko I, Haussler CA, Hsieh D, Hsieh JC, Jurutka PW. Molecular mechanisms of vitamin D action. *Calcif Tissue Int*. 2013, 92(2):77–98. doi: 10.1007/s00223-012-9619-0.

Hayes JD, Dinkova-Kostova AT. The Nrf2 regulatory network provides an interface between redox and intermediary metabolism. *Trends Biochem Sci*. 2014, 39(4):199–218. doi: 10.1016/j.tibs.2014.02.002.

He L, Wang M. Association of vitamin d receptor-a gene polymorphisms with coronary heart disease in Han Chinese. *Int J Clin Exp Med*. 2015, 8(4):6224–6229.

Heine G, Hoefer N, Franke A, Nöthling U, Schumann RR, Hamann L, Worm M. Association of vitamin D receptor gene polymorphisms with severe atopic dermatitis in adults. *Br J Dermatol*. 2013, 168(4):855–858. doi: 10.1111/bjd.12077.

Hewison M. Vitamin D and immune function: an overview. *Proc Nutr Soc*. 2012, 71(1):50–61. doi: 10.1017/S0029665111001650.

Hufnagl K, Jensen-Jarolim E. Vitamin A and D in allergy: from experimental animal models and cellular studies to human disease. *Allergo J Int*. 2018, 27(3):72–78. doi: 10.1007/s40629-018-0054-2.

Hussain T, Naushad SM, Ahmed A, Alamery S, Mohammed AA, Abdelkader MO, Alkhrm NAN. Association of vitamin D receptor TaqI and ApaI genetic polymorphisms with nephrolithiasis and end stage renal disease: a meta-analysis. *BMC Med Genet*. 2019, 20(1):193. doi: 10.1186/s12881-019-0932-6.

Imani D, Razi B, Motallebnezhad M, Rezaei R. Association between vitamin D receptor (VDR) polymorphisms and the risk of multiple sclerosis (MS): an updated meta-analysis. *BMC Neurol*. 2019, 19(1):339. doi: 10.1186/s12883-019-1577-y.

Imani D, Razi B, Khosrojerdi A, Lorian K, Motallebnezhad M, Rezaei R, Aslani S. Vitamin D receptor gene polymorphisms and susceptibility to urolithiasis: a meta-regression and meta-analysis. *BMC Nephrol*. 2020, 21(1):263. doi: 10.1186/s12882-020-01919-1.

Jeffery LE, Raza K, Hewison M. Vitamin D in rheumatoid arthritis-towards clinical application. *Nat Rev Rheumatol*. 2016, 12(4):201–210. doi: 10.1038/nrrheum.2015.140.

Jurutka PW, Remus LS, Whitfield GK, Thompson PD, Hsieh JC, Zitzer H, Tavakkoli P, Galligan MA, Dang HT, Haussler CA, Haussler MR. The polymorphic N terminus in human vitamin D receptor isoforms influences transcriptional activity by modulating interaction with transcription factor IIB. *Mol Endocrinol*. 2000, 14(3):401–420. doi: 10.1210/mend.14.3.0435.

Jurutka PW, Whitfield GK, Hsieh JC, Thompson PD, Haussler CA, Haussler MR. Molecular nature of the vitamin D receptor and its role in regulation of gene expression. *Rev Endocr Metab Disord.* 2001, 2(2):203–216. doi: 10.1023/a:1010062929140.

Kato S. The function of vitamin D receptor in vitamin D action. *J Biochem.* 2000, 127(5):717–22. doi: 10.1093/oxfordjournals.jbchem.a022662.

Keane JT, Elangovan H, Stokes RA, Gunton JE. Vitamin D and the liver-correlation or cause? *Nutrients.* 2018, 10(4):496. doi: 10.3390/nu10040496.

Köstner K, Denzer N, Müller CS, Klein R, Tilgen W, Reichrath J. The relevance of vitamin D receptor (VDR) gene polymorphisms for cancer: a review of the literature. *Anticancer Res.* 2009, 29(9):3511–3536.

Li YC, Qiao G, Uskokovic M, Xiang W, Zheng W, Kong J. Vitamin D: a negative endocrine regulator of the renin-angiotensin system and blood pressure. *J Steroid Biochem Mol Biol.* 2004, 89–90(1–5):387–392. doi: 10.1016/j.jsbmb.2004.03.004.

Lockau L and Atkinson SA. Vitamin D's role in health and disease: how does the present inform our understanding of the past? *Int J Paleopathol.* 2018, 23:6–14. doi: 10.1016/j.ijpp.2017.11.005.

LoPiccolo MC and Lim HW. Vitamin D in health and disease. *Photodermatol Photoimmunol Photomed.* 2010, 26(5):224–229. doi: 10.1111/j.1600-0781.2010.00524.x.

Lu M, Taylor BV, Körner H. Genomic Effects of the vitamin D receptor: potentially the link between vitamin d, immune cells, and multiple sclerosis. *Front Immunol.* 2018, 9:477. doi: 10.3389/fimmu.2018.00477.

Lu M, McComish BJ, Burdon KP, Taylor BV, Körner H. The association between vitamin D and multiple sclerosis risk: 1,25 (OH)2D3 induces super-enhancers bound by VDR. *Front Immunol.* 2019, 10:488. doi: 10.3389/fimmu.2019.00488.

Makoui MH, Imani D, Motallebnezhad M, Azimi M, Razi B. Vitamin D receptor gene polymorphism and susceptibility to asthma: Meta-analysis based on 17 case-control studies. *Ann Allergy Asthma Immunol.* 2020, 124(1):57–69. doi: 10.1016/j.anai.2019.10.014.

Mao S, Huang S. Association between vitamin D receptor gene BsmI, FokI, ApaI and TaqI polymorphisms and the risk of systemic lupus erythematosus: a meta-analysis. *Rheumatol Int.* 2014, 34(3):381–8. doi: 10.1007/s00296-013-2898-6.

McGinley MP, Goldschmidt CH, Rae-Grant AD. Diagnosis and treatment of multiple sclerosis: a review. *JAMA.* 2021 Feb 23;325(8):765–779. doi: 10.1001/jama.2020.26858. Erratum in: *JAMA.* 2021, 325(21): 2211.

Mocellin S and Nitti D. Vitamin D receptor polymorphisms and the risk of cutaneous melanoma. Cancer. 2008, 113:2398–2407. doi:https://doi.org/10.1002/cncr.23867.

Mohammadi A, Azarnezhad A, Khanbabaei H, Izadpanah E, Abdollahzadeh R, Barreto GE, Sahebkar A. Vitamin D receptor genetic polymorphisms and the risk of multiple sclerosis: a systematic review and meta-analysis. *Steroids.* 2020, 158:108615. doi: 10.1016/j.steroids.2020.108615.

Morales E, Sanchez-Solis M, Garcia-Marcos L. Vitamin D Metabolism Genes in Asthma and Atopy. *Mini Rev Med Chem.* 2015,15 (11):913–26. doi: 10.2174/1389557515666150519105944.

Mukhtar M, Sheikh N, Suqaina SK, Batool A, Fatima N, Mehmood R, Nazir S. Vitamin D Receptor Gene Polymorphism: an important predictor of arthritis development. *Biomed Res Int.* 2019, 2019:8326246. doi: 10.1155/2019/8326246.

Nagpal S, Na S, Rathnachalam R. Noncalcemic actions of vitamin D receptor ligands. *Endocr Rev.* 2005, 26(5):662–87. doi: 10.1210/er.2004-0002.

Nasri H, Behradmanesh S, Ahmadi A, Rafieian-Kopaei M. Impact of oral vitamin D (cholecalciferol) replacement therapy on blood pressure in type 2 diabetes patients; a randomized, double-blind, placebo controlled clinical trial. *J Nephropathol.* 2014, 3(1):29–33. doi: 10.12860/jnp.2014.07. Epub 2013 Dec 25.

Nikiphorou E, Uksila J, Sokka T. A cross-sectional study of vitamin D levels in a large cohort of patients with rheumatic diseases. *Clin Rheumatol.* 2018, 37(3):803–810. doi: 10.1007/s10067-017-3870-8.

Nikolac Gabaj N, Unic A, Miler M, Pavicic T, Culej J, Bolanca I, Herman Mahecic D, Milevoj Kopcinovic L, Vrtaric A. In sickness and in health: pivotal role of vitamin D. *Biochem Med (Zagreb).* 2020, 30(2):020501. doi: 10.11613/BM.2020.020501.

Nitsa A, Toutouza M, Machairas N, Mariolis A, Philippou A, Koutsilieris M. Vitamin D in cardiovascular disease. *In Vivo.* 2018, 32(5):977–981. doi: 10.21873/invivo.11338.

Norman AW. Sunlight, season, skin pigmentation, vitamin D, and 25-hydroxyvitamin D: integral components of the vitamin D endocrine system. *Am J Clin Nutr.* 1998, 67(6):1108–1110. doi: 10.1093/ajcn/67.6.1108.

Nunes IFOC, Cavalcante AACM, Alencar MVOB, Carvalho MDF, Sarmento JLR, Teixeira NSCCA, Paiva AA, Carvalho LR, Nascimento LFM, Cruz MSP, Rogero MM, Lima ACB, Carvalho CMRG. Meta-analysis of the association between the rs228570 vitamin D receptor gene polymorphism and arterial hypertension risk. *Adv Nutr.* 2020, 11(5):1211–1220. doi: 10.1093/advances/nmaa076. Erratum in: Adv Nutr. 2020 Sep 1;11(5): 1403.

Orlow I, Roy P, Reiner AS, Yoo S, Patel H, Paine S, Armstrong BK, Kricker A, Marrett LD, Millikan RC, Thomas NE, Gruber SB, Anton-Culver H, Rosso S, Gallagher RP, Dwyer T, Kanetsky PA, Busam K, From L, Begg CB, Berwick M; GEM Study Group. Vitamin D receptor polymorphisms in patients with cutaneous melanoma. *Int J Cancer.* 2012, 130(2):405–418. doi: 10.1002/ijc.26023.

Ozono K, Liao J, Kerner SA, Scott RA, Pike JW. The vitamin D-responsive element in the human osteocalcin gene. Association with a nuclear proto-oncogene enhancer. *J Biol Chem.* 1990, 265(35):21881–21888.

Patel TV, Singh AK. Role of vitamin D in chronic kidney disease. *Semin Nephrol.* 2009, 29(2):113–121. doi: 10.1016/j.semnephrol.2009.01.004.

Pfotenhauer KM, Shubrook JH. Vitamin D deficiency, its role in health and disease, and current supplementation recommendations. *J Am Osteopath Assoc.* 2017, 117(5):301–305. doi: 10.7556/jaoa.2017.055.

Pike JW and Christakos S. Biology and mechanisms of action of the vitamin D hormone. *Endocrinol Metab Clin North Am.* 2017, 46(4):815–843. doi: 10.1016/j.ecl.2017.07.001.

Reis GV, Gontijo NA, Rodrigues KF, Alves MT, Ferreira CN, Gomes KB. Vitamin D receptor polymorphisms and the polycystic ovary syndrome: a systematic review. *J Obstet Gynaecol Res.* 2017, 43(3):436–446. doi: 10.1111/jog.13250.

Rengarajan J, Szabo SJ, Glimcher LH. Transcriptional regulation of Th1/Th2 polarization. *Immunol Today*. 2000, 21:479–483.

Rochel N, Wurtz JM, Mitschler A, Klaholz B, Moras D. The crystal structure of the nuclear receptor for vitamin D bound to its natural ligand. *Mol Cell*. 2000, 5(1):173–179. doi: 10.1016/s1097-2765(00)80413-x.

Rucević I, Barisić-Drusko V, Glavas-Obrovac L, Stefanić M. Vitamin D endocrine system and psoriasis vulgaris-review of the literature. *Acta Dermatovenerol Croat*. 2009, 17(3):187–192.

Ruiz-Ballesteros AI, Meza-Meza MR, Vizmanos-Lamotte B, Parra-Rojas I, de la Cruz-Mosso U. Association of vitamin D metabolism gene polymorphisms with autoimmunity: evidence in population genetic studies. *Int J Mol Sci*. 2020, 21(24):9626. doi: 10.3390/ijms21249626.

Ruiz-Ojeda FJ, Anguita-Ruiz A, Leis R, Aguilera CM. Genetic Factors and Molecular mechanisms of vitamin D and obesity relationship. *Ann Nutr Metab*. 2018, 73(2):89–99. doi: 10.1159/000490669.

Saccone D, Asani F, Bornman L. Regulation of the vitamin D receptor gene by environment, genetics and epigenetics. *Gene*. 2015, 561(2):171–180.

Sameer AS, Banday MZ, Nissar S, Saeed SA. A Comparison of biomarkers in the assessment of glycemic control in diabetes: reviewing the evidence. *Curr Diabetes Rev*. 2019, 15(6):471–479.

Sassi F, Tamone C, D'Amelio P. Vitamin D: nutrient, hormone, and immunomodulator. *Nutrients*. 2018, 10(11):1656. doi: 10.3390/nu10111656.

Scazzone C, Agnello L, Bivona G, Lo Sasso B, Ciaccio M. Vitamin D and Genetic susceptibility to multiple sclerosis. *Biochem Genet*. 2021, 59(1):1–30. doi: 10.1007/s10528-020-10010-1.

Schug J, Overton GC. Modeling transcription factor binding sites with Gibbs Sampling and minimum description length encoding. *Proc Int Conf Intell Syst Mol Biol*. 1997, 5: 268–271.

Shaat N, Katsarou A, Shahida B, Prasad RB, Kristensen K, Planck T. Association between the rs1544410 polymorphism in the vitamin D receptor (VDR) gene and insulin secretion after gestational diabetes mellitus. *PLoS One*. 2020, 15(-5):e0232297. doi: 10.1371/journal.pone.0232297.

Shang M and Sun J. Vitamin D/VDR, probiotics, and gastrointestinal diseases. *Curr Med Chem*. 2017, 24(9):876–887. doi: 10.2174/0929867323666161202150008.

Shiozaki M, Yoshimura K, Shibata M, Koike M, Matsuura N, Uchiyama Y, Gotow T. Morphological and biochemical signs of age-related neurodegenerative changes in klotho mutant mice. *Neuroscience*. 2008, 152(4):924–941. doi: 10.1016/j.neuroscience.2008.01.032.

Song GG, Bae SC, Lee YH. Vitamin D receptor FokI, BsmI, and TaqI polymorphisms and susceptibility to rheumatoid arthritis: A meta-analysis. *Z Rheumatol*. 2016, 75(3):322–329. doi: 10.1007/s00393-015-1581-6. PMID: 26358095.

Swapna N, Vamsi UM, Usha G, Padma T. Risk conferred by FokI polymorphism of vitamin D receptor (VDR) gene for essential hypertension. *Indian J Hum Genet*. 2011, 17(3): 201–206. doi: 10.4103/0971-6866.92104.

Szymczak-Tomczak A, Krela-Kaźmierczak I, Kaczmarek-Ryś M, Hryhorowicz S, Stawczyk-Eder K, Szalata M, Skrzypczak-Zielińska M, Łykowska-Szuber L, Eder P, Michalak M, Dobrowolska A, Słomski R. Vitamin D receptor (VDR) TaqI polymorphism, vitamin D and bone mineral density in patients with inflammatory bowel diseases. *Adv Clin Exp Med*. 2019, 28(7):955–960. doi: 10.17219/acem/97376.

Tizaoui K, Kaabachi W, Hamzaoui A, Hamzaoui K. Association between vitamin D receptor polymorphisms and multiple sclerosis: systematic review and meta-analysis of case-control studies. *Cell Mol Immunol*. 2015, 12(2):243–252. doi: 10.1038/cmi.2014.47.

Trochoutsou AI, Kloukina V, Samitas K, Xanthou G. Vitamin-D in the immune system: genomic and non-genomic actions. *Mini Rev Med Chem*. 2015, 15(11):953–963. doi: 10.2174/1389557515666150519110830.

Uitterlinden AG, Fang Y, Van Meurs JB, Pols HA, Van Leeuwen JP. Genetics and biology of vitamin D receptor polymorphisms. *Gene*. 2004a, 338(2):143–156. doi: 10.1016/j.gene.2004.05.014.

Uitterlinden AG, Fang Y, van Meurs JB, van Leeuwen H, Pols HA. Vitamin D receptor gene polymorphisms in relation to Vitamin D related disease states. *J Steroid Biochem Mol Biol*. 2004b, 89–90(1–5):187–193. doi: 10.1016/j.jsbmb.2004.03.083.

Vahdat S. Vitamin D and kidney diseases: a narrative review. *Int J Prevent Med* 2020, 11: 195. https://doi.org/10.4103/ijpvm.IJPVM_54_19.

Valdivielso JM and Fernandez E. Vitamin D receptor polymorphisms and diseases. *Clin Chim Acta*. 2006, 371(1–2):1–12. doi: 10.1016/j.cca.2006.02.016.

Valdivielso JM, Fernandez E. Vitamin D receptor polymorphisms and diseases. *Clin Chim Acta*. 2006, 371(1–2):1–12. doi: 10.1016/j.cca.2006.02.016.

Vasilovici AF, Grigore LE, Ungureanu L, Fechete O, Candrea E, Trifa AP, Vişan S, Şenilă S, Cosgarea R. Vitamin D receptor polymorphisms and melanoma. *Oncol Lett*. 2019, 17(5):-4162–4169. doi: 10.3892/ol.2018.9733.

Wang S. Epidemiology of vitamin D in health and disease. *Nutr Res Rev*. 2009, 22(2):188–203. doi: 10.1017/S0954422409990151.

Wang Y, Sun Z. Current understanding of klotho. *Ageing Res Rev*. 2009, 8(1):43–51. doi: 10.1016/j.arr.2008.10.002.

Wang L, Ma J, Manson JE, Buring JE, Gaziano JM, Sesso HD. A prospective study of plasma vitamin D metabolites, vitamin D receptor gene polymorphisms, and risk of hypertension in men. *Eur J Nutr*. 2013, 52(7):1771–1779. doi: 10.1007/s00394-012-0480-8.

Whitfield GK, Remus LS, Jurutka PW, Zitzer H, Oza AK, Dang HT, Haussler CA, Galligan MA, Thatcher ML, Encinas Dominguez C, Haussler MR. Functionally relevant polymorphisms in the human nuclear vitamin D receptor gene. *Mol Cell Endocrinol*. 2001, 177(1–2):145–159. doi: 10.1016/s0303-7207(01)00406-3.

Yamamoto H, Miyamoto K, Li B, Taketani Y, Kitano M, Inoue Y, Morita K, Pike JW, Takeda E. The caudal-related homeodomain protein Cdx-2 regulates vitamin D receptor gene expression in the small intestine. *J Bone Miner Res*. 1999, 14(2):240–247. doi: 10.1359/jbmr.1999.14.2.240.

Yamamoto M, Clark JD, Pastor JV, Gurnani P, Nandi A, Kurosu H, Miyoshi M, Ogawa Y, Castrillon DH, Rosenblatt KP, Kuro-o M. Regulation of oxidative stress by the anti-aging hormone klotho. *J Biol Chem*. 2005, 280(45):38029–38034. doi: 10.1074/jbc.M509039200.

Yang L, Wu L, Fan Y, Ma J. Associations among four polymorphisms (BsmI, FokI, TaqI and ApaI) of vitamin D receptor gene and end-stage renal disease: a meta-analysis. *Arch Med Res*. 2015, 46(1):1–7. doi: 10.1016/j.arcmed.2014.11.017. Erratum in: Arch Med Res. 2015 Feb;46(2): 172.

Ye WZ, Reis AF, Velho G. Identification of a novel Tru9 I polymorphism in the human vitamin D receptor gene. *J Hum Genet*. 2000, 45(1):56–57. doi: 10.1007/s100380050011.

Yu F, Cui LL, Li X, Wang CJ, Ba Y, Wang L, Li J, Li C, Dai LP, Li WJ. The genetic polymorphisms in vitamin D receptor and the risk of type 2 diabetes mellitus: an updated meta-analysis. *Asia Pac J Clin Nutr.* 2016, 25(3):614–624. doi: 10.6133/apjcn.092015.12.

Zella LA, Kim S, Shevde NK, Pike JW. Enhancers located within two introns of the vitamin D receptor gene mediate transcriptional autoregulation by 1,25-dihydroxyvitamin D3. *Mol Endocrinol.* 2006, 20(6):1231–1247. doi: 10.1210/me.2006-0015.

Zhai N, Bidares R, Makoui MH, Aslani S, Mohammadi P, Razi B, Imani D, Yazdchi M, Mikaeili H. Vitamin D receptor gene polymorphisms and the risk of the type 1 diabetes: a meta-regression and updated meta-analysis. *BMC Endocr Disord.* 2020, 20(1):121. doi: 10.1186/s12902-020-00575-8.

Zhang D, Wang L, Zhang R, Li S. Association of vitamin D receptor gene polymorphisms and the risk of multiple sclerosis: a meta-analysis. *Arch Med Res*. 2019, 50(6):350–361. doi: 10.1016/j.arcmed.2019.10.007.

Zhang L, Zhang S, He C, Wang X. VDR Gene Polymorphisms and Allergic diseases: evidence from a meta-analysis. *Immunol Invest.* 2020a, 49(1–2):166–177. doi: 10.1080/08820139.2019.1674325.

Zhang WT, Jin TF, Chen L. Associations of four common VDR polymorphisms with rheumatoid arthritis and systemic lupus erythematosus: evidence from a meta-analysis. *Lupus*. 2020b, 29(4):364–370. doi: 10.1177/0961203320909432.

Zhou C, Lu F, Cao K, Xu D, Goltzman D, Miao D. Calcium-independent and 1,25(OH)2D3-dependent regulation of the renin-angiotensin system in 1alpha-hydroxylase knockout mice. *Kidney Int.* 2008, 74(2):170–179. doi: 10.1038/ki.2008.101.

Zhou TB, Jiang ZP, Huang MF. Association of vitamin D receptor BsmI (rs1544410) gene polymorphism with the chronic kidney disease susceptibility. *J Recept Signal Transduct Res.* 2015a, 35(1):54–57. doi: 10.3109/10799893.2014.926927.

Zhou TB, Jiang ZP, Huang MF, Su N. Association of vitamin D receptor Fok1 (rs2228570), TaqI (rs731236) and ApaI (rs7975232) gene polymorphism with the risk of chronic kidney disease. *J Recept Signal Transduct Res*. 2015b, 35 (1):58–62. doi: 10.3109/10799893.2014.926928.

Zhu B, Zhao HL, Ou C, Huang LS, Li PZ, Lao M. Association of vitamin D receptor BsmI gene polymorphism with the risk of type 2 diabetes mellitus. *J Recept Signal Transduct Res.* 2014, 34(6):458–462. doi: 10.3109/10799893.2014.919592.

Zhu YB, Li ZQ, Ding N, Yi HL. The association between vitamin D receptor gene polymorphism and susceptibility to hypertension: a meta-analysis. *Eur Rev Med Pharmacol Sci.* 2019, 23(20):9066–9074. doi: 10.26355/eurrev_201910_19309.

Zittermann A, Ernst JB. Calciotropic and phosphaturic hormones in heart failure. *Nutr Metab Cardiovasc Dis*. 2016, 26(11): 971–979. doi: 10.1016/j.numecd.2016.06.007.

Zittermann A, Trummer C, Theiler-Schwetz V, Lerchbaum E, März W, Pilz S. Vitamin D and Cardiovascular disease: an updated narrative review. *Int J Mol Sci*. 2021, 22(6):2896. doi: 10.3390/ijms22062896.

8

Genetic Polymorphisms in Fatty Acid Metabolism Genes and Cancer Susceptibility

Mosin S. Khan
Government Medical College Srinagar & Associated Hospitals, Karan Nagar, Srinagar, J&K, India
Government Medical College Baramulla & Associated Hospitals, Kanth Bagh, Baramulla, J&K, India

Sunia Faiz, Iqra Farooq, and Suhail S. Lone
Department of Biochemistry, Government Medical College Srinagar & Associated Hospitals, Karan Nagar, Srinagar, Jammu and Kashmir, India

Sabhiya Majid
Department of Biochemistry & Multidisciplinary Research Unit (MRU-ICMR), Government Medical College Srinagar & Associated Hospitals, Karan Nagar, Srinagar, Jammu and Kashmir, India

Waseem Qureshi
Registrar Academics, Government Medical College Srinagar & Associated Hospitals, Karan Nagar, Srinagar, Jammu and Kashmir, India

CONTENTS

DOI: 10.1201/9781003246244-8

ABSTRACT Fatty acids (FAs) form the following lipids: phospholipids, sphingolipids and triglycerides. They contain a carboxylic acid group and a hydrocarbon chain of varying carbon chain length and degree of desaturation. FAs play a crucial role in cancer cell membrane biosynthesis during uncontrolled cell divisions. Pathways in glucose metabolism, pentose phosphate pathway and glutamine metabolism have been reprogrammed in cancer cells to adapt to their rapid cell growth and proliferation. Reprogramming in FA metabolism provides energy, macromolecules for membrane synthesis and lipid signals during cancer development. Functional polymorphisms in FA metabolism genes such as *FASN, CYP2E1* and *LIPC* have been investigated in relation to many types of cancers. In this chapter, we summarize the polymorphisms of FA metabolism genes and their role in supporting tumorigenesis and cancer progression.

KEYWORDS: *Polymorphism; Genes; Cancer Genetics; Mutations; Colorectal Cancer; Breast Cancer; Prostate Cancer.*

8.1 Introduction

FAs are the long chains of lipid carboxylic acids. They are an important constituent of fats and oils and cell membranes where they are the components of phospholipids and glycolipids. They are a good source of energy and have important biological and structural roles. Their metabolism produces high energy. The muscles and heart obtain energy from the beta-oxidation of FAs. The human body can make many of these FAs, except some essential polyunsaturated FAs (PUFAs)—the linoleic acid (LA) and the alpha-linolenic acid (ALA). These two are present in different vegetable oils, but their metabolites are found mainly in the fish oil. The human body cannot make FAs that have odd number of carbon atoms; however, there were studies in which this type of FAs were found in plasma in least concentration (Tiuca et al., 2015). They are classified by length, by saturation vs. unsaturation, by even vs. odd carbon content and by linear vs. branched. Based on length, we have short-chain FAs (with aliphatic tails of five or fewer carbons, e.g., butyric acid), medium-chain FAs (with aliphatic tails of 6–12 carbons, which can form medium-chain triglycerides), long-chain FAs (with aliphatic tails of 13–21 carbons) and very long-chain FAs (with aliphatic tails of 22 or more carbons). Saturated FAs don't have C=C double bonds. Their general formula is $CH_3(-CH_2)_nCOOH$. Stearic acid ($n=16$) is an important saturated FA forming soap when neutralized with lye. Unsaturated FAs have one or more C=C double bonds, and therefore, they have cis or trans isomers. In cis form, two hydrogen atoms adjacent to the double bond are present on the same side of the chain. The firmness of the double bond freezes its confirmation and, in the case of the cis isomer, causes the chain to bend and restricts the conformational freedom of the FA. Cis isomer with more double bonds has the least flexibility. In trans isomer, the hydrogen atoms adjacent to the double bond are present on the opposite sides of the chain. As a result, they do not cause the chain to bend much, and their shape is similar to straight saturated FAs. Trans-FAs are rare in nature and are formed as a result of human processing, e.g., hydrogenation. Most FAs have even chains, e.g., stearic (C18) and oleic (C18) acids. Some FAs have odd numbers of carbon atoms and are called odd-chained FAs (OCFAs). The most common OCFAs are the saturated C15 and C17 derivatives, pentadecanoic acid and heptadecanoic acid, respectively, mostly found in dairy products (Pfeuffer et al., 2016). OCFAs are biosynthesized and metabolized differently from the even-chained ones at molecular level. FAs have very important functions such as biological signaling, energy storage, cellular fuel sources, composition of hormones and lipids and protein changes.

8.2 Fatty Acid Metabolism

Fatty acid (FA) metabolism involves both catabolic and anabolic processes that create biologically important molecules such as triglycerides, phospholipids, second messengers and ketone bodies using ATP molecules (Alves-Bezerra & Cohen, 20175). One important role of FAs is energy production in the form of ATP, and unlike other macronutrient classes, FAs produce the most ATP on an energy per gram basis through beta-oxidation and the citric acid cycle. FAs are the important components of the phospholipids that form the phospholipid bilayers of the plasma membrane. FAs form secondary messengers within the cell and local hormones in the immediate vicinity of the cell through cleavage from the chemical attachments in plasma membrane either wholly or partially.

8.2.1 Fatty Acid Catabolism

Adipose tissue releases fatty acids, and the released FAs are stored in the form of triglycerides as follows:

- **Lipolysis**: It is catalyzed by *lipases*. These lipases are activated by elevated concentrations of epinephrine and glucagon levels or norepinephrine secreted by sympathetic nerves in the adipose tissue (due to low blood glucose levels after meals, which decreases the insulin level in the blood) (Diwan, 2011) (Figure 8.1).
- FA molecules combine with ATP (which is broken down to AMP and inorganic pyrophosphate) to form fatty acyl-adenylate, which reacts with free coenzyme A (CoA) to form fatty acyl-coenzyme A (acyl-CoA) molecule (Figure 8.2).
- *Carnitine shuttle* carries acyl-CoA into the mitochondria (Longo et al., 2016) (Figure 8.3).

8.2.2 Beta-Oxidation

Beta-oxidation occurs in the mitochondrial matrix. It cleaves the long carbon chains of the FAs (in the form of acyl-CoA molecules) into a series of two-carbon (acetate) units, which then combine with CoA to form molecules of acetyl-CoA that condenses with oxaloacetate to form citrate at the initial steps of the citric acid cycle (Zechner et al., 2005). It is therefore convenient to think of this reaction as marking the "starting point" of this cycle, as this is when the fuel—acetyl-CoA—is

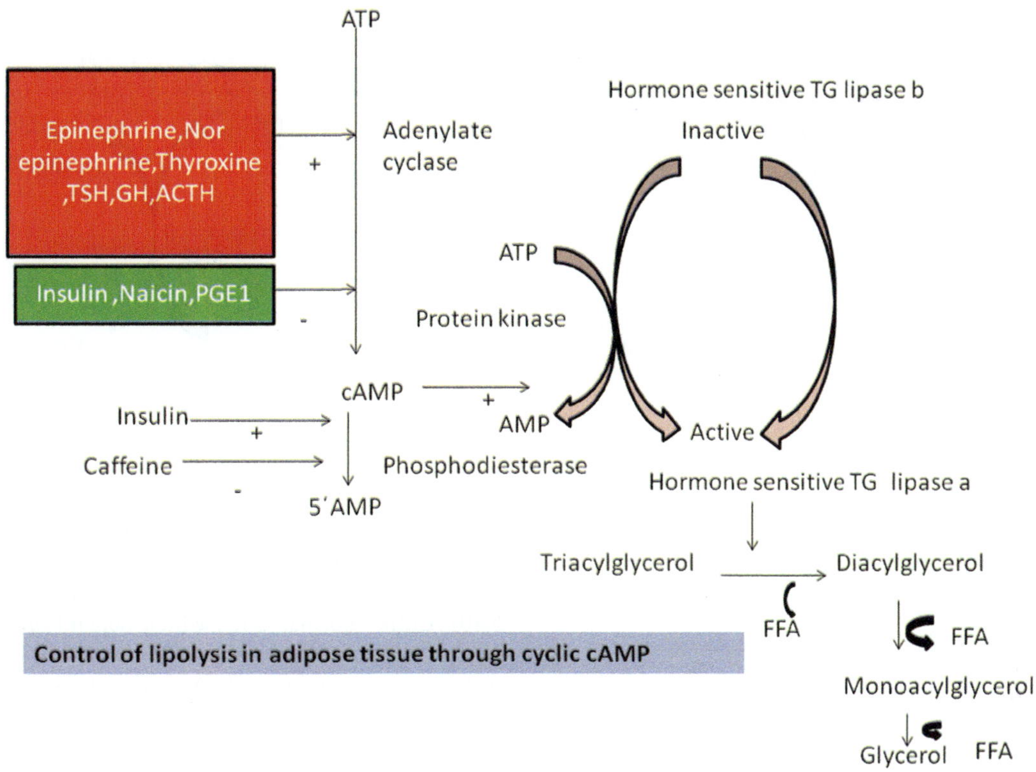

FIGURE 8.1 Control of lipolysis in adipose tissue through cAMP.

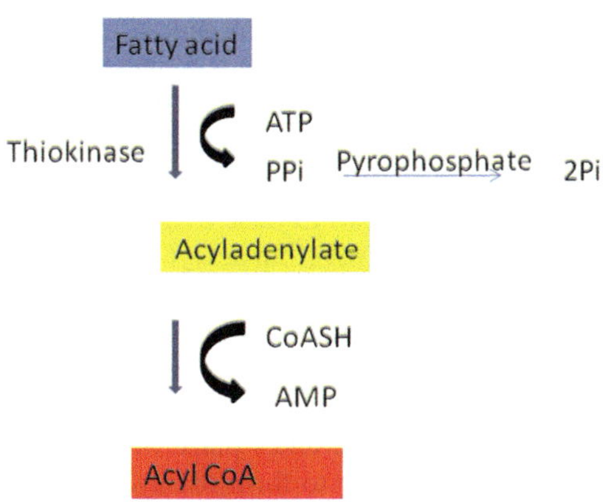

FIGURE 8.2 Activation of fatty acid to acyl-CoA by the enzyme thiokinase.

added to the cycle, which will be dissipated as CO_2 and H_2O with the release of high amount of ATP during the course of each turn of the cycle.

Briefly, the steps in beta-oxidation (the initial breakdown of free FAs into acetyl-CoA) are shown in Figure 8.4.

8.3 Fatty Acid Metabolism Genes

FA metabolism comprises the following genes: *FA synthase gene (FASN)* for the evaluation of de novo FA synthesis; *PPAR-gamma* gene; *FABP1* gene for the transport of FAs and other lipophilic substances; *CYP2E1* gene (cytochrome P450 2E1); *CYP1A1*, also known as AHH (aryl hydrocarbon hydroxylase); and *LIPC* (lipase C, hepatic type) (Poremski et al., 2008).

FA synthase is encoded by *FASN* gene. This enzyme comprises two similar 27 kDa multifunctional polypeptides, in which substrates are hanged from one functional domain to the next (Persson et al., 2009). It carries the synthesis of palmitate from acetyl-CoA and malonyl-CoA, in the presence of NADPH (Gipson et al., 2010).

PPAR-gamma gene encodes the peroxisome proliferator-activated receptor gamma or glitazone receptor (Park et al., 2017). This gene functions in the regulation of FA storage and glucose metabolism.

FABP1 gene codes for FA-binding proteins, a family of transport proteins for FAs and other lipophilic substances. They are highly conserved cytoplasmic proteins. These proteins can bind bile acids. *FABP1* gene also functions in mediating lipid metabolic pathways balancing the levels of intracellular FAs and regulating the expression of FA-responsive genes (Parmer et al., 2012).

CYP2E1 codes for cytochrome P450 2E1. It is a member of the cytochrome P450 oxidase system. This enzyme is involved in xenobiotics metabolism in human body. Liver highly expresses this protein, where it is composed of nearly 50% of the total hepatic cytochrome P450 mRNA (Westphal et al., 2011) and 07% of the hepatic cytochrome P450 protein (Fleming, 2014). This gene plays a crucial role in the metabolism of endogenous FAs such as the omega-1 hydroxylation of FAs such as arachidonic acid (Zhang et al., 2014).

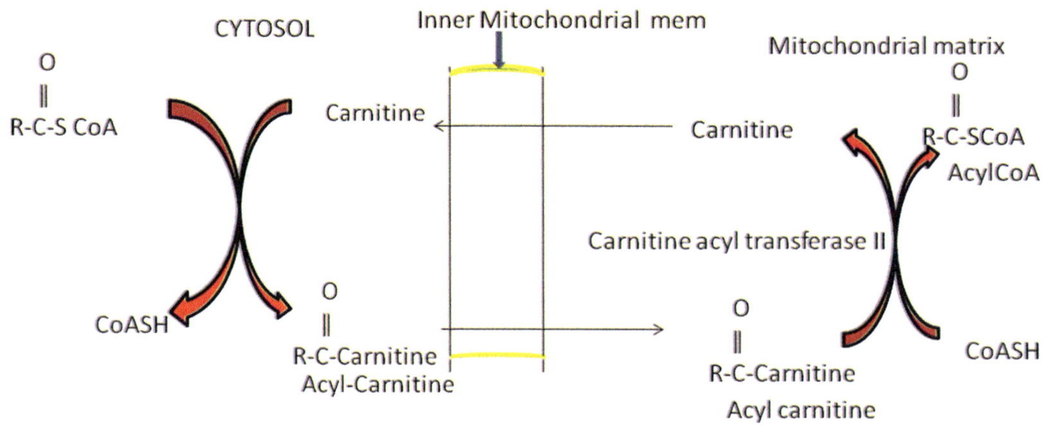

FIGURE 8.3 Carnitine shuttle for transport of activated fatty acid into mitochondria.

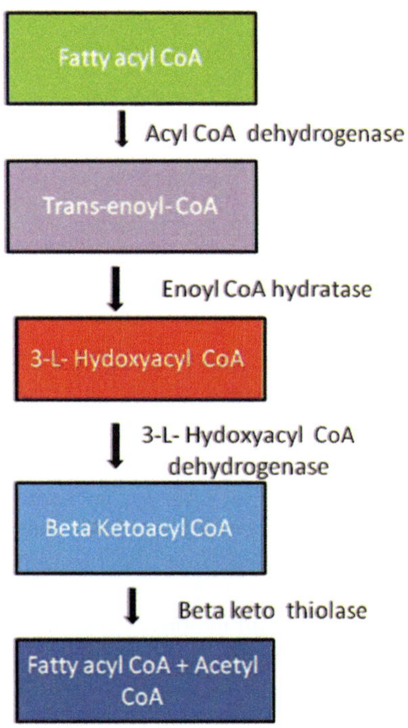

FIGURE 8.4 Steps in beta-oxidation.

The *LIPC* gene stimulates liver to produce hepatic lipase, which is secreted into the bloodstream where it helps with the conversion of fat-transporting molecules called very low-density lipoproteins (VLDLs) and intermediate-density lipoproteins (IDLs) to low-density lipoproteins (LDLs).

8.4 Fatty Acid Metabolism and Cancer

Cancer cells undergo cell proliferation, and their proliferation requires new biological membrane assembly, which is formed from lipids acquired through diet, i.e., exogenously or endogenously synthesized. Figure 8.5 depicts the FA metabolism and potential strategies for targeting lipid metabolism in cancer. Most of the lipids of adult tissues are synthesized through de novo synthesis in well-nourished individuals, as normal cells use exogenous FAs to satisfy their lipid necessity (Glatz et al., 2010). There are several proteins that facilitate the uptake of FAs from circulation, e.g., FABPs (Glatz et al., 2010). Unlike normal cells, which depend on the uptake of FAs, cancer cells reactivate de novo FA synthesis, irrespective of elevated concentration of FAs in blood (Metallo et al., 2011), which indicates their crucial role in tumorigenesis. FA synthesis is an anabolic process that converts nutrient-derived carbon produced from citric acid cycle into FAs. Notably, cancer cells exhibit metabolic plasticity in producing citrate for FA synthesis. During hypoxic conditions or in cancer cells with malfunctioning mitochondria, the TCA cycle is repressed, and citrate formation takes place from reductive carboxylation of glutamine-derived α-ketoglutarate by the NADPH-dependent isocitrate dehydrogenase (Mullen et al., 2011; Wise et al., 2011). Citrate is broken down into oxaloacetate and acetyl-CoA by ATP citrate lyase, which is the prime substrate for FA synthesis. Next, acetyl-CoA carboxylase (ACC), the rate-limiting enzyme, forms malonyl-CoA from acetyl-CoA. Malonyl-CoA is then used for FA synthesis and is used in elongation of FA through FASN until palmitate is formed. Additional changes in FAs are carried by elongases and desaturases to produce FAs at different carbon chain lengths and levels of saturation. The esterification then takes place in free FAs via the glycerol phosphate pathway to form the hydrophobic tails of phospholipids and sphingolipids, which together with the cholesterol form the key constituents of all biological membranes (Yen et al., 2008). FAs provide fuel for cellular bioenergetics through FA β-oxidation (FAO) by generating triacylglycerols (TGs) and cholesterol esters (CEs) for energy storage in the form of lipid droplets (Farese and Walther, 2009). Recent studies have demonstrated that many cancer cells also obtain FAs from their environment and that de novo synthesis and exogenous FAs lead to cancer progression (Pascual et al., 2017; Chen et al., 2018), which has substantial implications for cancer therapeutics.

During the process of metastasis, cells migrate from their primary sites of growth and enter blood in a systematic way using ATP for which FA oxidation is the most efficient process. The functional role of FAO in metastatic progress was seen in tumor metastasis to lymph nodes, where researchers found that lymph node metastasis requires tumor cells to

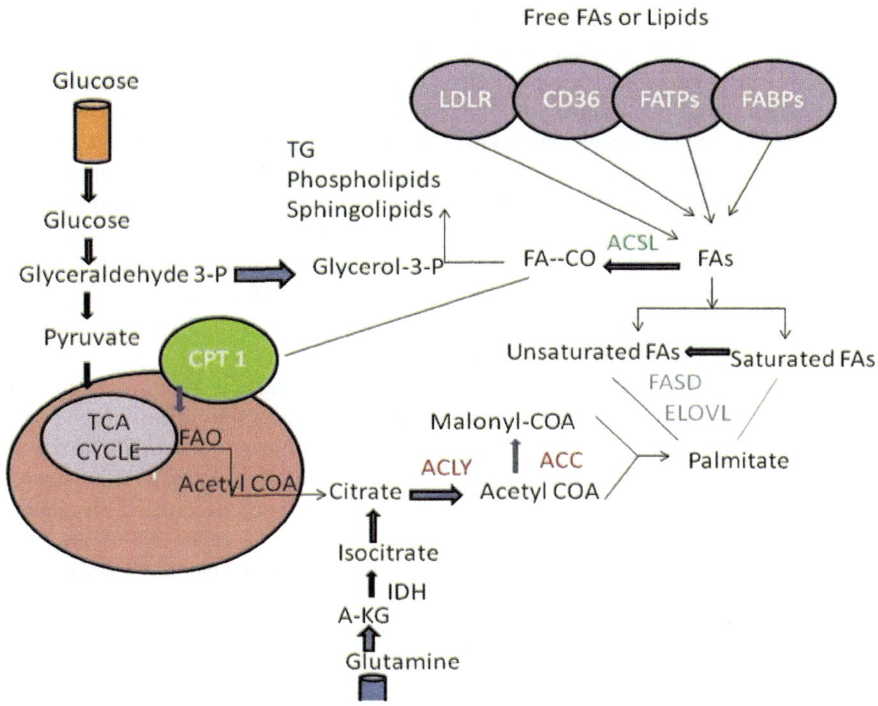

FIGURE 8.5 Fatty acid (FA) metabolism and possible strategies for targeting lipid metabolism in cancer.

undergo a metabolic shift toward FAO, with the transcription coactivator yes-associated protein (YAP) that carries out the upregulation of FAO-related metabolic genes (Lee et al., 2019). These studies lead to the conclusion that inhibition of the FAO metabolic pathway may merit exploration as a potential target for mitigating distant metastasis, which is the chief cause of cancer-related deaths.

Red shows protein targets in de novo FA synthesis; blue highlights protein targets involved in FA changes; white highlights protein targets having a role in FA uptake and degradation; and green highlights protein targets involved in FA activation.

8.5 Genetic Polymorphisms in Fatty Acid Metabolism Genes and Cancer

The important tool for revolution in medical genetics is to find variations within the DNA. These variations called polymorphisms occur in either exonic or intronic regions.

Polymorphisms signify sequence variations, which are present in the general population, and confer no negative effects. However, after the Human Genome Project, it became apparent that some "polymorphisms" were not entirely dangerous. This technique has located and identified genes for various disorders with a clear pattern of Mendelian inheritance, such as cystic fibrosis, the inherited muscular dystrophies and neurodegenerative disorders such as Huntington's disease. Methods that use genetic polymorphisms will also be significant for finding genes that incline people to more common conditions in which inheritance patterns are complex, such as diabetes, atherosclerosis and hypertension.

The set of FA metabolism genes involved in cancer are discussed below.

8.5.1 FASN Gene

FA synthase (FAS) is an enzyme that in humans is encoded by the *FASN* gene (Persson et al., 2009). This protein is a multi-enzyme subunit that catalyzes FA synthesis. It is composed of two 272 kDa multifunctional polypeptides, in which substrates move from one functional domain to the next (Smith et al., 2003). Its chief function is to catalyze the synthesis of palmitate (C16:0, a long-chain saturated FA) from acetyl-CoA and malonyl-CoA, using NADPH (Eggert et al., 2012). Its cytogenetic location is **17q25.3.**

8.5.1.1 Structure and Function

The type I FASN of fungi and animals are huge multifunctional polypeptides that combine all the FA synthesis steps into large macromolecular assemblies. FASN structure in mammals comprises two identical 270 kDa polypeptide chains, each comprising all seven required domains (i.e., beta-ketoacyl synthase, malonyl/acetyltransferase, dehydrase, enoyl reductase, ketoacyl reductase, acyl carrier protein and thioesterase) that assemble into homodimers for enzymatic activity (Maier et al., 2008). Condensation reaction occurs between malonyl group derived from malonyl-CoA and an acetyl group from acetyl-CoA at first. The reduction of the ketoacyl product includes the following three steps: ketoacyl reduction, hydroxyacyl dehydration and enoyl reduction, which lead to the saturated acyl derivative, which then acts as a primer for further elongation and reduction cycles to finally yield palmitoyl derivative. The latter undergoes hydrolysis catalyzed by thioesterase to free palmitate. Recent studies suggested that the FASN I enzymatic complex comprises seven separate enzymatic pockets as head-to-tail dimer with the beta-ketoacyl synthase and malonyl/acetyltransferase domains of one monomer working jointly with dehydrase, enoyl reductase, ketoacyl reductase, acyl carrier protein and thioesterase domains of the adjacent monomer (Hudgins et al., 2008). That is, two monomers of FASN are arranged in an antiparallel way allowing functional connections across the monomer interface. In 2005, Asturias et al. proposed another head-to-head model for FASN organization, which predicts that the beta-ketoacyl synthase and malonyl/acetyltransferase domains of both monomers lie close to the center of the FASN dimer, where they can access the acyl carrier protein of any subunit (Asturias et al., 2005). In 2006, Maier et al. reported the architecture of mammalian FASN from a 4.5 Å electron density map, in which most domains could be assigned, but no details were noticeable (Maier et al., 2006). Maier's FASN map fundamentally approved the Asturias' revised model and suggested that mammalian FASN is a scrambled dimer with a large dimerization interface through the body of the molecule, perpendicular to the interface projected in the classical scheme (Maier et al., 2006). The X-shape structure of mammalian FASN consists of full set of active sites present in each of two semicircular "reaction chambers" on both sides of the molecule. These authors revealed a mammalian FASN crystal structure at 3.2 Å resolution, which provides enough information to understand both the folding and the connectivity of, at least, five catalytic domains (Maier et al., 2008). Although the flexibly tethered acyl carrier protein and thioesterase domains remained unclear, the structure clearly revealed a complex architecture of alternating linkers and enzymatic domains. FASN architecture can now be segregated into two "arms"—a "selection/condensing arm" for adding new building blocks and a "modifying arm" for chemical processing of chain elongation intermediates (Smith and Sherman, 2008).

Moreover, the FASN structure fits perfectly for deletion and insertion of catalytic domains, mainly in the modifying arm. Surprisingly, Maier's FASN structure discovered two additional non-enzymatic domains, a pseudo-keto-reductase and a peripheral pseudo-methyltransferase that is perhaps a molecular relic of an ancestral methyltransferase domain maintained in some related polyketide synthases. As a result of its modular architecture resulting in differences in domain structure, the mammalian FASN should be viewed as a modular assembly line highly adjustable to new functions by achieving synthesis of various products. Its main function is to catalyze the synthesis of palmitate from acetyl-CoA and malonyl-CoA, in the presence of NADPH, into long-chain saturated FAs.

8.5.1.2 Reported SNPs in FASN Gene

The *C1483T* (rs2228305) substitution (missense mutation) of the *FASN* gene located within the peak of linkage to percentage of body fat on chromosome 17q25 appears to have a role in percentage body fat and body mass index, with increased substrate oxidation rate, and with a protective effect in leading to obesity in white boys, but not in girls. The *Val* for *Ile* change in the 1,483 position of the amino acid sequence is positioned within the interdomain region of the FASN protein that is close to two active centers of the FAS dimer. Recent studies evaluated adipose tissue FASN activity according to this *FASN* gene polymorphism and the role of this *FASN* gene variant in adiposity and insulin sensitivity in white adults. Preliminary analyses powerfully suggest that individuals with rs2228305 polymorphism in the *FASN* gene could have a lower risk of metabolic syndromes associated with obesity (Körner et al., 2007). Thus, reduced FASN activity in adipose tissue samples is seen from carriers of the *FASN C1483T* variant (rs2228305), and those individuals with decreased FASN activity linked to this polymorphism had increased insulin sensitivity (Körner et al., 2007).

Moreno-Navarrete et al. showed that the adipose tissue FAS activity was significantly higher in subjects with the *Val* variant in comparison with *Ile* variant-carrying individuals. Also, the *Val* allele in the FASN is associated with impaired glucose tolerance, visceral obesity, etc. (Kovacs et al., 2004; Moreno-Navarrete et al., 2009). An appealing finding in the present study was that the FASN was linked with depressive symptoms to the same degree as the 5-HTTLPR under the influence of perceived stress, suggesting that *C1483T (rs2228305)* polymorphism in the *FASN* gene can influence the vulnerability to depression and that the *Ile* variant may add to the vulnerability of psychological stress.

Several SNPs were discovered in bovine FASN and also their connection with fat related traits in beef and dairy cattle. Roy et al. identified two polymorphisms (G841C and A16024G) in Holstein Friesian and indicated that the allele frequencies of these two SNPs varied significantly between high and low milk-fat content groups (Roy et al., 2005). Abe et al. for the first time sequenced the full length of protein-coding regions (CDS) in FASN gene and detected 12 variants, including six non-synonymous SNPs in Japanese Black and Limousin families, and the AT haplotype of A16024G and C16039T was significantly related to the FA composition (Abe et al., 2009). Of the two alleles of the G841C, the G allele showed a major difference, being 40% higher than the C allele (Ordovás et al., 2008). This result suggested that the allele difference affects the expression of FASN gene and then would influence FA composition in the current study. In light of this gene function, the G841C would be the most promising candidate as one of the responsible mutations for bovine FA composition.

8.5.1.3 FASN Gene Polymorphisms and Cancer

Reduced FASN activity is found in normal cells. Although there is no regulation of FASN activity by allosteric effectors or covalent modification, the expression of FASN changes with the nutrition in lipogenic tissues. FASN-catalyzed endogenous FA biosynthesis in liver and adipose tissue is stimulated by a high-carbohydrate-containing diet, whereas FA suppresses it. FASN expression is also seen in hormone-sensitive cells. During the menstrual cycle, the expression of FASN in the human endometrium is closely associated with the expression of the proliferation antigen Ki-67, estrogen receptor and progesterone receptor, which means that there is a functional connection between FASN and the estradiol (E2)/estrogen receptor-dependent signaling in the normal control of endometrial cell propagation (Lupu and Menendez, 2006). In normal mammary glands, the stimulation of FASN expression and its activity in breast-feeding mothers is considered to be due to stimulatory effects of the following hormones: cortisol, prolactin and insulin, facilitated by reduced production of progestins. Four main mechanisms have been demonstrated for cancer-associated FASN expression: enhanced transcription of the FASN gene, gain of FASN gene copies, enhanced translation of FASN mRNA and increased stability of FASN protein. Molecular mechanisms regulating FASN expression in cancer cells are depicted in Figure 8.6. These modes of regulation of FASN, all of which are not mutually exclusive and may occur in parallel in tumor cells, protect it from responding to physiological regulators, thus resulting in constitutive tumor-associated FASN overexpression and hyperactivity. Figure 8.7 deduces the molecular control of endogenous FASN levels in normal and cancer cells.

In tumor cells, FASN carries out the synthesis of FAs to meet the excessive bioenergetic and structural demands of

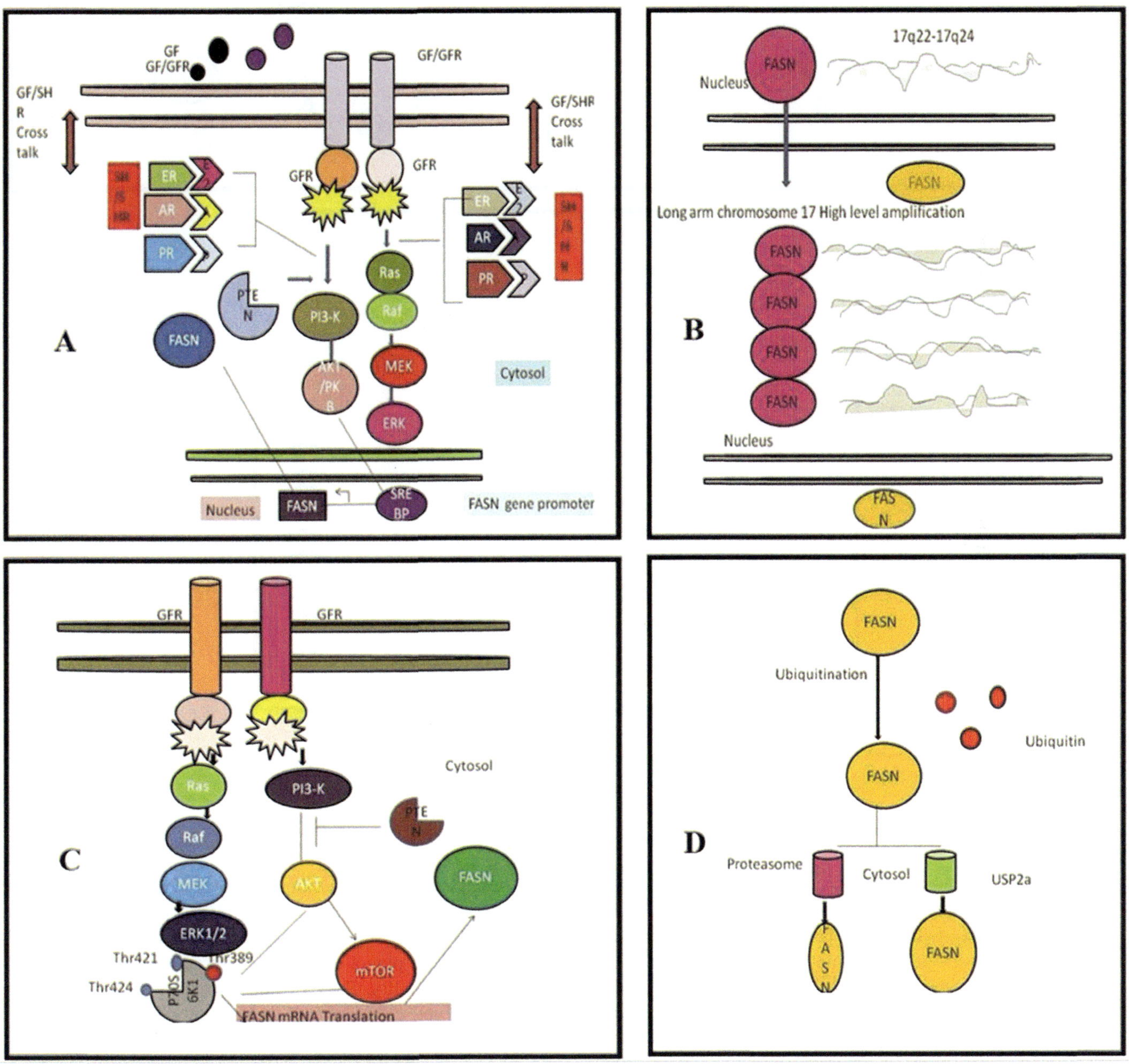

FIGURE 8.6 Molecular mechanisms regulating FASN expression in cancer cells.(a) Enhanced transcription of FASN gene. (b) The gain of FASN gene copies—17q22–17q24, the region to which the human FASN gene has been mapped—is prone to high-level amplification in tumor cells, and accordingly, the gain of FASN gene copy number has recently been shown in prostate cancer tissues. (c) Enhanced translation of FASN mRNA. (d) Increased stabilization of FASN protein.

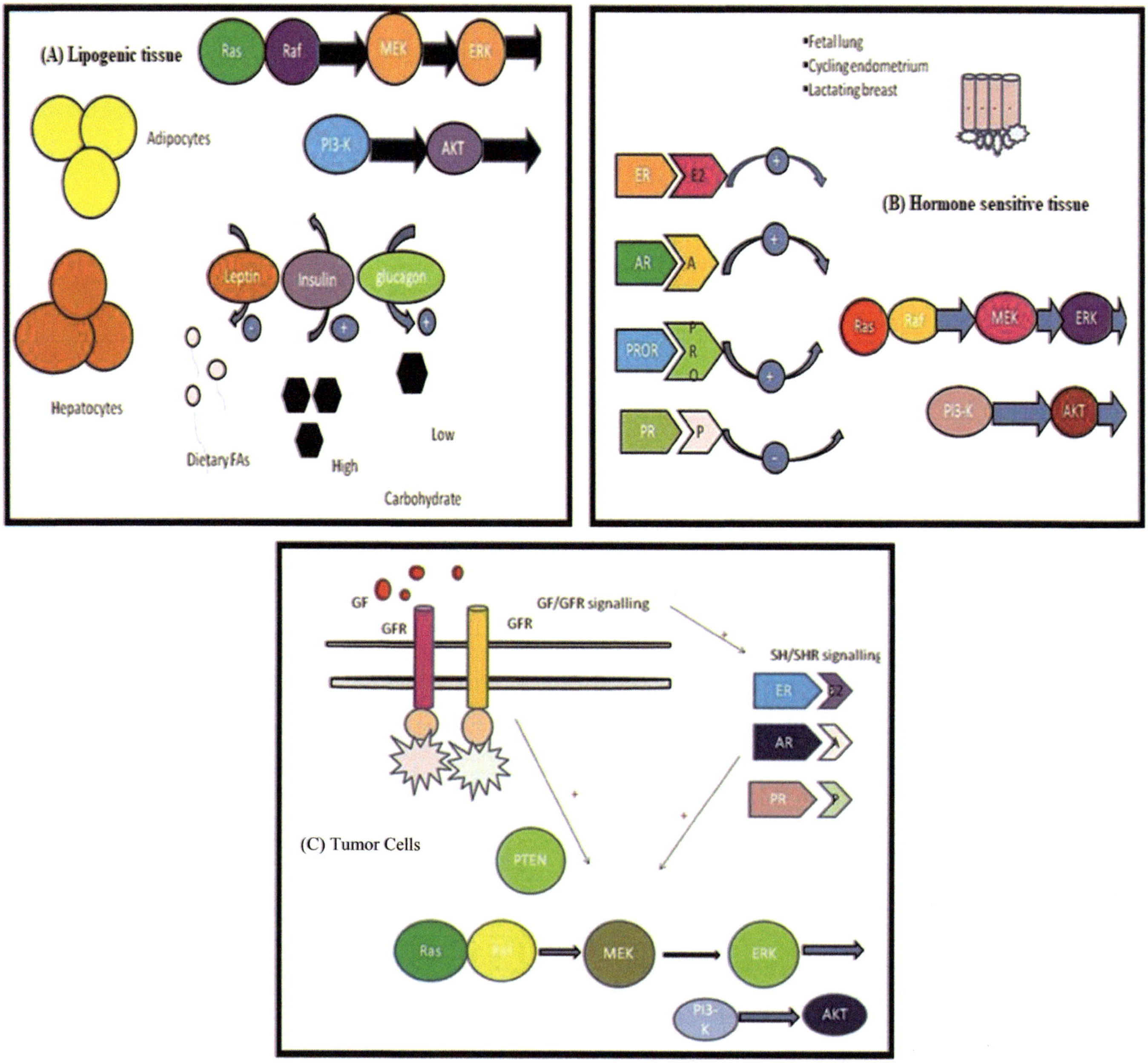

FIGURE 8.7 Molecular control of endogenous FASN levels in normal and cancer cells—similarities and differences. (a) Lipogenic tissue, (b) hormone-sensitive tissue and (c) tumor cells.

proliferation (Menendez and Lupu, 2007). In prostate tumor, FASN is highly expressed (Campa et al., 2009). Treatment with FASN inhibitors decreases prostate tumor cell proliferation and induces apoptosis (Ogino et al., 2008). Thus, FASN may act as oncogenic gene in epithelial tumors, including prostate cancer (PC) (Ogino et al., 2008). Obese patients are at high risk of PC, poor outcomes after prostatectomy (Freedland et al., 2004) or radiation (Freedland et al., 2004) and increased PC-specific deaths, but mechanisms causing these effects are not clear (Ma et al., 2008). Two studies link *FASN* polymorphisms with obesity in humans, and FASN inhibition quickly reduces fat stores in mice, clearly indicating that FASN functions in energy homeostasis (Ma et al., 2008). Links between *FASN* polymorphisms and obesity, and FASN expression and aggressive PC raise the chance that *FASN* could play a role in the interaction between obesity and PC lethality. Survival analysis among the patients with PC showed contrary associations between SNPs *c* and PC-specific deaths. Compared with the CC wild type of *rs4246444,* the hazard ratio (HR) for PC mortality for the homozygous AA variant was 0.40 (95% CI, 0.16–0.98), adjusting for age at diagnosis and baseline BMI, compared to those with GG wild type. Similarly, this SNP interacted with BMI for both the risk of advanced PC and PC–related death.

Gene	SNP	Function	Role in Disease
FASN	*rs2228305*	Appears to associate with lower percentage body fat and body mass index, with increased substrate oxidation rate, and with a shielding effect in leading to obesity in white boys, but not in girls. PC-specific deaths.	Impaired glucose tolerance and insulin sensitivity.
	rs4246444	Lowers the blood levels of myristoleic, palmitoleic and oleic acids. Also associated with a lower risk of death due to prostate cancer.	Prostate cancer.

8.5.2 PPARG Gene

Peroxisome proliferator-activated receptor gamma (PPAR-gamma or PPARG) belongs to type II nuclear receptor gene. It is positioned on chromosome 3 in humans and chromosome 6 in mice.

8.5.2.1 Structure and Function of PPARG

Similar to nuclear receptors, PPARG gene comprises the following domains: ligand-independent activation domain (AF-1 region and A/B domain), a DNA-binding domain (DBD) (C-domain), a hinge region (D-domain) and a ligand-dependent ligand-binding domain (LBD) (E/F-domain and AF-2 region) (Lewis et al., 2010).

PPARG1 and *PPARG2*, the two isoforms of this gene, vary by only 30 amino acids at the N-terminal end. These extra amino acids in PPARG2 isoforms lead to better potency and adipose selectivity, which makes this protein a participant of adipose differentiation (Tontonoz and Spiegelman, 2008). Its structure consists of 13 α-helices and 4 short β-strands (Lewis et al., 2010). It has the largest T-shaped binding pocket among all nuclear receptors (Pochetti et al., 2007), allowing for interaction with various ligands (Pochetti et al., 2007). LBD of *PPAR-gamma* has a helical sandwich shape so as to provide binding site for ligands and is composed of 250 amino acids (Lewis et al., 2010). Activation by full agonists results from hydrogen bond interaction between the S289, H323, Y473 and H449 residues of the PPARG-LBD (Pochetti et al., 2007). This activation leads to conformational changes in the LBD AF-2 region, important for recruiting coactivators.

PPARG maintains glucose and lipid homeostasis (Tontonoz and Spiegelman, 2008). This gene mediates adipocyte differentiation and alters insulin sensitivity, inflammatory processes and cell proliferation (Aprahamian et al., 2009). Ligand-dependent mechanisms involve inhibiting inflammatory cytokine production and macrophage activation (Aprahamian et al., 2009). *PPARG* controls the expression of many genes, especially FA metabolism genes.

8.5.2.2 Reported SNPs in PPARG

There are a number of polymorphisms found in *PPARG*, and they are associated with a variety of diseases. Several studies have shown that the SNPs in *PPARG* controls lipid and glucose metabolism (Sartor, 2006; Yang et al., 2009). Among them, the missense variant *C34G* (*rs1801282*), i.e., *Pro12Ala* in the exon b, leads to *proline (Pro)* to *alanine (Ala)* change and has been broadly reviewed in epidemiologic studies. Many case-control studies have revealed that this missense mutation can prevent from T2DM risk in East Asia (Japanese) (Aprahamian et al., 2009). Several meta-analysis studies have shown the role of this SNP in the risk of T2DM (Tong et al., 2012). Ludovico et al. showed a heterogeneous effect of the *Ala* allele in the development of T2DM in Europeans, Asians and North Americans (Ludovico et al., 2007). There are other polymorphisms found in PPARG, which are linked with T2DM. In a Singaporean population study, Tai et al. observed that the people carrying the CC homozygote polymorphism of *C34G* has a lower risk of diabetes (Tai et al., 2004). Considerable association between *PPARG c.C1347T (rs3586806)* polymorphism and amplified risk of cancer was also found among Asians and mixed populations, but not among Caucasians in a subgroup analysis by ethnicity. In a subgroup analysis by cancer type, *c.C1347T (rs3586806)* polymorphism was having a role in esophageal cancer and glioblastoma, but not in biliary tract, breast, colorectal, melanoma, ovarian and other cancers. *PPARG* mutation may boost CRC risk (Wan et al., 2011). The possible role of *PPARG rs1801282* polymorphism in CRC risk has widely been studied; however, the findings of those investigations were contradictory, especially in Asians.

8.5.2.3 PPARG Gene Polymorphisms in Cancer

Many studies have shown the role of PPARG in CRC risk. Understanding the polymorphisms in this gene associated with CRC susceptibility may be useful for better therapy for CRC. Recently, some case-control studies have stressed the association between *PPARG* polymorphisms and CRC. A common single nucleotide polymorphism in *PPARG* gene *C34G (Pro12Ala rs1801282)* has been established, which was linked with receptor activity, insulin sensitivity, body mass index (BMI) and risk of T2DM (Crous-Bou et al., 2012; Guerrero-Romero et al., 2010). Many studies have been done on the role of *PPARG C34G (rs1801282)* polymorphism in the risk of CRC. In a meta-analysis study done in Caucasians, *PPARG C34G* polymorphism was found to be associated with the reduced risk of CRC in Caucasians (Abuli et al., 2011). However, there were only three case-control studies with comparatively small sample sizes, which stressed the relationship between *PPARG C34G (rs1801282)* polymorphism and CRC in Asians (Wu et al., 2011). The results of the case-control study might be limited by sample size. To overcome this drawback, a comprehensive pooled analysis was subsequently carried out to determine the association of PPARG rs1801282 polymorphism with CRC risk. Several individual studies have deduced the positive role of *PPARG C1347T* polymorphism in cancer risk (Doecke et al., 2008); however, others observed no involvement. A meta-analysis reported that this polymorphism was allied with cancer risk among Asians and mixed populations (Xu et al., 2010). In *PPARG* exon 6, a C-to-T substitution is a synonymous polymorphism that codes for histidine with *PPARG C1347T* allele. The findings of earlier epidemiological studies showed a relationship of this polymorphism with metabolic disorders such as type 2 diabetes and atherosclerosis (Wang et al., 2015c). It is proposed that the C-to-T substitution may change the expression of this gene at transcription and translation levels. A tendency of increased risk was observed with *PPARG C1347T* polymorphism for NSCLC, and an increased risk was also found in the subsequent meta-analysis study (Xu et al., 2010). All these observations demonstrated that *PPARG C1347T* polymorphism might influence the development of cancer.

Gene	SNP	Function	Role and Disease
PPARG	*rs1801282*	This missense variant C12G (*rs1801282*), i.e., *Pro12Ala* in the exon b, results in the amino acid change from *proline (Pro)* to *alanine (Ala)*, which has broadly been reviewed in epidemiologic studies. Many case-control studies have revealed that this missense mutation can reduce T2DM risk in East Asia (Japanese).	T2DM
	rs3586806	A major association between rs35856806C/T polymorphism and obese and overweight patients with T2DM has been found.	Associated with the risk of esophageal cancer and glioblastoma, but not with biliary tract, breast, colorectal, melanoma, ovarian and other cancers.
	rs1801282 C>G	The functional SNPs were predicted to be correlated with the susceptibility of colorectal cancer.	Associated with colorectal cancer.

8.5.3 FABPI Gene

FABP1 gene codes for FABP1 (FA-binding protein 1). It is commonly known as liver-type FA-binding protein (LFABP). Its expression is high in the liver where it transports and metabolizes long-chain FAs (LCFAs), endocannabinoids, phytocannabinoids, synthetic cannabinoid receptor (CBR) agonists and antagonists and other hydrophobic molecules (Huang et al., 2016). Expressional changes in the protein have been linked to metabolic conditions, including obesity (Shi et al., 2012). The human *FABP1* gene is located at chr 2p12-q11 from base pair 88,122,982 to base pair 88,128,131 and is expressed in the intestine, liver, pancreas, stomach and kidney (Shi et al., 2012; Smathers and Petersen, 2011).

8.5.3.1 Structure and Function

FABP1 gene has a distinctive structure unlike other members of the FABP family, allowing it to bind multiple ligands (Wang et al., 2015a). It also has a wider solvent-accessible core compared to other FABPs, allowing more different substrate binding (Smathers and Petersen, 2011). The "portal hypothesis" has been put forward to elucidate the binding process of FABPs (Smathers and Petersen, 2011). It has been suggested that FAs enter the solvent-accessible area of the protein through a dynamic region comprising of α-helix II and turns between βC-βD and βE-βF loops (Atshaves et al., 2010). The FA is bound in the protein cavity for transport (Atshaves et al., 2010). The FABPs are a family of small, highly conserved proteins present in the cytoplasm and implicated in the binding of LCFAs. A high expression of this protein is found in the liver, where it accounts for 7%–11% of the total cytosolic protein, and it can also be found in the intestine, kidney, pancreas stomach and lungs (Thuijls et al., 2011). FABP1 is quite different from other hydrophobic ligands. It can bind bilirubin, monoglycerides, bile acids and fatty acyl-CoA (Thuijls et al., 2011). It has been proposed that FABP1 plays a significant role in preventing cytotoxicity by binding heme, FAs and other molecules that are potentially lethal when unbound (Obata et al., 2016).

8.5.3.2 Reported SNPs in FABP1 Gene

Several polymorphisms of FABP1 gene have been investigated. A threonine (T) to alanine (A) substitution of FABP1 at position 94 *(T94A rs2241883)* is associated with deranged lipid profile (increased serum cholesterol and triglycerides, which further increases the risk of cardiovascular disease, type 2 diabetes and the metabolic syndrome) (Fisher et al., 2007). Peng et al. reported that two genetic variants in *FABP1*, *rs2241883* and *rs1545224*, are associated with the vulnerability to nonalcoholic fatty liver disease (NAFLD) in a Chinese population (Peng et al., 2012). These observations clearly suggest that polymorphisms are possibly associated with diseases that are due to the change in lipid metabolism. The pathogenesis of AERD is related to inflammatory or anti-inflammatory lipid mediators, such as cysteinyl leukotrienes and lipoxins, whose precursor is arachidonic acid (Laidlaw and Boyce, 2013). Arachidonic acid is oxidized by 5-LO to form the unstable intermediate LTA4, which conjugates to reduced glutathione by the terminal enzyme LTC4 synthase (LTC4S) to form LTC4 (Cheong et al., 2011). FABP1 increases the half-life of LTA4 competitively to arachidonic acid and can bind to other arachidonic acid metabolites, hydroperoxyeicosatetraenoic acid (HPETE) and hydroxyeicosatetraenoic acid (HETE), with high affinity (Laidlaw and Boyce, 2013). FABP1 reduces 15-LO-induced oxygenation of linoleic acid and arachidonic acid (Cheong et al., 2011). In cooperation with 5-LO, 15-LO of epithelial cells is the key enzyme for the production of lipoxins, unstable anti-inflammatory derivatives of arachidonic acid (Lee and Stevenson, 2011). Various transcription factors regulate FABPs: PPARα/PPARγ (Lee and Stevenson, 2011), which are upregulated in various cells via the overproduction of eicosanoid metabolites, including 15-hydroxyeicosatetraenoic acid (15-HETE) and 15-deoxy-12,14-prostaglandin J2 (15d-PGJ2). Interestingly, a polymorphism in the gene encoding PPARγ is linked with aspirin hypersensitivity (Stephens et al., 2001).

FABP1 regulates various cellular processes, including inflammation, immunity, metabolism and energy homeostasis (Mukai et al., 2017). A number of recent studies have depicted its role in liver diseases. Mukai et al. found a reduction in liver weight and hepatic triglycerides in mice with *FABP1* knockdown and fed with a high-fat diet. They also observed that the levels of hepatic inflammation cytokines and chemokines (interleukin-6, tumor necrosis factor alpha and monocyte chemotactic protein 1) and an oxidative stress maker (heme oxygenase-1) were considerably least in this mouse model (Mukai et al., 2017). *FABP1 T94A* mutation induced increased hepatic lipid and changes in hepatic endocannabinoids system in males, which suggested the important role of *FABP1* in NAFLD (Martin et al., 2018).

8.5.3.3 FABP1 Gene Polymorphisms in Cancer

FABP1 is associated with many diseases, and variants in this gene can lead to various changes. *FABP1* is seen downregulated in HCC tissues and could therefore serve as a promising prognostic marker in HCC patients (Wang et al., 2014). *FABP1*-targeting microRNAs could appreciably reduce *FABP1* expression at translational level and ameliorate hepatocyte steatosis and injury (Jiang et al., 2013, 2015). The substitution of adenine to guanine at 2,919,872 position considerably reduced triglyceride concentration in serum and *FABP1* promoter activity (Peng et al., 2015). *T94A* variant in *FABP1* was associated with high plasma triglycerides, increased cholesterol accumulation, atherothrombotic cerebral infarction and NAFLD by affecting FABP1 structure, stability and conformational and functional response to fibrates. (Martin et al., 2013). *rs1545224* and *rs2241883* polymorphisms represent an A/G and a C/T single nucleotide polymorphism on human chromosome 2, respectively. The transversion results in an intronic variation of *rs1545224* polymorphism and missense mutation of *rs2241883* polymorphism. *rs1545224* A allele and *rs2241883* C allele were risk factors for NAFLD as shown by many studies and showed a cumulative outcome (Rashid et al., 2017). Increased levels of low-density lipoprotein and fasting plasma glucose were observed in individuals with *rs1545224* A allele and *rs2241883* C allele, respectively (Peng et al., 2012). A recent study conducted showed that *rs1545224* and *rs2241883* polymorphisms were not associated with hepatocellular carcinoma (HCC) risk in healthy individuals or chronic hepatitis B carriers. However, for patients with HCC, *rs1545224* A allele, AG and AA genotypes increased the risk of developing HCC to 1.33, 1.61 and 1.71 times, respectively (Tao et al., 2018). The variation in introns in *rs1545224* may change *FABP1* expression, and an increased expression of *FABP1* was seen in liver cirrhosis patients' plasma and HCC tissues than in those of healthy controls (Graupera et al., 2017). Although intron polymorphisms may have little effect on gene function, they may have an impact on transcription factor binding and even alter related protein expression (Tao et al., 2018).

8.5.4 CYP2E1 Gene

This gene codes for cytochrome P450 2E1 enzyme, a member of the cytochrome P450 mixed-function oxidase system. It carries out the metabolism of xenobiotics in the human body. The hepatic CYP2E1 enzyme plays a significant role in the metabolic activation of nitrosamines and other carcinogenic compounds (Hagiwara et al., 2005). This gene is present on chromosome 10 in humans.

Gene	SNP	Functions	Role and Disease
FABP	rs2241883	FABP T94A variant, the most common in the FABP family, has been associated with elevated liver triglyceride levels.	Cardiovascular diseases and nonalcoholic fatty liver disease in Chinese population.
	rs1545224	FABP rs15452224 polymorphisms might increase HCC risk in LC patients, indicating that FABP rs15452224 polymorphisms may be related to the process of developing HCC in LC patients.	Nonalcoholic fatty liver disease in Chinese population.
	T94A	It is found to play a more important role in the pathogenesis of PCOS. It provides evidence of an association between FABP1 genetic variants and PCOS women.	Related to high plasma triglycerides, increased cholesterol accumulation, atherothrombotic cerebral infarction and NAFLD.

8.5.4.1 Structure and Function

CYP2E1 is composed of 12 major α-helices and 4 β-sheets with short superseding helices interspersed in between (Porubsky et al., 2008). The active site of CYP2E1 contains an iron atom bound to the heme center, which mediates the electron transfer steps obligatory to carry out oxidation of its substrates. The active site of CYP2E1 is the smallest observed in human P450 enzymes, with its small capacity due to isoleucine at position 115. The side chain of this residue protrudes above the heme center, restricting active site volume compared to similar enzymes that have less massive residues at this position (Porubsky et al., 2008). T^{303} is important for substrate positioning. It also protrudes into the active site and is hence highly conserved by many cytochrome P450 enzymes (Porubsky et al., 2008). It donates hydrogen bond to acceptors from its hydroxyl group, and its methyl group has also been occupied in the positioning of FAs within the active site (Fischer et al., 2014). A number of residues adjacent to the active site, including L^{368}, create a hydrophobic access channel vital for determining the enzyme's specificity expressed in the liver, where it comprises nearly 50% of the total hepatic cytochrome P450 mRNA (Kitam et al., 2012) and 7% of the hepatic cytochrome P450 protein (Wagner et al., 2014). The liver is therefore the organ where most drugs undergo detoxification by CYP2E1, either directly or by facilitated excretion from the body. CYP2E1 metabolizes the small polar molecules, including toxic laboratory chemicals such as dimethylformamide, aniline and halogenated hydrocarbons. Several metabolic reactions are carried out by CYP2E1, including the conversion of ethanol to acetaldehyde and to acetate in humans (Wang et al., 2014) where it works alongside alcohol dehydrogenase and aldehyde dehydrogenase. In the conversion sequence of acetyl-CoA to glucose, CYP2E1 undergoes conversion of acetone via hydroxyacetone (acetol) into propylene glycol and methylglyoxal, the precursors of pyruvate, acetate and lactate (He et al., 2015).

(ω-1)-Hydroxylation of FAs such as arachidonic acid is also carried out by CYP2E1 gene, involving it in significant signaling pathways that may link it to diabetes and obesity (Zhang et al., 2014). Thus, it acts as a monooxygenase enzyme to metabolize arachidonic acid to 19-hydroxyeicosatetraenoic acid (19-HETE). However, it also acts as an epoxygenase to metabolize docosahexaenoic acid to epoxides, primarily 19*R*,20*S*-epoxyeicosapentaenoic acid and 19*S*,20*R*-epoxyeicosapentaenoic acid isomers (termed 19,20-EDP), and eicosapentaenoic acid to epoxides, primarily 17*R*,18*S*-eicosatetraenic acid and 17*S*,18*R*-eicosatetraenic acid isomers (termed 17,18-EEQ) (Zhang et al., 2014). 19-HETE inhibits 20-HETE, a largely active signaling molecule; for example, it constricts arterioles, increases blood pressure, promotes inflammation responses and stimulates the growth of a variety of tumor cells; however, the *in vivo* capability and importance of 19-HETE in inhibiting 20-HETE has not been confirmed. The EDP (epoxydocosapentaenoic acid) and EEQ (epoxyeicosatetraenoic acid) metabolites have numerous functions. In various animal models and *in vitro* studies on animal and human tissues, they decrease blood pressure and pain perception; suppress inflammation; inhibit angiogenesis, endothelial cell migration and endothelial cell proliferation; and reduce the growth and metastasis of human breast and prostate cancer cell lines (Fleming, 2014). It is suggested that the EDP and EEQ metabolites function in humans and animal models in the same way and that, as the products of omega-3 FAs, docosahexaenoic acid and eicosapentaenoic acid, the EDP and EEQ metabolites have many of the helpful effects attributed to dietary omega-3 FAs (Fleming, 2014). The EDP and EEQ metabolites have a short life span, being inactivated immediately after formation by epoxide hydrolases, particularly soluble epoxide hydrolase, and therefore act nearby. CYP2E1 could not contribute majorly to the formation of the cited epoxides (Westphal et al., 2011), but could act locally in certain tissues to do so.

8.5.4.2 Reported SNPs in CYP2E1

The oxidative activity of CYP2E1 is higher compared to other P450 cytochromes and oxidizes ethanol-producing highly toxic products, e.g., acetaldehyde and 1-hydroxyethyl. Therefore, CYP2E1 produces a high amount of reactive oxygen species (Zhang et al., 2011). Functional studies (Polonikov et al., 2011) showed that *C2* allele *(−1293C or −1053T)* of *CYP2E1* gene is responsible for more prominent activation of cytochrome in response to excessive alcohol consumption than *C1* allele *(−1293C or −1053T)*. Therefore, the link between functionally active allelic variant *−1293C* of *CYP2E1* gene and amplified risk of essential hypertension in men with alcohol abuse first detected in our experiments can be explained at the molecular basis. It can be hypothesized that the generation of highly reactive products of free radical oxidation resulting from potent cytochrome induction caused by excessive alcohol consumption in the carriers of *−1293C* allele of *CYP2E1* gene causes oxidative stress and endothelial dysfunction, serving as the pathogenetic foundation for essential hypertension.

rs3813867, rs2031920, rs2031921, rs3813870 and *rs2031922 (GTCAC)* SNPs in *CYP2E1* showed association with schizophrenia (Du et al., 2010). While haplotype analysis showed an important association with schizophrenia, none of the single SNPs in this haplotype showed positive association. This may be due to the fact that the haplotype analysis has sophisticated power than individual genetic markers in association analysis, given that the haplotype analysis underlines the correlation between the individual markers. The *RsaI* variant corresponding to a *C-1054T* mutation *(rs 2031920)* and the 96-bp insertion in the 5′-flanking region have drawn much attention with regard to having potential functionality (Hagiwara et al., 2005; Du et al., 2010).

Several restriction fragment length polymorphisms of the *CYP2E1* have been identified (Jemal et al., 2011; Feng et al., 2012), and a variant *C2* allele recognized by *RsaI* digestion in the 5′-flanking region of the gene is linked with decreased enzyme activity or non-inducibility (Martiny and Miteva, 2013; Feng et al., 2012). Some *in vitro* studies showed that the *C2* allele reduced the expression of a reporter gene construct (Feng et al., 2012; Ferlay et al., 2010).

8.5.4.3 CYP2E1 Gene Polymorphisms in Cancer

There are more than 100 variants of CYP2E1 gene. However, only common mutations were broadly studied as they might affect the activity of *CYP2E1* (Chong et al., 2016). *rs3813867 G > C* and *rs2031920 C > T* were two main SNPs among them, associated with *PstI* and *RsaI* restriction enzyme sites, respectively. Linkage disequilibrium of these SNPs also lead to the *CYP2E1*5* haplotype and result in three distinct genotypes—(1) *c1/c1 (RsaI+ PstI−)*, homozygous of normal alleles; (2) *c1/c2*, heterozygous; and (3) *c2/c2 (RsaI− PstI+)*, homozygous alleles after nucleotide exchange (Chong et al., 2016). Another variant, recognized by *DraI* restriction enzyme in intron 6 *(rs6413432)*, forms *CYP2E1*6* polymorphism and also results in three genotypes—*C/C, C/D and D/D* (Qin et al., 2013).

Various studies have shown the relationship between CYP2E1 gene polymorphism and bladder cancer offering conflicting results. A study conducted on patients having primary unilateral BC demonstrated lack of any association between *CYP2E1*5* polymorphisms and BC (Khedhaier et al., 2008). However, Wu et al. in their study on non-smoker and nonalcoholic women reported that individuals with the *c2/c2* genotype of *CYP2E1*5* had a lower BC risk than those with *c1/c1* (Wu et al., 2006). The most recent study by Chong *et al.* showed that the C1/C2 genotype or C2 allele carriers with *CYP2E1*5* polymorphism have almost 1.8-fold higher risk of BC (Chong et al., 2016). Only few studies have shown the association between the *RsaI* polymorphism of *CYP2E1* and colorectal cancer (CRC) risk, and their findings are rather conflicting. Individuals with the *RsaI C2* allele inclined to have lower risks of colon and rectal cancers in a population-based study conducted in Hawaii (Le Marchand et al., 2002), whereas a statistically considerable increase in CRC risk was reported for individuals with the *RsaI C2* allele in Hungary (Kiss et al., 2000) and for those homozygous for the *C2* allele in China (Gao et al., 2007). No connection was found between the *RsaI* polymorphism and the risk of CRC in Australia (Butler et al.,

2001) and the Netherlands (VanderLogt et al., 2006). The *RsaI C2* allele is rather rare in Caucasians, and results from studies of Caucasians are most likely subject to chance. Only one previous case-control study conducted in Hawaii examined the relation between the *96-bp insertion* polymorphism and CRC risk, wherein they showed a statistically significant risk of rectal cancer, but not of colon cancer, for individuals harboring at least one allele of the 96-bp insertion (VanderLogt et al., 2006). In a recent Japanese study on colorectal adenomas, the *96-bp insertion* allele was linked with an increased occurrence of large adenomas of ≥5 mm, but not of small adenomas (<5 mm) (Morita et al., 2008).

Gao *et al.* in their study showed the connection of *CYP2E1 RsaI* and *96-bp insertion* polymorphisms with CRC risk in the Fukuoka community of Japan with particular consideration to alcohol use, obesity and red meat intake (Gao et al., 2007), wherein they reported that *CYP2E1* is induced by alcohol consumption and obesity and it may differ by the *RsaI* and *96-bp insertion* polymorphisms (Gao et al., 2007).

CYP2E1 is an important phase I enzyme metabolizing alcohol and tobacco-generated N-nitrosamines. Change in its activity might be linked to the progress of BC (Ma et al., 2014). Although recent studies have investigated the role of *CYP2E1* gene polymorphisms in BC risk, the reported results were incompatible, with some studies reporting that *CYP2E1* gene polymorphisms may amplify the risk of BC, but not others (Cantor et al., 2010; Basma et al., 2013).

Gene	SNP	Function	Role and Disease
CYP2E1	*rs3813867* *rs2031920*	"T" allele carrier of the CYP2E1 rs23031920 C/T polymorphisms and "C" allele carrier of the rs3813867 G/C polymorphisms as protective factors for ESCC patients, especially in Asian populations.	Associated with schizophrenia.
	rs3813870 *rs2031922*	The haplotypes rs3813870 and rs2031922 were significantly associated with schizophrenia. Associated with *PstI* and *RsaI* restriction enzyme sites, respectively.	Decreased enzyme activity or non-inducibility. Breast cancer.

8.5.5 LIPC Gene

The *LIPC* gene provides instructions for making an enzyme called *hepatic lipase.* This enzyme is produced by liver cells and released into the circulation where it helps in the conversion of fat-transporting molecules called VLDLs and IDLs to LDLs. The enzyme also transports high-density lipoproteins (HDLs) that carry cholesterol and triglycerides from the blood to the liver, where the HDLs deposit these fats so they can reach other tissues or removed from the body. Hepatic lipase keeps balance of these fat-transporting molecules by regulating the formation of LDLs and the transport of HDLs. This gene is located on chromosome 15 in humans.

8.5.5.1 Structure and Function

The *LIPC* gene encodes a hepatic triglyceride lipase. The liver shows the high expression of this enzyme. It has the double function of triglycerides, phospholipids and lipoproteins hydrolase and binding factor for receptor-mediated lipoprotein uptake (Andrés-Blasco et al., 2016). LIPC is formed in the liver, after which it is secreted and bound to hepatocytes and hepatic endothelial surfaces via heparin sulfate proteoglycans (HSPGs). The active form of LIPC is a homodimer and has wide substrate specificity. It catalyzes the hydrolysis of fatty acyl chains at the SN-1 position of phospholipids and of mono-, di- and triacylglycerols associated with a number of lipoproteins, including HDLs. LIPC may also have a role in the binding and uptake of lipoproteins and selective uptake of cholesteryl esters from lipoproteins (Brown et al., 2004).

Hepatic lipase is categorized under hydrolase class of enzymes. Its function is to hydrolyze triacylglycerol to diacylglycerol and carboxylate (free FAs) by addition of water molecule (Sanan et al., 1997). The substrate, triacylglycerol, comes from IDLs, and the release of free FAs converts IDLs into LDLs (Chatterjee and Sparks, 2011). These remnants of LDLs are transported back to liver, where they can be stored for later use or broken down to produce energy. They can also be sent to peripheral cells because they contain cholesterol which can be used in anabolic pathways to assemble molecules that the cell needs such as hormones that include a cholesterol backbone (Chatterjee and Sparks, 2011).

8.5.5.2 Reported SNPs in LIPC Gene

LIPC gene has a strong role in cardiovascular diseases (Wang et al., 2015b; Lu et al., 2016) (see Table 8.1). Mägi et al. reported that the *rs2043085* SNP was the most considerable SNP in the *LIPC* locus linked with lipid traits and BMI. They also found that SNPs *rs2043085* and *rs1532085* were significantly associated with adiposity status (Mägi et al., 2017). *LIPC rs1800588* SNP present in the promoter *region of the gene is found to be strongly* associated with various phenotypes (Ahmad et al., 2011). *LIPC* is expressed only in the liver, and both *rs1532085* and *rs1800588* were associated with the expression of *LIPC* (Folkersen et al., 2010). *LIPC rs2043085* SNP exhibited a high linkage disequilibrium with *rs1532085*, and either of them has been shown to be the most important SNP in the *LIPC* locus associated with various phenotypes in several genome-wide association studies (Benner et al., 2016).

8.5.5.3 LIPC Gene Polymorphisms in Cancer

Several *LIPC* polymorphisms were found to be linked with the risk of cancers, mostly CRC, although no particular haplotype was related to CRC. The *rs12299484* SNP showed a role in CRC risk after the Bonferroni correction (Rebhan et al., 1998). A path analysis based on structural equation modeling showed a direct role of LIPC gene polymorphisms in colorectal carcinogenesis as well as an indirect effect mediated through serum lipid levels. So genetic polymorphisms in the *LIPC* gene have a possible role in colorectal carcinogenesis through the regulation of serum lipid levels. The *LIPC*

TABLE 8.1

Various SNPs Reported in LIPC Gene and Their Function and Relations to Diseases

Gene	SNP	Function	Role and Disease
LIPC	*rs2043085*	*rs2043085* SNP was the most important SNP in the *LIPC* locus linked with lipid traits and BMI. Expression of LIPC gene.	Adiposity.
	rs1532085	rs1532085 is known to a play role in lipid and lipoprotein metabolic pathways.	Colorectal cancer.
	rs1800588	rs1800588, a common variant in the LIPC gene, has been linked with higher concentration of HDL-C.	Increased risk of colorectal cancer and thyroid cancer.
	rs12299484	A path analysis based on structural equation modeling showed a direct efficient role of LIPC gene polymorphisms on colorectal carcinogenesis as well as an indirect effect mediated through serum lipid levels.	Role in CRC risk.
	rs3829462	rs3829462 is found to have a significant association with CRC risk.	Role in CRC.

gene codes for a hepatic triglyceride lipase, which is mostly expressed in the liver, and has the dual function of triglycerides, phospholipids and lipoproteins hydrolase and binding factor for receptor-mediated lipoprotein uptake (Rebhan et al., 1998). This gene has been broadly studied in relation to cardiovascular diseases (Chen et al., 2009). Hyperlipidemia, which is characterized by increased total cholesterol (TC), triglycerides, LDL and decreased HDL, has also been linked with cancer risk (Hoeft et al., 2010). Hyperlipidemia has been found due to several genetic polymorphisms in the lipid metabolism pathway (Aulchenko et al., 2009). Thus, the potential role of the *LIPC* gene in FA metabolism is of particular concern in carcinogenesis, probably mediated through hyperlipidemia. Several polymorphisms in *LIPC* have been found to be associated with CRC risk, although no particular haplotype was linked to CRC. The most important was the association of the G allele of *LIPC rs9652472* with the increased risk of CRC. The associations between the T allele of the *LIPC SNP rs1800588* and the higher concentration of serum HDL have been reported (Aulchenko et al., 2009; Fan et al., 2009). It is assumed that these associations are due to changes in the rate of transcription of the LIPC gene for the carriers of the T allele of this promoter polymorphism (Fan et al., 2009). Other studies have shown a high frequency of *LIPC SNP rs1800588* T allele in cases with higher TC and triglycerides, thereby increasing their risk of cancer (Gerber, 2009). Two studies analyzed the link between *LIPC* gene variants and CRC risk, but with conflicting results. While the study by Webb et al. reported an association between 10 polymorphisms of *LIPC* gene and CRC risk (Webb et al., 2006), including the *rs3829462* SNP, Frank et al. did not replicate these associations (Frank et al., 2008).

8.6 Conclusions

Cancer cells depend on FAs for all biological activities. This chapter provides an update on the extended role of FA metabolism genes in cancer as well as new opportunities to target this process. FA metabolism can be supposed as a network of pathways notable for its plasticity. This metabolic plasticity has key implications for new therapeutic strategies and subsequent drug resistance. The potential of cancer cells to switch between FA synthesis, FA uptake and degradation ensures their fitness when exposed to temporary fluctuations in the availability of nutrients and oxygen. The polymorphisms of the FA metabolism genes may play a vital role in evaluating the risk of cancer. There is a possibility for future research based on the effects of common polymorphisms on the phenotypic outcomes of cancer. Independent large-scale studies need to be done to authenticate the relationship among multiple polymorphisms of candidate genes, environment factors and cancer risk thereof. Although chemical inhibitors for specific metabolic pathways are available, successful therapies targeting FA metabolism may depend on an understanding of the specific metabolic plasticity associated with a particular type of cancer and may need targeting multiple sources of FAs simultaneously, and possibly in combination with stringent dietary regimens.

ABBREVIATIONS

ALA: Alpha-linoleic acid
ATP: Adenosine triphosphate
BC: Bladder cancer
BC: Breast cancer
BMI: Body mass index
CRC: Colorectal cancer
CYP2E1: Cytochrome P450 2E1
DNA: Deoxyribonucleic acid
EDP: Epoxydocosapentaenoic acid
EEQ: Epoxyeicosatetraenoic acid
FA: Fatty acid
FABP: FA-binding protein
FAS: FA synthase
HCC: Hepatocellular carcinoma
HDL: High-density lipoprotein
HETE: Hydroxyeicosatetraenoic acid
HSPGs: Heparin sulfate proteoglycans
GTP: Guanosine triphosphate
IDL: Intermediate-density lipoprotein
LA: Linoleic acid
LBD: Ligand-binding domain
LCFA: Long-chain FA
LDL: Low-density lipoprotein
LIPC: Lipase C, hepatic type
NAFLD: Nonalcoholic fatty liver disease
NSCLC: Non-small cell lung cancer

PPARG: Peroxisome proliferator-activated receptor gamma
PC: Prostate cancer
SCOPA: Software for correlated phenotype analysis
SNPs: Single nucleotide polymorphisms
T2DM: Type 2 diabetes mellitus

REFERENCES

Abe T, Saburi J, Hasebe H (2009) Novel mutations of the FASN gene and their effect on fatty acid composition in Japanese black beef. *Biochem Genet* 47:397–411.

Abuli A, Fernandez-Rozadilla C, Alonso-Espinaco V (2011) Case-control study for colorectal cancer genetic susceptibility in EPICOLON---previously identified variants and mucins. *BMC Cancer* 11:339.

Ahmad T, Chasman DI, Buring JE (2011) Physical activity modifies the effect of LPL, LIPC, and CETP polymorphisms on HDL-C levels and the risk of myocardial infarction in women of European ancestry. *Circ Cardiovasc Genet* 4:74–80.

Alves-Bezerra M, Cohen DE (2017) Triglyceride Metabolism in the Liver. *Compr Physiol 8*: 1–8.

Andrés-Blasco I, Vinué Á, Herrero-Cervera A, Martínez-Hervás S (2016) Hepatic lipase inactivation decreases atherosclerosis in insulin resistance by reducing LIGHT/lymphotoxin β-receptor pathway. *Thromb Haemost* 116:379–93.

Aprahamian T, Bonegio RG, Richez C (2009) The peroxisomal proliferator-activated receptor gamma agonist rosiglitazone ameliorates murine lupus by induction of adiponectin. *J. Immunol.* 182:340–6.

Asturias FJ, Chadick JZ, Cheung IK, Stark H (2005) Structure and molecular organization of mammalian FA synthase. *Nat Struct Mol Biol* 12:225–32.

Atshaves BP, McIntosh AL, Storey SM (2010) High dietary fat exacerbates weight gain and obesity in female liver FA binding protein gene-ablated mice. *Lipids* 45(2):97–110.

Aulchenko YS, Ripatti S, Lindqvist I (2009) Loci influencing lipid levels and coronary heart disease risk in 16 European population cohorts. *Nat Genet* 41:47–55.

Basma HA, Kobeissi LH, Jabbour ME (2013) CYP2E1 and NQO1 genotypes and bladder cancer risk in a Lebanese population. *Int J Mol Epidemiol Genet* 4:207–17.

Benner C, Spencer CC, Havulinna AS (2016) FINEMAP—efficient variable selection using summary data from genome-wide association studies. *Bioinformatics* 32:1493–501.

Brown RJ, Gauthier A, Parks RJ (2004) Severe hypoalphalipoproteinemia in mice expressing human hepatic lipase deficient in binding to heparan sulfate proteoglycan. *J Biol Chem* 279:42403–9.

Butler WJ, Ryan P, Roberts-Thomson IC (2001) Metabolic genotypes and risk for colorectal cancer. *J Gastroenterol Hepatol* 16:631–5.

Campa D, McKay J, Sinilnikova O (2009) Genetic variation in genes of the FA synthesis pathway and breast cancer risk. *Breast Cancer Res Treat* 118:565–74.

Cantor KP, Villanueva CM, Silverman DT (2010) Polymorphisms in GSTT1, GSTZ1, and CYP2E1, disinfection by-products, and risk of bladder cancer in Spain. *Environ Health Perspect* 118:1545–50.

Chatterjee C, Sparks DL (2011) Hepatic lipase, high density lipoproteins, and hypertriglyceridemia. *Am J Pathol* 178(4):1429–33.

Chen M, Zhang J, Sampieri K (2018) An aberrant SREBP dependent lipogenic program promotes metastatic prostate cancer. *Nat Genet* 50:206–18.

Chen SN, Cilingiroglu M, Todd J (2009) Candidate genetic analysis of plasma high-density lipoprotein-cholesterol and severity of coronary atherosclerosis. *BMC Med Genet* 10:111.

Cheong HS, Park SM, Kim MO (2011) Genome-wide methylation profile of nasal polyps---relation to aspirin hypersensitivity in asthmatics. *Allergy* 66:637–44.

Chiu HK, Qian K, Ogimoto K, Morton GJ (2010) Mice lacking hepatic lipase are lean and protected against diet-induced obesity and hepatic steatosis. *Endocrinology* 151:993–1001.

Chong ET, Goh LP, See EU, Chuah JA, Chua KH (2016) Association of CYP2E1, STK15 and XRCC1 polymorphisms with risk of breast cancer in Malaysian women. *Asian Pac J Cancer Prev* 17(2):647–53.

Crous-Bou M, Rennert G, Salazar R (2012) Genetic polymorphisms in FA metabolism genes and colorectal cancer. *Mutagenesis* 27:169–76.

Diwan, Joyce J (2011) *FA Synthesis.* Rensselaer Polytechnic Institute (RPI). Architecture, Business, Engineering, IT, Humanities, Science. Web. 30 Apr. 2011. <http://rpi.edu/dept/bcbp/molbiochem/MBWeb/mb2/part1/fasynthesis.htm

Doecke JD, Zhao ZZ, Stark MS (2008) Australian Cancer S. Single nucleotide polymorphisms in obesityrelated genes and the risk of esophageal cancers. *Cancer Epidemiol Biomarkers Prev* 17:1007–12.

Du J, Zhang A, Wang L, Xuan J, Yu L (2010) Relationship between response to risperidone, plasma concentrations of risperidone and CYP3A4 polymorphisms in schizophrenia patients. *J Psychopharmacol* 24:1115–20.

Eggert SL, Huyck KL, Somasundaram P (2012) Genome-wide linkage and association analyses implicate FASN in predisposition to uterine leiomyomata. *Am. J. Hum. Genet.* 91:621–8.

Fan YM, Raitakari OT, Kahonen M (2009) Hepatic lipase promoter C-480T polymorphism is associated with serum lipids levels, but not subclinical atherosclerosis---the Cardiovascular Risk in Young Finns Study. *Clin Genet* 76:46–53.

Farese RV Jr, Walther TC (2009) Lipid droplets finally get a little R-E-S-P-E-C-T. *Cell* 139:855–60.

Feng J, Pan X, Yu J (2012) Functional PstI/RsaI polymorphism in CYP2E1 is associated with the development, progression and poor outcome of gastric cancer. *PLoS One* 7:e44478.

Ferlay J, Shin H, Bray F (2010) Estimates of worldwide burden of cancer in 2008—GLOBOCAN 2008. *Int J Cancer* 127:2893–917.

Fischer R, Konkel A, Mehling H, Blossey K (2014) Dietary omega-3 FAs modulate the eicosanoid profile in man primarily via the CYP-epoxygenase pathway. *J Lipid Res* 55(6):1150–64.

Fisher E, Weikert C, Klapper M, Lindner I (2007) L-FABP T94A is associated with fasting triglycerides and LDL-cholesterol in women. *Mol Gen Met* 91:278–84.

Fleming I (2014) The pharmacology of the cytochrome P450 epoxygenase/soluble epoxide hydrolase axis in the vasculature and cardiovascular disease. *Pharmacol Rev* 66(4):1106–40.

Folkersen L, van't Hooft F, Chernogubova E (2010) Association of genetic risk variants with expression of proximal genes identifes novel susceptibility genes for cardiovascular disease. *Circ Cardiovasc Genet* 3:365–73.

Frank B, Burwinkel B, Bermejo JL, et al. (2008) Ten recently identified associations between nsSNPs and colorectal cancer could not be replicated in German families. *Cancer Lett* 271:153–7.

Freedland SJ, Aronson WJ, Kane CJ (2004) Impact of obesity on biochemical control after radical prostatectomy for clinically localized prostate cancer---A report by the Shared Equal Access Regional Cancer Hospital database study group. *J Clin Oncol* 22:446–53.

Gao CM, Takezaki T, Wu JZ (2007) CYP2E1 RsaI polymorphism impacts on risk of colorectal cancer association with smoking and alcohol drinking. *World J Gastroenterol* 13:5725–30.

Gerber M (2009) Background review paper on total fat, FA intake and cancers. *Ann Nutr Metab* 55:140–61.

Gipson P, Mills DJ, Wouts R (2010) Direct structural insight into the substrate-shuttling mechanism of yeast FA synthase by electron cryomicroscopy. *PNAS* 107(20):9164–9.

Glatz JF, Luiken JJ, Bonen A (2010) Membrane FA transporters as regulators of lipid metabolism: Implications for metabolic disease. *Physiol Rev* 90:367–417.

Graupera I, Coll M, Pose E (2017) Adipocyte fatty-acid binding protein is overexpressed in cirrhosis and correlates with clinical outcomes. *Sci Rep* 7:1829.

Guerrero-Romero F, Simental-Mendia LE, Gonzalez-Ortiz M (2010) The product of triglycerides and glucose, a simple measure of insulin sensitivity. Comparison with the euglycemic hyperinsulinemic clamp. *J Clin Endocrinol Metab* 95:3347–51.

Hagiwara T, Kono S, Yin G (2005) Genetic polymorphism in cytochrome P450 7A1 and risk of colorectal cancer---the Fukuoka Colorectal Cancer Study. *Cancer Res* 65:2979–82.

He J, Wang C, Zhu Y, Ai D (2015) Soluble epoxide hydrolase---A potential target for metabolic diseases. *J Diabetes* 8(3):305–13.

Hoeft B, Linseisen J, Beckmann L (2010) Polymorphisms in FA metabolism-related genes are associated with colorectal cancer risk. *Carcinogenesis* 3:466–72.

Huang H, McIntosh AL, Martin GG, Landrock D, Chung S, Landrock KK, Dangott LJ, Li S, Kier AB, Schroeder F (2016) FABP1: a novel hepatic endocannabinoid and cannabinoid binding protein. *Biochemistry* 55: 5243–55.

Hudgins LC, Baday A, Hellerstein MK (2008) The effect of dietary carbohydrate on genes for FA synthase and inflammatory cytokines in adipose tissues from lean and obese subjects. *J Nutr Biochem* 19:237–45.

Jemal R, Bray F, Center M (2011) Global cancer statistics. *CA Cancer J Clin* 61:69–90.

Jiang DK, Ma XP, Wu X (2015) Genetic variations in STAT4, C2, HLA-DRB1 and HLA-DQ associated with risk of hepatitis B virus-related liver cirrhosis. *Sci Rep* 5:16278.

Jiang DK, Sun J, Cao G (2013) Genetic variants in STAT4 and HLA-DQ genes confer risk of hepatitis B virus-related hepatocellular carcinoma. *Nat Genet* 45:72–5.

Khedhaier A, Hassen E, Bouaouina N, Gabbouj S, Ahmed SB, Chouchane L (2008) Implication of xenobiotic metabolizing enzyme gene (CYP2E1, CYP2C19, CYP2D6, mEH and NAT2) polymorphisms in breast carcinoma. *BMC Cancer* 8: 109.

Kiss I, Sa´ndor J, Pajkos G, Bogner B (2000) Colorectal cancer risk in relation to genetic polymorphism of cytochrome P450 1A1, 2E1, and glutathione-S-transferase M1 enzymes. *Anticancer Res* 20:519–22.

Kitam VO, Maksymchuk OV, Chashchyn MO (2012) The possible mechanisms of CYP2E1 interactions with HSP90 and the influence of ethanol on them. *BMC Struct Biol* 12(1):33.

Körner A, Ma L, Franks PW, Kiess W (2007) Sex-specific effect of the Val1483Ile polymorphism in the FA synthase gene (FAS) on body mass index and lipid profile in Caucasian children. *Int J Obes (Lond)* 31:353–8.

Kovacs P, Harper I, Hanson RL (2004) A novel missense substitution (Val1483Ile) in the FA synthase gene (FAS) is associated with percentage of body fat and substrate oxidation rates in nondiabetic Pima Indians. *Diabetes* 53:1915–9.

Laidlaw TM, Boyce JA (2013) Pathogenesis of aspirin-exacerbated respiratory disease and reactions. *Immunol Allergy Clin North Am* 33:195–210.

Le Marchand L, Donlon T, Seifried A, Wilkens LR (2002) Red meat intake, CYP2E1 genetic polymorphisms, and colorectal cancer risk. *Cancer Epidemiol Biomarkers Prev* 11:1019–24.

Lee CK, Jeong SH, Jang C (2019) Tumor metastasis to lymph nodes requires YAP-dependent metabolic adaptation. *Science* 363:644–9.

Lee RU, Stevenson DD (2011) Aspirin-exacerbated respiratory disease---evaluation and management. *Allergy Asthma Immunol Res* 3:3–10.

Lewis SN, Bassaganya-Riera J, Bevan DR (2010) Virtual screening as a technique for PPAR Modulatory Discovery. *PPAR Res* 2010:861238.

Longo N, Frigeni M, Pasquali M (2016) Carnitine transport and fatty acid oxidation. *Biochim Biophys Acta* 1863:2422–35.

Lu X, Huang J, Mo Z, He J, Wang L, Yang X (2016) Genetic susceptibility to lipid levels and lipid change over time and risk of incident hyperlipidemia in Chinese populations. *Circ Cardiovasc Genet* 9:37–44.

Ludovico O, Pellegrini F, Paola R, et al. (2007) Heterogeneous effect of peroxisome proliferator-activated receptor gamma 2 Ala12variant on type 2 diabetes risk. *Obesity* 15:1076–81.

Lupu R, Menendez JA (2006) Targeting FA synthase in breast and endometrial cancer---An alternative to selective estrogen receptor modulators. *Endocrinology* 147:4056–66.

Ma J, Li H, Giovannucci E (2008) Prediagnostic body-mass index, plasma C-peptide concentration, and prostate cancer-specific mortality in men with prostate cancer---A long-term survival analysis. *Lancet Oncol* 9:1039–47.

Ma L, Zhao J, Li T, He Y, Wang J, Xie L (2014) Association between tumor necrosis factor-alpha gene polymorphisms and prostate cancer risk---a meta-analysis. *Diagn Pathol* 9(1):74.

Mägi R, Suleimanov YV, Clarke GM (2017) SCOPA and META-SCOPA---software for the analysis and aggregation of genome-wide association studies of multiple correlated phenotypes. *BMC Bioinform* 11(18):25.

Maier T, Jenni S, Ban N (2006) Architecture of mammalian FA synthase at 4.5 Å resolution. *Science (Wash DC)* 311:1258–62.

Maier T, Leibundgut M, Ban N (2008) The crystal structure of a mammalian FA synthase. *Science (Wash DC)* 32:1315–22.

Martin GG, Landrock D, Dangott LJ (2018) Human liver FA binding protein-1 T94A variant, nonalcohol fatty liver disease, and hepatic endocannabinoid system. *Lipids* 53:27–40.

Martin GG, McIntosh AL, Huang H (2013) The human liver FA binding protein T94A variant alters the structure, stability, and interaction with fibrates. *Biochemistry* 52:9347–57.

Martiny VY, Miteva MA (2013) Advances in molecular modeling of human cytochrome P450 polymorphism. *J Mol Biol* 425:3978–92.

Menendez JA, Lupu R (2007) FA synthase and the lipogenic phenotype in cancer pathogenesis. *Nat Rev Cancer* 7:763–77.

Metallo CM, Gameiro PA, Bell EL (2011) Reductive glutamine metabolism by IDHI mediates lipogenesis under hypoxia. *Nature* 481:380–4.

Moreno-Navarrete JM, Botas P, Valdés S (2009) Val1483Ile in FASN gene is linked to central obesity. *J Affect Disord* 134:448–52.

Morita M, Tabata S, Tajima O, Yin G (2008) Genetic polymorphisms of CYP2E1 and risk of colorectal adenomas in the Self Defense Forces Health Study. *Cancer Epidemiol Biomarkers Prev* 17:1800–7.

Mukai T, Egawa M, Takeuchi T (2017) Silencing of FABP1 ameliorates hepatic steatosis, inflammation, and oxidative stress in mice with nonalcoholic fatty liver disease. *FEBS Open Bio* 7:1009–16.

Mullen AR, Wheaton WW, Jin ES (2011) Reductive carboxylation supports growth in tumour cells with defective mitochondria. *Nature* 481:385–8.

Obata Y, Kamijo-Ikemori A, Ichikawa D (2016) Clinical usefulness of urinary liver-type fatty-acid-binding protein as a perioperative marker of acute kidney injury in patients undergoing endovascular or open-abdominal aortic aneurysm repair. *J Anesth* 30(1):89–99.

Ogino S, Nosho K, Meyerhardt JA (2008) Cohort study of FA synthase expression and patient survival in colon cancer. *J Clin Oncol* 26:5713–20.

Ordovás L, Roy R, Pampín S, Zaragoza P (2008) The g.763G>C SNP of the bovine FASN gene affects its promoter activity via Sp-mediated regulation---implications for the bovine lactating mammary gland. *Physiol Genomics* 34:144–8.

Park YK, Wang L, Giampietro A (2017) Distinct roles of transcription factors KLF4, Krox20, and peroxisome proliferator-activated receptor γ in adipogenesis. *Mol Cell Biol* 37(2):18779–89.

Parmer MB, Venkatachalam AB, Wright JM (2012) The evolutionary relationship of the transcriptionally active FABP11a (intronless) and FABP11b genes of medaka with FABP11 genes of other teleost fishes. *FEBSJ* 279:2310–21.

Pascual G, Avgustinova A, Mejetta S (2017) Targeting metastasis-initiating cells through the FA receptor CD36. *Nature* 541:41–5.

Peng XE, Wu YL, Lu QQ (2012) Two genetic variants in FABP1 and susceptibility to non-alcohol fatty liver disease in a Chinese population. *Gene* 500:54–8.

Peng XE, Wu YL, Zhu YB (2015) Association of a human FABP1 gene promoter region polymorphism with altered serum triglyceride levels. *PloS One* 10:e0139417.

Persson B, Kallberg Y, Bray JE (2009) The SDR (short-chain dehydrogenase/reductase and related enzymes) nomenclature initiative. *Chem Biol Interact* 178(1–3):94–8.

Pfeuffer, Maria Jaudszus, Anke (2016) Pentadecanoic and heptadecanoic acids---multifaceted odd-chain FAs. *Int Rev* 7:730–4.

Pochetti G, Godio C, Mitro N (2007) Insights into the mechanism of partial agonism---crystal structures of the peroxisome proliferator-activated receptor gamma ligand-binding domain in the complex with two enantiomeric ligands. *JBC* 282:17314–24.

Polonikov AV, Ushachev DV, Shestakov AM (2011) *Kardiologiya* 51:33–8.

Poremski PR, Meneely KM, Scott EE (2008) Structures of human cytochrome P-450 2E1 insights into the binding of inhibitors and both small molecular weight and FA substrates. *The JBC* 283(48):33698–707.

Porubsky PR, Meneely KM, Scott EE (2008) Structures of human cytochrome P-450 2E1. Insights into the binding of inhibitors and both small molecular weight and fatty acid substrates. *JBC* 283: 33698–707.

Qin X, Peng Q, Chen Z, Deng Y, Huang S, Xu J (2013) The association between MTHFR gene polymorphisms and hepatocellular carcinoma risk---a meta-analysis. *PLoS One* 8(2):e56070.

Rashid N, Nigam A, Saxena P (2017) Association of IL-1beta, IL-1Ra and FABP1 gene polymorphisms with the metabolic features of polycystic ovary syndrome. *Inflamm Res* 66:621–36.

Rebhan M, Chalifa-Caspi V, Prilusky J (1998) Gene Cards—a novel functional genomics compendium with automated data mining and query reformulation support. *Bioinformatics* 14:656–64.

Roy R, Taourit S, Zaragoza P (2005) Genomic structure and alternative transcript of bovine fatty acid synthase gene FASN---comparative analysis of the FASN gene between monogenic and ruminant species. *Cytogenet Genome Res* 111:65–73.

Sanan DA, Fan J, Bensadoun A, Taylor JM (1997) Hepatic lipase is abundant on both hepatocyte and endothelial cell surfaces in the liver. *J Lipid Res* 38(5):1002–13.

Sartor RB (2006) Mechanism of disease---pathogenesis of Crohn' s disease and Ulcerative colitis. *Nature* 3:390–407.

Shi J, Zhang Y, Gu W, Cui B, Xu M (2012) Serum liver FA binding protein levels correlate positively with obesity and insulin resistance in Chinese young adults. *PLoS One* 7(11):e48777.

Smathers RL, Petersen DR (2011) The human FA-binding protein family---evolutionary divergences and functions. *Hum Genomics* 5:170–91.

Smith JL, Sherman DH (2008) Biochemistry---an enzyme assembly line. *Science (Wash DC)* 321:1304–5.

Smith S, Witkowski A, Joshi AK (2003) Structural and functional organization of the animal FA synthase. *Prog Lipid Res* 42:289–317.

Stephens M, Smith NJ, Donnelly P (2001) A new statistical method for haplotype reconstruction from population data. *Am J Hum Genet* 68:978–89.

Tai ES, Corella D, Deurenberg-Yapal (2004) Differential effects of the C1431T and Pro12Ala PPAR gamma gene variants on plasma lipids and diabetes risk in an Asian population. *J Lipid Res* 45:555–8.

Tao H, Zhou X, Xie Q (2018) SRR intronic variation inhibits expression of its neighbouring SMG6 gene and protects against temporal lobe epilepsy. *J Cell Mol Med* 22:1883–93.

Thuijls G, van Wijck K, Grootjans J (2011) Early diagnosis of intestinal ischemia using urinary and plasma FA binding proteins. *Ann Surg* 253(2):303–8.

Tiuca ID, Naggy K, Oprean R (2015) Development and optimization of a gas-chromatographic separation method of FAs in human serum. *World J Pharma Sci* 3:1713–9.

Tong J, Wang L, Yi Y-s, Pan X-h, Lü Y (2012) Relationship between PPARgamma2 Pro12Ala polymorphism and type 2 diabetes mellitus in Chinese Han population—A meta-analysis. *Zhonghua Yu Fang Yi Xue Za Zhi* 46:359–63.

Tontonoz P, Spiegelman BM (2008) Fat and beyond: The Diverse Biology of PPARG. *Annu Rev Biochem* 77:289–312.

VanderLogt EMJ, Bergevoet SM, Roelofs HMJ (2006) Role of epoxide hydrolase, NAD(P)H:quinone oxidoreductase, cytochrome P450 2E1 or alcohol dehydrogenase genotypes in susceptibility to colorectal cancer. *Mutat Res* 593:39–49.

Wagner K, Vito S, Inceoglu B, Hammock BD (2014) The role of long chain FAs and their epoxide metabolites in nociceptive signaling. *Prostaglandins & Other Lipid Mediators* 113–115:2–12.

Wan Z, Shi W, Shao B, Shi J, Shen A (2011) Peroxisome proliferator-activated receptor gamma agonist pioglitazone inhibits beta-catenin-mediated glioma cell growth and invasion. *Mol Cell Biochem* 349:1–10.

Wang B, Tao X, Huang CZ (2014) Decreased expression of liver-type FA -binding protein is associated with poor prognosis in hepatocellular carcinoma. *Hepato–Gastroenterology* 61:1321–6.

Wang G, Bonkovsky HL, de Lemos A, Burczynski FJ (2015a) Recent insights into the biological functions of liver FA binding protein 1. *Journal of Lipid Research* 56:2238–47.

Wang H, Zhang D, Ling J, Lu W, Zhang S (2015b) Gender specific effect of LIPC C-514T polymorphism on obesity and relationship with plasma lipid levels in Chinese children. *J Cell Mol Med* 19:2296–306.

Wang W, Shao Y, Tang S, Cheng X, Lian H (2015c) Peroxisome proliferator-activated receptor-gamma (PPAR gamma) Pro12Ala polymorphism and colorectal cancer (CRC) risk. *Int J Clin Exp Med* 8:4066–72.

Webb EL, Rudd MF, Sellick GS (2006) Search for low penetrance alleles for colorectal cancer through a scan of 1467 non-synonymous SNPs in 2575 cases and 2707 controls with validation by kin-cohort analysis of 14 704 first-degree relatives. *Hum Mol Genet* 15:3263–71.

Westphal C, Konkel A, Schunck WH (2011) CYP-eicosanoids—a new link between omega-3 FAs and cardiac disease. *Prostaglandins & Other Lipid Mediators* 96(1–4):99–108.

Wise DR, Ward PS, Shay JE (2011) Hypoxia promotes isocitrate dehydrogenase-dependent carboxylation of alpha-ketoglutarate to citrate to support cell growth and viability. *Proc Natl Acad Sci USA* 108:19611–6.

Wu MH, Chu CH, Chou YC, Chou WY (2011) Joint effect of peroxisome proliferator-activated receptor gamma genetic polymorphisms and estrogen-related risk factors on breast cancer risk—results from a case-control study in Taiwan. *Breast Cancer Res Treat* 127:777–84.

Wu SH, Tsai SM, Hou MF, Lin HS, Hou LA (2006) Interaction of genetic polymorphisms in cytochrome P450 2E1 and glutathione S-transferase M1 to breast cancer in Taiwanese woman without smoking and drinking habits. *Breast Cancer Res Treat* 100(1):93–8.

Xu W, Li Y, Wang X, Chen B, Liu S (2010) PPARgamma polymorphisms and cancer risk—a meta-analysis involving 32,138 subjects. *Oncol Rep* 24:579–85.

Yang Li, Hua Qi, Liu Rong Kun (2009) Two common polymorphisms of PPAR gamma gene and metabolic syndrome—Families in a Chinese population. *Arch Med Res* 40:89–96.

Yen CL, Stone SJ, Koliwad S (2008) Thematic review series Glycerolipids DGAT enzymes and triacylglycerol biosynthesis. *J Lipid Res* 49:2283–301.

Zechner R, Strauss JG, Haemmerle G (2005) Lipolysis—pathway under construction. *Curr Opin Lipidol* 16(3):333–40.

Zhang G, Kodani S, Hammock BD (2014) Stabilized epoxygenated FAs regulate inflammation, pain, angiogenesis and cancer. *Progress in Lipid Research* 53:108–23.

Zhang W, Lu D, Dong W (2011) Expression of CYP2E1 increases oxidative stress and induces apoptosis of cardiomyocytes in transgenic mice. *FEBS J* 278:1484–92.

9

Single Nucleotide Polymorphisms and Parkinson's Disease Susceptibility

Saniya Nissar
Department of Biochemistry, Government Medical College, Karan Nagar, Srinagar, Kashmir, India
Infinity Biochemistry Pvt. Ltd. Srinagar, Kashmir, India

Syed Sameer Aga
Basic Medical Sciences, College of Medicine, King Saud bin Abdulaziz University for Health Sciences (KSAU-HS), King Abdullah International Medical Research Center (KAIMRC), Ministry of National Guard Health Affairs (NGHA), Jeddah, Kingdom of Saudi Arabia
Infinity Biochemistry Pvt. Ltd. Srinagar, Kashmir, India

Mujeeb Zafar Banday
Department of Biochemistry, Government Medical College, Karan Nagar, Srinagar, Kashmir, India
Infinity Biochemistry Pvt. Ltd. Srinagar, Kashmir, India

Sabhiya Majid
Department of Biochemistry & Multidisciplinary Research Unit (MRU-ICMR), Government Medical College Srinagar & Associated Hospitals, Karan Nagar, Srinagar, Jammu and Kashmir, India

CONTENTS

DOI: 10.1201/9781003246244-9

ABSTRACT Parkinson's disease (PD) is one of the common neurodegenerative diseases, which lasts several years. The incidence is varied worldwide and increases with age in both males and females with lifetime risk being 1.5%. The pathogenesis of PD is complex and multifactorial owing to the interplay of genetic and environmental factors. However, with the advances in research, the study of risk factors has helped in better understanding the pathogenesis of PD. The genetics of PD is complex and involves various Mendelian factors such as genes (*PRKN*, *SNCA*, *LRRK2*, etc.) and non-Mendelian factors such as single nucleotide polymorphisms (SNPs), which play an important role in the disease susceptibility and pathogenesis and may influence diagnosis and treatment of the disease. This chapter aims to explore different SNPs in various genes involved in pathways such as dopamine metabolism, oxidative stress, detoxification metabolism and inflammatory processes and their impact on the risk of development of PD and to identify the relationship between SNPs and PD among different ethnic populations.

KEYWORDS: *Parkinson's Disease; Risk Factors; Single Nucleotide Polymorphism; Genome-Wide Association Studies.*

9.1 Introduction

Parkinson's disease (PD), also referred to as paralysis agitans, is the most prevalent neurodegenerative disease, which leads to movement disorder. PD is the second most common neurodegenerative disease after Alzheimer's dementia (Przedborski, 2005) and is generally diagnosed during the later stages, mostly in the sixth decade of life. Globally, it affects nearly 1%–2% of people above 65 years of age, and in the year 2020, around 9.4 million (M) (range 7.7–11.2 M) people were estimated to be living with this disease. The country-specific average numbers for PD are 930 thousand (K) in the USA (95% CI, 794 K–1.1 M); 344 K in Japan (95% CI, 278 K–425 K); 266 K in Germany (95% CI, 209 K–318 K); 157 K in France (95% CI, 127 K–192 K); 149 K in Italy (95% CI, 125 K–181 K); 142 K in the UK (95% CI, 117 K–174 K); and 120 K in Spain (95% CI, 97 K–150 K) (Maserejian et al., 2020) (Figure 9.1).

The epidemiological studies show age-, gender- and population-specific variations in the prevalence and incidence of PD and related disorders. In the general population of the high economic group countries, the prevalence is estimated to be 0.3% to 1.0% in people older than 60 years and 3.0% in people older than 80 years and the incidence rates range between 8 and 18/100,000 person-years (Lee et al., 2016). Recently, it has been shown that the prevalence of PD ranges between 41/100,000 in the age group of 40–49 years and 1903/100,000 in the age group >80 years. The prevalence of PD is more among males (134 per 100,000 individuals) in comparison with females (41 per 100,000 individuals), and the median age at onset (AAO) is 60 years. The incidence of PD also varies across different world regions, with less frequency found in Asia in comparison with Europe, Australia and North America among individuals in the age group of 70–79 years (Pringsheim et al., 2014). Another epidemiological study has reported the incidence of PD for individuals above 40 years of age to be 37.6 per 100,000 (for females) and 61.2 per 100,000 (for males) (Hirsch et al., 2016). The incidence increases with age in both males and females, and the lifetime risk is estimated at 1.5% (Searles et al., 2013). The prevalence in Taiwan has been reported to be 147.7 per 100,000 (Liu et al., 2016).

FIGURE 9.1 Estimation of the 2020 global population of Parkinson's disease. (N. Maserejian, L. Vinikoor-Imler, A. Dilley, Cambridge, MA.)

In Europe, the estimated prevalence and incidence rates for PD range between 65 and 12,500/100,000 and between 5 and 346/100,000 person-years, respectively.

With the increase in the aging population worldwide, there has been a significant rise in the prevalence of PD. Further, due to the high rate of disability associated with PD and long treatment courses as well as ineffective treatment and lack of early diagnosis, this disease has led to serious financial burden on individuals and their families and consequently tends to be a huge problem for society (Marshall and Willett, 2018).

PD is a complex disease influenced by several genetic and environmental factors. The basic clinical representation of PD involves the gradual impairment of voluntary motor control within an individual, and the prevalence tends to increase steadily with age. PD is represented clinically by motor features such as postural and gait impairments, which apparently occur due to the different levels of changes within the brain. Usually, the common representation of PD includes bradykinesia, rigidity, stooped posture, stiffness of limbs, poor balance and coordination as well as tremor while resting and finally extreme and worse level of bilateral vocal cord paralysis. These manifestations are utilized to assess the development and progression of disease and thereby response to the therapy (Frisardi et al., 2016; Mazzoni et al., 2012). In addition, there are multiple non-motor features mentioned further on in this chapter, which define the early stages of the disease. The characterization of these non-motor features is therefore important for early diagnosis of PD (Mahlknecht et al., 2015). Further, no individual biomarkers have been assigned for the diagnosis of PD; however, various diverse biomarkers cumulatively can effectively predict the status and progression of the disease. The Movement Disorders Society (MDS) has recently released the probability methodology and research criteria for the diagnosis of PD. Accordingly, there are three disease stages, and various clinical and imaging markers or tools including biomarkers have been assigned to each of these stages (Berg et al., 2015).

The disease stages include the following:

Prodromal PD is the stage where early non-motor symptoms and signs appear, but early motor features are only subtly detectable. The various symptoms, diagnostic markers or tools for this stage are mostly characterized by non-motor symptoms, including hyposmia (olfactory deficits or loss), visual abnormalities (oculomotor disturbances), urogenital dysfunction, constipation, other dysautonomic features, mood disorders, somnolence, erectile dysfunction and neuropsychiatric symptoms such as depression, anhedonia, anxiety, orthostatic hypotension and sleep disorders such as rapid eye movement (REM) sleep behavior disorder.

Preclinical PD is the stage where there is the presence of neurodegenerative synucleinopathy, but no clinical symptoms. The various diagnostic markers or tools for this stage include molecular imaging, dopamine transporter single-photon emission computed tomography (DAT-SPECT), PET scan, Fluorodopa (F-DOPA), optical resonance tomography, magnetic resonance imaging, transcranial sonography, biochemical biomarkers (CSF and blood) and genetic biomarkers.

Clinical PD is the stage where the diagnosis of PD can be made based on the presence of classical motor features or symptoms, including akinesia, bradykinesia, stooped posture, tremor at rest, postural instability, rigidity, hypokinesia, stiffness of arms, legs and trunk, gait impairment and poor balance and coordination.

The pathogenesis of PD including its development and progression is complex and involves multiple cellular pathways. Further, the factors that are responsible for the initiation of the disease remain largely unknown. However, various studies have shown that the initiation as well as the progression of the disease involves multiple genetic and environmental factors and, consequently, the expression of PD phenotypes is determined by the genetic and environmental factors as well as the complex interplay between the two. PD is characterized clinically by the neuronal degeneration with the substantia nigra pars compacta – one of the nuclei that constitute the basal ganglia in the brain. These neurons are involved in the transmission of dopamine to another basal ganglia nucleus and striatum and control the motor and cognitive functions of the brain. The progressive degeneration, thus, leads to the impairment of neuronal circuits inside the motor cortical areas and the basal ganglia. These impairments lead to movement abnormalities, which are the major manifestations of PD and adversely affect the quality of life in an individual (Mazzoni et al., 2012). The decrease in dopamine levels by 80% in the basal ganglia and substantia nigra (SN) results in the appearance of initial signs and symptoms of PD. The decrease in dopamine levels adversely affects the dopaminergic pathway, whereby dopamine plays an important role in the maintenance of balance through its interaction with the G protein-coupled receptors (GPCRs) family (Katritch et al., 2012; Gurevich et al., 2016). Another characteristic of PD is Lewy inclusion bodies, which are spherical abnormal insoluble fibril aggregates that develop in the cytoplasm of nerve cells (Cieri et al., 2017). These Lewy inclusion bodies are mainly composed of alpha-synuclein (ASN), a presynaptic protein, which is linked to PD both neuropathologically and genetically (Song et al., 2015).

In addition, increased oxidative stress and low anti-inflammatory responses, which are characteristic of continuously activated microglia, cause chronic inflammation in the brain and lead to neurodegeneration (Kalia and Lang, 2015). Furthermore, the aberrant functioning of other signaling pathways, which involve neurotransmitters such as serotonin, glutamate and NO, also contribute to PD pathogenesis (Politis and Niccolini, 2015; Zhang et al., 2019; Titova and Chaudhuri, 2017; Tiwari and Pal, 2017). Other neurotransmitters such as acetylcholine (ACh) and gamma-aminobutyric acid (GABA) do play a key role in the pathogenesis of PD. ACh functions together with dopamine as a postsynaptic excitatory neurotransmitter through cholinergic receptors (nicotinic and muscarinic receptors) and D1/D2 receptors, respectively. A reduced presynaptic inhibitory action is exerted by GABA via $GABA_A$ and $GABA_B$ receptors, whereas an excitotoxic effect plus a presynaptic inhibitory action through N-methyl-D-aspartate (NMDA) and metabotropic glutamate (glutaminergic) receptors (mGluRs) is exerted by glutamate (Werner and Covenas, 2015). Glutaminergic neurons strongly inhibit D2 dopaminergic neurons during presynapsis via NMDA receptors in the putamen – the outer part of the lentiform nucleus of the brain (Werner and Covenas, 2011). PD is a multi-neurotransmitter dysfunction disorder, involving impairment in the process of neuronal development, on the one hand, and cell death, on the other (Hama et al., 2015; Dauer and Przedborski, 2003).

PD, initially considered a sporadic disease, has now emerged as a disease in which multiple genes including *SNCA*, *PRKN*, *LRRK2*, *ATP13A2*, *PINK1*, *DJ1* and *PLA2G6*are involved in dwisease pathogenesis (Bekris et al., 2010; Wider and Wszolek, 2007). However, single genes involved in the heritable types of PD have not yet been identified and characterized. *SNCA* gene (chromosomal location, 4q22.1) encodes the ASN protein and corresponds to the *PARK1* and *PARK4* loci, and it has been reported that mutations in *SNCA* gene account for >1% of this disease in the general population (Blair and Dhillon, 2017). ASN is one of the highly conserved proteins that are abundantly found in neurons, especially presynaptic terminals. Its exact role has not yet been fully elucidated, but functions chiefly to control the vesicular neurotransmission. Additionally, human ASN plays a key role in regulating dopamine neurotransmission. ASN is the central component of Lewy bodies, and there is a twofold increase in its expression during triplication of the *SNCA* gene, which has been found to be responsible for orthostatic hypotension in PD patients. Thus, *SNCA* contributes to the etiology of PD (Doherty, 2013; Siddiqui et al., 2016). Another gene *LRRK2* (leucine-rich repeat kinase 2) (chromosomal location, 12q12) is the incomplete penetrance gene and accounts for 4% of hereditary PD. It has more than 100 mutations, but G2019S and R1441G/C/H have been characterized as the most common ones. G2019S substitution is found in higher frequency among Ashkenazi Jews, where it accounts for 30% of familial PD and 13% of sporadic PD (Ozelius et al., 2006). In contrast, in North African Arab-Berbers, G2019S substitution accounts for 37% of familial PD and 41% of sporadic PD (Lesage et al., 2006). Different mutations in *LRRK2* gene have been demonstrated to be linked with "classical" late-onset PD (LOPD), of which the age of onset is around the sixth decade of life. The clinical picture of LRRK2-linked Parkinsonism is much like the one seen in sporadic cases of PD (Ferreira and Massano, 2017). The *LRRK2* gene encodes a protein that belongs to the Roco protein family and consists of a Roc GTPase domain, a COR dimerization region and a protein kinase domain. Although the genetic component unequivocally contributes to familial PD, this type accounts for only a small percentage of overall PD, which is considered and mostly found to be sporadic in nature.

9.2 Risk Factors

Currently, the pathogenesis of PD is hypothesized to result from the complex cross-interdependence between genetic and environmental factors. Earlier, PD was considered sporadic in nature; however, numerous studies carried out over the past decade have highlighted the significant role of numerous genetic factors in the pathogenesis of this disease. Therefore, PD susceptibility involves both genetic and environmental factors (Ascherio and Schwarchild, 2016) and studying these factors has helped in the better understanding of the underlying mechanisms as well as phenotypic heterogeneity associated with PD (see Figure 9.2).

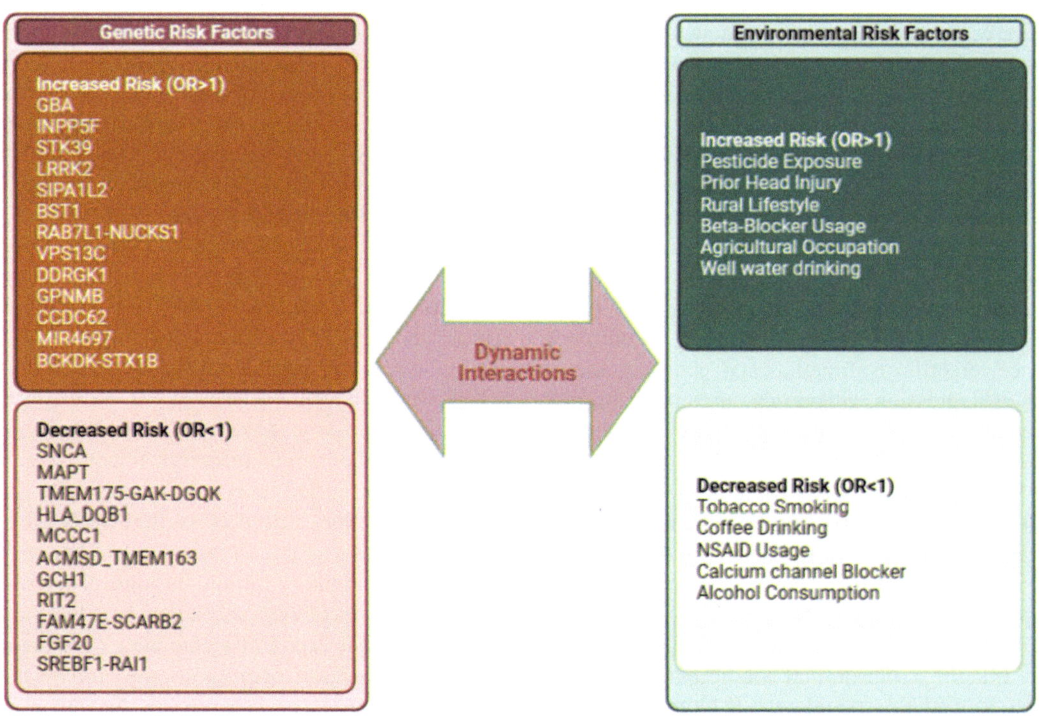

FIGURE 9.2 Diagrammatic representation of various genetic and environmental risk factors of Parkinson's disease and complex (dynamic) interactions between the two types of risk factors.

9.2.1 Environmental Risk Modifiers of Parkinson's Disease

9.2.1.1 *Pesticides*

The exposure to pesticides has been found to be associated with PD, mostly in people involved in farming and living in rural areas consuming pesticide-contaminated well water (Noyce et al., 2012). In France, PD is recognized as the occupational disease in farmers because of their chronic exposure to the pesticides. Contrarily, two recent meta-analyses point to the high heterogeneity and inconsistency among the studies (Breckenridge et al., 2016; Van Maele-Fabry et al., 2012). The exposure to insecticides such as organochlorine is found to be associated with late-onset PD in men (Elbaz et al., 2009). The use or exposure to rotenone and paraquat structurally related to 1-methyl-4-phenyl-1,2,3,6-tetrahydropyridine (MPTP) has been associated with PD in exposed individuals. In addition, the occupational welders who are exposed to heavy metals such as manganese and recreational ephedrine (methcathinone) users tend to develop PD (Selikhova et al., 2008). However, in the overall development of PD, these environmental toxins or factors seem unlikely to play more than a minor role and hence can be described as minor risk factors for PD (Noyce et al., 2012).

9.2.1.2 *Cigarette Smoking*

A number of studies have shown an opposite correlation between the risk of PD and smoking. For example, smoking has been found to be protective against PD (pooled OR=0.23–0.70) in two meta-analysis studies (Breckenridge et al., 2016; Ritz et al., 2007). Further, an inverse correlation has been demonstrated between the number of cigarettes smoked per day, the number of years of smoking as well as the number of years since quitting smoking and the risk of PD. In contrast, the findings of a more recent large-scale case-control study has shown that patients with PD are able to quit smoking more easily than controls, suggesting that the negative association with smoking could be due to a decreased responsiveness to nicotine during the prodromal phase of PD (Ritz et al., 2014). Further, smoking tends to be a risk factor for rapid eye movement (REM) sleep behavior disorder, but not for the conversion to PD even though a trend for fewer pack-years and current smokers in converters to PD has been observed (Postuma et al., 2015).

9.2.1.3 *Beverages*

A 25% reduction in the risk of PD for the coffee users has been reported. Caffeine acts as an adenosine receptor antagonist, which makes it a putative protective factor against the risk of PD (Costa et al., 2010). Further, regular tea intake has been demonstrated to decrease the risk of PD. A Chinese study has shown no significant association between green tea and PD; however, the consumption of black tea, which is another source of caffeine, is reported to lower the risk of PD in women (Tan et al., 2008).

9.2.1.4 *Urate*

PD patients have low plasma uric acid levels in comparison with healthy individuals, which may be attributed to its consumption during the state of increased oxidative stress in the body. Urate acts as a scavenger for reactive oxygen species (ROS) and peroxynitrate, due to which it reduces the cellular death and mitochondrial dysfunction in cell cultures exposed to MPTP. MPTP, a potent inhibitor of mitochondrial complex I, increases the levels of pro-oxidative stress iron (II) (Fe^{2+}) and promotes glutamate excitotoxicity (Meredith et al., 2009; Shen and Ji, 2013). All these processes have been associated with the risk of PD through their established roles in the pathogenesis of the disease. The increase in plasma antioxidant capacity has been reported to correlate inversely with the progression of clinical signs or disease stages in PD upon administration of inosine—the precursor of uric acid (Bhattacharyya et al., 2016).

9.2.1.5 *Head Injury*

The pooled observational data from multiple studies have shown that head injury is of modest significance in the risk of PD (Gardner et al., 2015). Further, there is increasing evidence that individuals with recurrent head injuries, especially sportspersons, have an increased risk of developing various degenerative neurological conditions, including Parkinsonism. However, the pathogenesis in such cases differs from that of typical PD (Mckee et al., 2013).

9.2.1.6 *Drugs*

A number of observational studies have reported that various medications including calcium channel blockers (CCBs), non-steroidal anti-inflammatory drugs (NSAIDs) and statins have negative associations with PD (Noyce et al., 2012). For example, isradipine (CCB) has been reported to impart protection against the neurotoxicity to the dopaminergic cells in both in vitro and in vivo models of PD. Isradipine tends to shift the pacemaking of these neurons to sodium-dependent mode from being calcium dependent (Chan et al., 2007). Ibuprofen (NSAID) has been shown to reduce the risk of PD. It acts as a ligand of peroxisome proliferator-activated receptor gamma (PPARγ) or PPARγ agonist and exerts both antioxidative and anti-apoptotic effects (Samii et al., 2009; Chaturvedi and Beal, 2008). Likewise, non-aspirin NSAIDs have been reported to have a protective effect on PD risk, whereas no such effect has been observed for aspirin usage (Gagne and Power, 2010). The use of statins for cholesterol imbalance has been found to be inversely associated with the risk of PD, which may be due to their anti-inflammatory or antioxidant effect (Bai et al., 2016). Another study has reported a dose-effect relationship between the statin use time duration and risk of development of PD (Wahner et al., 2008).

9.2.1.7 *Metabolic Risk Factors*

The metabolic risk factors include various disorders such as diabetes, anemia, obesity and affective disorders (psychiatric disorders or mood disorders) such as depression and anxiety. It has been observed that there is either a positive or a negative relationship between the diabetes and the risk of PD and the results are inconsistent and inconclusive (Cereda et al., 2011; Lu et al., 2014; Yue et al., 2016). It is noteworthy that the

TABLE 9.1
Various Genes Involved in the Pathogenesis of Parkinson's Disease

Mendelian Inherited Gene (Monogenic Forms)		Non-Mendelian Genetic Transmission
Autosomal Dominant Transmission	**Recessive Transmission**	**Single Nucleotide Polymorphism**
SNCA	PRKN	SNCA
LRRK2	PINK1	MAPT
VPS35	DJ1	LRRK2
CHCHD2	ATP13A2	HLA-DRB5
EIF4G1	PLA2G6	BST1
	FBXO7	GAK
	DNAJC6	ACMSD

anti-diabetic drugs have been observed to reduce the incidence of PD in diabetics (Brauer et al., 2015; Svenningsson et al., 2016). No association between BMI, obesity, overweight and PD risk has been found; however, PD patients usually have a lower weight and BMI in comparison with healthy individuals (van der Marck et al., 2012; Wang et al., 2015). In addition, anemia has been commonly reported in PD patients in comparison with healthy individuals (Savica et al., 2009). Anxiety and pessimism have been reported to increase the risk of PD (Bower et al., 2010). Depression either is considered as an early sign of PD or has common etiology as that of PD (Fang et al., 2010). Furthermore, estrogen has been suggested to have an effect on the risk of PD; however, the results from different studies are inconsistent (Miller and Cronin-Golomb, 2010; Liu and Dluzen, 2007). Females who have had oophorectomy at younger age have been found to have a strong risk of developing PD (Rocca et al., 2008).

9.2.2 Genetic Risk Modifiers (Contributors) of Parkinson's Disease

Numerous studies have been carried out to understand the role of genetic factors in the initiation and progression of PD, and several genes, which lead to monogenic familial forms of PD, have been identified. Further, genome-wide association studies (GWAS) have identified single nucleotide polymorphisms (SNPs) in several genes, which have a strong association with an elevated risk of PD. Such genetic studies have led to a better understanding of molecular pathogenesis of PD and have identified PD-associated genetic factors, which are normally involved in common intracellular functions.

Mendelian as well as non-Mendelian gene mutations have been implicated in familial forms of PD, and some polymorphisms tend to increase the risk of PD upon their interaction with environmental factors. Table 9.1 summarizes the genes that are involved in the pathogenesis of PD.

9.3 Genetic Polymorphisms in Parkinson's Disease

A number of recent meta-analysis studies have reported the association of polymorphisms in various genes with the risk of PD. These genes include those that function in several cellular pathways involved in dopamine metabolism, detoxification mechanisms, oxidative stress, inflammatory processes and mitochondrial dysfunction and include genes with putative relation.

9.3.1 Gene Polymorphisms Involved in Dopamine Metabolism

Dopamine serves as one of the key neurotransmitters, which is synthesized in both central nervous system and the periphery and is known to mediate its action via GPCRs. Dopamine receptors belong to the GPCR superfamily, which includes different types of G-proteins. GPCRs are functionally linked to the enzyme adenylyl cyclase (AC), which converts adenosine triphosphate (ATP) to cyclic adenosine monophosphate (cyclic AMP/cAMP), a second messenger. There are five dopamine GPCRs: D1, D2, D3, D4 and D5, which are subdivided into two major groups based upon their structure and pharmacological properties: D1-like receptors (D1Rs), which include D1 and D5, and D2-like receptors (D2Rs), which include D2, D3 and D4 (Baik, 2013). Since all dopamine receptors belong to the GPCR superfamily, they act as metabotropic in producing downstream second messengers, which trigger or block the activation of specific cell signaling pathways (Baik, 2013; Beaulieu et al., 2015).

D1Rs (D1 and D5 receptors) are functionally linked to stimulatory G-proteins, G_s ($G\alpha_s$ or $G\alpha_{olf}$), and activate the production of cAMP, whereas D2Rs (D2, D3 and D4) are functionally linked to inhibitory G-proteins, G_i ($G\alpha_{i/o}$), and inhibit the production of cAMP (Neve et al., 2004). The activation and inhibition of cAMP by D1Rs and D2Rs, respectively, provides a bidirectional mechanism to modulate the activity of this and several other related pathways and several protein targets. These include Akt-GSK3 pathway (Klein et al., 2019), PAR4 signaling pathway (Park et al., 2005), protein kinase A (PKA) and dopamine- and cAMP-regulated neuronal phosphoprotein, Mr 32 kDa (DARPP-32/phosphoprotein phosphatase 1 regulatory subunit 1B/PPP1R1B) (Greengard et al., 1999). DARPP-32 acts as a main modulator of dopamine receptor signaling. The dopamine receptors also elicit their functions through interactions with $G\alpha_q$ and $G\beta_\gamma$ receptors and through crosstalk with other pathways such as mitogen-activated protein kinase (MAPK-MEK–ERK). Dopaminergic (DAergic) signaling pathways are crucial for the maintenance of normal physiological processes and are intricately linked to several other cellular signaling pathways, and any sort of imbalance may lead to dysfunctions, which may manifest as neurodegenerative diseases (Klein et al., 2019). For instance, the activation of Akt/protein kinase B and ERK signaling pathways are involved in the prevention of dopaminergic neuronal death in PD.

Several molecules are involved in the synthesis, degradation as well as transportation of dopamine. For example, solute carrier family 18 member 1 (SLC18A1/vesicular monoamine transporter member 1/VMAT1) and solute carrier family 18 member 2 (SLC18A2/vesicular monoamine transporter member 2/VMAT2) are involved in the transportation of dopamine from cytosol into the synaptic vesicles where it is stored until release into synaptic cleft. The dopamine transportation into vesicles utilizes proton gradient, which is maintained across the vesicular membrane.

Several SNPs identified in mediators involved in dopamine metabolism pathway have been reported to affect the biosynthesis, secretion and functions of dopamine. Various SNPs in SLC18A2 (VMAT2) gene (chromosomal location, 10q25.3) have been studied for any possible association with the risk of PD. The SNP rs363371 has been reported to significantly correlate with the risk of PD in Han Chinese men (C vs. A allele: OR, 0.799, (95% CI, 0.665–0.959); *p* value, 0.016), whereas the rs363224 SNP has not shown any association with the risk of PD (Yang et al., 2015b).

Several SNPs in the genes encoding dopamine receptor D2 *(DRD2)* (chromosomal location, 11q23.2) and dopamine receptor D3 *(DRD3)* (chromosomal location, 3q13.31) have been reported to modulate the risk of PD. A meta-analysis based on 16 studies has shown a significant association between *DRD2* and *DRD3* gene SNPs and the risk of PD. The study consisted of 4,279 cases and 5,661 controls for *DRD2* gene study and 4,279 cases and 5,661 controls for *DRD3* gene study. *DRD2* gene rs1800497 SNP has been found to marginally increase the risk of PD in Europeans [OR, 1.13 (95% CI, 1.00–1.27); *p* value, 0.05], whereas *DRD3* gene rs2134655 SNP has shown a significant association with an increased risk of PD [OR, 1.17 (95% CI, 1.03–1.32); p value, 0.01] (Dai et al., 2014a). In addition, rs2652510 SNP in solute carrier family 6 member 3 *(SLC6A3)* gene (chromosomal location, 5p15.33) has shown a significant association with an increased risk of PD in Caucasians [allelic G: OR, 1.26 (95% CI, 1.04–1.54); *p* value, 0.018; genotypic GG: OR, 1.37 (95% CI, 1.03–1.84); *p* value, 0.032] (Zhai et al., 2014).

Several meta-analysis studies have been conducted to evaluate the role of SNPs in monoamine oxidase (MAO) genes in modulating the risk of PD. MAO belongs to a family of flavin MAO that catalyzes the oxidation of dopamine. They are located together on the short arm of the X chromosome, between bands Xp11.23 and Xp11.4, and are composed of 15 exons. MAOs can be divided into MAO-A and MAO-B based on their substrate selectivity and sensitivity to inhibitors. MAO-A is mainly involved in the catabolism (oxidation) of epinephrine, norepinephrine, serotonin, tryptamine, tyramine and dopamine, whereas MAO-B is mainly involved in the catabolism (oxidation) of benzylamine, phenylethylamine, tryptamine, tyramine and dopamine. MAO-A is inhibited irreversibly by clorgyline, whereas MAO-B is inhibited irreversibly by rasagiline and selegiline (*l*-deprenyl) (Finberg and Rabey, 2016).

Various SNPs in *MAOA* and *MAOB* genes have been studied for any possible association with the risk of PD in different populations. A significant association of *MAOA* gene rs1137070 SNP with an increased risk of PD has been reported. In a study, which included 735 cases and 951 controls, an overall increased risk has been reported in the allelic, recessive and homozygous models (M vs. W allele: OR, 1.19 (95% CI, 1.01–1.41); *p* value, 0.036; MM vs. WM+WW: OR, 2.19 (95% CI, 1.21–3.97); *p* value, 0.009; MM vs. WW: OR, 1.94 (95% CI, 1.35–2.79); *p* value, <0.001). In the subgroup analysis, a significant association of the SNP with the risk of PD has been reported in the dominant, recessive, homozygous and heterozygous models in Asians (WM+MM vs. WW: OR, 0.65 (95% CI, 0.45–0.94); *p* value, 0.021; MM vs. WM+WW: OR, 3.08 (95% CI, 1.32–7.21); *p* value, 0.009; MM vs. WW: OR, 2.25 (95% CI, 1.31–3.84); *p* value, 0.003; MM vs. WM: OR, 4.79 (95% CI, 1.01–22.75); *p* value, 0.049) and in the homozygous model in Caucasians (MM vs. WW: OR, 1.72 (95% CI, 1.05–2.81); *p* value, 0.031) (Sun et al., 2014).

A significant association of *MAOB* gene rs1799836 SNP with an increased risk of PD has also been reported. In another study, which included 2,846 cases and 3,508 controls, an increased risk of PD has been reported in the recessive model [OR, 1.32 (95% CI, 1.18–1.47), *p* value, <0.001] (Liu et al., 2014). On the contrary, in another meta-analysis study that involved 2,156 cases and 2,249 controls, rs1799836 SNP has shown a significant association with a decreased risk of PD in the allelic, recessive, homozygous and heterozygous models in the overall study population and in Asians [G allele vs. A allele: OR, 0.84 (95% CI, 0.72–0.97), *p* value, 0.021; GG vs. AA+AG: OR, 1.73 (95% CI, 1.25–2.39), *p* value, 0.001; GG vs. AA: OR, 1.56 (95% CI, 1.18–2.04), *p* value, 0.002; GG vs. AG: OR, 3.42 (95% CI, 1.86–6.30), *p* value, <0.001], but no such association has been reported in Caucasians (Sun et al., 2014).

Another study, which included 2,279 cases and 2,911 controls, showed that *MAOB* gene rs1799836 SNP tends to increase the risk of PD in Asians carrying the rs1799836A allele (OR, 1.182 (95% CI, 1.012–1.380), *p* value<0.05). A combination study on *MAOB* gene rs1799836 SNP/catechol-O-methyltransferase *(COMT)* gene rs4670 SNP has shown an overall increased risk of PD in the carriers of rs1799836A and rs4670LL genotypes [AA/A vs. GG+GA/G: OR, 1.610 (95% CI, 1.094–2.369); AA/A vs. GA: OR, 1.621 (95% CI, 1.004–2.619)] (Zhang et al., 2016). However, no association of *COMT* gene rs4680 SNP with the risk of PD has been found in another study, which included 9,719 cases and 14,634 controls [OR= 0.99 (0.94–1.04)] (JIménez-Jiménez et al., 2014).

COMT gene (chromosomal location, 22q11.21) encodes COMT enzyme, which is involved in the inactivation of dopamine in substantia nigra. The gene polymorphisms including SNPs associated with *COMT* gene influence the concentration and activity of the COMT enzyme and increase the risk of developing motor fluctuations in PD patients. A meta-analysis on the association of *COMT* gene rs4680 SNP with the risk of PD shows variations based on ethnicity. A positive association between *COMT* gene Val158Met polymorphism and PD risk has been reported in Japanese and Indian populations (Wang et al., 2019).

Brain-derived neurotrophic factor *(BDNF)* gene (chromosomal location, 11p14.1) encodes BDNF protein, which promotes dopamine uptake of midbrain dopaminergic neurons. A meta-analysis has reported a lack of any association between *BDNF* gene rs2049045 (Val66Met) SNP and the risk of PD (Mariani et al., 2015).

Methylenetetrahydrofolate reductase *(MTHFR)* gene (chromosomal location, 1p36.22) encodes MTHFR protein, an essential enzyme involved in folate metabolism. It is the key enzyme to catalyze 5,10-methylenetetrahydrofolate into 5-methyltetrahydrofolate – the dominant form of circulating folate. 5-Methyltetrahydrofolate serves as the donor of methyl group for the remethylation of homocysteine (Hcy), resulting in the synthesis of methionine. Certain *MTHFR* gene polymorphisms have been associated with decreased MTHFR enzyme activity and thus elevated plasma Hcy level (Yuan et al., 2009; Castro et al., 2003). The decrease in MTHFR enzyme activity is a potential risk factor for cognitive decline. It can exert other neurotoxic effects as well, such as DNA damage, basal ganglia disturbances and neurotransmitter imbalance in motor circuits, which increases the risk of neuronal cell movement dysfunctions and can ultimately lead to PD (Fong et al., 2011; Yuan et al., 2009; Castro et al., 2003). A study on *MTHFR* gene rs1801133 (C677T) SNP, which included 1,606 cases and 7,208 controls, and rs1801131 SNP, which involved 891 cases and 820 controls, was carried out by Wu et al. (2013b). The study has shown the association of *MTHFR* gene rs1801133 SNP with an increased risk of PD in the allelic, dominant and recessive models in Asians and Caucasians [T allele vs. C allele: OR, 1.24 (95% CI, 1.11–1.38); TT+CT vs. CC: OR, 1.27 (95% CI, 1.10–1.46); TT vs. CC: OR, 1.56 (95% CI, 1.22–1.98); TT vs. CT+CC: OR, 1.43 (95% CI, 1.14–1.79)]. However, no association of *MTHFR* gene rs1801131 SNP with the risk of PD has been reported. Another study, which included 1,820 cases and 7,530 controls, has shown the association of *MTHFR* gene rs1801133 SNP with an increased risk of PD. An increased risk of PD has been observed in the allelic model (T allele vs. C allele) in overall study population [OR, 1.212 (95% CI, 1.097–1.340); *p* value, 0.0002], in Asians [OR, 1.293 (95% CI, 1.058–1.580); *p* value, 0.012] and in Caucasians [OR, 1.187 (95% CI, 1.058–1.332); *p* value, 0.004] (Zhu et al., 2013). However, a study, which included 2,690 cases and 8,465 controls, has shown a lack of association of *MTHFR* gene rs1801133 SNP with the risk of PD among Asians in all genotypic models, but has reported an increased risk of PD among Caucasians in the allelic model [T allele vs. C allele, OR, 1.17 (95% CI, 1.04–1.31)]. The study has also reported the association of T allele with increased homocysteine levels (Zhu et al., 2015b).

9.3.2 Gene Polymorphisms Involved in Detoxification Metabolism

The detoxification metabolism is an important mechanism of eliminating both exogenous and endogenous toxins from the body, occurs in different phases and involves several genes. The polymorphisms including SNPs in the genes involved in detoxification metabolism may result in altered activity or complete loss of enzyme function and their ability to metabolize toxins including neurotoxins and may cause neuronal cell death, thus leading to the risk of development of PD.

The cytochrome *P450 (CYP)* gene family encodes enzymes that represent the phase I detoxification metabolism and play an important role in the biotransformation of toxins. There are number of isoforms of the CYP enzymes. The CYP enzyme debrisoquine 4-hydroxylase CYP encoded by CYP family 2 subfamily D member 6 (*CYP2D6*) gene (chromosomal location, 22q13.2) is expressed in neurons and metabolizes endogenous and environmental neurotoxins. Importantly, individuals with PD have lower *CYP2D6* expression in the brain (Mann et al., 2012). Likewise, glutathione S-transferases (GSTs) represent phase II detoxification metabolism and genetic polymorphisms (deletion) of *GSTM* and *GSTT* genes result in loss of catalytic activity of enzymes and thus may cause neuronal cell death, which may increase the risk of development of PD.

A meta-analysis based on 22 studies, which included 2,629 cases and 3,601 controls, has shown a significant association of *CYP2D6* gene polymorphism, *CYP2D6*4*, with the risk of PD. An increased risk of PD has been observed in the overall study population [A allele vs. G allele: OR, 1.28 (95% CI, 1.14–1.43); *p* value, 0.001] and in Caucasians [A allele vs. G allele: OR, 1.28 (95% CI, 1.13–1.44); *p* value=0.001], but no such association has been reported in all genetic models among Asians (Lu et al., 2013).

Another meta-analysis based on 16 studies, which included 3,296 cases and 5,014 controls, has shown a significant association of *GSTM1* gene null polymorphism with the risk of PD. An increased risk of PD has been observed in the overall study population [OR, 1.13 (95% CI, 1.03–1.24); *p* value, <0.05] and in Caucasians [OR, 1.16 (95% CI, 1.04–1.29); *p* value, <0.05], but not in Asians. However, a lack of association with the risk of PD for *GSTT1* gene null allele has been observed (Wang and Wang, 2014). Another meta-analysis has reported the association of *GSTM1* gene null allele with an increased risk of PD in the overall study population [OR, 1.11 (95% CI, 1.01–1.21); *p* value, <0.05] and in Caucasians [OR, 1.15 (95% CI, 1.05–1.27); *p* value, <0.05)], but not in Asians (Weikang et al., 2016). Further, a meta-analysis study has shown a significant association of *GSTT1* gene null polymorphism with an increased risk of PD in Caucasians under recessive model [OR, 1.20 (95% CI, 1.04–1.39), *p* value, 0.014], but no significant association has been observed in case of overall study population (Wang et al., 2014).

The *ALDH2* gene (chromosomal location, 12q24.12) encodes a biologically important active enzyme, aldehyde dehydrogenase 2 (ALDH2), which is involved in the metabolism and detoxification of aldehyde (Chen et al., 2016). The SNPs in *ALDH2* gene have been associated with the risk of neurodegenerative disorders. An association study on *ADH2* gene rs4767944 and rs441 SNPs and the risk of PD has shown the association of the *ALDH2* gene rs4767944 C>T SNP with an increased risk of PD in the homozygous model, but not in other genetic models. However, no significant association has been observed in case of *ADH2* gene rs441 SNP (Chen et al., 2019).

9.3.3 Polymorphisms of Genes Involved in Inflammatory Processes

The neuroinflammation, defined as inflammation of neurons, is one of the primary processes responsible for PD. Some of the important inflammatory mediators, which have been implicated in the pathogenesis of PD, include the triggering receptor expressed on myeloid cells 2 (TREM2), tumor necrosis factor alpha (TNF-α) and Toll-like receptor 4 (TLR4). These inflammatory mediators play different roles and use different mechanisms to cause neuroinflammation and subsequent neuronal degeneration. However, the progression of the disease mostly occurs because of inflammation-derived oxidative stress and cytokine-dependent toxicity. For example, TNF-α through its action on dopaminergic neurons directly leads to neuronal degeneration. TNF-α can also function as a mild mitochondrial uncoupler and cause uncoupling, a process in which the potential generated across mitochondrial membrane is dissociated from or no longer used for mitochondrial ATP synthesis. This process lowers the ATP levels, increases the production of ROS and ultimately results in neuronal cell death (Prajapati et al., 2015). The polymorphisms including SNPs in the genes, which encode these and various other inflammatory mediators, contribute to pro- or anti-inflammatory profile predominance mainly through alteration in cytokine production and thus may result in faster or diminished dopaminergic neuronal degeneration and may ultimately modulate or modify the risk of PD (Zhao et al., 2015; Lindenau et al., 2017).

A meta-analysis study, which included 2,858 cases and 2,907 controls, has reported a significant association of *TNF-α* gene (chromosomal location, 6p21.33) SNPs with the risk of PD. A significant association of *TNF-α* gene rs1799964 (–1031T/C) SNP with an increased risk of PD has been observed in the recessive [OR, 3.19 (95% CI, 1.66–613); *p* value, 0.0005] and the additive models [OR, 3.15 (95% CI, 1.63–3.51); *p* value, 0.0006]. However, no significant association has been observed in case of *TNF-α* gene rs1800629 (–308G/A) (Dai et al., 2014b). Another meta-analysis has reported a significant association of *TNF-α* gene rs1799964 SNP with an increased risk of PD in the overall study population [CC vs. TT: OR, 3.146 (95% CI, 1.631–6.070); *p* value, 0.008; CC vs. CT+TT: OR, 3.187 (95% CI, 1.657–6.128); *p* value, 0.008]. In the same study, a significant association of TNF-α gene rs1799964 SNP with an increased risk of PD has been reported in Asians [C allele vs. T allele: OR, 1.328 (95% CI, 1.053–1.675), *p* value, 0.034; CC vs. TT: OR, 3.207 (95% CI, 1.614–6.373), *p* value, 0.004; CC vs. CT+TT: OR, 3.238 (95% CI: 1.636–6.410), *p* value, 0.004] (Chu et al., 2012).

Various SNPs in the *TLR4* gene (chromosomal location, 9q33.1) and *TLR9* gene (chromosomal location, 3p21.2) have been reported to modulate the risk of PD. The rs1927911, rs1927914 and rs10116253 SNPs in *TLR4* gene and the rs352140 and rs187084 SNPs in *TLR9* gene have been studied for their association with the risk of PD. In a study on Han Chinese population, the *TLR4* gene rs1927914C allele has been associated with a decreased risk of PD (*p* value, 0.028) and early-onset PD or EOPD (*p* value, 0.023) and a decreased risk of PD has also been reported in the carriers of *TLR4* gene rs1927914C allele (*p* value, 0.032), especially males (*p* value, 0.034). The study has reported significant differences in males regarding the distribution of *TLR4* gene rs1927914 alleles (*p* value, 0.012) and *TLR4* gene rs1927914 genotypes (*p* value, 0.035) in comparison with healthy controls. The study indicated the protective role of *TLR4* gene rs1927914 SNP against sporadic PD and EOPD, especially in males. However, no significant association between *TLR4* gene rs1927911 and rs10116253 SNPs and the risk of PD has been reported (Zhao et al., 2015). In another study on Han Chinese population, which included 380 cases and 380 controls, a decreased risk of PD has been reported in carriers of *TLR9* gene rs352140T allele (TT+TC vs. CC, *p* value, 0.018). Further, the study has reported significant differences in females regarding the distribution of *TLR9* gene rs352140 alleles (*p* value, 0.040). The study indicated the protective role of *TLR9* gene rs352140 SNP against sporadic PD, especially in females. However, no significant association between *TLR9* gene rs187084 SNPs and the risk of PD has been reported in subgroup analysis regarding age and gender (Zhu et al., 2015).

A meta-analysis on rs75932628 SNP in *TREM2* gene (chromosomal location, 6p21.1), which included 39 studies, has reported a significant association of this SNP with an increased risk of PD among North Americans [OR, 3.59 (95% CI, 1.33–9.70); I^2, 0%]. However, no significant association between *TREM2* gene rs75932628 SNP and the risk of PD has been reported among Europeans [OR, 1.20 (95% CI, 0.83–1.72); I^2, 0%] (Zhou et al., 2019). Further, in another study on Han Chinese population, rs1801105 (Thr105Ile/314C/T) SNP in histamine N-methyltransferase *(HNMT)* gene (chromosomal location, 2q22.1) has shown a significant association with a decreased risk of PD [Ile vs. Thr allele: OR, 0.516 (95% CI, 0.318–0.838), *p* value, 0.007] (Yang et al., 2015).

9.3.4 Gene Polymorphisms Involved in Oxidative Stress

The characterization of chronic and progressive neurodegeneration of dopaminergic neurons has established oxidative stress as one of the most important processes in the etiology or pathogenesis of PD. The imbalance between the biological detoxification system and ROS production is a common factor underlying the neuronal cellular damage, which underlies the neurodegenerative disorder (Puspita et al., 2017).

Superoxide dismutase (SOD) gene acts as an important antioxidant defense mechanism against oxidative stress by catalyzing the dismutation of superoxide (O_2^{2-}) into oxygen and hydrogen peroxide. SOD genes SOD1 (chromosomal location, 21q22.11), SOD2 (chromosomal location, 6q25.3) and SOD3 (chromosomal location, 4p15.2) are highly polymorphic in nature with certain polymorphisms reported to be associated with serious diseases (Kim et al., 2015). SNPs in SOD genes include *SOD1* (*rs*2070424 A/G), *SOD2* (*rs*4880 T/C) and *SOD3* (*rs*1799895, C/G). A case-control study has reported a significant association of *SOD1* gene rs2070424 AG or GG genotypes and G allele with an increased risk of PD in comparison with *SOD1* gene rs2070424 AA genotype and A allele. Further, the study reported a significant association of *SOD2* gene rs4880 TC or CC genotypes and C allele with an increased risk of PD in comparison with *SOD2* gene rs4880 TT genotype and T allele. It has been reported that together, the carriers of *SOD1* gene rs2070424 G allele and *SOD2* gene rs4880C allele are at an enhanced genetic susceptibility toward PD (Liu et al., 2019). Another study has reported a significantly higher frequency of *SOD2* rs4880 CC genotype in PD cases in comparison with healthy controls. The study reported an increased risk of PD among the carriers of *SOD2* gene rs4880 CC genotype in the overall model [OR, 2.48 (95% CI, 1.03–5.99); *p* value, 0.04] as well as in the recessive model [OR, 2.61 (95% CI, 1.08–6.30); *p* value, 0.03] (Wu et al., 2021).

A meta-analysis on SNPs in nuclear factor erythroid-derived 2-like 2 *(NFE2L2)* gene (chromosomal location, 2q31.2) has reported a significant association of *NFE2L2* gene rs7557529 G/A, rs35652124 A/G, rs2886161 A/G and rs1806649 G/A SNPs with age at onset (AAO) of PD [(−1.0 year per A allele, *p* value, 0.042 for rs7557529 SNP), (−1.1 year per G allele, *p* value, 0.045 for rs35652124 SNP), (−1.2 year per G allele, *p* value, 0.021 for rs2886161 SNP) and (+1.2 year per A allele, *p* value, 0.029 for rs1806649 SNP)] (von Otter et al., 2014). However, no significant association between *NFE2L2* gene SNPs including rs35652124 A/G SNP and the risk of PD has been reported [G allele vs. A allele: OR, 1.001; *p* value, 0.986; GG vs. AA: OR, 1.026; *p* value, 0.785; GA vs. AA: OR, 1.023; *p* value, 0.786; GG+GA vs. AA: OR, 1.224; *p* value, 0.070; GG vs. GA+AA: OR, 0.994; *p* value, 0.926] (Zhu et al., 2016).

9.3.5 Gene Polymorphisms in Mitochondrial Function Pathway

The mitochondrial dysfunction is one of the important processes in the pathogenesis of PD (Bose and Beal, 2016). Various studies have provided novel insights into the role of disturbances in mitochondrial function in the pathogenesis of PD. The base pair 5178C/A transversion in the mitochondrial DNA has shown a significant association with the processes that promote longevity, including decreased production of mitochondrial ROS and reduced risk of several age-related diseases. In a study on Han Chinese population, which included 484 cases and 710 controls, the base pair 5178C/A transversion in the mitochondrial DNA has been associated with an increased risk of PD in males (*p* value, 0.011), but not in the overall study population (Gusdon et al., 2015). Besides rare genetic variants, which interfere with normal mitochondrial functions and confer substantial risk of development of typical PD, more common variants have been reported, for example, in the glucosylceramidase beta *(GBA)* or *LRRK2* genes, which may interfere with the mitochondrial homeostasis. Studies have reported a higher frequency of G2385R and R1628P variants of *LRRK2* gene among Asians, where these variants have been associated with two to three times increased risk of PD (Wu et al., 2013; Gopalai et al., 2014). With a frequency of 3%–5% in the Asian PD population, these two variants can be considered as the most common risk variants for the development of PD in East Asia (Pulkes et al., 2014).

9.3.6 Gene Polymorphisms of Trophic Factors

The trophic factors, produced by brain parenchyma, which includes neurons and glial cells, play a vital role in the development, differentiation, maintenance, regeneration and function of nervous tissue. The 22-member fibroblast growth factor (FGF) family includes some of the major trophic factors, which play crucial roles in the development and metabolism in brain cells. FGF20 is the major trophic factor expressed specifically in substantia nigra compacta region of the brain and is found to be responsible for the enhancement of the midbrain dopaminergic neuronal survival (Ohmachi et al., 2003). It serves as a neurotrophic factor within the dopaminergic cells via binding to its receptor, fibroblast growth factor receptor 1c (FGFR1c), and accelerates the phosphorylation of its specific cytoplasmic tyrosine residues. The phosphorylation of FGFR1c triggers the activation of the MAPK pathway, which plays important roles in the survival of dopaminergic neurons (Ohmachi et al., 2003) and seems to increase ASN protein levels in dopaminergic neurons (Rideout et al., 2003). Thus, genetic variations in *FGF20* gene may substantially influence the risk of PD. There are two known SNPs, rs1721100 and rs12720208 located in the 3′-UTR of *FGF20* gene (chromosomal location, 8p22), that have shown a significant association with the risk of PD. A meta-analysis study has shown a significant association between *FGF20* gene rs1721100 SNP and an increased risk of PD in the recessive model [GG versus CG/GG: OR, 1.15 (95% CI, 1.02–1.29), *p* value, 0.02], but not in the dominant model [CG/GG versus CC: OR, 1.03 (95% CI, 0.93–1.13), *p* value, 0.57]. The study therefore suggested that *FGF20* gene rs1721100GG genotype is a risk factor for PD (Zhu et al., 2014). Another meta-analysis study has shown a significant association of *FGF20* gene rs12720208 SNP with an increased risk of PD in allelic model [T allele vs. C allele: OR, 1.167 (95% CI, 1.020–1.335)] and in dominant model [TT+TC vs. CC: OR, 1.156 (95% CI, 1.001–1.335)] in Caucasians, but not in Asians (Wang et al., 2017). In addition, another study, which included 520 cases and 520 controls from Iran, has reported a significant association of the *FGF20* gene rs12720208 SNP with an increased risk of PD in allelic model [T allele vs. C allele: OR, 2.85 (95% CI, 1.80–4.51), *p* value, <0.0001] (Haghnejad et al., 2015). Further, the association of rs1058508 and rs1136201 SNPs in erb-b2 receptor tyrosine kinase 2 *(ERBB2/HER2)* gene (chromosomal location, 17q12) with the risk of PD has been evaluated. A significant association between the minor allele homozygous genotype 1170Ala/Ala of *ERBB2* gene rs1058508 SNP and an increased risk of PD, especially among females (*p* value, 0.014), has been reported. However, no significant association between *ERBB2* gene rs1136201 SNP and the risk of PD has been reported (Wang et al., 2013).

9.3.7 Other Gene Polymorphisms (Polymorphisms in Genes with Putative Relation)

Various genes expressed in the neuronal cell-rich regions of the brain are involved in the pathophysiology of several neurodegenerative diseases, including PD. One such neurotrophic factor is insulin-like growth factor 1 (IGF1). IGF1 enhances the phosphorylation levels of glycogen synthase kinase-3β (GSK3β) and activates the phosphatidylinositol-3-kinase (PI3K)/AKT pathway. It also improves the motor dysfunction and behavior abnormalities caused by 6-hydroxydopamine neurotoxicity, thus making it an important candidate gene in the modification of

PD risk (Cheng et al., 2011; Ayadi et al., 2016). A study on Han Chinese population, which included 512 sporadic PD cases and 535 controls, has reported a significant association of *IGF1* gene rs972936CC genotype with an increased risk of PD (*p* value, 0.009), especially in males (*p* value, 0.024) and with LOPD (*p* value, 0.013) (Xiao et al., 2017).

Vitamin D (1,25-dihydroxyvitamin D3) receptor (VDR) gene (chromosomal location, 12q13.11) is expressed in various body tissues, including brain (Eyles et al., 2005). It encodes a transcription factor, which is found to influence the calcium homeostasis, neurotrophic signaling, cellular growth as well as differentiation and regulation of immune system. It may also participate in the development of neurological disorders, including PD (Kato, 2000; Kesby et al., 2011). The outcomes of various studies, which investigated the association between *VDR* gene SNPs and the risk of PD, have been different. A study has reported a significant association between *VDR* gene rs2228570 SNP and an increased risk of PD in Asians [C allele vs. T allele: OR, 1.261 (95% CI, 1.080–1.472), *p* value, 0.003] (Gao et al., 2020).

The transcription factor paired-like homeodomain transcription factor 3 (PITX3) encoded by *PITX3* gene (chromosomal location, 10q24.32) belongs to pituitary homeobox subfamily and plays a vital role in the development of brain (Li et al., 2009). Precisely, PITX is essential for the differentiation and maintenance of dopaminergic neurons in midbrain (Hwang et al., 2009; Papanikolaou et al., 2009). The genetic polymorphisms including SNPs in the *PITX3* gene can lead to the loss of dopaminergic neurons, resulting in PD. A meta-analysis study based on ten studies, which included 5,172 cases and 7,290 controls, has reported a significant association of A allele of *PITX3* gene rs4919621 SNP with an increased risk of PD in Caucasians (*p* value, 0.04). Further, the C allele of *PITX3* gene rs2281983 SNP and the A allele of *PITX3* gene rs4919621 SNP have shown a significant association with the risk of EOPD (*p* value, ≤0.0001) (Tang et al., 2012).

9.4 Conclusions

Based on the results from different population studies, it has been suggested that there occur differences in the association of certain polymorphisms with the risk of Parkinson's disease. Thus, ethnicity seems to play an important role in the genetic association of PD. For example, there is an elevated risk of PD in Caucasians in relation to genes involved in detoxification processes, such as *GSTT1*, *GSTM1* and *CYP2D6*. Likewise, there is an increased risk of PD in Chinese Han population carrying the rs1564282 SNP in the cyclin G-associated kinase *(GAK)* gene in *PARK17* locus (chromosomal location, 4p16.3), while the same SNP lacks any sort of association in the Spanish population. Further, *COMT* gene Val158Met polymorphism is more frequent in the Japanese and Indian populations compared with other ethnicities. Moreover, the alpha-synuclein (*SNCA)* rs356220 polymorphism has been identified in both Caucasian and Asian populations. *PARK16* rs823128 variation and the PD risk has been shown to be significantly associated in Asian and Chilean populations, but not in Caucasians.

Several polymorphisms in genes associated with familial PD have been reported to modify the risk of PD. These include *PARK2* gene rs1801582 SNP (decreased PD risk); *PARK16* rs823128, rs947211, rs823156 and rs11240572 SNPs (decreased PD risk in Asians); *PARK5* rs5030732 (167G<A) SNP; Grb10-interacting GYF protein 2 (*GIGYF2*) gene 167G<A SNP; *GIGYF2* gene 3630A<G SNP; and *GBA* mutations (increased PD risk). Through different pathways, several genes have shown a positive association with PD risk, including genes involved in the dopamine metabolism, such as *DRD3*, *SLC6A3*, *MAOA*, *MAOB* and *MTHFR*; genes involved in oxidative stress, such as HFE gene; and the genes involved in inflammatory processes such as *TNFα* and *A2M* genes.

Among other acknowledged relations, an important association has been found between PD risk and estrogen receptor 1 *(ESR1)* gene rs2234693 (–397T>C) SNP, *ESR1* gene rs9340799 (–351A>G) SNP, *ESR2* gene *rs4986938* (1730G>A) SNP (increased risk), *VDR* gene rs2228570 SNP (increased risk) and granulin precursor *(GRN)* gene rs5848 SNP (increased risk). Moreover, microtubule-associated protein tau *(MAPT)* gene 238 bp polymorphism (decreased risk), saitohin *(STH)* gene Q7R polymorphism (decreased risk), semaphorin 5A *(SEMA5A)* gene rs7702187 SNP (decreased risk in Western populations), *GSK3β* gene rs334558 SNP and *GSK3β* gene rs6438552 (decreased risk in Asians) showed association with PD.

ABBREVIATIONS

AAO:	Age at onset
ACh:	Acetylcholine
ASN:	Alpha-synuclein
CCBs:	Calcium channel blockers
DNA:	Deoxyribonucleic acid
FGF:	Fibroblast growth factor
FGFR:	Fibroblast growth factor receptor
GABA:	Gamma-aminobutyric acid
GPCRs:	G protein-coupled receptors
GWAS:	Genome-wide association studies
K:	Thousand
LRRK2:	Leucine-rich repeat kinase 2
M:	Million
MAPK:	Mitogen-activated protein kinase
MDS:	Movement Disorders Society
NO:	Nitric oxide
NSAIDs:	Non-steroidal anti-inflammatory drugs
OR:	Odds ratio
PD:	Parkinson's disease
PPARγ:	Peroxisome proliferator-activated receptor gamma
PRKN:	Parkin gene
ROS:	Reactive oxygen species
rs:	Reference SNP cluster ID
SN:	Substantia nigra
SNCA:	Alpha-synuclein gene
SNP:	Single nucleotide polymorphism
SOD:	Superoxide dismutase

TLR4: Toll-like receptor 4
TNF-α: Tumor necrosis factor alpha
TREM2: Triggering receptor expressed on myeloid cells 2
UTR: Untranslated region
VDR: Vitamin D receptor

REFERENCES

Ascherio A, Schwarzschild MA. The epidemiology of Parkinson's disease: risk factors and prevention. *Lancet Neurol* 2016; 15(12):1257–72. doi: 10.1016/S1474-4422(16)30230-7.

Ayadi AE, Zigmond MJ, Smith AD IGF-1 protects dopamine neurons against oxidative stress: association with changes in phosphokinases. *Exp Brain Res* 2016;234:1863–73. doi: 10.1007/s00221-016-4572-1.

Bai S, Song Y, Huang X, et al. Statin use and the risk of Parkinson's disease: an updated meta-analysis. *PLoS One* 2016;11(3):e0152564

Baik JH. Dopamine signaling in reward-related behaviors. *Front Neural Circuits* 2013;7:152. doi: 10.3389/fncir.2013.00152.

Beaulieu JM, Espinoza S, Gainetdinov RR. Dopamine receptors - IUPHAR Review 13. *Br J Pharmacol.* 2015 Jan;172 (1):1–23. doi: 10.1111/bph.12906.

Bekris LM, Mata IF, Zabetian CP. The genetics of Parkinson's disease. *J Geriatr Psychiatry Neurol* 2010;23(4):228–42.

Berg D, Postuma RB, Adler CH, et al. MDS research criteria for prodromal Parkinson's disease. *Mov Disord* 2015;30:1600–11.

Bhattacharyya S, Bakshi R, Logan R, Ascherio A, Macklin EA, Schwarzs child MA. Oral inosine persistently elevates plasma antioxidant capacity in Parkinson's disease. *Mov Disord* 2016;31(3):417–21.

Blair HA, Dhillon S. Safinamide: a review in Parkinson's disease. *CNS Drugs* 2017;31:169–76. https://doi.org/10.1007/s40263-017-0408-1

Bose A, Beal MF. Mitochondrial dysfunction in Parkinson's disease. *J Neurochem* 2016;139 Suppl 1:216–31 doi: 10.1111/jnc.13731.

Bower JH, Grossardt BR, Maraganore DM, et al. Anxious personality predicts an increased risk of Parkinson's disease. *Mov Disord* 2010;25(13):2105–13.

Brauer R, Bhaskaran K, Chaturvedi N, Dexter DT, Smeeth L, Douglas I. Glitazone treatment and incidence of Parkinson's disease among people with diabetes: a retrospective cohort study. *PLoS Med* 2015;12(7):e1001854.

Breckenridge CB, Berry C, Chang ET, Sielken Jr RL, Mandel JS. Association between Parkinson's disease and cigarette smoking, rural living, well-water consumption, farming and pesticide use: systematic review and meta- analysis. *PLoS One* 2016;11(4):e0151841

Castro R, Rivera I, Ravasco P, et al. 5,10-Methylenetetrahydrofolate reductase 677C–NT and 1298A–NC mutations are genetic determinants of elevated homocysteine. *QJM* 2003;96:297–303.

Cereda E, Barichella M, Pedrolli C, et al. Diabetes and risk of Parkinson's disease: a systematic review and meta-analysis. *Diabetes Care* 2011;34(12):2614–23.

Chan CS, Guzman JN, Ilijic E, et al. 'Rejuvenation' protects neurons in mouse models of Parkinson's disease. *Nature* 2007;447(7148):1081–6.

Chaturvedi RK, Beal MF. PPAR: a therapeutic target in Parkinson's disease. *J Neurochem* 2008;106(2):506–18.

Chen J, Huang W, Cheng CH, Zhou L, Jiang GB, Hu YY. Association Between *Aldehyde dehydrogenase-2* polymorphisms and risk of Alzheimer's disease and Parkinson's disease: a meta-analysis based on 5,315 individuals. *Front Neurol* 2019;10:290. doi: 10.3389/fneur.2019.00290.

Chen CH, Joshi AU, Mochly-Rosen D. The role of mitochondrial aldehyde dehydrogenase 2 (ALDH2) in neuropathology and neurodegeneration. *Acta Neurol Taiwan* 2016;25:111–23.

Cheng B, Maffi SK, Martinez AA, Acosta YP, Morales LD, Roberts JL Insulin-like growth factor-I mediates neuroprotection in proteasome inhibition-induced cytotoxicity in SH-SY5Y cells. *Mol Cell Neurosci* 2011;47:181–90. doi: 10.1016/j.mcn.2011.04.002.

Chu K, Zhou X, Luo BY. Cytokine gene polymorphisms and Parkinson's disease: a meta-analysis. *Can J Neurol Sci* 2012;39(1):58–64. doi: 10.1017/s0317167100012695.

Cieri D, Brini M, Calì T. Emerging (and converging) pathways in Parkinson's disease: keeping mitochondrial wellness. *Biochem Biophys Res Commun* 2017;483:1020–30.

Costa J, Lunet N, Santos C, Santos J, Vaz- Carneiro A. Caffeine exposure and the risk of Parkinson's disease: a systematic review and meta-analysis of observational studies. *J Alzheimers Dis* 2010;20(Suppl 1):S221–38.

Dai D, Wang Y, Wang L, et al. Polymorphisms of DRD2 and DRD3 genes and Parkinson's disease: a meta-analysis. *Biomed Rep* 2014a;2:275–81.

Dai D, Lin P, Wang Y, et al. Association of NQO1 and TNF polymorphisms with Parkinson's disease: a meta-analysis of 15 genetic association studies. *Biomed Rep* 2014b;2:713–8.

Dauer W, Przedborski S. Parkinson's disease: mechanisms and models. *Neuron* 2003;39(6):889–909.

Doherty GH. Homocysteine and Parkinson's disease: a complex relationship. *J Neurol Disord* 2013;1:107.

Elbaz A, Clavel J, Rathouz PJ, et al. Professional exposure to pesticides and Parkinson disease. *Ann Neurol* 2009;66(4):494–504.

Eyles DW, Smith S, Kinobe R, Hewison M, McGrath JJ. Distribution of the vitamin D receptor and 1 alpha-hydroxylase in human brain. *J Chem Neuroanat* 2005;29:21–30.

Fang F, Xu Q, Park Y, et al. Depression and the subsequent risk of Parkinson's disease in the NIH-AARP Diet and Health Study. *Mov Disord* 2010;25(9):1157–62.

Ferreira M, Massano J. An updated review of Parkinson's disease genetics and clinicopathological correlations. *Acta Neurol Scand* 2017;135(3):273–84. doi: 10.1111/ane.12616. Epub 2016 Jun 8.

Finberg JP, Rabey JM. Inhibitors of MAO-A and MAO-B in Psychiatry and Neurology. *Front Pharmacol* 2016;7:340. doi:10.3389/fphar.2016.00340.

Fong CS, Shyu HY, Shieh JC, et al. Association of MTHFR, MTR, and MTRR polymorphisms with Parkinson's disease among ethnic Chinese in Taiwan. *Clin Chim Acta* 2011;412:332–8.

Frisardi V, Santamato A, Cheeran B. Parkinson's disease: new insights into pathophysiology and rehabilitative approaches. *Parkinson's Dis* 2016:2016:3121727.

Gagne JJ, Power MC. Anti-inflammatory drugs and risk of Parkinson disease: a meta-analysis. *Neurology* 2010;74(12):995–1002.

Gao J, Teng J, Liu Z, Cai M, Xie A. Association between vitamin D receptor polymorphisms and susceptibility to Parkinson's disease: an updated meta-analysis. *Neurosci Lett* 2020;720:134778. doi: 10.1016/j.neulet.2020.134778.

Gardner RC, Burke JF, Nettiksimmons J, et al. Traumatic brain injury in later life increases risk for Parkinson disease. *Ann Neurol* 2015;77:987–95.

Gopalai AA, Lim SY, Chua JY, Tey S, Lim TT, Mohamed Ibrahim N, Tan AH, Eow GB, Abdul Aziz Z, Puvanarajah SD, Viswanathan S, Looi I, Lim SK, Tan LP, Chong YB, Tan CT, Zhao Y, Tan EK, Ahmad-Annuar A. LRRK2 G2385R and R1628P mutations are associated with an increased risk of Parkinson's disease in the Malaysian population. *Biomed Res Int* 2014;2014:1–4.

Greengard P, Allen PB, Nairn AC. Beyond the dopamine receptor: the DARPP-32/protein phosphatase-1 cascade. *Neuron* 1999;23:435–47. doi: 10.1016/S0896-6273(00)80798-9.

Gurevich EV, Gainetdinov RR, Gurevich VV. G protein-coupled receptor kinases as regulators of dopamine receptor functions. *Pharmacol Res* 2016;111:1–6.

Gusdon AM, Fang F, Chen J, et al. Association of the mt-ND2 5178A/C polymorphism with Parkinson's disease. *Neurosci Lett* 2015;587:98–101.

Haghnejad L, Emamalizadeh B, Jamshidi J, et al. Variation in the miRNA-433 binding site of FGF20 is a risk factor for Parkinson's disease in Iranian population. *J Neurol Sci* 2015;355:72–4.

Hama Y, Yabe I, Wakabayashi K et al. Level of plasma neuregulin-1 SMDF is reduced in patients with idiopathic Parkinson's disease. *Neurosci Lett* 2015;587:17–21.

Hirsch L, Jette N, Frolkis A, Steeves T, Pring-sheim T. The incidence of Parkinson's disease: a systematic review and meta-analysis. *Neuroepidemiology* 2016;46(4):292–300.

Hwang DY, Hong S, Jeong JW, et al. Vesicular monoamine transporter 2 and dopamine transporter aremolecular targets of Pitx3 in the ventral midbrain dopamine neurons. *J Neurochem* 2009;111:1202–12.

Jiménez-Jiménez FJ, Alonso-Navarro H, García-Martín E, Agúndez JA. COMT gene and risk for Parkinson's disease: a systematic review and meta-analysis. *Pharmacogenet Genom* 2014;24:331–9.

Kalia LV, Lang AE. Parkinson's disease. *Lancet* 2015;386(9996): 896–912.

Kato S. The function of vitamin D receptor in vitamin D action. *J Biochem* 2000;127:717–22.

Katritch V, Cherezov V, Stevens RC. Diversity and modularity of G protein coupled receptor structures. *Trends Pharmacol Sci* 2012;33:17–27.

Kesby JP, Eyles DW, Burne TH, McGrath JJ. The effects of vitamin D on brain development and adult brain function. *Mol Cell Endocrinol* 2011;347:121–7.

Kim SH, Kim SH, Lee JH, Lee BH, Yoon HJ, Shin DH, Park SS, Jang SB, Park JS, Jee YK. Superoxide dismutase gene (SOD1, SOD2, and SOD3) polymorphisms and antituberculosis drug-induced hepatitis. *Allergy Asthma Immunol Res* 2015;7(1):88–91. doi: 10.4168/aair.2015.7.1.88.

Klein MO, Battagello DS, Cardoso AR, Hauser DN, Bittencourt JC, Correa RG. Dopamine: functions, signaling, and association with neurological diseases. *Cell Mol Neurobiol* 2019;39(1):31–59. doi: 10.1007/s10571-018-0632-3.

Lee A, Gilbert RM. Epidemiology of Parkinson disease. *Neurol Clin* 2016;34:955–65.

Lesage S, Dürr A, Tazir M, et al. LRRK2 G2019S as a cause of Parkinson's disease in North African Arabs. *N Engl J Med* 2006;354:422–3.

Li J, Dani JA, Le W. The role of transcription factor Pitx3 in dopamine neuron development and Parkinson's disease. *Curr Top Med Chem* 2009;9:855–9.

Lindenau JD, Altmann V, Schumacher-Schuh AF, Rieder CR, Hutz MH. Tumor necrosis factor alpha polymorphisms are associated with Parkinson's disease age at onset. *Neurosci Lett* 2017;658:133–6. doi: 10.1016/j.neulet.2017.08.049.

Liu B, Dluzen DE. Oestrogen and nigrostriatal dopaminergic neurodegeneration: animal models and clinical reports of Parkinson's disease. *Clin Exp Pharmacol Physiol* 2007;34(7):555–65.

Liu C, Fang J, Liu W. Superoxide dismutase coding of gene polymorphisms associated with susceptibility to Parkinson's disease. *J Integr Neurosci* 2019;18(3):299–303. doi: 10.31083/j.jin.2019.03.127.

Liu Y, Wang Z, Zhang B. The relationship between monoamine oxidase B (MAOB) A644G polymorphism and Parkinson disease risk: a meta-analysis. *Ann Saudi Med* 2014;34:12–7.

Liu WM, Wu RM, Lin JW, Liu YC, Chang CH, Lin CH. Time trends in the prevalence and incidence of Parkinson's disease in Taiwan: a nationwide, population-based study. *J Formos Med Assoc* 2016;115(7):531–8. https://doi.org/10.1016/j.jfma.2015.05.014

Lu L, Fu DL, Li HQ, Liu AJ, Li JH, Zheng GQ. Diabetes and risk of Parkinson's disease: an updated meta-analysis of case-control studies. *PLoS One* 2014;9(1):e85781.

Lu Y, Mo C, Zeng Z, et al. CYP2D6*4 allele polymorphism increases the risk of Parkinson's disease: evidence from meta-analysis. *PLoS One* 2013;8:e84413.

Mahlknecht P, Seppi K, Poewe W. The concept of prodromal Parkinson's disease. *J Parkinson's Dis* 2015;5:681–97.

Mann A, Miksys SL, Gaedigk A, Kish SJ, Mash DC, Tyndale RF. The neuroprotective enzyme CYP2D6 increases in the brain with age and is lower in Parkinson's disease patients. *Neurobiol Aging* 2012;33:2160–71.

Mariani S, Ventriglia M, Simonelli I, Bucossi S, Siotto M, R RS. Meta-analysis study on the role of bone-derived neurotrophic factor Val66Met polymorphism in Parkinson's disease. *Rejuvenation Res* 2015;18(1):40–7. doi: 10.1089/rej.2014.1612.

Marshall LJ, Willett C. Parkinson's disease research: adopting a more human perspective to accelerate advances. *Drug Discov Today* 2018;23:1950–61.

Maserejian N, Vinikoor-Imler L, Dilley A. Estimation of the 2020 global population of Parkinson's disease (PD) [abstract]. *Mov Disord* 2020;35(suppl 1).

Mazzoni P, Shabbott B, Corte´s JC. Motor control abnormalities in Parkinson's disease. *Cold Spring Harb Perspect Med* 2012;2:a009282.

McKee AC, Stein TD, Nowinski CJ, et al. The spectrum of disease in chronic traumatic encephalopathy. *Brain* 2013;136:43–64.

Meredith GE, Totterdell S, Beales M, Meshul CK. Impaired glutamate homeostasis and programmed cell death in a chronic MPTP mouse model of Parkinson's disease. *Exp Neurol* 2009;219(1):334–40. doi: 10.1016/j.expneurol.2009.06.005

Miller IN, Cronin-Golomb A. Gender differences in Parkinson's disease: clinical characteristics and cognition. *Mov Disord* 2010;25(16):2695–703.

Neve KA, Seamans JK, Trantham-Davidson H. Dopamine receptor signaling. *J Recept Signal Transduct Res* 2004;24:165–205. doi: 10.1081/RRS-200029981.

Noyce AJ, Bestwick JP, Silveira-Moriyama L, et al. Meta-analysis of early non-motor features and risk factors for Parkinson disease. *Ann Neurol* 2012;72:893–901.

Ohmachi S, Mikami T, Konishi M, Miyake A, Itoh N. Preferential neurotrophic activity of fibroblast growth factor-20 for dopaminergic neurons through fibroblast growth factor receptor-1c. *J Neurosci Res* 2003;72(4):436–43. doi: 10.1002/jnr.10592.

Ozelius LJ, Senthil G, Saunders-Pullman R, et al. LRRK2 G2019S as a cause of Parkinson's disease in Ashkenazi Jews. *N Engl J Med* 2006;354:424–5.

Papanikolaou T, Amano T, Lennington J, et al. In-vitro analysis of Pitx3 in mesodiencephalic dopaminergic neuron maturation. *Eur J Neurosci* 2009;29:2264–75.

Park SK, Nguyen MD, Fischer A, Luke MP, Affar el B, Dieffenbach PB, Tseng HC, Shi Y, Tsai LH. Par-4 links dopamine signaling and depression. *Cell* 2005;122:275–87. doi: 10.1016/j.cell.2005.05.031.

Politis M, Niccolini F. Serotonin in Parkinson's disease. *Behav Brain Res* 2015;277:136–45.

Postuma RB, Iranzo A, Hog IB, et al. Risk factors for neurodegeneration in idiopathic rapid eye movement sleep behavior disorder:a multi- center study. *Ann Neurol* 2015;77(5):830–9.

Prajapati P, Sripada L, Singh K, Bhatelia K, Singh R, Singh R. TNF-α regulates miRNA targeting mitochondrial complex-I and induces cell death in dopaminergic cells. *Biochim Biophys Acta* 2015;1852(3):451–61. doi: 10.1016/j.bbadis.2014.11.019.

Pringsheim T, Jette N, Frolkis A, Steeves TD. The prevalence of Parkinson's disease: a systematic review and meta-analysis. *Mov Disord* 2014;29(13):1583–90.

Przedborski S. Pathogenesis of nigral cell death in Parkinson's disease. *ParkinsonismRelat Disord* 2005;11:S3–7.

Pulkes T, Papsing C, Thakkinstian A, Pongpakdee S, Kulkantrakorn K, Hanchaiphiboolkul S, Tiamkao S, Boonkongchuen P. Confirmation of the association between LRRK2 R1628P variant and susceptibility to Parkinson's disease in the Thai population. *Park Relat Disord* 2014;20:1018–21.

Puspita L, Chung SY, Shim JW. Oxidative stress and cellular pathologies in Parkinson's disease. *Mol Brain* 2017;10(1):53. doi: 10.1186/s13041-017-0340-9.

Rideout HJ, Dietrich P, Savalle M, Dauer WT, Stefanis L. Regulation of alpha-synuclein by bFGF in cultured ventral midbrain dopaminergic neurons. *J Neurochem* 2003;84(4):803–13. doi: 10.1046/j.1471-4159.2003.01574.x.

Ritz B, Ascherio A, Checkoway H, et al. Pooled analysis of tobacco use and risk of Parkinson disease. *Arch Neurol* 2007;64(7):990–7.

Ritz B, Lee PC, Lassen CF, Arah OA. Parkinson disease and smoking revisited: ease of quitting is an early sign of the disease. *Neurology* 2014;83:1396–402.

Rocca WA, Bower JH, Maraganore DM, et al. Increased risk of parkinsonism women who underwent oophorectomy before menopause. *Neurology* 2008;70(3):200–9.

Samii A, Etminan M, Wiens MO, Jafari S. NSAID use and the risk of Parkinson's disease: systematic review and meta-analysis of observational studies. *Drugs Aging* 2009;26(9):769–79.

Savica R, Grossardt BR, Carlin JM, et al. Anemia or low hemoglobin levels preceding Parkinson disease: a case-control study. *Neurology* 2009;73(17):1381–7.

Searles Nielsen S, Bammler TK, Gallagher LG, et al. Genotype and age at Parkinson disease diagnosis. *Int J Mol Epidemiol Genet* 2013;4:61–9.

Selikhova M, Fedoryshyn L, Matviyenko Y, et al. Parkinsonism and dystonia caused by the illicit use of ephedrone—a longitudinal study. *Mov Disord* 2008;23:2224–31.

Shen L, Ji HF. Low uric acid levels in patients with Parkinson's disease: evidence from meta-analysis. *BMJ Open* 2013;3(11):e003620.

Siddiqui IJ, Pervaiz N, Abbasi AA. The Parkinson disease gene SNCA: evolutionary and structural insights with pathological implication. *Sci Rep* 2016;6.

Song LK, Ma KL, Yuan YH, et al. Targeted overexpression of a-synuclein by rAAV2/1 vectors induces progressive nigrostriatal degeneration and increases vulnerability to MPTP in mouse. *PLoS One* 2015;10:e0131281.

Sun YX, Wang XH, Xu AH, Zhao JH. Functional polymorphisms of the MAO gene with Parkinson disease susceptibility: a meta-analysis. *J Neurol Sci* 2014;345:97–105.

Svenningsson P, Wirdefeldt K, Yin L, et al. Reduced incidence of Parkinson's disease after dipeptidylpeptidase-4inhibitors – a nationwide case-control study. *Mov Disord* 2016;31(9):1422–3.

Tan LC, Koh WP, Yuan JM, et al. Differential effects of black versus green tea on risk of Parkinson's disease in the Singapore Chinese Health Study. *Am J Epidemiol* 2008;167(5):553–60.

Tang L, Zhao S, Wang M, Sheth A, Zhao Z, Chen L, Fan X, Chen L. Meta-analysis of association between PITX3 gene polymorphism and Parkinson's disease. *J Neurol Sci* 2012;317(1–2):80–6.

Titova N, Chaudhuri KR. Personalized medicine in Parkinson's disease: time to be precise. *Mov Disord* 2017;32(8):1147–54.

Tiwari PC, Pal R. The potential role of neuroinflammation and transcription factors in Parkinson disease. *Dialogues Clin Neurosci* 2017;19(1):71–80.

van der Marck MA, Dicke HC, Uc EY, et al. Body mass index in Parkinson's disease: a meta-analysis. *Parkinsonism Relat Disord* 2012;18(3):263–7.

Van Maele-Fabry G, Hoet P, Vilain F, Lison D. Occupational exposure to pesticides and Parkinson's disease: a systematic review and meta-analysis of cohort studies. *Environ Int* 2012;46:30–43.

von Otter M, Bergström P, Quattrone A, De Marco EV, Annesi G, Söderkvist P, Wettinger SB, Drozdzik M, Bialecka M, Nissbrandt H, Klein C, Nilsson M, Hammarsten O, Nilsson S, Zetterberg H. Genetic associations of

Nrf2-encoding NFE2L2 variants with Parkinson's disease – a multicenter study. *BMC Med Genet* 2014;15:131. doi: 10.1186/s12881-014-0131-4.

Wahner AD, Bronstein JM, Bordelon YM, Ritz B. Statin use and the risk of Parkinson disease. Neurology 2008;70(16Pt2):1418–22.

Wang V, Chuang TC, Kao MC, Shan DE, Soong BW, Shieh TM. Polymorphic Ala-allele carriers at residue 1170 of HER2 associated with Parkinson's disease. *J Neurol Sci* 2013;325:115–9.

Wang X, Sun X, Zhang X, Li H, Xie A. Quantitative assessment of the effect of FGF20 rs12720208 variant on the risk of Parkinson's disease: a meta-analysis. *Neurol Res* 2017;39(4): 374–80. doi: 10.1080/01616412.2017.1286542.

Wang T, Wang B. Association between glutathione S-transferase M1/glutathione stransferase T1 polymorphisms and Parkinson's disease: a meta-analysis. *J Neurol Sci* 2014;338:65–70.

Wang YL, Wang YT, Li JF, Zhang YZ, Yin HL, Han B. Body mass index and risk of Parkinson's disease:a dose-response meta-analysis of prospectivestudies. *PLoS One* 2015;10(6):e0131778.

Wang D, Zhai JX, Zhang LM, Liu DW. Null genotype of GSTT1 contributes to increased Parkinson's disease risk in Caucasians: evidence from a meta-analysis. *Mol Biol Rep* 2014;41:7423–30.

Wang YC, Zou YB, Xiao J, Pan CD, Jiang SD, Zheng ZJ, Yan ZR, Tang KY, Tan LM, Tang MS. COMT Val158Met polymorphism and Parkinson's disease risk: a pooled analysis in different populations. *Neurol Res* 2019;41(4):319–25.

Weikang C, Jie L, Likang L, Weiwen Q, Liping L. A meta-analysis of association between glutathione S-transferase M1 gene polymorphism and Parkinson's disease susceptibility. *Open Med (Wars)* 2016;11(1):578–83. doi: 10.1515/med-2016-0094.

Werner FM, Covenas R. Classical neurotransmitters and neuropeptides involved in generalized epilepsy: a focus on antiepileptic drugs. *Curr Med Chem* 2011;18:4933–48.

Werner FM, Covenas R. Classical neurotransmitters and neuropeptides involved in Parkinson's disease: focus on anti-Parkinsonian drugs. *Curr Drug Ther* 2015;10:66–81.

Wider C, Wszolek ZK. Clinical genetics of Parkinson's disease and related disorders. *Park Relat Disord* 2007;13(suppl 3):229–32.

Wu YR, Chang KH, Chang WT, Hsiao YC, Hsu HC, Jiang PR, Chen YC, Chao CY, Chang YC, Lee BH, Hu FJ, Chen WL, Lee-Chen GJ, Chen CM. Genetic variants of LRRK2 in Taiwanese Parkinson's disease. *PLoS One* 2013a;8:e82001.

Wu YR, Chang KH, Chao CY, Lin CH, Chen YC, Liu TW, Lee-Chen GJ, Chen CM. Association of SOD2 p.V16A polymorphism with Parkinson's disease: a meta-analysis in Han Chinese. *J Formos Med Assoc* 2021;120(1 Pt 2):501–7. doi: 10.1016/j.jfma.2020.06.023.

Wu YL, Ding XX, Sun YH, Yang HY, Sun L. Methylenetetrahydrofolate reductase (MTHFR) C677T/A1298C polymorphisms and susceptibility to Parkinson's disease: a meta-analysis. *J Neurol Sci* 2013b;335:14–21.

Xiao Y, Cen L, Mo M, Chen X, Huang S, Wei L, Li S, Yang X, Qu S, Pei Z, Xu P. Association of IGF1 gene polymorphism with Parkinson's disease in a Han Chinese population. *J Gene Med* 2017;19(4).

Yang X, Liu C, Zhang J, et al. Association of histamine N-methyltransferase Thr105Ile polymorphism with Parkinson's disease and schizophrenia in Han Chinese: a case-control study. *PLoS One* 2015a;10:e0119692.

Yang X, Xu P, Zhao Q, et al. Polymorphism in the vesicular monoamine transporter 2 gene decreases the risk of Parkinson's disease in Han Chinese Men. *Parkinsons Dis* 2015b;2015:903164. doi: 10.1155/2015/903164.

Yuan RY, Sheu JJ, Yu JM, et al. Methylenetetrahydrofolate reductase polymorphisms and plasma homocysteine in levodopa-treated and nontreated Parkinson's disease patients. *J Neurol Sci* 2009;287:64–8.

Yue X, Li H, Yan H, Zhang P, Chang L, Li T. Risk of Parkinson disease in diabetes mellitus an updated meta-analysis of population- based cohort studies. *Medicine (Baltimore)* 2016;95(18):e3549.

Zhai D, Li S, Zhao Y, Lin Z. SLC6A3 is a risk factor for Parkinson's disease: a meta-analysis of sixteen years' studies. *Neurosci Lett.* 2014 Apr 3;564:99–104. doi: 10.1016/j.neulet.2013.10.060. Epub 2013 Nov 7.

Zhang Y, Piao X, Wu J, Li Y, Liang Q. A meta-analysis on relationship of MAOB intron 13 polymorphisms interactions with smoking/COMT H158L polymorphisms with the risk of PD. *Int J Neurosci* 2016;126:400–7.

Zhang Z, Zhang S, Fu P, Lin K, Ko JK, Yung KK. Roles of glutamate receptors in Parkinson's disease. *Int J Mol Sci* 2019;20(18):4391.

Zhao J, Han X, Xue L, Zhu K, Liu H, Xie A. Association of TLR4 gene polymorphisms with sporadic Parkinson's disease in a Han Chinese population. *Neurol Sci* 2015;36(9):1659–65.

Zhou SL, Tan CC, Hou XH, et al. TREM2 variants and neurodegenerative diseases: a systematic review and meta-analysis. *J Alzheimers Dis* 2019:1171–84.

Zhu ZG, Ai QL, Wang WM, Xiao ZC. Meta-analysis supports association of a functional SNP (rs1801133) in the MTHFR gene with Parkinson's disease. *Gene* 2013;531:78–83.

Zhu K, Teng J, Zhao J, Liu H, Xie A. Association of TLR9 polymorphisms with sporadic Parkinson's disease in Chinese Han population. *Int J Neurosci* 2015a:1–5.

Zhu M, Zhou T, Zu G, Liang Z. The NFE2L2 rs35652124 polymorphism and the risk of Parkinson's disease: a systematic review and meta-analysis. *Neuroreport* 2016;27(12):901–5.

Zhu Y, Zhu RX, He ZY, Liu X, Liu HN. Association of MTHFR C677T with total homocysteine plasma levels and susceptibility to Parkinson's disease: a meta-analysis. *Neurol Sci* 2015b;36:945–51.

Zhu R, Zhu Y, Liu X, He Z. Fibroblast growth factor 20 (FGF20) gene polymorphism and risk of Parkinson's disease: a meta-analysis. *Neurol Sci* 2014;35(12):1889–94. doi: 10.1007/s10072-014-1853-y.

10

Genetic Polymorphisms, Role in Drug Metabolism and Effect on Clinical Pharmacy

Shagufta Sheikh
Department of Biochemistry, Government Degree College for Women, Anantnag, Jammu and Kashmir, India

Sanjeev Rana
Human Genomics Research Group, Shri Mata Vaishno Devi University (SMVDU), Katra, Jammu and Kashmir, India

Hilal A. Wani
Department of Biochemistry, Government Degree College, Sumbal, Jammu and Kashmir, India

Aarif Ali
Department of Clinical Biochemistry, School of Biological Sciences, University of Kashmir, Hazratbal, Jammu and Kashmir, India

Qazi Danish
Department of Biochemistry, Sri Pratap (SP) College, Cluster University, Srinagar, India

Sabhiya Majid and Nisar A. Naikoo
Department of Biochemistry & Multidisciplinary Research Unit (MRU-ICMR), Government Medical College Srinagar & Associated Hospitals, Karan Nagar, Srinagar, Jammu and Kashmir, India

Nouroz Sehar
Centre for Translational and Clinical Research, School of Chemical & Life Sciences, Jamia Hamdard, Hamdard Nagar, New Delhi, India

Amjad Faihan Binlebdah, Azher Arafah, and Muneeb U. Rehman
Department of Clinical Pharmacy, College of Pharmacy, King Saud University, Riyadh, Saudi Arabia

CONTENTS

DOI: 10.1201/9781003246244-10

ABSTRACT The polymorphism in drug-metabolizing enzymes has a significant impact on response as well as metabolism of drugs. Therefore, to uncover the genetic network that regulates a person's response to pharmacological treatment, pharmacogenomics has emerged as a developing discipline that employs genome-wide approaches in order to elucidate the genetic cause for disparities in therapeutic potency. Most drugs are widely known to have substantial inter-person variability in effectiveness and tolerability. Genetic differences in genes that encode drug-metabolizing enzymes, drug facilitators, and drug destinations contribute to these inter-individual variances in many drugs. The individuals who inherit clinical manifestations mu-st be allowed to be treated with significantly different doses of the medications. The majority of pharmacological effects are regulated by the interaction of multiple gene products that influence drug absorption, distribution, metabolism, and excretion as well as pharmacodynamics. The ultimate objective is to develop novel and revolutionary medication therapy optimization techniques closely related to every individual's genetic factors of pharmacological effectiveness and toxicity variables. This chapter provides insights into a person's genetic variations that tend to alter the expected response to therapeutic medications, as well as insights into the field's possible impact on pharmaceutical safety and efficacy.

KEYWORDS: *Polymorphism; Drug-Metabolizing Enzymes; Pharmacogenomics; Single Nucleotide Polymorphism; Cytochrome P450.*

10.1 Introduction

There is immense variation in DNA sequences among individuals, groups, and populations, and such variations are defined under the broad category called genetic polymorphisms. They include single nucleotide polymorphisms (SNPs), sequence repeats, deletions, insertions, and recombination (Novelletto & Ciminelli 2021). The interplay of multiple evolutionary mechanisms such as mutation migration, selection, and random genetic drift causes, maintains, and eliminates genetic variation in populations (Yeaman & Otto 2011). Polymorphism can be inherited, thus enabling its inheritance to be traced from one progeny to other; however, noncoding regions of DNA tend to have more polymorphisms (Rashid 2020). Thus, genetic polymorphism is considered as the inheritance of a trait regulated by a gene with two or more alleles in a population with a least common allele frequency of 1% (Witherspoon et al. 2007). It's interesting to note that although 99.9% of the DNA sequences of two different individuals are identical, still a significant number of differences lie within the variation of 0.1%. These variations can be attributed to phenotypic differences and susceptibility to pathology (Shastry 2002). Genetic polymorphism is one of the factors associated with sensitivity to common diseases, differences in phenotypic traits, and inter-person differences in drug response, according to an analysis of several comparative diseases (Shastry 2007). Based on this finding, it has been suggested that cataloguing genetic variations in different cultures and animals might lead to the formulation of patient-specific diet and therapy that is both safer and more successful. Differences in drug response mechanisms have long been recognized. Variability in drug effects may be caused by the kind and severity of the illness condition, as well as the individual's age, ethnicity, organ function, treatment methods, stage of disease, drug interactions, but most importantly, hereditary difference in drug metabolism, as well as genetic polymorphisms in drug targets (Carson et al. 1956). Clinical observation of inherited differences in drug metabolism was first made by Arno Motulsky in the year 1957, who proposed that "hereditary gene-controlled enzymatic factors determine why, with identical exposure, certain individuals become 'sick,' whereas others are not affected" (Motulsky 1957). The term "pharmacogenetics" was coined by Vogel to characterize the correlation between genetic determinants and pharmaceutical response (Vogel 1959). Under the title pharmacogenomics, it was eventually expanded to cover the complete genome in the search for connections with pharmacologic phenotypes. The ultimate objective of pharmacogenomics is to understand how genetic variations in therapeutic targets affect treatment response. The goal is to optimize drug therapy safety and bioefficacy through genetically guided, personalized treatment. The genetic factors involved in drug metabolism and its repercussions are rapidly being interpreted, thanks to advances in molecular biology, bioinformatics, genomics, and the availability of information from human genome projects. These findings are already being used to develop more rational and tailored pharmacological therapies (Evans & Relling 1999; McLeod & Evans 2001). This chapter attempts to decipher the network of genes that determines the potency of various pharmaceuticals and to use this knowledge to find new customized therapeutic targets and improve drug therapy for a more personalized experience.

10.2 Genetic Polymorphisms in Drug-Metabolizing Genes

Drug metabolism includes the conversion of compounds into hydrophilic metabolites for the simple removal from the biological system, as well as blockage of their biological and pharmacological activity (Lin & Lu 1997). The conversion of prodrugs into reactive molecules and the conversion of drugs into harmful metabolites are both possible outcomes of drug metabolism (Goodman et al. 2006). Drug-metabolizing enzyme polymorphism results in the formation of separate subgroups in the population that differs in their capacity to execute particular drug biotransformations. The genetic polymorphism in enzymes causes enhanced, reduced, or the absence of enzyme expression by various molecular pathways (Meyer & Zanger 1997). A range of metabolic processes are triggered by single alterations in one allele of a gene. The function of the enzyme is determined by the collection of alleles encoding the gene, and the overall function of the enzyme is the phenotype of the enzyme function (Zanger & Schwab 2013) as depicted in Figure 10.2. Genotyping can be used to discover DNA alterations in specific metabolic pathways that result in an abnormal phenotype. So a genetic test predicts the response of an individual to various prescription and non-prescription drugs. It examines for genes encoding for certain liver enzymes that have been activated, deactivated, or altered by medications. This separates patients into four categories according to their ability to metabolize drugs, and therefore, they have been classified into the following predictive phenotypes (Crew et al. 2012).

10.2.1 Poor Metabolizers

These kinds of individuals have two defective alleles, resulting in significantly reduced drug metabolism and, in certain cases, greater drug levels and a higher risk of severe drug responses (Demkow 2016). The CYP2D6*3, CYP2D6*4, CYP2D6*5, and CYP2D6*6 genotypes are found in the majority of poor metabolizers (Anzenbacher & Zanger 2012).

10.2.2 Intermediate Metabolizers

These individuals are heterozygous in nature with one wild-type allele and one defective-type allele. It must be noted that intermediate metabolizers have a lower enzymatic activity and can have two reduced function alleles (one deficient and one reduced function) (Zanger et al. 2004).

10.2.3 Extensive Metabolizers

These are supposed to carry two copies of wild-type alleles; these can be homozygous and heterozygous, e.g. CYP2D6*1/*2, CYP2D6*A/1a, and CYP2D6*1A/*5.

10.2.4 Ultra-rapid Metabolizers

In this type, individuals possess multiple copies of a gene resulting in increased drug metabolism (for example, CYP2D6*2/*3) (Mizutani 2003).

10.2.5 Normal Metabolizers

This category of individuals carry wild-type alleles (for example, CYP2D6*1/*3). It is to be noted that wild-type alleles encode genes for the enzymes with normal function. The poor metabolizers and ultra-rapid metabolizers (URMs) are the two ends of the continuum.

10.3 Inducers and Inhibitors

The effectiveness of a medicine is determined not only by the above metabolic states, but also by the type of drug administered. Drugs are mainly classified into two main groups: active drugs and prodrugs. Prodrugs are inactive until they are metabolized, while active drugs are active until they are metabolized (Lee & McLeod 2011). Increased substrate drug levels may occur from genetic changes that inactivate and limit enzyme function, yet genetic duplication may boost enzyme activity, leading to reduced substrate drug levels. Polymorphism influences drug interaction by changing the action of inhibitors and inducers on enzymes (Zhou et al. 2009). It might have an excessive or minimum influence on the substrate. An inhibitor is a molecule that binds to enzymes and reduces their activity, whereas a drug or molecule that increases the metabolic activity of an enzyme either by binding to the enzyme directly and activating it, or by increasing the expression of the gene coding for the enzyme is known as an activator. Inhibitors and inducers are classified into the following categories (Shargel 2010).

10.3.1 Extensive Metabolizer-Inhibitors

The amount of substrate medication is generally low due to the enzyme's fast action. An enzyme inhibitor will prevent extensive metabolism and generate significant increases in substrate drug levels. In an extensive metabolizer with high levels of substrate, the inhibitor's effect is substantially larger.

10.3.2 Poor Metabolizer-Inhibitors

Because the substrate's metabolism is substantially slower than normal, the amount of substrate drug remains high. When an inhibitor is introduced to the poor metabolizer, the extra-inhibition of metabolism is not substantially larger than what already exists. In a poor metabolizer, the inhibitor's action is reduced. It's possible that the medication interaction won't happen.

10.3.3 Extensive Metabolizer-Inducers

The level of substrate medication is lower than in a typical metabolizer due to rapid metabolism. Because the metabolism has already been considerably boosted, adding an inducer does not result in a bigger difference in the amount of substrate. It's possible that the drug interaction won't happen.

10.3.4 Poor Metabolizer-Inducers

Because of the slower metabolism of the substrate, the amount of medication in this case is higher than predicted in a typical metabolizer. The addition of inducer causes a considerable increase in substrate metabolism, resulting in a significantly lower amount of substrate than would be predicted in the usual metabolizer; drug interaction might be possible to some extent. Substrate levels equivalent to those of normal metabolizers may arise from drug interaction (Brunton et al. 2018).

Hence, we can say that the effect of inhibitor is great in extensive metabolizer than in poor metabolizer, while inducer impact is larger in poor metabolizers than in extensive metabolizers. When a substrate is processed by a number of enzyme systems and one or more enzymes are affected by polymorphism, there is a high possibility that complex drug interactions might occur. The substrate is first processed by a polymorphic enzyme, following which it forms an active metabolite. This active metabolite works as an inhibitor or inducer in the second system (Kivisto et al. 1995).

The other common explanations of drug metabolism include phase I oxidation activities, which are predominantly carried out by the liver's cytochrome P450 (CYP) enzymes, and phase II conjugations such as glucuronidation, sulphation, and acetylation (Dixit et al. 2017) as depicted in Figure 10.1. Both the processes transform lipid-soluble drugs into inert, water-soluble intermediates, enabling for more rapid elimination. Additional areas that have high metabolic activity include the GI tract, kidneys, and lungs.

The first drug-metabolizing polymorphism for acetylation was discovered over four decades ago. Two genes, NAT1 and NAT2, are now known to control N-acetyltransferase, with NAT2 A and B accounting for clinically significant metabolic variations (Summerscales & Josephy 2004). Caffeine, isoniazid, nitrazepam, and sulphonamides are examples of acetylated medicines (Shenfield 2004). After having received isoniazid or another marker drug and having their plasma or urine drug concentrations evaluated after a standard interval of time, individuals are divided into two groups as fast acetylators, who have only minimum parent drug concentrations remaining, and slow acetylators, who tend to have higher concentrations of parent drug left (Okumura et al. 1997).

10.4 Genetic Variability Associated with the Phase I Enzymes

Phase I processes, also known as non-synthetic reactions, are carried out by mixed function oxidases in the liver and involve oxidation, reduction, hydrolysis, cyclization, decyclization, and the addition or removal of oxygen or hydrogen. The most important category of enzymes that catalyse vast majority of phase I reactions include CYP enzyme superfamily, which biotransforms a wide range of exogenous and endogenous compounds, notably medications, toxins, and carcinogens (Debose-Byod 2007). They are found in the endoplasmic reticulum, and some are localized primarily in mitochondria. The CYP-dependent mixed function oxidase system catalyses the most prevalent alteration, i.e. hydroxylation (Gong, 2017). Omura and Sato coined the term "cytochrome P450" in 1962. It's described as a membrane-bound heme-containing protein with a 450 nm spectral peak being complexed with carbon

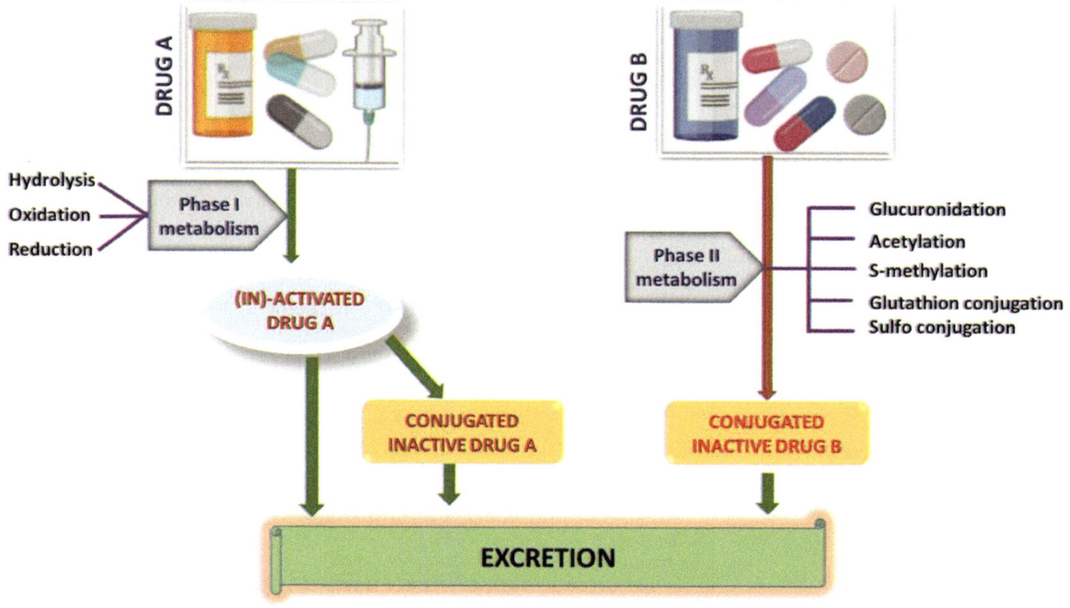

FIGURE 10.1 Phases of drug metabolism. Phase 1 drug-metabolizing enzymes mediate drug oxidation, reduction, or hydrolysis reactions, by which drugs may be activated or inactivated, followed by phase II reactions to further increase solubility and thereby facilitate excretion from the body.

monoxide (Omura & Sato 1962). Because of its peculiar spectrum characteristics, which included a typical absorption maximum of the reduced CO-bound complex at 450 nm, the term "cytochrome P450" was created. The terms/letters chrome, P, and 450 in cytochrome P450 stand for hemoprotein, pigment, and the absorption peak of the CO complex at 450 nm, respectively.

Based on amino acid sequence homology, CYP proteins are easily classified into numerous families and subfamilies. The human CYP gene family is a group of 57 genes that are divided into 18 families and 44 subfamilies (Hart et al. 2008; Gomes et al. 2009). Enzymes that exhibit 35%–40% similarity are given to the same family by an Arabic numeral, whereas those that show 55%–70% identity are allocated to a specific subgroup by a letter (Nebert & Russell 2002). For example, CYP2D6 refers to the family 2, subfamily D, and gene number 6 in the CYP2D gene family (Zahari et al. 2013). CYP polymorphism genotyping gives valuable genetic information that aids in understanding the impact of xenobiotics on the human body (Bozina et al. 2009). The most relevant polymorphisms associated with drug metabolism are mostly in the genes that encode for CYP2C9, CYP2C19, CYP2D6, and CYP3A4/5, which might result in therapeutic failure or severe adverse effects (Cooper et al. 1963). Around 70%–90% of the presently available prescription medicines are metabolized by these genes (Ingelman-Sundberg et al. 2007). The following sections detail the several gene polymorphisms found in some of the phase I XMEs in the research population, as well as their substrates and role in drug metabolism.

10.4.1 CYP2D6 (CYP 450 Family 2, Subfamily D, Polypeptide 6)

The CYP2D6 gene has locus on chromosome 22q13.1 and accounts for small percentage of all hepatic CYP. Its presence in other organs such as brain, gastrointestinal tract, and lung has also been confirmed. It is a 497-amino acid polypeptide that has nine exons and eight introns. It's one of the significant enzymes involved in the metabolism and elimination of xenobiotics via the process of hydroxylation, demethylation, and dealkylation. It's known to be involved in the metabolism of around 25% of commonly prescribed drugs. In the human population, the gene is extremely polymorphic, with more than 80 polymorphic CYP2D6 alleles identified. Most of the variants are quite rare, but some of the important ones include CYP2D6*9, CYP2D6*10, CYP2D6*17, CYP2D6*29, and CYP2D6*41, which possess decreased catalytic activity and hence form the intermediate metabolizer phenotype. CYP2D6*3, CYP2D6*8, and CY2D6*36 have no functional activity and thus form poor metabolizers. CYP2D6*1, CYP2D6*2, and CYP2D6*35 had enhanced functional capacity, due to gene duplication, and therefore form URM phenotype.

In the late 1980s, scientists discovered the molecular genetic basis for inherited drug metabolism variability after cloning a polymorphic human gene (CYP2D6) encoding the drug-metabolizing enzyme debrisoquin hydroxylase (Gonzalez et al. 1988). When it was discovered that a tiny proportion of people using the antihypertensive medicine debrisoquine observed a significant decline in blood pressure, it was discovered that this was due to unusually high plasma drug concentrations. Since the discovery of the CYP2D6 polymorphism, researchers have revealed that the enzyme CYP2D6 is involved in the metabolism of debrisoquine and over 100 additional medicines (Bertilsson et al. 2002). Many antidepressants, anti-arrhythmic medicines, beta-adrenergic blockers, opiates, neuroleptic agents, and other medications, including perhexiline and phenformin, are processed by CYP2D6 enzyme (Chamberlain 2018).

The CYP2D6 polymorphism and its impact on clinical outcomes associated with the use of tamoxifen in the treatment of oestrogen receptor-positive breast cancer is one of the most extensively studied models in pharmacogenetics (Olopade et al. 2008). The complex metabolism of tamoxifen is mediated by CYP2D6 via transformation of tamoxifen to endoxifen (4-hydroxy-N-desmethyltamoxifen), a compound having a far higher oestrogen receptor binding than the parent drug. Since CYP2D6 activity is so dynamic and is the crucial catalyst for endoxifen synthesis, it has piqued researchers' curiosity to investigate the effect of CYP2D6 polymorphism on response to tamoxifen therapy in women diagnosed with breast cancer. Despite the fact that results are inconsistent, it was observed that the patients having CYP2D6 poor metabolizer phenotype, treated with tamoxifen, had reduced circulating levels of endoxifen and thus increased chance of recurrence. Similarly, patients put on tamoxifen therapy and taking drugs such as paroxetine to alleviate tamoxifen-induced hot flashes that serve as powerful 2D6 inhibitors may also be at a higher risk of breast cancer recurrence. Routine CYP2D6 genotype testing in women with early breast cancer has not been generally used in clinical practice due to a lack of concordant, prospective evidence.

However, direct-to-consumer genetic testing for CYP2D6 variants is now accessible and might modify a patient's choice to begin tamoxifen medication. According to population surveys, PMs account for about 8% of Caucasians, but fewer than 1% of Asian races. The effects of different polymorphisms are similar to those of acetylation. PMs have greater plasma drug concentrations than usual, resulting in a higher rate of adverse drug responses (ADRs). They also lack metabolites that can be highly important, such as the active metabolite morphine, which is responsible for therapeutic effects when codeine is converted to morphine. Codeine has no analgesic effect in PMs CYP2D6 variants (Carranza-Leon et al. 2021). When conventional dosages of perhexiline, an antianginal drug, are administered for more than a short length of time, it causes severe and permanent peripheral neuropathy in PMs. Therefore, therapeutic monitoring of perhexiline is particularly critical in the early phases of therapy. Similarly, phenformin was pulled off the market due to a high alarming rate of life-threatening lactic acidosis, which mostly affected the PM segment of the population. EMs have a wide range of plasma concentrations for CYP2D6 target medicines, and they might be prone to inductive and repressive interactions with other medications. Inhibition can result in "phenocopying", when strong tranquillizers turn genetic EMs into phenotypic PMs. Antidepressants such as tricyclic antidepressants may be ineffective in the treatment of URMs (Shenfield & Gross 1999).

10.4.2 CYP2C19 (CYP 450 Family 2, Subfamily C, Polypeptide 19)

This gene is located on chromosome 10q24.1. It encodes liver enzyme, which is involved in the metabolism of 5%–10% of all drugs. Nearly 30 allelic variants have been identified so far (Sim et al. 2006). Some variants of CYP2C19 gene resulted in reduced activity (CYP2C19*5 and CYP2C19*8) or loss of functional activity (CYP2C19*2, CYP2C19*3, CYP2C19*4, CYP2C19*6, and CYP2C19*7) with poor metabolizer phenotype. The variant with high catalytic activity and extensive metabolizer phenotype is CYP2C19*17. S-mephenytoin has extensively been utilized as a metabolic test of CYP2C19 activity, and there has been a noticeable ethnic variation in the poor metabolizer phenotype. While the poor metabolizer phenotype is seen in just 2%–6% of Caucasians, it is found in up to 20% of Asians (Zhou et al. 2008). Fluoxetine and ketoconazole have been shown to inhibit the enzyme; with these drugs, homozygous extensive metabolizers (CYP2C19*17/*17) have the highest 2C19 inhibition, whereas homozygous poor metabolizers experience little to no inhibition. This phenomenon can be seen when drugs such as fluoxetine or cimetidine impede phenytoin metabolism, resulting in increased phenytoin exposure and severe side effects. In one of the studies carried out on approximately 1,500 patients having percutaneous coronary stent insertion, one or more CYP2C19*17 alleles exhibited a substantial increase in bleeding problems, a phenomenon associated with extensive metabolizers.

The effect of CYP2C19*2 (681G>A*)* allele has been investigated in several studies and a meta-analysis in patients with coronary artery disease treated with clopidogrel (antiplatelet agent). Clopidogrel is a prodrug, which is oxidized to its active metabolite largely by CYP2C19. Patients with CYP2C19*2 allele tend to have splicing defect and lead to formation of truncated protein, hence are unable to activate clopidogrel, and are at a higher risk of major cardiovascular events (Geisler et al. 2017).

Asians require lower doses and are more sensitive to adverse drug reactions from drugs such as diazepam. EMs are more prone to generate ADRs when the active component is a metabolite, for example proguanil, which is metabolized to cycloguanil. Helicobacter pylori eradication studies using a triple therapy of a proton pump inhibitor, clarithromycin, and amoxicillin have showed that eradication rates are higher in PMs than in EMs (especially homozygotes) (Zhao et al. 2008). The enzyme CYP2C19 methylates diazepam. Individuals who have the defective CYP2C19*2 allele have a longer plasma diazepam half-life than those who have the wild-type gene (Qin et al. 1999). Desmethyldiazepam metabolites demonstrated a longer half-life in CYP2C19-deficient individuals. Diazepam-induced toxicity can develop in Asians due to a slower metabolism; hence, dose should be carefully managed (Kim et al. 2004).

10.4.3 CYP2C9 (CYP 450 Family 2, Subfamily C, Polypeptide 9)

CYP2C9 is located on chromosomes 10q24.1 and constitutes 18% of CYP 450 protein in liver microsomes (Paine et al. 2006). This subfamily has a significant contribution to drug metabolism as well as cholesterol synthesis and other lipids. CYP2C9*2 (Arg144Cys) and CYP2C9*3 (Ile359Leu) are the two most widely studied genetic variants. The existence of two coding area polymorphisms resulting in the amino acid changes Arg144Cys and Ile359Leu was discovered in CYP2C9 cDNA sequences (Daly et al. 2018). Polymorphisms in the CYP2C9 gene that causes a lower enzyme activity are widespread, and this, combined with narrow therapeutic indices for numerous essential drug substrates, leads to some serious concerns about drug safety and efficacy (Daly et al. 2018). CYP2C9 is the primary enzyme involved in the metabolic clearance of a number of commonly used medicines with a limited therapeutic index (Fekete et al. 2021).

Evidence of a polymorphism affecting the metabolism of the antidiabetic medication tolbutamide was initially described in 1978, but it was assumed to be associated with the debrisoquine (CYP2D6) polymorphism at the time. Following research, it was discovered that the genetic control of tolbutamide metabolism differed from that of debrisoquine metabolism (Sangkuhl et al. 2021). Along with these investigations, an enzyme capable of oxidizing tolbutamide was discovered and called CYP2C9. These alterations were functionally significant, according to in vitro expression experiments. The functional significance of these SNPs was validated by genotyping of individuals receiving warfarin medication; patients requiring a lower maintenance dose of warfarin to achieve therapeutic anticoagulation were 6 times more likely to have one or more CYP2C9 variant alleles. Because of their well-established functional impacts, they've been widely researched in connection to a wide spectrum of medicines' metabolism, and they are not, however, the only clinically significant polymorphisms, nor are they even the most prevalent variations in some ethnic groups.

10.4.4 CYP3A4 (CYP 450 Family 3, Subfamily A, Polypeptide 4)

Human CYP3A enzymes are responsible in the oxidative, peroxidative, and reductive metabolism of a various endogenous chemicals, procarcinogens, and drugs. The variation in their expression between people may have a major impact on therapeutic treatment efficacy (Danielson 2002). It's found in the liver (30%) and small intestine (70%). CYP3A4 is one of the four isoforms of the human CYP3A subfamily, while other isoforms are CYP3A5, CYP3A7, and CYP3A43. In the human liver, these isoforms form up to 30% of the overall CYP enzyme make-up in various proportions (Qiu et al. 2008). On the 7q21–q22.1 chromosomal region, the CYP3A enzyme genes, as well as the CYP3AP1 and CYP3AP2 pseudogenes, are present in a 200 kb cluster. Each of the four CYP3A enzymes has 13 exons and are physically identical (between 71.5% and 84.1%).

For the CYP3A gene, more than 30 SNPs have been discovered. CYP3A4 does not have a deleted or null allele, unlike other P450s (Sun et al. 2017). The most frequent CYP3A4 variation, CYP3A4*1Bia, is an A392G transition within the promoter region known as the nifedipine response element. According to one research, the mutation is linked to cyclosporine clearance that is sluggish, but is a highly controversial finding.

10.4.5 CYP3A5 (CYP 450 Family 3, Subfamily A, Polypeptide 5)

CYP3A5 is located near CYP3A4 on chromosome 7 (Chr. 7q22.1). The CYP3A5*3 (A6986G) variant is caused by improper mRNA splicing and decreased functional protein translation (Fan et al. 2021). The extrahepatic CYP3A isoform, CYP3A5, has been linked to diseased risk and the metabolism of endogenous hormones or drugs in organs other than the liver. CYP3A5 has been connected to tacrolimus dosage required in solid organ transplant patients in order to maintain appropriate immunosuppression (Buck et al. 2019).

10.4.6 CYP3A7 (CYP 450 Family 3, Subfamily A, Polypeptide 7)

CYP3A7 is located on chromosome 7q22.1. It shares its 87% sequence with CYP3A4. During development, it's expressed in the foetal liver, although after the birth the hepatic expression is normally reduced (Domanski, 2001). The CYP3A7 protein has been discovered in certain people (Li & Lampe 2019). Increased CYP3A7 expression has been linked to the replacement of a 60-nucleotide fragment of the CYP3A7 promoter with the similar area form of the CYP3A4 promoter (CYP3A7*1C allele). The pregnane X receptor response element gene expression is boosted as a result of this promoter exchange (Bissig et al. 2018). PXR signalling regulates the expression of inducible CYP3A4 as well as a number of other drug-detoxifying genes. According to PXR polymorphisms, variation in CYP3A4 enzymatic activity may be due in part to inherited differences in the upstream essential proteins that influence gene expression induction.

10.4.7 Other P450 Enzymes

Several other P450 enzymes are expressed in human, which include CYP1A2, CYP2A6, CYP2B6, and CYP2E1, but in total only account for 5%–15% of drug metabolism (Boullata & Armenti 2004). The clinical significance of genetic polymorphisms associated with these enzymes remains unclear. However, CYP2B6 variant alleles (CYP2B6*6, CYP2B6*16, and CYP2B6*18) in some patients have shown reduced ability to activate the antineoplastic prodrug cyclophosphamide, leading to reduced anti-tumour efficacy, or to metabolize the antiretroviral efavirenz, leading to increased systemic toxicity. Since most of these are primarily small, single-centre genotype-phenotype association studies, they should be evaluated with care.

TABLE 10.1

Genetic Polymorphism of Genes (Phase II Enzymes) Involved in Drug Metabolism and Their Clinical Response

Enzymes	Substrates	Clinical Consequences
CYP1A1	Benzo(a)pyrene, phenacetin	Increased or decreased cancer risk
CYP1A2	Acetaminophen, amonafide, caffeine, paraxanthine, ethoxyresorufin, fluvoxamine	Decreased theophylline metabolism
CUP1B1	Oestrogen metabolites	Possible increased cancer risk
CYP2A6	Coumarin, nicotine, halothane	Decreased nicotine metabolism and cigarette addiction
CYP2B6	Cyclophosphamide, aflatoxin, mephenytoin	Significance not known
CYP2C8	Retinoic acid, paclitaxel	Significance not known
CYP2C9	Tolbutamide, warfarin, phenytoin, non-steroidal anti-inflammatories	Anticoagulant effect on warfarin
CYP2C19	Mephenytoin, omeprazole, hexobarbital, mephobarbital, propranolol, proguanil, phenytoin	Peptic ulcer response to omeprazole
CYP2D6	Beta-blockers, antidepressants, antipsychotics, codeine, debrisoquine, dextromethorphan, encainide, flecainide, fluoxetine, guanoxan, methoxyamphetamine, N-propylajmaline, perhexiline, phenacetin, phenformin, propafenone, sparteine	Tardive dyskinesia from antipsychotics, narcotic side effects, efficacy, dependence, imipramine dose requirement, beta-blocker effect
CYP2E1	N-nitrosodimethylamine, acetaminophen, ethanol	Possible effect on alcohol consumption
CYP3A4/3A5/3A7	Macrolides, cyclosporin, tacrolimus, calcium channel blockers, midazolam, terfenadine, lidocaine, dapsone, quinidine, triazolam, etoposide, teniposide, lovastatin, alfentanil, tamoxifen, steroids, benzo(a)pyrene	Tacrolimus dose requirement in paediatric cancer patients
Aldehyde dehydrogenase (ALDH2)	Cyclophosphamide, vinyl chloride	SCE frequency in lymphocytes
Alcohol dehydrogenase (ADH3)	Ethanol	Increased alcohol consumption and dependence
Dihydropyrimidine dehydrogenase	Fluorouracil	5-Fluorouracil neurotoxicity
NQO1 (DT-diaphorase)	Ubiquinone, menadione, mitomycin C	Menadione-associated urolithiasis, decreased tumour sensitivity to mitomycin C, possible increased cancer risk

10.4.8 Dihydropyrimidine Dehydrogenase

Dihydropyrimidine dehydrogenase (DPD) is a pyrimidine dehydrogenase enzyme located on chromosome 1p21.3. This enzyme is involved in the metabolism of endogenous pyrimidines thymine and uracil and, hence, is also known as dihydrothymine dehydrogenase or uracil reductase. It is also involved in the metabolism of antineoplastic agent fluorouracil, and around 80% metabolism of this anticancer drug 5-FU is performed by DPD. The 5-FU is catabolized into 5-fluoro-5,6-dihydrouracil (5-FUH2), which is then transformed into fluoro-beta-ureidopropionate (FUPA) and then fluoro-beta-alanine (FBAL) via dihydropyrimidinase (DPYS) and beta-ureidopropionase (UPB1) enzymes. This anabolic conversion of 5-FU into a metabolically active nucleotide form is critical to 5-FU's activity. DPD, on the other hand, lowers the availability of 5-FU. As a result, elevated DPD expression in tumours frequently leads to resistance to anticancer drugs such as 5-FU (Sharma et al. 2019). The lack of these enzymes, as well as any single nucleotide changes in the gene, may indicate a complete enzymatic deficiency. As a result, DPD polymorphisms will cause deficient phenotypes depending on the mutation and tying them to its polymorphic activity with a total frequency of 3%–5%. Insufficient synthesis of this enzyme leads to drug overaccumulation, which then leads to toxicity (5-FU).

There are over 50 polymorphisms in DPYD gene. Genetic aberrations such as exon skipping, missense mutations, and deletion in the DPYD gene cause the DPD to have a deficient phenotype. There are three primary DPYD genetic variants that have been reliably linked to the 5-FU's toxicity: DPYD*2A (rs3918290, IVS14+1G>A), 2*A (rs3918290 G>A) substitution, which causes an entire exon 14 to be skipped; DPYD*13 (rs5588602 T>G), which causes an Ile56Ser amino acid to be changed to a flavin-binding domain of the enzyme; and rs67376798 A>T, which causes an Asp949Val amino acid to be changed to an iron–sulphur motif (Ruzzo et al. 2017). The DPYD (IVS14+1G>A) mutation results in a 165-bp deletion in the mRNA, resulting in the DPYD*2A allele, which is followed by the newly shortened protein product; DPYD*2 A homozygous mutation in patients results in a total lack of this enzyme's function, whereas a heterozygous mutation results in a 50% loss of enzymatic activity, resulting in a life-threatening disease owing to the build-up of 5-FU in the body.

Table10.1 summarizes functional genetic variations in human drug-metabolizing enzymes, their substrates, and the clinical impact of polymorphisms in their genes. Almost every polymorphism that has been investigated so far has a varied frequency among ethnic and racial groups. The ideal dose of medicines may differ among distinct cultures due to significant ethnic variation in the frequency or kind of functional defects in drug-metabolizing enzymes, which is a major concern as drug research becomes more global.

10.5 Genetic Variability Associated with the Phase II Enzymes

Phase II enzymes are in charge of conjugating different molecules to create more hydrophilic metabolites that may be removed through the urine or faeces. Phenotypic variation and clinical consequences are also seen in phase II enzymes as listed in Table 10.2. Some of the important phase II enzymes are discussed below.

10.5.1 N-Acetyltransferase

The first drug-metabolizing polymorphism was described over 40years ago for acetylation. It is now known that N-acetyltransferase (NAT) is controlled by two genes (NAT1 and NAT2) and has a locus on chromosome 8p22. Compounds that

TABLE 10.2

Genetic Polymorphism of Genes (Phase II Enzymes) Involved in Drug Metabolism and Their Clinical Response

Enzymes	Substrate	Clinical Consequence
N-acetyltransferase (NAT1)	Aminosalicylic acids, aminobenzoic acid, sulfamethoxazole	Increased cancer risk, hypersensitivity to sulphonamides, amonafide toxicity; hydralazine-induced lupus, isoniazid neurotoxicity, and hepatitis
N-acetyl transferase (NAT2)	Isoniazid, hydralazine, procainamide, dapsone, caffeine	
Glutathione-transferase (GSTM1, M3, T1)	Busulfan, aminochrome, dopachrome, adrenochrome, noradrenochrome	Possible increased cancer risk, cisplatin-induced toxicity
Glutathione-transferase (GSTP1)	13-Cis retinoic acid, busulfan, ethacrynic, epirubicin	Possible increased cancer risk
Sulphotransferases	Steroids, acetaminophen, tamoxifen, oestrogens, dopamine	Increased or decreased cancer risk, clinical outcomes in women receiving tamoxifen for breast cancer
Catechol-O-methyltransferases	Oestrogens, levodopa, ascorbic acid	Decreased response to amphetamine substance abuse, levodopa response
Thiopurine methyltransferase	Mercaptopurine, thiguanine, azathioprine	Thiopurine toxicity and efficacy, risk of second cancers
UDP-glucuronosyltransferase (UGT1A1)	Irinotecan, troglitazone, bilirubin	Irinotecan, glucuronidation and toxicity, hyperbilirubinemia (Crigler-Najjar syndrome, Gilbert syndrome)
UDP-glucuronosyltransferase (UGT2B)	Opioids morphine, naproxen, ibuprofen, epirubicin	Significance unknown

are acetylated in animals are either aromatic amines or hydrazines, which are transformed to aromatic amides and aromatic hydrazides (Stanley 2017). For example, N-acetylation of isoniazid, a new anti-tuberculosis agent, to acetyl isoniazid (Hughes et al., 1954). It is also involved in the detoxification of hydralazine and endralazine (antihypertensive drugs), a number of sulphonamides (antibacterial drugs), procainamide (anti-arrhythmic drug), and nitrazepam (anti-inflammatory drug).

Individuals who are slow or rapid acetylators have experienced an uncommon side effect known as peripheral neuritis in some individuals treated with isoniazid. According to studies, patients who had a delayed transformation of isoniazid to acetyl isoniazid were more likely to develop peripheral neuritis. Individuals are either slow or fast acetylators. Slow acetylation was discovered to be a recessive trait after more research. There is ethnic diversity. Slow acetylation occurs in 10% of Japanese, 20% of Chinese, and 60% of Caucasians.

NAT1 and NAT2 are very similar to each other and share 80% identity at the amino acid level in these 2 proteins, which are each 290 amino acids long. NAT1 includes a functioning Cys-His-Asp catalytic triad, similar to cysteine proteases, like all mammalian arylamine N-acetyltransferase proteins. Acetyl-coenzyme A acetylates the catalytic Cys68 (AcCoA) followed by acetyl group transfer to the substrate. The three vital active site loop residues F125, Y127, and R129 have a significant impact on NAT1 substrate selection. The most prevalent NAT1 and NAT2 alleles found in some, but not all ethnic groups are NAT1*4 and NAT2*4. SNPs are recognized by assigning the number 1 to the "A" of the ATG translation start codon. Negative numbers represent SNPs upstream of the site, whereas positive values designate SNPs downstream of the site. The sluggish acetylator phenotype is related to NAT2 alleles carrying the G191A, T341C, A434C, G590A, and/or G857A missense-associated mutations. NAT2/NAT1 has been found in cancer cells, where it may not only play a role in cancer growth, but also contribute to some malignancies' resistance to cytotoxic drugs (Laurieri et al. 2018).

SNPs in the 3′-UTR of NAT1*10 have been linked to somewhat increased NAT1 activity in human bladder and colon and higher N-acetylation capacity in vivo. One of the SNPs in the 3′-UTR of NAT1*10 (1088T>A) changes a consensus polyadenylation signal (polyA-1), which supports the theory that greater activity is linked to a change in polyadenylation and hence improved mRNA stability (Sim & Laurieri 2018). Several investigations, however, have been unable to duplicate the higher catalytic activity linked to NAT1*10.

10.5.2 Thiopurine S-Methyltransferase (TPMT)

Thiopurine S-methyltransferase (TPMT) is an enzyme that in humans is encoded by the TPMT gene. A pseudogene for this locus is located on chromosome 18q (Katsanos & Tsianos 2010). Thiol metabolism is catalysed by two enzymes in humans, thiopurine S-methyltransferase (TPMT) and thiol methyltransferase (TMT). TMT is selective for aliphatic thiol substrates, whereas TPMT preferentially methylates thiopurine compounds. These enzymes have a wide range of activity in vivo, which causes toxicity in thiol-containing drugs like azathioprine that have been methylated. TPMT is a cytosolic enzyme that favours aromatic thiols like 6-mercaptopurine (6-MP) and 6-thioguanine over thiopurine compounds.

In humans, thiol methylation is a drug clearance mechanism. The effects of TPMT on the methylation of azathioprine, an immunosuppressant medication often used after organ donation, are a good example of this. Azathioprine is metabolized to 6-MP, which inhibits purine production and leads T cell apoptosis. TPMT catalyses methylation of the purine thiol, which is a significant route of drug clearance and detoxification for 6-MP.

In humans, the rate of thiol and thiopurine methylation is significantly varied. The TPMT genetic polymorphism is an excellent example of pharmacogenetics' therapeutic value. Thiopurine methylation is impacted by the presence of SNPs that reduce TPMT activity. Certain inactivating SNPs can cause a rise in circulating 6-MP levels, resulting in toxicity and undesired off-target effects. According to Ford and Berg, there are 29 different variant TPMT alleles, which may be associated with significant inter-individual heterogeneity in thiopurine drug toxicity and therapeutic efficacy. Genetic polymorphism allele frequencies are such that one in every 300 Caucasians is homozygous for a faulty allele or alleles for the characteristic of extremely low activity, while 11% are heterozygous and have intermediate activity. After therapy with a typical dosage of azathioprine, those who possess homozygous allele for low TPMT activity are at a greater risk of myelosuppression. In general, TPMT-deficient patients (homozygous mutant) can be treated with 6%–10% of the normal thiopurine dosage (Zhou et al. 2006). TPMT has the highest level of expression in the liver and kidney, and despite intensive research, the physiological purpose of this enzyme remains unknown.

10.5.3 UDP-glucuronosyltransferases (UGTs)

UDP-glucuronosyltransferase (UGT) is an enzyme that in humans is encoded by the UGT gene and is located in chromosome 15. UGTs consist of superfamily of enzymes that facilitate the covalent incorporation of sugars from nucleotide UDP-sugar donors to functional groups, typically hydroxyl, carboxyl, or amine, on a range of lipophilic substances (Bock, 2016). This biotransformation is vital in the removal of a variety of external compounds and endogenous metabolic by-products, as well as in modulating the quantities and distribution of various endogenous signalling molecules (Meech, 2019). These are commonly found in humans, plants, fungi, and bacteria.

In a wide range of animals, more than 35 different UGT gene products have been found. The superfamily in mammals is organized into four families as UGT1, UGT2, UGT3, and UGT8. UGT1 and UGT2 enzymes have critical roles in pharmacology and toxicology, regulating drug disposition and cancer risk. These UGTs are abundant in detoxification organs (e.g. the liver, kidney, and gut) and can be activated via pathways that detect detoxification demand and alter endobiotic signalling molecules. The functions of the UGT3 and UGT8 enzyme families have just recently been uncovered. In his renowned work analysing early important pharmacogenetic correlations, Motulsky addressed the mild hyperbilirubinemia of Gilbert syndrome, which is characterized by a failure to conjugate bilirubin, and hypothesized that variations in drug glucuronides could play a significant role (King et al. 2002).

In fact, impairments in UGT activity have been linked to both of these disorders. The UGT mutation that causes Gilbert has been cloned. UGT1A1*28 syndrome – a TA insertion in the promoter region of the gene causing reduced transcription – was obtained in 1996 (Sugatani et al. 2002). The presence of this variation has also been connected to irinotecan, an anti-cancer drug, which has caused significant haematologic and gastrointestinal side effects due to the lack of conjugation and, as a result, the delayed excretion of a hazardous metabolite (SN-38). Several UGT1A1 genotypes have been linked to the development of diabetes and serious toxicity with irinotecan, and the US FDA upgraded the warning in 2005 in the drug's label.

10.6 Clinical Implications of Other Drug Target Polymorphisms

In recent years, researchers have been intrigued by the influence of genetic variants in pharmacological targets on treatment success and toxicity (Carr et al. 2021). The purpose of the pharmacogenomics study is to discover the hereditary foundation for individual differences in medication response and toxicity, especially when pharmacokinetics fails to explain such kind of behaviour (Benjeddou & Peiró 2021). The published data on pharmacogenetics date back to the mid- to late 1990s, although the study on metabolic pharmacogenetics has been there since the 1950s. Furthermore, because most pharmacological effects are polygenic and high-throughput genotyping technologies are now available, the focus of pharmacogenomics is rapidly changing from single to polygenic genomics primarily because most pharmacological effects are polygenic in character, and techniques for efficient screening are now easily available.

The clinical implications of various drugs stated above are discussed in this chapter, as well as the logic and advantage of switching from a single gene procedure to a genomic approach to bring forth more therapeutically relevant data. Early pharmacogenetic drug targets research was limited to a single mutation in a single gene, whereas the gene is acting as the medication's direct target, e.g. the receptor. Tables 10.1 and 10.2 summarize the appropriate medication or drug class, as well as the drug response or adverse effects linked with the polymorphism. It provides examples of pharmacological target genetic polymorphisms that have been investigated to see whether they play a role in drug response variability.

According to the existing literature, drug target studies focusing on single polymorphism have established a substantial correlation with polymorphisms and projected response, but they have also proven disappointing in terms of the expected response. One of the most studied pharmacological targeted polymorphisms is the insertion/deletion (I/D) polymorphism in the angiotensin-converting enzyme (ACE) gene.

Among other ACE inhibitor treatment outcomes, the DD genotype was found to be linked to increased ACE activity, whereas the ACE I/D genotype has been linked to renal protective effect, hypertension reduction, reduced hypertrophy of left ventricle, and improved endothelial function. Nevertheless, these findings are contradictory in that some show no relationship between responsiveness and ACE I/D genotype, while others imply that the II genotype is connected to better treatment response. Some polymorphic genes are direct targets of psychiatric medicines, such as those found in dopamine receptor genes and 5-hydroxytryptamine (5HT)/serotonin receptor genes, and have also been the subject of several studies; for example, ACE and 2-β-adrenergic receptor gene polymorphisms and 5HT receptor genetic polymorphism have been linked to efficacy and toxicity in certain studies, but not all (Cacabelos 2020).

Although it may appear that relying on polymorphism in a drug's main target molecule is a sound strategy, it's simple to explain why this technique may yield ambiguous results, and why a genome approach could yield more reliable results, e.g. ACE inhibitors. The ACE inhibitors' direct protein target is ACE. The inhibition of ACE reduces the generation of angiotensin II as well as the decomposition of bradykinin and other vasodilator chemicals, all of which have their own signalling pathways. Inhibiting ACE inhibits the generation of angiotensin II as well as the breakdown of bradykinin and other vasodilator molecules, each of which operates through its own signalling cascade (Garvin et al. 2020). Angiotensin II binds to its G protein-coupled receptor (AT_1), beginning a signalling cascade that involves nearly about 20–30 proteins to achieve the required physiological effects (Pfleger et al. 2019). Considering that polymorphisms are expected to occur every 1,900 bp, each of these 20–30 proteins' genes is likely to contain one or two polymorphisms on average. As a result, it's easy to understand how a single ACE gene variation could not be enough to anticipate a patient's ACE drug response. Realizing that the majority of medications target enzymes or receptors with similar or more sophisticated signalling pathways, genomic methods will be required to better understand the genetic basis for variability in medication response. Many single gene polymorphic pharmacogenetic studies associated with altered disease risk have yielded consistent results. It's been observed that when a medicine known to induce an adverse event is combined with a genetic variation linked to the same adverse event, the chance of drug toxicity rises dramatically. In those individuals with prothrombin mutations, oral contraceptive use increases the risk of a thrombotic event substantially more than either the mutation or usage of oral contraceptives only. Long QT syndrome is connected to gene anomalies in cardiac potassium and sodium channels, which are in turn associated with an increased risk of drug-induced life-threatening heart rhythm disturbances known as "torsade de pointes". This might be a valuable model for recognizing patients who have the possibility of major drug-related side effects.

A well-known transcriptional inactivation mechanism is methylation of cytosines inside CpG-rich regions within promoters of genes in humans. Inter-individual variations in methylation of the O-6-methylguanine-DNA methyltransferase

(MGMT) promoter region have been linked to the responsiveness of malignant gliomas to carmustine treatment. Twelve of the nineteen patients with methylation in promoters of MGMT in their tumours exhibited responses, compared to just one of the 28 patients harbouring unmethylated promoters in MGMT. The mechanism proposed is that individuals with methylation in MGMT promoters had lower MGMT expression, due to the presence of enzyme that causes reversal of damage brought about by the process of alkylation in DNA and is crucial for the action of carmustine therapy. More research is required to confirm the findings from the preliminary investigation, in order to establish the exact mechanisms involved, and explain the genetic basis for inter-individual heterogeneity in methylation of MGMT promoter.

Polymorphisms associated with disease may have an impact on therapeutic effectiveness, in spite of the fact that significant proteins are not engaged directly in the pharmacologic activities of the drug. In this context, the HMG CoA-reductase inhibitors (statins) are quite fascinating. Statin studies in big populations have had mixed results. Polymorphic genes encoding the proteins such as cholesteryl ester transfer protein (CETP), apolipoprotein E, stromelysin-1, and fibrinogen have been examined in statin research in large populations with specific therapeutic goals (other than cholesterol changes) (Ordovas, 1995). The findings of these researches were strikingly comparable. The researchers first looked at a genetic variant that was thought to be or was proved to be related to an unfavourable clinical consequence, e.g. accelerated atherosclerosis progression, cardiovascular event, or even death, but not directly related to the medication's pharmacologic effect. The placebo arm of the investigation confirmed the genotype-clinical prognosis relationship in every case. According to the study, patients with the genotype linked to a higher risk of negative outcomes benefitted the most from statin medication, whereas those in a lower risk group benefitted little or not at all. These investigations also revealed that lipid changes linked with statin therapy are not genotype dependent, implying that the therapeutic advantages are somewhat independent of lipid reduction and hence of the medications' alleged direct pharmacological effect. Parallel findings were made with the polymorphisms associated with apolipoprotein E, efficacy and side effects of levodopa therapy in Parkinson's disease (Siest, 2002), and tacrine responsiveness in Alzheimer's disease (Farlow et al. 1998). Polymorphisms in genes of pathogens such as HIV and Helicobacter pylori are also important indicators of anti-infective drug response (Burgner et al. 2006). However, the details are beyond the scope of this chapter.

The aforementioned single polymorphism-single gene tests provide practical examples that diversity in medication response may be due to genetic variability. Despite the fact that these studies have shown that genetic variability can explain certain differences in pharmacological response, they do not always provide a clinically significant level of prediction. For example, patients without the APO ε4 allele responded well to tacrine in 83% of cases, but only 41% of individuals with the APO ε4 allele responded favourably. Even with these clear genetic response differences, restricting medication solely on depending on whether or not the APO ε4 allele is present would be unacceptable because a large number of patients harbouring this allele benefit from it. In fact, in this study, patients with the APO ε4 allele improved the most on tacrine (Ma & Lu 2011). Furthermore, unlike the tacrine-APO ε4 example, most of the other cases show a much lower difference in response across genotypes. Thus, in order to develop pharmacogenomics, more refined genetic approaches will be necessary, where genetic data may be utilized to estimate accurately the drug response, efficacy and toxicity, with future therapeutic-related decisions based on this understanding.

Regardless of the fact that there are few instances of genomic procedures correlating medication target polymorphisms to responsiveness or toxicity, research suggests that such kind of procedures will furnish more helpful information than the studies focusing on only single gene/polymorphism. Therefore, various polymorphisms in many genes that are known to play a crucial role in drug disposition are studied under "candidate gene approach" for their correlation with response and toxicity.

For example, a multiple candidate gene strategy was employed in a research of clozapine drug response in schizophrenia patients to provide an insight into the role of genetics in responding to clozapine (Lally et al. 2016). This investigation studied the correlation between clozapine sensitivity and polymorphisms in around ten genes; the combined effect of six genetic polymorphisms had the strongest correlation with response and had a positive predictive value of 76%, a poor prognostic value of 82%, and a sensitivity of 96% for recognizing patients affected with schizophrenia, who who improved after using clozapine drug, while a specificity of 38% was observed in patients who had a least response to clozapine. However, more study is needed to test the model in a larger patient population, to see whether it can speculate response to different antipsychotic medicines, and expand the gene and polymorphism repertoire that has been investigated. Even though the methodologies described above could be improved, they offer insights into a strategy that is far more likely to succeed in introducing pharmacogenomics into clinical settings.

The technique that requires no prior information of the drug's pharmacological actions is genome scanning approach, unlike other candidate gene procedures. The use of SNP maps to discover polymorphisms in genes implicated in pharmacokinetics is a pharmacogenomics method that several big pharmaceutical firms aim to adopt in their drug development process (Evans & Johnson 2001). While additional investigation into the mechanisms that cause such correlation is needed, once the relevant SNPs have been identified, the practical use of such technologies should assist in our knowledge of the genetic underpinnings for pharmaceutical efficacy and toxicity variations. To make this technique economical, either the expense of SNP detection must drop significantly, or the number of SNPs examined must be kept low (Sachidanandam et al. 2001) (Figure 10.2).

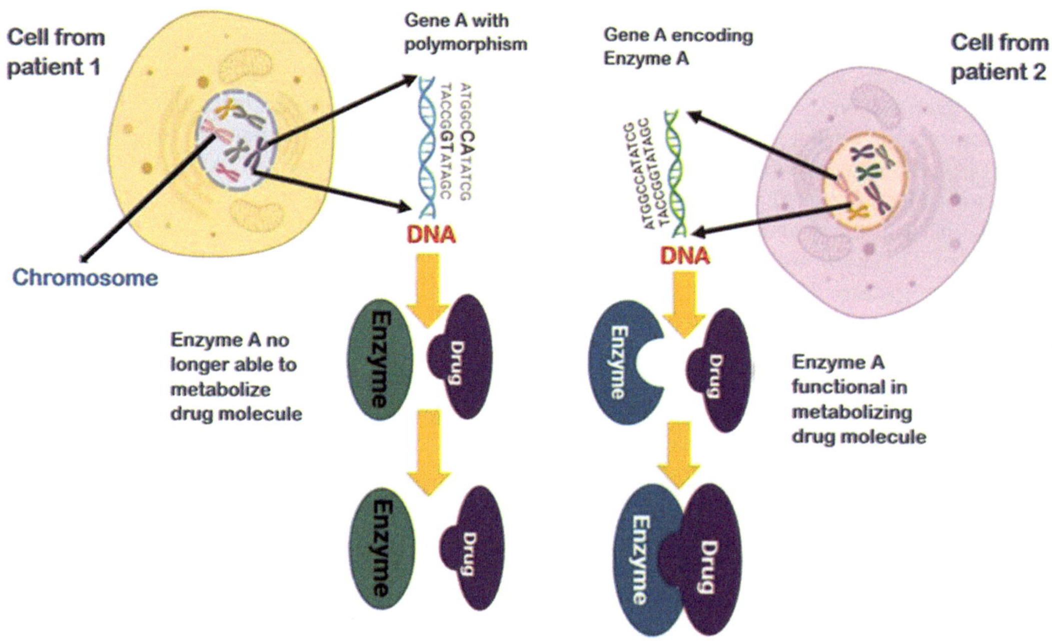

FIGURE 10.2 Genetic polymorphism can alter individual drug response by interfering with drug metabolism.

10.7 Conclusions and Future Aspects

With the evolution of personalized medicine, genetic biomarkers may inspire a shift away from empirical and population-based therapeutic interventions and toward selective pharmacological treatments based on superior patient results, putting an end to the "trial-and-error" treatment strategy. Even among people on the same treatment plan, ADRs might occur. Because adverse drug reactions are a primary cause of death in the USA, this uncertainty poses a serious safety risk. Finding genetic markers that put a patient at risk of an ADR would be extremely useful in preventing these responses. Genetic testing has the ability to predict therapy efficacy. Molecular screening tests should also be able to detect a variety of illnesses early on, when they are possibly preventive, treatable, or curable. Because genetic information is available long before many disease processes begin, early DNA testing may become critical in disease preventive measures. A high cholesterol level detected by a laboratory test, for example, might suggest a future risk of heart disease. On the other hand, a genetic test could be able to detect a mutation in the LDLR gene before cholesterol levels start to rise. Because delayed diagnosis is linked to greater morbidity and the need for more expensive medical treatments, genetic testing may be more cost-effective than phenotypic diagnostics. Advanced genetic screening tests may be able to predict the possibility of a recurrence of illness following a successful treatment. Furthermore, genomics-based prognostic tests may be able to reliably anticipate disease development, with the results helping doctors decide whether or not to utilize chemotherapy or radiation in cancer treatment. Only around half of the people are thought to have a positive reaction to their drugs. As a result, half of the patients may be under-medicated, or there may be therapeutic delays if patients need to switch medicines due to inefficacy. Pharmacogenetics and pharmacogenomics are important parts of the trend toward personalized medicine. Regardless of the fact that numerous variables other than genetics have a role in patient responses to drugs, medication has produced a mountain of evidence demonstrating that genetic polymorphism is a major cause of variation during the last half-century. At variable rates, adverse medication reactions and failures to deliver the intended therapeutic response occur. To attract the attention of both healthcare professionals and the general public, it was probably required to explain noteworthy pharmacogenomic examples, such as those represented by monogenic traits. In terms of translation into the clinic, genomic medical research has its own set of problems. However, there is little doubt that pharmacogenomics will become more extensively and thoroughly used in medicine, allowing us to tailor the usage of potent medications developed throughout the treatment process.

ABBREVIATIONS

5FU:	5-Fluorouracil
5HT:	5-Hydroxytryptamine
6-MP:	6-Mercaptopurine
ACE:	Angiotensin-converting enzyme
ADRs:	Adverse drug responses
APO E:	Apolipoprotein E
CETP:	Cholesteryl ester transfer protein
CpG:	5′-Cytosine-phosphate-guanine-3′
CYP:	Cytochrome P450
DPD:	Dihydropyrimidine dehydrogenase
FDA:	Food and Drug Administration
GI:	Gastrointestinal tract
GSTP:	Glutathione S-transferase pi gene
GWAS:	Genome-wide association study

HIV:	Human immunodeficiency virus
HMG CoA:	3-Hydroxy-3-methyl-glutaryl coenzyme A
I/D:	Insertion/deletion
IM:	Intermediate metabolizer
MGMT:	O-6-Methylguanine-DNA methyltransferase
NAT:	N-acetyltransferase
PMs:	Poor metabolizers
PXR:	Pregnane X receptor
SNP:	Single nucleotide polymorphism
TPMT:	Thiopurine S-methyltransferase
UDP:	Uridine diphosphate
UGTs:	UDP-glucuronosyltransferases
URMs:	Ultra-rapid metabolizers
XME:	Xenobiotic-metabolizing enzyme

REFERENCES

Anzenbacher P, Zanger UM, editors. *Metabolism of Drugs and Other Xenobiotics*. John Wiley & Sons; 2012. May 29. Weinhem Germany.

Benjeddou M, Peiró AM. Pharmacogenomics and prescription opioid use. *Pharmacogenomics*. 2021;22(04):235–45.

Bertilsson L, Dahl ML, Dalén P, Al-Shurbaji A. Molecular genetics of CYP2D6: clinical relevance with focus on psychotropic drugs. *British Journal of Clinical Pharmacology*. 2002;53(2):111–22.

Bissig KD, Han W, Barzi M, Kovalchuk N, Ding L, Fan X, Pankowicz FP, Zhang QY, Ding X. P450-humanized and human liver chimeric mouse models for studying xenobiotic metabolism and toxicity. *Drug Metabolism and Disposition*. 2018;46(11):1734–44.

Bock KW. The UDP-glycosyltransferase (UGT) superfamily expressed in humans, insects and plants: animal- plant arms-race and co-evolution. *Biochemical Pharmacology*. 2016;99:11–7.

Boullata JI, Armenti VT, editors. *Handbook of Drug-Nutrient Interactions*. Totowa, NJ: Humana Press; 2004.

Bozina N, Bradamante V, Lovrić M. Genetic polymorphism of metabolic enzymes P450 (CYP) as a susceptibility factor for drug response, toxicity, and cancer risk. *Arhiv za higijenu rada i toksikologiju*. 2009;60(2):217–42.

Brunton LL, Chabner B, Knollmann BC, editors. *Goodman & Gilman's the Pharmacological Basis of Therapeutics*. New York, NY: McGraw-Hill Education; 2018.

Buck E, Sprick M, Gaida MM, Grüllich C, Weber TF, Herpel E, Bruckner T, Koschny R. Tumor response to irinotecan is associated with CYP3A5 expression in colorectal cancer. *Oncology Letters*. 2019;17(4):3890–8.

Burgner D, Jamieson SE, Blackwell JM. Genetic susceptibility to infectious diseases: big is beautiful, but will bigger be even better? *The Lancet Infectious Diseases*. 2006;6(10):653–63.

Cacabelos R. Pharmacogenomics of Alzheimer's and Parkinson's diseases. *Neuroscience Letters*. 2020;726:133807.

Carr DF, Turner RM, Pirmohamed M. Pharmacogenomics of anticancer drugs: personalising the choice and dose to manage drug response. *British Journal of Clinical Pharmacology*. 2021;87(2):237–55.

Carranza-Leon D, Dickson AL, Gaedigk A, Stein CM, Chung CP. CYP2D6 genotype and reduced codeine analgesic effect in real-world clinical practice. *The Pharmacogenomics Journal*. 2021:1–7.

Carson PE, Flanagan CL, Ickes CE, Alving AS. Enzymatic deficiency in primaquine-sensitive erythrocytes. *Science*. 1956;124(3220):484–5.

Chamberlain J. *The Analysis of Drugs in Biological Fluids*. CRC press; 2018 Feb 6. Boca Raton.

Crews KR, Hicks JK, Pui CH, Relling MV, Evans WE. Pharmacogenomics and individualized medicine: translating science into practice. *Clinical Pharmacology & Therapeutics*. 2012;92(4):467–75.

Daly AK, Rettie AE, Fowler DM, Miners JO. Pharmacogenomics of CYP2C9: functional and clinical considerations. *Journal of Personalized Medicine*. 2018;8(1):1.

Danielson PÁ. The cytochrome P450 superfamily: biochemistry, evolution and drug metabolism in humans. *Current Drug Metabolism*. 2002;3(6):561–97.

DeBose-Boyd RA. A helping hand for cytochrome p450 enzymes. *Cell Metabolism*. 2007;5(2):81–3.

Demkow U. Next generation sequencing in pharmacogenomics. In *Clinical Applications for Next-Generation Sequencing* (pp. 217–40). Academic Press; 2016. Elsevier. USA Editor Urszula Demkow and Rafał Płoski. https://doi.org/10.1016/C2014-0-00597-7

Dixit VA, Lal LA, Agrawal SR. Recent advances in the prediction of non-CYP450-mediated drug metabolism. *Wiley Interdisciplinary Reviews: Computational Molecular Science*. 2017;7(6):e1323.

Domanski TL, Finta C, Halpert JR, Zaphiropoulos PG. cDNA cloning and initial characterization of CYP3A43, a novel human cytochrome P450. *Molecular Pharmacology*. 2001;59(2):386–92.

Evans WE, Johnson JA. Pharmacogenomics: the inherited basis for interindividual differences in drug response. *Annual Review of Genomics and Human Genetics*. 2001; 2(1):9–39.

Evans WE, Relling MV. Pharmacogenomics: translating functional genomics into rational therapeutics. *Science*. 1999;286(5439):487–91.

Fan Z, Zheng D, Wen X, Shen F, Lei L, Su S, Zhang S, Liu Q, Zhang X, Lu Y, Di L. CYP3A5* 3 polymorphism and age affect tacrolimus blood trough concentration in myasthenia gravis patients. *Journal of Neuroimmunology*. 2021;355:577571.

Farlow MR, Lahiri DK, Poirier J, Davignon J, Schneider L, Hui SL. Treatment outcome of tacrine therapy depends on apolipoprotein genotype and gender of the subjects with Alzheimer's disease. *Neurology*. 1998;50(3): 669–77.

Fekete F, Mangó K, Déri M, Incze E, Minus A, Monostory K. Impact of genetic and non-genetic factors on hepatic CYP2C9 expression and activity in Hungarian subjects. *Scientific Reports*. 2021;11(1):1–2.

Garvin MR, Alvarez C, Miller JI, Prates ET, Walker AM, Amos BK, Mast AE, Justice A, Aronow B, Jacobson D. A mechanistic model and therapeutic interventions for COVID-19 involving a RAS-mediated bradykinin storm. *Elife*. 2020;9:e59177.

Geisler T, Schaeffeler E, Schwab M. Platelets and Polymorphisms. In: Zirlik, A., Bode, C., Gawaz, M. (eds) Platelets, Haemostasis and Inflammation. Cardiac and Vascular Biology, vol 5. Springer, Cham. 2017. https://doi.org/10.1007/978-3-319-66224-4_17

Gomes AM, Winter S, Klein K, Turpeinen M, Schaeffeler E, Schwab M, Zanger UM. "Pharmacogenomics of human liver cytochrome P450 oxidoreductase: multifactorial analysis and impact on microsomal drug oxidation". 2009: 579–599.

Gong L, Zhang CM, Lv JF, Zhou HH, Fan L. Polymorphisms in cytochrome P450 oxidoreductase and its effect on drug metabolism and efficacy. *Pharmacogenetics and Genomics.* 2017;27(9):337–46.

Gonzalez FJ, Skodat RC, Kimura S, Umeno M, Zanger UM, Nebert DW, Gelboin HV, Hardwick JP, Meyer UA. Characterization of the common genetic defect in humans deficient in debrisoquine metabolism. *Nature.* 1988;331(6155):442–6.

Goodman LS, Gilman A, Brunton LL, Lazo JS, Parker K. *Goodman & Gilman's the Pharmacological Basis of Therapeutics*, 11th edn. McGraw-Hill; 2006. Editor: Laurence L brunton. United States of America.

Hart SN, Wang S, Nakamoto K, Wesselman C, Li Y, Zhong XB. Genetic polymorphisms in cytochrome P450 oxidoreductase influence microsomal P450-catalyzed drug metabolism. *Pharmacogenetics and Genomics.* 2008;18(1):11–24.

Hughes HB, Biehl JP, Jones AP, Schmidt LH. Metabolism of isoniazid in man as related to the occurrence of peripheral neuritis. *American Review of Tuberculosis.* 1954;70(2):266–73.

Ingelman-Sundberg M, Sim SC, Gomez A, Rodriguez-Antona C. Influence of cytochrome P450 polymorphisms on drug therapies: pharmacogenetic, pharmacoepigenetic and clinical aspects. *Pharmacology & Therapeutics.* 2007;116(3):496–526.

Katsanos K, Tsianos EV. Non-TPMT determinants of azathioprine toxicity in inflammatory bowel disease. *Annals of Gastroenterology.* 2010:95–104.

Kim K, Johnson JA, Derendorf H. Differences in drug pharmacokinetics between East Asians and Caucasians and the role of genetic polymorphisms. *The Journal of Clinical Pharmacology.* 2004;44(10):1083–105.

King RA, Rotter JI, Motulsky AG, editors. *The Genetic Basis of Common Diseases.* Oxford University Press; 2002. Oxford University Press. New York. Oxford monographs on medical genetics 44.

Kivisto KT, Kroemer HK, Eichelbaum M. The role of human cytochrome P450 enzymes in the metabolism of anticancer agents: implications for drug interactions. *British Journal of Clinical Pharmacology.* 1995;40(6):523–30.

Lally J, Gaughran F, Timms P, Curran SR. Treatment-resistant schizophrenia: current insights on the pharmacogenomics of antipsychotics. *Pharmacogenomics and Personalized Medicine.* 2016;9:117.

Laurieri N, Egleton JE, Russell AJ. Human arylamine N-acetyltransferase type 1 and breast cancer. In *Arylamine N-Acetyltransferases in Health and Disease: From Pharmacogenetics to Drug Discovery and Diagnostics* (pp. 351–384); 2018. World Scientific Publishing Co Pte Ltd. Singapore . Editors Nicola Laurieri edith Sim.

Lee SY, McLeod HL. Pharmacogenetic tests in cancer chemotherapy: what physicians should know for clinical application. *The Journal of Pathology.* 2011;223(1):15–27.

Li H, Lampe JN. Neonatal cytochrome P450 CYP3A7: a comprehensive review of its role in development, disease, and xenobiotic metabolism. *Archives of Biochemistry and Biophysics.* 2019;673:108078.

Lin JH, Lu AY. Role of pharmacokinetics and metabolism in drug discovery and development. *Pharmacological Reviews.* 1997;49(4):403–49.

Ma Q, Lu AY. Pharmacogenetics, pharmacogenomics, and individualized medicine. *Pharmacological Reviews.* 2011;63(2):437–59.

McLeod HL, Evans WE. Pharmacogenomics: unlocking the human genome for better drug therapy. *Annual Review of Pharmacology and Toxicology.* 2001;41(1):101–21.

Meech R, Hu DG, McKinnon RA, Mubarokah SN, Haines AZ, Nair PC, Rowland A, Mackenzie PI. The UDP-glycosyltransferase (UGT) superfamily: new members, new functions, and novel paradigms. *Physiological Reviews.* 2019;99(2):1153–222.

Meyer UA, Zanger UM. Molecular mechanisms of genetic polymorphisms of drug metabolism. *Annual Review of Pharmacology and Toxicology.* 1997;37(1):269–96.

Mizutani T. PM frequencies of major CYPs in Asians and Caucasians. *Drug Metabolism Reviews.* 2003;35(2–3):99–106.

Motulsky AG. Drug reactions, enzymes, and biochemical genetics. *Journal of the American Medical Association.* 1957;165(7):835–7.

Nebert DW, Russell DW. Clinical importance of the cytochromes P450. *The Lancet.* 2002;360(9340):1155–62.

Novelletto A, Ciminelli BM. DNA polymorphisms and genome structure: different scales of variation in the human genome. In *Forensic DNA Analysis* (pp. 3–22). Apple Academic Press; 2021. Boca Raton. Editor Jaiprash G Shewale and Ray H Liu Edition Ist. eBook ISBN9781003043027

Okumura K, Kita T, Chikazawa S, Komada F, Iwakawa S, Tanigawara Y. Genotyping of N-acetylation polymorphism and correlation with procainamide metabolism. *Clinical Pharmacology & Therapeutics.* 1997;61(5):509–17.

Olopade OI, Grushko TA, Nanda R, Huo D. Advances in breast cancer: pathways to personalized medicine. *Clinical Cancer Research.* 2008;14(24):7988–99.

Omura T, Sato R. A new cytochrome in liver microsomes. *Journal of Biological Chemistry.* 1962;237(4):PC1375–6.

Ordovas JM, Lopez-Miranda J, Perez-Jimenez F, Rodriguez C, Park JS, Cole T, Schaefer EJ. Effect of apolipoprotein E and A-IV phenotypes on the low density lipoprotein response to HMG CoA reductase inhibitor therapy. *Atherosclerosis.* 1995;113(2):157–66.

Paine MF, Hart HL, Ludington SS, Haining RL, Rettie AE, Zeldin DC. The human intestinal cytochrome P450 "pie". *Drug Metabolism and Disposition.* 2006;34(5):880–6.

Pfleger J, Gresham K, Koch WJ. G protein-coupled receptor kinases as therapeutic targets in the heart. *Nature Reviews Cardiology.* 2019;16(10):612–22.

Qin XP, Xie HG, Wang W, He N, Huang SL, Xu ZH, Ou-Yang DS, Wang YJ, Zhou HH. Effect of the gene dosage of CYP2C19 on diazepam metabolism in Chinese subjects. *Clinical Pharmacology & Therapeutics.* 1999;66(6):642–6.

Qiu H, Taudien S, Herlyn H, Schmitz J, Zhou Y, Chen G, Roberto R, Rocchi M, Platzer M, Wojnowski L. CYP3 phylogenomics: evidence for positive selection of CYP3A4 and CYP3A7. *Pharmacogenetics and Genomics.* 2008;18(1):53–66.

Rashid M. Human genome variations: databases and bioinformatics resources. *Polymorphism*. 2020;4:21–30.

Ruzzo A, Graziano F, Galli F, Galli F, Rulli E, Lonardi S, Ronzoni M, Massidda B, Zagonel V, Pella N, Mucciarini C. Dihydropyrimidine dehydrogenase pharmacogenetics for predicting fluoropyrimidine-related toxicity in the randomised, phase III adjuvant TOSCA trial in high-risk colon cancer patients. *British Journal of Cancer*. 2017;117(9):1269–77.

Sachidanandam R, Weissman D, Schmidt SC, Kakol JM, Stein LD, Marth G, Sherry S, Mullikin JC, Mortimore BJ, Willey DL, Hunt SE. A map of human genome sequence variation containing 1.42 million single nucleotide polymorphisms. *Nature*. 2001;409(6822):928–34.

Sangkuhl K, Claudio-Campos K, Cavallari LH, Agundez JA, Whirl-Carrillo M, Duconge J, Del Tredici AL, Wadelius M, Botton MR, Woodahl EL, Scott SA. PharmVar GeneFocus: CYP2C9. *Clinical Pharmacology & Therapeutics*. 2021. Sep;110(3):662–76.

Shargel, L. *Comprehensive Pharmacy Review*, 7th edn. Philadelphia: Lipincott William &Wilkins; 2010.

Sharma V, Gupta SK, Verma M. Dihydropyrimidine dehydrogenase in the metabolism of the anticancer drugs. *Cancer Chemotherapy and Pharmacology*. 2019;84(6): 1157–66.

Shastry BS. SNP alleles in human disease and evolution. *Journal of Human Genetics*. 2002;47(11):561–6.

Shastry BS. SNPs in disease gene mapping, medicinal drug development and evolution. *Journal of Human Genetics*. 2007;52(11):871–80.

Shenfield G, Gross A. The cytochrome P450 system and adverse drug reactions. *Adverse Drug Reaction Bulletin*. 1999;194(1):739–42.

Shenfield GM. Genetic polymorphisms, drug metabolism and drug concentrations. *The Clinical Biochemist Reviews*. 2004;25(4):203.

Siest G, Zaiou M, Visvikis S. Human apolipoprotein E concentration in response to diseases and therapeutic treatments. *Drug Development Research*. 2002;56(2):95–110.

Sim E, Laurieri N. Drug metabolism and pharmacogenetics then and now In *Arylamine N-Acetyltransferases in Health and Disease: From Pharmacogenetics to Drug Discovery and Diagnostics* (pp. 3–41); 2018.

Sim SC, Risinger C, Dahl ML, Aklillu E, Christensen M, Bertilsson L, Ingelman-Sundberg M. A common novel CYP2C19 gene variant causes ultrarapid drug metabolism relevant for the drug response to proton pump inhibitors and antidepressants. *Clinical Pharmacology & Therapeutics*. 2006;79(1):103–13.

Stanley LA. Drug metabolism. In *Pharmacognosy* (pp. 527–545). Academic Press; 2017. Elsevier. London editors Simone Badal and Rupika Delgode https://doi.org/10.1016/B978-0-12-802104-0.00027-5

Sugatani J, Yamakawa K, Yoshinari K, Machida T, Takagi H, Mori M, Kakizaki S, Sueyoshi T, Negishi M, Miwa M. Identification of a defect in the UGT1A1 gene promoter and its association with hyperbilirubinemia. *Biochemical and Biophysical Research Communications*. 2002;292(2):492–7.

Summerscales JE, Josephy PD. Human acetyl CoA: arylamine N-acetyltransferase variants generated by random mutagenesis. *Molecular Pharmacology*. 2004;65(1):220–6.

Sun B, Guo Y, Gao J, Shi W, Fan G, Li X, Qiu J, Qin Y, Liu G. Influence of CYP3A and ABCB1 polymorphisms on cyclosporine concentrations in renal transplant recipients. *Pharmacogenomics*. 2017;18(16):1503–13.

Vogel F. *Moderne probleme der humangenetik. InErgebnisse der inneren medizin und kinderheilkunde* (pp. 52–125). Berlin, Heidelberg: Springer; 1959.

Witherspoon DJ, Wooding S, Rogers AR, Marchani EE, Watkins WS, Batzer MA, Jorde LB. Genetic similarities within and between human populations. *Genetics*. 2007;176(1):351–9.

Yeaman S, Otto SP. Establishment and maintenance of adaptive genetic divergence under migration, selection, and drift. *Evolution: International Journal of Organic Evolution*. 2011;65(7):2123–9.

Zahari Z, Ismail R. Influence of Cytochrome P450, Family 2, Subfamily D, Polypeptide 6 (CYP2D6) polymorphisms on pain sensitivity and clinical response to weak opioid analgesics. *Drug Metabolism and Pharmacokinetics*. 2013:DMPK–13.

Zanger UM, Raimundo S, Eichelbaum M. Cytochrome P450 2D6: overview and update on pharmacology, genetics, biochemistry. *Naunyn-Schmiedeberg's Archives of Pharmacology*. 2004;369(1):23–37.

Zanger UM, Schwab M. Cytochrome P450 enzymes in drug metabolism: regulation of gene expression, enzyme activities, and impact of genetic variation. *Pharmacology & Therapeutics*. 2013;138(1):103–41.

Zhao F, Wang J, Yang Y, Wang X, Shi R, Xu Z, Huang Z, Zhang G. Effect of CYP2C19 genetic polymorphisms on the efficacy of proton pump inhibitor-based triple therapy for Helicobacter pylori eradication: a meta-analysis. *Helicobacter*. 2008;13(6):532–41.

Zhou SF, Ming Di Y, Chan E, Du YM, Chow VD, Xue CC, Lai X, Wang JC, Li CG, Tian M, Duan W. Clinical pharmacogenetics and potential application in personalized medicine. *Current Drug Metabolism*. 2008;9(8):738–84.

Zhou SF, Yang LP, Zhou ZW, Liu YH, Chan E. Insights into the substrate specificity, inhibitors, regulation, and polymorphisms and the clinical impact of human cytochrome P450 1A2. *The AAPS Journal*. 2009;11(3):481–94.

11

Role of Genetic Polymorphisms in Chronic Inflammatory Diseases

Saniya Nissar
Department of Biochemistry, Government Medical College, Karan Nagar, Srinagar, Kashmir, India
Infinity Biochemistry Pvt. Ltd. Srinagar, Kashmir, India

Syed Sameer Aga
Department of Basic Medical Sciences, College of Medicine, King Saud bin Abdulaziz University for Health Sciences (KSAU-HS), King Abdullah International Medical Research Center (KAIMRC), Ministry of National Guard Health Affairs (NGHA), Jeddah, Kingdom of Saudi Arabia
Infinity Biochemistry Pvt. Ltd. Srinagar, Kashmir, India

Mujeeb Zafar Banday
Department of Biochemistry, Government Medical College, Karan Nagar, Srinagar, Kashmir, India
Infinity Biochemistry Pvt. Ltd. Srinagar, Kashmir, India

Roohi Rasool
Department of Immunolgy and Molecular Medicine,
Sher-i-Kashmir Institute of Medical Sciences, Soura, Srinagar, Kashmir, India

CONTENTS

ABSTRACT Chronic inflammatory diseases (CIDs) are known to be the significant cause of mortality worldwide, with more than 50% of all deaths being attributed to CIDs, especially cancers, diabetes mellitus (DM), autoimmune disorders such as rheumatoid arthritis (RA), and neurodegenerative conditions such as multiple sclerosis (MS). Several genetic factors that result in the development of various CIDs have been identified. With the help of advanced techniques such as genome-wide association studies (GWAS), the functional significance and the impact of genetic polymorphisms, especially single nucleotide polymorphisms (SNPs), on the modulation of susceptibility to CID risk have widely been studied and reported all across the world in various populations. In addition, the role of SNPs in determining the disease progression is currently a hot field of research. Considering the strong evidence that inflammatory diseases occur because of inappropriate inflammatory responses that are regulated by both genetic as well as environmental factors, this chapter discusses the role of several important gene SNPs that have a direct bearing on risk modulation of different CIDs.

DOI: 10.1201/9781003246244-11

KEYWORDS: *Chronic Inflammatory Diseases; Polymorphisms; Single Nucleotide Polymorphism; SNP; Genome-Wide Association Studies; Inflammation; Rheumatoid Arthritis; Multiple Sclerosis; Diabetes Mellitus.*

11.1 Introduction

Chronic inflammatory diseases (CIDs) are the common diseases associated with inflammation. They are the most significant cause of death and greatest burden to human health because of life-long incapacitating illness, higher mortality, and expensive therapy and high care costs. Chronic inflammation is a process of slow and long-lasting inflammation, which prolongs from a few months to years. Usually, the impact of chronic inflammation depends upon the source of injury and the response of the human body toward such damage and its repairing ability to overcome it (National Centre for Chronic disease prevention, 2020).

There are a number of diseases mediated by chronic inflammation, which include diabetes, cardiovascular diseases, pulmonary disease, bowel diseases, cancer, arthritis and joint diseases, kidney, and liver disease (National Centre for Chronic disease prevention, 2020). The prevalence of CID tends to increase persistently. According to a survey (2014), it is estimated that nearly 60% of Americans have at least one chronic disease, 42% have more than one, and 12% of adults have 5 or more types of CIDs (Pahwa et al., 2021). Worldwide, 3 of 5 people die due to chronic inflammatory diseases such as stroke, heart diseases, cancer, and diabetes (Barcelos et al., 2019; Tsai et al., 2019).

Normally, there occurs temporary upregulation of inflammatory activity characterized by features such as vasodilation, increased blood flow, and capillary permeability and migration of neutrophils into the infected tissue through the capillary wall (diapedesis) during acute inflammation because of injury or damage, which resolve once the threat has passed (Kotas and Medzhitov, 2015; Straub RH, 2017; Yousuf et al., 2019). However, certain environmental factors prevent the resolution of acute inflammation and tend to promote it to higher grade, whereby the composition of the white blood cells starts to change and there occurs the infiltration of the primary inflammatory cells such as macrophages, lymphocytes, and plasma cells in the tissue site (Cutolo et al., 2019).

These cells produce inflammatory cytokines, growth factors, and enzymes and thus contribute to the progression of tissue damage and secondary repair together with fibrosis and granuloma formation (Needham et al., 2019; Milenkovic et al., 2019). Thus, because of a shift from short- to long-lived inflammatory response, the "immune tolerance" breaks down, which may cause alteration to normal cellular physiology, tissues, and finally major body organs (Straub, 2017). Thus, the risk of various diseases in both young and adult individuals increases (Fullerton and Gilroy, 2016; Miller and Raison 2016; Taniguchi and Karin, 2018).

11.2 Inflammatory Bowel Disease (IBD)

Inflammatory bowel disease (IBD) is a complex, multifunctional disorder, which is characterized by the chronic and recurrent inflammation of the digestive tract (Dudzinska et al., 2018). It includes two chronic idiopathic inflammatory diseases: ulcerative colitis (UC) and Crohn's disease (CD). Generally, both UC and CD etiologically arise in early adulthood, but they do have propensity to arise at any age from childhood onward. The diagnosis of these two diseases is often delayed despite their prominent signs and symptoms such as diarrhea, abdominal pain, and, in ulcerative colitis, peranal bleeding. CD affects the entire gastrointestinal tract (GIT) transmurally, beginning from the mouth till the anus, while UC mainly affects the colonic mucosa (Wehkamp et al., 2016).

The incidence and prevalence of IBDs are increasing day by day worldwide and are high in western countries (Sairenji et al., 2017). The highest annual incidence of UC in North America was reported to be 19.2 per 100,000 person-years, and the prevalence was 505 per 100,000 persons. For CD, the highest reported incidence was 20.2 per 100,000 person-years in North America and the prevalence was 322 per 100,000 persons (Molodecky et al., 2012). In Europe, the prevalence of CID is 12.7/100,000 among children and 24.3/100,000 in adults and the rates are increasing steadily among both young and adult population (Loddo and Romano, 2015). The age of onset for both CD and UC is the same, with the peak incidence in the second to fourth decade of life with no significant gender differences (Molodecky et al., 2012).

Being polygenic disorders, the risk is high if there is family history with relative risk of fivefold or greater in first-degree relatives developing IBD (Orholm et al., 1991). In addition, some high penetrative genes are common to both diseases, as cases of CD and UC can occur within the same family (Cho and Weaver, 2007). Earlier disease onset in children of parents (genetic predisposition) occurs in both diseases, but the tendency is stronger in UC (Bengtson et al., 2009; Lee et al., 1999). Environment factors also play a key trigger and primer in the manifestation of both of these diseases (Xavier and Podolsky, 2007). Dietary changes and frequent and unchecked use of antibiotics that affect the intestinal flora are also among the chief factors that play a role in the etiology of both the diseases (Eckburg and Relman, 2007).

The pathogenesis involves both genetic and environmental risk factors and appears as a complex interplay between host genetics, immune dysfunction, and microbial or environmental factors (Sairenji et al., 2017). The bowel inflammation is due to the chronic dysregulation of the immune system in response to the changes in commensal (non-pathogenic) gut flora (Abraham and Cho, 2009). Numerous genetic studies have demonstrated that the host-microbe interactions play a noticeable role in the pathogenesis of both UC and CD and involve genomic regions that regulate microbial defense and intestinal inflammation (Mangan et al., 2006).

11.2.1 Single Nucleotide Polymorphisms in IBD

Numerous SNPs in genes have been identified to be associated with CD and UC and may turn out either as harmful or as protective. Usually, a single SNP or several simultaneous SNPs within the critical gene are associated with the risk of developing IBD. It is reported that CD patients possess significantly more risk alleles as compared to controls, suggesting that individuals with fewer than six risk alleles are unlikely to develop CD (Weersema et al., 2009).

11.2.1.1 Human Leukocyte Antigen (HLA) Genes and IBD

The major histocompatibility complex (MHC) encodes different classes of human leukocyte antigen (HLA) genes and is located on chromosome 6p (6p21.31) covering an area of 7.6 Mb and containing more than 250 genes with different functions (Trowsdale and Knight, 2013). HLA genes have been reported to be significantly associated with IBD susceptibility, especially CD (van Heel et al., 2004; Ferguson et al., 2007), while HLA class II molecules DR and DQ have been reported to be associated with IBD susceptibility. Both HLA-DR and HLA-DQ are heterodimers composed of an alpha and beta chain, and both are cell surface receptors whose function is to present specific antigens to T lymphocytes. The DR α-chain is encoded by the DRA locus, and the DR β-chain is encoded by DRB loci. Similarly, DQA1 and DQB1 loci encode the α- and β-chains, respectively, of the DQ molecule. The DR and DQ loci are highly linked to each other and, to a lesser degree, to other HLA loci (Karp et al., 1990).

HLA class II genes DR2, DR9, and DR31*0103 have been reported to be highly susceptible genes for UC, whereas DR7, DRB3*0301, and DQ4 are susceptible genes for CD (Stokker et al., 1999). However, DR4 was found to be a protective gene for UC, whereas DR2 and DR3 were protective genes for CD (Stokker et al., 1999). The association between DR2 and UC can be attributed to the subtype DRB1*1502 (Trachtenberg et al., 2000). In particular, HLA-DR and HLA-DQ genes and HLA class II SNP rs9469220 were associated with CD (Wellcome Trust Case Control Consortium, 2007), whereas HLA DR-A SNP rs9268853 (Festen et al., 2010) and SNP rs2395185 (Juyal et al., 2011) were associated with UC. SNPs rs3763313 and rs2395185 in HLA butyrophilin-like 2 (HLA-BTNL2), which is implicated in T cell activation, were also found to be associated with UC (Juyal et al., 2011; Latiano et al., 2011; Valdes et al., 2011).

11.2.1.1.1 NK2 Transcription Factor-Related, Locus 3 (NKX2-3)

The NKX-3 gene located on chromosome 10q24.2 is a member of a family of genes that encode transcription factors containing homeodomains. This gene product is important for B cell- and T cell-dependent immune responses, and any mutation leads to altered gut migration of antigen-responsive lymphocytes, influencing inflammatory response (Weersma et al., 2009; Yu et al., 2009). The association between the NKX2-3 polymorphism and susceptibility to IBD was first reported in Caucasian patients (Wellcome Trust Case Control Consortium, 2007). SNP rs10883365 is linked to CD susceptibility with GG genotype being associated with CD compared with the wild-type AA genotype (Yu et al., 2009). Also NKX2-3 mRNA and protein expression are upregulated in CD patients, suggesting NKX2-3 is involved in CD pathogenesis (Yu et al., 2009). Moreover, SNP rs888208 is reported to be associated with UC, but the relation between SNP rs10883365 and UC remains unclear (Meggyesi et al., 2010; Arai et al., 2011).

A meta-analysis (17 studies; 17,329 cases and 18,029 controls) showed a significantly elevated risk of developing CD or UC in individuals carrying a G allele of SNP rs10883365 (A/G) compared to those having an A allele (OR = 1.226, 95% CI: 1.177–1.277 and OR = 1.274, 95% CI: 1.175–1.382, respectively). Also, a significantly increased CD risk was reported in both Europeans and Asians. For rs11190140 polymorphism (CT), there was a significant association (OR = 1.201, 95% CI: 1.136–1.269) with the development of CD risk (Lu et al., 2014).

Also, SNPs in the genes involved in T cell differentiation are found to increase the risk of developing UC. For example, the Runt domain transcription factor 3 (RUNX3) located on chromosome 1p36.11 has been found to be associated with the risk of susceptibility to UC. A study on SNP rs2236851 of RUNX3 shows the T allele is present with a higher frequency in UC patients compared with CD patients as well as controls (Weersma et al., 2009). Further, the prostaglandin E receptor 4 gene (PTGER4), located on 5p13.1, encodes a receptor that is involved in the initiation of immune responses, and SNP rs4613763 in PGTER4 has been found to be associated with UC (Latiano et al., 2011).

11.2.1.1.2 Tumor Necrosis Factor-Alpha (TNF-α)

TNF-α is one of the important cytokines which plays a pivotal role in the inflammatory response. TNF gene locus is found in the IBD3 region on chromosome 6p21, which has been found to be associated with an increased susceptibility to IBD. One of the important SNPs in the TNF-α promoter -308G/A (rs1800629) has been found to increase the risk of UC, while another SNP -857 C/T (rs1799724) has been found to decrease IBD risk in certain groups (Ferguson et al., 2008).

In one of the recent studies, TNF-α gene SNP 308G/A (rs1800629) has been associated with IBD risk in the Saudi population. The allele A and genotype GA were associated with susceptibility to IBD risk ($p < 0.001$), while the allele G and genotype GG were found to be protective against IBD ($p < 0.001$). Upon stratification, the results showed TNF-α (–308G/A) polymorphism is substantial for the risk of both UC and CD ($p < 0.001$) (Al-Meghaiseeb et al., 2016). Another study on TNF-α SNP rs1799964 (–1031T/C) revealed that the frequency of TNF-α-1031 C allele was considerably low, whereas the T allele conferred a significant risk of developing UC among patients with UC (Bonyadi et al., 2014).

Another study on TNF-α rs1800629 showed the significantly higher frequency of allele in UC patients compared with controls ($p = 0.0005$). Also, the dominant AA + AG vs GG ($p = 0.02$) and recessive models ($p = 0.0001$) were found to have a statistically significant association with UC. In the same way, an association of TNF-α rs1800629 variant was observed with IBD under the recessive model AA vs AG+GG ($p = 0.004$) (Senhaji et al., 2016).

The interaction between TNF-α and its receptor plays an important role in the development of inflammatory responses in IBD. The analysis of 23 SNPs indicated haplotypes GGC (2.9-fold more) and TAT (rs1061622/rs1061624/rs3397) were associated with CD, whereas TAC and TGT (0.47-fold less) were protective. Also, haplotypes TAC and GAT were associated with UC, whereas GAC and TGC were protective in Caucasians of New Zealand (Ferguson et al., 2009).

11.2.1.2 Non-Human Leukocyte Antigen (HLA) Susceptibility Genes

In addition to HLA genes, the genes that are located outside the HLA coding region (non-HLA genes) of the human genome have also been identified to be associated with autoimmune disorders. Several SNPs in such non-HLA genes, which are associated with CD and UC, have been identified. It is suggested that usually, there exists more than one risk allele in the single patient. For example, Weersma et al. found that it is likely that an individual with more than six risk alleles will develop CD. Several of the IBD susceptibility SNPs are discussed below.

11.2.1.2.1 Interleukin 17 (IL-17)

IL-17 cytokine has pleiotropic activities and plays an important role in immune responses (Liang et al., 2007). It is involved in the neutrophil proliferation, maturation, chemotaxis and T cell stimulation, and maturation of dendritic cells (Kolls and Lindén, 2004). IL-17 has six ligands (IL-17A to IL-17F) and five receptors, i.e., IL-17RA to IL-17RE (McGeachy et al., 2019).

IL-17A and IL-17F genes are usually expressed together and are located adjacent to each other on chromosome 6p12 (Moseley et al., 2003). They are the most extensively studied member of the IL-17 family. A study on IL-17F SNPs (rs763780, 7488T/C) and IL-17A SNP (rs2275913, G-197A) in Chinese Han population observed an increased risk of development of CD among subjects with variant IL-17F SNP, which affected some clinical features of UC and CD. However, IL-17A variant (AA genotype) had a weak association with the severity of UC. This study suggested a risk haplotype in IL-17A could elevate the UC risk (Zhang et al., 2013).

Also, a significant association was found in Japanese population between IL-17A SNP rs2275913 and IL-17F SNP rs763780 and the risk of UC (Arisawa et al., 2008). For IL-17F SNP (rs763780, 7488T/C), the frequency of wild genotype T/T was found to be high in comparison with other genotypes and significantly associated with an increased risk of development of UC (Arisawa et al., 2008). Furthermore, a number of studies showed a significant relationship between IL-17A SNP (rs2275913, G-197A) and UC susceptibility (Kim et al., 2011; Arisawa et al., 2008).

11.2.1.2.2 Interleukin 23 Receptor (IL-23R)

The IL-23 gene is located on chromosome 1p31.3. The functional IL-23R is composed of two subunits: IL-23R and IL-23Rß1 (Cho, 2008). The ligand IL-23 in combination with IL-6 and TGF-1 stimulates the naive CD^{4+}T cells and subverts T-helper cells (Th1 and Th2) differentiation for the generation of Th17 and promotes pro-inflammatory cytokine IL-17 secretion upon binding to its receptor (Veldhoen et al., 2006). This IL-23/IL-17 cytokine alliance mediates the development and progression of inflammation via Th17 cells (Rafa et al., 2013).

A meta-analysis (5,438 cases and 7,380 controls) observed the associations of the different SNPs in the IL-23R and their influence on the susceptibility to UC. The results showed that IL-23R SNP rs1120902 (dominant model), SNP rs7517847 (recessive model), and SNP rs11209032 (dominant model) were found to be associated with UC risk. It was also observed that the SNP rs11209026 and SNP rs7517847 in the IL-23R gene could have a protective effect, whereas the SNP rs11209032 was associated with a greater risk of UC in the Caucasians and the SNP rs1088967 was associated with a lower risk of UC among Asian populations. Moreover, there was no association between the four alleles of the rs2201841, rs1004819, rs1495965, and rs1343151 polymorphisms and the risk of developing UC in Caucasian and Asian populations (Liu et al., 2015b).

Another study by Glas et al. showed SNP rs1004819 is the major IL-23R variant associated with CD in German population, while the SNP rs11209026 variant is a protective marker for CD and UC (Glas et al., 2007). Further, IL-23R SNP rs11209026 was found to be associated with both CD (p value=0.0026, OR 0.54, 95% CI 0.36–0.81) and UC (p value=0.037, OR 0.66, 95% CI 0.45–0.98) (Roberts et al., 2007).

11.2.1.2.3 Caspase Activation and Recruitment Domain Genes (CARD15)

CARD15, also known as a nucleotide oligomerization domain 2 (NOD2), is located at chromosome 16 (Libioulle et al., 2007). It was the first CD susceptibility gene to be identified (Mitrovič and Potočnik, 2011) and functions to sense the components of pathogenic bacterial cell walls such as muramyl dipeptide and peptidoglycans. It is expressed primarily in macrophages and paneth cells and is involved in modulating the activity of nuclear factor-kB (NF-kB) (Ferguson et al., 2007). Several CARD15 gene polymorphisms (rs2066844, rs2066845, and rs2066847) were found to be associated with the susceptibility to developing CD (Hugot et al., 2001). These variant alleles result in altered structure of leucine-rich repeat domain of protein, resulting in overactivation of NF-kB on monocytes, which are involved in CD pathogenesis (Hugot et al., 2001). It was also found to be involved in increasing the risk of CD compared to UC and the development of either UC or CD in comparison with healthy Greeks (Gazouli et al., 2005). Further, CARD15 Leu1007fsinsC (OR=7.35), Gly908Arg (OR=5.70), and Arg702Trp (OR=2.45) were found to be associated with CD in Italian population (Bianchi et al., 2007). Higher rates of mutations in CARD15/NOD2 SNPs were also found in Ashkenazi Jews compared with Sephardic Jews (Karban et al., 2004).

Moreover, CARD4/NOD1, which is found on the short arm of chromosome 7 (7p14), does code for a pattern recognition receptor (PRR) activated on pathogen sensing, which leads to downstream signaling that activates innate immune responses. CARD4 156C>G SNP (rs2075818) had a significant effect on modulating the risk of CD and IBD, in general, but not UC, while other CARD4 SNPs, i.e., 27606C>T (rs2075822) and 45343G>A (rs2907748), have been reported to modulate the risk of UC, but not CD (Huebner et al., 2009).

11.2.1.2.4 Autophagy-Related 16 Like 1 (ATG16L1)

The ATG16L1 gene is located on chromosome 2, at 2q37.1, and plays an important role in the innate immune response and in the resistance to intracellular pathogens (Plantinga et al.,

2011). The ATG16L1 codes a protein that plays a role in the formation of autophagosomes during autophagy. Autophagy is a non-selective degradation system that has roles in starvation adaptation, intracellular protein and organelle clearance, development, anti-aging processes, elimination of microorganisms, cell death, tumor suppression, and antigen presentation (Mizushima, 2005). A common coding variant ATG16L1SNP (rs2241880), whereby a threonine-to-alanine substitution occurs at amino acid position 300 of the ATG16L1 protein (T300A), confers an increased risk of developing IBD.

Zhang et al. conducted an extensive meta-analysis involving 24 studies, totaling 13,022 cases and 17,532 controls, found a positive association between ATG16L1 (rs2241880) and CD in Caucasians, and suggested that the G allele of ATG16L1 (rs2241880) is a low-penetrance gene for developing CD in Caucasians. However, no significant association was found in Asians, suggesting that this difference is possibly related to ethnic variations. Another study also suggested the significance of ATG16L1 SNP (rs2241880) in CD susceptibility and stated the frequency of the allele G of the polymorphism rs2241880/T300A was higher in the group of CD patients in comparison with controls (Scolaro et al., 2014).

Another meta-analysis confirmed the positive association between the SNP rs2241880/T300A and susceptibility to CD among Asian, American, European, and Oceania populations (Cheng et al., 2010). Also, a significant association between ATG16L1 SNPs (rs11209026 and rs2241880) and CD susceptibility has been documented. For ATG16L1 rs11209026 SNP (p value=0.002, OR 0.54, 95% CI 0.36–0.81) and for SNP rs2241880 (p=0.0001, OR=1.41, 95% CI 1.18–1.67) was reported, suggesting their significance in developing risk of CD. In addition, the association of rs11209026 with UC was observed (p value=0.037, OR=0.66, 95% CI 0.45–0.98) (Roberts et al., 2007).

A GWAS of 735 individuals with CD and 368 controls, accounting for 19,779 non-synonymous SNPs, found a statistically significant interaction of ATG16L1 gene SNP rs2241880 with the incidence of CD in two studies involving German population and one study on British population . However, there was lack of any association with UC (Hampe et al., 2007).

11.2.1.2.5 Myosin IX B (MYO9B)

MYO9B gene located on chromosome 19p13.1 encodes a single-headed processive myosin class IX myosin family member (Inoue et al., 2002). This single motor protein has a Rho-GTPase-activating domain, which converts active Rho-GTP to inactive Rho-GDP and thus downregulates Rho-dependent signaling pathways (Müller et al., 1997). Rho-GTPase regulates the tight junctions and actin filament remodeling, which lead to enhanced paracellular epithelial permeability (Bruewer et al., 2004). Therefore, it may be responsible for the maintenance of intestinal barrier integrity. MYO9B has been found to be a candidate susceptibility gene for IBD risk (Rioux et al., 2000). There are five known SNPs in the MYO9B gene, of which SNPs rs962917, rs1545620, and rs2305764 were found to be linked with IBD, CD, and UC (Latiano et al., 2008). However, the SNPrs2305767 has been found to be associated with IBD and CD only (Cooney et al., 2009).

A meta-analysis including 8,975 cases and 9,482 controls reported that many SNPs of MY09B were significantly associated with the IBD, CD, or UC risk. The results suggest that SNP rs1457092 elevated the risk of UC and IBD, whereas SNP rs2305764 was significant in IBD and rs962917 increased the risk of IBD, UC, and CD in Caucasians (Wang et al., 2016).

Another meta-analysis revealed MYO9B rs1545620 (A/C) polymorphism was associated with both CD and UC susceptibility in Caucasians (OR=0.88, 95% CI=0.82–0.95, p=0.001; OR=0.82, 95% CI=0.76–0.89, p<0.001), but not in Chinese. The results also suggested that rs1457092 (G/T) and rs2305764 (C/T) polymorphisms are associated with UC susceptibility (OR=0.85, 95% CI=0.79–0.91, p<0.001; OR=0.88, 95% CI=0.83–0.93, p<0.001), but not with CD susceptibility in Caucasians (Li et al., 2016a).

11.3 Rheumatoid Arthritis (RA)

RA is one of the most common chronic autoimmune diseases with severe health impact (Iwaszko et al., 2016). It is characterized by the inflammation of synovial tissues and formation of rheumatoid pannus, which can erode adjacent cartilage and bone joints causing subsequent joint destruction and movement limitation (Chua et al., 2019). It is a long-standing disease, which mostly leads to the swelling and pain in the joints of hands and wrist (McInnes and Schett, 2017). Although the pathogenesis of this disease is poorly understood, it is characterized by synoviocyte hyperplasia, neo-vascularization, and partial infiltration of numerous T cells, inducing synovial inflammation, articular cartilage damages, bone, and joint capsules at different levels, and finally leads to joint deformity and functional incapacitation (Emery et al., 2015).

RA affects 0.5%–1% of world population, and its incidence rate is observed to be 5–50/100,000 annually, mostly affecting people in the age group of 35–50 years (Scott et al., 2010; Areskoug-Josefsson and Oberg, 2009). It is three times more common in females than in males (sex ratio 3:1) (Carmona et al., 2010; Oliver and Silman, 2009). The prevalence of RA is about 0.5%–1% in the normal population (Suzuki et al., 2013); 0.3%–1.1% among those of European origin; and 0.1%–0.5% among those of Asian ancestry (Kochi et al., 2010). It is a heritable disorder with a chance of high risk (60%) among family members, indicating that genetic factors could contribute to the main pathogenesis of the disease (MacGregor et al., 2000). The risk in monozygotic twins and siblings of affected individuals increased 12- to 62-fold and 2- to 17-fold, respectively (Seldin et al., 1999).

RA is classified into two subtypes based on the presence or absence of anti-citrullination protein antibodies (APCA): APCA-positive RA and APCA-negative RA. During the process of inflammation, the conversion of arginine amino acid to citrulline by the enzyme peptidylarginine deaminase (PAD) occurs and is known as citrullination. Both subtypes differ genetically, but the contribution of genetic factors appears to be equally high in both types of RA (van der Woude et al., 2009). Within the 3 years of follow-up, about 75% of ACPA with undifferentiated arthritis will eventually develop RA (van Gaalen et al., 2004).

The etiology of RA involves both genetic and environmental factors (Kharlamova et al., 2016). The genetic factors account for 50%–60% of all RA cases (Combe and Rincheval,

2015). Numerous RA-related susceptibility genes have been identified, including HLA, PADI4, and PTPN22. HLA class II molecules account for one-third of the total genetic susceptibility and are the most powerful genetic factors for RA across all populations (Newton et al., 2004). Also non-HLA genes, such as CTLA-4, IL-23R, PTPN22, PADI4, TNFAIP3, and MTHFR, have consistently been associated with RA predisposition (Flores-Borja et al., 2008; Plenge et al., 2007; Begovich et al., 2004; Matmati et al., 2011; Owen et al., 2013). Furthermore, other genes are still emerging to be implicated in the incidence of RA.

11.3.1 Single Nucleotide Polymorphisms in RA

Numerous case–control studies involving SNPs have helped in the determination of genetic risk factors in many autoimmune disorders, including RA. SNPs are much higher-density genetic markers than microsatellite markers that can be used for linkage studies and have led to the identification of the association of specific gene-population linkage. For example, SNPs in PADI4 gene are found in Asian RA cases (Suzuki et al., 2003) and PTPN22 in European (Begovich et al., 2004). More than 35 genetic regions have been reported as RA susceptibility loci with the help of GWAS using DNA array technologies (Gregersen, 2010). The most extensively studied locus of HLA region with respect to susceptibility to the risk of development of RA is HLA-DR locus, yet it only accounts for approximately one-third of the genetic susceptibility to RA. Recently, a trans-ethnic GWAS meta-analysis evaluated ~10 million SNPs in genes contributing to the RA disease and involved a total of 29,980 RA cases and 75,578 controls (Okada et al., 2014).

11.3.1.1 Human Leukocyte Antigen (HLA) Susceptibility Genes

The strongest genetic association in RA is observed in the HLA region, which is located on chromosome 6p21 and extends up to 3.6 Mb. It includes MHC classes I, II, and III and contains many genes that have immunoregulatory functions. HLA-DR4 is the oldest reported gene in RA (Stastny, 1978). Subsequently, the conserved sequence of the third hypervariable region in HLA-DRB1, termed "SE," has a strong association with the development of RA, and SE alleles vary among different ethnic groups. Table 11.1 summarizes the RA-associated HLA alleles (susceptible and protective) in different populations.

11.3.1.2 Non-Human Leukocyte Antigen (HLA) Genes

Several non-HLA genes are found to be associated with RA susceptibility and are well documented. Among the non-HLA genes associated with RA predisposition, IL-23R and PTPN22 genes have likely the strongest association (de Vries, 2011; Bax et al., 2011).

11.3.1.2.1 Peptidylarginine Deiminase 4 (PADI4)

PADI4 gene is located on chromosome 1p36 and is a highly expressed gene in inflamed synovium of patients with RA. PADI4 is a member of the PADI gene family and acts as an enzyme for protein citrullination, which converts peptidylarginine (arginine residue) into peptidylcitrulline (citrulline residue). Peptidylcitrulline acts as an antigen of ACPA and is provided only by PADs (translated protein from PADI genes), via modification of protein substrates (Worthington and John, 2003). PADI4 is highly expressed in bone marrow, macrophages, neutrophils, and monocytes (Chang et al., 2005). Numerous polymorphisms have been identified in PADI4 promoter, and SNP -92C/G (or rs874881) is one of the most extensively studied SNPs in RA (Suzuki et al., 2003; Iwamoto et al., 2006).

A meta-analysis including (8 studies; 1,351 cases and 1,585 controls) observed a significant association between the PADI4–92C/G polymorphism and RA in the Chinese population (G vs C: OR = 1.32, 95% CI = 1.02–1.71; GG + CG vs CC: OR = 1.75, 95% CI = 1.20–2.53) (Karami et al., 2019).

A study carried out in southern Mexican population showed the minor alleles of PADI4_89*G (rs11203366),

TABLE 11.1
RA and Associated Susceptible and Protective HLA Alleles in Different Populations

Ethnicity	Rheumatoid Arthritis Alleles							References
	HLA-A*	HLA-C*	HLA-B*	HLA-DRB1*	HLA-DQA1*	HLA-DQB1	HLA-DPB1*	
Caucasian			04:01, 04:04, 01:01, 04:05, 10:01					de Almeida et al. (2011)
			01:01, 01:02, [01:03], 04:01, 04:04, 04:05, 04:08, [07], 10:01, [12:01], [13:01], 14:02, [15:01]					Gough and Simmonds (2007)
			[01:03], [04:02], [13:01], [13:02]					van Drongelen and Holoshitz (2017)
Asian			01:01, [03:01], 04:01, [04:03], 04:05, [04:06], 04:10, [07:01], 10:01, [13:01], [14:05]					Jun et al. (2007)
African	No data							
Arabian			04:05, [06], 08, 10: 03:07, 03:08, 11, 15					Al-Swailem et al. (2006) and Alsaeid et al. (2002)

PADI4_90*T (rs11203367), and PADI4_92*G (rs874881) gene polymorphisms were found to be associated with RA susceptibility (p=0.04; p=0.03; p=0.04, respectively). The GTG haplotype was also significantly associated with RA (p=0.008) (Baños-Hernández et al., 2017).

Another meta-analysis (28 studies; 15,947 RA cases and 22,696 controls) showed a significant association between allele 2 of the PADI4_94 polymorphism and RA in the overall population (OR=1.155, 95% CI=1.069–1.249, $p=2.7\times10^{(-5)}$). The stratification analysis revealed that the PADI4_94 and PADI_104 polymorphisms are associated with the susceptibility to RA in Asians and Caucasians and that the PADI4_92 polymorphism is associated with susceptibility to RA in Asians only (Lee et al., 2015).

This meta-analysis demonstrates that PADI4-94G/A polymorphism is associated with susceptibility to RA in the overall population and in the Asian population. The PADI4-92C/G (rs874881) polymorphism confers susceptibility to RA in Africans, and the PADI4-90C/T (rs11203367) polymorphism was associated with RA in the Latin American population (Yang et al., 2015).

A meta-analysis including 27 studies suggests that PADI4 polymorphisms represent a significant risk factor for RA, especially in Asians. The results showed six SNPs (six gene loci: padi4_94, 104, 92, 90, 89, and 100) were significantly associated with RA in Asian populations. Three SNPs (PADI4_104, 90, and 89) showed significant associations, while the other three SNPs (PADI4_94, 92, and 100) exhibited no association in the European population (Hou et al., 2013).

11.3.1.2.2 Cytotoxic T Lymphocyte-Associated Protein 4 (CTLA-4)

CTLA-4 gene is located at chromosome 2q33.2 and is one of the widely studied non-HLA susceptibility genes of RA. It is mainly expressed on the surface of Treg cells and conventional T cells and suppresses self-reactive T cell responses via downregulating ligand availability for the co-stimulatory receptor CD28 to elicit inhibitory signals (Sansom, 2015). T cell activation is modulated through the interactions between the inhibitory CTLA-4 and CD28, as a stimulatory signal transducer, with their ligands. Both receptors are activated by the same ligands, namely CD80 and CD86, which are expressed particularly on the antigen-presenting cells (Borriello et al., 1997).

CTLA-4 plays a vital role in regulating T cell function, and RA is also a T cell-mediated autoimmune disease, suggesting its role in RA pathogenesis is most likely (Sansom, 2015; Yang et al., 2020a). SNPs in the CTLA-4 gene may contribute to RA, and the SNPs in three loci of CTLA-4, +49A/G (rs231775), −318C/T (rs5742909), and CT60 G/A(rs3087243), are the most often studied in RA subjects (Liu et al., 2013).

Mousavi et al. conducted a meta-analysis including 6 population studies (CTLA-4 rs231775 1,078 cases and 1,118 controls; rs5742909: 2,193 cases and 2,580 controls). The results indicated that G allele, GG and GA genotypes, and dominant model for rs231775, recessive model for rs5742909, and C allele, CC and CT genotypes were significantly associated with increased RA risk (Mousavi et al., 2021).

A meta-analysis (66 case–control studies; 21,681 cases and 23,457 controls) confirmed the relationship between CTLA-4 gene polymorphisms and RA. The results show that rs3087243 polymorphism has a significant association in both Asian (p=0.02) and Caucasian populations (p<0.00001). For rs231775 polymorphism, a significant association was observed in the Asians (p=0.001), but there was no effect in Caucasians. However, there was no association between rs5742909 polymorphism and RA (Zhou et al., 2021).

11.3.1.2.3 Signal Transducer and Activator of Transcription 4 (STAT4)

STAT4 gene plays an important role in cytokine signaling, particularly IL-12, and has a modest association with RA. They are a family of latent cytoplasmic transcription factors and are expressed in activated peripheral blood monocytes, dendritic cells, and macrophages (Levy and Darnell, 2002). STAT4 gene may play an important role in the pathogenesis of RA because the Th17 and Th1 (differentiation induced by STAT4) lineages are crucial effectors in chronic inflammatory disorders (Korman et al., 2008; McInnes and Schett, 2007). Polymorphisms in the STAT4 gene may increase susceptibility risk to RA in both positive and negative ACPA groups (Bax et al., 2011; Liang et al., 2012).

A meta-analysis on STAT4 rs7574865 (28 studies; 25,300 RA patients and 26,326 controls) was conducted for genotype TT versus GT+GG, GT+TT versus GG, TT versus GG and T allele. And it showed an association between RA and the TT genotype, GT+TT genotype, and T allele in all subjects. Also, according to ethnic subgroup analysis, it was shown that the TT genotype, GT+TT genotype, and T allele were significantly associated with RA in Europeans, Asians, Africans, and Latin Americans. A subgroup analysis according to the absence or presence of rheumatoid factor (RF) and anti-cyclic citrullinated peptide (anti-CCP) antibodies revealed that the association between the STAT4 rs7574865 polymorphism and RA may be independent of the presence of RF and anti-CCP antibodies (Gu et al., 2015).

Similarly, STAT4 rs7574865 polymorphism meta-analysis showed an association between STAT4 T allele and GT+TT and TT genotypes and RA susceptibility. The ethnic stratification of RA patients showed that STAT4 T allele and STAT4 GT+TT were significantly associated with RA in European, Asian, and African subjects, while STAT4 TT genotype was significantly associated with RA in European and Asian, but not in African subjects (Elshazli et al., 2015).

Moreover, a meta-analysis obtained similar results showing a significant association between rs7574865 polymorphism and RA risk in allele models in Caucasian and Asian ethnicities in five genetic models. There was also an association between rs7574865 and RA risk in Africans in the dominant model (GT+TT vs. GG: OR=2.77, 95% CI=1.95–3.93, p<0.01) and heterozygote model (GT vs. GG: OR=2.47, 95% CI=1.91–3.20, p<0.01) (Gao et al., 2020).

11.3.1.2.4 Protein Tyrosine Phosphatase Non-Receptor 22 (PTPN22)

PTPN22 gene located on chromosome 1p13 codes for lymphoid protein tyrosine phosphatase (LYP), which acts as a negative regulator of signaling pathway mediated via the T cell receptor and helps in the prevention of spontaneous activation and inactivation of TCR associated with dephosphorylation (Gregersen et al., 2006).

PTPN22 gene SNP rs2476601 (1858C/T) has been associated with susceptibility to risk of RA (Vang et al., 2007). Numerous single case–control studies worldwide have assessed the association of the PTPN22 gene rs2476601 SNP with susceptibility to RA.

A meta-analysis including 52 case–control studies (23,584 cases and 24,497 controls) observed a significantly strong association of rs2476601 polymorphism with the risk of RA in Caucasians and Africans under all genotype models. The results are as follows: dominant model (OR=1.69, 95% CI=1.55–1.84, p<0.001), recessive model (OR=2.50, 95% CI=2.06–3.05, p<0.001), allelic model (OR=1.80, 95% CI=1.60–2.2, p<0.001), TT versus CC model (OR=2.79, 95% CI=2.28–3.41, p<0.001), and CT versus CC model (OR=1.59, 95% CI=1.50–1.67, p<0.001) (Abbasifard et al., 2020).

Moreover, analyzing 43 case–control studies (containing 27,205 RA cases and 27,677 controls), Nabi et al. indicated that the PTPN22 gene rs2476601 SNP is associated with RA susceptibility in Caucasians. However, there was no significant association between this polymorphism and the risk of RA in the Asian population (Nabi et al., 2016). Similarly, another study revealed a significant association between the PTPN22 gene rs2476601 polymorphism and susceptibility to RA in European, but not in Asian and African subjects (Elshazli et al., 2015).

11.3.1.2.5 Interleukins

Since the role of interleukins in the pathogenesis of autoimmune disorders is already discussed in the first section of this chapter (IBD), here we will focus on SNPs in ILs and their importance in RA susceptibility.

Interleukin 23 receptor **(IL-23R)** SNPs are also found to be risk factors for RA. A meta-analysis (8 studies; 5,544 RA patients and 5,532 controls) suggested IL-23R gene rs2201841 polymorphism might be a susceptible factor for RA under G vs. A model, especially in Caucasian population (Zou et al., 2019). Further, a meta-analysis (37 case–control studies; 23,506 cases and 25,984 controls) analyzing the IL-23R, IL-21, and IL-1A gene polymorphisms and the risk of RA suggested that the minor A allele (rs1343151 SNP) significantly increased the risk of RA (log OR=0.085, 95% CI=0.008, 0.156). However, IL-21 and IL-1A gene polymorphisms were not found to be significant in the development of RA risk (Mohammadi et al., 2019).

The association between **interleukin 17 (IL-17)** and RA risk has been well studied, and the results suggest its role in the development of RA risk. A meta-analysis (19 studies; 5,298 cases and 5,675 healthy controls) showed a significant association between rs2275913 G allele and RA susceptibility (p<0.05). Also, the subgroup analysis by ethnicity indicated that rs763780C allele was closely related to Caucasian population (p<0.001), but not Mongolians. Also, the rs763780 and RA susceptibility in Caucasian population was found to be significant (p<0.05) (Shao et al., 2020).

A meta-analysis (25 studies; 7,474 cases and 10,628 controls) indicated that the IL-17A SNPs (rs2275913) and IL-17F (rs763780 T/C) may increase the RA risk. The results suggested that individuals who carry the rs2275913 GA genotype have a 20% or 41% increased risk compared to GG genotype and IL-17F rs763780 is significantly associated with the risk for RA (p< 0.005) (Eskandari-Nasab et al., 2017).

Interleukin 6 (IL-6) is a pro-inflammatory cytokine involved in RA susceptibility. It is reported that the IL-6 expression might in part be genetically modulated by polymorphisms located at the promoter region, i.e., rs1800795 (–174G/C), rs1800796 (-572G/C), and rs1800797 (–597G/A) (Arman et al., 2012). IL-6 promoter polymorphisms have been associated with susceptibility to RA, but with varied results among different populations (Liu et al., 2015b).

A study by Li et al. in the Chinese Han population showed a statistically significant association of IL-6 SNPs with RA. The observed ORs were as follows: for rs1800795, OR=4.823 (95% CI 1.338–17.379, p =0.016); for rs1800797, OR=5.905 (95% CI 1.270–27.448, p=0.024); and for rs1474347 SNP, OR=3.423 (95% CI 1.212–9.665, p=0.020) (Liu et al., 2015b).

Also, a significant association was found between IL-6 SNPs rs1800795 and rs1800796 and RA risk (p≤0.001) (Huang et al., 2007). Similarly, Li et al. observed the association of the rs1800795/ –174G/C variant with RA risk and studied circulating IL-6 levels also. The overall analysis showed that C allelic and GC+CC genotypes had 1.59-fold (95% CI: 1.19–2.14) and 1.63-fold (95% CI: 1.17–2.26) increased risk of developing RA, respectively. Asian populations showed a stronger association with 4.55-fold (95% CI: 1.62–12.75), 1.84-fold (95% CI: 1.13–2.99), and 4.69-fold (95% CI: 1.68–13.14) increased RA risk in carriers of –174C allelic, CC, and GC+CC genotypes, respectively (Li et al., 2016b).

Another meta-analysis including 13 studies (3,291 RA cases and 3,812 controls) analyzed the association of IL-6–174G>C SNP with the RA risk in different populations (Asian, Caucasian, and mixed). The results revealed the increased RA risk associated with IL-6–174G>C SNP in the overall population: allelic (OR=1.650, 95% CI=1.169–2.329, p=0.004); homozygous (OR=1.380, 95% CI=0.906–2.101, p=0.133); heterozygous (OR=1.559, 95% CI=1.001–2.428, p=0.049); dominant (OR=1.663, 95% CI=1.078–2.567, p=0.022); and recessive (OR=1.366, 95% CI=0.964–1.935, p=0.079). The subgroup analysis also showed this polymorphism to be associated with an increased RA risk in Asian population: allelic (OR=3.724, 95% CI=1.361–10.190, p=0.010), but not in Caucasian population. This meta-analysis shows that IL-6–174G>C SNP is significantly associated with an increased RA risk in the overall population, and specifically in the Asian population (Dar et al., 2017).

11.3.1.2.6 Tumor Necrosis Factor-Alpha-Inducible Protein 3 (TNFAIP3) Gene

TNFAIP3 gene is located at chromosome 6q23 and encodes ubiquitin-editing protein A20, which is an inhibitor of NF-kB activation in several signaling pathways such as those of TNF and Toll-like receptors. It deregulates NF-kB-dependent gene expression by deubiquitinating specific NF-kB signaling molecules (Lee et al., 2000). Several studies have investigated the associations between TNFAIP3 gene polymorphisms and RA susceptibility (Perkins et al., 2012; Vernerova et al., 2016).

A meta-analysis carried out to study two SNPs in this gene (rs6920220 and rs10499194) reported a significant association between the two alleles of rs6920220 and RA in Europeans (OR = 1.227). And a significant association between the RA and SNP rs10499194 was found in Asians (OR = 1.254), but not in Europeans. Furthermore, an association was found between the two alleles of another SNP rs2230926 and RA in all study subjects (OR = 1.390) (Lee et al., 2012). Also, TNFAIP3 rs719149 A allele is found to be a potential susceptibility locus in African Americans (p = 0.02) (Perkins et al., 2012).

Another meta-analysis including 10 case–control studies and involving 18,014 cases and 20,112 controls supports the association between TNFAIP3 gene rs2230926 and rs5029937 polymorphisms and RA risk. The stratification analysis of ethnicity indicated that rs5029937 polymorphism increased the risk of RA among Caucasians, while rs2230926 polymorphism increased the risk of RA among Asians and Caucasians (Zhang et al., 2017).

11.4 Multiple Sclerosis (MS)

Multiple sclerosis (MS) is a chronic autoimmune central nervous system (CNS) disease that is characterized by demyelination, inflammation, and neurodegeneration-axonal damage (Bjelobaba et al., 2017). MS is considered as a heterogeneous disease (Emamgholipour et al., 2013) with certain individuals experiencing a mild course with little disability progression, while others worsen insidiously and accumulate substantial disability over years. MS is the most common cause of progressive non-traumatic primary neurological disability in young adults (Marrie, 2004).

There are T cell, B cell, and macrophage infiltrations in active demyelination lesions in the brain of MS patients, and spinal cord atrophy is common (Frischer et al., 2009). The hallmarks of pathophysiology of MS are the formation of lesions ("plaques") in many areas in the brain and spinal cord because of infiltration/damage mediated by self-reactive T cell immune cells (Palmer, 2013; Baranzini and Oksenberg, 2017). It is suggested that auto-reactive T cells (CD4+ and CD8+ T-helper cells) initiate the cascade of inflammatory change. CD4+ cells recognize antigens that are presented by MHC molecules on specialized antigen-presenting cells and present them to specific T cell receptors (Haegert, 2011). This immune activation leads to the commencement of CD4+ cells-mediated pathological destruction of cells of self, particularly myelinated CNS neurons resulting in loss of myelin.

The most prominent candidates for autoantigens include the components of myelin sheath, such as myelin basic protein, neurofascin, myelin oligodendroglial glycoprotein, and proteolipid protein (Mathey et al., 2007). The B crystallin, a stress protein present in myelin sheath, is another candidate (Ousman et al., 2007). Lesions in the white matter (WM) characterize MS from the early phase, and the quantity and quality changes in the WM can be observed during the course of progression of the disease. The frequency and distribution of such lesions in the brain and spinal cord define the seriousness and variety of clinical symptoms of MS (Palmer, 2013).

A most extensive study on the epidemiology of MS shows that the number of people with MS has increased from 2.3 to 2.8 million globally from 2013 to 2020. The prevalence of the MS disease is 35.9 per 100,000 population globally. The incidence rate across 75 reporting countries is 2.1 per 100,000 persons/year, and the mean age of diagnosis is 32 years. The females are twice as likely to live with MS as males (Walton et al., 2020) (Figure 11.1).

Although the origin of inflammation is yet to be identified, the etiology of the MS remains elusive. However, there are both genetic and environmental complicating factors that play a role in susceptibility of MS (Liu et al., 2017).

The variation in HLA genes within the MHC is a strong and well-known risk factor for MS. However, GWAS have found more than 200 independent non-MHC variants that contribute to MS risk (International Multiple Sclerosis Genetics Consortium, 2019).

11.4.1 Single Nucleotide Gene Polymorphisms and MS

The genetic basis for the increase in the risk of MS has largely been determined by GWAS, indicating that a common variation in the regulatory regions of immune genes largely drives the variation in susceptibility to MS (International Multiple Sclerosis Genetics Consortium, 2019). Over 200 genes have now been identified (Patsopoulus, 2018), and of these, 110 non-MHC genetic loci have been detailed and 13 MHC loci have been identified (Moutsianas et al., 2015).

11.4.1.1 Human Leukocyte Antigen (HLA) Genes

International Multiple Sclerosis Genetic Consortium (IMSGC) completed the first MS GWAS and confirmed that rs3135391, with an SNP tagged for HLA-DRB1*1501, represented the strongest association with MS (International Multiple Sclerosis Genetic Consortium, 2007). The first MS risk gene variant, HLA-DRB1*15:01, causes a threefold increase in the risk (Olerup and Hillert, 1991). HLA-B was also found to be associated with MS. Table 11.2 summaries the HLA genes associated with MS risk in different ethnic populations.

11.4.1.2 Non-Human Leukocyte Antigen (HLA) Genes

Several studies have been carried out to evaluate the associations between MS and the candidate genes outside the HLA region, and GWAS have located more than 200 independent variants associated with the risk of MS. To date, these identified non-HLA MS risk alleles have shown a modest effect on disease development and progression and are mostly common in the general population. One of the most common non-HLA loci is found within the cluster of differentiation gene. Some of the non-HLA genes and their significance in the development of risk of MS are discussed as follows.

11.4.1.2.1 Cluster of Differentiation 24 (CD24)

CD24 is located on the chromosome 6q21 and is one of the most extensively studied factors. CD24 is considered to modify the risk of MS. It is a small protein molecule anchored by glycosylphosphatidylinositol (GPI) present inside the cell

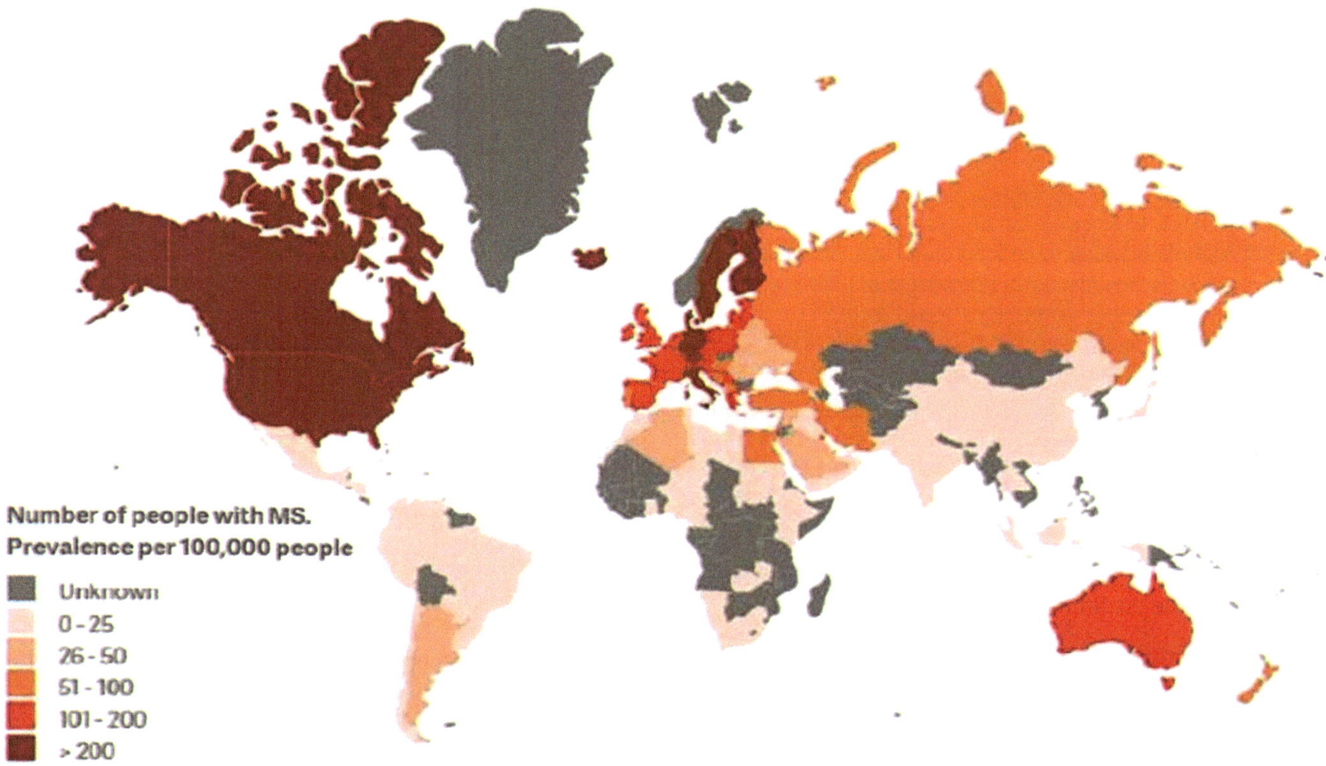

FIGURE 11.1 Map showing geographic variation in MS prevalence per 100,000 people by country in shades of orange and as per the key in red. Countries without prevalence data are shown in gray.
Country confidence scores were assigned using four variables: (1) size of the population covered by data source, from 0 if unknown to 5 if covering the whole country; (2) year of data collection by source, from 0 if prior to 2009 to 5 if in 2017–2019; (3) type of data source, from 0 if unknown to 5 if peer-reviewed journal publication; and (4) additional points were given for meeting certain methodological criteria: using the 2017 McDonald Criteria, performing a validation step, or using multiple consistent data sources for the estimate. Confidence ratings were assigned based on the total scores using the following thresholds: ⩽5 = very low, 6–10 = low, 11–15 = moderate, ⩾16 = high. (Wallter et al., *Rising Prevalence of Multiple Sclerosis Worldwide: Insights from the Atlas of MS*, 3rd edition, MultScler; 2020.)

TABLE 11.2

MS and Associated Susceptible and Protective HLA Alleles in Different Populations

Ethnicity	Multiple Sclerosis Alleles							References
	HLA-A*	HLA-C*	HLA-B*	HLA-DRB1*	HLA-DQA1*	HLA-DQB1	HLA-DPB1*	
Caucasian	[*02:01]		*37:01	03:01, 04:04, 13:03, 14:01, 15:01		06:02		Patsopoulos et al. (2013)
Asian				[09], 15				Qiu et al. (2011)
African	No data or research							
Arabian	2, 9, 19		5, 35, 40	15:01		02:01, [02:02], [05:03], 06:02, 06:03, [06:14]		Al Jumah et al. (2018)

surface of activated B cells, T cells, and other cells of CNS (Liu et al., 2007; Zhu et al., 2012). CD24 has also been identified as an important mediator in a CD28-independent co-stimulatory pathway in the activation of both CD8 and CD4 T cells (Saadatnia et al., 2012; Zhu et al., 2012). It also plays an important role in the binding of adhesion molecules such as the vascular cell adhesion molecule (VCAM1) and the very late activation antigen 4 (VLAA4), which are contribute to CNS of MS patients (Duggal et al., 2013).

A SNP rs52812045 (Ala57Val) in CD24 gene at codon 226 results in the replacement of an alanine amino acid with valine and has been confirmed to be associated with the susceptibility of MS. A meta-analysis by Yang et al. including 7 studies (2,079 cases and 2,295 controls) showed a significant association ($p<0.0032$) between CD24 Ala/Val polymorphism and susceptibility to MS. Also, the stratified analysis by area showed a significant association in Asians, but there was no association found in Europeans. This study suggested that the

CD24 Val allele was associated with an increased risk of MS (Yang et al., 2020b).

Moreover, Braliou et al. concluded in a meta-analysis (7 studies; 2,085 cases and 2,295 controls) that CD24 (Ala57Val) polymorphism is associated with MS according to the recessive mode of inheritance (OR = 1.75; 95% CI: 1.09–2.81) (Braliou et al., 2015). Also, a meta-analysis (11 studies; 2,466 cases and 2,650 controls) suggests that Val/Val genotypes of the CD24 Ala57Val polymorphism are associated with an increased risk of MS in Caucasians ($p<0.001$) (Huang et al., 2015).

11.4.1.2.2 Interleukin 2 Receptor α (IL-2RA)

IL-2RA, also known as CD25, is located at chromosome 10p15.1. CD25 is a central constituent of the trimeric IL-2 receptor complex that binds to IL. IL-2 has a wide range of effects that are essential for the balance between immune response and tolerance (Boyman and Sprent, 2012; Malek and Castro, 2010). IL-2 signaling is important for the survival and suppressive capacity of T_{reg} cells, and the expression of CD25 is required for the optimal development of TH1 cells (Boyman and Sprent, 2012; Hartmann et al., 2014). The alterations in the CD25 expression may influence the immune and inflammation signaling cascades, thus affecting CD4+ T cell differentiation and T_{reg} cells suppressive activity. IL-2RA SNPs rs2104286 and rs11256593 are associated with CD25 expression on CD4+ T cells and are associated with MS (Buhelt et al., 2019). The IL-2RA polymorphism regulates IL-2 responsiveness of naive Th cells and their propensity to develop into GM-CSF-producing memory Th cells. These results suggest that such nature of the Th cell's functionality can represent them to be an immunologically relevant risk factor for the development of MS (Hartmann et al., 2014).

This meta-analysis studied the association between IL-2RA rs2104286 (8 studies comprising 13,569 cases and 23,435 controls) and the susceptibility to MS. The results found that the T allele and the TT and TT+TC genotypes of the IL-2RA rs2104286 polymorphism were associated with MS. Also, the IL-2RA rs12722489 polymorphism (5 studies comprising 5,643 patients and 6,415 controls) revealed that the C allele and the CC genotype of the IL-2RA rs12722489 polymorphism were found to be associated with MS, but the CC+CT genotype was not. It was concluded that both IL-2RA gene polymorphisms, rs2104286 and rs12722489, were associated with increased susceptibility to MS (Wang et al., 2011). Similarly, it was shown that the gene polymorphisms at the loci of IL-2RA rs2104286 and rs12722489 may increase the risk of MS in the Han and Hui nationalities (Xia et al., 2018).

Another meta-analysis evaluated IL-2RA gene polymorphisms SNP rs2104286 (11 studies; 8,608 cases and 9,061 controls) and SNP rs12722489 (6 studies; 4,259 cases and 5,420 controls) and their association with the risk of MS. The results demonstrated that the A allele (rs2104286) was associated with an increased risk in both Caucasians and Asians ($p<0.001$) and the C allele (rs12722489) elevated the risk in Caucasians only. Thus, SNP rs2104286 was associated with susceptibility to MS in both Asians and Caucasians, whereas SNP rs2104286 was found only in Caucasians (Wang and Chen, 2018).

11.4.1.2.3 Interleukin 17 Receptor (IL-7R)

The IL-7R gene is located on chromosome 5p13 and consists of the gamma chain (IL-7Rγ) and alpha chain (IL-7RA) (Tavakolpour, 2016a). IL-7R regulates the T cell functioning and plays a main role in the development of functionally mature lymphocytes and thus influences the pathogenesis of MS (Traboulsee et al., 2014). IL-17 is one of the important interleukins that are involved in the immunopathogenesis of autoimmune disease, especially MS (Tavakolpour, 2016a). It plays an important role in the homeostasis of the memory T cells and is involved in the production of auto-reactive T cells in MS patients. Moreover, IL-7R is necessary for the maturity of gamma and delta T cells, the earliest T cells observed in the inflammatory lesions of MS (Jacobs et al., 2010).

The gene encoding IL-7RA is polymorphic, and SNPs in IL-7RA have recently been associated with a wide range of diseases (Hartling et al., 2015). Of all the identified SNPs in the IL-7RA gene, rs6897932 is the strongest SNP associated with MS (Alcina et al., 2008; Wu et al., 2016; Tavakolpour, 2016b).

A meta-analysis (17 studies; 15,270 cases and 17,971 controls) reported a positive role of SNP rs6897932 in the risk of development of MS ($p=0.001$). Upon stratification based on ethnicity, it was shown that this SNP is significant in MS among Europeans, but not in Asians (Wu et al., 2016). However, another meta-analysis (33 studies; 19,351 patients and 21,005 healthy controls) shows negative results, suggesting IL-7RA gene rs6897932 SNP decreases MS susceptibility in the overall population—Asians, Americans, and Europeans (Omraninava et al., 2021).

A meta-analysis on IL-7RA gene polymorphism SNP rs6897932 (including 11 case–control studies; 6,752 cases and 7,349 controls) and the risk of MS suggested a positive association of IL-7R rs6897932 (T244I) polymorphism with MS ($p<0.001$) (Tavakolpour, 2016b).

Sahami-Fard et al. conducted a meta-analysis (28 studies; 16,260 patients and 18,335 controls) and represented a significant association between the IL-7R T244I polymorphism and MS susceptibility in Europeans, but not in the Middle East population. The results include the following: recessive model: OR = 1.126, 95% CI 1.026–1.236, $p=0.012$; dominant model: OR = 1.172, 95% CI 1.024–1.341, $p=0.021$; homozygous model: OR = 1.213, 95% CI 1.038–1.417, $p=0.015$; and allelic model: OR = 1.109, 95% CI 1.025–1.200, $p=0.010$ (Sahami-Fard et al., 2020).

11.4.1.2.4 Vitamin D Receptor (VDR)

The VDR gene is one of the most studied genes related to vitamin D and is located on chromosome 12q13.11 (Narooie-Nejad et al., 2015). It consists of 9 exons and 8 introns and has numerous allelic variants. VDR is a member of the nuclear hormone receptor superfamily and a transcription factor (Holick, 2004). VDR is an intracellular receptor found in various tissues, including brain, immune cells, and peripheral blood monocytes (Elkama and Karahalil, 2018). Vitamin D plays a regulatory role within the immune system by binding to the nuclear VDR. Vitamin D interacts with the VDR and then binds to the retinoid X receptor (RXR) to form a heterodimer complex, which regulates gene transcription through vitamin D response elements (VDREs) in the promoter regions of vitamin D target genes, thereby modifying their expression (Jones et al., 1998). It is suggested that the vitamin D deficiency increases the risk of MS by putting the immune system in a more pro-inflammatory state (Harandi et al., 2014). The low levels of vitamin D in patients with MS suggest that vitamin D modulates the Th1-mediated immune response (Holick, 2004).

VDR gene variants can severely affect gene activation and disturb vitamin D and calcium metabolism, thereby increasing the risk of autoimmune diseases such as MS (Abdollahzadeh et al., 2016). There are four common SNPs of the VDR gene, namely FokI, BsmI, ApaI, and TaqI. FokI SNP in start codon produces a shorter protein with more activity, whereas ApaI, BsmI, and TaqI polymorphisms are in the 3′-end of VDR gene, which leads to silent mutation associated with increased VDR mRNA stability (Makoui et al., 2020). Among over 30 different VDR SNPs, the most studied are FokI (rs10735810), BsmI (rs1544410), TaqI (rs731236), and ApaI (rs7975232) in MS (Huang and Xie, 2012).

A meta-analysis (27 case–control studies; 4,879 MS patients and 5,402 controls) showed the association of the VDR TaqI polymorphism with MS risk, especially in Caucasians. Also, the ApaI and FokI polymorphisms correlated with MS risk among Asians, while BsmI polymorphisms showed no association with MS (Zhang et al., 2019).

Another meta-analysis (30 case–control studies) suggested a significant association between TaqI polymorphism and MS susceptibility (OR=1.27, 95% CI=1.01–1.59, random effect). Furthermore, BsmI polymorphism was associated with an increased risk, whereas ApaI polymorphism was associated with a decreased risk of MS in Asian populations (Imani et al., 2019).

Similarly, another meta-analysis (21 studies) suggested a significant association between the TaqI polymorphism and MS risk in the homozygous model (p=0.006) and indicated a significant protective effect of TT TaqI genotype (Chen et al., 2017).

The meta-analysis (13 case–control studies; 3,300 cases and 3,194 controls) revealed a significant association between the ApaI polymorphism and MS risk in the homozygous and codominant models (p=0.013 and p=0.031, respectively), suggesting that the AA ApaI genotype might be a significant MS risk factor. The AA ApaI and FF FokI genotypes are significant risk factors for MS. However, there were no significant associations found between TaqI and BsmI polymorphisms and MS risk (Tizaoui et al., 2015).

11.5 Type 1 Diabetes (T1D)

T1D, also known as insulin-dependent diabetes mellitus (IDDM), is an autoimmune disease characterized by the destruction of pancreatic islet β-cells, resulting in low insulin production and thus leading to hyperglycemia and other serious complications that cause many health issues such as ketoacidosis, kidney failure, heart disease, stroke, and shortened life expectancy and/or possibly death (Polychronakos and Li, 2011; Naser et al., 2013). The affected patients have lifetime requirements for exogenous insulin for survival (Lemos et al., 2018; Kiani et al., 2015) and have increased the life expectancy of T1D patients substantially in comparison with that three decades ago, although it is still shorter than that of healthy people. The etiology of T1D is very complex, involving both genetic and environmental factors. The interplay between genetic predisposition, environmental factors, and reprogramming of the immune system results in T1D. Although it can affect individuals at any age, it usually develops at a young age.

The T1D represents roughly 5%–10% of all diabetes cases, with an increasing prevalence rate (Miettinen et al., 2017). T1D inflicts mostly children under 15 years of age and almost 90,000 children are diagnosed each year, with almost 0.5 million children living with T1D (Diaz-Valencia et al., 2015). The incidence of T1D is also significantly different due to region and ethnicity (Reddy et al., 2011; Chellappan et al., 2018). Figure 11.2 shows the map of incidence of T1D in 0- to 14-year-olds worldwide. The data were summarized from studies published between January 1990 and December 2018 and from the 9th edition of International Diabetes Federation (IDF) Diabetes Atlas.

11.5.1 Single Nucleotide Gene Polymorphisms and T1D

GWAS have identified >60 loci within the human genome, which are marked by SNPs to be associated with the risk of developing T1D (Nyaga et al., 2018). These studies have given us a clearer picture of the allelic architecture of genetic susceptibility to T1D. It is suggested that the genetic variation in both pancreatic β-cells and immune cells is central in mediating T1D risk.

11.5.1.1 Human Leukocyte Antigen (HLA) Susceptibility Genes

The HLA region on chromosome 6p21 accounts for approximately 50% of the familial aggregation of T1D, and its association with T1D has been known since long. HLA-DR and HLA-DQ have the strongest association with T1D. The association of HLA genes with susceptibility risk of T1D is shown in Table 11.3.

11.5.1.2 Non-Human Leukocyte Antigen (HLA) Susceptibility Genes

A number of genes are located on non-HLA regions of the genome, and these predisposing genes result in the susceptibility to T1D. Besides the genes identified using candidate gene approaches, GWAS have provided a wealth of knowledge on the genetic basis of T1D, and over 50 loci have been associated with this disease (Redondo et al., 2018). Here, we will briefly explain the genes that show the strongest association with T1D risk.

11.5.1.2.1 Erb-b2 Receptor Tyrosine Kinase 3 (ERBB)

The ERRB gene is located on chromosome 12q13 and encodes erb-b2 receptor tyrosine kinase 3 (ErbB3), which is a member of the epidermal growth factor receptor (EGFR) family and regulates cell survival, differentiation, and proliferation in several cell types (Maziarz et al., 2015). ErbB3 interacts more specifically with PIK3 regulatory subunits, which transduce signals downstream to mTOR signaling pathway and regulate insulin production in beta-cells and subsequent glucose metabolism. The inhibition of the PI3K pathway reduces the regulatory T cell number in pancreatic cells, suggesting that the ERBB gene

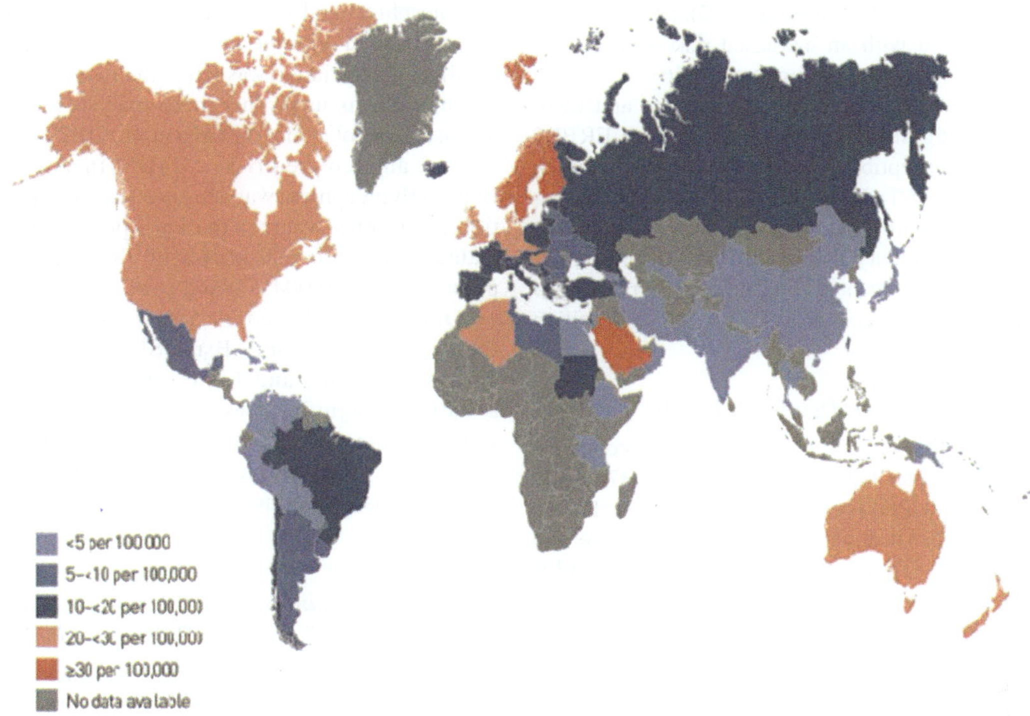

FIGURE 11.2 Age- and sex-standardized incidence rates (per 100,000) of type 1 diabetes in children aged under 15 years. (Patterson CC et al. *International Diabetes Federation Diabetes Atlas*, 9th edition; 2019.)

TABLE 11.3

T1D and Associated Susceptible and Protective HLA Alleles in Different Populations

	Type 1 Diabetes/Insulin-Dependent Diabetes Mellitus Alleles							
Ethnicity	**HLA-A***	**HLA-C***	**HLA-B***	**HLA-DRB1***	**HLA-DQA1***	**HLA-DQB1**	**HLA-DPB1***	**References**
Caucasian	30	03, 08, 16	18, 19, 44, 39	04:02, [04:03], 04:04, 04:05		[03:01], 03:02	02:02, 03:01, [04:02], [04:03]	Steck and Rewers (2011)
		03:03, 08:02, 16:01, 07:01, 07:02	18:01, 39:06, [44:03], [38:01]					Varade et al. (2018)
		02:01, 11:01, 24:02, 3201, 6601	07:02, 18:01, 35:02, 39:06, 44:03, [57:01]				[01:01], [02:02], [04:01], [04:02], [17:01]	Noble and Valdes (2011)
Asian	24:02, 24:03, [11:01]	01:02, 03:02, 15:02, 07:02	[13:01], 58:01	04:05		[01:01], [05:03]	02:02	Bugawan et al. (2002)
African				03:01, 04, 09	03	02, 03:02		Pirie et al. (2001)
Arabian				[04:01], 03:01:01, 04:01:01, [10:01:01], [11:01]		02:01, [03:01:01], 03:02, [05:01:01]		Al-Hussein et al. (2003) and Al-Harbi et al. (2004)

might play an important role in potential T1D pathogenesis (Wang et al., 2010; Kaur et al., 2016; Sun et al., 2016).

Several studies have suggested that ERBB3 may significantly contribute to the pathogenesis of T1DM through the modulation of APCs function, autoimmunity, and cytokine-induced beta-cell apoptosis. Therefore, several GWAS have identified ERBB3 gene as a susceptible locus for T1DM, with the strongest association signal being observed for the rs2292239 SNP in intron 7 of the gene (Lemos et al., 2018; Wang and Pan, 2019).

Sun et al. identified the SNP rs2292239 in ERBB3 gene was significantly associated with T1D. The results of the study show the frequency of the major G allele was significantly decreased in patients with T1D (68.8% in T1D vs 77.3% in controls, OR 0.65, 95% CI 0.53–0.79, p=0.02), and the minor allele T was associated with an increased risk of T1D (OR 1.55, 95% CI 1.26–1.90, p=0.02) (Sun et al., 2016).

A meta-analysis (9 studies; 5369 T1D patients and 6920 controls) showed a significant association between ERBB3 rs2292239 and T1D susceptibility in the overall population (A vs. C, OR=1.292, 95% CI=1.224–1.364, p value for the heterogeneity test, *P*H=0.450). Similar results were found in the subgroup analysis by ethnicity (Wang and Pan, 2019). Another study suggests that the A/A genotype of the ERBB3 rs2292239 SNP is associated with the risk of T1D in a white Brazilian population (Lemos et al., 2018).

11.5.1.2.2 Programmed Cell Death 1 (PDCD1 or PD1)

The PD1 gene, located on chromosome 2q37 encodes an immunoinhibitory receptor belonging to the CD28/B7 family (Prokunina et al., 2002). PD1 is expressed by myeloid cells and activated T and B cells. It downregulates their response and functions as a negative regulator of self-reactivity of T and B cells (Okazaki and Wang, 2005). It is suggested that PD1 may also regulate peripheral response, thereby playing a key role in autoimmune diseases. There are many known SNPs in the PD1 gene, and a number of studies have shown the role of such SNPs in the risk of T1D (Flores et al., 2010; Lee et al., 2015).

A study in China found that SNPs rs2227982 and rs2227981 were associated with T1D risk in a Chinese Han population of 3060 individuals (1019 T1D patients and 2041 healthy controls). Also there was a significant association of these SNPs with T1D risk in East Asians (OR=0.82, 95% CI 0.74–0.90 and OR=1.23, 95% CI 1.12–1.36, respectively), but not in Europeans (Gu et al., 2018).

Another meta-analysis (13,210 patients and 17,073 controls) revealed a significant association between T1D and the two alleles of the SNP rs111568821 (PD1.3) among the Europeans (p=0.005). It also revealed an association between the PD1 and PD1.9 polymorphism and T1D in Asian population (p<0.004) (Lee et al., 2015).

11.5.1.2.3 Vitamin D Receptor (VDR)

Vitamin D is a steroid molecule and employs its function through VDR. It is expressed by lymphocytes, macrophages, and pancreatic cells (Bikle, 2017). Being a member of the nuclear hormone receptor superfamily, VDR is linked to insulin sensitivity and secretion (Panierakis et al., 2009). Vitamin D deficiency leads to immune system reprogramming resulting in pre-sensitization to self-antigens (van Belle et al., 2011). It inhibits the activation and secretion of interleukins IL-1 and IL-6, INF-γ, and TNF—the key molecules in the pathogenesis of T1D (Trembleau et al., 1995; Riachy et al., 2006; Uitterlinden et al., 2004). Several studies have been carried out to observe the association of VDR SNPs with the risk of development of T1D, and some are listed below.

A meta-analysis including 79 studies (FokI: 22 studies; BsmI: 25 studies; ApaI: 17 studies; and TaqI: 15 studies) on T1DM reveals that BsmI SNP is associated with an increased risk of T1D in East Asian population (p=< 0.001). However, no significance of ApaI or TaqI and T1D was found (Wang et al., 2012). Similarly, in the Asian subgroup, only BsmI polymorphism and not FokI, ApaI, and TaqI polymorphisms was associated with the increased risk of T1DM, whereas no significant association was found in other populations for other polymorphisms in another meta-analysis including a total of 57 case–control studies (Zhang et al., 2012).

Also, an increased risk of T1DM in association with the BsmI polymorphism was observed in the East Asian population (p<0.001), whereas the FokI polymorphism was associated with an increased risk of T1DM in the West Asian population (p=0.004) (Wang et al., 2014). A meta-analysis by Qin et al. evaluated 39 case–control studies and found the association of only BsmI SNP with T1D risk among Asians, Africans, and Latino (Qin et al., 2014). Further, a statically significant association of FokI and TaqI VDR gene polymorphisms with T1DM (p<0.0001) was found in Kuwaiti Arab children. However, the ApaI and BsmI polymorphisms did not show any association with T1DM (Rasoul et al., 2019).

11.5.1.2.4 Protein Tyrosine Phosphatase Non-Receptor Type 2 and 22 (PTPN2 and PTPN22)

Both these genes encode signaling molecules tyrosine phosphatases (PTPs) that operate various cellular processes such as cell cycle, growth and differentiation, and oncogenic transformations (Estus et al., 2013). Changes in these genes tend to abnormal immune functioning resulting in diseases such as T1D.

PTPN2 located on chromosome 18 and PTPN22 gene located on chromosome 1p13 are members of the PTP family, which dephosphorylate the receptor tyrosine residues and regulate many signaling pathways (Cerosaletti and Buckner, 2012). PTPN22 negatively regulates T cell activation through dephosphorylation of kinases. Lck, Fyn, and ZAP70 are all involved in T cell signaling (Cerosaletti and Buckner, 2012). A SNP (rs2476601) is found to be associated with T1D. There occurs a single substitution of arginine for tryptophan in the encoded protein (R620W), leading to a decrease in T cell receptor and B cell receptor signaling (Menard et al., 2011; Espino-Paisan et al., 2011).

PTPN2 expression is a recognized factor involved in pancreatic beta-cell apoptosis (Santin et al., 2011). It regulates insulin signaling by inactivating its receptor through dephosphorylation of the insulin receptor β-chain in conjunction with the PTP1B, and any mutation leads to deficiency of PTPN22 expression leading to cytokine-induced pancreatic beta-cell apoptosis and thus impacts glucose homeostasis resulting in T1D (Cerosaletti and Buckner, 2012). A SNP rs2542151 in the PTPN2 gene is known to be associated with autoimmune diseases, including T1D (Espino-Paisan et al., 2011).

A significant association was found between PTPN22 SNP (rs1310182 and rs2476601) and T1D among the Emirati and Estonian populations (Sharma et al., 2021; Kisand and Uibo, 2012). Also, PTPN22 SNP rs2476601 was statistically significant in T1D among Colombian families (Rodríguez et al., 2015).

Another study found that the carriers of the T allele of PTPN22 were significantly more likely to develop T1D in Egyptian population (OR=2.2, 95% CI=1.2–4, p=0.01), especially females who carried the 1858T allele (Abdelrahman et al., 2016). Similarly, both the 1858CT and 1858TT genotypes and the 1858T allele were found more frequently in patients (32.5% and 18.7%, respectively) than in controls (10% and 5.0%, respectively), $p=0.013$ and $p=0.007$, respectively. Among females, the 1858T allele was more common in patients (18%) than in controls (2.6%), $p=0.014$ (El Fotoh et al., 2019). Further, a study in Chinese children found that the TT+TC genotype and the T allele of C1858T were more frequent in T1D patients (19.40% and 10.0%, respectively) than in healthy subjects (7.51% and 4.0%, respectively), and the difference was significant (both $p<0.001$) (Liu et al., 2015a).

A meta-analysis (24 studies; 8,129 cases and 9,641 controls) showed a significant association of PTPN22+1858C/T gene polymorphism in the overall population. When stratified by race, significance was observed in Europe and America, but not in Asia (Tang et al., 2012).

A study on PTPN2 SNP rs2542151 revealed a significant association with T1D in Chinese Han children (OR=0.6; 95% CI: 0.40–0.95; and $p=0.024$) (Peng et al., 2015). Also, the frequency of rs2542151*G carriers was significantly higher in the early-onset group compared with late-onset patients ($p=0.023$) and with controls (OR=1.61; 95% CI=1.14–2.26; $p=0.005$) (Espino-Paisan et al., 2011). Another study showed that PTPN2 SNP rs1893217 is significantly associated with T1D patients diagnosed age ≤18 (OR =0.80, 95% CI: 0.67–0.97, $p=0.02$) and may increase the risk of early-onset T1D (Chen et al., 2019).

11.5.1.2.5 Cytotoxic T lymphocyte-Associated Protein 4 (CTLA-4)

CTLA-4 gene (2q33), mapped to the IDDM 12 locus, encodes a T cell-specific transmembrane glycoprotein co-receptor (Polychronakos and Li, 2011). CTLA-4 is an important mediator of T cell activation in autoimmune diseases. The association of polymorphisms of CTLA gene with T1D has widely been reported (Khodaeian et al., 2015; Wang et al., 2017).

A meta-analysis (52 studies; 11,017 cases and 14,191 controls) demonstrated that the G allele of rs231775 of CTLA-4 is a risk factor associated with increased T1D susceptibility with an overall random effect (OR=1.41, 95% CI: 1.31–1.53, $p<10^{(-5)}$) for G allele versus A allele. Significant results were also observed for heterozygous (OR=1.29, 95% CI: 1.16–1.45, $p<10^{(-5)}$) and homozygous genetic models (OR=1.96, 95% CI: 1.66–2.31, $p<10^{(-5)}$) (Chen et al., 2013).

Another meta-analysis (19 case–control studies; 3,797 cases and 3,981 controls) reported that the G allele in CTLA-4 SNP rs231775 (+49A/G) had a 1.33-fold increased risk of developing T1D in comparison with children carrying the A allele; children who had a GG or GA genotype had a higher risk of developing T1D than children who had the AA genotype ($p<0.05$). Also, the ethnic stratification results showed that the risk was higher in Asians in the overall genetic model (for allelic: OR =1.60; for GG vs AA: OR =2.46; and for GA vs. AA, OR =1.58) in comparison with Caucasian population (for allelic, OR=1.24; for GG vs AA: OR=1.55; and for GA vs AA: OR=1.19), respectively (Wang et al., 2017).

Another meta-analysis (58 studies; 30,734 cases and 40,754 controls) demonstrated that the CTLA-4 SNPs rs231775 and rs3087243 are risk factors for developing T1D in Caucasian and Middle Eastern populations, but failed to show any significance in Africans. However, SNP rs231775 was found to be significantly associated with T1D among East Asians, while no associations were found for rs3087243 polymorphism, which was suggested to have a marginally significant association with T1D in Indian population (Wang et al., 2014).

11.6 Conclusions

SNPs are associated with susceptibility to CID progression, affecting the production or function of a protein and thus leading to the development of disease. Epidemiological and clinical studies suggest that the pathogenesis of inflammatory diseases is strongly linked to genetic predisposition.

HLA genes DR2, DR9, and DRB1*0103 are prone to UC risk, whereas DR7, DRB3*0301, and DQ4 are susceptibility genes for CD. HLA-DRB1*0103 is a significant risk allele for UC in American and European populations. Data from different studies have concluded that the genetic architecture in UC appears to be similar between Europeans and East Asians. The IL-23 pathway has shown stronger genetic effects in Japanese IBD cases, and the associations between genes in the epithelial barrier and the Th17-IL-23R pathways were similar in Japanese and European UC populations. Also, a positive association between ATG16L1 (rs2241880) and CD is found among Caucasians. SNPs in CARD15/NOD2 gene have an elevated susceptibility (40 times) to CD, and SNPs in IL-R23 and IL-10 have a significant association with UC. SNPs in the MY09B gene are associated with the risk of IBD, CD, or UC in Caucasian populations.

HLA-DR*15 and DQ*06 haplotypes are associated with MS in Europeans, and HLA-DR*03 and HLA-DQ*02 alleles are common in Sardinian population. HLA-DRB1*15:01 increases risk threefold. CD24 Ala57Val polymorphisms are associated with an increased risk of MS in Caucasians, and IL-7R polymorphism rs6897932 is significant in MS among Europeans. VDR TaqI polymorphism has been correlated with MS risk in Caucasians, whereas ApaI and FokI polymorphisms show significance in Asian populations.

It has been documented that shared epitope (SE) alleles, such as HLA-DRB1*01 and DRB1*04, and some HLA alleles such as HLA-DRB1*13 and DRB1*15 are involved in RA susceptibility. GWAS have identified at least 30 non-HLA susceptibility loci to be associated with the risk of RA. Among those alleles, PTPN22, IL-23R, TRAF1, STAT4, CD40, PADI4, IRF5, CCR6, and CTLA-4 play a major role in the pathogenesis. IL-23R and PTPN22 genes have likely the strongest association with the development of RA. PTPN22 rs2476601 is the most widely researched RA polymorphism that downregulates the TCR signaling.

HLA genotype (DR3/DR4-*DQB1**03:02) are the high-risk alleles for T1D. DR3/DR4 genotype is only found in approximately 2%–3% of the US population. DRB1*03:01-DQA1*05:01-DQB1*02:01 (DR3) and/or DRB1*04-DQA1*03:01-DQB1*03:02 (DR4) is strongly associated with susceptibility to T1D among Caucasians. CTLA-4 SNPs rs231775 and rs3087243 are risk factors for developing T1D in Caucasian and Middle Eastern populations. PTPN22 gene SNPs are significant risk factors for T1D in European and Middle Eastern populations.

The identification of SNPs in the individual genome can help to ascertain the patient's genotype and can improve the differential diagnosis and may optimize treatment efficacy. Also, the treatment plans can be individualized upon stratification of specific variants involved in the prognosis of patients. Therefore, with the help of pharmacogenetic approaches, it will be possible to increase the potential for personalized medicine in the future.

ABBREVIATIONS

CD:	Crohn's disease
CID:	Chronic inflammatory disease
GWAS:	Genome-wide association studies
HLA:	Human leukocyte antigen
IBD:	Inflammatory bowel disease
IDDM:	Insulin-dependent diabetes mellitus
IL:	Interleukin
INF:	Interferon
MHC:	Major histocompatibility complex
MS:	Multiple sclerosis
OR:	Odds ratio
p:	p value
RA:	Rheumatoid arthritis
SNP:	Single nucleotide polymorphism
T1D:	Type 1 diabetes
TNF:	Tumor necrosis factor
TNFRSF1B:	TNF-α receptor superfamily, member 1B
UC:	Ulcerative colitis
VDR:	Vitamin D receptor

REFERENCES

Abbasifard M, Imani D, Bagheri-Hosseinabadi Z. PTPN22 gene polymorphism and susceptibility to rheumatoid arthritis (RA): updated systematic review and meta-analysis. *J Gene Med.* 2020 Sep;22(9):e3204. doi: 10.1002/jgm.3204. Epub 2020 Jun 3.

Abdelrahman HM, Sherief LM, Abd Elrahman DM, Alghobashy A, Elsaadani HF, Mohamed RH. The association of PTPN22 (rs2476601) and IL2RA (rs11594656) polymorphisms with T1D in Egyptian children. *Hum Immunol.* 2016 Aug;77(-8):682–6. doi: 10.1016/j.humimm.2016.06.006. Epub 2016 Jun 8.

Abdollahzadeh R, Fard MS, Rahmani F, Moloudi K, Kalani BS, Azarnezhad A. Predisposing role of Vitamin D receptor (VDR) polymorphisms in the development of multiple sclerosis: a case-control study. *J Neurol Sci.* 2016 Aug 15;367:148–51. doi: 10.1016/j.jns.2016.05.053. Epub 2016 Jun 1.

Abraham C, Cho JH. Inflammatory bowel disease. *N Engl J Med.* 2009;361(21):2066–78.

Al Jumah M, Kojan S, Al Shehri AM, Al Balwi M, Al Abdulkarim I, Masuadi EM, Alhaidan Y, Alabdulrahman A, Fakhoury HM, Hajeer AH. HLA class II polymorphism in Saudi patients with multiple sclerosis. *HLA.* 2018 Jan;91(1):17–22. doi: 10.1111/tan.13173. Epub 2017 Dec 11.

Alcina A, Fedetz M, Ndagire D, Fernández O, Leyva L, Guerrero M, Arnal C, Delgado C, Matesanz F. The T244I variant of the interleukin-7 receptor-alpha gene and multiple sclerosis. *Tissue Antigens.* 2008 Aug;72(2):158–61. doi: 10.1111/j.1399-0039.2008.01075.x.

Al-Harbi EM, Abbassi AJ, Tamim H, al-Jenaidi F, Kooheji M, Kamal M, al-Mahroos S, al-Nasir F, Motala AA, Almawi WY. Specific HLA-DRB and -DQB alleles and haplotypes confer disease susceptibility or resistance in Bahraini type 1 diabetes patients. *Clin Diagn Lab Immunol.* 2004 Mar;11(2):292–6. doi: 10.1128/cdli.11.2.292-296.2004. PMID: 15013978.

Al-Hussein KA, Rama NR, Ahmad M, Rozemuller E, Tilanus MG. HLA-DPB1*0401 is associated with dominant protection against type 1 diabetes in the general Saudi population and in subjects with a high-risk DR/DQ haplotype. *Eur J Immunogenet.* 2003 Apr;30(2):115–9. doi: 10.1046/j.1365-2370.2003.00369.x.

Al-Meghaiseeb ES, Al-Robayan AA, Al-Otaibi MM, Arfin M, Al-Asmari AK. Association of tumor necrosis factor-α and -β gene polymorphisms in inflammatory bowel disease. *J Inflamm Res.* 2016 Jun 17;9:133–40. doi: 10.2147/JIR.S101225.

Alsaeid K, Haider MZ, Kamal H, Srivastva BS, Ayoub EM. Prevalence of human leukocyte antigen (HLA) DRB1 alleles in Kuwaiti children with juvenile rheumatoid arthritis. *Eur J Immunogenet.* 2002 Feb;29(1):1–5. doi: 10.1046/j.1365-2370.2002.00253.x.

Al-Swailem R, Al-Rayes H, Sobki S, Arfin M, Tariq M. HLA-DRB1 association in Saudi rheumatoid arthritis patients. *Rheumatol Int.* 2006 Sep;26(11):1019–24. doi: 10.1007/s00296-006-0119-2. Epub 2006 May 4.

Arai T, Kakuta Y, Kinouchi Y, Kimura T, Negoro K, Aihara H, Endo K, Shiga H, Kanazawa Y, Kuroha M, Moroi R, Nagasawa H, Shimodaira Y, Takahashi S, Shimosegawa T. Increased expression of NKX2.3 mRNA transcribed from the risk haplotype for ulcerative colitis in the involved colonic mucosa. *Hum Immunol.* 2011 Jul;72(7):587–91. doi: 10.1016/j.humimm.2011.03.023. Epub 2011 Apr 8.

Areskoug-Josefsson K, Oberg U. A literature review of the sexual health of women with rheumatoid arthritis. *Musculoskeletal Care.* 2009 Dec;7(4):219–26. doi: 10.1002/msc.152.

Arisawa T, Tahara T, Shibata T, Nagasaka M, Nakamura M, Kamiya Y, Fujita H, Nakamura M, Yoshioka D, Arima Y, Okubo M, Hirata I, Nakano H. The influence of polymorphisms of interleukin-17A and interleukin-17F genes on the susceptibility to ulcerative colitis. *J Clin Immunol.* 2008 Jan;28(1):44–9. doi: 10.1007/s10875-007-9125-8. Epub 2007 Sep 9.

Arman A, Coker A, Sarioz O, Inanc N, Direskeneli H. Lack of association between IL-6 gene polymorphisms and rheumatoid arthritis in Turkish population. *Rheumatol Int.* 2012 Jul;32(7):2199–201. doi: 10.1007/s00296-011-2057-x. Epub 2011 Jul 22.

Baños-Hernández CJ, Navarro-Zarza JE, Parra-Rojas I, Vázquez-Villamar M, Ramón Padilla-Gutiérrez J, Valle Y, Reyes-Castillo Z, Magdalena Torres-Carrillo N, García-Arellano S, Brennan-Bourdon LM, Muñoz-Valle JF. PADI4 polymorphisms and the functional haplotype are associated with increased rheumatoid arthritis susceptibility: a replication study in a Southern Mexican population. *Hum Immunol.* 2017 Sep;78(9):553–8. doi: 10.1016/j.humimm.2017.05.005. Epub 2017 May 25. PMID: 28551357.

Baranzini SE, Oksenberg JR. The genetics of multiple sclerosis: from 0 to 200 in 50 years. *Trends Genet.* 2017 Dec;33(-12):960–70. doi: 10.1016/j.tig.2017.09.004. Epub 2017 Oct 5.

Barcelos IP, Troxell RM, Graves JS. Mitochondrial dysfunction and multiple sclerosis. *Biology (Basel).* 2019 May 11;8(-2):37. doi: 10.3390/biology8020037.

Bax M, van Heemst J, Huizinga TW, Toes RE. Genetics of rheumatoid arthritis: what have we learned? *Immunogenetics.* 2011 Aug;63(8):459–66. doi: 10.1007/s00251-011-0528-6. Epub 2011 May 10.

Begovich AB, Carlton VE, Honigberg LA, Schrodi SJ, Chokkalingam AP, Alexander HC, Ardlie KG, Huang Q, Smith AM, Spoerke JM, Conn MT, Chang M, Chang SY, Saiki RK, Catanese JJ, Leong DU, Garcia VE, McAllister LB, Jeffery DA, Lee AT, Batliwalla F, Remmers E, Criswell LA, Seldin MF, Kastner DL, Amos CI, Sninsky JJ, Gregersen PK. A missense single-nucleotide polymorphism in a gene encoding a protein tyrosine phosphatase (PTPN22) is associated with rheumatoid arthritis. *Am J Hum Genet.* 2004 Aug;75(2):330–7. doi: 10.1086/422827. Epub 2004 Jun 18.

Bengtson MB, Solberg C, Aamodt G, Jahnsen J, Moum B, Sauar J, Vatn MH; Ibsen Study Group. Clustering in time of familial IBD separates ulcerative colitis from Crohn's disease. *Inflamm Bowel Dis.* 2009 Dec;15(12):1867–74. doi: 10.1002/ibd.20978. Epub 2009 May 11. PMID: 19434721.

Bianchi V, Maconi G, Ardizzone S, Colombo E, Ferrara E, Russo A, Tenchini ML, Porro GB. Association of NOD2/CARD15 mutations on Crohn's disease phenotype in an Italian population. *Eur J Gastroenterol Hepatol.* 2007 Mar;19(-3):217–23. doi: 10.1097/01.meg.0000250590.84102.12.

Bikle D. Vitamin D: production, metabolism, and mechanisms of action. 2017 Aug 11. In: Feingold KR, Anawalt B, Boyce A, Chrousos G, de Herder WW, Dhatariya K, Dungan K, Grossman A, Hershman JM, Hofland J, Kalra S, Kaltsas G, Koch C, Kopp P, Korbonits M, Kovacs CS, Kuohung W, Laferrère B, McGee EA, McLachlan R, Morley JE, New M, Purnell J, Sahay R, Singer F, Stratakis CA, Trence DL, Wilson DP, editors. *Endotext* [Internet]. South Dartmouth (MA): MDText.com, Inc.; 2000–2017.

Bjelobaba I, Savic D, Lavrnja I. Multiple sclerosis and neuroinflammation: the overview of current and prospective therapies. *Curr Pharm Des.* 2017;23(5):693–730. doi: 10.2174/1381612822666161214153108.

Bonyadi M, Abdolmohammadi R, Jahanafrooz Z, Somy MH, Khoshbaten M. TNF-alpha gene polymorphisms in Iranian Azari Turkish patients with inflammatory bowel diseases. *Saudi J Gastroenterol.* 2014 Mar–Apr;20(2):108–12. doi: 10.4103/1319-3767.129475.

Borriello F, Sethna MP, Boyd SD, Schweitzer AN, Tivol EA, Jacoby D, Strom TB, Simpson EM, Freeman GJ, Sharpe AH. B7-1 and B7-2 have overlapping, critical roles in immunoglobulin class switching and germinal center formation. *Immunity.* 1997 Mar;6(3):303–13. doi: 10.1016/s1074-7613(00)80333-7.

Boyman O, Sprent J. The role of interleukin-2 during homeostasis and activation of the immune system. *Nat Rev Immunol.* 2012 Feb 17;12(3):180–90. doi: 10.1038/nri3156.

Braliou GG, Pantavou KG, Kontou PI, Bagos PG. Polymorphisms of the CD24 gene are associated with risk of multiple sclerosis: a meta-analysis. *Int J Mol Sci.* 2015 Jun 1;16(6):12368–81. doi: 10.3390/ijms16061236.

Bruewer M, Hopkins AM, Hobert ME, Nusrat A, Madara JL. RhoA, Rac1, and Cdc42 exert distinct effects on epithelial barrier via selective structural and biochemical modulation of junctional proteins and F-actin. *Am J Physiol Cell Physiol.* 2004 Aug;287(2):C327–35. doi: 10.1152/ajpcell.00087.2004. Epub 2004 Mar 24.

Bugawan TL, Klitz W, Alejandrino M, Ching J, Panelo A, Solfelix CM, Petrone A, Buzzetti R, Pozzilli P, Erlich HA. The association of specific HLA class I and II alleles with type 1 diabetes among Filipinos. *Tissue Antigens.* 2002 Jun;59(6):452–69. doi: 10.1034/j.1399-0039.2002.590602.x.

Buhelt S, Søndergaard HB, Oturai A, Ullum H, von Essen MR, Sellebjerg F. Relationship between multiple sclerosis-associated IL2RA risk allele variants and circulating T cell phenotypes in healthy genotype-selected controls. *Cells.* 2019 Jun 25;8(6):634. doi: 10.3390/cells8060634.

Carmona L, Cross M, Williams B, Lassere M, March L. Rheumatoid arthritis. *Best Pract Res Clin Rheumatol.* 2010 Dec;24(6):733–45. doi: 10.1016/j.berh.2010.10.001. PMID: 21665122.

Cerosaletti K, Buckner JH. Protein tyrosine phosphatases and type 1 diabetes: genetic and functional implications of PTPN2 and PTPN22. *Rev Diabet Stud.* 2012 Winter;9(4):188–200. doi: 10.1900/RDS.2012.9.188. Epub 2012 Dec 28.

Chang X, Yamada R, Suzuki A, Sawada T, Yoshino S, Tokuhiro S, Yamamoto K. Localization of peptidylarginine deiminase 4 (PADI4) and citrullinated protein in synovial tissue of rheumatoid arthritis. *Rheumatology (Oxford).* 2005 Jan;44(1):40–50. doi: 10.1093/rheumatology/keh414. Epub 2004 Oct 5.

Chellappan DK, Sivam NS, Teoh KX, Leong WP, Fui TZ, Chooi K, Khoo N, Yi FJ, Chellian J, Cheng LL, Dahiya R, Gupta G, Singhvi G, Nammi S, Hansbro PM, Dua K. Gene therapy and type 1 diabetes mellitus. *Biomed Pharmacother.* 2018 Dec;108:1188–200. doi: 10.1016/j.biopha.2018.09.138. Epub 2018 Oct 2.

Chen S, Fan H, Feng Y, Zhang Y, Chen Y, Gu Y, Shi Y, Dai H, Zhang M, Xu X, Chen H, Yang T, Xu K. The association between rs1893217, rs478582 in PTPN2 and T1D risk with different diagnosed age, and related clinical characteristics in Chinese Han population. *Autoimmunity.* 2019 Mar;52(2):95–101. doi: 10.1080/08916934.2019.1608191. Epub 2019 Apr 29.

Chen XL, Zhang ML, Zhu L, Peng ML, Liu FZ, Zhang GX, Wang LM, Zhao J. Vitamin D receptor gene polymorphisms and the risk of multiple sclerosis: an updated meta-analysis. *Microb Pathog.* 2017 Sep;110:594–602. doi: 10.1016/j.micpath.2017.08.002. Epub 2017 Aug 3.

Chen Z, Fei M, Fu D, Zhang L, Ma Y, Wang Y, Zhang F, Xia Q, Wang X. Association between cytotoxic T lymphocyte antigen-4 polymorphism and type 1 diabetes: a meta-analysis. *Gene.* 2013 Mar 10;516(2):263–70. doi: 10.1016/j.gene.2012.12.030. Epub 2012 Dec 19.

Cheng JF, Ning YJ, Zhang W, Lu ZH, Lin L. T300A polymorphism of ATG16L1 and susceptibility to inflammatory bowel diseases: a meta-analysis. *World J Gastroenterol.* 2010 Mar 14;16(10):1258–66. doi: 10.3748/wjg.v16.i10.1258.

Cho JH. The genetics and immunopathogenesis of inflammatory bowel disease. *Nat Rev Immunol.* 2008 Jun;8(6):458–66. doi: 10.1038/nri2340.

Cho JH, Weaver CT. The genetics of inflammatory bowel disease. *Gastroenterology.* 2007 Oct;133(4):1327–39. doi: 10.1053/j.gastro.2007.08.032.

Chua JR, Jamal S, Riad M, Castrejon I, Malfait AM, Block JA, Pincus T. Disease burden in osteoarthritis is similar to that of rheumatoid arthritis at initial rheumatology visit and significantly greater six months later. *Arthritis Rheumatol.* 2019 Aug;71(8):1276–84. doi: 10.1002/art.40869. Epub 2019 Jul 3.

Combe B, Rincheval N. Early lessons from the recent-onset rheumatoid arthritis cohort ESPOIR. *Joint Bone Spine.* 2015 Jan;82(1):13–7. doi: 10.1016/j.jbspin.2014.07.003. Epub 2014 Sep 17.

Cooney R, Cummings JR, Pathan S, Beckly J, Geremia A, Hancock L, Guo C, Morris A, Jewell DP. Association between genetic variants in myosin IXB and Crohn's disease. *Inflamm Bowel Dis.* 2009 Jul;15(7):1014–21. doi: 10.1002/ibd.20885.

Cutolo M, Soldano S, Smith V. Pathophysiology of systemic sclerosis: current understanding and new insights. *Expert Rev Clin Immunol.* 2019 Jul;15(7):753–64.

Dar SA, Haque S, Mandal RK, Singh T, Wahid M, Jawed A, Panda AK, Akhter N, Lohani M, Areeshi MY, Rai G, Datt S, Bhattacharya SN, Ramachandran VG, Das S. Interleukin-6-174G>C (rs1800795) polymorphism distribution and its association with rheumatoid arthritis: a case-control study and meta-analysis. *Autoimmunity.* 2017 May;50(3):158–69. doi: 10.1080/08916934.2016.1261833. Epub 2016 Dec 23. PMID: 28010120.

de Almeida DE, Ling S, Holoshitz J. New insights into the functional role of the rheumatoid arthritis shared epitope. *FEBS Lett.* 2011 Dec 1;585(23):3619–26. doi: 10.1016/j.febslet.2011.03.035. Epub 2011 Mar 22.

de Vries R. Genetics of rheumatoid arthritis: time for a change! *Curr Opin Rheumatol.* 2011 May;23(3):227–32. doi: 10.1097/BOR.0b013e3283457524.

Diaz-Valencia PA, Bougnères P, Valleron AJ. Global epidemiology of type 1 diabetes in young adults and adults: a systematic review. *BMC Public Health.* 2015 Mar 17;15:255. doi: 10.1186/s12889-015-1591-y.

Dudzinska E, Gryzinska M, Ognik K, Krauze M, Listos P. Nonspecifcinfammatory bowel diseases and the risk of tumour growth. *Medycyna Weterynaryjna-Veterinary Medicine-Science and Practice.* 2018;74:228–32.

Duggal NA, Upton J, Phillips AC, Sapey E, Lord JM. An age-related numerical and functional deficit in CD19(+) CD24(hi) CD38(hi) B cells is associated with an increase in systemic autoimmunity. *Aging Cell.* 2013 Oct;12(5):873–81. doi: 10.1111/acel.12114. Epub 2013 Jul 19.

Eckburg PB, Relman DA. The role of microbes in Crohn's disease. *Clin Infect Dis.* 2007 Jan 15;44(2):256–62. doi: 10.1086/510385. Epub 2006 Dec 12.

El Fotoh WMMA, El RazekMidan DA, El Shalakany AH. Role of C1858T polymorphism of lymphoid tyrosine phosphatase in Egyptian children and adolescents with type 1 diabetes. *Curr Diabetes Rev.* 2019;16(1):73–79. doi: 10.2174/1573399814666180709102533.

Elkama A, Karahalil B. Role of gene polymorphisms in Vitamin D metabolism and in multiple sclerosis. *ArhHig Rada Toksikol.* 2018 Mar 1;69(1):25–31. doi: 10.2478/aiht-2018-69-3065.

Elshazli R, Settin A. Association of PTPN22 rs2476601 and STAT4 rs7574865 polymorphisms with rheumatoid arthritis: a meta-analysis update. *Immunobiology.* 2015 Aug;220(8):1012–24. doi: 10.1016/j.imbio.2015.04.003. Epub 2015 Apr 28.

Emamgholipour S, Eshaghi SM, Hossein-nezhad A, Mirzaei K, Maghbooli Z, Sahraian MA. Adipocytokine profile, cytokine levels and foxp3 expression in multiple sclerosis: a possible link to susceptibility and clinical course of disease. *PLoS One.* 2013 Oct 3;8(10):e76555. doi: 10.1371/journal.pone.0076555. Erratum in: *PLoS One.* 2013;8(11). doi:10.1371/annotation/-6de84c11-73f5-4ee4-aec7-ca83072401a4. Erratum in: *PLoS One.* 2014;9(1). doi:10.1371/annotation/8fc7e49e-2790-4cfa-900c-2a1e319d61b8.

Emery P, Kavanaugh A, Bao Y, Ganguli A, Mulani P. Comprehensive disease control (CDC): what does achieving CDC mean for patients with rheumatoid arthritis? *Ann Rheum Dis.* 2015 Dec;74(12):2165–74. doi: 10.1136/annrheumdis-2014-205302. Epub 2014 Aug 19.

Eskandari-Nasab E, Moghadampour M, Tahmasebi A. Meta-analysis of risk association between Interleukin-17A and F gene polymorphisms and inflammatory diseases. *J Interferon Cytokine Res.* 2017 Apr;37(4):165–74. doi: 10.1089/jir.2016.0088. Epub 2017 Feb 10. PMID: 28186427.

Espino-Paisan L, de la Calle H, Fernández-Arquero M, Figueredo MA, de la Concha EG, Urcelay E, Santiago JL. A polymorphism in PTPN2 gene is associated with an earlier onset of type 1 diabetes. *Immunogenetics.* 2011 Apr;63(4):255–8. doi: 10.1007/s00251-010-0500-x. Epub 2011 Jan 19.

Estus JL; Family Investigation of Nephropathy and Diabetes Research Group, Fardo DW. Combining genetic association study designs: a GWAS case study. *Front Genet.* 2013 Sep 27;4:186. doi: 10.3389/fgene.2013.00186.

Ferguson LR, Han DY, Huebner C, Petermann I, Barclay ML, Gearry RB, McCulloch A, Demmers PS. Tumor necrosis factor receptor superfamily, member 1B haplotypes increase or decrease the risk of inflammatory bowel diseases in a New Zealand caucasian population. *Gastroenterol Res Pract.* 2009;2009:591704. doi: 10.1155/2009/591704. Epub 2009 May 3.

Ferguson LR, Huebner C, Petermann I, Gearry RB, Barclay ML, Demmers P, McCulloch A, Han DY. Single nucleotide polymorphism in the tumor necrosis factor-alpha gene affects inflammatory bowel diseases risk. *World J Gastroenterol.* 2008 Aug 7;14(29):4652–61. doi: 10.3748/wjg.14.4652.

Ferguson LR, Shelling AN, Browning BL, Huebner C, Petermann I. Genes, diet and inflammatory bowel disease. *Mutat Res.* 2007 Sep 1;622(1–2):70–83. doi: 10.1016/j.mrfmmm.2007.05.011. Epub 2007 Jun 2.

Festen EA, Stokkers PC, van Diemen CC, van Bodegraven AA, Boezen HM, Crusius BJ, Hommes DW, van der Woude CJ, Balschun T, Verspaget HW, Schreiber S, de Jong DJ, Franke A, Dijkstra G, Wijmenga C, Weersma RK. Genetic analysis in a Dutch study sample identifies more ulcerative colitis susceptibility loci and shows their additive role in disease risk. *Am J Gastroenterol.* 2010 Feb;105(2):395–402. doi:

10.1038/ajg.2009.576. Epub 2009 Oct 27. Erratum in: *Am J Gastroenterol*. 2010 Feb;105(2):479.

Flores S, Beems M, Oyarzún A, Carrasco E, Pérez F. Polimorfismos del gen del receptor de muertecelularprogramada 1 (PDCD1) y diabetes tipo 1 en población Chilena [Programmed cell death 1 (PDCD1) gene polymorphisms and type 1 diabetes in Chilean children]. *Rev Med Chil*. 2010 May;138(5):543–50. Spanish. Epub 2010 Jul 12.

Flores-Borja F, Jury EC, Mauri C, Ehrenstein MR. Defects in CTLA-4 are associated with abnormal regulatory T cell function in rheumatoid arthritis. *Proc Natl Acad Sci U S A*. 2008 Dec 9;105(49):19396–401. doi: 10.1073/pnas.0806855105. Epub 2008 Nov 26.

Frischer JM, Bramow S, Dal-Bianco A, Lucchinetti CF, Rauschka H, Schmidbauer M, Laursen H, Sorensen PS, Lassmann H. The relation between inflammation and neurodegeneration in multiple sclerosis brains. *Brain*. 2009 May;132(Pt 5):1175–89. doi: 10.1093/brain/awp070. Epub 2009 Mar 31.

Fullerton JN, Gilroy DW. Resolution of inflammation: a new therapeutic frontier. *Nat Rev Drug Discov*. 2016 Aug;15 (8):551–67. doi: 10.1038/nrd.2016.39. Epub 2016 Mar 29.

Gao W, Dong X, Yang Z, Mao G, Xing W. Association between rs7574865 polymorphism in STAT4 gene and rheumatoid arthritis: an updated meta-analysis. *Eur J Intern Med*. 2020 Jan; 71:101–3. doi: 10.1016/j.ejim.2019.11.009. Epub 2019 Nov 19.

Gazouli M, Mantzaris G, Kotsinas A, Zacharatos P, Papalambros E, Archimandritis A, Ikonomopoulos J, Gorgoulis VG. Association between polymorphisms in the Toll-like receptor 4, CD14, and CARD15/NOD2 and inflammatory bowel disease in the Greek population. *World J Gastroenterol*. 2005 Feb 7;11(5):681–5. doi: 10.3748/wjg.v11.i5.681.

Glas J, Seiderer J, Wetzke M, Konrad A, Török HP, Schmechel S, Tonenchi L, Grassl C, Dambacher J, Pfennig S, Maier K, Griga T, Klein W, Epplen JT, Schiemann U, Folwaczny C, Lohse P, Göke B, Ochsenkühn T, Müller-Myhsok B, Folwaczny M, Mussack T, Brand S. rs1004819 is the main disease-associated IL23R variant in German Crohn's disease patients: combined analysis of IL23R, CARD15, and OCTN1/2 variants. *PLoS One*. 2007 Sep 5;2(9):e819. doi: 10.1371/journal.pone.0000819.

Gough SC, Simmonds MJ. The HLA region and autoimmune disease: associations and mechanisms of action. *Curr Genomics*. 2007 Nov;8(7):453–65. doi: 10.2174/138920207783591690.

Gregersen PK. Susceptibility genes for rheumatoid arthritis - a rapidly expanding harvest. *Bull NYU Hosp Jt Dis*. 2010;68(3):179–82.

Gregersen PK, Lee HS, Batliwalla F, Begovich AB. PTPN22: setting thresholds for autoimmunity. *Semin Immunol*. 2006 Aug;18(4):214–23. doi: 10.1016/j.smim.2006.03.009. Epub 2006 May 30.

Gu E, Lu J, Xing D, Chen X, Xie H, Liang J, Li L. rs7574865 polymorphism in signal transducers and activators of transcription 4 gene and rheumatoid arthritis: an updated meta-analysis of 28 case-control comparisons. *Int J Rheum Dis*. 2015 Jan;18(1):3–16. doi: 10.1111/1756-185X.12363. Epub 2014 Apr 19.

Gu Y, Xiao L, Gu W, Chen S, Feng Y, Wang J, Wang Z, Cai Y, Chen H, Xu X, Shi Y, Zhang M, Xu K, Yang T. rs2227982 and rs2227981 in PDCD1 gene are functional SNPs associated with T1D risk in East Asian. *Acta Diabetol*. 2018 Aug;55(8):813–9. doi: 10.1007/s00592-018-1152-9. Epub 2018 May 17.

Haegert DG. Multiple sclerosis: a disorder of altered T-cell homeostasis. *MultScler Int*. 2011;2011:461304. doi: 10.1155/2011/461304. Epub 2011 Sep 15.

Hampe J, Franke A, Rosenstiel P, Till A, Teuber M, Huse K, Albrecht M, Mayr G, De La Vega FM, Briggs J, Günther S, Prescott NJ, Onnie CM, Häsler R, Sipos B, Fölsch UR, Lengauer T, Platzer M, Mathew CG, Krawczak M, Schreiber S. A genome-wide association scan of nonsynonymous SNPs identifies a susceptibility variant for Crohn disease in ATG16L1. *Nat Genet*. 2007 Feb;39(2):207–11. doi: 10.1038/ng1954. Epub 2006 Dec 31.

Harandi AA, Harandi AA, Pakdaman H, Sahraian MA. Vitamin D and multiple sclerosis. *Iran J Neurol*. 2014;13(1):1–6.

Hartling HJ, Sørensen CJ, Rigas AS, Burgdorf KS, Nielsen SD, Pedersen OB, Petersen MS, Sørensen E, Kotzé S, Thørner LW, Hjalgrim H, Erikstrup C, Ullum H. No association between rs6897932 in the gene encoding interleukin-7 receptor α and low-grade inflammation or self-reported health - results from the Danish Blood Donor Study. *APMIS*. 2015 Feb;123(2):169–74. doi: 10.1111/apm.12334. Epub 2014 Nov 25.

Hartmann FJ, Khademi M, Aram J, Ammann S, Kockum I, Constantinescu C, Gran B, Piehl F, Olsson T, Codarri L, Becher B. Multiple sclerosis-associated IL2RA polymorphism controls GM-CSF production in human TH cells. *Nat Commun*. 2014 Oct 3;5:5056. doi: 10.1038/ncomms6056.

Holick MF. Sunlight and Vitamin D for bone health and prevention of autoimmune diseases, cancers, and cardiovascular disease. *Am J Clin Nutr*. 2004 Dec;80(6 Suppl):1678S–88S. doi: 10.1093/ajcn/80.6.1678S.

Hou S, Gao GP, Zhang XJ, Sun L, Peng WJ, Wang HF, Ge XJ, Huang W, Sun YH. PADI4 polymorphisms and susceptibility to rheumatoid arthritis: a meta-analysis. *Mod Rheumatol*. 2013 Jan;23(1):50–60. doi: 10.1007/s10165-012-0639-4. Epub 2012 May 3. Erratum in: *Mod Rheumatol*. 2013 Jan;23(1):61.

Huang J, Xie ZF. Polymorphisms in the Vitamin D receptor gene and multiple sclerosis risk: a meta-analysis of case-control studies. *J Neurol Sci*. 2012 Feb 15;313(1–2):79–85. doi: 10.1016/j.jns.2011.09.024. Epub 2011 Oct 26.

Huang J, Yang Y, Liang Z, Kang M, Kuang Y, Li F. Association between the CD24 Ala57Val polymorphism and risk for multiple sclerosis and systemic lupus erythematosus: a meta-analysis. *Sci Rep*. 2015 Apr 1;5:9557. doi: 10.1038/srep09557.

Huang XZ, Zhuang JH, Ren YG, Zhou LJ, Zhou Q. [Association of interleukin-6 and interleukin-18 gene polymorphism with rheumatoid arthritis in Guangdong Han population]. Nan Fang Yi Ke Da XueXue Bao. 2007 Nov;27(11):1661–4. Chinese. PMID: 18024284.

Huebner C, Ferguson LR, Han DY, Philpott M, Barclay ML, Gearry RB, McCulloch A, Demmers PS, Browning BL. Nucleotide-binding oligomerization domain containing 1 (NOD1) haplotypes and single nucleotide polymorphisms modify susceptibility to inflammatory bowel diseases in a New Zealand caucasian population: a case-control study. *BMC Res Notes*. 2009 Mar 27;2:52. doi: 10.1186/1756-0500-2-52.

Hugot JP, Chamaillard M, Zouali H, Lesage S, Cézard JP, Belaiche J, Almer S, Tysk C, O'Morain CA, Gassull M, Binder V, Finkel Y, Cortot A, Modigliani R, Laurent-Puig P, Gower-Rousseau C, Macry J, Colombel JF, Sahbatou M, Thomas G. Association of NOD2 leucine-rich repeat variants with susceptibility to Crohn's disease. *Nature.* 2001 May 31;411(6837):599–603. doi: 10.1038/35079107.

Imani D, Razi B, Motallebnezhad M, Rezaei R. Association between Vitamin D receptor (VDR) polymorphisms and the risk of multiple sclerosis (MS): an updated meta-analysis. *BMC Neurol.* 2019 Dec 26;19(1):339. doi: 10.1186/s12883-019-1577-y.

Inoue A, Saito J, Ikebe R, Ikebe M. Myosin IXb is a single-headed minus-end-directed processive motor. *Nat Cell Biol.* 2002 Apr;4(4):302–6. doi: 10.1038/ncb774.

International Multiple Sclerosis Genetics Consortium. Multiple sclerosis genomic map implicates peripheral immune cells and microglia in susceptibility. *Science.* 2019 Sep 27;365(6460):eaav7188. doi: 10.1126/science.aav7188.

International Multiple Sclerosis Genetics Consortium (IMSGC), Beecham AH, Patsopoulos NA, Xifara DK, Davis MF, Kemppinen A, Cotsapas C, Shah TS, Spencer C, Booth D, Goris A, Oturai A, Saarela J, Fontaine B, Hemmer B, Martin C, Zipp F, D'Alfonso S, Martinelli-Boneschi F, Taylor B, Harbo HF, Kockum I, Hillert J, Olsson T, Ban M, Oksenberg JR, Hintzen R, Barcellos LF; Wellcome Trust Case Control Consortium 2 (WTCCC2); International IBD Genetics Consortium (IIBDGC), Agliardi C, Alfredsson L, Alizadeh M, Anderson C, Andrews R, Søndergaard HB, Baker A, Band G, Baranzini SE, Barizzone N, Barrett J, Bellenguez C, Bergamaschi L, Bernardinelli L, Berthele A, Biberacher V, Binder TM, Blackburn H, Bomfim IL, Brambilla P, Broadley S, Brochet B, Brundin L, Buck D, Butzkueven H, Caillier SJ, Camu W, Carpentier W, Cavalla P, Celius EG, Coman I, Comi G, Corrado L, Cosemans L, Cournu-Rebeix I, Cree BA, Cusi D, Damotte V, Defer G, Delgado SR, Deloukas P, di Sapio A, Dilthey AT, Donnelly P, Dubois B, Duddy M, Edkins S, Elovaara I, Esposito F, Evangelou N, Fiddes B, Field J, Franke A, Freeman C, Frohlich IY, Galimberti D, Gieger C, Gourraud PA, Graetz C, Graham A, Grummel V, Guaschino C, Hadjixenofontos A, Hakonarson H, Halfpenny C, Hall G, Hall P, Hamsten A, Harley J, Harrower T, Hawkins C, Hellenthal G, Hillier C, Hobart J, Hoshi M, Hunt SE, Jagodic M, Jelčić I, Jochim A, Kendall B, Kermode A, Kilpatrick T, Koivisto K, Konidari I, Korn T, Kronsbein H, Langford C, Larsson M, Lathrop M, Lebrun-Frenay C, Lechner-Scott J, Lee MH, Leone MA, Leppä V, Liberatore G, Lie BA, Lill CM, Lindén M, Link J, Luessi F, Lycke J, Macciardi F, Männistö S, Manrique CP, Martin R, Martinelli V, Mason D, Mazibrada G, McCabe C, Mero IL, Mescheriakova J, Moutsianas L, Myhr KM, Nagels G, Nicholas R, Nilsson P, Piehl F, Pirinen M, Price SE, Quach H, Reunanen M, Robberecht W, Robertson NP, Rodegher M, Rog D, Salvetti M, Schnetz-Boutaud NC, Sellebjerg F, Selter RC, Schaefer C, Shaunak S, Shen L, Shields S, Siffrin V, Slee M, Sorensen PS, Sorosina M, Sospedra M, Spurkland A, Strange A, Sundqvist E, Thijs V, Thorpe J, Ticca A, Tienari P, van Duijn C, Visser EM, Vucic S, Westerlind H, Wiley JS, Wilkins A, Wilson JF, Winkelmann J, Zajicek J, Zindler E, Haines JL, Pericak-Vance MA, Ivinson AJ, Stewart G, Hafler D, Hauser SL, Compston A, McVean G, De Jager P, Sawcer SJ, McCauley JL. Analysis of immune-related loci identifies 48 new susceptibility variants for multiple sclerosis. *Nat Genet.* 2013 Nov;45(11):1353–60. doi: 10.1038/ng.2770. Epub 2013 Sep 29.

International Multiple Sclerosis Genetics Consortium, Hafler DA, Compston A, Sawcer S, Lander ES, Daly MJ, De Jager PL, de Bakker PI, Gabriel SB, Mirel DB, Ivinson AJ, Pericak-Vance MA, Gregory SG, Rioux JD, McCauley JL, Haines JL, Barcellos LF, Cree B, Oksenberg JR, Hauser SL. Risk alleles for multiple sclerosis identified by a genomewide study. *N Engl J Med.* 2007 Aug 30;357(9):851–62. doi: 10.1056/NEJMoa073493. Epub 2007 Jul 29.

Iwamoto T, Ikari K, Nakamura T, Kuwahara M, Toyama Y, Tomatsu T, Momohara S, Kamatani N. Association between PADI4 and rheumatoid arthritis: a meta-analysis. *Rheumatology (Oxford).* 2006 Jul;45(7):804–7. doi: 10.1093/rheumatology/kel023. Epub 2006 Jan 31.

Iwaszko M, Świerkot J, Kolossa K, Jeka S, Wiland P, Bogunia-Kubik K. Influence of CD94 and NKG2A variants on susceptibility to rheumatoid arthritis and efficacy of anti-TNF treatment. *Joint Bone Spine.* 2016 Jan;83(1):75–9. doi: 10.1016/j.jbspin.2015.06.010. Epub 2015 Oct 6.

Jacobs SR, Michalek RD, Rathmell JC. IL-7 is essential for homeostatic control of T cell metabolism in vivo. *J Immunol (Baltimore).* 2010;184:3461–9.

Jones G, Strugnell SA, DeLuca HF. Current understanding of the molecular actions of Vitamin D. *Physiol Rev.* 1998 Oct;78(4):1193–231. doi: 10.1152/physrev.1998.78.4.1193.

Jun KR, Choi SE, Cha CH, Oh HB, Heo YS, Ahn HY, Lee KJ. Meta-analysis of the association between HLA-DRB1 allele and rheumatoid arthritis susceptibility in Asian populations. *J Korean Med Sci.* 2007 Dec;22(6):973–80. doi: 10.3346/jkms.2007.22.6.973. PMID: 18162709; PMCID: PMC2694263.

Juyal G, Prasad P, Senapati S, Midha V, Sood A, Amre D, Juyal RC, BK T.An investigation of genomewide studies reported susceptibility loci for ulcerative colitis shows limited replication in north Indians. *PLoS One.* 2011;6:e16565.

Karami J, Aslani S, Jamshidi A, Garshasbi M, Mahmoudi M. Genetic implications in the pathogenesis of rheumatoid arthritis; an updated review. *Gene.* 2019 Jun 20;702:8–16. doi: 10.1016/j.gene.2019.03.033. Epub 2019 Mar 20.

Karban A, Waterman M, Panhuysen CI, Pollak RD, Nesher S, Datta L, Weiss B, Suissa A, Shamir R, Brant SR, Eliakim R. NOD2/CARD15 genotype and phenotype differences between Ashkenazi and Sephardic Jews with Crohn's disease. *Am J Gastroenterol.* 2004 Jun;99(6):1134-40. doi: 10.1111/j.1572-0241.2004.04156.x. PMID: 15180737.

Karp DR, Teletski CL, Jaraquemada D, Maloy WL, Coligan JE, Long EO. Structural requirements for pairing of alpha and beta chains in HLA-DR and HLA-DP molecules. *J Exp Med.* 1990 Mar 1;171(3):615–28. doi: 10.1084/jem.171.3.615.

Kaur S, Mirza AH, Brorsson CA, Fløyel T, Størling J, Mortensen HB, Pociot F; Hvidoere International Study Group. The genetic and regulatory architecture of ERBB3-type 1 diabetes susceptibility locus. *Mol Cell Endocrinol.* 2016 Jan 5;419:83–91. doi: 10.1016/j.mce.2015.10.002. Epub 2015 Oct 9.

Kharlamova N, Jiang X, Sherina N, Potempa B, Israelsson L, Quirke AM, Eriksson K, Yucel-Lindberg T, Venables PJ,

Potempa J, Alfredsson L, Lundberg K. Antibodies to porphyromonasgingivalis indicate interaction between oral infection, smoking, and risk genes in rheumatoid arthritis etiology. *Arthritis Rheumatol.* 2016 Mar;68(3):604–13. doi: 10.1002/art.39491.

Khodaeian M, Enayati S, Tabatabaei-Malazy O, Amoli MM. Association between genetic variants and diabetes mellitus in Iranian populations: a systematic review of observational studies. *J Diabetes Res.* 2015;2015:585917. doi: 10.1155/2015/585917. Epub 2015 Oct 26.

Kiani AK, John P, Bhatti A, Zia A, Shahid G, Akhtar P, Wang X, Demirci FY, Kamboh MI. Association of 32 type 1 diabetes risk loci in Pakistani patients. *Diabetes Res Clin Pract.* 2015 Apr;108(1):137–42. doi: 10.1016/j.diabres.2015.01.022. Epub 2015 Jan 21.

Kim SW, Kim ES, Moon CM, Park JJ, Kim TI, Kim WH, Cheon JH. Genetic polymorphisms of IL-23R and IL-17A and novel insights into their associations with inflammatory bowel disease. *Gut.* 2011 Nov;60(11):1527–36. doi: 10.1136/gut.2011.238477. Epub 2011 Jun 14.

Kisand K, UiboR. LADA and T1D in Estonian population - two different genetic risk profiles. *Gene.* 2012 Apr 15;497(2):-285–91. doi: 10.1016/j.gene.2012.01.089. Epub 2012 Feb 3.

Kochi Y, Suzuki A, Yamada R, Yamamoto K. Ethnogenetic heterogeneity of rheumatoid arthritis-implications for pathogenesis. *Nat Rev Rheumatol.* 2010 May;6(5):290–5. doi: 10.1038/nrrheum.2010.23. Epub 2010 Mar 16. PMID: 20234359.

Kolls JK, Lindén A. Interleukin-17 family members and inflammation. *Immunity.* 2004 Oct;21(4):467–76. doi: 10.1016/j.immuni.2004.08.018.

Korman BD, Kastner DL, Gregersen PK, Remmers EF. STAT4: genetics, mechanisms, and implications for autoimmunity. *Curr Allergy Asthma Rep.* 2008 Sep;8(5):398–403. doi: 10.1007/s11882-008-0077-8.

Kotas ME, Medzhitov R. Homeostasis, inflammation, and disease susceptibility. *Cell.* 2015 Feb 26;160(5):816–27. doi: 10.1016/j.cell.2015.02.010.

Latiano A, Palmieri O, Latiano T, Corritore G, Bossa F, Martino G, Biscaglia G, Scimeca D, Valvano MR, Pastore M, Marseglia A, D'Incà R, Andriulli A, Annese V. Investigation of multiple susceptibility loci for inflammatory bowel disease in an Italian cohort of patients. *PLoS One.* 2011;6(7):e22688. doi: 10.1371/journal.pone.0022688. Epub 2011 Jul 27.

Latiano A, Palmieri O, Valvano MR, D'Incà R, Caprilli R, Cucchiara S, Sturniolo GC, Bossa F, Andriulli A, Annese V. The association of MYO9B gene in Italian patients with inflammatory bowel diseases. *Aliment PharmacolTher.* 2008 Feb 1;27(3):241–8. doi: 10.1111/j.1365-2036.2007.03551.x. Epub 2007 Oct 15.

Lee EG, Boone DL, Chai S, Libby SL, Chien M, Lodolce JP, MA A (September 2000). Failure to regulate TNF-induced NF-kappaB and cell death responses in A20-deficient mice. Science. 289 (5488): 2350–2354. doi: 10.1126/science.289.5488.2350. PMC 3582399. PMID 11009421.

Lee JC, Bridger S, McGregor C, Macpherson AJ, Jones JE. Why children with inflammatory bowel disease are diagnosed at a younger age than their affected parent. *Gut.* 1999 Jun;44(6):808–11. doi: 10.1136/gut.44.6.808.

Lee YH, Bae SC, Choi SJ, Ji JD, Song GG. Associations between TNFAIP3 gene polymorphisms and rheumatoid arthritis: a meta-analysis. *Inflamm Res.* 2012 Jun;61(6):635–41. doi: 10.1007/s00011-012-0455-5. Epub 2012 Mar 9.

Lee YH, Bae SC, Kim JH, Song GG. Meta-analysis of genetic polymorphisms in programmed cell death 1. Associations with rheumatoid arthritis, ankylosing spondylitis, and type 1 diabetes susceptibility. *Z Rheumatol.* 2015 Apr;74(-3):230–9. doi: 10.1007/s00393-014-1415-y.

Lemos NE, Dieter C, Dorfman LE, Assmann TS, Duarte GCK, Canani LH, Bauer AC, Crispim D. The rs2292239 polymorphism in ERBB3 gene is associated with risk for type 1 diabetes mellitus in a Brazilian population. *Gene.* 2018 Feb 20;644:122–8. doi: 10.1016/j.gene.2017.11.009. Epub 2017 Nov 3.

Levy DE, Darnell JE Jr. Stats: transcriptional control and biological impact. *Nat Rev Mol Cell Biol.* 2002 Sep;3(9):651–62. doi: 10.1038/nrm909.

Li P, Yang XK, Wang X, Zhao MQ, Zhang C, Tao SS, Zhao W, Huang Q, Li LJ, Pan HF, Ye DQ. A meta-analysis of the relationship between MYO9B gene polymorphisms and susceptibility to Crohn's disease and ulcerative colitis. *Hum Immunol.* 2016a Oct;77(10):990–6. doi: 10.1016/j.humimm.2016.07.008. Epub 2016 Jul 18.

Li B, Xiao Y, Xing D, Ma XL, Liu J. Circulating interleukin-6 and rheumatoid arthritis: a Mendelian randomization meta-analysis. *Medicine (Baltimore).* 2016 Jun;95(-23):e3855. doi: 10.1097/MD.0000000000003855. Erratum in: *Medicine (Baltimore).* 2016b Jul 18;95(28):e0916.

Liang SC, Long AJ, Bennett F, Whitters MJ, Karim R, Collins M, Goldman SJ, Dunussi-Joannopoulos K, Williams CM, Wright JF, Fouser LA. An IL-17F/A heterodimer protein is produced by mouse Th17 cells and induces airway neutrophil recruitment. *J Immunol.* 2007 Dec 1;179(11):7791–9. doi: 10.4049/jimmunol.179.

Liang YL, Wu H, Shen X, Li PQ, Yang XQ, Liang L, Tian WH, Zhang LF, Xie XD. Association of STAT4 rs7574865 polymorphism with autoimmune diseases: a meta-analysis. *Mol Biol Rep.* 2012 Sep;39(9):8873–82. doi: 10.1007/s11033-012-1754-1. Epub 2012 Jun 20.

Libioulle C, Louis E, Hansoul S, Sandor C, Farnir F, Franchimont D, Vermeire S, Dewit O, de Vos M, Dixon A, Demarche B, Gut I, Heath S, Foglio M, Liang L, Laukens D, Mni M, Zelenika D, Van Gossum A, Rutgeerts P, Belaiche J, Lathrop M, Georges M. Novel Crohn disease locus identified by genome-wide association maps to a gene desert on 5p13.1 and modulates expression of PTGER4. *PLoS Genet.* 2007 Apr 20;3(4):e58. doi: 10.1371/journal.pgen.0030058. Epub 2007 Mar 5.

Liu CP, Jiang JA, Wang T, Liu XM, Gao L, Zhu RR, Shen Y, Wu M, Xu T, Zhang XG. CTLA-4 and CD86 genetic variants and haplotypes in patients with rheumatoid arthritis in Southeastern China. *Genet Mol Res.* 2013 Apr 25;12(-2):1373–82. doi: 10.4238/2013.April.25.8.

Liu H, Huang J, Dou M, Liu Y, Xiao B, Liu X, Huang Z. Variants in the IL7RA gene confer susceptibility to multiple sclerosis in Caucasians: evidence based on 9734 cases and 10436 controls. *Sci Rep.* 2017;7:1–10.

Liu HW, Xu RY, Sun RP, Wang Q, Liu JL, Ge W, Yu Z. Association of PTPN22 gene polymorphism with type 1 diabetes mellitus in Chinese children and adolescents. *Genet Mol Res.* 2015a Jan 15;14(1):63–8. doi: 10.4238/2015.January.15.8.

Liu JQ, Carl JW Jr, Joshi PS, RayChaudhury A, Pu XA, Shi FD, Bai XF. CD24 on the resident cells of the central nervous system enhances experimental autoimmune encephalomyelitis. *J Immunol.* 2007 May 15;178(10):6227–35. doi: 10.4049/jimmunol.178.10.6227.

Liu M, Zhu W, Wang J, Zhang J, Guo X, Wang J, Song J, Dong W. Interleukin-23 receptor genetic polymorphisms and ulcerative colitis susceptibility: a meta-analysis. *Clin Res Hepatol Gastroenterol.* 2015b Sep;39(4):516–25. doi: 10.1016/j.clinre.2014.10.009. Epub 2014 Nov 11.

Loddo I, Romano C. Infammatory bowel disease: genetics, epigenetics, and pathogenesis. *Frontiers in Immunology.* 2015;6:article 551.

Lu X, Tang L, Li K, Zheng J, Zhao P, Tao Y, Li LX. Contribution of NKX2-3 polymorphisms to inflammatory bowel diseases: a meta-analysis of 35358 subjects. *Sci Rep.* 2014 Jan 29;4:3924. doi: 10.1038/srep03924.

MacGregor AJ, Snieder H, Rigby AS, Koskenvuo M, Kaprio J, Aho K, Silman AJ. Characterizing the quantitative genetic contribution to rheumatoid arthritis using data from twins. *Arthritis Rheum.* 2000 Jan;43(1):30–7. doi: 10.1002/1529-0131(200001)43:1<30::AID-ANR5>3.0.CO;2-B.

Makoui MH, Imani D, Motallebnezhad M, Azimi M, Razi B. Vitamin D receptor gene polymorphism and susceptibility to asthma: meta-analysis based on 17 case-control studies. *Ann Allergy Asthma Immunol.* 2020 Jan;124(1):57–69. doi: 10.1016/j.anai.2019.10.014. Epub 2019 Oct 22.

Malek TR, Castro I. Interleukin-2 receptor signaling: at the interface between tolerance and immunity. *Immunity.* 2010 Aug 27;33(2):153–65. doi: 10.1016/j.immuni.2010.08.004.

Mangan PR, Harrington LE, O'Quinn DB, Helms WS, Bullard DC, Elson CO, Hatton RD, Wahl SM, Schoeb TR, Weaver CT . Transforming growth factor-beta induces development of the T(H)17 lineage. *Nature.* 2006;441(7090):231–4.

Marrie RA. Environmental risk factors in multiple sclerosis aetiology. *Lancet Neurol.* 2004 Dec;3(12):709–18. doi: 10.1016/S1474-4422(04)00933-0.

Mathey EK, Derfuss T, Storch MK, Williams KR, Hales K, Woolley DR, Al-Hayani A, Davies SN, Rasband MN, Olsson T, Moldenhauer A, Velhin S, Hohlfeld R, Meinl E, Linington C. Neurofascin as a novel target for autoantibody-mediated axonal injury. *J Exp Med.* 2007 Oct 1;204(10):2363–72. doi: 10.1084/jem.20071053. Epub 2007 Sep 10.

Matmati M, Jacques P, Maelfait J, Verheugen E, Kool M, Sze M, Geboes L, Louagie E, Mc Guire C, Vereecke L, Chu Y, Boon L, Staelens S, Matthys P, Lambrecht BN, Schmidt-Supprian M, Pasparakis M, Elewaut D, Beyaert R, van Loo G. A20 (TNFAIP3) deficiency in myeloid cells triggers erosive polyarthritis resembling rheumatoid arthritis. *Nat Genet.* 2011 Aug 14;43(9):908–12. doi: 10.1038/ng.874.

Maziarz M, Hagopian W, Palmer JP, Sanjeevi CB, Kockum I, Breslow N, Lernmark Å; Swedish Childhood Diabetes Register; Diabetes Incidence in Sweden Study Group; Type 1 Diabetes Genetics Consortium. Non-HLA type 1 diabetes genes modulate disease risk together with HLA-DQ and islet autoantibodies. *Genes Immun.* 2015 Dec;16(8):541–51. doi: 10.1038/gene.2015.43. Epub 2015 Oct 29.

McGeachy MJ, Cua DJ, Gaffen SL. The IL-17 family of cytokines in health and disease. *Immunity.* 2019 Apr 16;50(4):-892–906. doi: 10.1016/j.immuni.2019.03.021.

McInnes IB, Schett G. Cytokines in the pathogenesis of rheumatoid arthritis. *Nat Rev Immunol.* 2007 Jun;7(6):429–42. doi: 10.1038/nri2094.

McInnes IB, Schett G. Pathogenetic insights from the treatment of rheumatoid arthritis. *Lancet.* 2017 Jun 10;389(10086): 2328–37.doi: 10.1016/S0140-6736(17)31472-1.

Meggyesi N, Kiss LS, Koszarska M, Bortlik M, Duricova D, Lakatos L, Molnar T, Leniček M, Vítek L, Altorjay I, Papp M, Tulassay Z, Miheller P, Papp J, Tordai A, Andrikovics H, Lukas M, Lakatos PL. NKX2-3 and IRGM variants are associated with disease susceptibility to IBD in Eastern European patients. *World J Gastroenterol.* 2010 Nov 7;16(-41):5233–40. doi: 10.3748/wjg.v16.i41.5233..

Menard L, Saadoun D, Isnardi I, Ng YS, Meyers G, Massad C, Price C, Abraham C, Motaghedi R, Buckner JH, Gregersen PK, Meffre E. The PTPN22 allele encoding an R620W variant interferes with the removal of developing autoreactive B cells in humans. *J Clin Invest.* 2011 Sep;121(9):3635–44. doi: 10.1172/JCI45790. Epub 2011 Aug 1.

Miettinen ME, Smart MC, Kinnunen L, Harjutsalo V, Reinert-Hartwall L, Ylivinkka I, Surcel HM, Lamberg-Allardt C, Hitman GA, Tuomilehto J. Genetic determinants of serum 25-hydroxyvitamin D concentration during pregnancy and type 1 diabetes in the child. *PLoS One.* 2017 Oct 4;12(-10):e0184942. doi: 10.1371/journal.pone.0184942.

Milenkovic VM, Stanton EH, Nothdurfter C, Rupprecht R, Wetzel CH. The role of chemokines in the pathophysiology of major depressive disorder. *Int J Mol Sci.* 2019 May 09;20(9):2283. doi: 10.3390/ijms20092283.

Miller AH, Raison CL. The role of inflammation in depression: from evolutionary imperative to modern treatment target. *Nat Rev Immunol.* 2016 Jan;16(1):22–34. doi: 10.1038/nri.2015.5

Mitrovič M, Potočnik U. High-resolution melting curve analysis for high-throughput genotyping of NOD2/CARD15 mutations and distribution of these mutations in Slovenian inflammatory bowel diseases patients. *Dis Markers.* 2011;30(5):265–74. doi: 10.3233/DMA-2011-0783.

Mizushima N. The pleiotropic role of autophagy: from protein metabolism to bactericide. *Cell Death Differ.* 2005 Nov;12 Suppl 2:1535–41. doi: 10.1038/sj.cdd.4401728.

Mohammadi FS, Aslani S, Mostafaei S, Jamshidi A, Riahi P, Mahmoudi M. Are genetic variations in IL-21-IL-23R-IL-17A cytokine axis involved in a pathogenic pathway of rheumatoid arthritis? Bayesian hierarchical meta-analysis. *J Cell Physiol.* 2019 Aug;234(10):17159–71. doi: 10.1002/jcp.28495. Epub 2019 Mar 28.

Molodecky NA, Soon IS, Rabi DM, Ghali WA, Ferris M, Chernoff G, Benchimol EI, Panaccione R, Ghosh S, Barkema HW, Kaplan GG. Increasing incidence and prevalence of the inflammatory bowel diseases with time, based on systematic review. *Gastroenterology.* 2012 Jan;142(1):46–54.e42; quiz e30. doi: 10.1053/j.gastro.2011.10.001. Epub 2011 Oct 14.

Moseley TA, Haudenschild DR, Rose L, Reddi AH. Interleukin-17 family and IL-17 receptors. *Cytokine Growth Factor Rev.* 2003 Apr;14(2):155–74. doi: 10.1016/s1359-6101(03)00002-9.

Mousavi MJ, Shayesteh MRH, Jamalzehi S, Alimohammadi R, Rahimi A, Aslani S, Rezaei N. Association of the genetic polymorphisms in inhibiting and activating molecules of immune system with rheumatoid arthritis: a systematic review and meta-analysis. *J Res Med Sci.* 2021 Mar 31;26:22. doi: 10.4103/jrms.JRMS_567_20.

Moutsianas L, Jostins L, Beecham AH, Dilthey AT, Xifara DK, Ban M, Shah TS, Patsopoulos NA, Alfredsson L, Anderson CA, Attfield KE, Baranzini SE, Barrett J, Binder TMC, Booth D, Buck D, Celius EG, Cotsapas C, D'Alfonso S, Dendrou CA, Donnelly P, Dubois B, Fontaine B, Fugger L, Goris A, Gourraud PA, Graetz C, Hemmer B, Hillert J; International IBD Genetics Consortium (IIBDGC), Kockum I, Leslie S, Lill CM, Martinelli-Boneschi F, Oksenberg JR, Olsson T, Oturai A, Saarela J, Søndergaard HB, Spurkland A, Taylor B, Winkelmann J, Zipp F, Haines JL, Pericak-Vance MA, Spencer CCA, Stewart G, Hafler DA, Ivinson AJ, Harbo HF, Hauser SL, De Jager PL, Compston A, McCauley JL, Sawcer S, McVean G. Class II HLA interactions modulate genetic risk for multiple sclerosis. *Nat Genet.* 2015 Oct;47(10):1107–13. doi: 10.1038/ng.3395. Epub 2015 Sep 7.

Müller RT, Honnert U, Reinhard J, Bähler M. The rat myosin myr 5 is a GTPase-activating protein for Rho in vivo: essential role of arginine 1695. *Mol Biol Cell.* 1997 Oct;8(10):2039–53. doi: 10.1091/mbc.8.10.2039.

Nabi G, Akhter N, Wahid M, Bhatia K, Mandal RK, Dar SA, Jawed A, Haque S. Meta-analysis reveals PTPN22 1858C/T polymorphism confers susceptibility to rheumatoid arthritis in Caucasian but not in Asian population. *Autoimmunity.* 2016;49(3):197–210. doi: 10.3109/08916934.2015.1134514. Epub 2016 Jan 13. PMID: 26763276.

Narooie-Nejad M, Moossavi M, Torkamanzehi A, Moghtaderi A, Salimi S. Vitamin D receptor gene polymorphism and the risk of multiple sclerosis in South-Eastern of Iran. *J Mol Neurosci.* 2015 Jul;56(3):572–6. doi: 10.1007/s12031-015-0513-x. Epub 2015 Feb 18.

Naser SA, Thanigachalam S, Dow CT, Collins MT. Exploring the role of Mycobacterium avium subspecies paratuberculosis in the pathogenesis of type 1 diabetes mellitus: a pilot study. *Gut Pathog.* 2013 Jun 13;5:14. doi: 10.1186/1757-4749-5-14.

National Center for Chronic Disease Prevention and Health Promotion, Centers for Disease Control and Prevention Inflammatory Bowel Disease Prevalence. [Accessed on 27 November 2020]

Needham EJ, Helmy A, Zanier ER, Jones JL, Coles AJ, Menon DK. The immunological response to traumatic brain injury. *J Neuroimmunol.* 2019 Jul 15;332:112–25.

Newton JL, Harney SM, Wordsworth BP, Brown MA. A review of the MHC genetics of rheumatoid arthritis. *Genes Immun.* 2004 May;5(3):151–7. doi: 10.1038/sj.gene.6364045.

Noble JA, Valdes AM. Genetics of the HLA region in the prediction of type 1 diabetes. *Curr Diab Rep.* 2011 Dec;11(-6):533–42. doi: 10.1007/s11892-011-0223-x.

Nyaga DM, Vickers MH, Jefferies C, Perry JK, O'Sullivan JM. The genetic architecture of type 1 diabetes mellitus. *Mol Cell Endocrinol.* 2018 Dec 5;477:70–80. doi: 10.1016/j.mce.2018.06.002. Epub 2018 Jun 18.

Okada Y, Wu D, Trynka G, Raj T, Terao C, Ikari K, Kochi Y, Ohmura K, Suzuki A, Yoshida S, Graham RR, Manoharan A, Ortmann W, Bhangale T, Denny JC, Carroll RJ, Eyler AE, Greenberg JD, Kremer JM, Pappas DA, Jiang L, Yin J, Ye L, Su DF, Yang J, Xie G, Keystone E, Westra HJ, Esko T, Metspalu A, Zhou X, Gupta N, Mirel D, Stahl EA, Diogo D, Cui J, Liao K, Guo MH, Myouzen K, Kawaguchi T, Coenen MJ, van Riel PL, van de Laar MA, Guchelaar HJ, Huizinga TW, Dieudé P, Mariette X, Bridges SL Jr, Zhernakova A, Toes RE, Tak PP, Miceli-Richard C, Bang SY, Lee HS, Martin J, Gonzalez-Gay MA, Rodriguez-Rodriguez L, Rantapää-Dahlqvist S, Arlestig L, Choi HK, Kamatani Y, Galan P, Lathrop M; RACI consortium; GARNET consortium, Eyre S, Bowes J, Barton A, de Vries N, Moreland LW, Criswell LA, Karlson EW, Taniguchi A, Yamada R, Kubo M, Liu JS, Bae SC, Worthington J, Padyukov L, Klareskog L, Gregersen PK, Raychaudhuri S, Stranger BE, De Jager PL, Franke L, Visscher PM, Brown MA, Yamanaka H, Mimori T, Takahashi A, Xu H, Behrens TW, Siminovitch KA, Momohara S, Matsuda F, Yamamoto K, Plenge RM. Genetics of rheumatoid arthritis contributes to biology and drug discovery. *Nature.* 2014 Feb 20;506(7488):376–81. doi: 10.1038/nature12873. Epub 2013 Dec 25.

Okazaki T, Wang J. PD-1/PD-L pathway and autoimmunity. *Autoimmunity.* 2005 Aug;38(5):353–7. doi: 10.1080/08916930500124072.

Olerup O, Hillert J. HLA class II-associated genetic susceptibility in multiple sclerosis: a critical evaluation. *Tissue Antigens.* 1991 Jul;38(1):1–15. doi: 10.1111/j.1399-0039.1991.tb02029.x.

Oliver JE, Silman AJ. Why are women predisposed to autoimmune rheumatic diseases? *Arthritis Res Ther.* 2009;11(-5):252. doi: 10.1186/ar2825. Epub 2009 Oct 26.

Omraninava M, Mehranfar S, Vahedi P, Razi B, Imani D, Aslani S, Feyzinia S. Association between IL7 Receptor Alpha (Il7ra) gene rs6897932 polymorphism and the risk of Multiple Sclerosis: a meta-regression and meta-analysis. *Mult Scler Relat Disord.* 2021 Feb;48:102687. doi: 10.1016/j.msard.2020.102687. Epub 2020 Dec 15.

Orholm M, Munkholm P, Langholz E, Nielsen OH, Sørensen TI, Binder V. Familial occurrence of inflammatory bowel disease. *N Engl J Med.* 1991 Jan 10;324(2):84–8. doi: 10.1056/NEJM199101103240203.

Ousman SS, Tomooka BH, van Noort JM, Wawrousek EF, O'Connor KC, Hafler DA, Sobel RA, Robinson WH, Steinman L. Protective and therapeutic role for alphaB-crystallin in autoimmune demyelination. *Nature.* 2007 Jul 26;448(7152):474–9. doi: 10.1038/nature05935. Epub 2007 Jun 13.

Owen SA, Lunt M, Bowes J, Hider SL, Bruce IN, Thomson W, Barton A. MTHFR gene polymorphisms and outcome of methotrexate treatment in patients with rheumatoid arthritis: analysis of key polymorphisms and meta-analysis of C677T and A1298C polymorphisms. *Pharmacogenomics J.* 2013 Apr;13(2):137–47. doi: 10.1038/tpj.2011.42. Epub 2011 Sep 20.

Pahwa R, Goyal A, Bansal P, Jialal I. Chronic inflammation. [Updated 2021 Aug 11]. In: *StatPearls* [Internet]. Treasure Island (FL): StatPearls Publishing; 2021 Jan.

Palmer AM. Multiple sclerosis and the blood-central nervous system barrier. *Cardiovasc Psychiatry Neurol.* 2013;2013:530356. doi: 10.1155/2013/530356. Epub 2013 Jan 15. PMID: 23401746; PMCID: PMC3562587.

Panierakis C, Goulielmos G, Mamoulakis D, Petraki E, Papavasiliou E, Galanakis E. Vitamin D receptor gene polymorphisms and susceptibility to type 1 diabetes in Crete, Greece. *Clin Immunol.* 2009 Nov;133(2):276–81. doi: 10.1016/j.clim.2009.08.004. Epub 2009 Sep 5.

Patsopoulos NA. Genetics of Multiple Sclerosis: An Overview and New Directions. *Cold Spring Harb Perspect Med.* 2018 Jul 2;8(7):a028951. doi: 10.1101/cshperspect.a028951. PMID: 29440325; PMCID: PMC6027932.

Patsopoulos NA, Barcellos LF, Hintzen RQ, Schaefer C, van Duijn CM, Noble JA, Raj T; IMSGC; ANZgene, Gourraud PA, Stranger BE, Oksenberg J, Olsson T, Taylor BV, Sawcer S, Hafler DA, Carrington M, De Jager PL, de Bakker PI. Fine-mapping the genetic association of the major histocompatibility complex in multiple sclerosis: HLA and non-HLA effects. *PLoS Genet.* 2013 Nov;9(-11):e1003926. doi: 10.1371/journal.pgen.1003926. Epub 2013 Nov 21.

Patterson CC, Karuranga S, Salpea P, Saeedi P, Dahlquist G, Soltesz G, Ogle GD. Worldwide estimates of incidence, prevalence and mortality of type 1 diabetes in children and adolescents: results from the International Diabetes Federation Diabetes Atlas, 9th edition. *Diabetes Res Clin Pract.* 2019 Nov;157:107842. doi: 10.1016/j.diabres.2019.107842. Epub 2019 Sep 10.

Peng H, Li J, Chen X, Zhou X, Zhu W, Li F. Genetic variants of PTPN2 gene in Chinese children with type 1 diabetes mellitus. *Med Sci Monit.* 2015 Sep 7;21:2653–8. doi: 10.12659/MSM.893607.

Perkins EA, Landis D, Causey ZL, Edberg Y, Reynolds RJ, Hughes LB, Gregersen PK, Kimberly RP, Edberg JC, Bridges SL Jr; Consortium for the Longitudinal Evaluation of African Americans with Early Rheumatoid Arthritis Investigators. Association of single-nucleotide polymorphisms in CCR6, TAGAP, and TNFAIP3 with rheumatoid arthritis in African Americans. *Arthritis Rheum.* 2012 May;64(5):1355–8. doi: 10.1002/art.33464.

Pirie FJ, Hammond MG, Motala AA, Omar MA. HLA class II antigens in South African Blacks with type I diabetes. *Tissue Antigens.* 2001 Apr;57(4):348–52. doi: 10.1034/j.139 9-0039.2001.057004348.x.

Plantinga TS, Crisan TO, Oosting M, van de Veerdonk FL, de Jong DJ, Philpott DJ, van der Meer JW, Girardin SE, Joosten LA, Netea MG. Crohn's disease- associated ATG16L1 polymorphism modulates pro-inflammatory cytokine responses selectively upon activation of NOD2. *Gut.* 2011 Sep;60(9):1229–35. doi: 10.1136/gut.2010.228908. Epub 2011 Mar 15.

Plenge RM, Seielstad M, Padyukov L, Lee AT, Remmers EF, Ding B, Liew A, Khalili H, Chandrasekaran A, Davies LR, Li W, Tan AK, Bonnard C, Ong RT, Thalamuthu A, Pettersson S, Liu C, Tian C, Chen WV, Carulli JP, Beckman EM, Altshuler D, Alfredsson L, Criswell LA, Amos CI, Seldin MF, Kastner DL, Klareskog L, Gregersen PK. TRAF1-C5 as a risk locus for rheumatoid arthritis--a genomewide study. *N Engl J Med.* 2007 Sep 20;357(12):1199–209. doi: 10.1056/NEJMoa073491. Epub 2007 Sep 5.

Polychronakos C, Li Q. Understanding type 1 diabetes through genetics: advances and prospects. *Nat Rev Genet.* 2011 Oct 18;12(11):781–92. doi: 10.1038/nrg3069.

Prokunina L, Castillejo-López C, Oberg F, Gunnarsson I, Berg L, Magnusson V, Brookes AJ, Tentler D, Kristjansdóttir H, Gröndal G, Bolstad AI, Svenungsson E, Lundberg I, Sturfelt G, Jönssen A, Truedsson L, Lima G, Alcocer-Varela J, Jonsson R, Gyllensten UB, Harley JB, Alarcón-Segovia D, Steinsson K, Alarcón-Riquelme ME. A regulatory polymorphism in PDCD1 is associated with susceptibility to systemic lupus erythematosus in humans. *Nat Genet.* 2002 Dec;32(4):666–9. doi: 10.1038/ng1020. Epub 2002 Oct 28.

Qiu W, James I, Carroll WM, Mastaglia FL, Kermode AG. HLA-DR allele polymorphism and multiple sclerosis in Chinese populations: a meta-analysis. *Mult Scler.* 2011 Apr;17(4):382–8. doi: 10.1177/1352458510391345. Epub 2010 Dec 21.

Qin WH, Wang HX, Qiu JL, Huang XB, Huang Y, Wu NR, Liang HS. A meta-analysis of association of Vitamin D receptor BsmI gene polymorphism with the risk of type 1 diabetes mellitus. *J Recept Signal Transduct Res.* 2014 Oct;34(5):372–7. doi: 10.3109/10799893.2014.903420. Epub 2014 Mar 25.

Rafa H, Saoula H, Belkhelfa M, Medjeber O, Soufli I, Toumi R, de Launoit Y, Moralès O, Nakmouche M, Delhem N, Touil-Boukoffa C. IL-23/IL-17A axis correlates with the nitric oxide pathway in inflammatory bowel disease: immunomodulatory effect of retinoic acid. *J Interferon Cytokine Res.* 2013 Jul;33(7):355–68. doi: 10.1089/jir.2012.0063. Epub 2013 Mar 8.

Rasoul MA, Haider MZ, Al-Mahdi M, Al-Kandari H, Dhaunsi GS. Relationship of four Vitamin D receptor gene polymorphisms with type 1 diabetes mellitus susceptibility in Kuwaiti children. *BMC Pediatr.* 2019 Mar 7;19(1):71. doi: 10.1186/s12887-019-1448-0.

Reddy MV, Wang H, Liu S, Bode B, Reed JC, Steed RD, Anderson SW, Steed L, Hopkins D, She JX. Association between type 1 diabetes and GWAS SNPs in the southeast US Caucasian population. *Genes Immun.* 2011Apr;12(3):208–12. doi: 10.1038/gene.2010.70. Epub 2011 Jan 27.

Redondo MJ, Steck AK, Pugliese A. Genetics of type 1 diabetes. *Pediatr Diabetes.* 2018 May;19(3):346–53. doi:10.1111/pedi.12597.

Riachy R, Vandewalle B, Moerman E, Belaich S, Lukowiak B, Gmyr V, Muharram G, Kerr Conte J, Pattou F. 1,25-Dihydroxyvitamin D3 protects human pancreatic islets against cytokine-induced apoptosis via down-regulation of the Fas receptor. *Apoptosis.* 2006 Feb;11(2):151–9. doi: 10.1007/s10495-006-3558-z.

Rioux JD, Silverberg MS, Daly MJ, Steinhart AH, McLeod RS, Griffiths AM, Green T, Brettin TS, Stone V, Bull SB, Bitton A, Williams CN, Greenberg GR, Cohen Z, Lander ES, Hudson TJ, Siminovitch KA. Genomewide search in Canadian families with inflammatory bowel disease reveals two novel susceptibility loci. *Am J Hum Genet.* 2000 Jun;66(6):1863–70. doi: 10.1086/302913. Epub 2000 Apr 21.

Roberts RL, Gearry RB, Hollis-Moffatt JE, Miller AL, Reid J, Abkevich V, Timms KM, Gutin A, Lanchbury JS, Merriman TR, Barclay ML, Kennedy MA. IL23R R381Q and ATG16L1 T300A are strongly associated with Crohn's disease in a study of New Zealand Caucasians with inflammatory bowel disease. *Am J Gastroenterol.* 2007 Dec;102(12):2754–61. doi: 10.1111/j.1572-0241.2007.01525 .x. Epub 2007 Sep 25.

Rodríguez A, Alfaro JM, Balthazar V, Pineda Trujillo N. Association analysis of PTPN22, CTLA4 and IFIH1 genes with type 1 diabetes in Colombian families. *J Diabetes.* 2015 May;7(3):402–10. doi: 10.1111/1753-0407.12192. Epub 2014 Sep 10.

Saadatnia M, Najafi MR, Najafi F, Davoudi V, Keyhanian K, Maghzi AH. CD24 gene allele variation is not associated with oligoclonal IgG bands and IgG index of multiple sclerosis patients. *Neuroimmunomodulation*. 2012;19(3):195–9. doi: 10.1159/000332011. Epub 2012 Jan 18.

Sahami-Fard MH, Mozhdeh M, Izadpanah F, Kashani HH, Nezhadi A. Interleukin 7 receptor T244I polymorphism and the multiple sclerosis susceptibility: a meta-analysis. *J Neuroimmunol*. 2020 Apr 15;341:577166. doi: 10.1016/j.jneuroim.2020.577166. Epub 2020 Jan 22.

Sairenji T, Collins KL, Evans DV. An update on inflammatory bowel disease. *Prim Care*. 2017 Dec;44(4):673–92. doi: 10.1016/j.pop.2017.07.010. Epub 2017 Oct 5.

Sansom DM. IMMUNOLOGY. Moving CTLA-4 from the trash to recycling. *Science*. 2015 Jul 24;349(6246):377–8. doi: 10.1126/science.aac7888. Epub 2015 Jul 23.

Santin I, Moore F, Colli ML, Gurzov EN, Marselli L, Marchetti P, Eizirik DL. PTPN2, a candidate gene for type 1 diabetes, modulates pancreatic β-cell apoptosis via regulation of the BH3-only protein Bim. *Diabetes*. 2011 Dec;60(12):3279–88. doi: 10.2337/db11-0758. Epub 2011 Oct 7.

Scolaro BL, dos Santos E, Ferreira LE, França PH, Kleinubing H Jr, Kotze PG, PinhoMde S. T300A genetic polymorphism: a susceptibility factor for Crohn's disease? *Arq Gastroenterol*. 2014 Apr–Jun;51(2):97–101. doi: 10.1590/s0004-28032014000200005.

Scott DL, Wolfe F, Huizinga TW. Rheumatoid arthritis. *Lancet*. 2010 Sep 25;376(9746):1094–108. doi: 10.1016/S0140-6736(10)60826-4.

Seldin MF, Amos CI, Ward R, Gregersen PK. The genetics revolution and the assault on rheumatoid arthritis. *Arthritis Rheum*. 1999 Jun;42(6):1071–9. doi: 10.1002/1529-0131(199906)42:6<1071::AID-ANR1>3.0.CO;2-8.

Senhaji N, Serrano A, Badre W, Serbati N, Karkouri M, Zaid Y, Nadifi S, Martin J. Association of inflammatory cytokine gene polymorphisms with inflammatory bowel disease in a Moroccan cohort. *Genes Immun*. 2016 Jan–Feb;17(1):60–5. doi: 10.1038/gene.2015.52. Epub 2015 Dec 3.

Shao M, Xu S, Yang H, Xu W, Deng J, Chen Y, Gao X, Guan S, Xu S, Shuai Z, Pan F. Association between IL-17A and IL-17F gene polymorphism and susceptibility in inflammatory arthritis: a meta-analysis. *Clin Immunol*. 2020 Apr;213:108374. doi: 10.1016/j.clim.2020.108374. Epub 2020 Mar 5.

Sharma C, R Ali B, Osman W, Afandi B, Aburawi EH, Beshyah SA, Al-Mahayri Z, Al-Rifai RH, Al Yafei Z, ElGhazali G, Alkaabi J. Association of variants in *PTPN22*, *CTLA-4*, *IL2-RA*, and *INS* genes with type 1 diabetes in Emiratis. *Ann Hum Genet*. 2021 Mar;85(2):48–57. doi: 10.1111/ahg.12406. Epub 2020 Sep 24.

Stastny P. Association of the B-cell alloantigen DRw4 with rheumatoid arthritis. *N Engl J Med*. 1978 Apr 20;298(16):869–71. doi: 10.1056/NEJM197804202981602. PMID: 147420.

Stokkers PC, Reitsma PH, Tytgat GN, van Deventer SJ. HLA-DR and -DQ phenotypes in inflammatory bowel disease: a meta-analysis. *Gut*. 1999 Sep;45(3):395–401. doi: 10.1136/gut.45.3.395.

Straub RH. The brain and immune system prompt energy shortage in chronic inflammation and ageing. *Nat Rev Rheumatol*. 2017 Dec;13(12):743–51. doi: 10.1038/nrrheum.2017.172. Epub 2017 Oct 12.

Sun C, Wei H, Chen X, Zhao Z, Du H, Song W, Yang Y, Zhang M, Lu W, Pei Z, Xi L, Yan J, Zhi D, Cheng R, Luo F. ERBB3-rs2292239 as primary type 1 diabetes association locus among non-HLA genes in Chinese. *Meta Gene*. 2016 May 12;9:120–3. doi: 10.1016/j.mgene.2016.05.003.

Suzuki A, Yamada R, Chang X, Tokuhiro S, Sawada T, Suzuki M, Nagasaki M, Nakayama-Hamada M, Kawaida R, Ono M, Ohtsuki M, Furukawa H, Yoshino S, Yukioka M, Tohma S, Matsubara T, Wakitani S, Teshima R, Nishioka Y, Sekine A, Iida A, Takahashi A, Tsunoda T, Nakamura Y, Yamamoto K. Functional haplotypes of PADI4, encoding citrullinating enzyme peptidylarginine deiminase 4, are associated with rheumatoid arthritis. *Nat Genet*. 2003 Aug;34(4):395–402. doi: 10.1038/ng1206.

Suzuki T, Ikari K, Yano K, Inoue E, Toyama Y, Taniguchi A, Yamanaka H, Momohara S. PADI4 and HLA-DRB1 are genetic risks for radiographic progression in RA patients, independent of ACPA status: results from the IORRA cohort study. *PLoS One*. 2013;8(4):e61045. doi: 10.1371/journal.pone.0061045. Epub 2013 Apr 8.

Tang S, Peng W, Wang C, Tang H, Zhang Q. Association of the PTPN22 gene (+1858C/T, -1123G/C) polymorphisms with type 1 diabetes mellitus: a systematic review and meta-analysis. *Diabetes Res Clin Pract*. 2012 Sep;97(3):446–52. doi: 10.1016/j.diabres.2012.04.011. Epub 2012 May 7.

Taniguchi K, Karin M. NF-κB, inflammation, immunity and cancer: coming of age. *Nat Rev Immunol*. 2018 May;18(5):309–24. doi: 10.1038/nri.2017.142. Epub 2018 Jan 22. PMID: 29379212.

Tavakolpour S. Anti-interleukin and associated receptors monoclonal antibodies therapy in autoimmune diseases. *Receptor Clin Invest*. 2016a;3:e1173.

Tavakolpour S. Interleukin 7 receptor polymorphisms and the risk of multiple sclerosis: a meta-analysis. *Mult Scler Relat Disord*. 2016b Jul;8:66–73. doi: 10.1016/j.msard.2016.05.001. Epub 2016 May 2.

Tizaoui K, Kaabachi W, Hamzaoui A, Hamzaoui K. Association between Vitamin D receptor polymorphisms and multiple sclerosis: systematic review and meta-analysis of case-control studies. *Cell Mol Immunol*. 2015 Mar;12(2):243–52. doi: 10.1038/cmi.2014.47. Epub 2014 Jul 7.

Traboulsee AL, Bernales CQ, Ross JP, Lee JD, Sadovnick AD, Vilariño-Güell C. Genetic variants in IL2RA and IL7R a_ect multiple sclerosis disease risk and progression. *Neurogenetics*. 2014;15:165–9.

Trachtenberg EA, Yang H, Hayes E, Vinson M, Lin C, Targan SR, Tyan D, Erlich H, Rotter JI. HLA class II haplotype associations with inflammatory bowel disease in Jewish (Ashkenazi) and non-Jewish caucasian populations. *Hum Immunol*. 2000 Mar;61(3):326–33. doi: 10.1016/s0198-8859(99)00134-2.

Trembleau S, Germann T, Gately MK, Adorini L. The role of IL-12 in the induction of organ-specific autoimmune diseases. *Immunol Today*. 1995 Aug;16(8):383–6. doi: 10.1016/0167-5699(95)80006-9.

Trowsdale J, Knight JC. Major histocompatibility complex genomics and human disease. *Annu Rev Genomics Hum Genet*. 2013;14:301–23. doi: 10.1146/annurev-genom-091212-153455. Epub 2013 Jul 15.

Tsai DH, Riediker M, Berchet A, Paccaud F, Waeber G, Vollenweider P, Bochud M. Effects of short- and long-term exposures to particulate matter on inflammatory marker levels in the general population. *Environ Sci Pollut Res Int.* 2019 Jul;26(19):19697–704. doi: 10.1007/s11356-019-05194-y. Epub 2019 May 12.

Uitterlinden AG, Fang Y, Van Meurs JB, Pols HA, Van Leeuwen JP. Genetics and biology of Vitamin D receptor polymorphisms. *Gene.* 2004 Sep 1;338(2):143–56. doi: 10.1016/j.gene.2004.05.014.

Valdes AM, Styrkarsdottir U, Doherty M, , Morris DL, Mangino M, Tamm A, Doherty SA, Kisand K, Kerna I, Tamm A, Wheeler M, Maciewicz RA, Zhang W, Muir KR, Dennison EM, Hart DJ, Metrustry S, Jonsdottir I, Jonsson GF, Jonsson H, Ingvarsson T, Cooper C, Vyse TJ, Spector TD, Stefansson K, Arden NK. Large scale replication study of the association between HLA class II/BTNL2 variants and osteoarthritis of the knee in European-descent populations. *PLoS One.* 2011;6:e23371.

van Belle TL, Coppieters KT, von Herrath MG. Type 1 diabetes: etiology, immunology, and therapeutic strategies. *Physiol Rev.* 2011 Jan;91(1):79–118. doi: 10.1152/physrev.00003.2010.

van der Woude D, Houwing-Duistermaat JJ, Toes RE, Huizinga TW, Thomson W, Worthington J, van der Helm-van Mil AH, de Vries RR. Quantitative heritability of anti-citrullinated protein antibody-positive and anti-citrullinated protein antibody-negative rheumatoid arthritis. *Arthritis Rheum.* 2009 Apr;60(4):916–23. doi: 10.1002/art.24385.

van Drongelen V, Holoshitz J. Human leukocyte antigen-disease associations in rheumatoid arthritis. *Rheum Dis Clin North Am.* 2017 Aug;43(3):363–76. doi: 10.1016/j.rdc.2017.04.003.

van Gaalen FA, Linn-Rasker SP, van Venrooij WJ, de Jong BA, Breedveld FC, Verweij CL, Toes RE, Huizinga TW. Autoantibodies to cyclic citrullinated peptides predict progression to rheumatoid arthritis in patients with undifferentiated arthritis: a prospective cohort study. *Arthritis Rheum.* 2004 Mar;50(3):709–15. doi: 10.1002/art.20044.

van Heel DA, Fisher SA, Kirby A, Daly MJ, Rioux JD, Lewis CM; Genome Scan Meta-Analysis Group of the IBD International Genetics Consortium. Inflammatory bowel disease susceptibility loci defined by genome scan meta-analysis of 1952 affected relative pairs. *Hum Mol Genet.* 2004 Apr 1;13(7):763–70. doi: 10.1093/hmg/ddh090. Epub 2004 Feb 19.

Vang T, Miletic AV, Bottini N, Mustelin T. Protein tyrosine phosphatase PTPN22 in human autoimmunity. *Autoimmunity.* 2007 Sep;40(6):453–61. doi: 10.1080/08916930701464897.

Varade J, Wang N, Lim CK, Zhang T, Zhang Y, Liu X, Piehl F, Matell R, Cao H, Xu X, Hammarström L. Novel genetic loci associated HLA-B*08:01 positive myasthenia gravis. *J Autoimmun.* 2018 Mar;88:43–49. doi: 10.1016/j.jaut.2017.10.002. Epub 2017 Oct 14.

Veldhoen M, Hocking RJ, Atkins CJ, Locksley RM, Stockinger B. TGFbeta in the context of an inflammatory cytokine milieu supports de novo differentiation of IL-17-producing T cells. *Immunity.* 2006 Feb;24(2):179–89. doi: 10.1016/j.immuni.2006.01.001.

Vernerova L, Spoutil F, Vlcek M, Krskova K, Penesova A, Meskova M, Marko A, Raslova K, Vohnout B, Rovensky J, Killinger Z, Jochmanova I, Lazurova I, Steiner G, Smolen J, Imrich R. A Combination of CD28 (rs1980422) and IRF5 (rs10488631) polymorphisms is associated with seropositivity in rheumatoid arthritis: a case control study. *PLoS One.* 2016 Apr 19;11(4):e0153316. doi: 10.1371/journal.pone.0153316.

Walton C, King R, Rechtman L, Kaye W, Leray E, Marrie RA, Robertson N, La Rocca N, Uitdehaag B, van der Mei I, Wallin M, Helme A, Angood Napier C, Rijke N, Baneke P. Rising prevalence of multiple sclerosis worldwide: insights from the Atlas of MS, 3rd edition. *MultScler.* 2020 Dec;26(14):1816–21. doi: 10.1177/1352458520970841. Epub 2020 Nov 11.

Wang B, Du W, Jia Y, Zhang X, Ma G. Cytotoxic T-lymphocyte-associated protein 4 +49A/G polymorphisms contribute to the risk of type 1 diabetes in children: an updated systematic review and meta-analysis with trial sequential analysis. *Oncotarget.* 2017 Feb 7;8(6):10553–64. doi: 10.18632/oncotarget.14457.

Wang D, Pan G. The association between *rs2292239* polymorphism in *ERBB3* gene and type 1 diabetes: a meta-analysis. *Biomed Res Int.* 2019 Aug 6;2019:7689642. doi: 10.1155/2019/7689642.

Wang G, Zhang Q, Xu N, Xu K, Wang J, He W, Yang T. Associations between two polymorphisms (FokI and BsmI) of Vitamin D receptor gene and type 1 diabetes mellitus in Asian population: a meta-analysis. *PLoS One.* 2014 Mar 6;9(3):e89325. doi: 10.1371/journal.pone.0089325.

Wang H, Jin Y, Reddy MV, Podolsky R, Liu S, Yang P, Bode B, Reed JC, Steed RD, Anderson SW, Steed L, Hopkins D, Huang Y, She JX. Genetically dependent ERBB3 expression modulates antigen presenting cell function and type 1 diabetes risk. *PLoS One.* 2010 Jul 26;5(7):e11789. doi: 10.1371/journal.pone.0011789.

Wang J, Liu L, Ma J, Sun F, Zhao Z, Gu M. Common variants on cytotoxic T lymphocyte antigen-4 polymorphisms contributes to type 1 diabetes susceptibility: evidence based on 58 studies. *PLoS One.* 2014 Jan 23;9(1):e85982. doi: 10.1371/journal.pone.0085982.

Wang LM, Zhang DM, Xu YM, Sun SL. Interleukin 2 receptor α gene polymorphism and risk of multiple sclerosis: a meta-analysis. *J Int Med Res.* 2011;39(5):1625–35. doi: 10.1177/147323001103900505.

Wang MJ, Xu XL, Yao GL, Yu Q, Zhu CF, Kong ZJ, Zhao H, Tang LM, Qin XH. MYO9B gene polymorphisms are associated with the risk of inflammatory bowel diseases. *Oncotarget.* 2016;7(37):58862–75. https://doi.org/10.18632/oncotarget.11186.

Wang Q, Xi B, Reilly KH, Liu M, Fu M. Quantitative assessment of the associations between four polymorphisms (FokI, ApaI, BsmI, TaqI) of vitamin D receptor gene and risk of diabetes mellitus. *Mol Biol Rep.* 2012 Oct;39(10):9405–14. doi: 10.1007/s11033-012-1805-7. Epub 2012 Jul 20.

Wang XX, Chen T. Meta-analysis of the association of IL2RA polymorphisms rs2104286 and rs12722489 with multiple sclerosis risk. *Immunol Invest.* 2018 Jul;47(5):431–42. doi: 10.1080/08820139.2018.1425699. Epub 2018 Apr 12.

Weersma RK, Stokkers PC, Cleynen I, Wolfkamp SC, Henckaerts L, Schreiber S, Dijkstra G, Franke A, Nolte IM, Rutgeerts P, Wijmenga C, Vermeire S. Confirmation of multiple Crohn's disease susceptibility loci in a large Dutch-Belgian cohort. *Am J Gastroenterol.* 2009 Mar;104(3): 630–8. doi: 10.1038/ajg. 2008.112. Epub 2009 Jan 27.

Wehkamp J, Götz M, Herrlinger K, Steurer W, Stange EF. Inflammatory bowel disease. *Dtsch Arztebl Int.* 2016 Feb 5;113(5):72–82. doi: 10.3238/arztebl.2016.0072. PMID: 26900160.

Wellcome Trust Case Control Consortium. Genome-wide association study of 14,000 cases of seven common diseases and 3,000 shared controls. *Nature.* 2007 Jun 7;447(7145):661–78. doi: 10.1038/nature05911.

Worthington J, John S. Association of PADI4 and rheumatoid arthritis: a successful multidisciplinary approach. *Trends Mol Med.* 2003 Oct;9(10):405–7. doi: 10.1016/j.molmed.2003.08.007.

Wu S, Liu Q, Zhu JM, Wang MR, Li J, Sun MG. Association between the IL7R T244I polymorphism and multiple sclerosis risk: a meta analysis. *Neurol Sci.* 2016 Sep;37(9):1467–74. doi: 10.1007/s10072-016-2608-8. Epub 2016 May 17.

Xavier RJ, Podolsky DK. Unravelling the pathogenesis of inflammatory bowel disease. *Nature.* 2007 Jul 26;448(7152):427–34. doi: 10.1038/nature06005.

Xia ZL, Qin QM, Zhao QY. A genetic link between CXCR5 and IL2RA gene polymorphisms and susceptibility to multiple sclerosis. *Neurol Res.* 2018 Dec;40(12):1040–7. doi: 10.1080/01616412.2018.1517110.

Yang J, McGovern A, Martin P, Duffus K, Ge X, Zarrineh P, Morris AP, Adamson A, Fraser P, Rattray M, Eyre S. Analysis of chromatin organization and gene expression in T cells identifies functional genes for rheumatoid arthritis. *Nat Commun.* 2020a Sep 2;11(1):4402. doi: 10.1038/s41467-020-18180-7.

Yang W, Zhou W, Zhang BK, Kong LS, Zhu XX, Wang RX, Yang Y, Chen YF, Chen LR. Association between CD24Ala/Val polymorphism and multiple sclerosis risk: a meta analysis. *Medicine (Baltimore).* 2020bApr;99(15):e19530. doi: 10.1097/MD.0000000000019530.

Yang XK, Liu J, Liu J, Liang Y, Xu WD, Leng RX, Pan HF, Ye DQ. Associations between PADI4 gene polymorphisms and rheumatoid arthritis: an updated meta-analysis. *Arch Med Res.* 2015 May;46(4):317–25. doi: 10.1016/j.arcmed.2015.05.011. Epub 2015 Jun 1.

Yousuf A, Ibrahim W, Greening NJ, Brightling CE. T2 Biologics for chronic obstructive pulmonary disease. *J Allergy Clin Immunol Pract.* 2019 May–Jun;7(5): 1405–16.

Yu W, Lin Z, Kelly AA, Hegarty JP, Poritz LS, Wang Y, Li T, Schreiber S, Koltun WA. Association of a Nkx2-3 polymorphism with Crohn's disease and expression of Nkx2-3 is up-regulated in B cell lines and intestinal tissues with Crohn's disease. *J Crohns Colitis.* 2009 Sep;3(3):189–95. doi: 10.1016/j.crohns.2009.04.003. Epub 2009 Jun 10.

Zhang D, Wang L, Zhang R, Li S. Association of Vitamin D receptor gene polymorphisms and the risk of multiple sclerosis: a meta analysis. *Arch Med Res.* 2019 Aug;50(6):350–61. doi: 10.1016/j.arcmed.2019.10.007. Epub 2019 Oct 31.

Zhang J, Li W, Liu J, Wu W, Ouyang H, Zhang Q, Wang Y, Liu L, Yang R, Liu X, Meng Q, Lu J. Polymorphisms in the Vitamin D receptor gene and type 1 diabetes mellitus risk: an update by meta-analysis. *Mol Cell Endocrinol.* 2012 May 15;355(1):135–42. doi: 10.1016/j.mce.2012.02.003. Epub 2012 Feb 14.

Zhang L, Yuan X, Zhou Q, Shi J, Song Z, Quan R, Zhang D. Associations Between TNFAIP3 gene polymorphisms and rheumatoid arthritis risk: a meta-analysis. *Arch Med Res.* 2017 May;48(4):386–92. doi: 10.1016/j.arcmed.2017.08.003. Epub 2017 Sep 6.

Zhang X, Yu P, Wang Y, Jiang W, Shen F, Wang Y, Tu H, Yang X, Shi R, Zhang H. Genetic polymorphisms of interleukin 17A and interleukin 17F and their association with inflammatory bowel disease in a Chinese Han population. *Inflamm Res.* 2013 Aug;62(8):743–50. doi: 10.1007/s00011-013-0629-9. Epub 2013 May 8.

Zhou C, Gao S, Yuan X, Shu Z, Li S, Sun X, Xiao J, Liu H. Association between CTLA-4 gene polymorphism and risk of rheumatoid arthritis: a meta-analysis. *Aging (Albany NY).* 2021 Aug 2;13(15):19397–414. doi: 10.18632/aging.203349. Epub 2021 Aug 2.

Zhu J, He J, Liu Y, Simeone DM, Lubman DM. Identification of glycoprotein markers for pancreatic cancer CD24+CD44+ stem-like cells using nano-LC-MS/MS and tissue microarray. *J Proteome Res.* 2012 Apr 6;11(4):2272–81. doi: 10.1021/pr201059g. Epub 2012 Feb 28.

Zou Q, Zhao Y, Wang Y, Fang Y, Liu Y. Associations between IL-23R gene polymorphisms and the susceptibility of rheumatoid arthritis: a meta-analysis. *Artif Cells NanomedBiotechnol.* 2019 Dec;47(1):951–6. doi: 10.1080/21691401.2019.1579731.

12

Single Nucleotide Polymorphism-Related Computational Resources Part I: Databases

Manmeet Kaur
Department of Biotechnology, Guru Nanak Dev University, Amritsar, Punjab, India

Pallawi Kumari
Department of Bioscience & Biotechnology, Banasthali Vidyapith University, Tonk, Rajasthan, India

Nadiem Nazir Bhat
Division of Veterinary Biochemistry, Sher e Kashmir University of Agricultural Sciences & Technology of Kashmir Shuhama Alustang, Kashmir, Jammu and Kashmir, India

Showkat Ahmad Dar
Bioinformatics Center, CSIR Institute of Microbial Technology, Chandigarh, India

CONTENTS

DOI: 10.1201/9781003246244-12

ABSTRACT Single nucleotide polymorphisms (SNPs) are the predominant variations found in human genomes, and till now, more than a hundred million have been reported. These variations are found to be associated with various phenotypes, drug responses, gene expression, and disease biology, including cancers. Recently, there has been an exponential growth in the genomic field, mainly due to the progress in next-generation sequencing (NGS) technologies and enhancements in the speed and power of computers leading to the generation of enormous data, including these genomic variations. This has led to corresponding advances in the detection of such nucleotide variations via genome-wide association studies (GWAS) and their applications in disease treatment or as biomarkers. It has become necessary to store, process, retrieve, and link these data so that they can be optimally used by all researchers universally. This led to the development of specialized bioinformatics databases and algorithms via collaborations and projects worldwide, with the National Institutes of Health (the USA) being the frontrunner. Some projects that were milestones in the SNP-related research include Human Genome Project, The Cancer Genome Atlas Program, and the 1000 Genomes Project. In this chapter, we will discuss predominately the available SNP-related resources, along with many supplementary links redirecting to corresponding web servers and their biological relevance or functional aspects of diseases, especially cancer.

KEYWORDS: *Single Nucleotide Polymorphisms (SNPs); Computation; Bioinformatics; Web Servers; Databases; Resources; Cancer.*

12.1 Introduction

Single nucleotide polymorphisms (SNPs) are a class of genetic changes found throughout genomes of species and vary among individuals and populations. SNPs depict a change in a DNA base on a particular location at a single nucleotide level that may replace the purine with another purine or a pyrimidine or vice versa. In humans, their average frequency of occurrence is one out of every thousand base residues. About a hundred million SNPs have been characterized among the human populations globally (Al-Koofee and Mubarak, 2020). SNPs may be a characteristic of a disease, phenotype, or a particular ethnicity. SNPs often represent genetic biomarkers associated with disease-causing gene products. They may occur inside the main gene coding sequence or regulatory sequences associated with it, affecting the function, expression level, or transcription factor-binding activity. SNPs can be silent or mutagenic, affecting phenotype at varying levels of amplitude. They may help forecast drug response, gene heritability, and disease propensity risks that might be associated with them. SNPs have also been used in tracking diseased gene inheritance patterns across pedigree. The modern-day research is ever advancing to correlate more and more SNPs with phenotypes of complex and age-related diseases such as risks of cancer, diabetes, hypertension, Parkinson's, and Alzheimer's across multiethnic population groups.

After understanding the importance of genetic information (especially SNPs), a need to develop centralized repositories, software, or prediction models for managing SNP data in the scientific community emerged (Johnson, 2009). There were various initiatives in this endeavor to achieve the maximum possible benefits from the research performed all over the world. After the advancement of the next-generation sequencing (NGS) technologies and the informatics field, the generation of sequencing data increased humongously. It became a necessary challenge to store, retrieve, and process the genomic (and other related) data so that the entire scientific community can benefit from them. This led to the bioinformatics field advancement in general. The development of SNP-related resources got developed in particular. As of now, bioinformatics has become an essential component of research, especially in the genomics (SNPs) field. With the logarithmic growth in genomic scientific data, it has become essential to continuously update the resources and the software to achieve the processing or modeling of the (SNP) data. Also, with integration of the different aspects of the SNP research with other data types such as genome-wide association studies (GWAS), transcriptomics, proteomics, metabolomics, pharmacology, and the complex diseases such as cancer, more resources are being continuously developed in the field. In this chapter, we will discuss the projects that were started in the endeavor to act as stepping stones for the SNP-related research and describe the databases/resources that have the applicability to make the SNP-related data available or process the related data.

12.2 SNP Resources

The rate at which information progresses and data are exchanged is one of the key determinants in human development. Databases have become necessary to store, update, retrieve, and compare data in real time. Incorporating new findings in the current databases has become a critical aspect of managing research advancement. The rapid discovery of genes contributing to various diseases is a norm during current times. With advancement, there is a need for the development of databases to monitor, update, and store information related to such genetic variations. As mentioned earlier, SNPs are an important type of genetic alteration, and SNP-associated data have become the basis of an array of genetic research and databases. The other variations found within the genome include microsatellites, indels (insertions and deletions), epigenetic markers, and copy number variations (CNVs) (Johnson, 2009). Currently, SNPs are among the most widely studied variations in genetic studies. The patterns of linkage disequilibrium (LD) for common and uncommon SNPs essentially reflect historical recombinational events. The current chromosomal states are mosaics of ancestral chromosomal sections because of these recombinational events. This explains why haplotypes and LD patterns are shared by chromosomes that appear to be unrelated within a group and between populations. These findings provided the conceptual and factual information for creating the "HapMap" project – a human genome haplotype map (International HapMap Consortium, 2003). The map illustrates common patterns of genetic variation, including connections between SNPs. Its goal is to establish the changes in human genetic sequence patterns and provide this knowledge publicly for research.

The HapMap aids in the finding of sequence changes that affect usual diseases, as well as aids in creating diagnostic tools in order to increase our capacity to select treatment targets. This project aids researchers all around the world in discovering the genetic variables that influence disease susceptibility, disease protection, and medication response. It has offered a critical shortcut for conducting linkage-based, candidate gene, and genome-wide association research. The Database of Short Genetic Variations (dbSNP) was made after the HapMap project, and it was designed to assist this large-scale polymorphism discovery. We will discuss the dbSNP and other important resources in the following text, with the rest of the repositories in the form of the table because of the word limit of this chapter.

12.2.1 The Cancer Genome Atlas (TCGA)

Cancer is one of the complicated diseases with varying molecular biology across different types and subtypes. There are a variety of cancers that can be distinguished based on their molecular profiling. Every cancer type has different mechanistic biology and requires a unique, specific therapeutic plan, as is the goal of futuristic personalized genomic medicine. The Cancer Genome Atlas (TCGA) enhanced the understanding of cancer genetics by utilizing genome-analyzing technologies, helping to make novel therapies, detection, and preventive strategies related to cancer. TCGA is a collaborative project of "National Cancer Institute (NCI) and the National Human Genome Research Institute (NHGRI), both are associated with NIH", started in 2005 with the idea to "generate, quality control, merge, analyze, interpret cancer molecular profiles and other information". It started with the idea from the "National Cancer Advisory Board under Human Cancer Genome Project" to create and analyze the entire range of genomic variations that are involved in human cancers. Its architecture is properly planned and involves various cooperating centers, which manage the collection, processing of samples, and large-scale sequencing with advanced computational data processing and understanding.

The bioinformatics data analysis servers/tools include "Tissue Source Site (TSSs), Biospecimen Core Resource (BCR), Genome Sequencing Centers (GSCS), Cancer Genome Characterization Centers (GCCs), Proteome Characterization Centers (PCCs), Data Coordinating Center (DCC), Cancer Genomics Hub (CGHub), and Genome Data Analysis Centers (GDACs)". DCC further aims the data collected by the research system of the "atlas" and enters them into the public databases, which permit the researchers to retrieve the cancer datasets and to make innovations in cancer and linking technologies. Gradually, it collected the data systematically and by now it has characterized over 20,000 primary cancers in around 33 cancer types from over 11,000 patients because of the work of highly skilled researchers. These cancers include "glioblastoma, breast cancer, ovarian cancer, lung cancer, colon and rectal cancer, gastric adenocarcinoma, acute myeloid leukemia, clear cell renal cell carcinoma, endometrial carcinoma, cervical carcinoma, esophageal carcinoma, and pancreatic ductal adenocarcinoma".

Researchers at TCGA have accumulated a wide variety of individual cancer genomic data from different types and subtypes. The resource helps to understand the biology and pathology of every tumor, which results in better and more focused treatment methods. The main focus of the Pan-Cancer project of TCGA is to set up and run the novel exhaustive integrated analysis of genomic data among a variety of cancers. An increase in sample datasets of tumors resulted in an enrichment of the statistical power and, therefore, the ability to detect and analyze molecular defects prevalent in cancers. It provided scientists and researchers with a lot of data that involve similarities and dissimilarities across the various changes happening in tumors, helping them to group and make cancer group-related therapies. The Synapse platform http://sagebase.org/synapse/ shares the information and outcome of the project Pan-Cancer. The extent of data exceeds 2.5 petabytes that have been made universally available for research purposes (Bailey et al., 2018; Hutter and Zenklusen, 2018).

TCGA aims to find and categorize the genome variations that majorly cause cancer in order to develop an extensive directory comprising profiles related to the cancer genome. Researchers of TCGA have investigated a range of people having tumors by integrated multi-dimensional analyses and large-scale genome sequencing. The focus of this project is to assist various datasets in order to revamp the treatment levels and detection techniques in order to prevent cancer. The resource contains all the aspects of data, including but not limited to genomic, epigenomic, transcriptomic, proteomic, CNVs, and mutations (SNPs) datasets. It provides a bulk of knowledge regarding cancer cells, cancer biomarkers, cancer genetic and epigenetic profiles, etc. It also provides a glance at the diagnosis and therapeutics of cancer. The resource also aids in the development of personalized medicine and bioinformatics tools to pick significant signals from bulk data for investigative cancer research. The resource can be accessed via the "Genomic Data Commons Data Portal" at the URL: https://portal.gdc.cancer.gov. As per the new release in October 2021, it contains information on 3,599,319 mutations related to the cancer data. The web interface provides options to dig deep into the information based on projects, data exploration, analysis, and data repositories in it. For example, users can visualize the mutational landscape of the top 200 variations in TCGA data using the "OncoGrid", shown in Figure 12.1. We can see how intricately the connections between various aspects of the data are presented, shown in the data legend (and color-coding).

TCGA besides growth in the computational and other aspects of cancer-related data provides other interfaces too to dig data even deeper. Some of them are the following:

> (1) The Cancer Imaging Archive (TCIA) for radiological imaging data. (2) The Cancer Proteome Atlas Portal (TCPA), an integrative data portal for accessing, visualizing, and analyzing proteomic data. (3) The cBioPortal for Cancer Genomics, which provides visualization, analysis, and download of large-scale cancer genomics datasets. (4) Copy Number Portal for exploring copy number alterations of TCGA data. (5) The Integrative Genomics Viewer (IGV), a high-performance visualization tool. (6) The Mutation calling using a Markov Substitution model for Evolution.

12.2.2 The International Cancer Genome Consortium (ICGC)

The ICGC is a pan-institutional and multi-country collaborative project to create systematic data about different cancer types, supposedly 50 cancer types, similar to TCGA. The data can be accessed via its data portal via the URL: https://dcc.icgc.org. As per the 28th release on March 27, 2019, it contains information on 86 cancer projects with 22 primary cancer sites from over 24,000 donors and the number of simple somatic mutations touching around 82,000,000. The data types include the "simple somatic mutations, copy number alterations, structural rearrangements, gene expression, microRNAs, DNA methylation and exon junctions", and other associated metadata. The ICGC uses BioMart as the open-source data retrieval and management tool. It provides the central point for accessing the data. As multiple laboratories and disciplines are involved in the project, controlled reference datasets, models, and annotations are being updated continuously. Thus, the data in it agree with other universal databases such as COSMIC, Ensembl, KEGG, and Reactome (Zhang et al., 2011).

The ICGC web portal is easy to use and provides options for advanced searching, data analysis besides investigating cancer projects, data release, and repository information. Besides, the APIs for accessing data, other features like compound explorers, pathways are provided for research support. Ontologies, ICGC-Beacon searches, Pan-Cancer Analysis of Whole Genomes (PCAWG), etc. have been integrated along with good documentation. Some examples and associated data of the ICGC are shown in Figure 12.2.

12.2.3 Single Nucleotide Polymorphism Database (dbSNP)

The dbSNP is the most trusted, centralized online resource for the "short genetic variations", to complement GenBank. The resource is publicly available at the following URL (Uniform Resource Locator): https://www.ncbi.nlm.nih.gov/snp (Sherry et al., 2001). Its scope includes data related to single-base nucleotide substitutions, retroposable element insertions, small-scale multi-base indels, and microsatellite repeat variations. dbSNP offers detailed information on population-specific frequency, genotype data, molecular context, experimental settings, and mapping (physical and functional) information related to SNPs (Bianco et al., 2013). The scientific society quickly embraced and comprehensively used it for SNPs research and other variations such as microsatellites, indels, and non-polymorphic changes, for both neutral variants and symptomatic mutations. dbSNP is curated by the "National Center for Biotechnology Information" (NCBI, National Institute of Health – NIH), built after the Human Genome Project found a very high number of common genetic variants. The portal is the most exhaustive database of common genetic variation and, over the course of its existence, has grown at an exponential rate (both neutral variants and symptomatic mutations). The resource comprises frequency specific to population and data related to genotype, as well as experimental settings, molecular context, and mapping data.

The genome mapped with the dbSNP variations is featured in genome browsers (NCBI, EMBL, and UCSC). The tools also allow users to combine SNP data with other genome annotation features. snpBLAST (a tool in the suite) allows users to query sequences against dbSNP. Haplotype predictions are also available in the dbSNP (Johnson, 2009). Date-wise dbSNP was primarily conceived in mid-1999 by the collaboration of NCBI and "National Human Genome Research Institute (NHGRI)" and has now crossed over a billion unique SNPs with individual references called "RefSNP or rs (for dbSNP Reference SNP)". Over two decades, this resource has been the primary source of various other resources such as "ClinVar, OMIM (Online Mendelian Inheritance in Man), SNPedia, the Variant Annotation Integrator at UCSC (University of California, Santa Cruz), and the Variant Effect Predictor at EBI (European Bioinformatics Institute)". It has been cited in over 50,000 research articles and counting. Data have been made available in multiple formats and channels, with options of searching, browsing, and retrieving the required data. The data can be downloaded in JSON (JavaScript Object Notation) or VCF (Variant Call Format) file formats via the FTP (File Transfer Protocol) or specialized APIs (application programming interfaces) designed for the tasks. The database was modernized with new NCBI notation for variations called

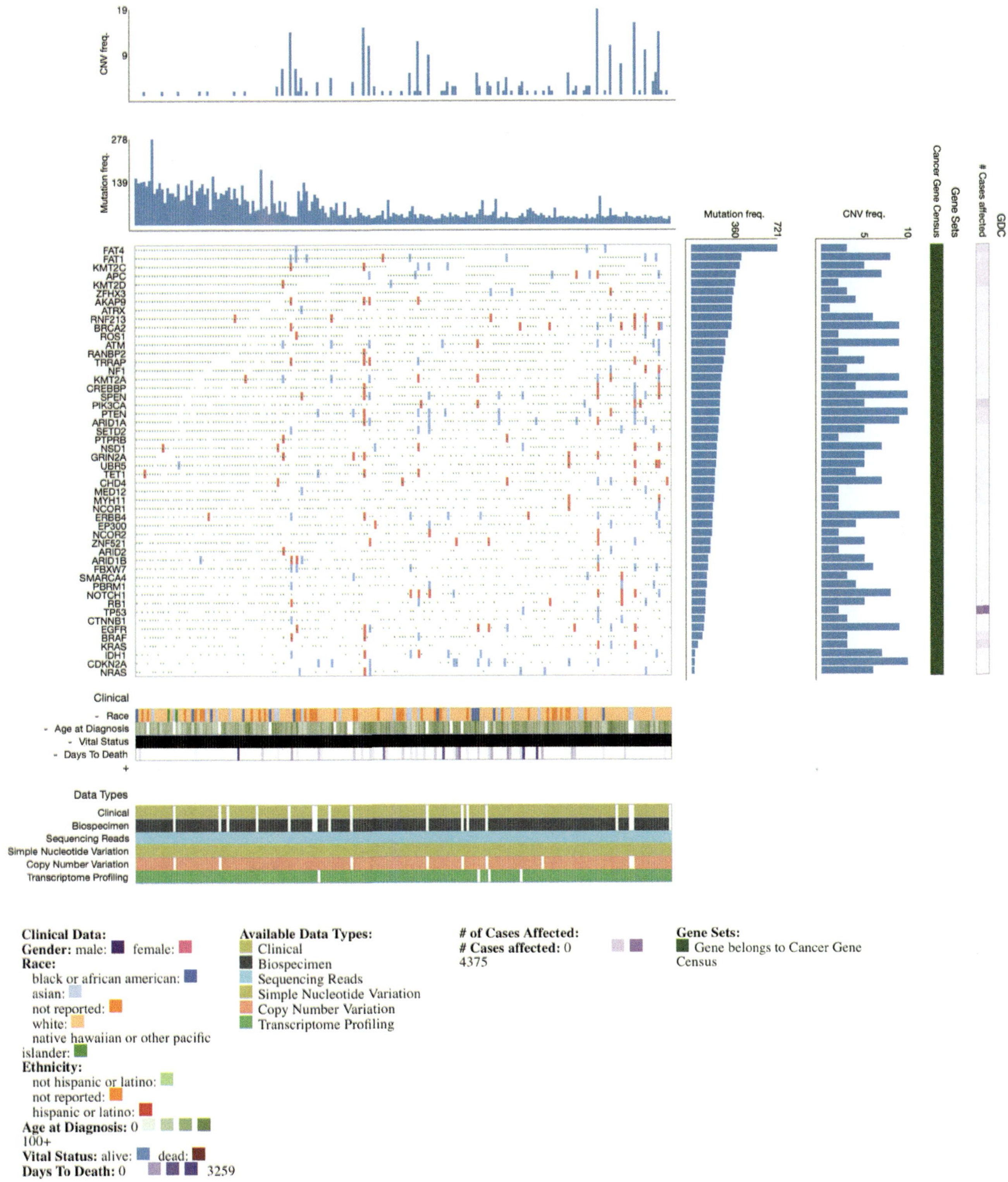

FIGURE 12.1 Example representation of the data from TCGA for simple somatic mutations (SSMs) with top 50 mutated genes with 200 most mutated cases. (Adapted from TCGA example dataset results online.)

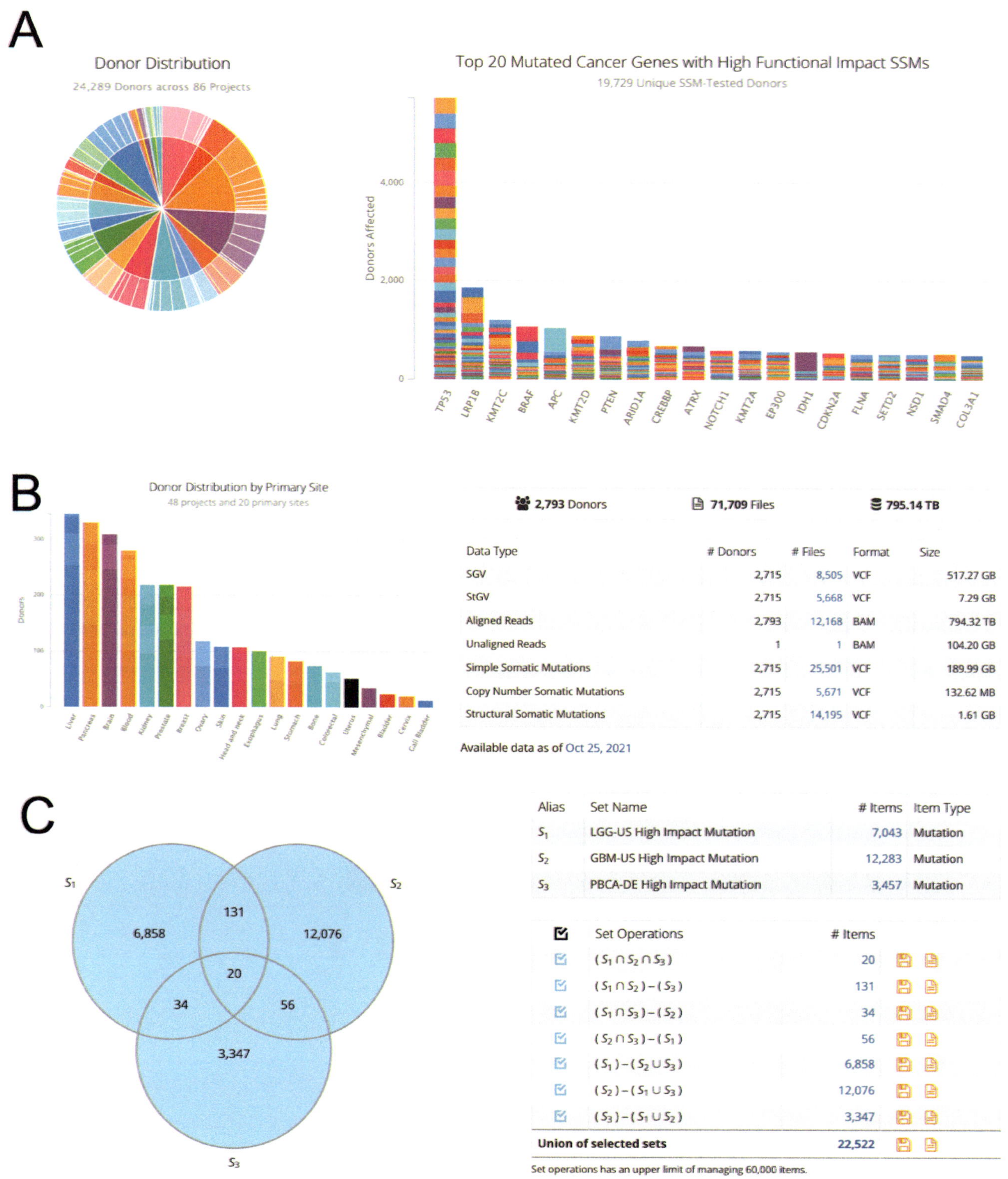

Data Type	# Donors	# Files	Format	Size
SGV	2,715	8,505	VCF	517.27 GB
StGV	2,715	5,668	VCF	7.29 GB
Aligned Reads	2,793	12,168	BAM	794.32 TB
Unaligned Reads	1	1	BAM	104.20 GB
Simple Somatic Mutations	2,715	25,501	VCF	189.99 GB
Copy Number Somatic Mutations	2,715	5,671	VCF	132.62 MB
Structural Somatic Mutations	2,715	14,195	VCF	1.61 GB

Alias	Set Name	# Items	Item Type
S_1	LGG-US High Impact Mutation	7,043	Mutation
S_2	GBM-US High Impact Mutation	12,283	Mutation
S_3	PBCA-DE High Impact Mutation	3,457	Mutation

Set Operations	# Items
$(S_1 \cap S_2 \cap S_3)$	20
$(S_1 \cap S_2) - (S_3)$	131
$(S_1 \cap S_3) - (S_2)$	34
$(S_2 \cap S_3) - (S_1)$	56
$(S_1) - (S_2 \cup S_3)$	6,858
$(S_2) - (S_1 \cup S_3)$	12,076
$(S_3) - (S_1 \cup S_2)$	3,347
Union of selected sets	**22,522**

FIGURE 12.2 Schematic representation of an example dataset from ICGC web server. (a) Pie diagram of the donors for the ICGC data portal (left panel) and top 20 mutated cancer genes with high functional impact SSMs (with 19,729 unique SSMs, right panel). (b) PCAWG – donor distribution by primary cancer site for 48 projects and 20 primary sites (right panel – colored bar plot) with corresponding information on files, sizes, and data types (in the right panel). (c) Venn diagram representation (left panel) of the three cancer types with the mutations and their overlap, with numerical data and other information in the right panel. Here, LGG-US is lower grade glioma (US data), GMB-US is glioblastomas (US data), and PBCA-DE represents pediatric brain cancer (Germany). (Adapted from ICGC example dataset results online.)

"Sequence Position Deletion Insertion or SPDI" and algorithm (Variant Overprecision Correction Algorithm or VOCA), to help in precisely "mapping, clustering and normalizing" the variants and other functions (Holmes et al., 2020).

The resource accepts the submission of the public data related to variation data with proper guidelines and how-to-do guides ("How to Submit to dbSNP" https://www.ncbi.nlm.nih.gov/snp/docs/submission/hts_launch_and_introductory_material). For further exploration of the resource, researchers can access the web link provided above or go through the resources, especially the NCBI bookshelf ID: NBK21088 https://www.ncbi.nlm.nih.gov/books/NBK21088/. An example of dbSNP query search and results is shown in Figure 12.3.

FIGURE 12.3 Schematic representation of (a) dbSNP query (*Homo sapiens* BRCA1) and its result as a list of all SNPs associated with the query. For further looking into single SNP, (b) data display of single SNP (rs12516) and its information with options for its details such as frequency, variant detail, and clinical significance. (c) Allele frequency and (d) genomic regions (in the variation viewer) are displayed here with SNPs (with rs-IDs). (Adapted from dbSNP web server online.)

(Continued)

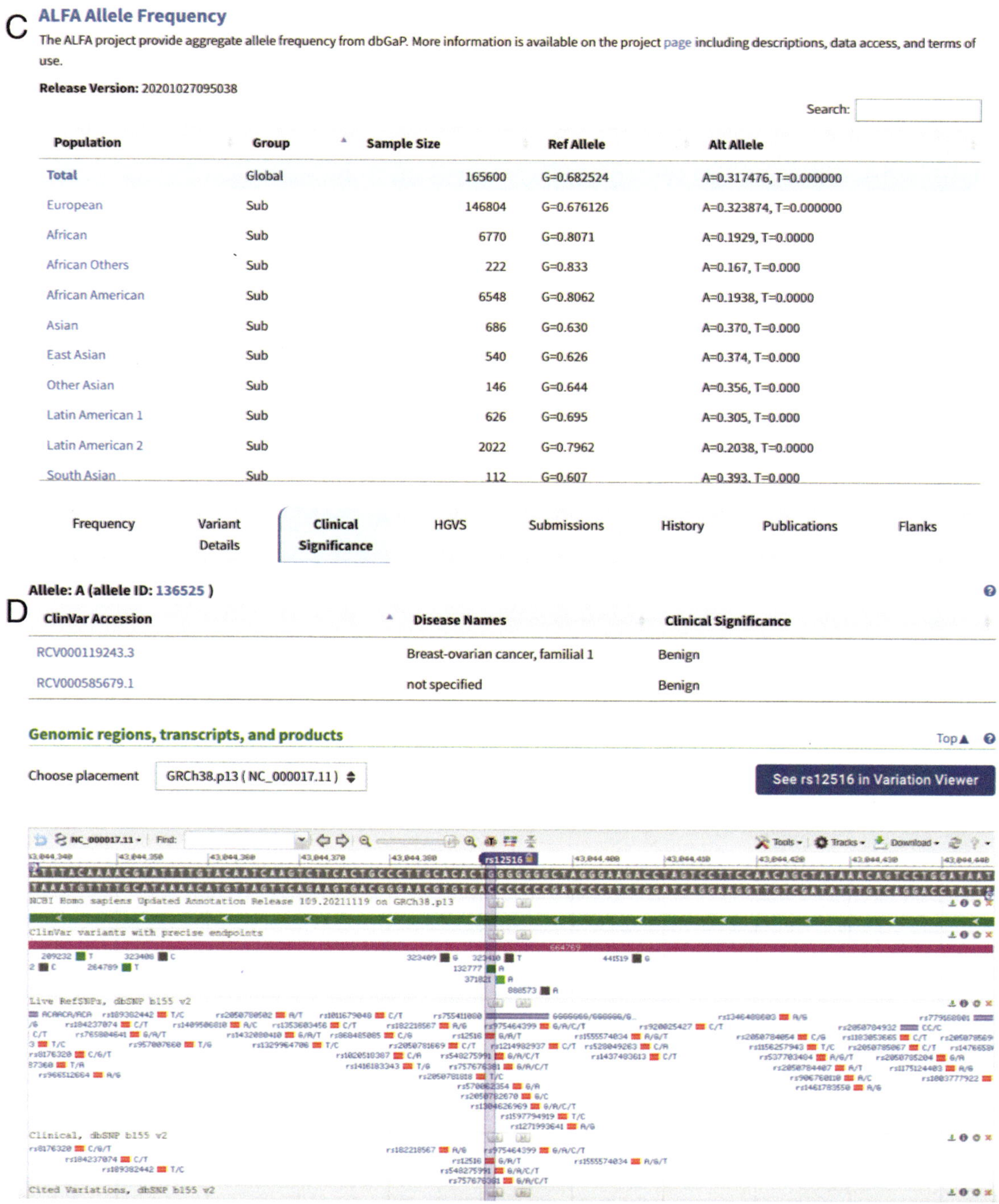

FIGURE 12.3 (*Continued*) Schematic representation of (a) dbSNP query (*Homo sapiens* BRCA1) and its result as a list of all SNPs associated with the query. For further looking into single SNP, (b) data display of single SNP (rs12516) and its information with options for its details such as frequency, variant detail, and clinical significance. (c) Allele frequency and (d) genomic regions (in the variation viewer) are displayed here with SNPs (with rs-IDs). (Adapted from dbSNP web server online.)

12.2.4 Database of Genotype and Phenotype (dbGaP)

The dbGaP (related database) was established in 2006 as an immense dataset of genomic and phenotypic support to research organizations around the world. It contains phenotype, genotype, exposure, sequencing data, and the relationships between them and was developed by NCBI (Tryka et al., 2014). The accessioned records, phenotypes, various molecular assay data, analyses, and documentation are all stored in a hierarchical structure. The dbGaP https://www.ncbi.nlm.nih.gov/gap/ provides available metadata about deposited researched data at the summary level and papers linked to the study (Tryka et al., 2014) (Wong et al., 2017).

12.2.5 1000 Genomes Project

The 1000 Genomes Project was a cooperative initiative by researchers from the USA, Germany, the UK, and China to create an exhaustive record of genetic variations in humans with the aim of assisting future medical research. The main purpose of the project was to create a deep characterization of the sequence variations in the human genome as a standard base for examining the connection between phenotype and genotype. The extended functionality included searching, genotype, and providing precise haplotype data on every form of human DNA polymorphisms across a wide variety of populations. Particularly, the aim was to characterize over 95% of variants that are present in the genomic area available to the present high-throughput sequencing technologies with at least 1% of allele frequency in each of five important population groups (East Asia, Europe, South Asia, the Americas, and West Africa). It extended the information from the International HapMap Project, which acted as a resource to search over diverse areas of the genome related to usual human diseases. The project manifested in presenting a library of variants, such as SNPs and structural variants, and their associated haplotype contexts. It allows GWAS to concentrate on most of the variants existing in the areas responsible for the disease. The project harbors an NGS-based genome repository of over 1000 individuals globally with free public access (1000 Genomes Project Consortium et al., 2010).

The 1000 Genomes Project has overshadowed the "International HapMap Project" or HapMap project. The latter was aimed at unfolding the haplotype map of human genomes to reveal the genetic variations for health and disease (International HapMap Consortium, 2003). The HapMap is no longer active, but the data can still be used by researchers freely. The data hosted have correlative implications in studies, such as an individual's response to various external factors such as drugs or chemicals.

12.2.6 Ensembl

The Ensembl genome database project by EBI is analogous to the "Human Genome Project" by NCBI. It aims to store, annotate, archive, and produce on demand the genomic and variation data (including SNPs) (Yates et al., 2016). It is an exhaustive repository of automated annotations of individual genomes, synteny, and orthology linkages between them available at the URL: http://www.ensembl.org/. The Ensembl specializes in analyzing genomic characteristics and sequences in order to best identify and annotate genes. It provides information on a wide range of species. For people interested in comparative genomic analysis studies, this is the most information-revealing tool. The Ensembl is a pioneer in automated gene annotation techniques and helps to simplify the difficult work of gene identification and characterization across varied species (Phillips, 2009). The Vertebrate Genome Annotation (VEGA) database, housed in the Ensembl interface, offers a variety of genomic analysis tools. The VEGA's major goal is to provide manual reasonable-quality annotation of the genome sequences in vertebrates. Besides, various other algorithms are associated with the interface, which include BLAST (Basic Local Alignment Search Tool)/BLAT (BLAST-Like Alignment Tool), BioMart, and a very essential tool called "Ensembl Variant Effect Predictor (VEP)" (McLaren et al., 2016). The Ensembl is tightly integrated and interconnected with other data repositories including UniProt/Swiss-Prot: https://www.ebi.ac.uk/uniprot (Phillips, 2009). As part of the gene build process, a variety of sequencing data are aligned to the genomic information. SNPs and other information are positioned using an externally given coordinate mapping. Regarding the SNP information of the gene, a "variational table" can be generated for a user-defined gene with information from dbSNP and other resources. The gene variational table displays various types of related information including the "ID", the unique RefSNP (rs-number) from dbSNP, followed by chromosome information, base pair information, allele or "global minor allele frequency" class of variation (such as SNP and indel), and clinical significance (ClinVar). Other information is also provided that links the variation data with the functional aspects such as sequence ontology and effect on pathogenicity. The algorithms used for the latter include SIFT (Sorting Intolerant From Tolerant), CADD (Combined Annotation Dependent Depletion), REVEL (rare exome variant ensemble learner), PolyPhen (Polymorphism Phenotyping), MEtaLR (predicts deleteriousness of missense variants), and Mutation Assessor. The Ensembl provides information to specially name and position the query attributes on the reference genome builds, with proper links to the primary information sources (Birney et al., 2004). The data from the database could be pulled down by using FTP and APIs. To further dig deep into the Ensembl and its use, a well-documented tutorial section has been provided on its website: http://useast.ensembl.org/info/website/tutorials/index.html. A representation for searching the SNP data in "Ensembl Variant Recorder" is shown in Figure 12.4.

12.2.7 ClinVar

The ClinVar is a database that compiles genetic variation data along with its impact on human clinical aspects or phenotypes, and it is maintained at NCBI (NIH). The resource is available at the web link: http://www.ncbi.nlm.nih.gov/clinvar (Landrum et al., 2018). It comprises over 8,25,000 records about germline and somatic variants. Majority of them fall under the category of SNPs. The data in the repository are collected and updated by the collective effort of the NIH genetic testing registry laboratories (approximately 402 in number). The ClinVar encompasses multiple data aspects such as the clinical effect of variation, its inheritance patterns, and the associated research with supporting metadata or information interconnecting the database with other resources. Each record is assigned an accession number, having three letters and nine numerals (e.g., VCV000000561). The three letters can be of three types: "SCV (Submitted record in ClinVar), RCV (Reference ClinVar record), and VCV (Variation ClinVar record)". Additionally, the Variation IDs ("identifiers specific to ClinVar") are three-digit numbers (e.g., 561) assigned to a record of the individual or set of variants. It has two classes: (I) directly interpreted and (II) included in the "context of the set of variants". All the ClinVar variations associated with a specific genomic location are included in the files. Variants, both directly interpreted and indirectly interpreted, are included in the VCF files (included variants) in compressed file formats. The digits of VCV correspond to the Variation ID, which is nine digits long because of the preceding zeros. The entire data

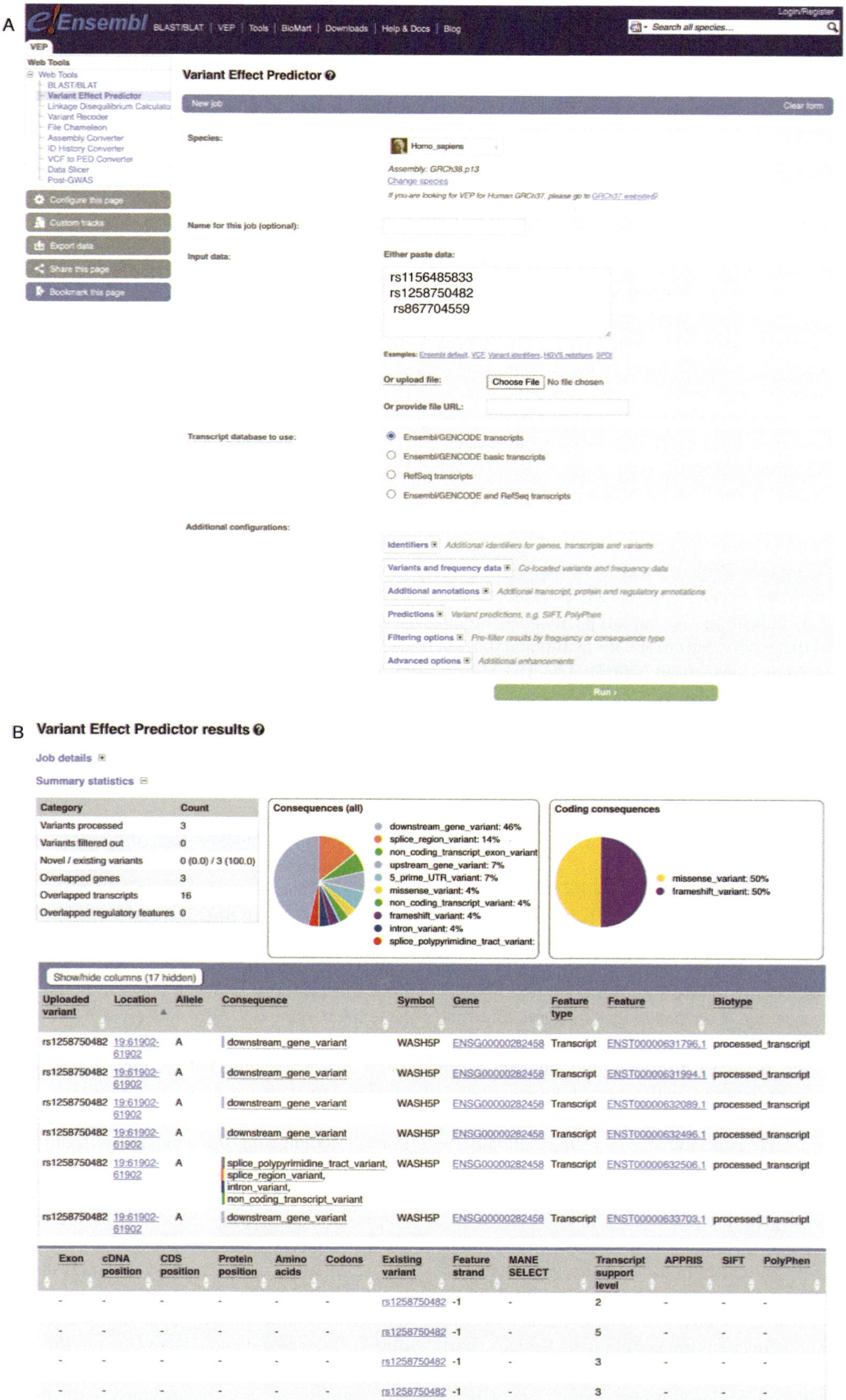

FIGURE 12.4 Schematic representation of (a) Ensembl variant recorder example search; to paste or upload the data, choose options for output and click the "Run" button (in light green color) and (b) the result of the queries in tabular format, each row for one query SNP. (Adapted from Ensembl example online.)

related to the particular disease are contained in a Variation ID (VCV record). The ID is used as the key in the VCF files, making it easier to compare the data.

The methods employed by ClinVar for acquiring the data include the "esearch, esummary, efetch, and elink" (the methods correspond with the E-utilities of NCBI's API). Users can execute the focused searches in ClinVar by limiting the area of the query phrase, as well as finding any search term. The Search option in the database can be used: https://www.ncbi.nlm.nih.gov/clinvar/advanced, to explore all options for indexed fields. Users can look up values, run, and bookmark the web links for revisiting in future (Landrum et al., 2020). Some of ClinVar's comprehensive search features include the following:

1. Variants can be searched depending on the "date" a ClinVar submitter rated their clinical significance.
2. HGNC IDs or GeneIDs can be used to search for variants in specific genes.
3. HPO IDs, such as "hp 0004322" [Trait identifier], can be used to search for variants associated with certain phenotypes.
4. Multiple other ways to filter the results include review status, clinical significance, the origin of allele, type of methods used, molecular consequence, type of variation, length of the variant, and variant gene relationship.

12.2.8 dbVar and DGVa – Both for Genomic Structural Variation (GSV)

"dbVar is NCBI's database of human genomic structural variations", usually for DNA size range up to 50 base pairs. The variation information provided regards duplications, insertions, deletions, multi-nucleotide substitutions, inversions, mobile element insertions, complex chromosomal rearrangements, and translocations. It recommends submitting variation data that are greater than 50 bp to dbVar and those less than this number to dbSNP. It was introduced by the NCBI in 2012 and can be accessed at https://www.ncbi.nlm.nih.gov/dbvar (Lappalainen et al., 2013). The "Database of Genomic Variants archive" (DGVa) stores and distributes the available structural variants of genes/proteins from every species. An EBI's counterpart of dbVAR can be found at http://www.ebi.ac.uk/dgva.

Both databases cumulatively offer archives of accessible data, for GSV (genomic structural variation) in a wide variety of species. The archives intercommunicate data and provide standard data format to the research community. If processed by DGVa, the accessioned objects are prefixed with e; if processed by dbVar, they are prefixed with n. Both databases have the same data architecture model that is intended to seize and represent the intricacy of GSV discovery, authorization, and genotype experiments. The resource assigns accession numbers to the studies, variant regions, and variant calls. The raw data from assays based on arrays or experiments related to sequencing are not stored in either of the databases. Instead, depositors are motivated to submit that data to a dedicated NCBI or EBI database. However, DGVa/dbVar submissions must include the accession numbers from these sources. The public can access the data from both archives in GVF (Reese et al., 2010). Germline and somatic variations are often represented as disjointed cases, cases where dbVar has remapped supplied information to contemporary versions of the assembly and could be downloaded for each taxonomic name and assembly. Data in tab-delimited files in XML format through dbVar are also available.

12.2.9 Database of Genomic Variants (DGV)

The DGV is a database with the goal to classify the best structural variant (SV) documented in the research in a way that molecular biologists and geneticists can understand. It is actively curated by the University of Toronto, Canada, and is freely accessible at http://dgv.tcag.ca/dgv/app/home. Researchers, as well as clinicians, use this resource regularly, either straight on the website, via data tracks presented on genome browsers, or through alternative interfaces via a wide range of commercial software applications (ChAS, BlueFuse Multi, and CytoSure) (MacDonald et al., 2014).

DGV studies are curated for accuracy and validity from accessioned records from central repositories, i.e., dbVar/DGVa. The DGV integrates these data of varying quality to structure a highly curated normal reference set of GSV in humans, using the range of various attributes provided, which include disease phenotypes (Lappalainen et al., 2013).

The DGVa, in collaboration with dbVar, formed a partnership to build a data store that permits the use of defined language and grants cross-platform unique identity numbers, providing users a smooth access to these data. Computational programs were created for the interchange of data between the archives, and every dataset representing structural variants in control samples of healthy individuals was transferred to DGV to organize, integrate, analyze and display. This design ensures that a consistent set of phrases and values represent the diverse features, allowing its curators to capture and preserve this information consistently and effectively. As each item has been documented in a logical and specific way, this makes it easy for users to compare data among various samples and studies (MacDonald et al., 2014). DGV contents are constantly included in other extensive genomic reference databases. Hence, they serve as a standard resource of information for the advanced development of databases and products, particularly in clinical laboratories employing CNV testing (MacDonald et al., 2014).

12.2.10 3-Dimensional Single Nucleotide Polymorphism (3DSNP)

3DSNP is an integrated resource for annotating noncoding variants in *Homo sapiens* by investigating their significance in distal gene and regulatory interactions in a three-dimensional context. It employs 1000 Genomes Project data on three-dimensional local chromatin signatures and interactions, in distinct cell types and LD as well as the genetic connections between variations displayed using 3DSNP's informative visualization tools. It is a useful database for annotating noncoding genomic sequences in humans and studying the effects of noncoding mutations on clinical characteristics (Lu et al., 2017;

TABLE 12.1

The Six-key Annotation Categories Available in 3DSNP along with the Available Tracks

Annotation Categories	Available Tracks Include
Population genetics	*F*ST, iHS, and XPNSL.
scATAC-seq	Contains profiles for 126 cell types/subtypes across 15 fetal tissues of human origin for chromatin accessibility.
Hi-C loops	Contain information for 49 human cell types regarding all SVs: altered loops and normal Hi-C loops.
Basic annotation	Repetitive genomic regions, DNase sites, and RefSeq genes are identified using Repeat Masker.
Variant pathogenicity	Provides information about ClinGen Triplosensitivity, ClinVar CNV, ClinVar SNV, ClinGen Validity, and ClinGen Haploinsufficiency.
Variant types	Provide information regarding SVs and SN from the dbSNP and HGSVC2.

Quan et al., 2021). Understanding human health and disease necessitates characterizing a wide spectrum of genetic variation. The structural variations such as indels and duplication areas on chromosomes are of immense importance in SNP biology as they lead to significant alterations in local chromatin structure.

In 2021, the database was upgraded with a newer version 3DSNP 2.0 and can be accessed at https://omic.tech/3dsnpv2/. The toolkit also helps to study noncoding variant annotation and structural variations in *Homo sapiens* species at a single-cell resolution. Thus, a need to upgrade the resource as a utility in single-molecule LRS and scATAC-seq is underway. The technologies act as paradigm-shifting biotechnologies used widely in genetics research, providing important data on noncoding variant annotations missing in pre-existing databases.

In 3DSNP 2.0, the former USCS genome browser interface scripts were replaced with the IGV.js plug-in to make it easier for the end-user to visualize local chromatin structure change and accessibility related to SNPs or SVs in question. Population genetics, scATAC-seq, Hi-C loops, basic annotation, variant pathogenicity, and variant types are the six lead annotation categories. The usage and features of these annotation categories are incorporated in Table 12.1.

12.2.11 RegulomeDB

The database incorporates disease association studies (GWAS) with variation annotation of SNPs in the intergenic areas of the human genome with familiar regulatory elements. It was developed in 2012 by Stanford University, USA, and is hosted on the web link: http://www.regulomedb.org/. There are options for data submission, for exploration of curated data, and for a comprehensive, customized data query. Users can explore the database via dbSNP IDs, single nucleotide information, or the chromosomal region in BED, VCF, or GFF3 file formats. RegulomeDB explains the role of the common variants resulting in altering gene expression. It ranks SNPs based on experimental data, including the SNP locus. These positions have been biochemically defined and annotated with other resources such as the Gene Expression Omnibus (GEO) and the ENCODE. The database covers around a thousand experimental datasets related to more than a hundred cell lines and tissues, characterizing about 60 million annotations. This project incorporates high-throughput studies with manual annotations and computationally inferred predictions, to decipher possible regulatory potential and functional disparity caused by these noncoding variants.

The tools included robustly score variants and help to significantly distinguish functional variants from a batch of candidates. The database can also identify the known linked functional variation via GWAS and offer a functional hypothesis (Boyle et al., 2012). Other features associated with the tool are highlighted below:

1. Information from ChIP-Seq for a range of essential regulatory factors.
2. Data on chromatin states in over 100 different cell types.
3. Expression quantitative trait loci (eQTLs) data linked with gene promoters.

It makes use of the subsequent techniques to generate computational predictions to assist with a useful annotation such as DNase footprinting exposing putative, precise protein-binding locations and provides clues about feasible nucleotide variation-based binding motif alterations.

12.2.12 MR-Base

The MR-Base (MR for Mendelian randomization) is a database and online platform that contains 11 billion SNPs associated with 17,000 GWAS. The resource automates 2SMR (2-sample Mendelian randomization) by combining a curated resource of summary of GWAS with an API, web app, and R packages. It was developed by "MRC Integrative Epidemiology Unit at the University of Bristol". It has developed several sensitivity assessments included in the software to analyze the effect of horizontal pleiotropy. The tool is available at http://www.mrbase.org (Hemani et al., 2018). PheWAS (phenome-wide association studies) relate known SNPs and other variants to multiple diverse phenotypes and uncover the probable origin of horizontal pleiotropy, as well as theory-free MR to better understand the effects of interference, which are among the applications supported by the MR-Base. These applications show that combining information and analytical techniques can lead to unique perceptions.

The MR-Base can leverage a variety of data sources to define exposure and outcome features:

1. For hundreds of characteristics, it has the entire GWAS summary data. The user can fish out the top hits based on their own criteria. Potential tools can also be found in the MRInstruments package, which contains independent SNP-trait associations with p-values less than $5e^{-8}$.
2. A variety of independent associations for thousands of phenotypes are generated by QTL investigations on gene expression, DNA methylation, protein, and metabolite levels. For simplicity of use within the MR-Base, the MRInstruments R package contains various omic QTLs.
3. There are various SNPs linked with a variety of complex attributes and disorders in the NHGRI-EBI GWAS Catalog. Within the MRInstruments R package, the list of potential tools are standardized and formatted for ease of usage.
4. Analysis can also be done on the user-provided data.

12.2.13 atSNP Search

atSNP Search is an exhaustive resource for analyzing motif matches to the human genome with both reference and variant alleles, as well as evaluating the overall relevance of variant modifications on motif matches. The resource currently contains over "37 billion variant-motif pairs" and is free to access at the URL: http://atsnp.biostat.wisc.edu/search (Shin et al., 2019). The primary components of the database include convenient search tools, using either SNPid list, SNPid window, genomic location, gene, or transcriptional factor. When paired with experimental validation, computational data from the atSNP Search may aid in the finding of the underlying disease processes.

The atSNP Search was created as a web resource in response to a requirement for

1. Finding more extensive motif-based regulatory variants.
2. Statistically accurate quantification of motif matches and associated variations.
3. Easy-to-understand graphical representation for motif-based data.

The outcome for every SNP-motif fusion having a significant score ($p<0.05$) is included in the atSNP Search. Plots which show motif matching with two gene copies are used to visually supplement the statistical quantifications.

12.2.14 OncoBase

OncoBase was created in order to decode tumor noncoding somatic mutations by base pair resolution estimation. OncoBase is located on the URL: http://www.oncobase.biols.ac.cn/. It is a database that annotates somatic variations in cancer types from various related projects by evaluating their functions in distal interactions between regulatory elements and target genes. Noncoding areas serve as the hub for over 95% of somatic variation in cancers. The noncoding area comprises many functional components based on one-dimensional epigenetic profiles (ENCODE and Roadmap Epigenomics). Combining three-dimensional (3D) spatial interactions datasets (such as Hi-C-Seq and ChIA-PET) can make precise enhancer and promoter pairs. Thus, merging somatic noncoding variations with 3D long-range interactions and 1D coordinated epigenetic profiles in particular types of cells or tissue will help in better understanding the underlying regulatory mechanism in human carcinogenesis.

OncoBase combines genotypic data, 3D chromatin interactions, phenotypic data, and essential genomic characteristics, which include histone modifications, chromatin states, gain or loss of TFBS motifs, and various QTLs among a wide variety of cell types. It also combines three-dimensional chromatin interactions, local chromatin signatures, and reconstructing enhancer-target networks using various algorithms. It assists the investigators by discovering functionally relevant noncoding somatic mutations in cancers based on their one- and three-dimensional genomic features. It is used for functionally annotating regulatory noncoding areas and orderly standardizing the regulatory impact of embedded noncoding somatic variations in cancer. OncoBase stores a collection of various somatic mutations from the TCGA or ICGC or COSMIC database and clinical mutations from the ClinVar database. The retrieved information in the database can be searched in the following ways: "somatic mutations", "gene symbols", "dbSNP ID", and "regulatory regions". OncoBase is an easy-to-use database that was created by merging jQuery and CodeIgniter, which supports flexible surfing and exploring functionalities. The data relating to annotation are kept as a MySQL database or flat files. The researchers have free access to the genetic data (Li et al., 2019).

12.2.15 Catalog of Somatic Mutations in Cancer (COSMIC)

COSMIC is an extensive resource providing information about somatic mutation effects in human cancers. This database mainly comprises the data with high precision and is manually curated by the team of researchers. It focuses on the experimentally known as well as probable cancer-related genes and mutations. It is available on the URL: https://cancer.sanger.ac.uk. The main source of data is derived manually by the curators from various scientific researches. It provides detailed, but focused information from the major cancer data portals and additional data, along with the research papers. These two factors merge to provide COSMIC with vast knowledge, thus making it the primary resource for exploring the cause and mutation in human cancer.

Not only does COSMIC cover the coding mutations, but also comprises each and every genetic mechanism that is involved in promoting cancer because of the various somatic mutations (drug resistance mutations, CNVs, gene fusions, etc.). COSMIC is manually curated by researchers and makes sure to provide quality, precision, and detailed description of the captured data. Its features are inclusive of novel steps that permit the hierarchization of important genes and disease processes rapidly and extensively to novel searches in the literature. The COSMIC database comprises more tools and databases such as the COSMIC Cell Lines Project, Cancer

Gene Census (CGC) and COSMIC-3D, which permit the exploration of mutations in the 3D structure of the protein, its functional and structural impact, and analysis of druggability (Tate et al., 2019).

12.2.16 SNP2APA

Alternative polyadenylation (APA) is an essential post-transcriptional regulatory mechanism that helps in recognizing various polyadenylation signals (PAS) on transcripts, which results in transcripts with distinct lengths of 3'-UTR, thus affecting various biological processes. Various researchers showed SNPs can also result in APA dysfunction and hence take part in disease development and can cause tumorigenesis. The SNP2APA web server has been created to investigate the outcome of SNPs on APA. Freely available for research and analysis at the URL: http://gong_lab.hzau.edu.cn/SNP2APA/#!/, it is used to browse, search, and download the data. Using the genotypic as well as APA data of various samples from TCGA and TC3A, it found SNPs that affect APA events across 32 types of cancer and described as APA quantitative trait loci (apaQTLs). The database has found the SNPs which affect the APA events using the genotypic data from The Cancer Genome Atlas: https://tcga-data.nci.nih.gov/tcga/ and PDUI data from The Cancer 3' UTR Atlas: http://tc3a.org/. The SNP2APA database provides cis-apaQTLs and trans-apaQTLs across 32 cancer types.

The SNP2APA provides an easy-to-use interface that enables the users to browse, search, and download the following datasets: survival-apaQTLs, *cis*-apaQTLs, GWAS-apaQTLs, and *trans*-apaQTLs. For exploring and visualizing a particular group, "Pancan-apaQTL" page was created.

A page named "PAS Predict" was designed for online prediction to see if a SNP can either create or destroy the polyadenylation signals of alternative polyadenylation. Polyadenylation signals are the most essential regulatory element throughout regulating events of alternative polyadenylation. The PAS Predict page was developed using Dragon PolyA Spotter to find the effect of SNP on polyadenylation signals (Yang et al., 2020).

12.2.17 Human Metabolome Database (HMDB)

The HMDB, developed by the Human Metabolome Project funded by Genome Canada, is a web-based metabolomics database that contains detailed data on metabolites, including biological functions, physiological concentrations, disease correlations, chemical processes, metabolic pathways, and reference spectra. It is a freely accessible web resource that contains complete information on the human metabolome at www.hmdb.ca (Wishart et al., 2018). Every entry related to metabolites in HMDB exhaustively describes the compound, information about the structure, associations of the diseases, data related to enzymes, information about the pathway, etc.; also, it has links to various other databases. There are other capabilities for extensive searching, relational querying, and data viewing. Additional data layers on SNP-metabolite interactions and the impact of medications on metabolite levels are being added and updated continuously, making it an important resource in the fields of clinical chemistry, medicine, nutrition, biochemistry, clinical genetics, and metabolomics science.

12.2.18 Human Mutation Analysis (HUMA)

The HUMA is a web resource that connects disease-relevant, structure, sequence, and variation data in a unified database. Easy accessibility and visualization are the main features of its user-friendly interface. Customized programmable data access is provided by the RESTful web API. Extensive analyses can be performed on the server because of the tools that have been included in the HUMA. Users can also add their own private altered datasets that are immediately mapped to public data for examination by its built-in tools. The resource can be found at https://huma.rubi.ru.ac.za (Brown and Tastan Bishop, 2018). The goal of HUMA is to create a platform for studying variance. The HUMA also includes several tools for analyzing differences in protein sequences and structures. The HUMA integrates and links a variety of information like Ensembl. The HUMA also stands out for its emphasis on protein variation, including protein-level analysis tools, and the option to add private datasets. Variants are plotted to protein and gene sequences in the HUMA database using chromosomal coordinates. When plotting variations to proteins, the coordinate ranges of the protein's coding sequence (CDS) must first be computed, which is a simple operation for genes. The retrieval of protein sequences from Ensembl and UniProt is plotted against each other based on 100% sequence identity; as a result, the protein dataset is non-redundant. The Ensembl provides exons and CDSs. The HUMA brings data from various other sources onto a single platform; some data sources the resource uses are listed in Table 12.2.

TABLE 12.2
Various Sources from Which the HUMA Gets the Data and Brings It to a Single Platform

UniProt and Ensembl	Secondary structure, binding sites, and charged residues, as well as names and descriptions, protein IDs, and associated literature. Ensembl is used to gather aiding sequences such as exons, transcripts, and CDS for proteins.
Pfam	The Pfam accession, name, and coordinates of the protein domain are used to relate data to proteins.
Protein Data Bank (PDB)	The required data, such as names, IDs, and UniProt accession numbers within the PDB files, and sequences of the decoded structures, are retrieved from the human PDB files.
Ensembl and the HUGO Gene Nomenclature Committee (HGNC)	Metadata and gene sequences are retrieved, including gene names, IDs, and signs along with the gene sequence's chromosomal coordinates. Genes linked with phenotypes are provided by Ensembl.
dbSNP and UniProt	Every human alteration in dbSNP is obtained from the dbSNP FTP site VCF. The file featured data tying variations to illnesses, despite the fact that UniProt variant data do not include chromosomal coordinates, and the majority of the variants overlap with those in dbSNP.
UniProt, ClinVar, and OMIM	Data related to diseases. ClinVar provided the IDs; OMIM data are utilized to provide the disease names.

12.2.19 miRdSNP

miRdSNP is a unique human-specific database of disease-associated SNPs (dSNPs) in 3'-UTRs that have been extensively curated from PubMed and can be found at the URL: http://mirdsnp.ccr.buffalo.edu/. miRdSNP associates genes substantiated experimentally to be the putative miRNA targets. TargetScan and PicTar tools included in the package help in the prediction of miRNA target sites as well. A comprehensive database of dSNPs and powerful tools for determining their reach on human genes' 3'-UTRs from miRNA target sites are being provided by miRdSNP. Gene names, miRBase IDs, target prediction algorithms, diseases, and any nucleotide gap between miRNA and dSNPs target sites may all be dynamically filtered in its comprehensive, flexible search options. The annotated positions for dSNPs and miRNA target sites on the full 3'-UTR succession can be examined at the sequence level. dSNPs may also be used in conjunction with the UCSC Genome Browser. All SNPs included on 3'-UTRs of separate genes allow the online interface to evaluate the connection of SNPs with miRNA and its dSNPs binding sites. The database annotates and curates experimentally proven miRNA-targeted genes from four independent curated databases.

For querying the database, miRdSNP provides an advanced search tool that allows users to look for miRNA targets and dSNPs in close vicinity. Gene names, miRbase IDs, target prediction algorithms, diseases, and any nucleotide gap between miRNA target sites and SNPs may all be used to refine searches. Users can additionally limit their search results to experimentally substantiated miRNA-targeted genes. Search outcomes are displayed in tabular format with exportability to CSV format, making the program output flexible to be used with other applications handling CSV data. Exporting data is flexibly context-sensitive, thus enabling outcome-relevant data for download. Users can search over 3 million records for exon distances between SNPs and miRNA target sites besides dSNPs. Separate search outcomes may also be seen at the sequence level, including annotated miRNA target sites, dSNPs, and SNPs. The succession views show the annotations of miRNA target sites from several forecast techniques and let the user switch between them for effortless comparison. Besides searching, the miRdSNP online interface allows users to explore RefSeq entry-based individual genes, by displaying all SNPs and miRNA targets (Bruno et al., 2012). The data are also available to download for big data and deep data analysis.

12.2.20 Protein ANalysis THrough Evolutionary Relationships (PANTHER)

PANTHER serves as an extensive program for analyzing genome-wide gene functions. It is freely available at http://pantherdb.org/ and was created intending to classify protein and gene functions. It has ever since expanded in the domain to be used for performing data analysis. PANTHER is one of the extensively used online resources. It is a member of the GO consortium and thus supports more genomes than other tools. The tool can be used in genome-wide gene function grouping. It employs experimental data, which include genomic, transcriptomic, and proteomic data for the process of gene function grouping. The genetic variation-based information layer has also been incorporated into its repository. It is used to examine genomic data, including SNPs, from experiments such as genome sequencing and GWAS. The layer incorporates various variants of a similar gene only once to avoid artifacts, and a single variant could be connected with just one gene (Mi et al., 2019).

12.2.21 Patrocles

The Patrocles database is a collection of "DNA sequence polymorphisms (DSPs)" that have the potential to disrupt microRNA-mediated regulation of genes. It is freely accessible on the URL: http://www.patrocles.org/.

Distinctive features of Patrocles include the following:

1. Seven vertebrate species are covered.
2. Sections involving the silencing procedure are covered (i.e., silencing machinery, targets, and precursors of microRNA).
3. It includes graphical data on eQTL and microRNA-target coexpression influence of genotype on target expression levels, which allows users to prioritize possible "Patrocles DSPs".
4. It includes CNV and eQTL data affecting miRNA precursors and genes encoding silencing machinery components.
5. Patrocles finder: the tool lets the user see if their selected DSP may cause microRNA-mediated regulation of genes of specific target sequences to be disrupted.

Patrocles is based on data from public databases and primary sources. It uses miRbase for miRNA catalogs and handles all of its mapping tasks in-house to ensure uniformity. Only the names, locations, and sequences of miRNAs are obtained from miRBase. The software design of Patrocles assures that every intake is plotted and every outcome is made using the similar releases of miRBase and Ensembl. DSPs are handled similarly to targets in both circumstances, with the exception that neither ancestrality nor conservation is considered. MicroRNA precursors and machinery genes could be analyzed besides CNVs using genomic intervals. Similarly, genes containing miRNA precursors in machinery and proteins could be compared to known human eQTLs/genes with allelic imbalance (Hiard et al., 2010).

12.2.22 SNPeffect

The SNPeffect database contains the information of SNVs in coding regions of proteins to their structural phenotype using sequence- and structure-based bioinformatics methods. Structural phenotyping combines "chaperone-binding prediction (LIMBO), protein stability analysis (FoldX), aggregation prediction (TANGO), and amyloid prediction (WALTZ)". The database was created with the motive of plotting the SNVs reaction on the protein homeostasis by combining various in-house established structural instruments. The tool provides mutant analyses that are calculated beforehand for over 60000 human coding protein variations, allowing for faster retrieval of the data. It also permits the generation of custom mutant groups and has options for meta-analysis of specified datasets, such as analyzing the proteostasis of a specific protein or protein family. It can be freely accessed on the URL: https://snpeffect.switchlab.org/sequences.

SNPeffect also includes information on affected enzymatic sites and several post-translational changes. Users can contribute custom protein variants for analysis using SNPeffect, which includes database-related automatic structure and modeling, using information from UniProt. The meta-analysis tool included allows visualization of the correlation values between phenotypic characteristics for a user-specified group of variants. It makes use of the FoldX force field to calculate realistic free energy changes following mutation (ΔG), resulting in more precise protein stability data. For the structures to be modeled, there should be at least 90% sequence identity with the template, since the quality of the structure is critical for the precision of ΔG predictions calculated by using FoldX (Buß et al., 2018).

Users can search for SNVs using the database interface. The interface includes filters for mutation type, molecular phenotypic effects, disease, dbSNP identifier, UniProt identification, and gene name. Filter application results in a list of variants that may be looked at in a protein-centered or variant-centered perspective. Each variant's wild-type protein is completely examined and referenced from the variant homepage. Self-explanatory histograms and bar plots are used to depict the effects. The Protein Data Bank (PDB) is used to get structural information. If sequence matches cannot be identified, models based on homology are constructed with a minimum of 90% identity to the original sequence. It provides visualization of both mutant residue and the wild-type environments in the protein structure if structural information is retrieved. A written and graphical comparison to the wild-type protein is provided with each phenotypic analysis (De Baets et al., 2012).

12.2.23 The Total Integrated Archive of Short-Read and Array (TIARA) Database

The TIARA genome database holds personal genomic data derived from ultra-high-resolution comparative genomic hybridization (CGH) arrays and NGS. This database enhances the correctness of recognizing SNPs, structural variants (SVs), and short indels in individuals' genomes. A user-friendly genome browser is available in TIARA at the URL: http://tiara.gmi.ac.kr. It provides log2 ratios and read depths (RDs) from CGH and NGS arrays, respectively. As more individuals undergo NGS or CGH array analysis, more personal genomes will be saved. More human sequencing data with greater depth make the TIARA genome database more comprehensive on rare human variations. In the genome browser, transcriptomic and genomic data are matched following TIARA's integrated method. The platform also provides tools to examine the transcriptome sequencing data, which include aligning reads, variations, and expression levels. Allele-specific gene expression, levels of gene expression, and editing of RNA or transcriptional base modifications (TBMs) could also be discovered by comparing genomic variants between transcriptome sequencing and whole genome sequencing for matched samples (Hong et al., 2011).

12.2.24 Genome Variation Map (GVM)

The GVM is a database of genome modifications that is open to the public. It is maintained by "the National Genomics Data Center (China)". The resource is available at the URL http://bigd.big.ac.cn/gvm/. Its goal is collecting and integrating genome variants for a variety of species, allowing contributions of various variation kinds globally. The resource is available free to the public to assist global research efforts. It contains 960 million variants from 41 species, including 64,819 unique genotypes and 260,393 high-quality genotype-to-phenotype relationships curated manually. Since its launch in 2017, the GVM has collected genetic variation data from 43,754 samples. The GVM issues valuable genome variants for a variety of species as a major resource in the NGDC and so plays a crucial role in both molecular breeding and functional genomics investigations. The resource has persistently been improved in recent years to deliver variant data and metadata of high-quality and external links to public repositories, including dbSNP (Li et al., 2021).

12.2.25 Variation Annotation Database for Human (VARAdb)

VARAdb is a complete variation annotation database for humans, specifically covering noncoding variants. The resource is available at the URL: http://www.licpathway.net/VARAdb/. It provides annotated data of 577,283,813 variations, ranked with the help of nine annotation groups, and facilitates pathways and downstream examination. It is unique as it incorporates genomic and epigenomic information into sections such as "Related genes, Variation information, Chromatin accessibility, Chromatin interaction, and Regulatory information". Motif changes, LD, risk SNPs, sequence conservation, eQTLs, somatic mutations, clinical variant-drug-gene pairs, super intensifiers, chromatin states, promoters, enhancers, transcription factors, chromatin accessibility regions, chromatin interactions, and histone modifications are all included in the detailed annotation information provided by the resource. VARAdb provides an easy-to-use interface to search, browse, view, or download variations and associated data. It provides insights into identifying putative functional variants and assessment of their implications on biological processes and diseases (Pan et al., 2021).

12.2.26 Catalog of Human Genome-Wide Association Studies (GWAS Catalog)

The NHGRI-EBI GWAS Catalog has been storing summary statistics files (aggregate p-values and association data) for every variant analyzed by mining published GWAS. The database resource is available at www.ebi.ac.uk/gwas/. Its interface features include ontology-assisted search for better curation. Recently, it has seen an increase in the frequency of data releases by automated curation and other improvements. All studies now have organized ancestry and recruitment information, expanding the variety of available data. The catalog's infrastructure enables it to scale for larger arrays, exomes, and studies related to sequencing.

It also includes published SNP-trait relationships. The catalog compiles a massive body of unorganized and diversified data from the literature into an easily searchable, organized, and quality-controlled resource. Various researchers use catalog data as a beginning point for additional research in finding usual variations, understanding mechanisms related to various diseases, and proposing targets for new therapeutics. The study of catalog data helps to discover possible common missense mutations and identify new targets for medications. Various resources, including Ensembl, PheGenI, the UCSC Genome Browser, GWASdb, and HuGE Navigator, use data from GWAS Catalog. Trained curators with experience in analyzing research design read each manuscript to establish the most scientifically accurate inferences understandably to be added to the GWAS Catalog (Buniello et al., 2019).

12.2.27 Mouse Genome Database (MGD)

MGD, commonly referred to as the mouse genome database, is a community GUI-based database for the experimental mice, available at the URL: http://www.informatics.jax.org. Mice are the animal model that is mostly used to investigate the genetic and molecular biology underpinning the phenotypes and disease mechanisms. It is a go-to resource for mouse gene information based on reference data. The International Cancer Screening Network (ICSN) for mice has classified MGD as the predominant resource for official gene, allele, and naming of the mouse strains. Its biocuration scientists collect data from biomedical research and major and minor datasets that researchers contribute. Its Multiple Genome Viewer tools allow users to explore the genomes of several mouse strains comprehensively. The database maintains its reputation as an essential community resource of laboratory mouse data to understand the genetics that is related to the corresponding human phenotypes. The database keeps track of mouse genes, genome characteristics, sequence data, and updated annotations. MGD helps in studies related to genotype-to-phenotype (G2P) annotations. The information in the resource is updated regularly, and its interface aids in easy data retrieval for gene identification with similar biological properties, as well as data mining for knowledge discovery (Blake et al., 2021).

12.2.28 dbNSFP – Database of Functional Predictions and Annotations for Human Non-synonymous and Splice-Site SNVs

dbNSFP is used to predict and annotate the functional effects of every possible non-synonymous single nucleotide variant (nsSNV) in the human genome. The database is available at the URL: http://database.liulab.science/dbNSFP. dbNSFP supports transcript-specific predictions and annotations. It has both nsSNVs and splice-site SNVs (ssSNVs) in it, which is centered on Gencode release 29/Ensembl version 94 currently. Its goal is to provide an open access for functional prognosis and annotations for the said SNVs, as well as to make refining and hierarchizing SNVs found in an exome sequencing study easier. There are currently 84,013,093 nsSNVs and ssSNVs in the database made available in contrast to human reference standards. Functional estimations and annotations for each SNV are gathered from a variety of approaches and resources. The database is simple to use with an easy interface for users with little or no bioinformatics experience (Liu et al., 2020).

12.2.29 Phenome Linking to the Genome (PheLiGe)

PheLiGe (https://phelige.com/about) enables users to analyze regional genotype-phenotype relationships across a wide range of phenotypes, providing information about the biological function affected by the variant in question. PheLiGe also allows users to compare geographical patterns of connection between various qualities. This method can determine whether a co-association is because of pleiotropy or linkage. It allows the researchers to explore regional and genome-wide associations, search phenotypes related to the genetic variant, and compare relationship patterns between distinct traits. Users can easily search for phenotypic association either directly or by a proxy variant in LD, which is noticed for a particular SNP. The tool employs the minor allele frequency (MAF) method to filter SNPs that can be represented by a dot at the highest resolution (Shashkova et al., 2021).

12.2.30 GWAS Atlas

GWAS Atlas is a database of documented genome-wide SNP-trait relationships in animals and plants, present on the URL: http://bigd.big.ac.cn/gwas/. It could be used for listing the degree of pleiotropy at the trait-associated locus, gene, or gene set levels. The resource can also distinguish the nature of trait-related variants to inspect genetic structure among a diversity of traits and domains in terms of trait polygenicity or SNP heritability. Most of the inspected traits are polygenic, with at least 0.1% of SNPs being causal. For all these associations, millions of individuals are needed for the identification of every significant SNP at the genome-wide level. Individuals with polygenic traits have a nearly specific set of risk alleles for a particular disease or trait. With the increase in polygenicity level and with a greater number of potential permutations of causative SNPs, the possible combinations of these increase logarithmically. The GWAS Atlas provides the most exhaustive summary of associated genetic areas, the extent of pleiotropy, and alteration in genetic structure among the traits (Watanabe et al., 2019).

12.2.31 ncRNA-eQTL

Genome-wide expression quantitative trait loci (eQTLs) are a tool used to evaluate the impact of SNPs on the expression of noncoding RNA (ncRNA) in various types of cancers. This tool helps us to know the relationship between risk alleles and tumorigenesis in the development of cancer. The ncRNA-eQTL web interface is located at http://ibi.hzau.edu.cn/ncRNA-eQTL/. The resource gives access to the ncRNA-eQTLs and their associated data. This database comprises mainly two sub-databases, i.e., lncRNA-eQTL home and miRNA-eQTL home. Each sub-database aims to serve four major datasets: trans-eQTLs, cis-eQTLs, GWAS-eQTLs, and survival-eQTLs.

Users can easily enter a particular SNP or gene by the quick search option present on the homepage (Li et al., 2020).

12.2.32 HACER

HACER is an acronym for Human ACtive Enhancer to interpret Regulatorly variants. The tool comprises data related to actively transcribed enhancers fostered from PRO-seq, GRO-seq, and CAGE data. The tool not only includes cell-type-specific enhancers, but also uses TF-enhancer binding and confirmed chromatin linkage data to connect enhancers into the transcriptional regulatory system. It is a one-of-a-kind repository for noncoding variations and enhancer-mediated regulatory research and is freely accessible on the URL: http://bioinfo.vanderbilt.edu/AE/HACER/.

It categorizes and annotates cell-type-specific enhancers transcribed in vivo and also places enhancers within their associated regulatory networks by combining ENCODE TF ChIP-Seq and approved chromatin interaction data. HACER serves as a beneficial resource for GWAS, enhancer-mediated regulation, and noncoding variants. It comprises tools that help to search the putative relationship mechanism between enhancer variants and the disease risk. Overall, the tool can be used to investigate the enhancer activity and associated gene regulation in different cell lines and to connect noncoding variants and disease genes for defining putative causal variants. It thus serves as an important resource for research on enhancer-mediated regulation and activity (Wang et al., 2019).

12.2.33 Genomic Information Portal on Cancer Cell Lines (GENIPAC)

GENIPAC was created particularly to host and display the genomic data. It is a specialized resource that hosts and displays genomic data of head and neck cancer (HNC) cell lines. It is an easy-to-access database as it allows the users to search very complicated datasets for developing mechanistic hypotheses based on the altered gene profiling and biological pathways. The database is used to study the various mechanisms which contribute to the development of cancer and precision therapies. This database uses head and neck-derived cell line data as the fundamental model. The tool also uses the oral cancer cell line data for various researches. GENIPAC could be accessed at https://genipac.cancerresearch.my/. This tool enables users to visualize the genomic alterations taking place within the pathways and aids the scientific community to have a better knowledge about the HNC and the evolution of novel precise therapies to treat HNC (Lee et al., 2018).

12.2.34 Database of Genomic Variants of Oral Cancer (dbGENVOC)

dbGENVOC hosted at http://research.nibmg.ac.in/dbcares/dbgenvoc/ is an easy-to-use web interface that permits the users to mine, store, retrieve, and analyze the variants of oral cancer, SNVs and indels that are identified by carrying out whole genome and exome sequencing from the patients suffering from oral cancer. It is an exhaustive database offering open access to the public for searching for variant information about oral cancer. It gives a holistic outlook to the biologists on large-scale cancer genomics at both individual and population levels. The resource is designed to access the variants that are particularly connected to Indian patients suffering from oral cancer. The database also incorporates the data from other sources globally and thus provides new avenues for competitive variant analysis geographically. Various data from TCGA manually curated by various researchers have also been added to the database for investigative comparison. It permits the investigators to query the database by a single gene, multiple variant sites, multiple genes, genomic region, pathway identities, or patient ID (Pradhan et al., 2021).

12.2.35 CancerSplicingQTL

CancerSplicingQTL is hosted at http://www.cancersplicingqtl-hust.com/#/. It is a database that uses SNPs to anticipate the impact of SNPs on alternative mRNA splicing. Alternative splicing (AS) is a molecular biology process in transcription that expands structural transcript variation and proteome diversity. It takes place in almost every multi-exon gene and plays an essential part in the cancer processes. Abnormal splicing patterns are often noticed in stages of cancer in the beginning, progression, diagnosis, and treatment. Various evidence has illustrated that AS events could experience adjustment by genetic variants. The recognition of splicing quantitative trait loci (sQTLs), genetic variants that influence AS events, may depict an essential step in completely understanding the involvement of genetic variants in the development of diseases. It is an easy-to-use database that allows users to browse, search, and download data easily. It acts as an exhaustive sQTL resource for categorizing the probable functions of SNPs that manage transcript isoforms in human cancers. There are six AS event types that are analyzed by this database, namely "exon skip (ES), alternate donor (AD) sites, alternate acceptor (AA) sites, retained intron (RI), alternate terminator (AT), and alternate promoter (AP)" (Tian et al., 2019).

12.2.36 SweGen

The SweGen project accessible at https://swefreq.nbis.se encompasses whole genome variation data frequencies for over a thousand people of Swedish origin. It serves as an extensive resource for genetic alteration in the Swedish population. This information hosted on the portal acts as a repository for clinical genetics and for research involving group-associated genetic variant frequencies. The project has accumulated data on the genetic architecture of the Swedish population by utilizing high-density SNP array information from a group of over 10,000 Swedish people. This database acts as an important resource for studying inheritable disorders and clinical sequencing. The portal helps in large cross-sectional population research for better understanding the disease trait expressivity of pathogenic variants and for interpreting and hierarchizing genetic alterations that might cause the disease (Ameur et al., 2017).

12.2.37 RMBase

RMBase is hosted at http://rna.sysu.edu.cn/rmbase/. This tool provides transcriptome-wide landscapes of RNA modifications. It is an exhaustive database that carries out the integration of epitranscriptome sequencing information for the exploration of post-transcriptional changes of RNAs and their associations with microRNA-binding affairs, dSNPs, and RNA-binding proteins (RBPs) in cancer. The components and tools incorporated in RMBase are as follows:

1. "MOTIF": offers the modified motifs' visualized logos and position weight matrices (PWMs).
2. "modRBP": used for investigating the connections between RBPs and RNA modifications.
3. "modMetagene": used for plotting the metagenes of RNA modification with a transcript model.

The database aims to help scientists investigate the potent roles and mechanisms of RNA modifications. It provides the interface for displaying the maps of RNA modifications in different cell types. Furthermore, by combining RBP-binding sites, microRNA targets, SNVs, and data from GWAS, the database can be utilized to search the associations among these RNA modification sites (Xuan et al., 2018).

12.2.38 SNP2TFBS

SNP2TFBS hosted at https://ccg.epfl.ch/snp2tfbs/ is a web interface for investigating alterations (SNPs/indels) in the human genome that affect the binding of transcription factors. The database comprises a stock of various files that aim to provide particular annotations for SNPs in humans. Specifically, the effect of SNP on transcription factor binding is evaluated based on the PWM model for assessing the binding specificity of the corresponding factor. The data files are revived regularly by an automated method that takes an extensive SNP catalog, a reference genome, and a group of PWMs as the input files. The database uses the PWM model for predicting the binding strength of transcription factors (Kumar et al., 2017).

12.2.39 RAAR – Regulators of Androgen Action Resource

The Regulators of Androgen Action Resource (RAAR) is hosted at http://www.lerner.ccf.org/cancerbio/heemers/RAAR/search/. It is a freely available, curated, non-redundant, and convenient web interface that enables the convenient and efficient investigation of androgen receptors and their prospective therapeutic targets. AR-related coregulators, including androgen transporters and androgen biosynthesis proteins, are getting popular as a therapeutic target for prostate cancer. However, such proteins are described by numerous transcriptional alternatives that tend to generate genetic changes and are expressed differentially in prostate cancer. The portal accumulates the review of diverse related datasets and databases, websites, and literature publications to determine their therapeutic potential. Mining and combining these databases is a time-consuming process, and RAAR tackled this problem. This database has been arranged in two layers. A "pre-receptor-level" database has information on androgen production, whereas a "post-receptor-level" database contains information on coregulatory genes (DePriest et al., 2016).

12.2.40 LincSNP 3.0

LincSNP 3.0 is accessible at http://bio-bigdata.hrbmu.edu.cn/lincsnp/. It is an up-to-date extensive database repository with the aim to store and annotate disease- or phenotype-related variants such as SNPs, LD SNPs, RNA editing and somatic mutations, in long noncoding RNAs (lncRNAs) and circular RNAs (circRNAs) or associated regulatory elements such as DNase I hypersensitive site (DHS), footprints, topologically associating domains (TADs), TFBSs, enhancers, and open chromatin areas. Also, the effects of SNP on methylated lncRNAs and circRNAs are included. In addition, four versatile web tools included functions to retrieve, display, and interpret data. The tools are named as Variant Mapper, Genome Browser, Functional Annotation, and Circos Plotter. Some other valuable tools are also offered to act as an intermediary research tool for functional variations and human circRNAs or lncRNAs, hence allowing a better analysis of lncRNA and circRNA functioning, especially their putative role in human diseases (Gao et al., 2021).

12.2.41 SomamiR

The SomamiR database http://compbio.uthsc.edu/SomamiR is a repository designed to supply extensive information for exploring the effects of germline and somatic variations on miRNA activity in cancer. It is widely known that many somatic variations distinguish the healthy and malignant genomes. While several germline variations in miRNAs and related targets have been found to change miRNA activity in malignancy, the effect of somatic variations on miRNA activity has gotten minimal attention. The tool thus caters to the same problem by including somatic and germline variations that can establish or disturb miRNA target sites. In addition, the repository contains a set of other somatic and germline alterations in miRNAs and related targets that have a demonstrative influence on cancer-related miRNA activity. SomamiR's cancer pathway analysis tools can find specific sections of pathways that frequently contain miRNA-related somatic alterations. The set of tools provided also includes a genome browser to allow browsing the genomic data (Bhattacharya et al., 2013).

12.2.42 MSDD PMID

The MiRNA SNP Disease Database (MSDD, http://bio-bigdata.hrbmu.edu.cn/msdd/home.jsp) is a hand-curated database containing extensive experimentally verified associations between miRNAs, SNPs, and human disorders. miRSNPs can cause dysfunction of miRNA and its target genes, increasing vulnerability to or causing the development of human diseases. For a complete knowledge of the processes and activities of miRSNPs, a curated compilation and overview of miRSNP-related disorders is required. Every relationship contains data on SNPs, miRNAs, miRNA target genes and diseases, SNP sites and alleles, and the miRNA malfunctioning pattern. It offers a convenient interface for submitting, browsing, retrieving, and downloading new data. MSDD acts as a relevant and beneficial resource as it considerably enhances our knowledge of miRNA malfunction in disease (Yue et al., 2018).

12.2.43 miRNASNP

miRNASNP-v3 is an online web interface tool (http://bioinfo.life.hust.edu.cn/miRNASNP/) with features to provide SNP exploration of traits associated with diseases. Cellular miRNASNP, biogenesis, and functions are often affected by SNP and variations such as SNP/SNV as well as miRNA target-binding sites influencing phenotypes. miRNASNP database could be extensively used for exploring disease-associated miRNA variations and mutations (Liu et al., 2021). The overview of functions provided by miRNASNP-v3 are as follows:

1. Identifying the effects of mutations over miRNA target binding.
2. Analysis of alteration arising by mutations (DRVs/SNPs) of miRNA target.
3. Identification of correlation among expression of miRNA and sensitivity of drugs.
4. Investigating miRNAs and their target expression profile in cancer.
5. Browsing secondary structure, modification, variation, and identifying changes defined by users on pre-miRNA- or miRNA-target binding sites.

12.3 Other Databases Related to SNP Research

SNP databases are important as they help to store, update, and compare related data in real time. The databases that are regularly updated prove to be significant as they keep on adding the updated relevant data, providing scientists and researchers the grant to a variety of SNP–relevant data. There are many more SNP databases, and some of them are enlisted in Table 12.3.

TABLE 12.3

Databases Related to SNPs and Their Web Links with Brief Descriptions

Database	Year	Links	Description
lncRNASNP	2015	http://bioinfo.life.hust.edu.cn/lncRNASNP/#!/	lncRNASNP is a database that contains a large number of SNPs in human and mouse lncRNAs. It includes lncRNA SNPs, lncRNA mutations, SNP impacts on lncRNA structure, and lncRNA:miRNA binding (Gong et al., 2015).
VariSNP	2015	http://structure.bmc.lu.se/VariSNP	Database package that includes variation datasets that may be used to build and test variant effect prediction systems. It is a collection of dbSNP datasets that have been screened for disease-related variants in ClinVar, PhenCode, and Swiss-Prot; thus, all variations are considered neutral/non-pathogenic. The datasets are organized according to the dbSNP functional classes (Schaafsma and Vihinen, 2015).
SorGSD	2015	https://ngdc.cncb.ac.cn/sorgsd/	Database containing 289 accessions, practical tools, phenotypes, resources, and key references for sorghum genomic variants (SNPs and indels) (Luo et al., 2016).
PeanutDB	2015	http://bioinfolab.muohio.edu/txid3818v1/	PeanutDB provides the research community with a convenient website that will surely aid genomics research and molecular breeding in an under-researched crop as a public genomic library that incorporates transcriptome information of peanuts from diverse sources (Duan et al., 2012).
DoGSD	2015	https://ngdc.cncb.ac.cn/idog/	The Dog Genome SNP Database (DoGSD) is a database that stores information about variation in dog and wolf genomes. It was created as an SNP detector and visualization tool to give the research community a great resource for studying the population, phenotype, evolution, and behavior of dogs (Bai et al., 2015).
GRASP	2014	http://grasp.nhlbi.nih.gov/	GRASP is designed to give a simple way to query reported SNP associations with human attributes, such as methylation and QTL studies (Leslie et al., 2014).
GWAS Central	2014	https://www.gwascentral.org/	A database of summary-level outcomes from both major and minor genetic association research. GWAS Central actively collects datasets from public domain projects and encourages community submissions (Beck et al., 2020).
rSNPBase	2014	http://rsnp3.psych.ac.cn/	SNP-related regulatory elements, SNP-based regulatory networks, and element-gene pairs are all available in rSNPBase. The SNP regulatory annotation databases rSNPBase and rVarBase have been upgraded to create this database (Guo and Wang, 2018).
GMTV database project	2014	http://mtb.dobzhanskycenter.org/	Clinical, microbiological, and epidemiological descriptions are combined with genome variations depending on whole genome sequencing data in the Genome-based Mycobacterium Tuberculosis Variation Database. It currently includes genomic data on 2819 mycobacterium TB strains (Sinkov et al., 2018).
Kazusa Marker DataBase	2014	http://marker.kazusa.or.jp/	The Kazusa Marker DataBase will be valuable in both basic and practical sciences such as genetics, genomics, and crop molecular breeding (Shirasawa et al., 2014).
PolymiRTS	2014	http://compbio.uthsc.edu/miRSNP/	A database of naturally occurring DNA polymorphisms in the seed areas and target sites of miRNAs. microRNAs bind to the transcripts of protein-coding genes, causing mRNA instability or translational suppression. SNPs and indels in miRNAs and their target locations may have an impact on miRNA-mRNA interaction, and hence miRNA-mediated gene regulation (Bhattacharya et al., 2014).

(Continued)

TABLE 12.3 (*Continued*)

Databases Related to SNPs and Their Web Links with Brief Descriptions

Database	Year	Links	Description
MirSNP	2012	http://bioinfo.bjmu.edu.cn/mirsnp/search/	MirSNP is a publicly accessible database that contains human SNPs in putative miRNA-mRNA binding sites (Liu et al., 2012).
SNPedia	2011	https://www.snpedia.com/index.php/SNPedia	SNPedia is a wiki site dedicated to the functional implications of human genetic diversity. SNPedia has concentrated on the phenotypic, medical, and genealogical relationships of SNPs (Cariaso and Lennon, 2012).
openSNP	2011	https://opensnp.org/	Database that allows direct-to-consumer genetic testing participants to post their phenotypic information and genetic data for free. It has become a resource for a wide range of investigations, including GWAS, as a result of this crowdsourcing initiative to collect phenotypic and genotypic information (Greshake et al., 2014).
PESNPdb	2010	http://bejerano.stanford.edu/pesnpdb/public/html/	PESNPdb is a database of SNPs that have been linked to pre-eclampsia, a pregnancy-related condition that can be fatal to both the mother and the child. All publications were hand-curated to contain specific information about each study, which is critical for data comparison and interpretation. PESNPdb seeks to aid research and bring fresh insights into pre-eclampsia genetic factors (Tuteja et al., 2012).
SNPnexus	2009	https://www.snp-nexus.org/v4/	A web-based resource for functional annotation of newly identified and publicly available SNPs (Oscanoa et al., 2020).
AutoSNPdb	2009	http://autosnpdb.appliedbioinformatics.com.au/	It features a versatile interface that allows for a wide range of queries. Users can look for SNPs in genes with expected functions or by comparing sequences to known genes. Additional levels of annotation and unique inquiries tailored to regions of interest can also be added. Plant cultivar information in the database allows for the detection of SNPs that differentiate between plant cultivars (Duran et al., 2009).
topoSNP	2004	http://sts.bioe.uic.edu/toposnp/	Database for assessing missense SNPs that may be mapped onto known three-dimensional protein structures. These include disease-related missense SNPs from OMIM database and additional missense SNPs from dbSNP (Stitziel et al., 2004).
YHRD	2003	https://yhrd.org	It generates estimation for Y-STR haplotypes, which result in its accurate frequency and may be utilized to assess matches quantitatively in forensic and kinship cases. Advanced tools and further materials for Y-STRs and Y-SNPs are also available (Willuweit et al., 2007).

12.4 Importance of SNP Data Resources in Research

Classic biology has associated the role of SNPs in identifying disease development genes alongside their association with variation in drug response, leading to the medical benefit relevance of SNPs to the genetic makeup and drug response of affected individuals in the form of developing molecular medicines aimed at individual SNP variations. SNPs also reveal information regarding mechanisms of sequence evolution and variation. Strong selective pressures are often significantly associated with modifying phenotypes. Substituted nucleotides fixed during evolution often tend to have positive selection advantages toward the overall fit for survival.

SNPs do not show a homogenous pattern and distribution throughout the genome; the frequency of SNPs is more in noncoding areas, generally where natural selection is acting and "fixing" the allele (eliminating other variants) of the SNP that constitutes the likely genetic adaptation. Mutation rate and genetic recombination are also known to determine SNP density and distribution (Barreiro et al., 2008). Specific SNPs in DNA sequences show an association with the development of diseases and response to chemicals, drugs, pathogens, and vaccines. SNPs thus have a big role to play in futuristic personalized medicine development (Laing et al., 2011)

The field of informatics biology plays a pivotal role in the identification and discovery of SNPs (Johnson, 2009). We sum up important computational methods used for the analysis of SNPs. Such servers also help in providing the information regarding if the SNPs are prospectively information about them to act as neutral or deleterious. In order to sum the information and resources for the scientific community working in fields related to SNP biology, a compendium of Internet-accessible tools/stand-alone tools have been enlisted. Such tools enable scientists/biologists to fetch SNPs based on data in the study of interest, their expression pattern distribution, and genetic/physical location within the genome.

Association studies of candidate genes are often used to associate SNPs with corresponding diseased phenotype, and the literature is full of SNP-diseased phenotype associations; for example, intronic SNP in PMS2 gene is a risk factor for male infertility (Ji et al., 2012). Similarly, a C/A SNP in FOXP3 is known to increase the predisposition of cells toward being cancerous (Ezzeddini et al., 2021). SNP Y402H in the CFH gene has been shown to have implications for the development of age-related macular degeneration (Landowski et al., 2019). SNP in PNPLA3 gene has been shown to be associated with non-alcoholic fatty liver disease.

A wide variety of SNPs have been shown to be associated with cancer; cancer-causing SNPs can be found in exons, introns, promoters, and untranslated regions; polymorphisms affect the gene expression by a variety of different mechanisms; and cancer-causing SNPs can be a primary cause of later occurring epigenetic modifications related to increased cancer susceptibility (You and Jones, 2012). SNPs in relation to cancer can be broadly classified as SNPs affecting disease susceptibility and SNPs affecting the disease outcome. SNPs in correlation to disease outcome show a relation to disease prognosis, survival, and response to therapies. A homozygous condition single gene transition G/A in myeloperoxidase (MPO) at 463 in the promoter region has been shown to decrease its gene expression and has been attributed to a significantly lower risk of lung cancer development (Yang et al., 2017). Similarly, polymorphisms in gene N-acetyltransferase (NAT2) and glutathione S-transferase M1 (GSTM1) have been found to be correlated to bladder, lung, and colon cancer risk (Zupa et al., 2009).

Variants because of SNPs are known to affect the outcome and thus have clinical relevance to them. For instance, CYP3A4 SNPs are found to affect the outcomes of prostate cancer (Han et al., 2015). Variations in the thiopurine S-methyltransferase (TPMT) gene are known to affect the outcome of therapeutic thiopurine drug response, commonly used for malignancies of hematopoietic origin (Krynetski and Evans, 2003). Similarly, polymorphisms in the CYP2D6 gene are classically known to affect the activity of various anticancer drugs by altering their pharmacokinetic properties (Carr et al., 2021).

Cancers often differ in their prevalence, incidence, and mortality among distinct ethnic groups. Prostate cancer incidence is known to be the highest among African American men in contrast to Japanese people, where the incidence of prostate cancer is the least (Shavers and Brown, 2002). Similarly, Native American and Hispanic women show a lower incidence of breast cancer in contrast to European descent non-Hispanic females (Fejerman and Ziv, 2008). Such population-specific variable effects are often attributed to variations within their genomes such as SNPs. For instance, 19 SNPs at 18 different loci in BRCA1 and BRCA2 have been associated with breast cancer susceptibility (Ingham et al., 2013). The field of ethnicity correlation with the disease is a vast community, and various database resources such as HapMap have made great advances in understanding allele-specific disease prevalence across various ethnicities (International HapMap Consortium, 2003).

12.5 Conclusions and Way Forward

With evolving NGS, GWAS, and data-based artificial intelligence prediction algorithms, discovery and correlational/consequential associations of SNPs have increased remarkably. Such technologies with affordable pricing have rationalized SNPs discovery in both model species and non-model species. The field has gained popularity in all areas of biology, be it physiological or evolutionary plant-animal studies, selection of desirable traits, or therapeutic efficiency of drugs. Alongside, the development of SNP arrays/chips has been emancipated as an assertive tool to study diversity patterns and marker associations in various species. Studies involving SNP biology are often initiatives aimed at large population-based studies. Therefore, SNPs find a high influence on the outcomes of large-scale epidemiology studies and their genetic reasons. The development of SNP maps in species has tremendous potential to contribute in every field, be it pathology, disease biology, diagnostics, breeding programs, or therapeutic efficacy.

The compendium of resources described in this review would help the broad scientific community about the databases and tools used in SNP research. The tools enlisted here may aid in deducing significant results from studies related to SNP biology, their important features/traits associated with the corresponding SNPs, and their probable effect on relevant biological and functional variables.

ABBREVIATIONS

AIMs: Ancestry informative markers
API: Application programming interface
CGAP: Cancer Genome Anatomy Project
CGH: Comparative genomic hybridization
CNV: Copy number variation
DRVs: Disease-related variations
dSNPs: Disease-associated single nucleotide polymorphisms
eQTLs: Expression quantitative trait loci
GEO: Gene expression omnibus
GSV: Genomic structure variation
GUI: Graphical user interface
GVF: Genome variation format
GWAS: Genome-wide association studies
LD: Linkage disequilibrium
LRS: Long-read sequencing
MAS: Marker-assisted selection
miSNPs: Single nucleotide polymorphisms in miRNA
NCBI: National Center of Biotechnology Information
nsSNVs: Non-synonymous single nucleotide variants
PCA: Principal component analysis
RAD: Restriction site-associated DNA
scATAC-seq: Single-cell assay for transposase accessible chromatin sequencing
scSNVs: Splicing consensus single nucleotide variants
SIFT: Sorting Intolerant From Tolerant
SNP: Single nucleotide polymorphism
SNV: Single nucleotide variant
ssSNVs: Splice-site SNVs
SVs: Structural variants
TBFs: Transcription factor-binding sites
URL: Uniform Resource Locator
VCF: Variant Call Format

REFERENCES

1000 Genomes Project Consortium, Abecasis, G. R., Altshuler, D., Auton, A., Brooks, L. D., Durbin, R. M., Gibbs, R. A., Hurles, M. E., and McVean, G. A. (2010). A map of human genome variation from population-scale sequencing. *Nature* 467, 1061–1073. doi:10.1038/nature09534.

Al-Koofee, A. F., and Mubarak, S. M. H. (2020). "Genetic polymorphisms," in *The Recent Topics in Genetic Polymorphisms*, eds. M. Çalışkan, O. Erol, and G. Cevahir Öz (IntechOpen). doi:10.5772/intechopen.88063.

Ameur, A., Dahlberg, J., Olason, P., Vezzi, F., Karlsson, R., Martin, M., Viklund, J., Kähäri, A. K., Lundin, P., Che, H., et al. (2017). SweGen: a whole-genome data resource of genetic variability in a cross-section of the Swedish population. *Eur. J. Hum. Genet.* 25, 1253–1260. doi:10.1038/ejhg.2017.130.

Bai, B., Zhao, W.-M., Tang, B.-X., Wang, Y.-Q., Wang, L., Zhang, Z., Yang, H.-C., Liu, Y.-H., Zhu, J.-W., Irwin, D. M., et al. (2015). DoGSD: the dog and wolf genome SNP database. *Nucleic Acids Res.* 43, D777–D783. doi:10.1093/nar/gku1174.

Bailey, M. H., Tokheim, C., Porta-Pardo, E., Sengupta, S., Bertrand, D., Weerasinghe, A., Colaprico, A., Wendl, M. C., Kim, J., Reardon, B., et al. (2018). Comprehensive characterization of cancer driver genes and mutations. *Cell* 173, 371–385.e18. doi:10.1016/j.cell.2018.02.060.

Barreiro, L. B., Laval, G., Quach, H., Patin, E., and Quintana-Murci, L. (2008). Natural selection has driven population differentiation in modern humans. *Nat. Genet.* 40, 340–345. doi:10.1038/ng.78.

Beck, T., Shorter, T., and Brookes, A. J. (2020). GWAS Central: a comprehensive resource for the discovery and comparison of genotype and phenotype data from genome-wide association studies. *Nucleic Acids Res.* 48, D933–D940. doi:10.1093/nar/gkz895.

Bhattacharya, A., Ziebarth, J. D., and Cui, Y. (2013). SomamiR: a database for somatic mutations impacting microRNA function in cancer. *Nucleic Acids Res.* 41, D977–982. doi:10.1093/nar/gks1138.

Bhattacharya, A., Ziebarth, J. D., and Cui, Y. (2014). PolymiRTS Database 3.0: linking polymorphisms in microRNAs and their target sites with human diseases and biological pathways. *Nucleic Acids Res.* 42, D86–91. doi:10.1093/nar/gkt1028.

Bianco, A. M., Marcuzzi, A., Zanin, V., Girardelli, M., Vuch, J., and Crovella, S. (2013). Database tools in genetic diseases research. *Genomics* 101, 75–85. doi:10.1016/j.ygeno.2012.11.001.

Birney, E., Andrews, T. D., Bevan, P., Caccamo, M., Chen, Y., Clarke, L., Coates, G., Cuff, J., Curwen, V., Cutts, T., et al. (2004). An overview of Ensembl. *Genome Res.* 14, 925–928. doi:10.1101/gr.1860604.

Blake, J. A., Baldarelli, R., Kadin, J. A., Richardson, J. E., Smith, C. L., Bult, C. J., and Mouse Genome Database Group (2021). Mouse Genome Database (MGD): knowledgebase for mouse-human comparative biology. *Nucleic Acids Res.* 49, D981–D987. doi:10.1093/nar/gkaa1083.

Boyle, A. P., Hong, E. L., Hariharan, M., Cheng, Y., Schaub, M. A., Kasowski, M., Karczewski, K. J., Park, J., Hitz, B. C., Weng, S., et al. (2012). Annotation of functional variation in personal genomes using RegulomeDB. *Genome Res.* 22, 1790–1797. doi:10.1101/gr.137323.112.

Brown, D. K., and Tastan Bishop, Ö. (2018). HUMA: a platform for the analysis of genetic variation in humans. *Hum. Mutat.* 39, 40–51. doi:10.1002/humu.23334.

Bruno, A. E., Li, L., Kalabus, J. L., Pan, Y., Yu, A., and Hu, Z. (2012). miRdSNP: a database of disease-associated SNPs and microRNA target sites on 3'UTRs of human genes. *BMC Genomics* 13, 44. doi:10.1186/1471-2164-13-44.

Buniello, A., MacArthur, J. A. L., Cerezo, M., Harris, L. W., Hayhurst, J., Malangone, C., McMahon, A., Morales, J., Mountjoy, E., Sollis, E., et al. (2019). The NHGRI-EBI GWAS catalog of published genome-wide association studies, targeted arrays and summary statistics 2019. *Nucleic Acids Res.* 47, D1005–D1012. doi:10.1093/nar/gky1120.

Buß, O., Rudat, J., and Ochsenreither, K. (2018). Foldx as protein engineering tool: better than random based approaches? *Comput. Struct. Biotechnol. J.* 16, 25–33. doi:10.1016/j.csbj.2018.01.002.

Cariaso, M., and Lennon, G. (2012). SNPedia: a wiki supporting personal genome annotation, interpretation and analysis. *Nucleic Acids Res.* 40, D1308–D1312. doi:10.1093/nar/gkr798.

Carr, D. F., Turner, R. M., and Pirmohamed, M. (2021). Pharmacogenomics of anticancer drugs: personalising the choice and dose to manage drug response. *Br. J. Clin. Pharmacol.* 87, 237–255. doi:10.1111/bcp.14407.

De Baets, G., Van Durme, J., Reumers, J., Maurer-Stroh, S., Vanhee, P., Dopazo, J., Schymkowitz, J., and Rousseau, F. (2012). SNPeffect 4.0: on-line prediction of molecular and structural effects of protein-coding variants. *Nucleic Acids Res.* 40, D935–D939. doi:10.1093/nar/gkr996.

DePriest, A. D., Fiandalo, M. V., Schlanger, S., Heemers, F., Mohler, J. L., Liu, S., and Heemers, H. V. (2016). Regulators of Androgen Action Resource: a one-stop shop for the comprehensive study of androgen receptor action. *Database (Oxford)* 2016. doi:10.1093/database/bav125.

Duan, X., Schmidt, E., Li, P., Lenox, D., Liu, L., Shu, C., Zhang, J., and Liang, C. (2012). PeanutDB: an integrated bioinformatics web portal for Arachis hypogaea transcriptomics. *BMC Plant Biol.* 12, 94. doi:10.1186/1471-2229-12-94.

Duran, C., Appleby, N., Clark, T., Wood, D., Imelfort, M., Batley, J., and Edwards, D. (2009). AutoSNPdb: an annotated single nucleotide polymorphism database for crop plants. *Nucleic Acids Res.* 37, D951–D953. doi:10.1093/nar/gkn650.

Ezzeddini, R., Somi, M. H., Taghikhani, M., Moaddab, S.-Y., Masnadi Shirazi, K., Shirmohammadi, M., Eftekharsadat, A. T., Sadighi Moghaddam, B., and Salek Farrokhi, A. (2021). Association of Foxp3 rs3761548 polymorphism with cytokines concentration in gastric adenocarcinoma patients. *Cytokine* 138, 155351. doi:10.1016/j.cyto.2020.155351.

Fejerman, L., and Ziv, E. (2008). Population differences in breast cancer severity. *Pharmacogenomics* 9, 323–333. doi:10.2217/14622416.9.3.323.

Gao, Y., Li, X., Shang, S., Guo, S., Wang, P., Sun, D., Gan, J., Sun, J., Zhang, Y., Wang, J., et al. (2021). LincSNP 3.0: an updated database for linking functional variants to human long non-coding RNAs, circular RNAs and their regulatory elements. *Nucleic Acids Res.* 49, D1244–D1250. doi:10.1093/nar/gkaa1037.

Gong, J., Liu, W., Zhang, J., Miao, X., and Guo, A.-Y. (2015). lncRNASNP: a database of SNPs in lncRNAs and their potential functions in human and mouse. *Nucleic Acids Res.* 43, D181–D186. doi:10.1093/nar/gku1000.

Greshake, B., Bayer, P. E., Rausch, H., and Reda, J. (2014). openSNP--a crowdsourced web resource for personal genomics. *PLoS ONE* 9, e89204. doi:10.1371/journal.pone.0089204.

Guo, L., and Wang, J. (2018). rSNPBase 3.0: an updated database of SNP-related regulatory elements, element-gene pairs and SNP-based gene regulatory networks. *Nucleic Acids Res.* 46, D1111–D1116. doi:10.1093/nar/gkx1101.

Han, J. H., Lee, Y. S., Kim, H. J., Lee, S. Y., and Myung, S. C. (2015). Association between cytochrome CYP17A1, CYP3A4, and CYP3A43 polymorphisms and prostate cancer risk and aggressiveness in a Korean study population. *Asian J. Androl.* 17, 285–291. doi:10.4103/1008-682X.133320.

Hemani, G., Zheng, J., Elsworth, B., Wade, K. H., Haberland, V., Baird, D., Laurin, C., Burgess, S., Bowden, J., Langdon, R., et al. (2018). The MR-Base platform supports systematic causal inference across the human phenome. *eLife* 7. doi:10.7554/eLife.34408.

Hiard, S., Charlier, C., Coppieters, W., Georges, M., and Baurain, D. (2010). Patrocles: a database of polymorphic miRNA-mediated gene regulation in vertebrates. *Nucleic Acids Res.* 38, D640–D651. doi:10.1093/nar/gkp926.

Holmes, J. B., Moyer, E., Phan, L., Maglott, D., and Kattman, B. (2020). SPDI: data model for variants and applications at NCBI. *Bioinformatics* 36, 1902–1907. doi:10.1093/bioinformatics/btz856.

Hong, D., Park, S.-S., Ju, Y. S., Kim, S., Shin, J.-Y., Kim, S., Yu, S.-B., Lee, W.-C., Lee, S., Park, H., et al. (2011). TIARA: a database for accurate analysis of multiple personal genomes based on cross-technology. *Nucleic Acids Res.* 39, D883–D888. doi:10.1093/nar/gkq1101.

Hutter, C., and Zenklusen, J. C. (2018). The cancer genome atlas: creating lasting value beyond its data. *Cell* 173, 283–285. doi:10.1016/j.cell.2018.03.042.

Ingham, S. L., Warwick, J., Byers, H., Lalloo, F., Newman, W. G., and Evans, D. G. R. (2013). Is multiple SNP testing in BRCA2 and BRCA1 female carriers ready for use in clinical practice? Results from a large Genetic Centre in the UK. *Clin. Genet.* 84, 37–42. doi:10.1111/cge.12035.

International HapMap Consortium (2003). The international hapmap project. *Nature* 426, 789–796. doi:10.1038/nature02168.

Ji, G., Long, Y., Zhou, Y., Huang, C., Gu, A., and Wang, X. (2012). Common variants in mismatch repair genes associated with increased risk of sperm DNA damage and male infertility. *BMC Med.* 10, 49. doi:10.1186/1741-7015-10-49.

Johnson, A. D. (2009). Single-nucleotide polymorphism bioinformatics: a comprehensive review of resources. *Circ. Cardiovasc. Genet.* 2, 530–536. doi:10.1161/CIRCGENETICS.109.872010.

Krynetski, E., and Evans, W. E. (2003). Drug methylation in cancer therapy: lessons from the TPMT polymorphism. *Oncogene* 22, 7403–7413. doi:10.1038/sj.onc.1206944.

Kumar, S., Ambrosini, G., and Bucher, P. (2017). SNP2TFBS - a database of regulatory SNPs affecting predicted transcription factor binding site affinity. *Nucleic Acids Res.* 45, D139–D144. doi:10.1093/nar/gkw1064.

Laing, R. E., Hess, P., Shen, Y., Wang, J., and Hu, S. X. (2011). The role and impact of SNPs in pharmacogenomics and personalized medicine. *Curr. Drug Metab.* 12, 460–486. doi:10.2174/138920011795495268.

Landowski, M., Kelly, U., Klingeborn, M., Groelle, M., Ding, J.-D., Grigsby, D., and Bowes Rickman, C. (2019). Human complement factor H Y402H polymorphism causes an age-related macular degeneration phenotype and lipoprotein dysregulation in mice. *Proc. Natl. Acad. Sci. USA* 116, 3703–3711. doi:10.1073/pnas.1814014116.

Landrum, M. J., Chitipiralla, S., Brown, G. R., Chen, C., Gu, B., Hart, J., Hoffman, D., Jang, W., Kaur, K., Liu, C., et al. (2020). ClinVar: improvements to accessing data. *Nucleic Acids Res.* 48, D835–D844. doi:10.1093/nar/gkz972.

Landrum, M. J., Lee, J. M., Benson, M., Brown, G. R., Chao, C., Chitipiralla, S., Gu, B., Hart, J., Hoffman, D., Jang, W., et al. (2018). ClinVar: improving access to variant interpretations and supporting evidence. *Nucleic Acids Res.* 46, D1062–D1067. doi:10.1093/nar/gkx1153.

Lappalainen, I., Lopez, J., Skipper, L., Hefferon, T., Spalding, J. D., Garner, J., Chen, C., Maguire, M., Corbett, M., Zhou, G., et al. (2013). DbVar and DGVa: public archives for genomic structural variation. *Nucleic Acids Res.* 41, D936–D941. doi:10.1093/nar/gks1213.

Lee, B. K. B., Gan, C. P., Chang, J. K., Tan, J. L., Fadlullah, M. Z., Abdul Rahman, Z. A., Prime, S. S., Gutkind, J. S., Liew, C. S., Khang, T. F., et al. (2018). GENIPAC: a genomic information portal for head and neck cancer cell systems. *J. Dent. Res.* 97, 909–916. doi:10.1177/0022034518759038.

Leslie, R., O'Donnell, C. J., and Johnson, A. D. (2014). GRASP: analysis of genotype-phenotype results from 1390 genome-wide association studies and corresponding open access database. *Bioinformatics* 30, i185–194. doi:10.1093/bioinformatics/btu273.

Li, C., Tian, D., Tang, B., Liu, X., Teng, X., Zhao, W., Zhang, Z., and Song, S. (2021). Genome Variation Map: a worldwide collection of genome variations across multiple species. *Nucleic Acids Res.* 49, D1186–D1191. doi:10.1093/nar/gkaa1005.

Li, J., Xue, Y., Amin, M. T., Yang, Y., Yang, J., Zhang, W., Yang, W., Niu, X., Zhang, H.-Y., and Gong, J. (2020). ncRNA-eQTL: a database to systematically evaluate the effects of SNPs on non-coding RNA expression across cancer types. *Nucleic Acids Res.* 48, D956–D963. doi:10.1093/nar/gkz711.

Li, X., Shi, L., Wang, Y., Zhong, J., Zhao, X., Teng, H., Shi, X., Yang, H., Ruan, S., Li, M., et al. (2019). OncoBase: a platform for decoding regulatory somatic mutations in human cancers. *Nucleic Acids Res.* 47, D1044–D1055. doi:10.1093/nar/gky1139.

Liu, C., Zhang, F., Li, T., Lu, M., Wang, L., Yue, W., and Zhang, D. (2012). MirSNP, a database of polymorphisms altering miRNA target sites, identifies miRNA-related SNPs in GWAS SNPs and eQTLs. *BMC Genomics* 13, 661. doi:10.1186/1471-2164-13-661.

Liu, C.-J., Fu, X., Xia, M., Zhang, Q., Gu, Z., and Guo, A.-Y. (2021). miRNASNP-v3: a comprehensive database for SNPs and disease-related variations in miRNAs and miRNA targets. *Nucleic Acids Res.* 49, D1276–D1281. doi:10.1093/nar/gkaa783.

Liu, X., Li, C., Mou, C., Dong, Y., and Tu, Y. (2020). dbNSFP v4: a comprehensive database of transcript-specific functional predictions and annotations for human nonsynonymous and splice-site SNVs. *Genome Med.* 12, 103. doi:10.1186/s13073-020-00803-9.

Lu, Y., Quan, C., Chen, H., Bo, X., and Zhang, C. (2017). 3DSNP: a database for linking human noncoding SNPs to their three-dimensional interacting genes. *Nucleic Acids Res.* 45, D643–D649. doi:10.1093/nar/gkw1022.

Luo, H., Zhao, W., Wang, Y., Xia, Y., Wu, X., Zhang, L., Tang, B., Zhu, J., Fang, L., Du, Z., et al. (2016). SorGSD: a sorghum genome SNP database. *Biotechnol. Biofuels* 9, 6. doi:10.1186/s13068-015-0415-8.

MacDonald, J. R., Ziman, R., Yuen, R. K. C., Feuk, L., and Scherer, S. W. (2014). The Database of Genomic Variants: a curated collection of structural variation in the human genome. *Nucleic Acids Res.* 42, D986–D992. doi:10.1093/nar/gkt958.

McLaren, W., Gil, L., Hunt, S. E., Riat, H. S., Ritchie, G. R. S., Thormann, A., Flicek, P., and Cunningham, F. (2016). The ensembl variant effect predictor. *Genome Biol.* 17, 122. doi:10.1186/s13059-016-0974-4.

Mi, H., Muruganujan, A., Ebert, D., Huang, X., and Thomas, P. D. (2019). PANTHER version 14: more genomes, a new PANTHER GO-slim and improvements in enrichment analysis tools. *Nucleic Acids Res.* 47, D419–D426. doi:10.1093/nar/gky1038.

Oscanoa, J., Sivapalan, L., Gadaleta, E., Dayem Ullah, A. Z., Lemoine, N. R., and Chelala, C. (2020). SNPnexus: a web server for functional annotation of human genome sequence variation (2020 update). *Nucleic Acids Res.* 48, W185–W192. doi:10.1093/nar/gkaa420.

Pan, Q., Liu, Y.-J., Bai, X.-F., Han, X.-L., Jiang, Y., Ai, B., Shi, S.-S., Wang, F., Xu, M.-C., Wang, Y.-Z., et al. (2021). VARAdb: a comprehensive variation annotation database for human. *Nucleic Acids Res.* 49, D1431–D1444. doi:10.1093/nar/gkaa922.

Phillips, C. (2009). SNP databases. *Methods Mol. Biol.* 578, 43–71. doi:10.1007/978-1-60327-411-1_3.

Pradhan, S., Das, S., Singh, A. K., Das, C., Basu, A., Majumder, P. P., and Biswas, N. K. (2021). dbGENVOC: database of GENomic Variants of Oral Cancer, with special reference to India. *Database (Oxford)* 2021. doi:10.1093/database/baab034.

Quan, C., Ping, J., Lu, H., Zhou, G., and Lu, Y. (2021). 3DSNP 2.0: update and expansion of the noncoding genomic variant annotation database. *Nucleic Acids Res.* doi:10.1093/nar/gkab1008.

Reese, M. G., Moore, B., Batchelor, C., Salas, F., Cunningham, F., Marth, G. T., Stein, L., Flicek, P., Yandell, M., and Eilbeck, K. (2010). A standard variation file format for human genome sequences. *Genome Biol.* 11, R88. doi:10.1186/gb-2010-11-8-r88.

Schaafsma, G. C. P., and Vihinen, M. (2015). VariSNP, a benchmark database for variations from dbSNP. *Hum. Mutat.* 36, 161–166. doi:10.1002/humu.22727.

Shashkova, T. I., Pakhomov, E. D., Gorev, D. D., Karssen, L. C., Joshi, P. K., and Aulchenko, Y. S. (2021). PheLiGe: an interactive database of billions of human genotype-phenotype associations. *Nucleic Acids Res.* 49, D1347–D1350. doi:10.1093/nar/gkaa1086.

Shavers, V. L., and Brown, M. L. (2002). Racial and ethnic disparities in the receipt of cancer treatment. *J. Natl. Cancer Inst.* 94, 334–357. doi:10.1093/jnci/94.5.334.

Sherry, S. T., Ward, M. H., Kholodov, M., Baker, J., Phan, L., Smigielski, E. M., and Sirotkin, K. (2001). dbSNP: the NCBI database of genetic variation. *Nucleic Acids Res.* 29, 308–311. doi:10.1093/nar/29.1.308.

Shin, S., Hudson, R., Harrison, C., Craven, M., and Keleş, S. (2019). atSNP Search: a web resource for statistically evaluating influence of human genetic variation on transcription factor binding. *Bioinformatics* 35, 2657–2659. doi:10.1093/bioinformatics/bty1010.

Shirasawa, K., Isobe, S., Tabata, S., and Hirakawa, H. (2014). Kazusa Marker DataBase: a database for genomics, genetics, and molecular breeding in plants. *Breed. Sci.* 64, 264–271. doi:10.1270/jsbbs.64.264.

Sinkov, V., Ogarkov, O., Mokrousov, I., Bukin, Y., Zhdanova, S., and Heysell, S. K. (2018). New epidemic cluster of pre-extensively drug resistant isolates of Mycobacterium tuberculosis Ural family emerging in Eastern Europe. *BMC Genomics* 19, 762. doi:10.1186/s12864-018-5162-3.

Stitziel, N. O., Binkowski, T. A., Tseng, Y. Y., Kasif, S., and Liang, J. (2004). topoSNP: a topographic database of non-synonymous single nucleotide polymorphisms with and without known disease association. *Nucleic Acids Res.* 32, D520–D522. doi:10.1093/nar/gkh104.

Tate, J. G., Bamford, S., Jubb, H. C., Sondka, Z., Beare, D. M., Bindal, N., Boutselakis, H., Cole, C. G., Creatore, C., Dawson, E., et al. (2019). COSMIC: the catalogue of somatic mutations in cancer. *Nucleic Acids Res.* 47, D941–D947. doi:10.1093/nar/gky1015.

Tian, J., Wang, Z., Mei, S., Yang, N., Yang, Y., Ke, J., Zhu, Y., Gong, Y., Zou, D., Peng, X., et al. (2019). CancerSplicingQTL: a database for genome-wide identification of splicing QTLs in human cancer. *Nucleic Acids Res.* 47, D909–D916. doi:10.1093/nar/gky954.

Tryka, K. A., Hao, L., Sturcke, A., Jin, Y., Wang, Z. Y., Ziyabari, L., Lee, M., Popova, N., Sharopova, N., Kimura, M., et al. (2014). NCBI's Database of genotypes and phenotypes: dbGaP. *Nucleic Acids Res.* 42, D975–D979. doi:10.1093/nar/gkt1211.

Tuteja, G., Cheng, E., Papadakis, H., and Bejerano, G. (2012). PESNPdb: a comprehensive database of SNPs studied in association with pre-eclampsia. *Placenta* 33, 1055–1057. doi:10.1016/j.placenta.2012.09.016.

Wang, J., Dai, X., Berry, L. D., Cogan, J. D., Liu, Q., and Shyr, Y. (2019). HACER: an atlas of human active enhancers to interpret regulatory variants. *Nucleic Acids Res.* 47, D106–D112. doi:10.1093/nar/gky864.

Watanabe, K., Stringer, S., Frei, O., Umićević Mirkov, M., de Leeuw, C., Polderman, T. J. C., van der Sluis, S., Andreassen, O. A., Neale, B. M., and Posthuma, D. (2019). A global overview of pleiotropy and genetic architecture in complex traits. *Nat. Genet.* 51, 1339–1348. doi:10.1038/s41588-019-0481-0.

Willuweit, S., Roewer, L., and International Forensic Y Chromosome User Group (2007). Y chromosome haplotype reference database (YHRD): update. *Forensic Sci. Int. Genet.* 1, 83–87. doi:10.1016/j.fsigen.2007.01.017.

Wishart, D. S., Feunang, Y. D., Marcu, A., Guo, A. C., Liang, K., Vázquez-Fresno, R., Sajed, T., Johnson, D., Li, C., Karu, N., et al. (2018). HMDB 4.0: the human metabolome database for 2018. *Nucleic Acids Res.* 46, D608–D617. doi:10.1093/nar/gkx1089.

Wong, K. M., Langlais, K., Tobias, G. S., Fletcher-Hoppe, C., Krasnewich, D., Leeds, H. S., Rodriguez, L. L., Godynskiy, G., Schneider, V. A., Ramos, E. M., et al. (2017). The dbGaP data browser: a new tool for browsing dbGaP controlled-access genomic data. *Nucleic Acids Res.* 45, D819–D826. doi:10.1093/nar/gkw1139.

Xuan, J.-J., Sun, W.-J., Lin, P.-H., Zhou, K.-R., Liu, S., Zheng, L.-L., Qu, L.-H., and Yang, J.-H. (2018). RMBase v2.0: deciphering the map of RNA modifications from epitranscriptome sequencing data. *Nucleic Acids Res.* 46, D327–D334. doi:10.1093/nar/gkx934.

Yang, W.-J., Wang, M.-Y., Pan, F.-Z., Shi, C., and Cen, H. (2017). Association between MPO-463G>A polymorphism and cancer risk: evidence from 60 case-control studies. *World J. Surg. Oncol.* 15, 144. doi:10.1186/s12957-017-1183-7.

Yang, Y., Zhang, Q., Miao, Y.-R., Yang, J., Yang, W., Yu, F., Wang, D., Guo, A.-Y., and Gong, J. (2020). SNP2APA: a database for evaluating effects of genetic variants on alternative polyadenylation in human cancers. *Nucleic Acids Res.* 48, D226–D232. doi:10.1093/nar/gkz793.

Yates, A., Akanni, W., Amode, M. R., Barrell, D., Billis, K., Carvalho-Silva, D., Cummins, C., Clapham, P., Fitzgerald, S., Gil, L., et al. (2016). Ensembl 2016. *Nucleic Acids Res.* 44, D710–D716. doi:10.1093/nar/gkv1157.

You, J. S., and Jones, P. A. (2012). Cancer genetics and epigenetics: two sides of the same coin? *Cancer Cell* 22, 9–20. doi:10.1016/j.ccr.2012.06.008.

Yue, M., Zhou, D., Zhi, H., Wang, P., Zhang, Y., Gao, Y., Guo, M., Li, X., Wang, Y., Zhang, Y., et al. (2018). MSDD: a manually curated database of experimentally supported associations among miRNAs, SNPs and human diseases. *Nucleic Acids Res.* 46, D181–D185. doi:10.1093/nar/gkx1035.

Zhang, J., Baran, J., Cros, A., Guberman, J. M., Haider, S., Hsu, J., Liang, Y., Rivkin, E., Wang, J., Whitty, B., et al. (2011). International Cancer Genome Consortium Data Portal--a one-stop shop for cancer genomics data. *Database (Oxford)* 2011, bar026. doi:10.1093/database/bar026.

Zupa, A., Sgambato, A., Bianchino, G., Improta, G., Grieco, V., LA Torre, G., Campisi, B., Traficante, A., Aieta, M., and Cittadini, A. (2009). GSTM1 and NAT2 polymorphisms and colon, lung and bladder cancer risk: a case-control study. *Anticancer Res.* 29, 1709–1714.

13

Single Nucleotide Polymorphism-Related Computational Resources Part II: Algorithms

Manmeet Kaur
Department of Biotechnology, Guru Nanak Dev University, Amritsar, Punjab, India

Pallawi Kumari
Department of Bioscience & Biotechnology, Banasthali Vidyapith University, Tonk, Rajasthan, India

Nadiem Nazir Bhat
Sher-e-Kashmir University of Agricultural Sciences & Technology of Kashmir
Shuhama Alustang, Kashmir, Jammu and Kashmir, India

Showkat Ahmad Dar
Bioinformatics Center, CSIR Institute of Microbial Technology, Chandigarh, India

CONTENTS

DOI: 10.1201/9781003246244-13

ABSTRACT The last twenty years have seen a high intervention of precision genomics that has led to the emergence of personalized genomics that aims for genetic feature-tailored health care. New investigative methodologies such as genome-wide association studies (GWAS) have led to a large-scale contribution to population genomic data of complex diseases. Similarly, in the predictive bioinformatics arena, every day new technologies and algorithms are being developed to make better and more accurate prediction models. Machine learning and pattern discovery algorithms applied to genetic data along with polygenic risk scoring have become methods of choice to design and code novel software for the detection of SNPs. Such algorithms have become the next big thing in predictive multidimensional data analysis in genomics. This chapter aims to review the most common predictors and web resources that can be used for genetic data-based SNP prediction, especially related to humans and diseases. We sum up and provide links for a range of web servers, predictors and applications in SNP prediction. The tools provided herein would act as a beneficial asset for the researchers targeting the domain of SNP biology and predictive exploratory genomic data analysis of nucleotide variations.

KEYWORDS: *Single nucleotide polymorphism (SNP); Computation; Bioinformatics; Web servers; Resources; Prediction algorithms; Machine learning; Statistical models.*

13.1 Introduction

Genomic data continue to grow exponentially, so do the requirements for new technologies to understand their functional aspects. In the current scenario, informatics-based, in silico, cost-effective data mining methods for single nucleotide polymorphism (SNP) discovery and analysis have gained global prominence. A comprehensive selection of tools and algorithms has been created to characterize and interpret the functional and molecular basis of these SNPs. Such tools and algorithms mainly focus on areas such as annotation of functional variation and predicting the functional effect of such mutations. This chapter is a compendium of the bioinformatics resources mainly used for predicting SNPs, along with their functional and structural relationship. The tools enlisted may help to provide phenotypic characteristics such as pathogenesis, gene expression, drug response, and disease correlations to be some areas associated with the SNPs. Combined approaches using bioinformatics-based predictions and wet lab validations are usually employed to detect informative SNP typing. Currently, because of advances in computational bioinformatics, a myriad of methods and tools have been created to investigate the interrelations between functional and structural characteristics of SNPs. Phenotypic effects because of SNPs show a wide domain of functional variations in protein features such as their dynamics, stability, and interactions. In silico tools for SNP prediction mainly employ training methodologies such as machine learning and artificial intelligence on genomic data followed by experimental cross-validations of predictions in laboratories. The combined approach of predictions and experimental validations is to enhance the reliability, precision, and reproducibility of the data generated (Ho et al., 2019).

13.2 Bioinformatics Approaches Used in Predicting SNPs

This chapter introduces online tools and packages for understanding the molecular and mutational aspects of SNP biology. Significant efforts have earlier been made in this direction, to understand the genetic drifts leading to variable phenotypic changes and diseases. These efforts have resulted in the development of a wide variety of web resources, tools, and databases to hypothesize about the molecular source and genetic basis of phenotypes. In order to identify functional sites, multiple databases for mutations related to SNP have been created that offer a range of genomic annotations. The underlying tools are often deployed with a web interface available online or as stand-alone software that can be installed locally. Let's go through them serially.

13.2.1 The UCSC Genome Browser

The UCSC Genome Browser is an online web resource that allows browsable presentation of selected sections of genomic sequences along with several parallel annotation features, for instance mRNAs associated, CpG islands, ESTs, predicted genes, known genes, assembly gaps, homologies, chromosomal bands, and assembly coverage (Figure 13.1). Each item hyperlinks to other pages redirecting to the links for repositories, such as GenBank, OMIM, Entrez, and PubMed. This browser is available at the URL: https://genome.ucsc.edu/

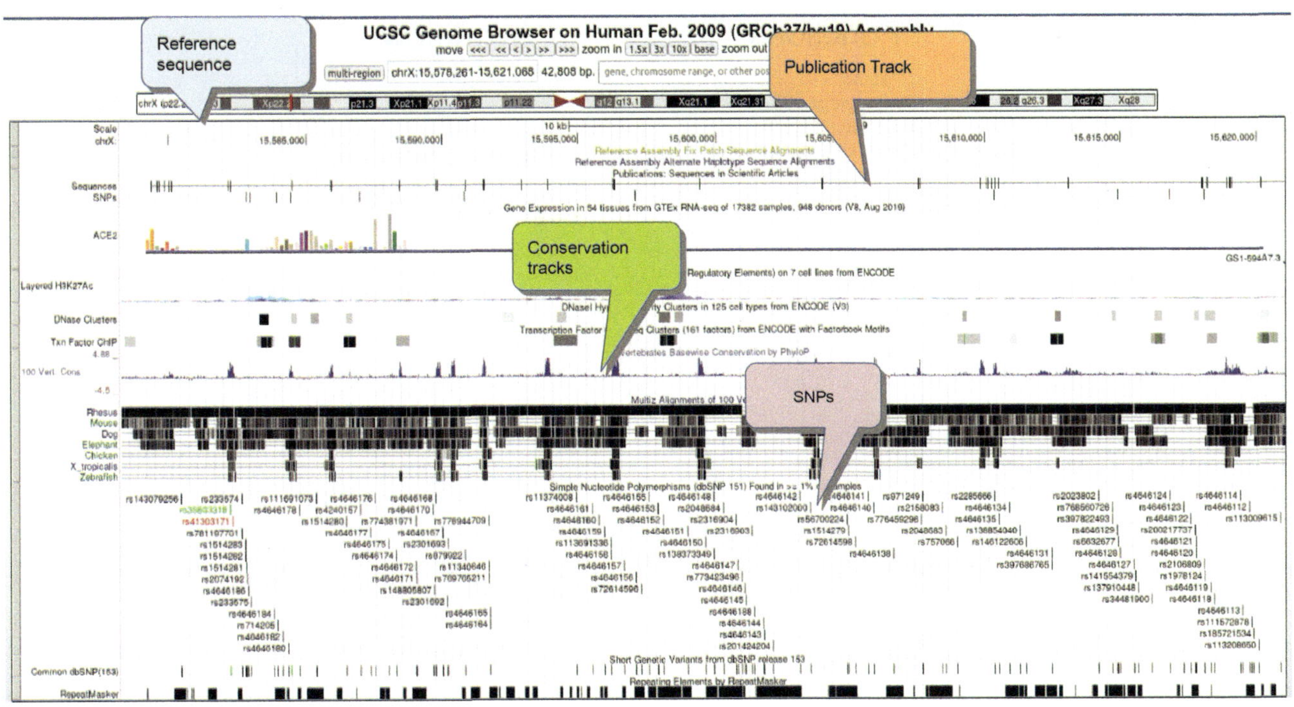

FIGURE 13.1 This figure shows the UCSC Genome Browser arranging and presenting the genomic information or reference genome sequence layout additional information as "annotation tracks" in the proper genomic position and providing circumstances to understand any genomic region. In this track (GRCh37/hg19), it shows human genome assembly. SNPs are shown with their rs number. Red: coding, non-synonymous; green: coding, synonymous; blue: untranslated regions; black: intronic regions. (Adopted from UCSC Genome Browser.)

(Mangan et al., 2014). The University of California Santa Cruz (UCSC) Cancer Genomics Browser (http://genome-cancer.ucsc.edu/) was created for extending and complementing the former by accelerating combined, collective, and flexible display, for exhaustive studies of cancer genetics and clinical data. The browser aims to visualize genome-wide pathway-directed experiment assessments for individuals and groups as heat maps. Clinical characteristics are coded mathematically and are exhibited in the distinct heat maps. Researchers can command, filter, collect, and visualize the data by chromosomal coordinates, data values, clinical features, user-defined gene-sets, or curated biological pathways. Statistical studies can be carried out and visualized via graphical representation through the browser. To preserve patients' confidentiality and restrict entry to unpublished information, a secure entry feature is provided as well. Researchers can use the web browser or use it as stand-alone software for local data visualization. The Cancer Genomics Browser, as an addition to the Genome Browser, intends to efficiently view and interpret cancer genomic and clinical data, offering a platform for the scientific community to display this information in an integrated system (Zhu et al., 2009).

13.2.2 GW-SEM

Genome-Wide Structural Equation Modeling (GW-SEM) is a computationally flexible and accessible algorithm designed to simplify the selection of models associated with multivariate genome-wide analysis studies. The latest version of GW-SEM 2.0 can be found on CRAN, and a GitHub link is provided here: https://jpritikin.github.io/gwsem/. GW-SEM enables users to correctly model the perfect function that is suitable for the analysis. GWAS are used to test the association in order to calculate the genetic correlations between SNPs and a single trait. GW-SEM provides a considerably deeper understanding of the genomic architecture for multiple GWAS methods. Its goal is to enable users to analyze the biomarkers, interconnected array of risk factors, comorbid disorders, and environmental antecedents, on genome-wide data by implementing structural equation modeling (Pritikin et al., 2021).

13.2.3 TransAT

TransAT is a versatile R language-based application available at the URL: https://github.com/ShihChingYu/TransAT. This tool provides genomic annotations for query transcripts, especially beneficial for physicians for personalized drugs (Figure 13.2). The tool takes transcript IDs and provides genomic positions with SNP ID and an SNP allele frequency from worldwide accessible population databases. The results generated from the TransAT can be easily represented understandably and can be downloaded easily (Shih et al., 2021).

13.2.4 GADGETS

GADGETS is a genetic algorithm-based tool and can be found at https://github.com/mnodzenski/epistasisGA. The tool is used to discover epistasis in SNP-sets related to case-parent or case-sibling studies. It analyzes the known SNPs datasets by evaluating the permutations and providing graphical representations.

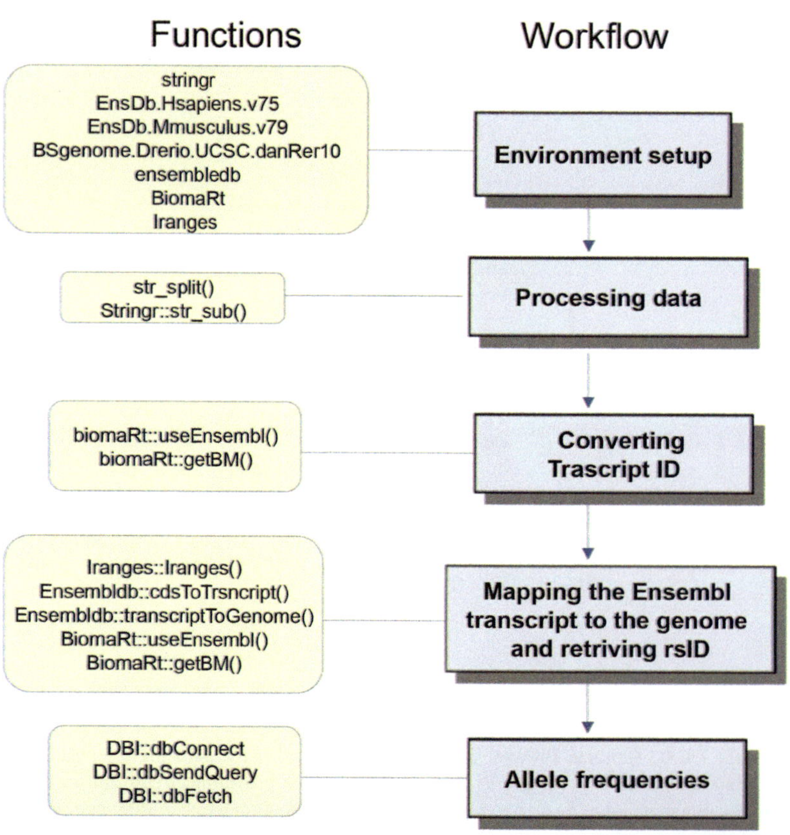

FIGURE 13.2 Representation of Transcript Annotation Tool (TransAT): steps and functions list. (This image has been adapted from Shih et al. (2021).)

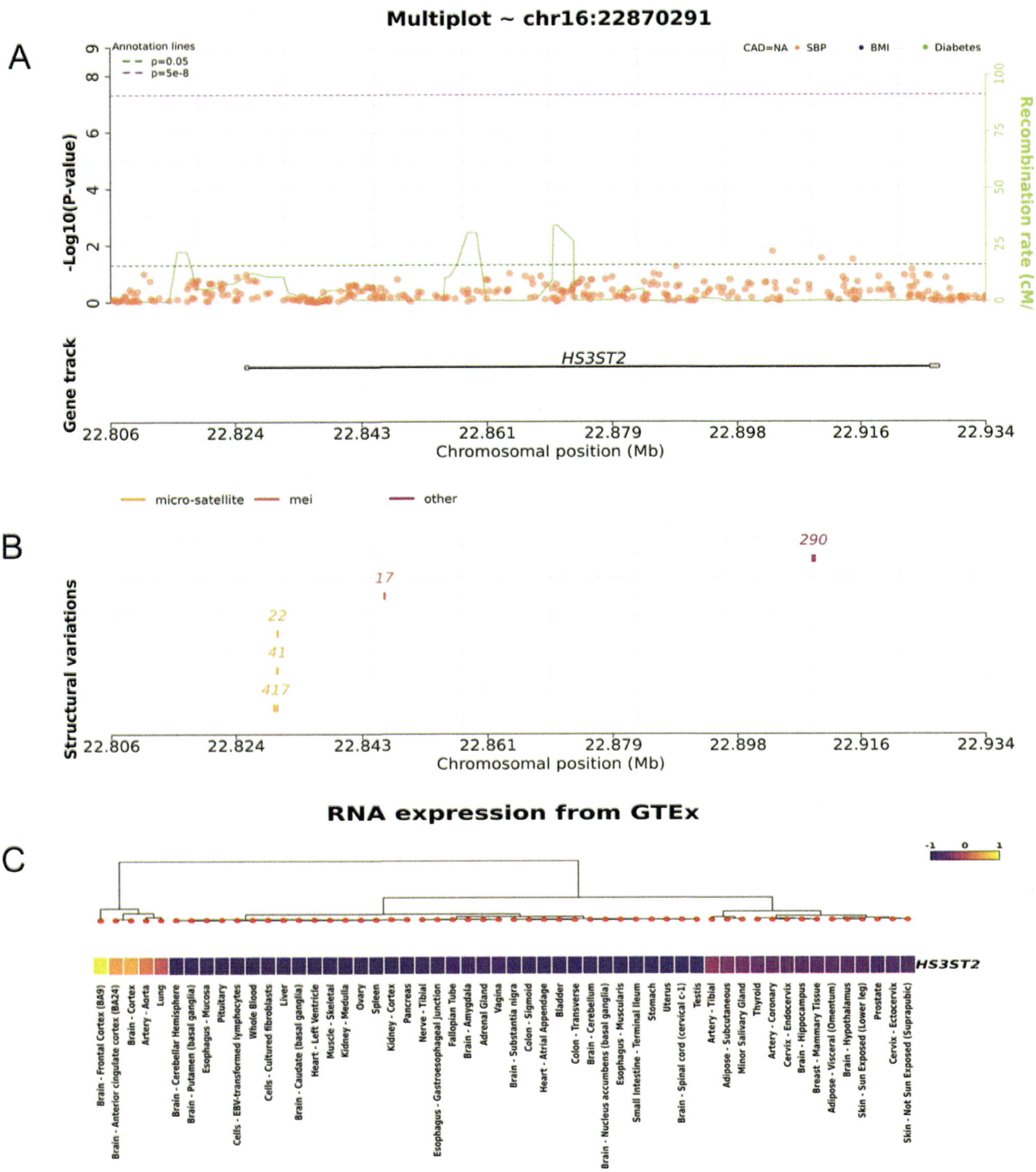

FIGURE 13.3 Some of the SnpXplorer visualization results representation. (a) Visual representation of interface with a statistical summary of several genetic studies with the p-value. (b) Structural variations within a colored segment. (c) Tissue-specific gene RNA expression.

GADGETS could correctly identify epistatic SNP-sets in realistically simulated case-parent triads (Nodzenski et al., 2021).

13.2.5 snpXplorer

The snpXplorer is a convenient web-based visualization tool used to explore functional SNPs and associated statistics (Figure 13.3). Its web server is available at the URL: https://snpxplorer.net, and its source code can be found at GitHub page https://github.com/TesiNicco/snpXplorer. The snpXplorer superimposes association statistics from several studies and shows information such as structural variation, SNP association, eQTL, recombination rates, gene expression, and linkage disequilibrium (LD) for a particular genomic region. The application can also interpret variant consequences by comparing associations among different traits from different GWAS. The tool can also perform gene-set enrichment analysis and variant to gene mapping from the list of SNPs, which can help further in identifying overrepresented molecular pathways (Tesi et al., 2021).

13.2.6 EpiGEN

EpiGEN is a freely available Python 3-based pipeline for epistasis simulation. It is available on GitHub at the URL: https://github.com/baumbachlab/epigen. It can generate LD patterns of SNP interactions as both quantitative and categorical phenotypes. Epistasis models can be specified in a risk model with parameters or with the relevant descriptions. The tool can also be used for observation bias simulation by selecting SNPs based on their minor allele frequencies. For GWAS, analysis simulators have a specific role in evaluating epistasis data (Blumenthal et al., 2020).

13.2.7 HiGwas

HiGwas is an R-based package used for the analysis of longitudinal GWAS data with the implementation of Lasso method for the identification of significant SNPs and calculating combined genetic effect. The source code for the package can be found at the URL: https://github.com/wzhy2000/higwas. The main areas of analysis included in the package could do single SNP analyses, pre-processing, and modeling for a high-dimensional set of SNPs. The tool has options for flexible significance-level adjustments. HiGwas also has functions for genetic and confidence interval estimation for these parameters (Wang et al., 2020b).

13.2.8 SEMpl

SEMpl is an open-source tool used to evaluate the association of SNPs with transcription factor-binding sites (TFBSs), observed from chromatin immunoprecipitation deep sequencing signals. The tool can be found at https://github.com/Boyle-Lab/SEM_CPP. The tool can predict the consequence of SNP on TF binding by classifying and ranking each potential alteration of the transcription factor-binding motif. Further, this information can discover probable disease-causing regulatory loci (Nishizaki et al., 2020).

13.2.9 SnpHub

SnpHub is a Shiny/R-based server configuration that can retrieve, analyze, and visualize large-scale genomic alteration data. Its source code can be found at the URL: http://guoweilong.github.io/SnpHub/. The framework can be easily built on any Linux server locally. The tool provides users with the functionality to retrieve gene- or locus-specific SNPs and small indels along with their annotated data. Users could also use haplotype networks, phylogenetic trees, geographical maps, and heat maps for exploring and visualizing genomic variants and results. The interface is also equipped with functions for post-processing and visualizing the genotyped data. It was specially created for retrieving and analyzing massive amounts of genomic alteration data easily and quickly. The tool also supports functions for interactive haplotype or genomic diversity analysis for a specific genetic locus (Wang et al., 2020a).

13.2.10 GeDi

The "Generalised suffix array-based Direct SNV caller, or GeDi" can find somatic SNVs in paired tumor-control NGS datasets, without the need for referencing or mapping. The source code of the tool is at the URL: https://github.com/izaak-coleman/GeDi. This tool uses a suffix array-based divergent approach to identify SNVs without using reference mappings or reading mapping coordinates to discover SNPs. It features great sensitivity and can detect SNVs at allele frequencies (<1%). The tool can also aid in locating SNPs at complicated hypermutation variant loci sites with precision (Coleman et al., 2020).

13.2.11 CKAT

CKAT, is an acronym for Composite Kernel Association Test for Pharmacogenetics Studies, can perform comprehensive simulation at the gene-level analysis of whole-genome association studies. It is an R-based software, free to use and hosted on the URL: https://cran.r-project.org/web/packages/CKAT/. The tool has an application in pharmacogenetics (PGx) as it can test the genomic and SNP interaction effects for SNP-sets embedded within randomized clinical trials. Recently, personalized drug development strategies have necessitated the importance of detecting new genomic markers. The relationship among collective effects of several SNPs in genomic regions and their collective associated drug response might help in such novel discoveries. Existing approaches do not have enough capability to analyze the whole genome for pharmacogenomics (PGx) studies (Zhang et al., 2020).

13.2.12 ICGRM

ICGRM is used to evaluate the genomic relationship matrix (GRM) through genomic SNPs. It was developed in C/C++, available at the URL: https://github.com/mingfang618/CLGRM. The tool provided genomic prediction methods, widely employed for evaluating genetic parameters and disease risk predictions in humans. The tool uses genome-wide SNP distribution data rather than pedigree data. GRM evaluation is the main step in such genomic prediction studies. The tool features the use of less computational power for calculating the GRM in such studies. The algorithm divides genomic SNPs into multiple segments and calculates a statistical summary of each GRM with minimum RAM sizes and then further integrates these statistical summaries into one GRM for whole-genome SNPs. This splitting process of SNPs reduces the required memory size by 15 times in simulation and makes it feasible for the computers with less computational power to efficiently handle big datasets as well (Jiang et al., 2019).

13.2.13 CoMutPlotter

CoMutPlotter is a tool used to compare the custom and ICGC/TCGA data. The tool is located at http://tardis.cgu.edu.tw/comutplotter (Figure 13.4). It is a web-based tool that aims to

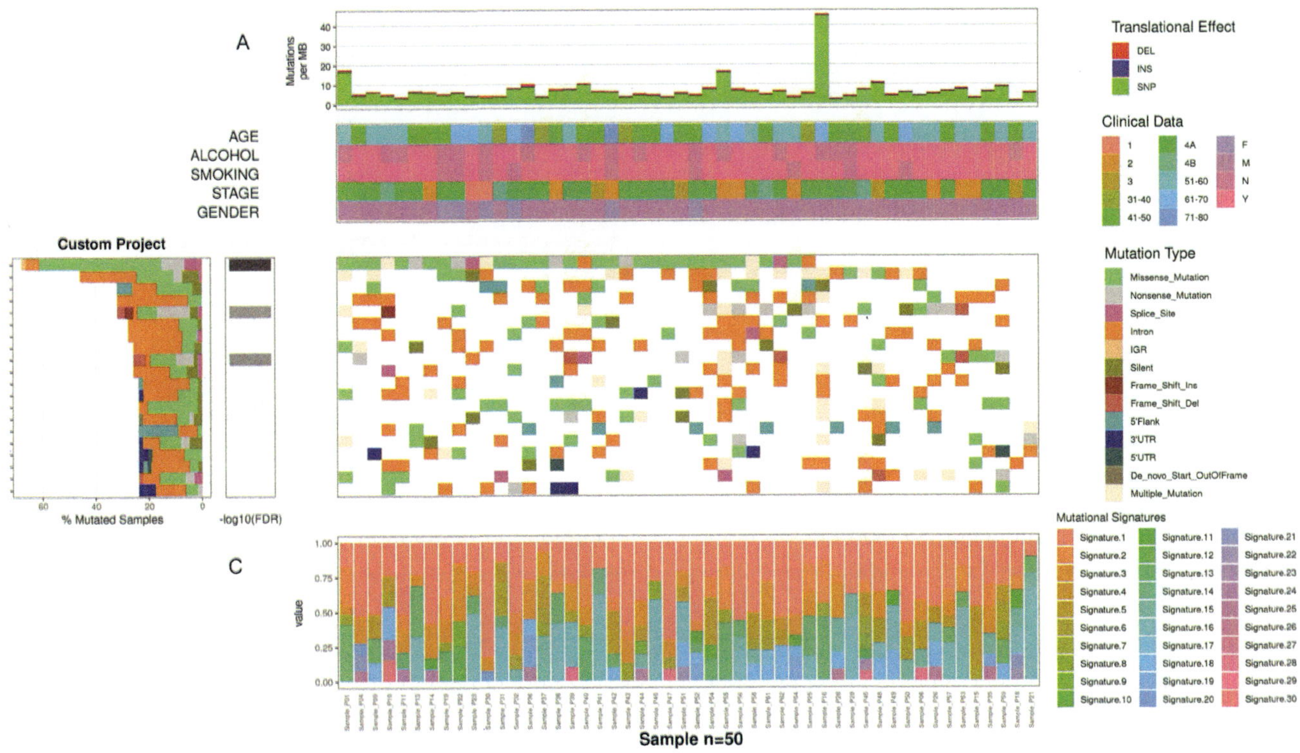

FIGURE 13.4 Representation of CoMutPlotter results: different panels showing (a) translational effect of clinical data, (b) mutation type, and (c) mutational signatures. (Adapted from CoMutPlotter web server.)

provide easy and automated graphical representations of the compared data. The tool breaks down hurdles between investigators and complicated cancer genomic data by offering accessibility to TCGA/ICGC data, along with unique cohort studies. This tool is primarily employed in cancer research and can also display a graphical representation of mutational regions in such groups. This tool could examine the frequency of mutation in genes and mutation in samples corroborating with their clinical descriptions, which acts as a preliminary step in studying the reappearance and coexistence of genetic variations. The tool features conversion of cancer mutational profiles into biological findings and therapeutic approaches and uses file formats such as MAF (Mutation Annotation Format), TSV, and VCF for the same. The peculiar feature of CoMutPlotter is its ability to compare mutation patterns between a customized cohort and the TCGA/ICGC study. The graphic overview from the tool also includes the inputs of COSMIC mutational profiles in selected samples, giving it a unique characteristic feature (Huang et al., 2019).

13.2.14 CERENKOV

CERENKOV is an acronym for Computational Elucidation of the REgulatory NonKOding Variome, and it is a computational machine-learning-based algorithm that can be used for detecting regulatory single nucleotide polymorphisms (rSNPs) within the noncoding genomes in GWAS. The tool can be found at https://github.com/ramseylab/cerenkov (Figure 13.5). The tool uses its own kind of CERENKOV data matrix to perform the analysis. It employs the XGBoost (Extreme Gradient Boosting) library to discover causative noncoding SNPs in locus recognized by GWAS. CERENKOV's accuracy of classification was tested with and without the new distance-based characteristics on the OSU18 SNP-set, and it was found that adding distance-based features enhances regulatory SNP identification evaluation as assessed by AVGRANK, AUPVR, and AUROC. Mainly, this software performs the extraction of the base feature matrix by computing likelihood ratio based on distance characteristics and evaluating putative SNPs responsible for disease occurrence (Yao et al., 2019).

13.2.15 gpart

The gpart is an R package used to discover LD structure in order to examine high-throughput genomic data generated by NGS technology. This package is available at https://bioconductor.org/packages/release/bioc/html/gpart.html and features functions for clustering algorithms that can be used for the analysis of SNP-based LD blocks. It can visualize gene position-based LD structures for up to 20,000 SNPs. The gpart package includes an imaging tool that displays the heat map associated with LD block, providing further insights into gene locations up to 20,000 SNPs. It can generate new SNP partition and LD blocks with computational efficiency and reasonable time. It employs a novel heuristic approach to increase the power and memory efficiency of building LD blocks (Kim et al., 2019).

FIGURE 13.5 Schematic representation of the Varanto web server, (a) with an input of test SNPs, with all variations, variation distance of 1,000 bp filters, with variation qualified impact as variation-annotation and cancer-gene neighborhood as gene-annotation function. (b) Heat map representation of SNPs and cancer-gene association and (c) karyotype/chromosome location of the SNPs. (Adapted from example dataset results from Varanto web server.)

13.2.16 CASMAP

Combinatorial association mapping or CASMAP is a tool used to determine the statistical relationship between high-order associations of genetic markers with their corresponding phenotypes. The tool can be found at https://github.com/BorgwardtLab/CASMAP. It is a software program that takes advantage of recent improvements in meaningful pattern mining to solve the computational and statistical constraints hindering the association mapping. The tool is employed to conduct association analysis specific to the area and find high-order epistatic associations of genomic variations. CASMAP enables the correction of categorical confounders such as gender or age, thus making it an ideal tool for GWAS-related studies (Llinares-López et al., 2019).

13.2.17 LD-annot

"LD-annot" is an automatic workflow that uses LD extent to identify the possible mutations in particular genomic areas of interest for a study. LD-annot provides identities, locations, genetic association, and gene annotation for every potential mutation in standard file formats and is located on the URL: https://github.com/ArnaudDroitLab/LD-annot. The tool focuses mainly on genotyping-by-sequencing GBS data and SNP arrays, as they are commonly used. However, it can also be used with any dataset that has genotypic information from a large number of variations and is structured in a VCF file. LD-annot analysis can be completed in little time, assisting research efforts aiming at identifying physiologically relevant genetic variants underlying phenotypic variance in a particular region/gene of interest (Prunier et al., 2019).

13.2.18 Multi-SNP Mediation Intersection-Union Test

SMUT or "multi-SNP mediation intersection-union test" package is designed in R language and is available for public use. The package is housed at the main CRAN repository, URL: https://CRAN.R-project.org/package=SMUT. It was introduced to analyze and link the SNPs with genomic noncoding GWAS variations. These kinds of variations require an understanding of their biological context in order to make sense of the coordinated or correlated biological effects on phenotype. To solve this problem, SMUT was designed and made available to the public for research use. Simulations have shown the reliability of SMUT over other similar methods as well as an increase in performance efficiency. The testing of this tool was done on the real dataset (METSIM, a randomly selected population of around 10,197 males from Finland). It was found that the tool was able to genuinely identify the significant mediation effect on chosen SNPs and genes (Zhong et al., 2019).

13.2.19 Varanto

Varanto is an R-based web app, designed in Shiny with a backend PostgreSQL database from http://bioinformatics.uef.fi/varanto. The GWAS identify relationships between genomic variations, for instance SNPs with particular diseases or traits. These SNPs can be further explored to identify their connections in biological processes, pathways, and disease. This tool accepts data from diverse resources to analyze, annotate, and visualize genomic alterations in humans. This tool uses their input variation to fetch their corresponding alterations and annotations by performing enrichment analysis and provides the visual representation of outputs (Timonen et al., 2019).

13.2.20 graph-GPA

The graph-GPA tool was developed as an R package "GGPA" with the combination of other web interfaces such as "DDNet". The tools are available on the following URLs: http://dongjunchung.github.io/GGPA/ and http://www.chunglab.io/ddnet/. This tool allows the identification of GWAS-based genomic variants and associating them with corresponding phenotypic studies to determine risk variants. This type of genotypic–phenotypic analysis remains challenging for GWAS datasets. The graph-GPA is an integrated Bayesian hierarchical model developed for analyzing multivariate GWAS data to enhance statistical power to find genomic variants with risk and determine pleiotropic architecture within a framework with the novel advancement (Kim et al., 2018).

13.2.21 SNPDelScore

SNPDelScore is available publicly at the URL: https://www.ncbi.nlm.nih.gov/research/snpdelscore/ (Figure 13.6). It quantifies deleterious effects of noncoding mutations with all known SNPs from 44 cell lines and predicts the "scores". These deleteriousness scores are supplemented by functional information from the "GWAS-Catalog, transcription factor-binding sites and genetic mutations". This tool helps in forecasting deleterious noncoding polymorphisms in genetics. The genome browser allows users to quickly locate consensus SNPs with the greatest negative scores, making them promising options for phenotype-causal alterations by presenting and comparing large collections of SNPs at a particular genomic location (Alvarez et al., 2018).

13.2.22 snpEnrichR

snpEnrichR is an R package available at https://github.com/kartiek/snpEnrichR, developed for the co-localization analysis of SNPs, especially the SNP trait related to regulatory regions (e.g., transcription factor-binding sites). It uses multiple tools to perform the analysis and produces random SNP profiles. That helps users by automating the analysis by creating a custom co-localization analysis of SNPs. The co-localization SNPs associated with the specific trait for a particular regulatory region or transcription factor-binding sites can provide a deep understanding of causal mechanisms. Since the exact causal effect of SNPs is unknown, the analysis is difficult; however, proxy SNPs might be reported in LD or in GWAS. For that reason, snpEnrichR, a comprehensive algorithm, links the co-localization analysis for SNP to find important information about their proxies and genotyped SNPs (Nousiainen et al., 2018).

13.2.23 WISH-R

The Weighted Interaction SNP Hub (WISH) technique combines epistatic associations and network-based research and is developed as an effective and user-friendly R package freely available on the URL: https://github.com/QSG-Group/WISH. Epistasis is challenging to analyze and comprehend because of data complexities and explanation problems. Computational restrictions are a barrier as the level of connection grows logarithmically with the increase in variations. Gene network-based techniques have proven to be effective in incorporating biological data and discovering essential core-genes and pathways involved with complicated characteristics. Based on this genomic information, the package estimated biological processes associations at a genome-wide level. It is a powerful tool for investigating genome-wide epistasis in complicated diseases and attributes. The software uses a parallel processing effective linear model and generalized linear models implemented to filter information based on LD and generate epistatic association coefficients among SNP pairings. Multiple testing difficulties are alleviated by analyzing normalized epistatic coefficients in a network architecture, which integrates biological signals to uncover modules and routes connected to complex features. It covers everything, including plans and procedures for assessing epistasis from data filtering up to interpreting final data (Carmelo et al., 2018).

13.2.24 FATHMM-XF (FATHMM with an eXtended Feature Set)

FATHMM-XF is a model for estimating harmful point mutations in the human genome. The tool is available at the URL:

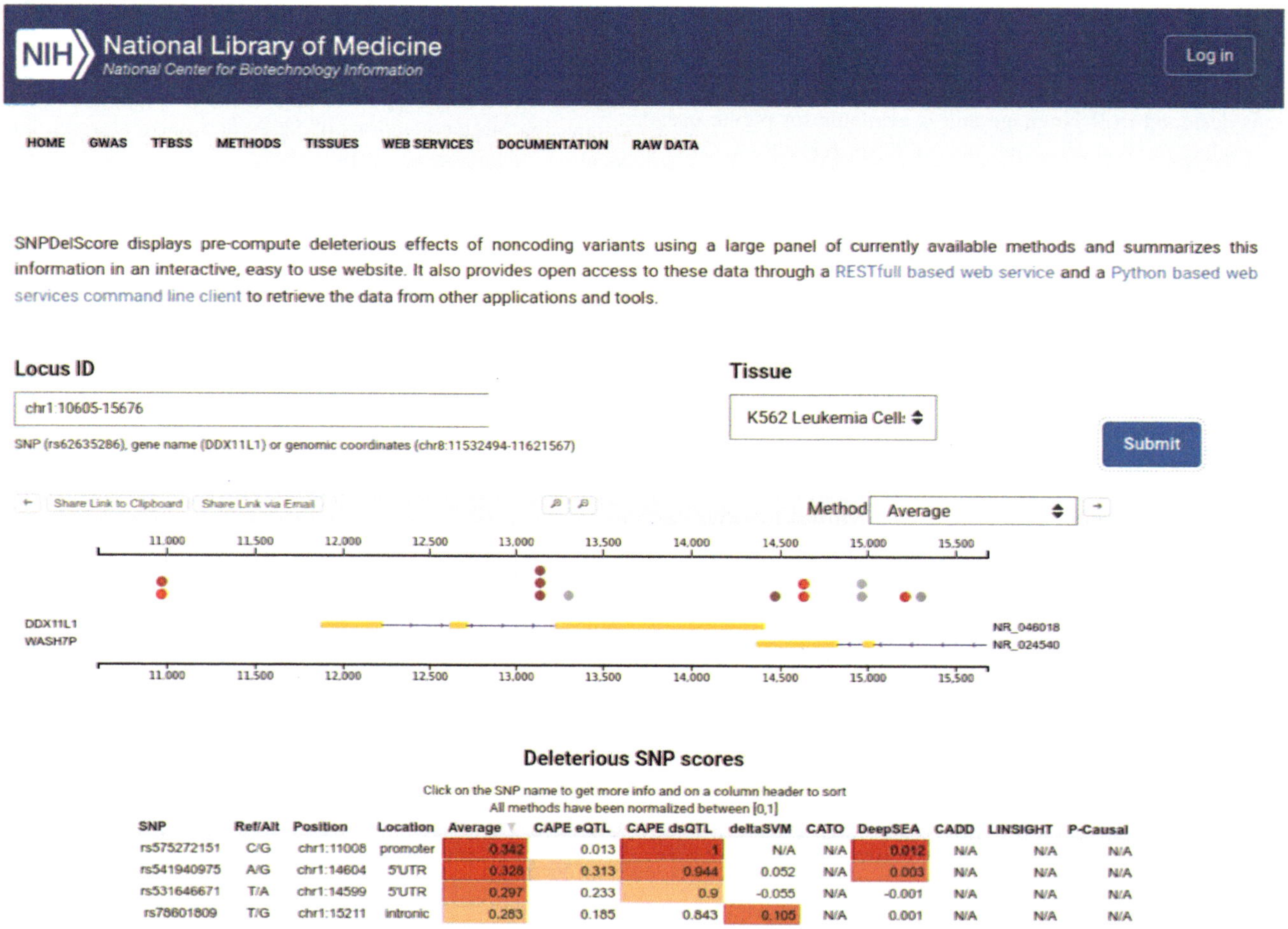

FIGURE 13.6 Representation of the SNPDelScore web server, for example SNP (rs62635286) on locus ID chr1:13115-13117 in the cancer cell line (K652 leukemia cell line), with deleterious scores predicted from multiple analysis methods. (Adapted from an example dataset on SNPDelScore web server.)

http://fathmm.biocompute.org.uk/fathmm-xf/ (Figure 13.7). It beats competitors on benchmarking evaluations because of its broad feature set and predictive results, especially in noncoding areas in which most of the pathogenic mutations are expected to be detected. It offers amazingly precise SNV estimations over the whole human genome. For easy evaluation and emphasis on the investigation of high-confidence predictions, FATHMM-XF provides a cutoff value (p-score) for each estimation. It matches or beats rival approaches in all testing, with promising results in noncoding areas (Rogers et al., 2018).

13.2.25 BSviewer

BSviewer is a software used for the visual representation of bisulfite sequencing data. The software is available at the URL: https://sunlab.cpy.cuhk.edu.hk/BSviewer/. It can be used for visualizing methylation positions at the nucleotide level while keeping the crucial details from the sequencing data. BSviewer uses aligned bisulfite sequence data to show a record of genomic areas given by the users. It first picks sequenced reads that come into the area for each genomic region and compares them to the reference genome. It can help to determine the methylation pattern on the CpG site and the nucleotides (SNPs). Finally, it gathers all the reads and produces the graphical output and can be very useful in variant-/allele-specific DNA methylation patterns at the nucleotide level (Sun et al., 2017).

13.2.26 Lep-MAP3

Lep-MAP3 software can map high-throughput WGS datasets with linkage mapping and is available online at the URL: https://sourceforge.net/projects/lep-map3/. It allows the genotyping of millions of SNPs for numerous different samples at a minimal cost. It can evaluate datasets with minimal coverage. Lep-algorithms can evaluate minimal datasets and simplify content categorization and filtration on any data. This results in plenty of markers in the resulting output with little technical inputs. Lep-MAP3 achieves excellent results on 5× sequence coverage and shows high performance on simulation data in terms of precision and pace (Rastas, 2017).

13.2.27 SCcaller

Defects related to cell rupture and whole-genome expansion can make mutation analysis in single-cell genome analysis difficult. "Single-cell multiple displacement amplification"

FIGURE 13.7 Visual representation of the FATHMM-XF web server. (a) Input data with example queries in the box provided and (b) its result with red color showing the pathogenic and green benign information aligned with the main columns of chromosome, position, variant (SNP), and scores (coding and noncoding). (Adopted from FATHMM-XF web server.)

(SCMDA) and SCcaller (https://github.com/biosinodx/SCcaller/) were developed to solve that problem. This tool can detect SNPs and indels in single-cell genomics data. This approach was confirmed as a valuable ground for the detection of the somatic mutation by correlating SCMDA-amplified cells with unamplified clones within a similar population (Dong et al., 2017).

13.2.28 TSGSIS

TSGSIS is a screening technique also called a grouped variable selection technique that helps to find the effect of SNP-SNP interactions on whole genomes. This tool is R-based and hosted at https://cran.r-project.org/web/packages/TSGSIS/index.html. The tool takes collective information about SNPs as input and concludes the candidate set's size in the model by detecting interaction effects. It is shown to perform well on simulated and real datasets when compared to other contemporary packages (Fang et al., 2017).

13.2.29 Heap

Heap is a tool that allows highly precise and reliable SNP calling, even with limited genomic coverage. The tool is available at the URL: https://github.com/meiji-bioinf/heap. Heap identifies genotyping and identifies SNPs at every site, except for reading end sites or carrying a minor allele backed by a single read, to minimize false-positive SNPs. Large-scale genomic resources have enabled GWAS and GP (genomic prediction) research to be carried out specifically with the data of NGS. The efficiency of GWAS and GP is determined by the amount and quality of variants used in the research, as well as the statistical models used. For higher SNP precision and specificity in NGS SNP calling, conventional technologies preferably demand more data. In the current NGS scenario, this tool has the potential to make GWAS and GP investigations more economical (Kobayashi et al., 2017).

13.2.30 DnaSP

DNA Sequence Polymorphism (DnaSP) is a software package freely available at http://www.ub.edu/dnasp. This software can analyze DNA polymorphism using multiple sequence alignment (MSA) data. The tool incorporates many innovative functionalities to evaluate large datasets. DnaSP 6 (current version 2021) implements other features as well, e.g., "different modules for reading as well as analyzing data from genomic partitioning, rapid methods extensible for high-throughput sequencing data". The tool also provides summary statistics for the analysis of population genetics data (Rozas et al., 2017).

13.2.31 HAPRAP

HAPRAP or HAPlotype Regional Association analysis Program is an "empirical iterative method" that allows exact mapping of distinctive reference groups using haplotype information and summary statistics. It is available on the URL: http://apps.biocompute.org.uk/haprap/. The output of HAPRAP is exceptionally consistent with the multiple regression, as shown by the simulation results with individual genotypes. For the identification of disease-associated loci, fine-mapping is one of the best approaches till now, which requires genotype from individuals and summary-level data from variants. Haplotypes are the appropriate representation of LD between different loci (Zheng et al., 2017).

13.2.32 CollapsABEL

CollapsABEL is a pipeline that contributes to statistical testing along with the graphical representation of plots for "Generalized Compound Double Heterozygosity (GCDH)". It is efficient in terms of computational power and efficiently filters out the compound heterozygosity (CH)-related copies of genes in high-throughput genomic data microarray or sequencing. CollapsABEL is written in R and available along with the source and binary package website (https://cran.r-project.org/web/packages/CollapsABEL). The presence of two recessive mutations in different forms is known as CH and is a fundamental part of absent heritability. The identification of CH-like effects might be important in describing missing heritability in GWAS (Zhong et al., 2016).

13.2.33 Monovar

Monovar is a tool for determining and genotyping SNPs in single-cell data using analytical techniques. The tool is available on the URL: https://bitbucket.org/hamimzafar/monovar. It was developed because present variant callers do not compensate for allelic dropouts, coverage nonuniformity, and false-positive errors; present variant callers are not adequate for single-cell DNA sequencing. It outperformed many traditional algorithms on assessments and in three distinct human tumor datasets for detecting driver gene mutations and clonal substructure delineation (Zafar et al., 2016).

13.2.34 NASP

NASP is a tool that was created as a standardized approach to handle a high number of WGS datasets that are frequently used in comparative genomics. The source code for the tool is housed at http://tgennorth.github.io/NASP/. Currently, WGS of isolated bacteria has become a normal procedure. Organic evolution and molecular epidemiology are two protocols used for WGS analysis, which mainly use SNPs as the evolutionary unit. NASP was compared to other methods in the study of two genuine bacterial genomics datasets and one simulation dataset. In comparison, the findings show NASP generates equal and comparable results. The tool is considerably more adaptable in terms of data input types, task management systems, tool diversity, and product formats. Different outcomes can be obtained from the tool depending on the reference genome used and whether phylogenies are inferred using concatenation SNPs or alignments with monomorphic locations (Sahl et al., 2016).

13.2.35 PredictSNP2

PredictSNP2 is a web-based platform located at http://loschmidt.chemi.muni.cz/predictsnp2. Its interface could be used for the accurate evaluation of SNP effects. To achieve this, it exploits different variant characteristics in distinct genomic regions. In order to provide precise personalized detection, diagnosis, and treatments of hereditary disorders, precise characterization of phenotypically causative variations is critical. Developers of the tool have employed three datasets encompassing multiple disease-related variations mainly grouped into five categories ("regulatory, splicing, missense, synonymous, and nonsense variants"). The data were used to generate threshold scores based on each category, employing methods such as "CADD, DANN, FATHMM, FitCons, FunSeq2, and GWAVA" for variant prioritization. The results obtained are used to provide the best consensus score and to find the best performing tool. PredictSNP2 is made as a publicly available convenient web interface with the aim of allowing quick access to tools to enable an in-depth variant evaluation and annotations from multiple databases (Bendl et al., 2016).

13.2.36 4Pipe4

4Pipe4 is a 454 sequencing platform-based data analysis pipeline that focuses on SNP discovery. The characteristic feature of this pipeline is that it does not use reference sequence or strain information. The pipeline is located at https://github.com/StuntsPT/4Pipe4. It features a command-line-based interface to run, analyze the output, and present the results. The variant-detecting functions are also included in the software. This platform is used to systematically examine sequencing input data, however, this tool is specifically designed 454 sequencing data, with information related to SNPs. It offers multiple HTML-formatted outputs that include statistics of many frequent assembly parameters. There is also a program for detecting potential simple sequence repeats (SSRs) in the data. This tool particularly benefits researchers who have 454 datasets from a panel of mixed people and aim to divulge SNPs-data from customized individualized genotyping. In contrast to other relative techniques, 4Pipe4 features the optimal ratio for the validation. This helps in obtaining a minimal group of SNPs with significantly less false-positive rates (Pina-Martins et al., 2016).

13.2.37 ALICE

ALICE or "Affymetrix Axiom single nucleotide polymorphism" (SNP) software is housed at http://hcyang.stat.sinica.edu.tw/software/ALICE.html. It has been developed as a convenient program to analyze SNP probe hybridization intensities and genotypes. The tool provides a high-throughput, high-density, and cost-effective genotyping solution for performing population-optimized analyses. This tool uses the latest population-optimized arrays to achieve the same. The tool houses integrated tools to compute parameters such as loss of heterozygosity (LOH), allele frequency (AF), "the long contiguous stretch of homozygosity (LCSH)", allelic imbalance (AI), and copy number variations (CNVs) (Huang et al., 2016).

13.2.38 Altools

Altools can discover CNVs, SNPs, and indels. The source code for the tool is located at https://sourceforge.net/projects/altools/. This tool employs the BWA/SAMtools/VarScan pipeline to identify SNPs and indels. It can also detect probable big deletions by using insert size data from matched end read alignments. A two-map technique used by the tool (BWA/BLASTn) detects accurate breakpoints while allowing for a quick explanation. Altools employs several procedures that provide more information about the genes affected by discovered mutations. The tool can also try to connect recognized variations and structural variants to their bioactivities, making it even more featureful. Currently, genotyping by resequencing is now a standard method used to estimate biodiversity, SNP diversity, and haplotype structure, as well as an effective process for addressing spatial community genomics in various model species. Most of the algorithms for the purpose map shorter reads to the reference genome to acquire key SNPs and indels to predict phylogenetic shapes of speciation. Variant calling is critical for establishing patterns in GWAS for QTLs, as well as determining populations and haplotype architectural styles based on SNPs, allowing for context-specific attribute and evolutionary research. This tool has been created to explore such variations and more complicated genomic adjustments, including CNVs, presence/absence variations, and massive deletions (Camiolo et al., 2016).

13.2.39 BAM-matcher

BAM-matcher is a software located at https://bitbucket.org/sacgf/bam-matcher/src/master/. This tool has been developed with the aim of using the genotype comparison in order to quickly assess whether two BAM files reflect specimens from a similar biological source. It has been created to be an easy-to-use software, with outcomes that are straightforward to read. It is best used at the beginning of data processing pipelines. By calculating genotype hits at frequent SNPs, a simple algorithm may determine whether two BAM documents contain readings sequenced from a similar specimen or patient. It is best for matching human data from WGS, WES, and RNA-Seq, but it can be modified to match panel or non-human data. The usual strategy for detecting wrongly labeled specimens by high-throughput genomic sequencing provision is to use independent high-density data related to SNP for the determination of the sample identity. It is now common to have various specimens sequenced from the same source, as for somatic variant assessment comparing corresponding tumor and normal specimens. We can use the genotype details in the sequence data to pair the specimen directly and avoid the requirement for extra laboratory testing (Wang et al., 2016).

13.2.40 SNP-sites

SNP-sites software can retrieve SNPs from multi-FASTA alignments quickly and in many forms for downstream analysis with minimal resources. It is available on the URL: https://www.sanger.ac.uk/tool/snp-sites/. SNPs can be retrieved quickly from the alignment files, making them suitable for use on low-cost machines. The total sample size used in prokaryotic surveys has increased proportionately as genome sequencing prices have decreased. Thus, the extraction of SNPs from a big whole genomic alignment is now a standard procedure, but current methods have struggled to keep up with the growing number of studies as they are sluggish, have memory deficit, and need inconvenient installation procedures. This tool was developed to cater to such problems (Page et al., 2016).

13.2.41 Ensembl Variant Effect Predictor (VEP)

Ensembl VEP is a powerful toolset to predict and analyze the functional impact of indels, SNPs, variance, regulatory regions, protein sequences, CNVs on genes, and transcripts. It is an open source and is freely available at the URL: https://grch37.ensembl.org/info/docs/tools/vep/index.html. It can be accessed via the web interface (http://useast.ensembl.org/Tools/VEP), command-line interface (GitHub or Docker file or zip file from website), and REST API (https://rest.ensembl.org/#VEP). All three are represented in Figure 13.8. It provides a huge collection of genomic annotations with different interfaces, as well as many simple options for configuring and broad analysis. The Ensembl VEP (SNP Effect Predictor) can speed up variant interpretation by simplifying a wide range of study designs (McLaren et al., 2016).

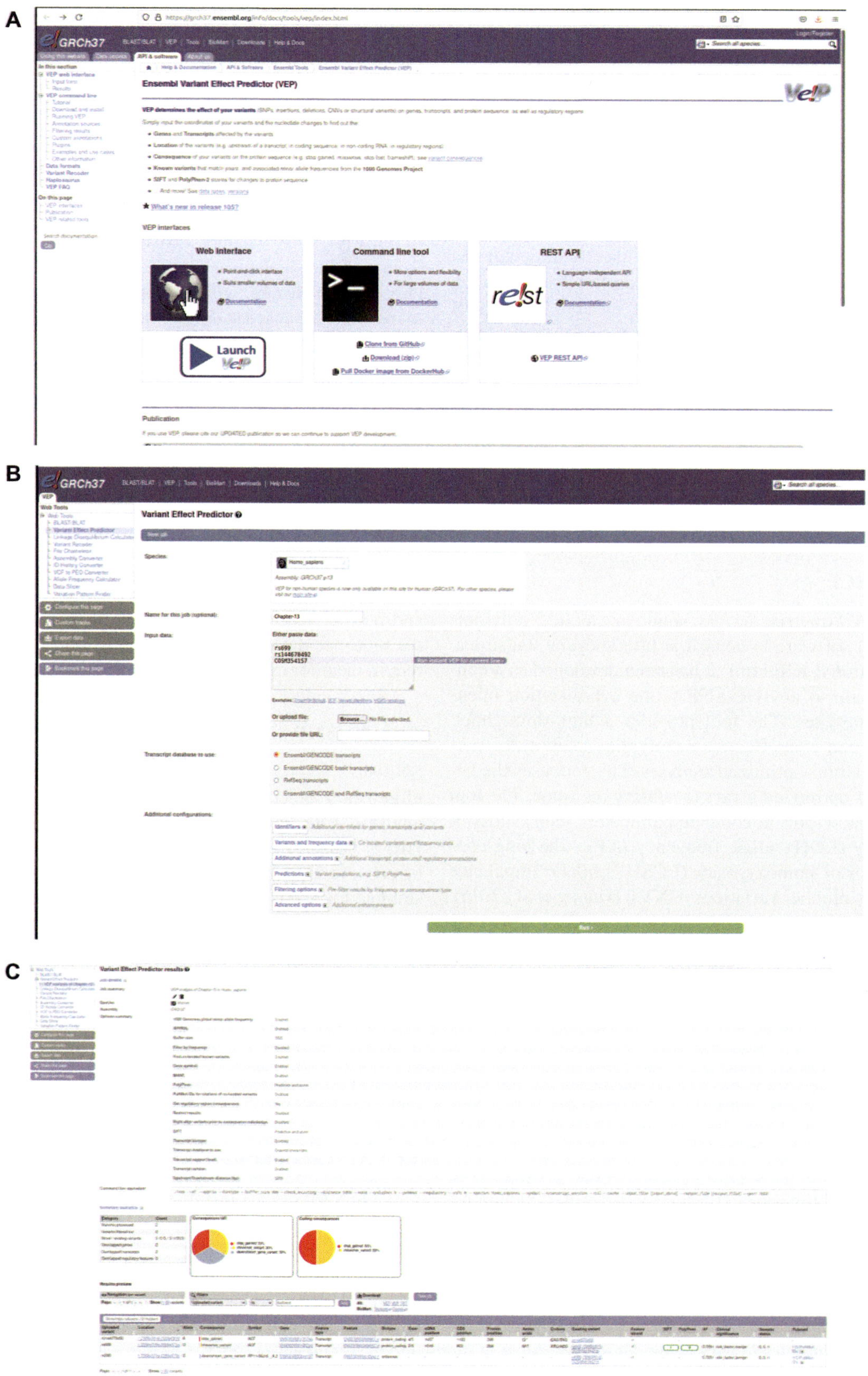

FIGURE 13.8 Visual representation for the usage of VEP web interface. (a) The main landing page with the three options (web interface, command line, and REST API) for using VEP tool. (b) Example query of SNPs to use in VEP web interface. (c) Result of the example query with information on query, command-line equivalent, summary statistics with plots, and results in tabular format. (Adapted from Ensembl VEP website.)

13.2.42 LAMPLINK

GWAS are one of the major techniques for decoding genetic links among several SNPs through high-order epistatic interaction. While existing tools lack significant interactions, LAMPLINK available at the URL: http://a-terada.github.io/lamplink/, serves an innovative approach with additional functions with PLINK and LAMP. The package executes analysis of case-control data using the chi-squared test or Fisher's exact test and identifies possible alterations of complex disease as well as statistically significant SNPs and can identify connective significant markers (Terada et al., 2016).

13.2.43 CINOEDV

CINOEDV or Co-Information based N-Order Epistasis Detector and Visualizer has been developed to detect and visualize epistatic interaction. The complete package is a freely available tool, is an R-based software package, and can be accessed from CRAN and SourceForge website at the URL: https://sourceforge.net/projects/cinoedv/files/. This tool has been created to study nonlinear epistatic interactions. Missing heritability is an unraveled mystery that has drawn scientists' attention toward depicting nonlinear interaction effects in bioinformatics. Epistatic interactions are one of them. However, computational and methodological challenges limit the studies of the related interaction. It comprises two stages: detection and visualization stages. In the first stage, n-order SNP phenotypic association is evaluated, where two search methods are used to detect epistatic interactions: a particle swarm optimization search and an exhaustive search. In the last visualization stage, epistatic interactions of order "n" are identified from the detection stage used to build a hypergraph; here, real and virtual vertices represent the main impact of SNP and interaction impact of an n-order epistatic interaction. Finally, this constructed hypograph can help to provide hidden information for understanding genetic composition that reveals clues for complex diseases (Shang et al., 2016).

13.2.44 JEPEGMIX

JEPEGMIX is a software package that can be used for detecting LD pattern groups with minimal error. To increase the gene-level analysis, most computational methods require input as a summary of statistical analysis from GWAS constructed with the help of LD. As an extension of JEPEG, JEPEGMIX can detect LD pattern groups without an error. Further, these LD estimates are used to (1) refer to the statistical summary of functional variants that are not measured and (2) to analyze the combined effects of all functional variations related to genes. The performance of this tool has been illustrated by GWAS meta-analysis from Multivariate Psychiatric Genome Consortium Schizophrenia stage 2 group. All related documentation and data such as software, database, annotation, and supporting examples can be accessed from http://dleelab.github.io/jepegmix/ (Lee et al., 2016).

13.2.45 snpGeneSets

The snpGeneSets tool can make annotation and interpretation of GWS data easier, and the source code for the package can be found at https://www.umc.edu/biostats_software/. The tool combines local copies of SNPs, genes, and gene-set knowledge bases with R language wrapper methods to provide open support for low databases for rapidly annotating huge genomic data. GWS of SNP correlations and differential gene expressions yield a plethora of findings, and NGS technology has increased the number of variants and genes discovered. Appropriate analysis necessitates a huge number of annotations and subsequent processing of these GWS, which is a computationally difficult operation (Mei et al., 2016). The package includes utilities for three different annotations:

1. Functional description for the set of genes and description of genome mappings for genes and SNPs.
2. Genes-SNPs, as well as genes-gene-sets, have bidirectional mappings.
3. Gene effect estimates are calculated from SNP relationships, and gene-set enrichment studies are used to uncover functioning pathways.

13.2.46 traseR

The traseR is an R-based package that helps in analyzing trait-associated SNPs with multiple options, including the incorporation of SNPs in LD regions. The source code for the tool can be found at https://Sbioconductor.org/packages/release/bioc/html/traseR.html. It is known that many significant sequence variants have been identified for common traits or associated with disease using GWAS. Several resources in genomic research collectively found loads of such SNPs associated with common or complex diseases. Some studies suggest that these trait-associated SNPs are phenotypically connected through a certain function which might be a disease-causing factor. The lack of computational tools to find the connection between phenotypic and corresponding genomic intervals inspired developing tools such as traseR (Chen and Qin, 2016).

13.2.47 SNPStats

SNPStats is an R-based Bioconductor application housed at the URL: https://www.bioconductor.org/packages/release/bioc/html/snpStats.html for GWAS. The focus of this analysis package is LD, descriptive analysis, and a test for Hardy-Weinberg equilibrium. This association analysis is based on logistic regression for binary disease status or linear regression for quantitative disease status. For single SNPs analysis, this package houses many functions and can perform multiple jobs such as interaction analysis for gene to environment or gene to gene and different genetic models such as dominant, codominant, over-dominant, log-additive, and recessive. This tool can also perform multiple SNPs analysis that includes evaluation of the relationship of haplotypes with counter-response and frequency estimation of haplotypes and interactions (Huber et al., 2015).

13.2.48 SNPsnap

SNPsnap software performs enrichment analysis on SNPs-data that helps in the etiology of complex biological systems, especially diseases based on the 1000 Genomes SNPs-data. It is freely available on the URL: http://www.broadinstitute.org/mpg/snpsnap/[Figure 13.9]. SNPsnap effectively matches querying SNPs to a random subset of SNPs based on their LD,

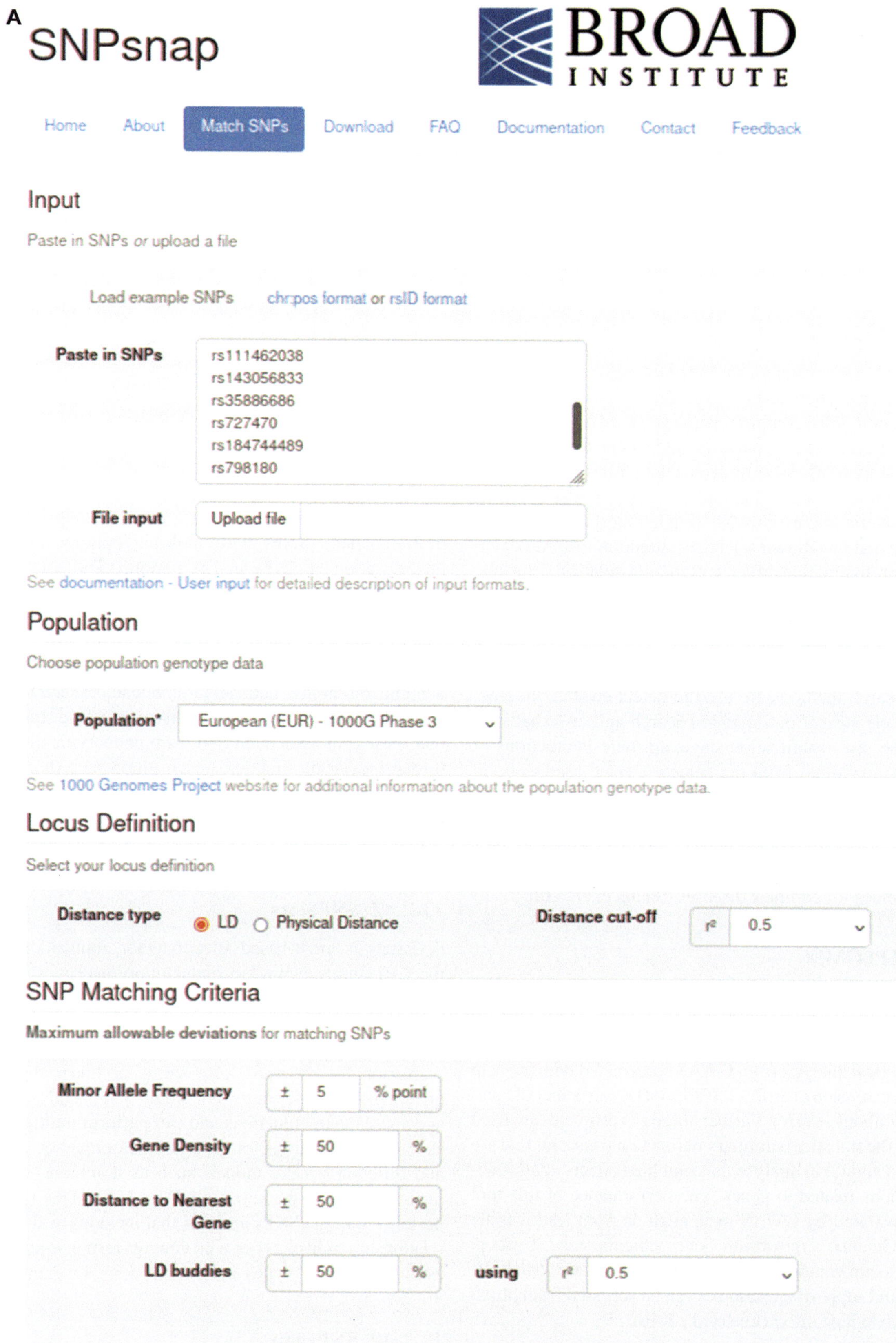

FIGURE 13.9 Visual representation of the SNPsnap website. (a) The main page with multiple options, including main function at the third position from the left "Match SNPs" with multiple input options to be filled including SNPs in the provided box or upload as file. (b) Options for the output file to be provided and (c) the results page, providing the link to the results page, followed by job status, SNPsnap score, and matching bias information with option to download the results information. (Adapted from SNPsnap website test data.)

(Continued)

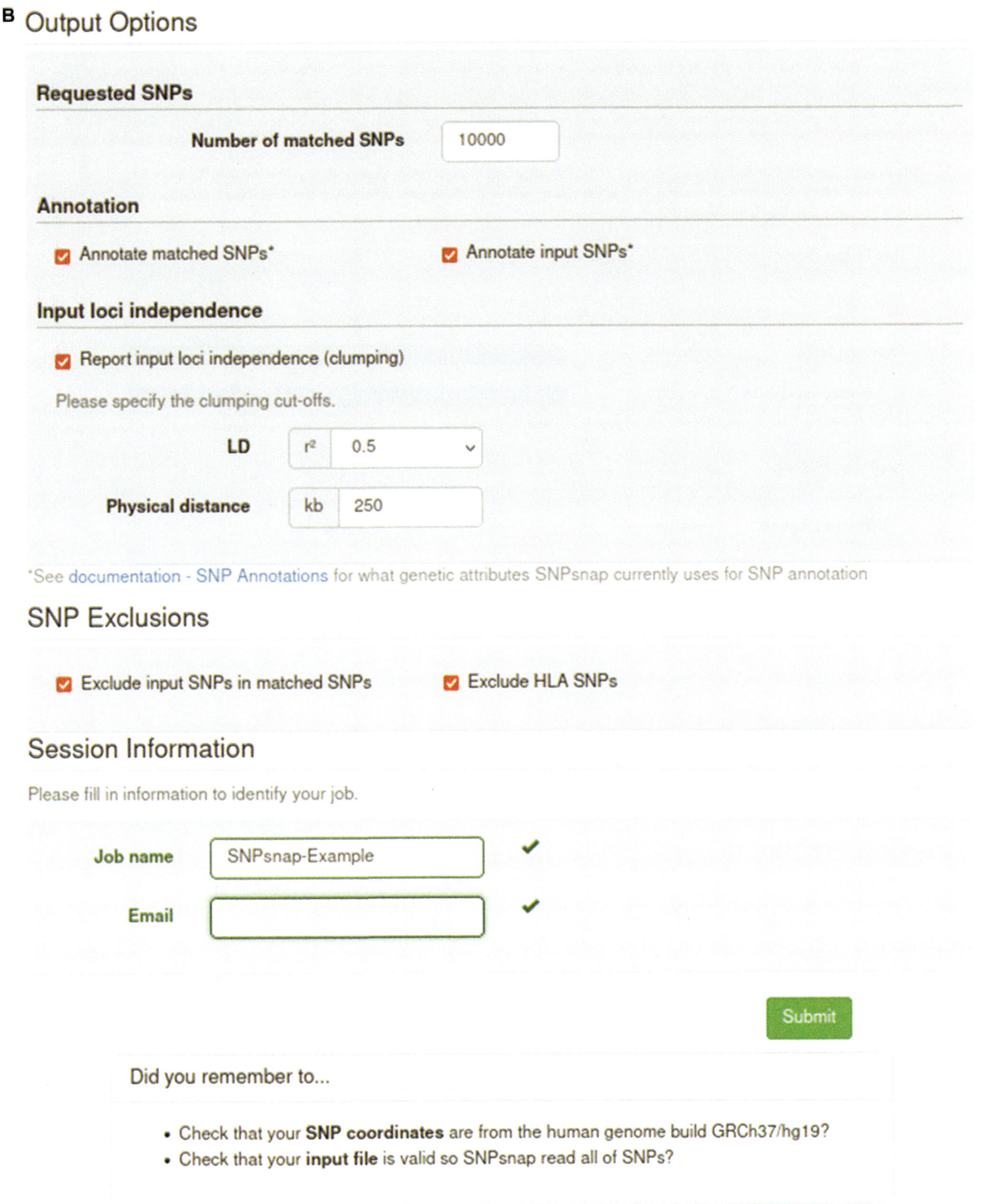

FIGURE 13.9 ***(Continued)*** Visual representation of the SNPsnap website. (a) The main page with multiple options, including main function at the third position from the left "Match SNPs" with multiple input options to be filled including SNPs in the provided box or upload as file. (b) Options for the output file to be provided and (c) the results page, providing the link to the results page, followed by job status, SNPsnap score, and matching bias information with option to download the results information. (Adapted from SNPsnap website test data.)

(Continued)

C

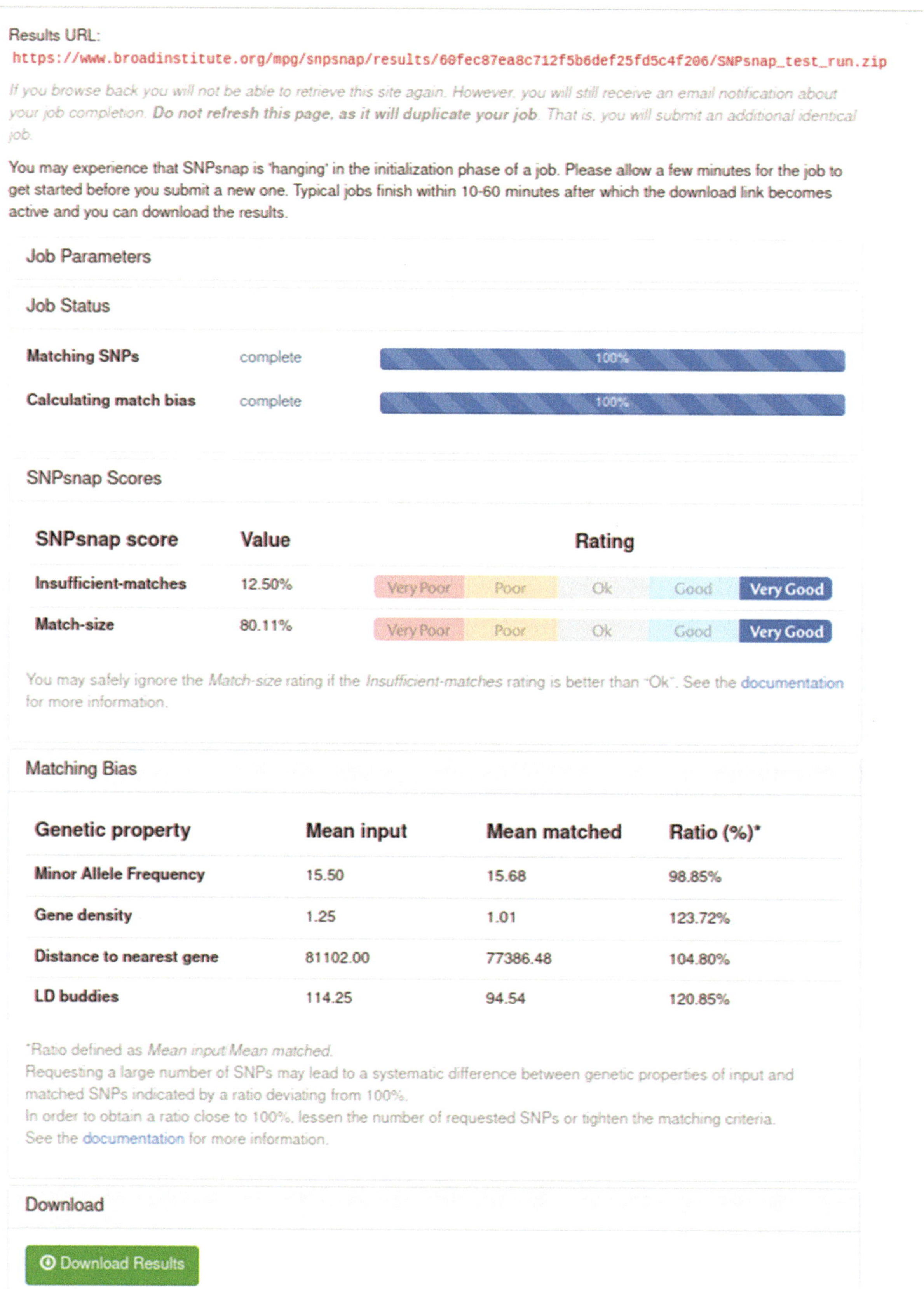

FIGURE 13.9 (*Continued*) Visual representation of the SNPsnap website. (a) The main page with multiple options, including main function at the third position from the left "Match SNPs" with multiple input options to be filled including SNPs in the provided box or upload as file. (b) Options for the output file to be provided and (c) the results page, providing the link to the results page, followed by job status, SNPsnap score, and matching bias information with option to download the results information. (Adapted from SNPsnap website test data.)

allele frequency, proximity to the closest gene, and density of the gene. The analysis of disease- or trait-associated SNPs with specific biological descriptions is an essential computational step after GWAS. SNP-based enhancement analysis must compensate for bias such as GWAS signal co-localization to gene-concentrated and high LD areas, along with gene sizes, placement, and functional correlations that are taken care of in this software. Furthermore, this pipeline uses paired SNP-sets to adjust backgrounds (Pers et al., 2015).

13.2.49 Phyre2

Phyre2 is a range of tools available at http://www.sbg.bio.ic.ac.uk/phyre2. It is used for the calculation and analysis of protein structure, mutations, and functions. The aim of this tool is to contribute to scientists with an intuitive and simple bioinformatics tool interface. This server uses advanced remote homology identification methods to predict ligand-binding sites, build 3D models, and analyze the effect of non-synonymous SNP (nsSNP) mutations. Phyre2 provides a user-friendly interface that walks users through the process of submitting a protein sequence for analysis of the secondary and tertiary structures. A set of subsidiary techniques have also been provided in order to discover protein structure by concurrently submitting huge sequences and leaving the software to do optimal analysis automatically (Kelley et al., 2015).

13.2.50 LDlink

LDlink is publicly available and could be reached at http://analysistools.nci.nih.gov/LDlink/. It is a group of different bioinformatics modules that generate interactive plots of haplotype tables from SNP's population from population groups. These modules have been developed with significance in query flexibility, proficiency of use, and interactive visualization of outputs. LDlink is customized for scientists involved in investigating common and uncommon abnormally sensitive loci by concentrating on result linking associated alleles and underlining hypothetical functional dissimilarity. Genomic identification of disease susceptibility in the mapping region and studying their population-specific genetic structure is important. LDlink is an investigative tool in this direction that helps in investigating the LD among ancestral populations in an area of interest (Machiela and Chanock, 2015).

13.2.51 PROVEAN

PROVEAN is a web-based tool for predicting the functional influence of single or multiple amino acid deletions, insertions, and substitutions. It can be accessed using the link http://provean.jcvi.org/index.php. The web server features a fast interpretation of protein variance from high-throughput analysis both at protein and genomic levels. PROVEAN community created a broad database by computing the pairwise alignments between a protein sequence and a large number of single-locus variants of other proteins using a pre-computed set of scores. The server features three primary functions: PROVEAN Protein Batches, PROVEAN Protein for All Organisms, and PROVEAN Genome Variants. Its data framework is based on an in-house database. The database comprises pre-computed PROVEAN scores, Ensembl gene annotation, and pre-computed supporting sequences. PROVEAN divides computing job submissions into disparate queues under the intended job size such that small computing jobs are handled promptly on a dedicated queue, resulting in time-saving quick result outputs (Choi and Chan, 2015).

13.2.52 SNP2GO

SNP2GO (http://www.cibiv.at/software/snp2go/manual.shtml) was developed to show that this conclusion is not always correct, as it uses a novel technique to appropriately check for potential SNP excess representations in biological pathway systems. GWAS can be aimed at finding the percentage of SNPs in genome sequences that are linked to a complicated trait. For the systematic study of GWAS, tactics built on the gene list enhancement idea are being used. The tool uses a notable excessive representation of potential genes linked with a biological system as a surrogate to predict the excessive representation of potential SNPs in the system (Szkiba et al., 2014).

13.2.53 SNPsea

SNPsea is a software located at https://github.com/slowkow/snpsea/releases. It is used to discover types of cells, tissues, and processes influenced by risk loci. It looks for enriched specificity to circumstances in trait-associated genomic regions. By comparing null SNP-sets to observed P-values, the tool employs a multivariate statistical technique. It shows datasets of SNPs linked with RBC count, multiple sclerosis, celiac disease, and HDL cholesterol as prototypes. SNPsea is a general technique that assesses the genes inside related regions for the enhancement of conditions specific to determine the conditions important to a trait. In an array of genes and conditions, it looks for the enhancement of condition specificity in genes associated with LD (Slowikowski et al., 2014).

13.2.54 PolyPhen-2

PolyPhen-2 is used to predict, annotate, and visualize SNPs by analyzing protein structures and multiple sequence alignments. This tool can be accessed at http://genetics.bwh.harvard.edu/pph2/. It uses three structure-based as well as eight sequence-based features that are selected automatically by its algorithm. Typically, the properties include wild-type allele differentiation, in contrast to the mutant allele, that is characteristically determined by amino acid replacements. It explains how well the two human alleles fit into the pattern of amino acid changes within the multiple sequence alignment of other homologous proteins, as well as how far the protein accommodates the first change from the human wild-type (normal) allele to the human protein or mutant originating from the hypermutable position. This pipeline chooses homologous sequences and refines them by clustering, constructing, and predicting an allele's functional significance (Adzhubei et al., 2013).

13.2.55 GWIS

As GWAS produce loads of likely SNP interactions, the data from such studies demand tools to analyze these data with accuracy. GWIS located at http://bioinformatics.research.nicta.com.au/gwis is a piece of software dedicated to finding all the possible epistatic interactions. Genome-Wide Interaction Search (GWIS) is a model-free tool, based on statistical test analysis of receiver operating characteristic (ROC) curve on GAS datasets. This tool is comparatively faster and time-efficient than other available tools (Goudey et al., 2013).

13.2.56 AnnTools

AnnTools is a flexible application for bioinformatics analysis and is especially created for diverse annotations of human genome modifications: unknown and well-known single nucleotide substitutions (SNV/SNP), structural variants or copy number variations (SVs/CNVs), and short insertions/deletions (indels). The tool is available at http://anntools.sourceforge.net/. AnnTools searches for disease-/trait-associated loci that overlap with regulatory elements, artifact-prone regions, and known segmental duplications. In addition, the tool can also perform functional characterization of coding variants. This tool accepts user-provided customized interpretation and provides flexibility in input data formats used for the analysis (Makarov et al., 2012).

13.2.57 SIFT

SIFT is a stand-alone software tool that offers to anticipate the effects of frameshifting indels and nsSNP on protein functions. SIFT is stand-alone software that can be accessed through http://sift-dna.org and helps to investigate the effectiveness of different mutations on protein roles (Sim et al., 2012).

13.2.58 GCTA

This tool has been created to assess the inherent correlation of genome scale variants, which can then infer variation described by SNPs using a mixed model approach. The tool features customizable flexibility for discovering diverse genetic models and separating genetic variance across the chromosomes. GCTA is used to estimate inherent associations from SNPs on the X chromosome, as well as to test the dose damage assumption. Data management, genomic connection estimate from SNPs, mixed linear model interpretation of SNP variation, LD with structure estimation, and GWAS simulation are the five main functionalities of GCTA. All functions focus on estimating the variation described by all SNPs on the X chromosome, as well as checking the dosage compensation assumption (Yang et al., 2011). The GCTA can be found at https://yanglab.westlake.edu.cn/software/gcta/#Overview.

13.2.59 adegenet

The adegenet (http://adegenet.r-forge.r-project.org/) is an R-based application used to investigate genomic SNPs. It comprises a set of tools and computational algorithms for the analysis of variants (Jombart and Ahmed, 2011).

13.2.60 SNPinfo

SNPinfo (https://snpinfo.niehs.nih.gov/) is a user-friendly web-based SNP selection tool to analyze clinical, mechanistic, and computerized data with the outputs of GWAS in order to annotate, organize, and select SNPs. This server can be used for small- and large-scale SNP selection and is specifically useful for association studies (Xu and Taylor, 2009).

13.2.61 SNAP

SNAP is a freely available query tool and can be accessed at http://www.broad.mit.edu/mpg/snap. The tool can discover and interpret SNPs that are adjacent to LD, according to HapMap. SNAP server features elucidative comparison and interpretation of GWAS and produces plots to visualize the results (Johnson et al., 2008).

13.2.62 WASP

WASP is an online tool housed at https://bioinfo.biotec.or.th/WASP. It is an allele-specific primer creation platform that can produce AS primers for human variants quickly. It implements a local SNP database with over 10 million SNPs from public databases, such as NCBI, dbSNP, HapMap, and JSNP, to aid in assembling important information regarding target polymorphisms. The tool allows users to do SNP design comprehensively and easily. WASP uses a local SNP database and design primer to search and get SNP information for multiple SNPs published in public databases. This tool uses flanking sequences to calculate allele-specific PCR assays for genotyping SNPs that aren't in the database. Primer-related factors can be used to tweak typical simulated PCR conditions in order to improve PCR results. The global uniqueness of the AS-PCR can be verified through its in silico PCR characteristic simulation (Wangkumhang et al., 2007).

13.2.63 PhD-SNP

PhD-SNP (https://snps.biofold.org/phd-snp/phd-snp.html) is a predictor based on support vector machine classifier, trained and tested on protein sequences to predict deleterious SNPs from humans. PhD-SNP predicts proteins that can be differentiated as neutral or as disease-related polymorphism, based on specific mutation. Protein sequence, position, and changed residue act as inputs for PhD-SNP. Predictions and multi-SVM are given as outputs in the form of a record table of the mutated spots in the protein sequence. The residue changes are then predicted as neutral or disease-causing changes by this predictor (Capriotti et al., 2006).

13.2.64 novoSNP

novoSNP is a tool that allows the identification of variations such as SNPs and small indels in resequencing projects. This program requires several sequencing files and reference sequences as input, and its output enlists potential modifications with a quality score. It provides an easy way to sort, filter, and check the variations identified. The results can be analyzed visually. The web link to access the tool is https://novosnp.bioinf.be/ (Weckx et al., 2005).

13.2.65 PolyBayes

PolyBayes is an automated analysis tool to report SNPs in duplicated DNA sequences. The program integrates three primary problems with sequence-based SNP detection. It employs multiple sequence alignments with paralog identification. The PolyBayes algorithm generates a list of putative polymorphic sites, each of which is connected with an SNP probability score. This allows the true-positive rate to be precisely estimated in invalidation trials. An adjustable cutoff score allows the user to strike a compromise between the accuracy of predictions. PolyBayes is available at http://clavius.bc.edu/~marth/PolyBayes/pages/main.html (Marth et al., 1999).

13.2.66 PolyPhred

PolyPhred is used to differentiate fluorescence-based sequences between outputs collected from various individuals, in order to determine heterogeneous spots for SNP. The functions of this tool are an integrated amalgam of three other programs: Phrap, Consed, and Phred. PolyPhred uses the base call metrics to identify potential heterozygotes. Peak instruction information is provided by Phred, and sequence alignment information is provided by Phrap. The tool also allows further potential heterogeneity analysis for fast investigation using the Consed tool with an accuracy rate of >99%. PolyPhred can be accessed at http://droog.gs.washington.edu/polyphred/ (Nickerson et al., 1997).

13.2.67 BGT

BGT is a quick command-line tool for querying whole-genome genotypes and frequency over thousands of samples. It is available at the URL: https://github.com/lh3/bgt. Many BGT functions can be performed with short scripts on multi-sample VCF; however, BGT offers a few advantages as a command-line tool.

1. It reduces the time to develop. Instead of using a script, one can use the command line to extract data from various files.
2. It reduces the time to process data. It's substantially quicker than VCF in a programming language such as Perl or Python because it uses high-performance C code at its heart.
3. BGT merge is faster than bcftools merge on two genotype-only data inputs because of the creation of a single non-reference allele.

The BGT server attempts to address a larger issue: data exchange. Rather than always supplying complete information in VCF, projects might use the BGT server to distribute data publicly, allowing users to choose a statistical summary of concern on the fly yet keeping samples as anonymous (Li, 2016) (Table 13.1).

TABLE 13.1

List of the Tools and Web Servers for the Functional SNPs and Mutations

Tools/Web Servers	Year	Links	Description
logicFS	2015	https://rdrr.io/bioc/logicFS/	Identify interactions of SNPs.
argyle	2015	https://github.com/andrewparkermorgan/argyle	Quality check and interpretation of genotype data including statistical and genomic analyses.
ASSIsT	2015	http://compbiotoolbox.fmach.it/assist/	Helps in the filtering of SNP markers; also analyzes various experimental data such as backcross and cross-pollination.
atSNP	2015	https://github.com/chandlerzuo/atsnp	Affinity test for regulatory SNP (atSNP) detection allows analysis of motif-SNP interaction by computing binding affinity, P-value.
SNiPlay3	2015	https://sniplay.southgreen.fr/cgi-bin/home.cgi	Manages SNPs identified from NGS technology, such as RNA-Seq, GBS, and WGRS. It allows users-specific options for comparing and filtering SNPs and detecting genotype data for further analysis.
BALL-SNP	2015	https://www.ccb.uni-saarland.de/software-old/ballsnp/	Very useful in structural bioinformatics and allows analysis of various mutations in nsSNPs in a protein.
SECA	2014	https://sites.google.com/site/qutsgel	SNP effect concordance analysis (SECA); inspects genetic overlap and pleiotropic effects between two phenotypes using GWA results.
MultiBLUP	2014	http://dougspeed.com/multiblup/	Best Linear Unbiased Prediction extended form of BLUP; it is used to build prediction models.
SuSPect	2014	http://www.sbg.bio.ic.ac.uk/suspect/	Predicts the probability of mutation linked to the disease.
AffyPipe	2014	https://github.com/nicolazzie/AffyPipe	Incorporates all Affymetrix tools and provides handy outputs with the ease of editing SNP probes.
CSI Phylogeny	2014	https://cge.cbs.dtu.dk/services/snpTree/	Web tool for automatic SNP analysis; constructs phylogeny with SNPs from WGS.
flashpca2	2014	https://github.com/gabraham/flashpca	Performs PCA from SNP data.

(Continued)

TABLE 13.1 (*Continued*)

List of the Tools and Web Servers for the Functional SNPs and Mutations

Tools/Web Servers	Year	Links	Description
Allele Workbench	2014	https://code.google.com/archive/p/allele-workbench/	Implemented in Java, it allows users to visualize variants, genes, libraries, transcripts, and exons. It also identifies allelic inequality at transcript and SNP levels.
Evaluation of Functional Impact of Non-synonymous SNPs (EFIN)	2014	http://paed.hku.hk/efin/	This tool helps in understanding prediction accuracy with improved significant values. It used an evolutionary method to cluster homologous sequences into multiple sections and identify the conservation of each cluster in improved prediction.
WS-SNPs&GO	2013	https://snps.biofold.org/snps-and-go/	These algorithms identify disease-related mutation by analyzing functional annotation of protein.
ChroMoS	2013	http://vm24141.virt.gwdg.de/services/chromos/	Chromatin-modified SNPs (ChroMoS); integrates epigenetic and genetic data and classifies SNPs and predicts functions.
linkSNPs	2013	http://www.sendou.soken.ac.jp/esb/innan/InnanLab/	This unique algorithm has been developed to perform haplotype phasing and SNP calling. It uses nearby SNPs from linkage information to remove incorrect haplotypes.
PHYLOViZ	2012	https://bitbucket.org/phyloviz/phyloviz-main/downloads/	Phylogenetic analysis and visual representation of large datasets, including SNPs.
AncestrySNPminer	2012	https://research.cchmc.org/mershalab/AncestrySNPminer/login.php	Fetches AIMs from genetic datasets; uses these data to link marker genes and annotation classes.
Bis-SNP	2012	http://people.csail.mit.edu/dnaase/bissnp2011/	Bisulfite SNP caller based on a model which predicts better SNP with better methylation estimation.
miRdSNP	2012	http://mirdsnp.ccr.buffalo.edu/	Provides a broad source of SNPs related to disease and a range of miRNA target sites.
SNPEVG	2012	https://animalgene.umn.edu/snpevg	SNPEVG is a versatile graphic representation tool for the visualization of local and global views from GWAS output. SNPEVG2 allows graphic visualization of several traits at the same time on the same graph with multiple options, while SNPEVG3 uses results from single-locus test outputs derived from epiSNP. Chromosome figure displays three (additive, dominance, and genotypic) genetic effects.
GenomeGems	2012	https://www.tau.ac.il/~nshomron/GenomeGems/	Compares SNPs from different datasets and helps in loading the data to UCSC Genome Browser for further exploration and visualization. This tool facilitates providing hierarchy in genetic SNP calling identified from deep sequencing.
ngs_backbone	2011	https://github.com/JoseBlanca/franklin	The package allows de novo assembly or alignment with the reference sequence and facilitates the annotation of SNPs, ORFs, GO, and SSRs.
SNP-Set Enrichment Analysis (SSEA)	2011	https://cbcl.ics.uci.edu/SSEA/	Enrichment method based on SNP pathway. It consists mainly of two steps: (1) identify all representative SNPs, perform average number calculation, and finally reselect statistically significant SNPs. (2) Rank significant SNPs and perform Kolmogorov-Smirnov testing of high-rank SNPs.
PoPoolation2	2011	https://sourceforge.net/projects/popoolation2/	Integrates a variety of different methods such as FST, Cochran-Mantel-Haenszel test, and Fisher's exact test, which are tested on SNPs, exons, and genes.
MicroSNiPer	2010	http://vm24141.virt.gwdg.de/services/microsniper/	Predicts the effect of SNP on hypothetical microRNA targets.
Atlas-SNP2	2010	https://www.hgsc.bcm.edu/software/atlas-2	Specific for the separation between proper SNP and indels from sequencing data.
i-GSEA4GWAS	2010	http://gsea4gwas.psych.ac.cn/	Upgraded versions of i-GSEA focus especially on biological pathways with highly significant genes. i-GSEA4GWAS allows deep understanding in complex disease studies.
CandiSNPer	2010	http://www2.hu-berlin.de/wikizbnutztier/software/CandiSNPer/	Detects LD region from GWAS nearby SNP with graphical representation.
HomozygosityMapper	2009	https://www.homozygositymapper.org/	This application helps in heterozygosity mapping. It has a database of marker data; when users upload their SNP genotype, it identifies homozygous regions and represents the output in an interactive graphical interface. This visualization can expand up to a single chromosome or a particular region or as per user interest. It can also find the candidate genes with the help of the GeneDistiller search engine.
VarScan	2009	http://varscan.sourceforge.net/	Detects variants/mutation from NGS data from the individual or pool of samples.
ADMIXTURE	2009	http://dalexander.github.io/admixture/	This is a very efficient stand-alone algorithm suitable for the population-based analysis to find SNP markers.

(*Continued*)

TABLE 13.1 (*Continued*)

List of the Tools and Web Servers for the Functional SNPs and Mutations

Tools/Web Servers	Year	Links	Description
AssociationViewer	2009	http://associationviewer.vital-it.ch/	Provides a graphical representation of genomic data analysis by GWAS; also provides analysis of SNP-related studies such as gene annotation and linkage disequilibrium.
SNPHarvester	2009	http://bioinformatics.ust.hk/SNPHarvester.html	Helps in the identification of SNP-SNP interactions by creating several paths of SNP groups of statistical significance and associated with disease.
Path	2009	https://genapha.hli.ubc.ca/PathTutorial/	Path provides an interface of multiple bioinformatics resources along with data associated with genetic studies to researchers. It identifies gene-gene interaction and helps in storing and visualizing data from GWA studies.
GSEA-SNP	2008	https://www.gsea-msigdb.org/gsea/index.jsp	Facilitates association mapping and identifies SNPs associated with disease and related pathways and helps in understanding their fundamental biological mechanisms. Helps in detecting markers which are undetectable in GWAS.
PolySearch	2008	http://polysearch.ca/	Web-based text-mining platform, identify the connection between disease, drug metabolites, genes, metabolic pathways, health effects (positive or negative) etc.
Snap2	2007	https://github.com/Rostlab/SNAP2	Uses neural networks to predict the effectiveness of SNVs in protein functions.
ArrayFusion	2006	http://microarray.ym.edu.tw/tools/arrayfusion	Annotates traditional CGH outputs and converts the data into a standard format.
QualitySNP	2006	https://www.bioinformatics.nl/tools/snpweb/	An algorithm to identify reliable SNPs, indels from EST data and stores in diploids and polyploid organisms using 3 filters. first, It screens all potential SNPs, second, cluster SNPs and identifies false SNPs coming from sequencing error, and third screen SNPs based on confidence scores.
SNP-VISTA	2005	https://genome.lbl.gov/vista/snpvista/	It provides deep insights into high-throughput SNP data to the users by representing the results in very interactive graphical representation and interactive exploration.
ALOHOMORA	2005	https://gmc.mdc-berlin.de/	Pedigree analysis tool with graphical representation facilitates gene mapping in genetic disease.
GeneCards 2002	2002	https://www.genecards.org/	A user-friendly, integrated, searchable database having data from different web resources, such as genetic, genomic, transcriptomic, transcriptomic, and functional information.

13.3 Conclusions

The population is experimentally proved (and predicted) to have vast numbers of SNPs, yet a small percentage of these genotypic polymorphisms are linked to phenotypic disease risks. Non-synonymous SNPs (nsSNPs) will probably have no effect since natural selection will have eliminated such variation effects in key places. The evaluation of non-neutral SNPs mostly depends on phylogeny evidence, which is supplemented to some extent by structural techniques. Much effort has been made to predict the likely effect on the phenotype of SNPs that induce amino acid alterations using various approaches. However, lately emphasis has been put on functional SNPs that influence regulatory areas or the splicing processes or even the noncoding areas of the genome. However, there is a clear proof that many human disease-associated traits are the consequence of exonic or noncoding changes influencing regulatory areas. Experimenting on the molecular mechanisms of diseases is lengthy and complex, and it's extremely difficult at the structural level, particularly when there are multiple missense mutations responsible for the disease. Theoretical and computational approaches in this direction can quickly provide relevant and accurate evidence regarding the influence of these variations on protein architecture and function. Computational algorithms and software predictions have been proven to be very useful to anticipate the biological effect of SNPs and their putative roles in gene regulation and phenotypic changes; the development of newer and efficient predictors and analysis tools thus would be of high applicability in futuristic molecular biology, precision medicine, population-level genotypic–phenotypic studies, and pharmacogenomics research, with the advancement of the SNP-related data as well in the application of machine learning and deep learning; and new algorithms and packages are being designed and tested to enhance our understanding in the field. These methods have also paved ways to integrate other large-scale datasets such as biological pathways and ontologies, making it more relevant for its application in medical sciences. However, we are still in the very beginning of such fields and we need more advancements to make them useful for humanity.

ABBREVIATIONS

AIMs:	Ancestry informative markers
API:	Application programming interface
AUPVR:	Area under the precision-vs-recall curve (higher is better)
AUROC:	Area under the receiver operating characteristic curve (higher is better)
AVGRANK:	Intralocus average score rank (lower is better)
CGAP:	Cancer Genome Anatomy Project
CGH:	Comparative genomic hybridization
CH:	Compound heterozygosity

CNV:	Copy number variation
dSNPs:	Disease-associated single nucleotide polymorphisms
eQTLs:	Expression quantitative trait loci
EST:	Expressed sequence tag
GEO:	Gene Expression Omnibus
GSV:	Genomic structural variation
GUI:	Graphical user interface
GVF:	Genome Variation Format
GWAS:	Genome-wide association studies
LD:	Linkage disequilibrium
LRS:	Long-read sequencing
MAF:	Minor allele frequency
MAS:	Marker-assisted selection
NCBI:	National Center for Biotechnology Information
nsSNVs:	Non-synonymous single nucleotide variants
PCA:	Principal component analysis
PGx:	Pharmacogenomics
RAD:	Restriction site-associated DNA
RNA-Seq:	RNA-sequencing
scATAC-seq:	Single-cell assay for transposase-accessible chromatin with sequencing
scSNVs:	Splicing consensus single nucleotide variants
SIFT:	Sorting Intolerant from Tolerant
SNPs:	Single nucleotide polymorphisms
SNVs:	single nucleotide variants
SSRs:	Simple sequence repeats
ssSNVs:	Splice site single nucleotide variants
SVs:	Structural variants
TSV:	Tab-separated values
URL:	Uniform Resource Locator
VCF:	Variant Call Format
WES:	Whole-exome sequencing
WGS:	Whole-genome sequencing

REFERENCES

Adzhubei, I., Jordan, D. M., and Sunyaev, S. R. (2013). Predicting functional effect of human missense mutations using PolyPhen-2. *Curr. Protoc. Hum. Genet.* Chapter 7, Unit7.20. doi:10.1002/0471142905.hg0720s76.

Alvarez, R. V., Li, S., Landsman, D., and Ovcharenko, I. (2018). SNPDelScore: combining multiple methods to score deleterious effects of noncoding mutations in the human genome. *Bioinformatics* 34, 289–291. doi:10.1093/bioinformatics/btx583.

Bendl, J., Musil, M., Štourač, J., Zendulka, J., Damborský, J., and Brezovský, J. (2016). Predictsnp2: a unified platform for accurately evaluating SNP effects by exploiting the different characteristics of variants in distinct genomic regions. *PLoS Comput. Biol.* 12, e1004962. doi:10.1371/journal.pcbi.1004962.

Blumenthal, D. B., Viola, L., List, M., Baumbach, J., Tieri, P., and Kacprowski, T. (2020). EpiGEN: an epistasis simulation pipeline. *Bioinformatics* 36, 4957–4959. doi:10.1093/bioinformatics/btaa245.

Camiolo, S., Sablok, G., and Porceddu, A. (2016). Altools: a user friendly NGS data analyser. *Biol. Direct* 11, 8. doi:10.1186/s13062-016-0110-0.

Capriotti, E., Calabrese, R., and Casadio, R. (2006). Predicting the insurgence of human genetic diseases associated to single point protein mutations with support vector machines and evolutionary information. *Bioinformatics* 22, 2729–2734. doi:10.1093/bioinformatics/btl423.

Carmelo, V. A. O., Kogelman, L. J. A., Madsen, M. B., and Kadarmideen, H. N. (2018). WISH-R-a fast and efficient tool for construction of epistatic networks for complex traits and diseases. *BMC Bioinformatics* 19, 277. doi:10.1186/s12859-018-2291-2.

Chen, L., and Qin, Z. S. (2016). traseR: an R package for performing trait-associated SNP enrichment analysis in genomic intervals. *Bioinformatics* 32, 1214–1216. doi:10.1093/bioinformatics/btv741.

Choi, Y., and Chan, A. P. (2015). PROVEAN web server: a tool to predict the functional effect of amino acid substitutions and indels. *Bioinformatics* 31, 2745–2747. doi:10.1093/bioinformatics/btv195.

Coleman, I., Corleone, G., Arram, J., Ng, H.-C., Magnani, L., and Luk, W. (2020). GeDi: applying suffix arrays to increase the repertoire of detectable SNVs in tumour genomes. *BMC Bioinformatics* 21, 45. doi:10.1186/s12859-020-3367-3.

Dong, X., Zhang, L., Milholland, B., Lee, M., Maslov, A. Y., Wang, T., and Vijg, J. (2017). Accurate identification of single-nucleotide variants in whole-genome-amplified single cells. *Nat. Methods* 14, 491–493. doi:10.1038/nmeth.4227.

Fang, Y.-H., Wang, J.-H., and Hsiung, C. A. (2017). TSGSIS: a high-dimensional grouped variable selection approach for detection of whole-genome SNP-SNP interactions. *Bioinformatics* 33, 3595–3602. doi:10.1093/bioinformatics/btx409.

Goudey, B., Rawlinson, D., Wang, Q., Shi, F., Ferra, H., Campbell, R. M., Stern, L., Inouye, M. T., Ong, C. S., and Kowalczyk, A. (2013). GWIS--model-free, fast and exhaustive search for epistatic interactions in case-control GWAS. *BMC Genomics* 14 Suppl 3, S10. doi:10.1186/1471-2164-14-S3-S10.

Ho, D. S. W., Schierding, W., Wake, M., Saffery, R., and O'Sullivan, J. (2019). Machine learning SNP based prediction for precision medicine. *Front. Genet.* 10, 267. doi:10.3389/fgene.2019.00267.

Huang, M.-C., Chuang, T.-P., Chen, C.-H., Wu, J.-Y., Chen, Y.-T., Li, L.-H., and Yang, H.-C. (2016). An integrated analysis tool for analyzing hybridization intensities and genotypes using new-generation population-optimized human arrays. *BMC Genomics* 17, 266. doi:10.1186/s12864-016-2478-8.

Huang, P.-J., Lin, H.-H., Lee, C.-C., Chiu, L.-Y., Wu, S.-M., Yeh, Y.-M., Tang, P., Chiu, C.-H., Lyu, P.-C., and Tsai, P.-C. (2019). CoMutPlotter: a web tool for visual summary of mutations in cancer cohorts. *BMC Med. Genomics* 12, 99. doi:10.1186/s12920-019-0510-y.

Huber, W., Carey, V. J., Gentleman, R., Anders, S., Carlson, M., Carvalho, B. S., Bravo, H. C., Davis, S., Gatto, L., Girke, T., et al. (2015). Orchestrating high-throughput genomic analysis with Bioconductor. *Nat. Methods* 12, 115–121. doi:10.1038/nmeth.3252.

Jiang, D., Xin, C., Ye, J., Yuan, Y., and Fang, M. (2019). ICGRM: integrative construction of genomic relationship matrix combining multiple genomic regions for big dataset. *BMC Bioinformatics* 20, 731. doi:10.1186/s12859-019-3319-y.

Johnson, A. D., Handsaker, R. E., Pulit, S. L., Nizzari, M. M., O'Donnell, C. J., and de Bakker, P. I. W. (2008). SNAP: a web-based tool for identification and annotation of proxy SNPs using HapMap. *Bioinformatics* 24, 2938–2939. doi:10.1093/bioinformatics/btn564.

Jombart, T., and Ahmed, I. (2011). adegenet 1.3-1: new tools for the analysis of genome-wide SNP data. *Bioinformatics* 27, 3070–3071. doi:10.1093/bioinformatics/btr521.

Kelley, L. A., Mezulis, S., Yates, C. M., Wass, M. N., and Sternberg, M. J. E. (2015). The Phyre2 web portal for protein modeling, prediction and analysis. *Nat. Protoc.* 10, 845–858. doi:10.1038/nprot.2015.053.

Kim, H. J., Yu, Z., Lawson, A., Zhao, H., and Chung, D. (2018). Improving SNP prioritization and pleiotropic architecture estimation by incorporating prior knowledge using graph-GPA. *Bioinformatics* 34, 2139–2141. doi:10.1093/bioinformatics/bty061.

Kim, S. A., Brossard, M., Roshandel, D., Paterson, A. D., Bull, S. B., and Yoo, Y. J. (2019). gpart: human genome partitioning and visualization of high-density SNP data by identifying haplotype blocks. *Bioinformatics* 35, 4419–4421. doi:10.1093/bioinformatics/btz308.

Kobayashi, M., Ohyanagi, H., Takanashi, H., Asano, S., Kudo, T., Kajiya-Kanegae, H., Nagano, A. J., Tainaka, H., Tokunaga, T., Sazuka, T., et al. (2017). Heap: a highly sensitive and accurate SNP detection tool for low-coverage high-throughput sequencing data. *DNA Res.* 24, 397–405. doi:10.1093/dnares/dsx012.

Lee, D., Williamson, V. S., Bigdeli, T. B., Riley, B. P., Webb, B. T., Fanous, A. H., Kendler, K. S., Vladimirov, V. I., and Bacanu, S.-A. (2016). JEPEGMIX: gene-level joint analysis of functional SNPs in cosmopolitan cohorts. *Bioinformatics* 32, 295–297. doi:10.1093/bioinformatics/btv567.

Li, H. (2016). BGT: efficient and flexible genotype query across many samples. *Bioinformatics* 32, 590–592. doi:10.1093/bioinformatics/btv613.

Llinares-López, F., Papaxanthos, L., Roqueiro, D., Bodenham, D., and Borgwardt, K. (2019). CASMAP: detection of statistically significant combinations of SNPs in association mapping. *Bioinformatics* 35, 2680–2682. doi:10.1093/bioinformatics/bty1020.

Machiela, M. J., and Chanock, S. J. (2015). LDlink: a web-based application for exploring population-specific haplotype structure and linking correlated alleles of possible functional variants. *Bioinformatics* 31, 3555–3557. doi:10.1093/bioinformatics/btv402.

Makarov, V., O'Grady, T., Cai, G., Lihm, J., Buxbaum, J. D., and Yoon, S. (2012). AnnTools: a comprehensive and versatile annotation toolkit for genomic variants. *Bioinformatics* 28, 724–725. doi:10.1093/bioinformatics/bts032.

Mangan, M. E., Williams, J. M., Kuhn, R. M., and Lathe, W. C. (2014). The UCSC genome browser: what every molecular biologist should know. *Curr. Protoc. Mol. Biol.* 107, 19.9. 1–36. doi:10.1002/0471142727.mb1909s107.

Marth, G. T., Korf, I., Yandell, M. D., Yeh, R. T., Gu, Z., Zakeri, H., Stitziel, N. O., Hillier, L., Kwok, P. Y., and Gish, W. R. (1999). A general approach to single-nucleotide polymorphism discovery. *Nat. Genet.* 23, 452–456. doi:10.1038/70570.

McLaren, W., Gil, L., Hunt, S. E., Riat, H. S., Ritchie, G. R. S., Thormann, A., Flicek, P., and Cunningham, F. (2016). The ensembl variant effect predictor. *Genome Biol.* 17, 122. doi:10.1186/s13059-016-0974-4.

Mei, H., Li, L., Jiang, F., Simino, J., Griswold, M., Mosley, T., and Liu, S. (2016). snpGeneSets: an R package for genome-wide study annotation. *G3 (Bethesda)* 6, 4087–4095. doi:10.1534/g3.116.034694.

Nickerson, D. A., Tobe, V. O., and Taylor, S. L. (1997). PolyPhred: automating the detection and genotyping of single nucleotide substitutions using fluorescence-based resequencing. *Nucleic Acids Res.* 25, 2745–2751. doi:10.1093/nar/25.14.2745.

Nishizaki, S. S., Ng, N., Dong, S., Porter, R. S., Morterud, C., Williams, C., Asman, C., Switzenberg, J. A., and Boyle, A. P. (2020). Predicting the effects of SNPs on transcription factor binding affinity. *Bioinformatics* 36, 364–372. doi:10.1093/bioinformatics/btz612.

Nodzenski, M., Shi, M., Krahn, J. M., Wise, A. S., Li, Y., Li, L., Umbach, D. M., and Weinberg, C. R. (2021). GADGETS: a genetic algorithm for detecting epistasis using nuclear families. *Bioinformatics*. doi:10.1093/bioinformatics/btab766.

Nousiainen, K., Kanduri, K., Ricaño-Ponce, I., Wijmenga, C., Lahesmaa, R., Kumar, V., and Lähdesmäki, H. (2018). snpEnrichR: analyzing co-localization of SNPs and their proxies in genomic regions. *Bioinformatics* 34, 4112–4114. doi:10.1093/bioinformatics/bty460.

Page, A. J., Taylor, B., Delaney, A. J., Soares, J., Seemann, T., Keane, J. A., and Harris, S. R. (2016). SNP-sites: rapid efficient extraction of SNPs from multi-FASTA alignments. *Microb. Genom.* 2, e000056. doi:10.1099/mgen.0.000056.

Pers, T. H., Timshel, P., and Hirschhorn, J. N. (2015). SNPsnap: a web-based tool for identification and annotation of matched SNPs. *Bioinformatics* 31, 418–420. doi:10.1093/bioinformatics/btu655.

Pina-Martins, F., Vieira, B. M., Seabra, S. G., Batista, D., and Paulo, O. S. (2016). 4Pipe4--A 454 data analysis pipeline for SNP detection in datasets with no reference sequence or strain information. *BMC Bioinformatics* 17, 41. doi:10.1186/s12859-016-0892-1.

Pritikin, J. N., Neale, M. C., Prom-Wormley, E. C., Clark, S. L., and Verhulst, B. (2021). GW-SEM 2.0: efficient, Flexible, and Accessible Multivariate GWAS. *Behav. Genet.* 51, 343–357. doi:10.1007/s10519-021-10043-1.

Prunier, J., Lemaçon, A., Bastien, A., Jafarikia, M., Porth, I., Robert, C., and Droit, A. (2019). LD-annot: a bioinformatics tool to automatically provide candidate SNPs with annotations for genetically linked genes. *Front. Genet.* 10, 1192. doi:10.3389/fgene.2019.01192.

Rastas, P. (2017). Lep-MAP3: robust linkage mapping even for low-coverage whole genome sequencing data. *Bioinformatics* 33, 3726–3732. doi:10.1093/bioinformatics/btx494.

Rogers, M. F., Shihab, H. A., Mort, M., Cooper, D. N., Gaunt, T. R., and Campbell, C. (2018). FATHMM-XF: accurate prediction of pathogenic point mutations via extended features. *Bioinformatics* 34, 511–513. doi:10.1093/bioinformatics/btx536.

Rozas, J., Ferrer-Mata, A., Sánchez-DelBarrio, J. C., Guirao-Rico, S., Librado, P., Ramos-Onsins, S. E., and Sánchez-Gracia, A. (2017). Dnasp 6: DNA sequence polymorphism analysis of large data sets. *Mol. Biol. Evol.* 34, 3299–3302. doi:10.1093/molbev/msx248.

Sahl, J. W., Lemmer, D., Travis, J., Schupp, J. M., Gillece, J. D., Aziz, M., Driebe, E. M., Drees, K. P., Hicks, N. D., Williamson, C. H. D., et al. (2016). NASP: an accurate, rapid method for the identification of SNPs in WGS datasets that supports flexible input and output formats. *Microb. Genom.* 2, e000074. doi:10.1099/mgen.0.000074.

Shang, J., Sun, Y., Liu, J.-X., Xia, J., Zhang, J., and Zheng, C.-H. (2016). CINOEDV: a co-information based method for detecting and visualizing n-order epistatic interactions. *BMC Bioinformatics* 17, 214. doi:10.1186/s12859-016-1076-8.

Shih, C.-Y., Chattopadhyay, A., Wu, C.-H., Tien, Y.-W., and Lu, T.-P. (2021). Transcript annotation tool (TransAT): an R package for retrieving annotations for transcript-specific genetic variants. *BMC Bioinformatics* 22, 350. doi:10.1186/s12859-021-04243-z.

Sim, N.-L., Kumar, P., Hu, J., Henikoff, S., Schneider, G., and Ng, P. C. (2012). SIFT web server: predicting effects of amino acid substitutions on proteins. *Nucleic Acids Res.* 40, W452–W457. doi:10.1093/nar/gks539.

Slowikowski, K., Hu, X., and Raychaudhuri, S. (2014). SNPsea: an algorithm to identify cell types, tissues and pathways affected by risk loci. *Bioinformatics* 30, 2496–2497. doi:10.1093/bioinformatics/btu326.

Sun, K., Lun, F. F. M., Jiang, P., and Sun, H. (2017). BSviewer: a genotype-preserving, nucleotide-level visualizer for bisulfite sequencing data. *Bioinformatics* 33, 3495–3496. doi:10.1093/bioinformatics/btx505.

Szkiba, D., Kapun, M., von Haeseler, A., and Gallach, M. (2014). SNP2GO: functional analysis of genome-wide association studies. *Genetics* 197, 285–289. doi:10.1534/genetics.113.160341.

Terada, A., Yamada, R., Tsuda, K., and Sese, J. (2016). LAMPLINK: detection of statistically significant SNP combinations from GWAS data. *Bioinformatics* 32, 3513–3515. doi:10.1093/bioinformatics/btw418.

Tesi, N., van der Lee, S., Hulsman, M., Holstege, H., and Reinders, M. J. T. (2021). snpXplorer: a web application to explore human SNP-associations and annotate SNP-sets. *Nucleic Acids Res.* 49, W603–W612. doi:10.1093/nar/gkab410.

Timonen, O., Särkkä, M., Fülöp, T., Mattsson, A., Kekäläinen, J., and Paananen, J. (2019). Varanto: variant enrichment analysis and annotation. *Bioinformatics* 35, 3154–3156. doi:10.1093/bioinformatics/btz046.

Wang, P. P. S., Parker, W. T., Branford, S., and Schreiber, A. W. (2016). BAM-matcher: a tool for rapid NGS sample matching. *Bioinformatics* 32, 2699–2701. doi:10.1093/bioinformatics/btw239.

Wang, W., Wang, Z., Li, X., Ni, Z., Hu, Z., Xin, M., Peng, H., Yao, Y., Sun, Q., and Guo, W. (2020a). SnpHub: an easy-to-set-up web server framework for exploring large-scale genomic variation data in the post-genomic era with applications in wheat. *Gigascience* 9. doi:10.1093/gigascience/giaa060.

Wang, Z., Wang, N., Wang, Z., Jiang, L., Wang, Y., Li, J., and Wu, R. (2020b). HiGwas: how to compute longitudinal GWAS data in population designs. *Bioinformatics* 36, 4222–4224. doi:10.1093/bioinformatics/btaa294.

Wangkumhang, P., Chaichoompu, K., Ngamphiw, C., Ruangrit, U., Chanprasert, J., Assawamakin, A., and Tongsima, S. (2007). WASP: a Web-based Allele-Specific PCR assay designing tool for detecting SNPs and mutations. *BMC Genomics* 8, 275. doi:10.1186/1471-2164-8-275.

Weckx, S., Del-Favero, J., Rademakers, R., Claes, L., Cruts, M., De Jonghe, P., Van Broeckhoven, C., and De Rijk, P. (2005). novoSNP, a novel computational tool for sequence variation discovery. *Genome Res.* 15, 436–442. doi:10.1101/gr.2754005.

Xu, Z., and Taylor, J. A. (2009). SNPinfo: integrating GWAS and candidate gene information into functional SNP selection for genetic association studies. *Nucleic Acids Res.* 37, W600–W605. doi:10.1093/nar/gkp290.

Yang, J., Lee, S. H., Goddard, M. E., and Visscher, P. M. (2011). GCTA: a tool for genome-wide complex trait analysis. *Am. J. Hum. Genet.* 88, 76–82. doi:10.1016/j.ajhg.2010.11.011.

Yao, Y., Liu, Z., Wei, Q., and Ramsey, S. A. (2019). CERENKOV2: improved detection of functional noncoding SNPs using data-space geometric features. *BMC Bioinformatics* 20, 63. doi:10.1186/s12859-019-2637-4.

Zafar, H., Wang, Y., Nakhleh, L., Navin, N., and Chen, K. (2016). Monovar: single-nucleotide variant detection in single cells. *Nat. Methods* 13, 505–507. doi:10.1038/nmeth.3835.

Zhang, H., Zhao, N., Mehrotra, D. V., and Shen, J. (2020). Composite Kernel Association Test (CKAT) for SNP-set joint assessment of genotype and genotype-by-treatment interaction in Pharmacogenetics studies. *Bioinformatics* 36, 3162–3168. doi:10.1093/bioinformatics/btaa125.

Zheng, J., Rodriguez, S., Laurin, C., Baird, D., Trela-Larsen, L., Erzurumluoglu, M. A., Zheng, Y., White, J., Giambartolomei, C., Zabaneh, D., et al. (2017). HAPRAP: a haplotype-based iterative method for statistical fine mapping using GWAS summary statistics. *Bioinformatics* 33, 79–86. doi:10.1093/bioinformatics/btw565.

Zhong, K., Karssen, L. C., Kayser, M., and Liu, F. (2016). CollapsABEL: an R library for detecting compound heterozygote alleles in genome-wide association studies. *BMC Bioinformatics* 17, 156. doi:10.1186/s12859-016-1006-9.

Zhong, W., Spracklen, C. N., Mohlke, K. L., Zheng, X., Fine, J., and Li, Y. (2019). Multi-SNP mediation intersection-union test. *Bioinformatics* 35, 4724–4729. doi:10.1093/bioinformatics/btz285.

Zhu, J., Sanborn, J. Z., Benz, S., Szeto, C., Hsu, F., Kuhn, R. M., Karolchik, D., Archie, J., Lenburg, M. E., Esserman, L. J., et al. (2009). The UCSC cancer genomics browser. *Nat. Methods* 6, 239–240. doi:10.1038/nmeth0409-239.

14

Role of Single Nucleotide Polymorphisms (SNPs) in Common Migraine

Syed Sameer Aga
Basic Medical Sciences, College of Medicine, King Saud bin Abdulaziz University for Health Sciences (KSAU-HS), King Abdullah International Medical Research Center (KAIMRC), Ministry of National Guard Health Affairs (NGHA), Jeddah, Kingdom of Saudi Arabia
Infinity Biochemistry Pvt. Ltd. Srinagar, Kashmir, India

Mujeeb Zafar Banday and Saniya Nissar
Department of Basic Medical Sciences, Bahria University Medical & Dental College, Karachi, Pakistan

Kulsum Akhtar
Department of Biochemistry, Sher-I-Kashmir Institute of Medical Sciences (SKIMS) Soura, Srinagar, India

Razaz Felemban and Moudi M. Alasmari
Department of Basic Medical Sciences, College of Medicine, King Saud bin Abdulaziz University for Health Sciences (KSAU-HS), King Abdullah International Medical Research Center (KAIMRC), Ministry of National Guard Health Affairs (NGHA), Jeddah, Kingdom of Saudi Arabia

Faheem Hyder Pottoo
Department of Pharmacology, College of Clinical Pharmacy, Imam Abdul Rahman Bin Faisal University, Dammam, Saudi Arabia

Sheikh A. Saeed
Department of Physiology, Bahria University Medical & Dental College, Karachi, Pakistan

CONTENTS

DOI: 10.1201/9781003246244-14

ABSTRACT Migraine is regarded as the most common disease affecting human functioning. Migraine is classified as the common type of primary headache disorder characterized by the "episodic attacks of severe unilateral throbbing pain." It is known to significantly reduce the quality of life of an individual and his/her capacity to do work. According to the International Classification of Headache Disorders, migraine is broadly classified into two types – migraine without aura (MO) and migraine with aura (MA), although both have a significant overlap of symptoms. The pathological mechanisms of migraine are very complex, but the role of various genetic polymorphisms have been reported in almost all populations of the world. In this chapter, we are going to elucidate the role played by various genetic polymorphisms of some critical genes (*PRDM, TRPM8, MTDH, MEF2D, MMP16, LRP, MTHFR, BDNF, NGF, TNF,* and *ESRs*) in the susceptibility to migraine in light of the available literature and meta-analyses.

KEYWORDS: *Polymorphism; Single Nucleotide Polymorphism; Migraine; Risk Factors; Aura; Meta-analyses; Susceptibility; Headache; Pain.*

14.1 Introduction

Migraine is a common disabling neurovascular disease that has a complex etiology and is usually characterized by episodic and debilitating unilateral attacks of headache for 1–3 days (IHS, 2018). It is often linked with photophobia, phonophobia, and autonomic dysfunction and/or nausea or vomiting (de Boer et al., 2019, 2020).

Overall burden of migraine is found in about 15% of the worldwide population, and women (18%) are found to be affected more than men (6%) (Kors et al., 2003). The higher prevalence of migraine in women is because of the estrogen effects (de Boer et al., 2019). Typically, migraines start during puberty or between the ages of 35 and 45 years (Stewart et al., 1994). Worldwide, it is categorized as the third most widespread disorder and can be easily diagnosed clinically by the proposed criteria of International Classification of Headache Disorders (ICHD, 2021; Arnold, 2018).

Migraine is a classic disabling disorder, which has a bearing on the day-to-day functioning of an individual. It is known to significantly reduce the quality of life of an individual and his/her capacity to do work (de Boer et al., 2019, 2020). Migraine is classified into various types depending upon the specific features and the severity of the symptoms (ICHD, 2021; Arnold, 2018). Furthermore, migraine is considered as the complex genetic disorder with a heritability of 0.34–0.81, and numerous mutations and polymorphisms of various pathway genes involved in the etiopathogenesis of the disease (Sutherland et al., 2019; Kaur et al., 2019a,b,c; de Vries et al., 2009).

14.2 Migraine

Migraine is classified as the common type of primary headache disorder characterized by the "*episodic attacks of severe unilateral throbbing pain*" (ICHD, 2021; Arnold, 2018). Usually, the secondary symptoms of photophobia and phonophobia together with nausea or vomiting are also associated with the disease. Migraine is ranked as the second most disabling disorder worldwide, which is often associated with other neurological or psychiatric disorders such as stroke, epilepsy, and depressive disorders (GBD, 2016; de Boer et al., 2019).

Although the pathophysiology of migraine is not well understood, some insights into the mechanism have been attributed to the events such as cortical spreading of depression (CSD) and activation of the trigeminovascular system together with its constituent neuropeptides (Sutherland et al., 2019; Goadsby et al., 2017; Ferrari et al., 2015). Since migraine is a brain-associated disorder, it is assumed to arise from its loss of homeostasis-control leading to the activation of nociceptors innervating the cranial blood vessels. This is eventually reflected in the increased release of vasoactive inflammatory mediators such as nitric oxide coupled with sensitization of the peripheral trigeminovascular neurons to dural stimuli both causing a signature throbbing pain of the disorder (Burstein et al., 2011).

Numerous external stimuli and factors have been identified that could trigger migraine, including chemicals, lack of sleep, stress, and skipping meals (Sutherland et al., 2019). Furthermore, in migraineurs, a chronic hypersensitivity to various sensory stimuli has also been found, which leads to the abnormal processing of the sensory information (Stankewitz et al., 2011).

14.3 Types of Migraine

According to the International Classification of Headache Disorders, migraine is broadly classified into two types – migraine without aura (MO) and migraine with aura (MA), although both have a significant overlap of symptoms (ICHD, 2021; de Boer et al., 2019; Arnold, 2018). Aura is a transient episode of focal neurologic dysfunction borne out of imbalance between excitatory and inhibitory neuronal activity at different levels in the central nervous system (CNS) (de Boer et al., 2019, 2020). A single episode of migraine usually follows four distinct phases: (1) prodrome, (2) aura, (3) headache (pain), and (4) postdrome, but not all migraineurs do experience all four phases.

14.3.1 Migraine without Aura (MO)

MO is usually defined as a recurrent headache disorder characterized by episodic attacks lasting between 4 and 72 hours and accompanying specific features and symptoms such as pulsating quality, unilateral location, moderate or severe pain intensity aggravation by routine physical activity, and association with nausea and/or photophobia and phonophobia (ICHD, 2021; Arnold, 2018).

The main criterion for a migraine to be classified as MO is that at least five attacks of headaches should occur within a time span of 4–72 hours (ICHD, 2021; de Boer et al., 2019; Arnold, 2018). The detailed diagnostic criteria are provided in Table 14.1.

14.3.2 Migraine with Aura (MA)

MA, also referred to as classical migraine, is defined as the recurrent headache that last for only minutes and is of unilateral, fully reversible visual, sensory, or other CNS symptoms. MA is known to develop gradually over 5–20 minutes and last less than 60 minutes, followed by headache and associated migraine symptoms (ICHD, 2021; Arnold, 2018).

The people who have MA develop temporary typical visual signs and symptoms such as blind spots (scotomas), zigzag lines across the field of vision, shimmering spots or stars, vision changes and flashes of light, or tingling in hand or face (Medscape, 2021). The detailed diagnostic criteria for MA are provided in Table 14.1.

MA is further divided into other categories depending upon the peculiar symptoms, the areas of brain they arise from, or involvement of motor weakness as migraine with brainstem aura, retinal migraine, or hemiplegic migraine (ICHD, 2021; Arnold, 2018). Hemiplegic migraine can be of either sporadic (SHM) or familial origin (FHM).

TABLE 14.1

Diagnostic Criteria for Different Categories of Migraine

1. Diagnostic criteria: Migraine without aura
   - A. At least five attacks[a] fulfilling criteria B–D
   - B. Headache attacks lasting 4–72 hours (untreated or unsuccessfully treated)[b,c]
   - C. Headache having at least two of the following four characteristics:
     1. Unilateral location
     2. Pulsating quality
     3. Moderate or severe pain intensity
     4. Aggravation by or causing avoidance of routine physical activity (e.g., walking or climbing stairs)
   - D. During headache, at least one of the following:
     1. Nausea and/or vomiting
     2. Photophobia and phonophobia
   - E. Not better accounted for by another ICHD-3 diagnosis.

[a] One or a few migraine attacks may be difficult to distinguish from symptomatic migraine-like attacks. Furthermore, the nature of a single or a few attacks may be difficult to understand. Therefore, at least five attacks are required. Individuals who otherwise meet criteria for migraine without aura, but have had fewer than five attacks should be coded as probable migraine without aura.

[b] When the patient falls asleep during migraine and wakes up without it, the duration of the attack is reckoned until the time of awakening.

[c] In children and adolescents (aged under 18 years), attacks may last 2–72 hours (the evidence for untreated durations of less than 2 hours in children has not been substantiated).

2. Diagnostic criteria: Migraine with aura
   - A. At least two attacks fulfilling criteria B and C
   - B. One or more of the following fully reversible aura symptoms:
     1. Visual
     2. Sensory
     3. Speech and/or language
     4. Motor
     5. Brainstem
     6. Retinal
   - C. At least three of the following six characteristics:
     1. at least one aura symptom spreads gradually over ≥5 minutes
     2. two or more aura symptoms occur in succession
     3. each individual aura symptom lasts 5–60 minutes[a]
     4. at least one aura symptom is unilateral[b]
     5. at least one aura symptom is positive[c]
     6. the aura is accompanied, or followed within 60 minutes, by headache
   - D. Not better accounted for by another ICHD-3 diagnosis.

[a] When, for example, three symptoms occur during an aura, the acceptable maximal duration is 3×60 minutes. Motor symptoms may last up to 72 hours.

[b] Aphasia is always regarded as a unilateral symptom; dysarthria may or may not be.

[c] Scintillations and pins and needles are positive symptoms of aura.

Sporadic hemiplegic migraine (SHM) occurs in individuals bearing no family history of the disease, while familial hemiplegic migraine (FHM) occurs in individuals where there is at least one first- or second-degree relative with HM disorder (ICHD, 2021; Torres-Ferrús et al., 2020; Burstein et al., 2015; Thomsen et al., 2002, 2003). In general population, nearly 67% of the HM cases are FHM as it follows the autosomal dominant pattern of inheritance with 70%–90% penetrance (de Boer et al., 2019; Silberstein & Dodick, 2013). FHM etiology is monogenic with genetic heterogeneity occurring because of the missense mutations in one of the three identified genes – *CACNA1A*, *ATP1A2*, and *SCN1A*. Accordingly, FHM is classified as FHM1 (MIM #141500), FHM2 (MIM #602481), or FHM3 (MIM #609634) depending upon whether the mutant gene is *CACNA1A*, *ATP1A2*, or *SCN1A*, respectively (Sutherland et al., 2019; de Boer et al., 2019).

14.4 Role of SNPs in Migraine

Out of various risk factors identified for the migraine, genetic predisposition stands out as the chief contributor to the disease. In the early 1990s, it was recognized that many patients diagnosed with migraine also have first-degree relative suffering from it as well (Russell & Olesen, 1996; Montagna, 2000; Wessman et al., 2004; de Vries et al., 2009; Anttila et al., 2018). With the advancement of sequencing techniques and completion of 1000 Genomes Project (1000 GPC, 2015), numerous genome-wide association studies (GWAS) and meta-analyses have furthermore corroborated the link between the various genetic variations and the risk of migraine – both sporadic and familial types (de Vries et al., 2009; Ligthart et al., 2011; Di Lorenzo et al., 2012; Gormley et al., 2016; Anttila et al., 2013, 2018; Sutherland & Griffiths, 2017; Sutherland et al., 2019).

For example, a rare form of migraine with motor aura called familial hemiplegic migraine (FHM) has been classified into three types – FHM1, FHM2, and FHM3, based on the type of genes which are mutated. For FHM1, *CACNA1A* gene (19p13) is found to be mutated; for FHM2, *ATP1A2* gene (1q23) and, for FHM3, *SCN1A* gene (2q24) are mutated (Montagna, 2000; Di Lorenzo et al., 2012; Sutherland & Griffiths, 2017).

Additionally, with various combined data, it has been estimated that the heritability of migraine surpasses 40%, establishing the genetic susceptibility to the disease with numerous metabolic, hormonal, vascular, inflammatory, and neurotrophic genes involved directly or indirectly with etiopathology (Montagna, 2000; Polderman et al., 2015; Gormley et al., 2016). The role of genetic polymorphisms in the modulation of the migraine risk has become evident with numerous GWAS and meta-analyses on different populations around the world (Ligthart et al., 2011; Freilinger et al., 2012; Gormley et al., 2016; Anttila et al., 2013; Polderman et al., 2015; Pickrell et al., 2016; Nyholt et al., 2017; Kaur et al., 2019a,b).

Various genetic polymorphisms currently identified can be divided into different pathways, based on their function as follows.

14.4.1 Neurological Genes

These are the group of genes that are directly involved in the expression of various ion channels on the cellular surfaces. These ion channels include candidate genes for calcium channel, voltage-dependent, P/Q type, alpha 1A subunit (CACNA1A), voltage-potassium intermediate/small conductance calcium-activated channel and subfamily N, member 3 (KCCN3), and Na+/K+-ATPase subunits. These genes and their products have a direct bearing on the synapse development, neurotransmission, neuroplasticity, and neurosensitivity (Sutherland & Griffiths, 2017; Sutherland et al., 2019).

14.4.2 Neurotransmission Genes

These genes (*MTDH*, *MEF2D*, *ASTN2*, and *FHL5*) are involved in the synthesis, release, and binding of neuropeptides (calcitonin gene-related peptide) or neurotransmitters (glutamate, GABA, dopamine, and serotonin) relevant to neuronal excitation, neuroplasticity, and neurosensitivity, and/or to nociception (Sutherland et al., 2019; de Boer et al., 2019, 2020).

14.4.3 Vascular Genes

This group of genes (*ACE*, *TGFBR2*, *PHACTR1*, *MTHFR*, *NOTCH3*, and *EDNRA*) are involved in the regulation of vascular hemostasis and metabolism, which play a role in endothelial cell function, vasoconstriction, and vasodilation. Many vascular genes associated with migraine also confer risk of stroke and heart disease. These functional variants in some of the vascular genes may cause migraine (Freilinger et al., 2012; Gormley et al., 2016; Bron et al., 2021).

14.4.4 Neurotrophic Genes

These include two important genes coding for brain-derived neurotrophic factor (BDNF) and nerve growth factor (NGF), both of which play an important role in synaptic plasticity, neurogenesis, neural growth, and differentiation (Coskun et al., 2016; Martins et al., 2017; Cai et al., 2017; Terrazzino et al., 2017).

14.4.5 Hormonal Genes

This group of genes are CYP450-dependent genes such as estrogen receptor 1 (ESR1), estrogen receptor 2 (ESR2), progesterone receptor (PGR), androgen receptor (AR), follicle-stimulating hormone receptor (FSHR), and nuclear receptor interacting protein 1 (NRIP1), which are involved in the metabolism of estrogen and progesterone especially relating to menstrual migraine (Gasparini et al., 2013; Freilinger et al., 2012; Gormley et al., 2016).

14.4.6 Inflammatory Genes

This group of genes include pro-inflammatory cytokine-coding genes (TNF-α; ILs; LTa; and LTb), which play a role in the activation of mast cells and macrophages resulting in neurogenic inflammation. Cytokines are one of the essential mediators of the immune and inflammatory pathways; moreover, their receptors are found to be widely expressed in the CNS by all types of cells, including neurons, indicating that they can act on neuronal receptors (Bruno et al., 2007; Anttila et al., 2013; de Boer et al., 2019, 2020).

14.5 Important Gene SNPs in Migraine

Here, we have attempted to discuss the most important genes that have been found to modulate the risk of migraine significantly in various populations around the world.

14.5.1 PRDM

Positive regulatory domain containing 16 (PRDM16) is an important regulatory protein encoded by *PRDM16* gene on chromosome 1 (p36.32) (Huret & Senon, 2004). PRDM16 belongs to the family of zinc finger-containing proteins, which play a key role in adipose biology by acting as a transcription co-regulator that determines the differentiation and development of brown adipose tissue (Kajimura, 2015; Chi & Cohen, 2016).

One of the key polymorphisms that are reported to affect the structure and functionality of PRDM16 protein is rs2651899. This is an intronic SNP, which is thought to affect the splicing pattern of *PRDM16* and its downstream regulatory elements, thereby affecting the expression of the protein which in turn results in the dysregulation of brown fat metabolism, resulting in higher BMIs (AlAmrani et al., 2018).

Numerous studies involving many diverse populations have reported on the association of *PRDM16* rs2651899 polymorphism with migraine (Chasman et al., 2011; Anttila et al., 2013; An et al., 2013b, 2017b; Fan et al., 2014; Ghosh et al., 2013; Sintas et al., 2015; Kaur et al., 2019a; Ebahimzadeh et al., 2021).

In a recent meta-analysis by Lee et al. (2020) involving six studies constituting 2,853 patients, it was concluded that *PRDM16* rs2651899 polymorphism was significantly associated with common migraine (both MA and MO). Moreover, Zhao et al. (2020) in their meta-analysis of 39 studies identified *PRDM16* gene as one of the six hub genes (*MEF2D, TSPAN2, PHACTR1, TRPM8*, and *PRDM16*) for its significant role in migraine susceptibility.

14.5.2 TRPM8

TRPM8 (transient receptor potential melastatin 8) is a member of TRP family coded by *TRMP8* gene located in chromosome 2 in humans (Bautista et al., 2007). It is also known as cold- and menthol-activated receptor 1 (CMR1), which functions as thermally regulated non-selective Ca^{2+}-permeable cation channel in most sensory neurons (Liu et al., 2020). TRPM8 channels have been demonstrated to get activated by temperatures <26°C in addition to cooling agents such as menthol and icilin (Dussor & Cao, 2016). Since cold temperature is one of the triggers of migraine and about 50% of patients exhibit cold allodynia, the role of TRPM8 in migraine susceptibility is a significant one (Burstein et al., 2000; Prince et al., 2004). Furthermore, numerous studies have implicated the increased expression of TRPM8 in cancers, urinary tract dysfunction, dry eye disease, obesity, and migraine; therefore, it serves as one of the novel biomarkers and a putative target of therapy as well (González-Muñiz et al., 2019; Liu et al., 2016).

Additionally, various studies have reported on the association of TRPM8 SNPs with migraine, out of which two are most important – SNP rs17862920 located within the first intron and SNP rs10166942 located 1 kb upstream of the transcription start of *TRPM8* gene (Chasman et al., 2011; Freilinger et al., 2012; Ling et al, 2019; Zhao et al., 2020).

In their study on 1904 Taiwanese individuals, Ling et al. (2019) reported that patients with *TRPM8* rs10166942 T allele were more likely to have chronic migraine than individuals with non-T allele carriers and concluded that it may contribute to the pathogenesis of chronic migraine. Zhao et al. also concluded TRPM8 SNP to be associated with the susceptibility to migraine in their own meta-analysis.

14.5.3 MTDH

Metadherin (MTDH), also known as astrocyte elevated gene-1 (AEG-1) or LYRIC (lysine-rich CEACAM1), is one of the key tumor-associated antigens (TAAs) (Britt et al., 2004). Two isoforms of MTDH are coded by the gene locus located on the chromosome 8 (8q22.1) (MTDH-Genecards, 2021). MTDH is a single domain transmembrane protein (64 kDa) expressed mainly in the endoplasmic reticulum and perinuclear space in the cell (Dhiman et al., 2019).

Furthermore, MTDH is known to mediate the downregulation of EAAT2 (excitatory amino acid transporter 2), which otherwise is a major glutamate transporter affecting the glutamate regulation at a synaptic level (Ran et al., 2018). In addition, MTDH activates the nuclear factor kappa B (NF-kB) transcription factor, promotes anchorage-independent growth, helps in cell invasion and metastasis, and reduces the effects of senescence (Dhiman et al., 2019; Kaur et al., 2019b). MTDH-mediated downregulation of EAAT2 results in the dearth of major glutamate transporter protein in neurons, resulting in the higher concentration of excitatory neurotransmitter glutamate in CNS (Goadsby, 2005). Additionally, migraineurs have been reported to have a higher level of glutamate and glutamine in the cerebrospinal fluids, which is correlated with the mean headache scores (Peres et al., 2004).

The importance of *MTDH* gene for migraine is because of its expression modulation by one of the important intergenic SNPs rs1835740 (Kaur et al., 2019b; de Boer et al., 2020). Ran et al. (2018) in their study on Swedish population concluded that *MTDH* rs1835740 SNP is associated with migraine as well as cluster headaches (CHs). Ligthart et al. (2011) had previously concluded the same in their own study on European migraine patients.

14.5.4 MEF2D

Myocyte-specific enhancer factor 2D (MEF2D) belongs to a family of specialized transcription factors, which play a pivotal role in the muscular and neuronal cell differentiation and development. They are tightly regulated by class II histone deacetylases (MEF2D-NCBI, 2021). *MEF2D* gene is found on chromosome 1 (1q22) (Chasman et al., 2014; Anttila et al., 2013).

One of the important SNPs present in the *MEFD2* gene, which is directly associated with the migraine susceptibility, is rs2274316 (Chasman et al., 2014; Anttila et al., 2013). Zhao et al. (2020) in their meta-analysis of 39 studies identified *MEF2D* gene as one of the six hub genes for its significant role in migraine susceptibility. Moreover, a study by An et al. (2017) identified *MEF2D* together with two other genes (*PRDM16* and *ASTN2*) to be associated with the susceptibility to migraine. However, a study by Fu et al. (2019) in Han Chinese population did not find any association between *MEF2D* rs2274316 SNP and migraine.

14.5.5 MMP16

Matrix metalloproteinases (MMPs) belong to a family of proteolytic enzymes that have the capability of disintegrating a variety of cell surface receptors, extracellular matrix, and numerous signaling molecules (Pei, 2005). Matrix metalloproteinase 16 (MMP16) is a membrane-type MMP that is encoded by *MMP* gene located at 8q21.3 (MMP16-NCBI, 2021). Although there is no plausible evidence for the actual mechanism of MMP16 in migraine, two important indirect functions do provide an insight. One of the key functions of MMP16 is the downstream cleavage of low-density lipoprotein receptor-related protein 1, LRP1 (Rozanov et al., 2004), and the other is its involvement in shedding of basal NgR1 in cortical neurons resulting in increasing axonal and synaptic plasticity (Chasman et al., 2011; Borrie et al., 2012; rs10504861-SNPedia, 2021).

One SNP located 200 Kb upstream of the *MMP16* gene has been found to affect the susceptibility to migraine (Anttila et al., 2013; Sintas et al., 2015). Anttila et al. (2013) had reported the association of *MMP16* rs10504861 SNP with migraine, but Sintas et al. (2015) in their study on 512 Spanish migraine patients reported only a nominal association; however, replicative studies by Lin et al. (2015) and Esserlind et al. (2016) failed to demonstrate any association.

14.5.6 LRP1

Low-density lipoprotein receptor-related protein 1 (LRP1) is one of the key transmembrane receptors that play multifunctional roles in the cell, including lipoprotein endocytosis and scavenging extracellular ligands, and regulates the metabolism and signaling receptor for integrins and tyrosine kinases (Potere et al., 2019; Qu et al., 2021). LRP1 is encoded by *LRP1* gene located on chromosome 12 (12q13.3) (Kaur et al., 2019b; Schürks, 2012).

LRP1's role in migraine is because of its interaction with the N-methyl-D-aspartate (NMDA) receptors affecting the glutamate pathway (Chasman et al., 2011). Additionally, one of its SNPs rs11172113 present in the first intron of the gene has been implicated in the migraine susceptibility (Kaur et al., 2019b; Chasman et al., 2011).

In their replicative study on Indian population, Ghosh et al. (2013) reported the significant protective association of *LRP1* rs11172113 SNPs with the migraine. Similar results were reported by Esserlind et al. (2013b) in their study on clinic-based Danish and Icelandic cohort. In contrast, Freilinger et al. (2012), An et al. (2013b), and Fan et al. (2014) in their respective studies reported a non-association of *LRP1* rs11172113 SNP with migraine.

14.5.7 MTHFR

5,10-Methylenetetrahydrofolate reductase (MTHFR) is one of the important enzymes in folate metabolism pathway, which converts 5,10-methylenetetrahydrofolate (5,10-MTHF) into 5-methyl-tetrahydrofolate (5-MTHF), which in turn serves as a co-substrate for homocysteine remethylation to synthesize methionine (Nazki et al., 2014). The enzyme is coded by *MTHFR* gene, located on chromosome 1p36.22 and composed of 11 exons. The gene codes for a 74.6 kDa protein of 656 amino acids (Saffroy et al., 2005).

Of the important functional polymorphisms that have been reported for *MTHFR* gene, two common SNPs –C677T (rs1801133) and A1298C (rs1801131) – have been studied ubiquitously around the world. *MTHFR* C677T SNP results in an alanine-to-valine substitution at codon 222 of the MTHFR protein and has a profound effect on the functionality of enzyme in reducing its thermal stability and activity (Frosst et al., 1995; Sameer et al., 2011). Individuals with the variant Val/Val genotype (TT) have no more than 30% of normal enzyme activity, which eventually results in causing lower levels of 5-methyl-tetrahydrofolate, an accumulation of 5,10-methylenetetrahydrofolate and an increased plasma homocysteine levels (Sameer et al., 2011; Kaur et al., 2018). *MTHFR* A1298C SNP occurs in the exon 7 at codon 429, causing the substitution of glutamine with alanine. This SNP results in the diminished activity of enzyme up to about 33% (Weisberg et al., 1998).

Numerous studies have been carried out on *MTHFR* SNPs and their association with migraine around the world. While many reported the positive association of *MTHFR* SNPs with migraine (Scher et al., 2006; Rubino et al., 2009; Stuart et al., 2012; An et al., 2013a; Kaur et al., 2018), others fail to draw any such conclusions (Todt et al., 2006; Kaunisto et al., 2006). Rubino et al. (2009) in their meta-analysis involving six major studies with 2,961 migraineurs from four different geographical regions concluded that the MTHFR TT genotype was associated with a higher risk of migraine with aura (MA) as compared with the CC genotype. Similarly, another meta-analysis by Liu et al. (2014) involving 17 different studies constituting of 8,903 migraineurs concluded that TT genotype was associated with a significantly higher risk of migraine with aura (MA) in Asians.

14.5.8 BDNF & NGF

Neurotrophic factors (NFs) play a key role in the functioning of the CNS, as well as in the development and modulation of pain (Martins et al., 2015; Merighi et al. 2008). Of the various NFs, two members have been implicated in the susceptibility to migraine i.e., brain-derived neurotrophic factor (BDNF) and nerve growth factor (NGF) (Martins et al., 2015, 2017). BDNF gene is located in the short arm of chromosome 11 (11p14.1) (BDNF-GeneCards, 2021), while NGF gene is located on the short arm of chromosome 1 (1p13.2) (NGF-GeneCards, 2021).

BDNF is the most common NF and plays an important role in synaptic plasticity, neurogenesis, neural growth, and survival of several neurons located in regions associated with behavior, mood, and pain modulation (Merighi et al., 2008). BDNF primarily functions at the central synapses in pain pathways and modulates glutamatergic (fast excitatory) and GABAergic/glycinergic (inhibitory) signals, as well as slow peptidergic neurotransmission in spinal cord (Martins et al., 2017; Merighi et al., 2008). NGF is demonstrated to be associated with peripheral nociception, especially during inflammatory conditions (Allen & Dawbarn, 2006). NGF plays a pivotal role in neuronal survival, neuroplasticity, and learning, which is crucial for reacting to stress (Yeh et al., 2015).

Many studies have investigated the role of various SNPs in the *BDNF* and *NGF* genes in modulating the risk of migraine. Of these, some important polymorphisms such as rs6265(-G/A), rs8192466(C/T), rs925946(G/T), rs2049046(A/T), and rs12273363(T/C) in the *BDNF* gene, and rs6330(C/T), rs11466112(C/T), rs11102930(C/A), and rs4839435(G/A) in the NGF gene have been reported in the literature (Yeh et al., 2015; Coskun et al., 2016; Cai et al., 2017; Terrazzino et al., 2017).

In their sentinel meta-analysis on *BDNF* SNPs, Terrazzino et al. (2017) included five studies with a total of 1,442 cases. They concluded that there was an increased risk of migraine associated with a dominant model of *BDNF* rs6265 SNP. Similar results were reported by Coskun et al. (2016) in their Turkish case-control study, in which they found BDNF rs6265 minor allele A to be borderline significant in migraineurs. In addition, they also reported that the minor TT genotype of rs6330 SNP of NGF is significantly higher in MA patients than in MO or healthy controls. Cai et al. (2017) also reported a significant association between *BDNF* rs6265 and rs2049046 (A/T) SNPs and migraine in allele, recessive, and additive models, but not in dominant model. They concluded that both *BDNF* rs6265 and rs2049046 (A/T) SNPs were associated with common migraine in Caucasian population.

14.5.9 TNF

Tumor necrosis factor alpha (TNF-α) is one of the predominant pro-inflammatory cytokines produced primarily by activated macrophages, lymphocytes, neutrophils, and neurons (Schwabe & Brenner, 2006). TNF-α plays a vital role in several processes, including autoimmunity, inflammation, cell growth, differentiation, proliferation, and apoptosis (Waters et al., 2013). TNF-α has a key role in migraine pathophysiology because it is known to activate the transcription of calcitonin gene-related peptide (CGRP) gene whose expression in eternal jugular vein is demonstrated to be significant in migraineurs (Durham, 2006; Goadsby et al., 1990).

The *TNF-α* gene is present on the long arm of chromosome 6 (6p21), within the major histocompatibility complex (MHC) class III area (Carroll et al., 1987). Among numerous SNPs present in *TNF-α* gene, the most common TNF-α polymorphism in general population is rs1800629 (–308G/A) SNP and is located in the promoter region of the gene at a position 308 bp from the transcriptional start site (Hajeer & Hutchinson, 2001). This SNP results in the enhanced constitutive and inducible expression TNF-α (Wilson et al., 1997). It has been reported to be associated with certain autoimmune, neoplastic, and infectious diseases and migraine as well (Wilson et al., 1995; Yilmaz et al., 2010; Gu et al., 2012).

Numerous studies and meta-analyses have reported the association of *TNF-α* rs1800629 (–308G/A) SNP with differences in susceptibility to migraine (Yilmaz et al., 2010; Ghosh et al., 2010; Gu et al., 2012). Ghosh et al. (2010) in their study on Indian population concluded that *TNF-α* rs1800629 (–308G/A) SNP was significantly associated with female migraineurs. Gu et al. (2012) in their meta-analysis of five studies from four countries concluded that it was associated with migraine risk in Asian population.

14.5.10 ESRs

Estrogen receptors (ESRs) are important receptors present in every cell, which are ligand-activated by estrogen (Enmark et al., 1997). On binding with the estrogen, they get activated to function as transcriptional activators of numerous metabolic genes by binding to their specific promoter region called hormone response elements, thereby increasing their expression (Enmark et al., 1997; Fuentes & Silveyra, 2019; Talarowska et al., 2019).

There are two kinds of ESRs – ES-α and ES-β – that are coded by two different *ESR* genes – *ESR-1* and *ESR-2*, respectively (Oterino et al., 2008; Fuentes & Silveyra, 2019); while *ESR-1* gene is located on the long arm of chromosome 6 (6q25.1), *ESR-2* gene is located on the long arm of chromosome 14 (14q23–24) (Enmark et al., 1997; Fuentes & Silveyra, 2019; Talarowska et al., 2019). The role played by *ESR* genes in migraine susceptibility is attributed to their function in translating the effects of sex hormones and the hormone treatment, especially estrogens (Gupta et al., 2007; Schürks et al., 2010).

Numerous polymorphisms have been identified in both *ESR-1* and *ESR-2* genes, and many of them have been identified to be a risk factor for migraine (Schürks et al., 2010; Joshi et al., 2010; Li et al., 2015; CoŞkun et al., 2016; An et al., 2017a). In their seminal study on Spanish migraineurs, Oterino et al. (2006) had demonstrated the significant association of *ESR1* rs1801132 (325G>C) polymorphism with migraine susceptibility; however, Kaunisto et al. (2006) failed to demonstrate the same in their replicative study in Finnish population. Additionally, Li et al. (2015) reported that genetic SNPs in *ESR1* gene may play a pivotal role in the pathogenesis of migraine in Caucasian populations, especially for *ESR1* rs1801132 (325G>C) and *ESR1* rs 2228480 (594G>A) SNPs. Similarly, Schürks et al. (2010) in their meta-analysis of four studies involving five different populations also reported that the two *ESR1* SNPs rs1801132 and rs 2228480 are associated with migraine susceptibility. Contrarily, Joshi et al. (2010) reported that *ESR1* rs2234693 (PvuII) is significantly associated with the risk of migraine in females but not two *ESR1*

SNPs rs1801132. These results were also reported by An et al. (2017a) by identifying *ESR1* rs2234693 (PvuII) as the crucial candidate SNP related to menstrual migraine susceptibility.

14.6 Conclusions

In this chapter, we tried to elucidate the role of essential genes that have extensively been linked to migraine susceptibility, in order to draw a broader picture for the genetics of migraine. However, as the field of genetics is an ever-expanding area, additional research with high-powered studies is warranted to delineate the association of various other less penetrative genes of numerous pathways with migraine across different populations of the world. Furthermore, more tractable functional studies to elucidate the commonalities between monogenic and polygenic migraine will allow in-depth information on mechanisms and treatment strategies for common migraine. Additionally, exhaustive genome-wide association studies (GWAS) on different populations would also help in identifying the common genetic SNPs involved in the pathogenesis of migraine, which could serve as easy biomarkers. These biomarkers would be helpful in revealing the phenotypic characteristics of migraineurs for monitoring the treatment response and in identifying new targets for pharmacotherapy and in future can lead to easy transition to personalized medicine in migraineurs.

ABBREVIATIONS

AEG-1: Astrocyte elevated gene-1
AR: Androgen receptor
BDNF: Brain-derived neurotrophic factor
CMR1: Cold- and menthol-activated receptor 1
CNS: Central nervous system
CSD: Cortical spreading of depression
DNA: Deoxyribonucleic acid
ESR: Estrogen receptor
FHM: Familial hemiplegic migraine
FSHR: Follicle-stimulating hormone receptor
GPs: Genetic polymorphisms
GVs: Genetic variations
GWAS: Genome-wide association studies
HM: Hemiplegic migraine
ICHD: International Classification of Headache Disorders
IHS: International Headache Society
LRP1: Low-density lipoprotein receptor-related protein 1
MA: Migraine with aura
MEF2D: Myocyte-specific enhancer factor 2D
MMPs: Matrix metalloproteinases
MO: Migraine without aura
MTHFR: 5,10-Methylenetetrahydrofolate reductase
NF-kB: Nuclear factor kappa B
NFs: Neurotrophic factors
NGF: Nerve growth factor
NRIP1: Nuclear receptor interacting protein 1
PGR: Progesterone receptor
PRDM16: Positive regulatory domain containing 16
SHM: Sporadic hemiplegic migraine
SNP: Single nucleotide polymorphism
TAAs: Tumor-associated antigens
TNF-α: Tumor necrosis factors alpha
TRPM8: Transient receptor potential melastatin 8
UTR: Untranslated region

REFERENCES

AlAmrani A, AbdelKarim M, AlZoghaibi M. PRDM16 gene polymorphism is associated with obesity and blood lipids profiles in Saudi population. *J Clin Med.* 2018;7(6):141.

Allen SJ, Dawbarn D. Clinical relevance of the neurotrophins and their receptors. *Clin Sci (Lond).* 2006;110(2):175–91. doi: 10.1042/CS20050161.

An X, Fang J, Lin Q, Lu C, Ma Q, Qu H. New evidence for involvement of ESR1 gene in susceptibility to Chinese migraine. *J Neurol.* 2017a;264(1):81–7. doi: 10.1007/s00415-016-8321-y.

An XK, Fang J, Yu ZZ, Lin Q, Lu CX, Qu HL, Ma QL. Multilocus analysis reveals three candidate genes for Chinese migraine susceptibility. *Clin Genet.* 2017b;92(2):143–9. doi: 10.1111/cge.12962.

An XK, Lu CX, Ma QL, Zhang XR, Burgunder JM, Lin Q, Qu HL. Association of MTHFR C677T polymorphism with susceptibility to migraine in the Chinese population. *Neurosci Lett.* 2013a;549:78–81. doi: 10.1016/j.neulet.2013.06.028.

An XK, Ma QL, Lin Q, Zhang XR, Lu CX, Qu HL. PRDM16 rs2651899 variant is a risk factor for Chinese common migraine patients. *Headache.* 2013b;53(10):1595–601. doi: 10.1111/head.12212.

Anttila V, Wessman M, Kallela M, Palotie A. Genetics of migraine. *Handb Clin Neurol.* 2018;148:493–503. doi: 10.1016/B978-0-444-64076-5.00031-4.

Anttila V, Winsvold BS, Gormley P, et al. Genome-wide meta-analysis identifies new susceptibility loci for migraine. *Nat Genet.* 2013;45:912–7.

Arnold M. Headache classification Committee of the International Headache Society (IHS) the international classification of headache disorders. *Cephalalgia.* 2018;38(1):1–211.

Bautista DM, Siemens J, Glazer JM, et al. The menthol receptor TRPM8 is the principal detector of environmental cold. *Nature.* 2007;448:204–8. doi: 10.1038/nature05910.

BDNF Gene. Gene Cards. https://www.genecards.org/cgi-bin/carddisp.pl?gene=BDNF. Accessed August 2021.

Borrie SC, Baeumer BE, Bandtlow CE. The Nogo-66 receptor family in the intact and diseased CNS. *Cell Tissue Res.* 2012;349(1):105–17. doi: 10.1007/s00441-012-1332-9.

Britt DE, Yang D-F, Yang D-Q, et al. Identification of a novel protein, LYRIC, localized to tight junctions of polarized epithelial cells. *Exp Cell Res.* 2004;300:134–48. doi: 10.1016/j.yexcr.2004.06.026.

Bron C, Sutherland HG, Griffiths LR. Exploring the hereditary nature of migraine. *Neuropsychiatr Dis Treat.* 2021;17:1183–94. doi: 10.2147/NDT.S282562.

Bruno PP, Carpino F, Carpino G, Zicari A. An overview on immune system and migraine. *Eur Rev Med Pharmacol Sci.* 2007;11(4):245–8.

Burstein R, Jakubowski M, Rauch SD. The science of migraine. *J Vestib Res.* 2011;21(6):305–14.

Burstein R, Noseda R, Borsook D. Migraine: multiple processes, complex pathophysiology. *J Neurosci.* 2015;35(17):6619–29. doi: 10.1523/JNEUROSCI.0373-15.2015.

Burstein R, Yarnitsky D, Goor-Aryeh I, Ransil BJ, Bajwa ZH. An association between migraine and cutaneous allodynia. *Ann Neurol*. 2000;47:614–62.

Cai X, Shi X, Zhang X, Zhang A, Zheng M, Fang Y. The association between brain-derived neurotrophic factor gene polymorphism and migraine: a meta-analysis. *J Headache Pain*. 2017;18(1):13. doi: 10.1186/s10194-017-0725-2.

Carroll MC, Katzman P, Alicot EM, Koller BH, Geraghty DE, Orr HT, Strominger JL, Spies T. Linkage map of the human major histocompatibility complex including the tumor necrosis factor genes. *Proc Natl Acad Sci U S A*. 1987;84(-23):8535–9. doi: 10.1073/pnas.84.23.8535.

Chasman DI, Anttila V, Buring JE, Ridker PM, Schürks M, Kurth T; International Headache Genetics Consortium. Selectivity in genetic association with sub-classified migraine in women. *PLoS Genet*. 2014 May 22;10(5):e1004366. doi: 10.1371/journal.pgen.1004366. Erratum in: PLoS Genet. 2015 Jun;11(6):e1005330. PMID: 24852292; PMCID: PMC4031047.

Chasman DI, Schürks M, Anttila V, et al. Genome-wide association study reveals three susceptibility loci for common migraine in the general population. *Nat Genet*. 2011;43(-7):695–8. doi: 10.1038/ng.856.

Chi J, Cohen P. The multifaceted roles of PRDM16: adipose biology and beyond. *Trends Endocrinol Metab*. 2016;27(1):11–23. doi: 10.1016/j.tem.2015.11.005.

Classification Committee of The International Headache Society (IHS). The international classification of headache disorders, 3rd edition. *Cephalalgia*. 2018;38(1):1–211.

Coskun S, Varol S, Ozdemir HH, Agacayak E, Aydın B, Kapan O, Camkurt MA, Tunc S, Cevik MU. Association of brain-derived neurotrophic factor and nerve growth factor gene polymorphisms with susceptibility to migraine. *Neuropsychiatr Dis Treat*. 2016;12:1779–85. doi: 10.2147/NDT.S108814.

CoŞkun S, Yůcel Y, Çim A, Cengiz B, Oztuzcu S, Varol S, Özdemir HH, Uzar E. Contribution of polymorphisms in ESR1, ESR2, FSHR, CYP19A1, SHBG, and NRIP1 genes to migraine susceptibility in Turkish population. *J Genet*. 2016;95(1):131–40. doi: 10.1007/s12041-016-0625-2.

de Boer I, Terwindt GM, van den Maagdenberg AMJM. Genetics of migraine aura: an update. *J Headache Pain*. 2020;21(1):64. doi: 10.1186/s10194-020-01125-2.

de Boer I, van den Maagdenberg AMJM, Terwindt GM. Advance in genetics of migraine. *Curr Opin Neurol*. 2019;32(3):413–21. doi: 10.1097/WCO.0000000000000687.

de Vries B, Frants RR, Ferrari MD, van den Maagdenberg AM. Molecular genetics of migraine. *Hum Genet*. 2009;126(-1):115–32. doi: 10.1007/s00439-009-0684-z.

Dhiman G, Srivastava N, Goyal M, Rakha E, Lothion-Roy J, Mongan NP, Miftakhova RR, Khaiboullina SF, Rizvanov AA, Baranwal M. Metadherin: a therapeutic target in multiple cancers. *Front Oncol*. 2019;9:349. doi: 10.3389/fonc.2019.00349.

Di Lorenzo C, Grieco GS, Santorelli FM. Migraine headache: a review of the molecular genetics of a common disorder. *J Headache Pain*. 2012;13(7):571–80. doi: 10.1007/s10194-012-0478-x.

Durham PL. Calcitonin gene-related peptide (CGRP) and migraine. *Headache*. 2006;46 Suppl 1(Suppl 1):S3–8. doi: 10.1111/j.1526-4610.2006.00483.x.

Dussor G, Cao YQ. TRPM8 and Migraine. *Headache*. 2016;56(9):1406–17. doi: 10.1111/head.12948.

Ebahimzadeh K, Gholipour M, Samadian M, Taheri M, Ghafouri-Fard S. A Comprehensive review on the role of genetic factors in the pathogenesis of migraine. *J Mol Neurosci*. 2021. doi: 10.1007/s12031-020-01788-1. Erratum in: *J Mol Neurosci*. 2021.

Enmark E, Pelto-Huikko M, Grandien K, Lagercrantz S, Lagercrantz J, Fried G, Nordenskjöld M, Gustafsson JA. Human estrogen receptor beta-gene structure, chromosomal localization, and expression pattern. *J Clin Endocrinol Metab*. 1997;82(12):4258–65. doi: 10.1210/jcem.82.12.4470.

Esserlind AL, Christensen AF, Le H, Kirchmann M, Hauge AW, Toyserkani NM, Hansen T, Grarup N, Werge T, Steinberg S, Bettella F, Stefansson H, Olesen J. Replication and meta-analysis of common variants identifies a genome-wide significant locus in migraine. *Eur J Neurol*. 2013;20(5):765–72. doi: 10.1111/ene.12055.

Esserlind AL, Christensen AF, Steinberg S, Grarup N, Pedersen O, Hansen T, Werge T, Hansen TF, Husemoen LL, Linneberg A, Budtz-Jorgensen E, Westergaard ML, Stefansson H, Olesen J. The association between candidate migraine susceptibility loci and severe migraine phenotype in a clinical sample. *Cephalalgia*. 2016;36(7):615–23. doi: 10.1177/0333102415570492.

Fan X, Wang J, Fan W, Chen L, Gui B, Tan G, Zhou J. Replication of migraine GWAS susceptibility loci in Chinese Han population. *Headache*. 2014;54(4):709–15. doi: 10.1111/head.12329.

Ferrari MD, Klever RR, Terwindt GM, Ayata C, van den Maagdenberg AM. Migraine pathophysiology: lessons from mouse models and human genetics. *Lancet Neurol*. 2015;14(1):65–80.

Freilinger T, Anttila V, de Vries B, International Headache Genetics Consortium, et al. Genome-wide association analysis identifies susceptibility loci for migraine without aura. *Nat Genet*. 2012;44(7):777–82. doi: 10.1038/ng.2307.

Frosst P, Blom HJ, Milos R, et al. A candidate genetic risk factor for vascular disease: a common mutation in methylenetetrahydofolatereductase. *Nat Genet*. 1995;10:111–3.

Fu X, Yang J, Wu X, Lin Q, Zeng Y, Xia Q, Cao L, Huang B, Huang G. Association between PRDM16, MEF2D, TRPM8, LRP1 gene polymorphisms and migraine susceptibility in the She ethnic population in China. *Clin Invest Med*. 2019;42(1):E21–30. doi: 10.25011/cim.v42i1.32389.

Fuentes N, Silveyra P. Estrogen receptor signaling mechanisms. *Adv Protein Chem Struct Biol*. 2019;116:135–70. doi: 10.1016/bs.apcsb.2019.01.001.

Gasparini CF, Sutherland HG, Griffiths LR. Studies on the pathophysiology and genetic basis of migraine. *Curr Genomics*. 2013;14(5):300–15. doi: 10.2174/13892029113149990007.

GBD 2016 Disease and Injury Incidence and Prevalence Collaborators. Global, regional, and national incidence, prevalence, and years lived with disability for 328 diseases and injuries for 195 countries, 1990–2016: a systematic analysis for the Global Burden of Disease Study 2016. *Lancet*. 2017;390(10100):1211–59. doi: 10.1016/S0140-6736(17)32154-2. Erratum in: *Lancet*. 2017;390(10106):e38.

Ghosh J, Joshi G, Pradhan S, Mittal B. Investigation of TNFA 308G>A and TNFB 252G>A polymorphisms in genetic susceptibility to migraine. *J Neurol*. 2010;257(6):898–904. doi: 10.1007/s00415-009-5430-x.

Ghosh J, Pradhan S, Mittal B. Genome-wide-associated variants in migraine susceptibility: a replication study from North India. *Headache*. 2013;53(10):1583–94. doi: 10.1111/head.12240.

Goadsby PJ. Migraine pathophysiology. *Headache*. 2005;45 (Suppl 1):S14–24.

Goadsby PJ, Edvinsson L, Ekman R. Vasoactive peptide release in the extracerebral circulation of humans during migraine headache. *Ann Neurol*. 1990;28(2):183–7. doi: 10.1002/ana.410280213.

Goadsby zPJ, Holland PR, Martins-Oliveira M, Hoffmann J, Schankin C, Akerman S. Pathophysiology of migraine: a disorder of sensory processing. *Physiol Rev*. 2017;97(2):553–622.

González-Muñiz R, Bonache MA, Martín-Escura C, Gómez-Monterrey I. Recent progress in TRPM8 modulation: an update. *Int J Mol Sci*. 2019;20(11):2618. doi: 10.3390/ijms 20112618.

Gormley P, Anttila V, Winsvold BS, et al., A. Meta-analysis of 375,000 individuals identifies 38 susceptibility loci for migraine. *Nat Genet*. 2016;48(8):856–66. doi: 10.1038/ng.3598. Erratum in: *Nat Genet*. 2016;48(10):1296.

Gu L, Yan Y, Long J, Su L, Hu Y, Chen Q, Xie J, Wu G. The TNF-α-308G/A polymorphism is associated with migraine risk: a meta-analysis. *Exp Ther Med*. 2012;3(6):1082–6. doi: 10.3892/etm.2012.533.

Gupta A, Mehrotra S, Villalón CM, Perusquía M, Saxena PR, MaasenVanDenBrink A. Potential role of female sex hormones in the pathophysiology of migraine. *Pharmacol Ther*. 2007;113:321–40.

Hajeer AH, Hutchinson IV. Influence of TNF alpha gene polymorphisms on TNFalpha production and disease. *Hum Immunol*. 2001;62(11):1191–9. doi: 10.1016/s0198-8859(01)00322-6.

Huret JL, Senon, S. PRDM16 (PR domain containing 16). *Atlas Genet Cytogenet Oncol Haematol*. 2004;8(2):76–7. Available at http://atlasgeneticsoncology.org/Genes/GC_PRDM16.html. Accessed 2021.

Joshi G, Pradhan S, Mittal B. Role of the oestrogen receptor (ESR1 PvuII and ESR1 325 C->G) and progesterone receptor (PROGINS) polymorphisms in genetic susceptibility to migraine in a North Indian population. *Cephalalgia*. 2010;30(3):311–20. doi: 10.1111/j.1468-2982.2009.01967.x.

Kajimura S. Promoting brown and beige adipocyte biogenesis through the PRDM16 pathway. *Int J Obes*. 2015;Suppl 5(Suppl 1):S11–4.

Kaunisto MA, Kallela M, Hämäläinen E, et al. Testing of variants ofthe MTHFR and ESR1 genes in 1798 Finnish individuals fails to confirm the association with migraine with aura. *Cephalalgia*. 2006;26:1462–72.

Kaur S, Ali A, Ahmad U, Pandey AK, Singh B. rs2651899 variant is associated with risk for migraine without aura from North Indian population. *Mol Biol Rep*. 2019a;46(1):1247–55. doi: 10.1007/s11033-019-04593-1.

Kaur S, Ali A, Ahmad U, Siahbalaei Y, Pandey AK, Singh B. Role of single nucleotide polymorphisms (SNPs) in common migraine. *Egypt J Neurol Psychiatry Neurosurg*. 2019b;55:47. https://doi.org/10.1186/s41983-019-0093–8

Kaur S, Ali A, Pandey AK, Singh B. Association of MTHFR gene polymorphisms with migraine in North Indian population. *Neurol Sci*. 2018;39(4):691–8. doi: 10.1007/s10072-018-3276-7.

Kaur S, Ali A, Siahbalaei Y, Ahmad U, Pandey AK, Singh B. Could rs4379368 be a genetic marker for North Indian migraine patients with aura?: Preliminary evidence by a replication study. *Neurosci Lett*. 2019c;712:134482. doi: 10.1016/j.neulet.2019.134482.

Kors E, Haan J, Ferrari M. Migraine genetics. *Curr Pain Headache Rep*. 2003;7:212–7.

Lee HH, Chen CC, Ong JR, Lin YF, Lee FP, Hu CJ, Wang YH. Association of rs2651899 polymorphism in the positive regulatory domain 16 and common migraine subtypes: a meta-analysis. *Headache*. 2020;60(1):71–80. doi: 10.1111/head.13670.

Li L, Liu R, Dong Z, Wang X, Yu S. Impact of ESR1 gene polymorphisms on migraine susceptibility: a meta-analysis. *Medicine (Baltimore)*. 2015;94(35):e0976. doi: 10.1097/MD.0000000000000976.

Ligthart L, de Vries B, Smith AV, et al. Meta-analysis of genome-wide association for migraine in six population-based European cohorts. *Eur J Hum Genet*. 2011;19(8):901–7. doi: 10.1038/ejhg.2011.48.

Lin QF, Fu XG, Yao LT, Yang J, Cao LY, Xin YT, Hou JX, Ye LF, Huang GB. Association of genetic loci for migraine susceptibility in the she people of China. *J Headache Pain*. 2015;16:553. doi: 10.1186/s10194-015-0553-1.

Ling YH, Chen SP, Fann CS, Wang SJ, Wang YF. TRPM8 genetic variant is associated with chronic migraine and allodynia. *J Headache Pain*. 2019;20(1):115. doi: 10.1186/s10194-019-1064-2.

Liu R, Geng P, Ma M, Yu S, Yang M, He M, Dong Z, Zhang W. MTHFR C677T polymorphism and migraine risk: a meta-analysis. *J Neurol Sci*. 2014;336(1–2):68–73. doi: 10.1016/j.jns.2013.10.008.

Liu Y, Mikrani R, He Y, Faran Ashraf Baig MM, Abbas M, Naveed M, Tang M, Zhang Q, Li C, Zhou X. TRPM8 channels: a review of distribution and clinical role. *Eur J Pharmacol*. 2020;882:173312. doi: 10.1016/j.ejphar.2020.173312.

Liu Z, Wu H, Wei Z, et al. TRPM8: a potential target for cancer treatment. *J Cancer Res Clin Oncol*. 2016;142:1871–81. doi: 10.1007/s00432-015-2112-1.

Martins LB, Duarte H, Ferreira AV, Rocha NP, Teixeira AL, Domingues RB. Migraine is associated with altered levels of neurotrophins. *Neurosci Lett*. 2015;587:6–10. doi: 10.1016/j.neulet.2014.12.022.

Martins LB, Teixeira AL, Domingues RB. Neurotrophins and migraine. *Vitam Horm*. 2017;104:459–73. doi: 10.1016/bs.vh.2016.10.003.

MEF2D. NCBI. https://www.ncbi.nlm.nih.gov/gene/4209. Accessed August 2021.

Merighi A, Salio C, Ghirri A, Lossi L, Ferrini F, Betelli C, Bardoni R. BDNF as a pain modulator. *Prog Neurobiol*. 2008;85(3):297–317. doi: 10.1016/j.pneurobio.2008.04.004.

Migraine with aura. Medscape. https://www.mayoclinic.org/diseases-conditions/migraine-with-aura/symptoms-causes/syc-20352072. 2021. Accessed August 2021.

MMP16. NCBI. https://www.ncbi.nlm.nih.gov/gene/4325. Accessed August 2021.

Montagna P. Molecular genetics of migraine headaches: a review. *Cephalalgia*. 2000;20(1):3–14. doi: 10.1046/j.1468-2982.2000.00003.x.

MTDH. Gene Cards. https://www.genecards.org/cgi-bin/carddisp.pl?gene=MTDH. Accessed August 2021.

Nazki FH, Sameer AS, Ganaie BA. Folate: metabolism, genes, polymorphisms, and the associated diseases. *Gene.* 2014;533(1):11–20. doi: 10.1016/j.gene.2013.09.063.

NGF Gene. Gene Cards. https://www.genecards.org/cgi-bin/carddisp.pl?gene=NGF. Accessed August 2021.

Nyholt DR, Borsook D, Griffiths LR. Migrainomics – identifying brain and genetic markers of migraine. *Nat Rev Neurol.* 2017;13(12):725–41. doi: 10.1038/nrneurol.2017.151.

Oterino A, Pascual J, Ruiz de Alegría C, Valle N, Castillo J, Bravo Y, González F, Sánchez-Velasco P, Cayón A, Leyva-Cobián F, Alonso-Arranz A, Muñoz P. Association of migraine and ESR1 G325C polymorphism. *Neuroreport.* 2006 Jan 23;17(1):61–. doi: 10.1097/01.wnr.0000192735.85287.f4. PMID: 16361951

Oterino A, Toriello M, Cayón A, Castillo J, Colas R, Alonson-Arranz A, Ruiz-Alegria C, Quintela E, Monton F, Ruiz-Lavilla N, Gonzalez F, Pascual J. Multilocus analyses reveal involvement of the ESR1, ESR2, and FSHR genes in migraine. *Headache.* 2008;48(10):1438–50. doi: 10.1111/j.1526-4610.2008.01294.x.

Pei D. Matrix metalloproteinases target protease-activated receptors on the tumor cell surface. *Cancer Cell.* 2005;7(3):207–8. doi: 10.1016/j.ccr.2005.02.011.

Peres MF, Zukerman E, SenneSoares CA, et al. Cerebrospinal fluid glutamate levels in chronic migraine. *Cephalalgia.* 2004;24(9):735–9.

Pickrell JK, Berisa T, Liu JZ, Ségurel L, Tung JY, Hinds DA. Detection and interpretation of shared genetic influences on 42 human traits. *Nat Genet.* 2016;48:709–17.

Polderman TJ, Benyamin B, de Leeuw CA, Sullivan PF, van Bochoven A, Visscher PM, Posthuma D. Meta-analysis of the heritability of human traits based on fifty years of twin studies. *Nat Genet.* 2015;47(7):702–9. doi: 10.1038/ng.3285.

Potere N, Del Buono MG, Mauro AG, Abbate A, Toldo S. Low density lipoprotein receptor-related protein-1 in cardiac inflammation and infarct healing. *Front Cardiovasc Med.* 2019;6:51. doi: 10.3389/fcvm.2019.00051.

Prince PB, Rapoport AM, Sheftell FD, Tepper SJ, Bigal ME. The effect of weather on headache. *Headache.* 2004;44:596–602.

Qu J, Fourman S, Fitzgerald M, Liu M, Nair S, Oses-Prieto J, Burlingame A, Morris JH, Davidson WS, Tso P, Bhargava A. Low-density lipoprotein receptor-related protein 1 (LRP1) is a novel receptor for apolipoprotein A4 (APOA4) in adipose tissue. *Sci Rep.* 2021;11(1):13289. doi: 10.1038/s41598-021-92711-0.

Ran C, Fourier C, Zinnegger M, Steinberg A, Sjöstrand C, Waldenlind E, Belin AC. Implications for the migraine SNP rs1835740 in a Swedish cluster headache population. *J Headache Pain.* 2018;19(1):100. doi: 10.1186/s10194-018-0937-0.

Rozanov DV, Hahn-Dantona E, Strickland DK, Strongin AY. The low density lipoprotein receptor-related protein LRP is regulated by membrane type-1 matrix metalloproteinase (MT1-MMP) proteolysis in malignant cells. *J Biol Chem.* 2004;279(6):4260–8. doi: 10.1074/jbc.M311569200.

rs10504861. SNPedia. https://www.snpedia.com/index.php/Rs10504861. Accessed August 2021.

Rubino E, Ferrero M, Rainero I, Binello E, Vaula G, Pinessi L. Association of the C677T polymorphism in the MTHFR gene withmigraine: a meta-analysis. *Cephalalgia.* 2009;29:818–25.

Russell MB, Olesen J. A nosographic analysis of the migraine aura in a general population. *Brain.* 1996;119:355–61.

Saffroy R, Lemoine A, Debuire B. MTHFR (5,10-Methylenetetrahydrofolate reductase). *Atlas Genet Cytogenet Oncol Haematol.* 2005. Available at http://AtlasGeneticsOncology.org/Genes/MTHFRID41448ch1p36.html. Accessed 2021.

Sameer AS, Shah ZA, Nissar S, Mudassar S, Siddiqi MA. Risk of colorectal cancer associated with the methylenetetrahydrofolate reductase (MTHFR) C677T polymorphism in the Kashmiri population. *Genet Mol Res.* 2011;10(2):1200–10. doi: 10.4238/vol10-2gmr1067.

Scher AI, Terwindt GM, Verschuren WM, et al. Migraine and-MTHFR C677T genotype in a population-based sample. *Ann Neurol.* 2006;59:372–5.36.

Schürks M. Genetics of migraine in the age of genome-wide association studies. *J Headache Pain.* 2012;13(1):1–9. doi: 10.1007/s10194-011-0399-0.

Schürks M, Rist PM, Kurth T. Sex hormone receptor gene polymorphisms and migraine: a systematic review and meta-analysis. *Cephalalgia.* 2010;30(11):1306–28. doi: 10.1177/0333102410364155.

Schwabe RF, Brenner DA. Mechanisms of liver injury. I. TNF-alpha-induced liver injury: role of IKK, JNK, and ROS pathways. *Am J Physiol Gastrointest Liver Physiol.* 2006;290(4):G583–9. doi: 10.1152/ajpgi.00422.2005.

Silberstein SD, Dodick DW. Migraine genetics: part II. *Headache.* 2013;53:1218–29.

Sintas C, Fernández-Morales J, Vila-Pueyo M, Narberhaus B, Arenas C, Pozo-Rosich P, Macaya A, Cormand B. Replication study of previous migraine genome-wide association study findings in a Spanish sample of migraine with aura. *Cephalalgia.* 2015;35(9):776–82. doi: 10.1177/0333102414557841.

Stankewitz A, Aderjan D, Eippert F, May A. Trigeminal nociceptive transmission in migraineurs predicts migraine attacks. *J Neurosci.* 2011;31(6):1937–43.

Stewart WF, Shechter A, Rasmussen BK. Migraine prevalence: a review of population-based studies. *Neurology.* 1994;44(6 Suppl 4):S17–23.

Stuart S, Cox HC, Lea RA, Griffiths LR. The role of the MTHFR gene in migraine. *Headache.* 2012;52(3):515–20. doi: 10.1111/j.1526-4610.2012.02106.x.

Sutherland HG, Albury CL, Griffiths LR. Advances in genetics of migraine. *J Headache Pain.* 2019;20(1):72. doi: 10.1186/s10194-019-1017-9.

Sutherland HG, Griffiths LR. Genetics of migraine: insights into the molecular basis of migraine disorders. *Headache.* 2017;57(4):537–69. doi: 10.1111/head.13053.

Talarowska ME, Szemraj J, Kuan-Pin S. Expression of ESR1 and ESR2 oestrogen receptor encoding gene and personality traits – preliminary study. *Prz Menopauzalny.* 2019;18(3):133–40. doi: 10.5114/pm.2019.90804.

Terrazzino S, Cargnin S, Viana M, Sances G, Tassorelli C. Brain-derived neurotrophic factor Val66Met gene polymorphism impacts on migraine susceptibility: a meta-analysis of case-control studies. *Front Neurol.* 2017;8:159. doi: 10.3389/fneur.2017.00159.

The International Classification of Headache Disorders. https://ichd-3.org/. Last Accessed August 2021.

Thomsen LL, Eriksen MK, Romer SF, et al. An epidemiological survey of hemiplegic migraine. *Cephalalgia.* 2002; 22:361–75.

Thomsen LL, Ostergaard E, Olesen J, Russell MB. Evidence for a separate type of migraine with aura: sporadic hemiplegic migraine. *Neurology.* 2003;60:595–601.

Todt U, Freudenberg J, Goebel I, et al. MTHFR C677T polymorphism and migraine with aura. *Ann Neurol.* 2006;60:621–2; author reply 2–3.39.

Torres-Ferrús M, Ursitti F, Alpuente A, et al. From transformation to chronification of migraine: pathophysiological and clinical aspects. *J Headache Pain.* 2020;21(1):42. doi: 10.1186/s10194-020-01111-8.

Waters JP, Pober JS, Bradley JR. Tumour necrosis factor in infectious disease. *J Pathol.* 2013;230(2):132–47. doi: 10.1002/path.4187.

Weisberg I, Tran P, Christensen B, Sibani S, Rozen R. A second genetic polymorphism in methylenetetrahydrofolate reductase (MTHFR) associated with decreased enzyme activity. *Mol Genet Metab.* 1998;64:169–72 doi: 10.1006/mgme.1998.2714.

Wessman M, Kaunisto MA, Kallela M, Palotie A. The molecular genetics of migraine. *Ann Med.* 2004;36(6):462–73. doi: 10.1080/07853890410018060.

Wilson AG, di Giovine FS, Duff GW. Genetics of tumour necrosis factor-alpha in autoimmune, infectious, and neoplastic diseases. *J Inflamm.* 1995;45(1):1–12.

Wilson AG, Symons JA, McDowell TL, McDevitt HO, Duff GW. Effects of a polymorphism in the human tumor necrosis factor alpha promoter on transcriptional activation. *Proc Natl Acad Sci U S A.* 1997;94(7):3195–9. doi: 10.1073/pnas.94.7.3195.

Yeh YW, Kuo SC, Chen CY, Liang CS, Ho PS, Yen CH, Chen TY, Shyu JF, Wan FJ, Lu RB, Huang SY. Harm avoidance involved in mediating the association between nerve growth factor (NGF) gene polymorphisms and antidepressant efficacy in patients with major depressive disorder. *J Affect Disord.* 2015;183:187–94. doi: 10.1016/j.jad.2015.05.012.

Yilmaz IA, Ozge A, Erdal ME, Edgünlü TG, Cakmak SE, Yalin OO. Cytokine polymorphism in patients with migraine: some suggestive clues of migraine and inflammation. *Pain Med.* 2010;11(4):492–7. doi: 10.1111/j.1526-4637.2009.00791.x.

Zhao Y, Zhu R, Xiao T, Liu X. Genetic variants in migraine: a field synopsis and systematic re-analysis of meta-analyses. *J Headache Pain.* 2020;21(1):13. doi: 10.1186/s10194-020-01087-5.

1000 Genomes Project Consortium (1000 GPC), Auton A, Brooks LD, Durbin RM, Garrison EP, Kang HM, Korbel JO, Marchini JL, McCarthy S, McVean GA, Abecasis GR. A global reference for human genetic variation. *Nature.* 2015;526(7571):68–74. doi: 10.1038/nature15393.

15

The Role of Vitamin D Single Nucleotide Polymorphisms in Cancer Susceptibility

Abdullah A. Awadh
Department of Basic Medical Sciences, College of Medicine,
King Saudi Bin Abdulaziz University for Health Sciences, Jeddah, Saudi Arabia

CONTENTS

ABSTRACT Vitamin D regulates calcium homeostasis and bone metabolism and modulates innate and adaptive immunological functioning, cardiovascular function, and normal and malignant keratinocyte proliferation and differentiation. The biologically active form of vitamin D, 1,25-dihydroxyvitamin D ($1,25(OH)_2D$), performs most of its actions via the widely dispersed nuclear vitamin D receptor (nVDR), which generates a heterodimer with the retinoid X receptor (RXR) upon stimulation with $1,25(OH)_2D$. The vitamin D receptor (VDR)/retinoid X receptor (RXR) complex regulates gene transcription by binding to DNA sequences known as vitamin D response elements in target genes. VDR signaling integrates with other intracellular signaling pathways to achieve its bioactivities. Despite the reported regulatory roles of vitamin D in several physiological activities, there are insufficient data to substantiate the significance of vitamin D and its receptors in the progression of cancers. Severe vitamin D deprivation or deletion of the VDR gene enhances cancer risk. In addition,

DOI: 10.1201/9781003246244-15

several studies have shown that animals injected with chemical analogs of the vitamin D hormone 1,25$(OH)_2$D have a lower risk of cancer, such as decreased tumor incidence or tumor size. This chapter discusses the function of vitamin D single nucleotide polymorphisms (SNPs) involvement in various forms of cancer.

KEYWORDS: *Vitamin D; Cancer; Vitamin D Receptor; Genomic Implications; Single Nucleotide Polymorphism.*

15.1 Introduction

Sir Edward Mellanby, a British physician, was apprehensive about the high prevalence of rickets in the UK, particularly in Scotland in the early twentieth century. Sir Mellanby was able to treat the sickness by administering cod liver oil; thus, he thought that vitamin A accounted for the disease's prevention (Mellanby, 1919). After this discovery, McCollum wanted to explore the notion that vitamin A was crucial for rickets healing. He observed that when he bubbled oxygen through cod liver oil, it degraded vitamin A and that this concoction could no longer avoid xerophthalmia and vitamin A deficiency, but it could still cure rickets (McCollum et al., 2002). McCollum et al. concluded that a new vitamin, vitamin D, is the agent that heals rickets.

Vitamin D and its metabolites are hormone precursors. Approximately 80% arises through UVB-induced photoconversion of 7-dehydrocholesterol to previtamin D3 in the skin, with the remaining portion coming from the diet and food supplements. Vitamin D3 is the natural form of vitamin D produced by UV irradiation of 7-dehydrocholesterol in the skin (Figure 15.1). Vitamin D and its metabolites are carried in the bloodstream by vitamin D-binding protein (VDBP), and once they enter their destination cells, they disassociate from VDBP and quickly penetrate the cells (Kochwvar et al., 2012; Holick, 2007; Searing et al., 2010). The significant bioactivity of 1,25$(OH)_2$D is to maintain calcium homeostasis, but it also regulates the maturation and recruitment of a vast number of cell types, including keratinocytes, endothelial cells, osteoblasts, and lymphocytes (Gallieni et al., 2009). The nuclear vitamin D receptor (nVDR), which works as a transcription factor modulating target gene transcription, is responsible for most of these physiological tasks (Corcoran et al., 2016). In healthy individuals, serum levels of 25(OH)D, the predominant circulating form of vitamin D, are employed to evaluate vitamin D deficiency, and the typical range is between 75 and 125 nmol/L (Holick, 2004). A blood level of 25(OH)D of less than 50 nmol/L is considered vitamin D deficiency (Tang et al., 2012). Vitamin D levels in the body vary, and fortification may be necessary for the aged, the disabled, and people who do not get enough

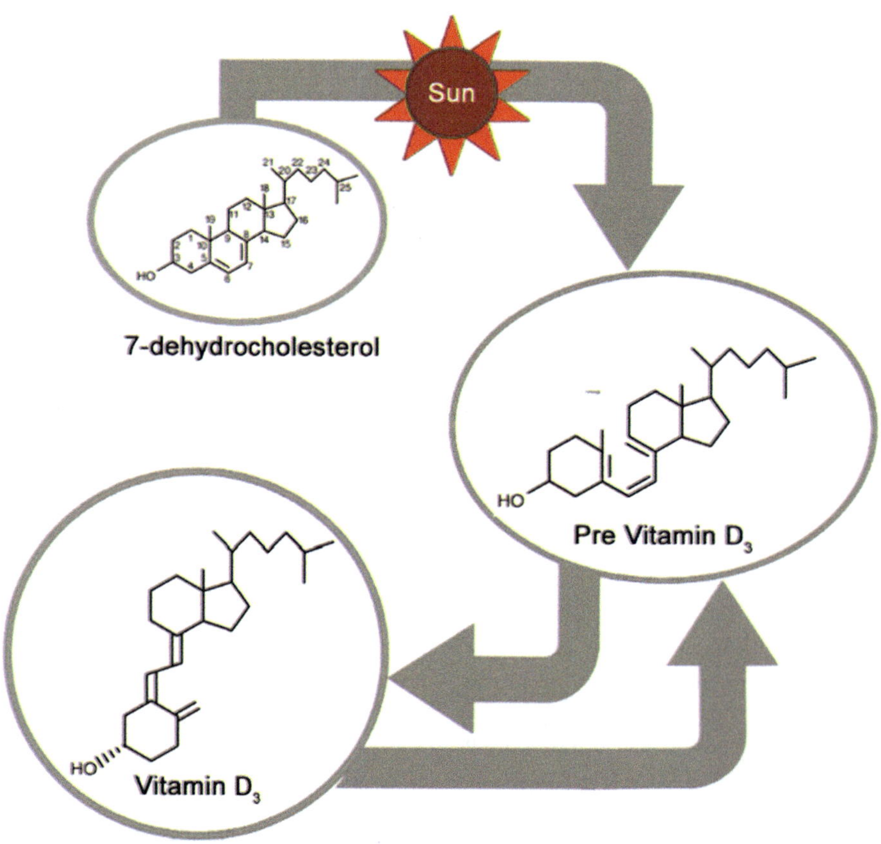

FIGURE 15.1 Conversion of 7-dehydrocholesterol to previtamin D3 in the presence of UV light.

sunlight (Dawson-Hughes et al., 2005). Multiple variables influence the blood level of physiologically active 1,25$(OH)_2D$, including the effectiveness of vitamin D precursor production in the skin, uptake from dietary sources, translation of 25(OH)D to 1,25$(OH)_2D$, and transportation and breakdown (Bringhurst et al., 2008).

Irrespective of whether 1,25$(OH)_2D$ is derived from the serum outside the target cell, 1,25$(OH)_2D$ binds vitamin D receptor (VDR) with maximum affinity for all vitamin D metabolites when observed at the molecular level. When ligated with 1,25$(OH)_2D$, the VDR predominantly forms heterodimers with the retinoid X receptor (RXR). The VDR and its unliganded RXR partner heterodimer form trans-acting complexes that bind to cis-acting vitamin D response elements (VDREs) in the genome when 1,25$(OH)_2D$ is present. The heterodimer engages with the transcriptional pathway, leading to 1,25$(OH)_2D$-regulated gene expression and positive or negative bioactivities. VDRE-like enhancer and repressor motifs can be detected a long way from the transcriptional initiation site of a vitamin D-regulated gene because of the three-dimensional "looping" characteristic of DNA-protein associations (Pike, 2011).

Inadequate vitamin D levels result in bone demineralization, which increases the risk of fractures. Vitamin D's significance in the modulation of cardiovascular system functions and its therapeutic influence on hypertension is the subject of much debate and research right now (De Luca et al., 2018; Moretti et al., 2018). In addition, vitamin D insufficiency has links to various neurological conditions. Severe vitamin D deprivation or deletion of the VDR gene enhances cancer risk (Sitrin et al., 1991; Zheng et al., 2012). In addition, many studies have shown that animals injected with chemical analogs of the vitamin D hormone 1,25$(OH)_2D$ have a lower risk of cancer, such as decreased tumor incidence or tumor size (Chen et al., 2003; Seubwai et al., 2010). While there is a widespread belief that vitamin D and its metabolites are vital for cancer prevention, the molecular basis for this shielding and the significance of vitamin D and its receptors in the progression of all cancer types are yet unknown. This chapter outlines how vitamin D metabolites regulate cancer development and discusses the function of vitamin D single nucleotide polymorphisms (SNPs) involvement in particular forms of cancer.

15.1.1 Transport of Vitamin D

The simple diffusion process allows a small amount of free or unbound 1,25$(OH)_2D$ to reach the cells. The bulk of systemic vitamin D and its metabolites are coupled to unique carrier proteins, DBP and albumin. The 1,25$(OH)_2D$-DBP complex permeates the cell via membrane proteins such as caveolin, cubilin, and megalin and simple diffusion. DBP is a steroid-binding protein produced in the liver and functions as a carrier molecule for 25(OH)D (Chun et al., 2012). It binds to 25(OH)D and 24,25$(OH)_2D$ more strongly than vitamin D or 1,25$(OH)_2D$. DBP is a reserve protein that concentrates 25(OH)D in the presence of 1,25$(OH)_2D$ and then unleashes it for bioactivity.

Patients with liver illnesses such as certain protein-losing enteropathy or nephrotic syndrome have impaired capacity for biosynthesis of sufficient levels of DBP and albumin. These individuals may have relatively lower DBP concentrations, leading to unattached vitamin D metabolites in their blood (Nussey et al., 2001). The liver is responsible for the internalization of vitamin D2- and D3-DBP complexes. Vitamin D is converted to 25(OH)D by hepatocytes. DBP-25(OH)D is taken up by receptor-mediated endocytosis in the renal tubules. Megalin and cubilin proteins are found in the proximal convoluted tube of the kidney and extrarenal tissues such as mammary cells.

15.1.2 Analogs of Vitamin D

Vitamin D supplementation has been restricted in clinical practice for many years due to the apparent risk of hypercalcemia. As a result, various organizations worldwide have looked into vitamin D analogs, which do not affect calcium levels, but still entail anti-cancer and immunomodulatory properties of vitamin D. Numerous synthetic analogs have been investigated so far, with impressive outcomes. Recent research has revealed that not only does 1,25$(OH)_2D3$ have biological action, but also its precursor, 25(OH)D3, can substantially suppress melanoma cell growth (Wasiewicz et al., 2018) or be utilized as a radio-sensitizing drug in the melanoma treatment (Podgorska et al., 2018).

Slominski and colleagues found that the active form of vitamin D (1,25$(OH)_2D3$) and one of the metabolites of CYP450ssc enzymatic activity, 20,23$(OH)_2D3$, share comparable but distinct chromosomal targets. This finding could elucidate, at least to some extent, why 20,23$(OH)_2D3$ has a lower impact on calcium levels in the blood than calcitriol (Slominski et al., 2018). Wasiewicz and co-workers investigated the anti-proliferative effects of 1,25$(OH)_2D3$, synthetic calcipotriol, and the short side chain vitamin D analog 21-hydoxypregnacalciferol (21(OH)pD) on three melanoma cell lines (Wasiewicz et al., 2018). The obtained results inferred that 21(OH)pD's anti-cancer activity is not entirely dependent on VDR expression. This finding could be significant because a lower level of VDR corresponds to the progression of the disease in melanoma, similar to its effect in many other malignancies (Slominski et al., 2018). A group of researchers investigated the impact of two vitamin D analogs, (24R)-1,24-dihydroxyvitamin D3 (PRI-2191) and 5,6-trans-isomer of calcipotriol (PRI-2205), on the solid tumor microenvironment and 4T1 mouse mammary gland cancer metastasis (Pawlik et al., 2018).

15.1.3 Vitamin D Receptors

The VDR can be found in nearly all human tissues and contains over 1,000 target genes (Maestro et al., 2016). Through a sequence of cell signaling events or as a ligand-activated transcription factor, 1,25$(OH)_2D3$ performs its influence with the support of high-affinity VDR (Rastinejad et al., 2013). The mechanism of action might be genomic or non-genomic, and it can be carried out with either nVDR or membrane VDR (mVDR) (Schwartz et al., 2002). In a cell, 1,25$(OH)_2D$ mediates its non-genomic impacts by attaching to a membrane receptor, triggering various signaling cascades.

15.1.3.1 Nuclear VDR

The VDR is extensively distributed and is not restricted to the conventional vitamin D target tissues. The VDR is a member of the nuclear receptor (NR) family. It regulates the expression of genes that regulate its bioactivity as a trans-activating transcription factor (Valdivielso et al., 2006). The ligand-activated transcription factors cause 1,25D-dependent genomic reactions in the nucleus, which are relatively gradual and may take several hours to days to elicit its response. The VDR is expressed by a large size gene (>100 kb) and is found on the long arm of chromosome 12q12 in humans. It comprises nine exons and has a complicated intron/exon architecture (Mangelsdorf et al., 1995). Exons 1a to 1f encode the VDR mRNA's 5′ untranslated region (UTR), which is variably spliced. Exons 2 and 3 contain the DNA-binding domain and encode the translation initiation site. This region, unique to the NRs, has two substantially conserved zinc-finger patterns, one for each region. The extensive ligand-binding domain (LBD) and robust heterodimerization domains are encoded by exons 4–9. The 3′ UTR is where exon 9 is found (Valdivielso et al., 2006).

The backbone of the VDR protein, which is essential in the stimulation of target genes, is a zinc-finger structure found in the DNA-binding domain of the VDR. When 1,25(OH)$_2$D3 binds to the stated receptor, two distinct protein interaction surfaces emerge on the VDR protein, one of which is required for targeted DNA binding and the other for recruiting massive coregulatory ensembles for gene modulation (Brown et al., 1998). The VDR then generates a heterodimer with the RXR. The VDRE is found in the promoter region of vitamin D target genes, and this heterodimer identifies it. The 12-helical LBD interacts with 1,25(OH)$_2$D3, which is found in the C-terminal region of the receptor (Garcia-Vallvé et al., 1998). The coregulatory proteins form a complex with 1,25(OH)$_2$D3 to either activate or suppress the transcriptional activity of the target gene. VDR trans-repression is characterized by various processes comprising the inhibition of gene expression by different transcription factors (Wei et al., 1998). Genes are transcribed, resulting in the zinc-finger transcription factor SLUG synthesis, which the VDR may use to downregulate genetic processes (Chen et al., 1998).

Polymorphisms in the nVDR gene are relatively common. Polymorphisms are modest sequence differences in a population that can be discovered utilizing techniques such as restriction fragment length polymorphisms (RFLPs). It is a strategy that makes use of differences in homologous DNA sequences. Polymorphisms in the VDR gene influence gene expression, changing protein levels and resulting in functional alterations (Uitterlinden et al., 2004). VDR gene variants have been linked to diabetes; colon and different types of cancers; immune disorders such as Crohn's disease, Grave's disease, multiple sclerosis and kidney ailments; and different neurological diseases such as schizophrenia (Huang et al., 2012). The cumulative impact of genomic and non-genomic processes is most definitely responsible for the functions of 1,25(OH)$_2$D3 (Hii and Ferrante, 2016).

15.1.3.2 Membrane VDR

The engagement of the 1,25(OH)$_2$D3 ligand with particular binding locations on the plasma membrane and vesicle membranes of target cells, termed mVDR, affects the fast, non-genomic action. mVDRs are also referred to as 1,25-membrane-associated rapid response steroid-binding proteins (1,25-MARRS), Pdia3 (protein disulfide isomerase family A member 3), ERp57, ERp60, and GRP58 (glucose-regulated protein) (Brown et al., 1999). Modified VDRs are hypothesized to be the source of these complexes. They are signal transduction systems-coupled chaperone proteins (Chen et al., 2013).

15.2 Vitamin D Signaling Pathways

This section emphasizes the signaling pathways of vitamin D, which majorly involves protein kinase C (PKC)-mediated calcium and phosphorus absorption (Farach-Carson and Nemere, 2003), phosphatidylinositol (PI) and phospholipase A2 (PLA2) expression (Sitrin et al., 1999), and the MAPK pathway (Chen et al., 1999). The coupling of 1,25(OH)$_2$D3 to 1,25-MARRS causes PKC activation in chondrocytes, resulting in a fast rise in PLA2 activity (Boyan et al., 2006). PLA2 stimulates PKC through a series of intermediate by-products, the most important being arachidonic acid. Arachidonic acid can activate PKC effectively, serving as a substrate for many enzymes. Lysophospholipids are arachidonic acid mediators that can activate phospholipase C, converting PI to diacylglycerol (DAG) and IP3 by hydrolyzation. IP3 interacts with receptors expressed on the smooth endoplasmic reticulum, prompting the flow of Ca^{2+}, which is a co-factor for PKC. PKC is activated and translocated to the plasma membrane when DAG interacts with it (Schwartz et al., 2003). Richard et al. were the first to show that 1,25-MARRS is involved in anti-cancer activities in breast cancer cells (Richard et al., 2010). The phosphorylation and expression of Akt, a kinase involved in cell survival and proliferation, is inhibited by 1,25(OH)$_2$D3 (Peehl et al., 1994). In squamous cell carcinoma cells, 1,25(OH)$_2$D3 stimulates the angiogenic factor interleukin 8 (IL-8); however, calcitriol suppresses prostate cancer cell migration and tube formation by interrupting the cascade of IL-8 signaling. In 1994, Hilliard and associates found that treating COS-1 cells with 1,25(OH)$_2$D3 increased the phosphorylation of serine residue 208 (S208) in the VDR, causing the coactivator to be recruited to form the nVDR-RXR heterodimer (Hilliard et al., 1994). However, in breast cancer cells, opposing actions of genomic and fast response pathways have been reported (Wu et al., 2010).

The genetic and quick signaling pathways engage with each other. Phosphorylation of the major proteins involved in the VDR-transcriptional complex may be responsible for the outlined process, which is frequently seen in cell cycle regulation. Relying on the cell or tissue type, the diverse signaling activities and crosstalk controlled by the 1,25(OH)$_2$D3-1,25-MARRS connection contribute to many gene transcriptional regulations. 1,25(OH)$_2$D3-1,25-MARRS facilitated fast activation of cytosolic kinases, according to Barletta and co-workers, and may be accountable for phosphorylating important coactivators engaged in a variety of pathobiological and physiological functions (Barletta et al., 2002). To some measure, the various biological impacts are caused by the target genes' genomic or non-genomic crosstalk and jargon actions. Several second messengers, particularly the rapidly accelerated fibrosarcoma (RAF)/mitogen-activated protein kinases (MAPK) pathway (Norman, 2006), may be involved with the nucleus to regulate gene expression.

15.3 Genomic and Non-genomic Implications of Vitamin D

Both genomic and non-genomic pathways mediate the activities of vitamin D and its analogs. $1,25(OH)_2D$ acts by pharmacologically distinct nuclear receptor-mediated and plasma membrane-initiated processes. $1,25(OH)_2D$ attaches to VDR in the cell nucleus, causing genomic implications, or caveolae in the plasma membrane, causing non-genomic impact. VDR is an intracellular nuclear receptor present in more than 30 human organs and is known to be activated by more than 60 genes in different cell lines. The VDR is expressed in both benign and malignant proliferating cells. Calcitriol generates heterodimers with the RXR and its ligand (9 cis-retinoic acids) when attached to VDR, and these dimers represent specific nucleotide sequences. In combination with several transcription factors, this complex resulted in the transcription of vitamin D-responsive genes (Vuolo et al., 2012).

Non-genomic systems can function together with the traditional genomic pathway. Non-genomic signaling is fast, does not require transcription, and indirectly influences transcription through bridging with other signaling pathways. Calcitriol's non-genomic effects cause fast calcium transport across intestinal mucosal membranes. The stimulation of one or more second messenger systems, such as phospholipase C, PKC, G protein-coupled receptors, or phosphatidylinositol-3-kinase, may occur when $1,25(OH)_2D$ binds to the plasma membrane. The activation of voltage-gated calcium or chloride channels and second messengers release are possible results. Some of these second messengers, especially the RAF/MAPK, may control crosstalk with the nucleus for gene expression regulation. The extracellular signal-regulated kinases (ERK) can boost VDR-transcriptional activity, while non-genomic activation of PKC may phosphorylate VDR and further stabilize it (Morelli et al., 2001).

$1,25(OH)_2D$ can indirectly cause apoptosis via the insulin-like growth factor (IGF) receptor and tumor necrosis factor alpha, or more directly via the Bcl-2 family system, including death receptors such as Fas and stress-activated protein kinase pathways involving Jun N-terminal kinase and p38. The suppression of tumor progression and metastasis has been established, with serine proteinases, metalloproteinases, and angiogenesis being among the mechanisms implicated (Seubwai et al., 2007). In vitro studies imply that $1,25(OH)_2D$ suppresses endothelial cell proliferation, while in vivo results proved the fact of lowering angiogenesis. On the other hand, calcitriol could not cause apoptosis and arrest the cell cycle in endothelial cells derived from normal tissues, but it could do so in tumor-derived endothelial cells (Trump et al., 2010). In vivo studies demonstrated that calcitriol given to mice with VEGF-overexpressing MCF-7 xenografts inhibited endothelial cell tube formation and tumor growth generated by VEGF (Bernardi et al., 2002; Chung et al., 2009). Calcitriol is known to promote the angiogenic factor IL-8 in squamous cell carcinoma cells, but it disrupts IL-8 signaling in prostate cancer cells, inhibiting endothelial cell motility and tube formation (Bao et al., 2006). Mouse models treated with calcitriol demonstrated inhibition of prostate and lung cancer metastasis (Nakagawa et al., 2005).

15.4 Role of Vitamin D and Its SNPs in Health and Diseases

Vitamin D insufficiency has an inextricable link to bone demineralization, which leads to an elevated risk of fractures. The expression of particular alleles of the FokI SNP in the VDR gene impacts cellular proliferation and inflammatory response. During the investigation, intervertebral disk cells from patients with discopathy or spondylolisthesis were exploited. Remarkably, the Ff variation of the FokI genotype of the VDR gene appeared to be more susceptible to vitamin D's anti-inflammatory actions and could be exploited as a diagnostic factor in discopathies (De Luca et al., 2018). In addition, the consequences of VDR SNPs in acute pancreatitis (AP) were studied. In comparison with alcohol-abuse controls, AP patients have more than threefold the frequency of a specific SNP (allele T in TaqI) (Cieślińska et al., 2018). As a result, changes in vitamin D signaling caused by unique SNPs in the VDR gene might have links to a higher risk of specific diseases.

The specific role of vitamin D in the control of cardiovascular system functions and its beneficial effect on hypertension is currently in the midst of heated controversy and research. Legarth et al. detailed the present understanding of the involvement of vitamin D in regulating blood pressure and the progression of hypertension (Legarth et al., 2018). Vitamin D significantly impacts the renin-angiotensin-aldosterone axis regulation and mutations in VDR, according to multiple in vivo experiments using VDR-null mutants and the fact that a low amount of vitamin D produces elevated blood pressure. Human trials or retrospective research, on the other hand, have yet to uncover a definite link between vitamin D levels and the incidence of hypertension, which was probably due to a flaw in the study design, but large-scale, randomized trials are still being undertaken to validate the potential benefits of vitamin D in the treatment of hypertension.

Vitamin D deficiency is frequently connected to several neurological disorders since the VDR localizes in several brain areas, including the hippocampus, hypothalamus, substantia nigra, and thalamus. Vitamin D regulates the expression of neurotrophins such as neural growth factor (NGF) and neurotransmitters (acetylcholine, dopamine, and gamma-aminobutyric acid). Morretti and colleagues presented the current state of knowledge about vitamin D's role in the treatment and cure of neurological illnesses, with a focus on multiple sclerosis, stroke, Alzheimer's disease, and Parkinson's disease (Moretti et al., 2018).

15.5 The Role of Vitamin D in Cancer

Cancer is one of the top reasons for death worldwide (Padhi et al., 2018; Verma et al., 2017). By 2040, the annual number of new cancer cases will predictably reach 29.5 million, with 16.4 million cancer-related deaths (cancer.gov/statistics). A decrease in age-adjusted mortality rates is the strongest predictor of effectiveness against cancer; however, other indicators, such as quality of life, are equally essential. Since the early 1990s, the overall cancer death rate in the USA has decreased dramatically. While death rates for many different cancers have decreased, incidences for a few malignancies have remained constant or even grown with time.

In men, the most common cancer is prostate cancer, followed by lung cancer, colorectal cancer, and liver cancer. Because of its elevated fatality rate, lung cancer is the most common cause of cancer death in men, usually accompanied by prostate cancer and liver cancer. In contrast to men, breast and cervical cancer are women's most commonly diagnosed cancers. The mortality profiles in females are more diverse than in males, with breast and cervical cancers being the leading causes of cancer death, followed by lung cancer (Sun et al., 2021).

The hallmarks of cancer mainly consist of eight fundamental biological capabilities that manifest during solid tumors' development. It provides a scientific framework for creating a logical approach to understanding the diversity of neoplastic diseases (Hanahan and Weinberg, 2011). The hallmarks of cancer include sustaining proliferative signals, evading the growth suppressors, resisting cell death, enabling replicative immortality, inducing angiogenesis, activating invasion and metastasis, increasing energy metabolism, and evading immune destruction. These biological activities result in genomic instability in cancer cells, leading to mutations in cells and the involvement of premalignant and malignant lesions, driven by cells of the immune system that promote tumor progression (Jones and Baylin, 2007; Dvorak, 1986).

15.5.1 Vitamin D and Cancer

Numerous studies show a relationship between vitamin D and cancer. Vitamin D levels in the blood are abnormally low, firmly linked to an increased risk of cancer, as demonstrated in vitro and epidemiological studies. High dosages of vitamin D compounds have been shown to inhibit tumor cell proliferation in many in vitro investigations. Figure 15.2 portrays the effect of activation of VDRs on tumorigenesis. Even though the relationship between vitamin D, inflammation, and cancer has until now been unknown, the fact that VDR is highly prevalent in the immune system has led to the speculation that vitamin D and its analogs may have immunomodulatory properties (Lucia and Torkko, 2004; Krishnan et al., 2007). Vitamin D regulates the function of various defense and immunological cells, including blood monocytes, macrophages, antigen-presenting cells, activated CD4 T cells, and epithelial cells (Veldman et al., 2000). T lymphocyte proliferation is modulated by $1,25(OH)_2D$. Dendritic cells and activated T lymphocytes could also synthesize $1,25(OH)_2D$ from sunlight-derived precursors, implying an immunological autocrine/paracrine action (Sigmundsdottir et al., 2007).

Calcitriol coupled to VDR forms heterodimers with the retinoic acid-related orphan receptor (ROR) and its ligand (9 cis-retinoic acids), which occupy specific nucleotide sequences known as VDREs. In concert with several transcription factors, this complex results in the transcription of vitamin D-responsive genes.

Despite the paucity of randomized clinical trial data, multiple epidemiological, controlled, clinical, preclinical, and

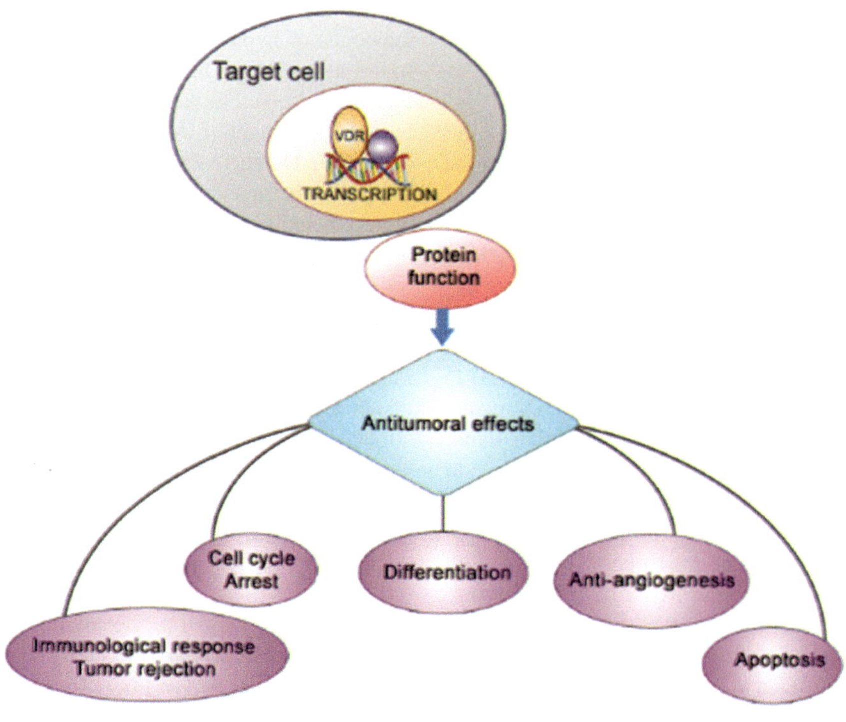

FIGURE 15.2 Effect of activation of the vitamin D receptor (VDR) on tumorigenesis. Calcitriol coupled to VDR forms heterodimers with the retinoid X receptor (RXR) and its ligand (9-cis-retinoic acid), which occupy certain nucleotide sequences known as vitamin D response elements (VDREs). This complex, in concert with a number of transcription factors, results in the transcription of vitamin D-responsive genes.

in vitro experimental studies show evidence that activating vitamin D signaling could be a potential method for cancer chemoprevention. As a result, numerous pharmaceutical techniques for cancer therapy have been explored and designed to address dysregulated vitamin D metabolism or action. However, there are predefined potential downsides of vitamin D-based cancer therapy that should be explored when better therapeutic techniques are developed. One possible drawback of systemic vitamin D signaling stimulation is the significant risk of hypercalcemia, which might have severe health consequences. Many endeavors are devoted to developing selective agonists of VDR that have limited influence on generating hypercalcemia while still having anti-cancer properties compared to calcitriol. Approximately 1,500 vitamin D analogs have been studied for these effects recently, but only a handful have the potential for further clinical studies in patients with different forms of cancer, such as leukemia, breast, prostate, and colon cancer.

Furthermore, the metabolites generated by the CYP11A1-driven alternate vitamin D metabolism pathway are a prejudiced agonist of VDR with reduced calcemic impact, but identical anti-proliferative characteristics in cancer to calcitriol. Because the alternate vitamin D metabolism pathway via CYP11A1 is still being studied, its involvement in cancer and the significant contributors of VDR and RXR are uncertain. As a result, there is anticipation for more studies into the additional vitamin D metabolism pathway and the successful implementation of this mechanism for cancer therapy. Furthermore, boosting local calcitriol concentrations in cancer cells might be harmful. CYP24A1 is a calcitriol-degrading enzyme commonly overexpressed in various malignancies; inhibiting CYP24A1 can boost localized calcitriol concentrations in cancer cells (Muindi et al., 2010). A recent research suggested that genetic knockdown or CYP24A1 inhibition dramatically increases the anti-tumor impact of calcitriol (Wang et al., 2020).

Numerous CYP24A1-specific inhibitors have been identified for clinical utilization; however, it is unclear whether they will be employed for cancer therapy with sufficient clinical efficacy and safety. Furthermore, since the expression levels of CYP27B1, VDR, CYP11A1, and ROR drop gradually in cancer cells as cancer progresses (Brożyna et al., 2020), the efficacy of vitamin D-based therapy may be confined to just early stages of cancer, rather than late stages, as previously indicated. As a result, researching and establishing novel diagnostic markers for forecasting the efficacy of vitamin D-based cancer therapy may be critical for its effective delivery. For comprehensive vitamin D-based cancer therapy, a better knowledge of the mechanisms through which cancer growth lowers the expression of those enzymes would be crucial.

Multiple malignant signaling pathways intersect on metabolic pathways via altering the expression and functioning of metabolic enzymes, according to recent discoveries in cancer metabolism research. Consequently, these carcinogenic signaling channels are highly likely to regulate vitamin D metabolism and functional enzymes expression or activity. Researchers will better understand how vitamin D metabolism and functionality are downregulated in cancer by unraveling such intricate systems integrating oncogenic and vitamin D metabolism pathways and offering new cancer treatments.

15.5.2 VDR Polymorphisms and Cancer

SNPs in the VDR gene could influence the risk of developing cancer (Duffy et al., 2017). Numerous studies have looked at VDR-affiliated SNPs, although their overall linkage with carcinogenesis is still debated. VDR polymorphisms have been linked to the breast (FokI, BsmI, and ApaI), prostate (FokI, BsmI, and TaqI), colorectal (FokI, BsmI, and TaqI), and skin cancer (FokI, BsmI, and TaqI) (Gandini et al., 2014). The CDX2 polymorphism (SNP rs11568820) of the VDR gene, for example, has been linked to the incidence of cancer risk with diversified outcomes. Several researchers conducted a meta-analysis to note a link between the VDR polymorphism and cancer risk (Lee et al., 2015; Lu et al., 2016; Serrano et al., 2016). Some cancer histotypes have been linked to the identification of specific SNPs in the VDR gene (VDR SNPs) and other chosen genes entangled in the vitamin D pathway, such as protein gene transport of vitamin D, CYP27B1, and CYP24A1, which encode the enzymes that generate and degrade $1,25(OH)_2D$, respectively (McCullough et al., 2009).

Around 200 distinct VDR polymorphisms have been identified, with FokI, BsmI, TaqI, ApaI, EcoRV, and CDX2 being the most connected with cancer, even though research in this field is often conflicting, which needs to be addressed. The RFLPs FokI and BsmI are the most researched SNPs in VDR. The FokI RFLP, which is found in the coding region of the VDR gene, results in the creation of a VDR protein three amino acids longer. The shorter the VDR variant, the more potent, although there are no substantial differences in ligand affinities for DNA binding.

The intronic BsmI gene is located at the 3-end of the gene. The BsmI does not affect the quantity, structure, or functionality of the final VDR protein, although it is tightly related to a poly(A) repeat and may impact the stability of VDR messenger RNA. Melanoma has the most robust connections with BsmI and FokI; prostate cancer has the most substantial ties with BsmI, FokI, and ApaI; and breast cancer has the strongest correlations with BsmI, FokI, and TaqI, as reported in a meta-analysis report (Chen et al., 2009; Kostner et al., 2009; Raimondi et al., 2009). The allelic variants BsmI (B)–ApaI (A)–TaqI (t) associate with a greater Gleason score than allelic variants BsmI (b)–ApaI (a)–TaqI (t) in prostate cancer, according to haplotype analysis (T).

Lowe et al. reported that females with low levels of 25(OH)D (20 ng/mL) and the BsmI bb genotype have a sixfold higher risk of getting breast cancer compared to females with sufficient amounts of 25(OH)D (>20 ng/mL) and the BB or Bb genotype (Lowe et al., 2005; Chen et al., 2009). A meta-analysis revealed the most vital link between the incidence of melanoma and the f allele of the polymorphism FokI and the B allele of the polymorphism BsmI (Gandini et al., 2009). Although it appears that the genotype ff is related to a faintly lesser risk of developing colorectal cancer than the FF genotype, a meta-analysis of the included studies on the incidence of colorectal cancer and VDR SNPs supports the substantiation of an association between BsmI polymorphism and a relatively low colorectal cancer risk rather than FokI, Poly(A), TaqI, CDX2, and ApaI VDR polymorphisms (McCullough et al., 2009; Touvier et al., 2011).

A study conducted in 172 patients with differentiated thyroid cancer implicated an association with VDR polymorphisms FokI, TaqI, and ApaI. In contrast, the alleles AA and FF of the ApaI and FokI VDR polymorphisms and the haplotype tABF bestowed the prevention from incidence of follicular carcinoma; the haplotype Tabf emerged to be correlated with an enhanced risk of follicular thyroid carcinoma. These research results need to be confirmed in large patient pools (Penna-Martinez et al., 2009). Dai et al. found that CDX2 mutations were significantly associated with an enhanced cancer risk of colorectal and ovarian cancer incidence in the homozygous model, particularly in African American patients pool, but not in Caucasians or Asians, in a meta-analysis involving 25 distinct researches comprising 34,018 individuals for noting the association of CDX2 polymorphism with cancer progression (Dai et al., 2015). In research studies among African American patients, data analysis of dominant, homozygous, and recessive comparative models revealed an essential link between cancer risk and CDX2 mutations. Data segmentation by cancer type resulted in a tangible link between VDR CDX2 polymorphisms and an increased risk of colorectal and ovarian cancers.

Gandini et al. explored the connection between VDR polymorphisms (FokI, BsmI, TaqI, ApaI, and CDX2) and various cancers, including ovarian cancer, renal cell carcinoma, thyroid carcinoma, hepatocellular carcinoma (HCC), sarcoma, skin cancer (melanoma and non-melanoma skin cancer), bladder cancer, non-Hodgkin lymphoma (Gandini et al., 2014). According to the findings, VDR polymorphisms were noted to be linked to the prostate (FokI, BsmI, and TaqI), breast (FokI, BsmI, and ApaI), colorectal (FokI, BsmI, and TaqI), and skin cancer (FokI, BsmI, and TaqI). A meta-analysis study of the affiliations between VDR polymorphisms (TaqI, ApaI, and CDX2) and the chances of various forms of cancer such as basal cell carcinoma, colorectal cancer, esophageal adenocarcinoma (EAC), HCC, head-and-neck cancer, non-Hodgkin's lymphoma, oral squamous cell carcinoma, renal cell carcinoma, squamous cell carcinoma, and thyroid carcinoma found that CDX2 gg versus GG was coupled with a significantly enhanced risk of cancer precipitation (Serrano et al., 2016). There were also substantial variations in frequency among ethnic groups. When VDR genotypes were pooled with particular haplotypes, related to further vitamin D pathway gene polymorphisms, or differentiated by cancer type or patient history, VDR genotypes may have greater importance. Nonetheless, in a handful of cases, there were obvious connections between the capability of VDR to operate appropriately and tumorigenesis.

15.5.3 Role of Vitamin D and Its Receptors in All Cancer Types

The function of vitamin D on cancer receives the most attention, but it is equally crucial to comprehend how cancer affects the vitamin D pathway. VDR expression is upregulated in numerous cancers, and it is regarded as a critical endogenous factor to tumor growth. VDR levels were lesser in normal colonic epithelial cells, enhanced in anomalous crypt foci, polyps, and well-differentiated tumor cells, and declined as a tumor cell de-differentiation component, according to Matusiak et al. (2005), demonstrating that the advancement of colon cancer can end up causing a diminished response to vitamin D action.

Overexpression of VDR in cholangiocarcinoma cell lines is associated with a favorable prognosis, suggesting that VDR modulates the anti-cancer properties of $1,25(OH)_2D$. VDR polymorphisms have been associated with a reduced risk of cancer (Figure 15.3). VDR overexpression appears to show an endogenous sensitivity to tumor development associated with an improved prognosis; in fact, overexpression of the receptor in tumor cell lines enhances the anti-proliferative impact of vitamin D (Pelczynska et al., 2005; Seubwai et al., 2007). It has also been proposed that malignant cells have a lower expression of the enzyme 25(OH)D-1-hydroxylase, albeit this has not been found uniformly in all malignant tumors.

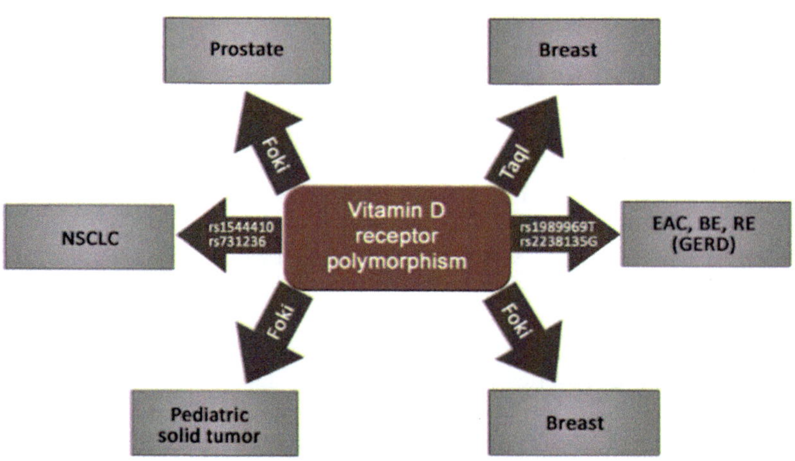

FIGURE 15.3 Association of vitamin D receptor (VDR) polymorphisms with reduced risk of cancer. BE, Barrett's esophagus; EAC, esophageal adenocarcinoma; GERD, gastroesophageal reflux disease; RE, reflux esophagitis; NSCLC, non-small cell lung cancer.

TABLE 15.1

List of the Research Works Associated with VDR Polymorphisms and the Incidence of Female Genital Tract Cancer

Study Type	Cancer Type	Polymorphism	Outcome	Reference
Meta-analysis	Breast cancer	FokI	FokI polymorphism was noted to be a risk factor in post-menopausal women, but not in premenopausal females.	Li et al. (2016)
Case-control study	Ovarian cancer	Fokl	FokI was found to be a risk factor for ovarian cancer in non-carriers of BRCA1/BRCA2 mutations of a Polish group of patients.	Mostowska et al. (2016)
Case-control study	Breast cancer	FokI and BsmI	Protective effect of the FokI polymorphism was conferred; however, BsmI polymorphism was observed to have no effect when studied in an Iranian population.	Talaneh et al. (2017)
Case-control study	Breast cancer	BsmI (rs1544410) and FokI (rs2228570)	BsmI polymorphism was observed to be associated with an increased risk of the said cancer when evaluated in a Pakistani population who were negative for BRCA1/ BRCA 2 germline mutations.	Rashid et al. (2015)
Meta-analysis	Breast cancer	CDX2, FokI, BsmI, ApaI, and TaqI	FokI and BsmI polymorphisms may be associated with an elevated risk of breast cancer.	Mun et al. (2015)

Friedrich et al. (2006) found that breast cancer cells have a higher expression of the enzyme 25(OH)D-1-hydroxylase than healthy breast cells.

Furthermore, the expression of the catabolic enzyme 24-hydroxylase was amplified in various tumor cell lines, such as breast and colorectal cancer cells (Townsend et al. and 2005; Beildeck). These studies support the theory that increased vitamin D levels safeguard against the beginning and tumor progression. Prostate cancer, breast cancer, melanoma, and colorectal cancer are the most common tumor models used to examine the link between vitamin D and cancer. Endocrine cancers are also an intriguing model to study as certain hormones may alter the vitamin D mechanisms.

15.5.3.1 Gynecological Malignancy

The VDR polymorphism has been implicated in female reproductive cancers such as breast, ovarian, cervical, endometrial, uterine, and vaginal cancers (Mun et al., 2015). VDR expression has been elevated in breast cancer and is inversely associated with tumor aggressiveness. Several case-control studies have revealed that the relationship between VDR polymorphism and breast cancer risk is race and ethnicity dependent. Additionally, the outcomes of these researches were different, with some indicating a link between VDR polymorphism and breast cancer and others indicating no link (Talaneh et al., 2017; Elzehery et al., 2017). No correlation was detected between VDR polymorphisms and breast cancer risk in Europeans or East Asians. However, this association was modified by the menopausal status in Europeans and East Asian females. In addition, in Europeans, variability by tumor histology was observed for the VDR polymorphisms (rs1544410, rs7967152, and rs2239186) (Shi et al., 2016). According to these case-control and meta-analysis investigations, the said polymorphism may be linked to an elevated risk of breast cancer. Moreover, Shaikh et al. investigated the influence of VDR gene polymorphisms FokI, BsmI, TaqI, ApaI, and poly(A) on the advancement of breast cancer and concluded that no concrete evidence links the risk of breast cancer with VDR polymorphisms owing to the variable outcomes of various studies (Shaikh et al., 2016). A list of the research works associated with VDR polymorphism and the incidence of female genital tract cancer is represented in Table 15.1.

15.5.3.2 Hematological Cancer

In a reported study, the FokI polymorphism was linked to increased vitamin D3-dependent tumor progression inhibition. The study's findings suggest that disrupting the VDR pathway's activation may have therapeutic value in improving responsiveness to vitamin D therapy in certain cancers (Gascoyne et al., 2017). Other research studies examined the relationship between VDR polymorphisms rs2228570 (FokI) and rs731236 (TaqI) and multiple myeloma susceptibility. The study revealed that the TT genotype of rs2228570 and the mutant C allele of rs731236 are substantially linked with an increased incidence of multiple myeloma. According to the study, the SNPs rs2228570 and rs731236 were critical genetic determinants in the predisposition to multiple myeloma (He et al., 2017). With a total of 133 patients and 300 controls, Esfahani et al. examined the correlation of VDR polymorphisms (FokI, BsmI, TaqI, and ApaI) with the precipitation of acute myeloid leukemia. They found a significant association between VDR polymorphism and acute myeloid leukemia and an association of Taql with clinical remission (Esfahani et al., 2017).

15.5.3.3 Urinary Tract Cancer

Lower and higher levels of circulating vitamin D have been associated with an increased risk of overall mortality in renal cell carcinoma (RCC) patients (Yang et al., 2016), suggesting that vitamin D levels may influence RCC pathophysiology and VDR may play a role in RCC carcinogenesis. VDR polymorphisms may be risk factors for the incidence of urological cancer, according to the findings of the research studies. Nevertheless, the proof for a clear relationship between urological cancer and VDR polymorphism was not definite due to the small number of research studies, and hence, more

TABLE 15.2
VDR Polymorphisms and the Incidence of Urinary Tract Cancer

Study Type	Polymorphism	Inference	Reference
Case-control study	FokI polymorphism	FokI polymorphisms ff genotype and f allele are linked to an enhanced incidence of urothelial bladder cancer.	Fradj et al. (2016)
Case-control study	*Apa*I, *Bsm*I, *Fok*I, and *Taq*I polymorphisms	*Apa*I genes are responsible for the progression of renal cell carcinoma in Chinese Han population.	Yang et al. (2016)

epidemiological investigations are necessary. Table 15.2 lists the relevant works associated with VDR polymorphism and the incidence of urinary tract cancer.

15.5.3.4 Pediatric Tumors

Vitamin D insufficiency has been linked to an elevated risk of cancer in children (Mohan et al., 2016). Vitamin D insufficiency plays a significant role in the inflammation and etiology of osteoporosis, osteoporosis, and bone homeostasis followed by metabolism. Additionally, osteosarcoma and Ewing sarcoma are linked to a much greater frequency of the Ff genotype for the FokI polymorphism, which shows that VDR polymorphism may function in bone malignancies (Rai et al., 2016). There was no link between the VDR polymorphism and Hodgkin's lymphoma in a pediatric sample. However, a moderate connection between the CT and CC FokI genotypes and a lower risk of pediatric solid tumor development has been documented in pediatric solid tumors (Bienertová-Vašků et al., 2016).

Furthermore, the correlation of VDR polymorphic variant rs1544410 with minor tumor recurrence at day 15 in juvenile B cell precursor acute lymphoblastic leukemia implied that the stated VDR polymorphism has both a prognostic and a pathogenic significance. In addition, the reported case-control studies revealed that the VDR polymorphism is linked to a lower risk of juvenile solid tumors (FokI), but not to Hodgkin's lymphoma and that it has the propensity to alter the bone mineral density, patient height, and overall survival rates. In the pediatric age group, vitamin D levels and VDR expression may determine solid tumors' development and overall survival (Tekgündüz et al., 2017). The impacts of VDR polymorphisms on tumorigenesis and other characteristics suggest that more investigation into the significance of VDR polymorphisms and their consequences on vitamin D metabolism in human physiology is warranted.

15.5.3.5 Prostate Cancer

Because there is a notion that a lack of vitamin D increases cancer risk, investigating the role of vitamin D in prostate cancer has progressed quickly. Most current clinical trials incorporate the injection of vitamin D analogs for cancer treatment (Petrioli et al., 2007). Prostate cancer is the most prevalent malignant tumor in men over 50 worldwide. Prostate cancer is becoming more common worldwide, with ethnicity, family history, hormonal status, dietary structure, age, and low vitamin D levels being the significant risk factors (Galunska et al., 2015). Additionally, studies have shown a correlation between VDR polymorphism and an elevated risk of prostate cancer (Cheteri et al., 2004). However, the findings of several investigations were inconclusive and conflicting (Yousaf et al., 2014).

A link was observed between the CC genotype (TaqI) and age (>58 years old) and the GG genotype (BsmI) along with a lower PSA level (10 ng/mL) (Nunes et al., 2016). There was also a 9.75-fold higher risk for prostate cancer in patients with reduced PSA level and GG/GG genotype (BsmI/ApaI). Several meta-analyses and case-control studies have been undertaken to determine whether there is a link between VDR polymorphisms (TaqI, FokI, CDX2, ApaI, BsmI, rs731236, rs7975232, rs731236, rs1544410, rs3782905, rs1544410, and rs2239185) and the incidence of prostate cancer (Mi et al., 2016; Wang et al., 2016b; Kang et al., 2016). The relationship differs by race (Asian, Caucasian, and African American) and VDR genotype-phenotype interaction. A case-control study looking into the potential interactions of VDR polymorphisms (FokI, BsmI, ApaI, and TaqI) and the cytochrome P450 17 alpha 1 (CYP17A1) gene (the MspA1I locus) with prostate cancer in a cohort of Lebanese men indicated that the overall polymorphism pattern of all genes entangled in prostate physiology is a good predictor than polymorphisms in individual genes (El Ezzi et al., 2017).

15.5.3.6 Melanoma

Vitamin D's preventive effect against UV-induced DNA damage and skin cancer progression and its prophylactic role in lowering skin cancer have been explored. Nevertheless, the link between vitamin D and the incidence of skin cancers, such as melanoma and keratinocyte carcinoma, is yet unknown. Vitamin D ingestion has been linked to a lower risk of basal cell carcinoma and melanoma and reduced squamous cell carcinoma and melanoma incidence (Park et al., 2016). Sunlight exposure and skin phototype and the antitumor and pro-carcinogenic impacts of vitamin D are the critical risk factors for skin cancer. VDR polymorphism, in addition to these parameters, is implicated in the pathogenesis of the said cancer and may enhance the risk of its aggressiveness (Schuckmann et al., 2016). Ombra et al. reviewed FokI, TaqI, ApaI, and BsmI polymorphisms with skin cancer (Ombra et al., 2017). A meta-analysis report found that the VDR polymorphism is linked to the incidence of keratinocyte cancer, but the recessive genotypes rs2228570, rs927650, and rs1544410 are linked to a lower risk of squamous cell carcinoma. The recessive genotypes of VDR rs7975232 and rs739837 were also correlated with a lower incidence of basal cell carcinoma (Schuckmann et al., 2016). Another research study discovered an association between the BsmI polymorphism and melanoma in the

TABLE 15.3

List of the Research Works Associated with VDR Polymorphisms and the Incidence of Gastric Tumors

Cancer Type	Study Type	Polymorphism	Outcome	Reference
CRC1	Meta-analysis	TaqI, ApaI, CDX2	TaqI was noted to be associated with a significant risk of CRC	Janmaat et al. (2015)
CRC	Case-control study (53)	rs2254210, rs1540339, rs2107301, rs11168267, rs11574113, rs731236, rs3847987, rs11574143	Limited association of VDR polymorphisms was observed with CRC progression	Budhathoki et al. (2016)
AP	Hospital-based study	TaqI, ApaI, FokI, BsmI	Sex-specific correlation of the adenomatous polyp was noted	Beckett et al. (2016)
CRC	Case-control study	35 different VDR types	Little effect of VDR polymorphisms on the risk of CRC was observed	Ashmore et al. (2015)
HCC	Case-control study	BsmI	No significant effect of polymorphism susceptibility on HCC incidence was observed	Youssef et al. (2016)
Pancreatic cancer	Cohort study	30 SNPs	No significant correlation was noted between the polymorphism and the said cancer type	Yuan et al. (2016)

CRC, colorectal cancer; AP, adenomatous polyp; HCC, hepatocellular carcinoma.

European population (Lee et al., 2015). Additionally, Orlow et al. discovered a substantial link between VDR polymorphism and melanoma-specific overall survival in a meta-analysis report (Orlow et al., 2016).

15.5.3.7 Gastric Cancer

Vitamin D has been demonstrated to shield cells against carcinogenesis in various solid tumors in several population-based studies; nevertheless, the current research on the probable adverse effects of high vitamin D status in EAC development has sparked alarm. A list of the research works associated with VDR polymorphisms and the incidence of gastric tumors is presented in Table 15.3.

In the Chinese Han population, a VDR polymorphism (rs10735810) indicating a T-to-C transition (ATG to ACG) in exon 2 of the VDR gene has been linked to the incidence of gastric cancer (Kong et al., 2015). VDR polymorphisms have thoroughly been examined concerning Barrett's esophagus and EAC risk. A profoundly dysplastic columnar, squamous epithelium around the gastroesophageal junction and a dysplastic esophageal mucosa define Barrett's esophagus. In a meta-analysis report, Janmaat et al. discovered that the VDR haplotype rs2238135G allele is linked to lower esophageal VDR expression and a lower likelihood of gastroesophageal reflux disease (GERD), Barrett's esophagus, and EAC incidences (Janmaat et al., 2015). In another research study, 15 haplotype SNPs in the VDR gene were compared to normal histological controls in patients with GERD, Barrett's esophagus, or EAC (12). When comparing Barrett's esophagus tissue to healthy control specimens, the researchers discovered that VDR gene expression was increased in Barrett's esophagus tissue. The VDR gene had a unique SNP, and patients with the said haplotype had a twofold lower incidence of GERD, Barrett's esophagus, and EAC.

A GATA-1-binding site was discovered to be caused by the mutant allele. GATA-1 inhibits transcription, which may elucidate the presence of a mutant VDR gene in that location which may have a role in suppressing inflammation and cellular proliferation (Kuo et al., 2017). Vitamin D and esophageal cancer have a link that is practically the polar opposite of that reported in other cancers. While tremendous and well-functioning vitamin D/VDR pathways have been shown to minimize cancer risk, robust vitamin D pathway activation has increased the risk associated with the prevalence of esophageal cancer and its associated precursor disorders.

A meta-analysis report revealed a tangible link between VDR SNPs and a lower risk of esophageal cancer, implying that a malfunctioning vitamin D circuit may benefit esophageal cancer (Zgaga et al., 2016). Insights into the mechanisms of action of the VDR haplotype in esophageal cancer could help determine people who would benefit from vitamin D cancer prevention. One possible reason for the adverse association between vitamin D and esophageal cancer is that VDR, in addition to its well-known activity as a vitamin receptor, has subsequently been identified as a bile acid receptor (Copple and Li, 2016).

Vitamin D plays an integral part in the etiology of colorectal cancer, and epidemiological studies have shown that it may also protect against the disease. Polymorphisms in the VDR gene and CYP27B2 and CYP24A1, implicated in the generation and suppression of vitamin D, have all been related to an increased risk of colorectal cancer (Vidigal et al., 2017). VDR genotype may significantly affect the preventative measures of advanced colorectal adenomas with vitamin D3 supplementation, according to a randomized clinical trial which enrolled a pool of 2,259 patients for 41 SNPs of seven genes, such as VDR, gene encoding vitamin D-binding protein, 7-dehydrocholesterol reductase (DHCR7), CYP2R1, CYP27B1, CYP24A1, and calcium-s (Barry et al., 2017). Additionally, one cohort study found that VDR gene variations had a minimal effect on the incidence of colorectal cancer (Ashmore et al., 2015).

Adenomatous polyps are a predecessor to colorectal cancer, and research indicated that FokI is linked to a reduced risk of adenomatous polyps in men and the BsmI/ApaI/TaqI haplotype in women (Beckett et al., 2016). The mixed results concerning the link between VDR polymorphisms and colorectal cancer risk point to the necessity for more research and broader cohort observational studies.

15.5.3.8 Other Types of Malignancies

VDR polymorphisms may involve the tumorigenesis of lung; thyroid cancers; cancers of the female genital tract, prostate, and gastrointestinal system; urological tumors; skin cancer; and pediatric tumors. A case-control study involving 426 patients with non-small cell lung cancer (NSCLC) and 445 healthy controls from China identified a statistically significant link between VDR polymorphisms rs1544410 and rs731236 and a lower risk of NSCLC (Wu et al., 2016). Additionally, a greater VDR mRNA expression has been linked to unfavorable prognostic variables in papillary thyroid carcinoma, suggesting that VDR may play a role in thyroid cancer (Choi et al., 2017).

VDR polymorphisms may play a role in thyroid cancer's pathophysiology and prognosis because vitamin D insufficiency is linked to thyroid illnesses and its metabolism is mediated by thyroid hormone (Mackawy et al., 2013). In lung cancer, a meta-analysis report discovered a clear and precise link between reduced baseline vitamin D status and weaker survival and between VDR genetic variants and progression-free survival. Therefore, no definite causal association could be identified from all controlled experiments included in the stated meta-analysis (Vaughan-Shaw et al., 2017).

15.6 Strength and Limitations of the Evidence of Vitamin D Role in Cancer Susceptibility

Multiple studies have shown that cancer patients who receive vitamin D supplements have a higher quality of life. The serum levels of 25(OH)D in oral cancer patients were noted to be lower when compared to healthy controls; however, administration with 1,000IU vitamin D twice daily restored the 25(OH)D levels (Anand et al., 2017). Compared to individuals who did not get vitamin D, these patients had minor chemoradiation-related side effects and improved their quality-of-life ratings. Similar findings were seen in esophageal cancer patients who self-reported taking vitamin D supplementation after esophagectomy (Wang et al., 2016a). Despite the lower amount of supplements, patients taking vitamin D supplementation experienced a higher quality of life, including functional ability, social functioning, weariness, and appetite loss. Vitamin D supplementation also increased disease-free survival in this research study.

While some researchers have confirmed that vitamin D supplementation reduces cancer risk, others have found that it has no impact or even increases cancer risk. There are various reasons for this variation, many of which can be traced back to variances in the study design of the clinical trials. For example, the doses of vitamin D supplied in investigations varied dramatically, ranging from hundreds of IUs to treatments in the tens of thousands of units (Hohaus et al., 2018). The research strategies also differed, ranging from self-reported vitamin D supplementation to trials assessing improvements in serum 25(OH)D levels (Cadaeu et al., 2016).

Genotypes of vitamin D genes must also be considered when determining whether vitamin D can lessen cancer risk. Vitamin D treatment at a dose of 1,000 IU/day significantly reduced the emergence of advanced colorectal adenomas in people with the VDR rs7968585 AA genotype, but raised the likelihood in those with either the AG or GG genotype (Barry et al., 2017). It is hard to comprehend data from observational clinical research on the relationship between vitamin D serum levels and the occurrence of any disorders. Associated risk factors such as older age, obesity, smoking, and insufficient physical activity are prevalent in both subjects with low 25(-OH)D and subjects with one or more diseases. Additionally, a low 25(OH)D in patients under surveillance may be a symptom instead of a significant cause of the disease (Christakos et al., 2016).

Study designs that concentrate on gender-selective malignancies, such as breast or prostate cancers, intrinsically introduce heterogeneity, impacting findings of vitamin D supplementation's efficiency, although it may not be related to the fact that cancer forms vary. According to studies, the gender of the cancer patient has an impact on the response to vitamin D therapy. Females had more significant improvements in serum 25(OH)D levels in response to vitamin D administration than males (Rees et al., 2016). Furthermore, it must be understood that vitamin D supplementation does not have to similarly affect all cancer kinds or phases of tumor growth. While the studies mentioned in the preceding sections looked at how vitamin D supplementation affects the risk of various cancers, there have not been comprehensive assessments to ascertain which cancers are more likely to be precluded or have an improved therapeutic outcome when supplemented with vitamin D.

15.7 Potential of Vitamin D and Its Receptors in Cancer Therapy

Various preclinical investigations and some epidemiologic outcomes prompted academic and pharmaceutical collaborations to develop vitamin D analogs with the best balance of in vivo qualities for application as chemoprevention or chemotherapeutic drug. Tuning vitamin D products for anti-cancer effects in vivo aims to strike an equilibrium between calcium mobilization properties, which causes hypercalcemia and boosts anti-cancer effects by engaging the VDR.

Numerous biopharmaceutical research groups took on this task and further created a slew of vitamin D mimics with many intriguing qualities, including resistance to metabolism and a modest effect on serum calcium levels (Posner, 1996; Sundaram et al., 2006; Flarakos et al., 2017). A few of these analogs have been used as chemical entities in the quest for illness scenarios where vitamin D compounds' anti-cancer properties can be used. Phase I studies, for instance, have been conducted in a variety of advanced cancers (Jain et al., 2011), leading to more focused phase II trials in pancreatic, liver, prostate, and breast malignancies (Amir et al., 2010). The schedules were well accepted in all cases; however, clinical outcomes were sometimes moderate.

A handful of researchers have examined treating males with localized prostate cancer prior to surgery and then investigating the prostate tumor following surgery to identify known VDR target genes. Beer and co-workers administered either 1,25$(OH)_2$D3 or placebo to roughly 40 patients diagnosed with localized prostate cancer for 4 weeks preceding radical prostatectomy (Beer et al., 2004). Variations in the expression of indicators of cellular proliferation in the VDR or recognized candidate VDR target genes were investigated. Other researchers used doxercalciferol instead of calciferol and found that transforming growth factor-beta receptor type 2 (TGFBR2) was significantly modulated. Microarray investigations of 1,25$(OH)_2$D3 sensitivity in isogenic breast cancer cell lines discovered that TGFBR2 was a critical mediator and marker of 1,25$(OH)_2$D3 sensitivity (Towsend et al., 2006). Other researchers looked at the therapeutic efficacy by raising the dose to see how well higher amounts of 1,25$(OH)_2$D3 could be tolerated (Wagner et al., 2013). These findings implied that higher dosages of 1,25$(OH)_2$D3 could be administered to prostate cancer patients, with detected alterations in VDR-dependent gene expression. VDR was downregulated in the therapy group, whereas the other genes, such as TGFBR2, remained constant.

Others have speculated that treatment with 1,25$(OH)_2$D3 could be enhanced by combining it with other cytotoxic drugs for a more significant therapeutic outcome. Composite studies were notoriously tricky; in the case of vitamin D, Novocea ended up taking a similar approach when developing DN-101, a novel formulation of 1,25$(OH)_2$D3, as a cancer therapy in combined effect with docetaxel for males with advanced prostate cancer and who had declined hormonal therapy, a condition known as castration-resistant prostate cancer.

According to Du et al., the interaction between 1,25$(OH)_2$D3 and the VDR can stimulate anti-tumor processes by inducing a sequence of gene regulation and cell signaling events (Du et al., 2017). Vitamin D and its equivalents are vital in the prevention of stomach adenocarcinomas. Increasing data suggest that preventing vitamin D deficiency is a cost-effective and prospective method to lower the risk of GC (Vyas et al., 2016). Wen et al. used immunohistochemical methods to investigate the expression of local VDR in normal, premalignant, and malignant gastric tissues (Wen et al., 2015). The VDR's connection with clinicopathological variables in GC patients was also investigated. The VDR was continually expressed in varied types of gastric tissues. VDR expression decreased from premalignant to gastric cancer tissues after treatment with 1,25$(OH)_2$D3, and VDR expression was lowest in moderately stratified tissues (Wen et al., 2015), suggesting that the VDR could be a potential predictive marker for GC.

15.7.1 Vitamin D and Proliferation

Numerous intracellular signal transmission networks are involved in the development and multiplication of hormone-dependent epithelial cells. The signaling channels involved may be insulin-like growth factor 1 or 2 activations (IGF-1 and IGF-2), EGF (epidermal growth factor), and pro-inflammatory cytokines, TNF (tumor necrosis factor), interleukin 2 (IL-2), and granulocyte-macrophage colony-stimulating factor (GM-CSF). When transcription signaling pathways are activated, multiple target genes that govern cell proliferation and genes that influence functions that facilitate the cellular transition in normal and proliferative tissues, such as inflammation, angiogenesis, cell mobility, and the capacity to metastasize, are modulated. Vitamin D predominantly affects the calcium and phosphate balance in the body. The complex of 1,25$(OH)_2$D3 and VDR interacts with the regulatory site in the promoter region of the p21 and p27 genes, increasing their expression and inhibiting cyclin-dependent kinases (CDKs), preventing retinoblastoma protein (Rb) phosphorylation, and arresting the cell cycle in the G1 phase (Feldman et al., 2014; Wade et al., 2002). Vitamin D inhibits cell proliferation by modulating intracellular kinase pathways p38 MAPK, ERK, and phosphoinositide 3-kinase (PI3K) and repressing the proto-oncogene Myc, vital for cell proliferation (Feldman et al., 2014).

15.7.2 Vitamin D and Metastasis

In vivo studies on prostate and bladder tumors in animals have indicated that 1,25$(OH)_2$D3 inhibits tumor invasiveness (Hansen et al., 1994; Konety et al., 2001). Vitamin D's "anti-invasive" role is thought to be mediated via inhibition of metalloproteinase and serine protease activities, an elevation in E-cadherin expression, and a decrease in integrins a6 and b4.

E-cadherin is a member of the calcium-dependent adhesion molecule superfamily. Modifications in these proteins' expression and modulation are inextricably connected to tumor invasiveness. The reduction in E-cadherin functioning has been linked to the clinical presentation of prostate cancer and its ability to spread and inferior overall patient survival (Prevarskaya et al., 2011; Cavallaro et al., 2002).

15.7.3 Vitamin D and Angiogenesis

Another anti-tumor strategy for 1,25$(OH)_2$D3 has been the inhibition of angiogenesis, for example, in prostate cancer, both effectively through the influence of tumor endothelial cells and indirectly through a decrease in COX-2-generated prostaglandin E2 (PGE2) levels (Feldman et al., 2014; Danilenko and Studzinski, 2004). IL-8 is one of the factors that promote angiogenesis. The chemotactic recruitment of neutrophils to the inflammatory sites and activation of their bactericidal characteristics are the two main activities of IL-8. Furthermore, IL-8 is an important stimulator of developing new blood vessels. 1,25$(OH)_2$D3 has been demonstrated to prevent the stimulation of the IL-8 gene transcription in prostate cancer cells, most probably through contact with the p65 subunit of nuclear factor-kB (NF-kB). Calcitriol also suppresses the expression of vascular endothelial growth factor (VEGF), angiopoietin 1, and platelet-derived growth factor (PDGF) through VDR, as well as transcriptional inhibition of hypoxia-inducible factor 1 alpha (HIF1) (Nabeshima and Imura, 2008).

15.8 Future Implications

These researches represent a shift toward therapeutic and preventive applications and reflect a growing understanding of the role of nutrition in the incidence and development of cancer and other aging illnesses. Identifying essential foods with vitamin D could be a simple preventive, therapeutic measure. Similar techniques, such as introducing folic acid to bread, have been effectively applied in the USA concerning the requirement for pregnant women to improve their intake. However, a comprehensive view of VDR signaling will be required for future advancements to emerge. Single nuclear receptors, such as the VDR, have previously been explored for their ability to regulate a small range of gene targets and impact cellular function. This method has yielded a wealth of information about many of these receptors precisely.

On the other hand, cell and organism function is dependent on the dynamic interplay of a multitude of receptors via the networks that connect them and inherent cellular programs such as those that drive genesis and differentiation. The limits of previously available approaches and tools to conduct such investigations might be linked to the existing absence of an integrated perspective of how these interconnections cause performance and dysfunction, for example, in aging patients. The integration of post-genomic approaches with bioinformatics and system biology methodology generates a holistic view, illuminating and characterizing the mechanisms through which cells, tissues, and organisms respond to environmental variables such as nutrition. As a result, it is generally unrealistic to believe that the VDR solely plays a vital and dominating role in cell and tissue functioning by operating alone, but rather that it is intricately coupled to the functions of associated nuclear receptors and co-factors. Similarly, the widely held belief is that the skin's overall signaling capacity is conserved in most cell types and represents a mix of VDR function and its interconnections with intrinsic transcriptional programs.

The aim is to construct the nuclear receptor network's spatiotemporal functions, particularly the magnitude to which the VDR imposes crucial regulation over the processes of transcription and translation. However, it requires a thorough comprehension of the chromatin design and background of promoter regions, genomic association, gene regulation hierarchies, and 1a,25(OH)$_2$D3-based metabolomic systems, all of which need to be carried out in the light of relevant cellular backgrounds. The ultimate treatment goal will be to deploy this understanding to techniques that target only subsets of VDR activities in distinct illness populations.

15.9 Conclusions

While 1,25(OH)$_2$D has historically been associated with reducing multiplication and activation of apoptosis in cancer, there is now ample proof that 1,25(OH)$_2$D affects additional cancer-relevant cellular activities. Multiple lines of evidence, for example, imply that vitamin D metabolites' immunoregulatory activities are worth investigating for cancer prevention. While several studies have shown that 1,25(OH)$_2$D can change transcript levels, it is unclear whether VDR directly regulates genes. Recent developments in global profiling of transcription factor-binding sites using techniques such as ChIP-Seq might help researchers distinguish between the direct and indirect effects of vitamin D in the context of cancer. Several investigations have suggested that VDR polymorphisms, namely FokI, ApaI, BsmI, and TaqI, play a critical role in the carcinogenesis of several cancer types by changing vitamin D metabolism and cellular responsiveness to vitamin D. The relationship between VDR polymorphisms and carcinogenesis varies according to age, sex, race, and ethnicity, as evidenced by the studies. These studies also imply that investigating VDR polymorphisms to improve our comprehension of the vitamin D pathway may offer more significant evidence for its cancer prevention therapeutic effects. VDR polymorphisms could predict cancer diagnosis, recurrence, prognosis, and survival. There have been numerous epidemiological studies on breast, prostate, and colorectal cancers, but few reports have shown the link between VDR polymorphisms and lung, thyroid, esophageal, ovarian, renal, and hepatocellular carcinoma. These publications found a link between the VDR polymorphism and elevated and lowered cancer risk. In order to demonstrate the substantial connections and causal links between VDR polymorphisms and cancer, further study with a larger cohort is needed. Moreover, while developing possible vitamin D supplementation and appropriate interventions, genetic diversity in the vitamin D circuit should be considered.

ABBREVIATIONS

1,25-MARRS:	1,25-membrane-associated rapid response steroid-binding proteins
1,25(OH)$_2$D:	1,25-dihydroxyvitamin D
AP:	Acute pancreatitis
CDK:	Cyclin-dependent kinase
DAG:	Diacylglycerol
EAC:	Esophageal adenocarcinoma
EGF:	Epidermal growth factor
ERK:	Extracellular signal-regulated kinase
GERD:	Gastroesophageal reflux disease
GM-CSF:	Colony-stimulating factor
GRP58:	Glucose-regulated protein
HCC:	Hepatocellular carcinoma
HIF1:	Hypoxia-inducible factor 1 alpha
MAPK:	Mitogen-activated protein kinase
mVDR:	Membrane vitamin D receptor
NF-kB:	Nuclear factor-kB
NGF:	Neural growth factor
nVDR:	Nuclear vitamin D receptor
PDGF:	Platelet-derived growth factor
Pdia3:	Protein disulfide isomerase family A member 3
PI:	Phosphatidylinositol
PI3K:	Phosphoinositide 3-kinase
PLC:	Phospholipase C
PLA2:	Phospholipase A2
PKC:	Protein kinase C
RAF:	Rapidly accelerated fibrosarcoma

Rb:	Retinoblastoma protein
RCC:	Renal cell carcinoma
RFLP:	Restriction fragment length polymorphism
ROR:	Retinoic acid-related orphan receptor
RXR:	Retinoid X receptor
SNP:	Single nucleotide polymorphism
TGFBR2:	Transforming growth factor-beta (TGF-β) receptor type 2
VDBP:	Vitamin D-binding protein
VDR:	Vitamin D receptor
VDRE:	Vitamin D response element
VEGF:	Vascular endothelial growth factor

REFERENCES

Amir E, Simmons CE, Freedman OC, Dranitsaris G, Cole DE, Vieth R, et al. A phase 2 trial exploring the effects of high-dose (10,000 IU/day) vitamin D(3) in breast cancer patients with bone metastases. *Cancer*. 2010;116(2):284–91. doi: 10.1002/cncr.24749.

Anand A, Singh S, Sonkar AA, Husain N, Singh KR, Singh S, et al. Expression of vitamin D receptor and vitamin D status in patients with oral neoplasms and effect of vitamin D supplementation on quality of life in advanced cancer treatment. *Contemporary Oncology*. 2017;21(2):145.

Ashmore JH, Gallagher CJ, Lesko SM, Muscat JE, Hartman TJ, Lazarus P. No association between vitamin D intake, VDR polymorphisms, and colorectal cancer in a population-based case–control study. *Cancer Epidemiology and Prevention Biomarkers*. 2015;24(10):1635–7.

Bao BY, Yao J, Lee YF. 1alpha, 25-dihydroxyvitamin D3 suppresses interleukin-8-mediated prostate cancer cell angiogenesis. *Carcinogenesis*. 2006;27(9):1883–93. doi: 10.1093/carcin/bgl041.

Barletta F, Freedman LP, Christakos S. Enhancement of VDR-mediated transcription by phosphorylation: correlation with increased interaction between the VDR and DRIP205, a subunit of the VDR-interacting protein coactivator complex. *Molecular Endocrinology*. 2002;16(2):301–14. doi: 10.1210/mend.16.2.0764.

Barry EL, Peacock JL, Rees JR, Bostick RM, Robertson DJ, Bresalier RS, et al. Vitamin D receptor genotype, vitamin D3 supplementation, and risk of colorectal adenomas: a randomized clinical trial. *JAMA Oncology*. 2017;3(5):628–35.

Beckett EL, Le Gras K, Martin C, Boyd L, Ng X, Duesing K, et al. Vitamin D receptor polymorphisms relate to risk of adenomatous polyps in a sex-specific manner. *Nutrition and Cancer*. 2016;68(2):193–200.

Beer TM, Myrthue A, Garzotto M, O'Hara M F, Chin R, Lowe BA, et al. Randomized study of high-dose pulse calcitriol or placebo prior to radical prostatectomy. *Cancer Epidemiology, Biomarkers & Prevention*. 2004;13(12):2225–32.

Bernardi RJ, Johnson CS, Modzelewski RA, Trump DL. Antiproliferative effects of 1alpha, 25-dihydroxyvitamin D(3) and vitamin D analogs on tumor-derived endothelial cells. *Endocrinology*. 2002;143(7):2508–14. doi: 10.1210/endo.143.7.8887.

Bienertová-Vašků J, Drábová K, Zlámal F, Tomandl J, Kýr M, Šplíchal Z, et al. Pre-treatment VD levels and VDR receptors as potential predictors of occurrence and overall survival in paediatric patients with solid tumours—a single institution pilot study. *Tumor Biology*. 2016;37(7):9209–19.

Boyan BD, Wang L, Wong KL, Jo H, Schwartz Z. Plasma membrane requirements for 1alpha, 25(OH)2D3 dependent PKC signaling in chondrocytes and osteoblasts. *Steroids*. 2006;71(4):286–90. doi: 10.1016/j.steroids.2005.09.018.

Bringhurst FR. Bone and mineral metabolism in health and disease. *Harrison's Principles of Internal Medicine*. 2008: 2365–77.

Brown A, Ritter C, Slatopolsky E, Muralidharan K, Okamura W, Reddy G. 1α, 25-dihydroxy-3-Epi-vitamin D3, a natural metabolite of 1α, 25-dihydroxyvitamin D3, is a potent suppressor of parathyroid hormone secretion. *Journal of Cellular Biochemistry*. 1999;73(1):106–13.

Brożyna AA, Kim T-K, Zabłocka M, Jóźwicki W, Yue J, Tuckey RC, et al. Association among vitamin D, retinoic acid-related orphan receptors, and vitamin D hydroxyderivatives in ovarian cancer. *Nutrients*. 2020;12(11):3541.

Budhathoki S, Yamaji T, Iwasaki M, Sawada N, Shimazu T, Sasazuki S, et al. Vitamin D receptor gene polymorphism and the risk of colorectal cancer: a nested case-control study. *PLoS One*. 2016;11(10):e0164648. doi: 10.1371/journal.pone.0164648.

Cavallaro U, Schaffhauser B, Christofori G. Cadherins and the tumour progression: is it all in a switch? *Cancer Letters*. 2002;176(2):123–8.

Chen A, Davis BH, Bissonnette M, Scaglione-Sewell B, Brasitus TA. 1,25-Dihydroxyvitamin D(3) stimulates activator protein-1-dependent Caco-2 cell differentiation. *Journal of Biological Chemistry*. 1999;274(50):35505–13. doi: 10.1074/jbc.274.50.35505.

Chen J, Doroudi M, Cheung J, Grozier AL, Schwartz Z, Boyan BD. Plasma membrane Pdia3 and VDR interact to elicit rapid responses to 1α, 25 (OH) 2D3. *Cellular Signalling*. 2013;25(12):2362–73.

Chen S, Wu J, Hsieh J-C, Whitfield GK, Jurutka PW, Haussler MR, et al. Suppression of ANP gene transcription by liganded vitamin D receptor: involvement of specific receptor domains. *Hypertension*. 1998;31(6):1338–42.

Cheteri MB, Stanford JL, Friedrichsen DM, Peters MA, Iwasaki L, Langlois MC, et al. Vitamin D receptor gene polymorphisms and prostate cancer risk. *Prostate*. 2004;59(4):409–18. doi: 10.1002/pros.20001.

Choi JY, Yi JW, Lee JH, Song R-Y, Yu H, Kwon H, et al. VDR mRNA overexpression is associated with worse prognostic factors in papillary thyroid carcinoma. *Endocrine Connections*. 2017;6(3):172–8.

Christakos S, Dhawan P, Verstuyf A, Verlinden L, Carmeliet G. Vitamin D: metabolism, molecular mechanism of action, and pleiotropic effects. *Physiological Reviews*. 2016;96(-1):365–408. doi: 10.1152/physrev.00014.2015.

Chun RF, Peercy BE, Adams JS, Hewison M. Vitamin D binding protein and monocyte response to 25-hydroxyvitamin D and 1, 25-dihydroxyvitamin D: analysis by mathematical modeling. *PLoS One*. 2012;7(1):e30773.

Chung I, Han G, Seshadri M, Gillard BM, Yu WD, Foster BA, et al. Role of vitamin D receptor in the antiproliferative effects of calcitriol in tumor-derived endothelial cells and tumor angiogenesis in vivo. *Cancer Research*. 2009;69(-3):967–75. doi: 10.1158/0008-5472.CAN-08-2307.

Cieślińska A, Kostyra E, Fiedorowicz E, Snarska J, Kordulewska N, Kiper K, et al. Single nucleotide polymorphisms in the vitamin D receptor gene (VDR) may have an impact on acute pancreatitis (AP) development: a prospective study in populations of AP patients and alcohol-abuse controls. *International Journal of Molecular Sciences.* 2018;19(7). doi: 10.3390/ijms19071919.

Copple BL, Li T. Pharmacology of bile acid receptors: evolution of bile acids from simple detergents to complex signaling molecules. *Pharmacological Research.* 2016;104:9–21.

Corcoran A, Nadkarni S, Yasuda K, Sakaki T, Brown G, Kutner A, et al. Biological evaluation of double point modified analogues of 1,25-dihydroxyvitamin D(2) as potential anti-leukemic agents. *International Journal of Molecular Sciences.* 2016;17(2). doi: 10.3390/ijms17020091.

Dai ZM, Fei YL, Zhang WG, Liu J, Cao XM, Qu QM, et al. Association of vitamin D receptor Cdx-2 polymorphism with cancer risk: a meta-analysis. *Medicine (Baltimore).* 2015;94(33):e1370. doi: 10.1097/MD.0000000000001370.

Danilenko M, Studzinski GP. Enhancement by other compounds of the anti-cancer activity of vitamin D3 and its analogs. *Experimental Cell Research.* 2004;298(2):339–58.

Dawson-Hughes B, Heaney RP, Holick MF, Lips P, Meunier PJ, Vieth R. Estimates of optimal vitamin D status. *Osteoporosis International.* 2005;16(7):713–6. doi: 10.1007/s00198-005-1867-7.

De Luca P, de Girolamo L, Perucca Orfei C, Vigano M, Cecchinato R, Brayda-Bruno M, et al. Vitamin D's effect on the proliferation and inflammation of human intervertebral disc cells in relation to the functional vitamin D receptor gene FokI polymorphism. *International Journal of Molecular Sciences.* 2018;19(7). doi: 10.3390/ijms19072002.

Du C, Yang S, Zhao X, Dong H. Pathogenic roles of alterations in vitamin D and vitamin D receptor in gastric tumorigenesis. *Oncotarget.* 2017;8(17):29474–86. doi: 10.18632/oncotarget.15298.

Duffy MJ, Murray A, Synnott NC, O'Donovan N, Crown J. Vitamin D analogues: potential use in cancer treatment. *Critical Reviews in Oncology/Hematology.* 2017;112:190–7. doi: 10.1016/j.critrevonc.2017.02.015.

Dvorak HF. Tumors: wounds that do not heal. Similarities between tumor stroma generation and wound healing. *New England Journal of Medicine.* 1986;315(26):1650–9. doi: 10.1056/NEJM198612253152606.

El Ezzi AA, Boyko VG, Baker MT, Zaidan WR, Hraiki KM, El Saidi MA, et al. Association of some polymorphisms in the VDR gene, CYP17 gene and SRD5A2 gene and prostate cancer among Lebanese men. *Asian Pacific Journal of Cancer Prevention.* 2017;18(1):93–100. doi: 10.22034/APJCP.2017.18.1.93.

Elzehery RR, Baiomy AA, Hegazy MA-F, Fares R, El-Gilany A-H, Hegazi R. Vitamin D status, receptor gene BsmI (A/G) polymorphism and breast cancer in a group of Egyptian females. *Egyptian Journal of Medical Human Genetics.* 2017;18(3):269–73.

Farach-Carson MC, Nemere I. Membrane receptors for vitamin D steroid hormones: potential new drug targets. *Current Drug Targets.* 2003;4(1):67–76. doi: 10.2174/1389450033347118.

Feldman D, Krishnan AV, Swami S, Giovannucci E, Feldman BJ. The role of vitamin D in reducing cancer risk and progression. *Nature Reviews Cancer.* 2014;14(5):342–57.

Flarakos CC, Weiskopf A, Robinson M, Wang G, Vouros P, Sasso GJ, et al. Metabolism of selective 20-epi-vitamin D3 analogs in rat osteosarcoma UMR-106 cells: isolation and identification of four novel C-1 fatty acid esters of 1alpha, 25-dihydroxy-16-ene-20-epi-vitamin D3. *Steroids.* 2017;119:18–30. doi: 10.1016/j.steroids.2016.12.013.

Fradj MKB, Kallel A, Gargouri MM, Chehida MAB, Sallemi A, Ouanes Y, et al. Association of FokI polymorphism of vitamin D receptor with urothelial bladder cancer in Tunisians: role of tobacco smoking and plasma vitamin D concentration. *Tumor Biology.* 2016;37(5):6197–203.

Gallieni M, Cozzolino M, Fallabrino G, Pasho S, Olivi L, Brancaccio D. Vitamin D: physiology and pathophysiology. *International Journal of Artificial Organs.* 2009;32(2):87–94. doi: 10.1177/039139880903200205.

Galunska B, Gerova D, Kosev P, Anakievski D, Hinev A. Serum 25-hydroxy vitamin D levels in Bulgarian patients with prostate cancer: a pilot study. Clinical Laboratory. 2015;61(-3–4):329–35. doi: 10.7754/clin.lab.2014.140802.

Gandini S, Gnagnarella P, Serrano D, Pasquali E, Raimondi S. Vitamin D receptor polymorphisms and cancer. *Advances in Experimental Medicine and Biology.* 2014;810:69–105. doi: 10.1007/978-1-4939-0437-2_5.

Garcia-Vallvé S, Palau J. Nuclear receptors, nuclear-receptor factors, and nuclear-receptor-like orphans form a large paralog cluster in Homo sapiens. *Molecular Biology and Evolution.* 1998;15(6):665–82.

Gascoyne DM, Lyne L, Spearman H, Buffa FM, Soilleux EJ, Banham AH. Vitamin D receptor expression in plasmablastic lymphoma and myeloma cells confers susceptibility to vitamin D. *Endocrinology.* 2017;158(3):503–15.

Hanahan D, Weinberg RA. Hallmarks of cancer: the next generation. *Cell.* 2011;144(5):646–74.

Hansen CM, Frandsen TL, Brünner N, Binderup L. 1α, 25-dihydroxyvitamin D 3 inhibits the invasive potential of human breast cancer cellsin vitro. *Clinical & Experimental Metastasis.* 1994;12(3):195–202.

He Y, Ou C, Pang W, Lin Y, He J, Li C, et al. Genetic association of VDR polymorphisms and multiple myeloma susceptibility: a case control study. *International Journal of Clinical and Experimental Pathology.* 2017;10(3):3538–42.

Hii C, Ferrante A. The non-genomic actions of vitamin D. *Nutrients.* 2016;8:135.

Hilliard GMt, Cook RG, Weigel NL, Pike JW. 1,25-dihydroxyvitamin D3 modulates phosphorylation of serine 205 in the human vitamin D receptor: site-directed mutagenesis of this residue promotes alternative phosphorylation. *Biochemistry.* 1994;33(14):4300–11. doi: 10.1021/bi00180a026.

Hohaus S, Tisi MC, Bellesi S, Maiolo E, Alma E, Tartaglia G, et al. Vitamin D deficiency and supplementation in patients with aggressive B-cell lymphomas treated with immunochemotherapy. *Cancer Medicine.* 2018;7(1):270–81.

Holick MF. Sunlight and vitamin D for bone health and prevention of autoimmune diseases, cancers, and cardiovascular disease. *American Journal of Clinical Nutrition.* 2004;80(6 Suppl):1678S–88S. doi: 10.1093/ajcn/80.6.1678S.

Holick MF. Vitamin D deficiency. *New England Journal of Medicine*. 2007;357(3):266–81. doi: 10.1056/NEJMra070553.

Huang J, Xie Z-F. Polymorphisms in the vitamin D receptor gene and multiple sclerosis risk: a meta-analysis of case–control studies. *Journal of the Neurological Sciences*. 2012;313(1–2):79–85.

Jain RK, Trump DL, Egorin MJ, Fernandez M, Johnson CS, Ramanathan RK. A phase I study of the vitamin D3 analogue ILX23-7553 administered orally to patients with advanced solid tumors. *Investigational New Drugs*. 2011;29(6): 1420–5. doi: 10.1007/s10637-010-9492-5.

Janmaat VT, Van De Winkel A, Peppelenbosch MP, Spaander MC, Uitterlinden AG, Pourfarzad F, et al. Vitamin D receptor polymorphisms are associated with reduced esophageal vitamin D receptor expression and reduced esophageal adenocarcinoma risk. *Molecular Medicine*. 2015;21:346–54. doi: 10.2119/molmed.2012.00336.

Jones PA, Baylin SB. The epigenomics of cancer. *Cell*. 2007;128(4): 683–92.

Kang S, Zhao Y, Liu J, Wang L, Zhao G, Chen X, et al. Association of vitamin D receptor Fok I polymorphism with the risk of prostate cancer: a meta-analysis. *Oncotarget*. 2016;7(47):-77878–89. doi: 10.18632/oncotarget.12837.

Konety BR, Lavelle JP, Pirtskalaishvili G, Dhir R, Meyers SA, NGUYEN T-ST, et al. Effects of vitamin D (calcitriol) on transitional cell carcinoma of the bladder in vitro and in vivo. *The Journal of Urology*. 2001;165(1):253–8.

Krishnan AV, Moreno J, Nonn L, Swami S, Peehl DM, Feldman D. Calcitriol as a chemopreventive and therapeutic agent in prostate cancer: role of anti-inflammatory activity. *Journal of Bone and Mineral Research*. 2007;22 Suppl 2:V74–80. doi: 10.1359/jbmr.07s213.

Kuo Y-Y, Chang Z-F. Correction for Kuo and Chang, GATA-1 and Gfi-1B interplay to regulate Bcl-xL transcription. *Molecular and Cellular Biology*. 2017;37(6):e00008–17.

Lee YH, Gyu Song G. Vitamin D receptor FokI, BsmI, TaqI, ApaI, and EcoRV polymorphisms and susceptibility to melanoma: a meta-analysis. *J BUON*. 2015;20(1):235–43.

Legarth C, Grimm D, Wehland M, Bauer J, Kruger M. The impact of vitamin D in the treatment of essential hypertension. *International Journal of Molecular Sciences*. 2018;19(2). doi: 10.3390/ijms19020455.

Lu D, Jing L, Zhang S. Vitamin D receptor polymorphism and breast cancer risk: a meta-analysis. *Medicine (Baltimore)*. 2016;95(18):e3535. doi: 10.1097/MD.0000000000003535.

Lucia MS, Torkko KC. Inflammation as a target for prostate cancer chemoprevention: pathological and laboratory rationale. *Journal of Urology*. 2004;171(2 Pt 2):S30–4; discussion: S5. doi: 10.1097/01.ju.0000108142.53241.47.

Mackawy AMH, Al-Ayed BM, Al-Rashidi BM. Vitamin D deficiency and its association with thyroid disease. *International Journal of Health Sciences*. 2013;7(3):267.

Maestro MA, Molnár F, Mouriño A, Carlberg C. Vitamin D receptor 2016: novel ligands and structural insights. *Expert Opinion on Therapeutic Patents*. 2016;26(11):1291–306.

Mangelsdorf DJ, Thummel C, Beato M, Herrlich P, Schütz G, Umesono K, et al. The nuclear receptor superfamily: the second decade. *Cell*. 1995;83(6):835.

McCollum E, Pitz W, Simmonds N, Becker J, Shipley P, Bunting R. The effect of additions of fluorine to the diet of the rat on the quality of the teeth. 1925. Studies on experimental rickets. XXI. An experimental demonstration of the existence of a vitamin which promotes calcium deposition. 1922. The effect of additions of fluorine to the diet of the rat on the quality of the teeth. 1925. *The Journal of Biological Chemistry*. 2002;277(19):E8.

Mellanby E. An experimental investigation of rickets. *Lancet*. 1919;1:407–12.

Mohan R, Mohan G, Scott JX, Rajendran A, Paramasivam V, Ravindran M. Vitamin D insufficiency among children with cancer in India. *Indian Journal of Medical and Paediatric Oncology*. 2016;37(1):14.

Morelli S, Buitrago C, Boland R, de Boland AR. The stimulation of MAP kinase by 1,25(OH)(2)-vitamin D(3) in skeletal muscle cells is mediated by protein kinase C and calcium. *Molecular and Cellular Endocrinology*. 2001;173(1–2):41–52. doi: 10.1016/s0303-7207(00)00435-4.

Moretti R, Morelli ME, Caruso P. Vitamin D in neurological diseases: a rationale for a pathogenic impact. *International Journal of Molecular Sciences*. 2018;19(8). doi: 10.3390/ijms19082245.

Mostowska A, Sajdak S, Pawlik P, Lianeri M, Jagodzinski PP. Polymorphic variants in the vitamin D pathway genes and the risk of ovarian cancer among non-carriers of BRCA1/BRCA2 mutations. *Oncology Letters*. 2016;11(2):1181–8. doi: 10.3892/ol.2015.4033.

Muindi JR, Yu W-D, Ma Y, Engler KL, Kong R-X, Trump DL, et al. CYP24A1 inhibition enhances the antitumor activity of calcitriol. *Endocrinology*. 2010;151(9):4301–12.

Mun M-J, Kim T-H, Hwang J-Y, Jang W-C. Vitamin D receptor gene polymorphisms and the risk for female reproductive cancers: a meta-analysis. *Maturitas*. 2015;81(2):256–65. doi: 10.1016/j.maturitas.2015.03.010.

Nabeshima Y-i, Imura H. α-Klotho: a regulator that integrates calcium homeostasis. *American Journal of Nephrology*. 2008;28(3):455–64.

Nakagawa K, Kawaura A, Kato S, Takeda E, Okano T. 1 alpha, 25-Dihydroxyvitamin D(3) is a preventive factor in the metastasis of lung cancer. *Carcinogenesis*. 2005;26(2):429–40. doi: 10.1093/carcin/bgh332.

Norman AW. Minireview: vitamin D receptor: new assignments for an already busy receptor. *Endocrinology*. 2006;147(-12):5542–8. doi: 10.1210/en.2006-0946.

Nunes SB, de Matos Oliveira F, Neves AF, Araujo GR, Marangoni K, Goulart LR, et al. Association of vitamin D receptor variants with clinical parameters in prostate cancer. *Springerplus*. 2016;5:364. doi: 10.1186/s40064-016-2009-8.

Nussey S, Whitehead S. *Endocrinology: An Integrated Approach*. Oxford, UK: BIOS Scientific; 2001.

Ombra MN, Paliogiannis P, Doneddu V, Sini MC, Colombino M, Rozzo C, et al. Vitamin D status and risk for malignant cutaneous melanoma: recent advances. *European Journal of Cancer Prevention*. 2017;26(6):532–41. doi: 10.1097/CEJ.0000000000000334.

Orlow I, Reiner AS, Thomas NE, Roy P, Kanetsky PA, Luo L, et al. Vitamin D receptor polymorphisms and survival in patients with cutaneous melanoma: a population-based study. *Carcinogenesis*. 2016;37(1):30–8. doi: 10.1093/carcin/bgv157.

Padhi S, Kapoor R, Verma D, Panda AK, Iqbal Z. Formulation and optimization of topotecan nanoparticles: in vitro characterization, cytotoxicity, cellular uptake and pharmacokinetic outcomes. *Journal of Photochemistry and Photobiology B: Biology*. 2018;183:222–32.

Park SM, Li T, Wu S, Li WQ, Qureshi AA, Cho E. Vitamin D intake and risk of skin cancer in US women and men. *PLoS One*. 2016;11(8):e0160308. doi: 10.1371/journal.pone.0160308.

Pawlik A, Anisiewicz A, Filip-Psurska B, Nowak M, Turlej E, Trynda J, et al. Calcitriol and its analogs establish the immunosuppressive microenvironment that drives metastasis in 4T1 mouse mammary gland cancer. *International Journal of Molecular Sciences*. 2018;19(7):2116.

Peehl DM, Skowronski RJ, Leung GK, Wong ST, Stamey TA, Feldman D. Antiproliferative effects of 1,25-dihydroxyvitamin D3 on primary cultures of human prostatic cells. *Cancer Research*. 1994;54(3):805–10.

Petrioli R, Pascucci A, Francini E, Marsili S, Sciandivasci A, De Rubertis G, et al. Weekly high-dose calcitriol and docetaxel in patients with metastatic hormone-refractory prostate cancer previously exposed to docetaxel. *BJU International*. 2007;100(4):775–9. doi: 10.1111/j.1464-410X.2007.07019.x.

Pike JW. Genome-wide principles of gene regulation by the vitamin D receptor and its activating ligand. *Molecular and Cellular Endocrinology*. 2011;347(1–2):3–10.

Podgorska E, Drzal A, Matuszak Z, Swakon J, Slominski A, Elas M, et al. Calcitriol and calcidiol can sensitize melanoma cells to low–LET proton beam irradiation. *International Journal of Molecular Sciences*. 2018;19(8):2236.

Prevarskaya N, Skryma R, Shuba Y. Calcium in tumour metastasis: new roles for known actors. *Nature Reviews Cancer*. 2011;11(8):609–18.

Rai V, Dietz NE, Dilisio MF, Radwan MM, Agrawal DK. Vitamin D attenuates inflammation, fatty infiltration, and cartilage loss in the knee of hyperlipidemic microswine. *Arthritis Research & Therapy*. 2016;18(1):1–17.

Rashid MU, Muzaffar M, Khan FA, Kabisch M, Muhammad N, Faiz S, et al. Association between the Bsm I polymorphism in the vitamin D receptor Gene and Breast cancer risk: results from a Pakistani case-control study. *PLoS One*. 2015;10(10):e0141562.

Rastinejad F, Huang P, Chandra V, Khorasanizadeh S. Understanding nuclear receptor form and function using structural biology. *Journal of Molecular Endocrinology*. 2013;51(3):T1–21.

Rees JR, Mott LA, Barry EL, Baron JA, Bostick RM, Figueiredo JC, et al. Lifestyle and other factors explain one-half of the variability in the serum 25-hydroxyvitamin D response to cholecalciferol supplementation in healthy adults. *The Journal of Nutrition*. 2016;146(11):2312–24.

Richard CL, Farach-Carson MC, Rohe B, Nemere I, Meckling KA. Involvement of 1,25D3-MARRS (membrane associated, rapid response steroid-binding), a novel vitamin D receptor, in growth inhibition of breast cancer cells. *Experimental Cell Research*. 2010;316(5):695–703. doi: 10.1016/j.yexcr.2009.12.015.

Schuckmann LA, Law MH, Montgomery GW, Green AC, JC VDP. Vitamin D pathway gene polymorphisms and keratinocyte cancers: a nested case-control study and meta-analysis. *Anticancer Research*. 2016;36(5):2145–52.

Schwartz Z, Shaked D, Hardin RR, Gruwell S, Dean DD, Sylvia VL, et al. 1alpha, 25(OH)2D3 causes a rapid increase in phosphatidylinositol-specific PLC-beta activity via phospholipase A2-dependent production of lysophospholipid. *Steroids*. 2003;68(5):423–37. doi: 10.1016/s0039-128x(03)00044-8.

Schwartz Z, Sylvia VL, Larsson D, Nemere I, Casasola D, Dean DD, et al. 1α, 25 (OH) 2D3 regulates chondrocyte matrix vesicle protein kinase C (PKC) directly via G-protein-dependent mechanisms and indirectly via incorporation of PKC during matrix vesicle biogenesis. *Journal of Biological Chemistry*. 2002;277(14):11828–37.

Searing DA, Leung DY. Vitamin D in atopic dermatitis, asthma and allergic diseases. *Immunology and Allergy Clinics of North America*. 2010;30(3):397–409. doi: 10.1016/j.iac.2010.05.005.

Serrano D, Gnagnarella P, Raimondi S, Gandini S. Meta-analysis on vitamin D receptor and cancer risk: focus on the role of TaqI, ApaI, and Cdx2 polymorphisms. *European Journal of Cancer Prevention*. 2016;25(1):85–96. doi: 10.1097/CEJ.0000000000000132.

Seubwai W, Wongkham C, Puapairoj A, Khuntikeo N, Wongkham S. Overexpression of vitamin D receptor indicates a good prognosis for cholangiocarcinoma: implications for therapeutics. *Cancer*. 2007;109(12):2497–505. doi: 10.1002/cncr.22716.

Seubwai W, Wongkham C, Puapairoj A, Okada S, Wongkham S. 22-Oxa-1, 25-dihydroxyvitamin D3 efficiently inhibits tumor growth in inoculated mice and primary histoculutre of cholangiocarcinoma. *Cancer*. 2010;116(23):5535–43.

Shaikh F, Baig S, Jamal Q. Do VDR gene polymorphisms contribute to breast cancer? *Asian Pacific Journal of Cancer Prevention*. 2016;17(2):479–83.

Shi J, Grundy A, Richardson H, Burstyn I, Schuetz JM, Lohrisch CA, et al. Genetic variation in vitamin D-related genes and risk of breast cancer among women of European and East Asian descent. *Tumor Biology*. 2016;37(5):6379–87.

Sigmundsdottir H, Pan J, Debes GF, Alt C, Habtezion A, Soler D, et al. DCs metabolize sunlight-induced vitamin D3 to 'program' T cell attraction to the epidermal chemokine CCL27. *Nature Immunology*. 2007;8(3):285–93. doi: 10.1038/ni1433.

Sitrin MD, Bissonnette M, Bolt MJ, Wali R, Khare S, Scaglione-Sewell B, et al. Rapid effects of 1,25(OH)2 vitamin D3 on signal transduction systems in colonic cells. *Steroids*. 1999;64(1–2):137–42. doi: 10.1016/s0039-128x(98)00102-0.

Sitrin MD, Halline AG, Abrahams C, Brasitus TA. Dietary calcium and vitamin D modulate 1, 2-dimethylhydrazine-induced colonic carcinogenesis in the rat. *Cancer Research*. 1991;51(20):5608–13.

Slominski AT, Kim T-K, Janjetovic Z, Brożyna AA, Żmijewski MA, Xu H, et al. Differential and overlapping effects of 20, 23 (OH) 2D3 and 1, 25 (OH) 2D3 on gene expression in human epidermal keratinocytes: identification of AhR as an alternative receptor for 20, 23 (OH) 2D3. *International Journal of Molecular Sciences*. 2018;19(10):3072.

Sundaram S, Beckman MJ, Bajwa A, Wei J, Smith KM, Posner GH, et al. QW-1624F2-2, a synthetic analogue of 1,25-dihydroxyvitamin D3, enhances the response to other deltanoids and suppresses the invasiveness of human metastatic breast tumor cells. *Molecular Cancer Therapeutics*. 2006;5(11):2806–14. doi: 10.1158/1535-7163.MCT-06-0092.

Talaneh S, Ghorbani A, Oghabi Bakhshaiesh T, Jafari B. FokI and BsmI Polymorphisms of the VDR gene and breast cancer risk. *Multidisciplinary Cancer Investigation.* 2017;1(1):21–5.

Tang JY, Fu T, Lau C, Oh DH, Bikle DD, Asgari MM. Vitamin D in cutaneous carcinogenesis: part II. *Journal of the American Academy of Dermatology.* 2012;67(5):817 e1–11; quiz 27–8. doi: 10.1016/j.jaad.2012.07.022.

Tekgündüz SA, Yeşil Ş, Ören AC, Tanyildiz HG, Çandir MO, Bozkurt C, et al. Vitamin D receptor (VDR) polymorphisms in pediatric patients presenting with hodgkin's lymphoma. *Journal of Pediatric Hematology/Oncology.* 2017;39(2):e59–61.

Towsend K, Trevino V, Falciani F, Stewart PM, Hewison M, Campbell MJ. Identification of VDR-responsive gene signatures in breast cancer cells. *Oncology.* 2006;71(1–2):111–23. doi: 10.1159/000100989.

Trump DL, Deeb KK, Johnson CS. Vitamin D: considerations in the continued development as an agent for cancer prevention and therapy. *Cancer Journal.* 2010;16(1):1–9. doi: 10.1097/PPO.0b013e3181c51ee6.

Valdivielso JM, Fernandez E. Vitamin D receptor polymorphisms and diseases. *Clinica Chimica Acta.* 2006;371(1–2):1–12.

Vaughan-Shaw P, O'sullivan F, Farrington SM, Theodoratou E, Campbell H, Dunlop MG, et al. The impact of vitamin D pathway genetic variation and circulating 25-hydroxyvitamin D on cancer outcome: systematic review and meta-analysis. *British Journal of Cancer.* 2017;116(8):1092–110.

Veldman CM, Cantorna MT, DeLuca HF. Expression of 1,25-dihydroxyvitamin D(3) receptor in the immune system. *Archives of Biochemistry and Biophysics.* 2000;374(-2):334–8. doi: 10.1006/abbi.1999.1605.

Verma D, Thakur PS, Padhi S, Khuroo T, Talegaonkar S, Iqbal Z. Design expert assisted nanoformulation design for co-delivery of topotecan and thymoquinone: optimization, in vitro characterization and stability assessment. *Journal of Molecular Liquids.* 2017;242:382–94.

Vidigal VM, Silva TD, de Oliveira J, Pimenta CAM, Felipe AV, Forones NM. Genetic polymorphisms of vitamin D receptor (VDR), CYP27B1 and CYP24A1 genes and the risk of colorectal cancer. *The International Journal of Biological Markers.* 2017;32(2):224–30.

Vuolo L, Di Somma C, Faggiano A, Colao A. Vitamin D and cancer. *Frontiers in Endocrinology (Lausanne).* 2012;3:58. doi: 10.3389/fendo.2012.00058.

Vyas N, Companioni RC, Tiba M, Alkhawam H, Catalano C, Sogomonian R, et al. Association between serum vitamin D levels and gastric cancer: a retrospective chart analysis. *World Journal of Gastrointestinal Oncology.* 2016;8(-9):688–94. doi: 10.4251/wjgo.v8.i9.688.

Wade WN, Willingham MC, Koumenis C, Cramer SD. p27Kip1 is essential for the antiproliferative action of 1, 25-dihydroxyvitamin D3 in primary, but not immortalized, mouse embryonic fibroblasts. *Journal of Biological Chemistry.* 2002;277(40):37301–6.

Wagner D, Trudel D, Van der Kwast T, Nonn L, Giangreco AA, Li D, et al. Randomized clinical trial of vitamin D3 doses on prostatic vitamin D metabolite levels and ki67 labeling in prostate cancer patients. *Journal of Clinical Endocrinology and Metabolism.* 2013;98(4):1498–507. doi: 10.1210/jc.2012-4019.

Wang L, Wang C, Wang J, Huang X, Cheng Y. Longitudinal, observational study on associations between postoperative nutritional vitamin D supplementation and clinical outcomes in esophageal cancer patients undergoing esophagectomy. *Scientific Reports.* 2016a;6(1):1–10.

Wang K, Wu G, Li J, Song W. Role of vitamin D receptor gene Cdx2 and Apa1 polymorphisms in prostate cancer susceptibility: a meta-analysis. *BMC Cancer.* 2016b;16(1):674. doi: 10.1186/s12885-016-2722-2.

Wang P, Xu J, You W, Hou Y, Wang S, Ma Y, et al. Knockdown of CYP24A1 aggravates 1α, 25 (OH) 2D3-inhibited migration and invasion of mouse ovarian epithelial cells by suppressing EMT. *Frontiers in Oncology.* 2020;10:1258.

Wasiewicz T, Piotrowska A, Wierzbicka J, Slominski AT, Zmijewski MA. Antiproliferative activity of non-calcemic vitamin D analogs on human melanoma lines in relation to VDR and PDIA3 receptors. *International Journal of Molecular Sciences.* 2018;19(9):2583.

Wei P, Inamdar N, Vedeckis WV. Transrepression of c-jun gene expression by the glucocorticoid receptor requires both AP-1 sites in the c-jun promoter. *Molecular Endocrinology.* 1998;12(9):1322–33.

Wen Y, Da M, Zhang Y, Peng L, Yao J, Duan Y. Alterations in vitamin D signaling pathway in gastric cancer progression: a study of vitamin D receptor expression in human normal, premalignant, and malignant gastric tissue. *International Journal of Clinical and Experimental Pathology.* 2015;8(10):13176–84.

Wu W, Beilhartz G, Roy Y, Richard CL, Curtin M, Brown L, et al. Nuclear translocation of the 1,25D3-MARRS (membrane associated rapid response to steroids) receptor protein and NFkappaB in differentiating NB4 leukemia cells. *Experimental Cell Research.* 2010;316(7):1101–8. doi: 10.1016/j.yexcr.2010.01.010.

Wu X, Cheng J, Yang K. Vitamin D-related gene polymorphisms, plasma 25-hydroxy-vitamin D, cigarette smoke and non-small cell lung cancer (NSCLC) risk. *International Journal of Molecular Sciences.* 2016;17(10):1597.

Yang C, Jia L, Li Y, Wu D, Sui C, Jiang Y, et al. The vitamin D receptor gene ApaI polymorphism is associated with increased risk of renal cell carcinoma in Chinese population. *Scientific Reports.* 2016;6(1):1–6.

Yousaf N, Afzal S, Hayat T, Shah J, Ahmad N, Abbasi R, et al. Association of vitamin D receptor gene polymorphisms with prostate cancer risk in the Pakistani population. *Asian Pacific Journal of Cancer Prevention.* 2014;15(22):10009–13. doi: 10.7314/apjcp.2014.15.22.10009.

Youssef E, Mohamed F, Edreis A, Sedik W, Alblihed M, Soliman A, et al. Evaluation of OPN level and VDR gene polymorphism in patients with hepatocellular carcinoma. *Research in Cancer and Tumor.* 2016;5:10–6.

Yuan C, Qian ZR, Babic A, Morales-Oyarvide V, Rubinson DA, Kraft P, et al. Prediagnostic plasma 25-hydroxyvitamin D and pancreatic cancer survival. *Journal of Clinical Oncology.* 2016;34(24):2899.

Zgaga L, O'Sullivan F, Cantwell MM, Murray LJ, Thota PN, Coleman HG. Markers of vitamin D exposure and esophageal cancer risk: a systematic review and meta-analysis. *Cancer Epidemiology and Prevention Biomarkers.* 2016;25(6):877–86.

Zheng W, Wong KE, Zhang Z, Dougherty U, Mustafi R, Kong J, et al. Inactivation of the vitamin D receptor in APCmin/+ mice reveals a critical role for the vitamin D receptor in intestinal tumor growth. *International Journal of Cancer.* 2012;130(1):10–9.

16

Role of Genetic Polymorphisms in Lipid Metabolism Pathways in Modulating Cancer Promotion

Racha Kerek
Faculty of Sciences, Department of Biochemistry, Jinan University, Tripoli, Lebanon

CONTENTS

ABSTRACT During carscinogenesis, neoplastic cell formation relies on fatty acid anabolic pathways for regulatory signaling molecules synthesis. Lipids rearrangement during tumor cell reprogramming includes fatty acids incorporation in cell membranes. This process is activated by the acyl-coenzyme A synthetase long-chain family of enzymes. Lipid uptake by neoplastic cells also involves several lipase class enzymes – lipoprotein lipase and hepatic lipase – and their regulatory proteins – apolipoprotein E and peroxisome proliferator-activated receptor gamma. In addition, polyunsaturated fatty acids metabolism releases bioactive molecules inducing carcinogenesis either by regulating the action of the aforementioned regulators, or by acting as signaling molecules activating pro-oncogenic oxidative pathways. Arachidonic acid may be transformed into one of several pro-tumorigenic metabolites through activation of enzymes of the prostaglandin-endoperoxide synthases or the cytochrome P450. Differential gene expression coding for these regulators results in the pathophysiological state of cancer. Genetic polymorphisms are involved in the activation of such processes. Certain variants of enzymes involved in lipid metabolism increase susceptibility for certain cancers, depending on the population. Therefore, studying genetic variants is crucial for providing a better prognosis for cancer patients. We'll review the role of single nucleotide polymorphisms in six genes of interest implicated in fatty acid metabolism and their association with cancer prognosis.

KEYWORDS: *Lipid Metabolism; De Novo Lipid Synthesis; Polyunsaturated Fatty Acids; Cancer Genetics; Cancer Susceptibility; Polymorphisms; Metabolic Disorders.*

16.1 Introduction

Cancer progression is linked to metabolic reprogramming, known as a set of metabolic changes by which cancer cells adapt to their microenvironment. Lipid metabolism is emerging as one of these key changes involved in cancer initiation and progression. Rapidly proliferating cells rely on fatty acid anabolic pathways for synthesis of both membranes and signaling molecules involved in neoplastic cells, in addition to providing a source for energy production. Expression of genes involved in these pathways, along with fatty acids' exogenous and endogenous synthesis, affects prognosis in different types of cancer. Gene-environment interaction has been a topic of interest in the latest clinical and epidemiological studies. We therefore review the impact of key genes involved in lipid metabolism, their polymorphisms, the additivity of genetic variant interactions, and dietary factors on tumorigenesis outcome.

DOI: 10.1201/9781003246244-16

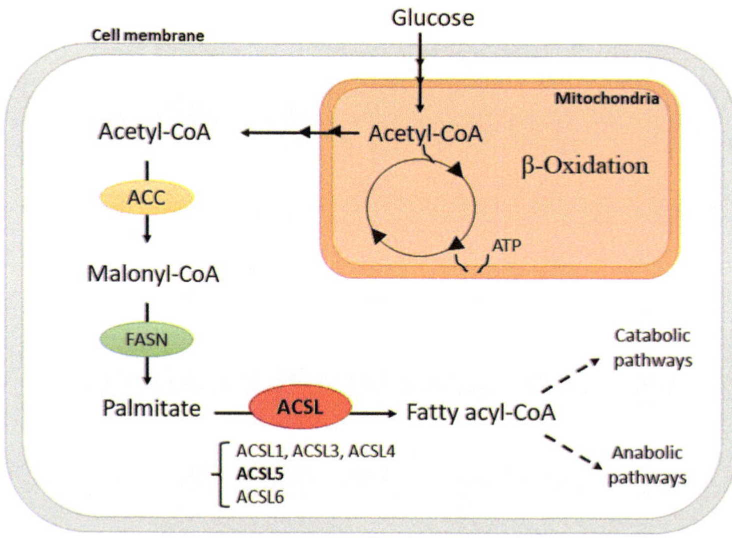

FIGURE 16.1 Overview of de novo long-chain fatty acid synthesis.
Acetyl-CoA carboxylase (ACC, rs11871275), fatty acid synthase (FASN, rs4247357), acyl-CoA synthetase long-chain family (ACSL).

16.2 *De novo* Lipid Synthesis

Throughout the recent years, the reprogramming of lipid metabolism has become widely accepted and established as having a major role in cancer development, its progression, and in its aggressiveness. *De novo* lipid synthesis is activated in tumor cells compared to normal cells, as evidenced by the significant increase in the levels of expression of different enzymes involved in the regulation of this pathway. Evidence is accumulating in this regard. As a recent example, a study by Xu et al. (2021) on clear cell renal cell carcinoma (ccRCC) has shown a significant increase in mRNA expression of the gene coding for fatty acid synthase (FASN) enzyme implicated in the *de novo* synthesis of long-chain fatty acids (LCFAs). The increased expression is also significantly correlated to advanced clinicopathological characteristics of ccRCC patients and poor survival rates in two different cohorts (Xu et al., 2021). FASN enzyme, encoded by the FASN gene located on the long arm of chromosome 17 (17q25.3), acts to catalyze the transformation of malonyl-CoA into palmitate. And, at the cellular level, the regulation of metabolic disorders in tumor cells microenvironment seems disturbed with *FASN* overexpression inducing cancer cell proliferation and inhibiting apoptosis in several types of cancers (Xu et al., 2021; Ezzeddini et al., 2019).

Another important gene is the acetyl-coA carboxylase (*ACC*) also mapped to chromosome 17 (17q12). The encoded enzyme is central in the de novo lipid synthesis pathway and is essential for the catalysis of acetyl-CoA carboxylation to malonyl-CoA in a biotin-dependent reaction. This reaction is key in fatty acid synthesis as it constitutes the rate-limiting step as it is allosterically regulated by the formation of palmitoyl-CoA (Figure 16.1).

In a similar manner to that observed in the case of an increase in *FASN* expression, a simultaneous increase in *ACC* leads to the spontaneous development of hepatocellular carcinoma (HCC) independently of nonalcoholic fatty liver disease (Li et al., 2019). This has been further elaborated by loss of functions study, which has shown that using ACC inhibitors or siRNA knockdown inhibited human glioblastoma cell proliferation, among other cancers, and was associated with clinical benefits regarding this disease (Indraccolo et al., 2020).

The study of genetic variations of these enzymes implicated in *de novo* lipid synthesis indicates certain associations with some types of cancers. In this regard, a case-control study conducted on 1,331 prostate cancer patients showed that four of five single nucleotide polymorphisms (SNPs) studied in the *FASN* gene were significantly related to increased cancer incidence and mortality in the USA (Nguyen et al., 2010). However, data from European studies show that, although some correlation exists between genetic variation in *FASN* and cancer, it doesn't reach significance neither in prostate cancer males, nor in breast cancer females (Campa et al., 2011). Another example in which ethnicity seems to be relevant regarding *FASN* variants and their association with cancer risk is that related to rs4247357 and the risk of uterine leiomyomata. Results show that while this SNP increases the risk of this form of cancer in women of European ancestry, however, it is not the case for patients of African ancestry (Wise et al., 2016; Eggert et al., 2012). In a similar manner, genetic variants in *ACC* may be positively correlated or at least influence the incidence of certain type of malignancies. Results vary depending on cancer type. For example, one study shows that rs11871275, an ACC variant at the site of binding of transcription factor, has no correlation to recurrence and survival of lung cancer (Jin et al., 2014). However, this same rs11871275 variation is being investigated as a prospective biomarker in HCC for the prediction of clinical outcomes in affected patients (Jiang et al., 2015).

Given the examples presented above, this proves the importance of gene variants of the enzymes implicated in lipid synthesis and their relevance to cancer incidence, progression, and aggressiveness. However, the expected correlations depend not only on the gene, but also on the specific mutation, different ethnicities, and the site of cancer. Among these different genes, we will detail the long-chain acyl-CoA synthetase family (ACSL) gene, its isoforms, and gene variants' correlation to all-cancer incidence.

16.3 ACSL

Metabolic reprograming during carcinogenesis is induced by lipids rearrangement and increase in fatty accumulation (-Gong et al., 2020). They also implicate alterations in the lipid metabolism to sustain increased phospholipids biosynthesis for membrane incorporation (Mika et al., 2020; Ventura et al., 2015), to regulate transcription (Blücher et al., 2020; Ventura et al., 2015), and to activate oncogenic and anti-apoptotic signaling pathways (Pang et al., 2021). The results of a recent transcriptomic study have also revealed the correlation of genes implicated in the formation and desaturation of LCFAs with some cancers, depending on tissue and location (Uhlen et al., 2017).

Among different genes of the lipid metabolic pathways, specifically those involved in LCFAs synthesis, accumulating evidence shows the roles of ACSL misexpression during cancer. ACSL is a family of lipid synthetase enzymes present in all organisms such as fungi and humans and seems implicated in both anabolic and catabolic pathways of lipid metabolism (Guo et al., 2014; Zhao et al., 2019). They act by establishing a thioester bond with the coenzyme A, an initial step necessary for activating fatty acid degradation, *de novo* lipid synthesis, and cell membrane renewal. In these ways, activated fatty acyl may be either oxidized by β-oxidation or incorporated into phospholipids depending on the body's physiological needs. Here, we discuss the function of the ACSL enzymes in activating the long-chain acyl-CoA and the role they play in cancers and other diseases.

Five different isoforms, ACSL1, ACSL3, ACSL4, ACSL5, and ACSL6, make up the ACSL family. Genes coding for these different isoforms are mapped to different chromosomes: *ACSL1* (4q35.1), *ACSL3* (2.q36.1), *ACSL4* (X.q23), *ACSL5* (10.q25.2), and *ACSL6* (5.q31.1). However, all isoforms share a common function of activation of fatty acids with chain length between 12 and 22 carbons. They are distinguished based on their differential expression in different tissues and their distribution at the subcellular level (Cooper et al., 2015). However, the main functional difference between these five isoenzymes relates to the length of their preferred substrate. Comparatively, ASCL1's preferred substrates seem to be shorter than those of ASCL4 (Klett et al., 2017). Extensive studies have also been done to show that ACSL4 preferentially acts on polyunsaturated fatty acids (PUFAs) incorporation, mainly arachidonic acid (AA) and eicosapentaenoic acid (EPA) (Shimbara-Matsubayashi et al., 2019). At the subcellular level, its localization in the cytoplasm, both in the vicinity of the nucleus and in the endoplasmic reticulum (ER), contributes to its pivotal role in the biosynthesis of highly unsaturated fatty acids (HUFAs) (Kuwata et al., 2019).

For all these reasons, each ACSL isoform seems to be implicated differently in distinct types of cancers (Chen et al., 2016; Tang et al., 2018; Rossi Sebastioano and Konstantinidou, 2019). The involvement of individual forms of ACSL in cancer prognosis is summarized in Table 16.1. Nonetheless, links between different isoforms have also been established. For instance, ASCL1 and ASCL4, in addition to another enzyme of the lipid metabolism the stearoyl-CoA desaturase (SCD), form a network acting together to increase the invasiveness of cells in colorectal cancer (Sánchez-Martínez et al., 2017). Another example is the study of the role of the androgen receptor-mediated signaling in prostate cancer cells, which demonstrates contrasting, but complementary ramifications on the expression levels of both ACSL3 and ACSL4, leading to accelerated tumor progression (Ma et al., 2021). Generally speaking, ASCL1 and ASCL4 seem to be overexpressed in certain types of cancer where ASCL5 expression is decreased (Tang et al., 2018).

16.3.1 ACSL5

As seen in Table 16.1, ACSL5 is distinguished from other isoforms in the fact that it mostly seems to have a protective influence against cancer prognosis, with the exception in relation to gastric cancer, where *ACSL5* induction promotes cancer cell progression (Seo et al., 2020). To further understand this, we focus here on the functions of *ACSL5* variants in cancers and other diseases.

In humans, the *ASC5L* gene, through different splicing schemes, encodes two different proteins: the full-length ASCL5 and the short ASCL5 protein, lacking exon 20. ACSL5, like the other isoenzymes of the ACSL family, generates activated fatty

TABLE 16.1

ACSL Isoforms' Expression in Different Types of Cancer

	Myeloid Leukemia	Lung Cancer	Breast Cancer	Liver Cancer	Gastric Cancer	Colorectal Cancer	Prostate Cancer	Ovarian Cancer	Bladder Cancer
ACSL1			Dark grey			Dark grey			
ACSL3				Dark grey					
ACSL4		Dark grey	Light grey	Dark grey			Dark grey		
ACSL5		Light grey	Light grey		Dark grey	Light grey		Light grey	Light grey
ACSL6	Light grey								

Light grey cases show cancers where the correspondent ACSL isoform expression is decreased, and dark grey cases show cancers where the correspondent ACSL isoform expression is increased.

acyls: fatty acids esterified to coenzyme A. This initial activating step is essential for subsequent reactions involved in cell proliferation and underlying fat absorption and energy regulation (Bowman et al., 2016; Pérez-Núñez et al., 2019). Expression of the full-length ASCL5 has a protective effect against cancer by acting on the balance of anti- and pro-apoptotic proteins (Gassler et al., 2007). In a study conducted to investigate the links between genetic factors and migraine, an *ACSL5* G2079+7A (-rs2256368) functional variant was identified within the 20th intron of the *ASCL5* gene. The author has shown that this single nucleotide substitution was responsible for exon 20 skipping encoding the short ASCL5 protein (p.Val671_Val694del) (Matesanz et al., 2016). At the functional level, rs2256368 not only led to a shorter protein, but also led to a decrease in the expression of this form of ASCL5 due to inefficient translation. Nevertheless, even with the diminished level of expression, the spliced protein was correlated to an inhibition of cell growth and excessive stress related to reactive oxygen species production (Pérez-Núñez et al., 2019). According to these results, *ASCL5* variants need to be better explored to understand their relevance to cancers and other diseases in which a shift in lipid metabolism is implicated.

In addition to this variation, 60 other polymorphisms are reported across the region of DNA comprising the *ACSL5* gene, but their functional and/or physiological significance has not yet been elucidated (Adamo et al., 2007). However, the authors have shown that another variant, the *ACSL5* +C986T (rs2419621), showed association with weight loss and adiposity in a Canadian population and may be used as a predictor to the carriers' responsiveness to diet. Similar results have also been confirmed on a Spanish population (Izaola Jáuregui et al., 2020). At the cellular level, this polymorphism is associated with an increase in the expression of *ACSL5* mRNA, mostly among homozygote individuals (Rajkumar et al., 2016; Adamo et al., 2007). Results also suggest the presence of a link between *ACSL5* rs2419621 and the peroxisome proliferator-activated receptor (*PPAR*) gamma 2 Pro12Ala allele variant, a relationship worth exploring, especially for the role of PPARs in lipid metabolism (Adamo et al., 2007). Although the results of this study focus on the significance of these polymorphisms in relation to diet, their influence on lipid metabolism and fatty acid uptake and oxidation makes them possible targets in the context of study of other metabolic disorders and cancer. Further investigation has hence demonstrated that rs2419621 T allele carriers exhibit an increased expression of the short ACSL5 protein in myocytes, similar to the previously reported genotype (Rajkumar et al., 2018). In contrast, HepG2 cell characteristics of liver cell carcinoma show expression of both long and short isoforms. At the cellular and subcellular levels, the presence of this truncated protein was associated with an increase in fatty oxidation in myocytes, which could be partly explained by the protein's localization within the mitochondria (Rajkumar et al., 2018).

16.4 PPARG

Peroxisome proliferator-activated receptor gamma (PPARG) is a member of a subfamily of nuclear receptors encoded by a gene located on the short arm of chromosome 3 (3p25.2). PPARs are a class of transcription factors superfamily that interact with retinoid X nuclear receptors (RXR) inducing target genes' expression by binding to the peroxisome proliferator response elements (PPREs) in their promoter region (Robbins and Nie, 2012). There are three isoforms of PPAR. Among these, PPARβ/γ controls the expression of genes used in beta-oxidation, while PPARG (also referred to as PPARγ) is involved in adipocyte differentiation. Structurally, two major isoforms of PPARG (PPARG1 and PPARG2) may be generated through alternative promoter usage, but all isoforms are involved in adipocyte differentiation and regulation of genes used in lipid storage (Al Haj et al., 2020). Indeed, the increase in lipogenesis is concomitant with the expression of certain transcription factors for lipogenesis enzymes, especially PPARs, in addition to the sterol regulatory element-binding protein (SREBP) 1 and 2 (Ezzeddini et al., 2019; Robbins and Nie, 2012). The key target genes of *PPARG* are regulators of fatty acid synthesis (Figure 16.2), such as *ACSL*, *SCD*, *FASN*, lipoprotein lipase (LPL), leptin, and fatty acid-binding protein 4 (*FABP-4*) (Moseti et al., 2016; Blanchard et al., 2016;

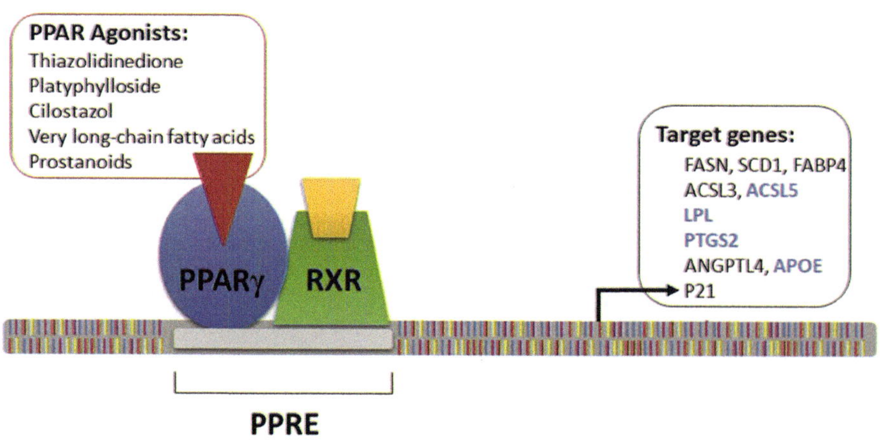

FIGURE 16.2 PPARG ligands and target genes.
Genes marked in blue are the genes of interest reviewed in this chapter.

Mansour et al., 2011). Antidiabetics of the thiazolidinedione family are synthetic agonists of PPARG, while very long-chain fatty acids and prostanoids such as prostaglandin 2 (PG2) act as diet-derived biological ligands (Robbins and Nie, 2012). PPARG is implicated in controlling cell proliferation as well as differentiation and might in this way affect carcinogenesis and other metabolic diseases (Robbins and Nie, 2012). The activation of PPARG by one of its synthetic or biological agonists represses the proliferation and differentiation of glioblastoma and bladder cancer cells, engenders apoptosis of prostate cancer cells, and upregulates p21 expression inducing cell cycle arrest in colon cancer (Mehus et al., 2020; Sabatino et al., 2019; Mansour et al., 2011).

On another note, epidemiological studies, especially in terms of studying *PPARG* gene variants, allow for a better understanding of this transcription factor's association with cancer. This is mostly relevant because of the association of *PPARG* induction with bioactive lipids, and the fact that different cells by site of cancer respond differently to their lipid microenvironment. In this regard, *PPARG* C34G, also referred to as *PPARG* Pro12Ala (rs1801282) polymorphism, has been found to improve LPL enzyme functioning in terms of traits related to insulin resistance (López-Alarcón et al., 2012). *PPARG* and *LPL* variants act in a combinatory manner as risk factors for the metabolic syndrome, and their study is thus important for the prevention of metabolic diseases (Bozina et al., 2013). No similar association in the two gene variants was significant in relation to cancer risk. However, computational analysis has identified 297 non-synonymous SNPs and 163 coding SNPs in the *PPARG* gene. Among them, 5 SNPs were identified as benign, whereas 12 SNPs were identified as probable influencers for obesity, diabetes, and cancer (Stalin et al., 2020).

PPARG variants are indeed widely studied for their implication with diabetes incidence. A meta-analysis reviewed 73 studies between 1990 and 2017 (including 62,250 cases and 69,613 controls), and the results showed a protective influence of the minor G allele (12Ala) with diminished risk of type II diabetes mellitus (T2DM) among its carriers (Sarhangi et al., 2020). Different ethnicities showed different percentages, and the effect was the strongest among the Europeans. Interestingly, however, although the 12Ala allele seems to protect against T2DM, the minor allele presents a risk factor for the metabolic syndrome as concluded in a meta-analysis among 13 studies, in a mechanism probably involving lipid metabolism and weight accumulation (Povel et al., 2011). In this context, PPARG is studied as a candidate gene and a predisposing factor to conditions involved in lipid metabolism and dyslipidemia. A meta-analysis including 74 studies and 54,953 subjects was carried out to clarify the relationship between three PPARG variants (Pro12Ala, C1347T, and C161T) and different lipid components levels in the serum (Li et al., 2015). A significant association has been found only for the Pro12ala with increased detection of lipid components (cholesterol, lipoproteins, and triglycerides) in the serum of carriers.

As lipid metabolism is reprogrammed during cancer progression, scientists looked for similar association in the context of cancer. In this regard, an investigation assessing published meta-analyses and genome-wide association studies (GWAS) on links between candidate genes/variants and the prospect of gastric cancer has demonstrated that *PPARG* Pro12Ala (rs1801282) variant presented strong corroboration for a significant association with cancer incidence (Tian et al., 2019). *PPARG* Pro12Ala (rs1801282) was indeed identified as one of the nine variants on nine genes that presented the higher risk among 61 total variants. A meta-analysis studying *PPARG* Pro12Ala (rs1801282) variant in 51 case-control and cohort studies (including 14 types of cancer, different ethnicities, and 16,844 cases and 23,736 controls) showed no interrelation with overall cancer risk (Wang et al., 2015a). Disparities, however, are noticeable when screening for cancer site and populations. Meta-analyses identify *PPARG* Pro12Ala (rs1801282) polymorphism as significantly associated with susceptibility to polycystic ovary syndrome (PCOS) and gastric cancer in the overall population (Liang et al., 2019; Tian et al., 2019). Another polymorphism, *PPARG* C1347T (rs3856806) was considered as a genetic predisposition factor to PCOS only in Asian ethnicity (Liang et al., 2019). However, in another type of cancer, a meta-analysis on 9 case-control and cohort studies (including a total of 3,715 cases and 5,166 controls) identified *PPARG* Pro12Ala (rs1801282) variants as protective for breast cancer in Caucasians, while *PPARG* rs3856806 polymorphism showed no association (Tang et al., 2015). Case-control studies show different results based on ethnicity, cancer, and variation site (Table 16.2).

Lin et al. (2019) in their study of six *PPARG* SNPs and their relationship with the risk of colorectal cancer development

TABLE 16.2

Prominent *PPARG* Variants (rs3856806 and rs1801282) in Relation to Cancer Risk

Polymorphism	Country or Population	Case/Control	Type of Cancer	Risk Alleles	References
C1347T, rs3856806	Chinese Han	1003/1303	Colorectal	T allele	Lin et al. (2019)
C1347T, rs3856806	Korean	971/658	Colorectal	T allele	Kim et al. (2019)
Pro12Ala (C>G), rs1801282	Multi-ethnic USA	553 (cohort study)	Breast	No correlation	Duggan et al. (2018)
C1347T, rs3856806 Pro12Ala (C>G), rs1801282	Chinese Han	584/923	Hepato-cellular	T allele (rs3856806) C allele (rs1801282)	Zhang et al. (2018)
Pro12Ala (C>G), rs1801282	Chinese Han	387/536	Colorectal	C allele	Jiang et al. (2017)
C1347T, rs3856806 Pro12Ala (C>G), rs1801282	Chinese Han	507/1496	Esophageal squamous cell	No correlation (rs3856806) G allele (rs1801282)	Qiu et al. (2017a)
Pro12Ala (C>G), rs1801282	Pakistani	200/200	Urothelial bladder	G allele	Ali et al. (2017)

perceived that only the *PPARG* 3856806 polymorphism increased the risk of cancer incidence. This association based on genotyping analysis for 1,003 cases and 1,303 controls in Chinese Han population was found to be stronger in men aged 61 and above. In the same population, but regarding HCC risk, Zhang et al. (2018) revealed that two *PPARG* single nucleotide polymorphisms influenced cancer risk in their case-control study (including 584 cases and 923 controls). Interestingly, however, while one variant, *PPARG* Pro12Ala (rs1801282) polymorphism, was considered as a protective factor against HCC, another one, *PPARG* C1347T (rs3856806) polymorphism, was correlated with increased susceptibility to HCC. In Pakistani population, a study of 17 candidate genes in 200 urothelial bladder carcinoma patients and 200 controls showed the association of *PPARG* variants with high-grade cancer incidence (Ali et al., 2017). Furthermore, *PPARG* variants were also studied in the Indian population in relation to PCOS, a disease which is known to predispose women to cancer risk. A case-control study on PCOS, including 169 patients and 169 control subjects, was conducted to look for an association between *PPARG* Pro112 Ala (rs1801282) and *PPARG* C1347 (rs3856806) and PCOS. Statistical analyses showed that the minor allele frequency for both variants, regarding patients and control women, was not significantly different and therefore exhibited no association with PCOS risk (Thangavelu et al., 2017). Results in Indian population were confirmed in another study that came out in the same year by Reddy et al. (2017), which also showed no significant correlation between *PPARG* Pro12Ala gene variant and PCOS risk. In Caucasians, based on research studies done in the last 10 years, *PPARG* variants lack associations with colorectal or gastric cancer risks (Kopp et al., 2013; Crous-Bou et al., 2012).

16.5 Lipoprotein Lipase and Hepatic Lipase

In the liver, triglycerides resulting from *de novo* lipid synthesis are packaged along with phospholipids, cholesterol, and apoproteins into very low-density lipoproteins (VLDLs). VLDLs go into the bloodstream to deliver triglycerides to different tissues for storage or use as energy sources. For this process, and upon entering circulation, lipoproteins release free fatty acids through lipoprotein lipase-catalyzed lipolysis.

Lipoprotein lipase-coding gene is found on chromosome 8 (8p21.3), and LPL enzyme is expressed as a homodimer in different tissues involved with lipid metabolism, mainly the adipose tissue in order to facilitate lipoprotein uptake. Its differential expression in some cells, especially during tumorigenesis-mediated lipid reprogramming, might be a prognostic marker in some cancers. Studies have shown an increase in LPL expression in breast, liposarcoma, and prostate cancer tissues (Kuemmerle et al., 2011). The angiopoietin-like 4 (ANGPTL4) protein, upregulated essentially in the liver and adipose tissue in a PPARG-dependent manner, promotes cancer development and progression through LPL activity inhibition (Blücher et al., 2020; La Paglia et al., 2017). Indeed, PPARG agonists, such as the platyphylloside, induce lipogenesis in a mechanism involving downregulation of gene promoters specific to differentiated adipocytes such as *LPL* (Lee and Sung, 2016). *LPL*'s differential expression during cancer-mediated lipid metabolism reprogramming may be implicated in tumor formation and progression (Dong et al., 2017). In clinical studies, *LPL* has been established as the strongest prognostic factor for chronic lymphocytic leukemia (CLL) based on mRNA expression and it is also correlated with decreased overall survival in these patients (Kaderi et al., 2011). *LPL* deletion affects therapeutic response in CLL patients as demonstrated by Liu et al. (2020) in their study of the 17p deletion related to *LPL* deletion (*LPL*-del) in 176 CLL patients.

Different tissues express specific type of lipase. Hepatocytes synthesize and secrete the hepatic lipase (*LIPC*, also known as *HL*) through differential expression of the corresponding gene located at chromosome 15 (15q21.3). LIPC is mainly responsible for triglycerides and phospholipids hydrolysis at this site (Kobayashi et al., 2015). The LIPC attached to endothelial cells of the liver is inactive. When secreted, LIPC circulating in bloodstream metabolizes VLDL and HDL substrates, releasing fatty acids and other bioactive lipid compounds (Chatterjee and Sparks, 2011). *LIPC*'s present, but differential expression has also been demonstrated in several types of cancers. Immunohistochemical LIPC expression analysis in 324 primary gastric cancer specimens, 346 breast cancer tissues, and esophageal adenocarcinoma cell lines confirmed the protein's upregulation and its possible role as a prognostic factor for cancer. Furthermore, individuals with increased LIPC expression had a higher rate of metastasis and exhibited poor overall survival rates, regardless of the type of cancer (Huang et al., 2020; Ishimine et al., 2016).

16.5.1 LPL and LIPC SNPs

Investigations on the relationship between LPL SNPs and cancer are scarce compared with other genes related to lipid metabolism. However, due to LPL's function in the body, associations center more around its involvement in gene-environment interactions, especially the effect of dietary intake, metabolic markers, and the risk of metabolic diseases. Indeed, a large GWAS including individuals of African ancestry stratified by gender has identified a novel variant *LPL* rs149352150 for BMI correlations in males (Ng et al., 2017). A recent large-scale study applied statistical approach to analyze data from different meta-analyses (66,430 men and women, aged 19–80 years, and European ancestry) and to search for what they referred to as gene-lipid associations. During their investigations regarding the implication of rare variants or those who appear in lox frequency, the authors looked at thirteen previously identified lipid loci (21 variants), and amid other results, *LPL* variants showed associations with high- and low-density lipoprotein plasma levels (Wang et al., 2020). Similarly, in a population study of 25,074 subjects from the UK, C allele carrier of *LPL* C1421A (rs328) variant had higher levels of plasma triglycerides concentrations and scored a higher risk in relation to the metabolic syndrome phenotypes (Verbeek et al., 2019). These phenotypes include the risk of stroke, cardiovascular diseases, and cancer, all conditioned by the lipid status in the cell. In this regard, a meta-analysis (13 studies; 4,681 cases and 8,516 controls) looking for a correlation between *LPL* polymorphisms and ischemic strokes indicated a protective role for *LPL* Ser447Ter, which was significant against atherosclerotic stroke risk (Wang et al., 2011). In a similar context, an association study (14 studies; 3,592 cases and 4,643 controls) on *LPL* variants and hypertension also showed the effect of LPL Ser447Ter in protection against hypertension (Wang et al., 2017).

Concerning carcinogenic associations, based on a scientific review of the published data in the last 10 years, *LPL* variants are mostly studied in relation to leukemia. *LPL* G644A (rs118204057) polymorphism has been identified in one survivor of acute lymphoblastic leukemia, compared to none in control subjects (Morel et al., 2017). This SNP is known to result in a missense mutation causing LPL deficiency, and elevated plasma triglycerides levels (Morel et al., 2017). Rombout et al. (2015) investigated three *LPL* polymorphisms T1139+43C (rs301), C1421G (rs328), and T1661C (rs13702) and their association with chronic lymphocytic leukemia (248 cases, Caucasians). Results showed that two variants, rs301 and rs13702, were significantly linked in a gene-gene sort of interaction, and wild-type allele in both was individually correlated to shorter overall survival. These SNPs mediated their effects by acting on the amount of LPL mRNA expressed.

In a similar manner, *LIPC* variants influence metabolic outcome in a plasma lipids content-dependent manner. A meta-analysis has shown the gender-based association between *LIPC* C514T (rs1800588) and obesity (Wang et al., 2015b). Additionally, according to the study, there is a significant relationship between rs1800588 and cholesterol (total and high-density lipoproteins), regardless of the obese phenotype. *LIPC* G250A (rs2070895) was found to be a marker for increased hypertension risk according to a meta-analysis including 2,599 participants (Zhao et al., 2018). Concerning the *LIPC* polymorphisms and cancer risk, although the relationship has not been well reported, as expected, the existing data report liver cancer. Indeed, four common variants have been discerned in the promoter region of the gene coding for *LIPC*: G250A (rs2070895), C514T (rs1800588), T710C (rs1532085), and A763G (rs1077834). These variants all affect the circulating lipoproteins levels and are thus linked to phenotypic changes. In a case-control study on 279 Chinese HCC patients vs 200 healthy controls, genotyping experiments revealed that the allelic frequency and distribution in the G250A (rs2070895) and C514T (rs1800588) genotypes were significantly different between the two groups. For the G250A (rs2070895), GG wild-type carriers seem to be predisposed to HCC, while the T allele carriers in C514T (rs1800588) presented the higher risk of cancer (Niu et al., 2018).

16.6 ApoE

Two post-translational mechanisms of control of LPL activity exist. These involve, on the one hand, the ANGPTL family of angiopoietin-like proteins such as ANGPTL4 (La Paglia et al., 2017), which function in their transport to capillary endothelium of target tissues. On the other hand, another group of proteins, the apolipoproteins produced by the liver, affects the LPL activity. Apolipoprotein E (APOE), a member of this family, reduces triglycerides clearance mediated by LPL activity. *APOE* gene is mapped to the 19q13.32 location on the 19th chromosome, a region which also contains other isoforms of the APO family, APOC1 and APOC2. The encoded protein contains a binding site to the low-density lipoprotein receptor (Phillips, 2014). The low-density lipoprotein receptor-related protein 1 (LRP1) binds to several targets implicated in lipid metabolic reactions, namely our three genes of interest for triglycerides metabolism *LPL*, *LIPC*, and *APOE*. In HepG2 liver cancer cell lines, treatment with cilostazol, a PPARγ agonist, increased APOE production, activated LPL enzyme, and enhanced clearance of lipoprotein (Kim et al., 2014).

Variations in the *APOE* gene have mostly been described in relation to metabolic disorders and Alzheimer's disease. Studies on the association with cancer are limited, but the link is well established in some types of cancer. The human *APOE* gene is polymorphic, and variants consist mainly of a combination of two SNPs (rs429358/rs7412), which results in three different isoforms: the *APOE* ε2/ε3/ε4. The presence of one or a combination of these different alleles may predispose to different health risks, including cancer (Table 16.3).

TABLE 16.3

***APOE* Genetic Polymorphisms in Relation to Cancer**

Polymorphism	Country or Population	Case/Control	Type of Cancer	Risk Isoform	References
ε2/ε3, rs7412; ε2/ε4, rs429358	Lithuania	142/256	Pituitary adenoma	ε2/ε3 for active adenoma; ε4 for recurrent adenoma	Sidaraite et al. (2019)
ε2/ε3, rs7412	European	80/22	Lymphoblastic leukemia	No correlation	Morel et al. (2017)
ε2/ε3, rs7412 ε2/ε4, rs429358	Chinese Han	550/550	Gastric	ε2	Kang et al. (2016)
C219A, r405509	Chinese Han	81/80	Renal cell	No correlation	Lv et al. (2015)
ε2/ε3, rs7412 ε2/ε4, rs429358	Russia	186/1110	Endometrial	ε2	Ivanova et al. (2015)
ε2/ε3, rs7412 ε2/ε4, rs429358	Brazil	47/165	Breast	ε4	Cibeira et al. (2014)
ε2/ε3, rs7412 ε2/ε4, rs429358	Italy	156/444	Gastric	ε3	De Feo et al. (2012)
ε2/ε3, rs7412 ε2/ε4, rs429358	Taiwan	306/300	Breast	ε4	Wu et al. (2012)

A meta-analysis by Liu et al. (2016) reviewed 12 articles (including 2,074 cases and 2,372 controls) on the association between the *APOE* rs429358/rs7412 polymorphism referred to as the *APOE* ε2/ε3/ε4 polymorphism and breast cancer risk. Results showed that ε2 and ε4 alleles don't demonstrate a significant association with breast cancer risk. However, after stratification by ethnicity, the authors established a possible correlation only for the ε4 allele in Asian populations regarding cancer at the breast tissue site (Liu et al., 2016). Another meta-analysis on the overall cancer risk (26 articles; 13,478 cases and 77,592 controls) showed the same results concluding with the *APOE* gene ε4 allele's effect also in the Asian population, but this time on the overall cancer risk (Yang et al., 2015). Alternately, the same allele had a tendency to lower the risk of developing proximal colorectal neoplasm according to a meta-analysis including 8 studies (4,310 cases and 9,243 controls) on the association between *APOE* gene polymorphism and colorectal neoplasia (Tian et al., 2014). In addition, in a cohort study following 5,117 participants for over 36–60 years, Kulminski et al. (2013) showed that in each generation, carrying the *APOE* ε4 allele increased the expectancy of life without cancer.

More intriguingly, in the context of our genes of interest, when we looked for a possible cumulative effect of *APOE* and *LPL* polymorphisms on lipid metabolism reprogramming and on cancer risk, several studies have been conducted. Shatwan et al. (2018) analyzed the seven most common genetic variations of *APOE* in a population-based study including 660 individuals (European ancestry, aged 60–74). Among these, cases genotyped to carry the ε4 and GG allele of the *APOE* rs429358 and rs445925, respectively, had significantly higher association with total cholesterol. Concerning the *LPL* polymorphisms, two variants were analyzed: rs328 and rs320. Carriers of the 'G' minor allele of both variants were diagnosed with increased amounts of HDL. No direct interaction could, however, be established between the two genes polymorphisms. *APOE* genotype influences the chronic lymphocytic leukemia survival rate, and response to therapy yet exposes them to a higher risk of dyslipidemia (Morel et al., 2017). Women carrier of the *APOE* gene ε4 allele showed a long survival rate compared to non-carrier patients.

16.7 Polyunsaturated Fatty Acids Metabolism

Certain fatty acids are essential to humans, and we must acquire them from our diet. Essential fatty acids are those belonging to the omega-3 or α-linolenic acid and omega-6 or linoleic acid polyunsaturated fatty acid families, because humans lack the activity of desaturase enzymes at the corresponding positions (Spector and Kim, 2015). Other PUFAs are also acquired from the diet, eicosapentaenoic acid and docosahexaenoic acid (DHA) mainly found in fish. Once acquired by the body, humans, however, may use these precursors in series of elongation and desaturation reactions to produce other PUFAs, mainly AA (Fuentes et al., 2018).

AA, along with other PUFAs, may be esterified to glycerol molecules and is incorporated in cell membranes as part of their phospholipids' components. Their presence contributes to the fluidity of the membranes, and their exact composition affects the physical properties of cells and their specificity in certain cells and tissues. Lipids and PUFAs are the building blocks for membrane synthesis, and those inside the leaflet may be metabolized through different pathways. The resulting metabolites are chemically derived bioactive molecules, often recognized as lipid signaling molecules (Fuentes et al., 2018). This metabolic pathway is initiated by PUFA disintegration from phospholipids forming the membranes, a reaction catalyzed by phospholipase A2 enzymes.

Cytochrome P450 (CYP) are key to the internal turnover of many endogenous entities, and many of them are lipids and lipid derivatives. They thus drive the biosynthesis of bile acids and steroid hormones from cholesterol through oxidation. Enzymatically oxidized lipids also include oxygenated polyunsaturated fatty acids, mainly AA. PUFAs containing 20 carbons such as the AA can be broken down into prothrombin and leukotrienes, collectively known as eicosanoids (Figure 16.3).

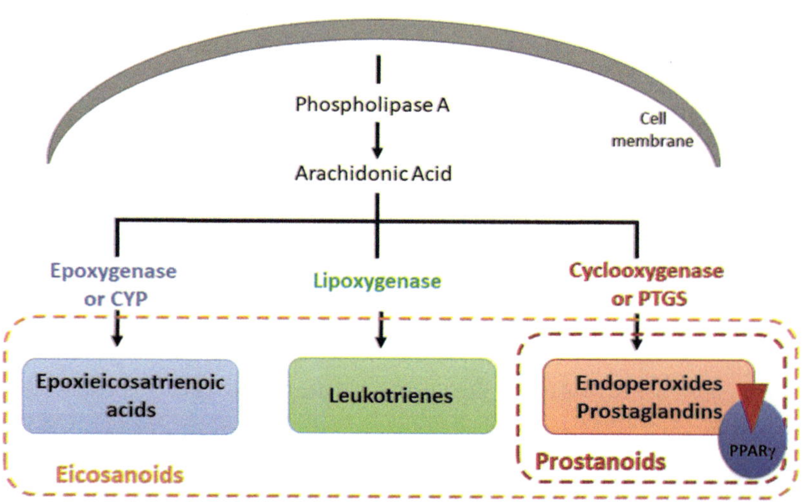

FIGURE 16.3 Overview of arachidonic acid metabolism.

They are produced by different tissues in our body and thus act as local mediators to several tissue-specific metabolic reactions. These reactions are facilitated by two types of enzymes, the lipoxygenase (LOX) and the prostaglandin-endoperoxide synthase (PTGS) enzymes also known as cyclooxygenase (COX), in addition to the CYP450, also known as the epoxygenase enzymes (Guengerich, 2017).

16.8 CYP

Environmental pollutants or xenobiotics that have entered the body are metabolized and then may be eliminated by either phase I or II enzymes encoded by what is known as the xenobiotic metabolism genes (Omiecinski et al., 2011). Phase I corresponds to the functionalization reactions, mediating the modification of the xenobiotic by a functional group in preparation for detoxification. These reactions are catalyzed mainly by monooxygenase CYP450 family enzymes among other enzymes including the aldehyde dehydrogenase and the superoxide dismutase (Rendic and Guengerich, 2012). Phase II corresponds to conjugation reactions. The enzymes involved in phase II are mainly transferases such as the arylamine N-acetyltransferases, the glutathione S-transferases, and the catechol-O-methyltransferase (Table 16.4).

CYP form an array of intrinsic integral proteins that are expressed in several species and are mainly located at the inner membrane of the mitochondria. CYP enzymes have a wide substrate specificity explaining their interplay both in the generation of cellular by-products and in the processing of xenobiotics (Guengerich, 2017). Enzymes belonging to the same subfamily also share some similarities at the level of their substrate specificity. For instance, CYP1A1 and CYP1A2 share a common flanking region on chromosome 15 and evidence shows that they have some, although minor, overlapping substrate specificities (Kwon et al., 2021).

In addition to endogenous molecules metabolism, CYPs also have a role in the detoxification of exogenous molecules. Hence, CYPs have a predominant role in the bioactivation of procarcinogens. Indeed, these enzymes are accountable for the generation of highly reactive by-products that could form an attachment to cellular macromolecules such as DNA (Rendic and Guengerich, 2012). Among the enzymes behind the triggering of procarcinogenic molecules are the CYP3A4, @c, 1A1/2, and 2E1, predominant in the liver in humans (Guengerich, 2017). Likewise, these isoforms are primarily the key for the internal processing of 90% of drugs and are largely involved in the generation of reactive oxygen entities within the mitochondria and the ER. Polycyclic aromatic hydrocarbons (PAHs) and heterocyclic amines are among the substrates to the CYP1 enzyme family. In the next section, we focus on the carcinogen-metabolizing enzyme, the CYP1A2, and its interplay between lipid metabolism and cancer.

16.8.1 CYP1A2

CYP1A2 gene is mapped to chromosomal region 15q24.1 in humans encoding one of the members of the cytochrome P450 superfamily of enzymes. CYP1A2, a 7-exon protein, is localized to the ER and has been proven to be involved in linoleic acid metabolism and its conversion to form 11-hydroxyoctadecadienoic acid (Brignac-Huber et al., 2011). AA, EPA, and DHA are other endogenous substrates to CYP1A2 that mediate their epoxygenation and conversion to one of the different hydroxyeicosatetraenoic acids and epoxyeicosatrienoic fatty acid isomers (Lucas et al., 2010). This conversion is tissue specific and has the highest activity in the liver (El-Sherbeni and El-Kadi, 2014). It is well established that AA hydroxy and epoxy metabolites are involved in carcinogenesis (Sausville et al., 2019). These biological products are considered pro-oncogenic lipid derivatives.

CYP1A2 gene can be regulated by the aryl hydrocarbon receptor (AHR), which may control the body's response to xenobiotics (Sundberg and Hankinson, 2019). This transcriptional regulation may induce an apoptotic effect on cancer cells (Figure 16.4). This effect has been proven on human colorectal cancer cells and esophageal squamous cell carcinoma, among others (Ronnekleiv-Kelly et al., 2016; Zhu et al., 2020). AHR expression in these cells, modulated through 3,3′-diindolylmethane treatment, led to a decrease in cell invasiveness. This effect was mediated through reprogramming of the lipid metabolism in these cells via inhibition of the cyclooxygenase/-prostaglandin pathway COX2/PGE2 regulated by NF-κB expression (Zhu et al., 2020). In another study on human lung

TABLE 16.4
Contribution of Different Metabolizing Enzymes to Procarcinogens Activation

Microsomal flavin-containing monooxygenase (FMO)	1%
N-acetyltransferase (NAT)	7%
Sulfotransferase (SULT)	13%
Aldo-keto reductase (AKR)	8%
Cyclooxygenase/prostaglandin synthase (COX/PTGS)	2%
Cytochrome P450	66%
	CYP1A1/2 – 37%
	CYP1B1 – 11%
	CYP2A6 – 8%
	CYP2E1 – 11%
	CY3A4 – 10%
	Others – 23%
Others	4%

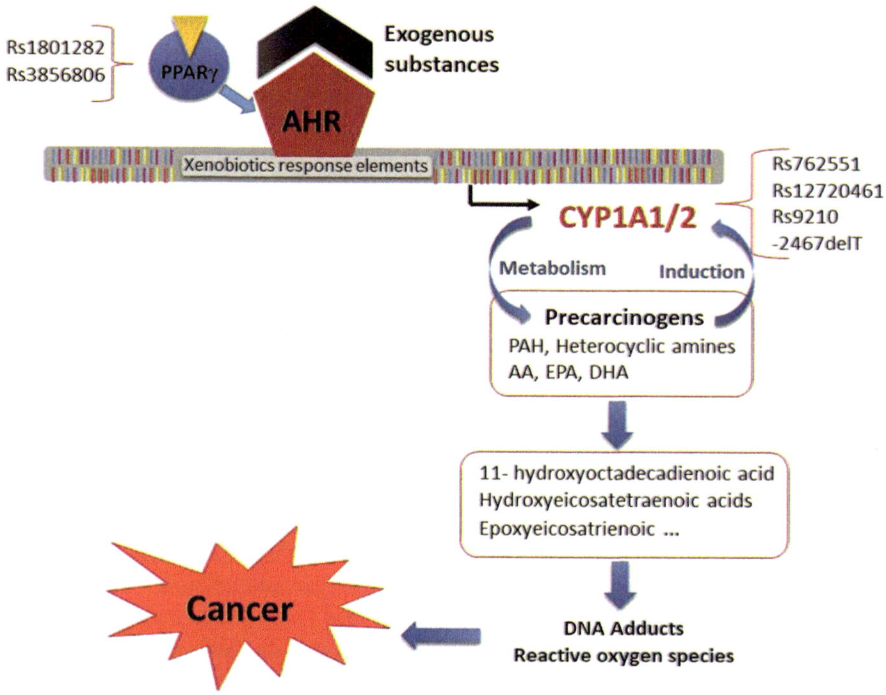

FIGURE 16.4 CYP1A2 expression and effect on cancer.

TABLE 16.5

Association of *CYP1A2* rs762551 Genetic Polymorphism with Cancer Risk

Polymorphism	Country	Case/Control	Type of Cancer	Risk Alleles	References
rs762551: C>A	Jordan	221/218	Breast	AC+CA genotype	Al-Eitan et al. (2019)
rs762551: C>A	Japan	351/870	Prostate	A allele	Koda et al. (2017)
rs762551: C>A	India	404/404	Esophageal squamous cell	AA genotype	Shah et al. (2018)
rs762551: C>A	China	212/200	Bladder	AA genotype	Song et al. (2016)
rs762551: C>A	Sweden	42/200	Breast	CC genotype	Simonsson et al. (2016)

cancer cells, AHR activated fatty acid accumulation, this time via the PPARG pathway (Ye et al., 2021). Evidence also shows that AHR ligands impact tumor growth primarily via PUFA epoxides and *CYP1A* genes overexpression (Huerta-Yepez, 2020). In this regard, it is a growing interest to study the effect of CYP1A2 metabolism, activity, and *CYP1A2* polymorphisms in tumor promotion and growth.

Different *CYP* genes have varying impact on cancer promotion, with ethnicities being a major factor explaining inter-individual and inter-population variations regarding cancer risk (Zhou et al., 2017). In this study, the authors analyzed genetic variability within 12 *CYP* genes accounting for 176 alleles. The analyzed isoforms collectively account for the majority of phase 1 metabolizing enzymes in humans. The study focuses on functional variations by determining frequencies of allelic variants related to phenotypic changes among *CYP450* genes. Data were retrieved from the Exome Aggregation Consortium based on large-scale projects in different ethnicities. Apart from *CYP2 – A6, B6, C8, C9, C19, D6* – and *CYP3 – A4* and *A5* – which hold the most significant pharmacogenetic consequences, *CYP1A2* came up as one of the next important genes on the functional level. According to the study, two CYP1A2 genetic substitutions located in the 5′-UTR, C154A (rs762551) and C113T (rs12720461), held phenotypic consequences (Table 16.5). Ten other missense variants were analyzed with no significant outcome other than change in the enzyme's activity. Regarding rs762551, allelic frequencies were higher among European and American populations, while rs12720461 allelic frequencies were higher among Africans and South-Asians (Zhou et al., 2017).

An updated GWAS that evaluated candidate genetic markers for breast cancer risk in women of European and Asian ancestry, 228 single nucleotide variants were analyzed. Genotyping results suggested the association of 14 variants with the risk for breast cancer, two of them being *CYP1A2* variants: the rs762551 and a novel T2544A (rs9210) (Yang et al., 2019).

Results, however, depend on the population and the type of cancer. For instance, in terms of the effect of CYP1A2 in the metabolization of exogenous substances, a study was conducted to test the contribution of cigarette smoking to head and neck squamous cell cancer risk. Four *CYP1A2* gene variants were analyzed, and no significant association was found in regard to this cancer type in Americans from different ethnicities (Jorge-Nebert et al., 2016). *CYP1A2* polymorphisms tailor patients' response to anti-cancer drugs and the occurrence of adverse drug reactions during the treatment for osteosarcoma (Trujillo-Paolillo et al., 2020). In a Dutch population, only patients carrying the double CC genotype of the rs762551 *CYP1A2* variant were slow metabolizers and required dose reduction in the daily dose of 400 mg of imatinib, an anti-cancer drug used for gastrointestinal stromal tumor treatment (Verboom et al., 2019). However, rs762551 *CYP1A2* polymorphism has no correlation in regard to imatinib treatment of gastrointestinal stromal tumor in a Chinese population (Qiu et al., 2017b). On another note, in a study conducted on four genetic variants of *CYP1A2* in a Korean population, only the -2467delT showed correlation with metabolites concentrations of tamoxifen used as treatment in breast cancer patients. Both heterozygous and homozygous patients were slow metabolizers compared to wild-type patients (Woo et al., 2017).

In conclusion, *CYP1A2* polymorphisms may affect cancer prognosis either positively or negatively. To elaborate, we refer the reader to Table 16.5; when we focus on the rs762551, the minor C allele is associated with breast cancer risk, especially in its homozygous CC form. On the other hand, this variant seems to have a protective effect in prostate and bladder cancers. Previous meta-analysis has implied the lack of significant association between breast cancer and *CYP1A2* rs762551 (Vukovic et al., 2016). In terms of these researches, further analysis is recommended on larger samples and in multi-ethnic populations.

16.9 Prostaglandin-Endoperoxide Synthases (PTGS)

Upon AA release from membranes, it may also be metabolized by other enzymes than the CYP families. AA may be oxidized by PTGS to form prostaglandins. Two isoforms of this enzyme exist, the PTGS1 and PTGS2, also known as COX-1 and COX-2. In general, bioactive metabolites of the PTGS enzymes are referred to as prostanoids (Christie and Harwood, 2020).

Among these two isoforms, studies mostly focus on the inducible form of the enzyme, PTGS2. The gene coding for this enzyme constitutes of 10 exons located on the long arm of chromosome 1 (1q31.1). PTGS2 expression is stimulated under inflammatory conditions as the cell's adaptive mechanism is employed to restore normal physiological conditions. These conditions include, among others, secretion of cytokines, tumor promoters, and other lipid metabolites (Christie and Harwood, 2020; Fuentes et al., 2018). The long-term exposure of cells to these conditions activates oncogenic pathways and promotes oxidative and DNA damage, leading to tumor formation (Ferreri et al., 2020). In this regard, the effect of diet, mainly in regard to inflammatory components, and PUFA content in regard to cancer prognosis have been investigated in a prospective study on 42,511 individuals between 40 and 60 years of age. Healthy individuals have been followed up for 15 years, with regular measurements for dietary information, inflammatory markers, and cancer incidence. Results show the possible effect of certain dietary components known as 'inflammatory food' on increased cancer risk, mainly at the sites of the lungs and the stomach (Bodén et al., 2019). PTGS2 expression seems central within this inflammation-induced carcinogenesis. Indeed, PTGS2 expression modulates PUFA's anti-neoplastic activities in human and mouse colorectal cancer cells exposed to EPA (Volpato et al., 2021; Huang et al., 2019). As mentioned before, prostaglandin (PG) released under the enzymatic activity of PTGS2 acts as a signaling molecule. Studies have shown that PG2 binding to their specific receptor, mainly EP4 receptor, and its activation induces urothelial tumorigenesis and affects cancer cells survival via modulation of pro-oncogenic pathway activation and/or oxidative stress (Huang et al., 2019; Kashiwagi et al., 2018).

Moreover, exploration of genetic variants of *PTGS2* is important in a pharmaceutical context. Research in this regard might be relevant for a personalized therapy approach, where results suggest that the dose of nonsteroidal anti-inflammatory drugs used in some types of cancers might need to be adjusted according to *PTGS2* genotypes (Wang et al., 2021). Among different *PTGS2* variants, those that have an effect on enzyme expression and activity are most likely to affect response to treatment (Lee et al., 2017). As seen in Table 16.6, case-control

TABLE 16.6

The Most Common *PTGS2* Genetic Polymorphisms in Relation to Cancer Risk

Polymorphism	Country	Case/Control	Type of Cancer	Risk Alleles	References
A3809G, rs689466	Portugal	260/476	Gastric	GG genotype	Lopes et al. (2021)
G1195A	China	114/56	Naso-pharyngeal	A allele	Sun et al. (2019)
G765C, rs20417 A3809G, rs689466	Turkey	114/165	Oral squamous cell carcinoma	CC genotype AA genotype	Cacina et al. (2018)
T8473C, rs5275	China	168/195	Naso-pharyngeal	C allele	Fu et al. (2018)
T8473C, rs5275 G765C, rs20417 A3809G, rs689466	Turkey	96/83	Thyroid	CC genotype No correlation No correlation	Uçan et al. (2017)
G765C, rs20417	Portugal	25/60	Colorectal	C allele	Cossiolo et al. (2017)
Rs2745557: A>G	Egypt	112/120	Prostate	G allele	Fawzy et al. (2016)

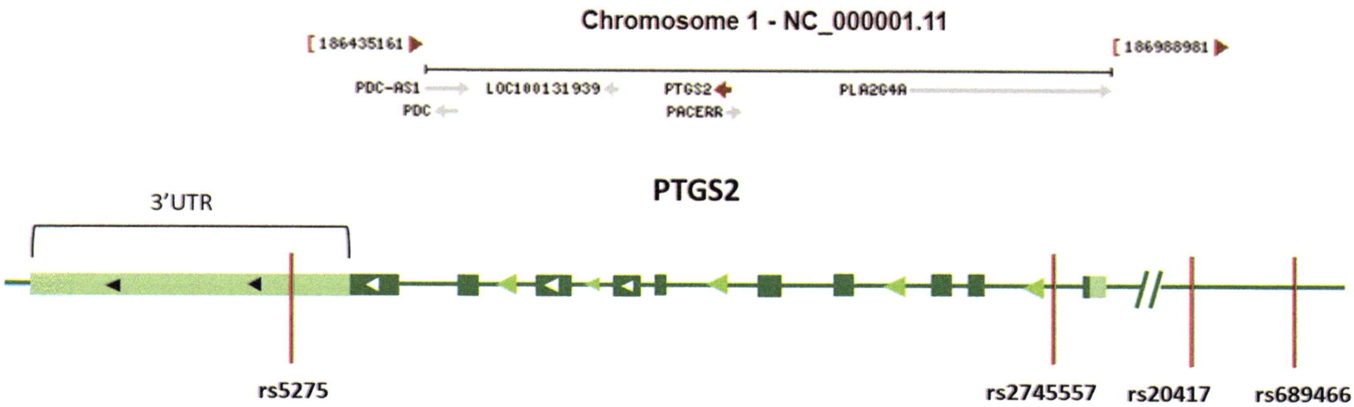

FIGURE 16.5 Schematic representation of the functional and most common *PTGS2* polymorphisms and their position relative to the *PTGS2* gene on chromosome 1 (1q31.1).

studies in several populations and classified by cancer type show an implication of genetic polymorphisms of *PTGS2* with an increased risk of cancer incidence.

In this table, we summarize the results of the most recent studies investigating this association. As it is usually the case in polymorphisms study, results depend on the population. The minor alleles of *PTGS2* gene rs5275 and rs689466 polymorphisms (Figure 16.5) disrupt miR-542-3p function or act on the promoter region, respectively, thus promoting PTGS2 overexpression (Moore et al., 2012; Piranda et al., 2017).

In other polymorphisms where the functional impact of the variations has not yet been fully elucidated, it has also been implied that it might be the change in enzyme activity (Fawzy et al., 2016). Indeed, different variants of the *PTGS2* SNP act by different mechanisms, leading to accumulation of pro-inflammatory *PTGS2* mRNA inducing carcinogenesis (-Vogel et al., 2014). Two recent meta-analyses studying the effect of the polymorphic form of *PTGS2* rs20417 and vulnerability to gastric cancer in different ethnicities demonstrated that this SNP only showed risk in Asians and not in Caucasians (Chen et al., 2019; Luo et al., 2019). On another note, Cox2-1195G>A showed significant association in terms of susceptibility to gastric cancer. This correlation was still significant even after stratification by different ethnicities (Luo et al., 2019).

Furthermore, investigations also imply that while some variants are linked to increased susceptibility to certain cancers, in defined ethnic groups and populations, other *PTGS2* polymorphisms seem to have a protective effect. A meta-analysis found that rs5275 and rs689466 are correlated to reduced malignancy at the level of the lungs (Li J et al., 2018). Another polymorphism, *PTGS2* 8473 T>C, exhibited no effect in terms of all-cancer threat. However, when stratified by type of cancer, this SNP was significantly correlated to the likelihood of malignancies at the esophagus and skin sites (Li Q et al., 2018). Overall, this indicates that genetic heterogeneity may exist in different populations according to *PTGS2* variant and cancer site. Furthermore, investigations also showed that even in the absence of association with the risk of cancer incidence, in certain populations, *PTGS2* polymorphisms might also be associated with a lower chance of cancer survival or increased cancer recurrence risk (Freitas-Alves et al., 2018; Butkiewicz et al., 2015).

16.10 Conclusions

A case-control study (including 3644 participants, Caucasian ethnicity, 17,80 cases and 1,864 controls) looked for an association of the eight genes of interest in this chapter (*ACSL5*, *PPARG*, *LPL*, *LIPC*, *APOE*, *CYP1A2*, and *PTGS2*), related to lipid metabolism, with the risk of colorectal cancer (Crous-Bou et al., 2012). Among the 30 researched SNPs, four *LIPC* variants, *PPARG* Pro12Ala rs1801282, and *PPARG* rs3856806 have indicated significant associations with colorectal cancer. In terms of these results and based on our report here presented, further characterization of these gene variants in different cancer types and populations and their association with different metabolic and inflammatory markers, diet choices, and clinicopathological data are highly encouraged.

It is proposed that an individual's genetic profile might influence his dietary choices, especially in regard to lipid intake. Epidemiological reports highlighted in this chapter and other studies involving variants implicated in the gene-environment or gene-diet relationships show why pharmacogenomics and nutrigenomics are rapidly gaining scientific interest. When we focus on the links between lipid polymorphism, genetic susceptibility, and cancer risk, in vitro, in vivo, and population studies show differences depending on cancer by site and grade. In addition, cancer incidence and outcome diverge based on populations and ethnicities. This chapter goes over how different bioactive lipids and, more importantly, how the variation in genes and enzymes responsible for their metabolism may be attributed to modified cancer risks. Dietary choices and lipids, in particular, are thus mediators of disease risk and cancer incidence.

ABBREVIATIONS

AA: Arachidonic acid
ACC: Acetyl-CoA carboxylase
ACSL: Acyl-CoA synthetase long-chain family
AHR: Aryl hydrocarbon receptor
ANGPTL4: Angiopoietin-like 4
APOE: Apolipoprotein E

ccRCC: Clear cell renal cell carcinoma
CLL: Chronic lymphocytic leukemia
COX: Cyclooxygenase
CYP: Cytochrome P450
DHA: Docosahexaenoic acid
EPA: Eicosapentaenoic acid
FABP: Fatty acid-binding protein
FASN: Fatty acid synthase
GWAS: Genome-wide association studies
HCC: Hepatocellular carcinoma
HUFAs: Highly unsaturated fatty acids
LCFA: Long-chain fatty acids
LIPC: Hepatic lipase
LOX: Lipoxygenase
LPL: Lipoprotein lipase
LRP1: Low-density lipoprotein receptor-related protein 1
PAHs: Polycyclic aromatic hydrocarbons
PGE: Prostaglandin
PCOS: Polycystic ovary syndrome
PPAR: Peroxisome proliferator-activated receptor
PPREs: Peroxisome proliferator response elements
PTGS: Prostaglandin-endoperoxide synthases
PUFAs: Polyunsaturated fatty acids
RXR: Retinoid X receptor
SCD: Stearoyl-CoA desaturase
SNP: Single nucleotide polymorphism
SREBP: Sterol regulatory element-binding protein
T2DM: Type II diabetes mellitus
VLDL: Very low-density lipoprotein

REFERENCES

Adamo KB, Dent R, Langefeld CD, Cox M, Williams K, Carrick KM, et al. Peroxisome proliferator-activated receptor gamma 2 and acyl-CoA synthetase 5 polymorphisms influence diet response. *Obesity (Silver Spring)*. 2007 May; 15(5):1068–75. doi: 10.1038/oby.2007.630.

Al-Eitan LN, Rababa'h DM, Alghamdi MA, Khasawneh RH. Association of CYP gene polymorphisms with breast cancer risk and prognostic factors in the Jordanian population. *BMC Med Genet*. 2019 Sep 2;20(1):148. doi: 10.1186/s12881-019-0884-x.

Al Haj G, Rey F, Giallongo T, Colli M, Marzani B, Giuliani G, et al. New selective PPARγ modulator inhibits triglycerides accumulation during murine adipocytes' and human adipose-derived mesenchymal stem cells differentiation. *Int J Mol Sci*. 2020 Jun 21;21(12):4415. doi: 10.3390/ijms21124415.

Ali SHB, Bangash KS, Rauf A, Younis M, Anwar K, Khurram R, et al. Identification of novel potential genetic predictors of urothelial bladder carcinoma susceptibility in Pakistani population. *Fam Cancer*. 2017 Oct;16(4):577–94. doi: 10.1007/s10689-017-9991-z.

Blanchard PG, Turcotte V, Côté M, Gélinas Y, Nilsson S, Olivecrona G, et al. Peroxisome proliferator-activated receptor γ activation favours selective subcutaneous lipid deposition by coordinately regulating lipoprotein lipase modulators, fatty acid transporters and lipogenic enzymes. *Acta Physiol (Oxf)*. 2016 Jul;217(3):227–39. doi: 10.1111/apha.12665.

Blücher C, Iberl S, Schwagarus N, Müller S, Liebisch G, Höring M, et al. Secreted factors from adipose tissue reprogram tumor lipid metabolism and induce motility by modulating PPARα/ANGPTL4 and FAK. *Mol Cancer Res*. 2020 Dec;18 (12):1849–62. doi: 10.1158/1541-7786.MCR-19-1223.

Bodén S, Myte R, Wennberg M, Harlid S, Johansson I, Shivappa N, et al. The inflammatory potential of diet in determining cancer risk; A prospective investigation of two dietary pattern scores. *PLoS One*. 2019 Apr 12;14(4):e0214551. doi: 10.1371/journal.pone.0214551.

Bowman TA, O'Keeffe KR, D'Aquila T, Yan QW, Griffin JD, Killion EA, et al. Acyl CoA synthetase 5 (ACSL5) ablation in mice increases energy expenditure and insulin sensitivity and delays fat absorption. *Mol Metab*. 2016 Jan 11;5(3): 210–20. doi: 10.1016/j.molmet.2016.01.001.

Bozina T, Simić I, Lovrić J, Pećin I, Jelaković B, Sertić J, et al. Effects of lipoprotein lipase and peroxisome proliferator-activated receptor-gamma gene variants on metabolic syndrome traits. *Coll Antropol*. 2013 Sep;37(3):801–8.

Brignac-Huber L, Reed JR, Backes WL. Organization of NADPH-cytochrome P450 reductase and CYP1A2 in the endoplasmic reticulum--microdomain localization affects monooxygenase function. *Mol Pharmacol*. 2011 Mar;79 (3):549–57. doi: 10.1124/mol.110.068817.

Butkiewicz D, Krześniak M, Drosik A, Giglok M, Gdowicz-Kłosok A, Kosarewicz A, et al. The VEGFR2, COX-2 and MMP-2 polymorphisms are associated with clinical outcome of patients with inoperable non-small cell lung cancer. *Int J Cancer*. 2015 Nov 15;137(10):2332–42. doi: 10.1002/ijc.29605.

Cacina C, Kaşarci G, Bektaş K, Unur M, Cakmakoglu B. The COX2 genetic variants in oral squamous cell carcinoma in Turkish population. *Cell Mol Biol (Noisy-le-grand)*. 2018 Nov 30;64(14):96–100.

Campa D, Hüsing A, Chang-Claude J, Dostal L, Boeing H, Kröger J, et al. Genetic variability of the fatty acid synthase pathway is not associated with prostate cancer risk in the European Prospective Investigation on Cancer (EPIC). *Eur J Cancer*. 2011 Feb;47(3):420–7. doi: 10.1016/j.ejca.2010.09.029.

Chatterjee C, Sparks DL. Hepatic lipase, high density lipoproteins, and hypertriglyceridemia. *Am J Pathol*. 2011 Apr;178 (4):1429–33. doi: 10.1016/j.ajpath.2010.12.050.

Chen S, Chen L, Tan Y, Wang J. Association between rs20417 polymorphism in cyclooxygenase-2 and gastric cancer susceptibility: evidence from15 case-control studies. *Medicine (Baltimore)*. 2019 May;98(18):e15468. doi: 10.1097/MD.0000000000015468.

Chen WC, Wang CY, Hung YH, Weng TY, Yen MC, Lai MD. Systematic analysis of gene expression alterations and clinical outcomes for long-chain acyl-coenzyme a synthetase family in cancer. *PLoS One*. 2016 May 12;11(5):e0155660. doi: 10.1371/journal.pone.0155660.

Christie WW, Harwood JL. Oxidation of polyunsaturated fatty acids to produce lipid mediators. *Essays Biochem*. 2020 Sep 23;64(3):401–21. doi: 10.1042/EBC20190082.

Cibeira GH, Giacomazzi J, Aguiar E, Schneider S, Ettrich B, DE Souza CI, et al. Apolipoprotein E genetic polymorphism, serum lipoprotein levels and breast cancer risk: a case-control study. *Mol Clin Oncol*. 2014 Nov;2(6):1009–15. doi: 10.3892/mco.2014.369.

Cooper DE, Young PA, Klett EL, Coleman RA. Physiological consequences of compartmentalized acyl-CoA metabolism.

J Biol Chem. 2015 Aug 14;290(33):20023–31. doi: 10.1074/jbc.R115.663260.

Cossiolo DC, Costa HCM, Fernandes KBP, Laranjeira LLS, Fernandes MTP, Poli-Frederico RC. Polymorphism of the COX-2 gene and susceptibility to colon and rectal cancer. *Arq Bras Cir Dig*. 2017 Apr–Jun;30(2):114–7. doi: 10.1590/0102-6720201700020008.

Crous-Bou M, Rennert G, Salazar R, Rodriguez-Moranta F, Rennert HS, Lejbkowicz F, et al. Genetic polymorphisms in fatty acid metabolism genes and colorectal cancer. *Mutagenesis*. 2012 Mar;27(2):169–76. doi: 10.1093/mutage/ger066.

De Feo E, Simone B, Persiani R, Cananzi F, Biondi A, Arzani D, et al. A case-control study on the effect of Apolipoprotein E genotypes on gastric cancer risk and progression. *BMC Cancer*. 2012 Oct 25;12:494. doi: 10.1186/1471-2407-12-494.

Dong W, Gong H, Zhang G, Vuletic S, Albers J, Zhang J, et al. Lipoprotein lipase and phospholipid transfer protein overexpression in human glioma cells and their effect on cell growth, apoptosis, and migration. *Acta Biochim Biophys Sin (Shanghai)*. 2017 Jan;49(1):62–73. doi: 10.1093/abbs/gmw117.

Duggan C, Baumgartner RN, Baumgartner KB, Bernstein L, George S, Ballard R, et al. Genetic variation in TNFα, PPARγ, and IRS-1 genes, and their association with breast-cancer survival in the HEAL cohort. *Breast Cancer Res Treat*. 2018 Apr;168(2):567–76. doi: 10.1007/s10549-017-4621-x.

Eggert SL, Huyck KL, Somasundaram P, Kavalla R, Stewart EA, Lu AT, et al. Genome-wide linkage and association analyses implicate FASN in predisposition to Uterine Leiomyomata. *Am J Hum Genet*. 2012 Oct 5;91(4):621–8. doi: 10.1016/j.ajhg.2012.08.009.

El-Sherbeni AA, El-Kadi AO. Characterization of arachidonic acid metabolism by rat cytochrome P450 enzymes: the involvement of CYP1As. *Drug Metab Dispos*. 2014 Sep;42(9):1498–507. doi: 10.1124/dmd.114.057836.

Ezzeddini R, Taghikhani M, Somi MH, Samadi N, Rasaee MJ. Clinical importance of FASN in relation to HIF-1α and SREBP-1c in gastric adenocarcinoma. *Life Sci*. 2019 May 1;224:169–76. doi: 10.1016/j.lfs.2019.03.056.

Fawzy MS, Elfayoumi AR, Mohamed RH, Fatah IRA, Saadawy SF. Cyclooxygenase 2 (rs2745557) polymorphism and the susceptibility to benign prostate hyperplasia and prostate cancer in Egyptians. *Biochem Genet*. 2016 Jun;54(3):326–36. doi: 10.1007/s10528-016-9722-4.

Ferreri C, Sansone A, Ferreri R, Amézaga J, Tueros I. Fatty acids and membrane lipidomics in oncology: a cross-road of nutritional, signaling and metabolic pathways. *Metabolites*. 2020 Aug 25;10(9):345. doi: 10.3390/metabo10090345.

Freitas-Alves DR, Vieira-Monteiro HA, Piranda DN, Sobral-Leite M, da Silva TSL, Bergmann A, et al. *PTGS2* polymorphism rs689466 favors breast cancer recurrence in obese patients. *Endocr Relat Cancer*. 2018 Mar;25(3):351–65. doi: 10.1530/ERC-17-0374.

Fu J, Li Z, Li N. The association between COX-2 gene rs5275 polymorphism and Nasopharyngeal carcinoma risk. *Pathol Res Pract*. 2018 Oct;214(10):1579–82. doi: 10.1016/j.prp.2018.07.028.

Fuentes NR, Kim E, Fan YY, Chapkin RS. Omega-3 fatty acids, membrane remodeling and cancer prevention. *Mol Aspects Med*. 2018 Dec;64:79–91. doi: 10.1016/j.mam.2018.04.001.

Gassler N, Roth W, Funke B, Schneider A, Herzog F, Tischendorf JJ, et al. Regulation of enterocyte apoptosis by acyl-CoA synthetase 5 splicing. *Gastroenterology*. 2007 Aug;133(2):587–98. doi: 10.1053/j.gastro.2007.06.005.

Gong J, Lin Y, Zhang H, Liu C, Cheng Z, Yang X, et al. Reprogramming of lipid metabolism in cancer-associated fibroblasts potentiates migration of colorectal cancer cells. *Cell Death Dis*. 2020 Apr 23;11(4):267. doi: 10.1038/s41419-020-2434-z.

Guengerich FP. Intersection of the roles of cytochrome P450 enzymes with xenobiotic and endogenous substrates: relevance to toxicity and drug interactions. *Chem Res Toxicol*. 2017 Jan 17;30(1):2–12. doi: 10.1021/acs.chemrestox.6b00226.

Guo X, Jiang M, Wan X, Hu C, Gong Y. Identification and biochemical characterization of five long-chain acyl-coenzyme A synthetases from the diatom Phaeodactylum tricornutum. *Plant Physiol Biochem*. 2014 Jan;74:33–41. doi: 10.1016/j.plaphy.2013.10.036.

Huang H, Aladelokun O, Ideta T, Giardina C, Ellis LM, Rosenberg DW. Inhibition of PGE_2/EP4 receptor signaling enhances oxaliplatin efficacy in resistant colon cancer cells through modulation of oxidative stress. *Sci Rep*. 2019 Mar 20;9(1):4954. doi: 10.1038/s41598-019-40848-4.

Huang JY, Zhang WL, Xing YN, Hou WB, Yin SC, Wang ZN, et al. Increased expression of LIPC is associated with aggressive phenotype of Borrmann type 4 gastric cancer. *J Gastrointest Surg*. 2020 Mar 10. doi: 10.1007/s11605-020-04550-5.

Huerta-Yepez S, Tirado-Rodriguez A, Montecillo-Aguado MR, Yang J, Hammock BD, Hankinson O. Aryl Hydrocarbon receptor-dependent inductions of omega-3 and omega-6 polyunsaturated fatty acid metabolism act inversely on tumor progression. *Sci Rep*. 2020 May 12;10(1):7843. doi: 10.1038/s41598-020-64146-6.

Indraccolo S, De Salvo GL, Verza M, Caccese M, Esposito G, Piga I, et al. Phosphorylated acetyl-CoA carboxylase is associated with clinical benefit with regorafenib in relapsed glioblastoma: REGOMA trial biomarker analysis. *Clin Cancer Res*. 2020 Sep 1;26(17):4478–84. doi: 10.1158/1078-0432.CCR-19-4055.

Ishimine H, Zhou R, Sumitomo K, Ito Y, Seki Y, Yoshida Y, et al. Lipase member H frequently overexpressed in human esophageal adenocarcinomas. *Tumour Biol*. 2016 Feb;37(2):2075–81. doi: 10.1007/s13277-015-3985-y.

Ivanova TI, Krikunova LI, Ryabchenko NI, Mkrtchyan LS, Khorokhorina VA, et al. Association of the apolipoprotein E 2 allele with concurrent occurrence of endometrial hyperplasia and endometrial carcinoma. *Oxid Med Cell Longev*. 2015;2015:593658. doi: 10.1155/2015/593658.

Izaola Jáuregui O, López Gómez JJ, Primo Martín D, Torres Torres B, Gómez Hoyos E, Ortolá Buigues A, et al. ACYL-CoA synthetase long-chain 5 polymorphism is associated with weight loss and metabolic changes in response to a partial meal-replacement hypocaloric diet. *Nutr Hosp*. 2020 Aug 27;37(4):757–62.

Jiang H, Dai J, Huang X, Chen Y, Qu P, Li J, et al. Genetic variants in de novo lipogenic pathway genes predict the prognosis of surgically-treated hepatocellular carcinoma. *Sci Rep*. 2015 Mar 26;5:9536. doi: 10.1038/srep09536.

Jiang J, Xie Z, Guo J, Wang Y, Liu C, Zhang S, et al. Association of *PPARG* rs 1801282 C>G polymorphism with risk of colorectal cancer: from a case-control study to a meta-analysis. *Oncotarget*. 2017 Aug 10;8(59):100558–69. doi: 10.18632/oncotarget.20138.

Jin X, Zhang KJ, Guo X, Myers R, Ye Z, Zhang ZP, et al. Fatty acid synthesis pathway genetic variants and clinical outcome of non-small cell lung cancer patients after surgery. *Asian Pac J Cancer Prev*. 2014;15(17):7097–103. doi: 10.7314/apjcp.2014.15.17.7097.

Jorge-Nebert LF, Zhang G, Wilson KM, Jiang Z, Butler R, Gluckman JL, et al. Head-and-neck squamous cell carcinoma risk in smokers: no association detected between phenotype and AHR, CYP1A1, CYP1A2, or CYP1B1 genotype. *Hum Genomics*. 2016 Nov 28;10(1):39. doi: 10.1186/s40246-016-0094-y.

Kaderi MA, Kanduri M, Buhl AM, Sevov M, Cahill N, Gunnarsson R, et al. LPL is the strongest prognostic factor in a comparative analysis of RNA-based markers in early chronic lymphocytic leukemia. *Haematologica*. 2011 Aug;96 (8):1153–60. doi: 10.3324/haematol.2010.039396.

Kang R, Li P, Wang T, Li X, Wei Z, Zhang Z, et al. Apolipoprotein E epsilon 2 allele and low serum cholesterol as risk factors for gastric cancer in a Chinese Han population. *Sci Rep*. 2016 Jan 28;6:19930. doi: 10.1038/srep19930.

Kashiwagi E, Inoue S, Mizushima T, Chen J, Ide H, Kawahara T, et al. Prostaglandin receptors induce urothelial tumourigenesis as well as bladder cancer progression and cisplatin resistance presumably via modulating PTEN expression. *Br J Cancer*. 2018 Jan;118(2):213–23. doi: 10.1038/bjc.2017.393.

Kim HJ, Moon JH, Kim HM, Yun MR, Jeon BH, Lee B, et al. The hypolipidemic effect of cilostazol can be mediated by regulation of hepatic low-density lipoprotein receptor-related protein 1 (LRP1) expression. *Metabolism*. 2014 Jan;63 (1):112–9. doi: 10.1016/j.metabol.2013.09.006.

Kim NH, Seol JE, Kim J, Lee BH, Hwang DY, Jeong J, et al. Red meat intake, CYP2E1 and PPARγ polymorphisms, and colorectal cancer risk. *Eur J Cancer Prev*. 2019 Jul;28 (4):304–10. doi: 10.1097/CEJ.0000000000000462.

Klett EL, Chen S, Yechoor A, Lih FB, Coleman RA. Long-chain acyl-CoA synthetase isoforms differ in preferences for eicosanoid species and long-chain fatty acids. *J Lipid Res*. 2017 May;58(5):884–94. doi: 10.1194/jlr.M072512.

Kobayashi J, Miyashita K, Nakajima K, Mabuchi H. Hepatic lipase: a comprehensive view of its role on plasma lipid and lipoprotein metabolism. *J Atheroscler Thromb*. 2015;22 (10):1001–11. doi: 10.5551/jat.31617.

Koda M, Iwasaki M, Yamano Y, Lu X, Katoh T. Association between NAT2, CYP1A1, and CYP1A2 genotypes, heterocyclic aromatic amines, and prostate cancer risk: a case control study in Japan. *Environ Health Prev Med*. 2017 Oct 24;22(1):72. doi: 10.1186/s12199-017-0681-0.

Kopp TI, Friis S, Christensen J, Tjønneland A, Vogel U. Polymorphisms in genes related to inflammation, NSAID use, and the risk of prostate cancer among Danish men. *Cancer Genet*. 2013 Jul–Aug;206(7–8):266–78. doi: 10.1016/j.cancergen.2013.06.001.

Kuemmerle NB, Rysman E, Lombardo PS, Flanagan AJ, Lipe BC, Wells WA, et al. Lipoprotein lipase links dietary fat to solid tumor cell proliferation. *Mol Cancer Ther*. 2011 Mar;10 (3):427–36. doi: 10.1158/1535-7163.MCT-10-0802.

Kulminski AM, Culminskaya I, Arbeev KG, Ukraintseva SV, Arbeeva L, Yashin AI. Trade-off in the effect of the APOE gene on the ages at onset of cardiocascular disease and cancer across ages, gender, and human generations. *Rejuvenation Res*. 2013 Feb;16(1):28–34. doi: 10.1089/rej.2012.1362.

Kuwata H, Nakatani E, Shimbara-Matsubayashi S, Ishikawa F, Shibanuma M, Sasaki Y, et al. Long-chain acyl-CoA synthetase 4 participates in the formation of highly unsaturated fatty acid-containing phospholipids in murine macrophages. *Biochim Biophys Acta Mol Cell Biol Lipids*. 2019 Nov;1864(11):1606–18. doi: 10.1016/j.bbalip.2019.07.013.

Kwon YJ, Shin S, Chun YJ. Biological roles of cytochrome P450 1A1, 1A2, and 1B1 enzymes. *Arch Pharm Res*. 2021 Jan;44(1):63–83. doi: 10.1007/s12272-021-01306-w.

La Paglia L, Listì A, Caruso S, Amodeo V, Passiglia F, Bazan V, et al. Potential role of ANGPTL4 in the cross talk between metabolism and cancer through PPAR signaling pathway. *PPAR Res*. 2017;2017:8187235. doi: 10.1155/2017/8187235.

Lee M, Sung SH. Platyphylloside isolated from *Betula platyphylla* inhibit adipocyte differentiation and induce lipolysis via regulating adipokines including PPARγ in 3T3-L1 cells. *Pharmacogn Mag*. 2016 Oct–Dec;12(48):276–81. doi: 10.4103/0973-1296.192208.

Lee SJ, Park MK, Shin DS, Chun MH. Variability of the drug response to nonsteroidal anti-inflammatory drugs according to cyclooxygenase-2 genetic polymorphism. *Drug Des Devel Ther*. 2017 Sep 13;11:2727–36. doi: 10.2147/DDDT.S143807.

Li J, Lu X, Zou X, Jiang Y, Yao J, Liu H, et al. COX-2 rs5275 and rs689466 polymorphism and risk of lung cancer: a PRISMA-compliant meta-analysis. *Medicine (Baltimore)*. 2018 Aug;97(35):e11859. doi: 10.1097/MD.0000000000011859.

Li Q, Chen R, Bie L, Zhao D, Huang C, et al. Association of the variants in the PPARG gene and serum lipid levels: a meta-analysis of 74 studies. *J Cell Mol Med*. 2015 Jan;19 (1):198–209. doi: 10.1111/jcmm.12417.

Li Q, Ma C, Zhang Z, Chen S, Zhi W, Zhang L, et al. Association between cyclooxygenase-2 (COX-2) 8473 T>C polymorphism and cancer risk: a meta-analysis and trial sequential analysis. *BMC Cancer*. 2018 Aug 24;18(1):847. doi: 10.1186/s12885-018-4753-3.

Li T, Weng J, Zhang Y, Liang K, Fu G, Li Y, et al. mTOR direct crosstalk with STAT5 promotes de novo lipid synthesis and induces hepatocellular carcinoma. *Cell Death Dis*. 2019 Aug 14;10(8):619. doi: 10.1038/s41419-019-1828-2.

Liang J, Lan J, Li M, Wang F. Associations of leptin receptor and peroxisome proliferator-activated receptor gamma polymorphisms with polycystic ovary syndrome: a meta-analysis. *Ann Nutr Metab*. 2019;75(1):1–8. doi: 10.1159/000500996.

Lin J, Chen Y, Tang WF, Liu C, Zhang S, Guo ZQ, et al. *PPARG* rs3856806 C>T polymorphism increased the risk of colorectal cancer: a case-control study in Eastern Chinese Han population. *Front Oncol*. 2019 Feb 19;9:63. doi: 10.3389/fonc.2019.00063.

Liu W, Burger JA, Xu J, Tang Z, Toruner G, Khanlari M, et al. LPL deletion is associated with poorer response to ibrutinib-based treatments and overall survival in TP53-deleted chronic lymphocytic leukemia. *Ann Hematol*. 2020 Oct;99(10):2343–9. doi: 10.1007/s00277-020-04223-y.

Liu YL, Zhang HM, Pan HM, Bao YH, Xue J, Wang TC, et al. The relationship between apolipoprotein E gene ε2/ε3/ε4 polymorphism and breast cancer risk: a systematic review and meta-analysis. *Onco Targets Ther*. 2016 Mar 7;9:1241–9. doi: 10.2147/OTT.S94228.

Lopes C, Pereira C, Farinha M, Medeiros R, Dinis-Ribeiro M. Genetic variations in Prostaglandin E2 pathway identified as susceptibility biomarkers for gastric cancer in an intermediate risk European country. *Int J Mol Sci*. 2021 Jan 11;22 (2):648. doi: 10.3390/ijms22020648.

López-Alarcón M, Rodríguez-Cruz M, Vital-Reyes VS, Zavala-Ortega MI, Hinojosa-Cruz JC, Canizales-Quinteros S, et al. PPARγ2 Pro12Ala polymorphism is associated with improved lipoprotein lipase functioning in adipose tissue of insulin resistant obese women. *Gene*. 2012 Dec 15;511(2): 404–10. doi: 10.1016/j.gene.2012.09.057.

Lucas D, Goulitquer S, Marienhagen J, Fer M, Dreano Y, Schwaneberg U, et al. Stereoselective epoxidation of the last double bond of polyunsaturated fatty acids by human cytochromes P450. *J Lipid Res*. 2010 May;51(5):1125–33. doi: 10.1194/jlr.M003061.

Luo MX, Long BB, Li F, Zhang C, Pan MT, Huang YQ, et al. Roles of Cyclooxygenase-2 gene -765G > C (rs20417) and -1195G > A (rs689466) polymorphisms in gastric cancer: a systematic review and meta-analysis. *Gene*. 2019 Feb 15;685:125–35. doi: 10.1016/j.gene.2018.10.077.

Lv C, Bai Z, Liu Z, Luo P, Zhang J. Renal cell carcinoma risk is associated with the interactions of APOE, VHL and MTHFR gene polymorphisms. *Int J Clin Exp Pathol*. 2015 May 1;8(5):5781–6.

Ma Y, Zhang X, Alsaidan OA, Yang X, Sulejmani E, Zha J, et al. Long-Chain acyl-CoA synthetase 4-mediated fatty acid metabolism sustains androgen receptor pathway-independent prostate cancer. *Mol Cancer Res*. 2021 Jan;19(1):124–35. doi: 10.1158/1541-7786.MCR-20-0379.

Mansour M, Schwartz D, Judd R, Akingbemi B, Braden T, Morrison E, et al. Thiazolidinediones/PPARγ agonists and fatty acid synthase inhibitors as an experimental combination therapy for prostate cancer. *Int J Oncol*. 2011 Feb;38 (2):537–46. doi: 10.3892/ijo.2010.877.

Matesanz F, Fedetz M, Barrionuevo C, Karaky M, Catalá-Rabasa A, Potenciano V, et al. A splice variant in the ACSL5 gene relates migraine with fatty acid activation in mitochondria. *Eur J Hum Genet*. 2016 Nov;24(11):1572–7. doi: 10.1038/ejhg.2016.54.

Mehus AA, Bergum N, Knutson P, Shrestha S, Zhou XD, Garrett SH, et al. Activation of PPARγ and inhibition of cell proliferation reduces key proteins associated with the basal subtype of bladder cancer in As3+-transformed UROtsa cells. *PLoS One*. 2020 Aug 21;15(8):e0237976. doi: 10.1371/journal. pone.0237976.

Mika A, Pakiet A, Czumaj A, Kaczynski Z, Liakh I, Kobiela J, et al. Decreased triacylglycerol content and elevated contents of cell membrane lipids in colorectal cancer tissue: a lipidomic study. *J Clin Med*. 2020 Apr 12;9(4):1095. doi: 10.3390/jcm9041095.

Moore AE, Young LE, Dixon DA. A common single-nucleotide polymorphism in cyclooxygenase-2 disrupts microRNA-mediated regulation. *Oncogene*. 2012 Mar 22;31(12):1592–8. doi: 10.1038/onc.2011.349.

Morel S, Leahy J, Fournier M, Lamarche B, Garofalo C, Grimard G, et al. Lipid and lipoprotein abnormalities in acute lymphoblastic leukemia survivors. *J Lipid Res*. 2017 May;58 (5):982–93. doi: 10.1194/jlr.M072207.

Moseti D, Regassa A, Kim WK. Molecular regulation of adipogenesis and potential anti-adipogenic bioactive molecules. *Int J Mol Sci*. 2016 Jan 19;17(1):124. doi: 10.3390/ijms17010124.

Ng MCY, Graff M, Lu Y, Justice AE, Mudgal P, Liu CT, Young K, et al. Discovery and fine-mapping of adiposity loci using high density imputation of genome-wide association studies in individuals of African ancestry: African Ancestry Anthropometry Genetics Consortium. *PLoS Genet*. 2017 Apr 21;13(4):e1006719. doi: 10.1371/journal.pgen.1006719.

Nguyen PL, Ma J, Chavarro JE, Freedman ML, Lis R, Fedele G, et al. Fatty acid synthase polymorphisms, tumor expression, body mass index, prostate cancer risk, and survival. *J Clin Oncol*. 2010 Sep 1;28(25):3958–64. doi: 10.1200/JCO.2009.27.0793.

Niu CZ, Zhang FH, Li Y, Liu JJ, Bao CX. The -250G/A and -514C/T polymorphisms in hepatic lipase gene promoter confers an increased risk of hepatocellular carcinoma in a Chinese population. *Ann Hepatol*. 2018 Oct 16;17(6):992–1000. doi: 10.5604/01.3001.0012.7199.

Omiecinski CJ, Vanden Heuvel JP, Perdew GH, Peters JM. Xenobiotic metabolism, disposition, and regulation by receptors: from biochemical phenomenon to predictors of major toxicities. *Toxicol Sci*. 2011 Mar;120 Suppl 1(Suppl 1):S49–75. doi: 10.1093/toxsci/kfq338.

Pang BO, Zhang J, Zhang XI, Yuan J, Shi Y, Qiao L. Inhibition of lipogenesis and induction of apoptosis by valproic acid in prostate cancer cells via the C/EBPα/SREBP-1 pathway. *Acta Biochim Biophys Sin (Shanghai)*. 2021 Jan 20:gmab002. doi: 10.1093/abbs/gmab002.

Pérez-Núñez I, Karaky M, Fedetz M, Barrionuevo C, Izquierdo G, Matesanz F, Alcina A. Splice-site variant in ACSL5: a marker promoting opposing effect on cell viability and protein expression. *Eur J Hum Genet*. 2019 Dec;27 (12):1836–44. doi: 10.1038/s41431-019-0414-5.

Phillips MC. Apolipoprotein E isoforms and lipoprotein metabolism. *IUBMB Life*. 2014 Sep;66(9):616–23. doi: 10.1002/iub.1314.

Piranda DN, Abreu RBV, Freitas-Alves DR, de Carvalho MA, Vianna-Jorge R. Modulation of the prostaglandin-endoperoxide synthase 2 gene expression by variant haplotypes: influence of the 3'-untranslated region. *Braz J Med Biol Res*. 2017 Nov 30;51(2):e6546. doi: 10.1590/1414-431X20176546.

Povel CM, Boer JM, Reiling E, Feskens EJ. Genetic variants and the metabolic syndrome: a systematic review. *Obes Rev*. 2011 Nov;12(11):952–67. doi: 10.1111/j.1467-789X.2011.00907.x.

Qiu H, Wang Y, Kang M, Ding H, Liu C, Tang W, et al. The relationship between IGF2BP2 and *PPARG* polymorphisms and susceptibility to esophageal squamous-cell carcinomas in the Eastern Chinese Han population. *Onco Targets Ther*. 2017a Nov 21;10:5525–32. doi: 10.2147/OTT.S145776.

Qiu H, Zhuang W, Wang X, Huang M, Zhou Z. [Association between genetic polymorphisms and variation of imatinib pharmacokinetics in gastrointestinal stromal tumors]. *Zhonghua Wei Chang Wai Ke Za Zhi*. 2017b Sep 25;20(9):-1031–34. Chinese.

Rajkumar A, Lamothe G, Bolongo P, Harper ME, Adamo K, Doucet É, et al. Acyl-CoA synthetase long-chain 5 genotype is associated with body composition changes in response to lifestyle interventions in postmenopausal women with overweight and obesity: a genetic association study on cohorts Montréal-Ottawa New Emerging Team, and Complications

Associated with Obesity. *BMC Med Genet.* 2016 Aug 11;17(1):56. doi: 10.1186/s12881-016-0320-4.

Rajkumar A, Liaghati A, Chan J, Lamothe G, Dent R, Doucet É, et al. ACSL5 genotype influence on fatty acid metabolism: a cellular, tissue, and whole-body study. *Metabolism.* 2018 Jun;83:271–9. doi: 10.1016/j.metabol.2018.03.019.

Reddy BM, Kommoju UJ, Dasgupta S, Rayabarapu P. Association of type 2 diabetes mellitus genes in polycystic ovary syndrome aetiology among women from southern India. *Indian J Med Res.* 2016 Sep;144(3):400-408. doi: 10.4103/0971-5916.198678.

Rendic S, Guengerich FP. Contributions of human enzymes in carcinogen metabolism. *Chem Res Toxicol.* 2012 Jul 16;25 (7):1316–83. doi: 10.1021/tx300132k. Epub 2012 May 10. PMID: 22531028; PMCID: PMC3398241.

Robbins GT, Nie D. PPAR gamma, bioactive lipids, and cancer progression. Front Biosci (Landmark Ed). 2012 Jan 1;17: 1816–34. doi: 10.2741/4021.

Rombout A, Stamatopoulos B, Lagneaux L, Lust S, Offner F, Naessens E, et al. Lipoprotein lipase SNPs rs13702 and rs301 correlate with clinical outcome in chronic lymphocytic leukemia patients. *PLoS One.* 2015 Mar 26;10(3):e0121526. doi: 10.1371/journal.pone.0121526. Erratum in: PLoS One. 2015;10(6):e0131029.

Ronnekleiv-Kelly SM, Nukaya M, Díaz-Díaz CJ, Megna BW, Carney PR, Geiger PG, et al. Aryl hydrocarbon receptor-dependent apoptotic cell death induced by the flavonoid chrysin in human colorectal cancer cells. *Cancer Lett.* 2016 Jan 1;370 (1):91–9. doi: 10.1016/j.canlet.2015.10.014.

Rossi Sebastiano M, Konstantinidou G. Targeting long chain acyl-CoA synthetases for cancer therapy. *Int J Mol Sci.* 2019 Jul 24;20(15):3624. doi: 10.3390/ijms20153624.

Sabatino L, Ziccardi P, Cerchia C, Muccillo L, Piemontese L, Loiodice F, et al. Chiral phenoxyacetic acid analogues inhibit colon cancer cell proliferation acting as PPARγ partial agonists. *Sci Rep.* 2019 Apr 1;9(1):5434. doi: 10.1038/s41598-019-41765-2.

Sánchez-Martínez R, Cruz-Gil S, García-Álvarez MS, Reglero G, Ramírez de Molina A. Complementary ACSL isoforms contribute to a non-Warburg advantageous energetic status characterizing invasive colon cancer cells. *Sci Rep.* 2017 Sep 11;7(1):11143. doi: 10.1038/s41598-017-11612-3.

Sarhangi N, Sharifi F, Hashemian L, Hassani Doabsari M, Heshmatzad K, Rahbaran M, et al. PPARG (Pro12Ala) genetic variant and risk of T2DM: a systematic review and meta-analysis. *Sci Rep.* 2020 Jul 29;10(1):12764. doi: 10.1038/s41598-020-69363-7.

Sausville LN, Williams SM, Pozzi A. Cytochrome P450 epoxygenases and cancer: a genetic and a molecular perspective. *Pharmacol Ther.* 2019 Apr;196:183–94. doi: 10.1016/j.pharmthera.2018.11.009.

Seo EH, Kim HJ, Kim JH, Lim B, Park JL, Kim SY, et al. ONECUT2 upregulation is associated with CpG hypomethylation at promoter-proximal DNA in gastric cancer and triggers ACSL5. *Int J Cancer.* 2020 Jun 15;146(12):3354–68. doi: 10.1002/ijc.32946.

Shah IA, Mehta P, Lone MM, Rasool MT, Lone GN, Gulzar GM, et al. CYP1A2*1F gene variant, alkaline salt tea intake and risk of esophageal squamous cell carcinoma. *Nutr Cancer.* 2018 Jan;70(1):146–52. doi: 10.1080/01635581.2018.1412482.

Shatwan IM, Winther KH, Ellahi B, Elwood P, Ben-Shlomo Y, Givens I, et al. Association of apolipoprotein E gene polymorphisms with blood lipids and their interaction with dietary factors. *Lipids Health Dis.* 2018 Apr 30;17(1):98. doi: 10.1186/s12944-018-0744-2.

Shimbara-Matsubayashi S, Kuwata H, Tanaka N, Kato M, Hara S. Analysis on the substrate specificity of recombinant human acyl-CoA synthetase ACSL4 variants. *Biol Pharm Bull.* 2019;42(5):850–5. doi: 10.1248/bpb.b19-00085.

Sidaraite A, Vilkeviciute A, Glebauskiene B, Kriauciuniene L, Zaliuniene D, Liutkeviciene R. Association of ApoE haplotype with clinical evidence of pituitary adenoma. *Gene.* 2019 Jul 20;706:154–61. doi: 10.1016/j.gene.2019.04.090.

Simonsson M, Veerla S, Markkula A, Rose C, Ingvar C, Jernström H. CYP1A2--a novel genetic marker for early aromatase inhibitor response in the treatment of breast cancer patients. *BMC Cancer.* 2016 Mar 31;16:256. doi: 10.1186/s12885-016-2284-3.

Song YL, Wang L, Ren JC, Xu ZH. CYP1A2-163C/A (rs762551) polymorphism and bladder cancer risk: a case-control study. *Genet Mol Res.* 2016 Apr 26;15(2). doi: 10.4238/gmr.15026298

Spector AA, Kim HY. Discovery of essential fatty acids. *J Lipid Res.* 2015 Jan;56(1):11–21. doi: 10.1194/jlr.R055095.

Stalin A, Lin D, Josephine Princy J, Feng Y, Xiang H, Ignacimuthu S, et al. Computational analysis of single nucleotide polymorphisms (SNPs) in PPAR gamma associated with obesity, diabetes and cancer. *J Biomol Struct Dyn.* 2020 Oct 22:1–15. doi: 10.1080/07391102.2020.1835724.

Sun N, Wang XQ. Correlation between COX-2 gene polymorphism and susceptibility to nasopharyngeal carcinoma. *Eur Rev Med Pharmacol Sci.* 2019 Jul;23(13):5770–8. doi: 10.26355/eurrev_201907_18315.

Sundberg CD, Hankinson O. A CRISPR/Cas9 whole-genome screen identifies genes required for aryl hydrocarbon receptor-dependent induction of functional CYP1A1. *Toxicol Sci.* 2019 Aug 1;170(2):310–9. doi: 10.1093/toxsci/kfz111.

Tang W, Chen Y, Wang Y, Gu H, Chen S, Kang M. Peroxisome proliferator-activated receptor gamma (PPARG) polymorphisms and breast cancer susceptibility: a meta-analysis. *Int J Clin Exp Med.* 2015 Aug 15;8(8):12226–38.

Tang Y, Zhou J, Hooi SC, Jiang YM, Lu GD. Fatty acid activation in carcinogenesis and cancer development: essential roles of long-chain acyl-CoA synthetases. *Oncol Lett.* 2018 Aug;16(2):1390–6. doi: 10.3892/ol.2018.8843.

Thangavelu M, Godla UR, Paul Solomon FD, Maddaly R. Single-nucleotide polymorphism of INS, INSR, IRS1, IRS2, PPAR-G and CAPN10 genes in the pathogenesis of polycystic ovary syndrome. *J Genet.* 2017 Mar;96(1):87–96. doi: 10.1007/s12041-017-0749-z.

Tian J, Liu G, Zuo C, Liu C, He W, Chen H. Genetic polymorphisms and gastric cancer risk: a comprehensive review synopsis from meta-analysis and genome-wide association studies. *Cancer Biol Med.* 2019 May;16(2):361–89. doi: 10.20892/j.issn.2095-3941.2018.0290.

Tian Y, Wang J, Ye Y, Sun L, Fan Y, Wang L, Li J, Wang Z, Wang K. Apolipoprotein E polymorphism and colorectal neoplasm: results from a meta-analysis. *PLoS One.* 2014 Jul 16;9(7):e102477. doi: 10.1371/journal.pone.0102477.

Trujillo-Paolillo A, Salinas-Souza C, Dias-Oliveira I, Petrilli AS, Toledo SRC. *CYP* Genotypes are associated with toxicity and survival in osteosarcoma patients. *J Adolesc Young Adult Oncol.* 2020 Dec;9(6):621–7. doi: 10.1089/jayao.2019.0180.

Uçan B, Özbek M, Şahin M, Kızılgül M, Çakal E. Cyclooxygenase-2 (COX-2) gene polymorphism in patients with differentiated thyroid carcinomas in the Turkish population. *Turk J Med Sci*. 2017 Dec 19;47(6):1848–53. doi: 10.3906/sag-1708-49.

Uhlen M, Zhang C, Lee S, Sjöstedt E, Fagerberg L, Bidkhori G, et al. A pathology atlas of the human cancer transcriptome. *Science*. 2017 Aug 18;357(6352):eaan2507.

Ventura R, Mordec K, Waszczuk J, Wang Z, Lai J, Fridlib M, et al. Inhibition of de novo palmitate synthesis by fatty acid synthase induces apoptosis in tumor cells by remodeling cell membranes, inhibiting signaling pathways, and reprogramming gene expression. *EBioMedicine*. 2015 Jul 2;2(8):808–24. doi: 10.1016/j.ebiom.2015.06.020.

Verbeek R, Oldoni F, Surendran RP, Zwinderman AH, Khaw KT, Stroes ESG, et al. A 3-SNP gene risk score and a metabolic risk score both predict hypertriglyceridemia and cardiovascular disease risk. *J Clin Lipidol*. 2019 May–Jun;13 (3):492–501. doi: 10.1016/j.jacl.2019.02.005.

Verboom MC, Kloth JSL, Swen JJ, Sleijfer S, Reyners AKL, Steeghs N, et al. Genetic polymorphisms in ABCG2 and CYP1A2 are associated with imatinib dose reduction in patients treated for gastrointestinal stromal tumors. *Pharmacogenomics J*. 2019 Oct;19(5):473–9. doi: 10.1038/s41397-019-0079-z.

Vogel LK, Sæbø M, Høyer H, Kopp TI, Vogel U, Godiksen S, Frenzel FB, Hamfjord J, Bowitz-Lothe IM, Johnson E, Kure EH, Andersen V. Intestinal PTGS2 mRNA levels, PTGS2 gene polymorphisms, and colorectal carcinogenesis. *PLoS One*. 2014 Aug 28;9(8):e105254. doi: 10.1371/journal.pone.0105254.

Volpato M, Ingram N, Perry SL, Spencer J, Race AD, Marshall C, et al. Cyclooxygenase activity mediates colorectal cancer cell resistance to the omega-3 polyunsaturated fatty acid eicosapentaenoic acid. *Cancer Chemother Pharmacol*. 2021 Feb;87(2):173–84. doi: 10.1007/s00280-020-04157-2.

Vukovic V, Ianuale C, Leoncini E, Pastorino R, Gualano MR, Amore R, et al. Lack of association between polymorphisms in the CYP1A2 gene and risk of cancer: evidence from meta-analyses. *BMC Cancer*. 2016 Feb 10;16:83. doi: 10.1186/s12885-016-2096-5.

Wang C, Sun T, Li H, Bai J, Li Y. Lipoprotein lipase Ser447Ter polymorphism associated with the risk of ischemic stroke: a meta-analysis. *Thromb Res*. 2011 Nov;128(5):e107–12. doi:10.1016/j.thromres.2011.07.016.

Wang F, Wang M, Yin H, Long Z, Zhu L, Yu H, et al. Association between plasma prostaglandin E2 level and colorectal cancer. *Eur J Cancer Prev*. 2021 Jan;30(1):59–68. doi: 10.1097/CEJ.0000000000000583.

Wang H, Zhang D, Ling J, Lu W, Zhang S, Zhu Y, et al. Gender specific effect of LIPC C-514T polymorphism on obesity and relationship with plasma lipid levels in Chinese children. *J Cell Mol Med*. 2015b Sep;19(9):2296–306. doi: 10.1111/jcmm.12663.

Wang J, Du S, Wang J, Zhu M, Wen X, Yang W. Association of the lipoprotein lipase gene Ser447Ter polymorphism with hypertension and blood pressure variation: evidence from an updated meta-analysis. *Clin Exp Hypertens*. 2017;39(7):-655–64. doi: 10.1080/10641963.2017.1313848.

Wang Y, Chen Y, Jiang H, Tang W, Kang M, Liu T, et al. Peroxisome proliferator-activated receptor gamma (PPARG) rs1801282 C>G polymorphism is associated with cancer susceptibility in Asians: an updated meta-analysis. *Int J Clin Exp Med*. 2015a Aug 15;8(8):12661–73.

Wang Z, Chen H, Bartz TM, Bielak LF, Chasman DI, Feitosa MF, et al. Gene-lifestyle interactions working group. Role of rare and low-frequency variants in gene-alcohol interactions on plasma lipid levels. *Circ Genom Precis Med*. 2020 Aug;13(-4):e002772. doi: 10.1161/CIRCGEN.119.002772.

Wise LA, Palmer JR, Rosenberg L, Haddad SA, Ruiz-Narváez EA. FASN, dietary fat intake, and risk of uterine leiomyomata in the Black Women's Health Study. *Fertil Steril*. 2016 Oct;106(5):1136–41. doi: 10.1016/j.fertnstert.2016.06.019.

Woo HI, Lee SK, Kim J, Kim SW, Yu J, Bae SY, et al. Variations in plasma concentrations of tamoxifen metabolites and the effects of genetic polymorphisms on tamoxifen metabolism in Korean patients with breast cancer. *Oncotarget*. 2017 Nov 1;8(59):100296–311. doi: 10.18632/oncotarget.22220.

Wu CT, Cheng YH, Chen FN, Chen DR, Wei MF, Chang NW. Combined effects of peroxisome proliferator-activated receptor alpha and apolipoprotein E polymorphisms on risk of breast cancer in a Taiwanese population. *J Investig Med*. 2012 Dec;60(8):1209–13. doi: 10.2310/JIM.0b013e31826be86a.

Xu W, Hu X, Anwaier A, Wang J, Liu W, Tian X, et al. Fatty acid synthase correlates with prognosis-related abdominal adipose distribution and metabolic disorders of clear cell renal cell carcinoma. *Front Mol Biosci*. 2021 Jan 25;7:610229. doi: 10.3389/fmolb.2020.610229.

Yang C, Tian G, Mi J, Wei X, Li X, Li X, et al. Causal relevance of circulating high-density lipoprotein cholesterol with cancer: a Mendelian randomization meta-analysis. *Sci Rep*. 2015 Mar 30;5:9495. doi: 10.1038/srep09495.

Yang Y, Shu X, Shu XO, Bolla MK, Kweon SS, Cai Q, et al. Re-evaluating genetic variants identified in candidate gene studies of breast cancer risk using data from nearly 280,000 women of Asian and European ancestry. *EBioMedicine*. 2019 Oct;48:203–11. doi: 10.1016/j.ebiom.2019.09.006.

Ye G, Gao H, Zhang X, Liu X, Chen J, Liao X, et al. Aryl hydrocarbon receptor mediates benzo[a]pyrene-induced metabolic reprogramming in human lung epithelial BEAS-2B cells. *Sci Total Environ*. 2021 Feb 20;756:144130. doi: 10.1016/j.scitotenv.2020.144130.

Zhang S, Jiang J, Chen Z, Wang Y, Tang W, Chen Y, et al. Relationship of *PPARG*, *PPARGC1A*, and PPARGC1B polymorphisms with susceptibility to hepatocellular carcinoma in an Eastern Chinese Han population. *Onco Targets Ther*. 2018 Aug 8;11:4651–60. doi: 10.2147/OTT.S168274.

Zhao C, Liu G, Shang S, Wei Q, Zhang L, Xia T, Yang X, Sun G, Zhang H. Adaptive evolution of the ACSL gene family in Carnivora. *Genetica*. 2019 Apr;147(2):141–8. doi: 10.1007/s10709-019-00057-3.

Zhao X, Ren Y, Li H, Wu Y. Association of LIPC -250G/A and -514C/T polymorphisms and hypertension: a systematic review and meta-analysis. *Lipids Health Dis*. 2018 Oct 15;17(1):238. doi: 10.1186/s12944-018-0884-4.

Zhou Y, Ingelman-Sundberg M, Lauschke VM. Worldwide distribution of cytochrome P450 alleles: a meta-analysis of population-scale sequencing projects. *Clin Pharmacol Ther*. 2017 Oct;102(4):688–700. doi: 10.1002/cpt.690.

Zhu P, Yu H, Zhou K, Bai Y, Qi R, Zhang S. 3,3'-Diindolylmethane modulates aryl hydrocarbon receptor of esophageal squamous cell carcinoma to reverse epithelial-mesenchymal transition through repressing RhoA/ROCK1-mediated COX2/PGE_2 pathway. *J Exp Clin Cancer Res*. 2020 Jun 16;39(1):113. doi: 10.1186/s13046-020-01618-7.

17

Toll-Like Receptor (TLR) Polymorphisms in Prostate Cancer

Payam Behzadi
Department of Microbiology, Shahr-e-Qods Branch, Islamic Azad University, Tehran, Iran

CONTENTS

ABSTRACT Prostate cancer (PCa) is a huge concern among men worldwide, especially in global industrial regions. In recent years, the incidence of PCa has been growing up in Asian countries, too. Although the PCa is a multifactorial malignancy, the immunogenetic factors are highlighted as considerable items. Due to this fact, I as the author of this chapter will have a deep look at the immunogenetic characteristics of the PCa.

This chapter is prepared on the basis of an effective search in association with academic English language papers published during the last two decades. The search was achieved through the Internet by recruitment of versatile search engines, databases and websites comprising Google Scholar, MEDLINE and Scopus. For papers selection, the quality of the collected publications was determined by iCite bibliometrics (https://icite.od.nih.gov/analysis).

In recent years, genome-wide association investigations have revealed the importance of the occurrence of single nucleotide polymorphisms (SNPs) in different genes. Hence, SNPs may target the genes of immunogenetic glycoproteins of Toll-like receptors (TLRs). This feature may result in a variety of disorders, malignancies and cancers such as PCa. In this regard, it is aimed to investigate the effects of SNPs that appear in TLR genes relating to PCa throughout this chapter.

KEYWORDS: *Prostate Cancer; Risk Factors; Single Nucleotide Polymorphism; Toll-Like Receptors; Signaling Pathways; Immunogenetics; Molecular Biology.*

17.1 Introduction

Prostate cancer (PCa) is known as the first-ranked non-cutaneous malignancy and simultaneously recognized as the second-ranked malignant cancerous tumor after lung cancer and the fifth cancerous death reason in men worldwide. To date, PCa has been known as the third cancerous mortality reason in the USA, and in accordance with the recent reports, the incidence of this malignant tumor has been growing up in Asian countries, too (Castillo et al. 2019; Ma 2019; Merriel et al. 2018). However, the incidence of PCa in Europe and the USA is higher than in Asian countries such as Iran (Basiri et al. 2020). PCa is known as a serious male malignancy, which directly affects the urinary system and sexual activities and results in a poor and unfavorable life quality (Chipman et al. 2014; Ma 2019). In this regard, men in the age range >50 with lower urinary tract symptoms, e.g., retention or urgency or erectile dysfunction or apparent hematuria, nocturia and urinary frequency, should be suspected for PCa (Merriel et al. 2018; NICE 2015).

DOI: 10.1201/9781003246244-17

From the molecular and genomic aspects, the sequencing technologies indicate that PCa is a heterogeneous disease both biologically and clinically. Because of the significant molecular heterogeneity of PCa, the molecular subtypes of PCa may help us to have effective outcomes in prognostic, diagnostic and therapeutic issues (Faisal and Lotan 2020; Vlajnic and Bubendorf 2020).

Indeed, PCa is a multifactorial feature that is strongly associated with the immunogenetic characteristics of individuals. The immune system functions with double-edged sword mechanisms in contrast to cancers such as PCa, as it does by the appearance of single nucleotide polymorphisms (SNPs) in the genes encoding the immune glycoproteins of Toll-like receptors (TLRs). This feature is defined in the field of immunogenetics (Angrini et al. 2020; Behzadi 2021; Dreussi et al. 2018). TLRs are important biomolecules that bridge the innate and adaptive immune systems together. The TLRs are pivotal immune glycoproteins that contribute to the host defense system against causative microbial agents of infectious diseases and cancers (Behzadi and Behzadi 2016; Behzadi 2021; Behzadi et al. 2019, 2021; Gajdács and Behzadi 2020; Hirbod-Mobarakeh et al. 2020). Hence, this chapter focuses on the role of SNPs in *TLR* genes associated with PCa.

17.2 Urogenital System: Prostate Anatomy, Structure and Characteristics

The term "prostate" has its root in the Greek term/word *prohistanai*, meaning "placed in front of" (Prostate 2020; Sharma et al. 2017). The prostate as a compartment of the male urogenital system (UGS) is situated between the urinary bladder and the penis (in pelvic cavity), which is determined as the king accessory and exocrine sex gland that is highly involved in men's reproductive system. In other words, the function of all the male UGS components and their effective interactions may lead to a successful fertility in men. Prostate acts as a double-edged sword: on the one hand, this glandular organ is necessary for men's fertility, and on the other hand, it is responsible for important malignancies such as prostatitis and PCa. As a huge number of proteins are present in seminal plasma, the semen fluid is a great treasure of protein biomarkers to trace different diseases and disorders in men's UGS. The prostatic fluid contains kallikreins (KLKs) (a subfamily of serine proteases with 15 members; the prostate-specific antigen (PSA/KALK-3) is encoded by the *klk-3* gene), Zn^{2+} ions and citrate. KLK-3 (PSA) is known as the pivotal protein biomarker of PCa (Franz et al. 2013; Gilany et al. 2015; Kalinska et al. 2016; Sharma et al. 2017; Tonry et al. 2020; Verze et al. 2016).

The human prostate in adult men involves a spatial space of 4 cm (width) by 3 cm (high) with a weight of ~20 g, capturing a size of walnut which is enveloped by a capsule of collagen, Denonvilliers' fascia and plexuses. This glandular organ develops from urogenital sinus with embryonic origin. The prostate in adult men consists of a stromal tissue enclosing the epithelial bilayer section. However, the epithelial and stromal sections are away from the basement membrane of basal lamina. Simultaneously, the human prostate gland encompasses three glandular zones including central (involving 20%–25% of the prostate tissue), peripheral (comprising up to 70% of the prostate tissue) and transition (containing 5%–10% of the prostate tissue) lobes and an anterior section involving a fibromuscular stromal portion with no glandular tissue. The stromal portion of prostate guarantees suitable microenvironment for epithelial section. It is also composed of collagen- and laminin-enriched extracellular matrix, fibroblasts, immune cells, myofibroblasts, smooth muscle cells armed by androgen receptors (ARs) and nervous and blood vessel plexuses. In other words, the stroma is the cornerstone of prostate health and homeostasis. It is reported that the stroma has a key role in inflammatory responses of prostate. In contrast, the epithelial compartment of prostate produces prostatic fluid, which covers about 20%–30% of the seminal fluid. The pH of the prostatic fluid is estimated in the acidic range of ~6.4 (Franz et al. 2013; Nieto et al. 2014; Saranyutanon et al. 2020; Sharma et al. 2017; Tonry et al. 2020; Verze et al. 2016; Whitaker et al. 2020; Xin 2019).

Anatomically, the central lobe of prostate gland grows in the form of a cone which surrounds the ejaculatory ducts immediately below the urinary bladder. The peripheral lobe, which constitutes the major section of the reproductive gland of prostate, forms the lateral and posterior zones. Chronic prostatitis and carcinoma often appear in the peripheral zone. The transition section covers the anterior minor segment of the prostate. The central and transition sections are the main zones in which the benign prostatic hyperplasia (BPH) origins and develops. Besides, the central section is normally regarded with adenocarcinoma. The prostate fluid is produced by different zones of prostate gland and drains into the ejaculatory ducts and in consequence into the urethra. Previous studies show that peripheral and transition lobes have the highest rate with endodermal origin (Franz et al. 2013; Henry et al. 2018; Nieto et al. 2014; Saranyutanon et al. 2020; Sharma et al. 2017; Verze et al. 2016; Whitaker et al. 2020; Xin 2019).

Moreover, Henry et al. (2018) discovered two new types of epithelial cells including secretoglobin family 1A member 1^+ $(SCGB1A1)^+$ and keratin 13^+ ($KRT13^+$) in urethral section and the related ducts located in the proximal zone of transition lobe in prostate. The $SCGB1A1^+$ cells in prostate have close similarity to respiratory tract epithelial club cells in shape and morphology. The prostate and lung epithelial club cells are involved in immunomodulatory system. Besides, the $KRT13^+$ cells in prostate are similar to the lung hillock basal cells in shape and morphology. Both cells of $SCGB1A1^+$ and $KRT13^+$ express the prostate stem cell antigens (Henry et al. 2018; Montoro et al. 2018).

According to previous studies, the peripheral zone as the major section of prostate gland involves the majority of glandular compartments and is located in the neighborhood of the rectum. This section is encapsulated by a fibrous layer. The reported results show that 68%–70% of the PCa cases are in association with peripheral section of the glandular organ of prostate (Franz et al. 2013; Whitaker et al. 2020).

Indeed, the epithelial bilayer, which consists of a secretory luminal layer and a basal layer, is situated within a fibromuscular matrix. The luminal layer of the epithelium is composed of long and columnar secretory (luminal acinar) cells. The major populated secretory cells are characterized by the expression of heterogeneous antigens comprising ARs, cytokeratins 8 and 18,

NKX3.1 and CD38 and CD57 molecules as cell surface markers. These cells also contribute to the production of secretory proteins such as KLK-2 and glycoprotein of PSA as constitutional and liquefying structure of semen. The glycoprotein of PSA, which is recognized in blood circulation, is known as a PCa biomarker. The rate of circulating PSA in blood is directly associated with the size of prostate. Furthermore, the AR is composed of N-terminal (NTD)-, DNA binding (DNABD)-, joint- and ligand-linking domains. In addition, it is known as a transcription factor situated on X chromosome, which can be activated by the related ligands. The occurrence of dysregulation in AR function is relating to PCa (Adamowicz et al. 2017; Franz et al. 2013; Nieto et al. 2014; Saranyutanon et al. 2020; Sharma et al. 2017; Verze et al. 2016; Whitaker et al. 2020; Xin 2019).

The cytoplasm of epithelial cells is the main storage of Zn^{2+} ions and citrate (an intermediate product from the metabolic pathway of Krebs). The Zn^{2+} ions are the main molecules that contribute to the metabolic pathway of prostate cells. The imbalanced concentration of intracellular Zn^{2+} ions may lead to PCa. The concentration of Zn^{2+} ions within the cells' cytoplasm is regulated by two zinc transporter families of SLC30 and SLC39. Due to this fact, the prostate epithelium has a direct association with citrate and Zn^{2+} ions. Moreover, the epithelial cells of prostate as the only human cells are able to provide energy throughout the glycolysis metabolism pathway in the normal situation. However, this property acts as a double-edged sword, because this energy can also support the proliferation of cancerous cells and tumors in prostate (Franz et al. 2013; Kalinska et al. 2016; Nieto et al. 2014; Verze et al. 2016; Whitaker et al. 2020).

The prostate organ in men bears the maximal content of Zn^{2+} ions in comparison with other soft tissues in human body because of the presence of epithelium section. The epithelial secretory cells possess mitochondria, ribosomes and endoplasmic reticulum in their cytoplasmic areas, and these cells contribute to signaling pathways of endocrine and paracrine systems. The basal layer, which supports the secretory luminal layer, consists of two groups of cells including neuroendocrine cells and the polygonal and cuboidal basal cells. The interspersed and sporadic spread neuroendocrine cells and basal cells have no ARs. According to a previous study, the neuroendocrine cells are divided into two groups of open and closed types. The open neuroendocrine cells join to lumen, while the closed type does not. The neuroendocrine cells are recognized by the marker molecules of bombesin, calcitonin, neuropeptides of chromogranin A and somatostatin, enolase (neuron specific), serotonin, etc. The basal cells bear endoplasmic reticulum, the Golgi apparatus, mitochondria and ribosomes within their cytoplasmic zones. The basal cells phenotypically have a significant heterogeneity and are therefore characterized by the expression of a wide range of heterogeneous cell marker molecules. These markers involve high weight molecules of cytokeratins 5 and 14, the cell surface marker of Trop-2, CD133, CD44 CD49f, EpCAM, c-Kit and the P63 (Adamowicz et al. 2017; Franz et al. 2013; Kalinska et al. 2016; Nieto et al. 2014; Sharma et al. 2017; Verze et al. 2016; Whitaker et al. 2020; Xin 2019).

In addition to the mentioned cell types, the stem cells are important cell populations that are recognized in prostate cell structure. Due to this fact, the stem cells support the development of prostate epithelium. The stem cells are characterized by their self-renewal process throughout symmetrical cell division. This property maintains the stem cells population. On the other hand, the achievement of asymmetrical stem cell division leads to the occurrence of differentiated daughter cells of intermediate or transit-amplifying cells. The stem cells are recognized by their specific marker molecules including $\alpha_2\beta_1$ integrin, ATP-binding cassette subfamily G member 2 (ABCG-2), breast cancer reversed protein-1 (BCRP-1), CD44, CD133, cytokeratins 5/14 and telomerase reverse transcriptase (TERT) (Sharma et al. 2017).

The intermediate cells respond to androgen and have the ability of differentiation from basal into luminal cells. The differentiation process is performed by cytokeratins 5 and 14 downregulation and cytokeratin 18 upregulation. Therefore, the epithelium portion of prostate has a reciprocal reaction to androgen because of the presence of transit-amplifying cells. In the presence of androgen, these cells can be differentiated into basal and/or luminal cell linages. Thus, this property enables intermediate cells to represent the both groups of marker molecules in association with luminal and basal cells (Adamowicz et al. 2017).

17.3 Prostate Cancer

PCa – a known human male malignancy – appeared because of different factors such as inheritance. In this regard, the PCa natural process is significantly long and may take decades. Due to this fact, the tumorigenic process and PCa may often begin restricted and inertly in life commencement with slow rate. However, the cellular origin of PCa is not completely understood (Arora and Barbieri 2018; Gandhi et al. 2018; Packer and Maitland 2016; Saranyutanon et al. 2020; Stoyanova et al. 2013; Wang et al. 2014; Whitaker et al. 2020).

Previous reports indicate different types of stress factors including prostatic inflammation and luminal cell apoptosis (in the absence of E-cadherin) (anoikis) as important factors that stimulate prostatic basal cells to produce luminal cells or basal-to-luminal cells differentiation in mice. Furthermore, the occurrence of acute and chronic inflammations (prostatitis) in human caused by different agents such as pathogenic bacteria and viruses, hormones, traumas (chemical and physical), autoimmune responses and dietary regime may lead to basal-to-luminal differentiation in the aged men, which may lead to tissue damage (Henry et al. 2018; Kwon et al. 2014; Toivanen et al. 2016; Tonry et al. 2020; Verze et al. 2016).

The occurrence of tissue damage in epithelial and stromal cells of prostate results in remodeling of prostate tissue, and in consequence, it may lead to expansion of the prostate. The prostate enlargement is the direct outcome of testosterone reduction and estrogen promotion. The extension of prostate is achieved in parallel with production and secretion of immune molecules such as interleukins (ILs) and growth factors. The enlargement of prostate pressures the urinary system compartments such as urinary bladder and urethra and may lead to clinical symptoms comprising frequent urination, urgent urination and unaccommodating initiation in micturition. These symptoms are common between BPH and early stages of PCa. The enlargement of prostate and the related remodeling of its tissues are the cornerstone of benign or malignant PCa (Ficarra et al. 2014; Gandaglia et al. 2013; Parsons 2010; Tonry et al. 2020; Verze et al. 2016).

In accordance with previous reports, PCa is a multifocal carcinoma with versatile size and cell types in tumors. The recognition of heterogeneous tumors in the same or different foci is known as a typical characteristic of PCa, which facilitates the metastasis feature in this regard (Bolla et al. 2012; Boyd et al. 2012; Rycaj et al. 2016; Tonry et al. 2020).

It seems that the premalignant trauma as the origin of prostate neoplasia may first lead to low-grade prostatic intraepithelial neoplasia (LG-PIN) as an asymptomatic phase of adenocarcinoma and then transforms into high-grade prostatic intraepithelial neoplasia (HG-PIN), which is recognized as symptomatic adenocarcinoma. The HG-PIN can be recognized by the identification of cytologic abnormalities in prostatic glandular zones or ducts. In accordance with previous reports, the HG-PIN bridges the benign epithelial prostate and invasive form of PCa. Indeed, PCa is related to several genomic alterations which occur by the time and get significant molecular and genomic complications which may result in metastasis. The genomic profile of tumors is directly related to their characteristics. Due to this fact, tumors with transition zone origin are larger in volume and produce higher amount of PSA. Sometimes, the appearance of mutations in prostatic tumors leads to appearance of resistance in cells against therapeutics and leads to metastasis. As PCa is recognized as a type of adenocarcinoma, it resembles breast and colon cancers (Al Olama et al. 2014; Arora and Barbieri 2018; Gandhi et al. 2018; Kumar et al. 2016; Packer and Maitland 2016; Rycaj et al. 2016; Saranyutanon et al. 2020; Tonry et al. 2020).

It should be noticed that a significant number of molecular and genomic alterations in the beginning of PCa occur in androgen-dependent genes. In this regard, the AR as a member of the nuclear hormone receptor family is in association with transcription factors. The AR locus is situated on Xq11–12 of men's chromosomal genome. Although the AR is a pivotal protein and main transcriptional regulator in normal prostate gland, the abnormal activity of AR leads to the occurrence of PCa. Due to this fact, the linkage of AR with the related ligands such as testosterone results in the recognition of promoter regions of the target genes by AR. Thus, the AR together with its co-activators turns on the target genes throughout the transcription process. In the beginning of PCa, the androgen-dependent genes as the main targets of AR involve the most molecular and genomic alterations which result in a considerable alteration in signaling pathways (Arora and Barbieri 2018; Compérat et al. 2020; Koryakina et al. 2014; Robinson et al. 2015; Schweizer and Evan 2015).

Based on different characteristics obtained from epigenetic, genomic, metabolomic, proteomic and transcriptomic data, PCa is divided into three molecular subtypes comprising localized PCa, metastatic PCa and castration-resistant PCa (Arora and Barbieri 2018). However, histologically the majority of PCa cases (90%) in humans are associated with acinar adenocarcinoma, while 5%–10% of the PCa cases are classified as non-acinar carcinoma (Humphrey 2012; Xin 2019).

17.4 Classification and Grading of Prostate Cancer

There are two grading systems for PCa involving Gleason and TNM. The Gleason grading system is based on histologic architecture and cell morphology. The scoring range in Gleason system ranges from one to five. The scoring system consists of 1, 2, 3 (3A, 3B, 3C), 4 (4A, 4B) and 5 (5A and 5B). The scoring levels of 1, 2 and 3A involve medium size of the gland. 3B includes small to very small, while 3C relates to medium to large size of the gland. 4A and 4B both depict small, medium and large size levels of the gland. The 5A score involves different levels of gland size, while the 5B score includes only the small gland size. The more aggressive the tumorigenesis, the higher the score gets (Cimadamore et al. 2020; Franz et al. 2013; Tonry et al. 2020).

The tumor, node, metastasis (TNM) staging system is based on determined criteria which can be modified in each edition. Thus, the American Joint Committee on Cancer (AJCC) represents the TNM classification for genitourinary malignancies such as prostate. In accordance with TNM classification for PCa, the T, N and M are categorized into T2 (T2a, T2b, T2c), NX, N0 and N1 and M0, M1 (M1a, M1b, M1c) subgroups, alternatively (Behzadi 2021; Cimadamore et al. 2020; Cornejo et al. 2020).

In accordance with the World Health Organization (WHO) grading system edition 2016, which has similar considerations of WHO/International Society of Urological Pathology (ISUP) 1998 and WHO 2004 classifications, the urinary system tumors with pTa and pT1 grades are recognized as low and high grades (Behzadi 2021; Compérat et al. 2019).

17.5 Epidemiology of Prostate Cancer

In accordance with the latest reports by GLOBOCAN (November 2020) (https://gco.iarc.fr/today/data/factsheets/populations/900-world-fact-sheets.pdf), the incidence of PCa among new cases recognized in 2020 (for both sexes/with age range 0–85+) ranks fourth with the percentage of 7.3% (breast (11.7%), lung (11.4%) and colorectal (10%) cancers incidences rank first to third). Moreover, the incidence of PCa among male populations with regard to new cases in 2020 ranks second with the percentage of 14.1%, while the lung cancer ranks first with the percentage of 14.3% (https://gco.iarc.fr/today/data/factsheets/populations/900-world-fact-sheets.pdf). In accordance with the International Agency for Research on Cancer (IARC) (https://gco.iarc.fr/tomorrow/en/dataviz/isotype?cancers=27&single_unit=50000) reports, the new cases of PCa from 2020 (by 1,410,000 identified individuals) will increase by up to 2,430,000 individuals in 2040. The estimated age-standardized rate (ASR) in 2020 for male in all ages in Europe is highest in Ireland (110.7 per 100,000 individuals), followed by Estonia (102.1 per 100,000 individuals) and Sweden (100.4 per 100,000 individuals). The estimated age-standardized rates (ASRs) in 2020 for male in all ages in other countries are as follows: the USA (72.0 per 100,000 individuals), Canada (80.4 per 100,000 individuals), Brazil (78.0

per 100,000 individuals), Colombia (49.8 per 100,000 individuals), Chile (56.7 per 100,000 individuals), Zimbabwe (70.6 per 100,000 individuals), Zambia (69.2 per 100,000 individuals), South Africa (68.3 per 100,000 individuals), Egypt (13.9 per 100,000 individuals), Iran (21.2 per 100,000 individuals), Iraq (12.9 per 100,000 individuals), Kuwait (19.6 per 100,000 individuals), Bahrain (13.3 per 100,000 individuals), Qatar (21.1 per 100,000 individuals), United Arab Emirates (UAE) (13.4 per 100,000 individuals), Oman (13.8 per 100,000 individuals), Kingdom of Saudi Arabia (KSA) (7.0 per 100,000 individuals), Russian Federation (43.7 per 100,000 individuals), India (5.5 per 100,000 individuals), China (10.2 per 100,000 individuals), Japan (51.8 per 100,000 individuals), Republic of Korea (South Korea) (27.3 per 100,000 individuals), Indonesia (11.6 per 100,000 individuals), Australia (72.5 per 100,000 individuals) and New Zealand (92.9 per 100,000 individuals). The first three countries with the highest estimated age-standardized rate (ASR) in 2020 for male in all ages worldwide are Guadeloupe (France) (183.6 per 100,000 individuals), Martinique (France) (168.2 per 100,000 individuals) and Barbados (110.3 per 100,000 individuals). In contrast, the first three countries with the lowest estimated age-standardized rate (ASR) in 2020 for male in all ages worldwide are Bhutan (0.90 per 100,000 individuals), Yemen (2.8 per 100,000 individuals) and Democratic Republic of Korea (North Korea) (3.6 per 100,000 individuals) (Cancer Today 2020). The estimated incidence and mortality rates of PCa between the years of 2020 and 2040 in different continents and Asian region countries are shown in Tables 17.1 and 17.2 (https://gco.iarc.fr/tomorrow).

17.6 Prostate Cancer Risk Factors

Although there are numerous risk factors that predispose individuals to PCa, the main parameters include age, race and ethnicity, inborn genetics, inheritance and family history, lifestyle, diet and gut microbiota (Barsouk et al. 2020; Bratt et al. 2016; Jian et al. 2018; Kucera et al. 2020; Peisch et al. 2017; Rawla 2019; Rebbeck 2017; Whitaker et al. 2020).

TABLE 17.1

The Estimated Incidence and Mortality Rates of New Cases of PCa in the Years 2020 and 2040 in Different Continents in Both Sexes of Male and Female, with the Age Range of 65–85+

	Incidence		Mortality		
Continent	**2020**	**2040**	**2020**	**2040**	**Reference**
Africa	62,443	133,761 **(+114.2%)**[a]	37,153	82,108 **(+121.0%)**[b]	https://gco.iarc.fr/tomorrow
Asia	292,235	541,526 **(+85.3%)**[c]	105,394	237,777 **(+125.6%)**[a]	
Europe	337,539	470,664 (+39.4%)	98,993	158,377 (+60.0%)	
Latin America and the Caribbean	145,160	289,722 **(+99.6%)**[b]	51,530	111,406 **(+116.2%)**[c]	
North America	139,748	197,030 (+41%)	33,592	64,882 (+93.1%)	
Oceania	14,976	23,383 (+56.1%)	4,409	8,827 (+100.2%)	

[a] The first-ranked percentage.
[b] The second-ranked percentage.
[c] The third-ranked percentage.

TABLE 17.2

The Estimated Incidence and Mortality Rates of New Cases of PCa in the Years 2020 and 2040 in Different Asian Region Countries in Both Sexes of Male and Female, with the Age Range of 65–85+

	Incidence		Mortality		
Asian Region Countries	**2020**	**2040**	**2020**	**2040**	**Reference**
Central Asia Countries[d]	1,559	3,646 **(+133.9%)**[c]	838	1,970 **(+135.1%)**[c]	https://gco.iarc.fr/tomorrow
East Asia Countries[e]	199,915	337,131 (+68.6%)	62,966	140,245 (+122.7%)	
South Asia Countries[f]	32,258	64,471 (+99.9%)	16,250	33,032 (+103.3%)	
Southeast Asia Countries[g]	29,614	69,920 **(+136.1%)**[b]	13,465	34,554 **(+156.6%)**[a]	
The Persian Gulf Region Countries[h]	8,851	24,191 **(+173.3%)**[a]	4,474	11,104 **(+148.2%)**[b]	

[a] The first-ranked percentage.
[b] The second-ranked percentage.
[c] The third-ranked percentage.
[d] Kazakhstan, Kyrgyzstan, Tajikistan, Turkmenistan and Uzbekistan (Wikipedia.org).
[e] China (+Taiwan + Hong Kong), Japan, Korea (North), Korea (South), Mongolia and Macau (Wikipedia.org). Patients belonging to Hong Kong, Macau and Taiwan are not calculated.
[f] Afghanistan, Bangladesh, Bhutan, India, the Maldives, Nepal, Pakistan, Sri Lanka (Wikipedia.org).
[g] Brunei, Cambodia, Indonesia, Laos, Malaysia, Myanmar, the Philippines, Singapore, Thailand, Timor (East) and Vietnam (Wikipedia.org).
[h] Bahrain, Iran, Iraq, Kingdom of Saudi Arabia (KSA), Kuwait, Oman, Qatar and United Arab Emirates (UAE) (Wikipedia.org).

- **Age:** Age is one of the most important predisposing factors with regard to PCa (Tables 17.1 and 17.2). The age range ≥40 is determined as an important risk factor for PCa in both black and white populations with family history regarding PCa (Barsouk et al. 2020; Culp et al. 2020; Rawla 2019). The age factor is known as a non-modifiable predisposing factor in association with PCa. However, choosing a suitable lifestyle may decrease the acceleration of aging (Kucera et al. 2020).
- **Race and Ethnicity:** The race and ethnicity as another non-modifiable predisposing factor has a significant role in PCa occurrence (Kucera et al. 2020). Among Asian countries, the Southeast Asia countries rank first, followed by the Persian Gulf Region countries and the Central Asia countries ranking second and third, respectively (Table 17.2).

  The African ethnicity is considered as racial predisposition for PCa, which can be related to the genetic characteristics and biology of tumorigenesis (Culp et al. 2020). However, as Table 17.1 shows, the Asian racial factor is detected as the most important predisposing racial factor for PCa.
- **Inborn Genetics, Inheritance and Family History:** The inborn genetics, inheritance and family history are regarded as non-modifiable predisposing factors (Kucera et al. 2020). The inborn genetics and inheritance involve ~5% of the PCa risk, while 20% of patients with PCa have a positive family history (Barsouk et al. 2020; Rawla 2019). A family history with positive PCa for father and brother increases the possibility of PCa occurrence by 2.3 times (Kucera et al. 2020).
- **Lifestyle:** Body mass index (BMI; kg/m^2), Human Development Index (HDI), physical activity, smoking, sleep disorders, occupation and sexual activity are the most relevant cases to lifestyle that are recognized as modifiable predisposing factors. High BMI and obesity have a direct relation with PCa mortality. An individual with a BMI ≥30 kg/m^2 is predisposed two times more than a man with normal BMI for PCa mortality. Indeed, the obesity is involved in the malfunction of insulin hormone and cytokine signaling (inflammatory immune system) pathways and the triggering of the DNA-damaging oxidative stress (Barsouk et al. 2020; Kucera et al. 2020; Peisch et al. 2017; Rawla 2019; Sadeghi-Gandomani et al. 2017).
- Sexual activities with the occurrence risk of sexually transmitted diseases (STDs) are associated with PCa incidence and mortality. The human papilloma virus (HPV) and herpes simplex virus (HSV) are important and common viral agents of STD that may lead to PCa (Behzadi et al. 2015; Kucera et al. 2020; Rawla 2019; Sadeghi-Gandomani et al. 2017). The bacterial prostatitis and prostatic atrophy caused by bacterial pathogens such as *Neisseria gonorrhoeae* lead to PCa (Kucera et al. 2020; Rawla 2019; Sadeghi-Gandomani et al. 2017). Trichomoniasis as a protozoan infection is an important risk factor for aggressive PCa. Patients with PCa and trichomoniasis are threefold more predisposed to die. Indeed, the presence of *Trichomonas vaginalis* may trigger the prostatic epithelial cells to activate the inflammatory signaling pathway (Han et al. 2019; Kucera et al. 2020; Sadeghi-Gandomani et al. 2017).
- **Diet:** Dietary compartments include proteins, plants, diary and vitamins. The egg (choline intake) can be regarded as a predisposing factor of PCa, while fish consumption (because of the presence of omega-3) can reduce the risk of PCa. The skinless poultry consumption is not recognized as a predisposing factor of PCa. However, the grilled, pan-fried, oven-broiled poultry contain a high level of heterocyclic amines that are recognized as a PCa risk factor. The processed red meat comprises hot dogs, sausage, salami, etc. In accordance with the recent classification represented by WHO, the processed meat are recognized as "carcinogenic to human" (Group 1), and due to this fact, the red meat are classified as "probably carcinogenic to human" (Group 2A) (Kucera et al. 2020; Peisch et al. 2017; Rawla 2019; Richman et al. 2011, 2012).
- Plants dietary including tomato and lycopene, green tea, pomegranate and garlic consumption may reduce the PCa risk. Besides, the cruciferous vegetables consumption may lead to the detoxification of carcinogenic substances and reduce the risk of PCa. The consumption of coffee as a strong antioxidant is recommended to reduce the risk of PCa (Ferrís-Tortajada et al. 2012; Kucera et al. 2020; Livingstone et al. 2019; Peisch et al. 2017; Rawla 2019; Sadeghi-Gandomani et al. 2017).

  The A, C, D, E and K vitamins also have reductive effects on PCa with different types of mechanisms (Kucera et al. 2020).
- **Gut Microbiota:** The type and composition of gut microbiota have a direct and influent effect on tumorigenic processes. The presence of different microbial populations can reduce or increase the risk of PCa. However, more investigations with larger number of cases are needed. Furthermore, the gut microbiota has a direct association with the host immune system (Katongole et al. 2020; Kucera et al. 2020; Porter et al. 2018).

17.7 Single Nucleotide Polymorphisms and Cancers

The single nucleotide polymorphism (SNP) is recognized as one of the most significant genomic alterations in human. The SNPs may occur in different gene sets and clusters such as those that are associated with immune system. It is attested that the presence of SNPs in immune system-related genes predisposes individuals to cancer. Besides, the SNPs are known as worthy biomarkers for the detection and identification cancers (Deng et al. 2017; Ding et al. 2017; Priyadarshini et al. 2013; Sameer and Nissar 2021a; Shui et al. 2012).

Previous studies have revealed that the SNPs can occur in different parts of genes. The SNPs involve exonal and intronal regions of the genes as well as promoters and untranslated regions (UTRs) of 5′ and 3′. The presence of SNPs in each section of the genes predisposes individuals to different types of cancers, e.g., PCa (Akhter et al. 2012; Behzadi et al. 2014; Deng et al. 2017; Nissar et al. 2017; Shao et al. 2014). A promoter is the main gate for gene transcription and expression. Due to this fact, the cis-acting elements and trans-acting factors are the keys of the gate of promoters. With the occurrence of SNPs in promoters and the associated structures, the cis-acting elements, which act as specific protein-binding sites, get affected. Hence, the active proteins that contribute to the initiation and modulation of the gene transcription get alterations in their functions. Following this, the transcription process, mRNA molecules characteristics and translation process get alterations. The proteins expressed by the genes containing SNPs predispose humans to different disorders and diseases such as cancer (Behzadi et al. 2014; Deng et al. 2017; Hernandez-Garcia and Finer 2014). The TATA box, which is recognized as a core promoter, can be a target sequence for SNPs. The reported results show that the occurrence of SNPs in the TATA box as the core promoter region hinders the promoter functions and in consequence reduces the transcription process (Deng et al. 2017; Hernandez-Garcia and Finer 2014). The exonal SNPs are categorized into synonymous coding SNPs (scSNPs) and non-synonymous coding SNPs (nscSNPs). Indeed, the scSNPs are usually involved in the alteration of proteins expressions, their spatial structures and conformations (e.g., changing the α-helix and β-sheet configurations in secondary structures) and activities, indirectly. The scSNPs affect and alter the expression of adjacent genes of the target gene. This feature alters the expression of the target gene and in consequence changes its function and activity, too. The reports reveal that the scSNPs alter the splicing of mRNAs, the stability of the mRNAs and their structures. Besides, the scSNPs are able to alter the protein folding. In contrast, the nscSNPs vary the sequence of expressed protein amino acids. These sequential alterations caused by nscSNPs may lead to an increase or reduction in the hydrogen linkages number and phosphorylation feature in the expressed proteins. These features result in effective changes in secondary structure of the encoded proteins and in consequence vary the proteins activities and functions. Therefore, the nscSNPs are able to alter the proteins stability, to change the related modifications in post-translation process and to vary the proteins interactions. The final outcome of the occurrence of nscSNPs within exons is changes in signaling pathways and the proteins which act as tumor suppressors (Deng et al. 2017; Tennessen et al. 2012; Yates and Sternberg 2013).

Introns as important genetic elements that contribute to the upregulation of gene expression involve the cis-acting elements including enhancers, insulators, silencers and transcription factors. Therefore, the presence of SNPs in introns predispose individuals to genetic malignancies comprising cancers and tumorigenesis. The occurrence of SNPs in introns affects the protein production and gene expression through mRNA splicing and long noncoding RNAs, respectively (lncRNAs) (Behzadi et al. 2014; Deng et al. 2017; Zhang et al. 2014a).

The appearance of SNPs in the UTRs of 5′ and 3′ of mRNA molecules leads to changes and variations in the related sequences pertaining to these critical regions as translation process controller; hence, the SNPs have a direct effect on translation efficacy. The 5′-UTR is involved in transcription, transmission, processing, stability and translation processes of mRNA molecules. In other words, the 5′-UTR is the cornerstone of translation initiation. In addition to the 5′-UTR, the translation initiation site, the start codons of AUG (upstream), the open reading frame (ORF) (upstream), the ribosomal linking sites and the related spatial configuration of the mentioned structures determine the process of translation. Therefore, the presence of SNPs in the 5′-UTR affects all the related structures involved in translation, and due to this fact, it may lead to a variety of cancers in humans (Figure 17.1). On the other hand,

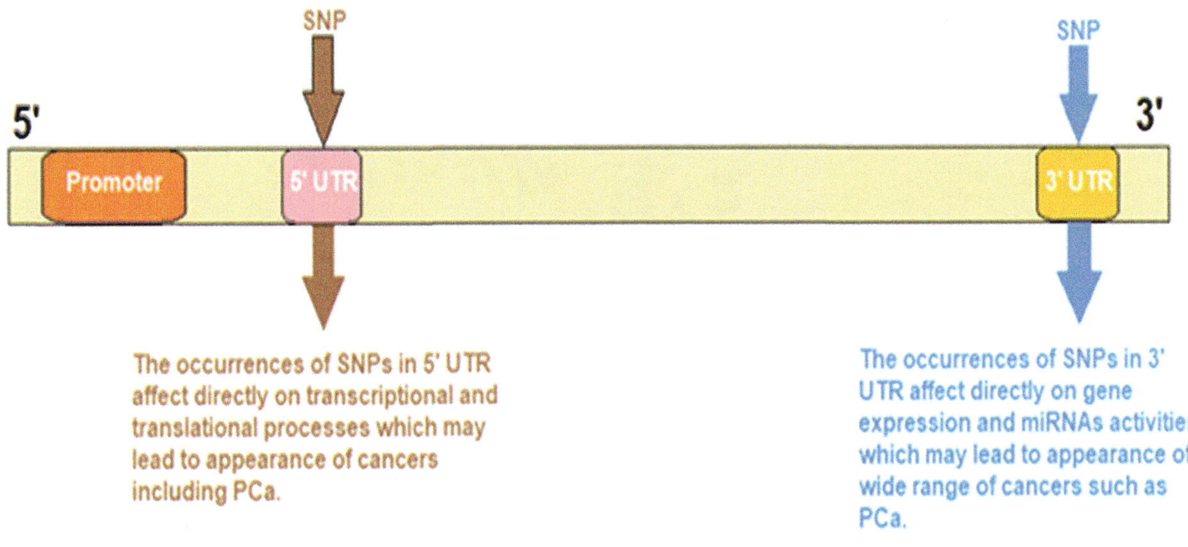

FIGURE 17.1 The concept of SNPs affecting the 5′- and 3′-UTRs.

the 3′-UTR governs the stability and degradation of mRNA molecules by modulating the polyadenylation of the mRNAs' tails and it mediates the end of translation process. Thus, the 3′-UTR is directly involved in gene expression. Furthermore, the 3′-UTR possesses microRNAs (miRNAs)-binding sites. The miRNA molecules are natural inhibitors of translation process and effective destabilizers of the related target molecules of mRNAs. The miRNA molecules activities can be achieved through their linkages to the 3′-UTR binding sites located on the target mRNA molecules. Hence, the occurrence of SNPs in the 3′-UTR may lead to appearance of a wide range of cancers (Figure 17.1). Simultaneously, the SNPs can occur in miRNA molecules, which results in the changes in the associated binding sites, prevention of miRNA processing, and inhibition of the molecule's targeting. In consequence of this feature, the synthesis of proteins and expression of the genes get inactivated (Behzadi et al. 2014; Deng et al. 2017; Fan et al. 2013; Ruiz-Ballesteros et al. 2020; Shui et al. 2012).

17.8 Introduction to Toll-Like Receptors, Structures, Locations and Functions in Normal Health

TLR1

The TLR1 molecules in human are encoded by the gene cluster of *TLR6-TLR1-TLR10* on the genomic chromosome of 4p14 within a region composed of 54,000 bases. *TLR1* gene transcript with 8,518 bases involves four exons (Behzadi and Behzadi 2016; Behzadi et al. 2021; Dannemann et al. 2016; Sameer and Nissar 2021b; Sun et al. 2005). The TLR1 molecules as ubiquitous glycoproteins are expressed by both innate and adaptive (B cells) immune cells and tumor cells of myeloma, too. The TLR1 is activated through dimerization with TLR2 to create TLR1/TLR2 heterodimer in the presence of lipoprotein ligands. Both the molecules of TLR1 and TLR2 are known as cell surface molecules which encompass horseshoe-like extracellular domain. Because of their molecular structure, the TLR1/TLR2 heterodimers appear as the English letter of m (Figure 17.2). The TLR1/TLR2 homodimers have anti-tumoral effects on colon cancer (through the increase of apoptotic effect on tumor cells and reduction in proliferation of tumor cells), hepatocellular carcinoma and lung cancer (by the increase in cytotoxic lymphocytes and reduction in immunosuppressive immune cells such as T_{regs}) (Behzadi and Behzadi 2016; Behzadi 2021; Behzadi et al. 2019, 2021; Botos et al. 2011; Dajon et al. 2017; de Oliviera Nascimento et al. 2012; Gomaz et al. 2012; Ohadian Moghadam and Nowroozi 2019; Qiu et al. 2013; Sameer and Nissar 2021; Su et al. 2019).

TLR2

The cell surface of TLR2 is an interesting molecule with a variety of co-receptors including TLR1 (Figure 17.2), TLR2, TLR6 and TLR10, which interact with a wide range of molecules including bacterial, fungal, protozoan and viral ligands. The three-exonal gene of *TLR2* is located on chromosomal genome (4q32) (Behzadi and Behzadi 2016; Behzadi 2021; Behzadi et al. 2019, 2021; Botos et al. 2011; de Oliviera Nascimento et al. 2012; Gomaz et al. 2012; Ohadian Moghadam and Nowroozi 2019; Qiu et al. 2013; Su et al. 2019; Wang et al. 2020).

The TLR2 biomolecules are expressed by both the innate and adaptive immune cells comprising. Besides, TLR2 glycoproteins are expressed in a wide range of tumor cells in different types of cancers including oral squamous cell carcinomas, myelogenous leukemia, colorectal cancer, brain cancer, pancreatic cancer, laryngeal carcinoma, gastrointestinal carcinoma, breast cancer and hepatocellular carcinoma. The stimulation and/or expression of TLR2 have pro-tumoral effects on breast cancer, head-and-neck cancers, hepatocellular carcinoma, intestinal cancer and lymphoma. Moreover, the overexpression of *TLR2* gene is associated with the development of some

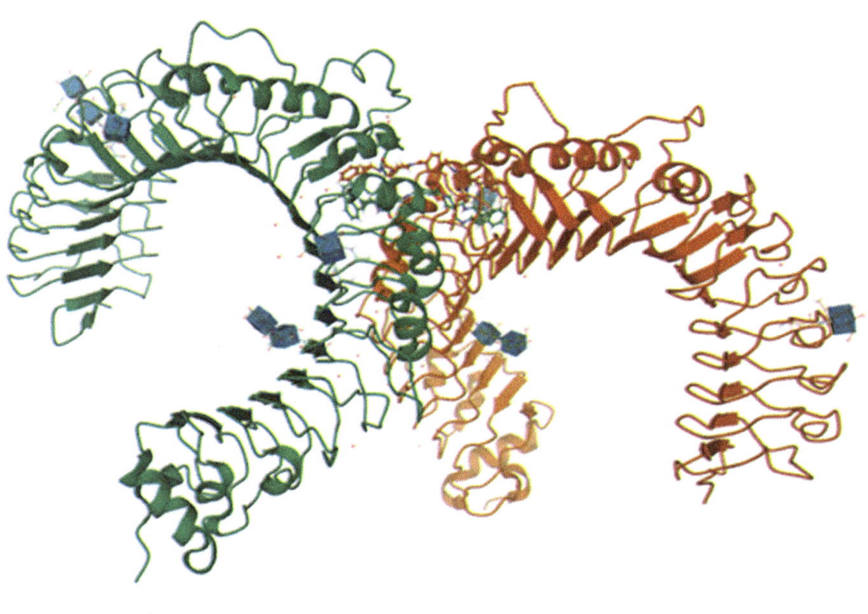

FIGURE 17.2 The m-letter-like structure of TLR1/TLR2 heterodimer (6NIG PBD file (Su et al. 2019)).

cancers, e.g., colorectal cancer, gastric cancer, ovarian cancer, melanoma, pancreatic cancer, lung cancer, oral squamous cell carcinomas, hepatocellular carcinoma, adrenocortical carcinoma and lung cancer (Ayala-Cuellar et al. 2019; Behzadi 2021; Behzadi et al. 2021; Dajon et al. 2017; Mokhtari et al. 2020; Ohadian Moghadam and Nowroozi 2019; Saghazadeh and Rezaei 2020).

TLR3

The endosomal receptors of TLR3 are functional molecules in the form of homodimers. The extracellular domain of TLR3 has a horseshoe-like structure, which in TLR3 homodimer form has the structure of the letter m (Figure 17.3). The *TLR3* gene is composed of five exons and located on 4q35 genomic chromosome. In contrast to other TLRs that use MyD88-dependent signaling pathway (with the exception of endosomal TLR4), TLR3 applies the TRIF-dependent signaling pathway (Figures 17.4 and 17.5). The well-known ligand of TLR3/TLR3 homodimer is the viral double-stranded RNAs (dsRNAs). mRNAs and tRNAs are other target ligands for TLR3/TLR3 homodimers (Behzadi and Behzadi 2016; Behzadi 2021; Behzadi et al. 2019, 2021; Botos et al. 2011; Gomaz et al. 2012; Ohadian Moghadam and Nowroozi 2019; Sameer and Nissar 2021; Zhao et al. 2014).

The TLR3 molecules are expressed by innate and adaptive immune cells. Moreover, these molecules are expressed in tumor cells regarding a wide range of cancers such as prostate cancer, cervical carcinoma, breast cancer, ovarian carcinoma, laryngeal and pharyngeal carcinoma, myelogenous leukemia, colorectal carcinoma, gastric cancer, hepatocellular carcinoma, melanoma, lung cancer, esophageal squamous cell

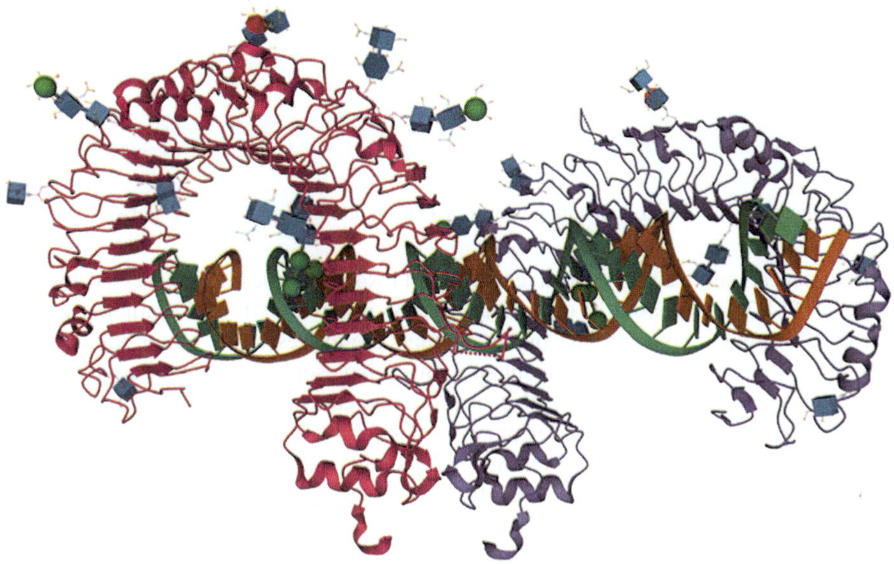

FIGURE 17.3 The mouse m-letter-like structure of TLR3/TLR3 homodimer with a ligand of dsRNA (3CIY PBD file (Liu et al. 2008)).

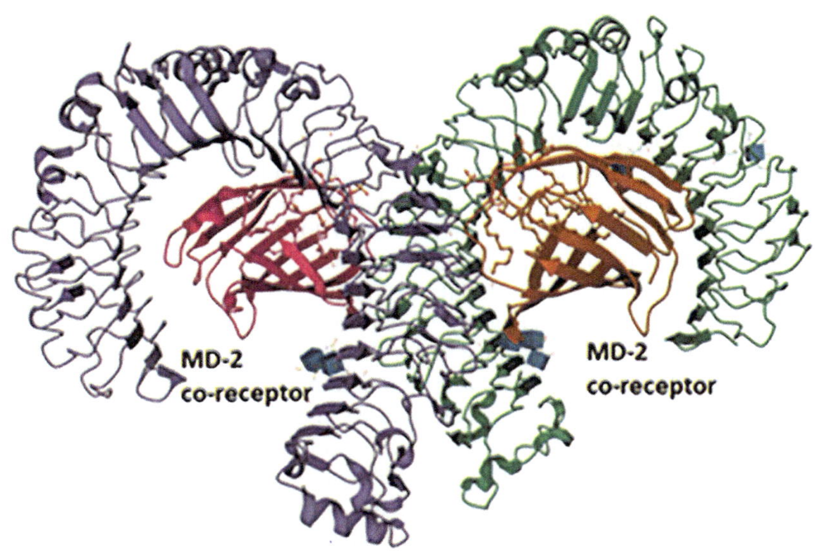

FIGURE 17.4 The m-letter-like structure of TLR4/TLR4 homodimer together with MD-2 co-receptor and LPS ligand (3VQ2 PBD file (Ohto et al. 2012)).

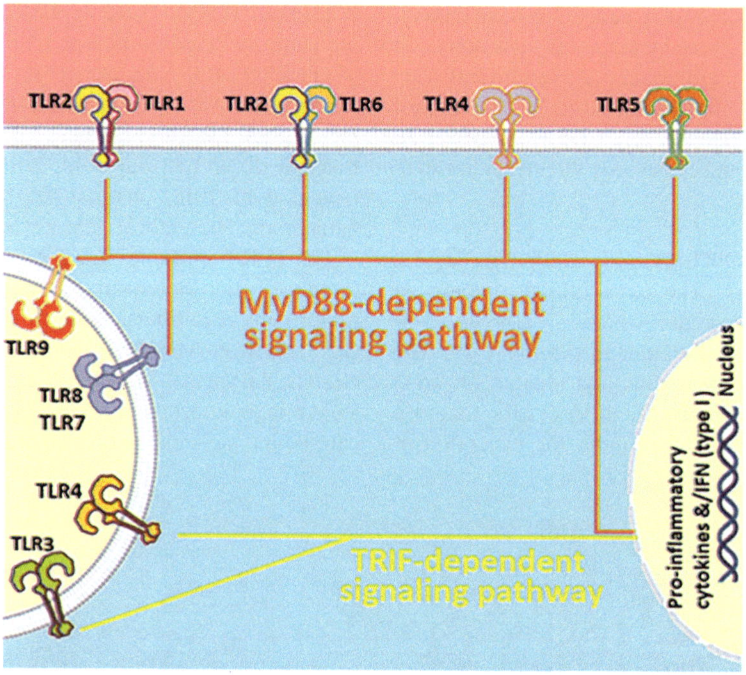

FIGURE 17.5 A simple illustration regarding the TLR signaling pathways. (This figure is prepared by the software tool service by Biorender.com.)

carcinoma, head-and-neck cancer and neuroblastoma cells. Moreover, the expression of TLR3 is increased in colon and ovarian cancers. According to previous reports, the overexpression of TLR3 is in association with the development of breast cancer, melanoma, head-and-neck squamous cell carcinoma, lung cancer, oral squamous cell carcinomas, urinary bladder cancer, prostate cancer, hepatocellular carcinoma and esophageal cancer (Ayala-Cuellar et al. 2019; Behzadi and Behzadi 2016; Behzadi 2021; Behzadi et al. 2021; Dajon et al. 2017; González-Reyes et al. 2011; Mokhtari et al. 2020; Ohadian Moghadam and Nowroozi 2019; Saghazadeh and Rezaei 2020; Zhao et al. 2014).

TLR4

The TLR4 molecule as a bi-zonal immunoglycoprotein is identified on both places of cell surfaces and endosomes. Hence, the TLR4 is the only TLR biomolecule which recruits both the MyD88- and TRIF-dependent pathways to express pro-inflammatory cytokines and type I IFN molecules through the ligand-ectodomain linkage. Moreover, because of the specific characteristics of TLR4, it is the only TLR biomolecule that employs all the types of adaptors along with MyD88- and TRIF-dependent signaling pathways (Figure 17.5). LPS is the well-known ligand of TLR4. However, a variety of bacterial, fungal, viral and protozoan ligands are known as TLR4 ligands. The gene of the *TLR4* is composed of four exons and situated on 9q32–33 of genomic chromosome. The functional TLR4 acts in the form of TLR4/TLR4 homodimers and should simultaneously be incorporated with CD14 and MD-2 glycoproteins. The TLR4/TLR4 homodimer creates an m-letter-like structure (Figure 17.4). The TLR4 is able to heterodimerize with TLR6 to act as TLR4/TLR6 heterodimer (Behzadi and Behzadi 2016; Behzadi 2020; Behzadi et al. 2019, 2021; Botos et al. 2011; Gomaz et al. 2012; Kawasaki and Kawai 2014; Luo et al. 2020; Ohadian Moghadam and Nowroozi 2019; Ohto et al. 2012).

TLR4 molecules are expressed by innate and adaptive (B cells) immune cells and epithelial cells, too. TLR glycoproteins are expressed by tumor cells in a wide range of cancers involving esophageal squamous cell carcinoma, breast cancer, ovarian cancer, prostate cancer, cervical cancer, adrenocortical carcinoma, lung cancer, colorectal cancer, myelogenous leukemia, intestinal cancer, laryngeal cancer, gastric cancer, neuroblastoma, pancreatic cancer, hepatocellular carcinoma, melanoma, skin cancer, brain cancer and head-and-neck squamous cell carcinoma. According to previous results, the expression of TLR4 increases in colon cancer, esophageal adenocarcinoma and ovarian cancer. Besides, the overexpression of TLR4 is associated with the development of various cancers comprising adrenocortical carcinoma, bladder cancer, head-and-neck squamous cell cancer, colorectal tumor, lung cancer, esophageal cancer, oral squamous cell carcinoma, colon cancer, pancreatic cancer, gastric cancer, melanoma, ovarian cancer and breast cancer (Ayala-Cuellar et al. 2019; Behzadi 2021; Dajon et al. 2017; Mokhtari et al. 2020; Ohadian Moghadam and Nowroozi 2019; Saghazadeh and Rezaei 2020).

TLR5

The TLR5 is encoded by the six-exonal *TLR5* gene, which is located on genomic chromosome of 1q33.3. Bacterial flagellin is the main ligand for the active form of TLR5/TLR5 homodimers. TLR5 glycoproteins are known as cell surface TLRs and are expressed by both the innate and adaptive (T cells) immune cells. The TLR5 is also expressed by tumor cells in ovarian cancer, breast cancer, gastric cancer, intestinal cancer, colon cancer and squamous cell carcinomas (Behzadi and Behzadi 2016; Behzadi 2021; Behzadi et al. 2019, 2021; Dajon et al. 2017; Gomaz et al. 2012; Ohadian Moghadam and Nowroozi 2019; Saghazadeh and Rezaei 2020).

The expression of TLR5 glycoproteins increases in cervical and ovarian cancers, while the TLR5 expression reduces in esophageal adenocarcinoma. Furthermore, there is a correlation between the overexpression of TLR5 molecules and the risk of development of ovarian cancer, gastric cancer, oral squamous cell carcinomas (Ayala-Cuellar et al. 2019; Mokhtari et al. 2020).

TLR6

The TLR6 is a cell surface TLR which can be functional through homodimerization with TLR2 (TLR2/TLR6) or heterodimerization with TLR4 (TLR4/TLR6) (in Alzheimer's disease). Lipoproteins are important ligands of the TLR2/TLR6 heterodimers. The *TLR6* gene bearing a transcript with two exons and 2,753 bases is involved in the gene cluster of *TLR6-TLR1-TLR10* and located on 4p14 chromosome (Behzadi and Behzadi 2016; Behzadi 2021; Behzadi et al. 2019, 2021; Gomaz et al. 2012; Kawasaki and Kawai 2014; Sun et al. 2005; Wang et al. 2020). The TLR6 molecules are expressed by both the innate and adaptive (B cells) cells of the immune system. The TLR6 is also expressed by tumor cells in hepatocellular carcinoma. The overexpression of TLR6 is associated with the risk of development of hepatocellular carcinoma (Dajon et al. 2017; Mokhtari et al. 2020; Ohadian Moghadam and Nowroozi 2019; Saghazadeh and Rezaei 2020).

TLR7

The *TLR7* gene is composed of three exons and situated on Xp22.3 of chromosomal genome. The biomolecule of TLR7 – resembling TLR4 – is a bi-zonal glycoprotein that can be detected in two different sites: on the cell surfaces and within the endosomes. From the point of structure and function, TLR7 is very close to TLR8. Both of them are recognized as the receptors of the viral single-stranded RNAs (ssRNAs). The active form of TLR7 is as TLR7/TLR7 homodimer (Behzadi and Behzadi 2016; Behzadi 2021; Behzadi et al. 2019, 2021; Wang et al. 2020).

TLR7 immunoproteins are expressed by innate and adaptive (B cells) immune cells. The TLR7 is expressed by tumor cells in lung cancer, myeloma, pancreatic ductal adenocarcinoma, esophageal squamous carcinoma, chronic lymphocytic leukemia and colorectal cancer, too. An overexpression of TLR7 is detected in myeloma cells, lung cancer and esophageal cancer, which has a correlation with the risk of cancer development (Dajon et al. 2017; Mokhtari et al. 2020; Ohadian Moghadam and Nowroozi 2019; Saghazadeh and Rezaei 2020).

TLR8

The intracellular TLR8 is encoded by the TLR8 gene, which consists of two exons and, resembling TLR7, has a chromosomal position on Xp22. The functional TLR8 acts as TLR8/TLR8 homodimers. The sequence, structure and genomic position of TLR8 glycoproteins are similar to the TLR7 molecules. As mentioned before, the main ligand of TLR8 homodimer is viral ssRNA (Behzadi and Behzadi 2016; Behzadi 2021; Behzadi et al. 2019, 2021; Wang et al. 2020). The TLR8 is expressed by the innate and adaptive immune cells (like T cells and T_{regs}). The biomolecule of TLR8 is also expressed by tumor cells in lung cancer, breast cancer and colorectal cancer. The overexpression of TLR8 is associated with the risk of lung cancer development (Dajon et al. 2017; Ohadian Moghadam and Nowroozi 2019).

TLR9

The intracellular TLR9 is encoded by the two-exonal *TLR9* gene, which is situated on 3p21.3 of the genomic chromosome. The TLR9 is the receptor of the unmethylated cytosine-phosphate-guanine (CpG) DNA molecule. The unmethylated CpG DNA acts as the main ligand of the functional homodimer of TLR9/TLR9 (Behzadi and Behzadi 2016; Behzadi 2021; Behzadi et al. 2019, 2021; Wang et al. 2020). The TLR9 glycoproteins are expressed by the innate and adaptive immune cells. Furthermore, the TLR9 molecules are expressed by tumor cells in myeloma, myelogenous leukemia, cervical squamous cell carcinoma, breast cancer, prostate cancer, lung cancer, hepatocellular carcinoma, gastric cancer, esophageal squamous cell carcinoma, colorectal cancer, renal cell carcinoma and ovarian cancer. The expression of TLR9 biomolecules increases in the early stages of breast cancer (while decreases in advanced stages), squamous cell carcinoma, pancreatic cancer, cervical cancer, prostate cancer, renal cancer, mucoepidermoid salivary gland carcinoma and neuroblastoma (while decreases in advanced stages). In addition, there is a correlation between the overexpression of TLR9 and the risk of cancer development in breast cancer, prostate cancer, myeloma, gastric cancer, lung cancer, pancreatic cancer and esophageal cancer (Ayala-Cuellar et al. 2019; Dajon et al. 2017; Mokhtari et al. 2020; Ohadian Moghadam and Nowroozi 2019; Saghazadeh and Rezaei 2020).

TLR10

As mentioned before, the *TLR10* gene with three exons is located within the gene cluster of *TLR6-TLR1-TLR10* on 4p14 chromosome. The *TLR10* gene transcript consists of 3,270 bases in three exons. The TLR10 dimers whether homodimers of TLR10/TLR10 or heterodimers of TLR1/TLR10 and TLR2/TLR10 are formed and activated by lipoproteins. The TLR10 is a bi-zonal molecule that can be detected as cell surface and endosomal TLR molecule (Behzadi and Behzadi 2016; Behzadi 2021; Behzadi et al. 2019, 2021; Gomaz et al. 2012; Sun et al. 2005; Wang et al. 2020).

17.8.1 Toll-Like Receptors, Cascade of TLR Signaling Pathways and the Related Adaptors

TLRs in humans are a ten-membered family of glycoproteins which can easily be identified by their horseshoe-structured ectodomains (Behzadi 2021; Botos et al. 2011; Kawasaki and Kawai 2014). As mentioned before, the extracellular TLRs (e.g., TLRs of 1, 2, 4, 5, 6, 7 and 10) and the intracellular TLRs (e.g., TLRs of 3, 4, 7, 8, 9 and 10) activate signaling pathways by different mechanisms (Figure 17.3). In all molecules of TLRs, the intracellular domain of TIR, which often consists of ~150 amino acids, starts the process of signaling pathway by the linkage between the extracellular domain of TLR and the related ligand. This linkage results in alteration in the receptor spatial configuration, which may lead to TIR signaling activation. For the initiation of TLR signaling pathway, the TIR domain participates in TLR dimerization in both forms of hetero- and homodimerization. Excluding TLR3 and the endosomal TLR4, which apply with TIR domain-containing adaptor-inducing IFN-β (TRIF)-dependent signaling pathway, the other TLR molecules exploit myeloid differentiation primary response protein 88 (MyD88)-dependent signaling pathway. Indeed, a variety of adaptors which bear TIR domain such as B-cell adaptor for phosphoinositide (BCAP), MyD88 adaptor-like (MAL/TIR domain-containing adapter molecule (TIRAP)), MyD88, sterile-α- and armadillo motif-containing protein (SARM), SLP65/76 and CSK-interacting membrane protein (SCIMP), TRIF and TRIF-related adaptor molecule induce the cascade of signaling pathways (Behzadi 2021; Behzadi et al. 2019; Berglund et al. 2015; De Nardo 2015; Kawasaki and Kawai 2014; Keshavarz et al. 2021; Luo et al. 2020; Mokhtari et al. 2020; Wang et al. 2020).

In this regard, the TLR2/TLR2 homodimers and TLR2 heterodimers (TLR1/TLR2 and TLR2/TLR6) induce the cascade of signaling pathway through the adaptors of BCAP, MAL, MyD88 and SCIMP (Behzadi 2021; Behzadi et al. 2021; Luo et al. 2020; Sameer and Nissar 2021; Sepehri et al. 2016). The homodimers of TLR/3TLR3 utilize the TIR adaptors of TRIF, SARM and SCIM to activate their signaling pathway. Among TLR molecules, the homodimers of TLR4/TLR4 (extracellular and intracellular molecules) employ the highest number of TIR adaptors comprising BCAP, MAL, MyD88, TRAM, TRIF, SARM and SCIMP. The homodimers of TLR5/TLR5 recruit the adaptors of MyD88 and TRIF for signal transduction. The TLR9/TLR9 homodimers use MAL, MyD88 and SCIMP adaptors to trigger the cascade of signaling. The other homodimers of TLR7/TLR7, TLR8/TLR8 and TLR10/TLR10 employ only the adaptor of MyD88. The endosomal/lysosomal/phagosomal TLRs (ligand activated) employ the canonical adaptors of TRAM and TRIF for signaling through invoking them to the endosomal membrane, while the cell surface TLRs recruit the canonical adaptors of MAL and MyD88 to induce signaling via invoking them to the cell surface (Figure 17.3) (Behzadi 2021; Behzadi et al. 2021; Luo et al. 2020; Sameer and Nissar 2021).

17.8.2 Toll-Like Receptors and Tumor Biology

As mentioned before, the TLRs act as a double-edged sword. In some cases, the TLRs are highly expressed in tumor cells than in normal cells. Even, in accordance with the condition of cancer (relating to stages and types), the expression of TLRs differs from one to another. The expression and overexpression of TLRs in some tissues directly depict the invasive and aggressive condition of the tumor and carcinoma. Moreover, the high expression of some TLRs in tumors indicates the enlargement of the tumor, the presence of malignancies and cancerous metastasis probability. In this regard, the occurrence of some SNPs is associated with different types of cancers, while some others have protective effects against cancers (Angrini et al. 2020; Chatterjee et al. 2014; De Re et al. 2019; Fan et al. 2019; Gast et al. 2011; González-Reyes et al. 2010; Gu et al. 2018; Li et al. 2014; Mäkinen et al. 2015; Messaritakis et al. 2018; Minmin et al. 2011; Slattery et al. 2012; Theodoropoulos et al. 2012; Tian et al. 2018; Wang et al. 2018; Yang et al. 2012; Yuan et al. 2013; Zeng et al. 2011; Zhang et al. 2014b).

17.9 Immunogenetics and Toll-Like Receptors

The SNP feature can occur in different genes including immunomolecules and glycoproteins, e.g., TLRs. Due to this fact, the SNPs can be detected as significant biomarkers for different types of cancers with different stages. TLRs are a member of pathogen recognition receptors (PRRs), which contribute to the innate immune system and simultaneously link the innate immune system to the adaptive immune system. In other words, the TLRs are vital immune biomolecules that form an effective and functional network consisting of chemokines, ILs and cytokines. Indeed, the stimulation of TLR molecules results in the release of a wide range of cytokines and chemokines. All these molecules are involved in host defense system against infectious diseases, autoimmune diseases and cancers (Behzadi and Behzadi 2016; Behzadi 2021; Behzadi et al. 2019, 2021; Boozari et al. 2019; Dreussi et al. 2018; Gajdács and Behzadi 2020; Heidarzadeh et al. 2020; Hosseini et al. 2015; Sameer and Nissar 2021).

The TLRs are known as evolutionarily conserved glycoproteins, the oldest of which is borne by nematodes such as *Caenorhabditis elegans*. Each TLR molecule consists of three distinct domains including an extracellular domain of leucine-rich repeat (LRR), a transmembrane domain and the intracellular section of the Toll/IL-1R (TIR)-like domain (Figure 17.6). These domains modulate the binding of ligands to TLR (via LRR) and the induction of signaling pathways (through the transmembrane and TIR domains), which trigger the activation of a variety of immune responses in a cascade manner (Behzadi and Behzadi 2016; Behzadi 2021; Behzadi et al. 2016, 2019, 2021; Sameer and Nissar 2021; Vijay 2018).

TLRs as a family of transmembrane glycoprotein receptors are able to identify and bind to different types of ligands comprising damage-/danger-associated molecular patterns (DAMPs), microbial-/pathogen-associated molecular patterns (MAMPs/PAMPs) (with bacterial, fungal, parasital and viral origins) and xenobiotic-associated molecular patterns (XAMPs), which may lead to inducing the cascades of signaling pathways (Behzadi 2021; Behzadi et al. 2021; Fleshner 2013; Heidarzadeh et al. 2020; Hirbod-Mobarakeh et al. 2020; Prasad and Kumar 2020; Sameer and Nissar 2021; Wang et al. 2020; Zhao et al. 2014).

According to our knowledge, we now know that TLRs involve the simplest organisms such as nematodes to advanced organisms such as humans and these biomolecules act as immunogenetic molecules that the appearance of mutations and polymorphisms within their genes may lead to serious malignancies such as PCa. The SNPs can appear in different situations of the genes. Due to this fact, Sato et al. shown the occurrence of rs11536889 (G>C) SNP in the 3′-UTR of *TLR4* gene. This SNP alters the expression of *TLR4* gene by affecting the translation process, perhaps through linking to molecules of miRNAs (Behzadi 2021; Dreussi et al. 2018; Sato et al. 2012; Vijay 2018).

The immunogenetic molecules of TLRS in humans consist of ten members comprising TLR1–TLR10. Some of them are located upon the cell membrane and identified as extracellular TLRs (e.g., TLRs of 1, 2, 4, 5, 6, 7 and 10), while others belong

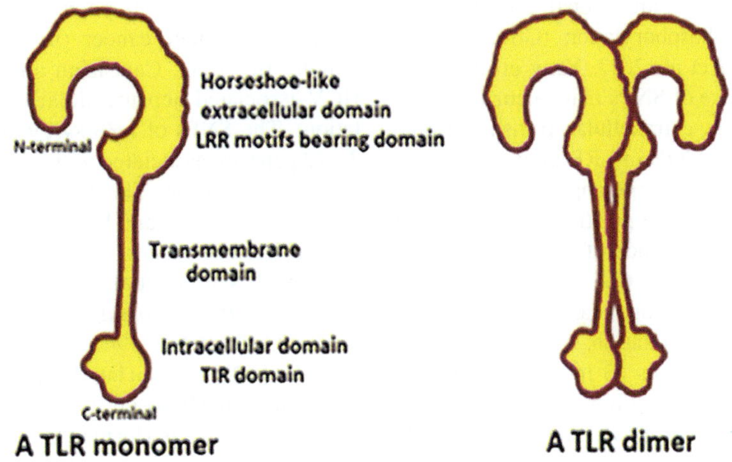

FIGURE 17.6 The structure of TLR and the related domains. Mono- and dimer forms are shown. (This figure is prepared through the software tool service by Biorender.com.)

to endosomes and lysosomes and are known as intracellular TLRs (e.g., TLRs of 3, 4, 7, 8, 9 and 10) (Figure 17.5). However, some of the TLRs are determined as bi-zonal TLRs including the TLRs of 4, 7 and 10, which can appear as extracellular and intracellular TLRs (Behzadi 2021; Kawasaki and Kawai 2014; Wang et al. 2020). Nevertheless, the biosynthesis of the human TLRs occurs within the membranous web of endoplasmic reticulum. Following this, the TLRs are translocated into the flat sacs of the Golgi apparatus. The glycosylation of TLRs is achieved in the N-terminal of TLRs' ectodomains, and this happens for the TLRs' co-receptors, too. Each TLR should be glycosylated for performing its function, interaction and signaling pathways. The glycosylation feature is a post-translational process that modifies lipids and proteins for the most. The process of glycosylation of TLRs and the associated co-receptors is unavoidable because it affects proteins' functions, configuration (folding), transfer and localization. Hence, the glycoproteins such as TLRs contribute to the important vital activities including induction of signaling pathways and immune responses (Kawasaki and Kawai 2014; Leifer and Medvedev 2016; Lin et al. 2020; Vijay 2018).

The highest level of TLR molecules' expression is found in a wide range of innate and adaptive immune cells comprising antigen-presenting cells (APCs). However, the APCs are divided into two groups of professional and non-professional cells. The non-professional APCs involve hepatocytes and fibroblasts, while the professional APCs include B cells, dendritic cells (DCs) and macrophages (MΦ)/monocytes (Angrini et al. 2020; Eiz-Vesper and Schmetzer 2020; Hirbod-Mobarakeh et al. 2020; Lin and Loré 2017; Zhao et al. 2014).

The TLR recognition of the related ligand leads to activation and maturation of APCs, which results in APCs translocation into the lymph nodes. As a cascade process, the APCs secrete a variety of pro-inflammatory cytokines consisting of interferons (IFNs), tumor necrosis factor-α (TNF-α), etc. The IFNs, in turn, induce the lymphocytes of cytotoxic T and natural killer (NK) cells (Angrini et al. 2020).

Zhao et al. (2014) described the role of TLRs and their signaling pathway in PCa. The activation of TLRs of 3, 4, 5 and 7 in NK cells situated in cancer cells, tumors and tumor microenvironment may result in immune responses by cytotoxic T and T helper 1 (T_{h1}) cells, which may lead to tumor inhibition. Moreover, the activation of TLRs of 3, 4, 5 and 7 within the APCs including B cells, MΦs and DCs disseminated in tumors and tumor microenvironment induces T cells to mediate the immune responses via T_{h1} and cytotoxic T cells. In consequence, this process leads to tumor inhibition, too (this scenario is the same as the one that occurs by the activation of TLRs of 3, 4, 5 and 7 within NK cells). In contrast, the activation of TLRs of 2, 4 and 9 triggers T cells to mediate the immune responses via T_{h2} and regulatory T cells (T_{regs}), which may lead to tumor progression. Hence, the TLR molecules function as a double-edged sword that, on the one hand, inhibit the growth and development of tumors (anti-tumorigenic effect) and, on the other hand, support the process of tumorigenesis (pro-tumorigenic effect) (Angrini et al. 2020; Behzadi 2021; Gomaz et al. 2012; Zhao et al. 2014).

17.10 Toll-Like Receptors: Single Nucleotide Polymorphisms and Cancers

The appearance of SNPs in genes may lead to structural and functional changes in protein products. Besides, it can affect the affinity of the ligands (proteins) and the related receptors. In accordance with previous reports, the occurrence of SNPs in *TLR* genes often results in different types of cancers. In this regard, the highest risks of cancers are detected (Gomaz et al. 2012; Nissar et al. 2017).

According to previous studies, the polymorphisms in *TLR2* and *TLR4* genes have the highest association with cancers, respectively. SNPs are detected in different genomic elements including *TLR* genes. In many cases, the presence of SNPs in *TLR* genes may be ineffective and neutral. However, in some cases, the appearance of SNPs in *TLR* genes results in alterations in amino acid, protein structure, activity and stability. Besides, the SNPs features may be affecting the process of gene expression comprising mRNA splicing, life length of mRNA molecules and the post-translational modifications such as glycosylation and phosphorylation (Gomaz et al. 2012; Lin et al. 2020; Nissar et al. 2017; Yang et al. 2019). Due to this fact, the occurrence of SNPs may occur in different parts of TLRs including the extracellular, transmembrane and intracellular domains. The SNPs in LRR alter the affinity between ligands and receptor, while the appearance of SNPs in transmembrane domain affects the position and movement of the receptor. Moreover, the presence of SNPs in intracellular domain of the TLR molecules (the TIR section) may lead to disruption in TLR dimerization, adaptors employment and TLR (receptor-ligand) internalization. The TIR domain related to the *TLR* gene encompasses 13 introns that can be effective targets for different types of polymorphisms. Now we know that polymorphisms involving SNPs may affect directly the strength or weakness of the immune system against infectious diseases, autoimmune diseases and cancers (Gomaz et al. 2012; Medvedev 2013; Vijay 2018).

The TLRs and the associated signaling pathways are activated by the exposure of DAMPs, MAMPs/PAMPs and XAMPs. Hence, in the following, I describe the structure and function of each TLR, the related SNPs and the associated signaling pathways. Now, I am going to discuss about the TLR SNPs and their roles associated with PCa.

17.10.1 Toll-Like Receptors, Single Nucleotide Polymorphism in Toll-Like Receptors and Prostate Cancer

TLR1 SNPs and their role

The occurrence of SNPs in the gene cluster of *TLR6-TLR1-TLR10* is directly associated with PCa (Gomaz et al. 2012; Vijay 2018). In some cases, the occurrence of SNPs in TLR1 (e.g., rs7696175 associated with the increased risk of breast cancer in European and Chinese populations; rs4833095 associated with the increased risk of breast cancer in Chinese population; rs4833095 associated with the development of gastric cancer in patients infected by *Helicobacter pylori*) has association with other cancers rather than PCa (Behzadi 2021; Semlali et al. 2018).

TLR1 SNPs in Prostate Cancer

The SNPs of rs5743604, rs5743551 and rs5743556 belonging to *TLR1* allele (–833C/T (C allele/on intron 3), –7202A/G (G allele/on promoter) and 6399C/T (C allele/on promoter)) increase the risk of PCa among Swedes. As the *TLR1* gene is a part of *TLR6-TLR1-TLR10* gene cluster, the occurrence of SNPs in any member of this gene cluster (*TLR6* and *TLR10*) is involved in PCa development (Kazma et al. 2012; Khan et al. 2016; Kutikhin and Yuzhalin 2012; Sun et al. 2005; Vijay 2018). The SNP of rs4833095 (AG/GG) belonging to *TLR1* gene is associated with advanced PCa risk in African American and Caucasian populations. Furthermore, the SNP of rs5743618 (TT) is recognized as an advanced PCa risk factor in African Americans (Kazma et al. 2012). The SNPs of rs4833095 and rs5743596 belonging to *TLR1* allele (248S and S26L) decrease the risk of PCa (Khan et al. 2016; Mukherjee et al. 2019).

TLR2 SNPs and their role

The TLR2 SNPs may lead to different types of cancers. Due to this fact, the "del" allele (–196 to –174) of *TLR2* gene (on exon 1; 5′-UTR) (Δ22) is related to the increased risk of breast cancer (between Greek population), cervical cancer (in North India), gallbladder cancer (in North India), gastric cancer (among Brazilian, Caucasian and Japanese populations) and HCV-associated hepatocellular carcinoma (in German population). The SNPs of rs3804099 and rs3804100 pertaining to *TLR2* gene are associated with papillary thyroid cancer among Korean population. The SNPs of rs3804099 and rs3804100 predispose Chinese and Brazilian populations to the increased risk of gastric cancer, respectively (Behzadi 2021; Gomaz et al. 2012; Semlali et al. 2018).

TLR2 SNPs in Prostate Cancer

The "del" allele (–196 to –174) of *TLR2* gene (on exon 1; 5′-UTR) (Δ22) is associated with PCa among people who live in North India (Behzadi 2021; Gomaz et al. 2012). The result of another study performed by Rogers et al. (2013) showed the significant association between the SNP of *TLR2* gene (rs1898830) and the risk of PCa among African Americans (Khan et al. 2016). However, there is a complex interaction between the rs1898830 in *TLR2* gene and rs4251545 in *IRAK4* gene, which can be recognized as a predictable risk of PCa among African Americans (Rogers et al. 2013).

TLR3 SNPs and their role

The SNPs of *TLR3* gene have some significant roles in the appearance and development of different cancers. The SNP of 829A>C in *TLR3* gene is in association with the risk of nasopharyngeal carcinoma (Gomaz et al. 2012). The SNP of rs10025405 (G allele) in the 3′-UTR of *TLR3* gene is related to breast cancer. The SNP of rs3775291 (A allele) on the fourth exon of *TLR3* gene is related to breast cancer recurrence among Chinese population. Simultaneously, the occurrence of this SNP may increase the survival of patients with colorectal carcinoma (stage II) among German population. The result of an investigation (Pocnet et al. 2017) indicated no considerable relationship between cervical cancer among women in North India and the *TLR3* gene SNP of rs3775290; however, the result of another study (Zidi et al. 2016) showed a significant relationship between the *TLR3* gene SNP of rs3775290 (+1377C>T) and the evolution of cervical cancer among Tunisian women (Behzadi 2021; Khan et al. 2016; Semlali et al. 2018). The SNPs of rs3775292, rs11721827 and rs7668666 in *TLR3* gene are associated with colon cancer, rectal cancer and melanoma-specific death, respectively (Khan et al. 2016). The rs3775290 SNP detected in *TLR3* gene has an association with the risk of urinary bladder cancer among Indian population (Behzadi 2021; Cheng et al. 2014).

TLR3 SNPs in Prostate Cancer

The rs3775290 SNP detected in *TLR3* gene has an association with the risk of PCa among Indian population, too (Behzadi 2021; Cheng et al. 2014).

TLR4 SNPs and their role

In the study done by Semlali et al., the results show a significant correlation between the *TLR4* SNP of rs10759931 and a strong risk of breast cancer among Saudi Arabian population (Semlali et al. 2017, 2018). An investigation carried out by Nissar et al. indicated that the *TLR4* gene polymorphism is not a pivotal biomarker for the development of colorectal cancer among Kashmir population (Nissar et al. 2017). The results of another survey performed by Castaño-Rodríguez et al. (2014) showed a significant correlation between the *TLR4* SNP of rs11536889 and gastric cancer in a Chinese population.

In contrast, the *TLR4* SNPs of rs10116253, rs1927911 and rs10759931 are reducing the risk factors for gastric cancers in a Chinese population (Castano-Rodriguez et al. 2014; Khan et al. 2016). Moreover, the SNPs of rs11536865 (–729G/C) (C allele) and rs11536889 (GC and CC genotypes) (C allele) in *TLR4* gene have a correlation with the risk of urinary bladder cancer among Chinese population (Behzadi 2021; Pandey et al. 2018).

TLR4 SNPs in Prostate Cancer

In accordance with the results obtained by Kim et al. (2012), the *TLR4 gene* SNP of rs11536889, which occurs in the 3′-UTR, has a significant association with the risk of PCa in Korean men. Another survey reported that there is a significant correlation between the occurrence of *TLR4* SNP of rs11536858 G/A (rs10759931) (in the 5′-UTR) (GG genotype) and rs1927911 (on intron 1) (TC genotype) and the increased risk of PCa development in Korean patients (Gomaz et al. 2012; Pandey et al. 2018). Moreover, in another study among Swedish population, the results showed that the variant of 11381G/C pertaining to the *TLR4 gene* SNP of rs11536889, which occurs in the 3′-UTR, has a significant correlation with the risk of PCa, too. Qiu et al. showed that there is a significant relationship between the *TLR4 gene* SNP of rs4986791 and the increased risk of BPH (Kim et al. 2012; Qiu et al. 2019). In a study, the reported results indicated a significant correlation between the occurrence of *TLR4 gene* SNP of rs10759932 T/C (CC genotype) in the 5′-UTR and the risk of PCa among African Americans and Caucasian population. In an investigation which was carried out among US population involving 97% Caucasians, eight *TLR4* gene SNPs comprising rs2149356, rs6478317, rs1927914, rs7873784, rs11536891, rs10116253, rs11536898 and rs1927911 were detected. All of these eight *TLR4 gene* SNPs have a significant correlation with a lower risk of PCa development. In another study which was done among US population containing 2% of African Americans and 98% of Caucasians, two *TLR4 gene* SNPs of rs10116253 T/C in the 5′-UTR and rs4986790 A/G were detected. Both of these *TLR4* gene SNPs have a significant correlation with the risk of PCa (Kutikhin and Yuzhalin 2012; Pandey et al. 2018). These invaluable results indicate that the majority of *TLR4 gene* SNPs occur in the 3′- and 5′-UTRs. As we know, the 3′- and 5′-UTRs are directly in relationship with translation initiation (ribosome recognition), translation termination and post-translational modifications. The appearance of *TLR4* gene SNPs in the 3′- and 5′-UTRs directly affect the expression and the activity of TLR4 glycoprotein. Hence, the occurrence of *TLR4* gene SNPs in the 3′- and 5′-UTRs has a significant relationship with different conditions of PCa (Barrett et al. 2012; Pandey et al. 2018).

TLR5 SNPs and their role

Some SNPs regarding *TLR5* gene including rs5744174 (1846 T>C, Phe616Leu) on the sixth exon (pertaining to LRRCT) and rs2072493 have a significant correlation with gastric cancer and colon cancer among Chinese population (Gomaz et al. 2012; Semlali et al. 2018).

TLR5 SNPs in Prostate Cancer

The *TLR5* gene SNPs of rs2072493 and rs5744174 may be in correlation with BPH (Seok et al. 2013).

TLR6 SNPs and their role

The *TLR6* gene SNPs of rs5743815 (1280 T>C, Val427Ala) on the second exon region pertaining to LRR section of the TLR molecule and the rs7696175 (detected in TLR1 gene, too) are correlated with non-Hodgkin's lymphoma among Caucasians and the increasing risk of breast cancer among Chinese and European populations, respectively (Gomaz et al. 2012; Khan et al. 2016; Semlali et al. 2018).

TLR6 SNPs in Prostate Cancer

The *TLR6* gene SNPs of rs5743795 (–1401A/G, A allele, on promoter) and rs5743806 (–673C/T, T allele, on promoter) are in association with PCa risk (Kutikhin and Yuzhalin 2012; Sun et al. 2005).

TLR7 SNPs and their role

The *TLR7* gene SNP of rs17900 (T allele) is associated with Hodgkin's disease (Kutikhin 2011).

TLR7 SNPs in Prostate Cancer

To date, no clear meta-analysis has been done to show a significant correlation between *TLR7* gene SNPs and PCa.

TLR8 SNPs and their role

To date, the SNPs of *TLR8* gene have not been reported.

TLR8 SNPs in Prostate Cancer

To date, no *TLR8* gene SNPs associated with PCa or other cancers have been reported.

TLR9 SNPs and their role

Several SNPs regarding TLR9 gene have been detected. The SNPs of rs187084 (–1486 T>C) (allele C) in the 5′-UTR and rs352140 (2848 C>T, Pro545Pro) (CT and TT genotypes) on exon 2 pertaining to LRR domain have a correlation with the increased cancer risk and cervical cancer. Besides, the SNPs of rs352140 (2848 A>G, Pro545Pro) (allele A) on exon 2 belonging to LRR domain and rs5743836 (–1237 T>C) (allele C) in the 5′-UTR have a relationship with the increased cancer risk and Hodgkin's lymphoma (Behzadi 2021; Gomaz et al. 2012). In another study, it was reported that the SNPs of rs352139 (A/G) and rs352144 (A/C) detected in the promoter region in *TLR9* gene are significantly correlated with the risk of cancer development and colorectal cancer (Behzadi 2021; Semlali et al. 2018).

TLR9 SNPs in Prostate Cancer

No significant relationship between the occurrence of SNPs in *TLR9* gene and PCa has been reported to date.

TLR10 SNPs and their role

In an investigation, it was reported that there may be a significant correlation between the SNP of rs11466657 in *TLR10* gene and the reduction in cancer risk regarding meningioma, while the SNP of rs7696175 in *TLR10* gene (allele T) has a significant relationship with the increased risk of breast cancer (Kutikhin 2011).

TLR10 SNPs in Prostate Cancer

The SNPs of rs11096955 (G1031T), rs11096957 (N241H), rs4274855 and rs4129009 in *TLR10* gene are in association with the risk of PCa (Khan et al. 2016; Kutikhin and Yuzhalin 2012; Mukherjee et al. 2019).

17.11 Conclusions

PCa is a significant global concern among men, which may result in serious morbidity and mortality. This cancer is the outcome of a wide range of risk factors including age, race and ethnicity, inborn genetics, inheritance and family history, lifestyle, diet and gut microbiota. Due to this knowledge, each type of mentioned risk factors may simultaneously predispose to different types of mutations in men's genomic pools, e.g., occurrence of SNPs in *TLR* genes. TLRs are important glycoproteins that can be determined as pivotal biomolecules which participate in different parts of immune system and human body. They connect innate and adaptive immune systems into a unite immune system. Each TLR is expressed by its own chromosomal gene. These immunoglycoproteins may be induced to initiate a signaling pathway which may lead to the expression of pro-inflammatory cytokines IFN type I. The expression of TLRs may decrease or increase in different conditions and in different parts of human body. This characteristic of TLRs makes them worthy molecules to be recognized as biomarkers. Thus, the TLRs are invaluable biomarkers that can help us to detect abnormal situations, malignancies and diseases. In this regard, a rigorous understanding of polymorphisms, e.g., the occurrence of SNPs in *TLR* genes, may help us to use these immunogenomic features as important biomarkers for different procedures of prognosis, diagnosis and treatment of cancers.

Acknowledgments

I have special thanks to Prof. Syed Sameer Aga, the editor of the present book, for his sincere guidance, help and support.

ABBREVIATIONS

AJCC:	American Joint Committee on Cancer
AR:	Androgen receptor
BCAP:	B-cell adaptor for phosphoinositide
BPH:	Benign prostatic hyperplasia
DAMP:	Damage-/danger-associated molecular pattern
GLOBOCAN:	Global cancer
HPV:	Human papilloma virus
HSV:	Herpes simplex virus
IARC:	International Agency for Research on Cancer
IFN:	Interferon
IL:	Interleukin
ISUP:	International Society of Urological Pathology
KLK:	Kallikrein
KRT13:	Keratin 13
LRR:	Leucine-rich repeat
MAL:	MyD88 adaptor-like
MAMP:	Microbial-associated molecular pattern
MD-2:	Myeloid differentiation factor 2
MyD88:	Myeloid differentiation primary response protein 88
PAMP:	Pathogen-associated molecular pattern
PCa:	Prostate cancer
PIN:	Prostatic intraepithelial neoplasia
PRR:	Pathogen recognition receptor
PSA:	Prostate-specific antigen
SARM:	Sterile-α- and armadillo motif-containing protein
SCGB1A1:	Secretoglobin family 1A member 1
SCIMP:	SLP65/76 and CSK-interacting membrane protein
SNP:	Single nucleotide polymorphism
STD:	Sexually transmitted disease
TERT:	Telomerase reverse transcriptase
TIR:	Toll/IL-1R
TIRAP:	TIR domain-containing adapter molecule
TLR:	Toll-like receptor
TNF-α:	Tumor necrosis factor-α
TNM:	Tumor, node, metastasis
TRIF:	TIR domain-containing adaptor-inducing IFN-β
UGS:	Urogenital system
UTR:	Untranslated region
WHO:	World Health Organization
XAMP:	Xenobiotic-associated molecular pattern

REFERENCES

Adamowicz J, Pakravan K, Bakhshinejad B, Drewa T, Babashah S (2017) Prostate cancer stem cells: from theory to practice. *Scandinavian Journal of Urology* 51:95–106.

Akhter Q et al. (2012) Polymorphisms in the 3′UTR of the human leptin gene and their role in hypertension. *Molecular Medicine Reports* 5:1058–1062.

Al Olama AA et al. (2014) A meta-analysis of 87,040 individuals identifies 23 new susceptibility loci for prostate cancer. *Nature Genetics* 46:1103–1109.

Angrini M, Varthaman A, Cremer I (2020) Toll-like receptors (TLRs) in the tumor microenvironment (TME): a dragon-like weapon in a non-fantasy game of thrones. In: Alexander Birbrair (ed.), *Tumor Microenvironment*. Switzerland: Springer, pp. 145–173.

Arora K, Barbieri CE (2018) Molecular subtypes of prostate cancer. *Current Oncology Reports* 20:58.

Ayala-Cuellar AP, Cho J, Choi KC (2019) Toll-like receptors: a pathway alluding to cancer control. *Journal of Cellular Physiology* 234:21707–21715.

Barrett LW, Fletcher S, Wilton SD (2012) Regulation of eukaryotic gene expression by the untranslated gene regions and other non-coding elements. *Cellular and Molecular Life Sciences* 69:3613–3634.

Barsouk A et al. (2020) Epidemiology, staging and management of prostate cancer. *Medical Sciences* 8:28.

Basiri A, Eshrati B, Zarehoroki A, Golshan S, Shakhssalim N, Khoshdel A, Kashi AH (2020) Incidence, gleason score and ethnicity pattern of prostate cancer in the multi-ethnicity country of Iran during 2008–2010. *Urology Journal* 17(6):602–606.

Behzadi P (2020) Classical chaperone-usher (CU) adhesive fimbriome: uropathogenic Escherichia coli (UPEC) and urinary tract infections (UTIs). *Folia Microbiologica* 65:45–65.

Behzadi P (2021) The role of toll-like receptor (TLR) polymorphisms in urinary bladder cancer. In: Sameer A.S. BMZ, Nissar S. (eds) *Genetic Polymorphism and Cancer*

Susceptibility, vol I. Springer, Singapore, pp. 281–317. doi: https://doi.org/10.1007/978-981-33-6699-2_9.

Behzadi E, Behzadi P (2016) The role of toll-like receptors (TLRs) in urinary tract infections (UTIs). *Central European Journal of Urology* 69:404–410.

Behzadi P, Behzadi E, Pawlak-Adamska EA (2019) Urinary tract infections (UTIs) or genital tract infections (GTIs)? It's the diagnostics that count. *GMS Hygiene and Infection Control* 14:1–12.

Behzadi P, Behzadi E, Ranjbar R (2014) *Basic Modern Molecular Biology*. 1st edn. Persian Science & Research Publisher, Tehran, Iran.

Behzadi P, Behzadi E, Ranjbar R (2016) IL-12 family cytokines: general characteristics, pathogenic microorganisms, receptors, and signalling pathways. *Acta microbiologica et immunologica Hungarica* 63:1–25.

Behzadi P, Behzadi E, Ranjbar R, Saberfar E, Morovvati S (2015) Herpes simplex virus and pregnant women. *Ginecologia'ro* 10:32–34.

Behzadi P, García-Perdomo HA, Karpiński TM (2021) Toll-like receptors: general molecular and structural biology. *Journal of Immunology Research* 2021:9914854.

Berglund NA, Kargas V, Ortiz-Suarez ML, Bond PJ (2015) The role of protein–protein interactions in toll-like receptor function. *Progress in Biophysics and Molecular Biology* 119:72–83.

Bolla M et al. (2012) Postoperative radiotherapy after radical prostatectomy for high-risk prostate cancer: long-term results of a randomised controlled trial (EORTC trial 22911). *The Lancet* 380:2018–2027.

Boozari M, Butler AE, Sahebkar A (2019) Impact of curcumin on toll-like receptors. *Journal of Cellular Physiology* 234:12471–12482.

Botos I, Segal DM, Davies DR (2011) The structural biology of toll-like receptors. *Structure* 19:447–459.

Boyd LK, Mao X, Lu Y-J (2012) The complexity of prostate cancer: genomic alterations and heterogeneity. *Nature Reviews Urology* 9:652.

Bratt O, Drevin L, Akre O, Garmo H, Stattin P (2016) Family history and probability of prostate cancer, differentiated by risk category: a nationwide population-based study. *Journal of the National Cancer Institute* 108:djw110.

Cancer Today (2020) GLOBOCAN 2020, International Agency for Research on Cancer 2020, World Health Organization (WHO). https://gco.iarc.fr/today/. Accessed 30 December 2020.

Castaño-Rodríguez N, Kaakoush NO, Pardo AL, Goh K-L, Fock KM, Mitchell HM (2014) Genetic polymorphisms in the toll-like receptor signalling pathway in Helicobacter pylori infection and related gastric cancer. *Human Immunology* 75:808–815.

Castillo TJM, Starmans M, Niessen W, Schoots I, Klein S, Veenland J (2019) Classification of prostate cancer: high grade versus low grade using a radiomics approach. *2019 IEEE 16th International Symposium on Biomedical Imaging (ISBI 2019)*, 2019, pp. 1319–1322, doi: 10.1109/ISBI.2019.8759217.

Chatterjee S et al. (2014) TLR7 promotes tumor progression, chemotherapy resistance, and poor clinical outcomes in non–small cell lung cancer. *Cancer Research* 74:5008–5018.

Cheng D, Hao Y, Zhou W, Ma Y (2014) Association between toll-like receptor 3 polymorphisms and cancer risk: a meta-analysis. *Tumor Biology* 35:7837–7846.

Chipman JJ et al. (2014) Measuring and predicting prostate cancer related quality of life changes using EPIC for clinical practice. *The Journal of Urology* 191:638–645.

Cimadamore A et al. (2020) Prostate cancer pathology: what has changed in the last 5 years. *Urologia Journal* 87:3–10.

Compérat E, Wasinger G, Oszwald A, Kain R, Cancel-Tassin G, Cussenot O (2020) The genetic complexity of prostate cancer. *Genes* 11:1396.

Compérat EM et al. (2019) Grading of urothelial carcinoma and the new "World Health Organisation classification of Tumours of the urinary system and male genital organs 2016". *European Urology Focus* 5:457–466.

Cornejo KM, Rice-Stitt T, Wu C-L (2020) Updates in staging and reporting of genitourinary malignancies. *Archives of Pathology & Laboratory Medicine* 144:305–319.

Culp MB, Soerjomataram I, Efstathiou JA, Bray F, Jemal A (2020) Recent global patterns in prostate cancer incidence and mortality rates. *European Urology* 77:38–52.

Dajon M, Iribarren K, Cremer I (2017) Toll-like receptor stimulation in cancer: a pro-and anti-tumor double-edged sword. *Immunobiology* 222:89–100.

Dannemann M, Andrés AM, Kelso J (2016) Introgression of Neandertal-and Denisovan-like haplotypes contributes to adaptive variation in human toll-like receptors. *The American Journal of Human Genetics* 98:22–33.

De Nardo D (2015) Toll-like receptors: activation, signalling and transcriptional modulation. *Cytokine* 74:181–189.

de Oliviera Nascimento L, Massari P, Wetzler LM (2012) The role of TLR2 in infection and immunity. *Frontiers in Immunology* 3:79.

De Re V et al. (2019) Polymorphism in toll-like receptors and Helicobacter pylori motility in autoimmune atrophic gastritis and gastric cancer. *Cancers* 11:648.

Deng N, Zhou H, Fan H, Yuan Y (2017) Single nucleotide polymorphisms and cancer susceptibility. *Oncotarget* 8: 110635.

Ding L, Jiang Q, Li G, Shen J, Du J, Lu X, Xiong X (2017) Comprehensive assessment of association between TLR4 gene polymorphisms and cancer risk: a systematic meta-analysis. *Oncotarget* 8:100593.

Dreussi E, Ecca F, Scarabel L, Gagno S, Toffoli G (2018) Immunogenetics of prostate cancer: a still unexplored field of study. *Pharmacogenomics* 19:263–283.

Eiz-Vesper B, Schmetzer HM (2020) Antigen-presenting cells: potential of proven und new players in immune therapies. *Transfusion Medicine and Hemotherapy* 47:429–431.

Faisal FA, Lotan TL (2020) The genomic and molecular pathology of prostate cancer: clinical implications for diagnosis, prognosis, and therapy. *Advances in Anatomic Pathology* 27:11–19.

Fan L, Zhou P, Hong Q, Chen A-X, Liu G-Y, Yu K-D, Shao Z-M (2019) Toll-like receptor 3 acts as a suppressor gene in breast cancer initiation and progression: a two-stage association study and functional investigation. *Oncoimmunology* 8:e1593801.

Fan Q et al. (2013) Functional polymorphism in the 5′-UTR of CR2 is associated with susceptibility to nasopharyngeal carcinoma. *Oncology Reports* 30:11–16.

Ferrís-Tortajada J, Berbel-Tornero O, García-Castell J, Ortega-García J, López-Andreu J (2012) Dietetic factors associated with prostate cancer. Protective effects of editerranean diet. *Actas Urológicas Españolas (English Edition)* 36:239–245.

Ficarra V et al. (2014) The role of inflammation in lower urinary tract symptoms (LUTS) due to benign prostatic hyperplasia (BPH) and its potential impact on medical therapy. *Current Urology Reports* 15:463.

Fleshner M (2013) Stress-evoked sterile inflammation, danger associated molecular patterns (DAMPs), microbial associated molecular patterns (MAMPs) and the inflammasome. *Brain, Behavior, and Immunity* 27:1–7.

Franz M-C, Anderle P, Bürzle M, Suzuki Y, Freeman M, Hediger M, Kovacs G (2013) Zinc transporters in prostate cancer. *Molecular Aspects of Medicine* 34:735–741.

Gajdács M, Behzadi P (2020) Introductory chapter: cytokines-the diamonds and pearls of biological systems cytokines:1. doi: 10.5772/intechopen.93197.

Gandaglia G et al. (2013) The role of chronic prostatic inflammation in the pathogenesis and progression of benign prostatic hyperplasia (BPH). *BJU International* 112:432–441.

Gandhi J et al. (2018) The molecular biology of prostate cancer: current understanding and clinical implications. *Prostate Cancer and Prostatic Diseases* 21:22–36.

Gast A et al. (2011) Association of inherited variation in toll-like receptor genes with malignant melanoma susceptibility and survival. *PLoS One* 6:e24370.

Gilany K, Minai-Tehrani A, Savadi-Shiraz E, Rezadoost H, Lakpour N (2015) Exploring the human seminal plasma proteome: an unexplored gold mine of biomarker for male infertility and male reproduction disorder. *Journal of Reproduction & Infertility* 16:61.

Gomaz A, Pavelić J, Matijević Glavan T (2012) The polymorphisms in toll-like receptor genes and cancer risk. *Periodicum Biologorum* 114:461–469.

González-Reyes S et al. (2010) Study of TLR3, TLR4 and TLR9 in breast carcinomas and their association with metastasis. *BMC Cancer* 10:1–9.

González-Reyes S et al. (2011) Study of TLR3, TLR4, and TLR9 in prostate carcinomas and their association with biochemical recurrence. *Cancer Immunology, Immunotherapy* 60:217–226.

Gu J, Liu Y, Xie B, Ye P, Huang J, Lu Z (2018) Roles of toll-like receptors: from inflammation to lung cancer progression. *Biomedical Reports* 8:126–132.

Han IH, Kim JH, Jang KS, Ryu JS (2019) Inflammatory mediators of prostate epithelial cells stimulated with Trichomonas vaginalis promote proliferative and invasive properties of prostate cancer cells. *The Prostate* 79:1133–1146.

Heidarzadeh M, Roodbari F, Hassanpour M, Ahmadi M, Saberianpour S, Rahbarghazi R (2020) Toll-like receptor bioactivity in endothelial progenitor cells. *Cell and Tissue Research*:1–8.

Henry GH et al. (2018) A cellular anatomy of the normal adult human prostate and prostatic urethra. *Cell Reports* 25:-3530–3542. e3535.

Hernandez-Garcia CM, Finer JJ (2014) Identification and validation of promoters and cis-acting regulatory elements. *Plant Science* 217:109–119.

Hirbod-Mobarakeh A et al. (2020) Immunogenetics of cancer. In: Nima Rezaei (ed.), *Cancer Immunology*. Switzerland: Springer, pp. 417–478.

Hosseini AM, Majidi J, Baradaran B, Yousefi M (2015) Toll-like receptors in the pathogenesis of autoimmune diseases. *Advanced Pharmaceutical Bulletin* 5:605.

Humphrey PA (2012) Histological variants of prostatic carcinoma and their significance. *Histopathology* 60:59–74.

Jian Z, Ye D, Chen Y, Li H, Wang K (2018) Sexual activity and risk of prostate cancer: a dose–response meta-analysis. *The Journal of Sexual Medicine* 15:1300–1309.

Kalinska M, Meyer-Hoffert U, Kantyka T, Potempa J (2016) Kallikreins–the melting pot of activity and function. *Biochimie* 122:270–282.

Katongole P, Sande OJ, Joloba M, Reynolds SJ, Niyonzima N (2020) The human microbiome and its link in prostate cancer risk and pathogenesis. *Infectious Agents and Cancer* 15:1–8.

Kawasaki T, Kawai T (2014) Toll-like receptor signaling pathways. *Frontiers in Immunology* 5:461.

Kazma R et al. (2012) Association of the innate immunity and inflammation pathway with advanced prostate cancer risk. *PLoS One* 7:e51680.

Keshavarz A, Pourbagheri-Sigaroodi A, Zafari P, Bagheri N, Ghaffari SH, Bashash D (2021) Toll-like receptors (TLRs) in cancer; with an extensive focus on TLR agonists and antagonists. *IUBMB Life* 73:10–25.

Khan A, Khan Z, Warnakulasuriya S (2016) Cancer-associated toll-like receptor modulation and insinuation in infection susceptibility: association or coincidence? *Annals of Oncology* 27:984–997.

Kim HJ et al. (2012) Sequence variants of toll-like receptor 4 (TLR4) and the risk of prostate cancer in Korean men. *World Journal of Urology* 30:225–232.

Koryakina Y, Ta HQ, Gioeli D (2014) Androgen receptor phosphorylation: biological context and functional consequences. *Endocrine-Related Cancer* 21:T131–T145.

Kucera R, Pecen L, Topolcan O, Dahal AR, Costigliola V, Giordano FA, Golubnitschaja O (2020) Prostate cancer management: long-term beliefs, epidemic developments in the early twenty-first century and 3PM dimensional solutions. *EPMA Journal*:1–20.

Kumar A et al. (2016) Substantial interindividual and limited intraindividual genomic diversity among tumors from men with metastatic prostate cancer. *Nature Medicine* 22:369–378.

Kutikhin AG (2011) Association of polymorphisms in TLR genes and in genes of the toll-like receptor signaling pathway with cancer risk. *Human Immunology* 72:1095–1116.

Kutikhin AG, Yuzhalin AE (2012) Are toll-like receptor gene polymorphisms associated with prostate cancer? *Cancer Management and Research* 4:23.

Kwon O-J, Zhang L, Ittmann MM, Xin L (2014) Prostatic inflammation enhances basal-to-luminal differentiation and accelerates initiation of prostate cancer with a basal cell origin. *Proceedings of the National Academy of Sciences* 111:E592–E600.

Leifer CA, Medvedev AE (2016) Molecular mechanisms of regulation of toll-like receptor signaling. *Journal of Leukocyte Biology* 100:927–941.

Li D, Jin Y, Sun Y, Lei J, Liu C (2014) Knockdown of toll-like receptor 4 inhibits human NSCLC cancer cell growth and inflammatory cytokine secretion in vitro and in vivo. *International Journal of Oncology* 45:813–821.

Lin A, Loré K (2017) Granulocytes: new members of the antigen-presenting cell family. *Frontiers in Immunology* 8:1781.

Lin B, Qing X, Liao J, Zhuo K (2020) Role of protein glycosylation in host-pathogen interaction. *Cells* 9:1022.

Liu L, Botos I, Wang Y, Leonard JN, Shiloach J, Segal DM, Davies DR (2008) Structural basis of toll-like receptor 3 signaling with double-stranded RNA. *Science* 320:379–381.

Livingstone TL, Beasy G, Mills RD, Plumb J, Needs PW, Mithen R, Traka MH (2019) Plant bioactives and the prevention of prostate cancer: evidence from human studies. *Nutrients* 11:2245.

Luo L, Lucas RM, Liu L, Stow JL (2020) Signalling, sorting and scaffolding adaptors for toll-like receptors. *Journal of Cell Science* 133. doi: 10.1242/jcs.239194.

Ma X (2019) The microbiome of prostate fluid is associated with prostate cancer. *Frontiers in Microbiology* 10:1664.

Mäkinen LK et al. (2015) Predictive role of toll-like receptors 2, 4, and 9 in oral tongue squamous cell carcinoma. Oral oncology 51:96–102.

Medvedev AE (2013) Toll-like receptor polymorphisms, inflammatory and infectious diseases, allergies, and cancer. *Journal of Interferon & Cytokine Research* 33:467–484.

Merriel SW, Funston G, Hamilton W (2018) Prostate cancer in primary care. *Advances in Therapy* 35:1285–1294.

Messaritakis I et al. (2018) Evaluation of the detection of toll-like receptors (TLRs) in cancer development and progression in patients with colorectal cancer. *PLoS One* 13:e0197327.

Minmin S et al. (2011) Single nucleotide polymorphisms of toll-like receptor 4 decrease the risk of development of hepatocellular carcinoma. *PLoS One* 6:e19466.

Mokhtari Y, Pourbagheri-Sigaroodi A, Zafari P, Bagheri N, Ghaffari SH, Bashash D (2020) Toll-like receptors (TLRs): an old family of immune receptors with a new face in cancer pathogenesis. *Journal of Cellular and Molecular Medicine* 25(2):639–651.

Montoro DT et al. (2018) A revised airway epithelial hierarchy includes CFTR-expressing ionocytes. *Nature* 560:319–324.

Mukherjee S, Huda S, Sinha Babu SP (2019) Toll-like receptor polymorphism in host immune response to infectious diseases: a review. *Scandinavian Journal of Immunology* 90:e12771.

Nieto CM, Rider LC, Cramer SD (2014) Influence of stromal–epithelial interactions on androgen action. *Endocrine-Related Cancer* 21:T147–T160.

Nissar S, Sameer AS, Rasool R, Qadri Q, Chowdri NA, Rashid F (2017) Role of TLR4 gene polymorphisms in the colorectal cancer risk modulation in ethnic Kashmiri population–a case–control study. *Egyptian Journal of Medical Human Genetics* 18:53–59.

Ohadian Moghadam S, Nowroozi MR (2019) Toll-like receptors: the role in bladder cancer development, progression and immunotherapy. *Scandinavian Journal of Immunology* 90:e12818.

Ohto U, Fukase K, Miyake K, Shimizu T (2012) Structural basis of species-specific endotoxin sensing by innate immune receptor TLR4/MD-2. *Proceedings of the National Academy of Sciences* 109:7421–7426.

Packer JR, Maitland NJ (2016) The molecular and cellular origin of human prostate cancer. *Biochimica et Biophysica Acta (BBA)-Molecular Cell Research* 1863:1238–1260.

Pandey N, Chauhan A, Jain N (2018) TLR4 polymorphisms and expression in solid cancers. *Molecular Diagnosis & Therapy* 22:683–702.

Parsons JK (2010) Benign prostatic hyperplasia and male lower urinary tract symptoms: epidemiology and risk factors. *Current Bladder Dysfunction Reports* 5:212–218.

Peisch SF, Van Blarigan EL, Chan JM, Stampfer MJ, Kenfield SA (2017) Prostate cancer progression and mortality: a review of diet and lifestyle factors. *World Journal of Urology* 35:867–874.

Pocnet C, Dupuis M, Congard A, Jopp D (2017) Personality and its links to quality of life: mediating effects of emotion regulation and self-efficacy beliefs. *Motivation and Emotion* 41:196–208.

Porter CM, Shrestha E, Peiffer LB, Sfanos KS (2018) The microbiome in prostate inflammation and prostate cancer. *Prostate Cancer and Prostatic Diseases* 21:345.

Prasad SB, Kumar R (2020) Role of toll-like receptor (TLR)-signaling in cancer progression and treatment. In: Bhawana Singh (ed.), *Cell Interaction-Regulation of Immune Responses, Disease Development and Management Strategies*. Croatia: IntechOpen.

Priyadarshini A, Chakraborti A, Mandal AK, Singh SK (2013) Asp299Gly and Thr399Ile polymorphism of TLR-4 gene in patients with prostate cancer from North India. Indian *Journal of Urology: IJU: Journal of the Urological Society of India* 29:37.

Prostate. (2020) Farlex. https://www.thefreedictionary.com/prostate. Accessed December 21 2020.

Qiu Y, Ding Y, Zou L, Tan Z, Liu T, Fu X, Xu W (2013) Divergent roles of amino acid residues inside and outside the BB loop affect human toll-like receptor (TLR) 2/2, TLR2/1 and TLR2/6 responsiveness. *PLoS One* 8:e61508.

Qiu Y, Zheng J, Yang J, Li F, Zhou X, Song X (2019) The predictive role of toll-like receptor-4 genetic polymorphisms in susceptibility to and prognosis of prostatic hyperplasia. *Iranian Journal of Basic Medical Sciences* 22:86.

Rawla P (2019) Epidemiology of prostate cancer. *World Journal of Oncology* 10:63.

Rebbeck TR (2017) Prostate cancer genetics: variation by race, ethnicity, and geography. In: Jason Efstathiou (ed.), *Seminars in Radiation Oncology*, vol 1. Elsevier, pp. 3–10.

Richman EL, Kenfield SA, Stampfer MJ, Giovannucci EL, Chan JM (2011) Egg, red meat, and poultry intake and risk of lethal prostate cancer in the prostate-specific antigen-era: incidence and survival. *Cancer Prevention Research* 4:2110–2121.

Richman EL, Kenfield SA, Stampfer MJ, Giovannucci EL, Zeisel SH, Willett WC, Chan JM (2012) Choline intake and risk of lethal prostate cancer: incidence and survival. *The American Journal of Clinical Nutrition* 96:855–863.

Robinson D et al. (2015) Integrative clinical genomics of advanced prostate cancer. *Cell* 161:1215–1228.

Rogers EN et al. (2013) Toll-like receptor-associated sequence variants and prostate cancer risk among men of African descent. *Genes & Immunity* 14:347–355.

Ruiz-Ballesteros AI, Meza-Meza MR, Vizmanos-Lamotte B, Parra-Rojas I, de la Cruz-Mosso U (2020) Association of vitamin D metabolism gene polymorphisms with autoimmunity: evidence in population genetic studies. *International Journal of Molecular Sciences* 21:9626.

Rycaj K et al. (2016) Longitudinal tracking of subpopulation dynamics and molecular changes during LNCaP cell castration and identification of inhibitors that could target the PSA–/lo castration-resistant cells. *Oncotarget* 7:14220.

Sadeghi-Gandomani H et al. (2017) The incidence, risk factors, and knowledge about the prostate cancer through worldwide and Iran. *World Cancer Research Journal* 4:1–8.

Saghazadeh A, Rezaei N (2020) Introductory chapter: toll-like receptors:1. doi: 10.5772/intechopen.88493.

Sameer A.S. BMZ, Nissar S (2021a) Mutations and polymorphisms: what is the difference? In: Sameer A.S. BMZ, Nissar S (eds) *Genetic Polymorphism and Cancer Susceptibility*, vol I. 1st edn. Springer, Singapore, pp. 1–21. doi: 10.1007/978–981-33-6699-2_1.

Sameer AS, Nissar S (2021b) Toll-like receptors (TLRs): structure, functions, signaling, and role of their polymorphisms in colorectal cancer susceptibility. *BioMed Research International* 2021:1157023.

Saranyutanon S, Deshmukh SK, Dasgupta S, Pai S, Singh S, Singh AP (2020) Cellular and molecular progression of prostate cancer: models for basic and preclinical research. *Cancers* 12:2651.

Sato K et al. (2012) A single nucleotide polymorphism in 3′-untranslated region contributes to the regulation of toll-like receptor 4 translation. *Journal of Biological Chemistry* 287:25163–25172.

Schweizer MT, Evan YY (2015) Persistent androgen receptor addiction in castration-resistant prostate cancer. *Journal of Hematology & Oncology* 8:1–14.

Semlali A, Alnemari R, Almalki E, Alrashed R, Alanazi M (2018) Toll-like receptors gene polymorphism and susceptibility to cancer development. *Genetic Diversity and Disease Susceptibility* 57:57–72.

Semlali A et al. (2017) Toll-like receptor 4 as a predictor of clinical outcomes of estrogen receptor-negative breast cancer in Saudi women. *OncoTargets and Therapy* 10:1207.

Seok H, Lee B-C, Kim YO, Chung J-H (2013) Association between exonic SNPs of TLR5 and benign prostate hyperplasia in Korean population. 스트레스研究 21:331–337.

Sepehri Z, Kiani Z, Nasiri AA, Kohan F (2016) Toll-like receptor 2 and type 2 diabetes. *Cellular & Molecular Biology Letters* 21:1–9.

Shao N, Li J, Xu B, Wang Y, Lu X, Feng N (2014) Role of the functional variant (– 652T> G) in the XRCC4 promoter in prostate cancer. *Molecular Biology Reports* 41: 7463–7470.

Sharma M, Gupta S, Dhole B, Kumar A (2017) The prostate gland. In: Anand Kumar, Mona Sharma (eds.), *Basics of Human Andrology*. Switzerland: Springer, pp. 17–35.

Shui IM et al. (2012) Genetic variation in the toll-like receptor 4 and prostate cancer incidence and mortality. *The Prostate* 72:209–216.

Slattery ML, Herrick JS, Bondurant KL, Wolff RK (2012) Toll-like receptor genes and their association with colon and rectal cancer development and prognosis. *International Journal of Cancer* 130:2974–2980.

Stoyanova T et al. (2013) Prostate cancer originating in basal cells progresses to adenocarcinoma propagated by luminal-like cells. *Proceedings of the National Academy of Sciences* 110:20111–20116.

Su L et al. (2019) Structural basis of TLR2/TLR1 activation by the synthetic agonist diprovocim. *Journal of Medicinal Chemistry* 62:2938–2949.

Sun J et al. (2005) Sequence variants in toll-like receptor gene cluster (TLR6-TLR1-TLR10) and prostate cancer risk. *Journal of the National Cancer Institute* 97:525–532.

Suspected cancer: recognition and referral (2015) https://www.nice.org.uk/guidance/ng12.

Tennessen JA et al. (2012) Evolution and functional impact of rare coding variation from deep sequencing of human exomes. *Science* 337:64–69.

Theodoropoulos GE et al. (2012) Toll-like receptors gene polymorphisms may confer increased susceptibility to breast cancer development. *The Breast* 21:534–538.

Tian S et al. (2018) The associations between toll-like receptor 9 gene polymorphisms and cervical cancer susceptibility. *Mediators of Inflammation* 2018. doi: 10.1155/2018/9127146.

Toivanen R, Mohan A, Shen MM (2016) Basal progenitors contribute to repair of the prostate epithelium following induced luminal anoikis. *Stem Cell Reports* 6:660–667.

Tonry C, Finn S, Armstrong J, Pennington SR (2020) Clinical proteomics for prostate cancer: understanding prostate cancer pathology and protein biomarkers for improved disease management. *Clinical Proteomics* 17:1–31.

Verze P, Cai T, Lorenzetti S (2016) The role of the prostate in male fertility, health and disease. *Nature Reviews Urology* 13:379.

Vijay K (2018) Toll-like receptors in immunity and inflammatory diseases: past, present, and future. *International Immunopharmacology* 59:391–412.

Vlajnic T, Bubendorf L (2020) Molecular pathology of prostate cancer: a practical approach. *Pathology* 53(1):36–43.

Wang Q, Zhang X, Xiao T, Pan C, Liu X, Zhao Y (2018) Prognostic role of toll-like receptors in cancer: a meta-analysis. *Therapeutics and Clinical Risk Management* 14:1323.

Wang Y et al. (2020) Small-molecule modulators of toll-like receptors. *Accounts of Chemical Research* 53(5):1046–1055.

Wang ZA, Toivanen R, Bergren SK, Chambon P, Shen MM (2014) Luminal cells are favored as the cell of origin for prostate cancer. *Cell Reports* 8:1339–1346.

Whitaker H, Tam JO, Connor MJ, Grey A (2020) Prostate cancer biology & genomics. *Translational Andrology and Urology* 9:1481.

Xin L (2019) Cells of origin for prostate cancer. In: Scott M. Dehm, Donald J. Tindall (eds.), *Prostate Cancer*. Cham: Springer, pp. 67–86.

Yang Y et al. (2019) AWESOME: a database of SNPs that affect protein post-translational modifications. *Nucleic Acids Research* 47:D874–D880.

Yang ZH, Dai Q, Gu YJ, Guo QX, Gong L (2012) Cytokine and chemokine modification by toll-like receptor polymorphisms is associated with nasopharyngeal carcinoma. *Cancer Science* 103:653–658.

Yates CM, Sternberg MJ (2013) The effects of non-synonymous single nucleotide polymorphisms (nsSNPs) on protein–protein interactions. *Journal of Molecular Biology* 425:3949–3963.

Yuan X et al. (2013) Activation of TLR4 signaling promotes gastric cancer progression by inducing mitochondrial ROS production. *Cell Death & Disease* 4:e794–e794.

Zeng H-M et al. (2011) Genetic variants of toll-like receptor 2 and 5, helicobacter pylori infection, and risk of gastric cancer and its precursors in a Chinese population. *Cancer Epidemiology and Prevention Biomarkers* 20:2594–2602.

Zhang X, Bailey SD, Lupien M (2014a) Laying a solid foundation for Manhattan–'setting the functional basis for the post-GWAS era'. *Trends in Genetics* 30:140–149.

Zhang Y, Wang Q, Ma A, Li Y, Li R, Wang Y (2014b) Functional expression of TLR9 in esophageal cancer. *Oncology Reports* 31:2298–2304.

Zhao S, Zhang Y, Zhang Q, Wang F, Zhang D (2014) Toll-like receptors and prostate cancer. *Frontiers in Immunology* 5:352.

Zidi S, Sghaier I, Gazouani E, Mezlini A, Yacoubi-Loueslati B (2016) Evaluation of toll-like receptors 2/3/4/9 gene polymorphisms in cervical cancer evolution. *Pathology & Oncology Research* 22:323–330.

18

Role of Single Nucleotide Polymorphism of Cytochrome P450 *Genes in Cancer Susceptibility*

Sachin Kumar and Rajiv Kumar Tonk
Department of Pharmacology, Delhi Pharmaceutical Sciences and Research University, Govt. of NCT of Delhi, New Delhi, India

Faheem Hyder Pottoo
Department of Pharmacology, College of Clinical Pharmacy, Imam Abdul Rahman Bin Faisal University, Dammam, Saudi Arabia

CONTENTS

ABSTRACT Genetic polymorphisms in *cytochrome P450 (CYP)* can have a leading role in disease development besides the usual capability of metabolizing drugs. The members of the CYP enzyme family vary in their primary structures, tissue distribution, substrate specificities, and immunological cross-reactivities. Mutations, for instance, single nucleotide polymorphisms (SNPs), in a *CYP* gene can modify the activity of the expressed CYP enzyme; for example, its functionality of rendering the lipophilic xenobiotics more hydrophilic may increase or decrease. The alteration in the genotypes can result in four possible phenotypes other than the normal, i.e., poor metabolizer, intermediate metabolizer, extensive metabolizer, and ultra-rapid metabolizer. Another aspect of the vulnerability of the SNP of *CYP* as a risk factor for cancer is the disposition of a particular SNP in a population and the environment of the population. In this chapter, we discuss some of the CYP isoforms associated with cancer. Moreover, we discuss the role of SNPs in *CYP* in lung, breast, liver, gastric, colorectal, prostate, and head-and-neck cancers.

18.1 Introduction

Humans are constantly exposed to different sorts of chemicals, food additives, drugs, and environmental pollutants that could eventually impair the homeostasis of the body and consequently produce undesirable health issues. Among various defense systems present in the human body, cytochrome P450 (CYP) enzymes have a pertinent role in the protection against toxic chemicals (Guengerich et al. 2016; Zanger and Schwab 2013). CYP constitutes a superfamily of hemeproteins comprising 400–500 amino acid residues and a heme prosthetic group at the active site. The CYP enzymes are membrane-bound proteins that are ubiquitously found in bacteria, fungi, plants, and animals. It has been documented that, in human beings, at least 57 different *CYP* genes and 58 pseudogenes translate for 18 different families and 44 subfamilies of CYP enzymes for performing essential roles in cellular metabolism and homeostasis besides the detoxification of xenobiotics (Rendic and Guengerich 2015).

DOI: 10.1201/9781003246244-18

18.2 Classification of CYP Enzymes

The accepted nomenclature of CYP enzymes has been established by CYP Nomenclature Committee (Nelson et al. 1996, 2004). The term CYP denotes a heme-containing protein having maximum absorption, in the reduced state, at the wavelength of 450 nm in carbon monoxide environment (Omura 1999). The CYP family is represented by an Arabic numeral following "CYP" (e.g., CYP1, CYP2, and CYP3). The Arabic numeral is accompanied by a letter and another Arabic numeral, which represent the subfamily (e.g., CYP1A) and the individual gene/isoenzyme/isozyme/isoform (e.g., CYP1A1), respectively. While the gene and the cDNA are written in italics (*CYP1A1*), the mRNA and the protein are denoted in the regular uppercase font (CYP1A1).

Different members of the CYP family vary in their primary structures, tissue distribution, substrate specificities, and immunological cross-reactivities. The homology of amino acid sequences between the family and subfamily members is 40% and 55%, respectively (e.g., CYP2A6 and CYP2A7) (Guengerich et al. 2016; Nelson et al. 1996, 2004). The enzymes in the families 1–4 (CYP1A2, 2C9, 2C19, 2D6, 2E1, and 3A4) are effectual in xenobiotics metabolism, while the other families have exhibited association with significant endogenous functions (Guengerich et al. 2016; Guengerich 2008; Rendic and Guengerich 2015) (Table 18.1).

In mammals, however, CYP enzymes are found in all tissues, but are highly concentrated in the small intestine and liver. The membrane-bound enzymes, abundant in the microsomal fraction of the liver, perform essential functions such as bile acid biosynthesis and metabolism of exogenous compounds such as drugs, procarcinogens, carcinogens, and ecological pollutants (Guengerich et al. 2016; Hasler et al. 1999). CYP enzymes present on the inner mitochondrial membrane of steroidogenic tissues such as the breast, ovary, testis, adrenal cortex, and placenta have a role in the production and breakdown of endogenous steroid hormones. Furthermore, CYP enzymes are required for cholesterol synthesis, oxidation of unsaturated fatty acids, and metabolism of vitamins (Hasler et al. 1999; Omura 1999). It has been documented that CYP enzymes regulate the functioning of $GABA_A$ receptor, cholesterol homeostasis, and retinoids elimination in the brain (Hedlund et al. 2001). Thus, CYP enzymes have a crucial role in the regulation of cell metabolism and homeostasis (Hasler et al. 1999; Hedlund et al. 2001; Omura 1999).

There are different viewpoints to classify CYP enzymes. In the typical sense, CYP enzymes are assembled into two major groups – they either carry out the detoxification of xenobiotics, or effectuate the biosynthesis of endogenous molecules. From the solubility point of view, CYP enzymes are also categorized into insoluble membrane-bound kinds that are customarily observed in eukaryotes and solubilized types that are spotted in prokaryotes (Manikandan and Nagini 2017; Porter and Coon 1991). Another perspective for classification utilizes the localization of CYP enzymes and the electron transfer process as the basis. In eukaryotes, CYP enzymes that are predominantly situated in the endoplasmic reticulum and a membrane-bound CYP reductase, comprised of FAD/FMN, carry out the reduction. On the contrary, in prokaryotes and mitochondrial CYP enzymes, the electron transfer process is executed by a soluble two-component reductase setup having FAD and iron-sulfur protein (McLean et al. 2005). Lately, CYP enzymes have been classified based on diversity in their functionality as moonlighting and non-moonlighting proteins. A *moonlighting protein* accomplishes several autonomous and oft-times unrelated functions such as the biosynthesis of endogenous compounds. Non-moonlighting proteins are those that carry out the xenobiotics metabolism. The moonlighting CYP enzymes play diverse roles in divergent tissues (*e.g.*, CYP7B1 and CYP17) or have manifold catalytic sites (CYP170A1) (Jeffery 1999; Zhao and Waterman 2011). For example, CYP7B1 is an archetype moonlighting enzyme that is associated with the steroidal hormone metabolism, synthesis of bile salts, regulation of estrogen receptor ligand, and production of antibodies in the brain, liver, prostate, and immune cells, respectively. In contrast, CYP170A1, a dual functional protein, has two categorical active sites for catalyzing monooxygenase and synthesizing terpene (Jeffery 1999; Zhao and Waterman 2011).

18.3 Mechanism of Working of CYP Enzymes

The general structure of CYP enzymes bears a resemblance to an inverted triangle (Elfaki et al. 2018). The CYP enzyme comprises a protein moiety and a prosthetic group, iron protoporphyrin IX or heme (Ortiz de Montellano 2010). CYP enzymes catalyzed the metabolism of the lipophilic xenobiotics or endogenous molecules by converting them into polar or hydrophilic compounds so as to facilitate their excretion through the kidneys (Chang and Kam 1999). The substrate interacts with the active site of the CYP enzyme, which consists of heme iron. The heme iron is reduced to the ferrous state in the presence of an electron donor, that is reduced NADPH (Chang and Kam 1999). The ferrous state heme temporarily accepts molecular oxygen (Zanger and Schwab 2013), and later, one atom of oxygen atom is incorporated into the substrate molecule, while the other atom is combined with hydrogen to form water. Therefore, CYP enzymes are categorized as monooxygenases as they combined only one atom of the molecular oxygen with their substrates (Ortiz de Montellano 2010). The reaction for the monooxygenase is presented in Figure 18.1, where RH is the substrate that can accept the hydroxyl group.

In particular, CYP families 1–4 catalyze reactions associated with the detoxification of drugs and xenobiotics. However, phase I CYP enzymes catalyze the transformation of xenobiotics or procarcinogens into electrophilic moieties (an eventual carcinogen) and the phase II reactions convert these electrophilic molecules into non-toxic compounds with help of other enzymes such as glutathione S-transferases and quinone reductase. Nevertheless, the reactive electrophilic intermediates that slip away from phase

$$\mathrm{RH + NAD(P)H + O_2 + H^+ \longrightarrow ROH + NAD(P) + H_2O}$$

FIGURE 18.1 The reaction for the monooxygenase.

TABLE 18.1

Different CYP Enzymes with Their Substrates, Inhibitors, and Effects

CYP	Endogenous Substrate	Drug Substrate	Inhibitors	Procarcinogen Activation	Effects	References
1A1	Melatonin, steroid hormone, polyunsaturated fatty acids	Acetaminophen, granisetron, clozapine, warfarin	Quercetin, resveratrol	Yes	Anti-estrogenic, metabolism of arachidonic acids and eicosanoids	Mescher and Haarmann-Stemmann (2018)
1A2	Melatonin, bilirubin, estrone, estradiol, uroporphyrinogen	Olanzapine, clozapine, haloperidol	Cimetidine, amiodarone, fluoroquinolones, fluvoxamine, interferon	Yes		Gunes and Dahl (2008), Na Takuathung et al. (2019)
2A6	Trans-retinoic acid, testosterone, estradiol	Nicotine, cotinine, tegafur, letrozole, efavirenz, valproic acid, pilocarpine, artemisinin	Methoxsalen, selegiline, tranylcypromine, menthol, grapefruit juice, caffeic acid, p-coumaric acid, quercetin, cinnamaldehyde	Yes	May contribute to short- and long-term toxicity due to the metabolites	Di et al. (2009), Raunio and Rahnasto-Rilla (2012), Tanner and Tyndale (2017)
2A13		Nicotine, coumarin, phenacetin, theophylline	Bacterial endotoxin lipopolysaccharide, kaempferol	Yes	Expression is markedly increased in non-small cell lung carcinomas, especially adenocarcinoma	Boonruang et al. (2020), Goto et al. (2010), Wu et al. (2013)
2B6	Testosterone	Cyclophosphamide, ifosfamide, propofol, ketamine, pethidine, nevirapine, sertraline, efavirenz, methadone, bupropion	Orphenadrine, 9-ethynylphenanthrene, methoxychlor	Yes	Affect drug metabolism and alter responses to certain drugs	Ekins and Wrighton (1999), Hedrich et al. (2016), Hodgson and Rose (2007)
2C8	Arachidonic acid, eicosapentaenoic acid	Amodiaquine, cerivastatin, dasabuvir, enzalutamide, imatinib, loperamide, montelukast, paclitaxel, pioglitazone, repaglinide, rosiglitazone	Gemfibrozil, montelukast, candesartan, cilexetil, zafirlukast, clotrimazole, felodipine, mometasone furoate	Yes	Highly expressed in the human liver and known to metabolize more than 100 drugs	Backman et al. (2016), van Eijk et al. (2019)
2C9	Steroids, melatonin, retinoids, arachidonic acid	Tolbutamide, glyburide, diclofenac, celecoxib, torasemide, phenytoin, losartan, S-warfarin	Histamine, tienilic acid, suprofen, silybin	No	Decreased enzyme activity along with drugs having narrow therapeutic indices results in safety and efficacy concerns	Zhou et al. (2009)
2C19	Estradiol, progesterone, testosterone, estrone	Clopidogrel, diazepam, omeprazole, citalopram, amitriptyline, clomipramine, doxepin, imipramine, trimipramine	Cannabidiol, ketoconazole, omeprazole, indomethacin	No	Isoform of the enzyme might alter the safety and efficacy of the substrate drugs	Elfaki et al. (2018)
2D6	Hydroxytryptamines, neurosteroids	Tamoxifen, sparteine, dextromethorphan, debrisoquine	Bupropion	No	The magnitude of drug-drug interaction and drug-gene interaction depends on the number of functional CYP alleles	Bahar et al. (2017)
2E1	Acetone, fatty acids	Ethanol, nicotine, acetaminophen, aspartame, chloroform, chlorzoxazone, tetrachloride, phenobarbital, isoflurane, enflurane	Diallyl sulfide, phenethyl isothiocyanate, chlormethiazole	Yes	Involved in the bioactivation and detoxification of several chemical carcinogens	Kessova and Cederbaum (2003), Peter Guengerich and Avadhani (2018)

(Continued)

TABLE 18.1 (*Continued*)

Different CYP Enzymes with Their Substrates, Inhibitors, and Effects

CYP	Endogenous Substrate	Drug Substrate	Inhibitors	Procarcinogen Activation	Effects	References
2F1		Naphthalene, benzene, styrene	3-Methylindole	Yes	Selectively expressed in lung tissue and involved in the metabolism of various pneumotoxicants with potential carcinogenic effects	Kartha and Yost (2008), Tournel et al. (2007)
2J2	Arachidonic acid	Astemizole, terfenadine	Danazol, telmisartan	Yes	Upregulated in hematological malignancies	Karkhanis et al. (2017)
2R1	Vitamin D			No	Microsomal vitamin D 25-hydroxylase	Chandler et al. (2018), Christakos et al. (2015), Kopp et al. (2020)
2S1	Retinoic acid, prostaglandin G2, prostaglandin H2, thromboxane A2, hydroperoxyeicosatetraenoic acids, 13S-hydroperoxy-octadecadienoic acid	Naphthalene			Metabolize some of the key intermediate eicosanoids of the cyclooxygenase and lipoxygenase pathways and thus might play an important role in modulating the inflammatory process	Bui et al. (2011), Saarikoski et al. (2005)
2W1	Arachidonic acid, trans-retinol, farnesol, lysophosphatidylcholines, lysophospholipids, free fatty acids	Benzothiazole, fluorobenzothiazole, duocarmycin, aflatoxin B1, aromatic amines, polycyclic aromatic hydrocarbons		Yes	Serve as a target in the prodrug-based cancer therapy whereby an inert prodrug can be converted by CYP2W1 into cytotoxic metabolites, harmful only for cancer cells expressing CYP2W1	Guo et al. (2016)
3A4	Cholesterol, cortisol, lithocholic acid, testosterone	Albendazole, amiodarone, caffeine, cyclophosphamide, doxorubicin, etoposide, ketoconazole, omeprazole, clopidogrel, methadone	Ketoconazole, clarithromycin, diltiazem, isoniazid, mifepristone, nicardipine, ritonavir, verapamil, voriconazole	Yes	CYP3A4 is expressed in the liver and small intestine and contribute to the pre-systemic and systemic metabolism of approximately 30% of all drugs	Ou et al. (2019), Waring (2020), Werk and Cascorbi (2014), Zanger and Schwab (2013)
3A5	Cholesterol, cortisol, lithocholic acid, testosterone	Cyclosporine, statins, taxanes, macrolide antibiotics	Itraconazole, fluconazole, voriconazole, miconazole, dronedarone, N-desbutyl dronedarone, mifepristone, nicardipine, ritonavir, verapamil	No	Involved in the metabolism of a number of drugs	Hong et al. (2016), Niwa et al. (2014), Zanger and Schwab (2013)

(*Continued*)

TABLE 18.1 *(Continued)*

Different CYP Enzymes with Their Substrates, Inhibitors, and Effects

CYP	Endogenous Substrate	Drug Substrate	Inhibitors	Procarcinogen Activation	Effects	References
3A7	Dehydroepiandrosterone sulfate, bile acid, progesterone, estrone, estradiol, trans-retinoic acid	Midazolam, tacrolimus, clarithromycin, nifedipine, sildenafil, carbamazepine, cyclosporine, glyburide, aflatoxin B1, meperidine	Triazolam, indinavir, saquinavir, astemizole, nicardipine, terfenadine, verapamil	No	CYP3A7 is by far the most abundant CYP isoform and accounts for up to 50% of the total P450 in embryonic, fetal, and neonatal livers	Lee et al. (2010), Li and Lampe (2019)
3A43	Testosterone	Olanzapine		No	The enzyme is involved in the metabolism of testosterone and might have a role in prostate cancer	Maksymchuk and Kashuba (2020), Stone et al. (2005)
4F2	Arachidonic acid, leukotriene B4, α-tocopherylquinone	Coumarin	Ketoconazole	No	CYP4F2 might influence warfarin dose requirements, possibly through vitamin K1 metabolism, because vitamin E, structurally analogous to vitamin K1, is also catalyzed by CYP4F	Danese et al. (2019), Park et al. (2019), Stec et al. (2007), Taylor et al. (2018)
4A11	Arachidonic acid	Aplidine	Plumbagin, isoliquiritigenin	No	Regulate essential hypertension	Brandon et al. (2007), Choi et al. (2018), Gainer et al. (2005), Wang et al. (2019a)
4A22	-	-	-	-	CYP4A22 is an orphan CYP with unknown function	Durairaj et al. (2019)
4B1	Arachidonic acid	2-Aminofluorene, 2-naphthylamine, benzidine, 4-ipomeanol, valproic acid	-	Yes	CYP4B1 catalyzes the bioactivation of several bladder carcinogens	Jarrar and Lee (2019), Jennings et al. (2018), Kowalski et al. (2019), Sasaki et al. (2008)
17A1	Pregnenolone, 17α-hydroxypregnenolone, progesterone, 17α-hydroxyprogesterone	Aminopyrine, 7,12-dimethylbenze[a] anthracene, aflatoxin B1	Ketoconazole, galeterone, orteronel, abiraterone acetate	Yes	Catalyzes both 17α-hydroxylase and 17,20-lyase activities and is essential for the production of both androgens and glucocorticoids	Kmeťová Sivoňová et al. (2017), Sushko et al. (2014)
19A1	Androstenedione, testosterone, 16α-hydroxydehydroepiandrosterone		Gossypol, apigenin, resveratrol, nicotine, tebuconazole, phthalates, triclosan, methoxychlor	Yes	Converts androgens into estrogens and is required for female sexual development and growth and development in both sexes	Parween et al. (2020), Wang et al. (2019b)

(Continued)

TABLE 18.1 (*Continued*)

Different CYP Enzymes with Their Substrates, Inhibitors, and Effects

CYP	Endogenous Substrate	Drug Substrate	Inhibitors	Procarcinogen Activation	Effects	References
21A2	17α-hydroxyprogesterone		Abiraterone	No	Deficiency of the enzyme is a major reason for congenital adrenal hyperplasia	Malikova et al. (2017), Sumińska et al. (2020)
26A1	Trans-retinoic acid	Tazarotenic acid	Ketoconazole, clotrimazole, miconazole, metyrapone, azolyl retinoids	No	Catalyze hydroxylation of retinoic acid, and the gene expression was inducible by exposure to exogenous retinoic acid	Foti et al. (2016), Isoherranen and Zhong (2019), Njar (2005)

II detoxification arrangements react with macromolecules such as RNA, DNA, and proteins, ultimately leading to adverse drug reactions (ADRs) and pathogenesis (Hollenberg et al. 2008; Manikandan and Nagini 2017). Polycyclic aromatic hydrocarbons (PAHs), chemicals comprising two or more fused benzene rings without any heteroatoms, are common environmental procarcinogens that can stimulate carcinogenesis upon activation by CYP enzymes and other enzymes, for example aldo-keto reductase and epoxide hydrolase (Moorthy et al. 2015; Shimada et al. 2008). The PAHs' instigation leads to the generation of reactive and redox-active diolepoxides, o-quinones, or radical cations (Moorthy et al. 2015), which consequently form DNA adducts upon reacting with DNA. These DNA adducts eventually become the locus for the commencement of carcinogenesis (Moorthy et al. 2015). Reports are available that instigate CYP1B1, expressed in pulmonary and other non-hepatic tissues, for the metabolic transformation of the PAHs (Shimada and Fujii-Kuriyama 2004). The prominent PAHs that are activated by the CYP1B1 and CYP1A1 are 5-methylchrysene, benzo[a]pyrene, benzo[a]pyrene, and 7,12-dimethylbenz[a]anthracene (Shimada et al. 2001).

18.4 Expression and Functioning of CYP Enzymes

A range of features impact the expression and functioning of CYP enzymes. These could be of physiological origin, such as gender, age, hormones, location, epigenetic and genetic polymorphisms, or pathophysiological nature such as inflammation and cancer (Congiu et al. 2009; Harvey and Morgan 2014; Zhang et al. 2011). Other factors such as drug-drug interactions, enzyme induction, or enzyme inhibition also confer the toxicities associated with the CYP-catalyzed reactions (Kumar et al. 2012; Shen and Meng 2012). Enhanced CYP enzymes induction is recognized to cause serious toxicological consequences as a result of the altered rate of xenobiotic metabolism, endocrine interference, drug-drug interactions, and activation of procarcinogens or toxic metabolites (Manikandan and Nagini 2017).

Mutations in a *CYP* gene can modify the activity of the expressed enzymes; for example, it may have increased or decreased functionality. If a particular mutant allele, in a population, exists at a frequency of no less than 1%, it is denoted as a pharmacogenetic polymorphism (Hasler et al. 1999).

The last two decades brought attention to the genetic disparity in *CYP*, which has revealed the frequency of expression of CYP isoenzyme polymorphism in different populations. Several genetic variants of *CYP* have been recognized in humans, and these variants have been assorted into different groups such as copy number variants (CNVs), insertions, inversion, deletion, and single nucleotide variants (SNVs) (Polimanti et al. 2012).

As revealed, more or less all the xenobiotic-metabolizer CYPs are polymorphic, and the allelic variants are abstracted and cataloged on the home page of the Human CYP Allele Nomenclature Committee (see www.PharmVer.org). The existence of different genotypes results in five principal phenotypes (Table 18.2). Particularly, considerable information exists about single nucleotide polymorphisms (SNPs) due to SNVs, which are the prime constituent of human genome variation. All these advanced genetic data could empower us to engineer a personalized medicine strategy based on individual genetic details (Fernald et al. 2011). Therefore, an optimal plan of action for achieving preventive and therapeutic objectives in an individual could be devised by analyzing the specific enzymatic systems.

Excessive polymorphism is observed in CYP families 1–3, having 22 diverse P450 isoforms, as compared to CYP families 4–51, largely accountable for anabolism and catabolism of endogenous compounds. Apropos penetrance, the xenobiotic-metabolizer polymorphic CYP enzymes can be grouped into two (Table 18.2) (Rodriguez-Antona and Ingelman-Sundberg 2006). The case of CYP1B1 is notable, and several of its defective alleles are linked to glaucoma (Libby et al. 2003). The *CYP2C9, CYP2C19*, and *CYP2D6* genes are recognized as the major polymorphic forms responsible for 40% of oxidative drug metabolism (Sadee et al. 2011). Recently, more than 350 functionally polymorphic CYPs have been presented at the CYP allele website. However, the highest numbers of variant alleles are described for *CYP2D6* (63 alleles), *CYP2B6* (28 alleles), *CYP1B1* (26 alleles), and *CYP2A6* (22 alleles) (Ingelman-Sundberg et al. 2007). *CYP* polymorphisms have been associated with the risk of disease development as well as the ability to metabolize drugs (Ingelman-Sundberg et al. 2007; Ingelman-Sundberg 2004; van Schaik 2008). Consequently, the genetic diversity of the Group I P450s does not convert into diverse phenotypes assigning more or less susceptibility to individuals toward environmentally induced

TABLE 18.2

Different Phenotypes of Cytochrome P450 Enzymes Expressed in the Human Body

Phenotype of CYP	Genotype of CYP	Effect
Poor metabolizer	Carrying two defective alleles	Completely lack a certain enzyme
Intermediate metabolizer	Heterozygous for a defect allele and a wild-type allele	Enzyme with a decreased rate of activity
Normal metabolizer	Carrying two wild-type functional alleles	Enzyme with normal rate of activity
Extensive metabolizer	Heterozygous for a wild-type functional allele and a highly active allele	Enzyme with an increased rate of activity
Ultra-rapid metabolizer	Carrying two highly active alleles or more than two active gene copies	Enzyme with the highest rate of activity

TABLE 18.3

Categorization of Cytochrome P450 Enzymes Based on Polymorphism Status and Metabolism of Environmental Procarcinogens

Group	CYPs	Polymorphic Status	Metabolism of Procarcinogen
I	CYP1A1, CYP1A2, CYP2E1, and CYP3A4	No significant functional polymorphism	Active in the metabolism of procarcinogens as well as drugs
II	CYP2A6, CYP2B6, CYP2C9, CYP2C19, and CYP2D6	Extremely polymorphic	Significant for the metabolism of drugs, but not concerned with procarcinogens

TABLE 18.4

Single Nucleotide Polymorphic Forms of *CYP* Associated with Cancers (pharmver.org)

Gene	SNP	rs Identity Number	Nucleotide Change (Reference > Alternate)	Amino Acid Change (Reference Position Alternate)	Gene Variant	Related Cancer
CYP1A1	*2A	4646903	11229T>C	500 b downstream variant	500 b downstream variant	Lung cancer
	*2C	1048943	9885A>G	Isoleucine462Valine	Missense variant	Lung cancer
	*13	4646422	7565G>C or 7565G>A	Glycine45Aspartic acid or Glycine45Alanine	Missense variant	Gastric cancer
CYP2A6	*9 or T-48G	28399433	4974T>G	2 Kb upstream variant	2 Kb upstream variant	Lung cancer
CYP2A13	Exon 5 C/T variant	8192789	8384C>T	Arginine257Cysteine	Missense variant	Lung cancer
	*4	148044792	5588G>A	Arginine101Glycine	Missense variant	Lung cancer
CYP1B1	*2 or l432V	1056836	10042C>G	Leucine432Valine	Missense variant	Lung cancer
CYP1A2	*1F or −163c/c	762551	32035C>A	Intron variant	Intron variant	Breast cancer
CYP19A1	*4	700519	127828C>T	Intron variant	Intron variant	Breast cancer
	-	7176005	4517G>A	2 Kb upstream variant	2 Kb upstream variant	Breast cancer
	-	6493497	4961C>T	2 Kb upstream variant	2 Kb upstream variant	Breast cancer
CYP2C19	*17 or −806C>T	12248560	4220C>T	2 Kb upstream variant	2 Kb upstream variant	Breast cancer
CYP2D6	*10	1065852	4300C>T	Intron variant	Intron variant	Hepatic cancer
CYP2E1	*5 or RsaI	2031920	3979C>T	2 Kb upstream variant	2 Kb upstream variant	Colorectal cancer, gastric cancer

cancer. Alternatively, there are tangible prospects of pinpointing CYP-based genetic biomarkers to forecast therapy outcome taking into consideration ADRs besides therapeutic non-performance due to rapid degradation (Johansson and Ingelman-Sundberg 2011) (Table 18.3).

CYP enzymes are correlated with the etiology of different types of cancer through diverse mechanisms. First, CYP enzymes catalyze the transformation of environmental procarcinogens into highly carcinogenic conversion compounds. Second, they contribute to the generation of reactive oxygen species that could stimulate carcinogenesis. Third, CYP enzymes can alter the pharmacokinetics of various anticancer drugs, which can influence the therapeutic outcomes of the treatment. Fourth, the SNP in the *CYP* gene could result in diverse phenotypes of CYP enzymes, which can increase the production of carcinogens or decrease the elimination of carcinogens. Thus, the polymorphism of the *CYP* gene has potential implications in the etiology of cancer. In this chapter, we discuss the association of SNPs in *CYP* genes with the worldwide death-causing cancers (Table 18.4).

18.5 Role of Cytochrome P450 in Cancers

18.5.1 Lung Cancer

Lung cancer is the foremost reason for cancer-related deaths in men and women, accounting for 1.76 million deaths worldwide in 2018 (Globocan Observatory 2019). In nearly 90% of the cases of lung cancer, tobacco smoking is the sustaining cause, while secondhand smoke and air pollution account for 1%–2% of the cases (Moorthy et al. 2015). While nicotine is the chief psychoactive component of tobacco, other procarcinogens such as nitrosamines are also available in tobacco and tobacco smoke (Figure 18.2). Compounds such as 4-(methylnitrosamino)-1-(3-pyridyl)-1-butanone (NNK) have been acknowledged for being linked with lung cancer. Nicotine dependence extends the carcinogenic exposure, predominantly via CYP2As-mediated metabolic activation of procarcinogens present in tobacco and its smoke, and could stimulate tumorigenesis. CYP2A6 and CYP2A13 (expressed in the respiratory tract) are associated with nicotine metabolism and metabolic activation of NNK (Boonruang et al. 2017).

Various polymorphic forms of the *CYP2A6* gene have been recognized to date, and three alleles, *1, *4, and *9, have been reported to be the chief polymorphisms in the Japanese. *CYP2A6*1* is considered a wild-type allele that expresses a protein with a normal rate of activity. *CYP2A6*4* is a deletion type of the *CYP2A6* gene (Nunoya et al. 1998, 1999). *CYP2A6*9* has an SNP in TATA box, T-48G substitution, which impedes the transcriptions (Pitarque et al. 2001) and, in consequence, the enzyme activity (Kiyotani et al. 2003; Yoshida et al. 2003). Kubota et al. documented, in the Japanese population, that the normal and intermediate metabolizer phenotypes of CYP2A6 are associated with the low blood level of nicotine and serious smoking withdrawal symptoms, which could manifest the nicotine addiction (Kubota et al. 2006). Therefore, an individualized smoking cessation program based on the pharmacogenetic data could be an innovative strategy for reducing the incidence of lung cancer related to smoking.

An SNP of *CYP2A13*, having C/T variation in exon 5, has been recognized, and this SNP resulted in Arg257Cys amino acid change. This 257Cys variant, compared to wild-type 257Arg variant, has been reported to have reduced. Functional analysis showed that the 257Cys variant has significantly moderate activity toward its substrates such as NNK, indicating that this SNP might have a protective role against genotoxicity and carcinogenicity in individuals carrying the *257Cys* allele (Zhang et al. 2002). Also, another report communicated that the poor metabolizer phenotype of *CYP2A13*4* metabolizes NNK and is associated with reduced risk of lung adenocarcinoma, but does not affect the incidence of squamous cell carcinoma (Wang et al. 2003). The conflicting results might be due to overlying substrate specificity and tissue expression of CYP2A6 and CYP2A13 and their communication with other significant metabolizing CYP enzymes.

Other studies have evaluated the influence of genetic polymorphisms in CYP1A1, an enzyme that can metabolize procarcinogens available in tobacco, in the Chinese population (Lin et al. 2014; Song et al. 2001). Various studies have concluded that genotypes *CYP1A1*2A* and *CYP1A1*2C* are positively associated with tobacco smoking-fostered lung cancer (Hussein et al. 2014; Lin et al. 2014; Song et al. 2001).

It has been reported that CYP1B1 is induced by tobacco and can metabolize benzo[a]pyrene (B[a]P), a procarcinogen present in tobacco, into carcinogenic intermediates (Port et al. 2004). A polymorphic form of CYP1B1, in which leucine is replaced by valine at codon 432 (CYP1B1 Leu432Val), has been correlated with increased levels of DNA adducts in white blood cells (Schoket et al. 2001). A research concluded the increased expression of CYP1B1 in the peripheral leukocytes of a lung cancer patient when compared

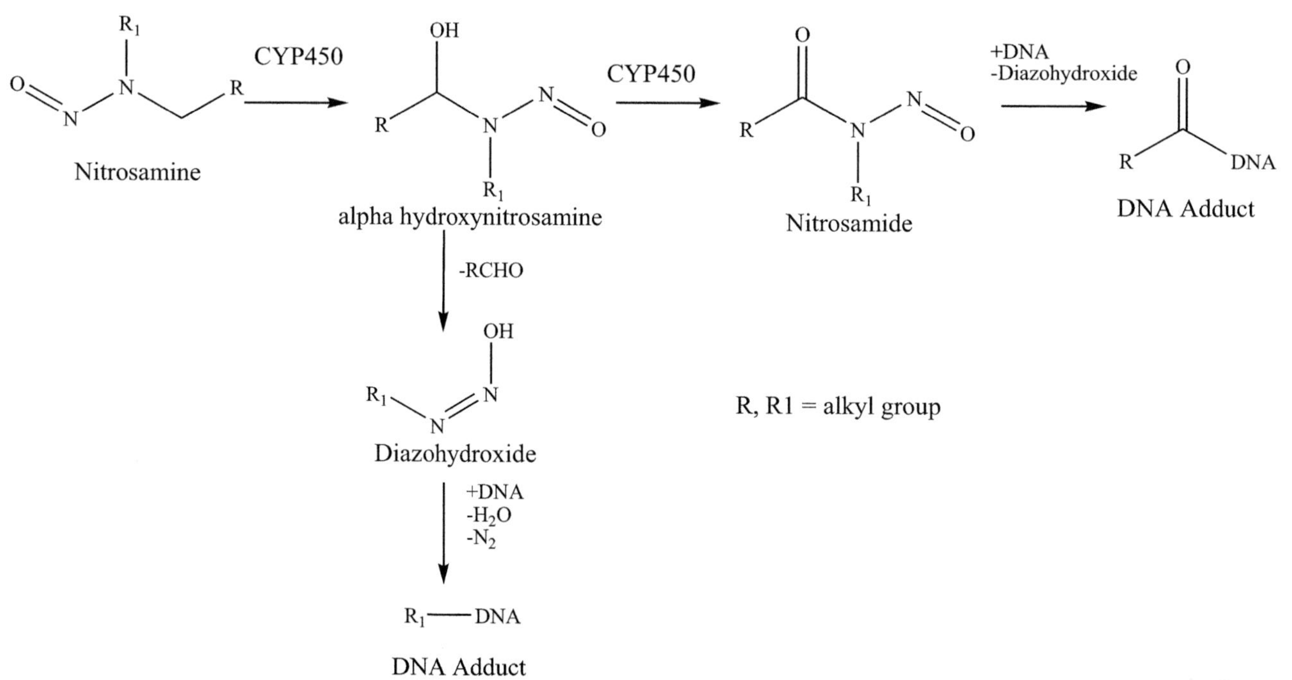

FIGURE 18.2 Metabolism of nitrosamine.

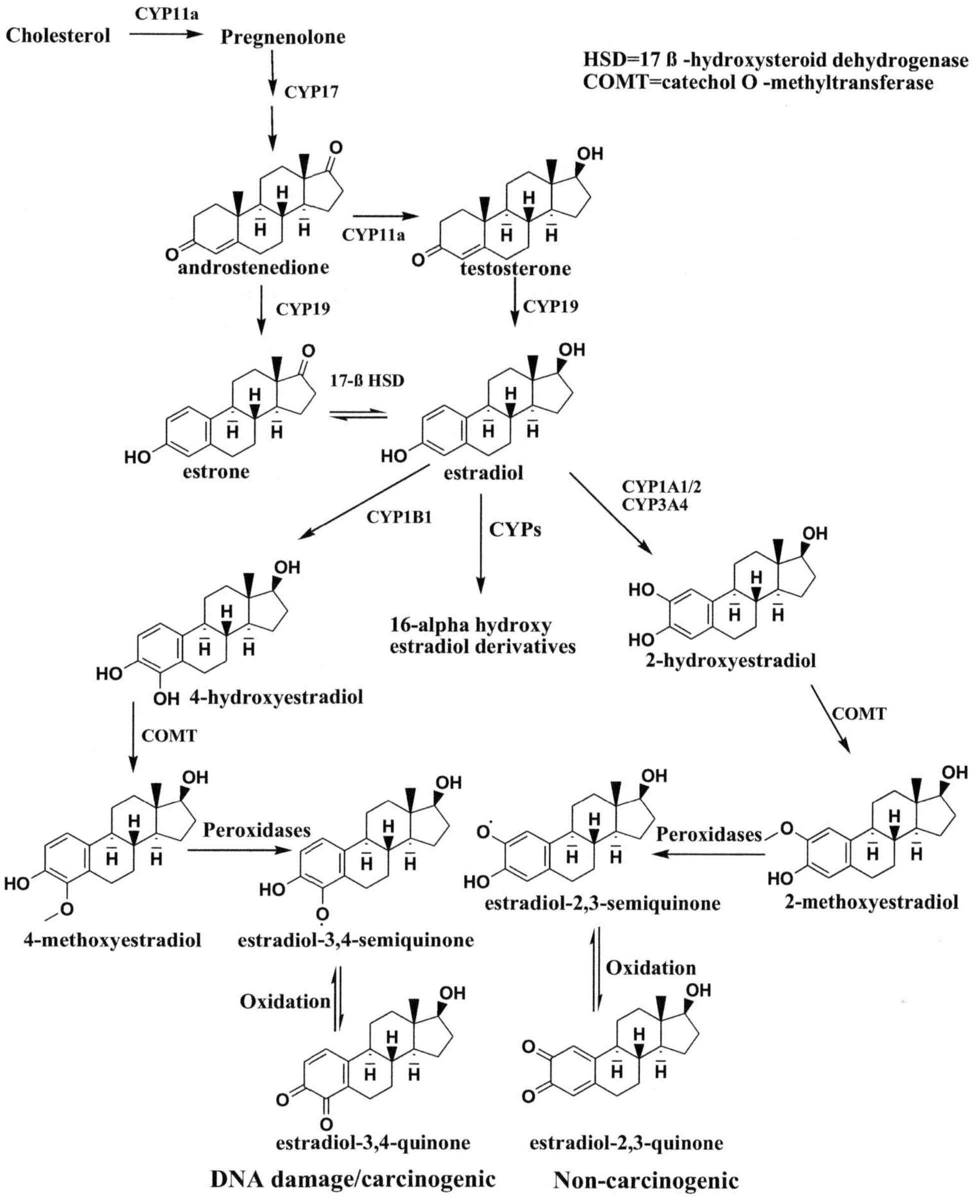

FIGURE 18.3 Hydroxylation of estradiol.

to the normal population (Wu et al. 2004). Wenzlaff et al. studied the correlation between the polymorphic CYP1B1 Leu432Val and lung cancer among the community that is exposed to environmental tobacco smoke. The study found a positive correlation in the Caucasian population, while no correlation was found among the African-American population (Wenzlaff et al. 2005). Therefore, CYP genes may vary in their impact on the pathogenesis of the disease as inherited SNPs, at the individual as well as the population level, might have interethnic variations in carcinogenesis and its treatment response.

18.5.2 Breast Cancer

Globally, breast cancer is the most predictable form of malignancy among women. Breast cancer is established as a multifactorial disease as it is affected by hormones, environment, lifestyle, and genetics (Fahad Ullah 2019).

Persistent high circulating level of estrogens and oxidative metabolites thereof is crucial for the growth and advancement of breast cancer. CYP1A1 and CYP1A2 are reckoned as key CYP enzymes involved in the 2-hydroxylation of estrone and estradiol in breast tissue (Zhou et al. 2010). Another exclusive pathway for estrogen metabolism includes 16α-hydroxylation (Figure 18.3). While 2-hydroxyestrone has weak estrogen or anti-estrogen-like effects, 16α-hydroxyestrone has stronger estrogen-like effects on estrogen receptors (Bageman et al. 2008), and increased blood level of 16α-hydroxyestrone is associated with an increased risk of breast carcinogenesis (Kabat et al. 1997; Liehr and Ricci 1996). CYP1A2 reportedly has the highest rate of transforming 17β-estradiol and estrone into 2-hydroxyestradiol and 2-hydroxyestrone, respectively, and further, CYP1A2 can perform their 4-hydroxylation at a considerable rate (Lee et al. 2003). It is reported that the SNP *CYP1A2*1F* has a lower inducibility and has C to A polymorphism in intron 1 (Sachse et al. 1999). A multi-ethnic cohort study done by Le Marchand et al. concluded the negative correlation between *CYP1A2–163C/C* (i.e., *CYP1A2*1F*) genotype and breast cancer risk, although this phenotype might have a role in ovarian and endometrial cancers (Le Marchand et al. 2005; Mikhailova et al. 2006). It is also reported that genotype *-163C/C* in premenopausal women is associated with a reduced blood level of estradiol in comparison with *–163 A/A* and *–163 A/C* genotypes (Haiman et al. 2007).

The *CYP19A1* gene encodes for aromatase enzyme, which is critical for the biosynthesis of estrogen and is targeted by aromatase inhibitors in breast cancer therapy. As for *CYP19A1* polymorphisms rs6493497 and rs7176005, aromatase inhibitor produced diverse effects in initial-stage breast cancer patients, whereas other studies revealed that the biosynthesis of estrogen is not increased in the presence of these polymorphs in postmenopausal women (Straume et al. 2012; Wang et al. 2010). In a meta-analysis of over 20,000 cases and controls, it was found that *CYP19A1* polymorphisms rs10046 on their own did not increase breast cancer risk (Pineda et al. 2013). A systematic review done by Artigalas et al. conveyed that the *CYP19A1* SNP rs4646 extends a beneficial effect in metastatic breast cancer by increasing the progression time (Artigalás et al. 2015). Also, conflicting reports are available about the rs700519 SNP of the *CYP19A1*, whereas one study found no significant association between the gene and breast cancer in any ethnic population, but another meta-analysis observed a correlation between the rs700519 SNP of *CYP19A1* and susceptibility to breast cancer in the Han Chinese population (Ma et al. 2010; Sun et al. 2015). In contrast, El Aitan et al. reported that in the Jordanian population, breast cancer is significantly associated with the *CYP19A1* SNPs rs7176005 and rs6493497 (Al-Eitan et al. 2019). The consistent variations in the expression of the SNPs of *CYP19A1* in different ethnicities indicate that stratification of patients on the pharmacogenomics basis can help in designing personalized medicines meant for individuals.

Justenhoven et al. investigated the correlation of three polymorphisms of *CYP2C19*, namely *CYP2C19*2*, *CYP2C19*3*, and *CYP2C19*17*, with breast cancer susceptibility. The study reported that women having hormone therapy for 10 or more years and carrying the *CYP2C19*17* were having a substantially decreased risk of breast cancer (Justenhoven et al. 2009). Sim et al. reported that *CYP2C19*17* is having two SNPs, *–806C>T* and *–3402C>T*, in 5′-flanked region (Sim et al. 2006). Thus, the protective role of *CYP2C19*17* could be attributed to the fact that the ultra-rapid metabolizer phenotype leads to the lower blood level of estrogen and consequently reduced the risk of breast cancer.

The metabolic activities of CYP enzymes not only affect the pathophysiology per se, but also influence the outcome of a particular treatment and consequently choice of the treatment. Tamoxifen, an estrogen receptor antagonist, is commonly used in the therapeutics of the estrogen receptor- and/or progesterone receptor-positive breast cancer (Abe et al. 2005). It is metabolically transformed into two significantly potent metabolites, i.e., 4-hydroxytamoxifen (4-OHT) generated via 4-hydroxylation of tamoxifen and 4-hydroxy-N-desmethyltamoxifen (endoxifen) resulting from the N-demethylation of 4-OHT. These metabolites have 100 times better affinity for the estrogen receptor and are 30–100 times potent than tamoxifen (Johnson et al. 2004). Most of the endoxifen is biosynthesized via an alternate pathway that requires CYP3A4/5 and CYP2D6 for the formation of N-desmethyltamoxifen (NDM) and 4-hydroxylation of NDM, respectively (Stearns et al. 2003). An investigative study done by Lim et al. concluded that *CYP2D6*5/*10* and **10/*10* genotypes, in Asians with breast cancer, are significantly linked with the low circulatory levels of endoxifen (Lim et al. 2011). This information is clinically vital because initial-stage breast cancer patients, who are poor metabolizers of tamoxifen, are therapeutically better managed with alternate adjuvant endocrine therapy, i.e., aromatase inhibitors.

18.5.3 Liver cancer

Hepatocellular carcinoma (HCC) is the most prevalent type of liver cancer and a prominent determinant of liver cancer-related deaths worldwide. Multiple factors are involved in the pathogenesis of HCC, such as a chronic viral infection, especially the hepatitis B and C virus (Zhou et al. 2016). Reduced metabolism of carcinogens and enhanced transformation of procarcinogens into carcinogens have also been reported as risk factors along with the transition in the metabolism of environmental toxins that arise as a result of *CYP* polymorphism (Guengerich et al. 1996).

Various studies have reported that there is a significant association between the allelic frequency distributions of *CYP2D6*10* and HCC. The mutant homozygote, i.e., *CYP2D6*10 T/T* type, is more prevalent in healthy individuals compared to the patients with HCC, which favors either *C/C* or *C/T* carriers (Agundez et al. 1995; Silvestri et al. 2003; Zhou et al. 2016). The study done by Zhou et al. also showed that smoking may not be an influencing factor in HCC in the Chinese Han population (Zhou et al. 2016).

Aflatoxin B1 is one of the potent mutagenic chemicals among aflatoxins. Aflatoxin B1 is transformed into aflatoxin B1–8,9-epoxide, a cytotoxic DNA-binding intermediate, with the help of CYP (Gallagher et al. 1994, 1996; Macé et al. 1997). Different studies have acknowledged the considerable variation in the production and metabolism of aflatoxin B1–8,9-epoxide (Doi et al. 2002; Heinonen et al. 1996), and these divergences play a critical role in predisposing individuals to HCC. Shreds of evidence have revealed that CYP1A2, a high-affinity P450 isoenzyme, is associated with aflatoxin B1 metabolism and is responsible for more than 95% of aflatoxin B1 biotransformation in the human liver (Gallagher et al. 1994, 1996; Macé et al. 1997). Besides, studies have indicated that much of the chemopreventive outcome of oltipraz, a recognized chemopreventive agent particularly for aflatoxin B1-related HCC, is effectuated by CYP1A2 inhibition (Chen et al. 2006; Langouët et al. 1995; Sofowora et al. 2001).

CYP2E1, another subtype of CYP, is popular for expediting the xenobiotics metabolism including ethanol and carcinogenic precursors (Dey 2013), and ethanol is said to be instrumental in the induction of the CYP2E1 (Guengerich et al. 1991). Plentiful investigations' indications also endorse the significance of CYP2E1 in the development of liver cancer (Zhang et al. 2014). Thus, it is plausible that *CYP2E1* polymorphisms may have a role in the advancement of liver carcinogenesis. The *CYP2E1* gene is positioned at 10q24.3 and comprises nine exons and eight introns with the possibility of SNP at more than 279 SNP sites. Numerous sites near the 5′-end regulatory region can alter *CYP2E1* expression. The most common polymorphisms of *CYP2E1* are *PstI* 1239 G>C (rs3813867), *RsaI*-1055 C>T (rs2031920), and *DraI* (rs6413432) (Guengerich et al. 1991). On the one hand, various studies have shown that the *PstI/RsaI* SNP of the CYP2E1 is indistinguishably linked to liver cancer incidents (Bose et al. 2013; Huang et al. 2014; Ladero et al. 1996). However, on the other hand, few studies have said that the interconnection between *CYP2E1* polymorphisms and liver carcinogenesis might be implausible or even non-existent (Lee et al. 1997). To establish the role of *CYP2E1* polymorphism in liver cancer, Fu et al. conducted a meta-analysis of 4,862 cases (1,820 liver cancer cases and 3,042 controls) and contemplated the lack of any potential correlation between *CYP2E1* polymorphisms and liver cancer in pooled population studies (Fu et al. 2017).

An interesting notion had been obtained that *CYP2E1* polymorphism, in alcoholics, might have a dominant role in liver carcinogenesis because of the result of the union of different SNPs in specific populations (Fu et al. 2017). Aldehyde dehydrogenase 2 (ALDH2), a mitochondrial enzyme, is instrumental in the metabolism of alcohol. It catalyzes the rate-limiting step of converting acetaldehyde into acetic acid. Sakamoto et al. proposed that *ALDH2* rs671, wherein a point mutation at residue 487 resulted in a glutamic acid-to-lysine switching at ALDH2 protein (expressed by *ALDH2**2*), might revamp the risk of instituting HCC (Lee et al. 2017; Sakamoto et al. 2006). But, a recent study concluded that the association of *CYP2E1* rs2031920 and *ALDH2* rs671 has no significant effect on the development of HCC in the East Asian population (Chen et al. 2020).

18.5.4 Gastric Cancer

Gastric cancer continues to be a health burden worldwide as it is a leading cause of cancer-related deaths. The principal risk factor for gastric cancer is *Helicobacter pylori* (*H. pylori*) infection. Other risk factors include age, sex, smoking, geographical location, socioeconomic status, consuming salty meats, and genetics. Gastric cancer is generally segregated into two: cardia and non-cardia gastric cancer. Cancer near about the gastroesophageal junction is termed as cardia gastric cancer, whereas non-cardia gastric cancer usually affects the lower section of the stomach. Universally, non-cardia gastric cancer is more ubiquitous, particularly in the East Asian region. Non-cardia gastric cancers are further subcategorized into intestinal and diffuse; while males and older adults are prone to intestinal non-cardia gastric cancer, diffuse-type non-cardia gastric cancer is prevalent in both the genders and in all age groups. Comparatively, diffuse non-cardia gastric cancer has a faster progression and more adverse prognosis. The incident rate of non-cardia gastric cancer is approximately twice in men than in women (Lyons et al. 2019). Studies have reported that despite the high incidence rate of *H. pylori* infection worldwide, only a smaller population developed gastric cancer, particularly in Asian and African regions (Ghoshal et al. 2014). This might be because the genetic makeup of an individual, in addition to *H. pylori*, does have a significant contribution to gastric carcinogenesis.

Out of many investigated genetic polymorphisms in the *CYP2E1* gene, the 96-bp insertion in the 5′-flanking region has attracted much curiosity because of its probable functionality. As per the accepted nomenclature, the insertion allele in the 96-bp region is called *CYP2E1**1D, while the non-insertion allele is *CYP2E1**1C (Silva et al. 2012). The study done by Ghoshal et al. concluded that the presence of *CYP2E1* (96-bp insertion) increased enzyme activity and was related to the twofold increased risk of gastric cancer even in the absence of *H. pylori* (Ghoshal et al. 2014). The explanations could be elucidated by the fact that *H. pylori* infection is recognized to decrease the CYP enzyme activity (Giannini et al. 2003). Therefore, *CYP2E1* (96-bp insertion) might have higher activity in the absence of *H. pylori* infection.

Elingarami et al. piloted a hospital-based case-control study to assess the possible association between *CYP2E1* polymorphisms (rs2031920 and rs2480256) and predisposition to gastric carcinogenesis in a Northern Jiangsu population. The study showed that subjects with the *CYP2E1 PstI* polymorphism (rs2031920), having a mutant *TT* allele, are significantly more prone to develop gastric cancer in contrast to subjects with the heterozygous *C/T* allele (Elingarami et al. 2015). Several studies have indicated that smoking can considerably escalate the rate of chlorzoxazone metabolism and amplify the activity of CYP2E1, which can evidently activate several procarcinogens and consequently increase the risk of gastric carcinoma in long-term smokers (Benowitz et al. 2003; Howard et al. 2001). However, the study done by Elingarami et al. remains inconclusive about the influence of smoking in developing gastric cancer via *CYP2E1* polymorphism in the Northern Jiangsu population (Elingarami et al. 2015).

A prospective study has shown that higher concentrations of 1-hydroxypyrene glucuronide, a PAH metabolite, were related to a higher risk of gastric cancer in the Chinese population. CYP1A1 along with other enzymes is associated with the metabolism of PAHs. Hidaka et al. conducted a nested case-control study in the Japanese population to investigate the association between SNP *CYP1A1* (rs4646421, rs4646422, and rs1048943) and gastric cancer risk. The study concluded that SNP *CYP1A1* (rs4646422) allele carriers, compared to homozygous wild-type genotype, had a significantly higher risk of gastric cancer even after the adjustment of potential confounding factors. The interaction between the SNP *CYP1A1* (rs4646422) and gastric cancer risk was also veritable among never-smokers (Hidaka et al. 2016).

18.5.5 Colorectal Cancer (CRC)

CRC is the fourth most commonly identified malignancy and is among the top five fatal cancers worldwide (Globocan Observatory 2019). Prominently, CRC is more prevalent in people aged 50 years or older, and age is also a risk factor. Reports have indicated that about 6%–7% of CRC cases have a genetic predisposition. Further, certain hereditary syndromes such as adenomatous polyposis syndrome and Lynch syndrome can also increase the risk of CRC genesis. Besides, other lifestyle factors such as smoking, alcohol consumption, high-fat diet consumption, physical inactivity, and consumption of red meat or processed meat are also accounted as risk factors for

CRC (Doaei et al. 2018). Studies reported that genetic polymorphism can influence the effects of these lifestyle factors (Akbari et al. 2017; Kalantari et al. 2016), potentially putting certain individuals at higher risk of developing CRC.

Le Marchand et al. conducted a population-based case-control study in Hawaii to test the interaction of two functional SNPs in *CYP2E1* (the G1259C *RsaI* substitution and a 5′-end 96-bp insertion variant) with CRC, along with their tailoring effects on the interdependence of CRC and red meat and processed meat. The study concluded, with the lack of statistically significant results, that subjects with the high-activity alleles were at a higher risk of CRC development, although this interrelation was much favored for the insert polymorphism instead of *RsaI*. Fascinatingly, the highest risk of CRC was found in individuals who have the insert and increased consumption of red meat or processed meat (Le Marchand et al. 2002). Another study, Fukuoka Colorectal Cancer Study, a community-based case-control study in Japan, explored the association of SNPs of *CYP2E1* (*RsaI* and 96-bp insertion) with CRC, with specific consideration to the outcome of these SNPs along with alcoholism, obesity, and consumption of red meat. The study observed that the risk of CRC was reduced in subjects with the *CYP2E1 RsaI c2* allele and was due to the *CYP2E1* 96-bp insertion allele (Morita et al. 2009).

18.5.6 Prostate Cancer

Prostate cancer (PC) is the most prevalent non-cutaneous cancer among men globally, with an estimated annual incidence of 1.6 million and 366,000 deaths (Torre et al. 2015). The malignant alteration of the prostate is a several-step sequential process that initiates as prostatic intraepithelial neoplasia trailed by localized PC and then advanced prostate adenocarcinoma with local invasion, concluding in metastatic PC (Shen and Abate-Shen 2010). A salient feature of PC is its hormone responsiveness that established the androgen deprivation therapy (ADT), initially achieved by castration, as the standard care for PC. Resistance to ADT results in primary castration-resistant PC or metastatic castration-resistant PC. Recently, cases of PC variants with neuroendocrine features (NEPC) or small cell features having reduced expression of androgen receptor (AR-low) or complete absence of expression of androgen receptor (AR–) have increased in the clinic, which may correlate with the use of potent androgen receptor antagonists. Additionally, a subset of androgen receptor-independent tumors devoid of markers of neuroendocrine differentiation that are entirely unresponsive to ADT might have emerged from AR-low or AR– clones and/or the transdifferentiation of androgen receptor-expressing adenocarcinoma into AR-low and AR– tumors (Shen and Abate-Shen 2010; Zou et al. 2017).

The CYP3A5 isoform is particularly expressed in the prostate and transforms testosterone by 6β-hydroxylation into a physiologically less active form for expedited elimination (mitra2015). SNP in the intron of *CYP3A5*3*, having A to G transition in intron 3 at position 6986 (*CYP3A5*3*, rs776746 A>G), is the most frequent and functional polymorphism, inactivates *CYP3A5*3*, and affects the metabolism of testosterone (Agundez 2005). Liang et al. conducted a meta-analysis of six studies to explore the correlation between CYP3A5*3 SNP and PC risk and found a significant association between the two. The study done by Liang et al. concluded that heterogeneity in cancer incidences among the different populations could be because of ethnic variation and SNP CYP3A5*3 could be a risk factor for PC in Africans, while this may not be the case in Caucasians and Asians (Liang et al. 2018).

The CYP1A1 Ile462Val (A4889G, rs1048943) SNP is a point mutation at position 4889 in exon 7, which results in the substitution of isoleucine by valine in the primary structure of the enzyme (Murata et al. 2001). CYP1A1 contributes to the transformation of estrogen into 2-hydroxyestradiol and 4-hydroxyestradiol, and studies have proposed these metabolites of estrogens as risk factors for PC (Anwar-Mohamed et al. 2009; Huang et al. 2012). Han et al. concluded through a meta-analysis that CYP1A1 Ile462Val SNP is a potential risk factor for PC in the Asian population, whereas Koda et al. found a similar association between CYP1A1 Ile462Val and PC among Japanese (Han et al. 2013; Koda et al. 2017).

18.5.7 Head-and-Neck Cancer

As per the National Cancer Institute, the USA, globally, head-and-neck cancers usually originate in the squamous cells of the mucosal surfaces of the head and neck and can further be categorized based on the area of their originations. Cancers of the oral cavity – including lips, tongue, gums, and inside the cheeks; pharynx (nasopharynx, oropharynx, and hypopharynx); larynx; nasal cavity; and salivary glands constitute the group head-and-neck cancers (https://www.cancer.gov/types/head-and-neck/head-neck-fact-sheet).

Oral cancer is one of the foremost cancers embracing 2% of all cancer cases (14.1 million cases in 2012) and has an approximately 50% mortality rate (Li et al. 2020). Internationally, the exorbitant incidences of oral cancer are found in South Asian countries such as India, Sri Lanka, and Taiwan, which could be accredited to the excessive cigarette smoking and chewing of areca nut in these countries (Warnakulasuriya 2010). Each year, 48,330 cases (3% of all cancer) of oral and oropharyngeal cancer are diagnosed in the USA (Jemal et al. 2009). Oral cancer is placed in the eighth position among leading cancers in men, having approximately 70% of male patients (Siegel et al. 2016).

A study done by Kao et al. indicated a correlation between CYP1A1 exon 7 polymorphisms and susceptibility to oral squamous cell carcinoma and oral precancerous lesion. It was observed in the study that the period of betel use and smoking in patients with oral precancerous lesions was shorter when compared to the oral squamous cell carcinoma patients. The study reported that individuals with SNP CYP1A1 exon 7 earlier had an earlier onset of oral squamous cell carcinoma even with the shorter period of tobacco (Kao et al. 2002).

Zhou et al. carried out a meta-analysis to assess the association between CYP1A1 Ile462Val SNP and oral cancer. The meta-analysis included 13 studies with 1,515 oral cancer patients and 2,233 controls. It was found that there is no association between CYP1A1 Ile462Val SNP and oral cancer, but in the subgroup analysis by ethnicity, Asians with CYP1A1 Ile462Val are at a high risk of developing oral cancer (Zhuo et al. 2012).

Maurya et al. investigated the association of alcohol consumption, smoking, and tobacco chewing with polymorphic alcohol- or tobacco-metabolizing CYP enzymes (CYP1A1, 1B1, and 2E1) as risk factors for cancers of the larynx, pharynx, and oral cavity. The study found that SNPs CYP1A1*2A, CYP1B1*2, and CYP2E1*5B or CYP2E1*6 along with tobacco smoking significantly increased the risk of oral cavity cancer, laryngeal cancer, and pharyngeal cancer. Similarly, tobacco chewing in genotypes CYP1B1*3 and CYP2E1*5B increased the risk of oral, laryngeal, and pharyngeal cancers. Also, genotypes CYP1A1*2A, CYP1A1*2C, CYP1B1*2, CYP1B1*3, and CYP2E1*5B were found to be having increased risk of developing laryngeal cancer when they consumed alcohol (Maurya et al. 2015).

18.6 Conclusions

The superfamily of CYP enzymes has a germane function of protecting the cells from the effects of toxic chemicals by transforming the lipophilic toxicants into hydrophilic and excretable moieties. The significance of the CYP enzymes system could be ascertained from the fact that at least 57 different CYP genes and 58 pseudogenes translate for 18 different families and 44 subfamilies of CYP enzymes. Mutations in a *CYP* gene can reorient the activity of the expressed enzymes either positively or negatively. The SNP of the *CYP* gene has the aptitude to govern not only the etiology of cancer, but the therapy as well. At the individual and population levels, CYP gene SNPs have influenced the interethnic variations in different cancers and the outcome of the therapy employed for the treatment of cancer. Thus, stratification of patients on a pharmacogenomics basis could be valuable for formulating personalized medicines or moving to a better alternative. CYP SNPs may contribute to carcinogenesis, especially in individuals or populations with a particular lifestyle or consumption habit, and this could give rise to variations in different individuals with different behaviors and eating habits. Therefore, lifestyle or dietary exhortations for the prevention and control of particular cancer should be adjusted according to the genotype of different individuals.

ABBREVIATIONS

ADRs: Adverse drug reactions
ADT: Androgen deprivation therapy
AR: Androgen receptor
B[a]P: Benzo[a]pyrene
cDNA: Complementary deoxyribonucleic acid
CNVs: Copy number variants
CYP: Cytochrome P450
FAD: Flavin adenine dinucleotide
FMN: Flavin adenine mononucleotide
GABA: Gamma-aminobutyric acid
HCC: Hepatocellular carcinoma
H. pylori: *Helicobacter pylori*
NADPH: Nicotinamide adenine dinucleotide
NDM: N-desmethyltamoxifen
NEPC: Neuroendocrine prostate cancer
NNK: Nicotine-derived nitrosamine
4-OHT: 4-Hydroxytamoxifen
PAHs: Polycyclic aromatic hydrocarbons
RNA: Ribonucleic acid
SNP: Single nucleotide polymorphism
SNVs: Single nucleotide variants

BIBLIOGRAPHY

Abe O, Abe R, Enomoto K, Kikuchi K, Koyama H, Masuda H, et al. Effects of chemotherapy and hormonal therapy for early breast cancer on recurrence and 15-year survival: An overview of the randomised trials. *Lancet*. 2005;365(9472):1687–717.

Agundez J. Cytochrome P450 gene polymorphism and cancer. *Curr. Drug Metab*. 2005;5(3):211–24.

Agundez JAG, Ledesma MC, Benitez J, Ladero JM, Rodríguez-Lescure A, Diaz-Rubio E, et al. CYP2D6 genes and risk of liver cancer. *Lancet*. 1995;345:830–1.

Akbari ME, Gholamalizadeh M, Doaei S, Mirsafa F. FTO gene affects obesity and breast cancer through similar mechanisms: A new insight into the molecular therapeutic targets. *Nutr. Cancer*. Taylor & Francis. 2017;70(1):30–6.

Al-Eitan LN, Rababa'H DM, Alghamdi MA, Khasawneh RH. Association of CYP gene polymorphisms with breast cancer risk and prognostic factors in the Jordanian population. *BMC Med. Genet*. 2019;20(1):1–12.

Anwar-Mohamed A, Elbekai RH, El-Kadi AOS. Regulation of CYP1A1 by heavy metals and consequences for drug metabolism. *Expert Opin. Drug Metab. Toxicol*. 2009;5(5):501–21.

Artigalás O, Vanni T, Hutz MH, Ashton-Prolla P, Schwartz IV. Influence of CYP19A1 polymorphisms on the treatment of breast cancer with aromatase inhibitors: A systematic review and meta-analysis. *BMC Med*. 2015;13(1):1–10.

Backman JT, Filppula AM, Niemi M, Neuvonen PJ. Role of cytochrome P450 2C8 in drug metabolism and interactions. *Pharmacol. Rev*. 2016;68(1):168–241.

Bageman E, Ingvar C, Rose C, Jernström H. Coffee consumption and CYP1A2*1F genotype modify age at breast cancer diagnosis and estrogen receptor status. *Cancer Epidemiol. Biomarkers Prev*. 2008;17(4):895–901.

Bahar MA, Setiawan D, Hak E, Wilffert B. Pharmacogenetics of drug-drug interaction and drug-drug-gene interaction: A systematic review on CYP2C9, CYP2C19 and CYP2D6. *Pharmacogenomics*. 2017;18(7):701–39.

Benowitz NL, Peng M, Jacob P. Effects of cigarette smoking and carbon monoxide on chlorzoxazone and caffeine metabolism. *Clin. Pharmacol. Ther*. 2003;74(5):468–74.

Boonruang S, Prakobsri K, Pouyfung P, Prasopthum A, Rongnoparut P, Sarapusit S. Structure–activity relationship and in vitro inhibition of human cytochrome CYP2A6 and CYP2A13 by flavonoids. *Xenobiotica* [Internet]. Taylor & Francis. 2020;50(6):630–9. Available from: http://dx.doi.org/10.1080/00498254.2019.1675101

Boonruang S, Prakobsri K, Pouyfung P, Srisook E, Prasopthum A, Rongnoparut P, et al. Inhibition of human cytochromes P450 2A6 and 2A13 by flavonoids, acetylenic thiophenes and sesquiterpene lactones from Pluchea indica and Vernonia cinerea. *J. Enzyme Inhib. Med. Chem*. [Internet]. Informa UK Ltd. 2017;32(1):1136–42. Available from: https://doi.org/10.1080/14756366.2017.1363741

Bose S, Tripathi DM, Sukriti, Sakhuja P, Kazim SN, Sarin SK. Genetic polymorphisms of CYP2E1 and DNA repair genes HOGG1 and XRCC1: Association with hepatitis B related advanced liver disease and cancer. *Gene* [Internet]. Elsevier B.V. 2013;519(2):231–7. Available from: http://dx.doi.org/10.1016/j.gene.2013.02.025

Brandon EFA, Sparidans RW, Van Ooijen RD, Meijerman I, Lazaro LL, Manzanares I, et al. In vitro characterization of the human biotransformation pathways of aplidine, a novel marine anti-cancer drug. *Invest. New Drugs*. 2007; 25(1):9–19.

Bui P, Imaizumi S, Beedanagari SR, Reddy ST, Hankinson O. Human CYP2S1 metabolizes cyclooxygenase- and lipoxygenase-derived eicosanoids. *Drug Metab. Dispos.* 2011;39(2):180–90.

Chandler PD, Tobias DK, Wang L, Smith-Warner SA, Chasman DI, Rose L, et al. Association between vitamin D genetic risk score and cancer risk in a large cohort of U.S. women. *Nutrients*. 2018;10(1):1–11.

Chang GWM, Kam PCA. The physiological and pharmacological roles of cytochrome P450 isoenzymes. *Anaesthesia*. 1999;54(1):42–50.

Chen J, Pan W, Chen Y, Wen L, Tu J, Liu K. Relationship of ALDH2 rs671 and CYP2E1 rs2031920 with hepatocellular carcinoma susceptibility in East Asians: A meta-analysis. *World J. Surg. Oncol.* 2020;18(1):1–12.

Chen X, Wang H, Xie W, Liang R, Wei Z, Zhi L, et al. Association of CYP1A2 genetic polymorphisms with hepatocellular carcinoma susceptibility: A case-control study in a high-risk region of China. *Pharmacogenet. Genom.* 2006;16(3): 219–27.

Choi YJ, Zhou Y, Lee JY, Ryu CS, Kim YH, Lee K, et al. Cytochrome P450 4A11 inhibition assays based on characterization of lauric acid metabolites [Internet]. *Food Chem. Toxicol.* Elsevier Ltd. 2018. Available from: https://doi.org/10.1016/j.fct.2017.12.063

Christakos S, Dhawan P, Verstuyf A, Verlinden L, Carmeliet G. Vitamin D: Metabolism, molecular mechanism of action, and pleiotropic effects. *Physiol. Rev.* 2015;96(1): 365–408.

Congiu M, Mashford ML, Slavin JL, Desmond P V. Coordinate regulation of metabolic enzymes and transporters by nuclear transcription factors in human liver disease. *J. Gastroenterol. Hepatol.* 2009;24(6):1038–44.

Danese E, Raimondi S, Montagnana M, Tagetti A, Langaee T, Borgiani P, et al. Effect of CYP4F2, VKORC1, and CYP2C9 in influencing coumarin dose: A single-patient data meta-analysis in more than 15,000 individuals. *Clin. Pharmacol. Ther.* 2019;105(6):1477–91.

Dey A. Cytochrome P450 2E1: Its role in disease and drug metabolism [Internet]. 2013. Available from: http://link.springer.com/10.1007/978-94-007-5881-0

Di Y, Chow V, Yang L-P, Zhou S-F. Structure, function, regulation and polymorphism of human cytochrome P450 2A6. *Curr. Drug Metab.* 2009;10(7):754–80.

Doaei S, Hajiesmaeil M, Aminifard A, Mosavi-Jarrahi SA, Akbari ME, Gholamalizadeh M. Effects of gene polymorphisms of metabolic enzymes on the association between red and processed meat consumption and the development of colon cancer; A literature review. *J. Nutr. Sci.* 2018 (7), e26, 1–7.

Doi AM, Patterson PE, Gallagher EP. Variability in aflatoxin B1-macromolecular binding and relationship to biotransformation enzyme expression in human prenatal and adult liver. *Toxicol. Appl. Pharmacol.* 2002;181(1):48–59.

Durairaj P, Fan L, Machalz D, Wolber G, Bureik M. Functional characterization and mechanistic modeling of the human cytochrome P450 enzyme CYP4A22. *FEBS Lett.* 2019;593(16):2214–25.

van Eijk M, Boosman RJ, Schinkel AH, Huitema ADR, Beijnen JH. Cytochrome P450 3A4, 3A5, and 2C8 expression in breast, prostate, lung, endometrial, and ovarian tumors: Relevance for resistance to taxanes. *Cancer Chemother. Pharmacol.* [Internet]. Springer: Berlin Heidelberg. 2019;84(3):487–99. Available from: https://doi.org/10.1007/s00280-019-03905-3

Ekins S, Wrighton SA. The role of CYP2B6 in human xenobiotic metabolism. *Drug Metab. Rev.* 1999;31(3):719–54.

Elfaki I, Mir R, Almutairi FM, Abu Duhier FM. Cytochrome P450: Polymorphisms and roles in cancer, diabetes and atherosclerosis. *Asian Pacific J. Cancer Prev.* 2018;19(8):2057–70.

Elingarami S, Liu H, Kalinjuma AV, Hu W, Li S, He N. Polymorphisms in NEIL-2, APE-1, CYP2E1 and MDM2 genes are independent predictors of gastric cancer risk in a Northern Jiangsu population (China). *J. Nanosci. Nanotechnol.* 2015;15(7):4815–28.

Fahad Ullah M. Breast cancer: Current perspectives on the disease status. In: Ahmad A, editor. *Adv. Exp. Med. Biol.* Breast Cancer Metastasis Drug Resist. Challenges Prog. [Internet]. 2nd ed. Springer Nature: Switzerland AG. 2019. pp. 51–64. Available from: http://link.springer.com/10.1007/978-3-030-20301-6_4

Fernald GH, Capriotti E, Daneshjou R, Karczewski KJ, Altman RB. Bioinformatics challenges for personalized medicine. *Bioinformatics*. 2011;27(13):1741–8.

Foti RS, Isoherranen N, Zelter A, Dickmann LJ, Buttrick BR, Diaz P, et al. Identification of tazarotenic acid as the first xenobiotic substrate of human retinoic acid hydroxylase CYP26A1 and CYP26B1. *J. Pharmacol. Exp. Ther.* 2016;357(2):281–92.

Fu P, Yang F, Li B, Zhang B, Guan L, Sheng J, et al. Meta-analysis of CYP2E1 polymorphisms in liver carcinogenesis. *Dig. Liver Dis.* 2017;49(1):77–83.

Gainer JV, Bellamine A, Dawson EP, Womble KE, Grant SW, Wang Y, et al. Functional variant of CYP4A11 20-hydroxyeicosatetraenoic acid synthase is associated with essential hypertension. *Circulation*. 2005;111(1):63–9.

Gallagher EP, Kunze KL, Stapleton PL, Eaton DL. The kinetics of aflatoxin B1 oxidation by human cDNA-expressed and human liver microsomal cytochromes P450 1A2 and 3A4. *Toxicol. Appl. Pharmacol.* 1996;141(2):595–606.

Gallagher EP, Wienkers LC, Stapleton PL, Kunze KL, Eaton DL. Role of human microsomal and human complementary DNA-expressed cytochromes P4501A2 and P4503A4 in the bioactivation of aflatoxin B11. *Cancer Res.* 1994; 54(1):101–8.

Ghoshal U, Tripathi S, Kumar S, Mittal B, Chourasia D, Kumari N, et al. Genetic polymorphism of cytochrome P450 (CYP) 1A1, CYP1A2, and CYP2E1 genes modulate susceptibility to gastric cancer in patients with Helicobacter pylori infection. *Gastric Cancer*. 2014;17(2):226–34.

Giannini E, Fasoli A, Botta F, Romagnoli P, Malfatti F, Chiarbonello B, et al. Helicobacter pylori infection is associated with greater impairment of cytochrome P-450 liver metabolic activity in anti-HCV positive cirrhotic patients. *Dig. Dis. Sci.* 2003;48(4):802–8.

Globocan Observatory W. Cancer Today – World. *Int. Agency Res. Cancer [Internet].* 2019;876:2018–9. Available from: https://gco.iarc.fr/today/data/factsheets/populations/900-world-fact-sheets.pdf

Goto T, Moriuchi H, Fu X, Ikegawa T, Matsubara T, Chang G, et al. The effects of single nucleotide polymorphisms in CYP2A13 on metabolism of 5-methoxypsoralen. *Drug Metab. Dispos.* 2010;38(12):2110–6.

Guengerich FP. Cytochrome p450 and chemical toxicology. *Chem. Res. Toxicol.* 2008;21(Table 1):70–83.

Guengerich FP, Johnson WW, Ueng YF, Yamazaki H, Shimada T. Involvement of cytochrome P450, glutathione S-transferase, and epoxide hydrolase in the metabolism of aflatoxin B1 and relevance to risk of human liver cancer. *Environ. Health Perspect.* 1996;104(SUPPL. 3):557–62.

Guengerich FP, Kim DH, Iwasaki M. Role of human cytochrome P-450 IIE1 in the oxidation of many low molecular weight cancer suspects. *Chem. Res. Toxicol.* 1991;4(2):168–79.

Guengerich FP, Waterman MR, Egli M. Recent structural insights into cytochrome P450 function. *Trends Pharmacol. Sci.* [Internet]. Elsevier Ltd. 2016;37(8):625–40. Available from: http://dx.doi.org/10.1016/j.tips.2016.05.006

Gunes A, Dahl ML. Variation in CYP1A2 activity and its clinical implications: Influence of environmental factors and genetic polymorphisms. *Pharmacogenomics.* 2008;9(5):625–37.

Guo J, Johansson I, Mkrtchian S, Ingelman-Sundberg M. The CYP2W1 enzyme: Regulation, properties and activation of prodrugs. *Drug Metab. Rev.* 2016;48(3):369–78.

Haiman CA, Dossus L, Setiawan VW, Stram DO, Dunning AM, Thomas G, et al. Genetic variation at the CYP19A1 locus predicts circulating estrogen levels but not breast cancer risk in postmenopausal women. *Cancer Res.* 2007;67(5):1893–7.

Han G, Ma Y, Liu P, Wei X, Zhang X, Zhu F. Quantitative synthesis of the association between the cytochrome P450 1A1 Ile462Val polymorphism and prostate cancer risk. *Tumor Biol.* 2013;34(3):1511–6.

Harvey RD, Morgan ET. Cancer, inflammation, and therapy: Effects on cytochrome P450-mediated drug metabolism and implications for novel immunotherapeutic agents. *Clin. Pharmacol. Ther.* 2014;96(4):449–57.

Hasler JA, Estabrook R, Murray M, Pikuleva I, Waterman M, Capdevila J, et al. Human cytochromes P450. *Mol. Aspects Med.* 1999;20(1–2):1–137.

Hedlund E, Gustafsson J, Warner M. Cytochrome P450 in the brain; A review. *Curr. Drug Metab.* [Internet]. 2001;2(3):245–63. Available from: http://www.eurekaselect.com/openurl/content.php?genre=article&issn=1389-2002&volume=2&issue=3&spage=245

Hedrich WD, Hassan HE, Wang H. Insights into CYP2B6-mediated drug-drug interactions. *Acta Pharm. Sin. B* [Internet]. Elsevier. 2016;6(5):413–25. Available from: http://dx.doi.org/10.1016/j.apsb.2016.07.016

Heinonen JT, Fisher R, Brendel K, Eaton DL. Determination of aflatoxin b1 biotransformation and binding to hepatic macromolecules in human precision liver slices. *Toxicol. Appl. Pharmacol.* 1996;136(1):1–7.

Hidaka A, Sasazuki S, Matsuo K, Ito H, Charvat H, Sawada N, et al. CYP1A1, GSTM1 and GSTT1 genetic polymorphisms and gastric cancer risk among Japanese: A nested case–control study within a large-scale population-based prospective study. *Int. J. Cancer.* 2016;139(4):759–68.

Hodgson E, Rose RL. The importance of cytochrome P450 2B6 in the human metabolism of environmental chemicals. *Pharmacol. Ther.* 2007;113(2):420–8.

Hollenberg PF, Kent UM, Bumpus NN. Mechanism-based inactivation of human cytochromes P450s: Experimental characterization, reactive intermediates, and clinical implications. *Chem. Res. Toxicol.* 2008;21(1):189–205.

Hong Y, Chia YMF, Yeo RH, Venkatesan G, Koh SK, Chai CLL, et al. Inactivation of human cytochrome P450 3A4 and 3A5 by dronedarone and n-desbutyl dronedarone. *Mol. Pharmacol.* 2016;89(1):1–13.

Howard LA, Micu AL, Sellers EM, Tyndale RF. Low doses of nicotine and ethanol induce CYP2E1 and chlorzoxazone metabolism in rat liver. *J. Pharmacol. Exp. Ther.* 2001;299(2):542–50.

Huang Q, Deshmukh RS, Ericksen SS, Tu Y, Szklarz GD. Preferred binding orientations of phenacetin in CYP1A1 and CYP1A2 are associated with isoform-selective metabolism. *Drug Metab. Dispos.* 2012;40(12):2324–31.

Huang APH, Tsai JC, Kuo LT, Lee CW, Lai HS, Tsai LK, et al. Clinical application of perfusion computed tomography in neurosurgery: Clinical article. *J. Neurosurg.* 2014;120(2):473–88.

Hussein AG, Pasha HF, El-Shahat HM, Gad DM, Toam MM. CYP1A1 gene polymorphisms and smoking status as modifier factors for lung cancer risk. *Gene* [Internet]. Elsevier B.V. 2014;541(1):26–30. Available from: http://dx.doi.org/10.1016/j.gene.2014.03.003

Ingelman-Sundberg M. Pharmacogenetics of cytochrome P450 and its applications in drug therapy: The past, present and future. *Trends Pharmacol. Sci.* 2004;25(4):193–200.

Ingelman-Sundberg M, Sim SC, Gomez A, Rodriguez-Antona C. Influence of cytochrome P450 polymorphisms on drug therapies: Pharmacogenetic, pharmacoepigenetic and clinical aspects. *Pharmacol. Ther.* 2007;116(3):496–526.

Isoherranen N, Zhong G. Biochemical and physiological importance of the CYP26 retinoic acid hydroxylases. *Pharmacol. Ther.* [Internet]. Elsevier Inc. 2019;204:107400. Available from: https://doi.org/10.1016/j.pharmthera.2019.107400

Jarrar YB, Lee SJ. Molecular functionality of cytochrome P450 4 (CYP4) genetic polymorphisms and their clinical implications. *Int. J. Mol. Sci.* 2019;20(17) 4274.

Jeffery CJ. Moonlighning proteins. *Trends Biochem. Sci.* 1999; 24(1):8–11.

Jemal A, Siegel R, Ward E, Hao Y, Xu J, Thun MJ. Cancer statistics, 2009 both sexes female both sexes estimated deaths. *CA Cancer J Clin.* 2009;59(4):1–25.

Jennings GK, Hsu MH, Shock LS, Johnson EF, Hackett JC. Noncovalent interactions dominate dynamic heme distortion in cytochrome P450 4B1. *J. Biol. Chem.* 2018;293(29): 11433–46.

Johansson I, Ingelman-Sundberg M. Genetic polymorphism and toxicology-with emphasis on cytochrome P450. *Toxicol. Sci.* 2011;120(1):1–13.

Johnson MD, Zuo H, Lee KH, Trebley JP, Rae JM, Weatherman R V., et al. Pharmacological characterization of 4-hydroxy-Ndesmethyl tamoxifen, a novel active metabolite of tamoxifen. *Breast Cancer Res. Treat.* 2004;85(2):151–9.

Justenhoven C, Hamann U, Pierl CB, Baisch C, Harth V, Rabstein S, et al. CYP2C19*17 is associated with decreased breast cancer risk. *Breast Cancer Res. Treat.* 2009;115(2):391–6.

Kabat GC, Chang CJ, Sparano JA, Sepkovic DW, Hu XP, Khalil A, et al. Urinary estrogen metabolites and breast cancer: A case-control study. *Cancer Epidemiol. Biomarkers Prev.* 1997;6(7):505–9.

Kalantari N, Doaei S, Keshavarz-Mohammadi N, Gholamalizadeh M, Pazan N. Review of studies on the fat mass and obesity-associated (FTO) gene interactions with environmental factors affecting on obesity and its impact on lifestyle interventions. *ARYA Atheroscler.* 2016;12(6):281–90.

Kao SY, Wu CH, Lin SC, Yap SK, Chang CS, Wong YK, et al. Genetic polymorphism of cytochrome P4501A1 and susceptibility to oral squamous cell carcinoma and oral precancer lesions associated with smoking/betel use. *J. Oral Pathol. Med.* 2002;31(9):505–11.

Karkhanis A, Hong Y, Chan ECY. Inhibition and inactivation of human CYP2J2: Implications in cardiac pathophysiology and opportunities in cancer therapy. *Biochem. Pharmacol.* [Internet]. Elsevier Inc. 2017;135:12–21. Available from: http://dx.doi.org/10.1016/j.bcp.2017.02.017

Kartha JS, Yost GS. Mechanism-based inactivation of lung-selective cytochrome P450 CYP2F enzymes. *Drug Metab. Dispos.* 2008;36(1):155–62.

Kessova I, Cederbaum AI. CYP2E1: Biochemistry, toxicology, regulation and function in ethanol-induced liver injury. *Curr. Mol. Med.* 2003;3(6):509–18.

Kiyotani K, Yamazaki H, Fujieda M, Iwano S, Matsumura K, Satarug S, et al. Decreased coumarin 7-hydroxylase activities and CYP2A6 expression levels in humans caused by genetic polymorphism in CYP2A6 promoter region (CYP2A6*9). *Pharmacogenetics.* 2003;13(11):689–95.

Kmeťová Sivoňová M, Jurečeková J, Tatarková Z, Kaplán P, Lichardusová L, Hatok J. The role of CYP17A1 in prostate cancer development: Structure, function, mechanism of action, genetic variations and its inhibition. *Gen. Physiol. Biophys.* [Internet]. 2017;36(05):487–99. Available from: http://www.elis.sk/index.php?page=shop.product_details&flypage=flypage.tpl&product_id=5471&category_id=132&option=com_virtuemart

Koda M, Iwasaki M, Yamano Y, Lu X, Katoh T. Association between NAT2, CYP1A1, and CYP1A2 genotypes, heterocyclic aromatic amines, and prostate cancer risk: A case control study in Japan. *Environ. Health Prev. Med.* 2017;22(1):72.

Kopp TI, Vogel U, Andersen V. Associations between common polymorphisms in CYP2R1 and GC, Vitamin D intake and risk of colorectal cancer in a prospective case-cohort study in Danes. *PLoS One.* 2020;15(2):1–14.

Kowalski JP, Mcdonald MG, Whittington D, Guttman M, Scian M, Girhard M, et al. Structure-activity relationships for CYP4B1 bioactivation of 4-ipomeanol congeners: Direct correlation between cytotoxicity and trapped reactive intermediates. *Chem. Res. Toxicol.* 2019;32(12):2488–98.

Kubota T, Nakajima-Taniguchi C, Fukuda T, Funamoto M, Maeda M, Tange E, et al. CYP2A6 polymorphisms are associated with nicotine dependence and influence withdrawal symptoms in smoking cessation. *Pharmacogenomics J.* 2006;6(2):115–9.

Kumar S, Sharma R, Roychowdhury A. Modulation of cytochrome-P450 inhibition (CYP) in drug discovery: A medicinal chemistry perspective. *Curr. Med. Chem.* 2012;19(21):3605–21.

Ladero JM, Agúndez JAG, Rodríguez-Lescure A, Diaz-Rubio M, Benítez J. RsaI polymorphism at the cytochrome P4502E1 locus and risk of hepatocellular carcinoma. *Gut.* 1996;39(2):330–3.

Langouët S, Coles B, Morel F, Becquemont L, Beaune P, Guengerich FP, et al. Inhibition of CYP1A2 and CYP3A4 by oltipraz results in reduction of aflatoxin Bt metabolism in human hepatocytes in primary culture. *Cancer Res.* 1995;55(23):5574–9.

Lee AJ, Cai MX, Thomas PE, Conney AH, Zhu BT. Characterization of the oxidative metabolites of 17β-estradiol and estrone formed by 15 selectively expressed human cytochrome P450 isoforms. *Endocrinology.* 2003;144(8):3382–98.

Lee SS, Jung HJ, Park JS, Cha IJ, Cho DY, Shin JG. Identification of a null allele of cytochrome P450 3A7: CYP3A7 polymorphism in a Korean population. *Mol. Biol. Rep.* 2010;37(1):213–7.

Lee DJ, Lee HM, Kim JH, Park IS, Rho YS. Heavy alcohol drinking downregulates ALDH2 gene expression but heavy smoking up-regulates SOD2 gene expression in head and neck squamous cell carcinoma. *World J. Surg. Oncol.* 2017;15(1):1–8.

Lee HS, Yoon JH, Kamimura S, Iwata K, Watanabe H, Kim CY. Lack of association of cytochrome P450 2E1 genetic polymorphisms with the risk of human hepatocellular carcinoma. *Int. J. Cancer.* 1997;71(5):737–40.

Li H, Lampe JN. Neonatal cytochrome P450 CYP3A7: A comprehensive review of its role in development, disease, and xenobiotic metabolism. *Arch. Biochem. Biophys.* [Internet]. Elsevier. 2019;673(August):108078. Available from: https://doi.org/10.1016/j.abb.2019.108078

Li CC, Shen Z, Bavarian R, Yang F, Bhattacharya A. Oral cancer: Genetics and the role of precision medicine. *Surg. Oncol. Clin. N. Am.* 2020;29(1):127–44.

Liang Y, Han W, Yan H, Mao Q. Association of CYP3A5*3 polymorphisms and prostate cancer risk: A meta-analysis. *J. Cancer Res. Ther.* 2018;14(9):S463–7.

Libby RT, Smith RS, Savinova OV, Zabaleta A, Martin JE, Gonzalez FJ, et al. Modification of ocular defects in mouse developmental glaucoma models by tyrosinase. *Science.* 2003;299(5612):1578–81.

Liehr JG, Ricci MJ. 4-Hydroxylation of estrogens as marker of human mammary tumors. *Proc. Natl. Acad. Sci. U. S. A.* 1996;93(8):3294–6.

Lim JSL, Chen XA, Singh O, Yap YS, Ng RCH, Wong NS, et al. Impact of CYP2D6, CYP3A5, CYP2C9 and CYP2C19 polymorphisms on tamoxifen pharmacokinetics in Asian breast cancer patients. *Br. J. Clin. Pharmacol.* 2011;71(5):737–50.

Lin J, He B, Cao L, Zhang Z, Liu H, Rao J, et al. CYP1A1 Ile462Val polymorphism and the risk of non-small cell lung cancer in a Chinese population. *Tumori.* 2014;100(5):547–52.

Lyons K, Le LC, Pham YTH, Borron C, Park JY, Tran CTD, et al. Gastric cancer: Epidemiology, biology, and prevention: A mini review. *Eur. J. Cancer Prev.* 2019;28(5):397–412.

Ma X, Qi X, Chen C, Lin H, Xiong H, Li Y, et al. Association between CYP19 polymorphisms and breast cancer risk: Results from 10,592 cases and 11,720 controls. *Breast Cancer Res. Treat.* 2010;122(2):495–501.

Macé K, Aguilar F, Wang JS, Vautravers P, Gómez-Lechón M, Gonzalez FJ, et al. Aflatoxin B1-induced DNA adduct formation and p53 mutations in CYP450-expressing human liver cell lines. *Carcinogenesis*. 1997;18(7):1291–7.

Maksymchuk O V., Kashuba VI. Altered expression of cytochrome P450 enzymes involved in metabolism of androgens and vitamin D in the prostate as a risk factor for prostate cancer. *Pharmacol. Reports* [Internet]. Springer International Publishing. 2020;72(5):1161–72. Available from: https://doi.org/10.1007/s43440-020-00133–y

Malikova J, Brixius-Anderko S, Udhane SS, Parween S, Dick B, Bernhardt R, et al. CYP17A1 inhibitor abiraterone, an anti-prostate cancer drug, also inhibits the 21-hydroxylase activity of CYP21A2. *J. Steroid Biochem. Mol. Biol.* [Internet]. Elsevier Ltd. 2017;174:192–200. Available from: http://dx.doi.org/10.1016/j.jsbmb.2017.09.007

Manikandan P, Nagini S. Cytochrome P450 structure, function and clinical significance: A review. *Curr. Drug Targets*. 2017;19(1):38–54.

Le Marchand L, Donlon T, Kolonel LN, Henderson BE, Wilkens LR. Estrogen metabolism-related genes and breast cancer risk: The multiethnic cohort study. *Cancer Epidemiol. Biomarkers Prev*. 2005;14(8):1998–2003.

Le Marchand L, Donlon T, Seifried A, Wilkens LR. Red meat intake CYP2E1 genetic polymorphisms, and colorectal cancer risk. *Cancer Epidemiol. Biomarkers Prev*. 2002;11(10 I):1019–24.

Maurya SS, Katiyar T, Dhawan A, Singh S, Jain SK, Pant MC, et al. Gene-environment interactions in determining differences in genetic susceptibility to cancer in subsites of the head and neck. *Environ. Mol. Mutagen.* [Internet]. 2015;56(3):313–21. Available from: http://dx.doi.org/10.1002/em.20575

McLean KJ, Sabri M, Marshall KR, Lawson RJ, Lewis DG, Clift D, et al. Biodiversity of cytochrome P450 redox systems. *Biochem. Soc. Trans*. 2005;33(4):796–801.

Mescher M, Haarmann-Stemmann T. Modulation of CYP1A1 metabolism: From adverse health effects to chemoprevention and therapeutic options [Internet]. *Pharmacol. Ther.* Elsevier Inc. 2018. Available from: https://doi.org/10.1016/j.pharmthera.2018.02.012

Mikhailova ON, Gulyaeva LF, Prudnikov A V., Gerasimov A V., Krasilnikov SE. Estrogen-metabolizing gene polymorphisms in the assessment of female hormone-dependent cancer risk. *Pharmacogenomics J*. 2006;6(3):189–93.

Moorthy B, Chu C, Carlin DJ. Polycyclic aromatic hydrocarbons: From metabolism to lung cancer. *Toxicol. Sci.* 2015;145(1):5–15.

Morita M, Le Marchand L, Kono S, Yin G, Toyomura K, Nagano J, et al. Genetic polymorphisms of CYP2E1 and risk of colorectal cancer: The fukuoka colorectal cancer study. *Cancer Epidemiol. Biomarkers Prev*. 2009;18(1):235–41.

Murata M, Watanabe M, Yamanaka M, Kubota Y, Ito H, Nagao M, et al. Genetic polymorphisms in cytochrome P450 (CYP) 1A1, CYP1A2, CYP2E1, glutathione S-transferase (GST) M1 and GSTT1 and susceptibility to prostate cancer in the Japanese population. *Cancer Lett*. 2001;165(2):171–7.

Na Takuathung M, Hanprasertpong N, Teekachunhatean S, Koonrungsesomboon N. Impact of CYP1A2 genetic polymorphisms on pharmacokinetics of antipsychotic drugs: A systematic review and meta-analysis. *Acta Psychiatr. Scand*. 2019; 139(1):15–25.

Nelson DR, Koymans L, Kamataki T, Stegeman JJ, Feyereisen R, Waxman DJ, et al. P450 superfamily: Update on new sequences, gene mapping, accession numbers and nomenclature. *Pharmacogenetics*. 1996:1–42.

Nelson DR, Zeldin DC, Hoffman SMG, Maltais LJ, Wain HM, Nebert DW. Comparison of cytochrome P450 (CYP) genes from the mouse and human genomes, including recommendations for genes, pseudogenes and alternative-splice variants. *Pharmacogenetics*. 2004;14(1):1–18.

Niwa T, Imagawa Y, Yamazaki H. Drug interactions between nine antifungal agents and drugs metabolized by human cytochromes P450. *Curr. Drug Metab*. 2014;15(7): 651–79.

Njar V. Cytochrome P450 retinoic acid 4-hydroxylase inhibitors: Potential agents for cancer therapy. *Mini-Reviews Med. Chem*. 2005;2(3):261–9.

Nunoya KI, Yokoi T, Kimura K, Inoue K, Kodama T, Funayama M, et al. A new deleted allele in the human cytochrome (CYP2A6) gene found in individuals showing poor metabolic capacity to coumarin and (+)-cis-3,5-dimethyl-2-(3_pryidyl)thiazolidin-4-one hydrochloride (SM-12502). *Pharmacogenetics*. 1998;8:239–49.

Nunoya KI, Yokoi T, Kimura K, Kainuma T, Satoh K, Kinoshita M, et al. A new CYP2A6 gene deletion responsible for the in vivo polymorphic metabolism of (+)-cis-3,5-dimethyl-2-(3-pyridyl)thiazolidin-4-one hydrochloride in humans. *J. Pharmacol. Exp. Ther*. 1999;289(1):437–42.

Omura T. Forty years of cytochrome P450. *Biochem. Biophys. Res. Commun*. 1999;266(3):690–8.

Ortiz de Montellano PR. Hydrocarbon hydroxylation by cytochrome P450 enzymes. *Chem. Rev.* 2010;110:932–48.

Ou Y, Xu Y, Gore L, Harvey RD, Mita A, Papadopoulos KP, et al. Physiologically-based pharmacokinetic modelling to predict oprozomib CYP3A drug–drug interaction potential in patients with advanced malignancies. *Br. J. Clin. Pharmacol.* 2019;85(3):530–9.

Park JW, Kim KA, Park JY. Effects of ketoconazole, a CYP4F2 inhibitor, and CYP4F2*3 genetic polymorphism on pharmacokinetics of vitamin K1. *J. Clin. Pharmacol.* 2019;59(11):1453–61.

Parween S, Fernández-Cancio M, Benito-Sanz S, Camats N, Rojas Velazquez MN, López-Siguero J-P, et al. Molecular basis of CYP19A1 deficiency in a 46,XX patient with R550W mutation in POR: Expanding the PORD phenotype. *J. Clin. Endocrinol. Metab.* [Internet]. 2020;105 (4):e1272–90. Available from: https://academic.oup.com/jcem/article/105/4/e1272/5736381

Peter Guengerich F, Avadhani NG. Roles of cytochrome P450 in metabolism of ethanol and carcinogens. *Adv. Exp. Med. Biol.* 2018;1032:15–35.

Pineda B, García-Pérez MÁ, Cano A, Lluch A, Eroles P. Associations between aromatase CYP19 rs10046 polymorphism and breast cancer risk: From a case-control to a meta-analysis of 20,098 subjects. *PLoS One*. 2013;8(1):1–9.

Pitarque M, Von Richter O, Oke B, Berkkan H, Oscarson M, Ingelman-Sundberg M. Identification of a single nucleotide polymorphism in the TATA box of the CYP2A6 gene: Impairment of its promoter activity. *Biochem. Biophys. Res. Commun*. 2001;284(2):455–60.

Polimanti R, Piacentini S, Manfellotto D, Fuciarelli M. Human genetic variation of CYP450 superfamily: Analysis of functional diversity in worldwide populations. *Pharmacogenomics*. 2012;13(16):1951–60.

Port JL, Yamaguchi K, Du B, De Lorenzo M, Chang M, Heerdt PM, et al. Tobacco smoke induces CYP1B1 in the aerodigestive tract. *Carcinogenesis*. 2004;25(11):2275–81.

Porter TD, Coon MJ. Cytochrome P-450. Multiplicity of isoforms, substrates, and catalytic and regulatory mechanisms. *J. Biol. Chem.* 1991;266(21):13459–72.

Raunio H, Rahnasto-Rilla M. CYP2A6: Genetics, structure, regulation, and function. *Drug Metabol. Drug Interact.* 2012;27(2):73–88.

Rendic S, Guengerich FP. Survey of human oxidoreductases and cytochrome P450 enzymes involved in the metabolism of xenobiotic and natural chemicals. *Chem. Res. Toxicol.* 2015; 28(1):38–42.

Rodriguez-Antona C, Ingelman-Sundberg M. Cytochrome P450 pharmacogenetics and cancer. *Oncogene.* 2006;25(11): 1679–91.

Saarikoski ST, Rivera SP, Hankinson O, Husgafvel-Pursiainen K. CYP2S1: A short review. *Toxicol. Appl. Pharmacol.* 2005;207(2 SUPPL.):62–9.

Sachse C, Brockmöller J, Bauer S, Roots I. Functional significance of a C→A polymorphism in intron I of the cytochrome P450 CYP1A2 gene tested with caffeine. *Br. J. Clin. Pharmacol.* 1999;47(4):445–9.

Sadee W, Wang D, Papp AC, Pinsonneault JK, Smith RM, Moyer RA, et al. Pharmacogenomics of the RNA world: Structural RNA polymorphisms in drug therapy. *Clin. Pharmacol. Ther.* [Internet]. Nature Publishing Group. 2011;89(3):355–65. Available from: http://dx.doi.org/10.1038/clpt.2010.314/nature06264

Sakamoto T, Hara M, Higaki Y, Ichiba M, Horita M, Mizuta T, et al. Influence of alcohol consumption and gene polymorphisms of ADH2 and ALDH2 on hepatocellular carcinoma in a Japanese population. *Int. J. Cancer.* 2006;118(6):1501–7.

Sasaki T, Horikawa M, Orikasa K, Sato M, Arai Y, Mitachi Y, et al. Possible relationship between the risk of Japanese bladder cancer cases and the CYP4B1 genotype. *Jpn. J. Clin. Oncol.* 2008;38(9):634–40.

van Schaik RHN. CYP450 pharmacogenetics for personalizing cancer therapy. *Drug Resist. Updat.* 2008;11(3):77–98.

Schoket B, Papp G, Lévay K, Mracková G, Kadlubar FF, Vincze I. Impact of metabolic genotypes on levels of biomarkers of genotoxic exposure. *Mutat. Res. – Fundam. Mol. Mech. Mutagen.* 2001;482(1–2):57–69.

Shen MM, Abate-Shen C. Molecular genetics of prostate cancer: New prospects for old challenges. *Genes Dev.* 2010;24(18): 1967–2000.

Shen C, Meng Q. Prediction of cytochrome 450 mediated drug-drug interactions by three-dimensional cultured hepatocytes. *Mini-Reviews Med. Chem.* 2012;12(10):1028–36.

Shimada T, Fujii-Kuriyama Y. Metabolic activation of polycyclic aromatic hydrocarbons to carcinogens by cytochromes P450 1A1 and 1B1. *Gann Monogr. Cancer Res.* 2004;52(1): 109–24.

Shimada T, Murayama N, Tanaka K, Takenaka S, Imai Y, Hopkins NE, et al. Interaction of polycyclic aromatic hydrocarbons with human cytochrome P450 1B1 in inhibiting catalytic activity. *Chem. Res. Toxicol.* 2008;21(12):2313–23.

Shimada T, Oda Y, Gillam EMJ, Guengerich FP, Inoue K. Metabolic activation of polycyclic aromatic hydrocarbons and other procarcinogens by cytochromes P450 1A1 and P450 1B1 allelic variants and other human cytochromes P450 in Salmonella Typhimurium NM2009. *Drug Metab. Dispos.* 2001;29(9):1176–82.

Siegel RL, Miller KD, Jemal A. Cancer statistics, 2016. *CA Cancer J. Clin.* 2016;66(1):7–30.

Silva TD, Felipe AV, Pimenta CA, Barão K, Forones NM. CYP2E1 RsaI and 96-bp insertion genetic polymorphisms associated with risk for colorectal cancer. *Genet. Mol. Res.* 2012;11(3):3138–45.

Silvestri L, Sonzogni L, De Silvestri A, Gritti C, Foti L, Zavaglia C, et al. CYP enzyme polymorphisms and susceptibility to HCV-related chronic liver disease and liver cancer. *Int. J. Cancer.* 2003;104(3):310–7.

Sim SC, Risinger C, Dahl ML, Aklillu E, Christensen M, Bertilsson L, et al. A common novel CYP2C19 gene variant causes ultrarapid drug metabolism relevant for the drug response to proton pump inhibitors and antidepressants. *Clin. Pharmacol. Ther.* 2006;79(1):103–13.

Sofowora GG, Choo EF, Mayo G, Shyr Y, Wilkinson GR. In vivo inhibition of human CYP1A2 activity by oltipraz. *Cancer Chemother. Pharmacol.* 2001;47(6):505–10.

Song N, Tan W, Xing D, Lin D. CYP 1A1 polymorphism and risk of lung cancer in relation to tobacco smoking: A case-control study in China. *Carcinogenesis.* 2001;22(1):11–6.

Stearns V, Johnson MD, Rae J, Morocho A, Novielli A, Bhargava P, et al. Active tamoxifen metabolite plasma concentrations after coadministration of tamoxifen and the selective serotonin reuptake inhibitor paroxetine. *J. Natl. Cancer Inst.* 2003;95(23):1758–64.

Stec DE, Roman RJ, Flasch A, Rieder MJ. Functional polymorphism in human CYP4F2 decreases 20-HETE production. *Physiol. Genomics.* 2007;30(1):74–81.

Stone A, Ratnasinghe LD, Emerson GL, Modali R, Lehman T, Runnells G, et al. CYP3A43 Pro 340 Ala polymorphism and prostate cancer risk in African Americans and Caucasians. *Cancer Epidemiol. Biomarkers Prev.* 2005;14(5):1257–61.

Straume AH, Knappskog S, Lønning PE. Effect of CYP19 rs6493497 and rs7176005 haplotype status on in vivo aromatase transcription, plasma and tissue estrogen levels in postmenopausal women. *J. Steroid Biochem. Mol. Biol.* [Internet]. Elsevier Ltd. 2012;128(1–2):69–75. Available from: http://dx.doi.org/10.1016/j.jsbmb.2011.08.015

Sumińska M, Bogusz-Górna K, Wegner D, Fichna M. Non-classic disorder of adrenal steroidogenesis and clinical dilemmas in 21-hydroxylase deficiency combined with backdoor androgen pathway. Mini-review and case report. *Int. J. Mol. Sci.* 2020;21(13):1–16.

Sun MY, Du HY, Zhu AN, Liang HY, de Garibay GR, Li FX, et al. Genetic polymorphisms in estrogen-related genes and the risk of breast cancer among Han Chinese women. *Int. J. Mol. Sci.* 2015;16(2):4121–35.

Sushko T, Gilep A, Usanov S. Genetics, structure, function, mode of actions and role in cancer development of CYP17. *Anticancer. Agents Med. Chem.* 2014;14(1):66–76.

Tanner JA, Tyndale RF. Variation in CYP2A6 activity and personalized medicine. *J. Pers. Med.* 2017;7(4):1–29.

Taylor L, Krueger N, Malysheva O, Atkinson J, Parker RS. ω-Hydroxylation of α-tocopheryl quinone reveals a dual function for cytochrome P450–4F2 in vitamin E metabolism. *Bioorganic Med. Chem.* [Internet]. Elsevier. 2018;26

(20):5555–65. Available from: https://doi.org/10.1016/j.bmc.2018.10.002

Torre LA, Bray F, Siegel RL, Ferlay J, Lortet-Tieulent J, Jemal A. Global cancer statistics, 2012. *CA Cancer J. Clin.* 2015;65(2):87–108.

Tournel G, Cauffiez C, Billaut-Laden I, Allorge D, Chevalier D, Bonnifet F, et al. Molecular analysis of the CYP2F1 gene: Identification of a frequent non-functional allelic variant. *Mutat. Res. – Fundam. Mol. Mech. Mutagen.* 2007;617(1–2):79–89.

Wang C, Chen Y, Wang Y, Liu X, Liu Y, Li Y, et al. Inhibition of COX-2, mPGES-1 and CYP4A by isoliquiritigenin blocks the angiogenic Akt signaling in glioma through ceRNA effect of miR-194–5p and lncRNA NEAT1. *J. Exp. Clin. Cancer Res.* 2019a;38(1):1–14.

Wang L, Ellsworth KA, Moon I, Pelleymounter LL, Eckloff BW, Martin YN, et al. Functional genetic polymorphisms in the aromatase gene CYP19 vary the response of breast cancer patients to neoadjuvant therapy with aromatase inhibitors. *Cancer Res.* 2010;70(1):319–28.

Wang Y, Pan P, Li X, Zhu Q, Huang T, Ge RS. Food components and environmental chemicals of inhibiting human placental aromatase. *Food Chem. Toxicol.* [Internet]. Elsevier. 2019b;128(January):46–53. Available from: https://doi.org/10.1016/j.fct.2019.03.043

Wang H, Tan W, Hao B, Miao X, Zhou G, He F, et al. Substantial reduction in risk of lung adenocarcinoma associated with genetic polymorphism in CYP2A13, the most active cytochrome P450 for the metabolic activation of tobacco-specific carcinogen NNK. *Cancer Res.* 2003;63(22):8057–61.

Waring RH. Cytochrome P450: Genotype to phenotype. *Xenobiotica [Internet].* Taylor & Francis. 2020;50(1):9–18. Available from: https://doi.org/10.1080/00498254.2019.1648911

Warnakulasuriya S. Living with oral cancer: Epidemiology with particular reference to prevalence and life-style changes that influence survival. *Oral Oncol.* [Internet]. Elsevier Ltd. 2010;46(6):407–10. Available from: http://dx.doi.org/10.1016/j.oraloncology.2010.02.015

Wenzlaff AS, Cote ML, Bock CH, Land SJ, Santer SK, Schwartz DR, et al. CYP1A1 and CYP1B1 polymorphisms and risk of lung cancer among never smokers: A population-based study. *Carcinogenesis.* 2005;26(12):2207–12.

Werk AN, Cascorbi I. Functional gene variants of CYP3A4. *Clin. Pharmacol. Ther.* 2014;96(3):340–8.

Wu H, Liu Z, Ling G, Lawrence D, Ding X. Transcriptional suppression of CYP2A13 expression by lipopolysaccharide in cultured human lung cells and the lungs of a CYP2A13-humanized mouse model. *Toxicol. Sci.* 2013; 135(2):476–85.

Wu MF, Wu WJ, Chang GC, Chen CY, Hu SW, Tsai WT, et al. Increased expression of cytochrome P4501B1 in peripheral leukocytes from lung cancer patients. *Toxicol. Lett.* 2004;150(2):211–9.

Yoshida R, Nakajima M, Nishimura K, Tokudome S, Kwon JT, Yokoi T. Effects of polymorphism in promoter region of human CYP2A6 gene (CYP2A6*9) on expression level of messenger ribonucleic acid and enzymatic activity in vivo and in vitro. *Clin. Pharmacol. Ther.* 2003;74(1):69–76.

Zanger UM, Schwab M. Cytochrome P450 enzymes in drug metabolism: Regulation of gene expression, enzyme activities, and impact of genetic variation. *Pharmacol. Ther.* [Internet]. Elsevier Inc. 2013;138(1):103–41. Available from: http://dx.doi.org/10.1016/j.pharmthera.2012.12.007

Zhang Y, Klein K, Sugathan A, Nassery N, Dombkowski A, Zanger UM, et al. Transcriptional profiling of human liver identifies sex-biased genes associated with polygenic dyslipidemia and coronary artery disease. *PLoS One.* 2011;6(8) e23506.

Zhang X, Su T, Zhang Q-Y, Gu J, Caggana M, Li H, et al. Genetic polymorphisms of the human CYP2A13 gene: Identification of single-nucleotide polymorphisms and functional characterization of an Arg257Cys variant. *J. Pharmacol. Exp. Ther.* [Internet]. 2002;302(2):416–23. Available from: http://jpet.aspetjournals.org/lookup/doi/10.1124/jpet.302.2.416

Zhang C, Wei Q, Jiang T, Shou X, Li ZQ, Wen H. Liver cancer-related gene CYP2E1 expression in HBV transgenic mice with acute liver injury. *Tumor Biol.* 2014;35(4):3671–7.

Zhao B, Waterman MR. Moonlighting cytochrome P450 monooxygenases. *IUBMB Life.* 2011;63(7):473–7.

Zhou SF, Wang B, Yang LP, Liu JP. Structure, function, regulation and polymorphism and the clinical significance of human cytochrome P450 1A2. *Drug Metab. Rev.* 2010;42(2): 268–354.

Zhou J, Wen Q, Li SF, Zhang YF, Gao N, Tian X, et al. Significant change of cytochrome P450s activities in patients with hepatocellular carcinoma. *Oncotarget.* 2016;7(31):50612–23.

Zhuo X, Zhao H, Chang A, Ye H, Zhou Y, Song Y, et al. Cytochrome P450 1A1 Ile462Val polymorphism and oral carcinoma risk: An updated meta-analysis including 1,515 cases and 2,233 controls. *Tumour Biol.* 2012;33(6):2079–89.

Zhou S-F, Zhou Z-W, Yang L-P, Cai J-P. Substrates, Inducers, inhibitors and structure-activity relationships of human cytochrome P450 2C9 and implications in drug development. *Curr. Med. Chem.* 2009;16(27):3480–675.

Zou M, Toivanen R, Mitrofanova A, Floch N, Hayati S, Sun Y, et al. Transdifferentiation as a mechanism of treatment resistance in a mouse model of castration-resistant prostate cancer. *Cancer Discov.* 2017;7(7):736–49.

19

Role of Single Nucleotide Polymorphisms of Inflammatory Molecules in Susceptibility to Cardiovascular Diseases (CVDs)

Vishal Kumar Deb and Ahmad Rabbani
Department of Microbiology, School of Life and Allied Sciences (SLAS), Baba Farid Institute of Technology (BFIT), Hemvati Nandan Bahuguna Garhwal Central University, Uttarakhand, India

Sabhiya Majid
Department of Biochemistry & Multidisciplinary Research Unit (MRU-ICMR), Government Medical College Srinagar & Associated Hospitals, Karan Nagar, Srinagar, Jammu and Kashmir, India

Amir Abushouk
Department of Basic Medical Sciences, College of Medicine, King Saud bin Abdulaziz University for Health Sciences (KSAU-HS), King Abdullah International Medical Research Center (KAIMRC), Ministry of National Guard Health Affairs (NGHA), Jeddah, Kingdom of Saudi Arabia

Johra Khan
Department of Medical Laboratory Sciences, College of Applied Medical Sciences, Majmaah University, Majmaah, Saudi Arabia

Bader Al Khawaji
Basic Medical Sciences, College of Medicine, King Saud bin Abdulaziz University for Health Sciences (KSAU-HS), King Abdullah International Medical Research Center (KAIMRC), Ministry of National Guard Health Affairs (NGHA), Jeddah, Kingdom of Saudi Arabia

Ruqaya Aziz
Department of Biochemistry, SKIMS Medical College & Associated Hospital, Srinagar, India

Mujeeb Zafar Banday and Saniya Nissar
Department of Biochemistry, Government Medical College, Karan Nagar, Srinagar, Kashmir, India
Infinity Biochemistry Pvt. Ltd. Srinagar, Kashmir, India

Nusrath Yasmeen
Department of Biotechnology, Amity Institute of Biotechnology, Amity University Rajasthan, Jaipur, India

CONTENTS

DOI: 10.1201/9781003246244-19

ABSTRACT Cardiovascular disease (CVD) is a chronic pathological condition of the heart that is caused due to the deposition of the fats inside the artery that increases the chances of the clotting of blood. Numerous complications arising from the CVDs such as stroke, coronary heart disease, congenital heart disease or atherosclerotic heart disease, rheumatic heart disease, and peripheral arterial disease are reported around the world. Genetic polymorphisms, in distinct types of inflammatory molecule-coding genes, are one of the important factors increasing the risk of CVDs. Inflammation is the protective biological response to any type of injury, damage, and also an infection that is associated with pathogens. The molecules that take part in responding to any unfavorable condition of the body in the form of inflammation are known as inflammatory molecules. Many clinical-based epidemiological studies have reported that the variations in gene sequence, especially single nucleotide polymorphisms (SNPs), in various inflammatory molecule-coding genes are associated with CVD risk. In this chapter, we describe the general role of inflammatory molecules in CVDs and the role of SNPs in inflammatory molecule-coding genes (IL-1, IL-6, IL-10, TNF-α, TGF-β, and CRP) in the initiation and development of CVDs.

KEYWORDS: *Cardiovascular Diseases; Risk Factors; Single Nucleotide Polymorphisms; Inflammatory Molecules and C-reactive Proteins.*

19.1 Introduction

One-third of global deaths are caused due to cardiovascular diseases (CVDs) (WHO–CVDs, 2017). CVDs are one of the biggest health issues among the non-communicable diseases, which account for about 46% of deaths due to stroke and ischemic heart disease (WHO Press, 2014; Fiatal and Adany, 2018). A large number of factors are responsible for CVDs, including genetic factors and environmental factors (Fiatal and Adany, 2018). In different case studies of CVD patients, it has been identified that the cardiovascular traits are associated with a set of variations in genes (Fiatal and Adany, 2018). CVD is a chronic pathological condition of the heart that is caused due to the deposition of the fats inside the arteries, thereby increasing the chances of the clotting of blood. The death rate due to CVDs is increasing day by day, and it has been projected that the number of deaths will reach >23.3 million by 2023 annually (Capewell et al., 2010; Kumar, 2014). The genetic variation or polymorphism is the risk factor for 30%–50% of the risk of hypertension and 50%–60% of coronary artery disease (CAD, a type of CVDs) (Butler, 2010; Dai et al., 2016; Luft, 2017; Fiatal and Adany, 2018).

Genetic polymorphisms (variation in gene sequence), in distinct types of inflammatory molecule-coding genes, are one of the key risk factors for CVDs. The inflammatory molecules such as interleukins (ILs), interferons (IFNs), transforming growth factors (TGFs), tumor necrosis factors (TNFs), and C-reactive protein (CRP) play roles in protecting the host body from the injury by activating the host (human) body immune system. In the case of CVDs, inflammatory molecules (cytokines) modulate the risk of CVDs progression by influencing the transcription of genes (Rao et al., 2007). Numerous single nucleotide polymorphisms (SNPs) in inflammatory molecules (cytokines)-coding genes have been reported to be responsible for CVDs (Javor et al., 2007).

19.1.1 Cardiovascular Disease (CVD) and Its Types

CVD is the largest cause of death worldwide. CVD is not a single form of disease; it is a group of diseases in heart and blood vessels (known as cardiovascular system) (WHO–CVDs, 2017). The onset of this disease occurs in the age group of 30–44 years (Crespo et al., 2019). Stroke and other cardio-related health problems such as ischemic heart attack are the reason for blocking of the supply of blood to the brain that leads to cerebrovascular disease, a type of CVDs. The blockage of the blood supply may lead to damage of the brain tissue as well (Berry et al., 2019). There are many other distinct types of CVDs that are also responsible for fatalities. These include stroke, angina, congenital heart disease, rheumatic heart disease, and coronary heart disease (CHD). Stroke is a serious neurological damage where blood supply gets interrupted and the rupture of artery and hemorrhage happens. This is the reason for which artery cannot get the oxygen properly and brain cell will start to die (necrotic) and it is the other type of active causing CVDs which is the result of deposition of the atheromatous plaques inside the wall of an artery which supplies to the muscle of the heart.

The advanced level of the CHD and its associated pain is known as angina, which gets expressed in the form of symptoms such as pain in the arm, chest, and jaw. The abnormalities in the form of structure or functions of heart that developed before the birth of individuals are known as congenital heart disease. There are other special types of CVDs that are associated with bacterial pathogenesis. The streptococcus pathogenesis causes rheumatic fever and leads to damage of the heart valves and is known as rheumatic heart disease (RHD).

RHD is a heart problem where connective tissue (in the heart) gets affected due to rheumatic fever causing inflammation (Boyarchuk, 2020).

19.1.2 Risk Factors

CVD is responsible for causing the disease not only in developing societies, but also in developed societies, according to the report of Gaziano et al. (2010). Several risk factors are associated with CVDs, which are categorized as modifiable and non-modifiable. The risk factors that are modifiable include lipid-related risk factors, inflammation-related risk factors, and lifestyle-based and behavior-based risk factors. The lipid-related risk factors (an example of modifiable risk factors) are classified into low-density lipoprotein (LDL) cholesterol (the low-density lipoprotein is related to traits of gene and other major genes that is unknown and the formation of LDL occurred due to the sequential exchange of lipids between LDL and triglyceride-rich lipoproteins) and high-density lipoprotein (HDL) where the risk of CVD get an increase when the levels of HDL-c get decrease to find (Vergeer et al., 2010). Inflammation-related risk factors are due to the deposition of the fatty acids inside the wall of the artery. Various stimulations get generated through the lymphocytes and monocytes that generate inflammation in atherosclerosis or other types of CVDs (Rafieian-Kopaei et al., 2014). Lifestyle-based and behavior-based risk factors are further divided into dietary factors and tobacco (Rippe, 2019) (Figure 19.1).

19.2 Cardiovascular Disease-Associated Inflammatory Pathways

19.2.1 HDL (High-Density Lipoprotein) and Apolipoprotein-Associated Inflammatory Pathway

There are distinct types of factors that are responsible for causing CVDs; among them, high-density lipoprotein cholesterol or HDL-C in low level is a risk factor for CVD (Ali et al., 2012). Regarding the structure of the HDL, it has an esterified cholesterol in the core central region covered by a single layer

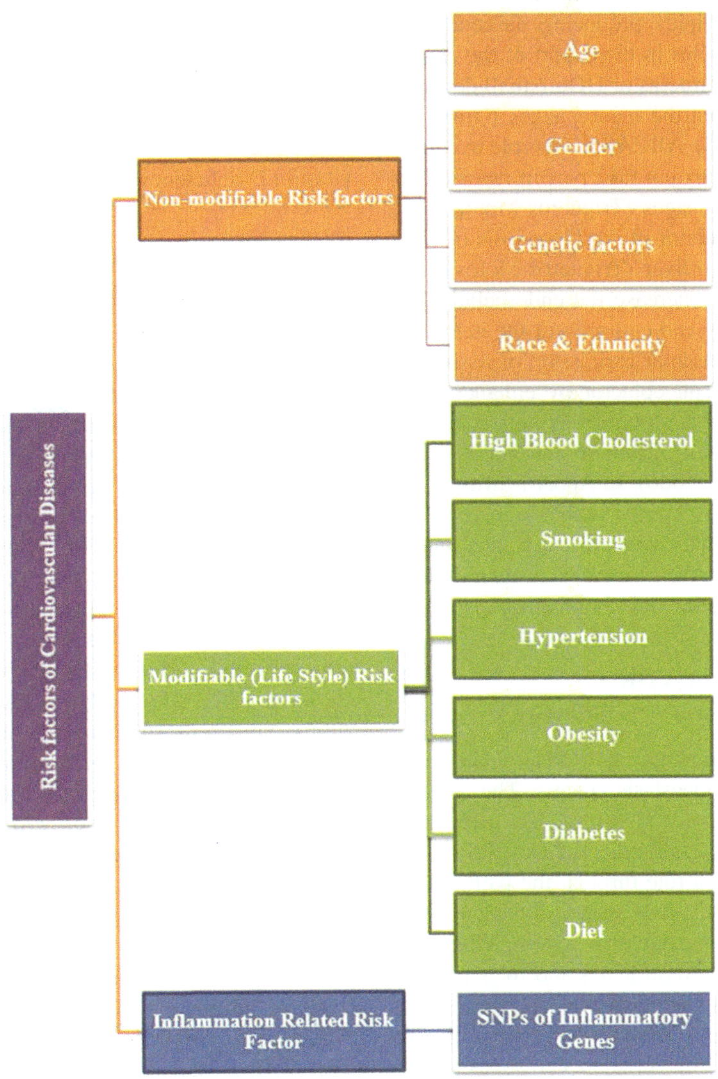

FIGURE 19.1 Three types of risk factors responsible for cardiovascular disease development.

of free cholesterol, phospholipids, and also apolipoproteins (apoA-I and apoA-II) (Heinecke, 2009; Badimon and Vilahur, 2014). Various enzymes help in dynamic remodeling as well as biogenesis of the HDL, and these enzymes are CETP (cholesteryl ester transfer protein), hepatic lipase, lecithin–cholesterol acyltransferase, etc. Based on size, HDL is divided into two particles, i.e., smaller density particles (HDL3) and large buoyant particles (HDL2) (Pirillo et al., 2013). CETP is a plasma protein that helps in the HDL-cholesteryl ester clearance pathway where ester gets uptake by the steroidogenic tissues or by liver directly and transfer it to lipoprotenis that has apoB and the process of taking up ester by the steroidogenic tissues done by the scavenger receptor BI and transfer to lipoprotein done by assisting nature of the CETP. After this process, the ester-free or cholesteryl ester-free HDL particle that has only the apoA-I gets disassociated and further enters the recycle process (Hoekstra et al., 2010).

The synthesis process of the HDL occurs in the liver and also in intestine, and in both cases, they secret apoA-I. After this process, apoA-I is able to interact with the ATP-binding cassette A1 (ABCA1—a cholesterol–phospholipid transporter that is expressed by macrophages, hepatocytes, and other cells) to transfer or exchange the lipids producing nascent HDL (Parks et al., 2012). ABCA1 helps in the export of the cholesterol to the lipid-poor apolipoproteins or HDL (small dense) in the cholesterol efflux step from the macrophages (Oram and Heinecke, 2005; Du et al., 2015). ABCA8 has its role in cholesterol efflux to apoA-I, but it may not take part in reverse cholesterol transport from macrophages (Trigueros-Motos et al., 2017). ABCA8 acts as a stimulator in the biogenesis of HDL by interacting with ABCA1 in liver (Trigueros-Motos et al., 2017). HDL has the ability to increase the endothelial nitric oxide (NO) and bioavailability and can prevent the activation of the platelet to reduce the vascular expression of tissue factor and activity of anticoagulant cofactor by enhancing the APC (activated protein C), that is protein S. Protein S helps in the reduction in the generation of thrombin and fibrin formation's precursors (Mineo et al., 2006). In ApoA-I-infused animals, it has been shown that the ApoA-I is able to increase the re-endothelialization post-percutaneous coronary intervention by increasing the recruitment of the endothelial progenitor cells and by inhibiting the inflammation or inflammatory responses.

19.2.2 Vitamin D Receptor(VDR)-Associated Inflammatory Pathway

Vitamin D is a fat-soluble vitamin that acts as a steroid hormone and participates in mineral homeostasis process. The diseases that occur due to the deficiency of vitamin D are osteomalacia (in adults) and rickets (in children) (Holick, 2007). The metabolites of vitamin D play a role through the pathways in the function of cardiovascular diseases such as thrombosis and inflammation and in the renin–angiotensin system. Vitamin D receptor (VDR) regulates the several genes and their promoters that have the vitamin D response elements, and these genes are also associated with various functions or regulatory process such as cell proliferation, cell differentiation, cell stress regulation, cell apoptosis process regulation, and regulatory processes of potential relevance to CVD (Bouillon et al., 2008). From structural point of view, VDR has two domains—an N-terminal domain, a DNA-binding domain that has two zinc fingers and a lipophilic 1,25(OH)2D-binding domain, and a C-terminal domain (a coactivator-binding domain) (Rosen et al., 2012). In the presence of UV (ultraviolet) light, vitamin D gets produced in the skin and metabolizes the enzymes such as CYP24A1, CYP27B1, CYP2R1, and CYP27B1, and theses enzymes are also known as cytochrome P450 enzymes. After the production of the vitamin D in the skin, it moves to liver through the bloodstream and binds with VD-binding proteins. To form the biologically active vitamin D (calcitriol (1,25(OH)2D)), the synthesized vitamin D is hydrolyzed with the Cyp2R1 forming calcidiol (25(OH)D) that is further hydrolyzed with Cyp27B1 to form the biologically active vitamin D in kidney.

When the production of the vitamin D get reduced, it leads to deficiency of the vitamin D that acts as a risk factor for CVDs such as myocardial infarction (MI) (Aleksova et al., 2015; Anderson et al., 2010), cardiac fibrosis (Charytan et al., 2015; Dorsch et al., 2014), peripheral arterial disease (PAD) (Nsengiyumva et al., 2015; Oh et al., 2016), and CAD (Chen et al., 2016). According to the research studies of Assalin et al. (2013) in animals, it has been reported that the vitamin D deficiency is associated with cardiac oxidative stress, metabolic changes of heart, and inflammation. Deficiency of the vitamin D decreased the VDR expression and its activity that increases the fibrosis and inflammation through the inflammatory cytokines such as TNF-α and IFN-γ. These are the ways by which vitamin D deficiency plays a key role in CVD pathogenesis.

19.2.3 Adiponectin-Associated Inflammatory Pathway

Adiponectin is an adipose tissue-secreted cytokine that is referred to as ADIPOQ and ACRP30 (Berg and Scherer, 2005; Shibata et al., 2009) and present in human's plasma (Berg and Scherer, 2005; Shibata et al., 2009). The adipose tissue-secreted cytokines are plasminogen activator inhibitor type 1, interleukin 10, tumor necrosis factor-α, adiponectin, etc., which have a direct association with obesity-mediated disorders (Berg and Scherer, 2005; Shibata et al., 2009). Structurally, adiponectin has two types of domains—N-terminal domain and C-terminal domain (a globular domain that is present in C-terminal). The relation between the CAD and the plasma level of adiponectin is very controversial as the risk of CAD is very high when the level of the adiponectin is low in plasma (Schulze et al., 2005). Numerous studies have suggested that increased levels of adipose tissue-derived inflammatory cytokines such as IL-6, VCAM-1 (vascular cell adhesion molecule 1), fibrinogen, and CRP are associated with increased BMI (body mass index) (Berg and Scherer, 2005), and CRP is the best example of the clinical inflammatory marker. Although the contribution of the adipose tissue to CRP levels is not so more, adipocyte-mediated factors have more and strong impact on the CRP production. CRP is also responsible for the elevated levels of interleukin 18, interleukin 6, SAA3, and ICAM-1 (Vasan et al., 2003; Willerson and Ridker, 2004). An elevated level of IL-6 leads to CVD (congestive heart failure) (Vasan et al., 2003). Elevated levels of CRP are also associated with the induction of PAI-1, MCP-1 (monocyte chemoattractant protein-1), and endothelin-1, and the inhibition of the NO signaling (Labarrere and Zaloga, 2004). Hypertension and improper vascular relaxation-contraction start due to the endothelin-1- and angiotensin II-mediated increased vascular contractile signals. These molecular mechanisms of the CRP lead to the risk of CVDs.

19.2.4 General Role of Inflammatory Molecules in Cardiovascular Diseases

The inflammatory molecules have an important role in cardiovascular risks such as MI (Hansson, 2005), myocarditis (Elamm et al., 2012), heart failure (Braunwald, 2008), arrhythmias (Stuart et al., 2016), sepsis-induced cardiomyopathy (Kakihana et al., 2016), and pericarditis (Imazio and Gaita, 2015; Szekely and Arbel, 2018). The cardiovascular events and atherosclerosis risk are associated with IL-1β contribution, which is found in different clinical studies (Herder et al., 2015, 2017). Different studies on CVD have shown that in the coronary arteries of a healthy human, the low level of atherosclerotic plaques gets progress by the expression of IL-1α and IL-1β which further turned into complicated plaques with high level (Van Tassell et al., 2013). IL-6 is an important cytokine that has a key role in the inflammation of the heart, especially in atherosclerosis as well as in other organs (Bacchiega et al., 2017). In case of mortality related to CVD, 50% are increased due to the pathogenesis of rheumatoid arthritis (RA) (DeMizio and Geraldino-Pardilla, 2020) and IL-6 has a role in activating the CRP by the maturation of the B lymphocytes and production of the auto-antibodies that plays a role in the RA disease progression (Turiano et al., 2013; Srirangan and Choy, 2010).

Another important interleukin is IL-10, which is considered as an anti-atherosclerotic cytokine and also has other features such as anti-thrombotic feature (Stenvinkel et al., 2005; Yilmaz et al., 2014). In the cardiac fibrosis and the cardiac injury, the role of IL-10 is rather indirect and downstream. STAT3 signals mediated by IL-10 are critical to recruit the endothelial progenitor cells from the bone marrow to the part of the heart injury (Krishnamurthy et al., 2011; Rajasingh et al., 2011; Steen et al., 2020). In CAD, atherosclerosis, and acute coronary syndrome (ACS), TGF-β acts as a predictor for disease progression in patients (Wang et al., 2019). TGF-β has an activity in the development of fibrosis in different tissues such as heart, lung, liver, and kidney.

Due to the high amount of deposition of extracellular matrix protein in the heart through the activity of TGF-β, it turns into cardiac fibrosis that is the most important pathological condition of CVDs (Biernacka et al., 2011; Pohlers et al., 2009). The more amount of extracellular matrix protein deposition can occur due to the high consumption of alcohol (Voskoboinik et al., 2019; El Hajj et al., 2014), aging (Parichatikanond et al., 2020), hypertension (Parichatikanond et al., 2020), and also diabetes (Parichatikanond et al., 2020; Yue et al., 2017), and these all are responsible for myocardial stiffness and cardiac fibrosis. Cardiac fibrosis that is induced by the TGF-β gets involved in arrhythmia pathogenesis by the electrical signal conduction disturbance. In normal heart function, the fibroblast is abundant, but pro-fibrotic mediators such as TGF-β help in myofibroblasts differentiation from the fibroblasts (Van Putten et al., 2016; Travers et al., 2016), which is the main source of the fibrosis in the heart (Parichatikanond et al., 2020).

In 1982, it was first suggested that CRP has the pro-atherogenic activity through the discovery of the binding of the VLDL and LDL with CRP in the atherosclerotic plaque (Casas et al., 2008). Atherosclerosis is an important example of CVD where in every stage of disease development, CRP is involved to complement activation, vascular endothelial cell activation, recruitment of the monocyte, and accumulation the lipid, and also, it is involved in thrombosis (Shrivastava et al., 2015). CRP is also able to facilitate innate immunity by the activation of the classical pathway (complement system) that assists in the progress of the CHD. During the inflammatory state of this disease, CRP also helps to activate the macrophages that further produce the tissue factor that helps to disseminate intravascular coagulation and leads to thrombosis. CRP has a role in increasing the activity of the protease inhibitor such as PAI-1 that inhibits the activator of tissue plasminogen and promotes fibrinolysis to cause atherogenesis (Shrivastava et al., 2015; Davis et al., 2012).

19.2.4.1 Relation between Cardiovascular Diseases and Inflammation

CVD has made a critical situation by causing death in several entities such as peripheral arterial disease and CAD (Moran et al., 2014). CVDs, especially stroke and CAD, are responsible for the deaths of 75% females and 80% males (Moran et al., 2014). Atherosclerosis is also a serious type of CVD, among others, where cholesterol plays a major role in developing this disease (Linton et al., 2019). There is also another major factor that is responsible for causing the atherogenesis, i.e., inflammatory molecules (cytokines), which have a key role in human arterial endothelial cells to promote atherogenesis (Fatkhullina et al., 2016). TNF-α, IL-6, and IL-1 are the inflammatory cytokines that are produced due to the impairment of endothelial metabolism and vascular endothelial cells and accumulation of macrophage in that specific area where CVDs (atherogenesis) develop (Tousoulis et al., 2016). RA, an example of inflammatory joint disease, leads to CVD as an independent risk factor, analogous to diabetes mellitus (Peters et al., 2009). The relation of inflammatory molecules with the disease condition may be addressing a research prospect toward the therapeutic aspect (Zhu et al., 2018).

19.3 Genetic Polymorphisms and Their Role in Inflammatory Molecule-Coding Gene in CVDs

CVDs are a combination of different heart diseases, which are very complex as well as multifactorial diseases such as CHD. These diseases are having distinct types of risk factors, which have already been discussed briefly in this chapter. Among the several types of factors that are responsible for causing CVDs, multiple genetic factors or genetic variations of inflammatory molecules have an important role in the risk of disease. Atherosclerosis is a common type of CVD where lots of evidence suggested that inflammation plays a key role in the initiation of it as an inflammatory disorder (Weber and Noels, 2011; Taube et al., 2012; Zhou et al., 2012; Shanker and Kakkar, 2010). In CVD-related diseases, various inflammatory cells are associated in the part of the artery where the inflammatory cells take part in plaque formation, vessel thrombosis that develops the lesions of the artery (Shanker and Kakkar, 2010). It was seen from various studies that SNPs in the inflammatory molecule-coding genes are able to enhance the risk of CVDs. In the gene activity, gene regulation, and function, the allele frequency in high level for variants in the constitution of the haplotypes has the maximum impact on CVDs (Terry et al., 2000) and also increases the clinical risk (Shanker and Kakkar, 2010; Humphries et al., 2008) (Figure 19.2 and Table 19.1).

19.3.1 Role of SNPs in IL-1-Coding Gene in Cardiovascular Disease

IL-1 is an important member of the interleukins or cytokines family as it has a similar function of TLR (Toll-like receptor). The broad-spectrum responses in immunological or inflammatory aspects are held on by the interleukin 1 (IL-1) cytokines and their receptors (Garlanda et al., 2013). Based on consensus sequence and ligand-binding receptors (primary) of 11 members of the interleukin 1 family, they are divided into three (3) subfamilies. They are present in the cytoplasm in diffuse form containing precursors, and each precursor has three amino acids. Interleukin 10 has three conserved consequences named as A-X-D. Different conserved sequences have different amino acids. For example, the conserved sequence A has aliphatic amino acids, X has any type of amino acids, and D has the aspartic acid (Towne et al., 2011). When IL-1 binds to its receptor, it forms a barrel shape due to the presence of the nine amino acids of IL-1 N-terminal consequences (Dinarello, 2018).

IL-1 is an important cytokine, and its polymorphism in coding gene plays an important role as a risk factor for inflammatory disease of several cardiovascular events such as Kawasaki disease (Fu et al., 2019), atherosclerotic inflammation (Rai et al., 2016), and coronary atherosclerotic heart disease. IL-1β is a family member of the IL-1 family cytokine that is also involved in inflammatory reactions through the activity of multiple cascades (Fang et al., 2018). The SNP in the IL-1β and its activity have been reported. The SNP in IL-1βC-31T (rs1143627) has a role in reducing the activity of the gene promoter, and IL-1βC-511T (rs16944) increases the transcriptional activity and increases the risk of the Kawasaki disease (Fu et al., 2019). It was also reported that the SNPs in IL-1β T+3954C (rs1143634) and IL-1αC-889T (rs1800587) have a role in the susceptibility to the Kawasaki disease in Iranian (Assari et al., 2018), Taiwanese (Weng et al., 2010), and Korean (Kim et al., 2011) populations (Fu et al., 2019). Atherosclerosis and cardiovascular autoimmune disease may

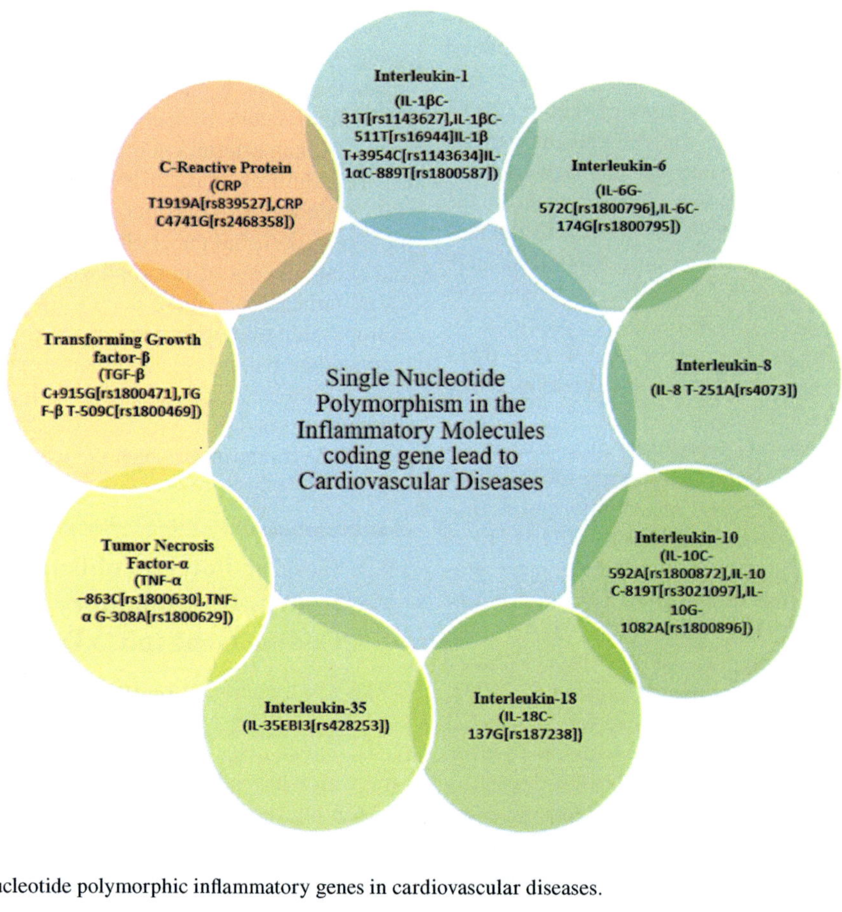

FIGURE 19.2 Single nucleotide polymorphic inflammatory genes in cardiovascular diseases.

TABLE 19.1
Molecular Weight and Role of Inflammatory Molecules in Cardiovascular Diseases

Inflammatory Molecules		Molecular Weight (mol. wt in kDa)	Role in Cardiovascular Diseases
Interleukin 1 (IL-1)	IL-1α	31	• Pro-inflammatory cytokine • Atherosclerosis risk • Complicated plaques formation in artery
	IL-1β	31	
Interleukin 6 (IL-6)		21–28	• Pro-inflammatory and anti-inflammatory cytokine • Pathogenesis of rheumatoid arthritis • Maturation of B lymphocytes • Production of auto-antibodies • Activates CRP
Interleukin 10 (IL-10)		36	• Anti-atherosclerotic cytokine • Anti-thrombotic cytokine • In the cardiac fibrosis and the cardiac injury, STAT3 signals mediated by IL-10 are critical to recruiting the endothelial progenitor cells from the bone marrow to the part of the heart injury
Transforming growth factor-β (TGF-β)		25 (prototype TGF-β1)	• Development of fibrosis in different tissues • Helps in myofibroblasts differentiation from the fibroblasts • Cardiac fibrosis
Tumor necrosis factor-α (TNF-α)		27 (proteolytically cleaved to 17 kDa)	• Integrates the innate immunity and adaptive immunity • Plays an important role in injury in the form of inflammatory molecules in cardiovascular diseases
C-reactive protein (CRP)		20–30	• Facilitates the innate immunity by the activation of the classical pathway of complement system • Role in increasing the activity of the protease inhibitors such as PAI-1 that inhibits the activator of tissue plasminogen and promotes the fibrinolysis and turned into atherogenesis

also initiate due to the IL-1β gene polymorphism, that is SNP in the position of the IL-1β C+3954T[rs1143634] (Fang et al., 2018). From other studies, it has also been reported that the SNP of the IL-1βC-511T (rs16944) can also increase the risk of atherosclerosis (Iacoviello et al., 2005; Licastro et al., 2004).

In 2006, Zhang et al. also suggested that the polymorphism (SNP) in the promoter part of the IL-1β gene at C/T locus is also associated with CAD and plays a role in dyslipidemia and aggravation of inflammation (Zhang et al., 2006). In the population of the Japanese, −511 T allele is a risk factor, which plays a role in atherosclerosis disease progression, according to the study of Oda et al. (2007). The MI is also linked with the genotype of IL-1β-511 CC, found by Iacoviello et al. (2005). Genetic variation or polymorphism (SNPs) in an IL-1 family member that acts as a pro-inflammatory molecule (IL-1alpha and IL-1beta) or as an anti-inflammatory molecule (interleukin 1 receptor antagonist) (Fearon and Fearon, 2008; Dinarello et al., 2012; Duff, 2007) affects the regulation of gene (Chen et al., 2006) and acts as an elevated level of inflammatory molecules or mediators (pro-inflammatory) in CVD progression (Berger et al., 2002; Iacoviello et al., 2005; Reiner et al., 2008; Rogus et al., 2008). Due to SNPs in IL-1, the regulation of that gene is not able to give responses in the normal way (Tsimikas et al., 2014).

19.3.2 Role of SNPs in IL-6-Coding Gene in Cardiovascular Disease

IL-6 is having multifunctional activities that can make an impact on the homeostatic process, and the most important feature of the IL-6 is that it has anti-inflammatory as well as pro-inflammatory roles. IL-6 and its receptors, i.e., IL-6RA that named as CD126 and IL-6RB that named as CD130 that help to make a hexameric complex structure with dimerization of gp130 induction and gp130 protein and this complex structure have the capacity for signaling. The IL-6R found in the body fluids in the form of soluble protein makes the signal transduction after binding to the IL-6 molecule (Tanaka et al., 2014). The molecular weight of the ligand-binding part of the IL-6R is 80–kDa, which helps it to bind with interleukin 6 (Uciechowski and Dempke, 2020; Scheller et al., 2011). The gp130 molecule has an important role in the signal transduction mechanism, mediated by the IL-6 and IL-6R, IL-6 cytokines (Uciechowski and Dempke, 2020; Scheller et al., 2011).

IL-6 is one type of pro-inflammatory cytokine that has a role in CVD pathogenesis. The genetic variation or polymorphism or SNP in IL-6 gene and its receptor is also associated with the immune response in CAD pathogenesis. The regulation of IL-6 in inflammatory reaction in the wall of vessel is censorious to thrombosis and atherogenesis (Jia et al., 2007; Fox and Kahn, 2005; Mahemuti et al., 2012). Lots of genetic as well as environmental factors are present that affect the activity of the IL-6 and its elevated level that is susceptible to VTE (venous thromboembolism) (Yadav et al., 2015).

It was indicated from the different empirical data that the development of VTE is linked with genetic factor and variation (Yadav et al., 2015; Fan et al., 2011). Different studies on IL-6 gene SNPs such as IL-6G-572C (rs1800796) and IL-6C-174G[rs1800795], which are in the IL-6 promoter region, have shown the susceptibility to the pathogenesis of CAD

(Tyagi et al., 2018), and in vitro studies also showed the same (Song et al., 2015; Yan et al., 2015; Ma et al., 2016; Mitrokhin et al., 2017), but reasons are unclear (Tyagi et al., 2018). The further investigation of the reason for that study was expanded for the identified additional SNPs in the genes of IL-6 and its receptor (IL-6R) in dbSNP database (Mitrokhin et al., 2017).

19.3.3 Role of SNPs in IL-8-Coding Gene in Cardiovascular Disease

Interleukin 8 is produced by the macrophages that act as a pro-inflammatory mediator (Danikiewicz et al., 2017). The increased level of IL-8 has been shown in CAD patients (Danikiewicz et al., 2017). The elevated levels of IL-8 in serum also act as a predictor for acute coronary syndrome and acute MI (Cavusoglu et al., 2015). Zhang et al. reported that the SNP in the promoter region of the IL-8T-251A (rs4073) is associated with the susceptibility to acute coronary syndrome (Zhang et al., 2011). Zhang et al. also reported from their research study that patients having the AT and AA genotypes have more chances of risk of CAD than the patients with TT genotypes (Zhang et al., 2017).

19.3.4 Role of SNPs in IL-10-Coding Gene in Cardiovascular Disease

IL-10 is another interleukin that is located on chromosome 1 and has the role as an important cytokine in the form of an anti-inflammatory molecule. IL-10 coding gene comprises five exons and has a total of 5.1 kbp (Sabat et al., 2010). IL-10 is a homodimer having 35 kDa molecular weight, which consists of two non-covalently bonded monomers. The homodimer of the IL-10 has two V-shaped domains with six helices, and out of the six helices, four helices are named as A, B, C, and D and two are named as E0 and F0, and they have a role in maintaining the structure and cytokine activity. Interleukin 10 receptor has IL-10R1 and IL-10R2, two distinct chains that belong to the cytokine class II receptor family that is transmembrane glycoproteins. The extracellular domain of the cytokine class II receptor has 210 amino acids, and the intracellular domains are different in length (Sabat et al., 2010).

IL-10 is a well-known anti-inflammatory cytokine that suppresses the inflammatory expression by inactivating the regulation of the Th1 cell and also by suppressing the activity of the IL-1, TNF-α, TGF-β, and IL-6 (the pro-inflammatory cytokines) (Mosser and Zhang, 2008). By preserving the fibrous cap, IL-10 promotes plaque stability (Apostolakis et al., 2007), which also acts as a factor to develop CHD among the people and also helps to develop atherosclerosis by blocking the activity of the metalloproteinase (Mousavi et al., 2018). Atherosclerosis is a developed form of CAD that develops due to the deposition of the plaques in the walls of arteries (Trégouët et al., 2009). The expression of the IL-10 is influenced by the polymorphisms of the IL-10 gene or the variable sites (microsatellites) of the IL-10 (Mousavi et al., 2018). The transcription activity of the IL-10 gene and the increasing level of the IL-10 in the plasma are influenced by the SNPs in the promoter region of the gene IL-10, that is IL-10C-592A[rs1800872], IL-10 C-819T (rs3021097), and IL-10G-1082A (rs1800896) (Zuo et al., 2014; Fragoso et al., 2011), and these SNPs are also involved in the thrombotic disease of the artery (Xuan et al., 2016).

19.3.5 Role of SNPs in IL-18-Coding Gene in Cardiovascular Disease

Interleukin 18 has a direct association with the development of unstable atherosclerotic plaques as it acts as a pro-atherogenic and pro-inflammatory cytokine (Tiret et al., 2005; Hernesniemi et al., 2009) (Table 19.2). The factors such as MMPs (matrix

TABLE 19.2

Polymorphic Genes and Cardiovascular Diseases

Inflammatory Molecules	Polymorphisms (SNPs)	Cardiovascular Diseases
IL-1	IL-1βC-31T[rs1143627]	Kawasaki disease
	IL-1βC-511T[rs16944]	
	IL-1βT+3954C[rs1143634]	
	IL-1αC-889T[rs1800587]	
IL-6	IL-6G-572C[rs1800796]	Pathogenesis of coronary artery disease
	IL-6C-174G[rs1800795]	
IL-8	IL-8T-251A[rs4073]	Coronary artery disease
IL-10	IL-10C-592A[rs1800872]	Thrombotic disease of artery/atherosclerosis
	IL-10C-819T[rs3021097]	
	IL-10G-1082A[rs1800896]	
IL-18	IL-18C-137G[rs187238]	Myocardial infarction
IL-35	IL-35EBI3[rs428253]	Coronary heart disease
TGF-β	TGF-βC+915G[rs1800471]	Cardiac fibrosis/coronary artery disease
	TGF-βT-509C[rs1800469]	
TNF-α	TNF-α −863C[rs1800630]	Coronary artery disease
	TNF-αG-308A[rs1800629]	
CRP	CRP T1919A[rs839527]	Coronary heart disease
	CRP C4741G[rs2468358]	

metalloproteinases) and IFN-γ (interferon-gamma) are associated with the development of atherosclerotic plaques, and these factors are secreted by the IL-18 cytokine (Hernesniemi et al., 2009; Gerdes et al., 2002). These developed unstable plaques lead to an increase in the risk of ASEs (acute cardiac events) to develop coronary stenosis (Hernesniemi et al., 2009). The SNP in the promoter region of the IL-18C-137G (rs187238) may increase the association of the acute fatal coronary events and MI and SCD (Mallat et al., 2002; Hernesniemi et al., 2009), but the mechanism of the IL-18 cytokine SNP-mediated SCD occurrence is not clear. Hernesniemi et al. (2009) concluded from their research study that the development of CAD and SCD and atherosclerotic plaques is modulated by the SNP of IL-18C2137G promoter region.

19.3.6 Role of SNPs in IL-35-Coding Gene in Cardiovascular Disease

IL-35 is a member of the family of IL-12 that is also an anti-inflammatory cytokine, and structurally, it is a heterodimer having two subunits, i.e., p35 and EBI3 (Huang et al., 2013; Vignali and Kuchroo, 2012). Two studies have been conducted on the relationship between gene polymorphisms of IL-35 and its susceptibility to CHD (Posadas-Sánchez et al., 2017; Lin et al., 2018). Hu et al. (2020) reported that the smokers having rs428253 GC/CC polymorphism has a high risk of CHD than the non-smokers who are having the IL-35EBI3 (rs428253) genotype polymorphism in IL-35-coded gene. Recent studies on the IL-35 polymorphism have suggested that the C allele polymorphism of IL-35EBI3 (rs428253) in the IL-35 gene acts as a risk factor for CHD (Hu et al., 2020).

19.3.7 Role of SNPs in IL-37-Coding Gene in Cardiovascular Disease

IL-37 (IL-1F7) belongs to the IL-1 family that modulates the immune responses and inflammatory responses (Kumar et al., 2000; Veerdonk and Netea, 2013). The translocation of the IL-37 to the nucleus occurs after the processing of the IL-37 caspase-1 site, and this translocated IL-37 reduces the production of the pro-inflammatory cytokines in murine RAW cells, meaning that IL-37 works as an anti-inflammatory cytokine (Sharma et al., 2008; Nold et al., 2010). Based on their research study, Yin et al. (2017) reported that SNPs of IL-37, that is SNP in the minor allele A of rs3811047 in the IL-37 gene, are associated with CAD as a risk factor.

19.3.8 Role of SNPs in TNF-α-Coding Gene in Cardiovascular Disease

TNF-α plays an important role in injury in the form of inflammatory molecules, and by the same way, it gives responses to CVD. Csiszar et al. (2004, 2007) suggested from their studies on young rats and its carotid arteries that TNF-α (that is recombinant) elicits oxidative stress, endothelial dysfunction, and also gene expression of the pro-inflammatory molecules. Due to the TNF-α induction, the endothelial cytoskeleton gets destabilized and forms the intracellular gap that allows the movement of macromolecules across the monolayer site of endothelial cells (Urschel and Cicha, 2015). TNF-α does not take part in the endothelial apoptosis, and it just acts as an endothelial cells' potent activator (Urschel and Cicha, 2015).

In CVDs such as CAD and MI, TNF-α takes part as a risk factor (Chu et al., 2012). Different experimental studies on SNPs of TNF-α-coding gene and in its promoter region and their associated CVDs in different populations of different countries have been done. In Pakistani population, the SNPs in the promoter region of TNF-α gene were studied and it was observed that the SNPs in −863C>A, that is TNF-α −863C (rs1800630) (promoter region of TNF-α gene), are associated with the CAD pathogenesis (Asifa et al., 2013). TNF-α level in serum is also linked with the polymorphism or genetic variation in TNF-α gene promoter region (Asifa et al., 2013; Bennet et al., 2006; Sharma et al., 2006; Boehm et al., 2011). Hongxia Chu et al. also suggested through their study on the TNF-α G-308A (rs1800629) polymorphism that TNF-α G-308A (rs1800629) has an association with the CAD and MI risk (Chu et al., 2012).

19.3.9 Role of SNPs in TGF-β-Coding Gene in Cardiovascular Disease

The TGF-β consists of a pro-domain with 250 residues and a domain of growth factor, having 110 residues, containing a single peptide that is encoded by 33 genes of TGF-β (Hinck et al., 2016). TGF-β, a multifunctional cytokine, is able to bind to serine/threonine kinase complex-containing membrane receptors (Hanafy and El Kemary, 2019). Smad2 and Smad3 are the transcription factors that are phosphorylated by the binding of TGF-β to receptors and then further bind to the Smad4 and, after that, start to regulate the transcription of a gene which is targeted in the nucleus (Hanafy, 2018; Hanafy et al., 2018). There are three types of isoforms of TGF-β, TGF-β1, TGF-β2, and TGF-β3, and these have highly conserved regions and also have three domains—an N-terminal domain that is extracellular, a transmembrane domain, and a C-terminal domain that contains serine/threonine kinase. The molecular weight of the TGF-β receptors is 85–110 kDa for type II receptors and 65–70 kDa of type I receptors (Hanafy and El Kemary, 2019).

The TGF-β has its putative signaling role in CHD, as suggested from the data of gene studies (Manginas et al., 2008; Sie et al., 2006). It was described that the TGF-β and its isoforms have a role in allowing the development of the fibrous plaque. The TGF-β1 has a fundamental role in the upregulation of the production of the extracellular matrix that makes inhibition of the proliferation of the vascular cells and reduces the division of the endothelial cells (Morris et al., 2012). The risk of the devilment of the several common types of CVDs is associated with the dysregulation of the signaling pathways of TGF-β, and it occurs due to the genetic variation or polymorphism, especially the SNPs in the TGF-β coding gene (Morris et al., 2012; Biros et al., 2011).

Lots of studies have been done on the relationship between the SNPs of TGF-β and CVD; among them, some genes have a negative link with CHD (Manginas et al., 2008; Sie et al., 2006;

TABLE 19.3
Receptors of C-reactive Protein and Their Function

Receptors of C-Reactive Protein	FcγRIII	FcγRIIA	FcγRIIB	FcγRI
Cluster of differentiation (other name of CRP receptors)	CD16	CD32A	CD32B	CD64
Ligands of C-reactive protein	mCRP	pCRP	mCRP/Pcrp	pCRP
Function	• LPL expression • Neutrophil binding promotes by it	• Phagocytosis process mediator • Platelets binding inhibition to neutrophils	• Inhibition of eNOS activation that is mediated by bradykinin and insulin	• Phagocytosis process mediator
References	Shih et al. (2009), Fujita et al. (2011), Boncler et al. (2011)	Shih et al. (2009), Fujita et al. (2011)	Hein et al. (2014)	Shih et al. (2009)

Morris et al., 2012). Some confirmed SNPs of the TGF-β have reported that they have a role in the CHD, and these SNPs are TGF-β C+915G (rs1800471) and TGF-β T-509C[rs1800469] (Morris et al., 2012). The SNP in the locus of TGF-β1 (+915 G → C at codon 25 and +869 T → C at codon 10) is associated with the development of the atherosclerotic disease and cardiac death (Rao et al., 2004).

19.3.10 Role of SNPs in CRP (C-Reactive Protein)-Coding Gene in Cardiovascular Disease

CRP is one type of protein that is found in the plasma, and it has two distinct types of conformational forms, i.e., monomeric isoform of CRP and pentameric isoform, which have different biological activities (Huang et al., 2009; Salazar et al., 2014).

Lots of evidence have suggested the association of the inflammation with the CVD (Markt et al., 2014; Hage and Szalai, 2009). The expression of the inflammatory molecules such as CRP as a pro-inflammatory molecule in plasma acts as an inflammatory biomarker. The immune cells also induce several chemokines and cytokines such as IL-6, which also induce the production of the CRP molecule (Berezin, 2018). In the stress condition of the liver cells and adipose tissue, CRP starts to secrete (Berezin, 2018). Pai et al. reported that the SNPs of CRP-coding gene—CRP T1919A (rs839527) and CRP C4741G (rs2468358)—are linked to the risk of CHD (Pai et al., 2008) (Table 19.3).

19.4 Conclusions

CVDs have been found as serious death-causing health issues that are not only associated with a single factor. These diseases have multiple factors such as lipids, inflammation-based risk factors, and behavioral and lifestyle-based risk factor, which are also divided into various sub-branches in the form of risk factors. After several studies on CVDs, it has been projected that the number of deaths will reach >23.3 million by 2023 annually and CVDs are arising as a thinkable problem in developing as well as developed countries. As CVDs have become a continuously increasing problem, it has come out from various studies that the polymorphisms, especially SNPs in the inflammation-causing molecule's genes, are also a major factor to promote this disease. The SNPs in the distinct types of inflammatory cytokines have a role in developing several problems such as dysregulation of the normal signaling pathway, stress condition in tissues, increase in the chance of other autoimmune diseases, and deposition of plaques in the walls of arteries, but the mechanism(s) that are associated with SNPs to develop CVDs are not fully understood. After explaining the different types of inflammatory molecule-coding genes' SNPs, the SNP in the CRP gene and other inflammatory molecule-coding genes may act as a predictive biomarker for the cardiovascular disease diagnosis.

ABBREVIATIONS

ABCA1: ATP-binding cassette A1
ACS: Acute coronary syndrome
APC: Activated protein C
ApoA-I: Apolipoprotein A-I
BMI: Body mass index
CAD: Coronary artery disease
CD: Cluster of differentiation
CETP: Cholesteryl ester transfer protein
CHD: Coronary heart disease
CRP: C-reactive protein
CVDs: Cardiovascular diseases
HDL: High-density lipoprotein
HDL-C: High-density lipoprotein cholesterol
ICAM-1: Intercellular adhesion molecule 1
IFN: Interferon
IL: Interleukin
kDa: Kilodaltons
LDL: Low-density lipoprotein
MI: Myocardial infarction
NO: Nitrogen monoxide
PAD: Peripheral arterial disease
PAI: Plasminogen activator inhibitor
RHD: Rheumatic heart disease
SAA3: Serum amyloid A

SNPs: Single nucleotide polymorphisms
STAT: Signal transducers and activators of transcription
TGF: Transforming growth factor
TLR: Toll-like receptor
TNF: Tumor necrosis factor
VCAM: Vascular cell adhesion molecule
VDR: Vitamin D receptor

REFERENCES

Aleksova A, Belfiore R, Carriere C, Kassem S, La Carrubba S, Barbati G et al. (2015) Vitamin D deficiency in patients with acute myocardial infarction – an Italian single-center study. *International Journal for Vitamin and Nutrition Research*. 85(1–2):23. doi: 10.1024/0300–9831/a000220.

Ali KM, Wonnerth A, Huber K, Wojta J. (2012) Cardiovascular disease risk reduction by raising HDL cholesterol–current therapies and future opportunities. *British Journal of Pharmacology*. 167(6):1177. doi: 10.1111/j.1476–5381.2012.02081.x.

Anderson JL, May HT, Horne BD, Bair TL, Hall NL, Carlquist JF et al. (2010) Relation of vitamin D deficiency to cardiovascular risk factors, disease status, and incident events in a general healthcare population. *The American Journal of Cardiology*. 106(7):963. doi: 10.1016/j.amjcard.2010.05.027.

Apostolakis S, Baritaki S, Kochiadakis GE, Igoumenidis NE, Panutsopulos D, Spandidos DA. (2007) Effects of polymorphisms in chemokine ligands and receptors on susceptibility to coronary artery disease. *Thrombosis Research*. 119(1):63. doi: 10.1016/j.thromres.2005.12.016.

Asifa GZ, Liaquat A, Murtaza I, Kazmi SAR and Javed Q. (2013) Tumor necrosis factor-alpha gene promoter region polymorphism and the risk of coronary heart disease. *The Scientific World Journal*. doi: 10.1155/2013/203492.

Assalin HB, Rafacho BP, Santos PPD, Ardisson LP, Roscani MG, Chiuso-Minicucci F et al. (2013). Impact of the length of vitamin D deficiency on cardiac remodeling. *Circulation: Heart Failure*. 6(4):809. doi: 10.1161/CIRCHEARTFAILURE.112.000298.

Assari R, Aghighi Y, Ziaee V, Sadr M., Rahmani F, Rezaei A et al. (2018) Pro-inflammatory cytokine single nucleotide polymorphisms in Kawasaki disease. *International Journal of Rheumatic Diseases*. 21(5):1120. doi: 10.1111/1756-185X.12911.

Bacchiega BC, Bacchiega AB, Usnayo MJG, Bedirian R, Singh G, Pinheiro GDRC. (2017) Interleukin 6 inhibition and coronary artery disease in a high-risk population – a prospective community-based clinical study. *Journal of the American Heart Association*. 6(3):e005038. doi: 10.1161/JAHA.116.005038.

Badimon L, Vilahur G. (2014) HDL particles—more complex than we thought. *Thrombosis and Haemostasis*. 112–857.

Bennet AM, Van Maarle MC, Hallqvist J, Morgenstern R, Frostegård J, Wiman B et al. (2006) Association of TNF-α serum levels and TNFA promoter polymorphisms with risk of myocardial infarction. *Atherosclerosis*. 187(2):408. doi: 10.1016/j.atherosclerosis.2005.09.022.

Berezin A. (2018) The single nucleotide polymorphisms in the C-reactive protein gene – are they biomarkers of cardiovascular risk? *International Biological and Biomedical Journal*. 4(2):122.

Berg AH, Scherer PE. (2005) Adipose tissue, inflammation, and cardiovascular disease. *Circ Res*. May 13;96(9): 939–949. doi: 10.1161/01.RES.0000163635.62927.34. PMID: 15890981.

Berger P, McConnell JP, Nunn M, Kornman KS, Sorrell J, Stephenson K et al. (2002) C-reactive protein levels are influenced by common IL-1 gene variations. *Cytokine*. 17(-4):171. doi: 10.1006/cyto.2001.0974.

Berry C, Sidik N, Pereira AC, Ford TJ, Touyz RM, Kaski JC et al. (2019) Small-vessel disease in the heart and brain – current knowledge, unmet therapeutic need, and future directions. *Journal of the American Heart Association*. 8(3):e011104. doi: 10.1161/JAHA.118.011104.

Biernacka A, Dobaczewski M, Frangogiannis NG. (2011) TGF-β signaling in fibrosis. *Growth Factors*. 29(5):196. doi: 10.3109/08977194.2011.595714.

Biros E, Norman PE, Jones GT, van Rij AM, Yu G, Moxon JV et al. (2011) Meta-analysis of the association between single nucleotide polymorphisms in TGF-β receptor genes and abdominal aortic aneurysm. *Atherosclerosis*. 219(1):218–223. doi: 10.1016/j.atherosclerosis.2011.07.105.

Boehm J, Hauner K, Grammer J, Dietrich W, Wagenpfeil S, Braun S et al. (2011) Tumor necrosis factor-α–863 C/A promoter polymorphism affects the inflammatory response after cardiac surgery. *European Journal of Cardio-Thoracic Surgery*. 40(1):e50–4. doi: 10.1016/j.ejcts.2011.01.084.

Boncler M, Rywaniak J, Szymański J, Potempa LA, Rychlik B, Watała C. (2011) Modified C-reactive protein interacts with platelet glycoprotein Ibα. *Pharmacological Reports*. 63(2):464.

Bouillon R, Carmeliet G, Verlinden L, van Etten E, Verstuyf A, Luderer HF et al. (2008) Vitamin D and human health – lessons from vitamin D receptor null mice. *Endocrine Reviews*. 29(6):726. doi: 10.1210/er.2008–0004.

Boyarchuk O. (2020) Features of heritable disorders of connective tissue in children with acute rheumatic fever and rheumatic heart disease. *Reumatologia*. 58(1):21.

Braunwald E. (2008) Biomarkers in heart failure. *New England Journal of Medicine*. 358(20):2148.

Butler MG. (2010) Pediatric hypertension – genetics of hypertension. Current status. *Le Journal medical libanais. The Lebanese Medical Journal*. 58(3):175.

Capewell S, Ford ES, Croft JB, Critchley JA, Greenlund KJ, Labarthe DR. (2010) Cardiovascular risk factor trends and potential for reducing coronary heart disease mortality in the United States of America. *Bulletin of the World Health Organization*. 88:120. doi: 10.2471/BLT.08.057885.

Casas JP, Shah T, Hingorani AD, Danesh J, Pepys MB. (2008) C-reactive protein and coronary heart disease – a critical review. *Journal of Internal Medicine*. 264(4):295. doi: 10.1111/j.1365–2796.2008.02015.x.

Cavusoglu E, Marmur JD, Yanamadala S, Chopra V, Hegde S, Nazli A et al. (2015) Elevated baseline plasma IL-8 levels are an independent predictor of long-term all-cause mortality in patients with acute coronary syndrome. *Atherosclerosis*. 242(2):589. doi: 10.1016/j.atherosclerosis.2015.08.022.

Charytan DM, Padera RF, Helfand AM, Zeisberg EM. (2015) Association of activated vitamin D use with myocardial fibrosis and capillary supply – results of an autopsy study.

Renal Failure. 37(6):1067. doi: 10.3109/0886022X.2015.1040704.

Chen S, Swier VJ, Boosani CS, Radwan MM, Agrawal DK. (2016). Vitamin D deficiency accelerates coronary artery disease progression in swine. *Arteriosclerosis, Thrombosis, and Vascular Biology.* 36(8):1651. doi: 10.1161/ATVBAHA.116.307586.

Chen H, Wilkins LM, Aziz N, Cannings C, Wyllie DH, Bingle C et al. (2006) Single nucleotide polymorphisms in the human interleukin-1B gene affect transcription according to haplotype context. *Human Molecular Genetics.* 15(4):519. doi: 10.1093/hmg/ddi469.

Chen H, Xiao L, Zhang H, Liu N, Liu T, Liu L et al. (2011) The involvement of β-actin in the signaling of transmembrane TNF-α-mediated cytotoxicity. *Journal of Leukocyte Biology.* 89(6):917. doi: 10.1189/jlb.1209812.

Chu H, Yang J, Mi S, Bhuyan SS, Li J, Zhong L et al. (2012) Tumor necrosis factor-alpha G-308 A polymorphism and risk of coronary heart disease and myocardial infarction – a case–control study and meta-analysis. *Journal of Cardiovascular Disease Research.* 3(2):84. doi: 10.4103/0975–3583.95359.

Crespo CJ, Di Angelantonio E, WHO CVD Risk Chart Working Group. (2019) World Health Organization cardiovascular disease risk charts – revised models to estimate risk in 21 global regions. *The Lancet.* 2019;7.

Csiszar A, Labinskyy N, Smith K, Rivera A, Orosz Z, Ungvari Z. (2007) Vasculoprotective effects of anti-tumor necrosis factor-α treatment in aging. *The American Journal of Pathology.* 170(1):388. doi: 10.2353/ajpath.2007.060708.

Csiszar A, Ungvari Z, Koller A, Edwards JG, Kaley G. (2004) Proinflammatory phenotype of coronary arteries promotes endothelial apoptosis in aging. *Physiological Genomics.* 17(1):21.doi: 10.1152/physiolgenomics.00136.2003.

Dai X, Wiernek S, Evans JP, Runge MS. (2016) Genetics of coronary artery disease and myocardial infarction. *World Journal of Cardiology.* 8(1):1. doi: 10.4330/wjc. v8.i1.1.

Danikiewicz A, Szkodzinski J, Hudzik B, Korzonek-Szlacheta I, Gąsior M, Zubelewicz-Szkodzińska B. (2017) Effects of trimetazidine on interleukin-2 and interleukin-8 concentrations in patients with coronary artery disease. *Canadian Journal of Physiology and Pharmacology.* 2017;95:759. doi: 10.1139/cjpp–2016–0424.

Davis FJ, Vidyasagar S, Maiya GA. (2012) C-reactive protein and coronary heart disease-risk marker or risk factor. *Journal of Clinical and Scientific Research.* 1:178.

DeMizio DJ, Geraldino-Pardilla LB. (2020) Autoimmunity and inflammation link to cardiovascular disease risk in rheumatoid arthritis. *Rheumatology and Therapy.* 7(1):19. doi: 10.1007/s40744-019-00189–0.

Dinarello CA. (2018) Overview of the IL-1 family in innate inflammation and acquired immunity. *Immunological Reviews.* 281(1):8–27. doi: 10.1111/imr.12621.

Dinarello CA, Simon A, Van Der Meer JW. (2012) Treating inflammation by blocking interleukin-1 in a broad spectrum of diseases. *Nature Reviews Drug Discovery.* 11(8):633.

Dorsch MP, Nemerovski CW, Ellingrod VL, Cowger JA, Dyke DB, Koelling TM et al. (2014) Vitamin D receptor genetics on extracellular matrix biomarkers and hemodynamics in systolic heart failure. *Journal of Cardiovascular Pharmacology and Therapeutics.* 19(5):439. doi: 10.1177/1074248413517747.

Du XM, Kim MJ, Hou L, Le Goff W, Chapman MJ, Van Eck M et al. (2015). HDL particle size is a critical determinant of ABCA1-mediated macrophage cellular cholesterol export. *Circulation Research.* 116(7):1133. doi: 10.1161/CIRCRESAHA.116.305485.

Duff GW. (2007) Influence of genetics on disease susceptibility and progression. *Nutrition Reviews.* 65(suppl_3):S177–S181. doi: 10.1301/nr.2007.dec.S177–S181.

Elamm C, Fairweather D, Cooper LT. (2012) Pathogenesis and diagnosis of myocarditis. *Heart.* 98(11):835. doi: 10.1136/heartjnl–2012–301686.

El Hajj EC, El Hajj MC, Voloshenyuk TG, Mouton AJ, Khoutorova E, Molina PE et al. (2014) Alcohol modulation of cardiac matrix metalloproteinases (MMPs) and tissue inhibitors of MMP s favors collagen accumulation. *Alcoholism – Clinical and Experimental Research.* 38 (2):448. doi: 10.1111/acer.12239.

Fan WH, Liu DL, Xiao LM, Xie CJ, Sun SY, Zhang JC. (2011) Coronary heart disease and chronic periodontitis – is polymorphism of interleukin-6 gene the common risk factor in a Chinese population? *Oral Diseases.* 17(3):270. doi: 10.1111/j.1601–0825.2010.01736.x.

Fang Y, Xie H, Lin Z. (2018) Association between IL-1β+3954C/T polymorphism and myocardial infarction risk – a meta-analysis. *Medicine.* 97(30). doi: 10.1097/MD.0000000000011645.

Fatkhullina AR, Peshkova IO, Koltsova EK. (2016) The role of cytokines in the development of atherosclerosis. *Biochemistry (Moscow).* 81(11):1358.doi: 10.1134/S0006297916110134.

Fearon WF, Fearon DT. (2008) Inflammation and cardiovascular disease – role of the interleukin-1 receptor antagonist. *Circulation.* 117:2577. doi: 10.1161/CIRCULATIONAHA.108.772491.

Fiatal S, Adany R. (2018) Application of single-nucleotide polymorphism-related risk estimates in identification of increased genetic susceptibility to cardiovascular diseases – a literature review. *Frontiers in Public Health.* 5:358. doi: 10.3389/fpubh.2017.00358.

Fox EA, Kahn SR. (2005) The relationship between inflammation and venous thrombosis. *Thrombosis and Haemostasis.* 94 (2):362. doi: 10.1160/TH05–04–0266.

Fragoso JM, Vallejo M, Alvarez-León E, Delgadillo H, Peña-Duque MA, Cardoso-Saldaña G et al. (2011) Alleles and haplotypes of the interleukin 10 gene polymorphisms are associated with risk of developing acute coronary syndrome in Mexican patients. *Cytokine.* 55(1):29–33. doi: 10.1016/j.cyto.2011.03.021.

Fu LY, Qiu X, Deng QL, Huang P, Pi L, Xu Y, et al. (2019) The IL-1B gene polymorphisms rs16944 and rs1143627 contribute to an increased risk of coronary artery lesions in Southern Chinese Children with Kawasaki disease. *Journal of Immunology Research.* 2019. doi: 10.1155/2019/4730507.

Fujita Y, Yamaguchi S, Kakino A, Iwamoto S, Yoshimoto R, Sawamura T. (2011) Lectin-like oxidized LDL receptor 1 is involved in CRP-mediated complement activation. *Clinical Chemistry.* 57(10):1398. doi: 10.1373/clinchem.2011.168625.

Garlanda C, Dinarello CA, Mantovani A. (2013) The interleukin-1 family – back to the future. *Immunity.* 39(6):1003. doi: 10.1016/j.immuni.2013.11.010.

Gaziano TA, Bitton A, Anand S, Abrahams-Gessel S, Murphy A. (2010) Growing epidemic of coronary heart disease in low-and middle-income countries. *Current Problems in Cardiology.* 35(2):72. doi: 10.1016/j.cpcardiol.2009.10.002.

Gerdes N, Sukhova GK, Libby P, Reynolds RS, Young JL, Schönbeck U. (2002) Expression of interleukin (IL)-18 and functional IL-18 receptor on human vascular endothelial cells, smooth muscle cells, and macrophages – implications for atherogenesis. *The Journal of Experimental Medicine.* 195(2):245. doi: 10.1084/jem.20011022.

Hage FG, Szalai AJ. (2009) The role of C-reactive protein polymorphisms in inflammation and cardiovascular risk. *Current Atherosclerosis Reports.* 11(2):124. doi: 10.1007%2Fs11883-009-0020–z.

Hanafy NAN. (2018) The growth of hepatocellular carcinoma can be inhibited by encapsulation of TGF ß1 antagonists. *Pharmacology & Toxicology.* 1(1):11.

Hanafy NA, El Kemary M. (2019) TGFβ1 as a Good and Bad Biological Molecule – structure and function. *Biomedical Journal of Scientific & Technical Research.* 17;(2):2019. doi: 10.26717/BJSTR.2019.17.002966.

Hanafy NAN, El-Kemary M, Leporatti S. (2018) Understanding TGF β1 signalling pathway is well strategy to use its encapsulated antagonist as nano therapeutic molecules. *Journal of Translational Science.* doi: 10.15761/JTS.1000264.

Hansson GK. (2005) Inflammation, atherosclerosis, and coronary artery disease. *New England Journal of Medicine.* 352 (16):1685. doi: 10.1056/NEJMra043430.

Hein TW, Qamirani E, Ren Y, Xu X, Thengchaisri N, Kuo L. (2014) Selective activation of lectin-like oxidized low-density lipoprotein receptor-1 mediates C-reactive protein–evoked endothelial vasodilator dysfunction in coronary arterioles. *Circulation Research.* 114(1):92. doi: 10.1161/CIRCRESAHA.114.301763.

Heinecke JW. (2009) The HDL proteome – a marker–and perhaps mediator–of coronary artery disease. *Journal of Lipid Research.* 50(Supplement):S167. doi: 10.1194/jlr.R800097-JLR200.

Herder C, Dalmas E, Böni-Schnetzler M, Donath MY. (2015) The IL-1 pathway in type 2 diabetes and cardiovascular complications. *Trends in Endocrinology & Metabolism.* 26(-10):551. doi: 10.1016/j.tem.2015.08.001.

Herder C, de Las Heras Gala T, Carstensen-Kirberg M, Huth C, Zierer A, Wahl S et al. (2017) Circulating levels of interleukin 1-receptor antagonist and risk of cardiovascular disease – meta-analysis of six population-based cohorts. *Arteriosclerosis, Thrombosis, and Vascular Biology.* 37(-6):1222. doi: 10.1161/ATVBAHA.117.309307.

Hernesniemi JA, Karhunen PJ, Oksala N, Kähönen M, Levula M, Rontu R et al. (2009) Interleukin 18 gene promoter polymorphism – a link between hypertension and pre-hospital sudden cardiac death – the Helsinki Sudden Death Study. *European Heart Journal.* 30(23):2939. doi: 10.1093/eurheartj/ehp316.

Hinck AP, Mueller TD, Springer TA. (2016) Structural biology and evolution of the TGF-β family. *Cold Spring Harbor Perspectives in Biology.* 8(12):a022103. doi: 10.1101/cshperspect.a022103.

Hoekstra M, Van Berkel TJ, Van Eck M. (2010) Scavenger receptor BI – a multi-purpose player in cholesterol and steroid metabolism. *World Journal of Gastroenterology – WJG.* 16(47):5916. doi: 10.3748/wjg.v16.i47.5916.

Holick MF. (2007) Vitamin D deficiency. *New England Journal of Medicine.* 357(3):266. doi: 10.1056/NEJMra070553.

Hu LI, Ying-Xue LIU, Huang JY, Zhu YF, Wang K. (2020). Analysis for interaction between interleukin-35 gene polymorphisms and risk factors on susceptibility to coronary heart disease in the Chinese Han population. *BMC Cardiovascular Disorders.* doi: 10.21203/rs.3.rs–31682/v1.

Huang Y, Lin YZ, Shi Y, Ji QW. (2013) IL-35 – a potential target for the treatment of atherosclerosis. *Die Pharmazie-An International Journal of Pharmaceutical Sciences.* 68 (10):793. doi: 10.1691/ph.2013.2928.

Huang G, Luo C, Gu X, Wu Z, Wang Z, Du Z et al. (2009) Mechanical strain induces expression of C-reactive protein in human blood vessels. *Journal of Pharmacology and Experimental Therapeutics.* 330(1):206. doi: 10.1124/jpet.109.150961.

Humphries SE, Yiannakouris N, Talmud PJ. (2008) Cardiovascular disease risk prediction using genetic information (gene scores) – is it really informative? *Current Opinion in Lipidology.* 19(2):128. doi: 10.1097/MOL.0b013e3282f5283e.

Iacoviello L, Di Castelnuovo A, Gattone M, Pezzini A, Assanelli D, Lorenzet R et al. (2005) Polymorphisms of the interleukin-1β gene affect the risk of myocardial infarction and ischemic stroke at young age and the response of mononuclear cells to stimulation in vitro. *Arteriosclerosis, Thrombosis, and Vascular Biology.* 25(1):222–227. doi: 10.1161/01.ATV.0000150039.60906.02.

Imazio M, Gaita F. (2015) Diagnosis and treatment of pericarditis. *Heart.* 101(14):1159–1168.

Javor J, Bucova M, Ferencik S, Grosse-Wilde H, Buc M. (2007) Single nucleotide polymorphisms of cytokine genes in the healthy Slovak population. *International Journal of Immunogenetics.* 34(4):273. doi: 10.1111/j.1744-313X.2007.00693.x.

Jia XW, Tian YP, Wang Y, Deng XX, Dong ZN. (2007) Correlation of polymorphism in IL-6 gene promoter with BMI, inflammatory factors, and pathogenesis and progression of CHD. *Zhongguoshi yan xue ye xue za zhi.* 15(6):1270.

Kakihana Y, Ito T, Nakahara M, Yamaguchi K, Yasuda T. (2016) Sepsis-induced myocardial dysfunction – pathophysiology and management. *Journal of Intensive Care.* 4(1):1. doi: 10.1186/s40560-016-0148–1.

Kim SK, Kang SW, Chung JH, Lee JS, Park HK, Yoon KL et al. (2011) Coding single-nucleotide polymorphisms of interleukin-1 gene cluster are not associated with Kawasaki disease in the Korean population. *Pediatric Cardiology.* 32(4):381. doi: 10.1007/s00246-010-9858–7.

Krishnamurthy P, Thal M, Verma S, Hoxha E, Lambers E, Ramirez V et al. (2011) Interleukin-10 deficiency impairs bone marrow–derived endothelial progenitor cell survival and function in ischemic myocardium. *Circulation Research.* 109(11):1280. doi: 10.1161/CIRCRESAHA.111.248369.

Kumar A. (2014) Cardiovascular diseases – are we overlooking some cardiovascular disease risk factors/markers? *Journal of Biomedical Sciences.* 2014;3(1). doi: 10.3823/1021.

Kumar S, McDonnell PC, Lehr R, Tierney L, Tzimas MN, Griswold DE et al. (2000) Identification and initial characterization of four novel members of the interleukin-1 family. *Journal of Biological Chemistry.* 275(14):10308. doi: 10.1074/jbc.275.14.10308.

Labarrere CA, Zaloga GP. (2004) C-reactive protein – from innocent bystander to pivotal mediator of atherosclerosis. *The*

American Journal of Medicine. 117(7):499. doi: 10.1016/j.amjmed.2004.03.039.

Licastro F, Chiappelli, M, Caldarera CM, Tampieri C, Nanni S, Gallina M et al. (2004) The concomitant presence of polymorphic alleles of interleukin-1β, interleukin-6 and apolipoprotein E is associated with an increased risk of myocardial infarction in elderly men – results from a pilot study. *Mechanisms of Ageing and Development.* 125 (8):575–579. doi: 10.1016/j.mad.2004.06.008.

Lin Y, Xue Y, Huang X, Lu J, Yang Z, Ye J et al. (2018). Association between interleukin-35 polymorphisms and coronary heart disease in the Chinese Zhuang population – a case–control study. *Coronary Artery Disease.* 29(5):423. doi: 10.1097/MCA.0000000000000635.

Linton MF, Yancey PG, Davies SS, Jerome WG, Linton EF, Song WL, et al. (2019) The role of lipids and lipoproteins in atherosclerosis. In *Endotext* [Internet]. MDText. com, Inc.

Luft FC. (2017) What have we learned from the genetics of hypertension? *Medical Clinics of North America.* 101:195–206. doi: 10.1016/j.mcna.2016.08.015.

Ma H, Sun G, Wang W, Zhou Y, Liu D, Tong Y et al. (2016) Association between interleukin-6–572 C> G and-174 G> C polymorphisms and hypertension – a meta-analysis of case-control studies. *Medicine.* 95(2). doi: 10.1097/MD.0000000000002416.

Mahemuti A, Abudureheman K, Aihemaiti X, Hu XM, Xia YN, Tang BP et al. (2012) Association of interleukin-6 and C-reactive protein genetic polymorphisms levels with venous thromboembolism. *Chinese Medical Journal.* 125(-22):3997. doi: 10.3760/cma.j.issn.0366–6999.2012.22.016.

Mallat Z, Henry P, Fressonnet R, Alouani S, Scoazec A, Beaufils P et al. (2002) Increased plasma concentrations of interleukin-18 in acute coronary syndromes. *Heart.* 88(5):467. doi: 10.1136/heart.88.5.467.

Manginas A, Tsiavou A, Chaidaroglou A, Giamouzis G, Degiannis D, Panagiotakos D et al. (2008) Inflammatory cytokine gene variants in coronary artery disease patients in Greece. *Coronary Artery Disease.* 19(8):575. doi: 10.1097/MCA.0b013e32831286e8.

Markt SC, Rider JR, Penney KL, Schumacher FR, Epstein MM, Fall K et al. (2014) Genetic variation across C-reactive protein and risk of prostate cancer. *The Prostate.* 74(10):1034. doi: 10.1002/pros.22820.

Mineo C, Deguchi H, Griffin JH, Shaul PW. (2006) Endothelial and antithrombotic actions of HDL. *Circulation Research.* 98(11):1352. doi: 10.1161/01.RES.0000225982.01988.93.

Mitrokhin V, Nikitin A, Brovkina O, Khodyrev D, Zotov A, Vachrushev N et al. (2017) Association between interleukin-6/6R gene polymorphisms and coronary artery disease in Russian population – influence of interleukin-6/6R gene polymorphisms on inflammatory markers. *Journal of Inflammation Research.* 10:151. doi: 10.2147/JIR.S141682.

Moran AE, Roth GA, Narula J, Mensah GA. (2014) 1990–2010 global cardiovascular disease atlas. *Global Heart.* 9(1):3–16. doi: 10.1016/j.gheart.2014.03.1220.

Morris DR., Moxon JV, Biros E, Krishna SM, Golledge J. (2012) Meta-analysis of the association between transforming growth factor-beta polymorphisms and complications of coronary heart disease. *PLoS One.* 7(5):e37878. doi: 10.1371/journal.pone.0037878.

Mosser DM, Zhang X. (2008) Interleukin-10 – new perspectives on an old cytokine. *Immunological Reviews.* 226(1):205. doi: 10.1111/j.1600–065X.2008.00706.x.

Mousavi SZ, Salehi A, Jorjani E, Manzari RS, Farazmandfar T, Shahbazi M. (2018) Association assessment of Interleukine-10 gene polymorphism and its expression status with susceptibility to coronary artery disease in Iran. *Egyptian Journal of Medical Human Genetics.* 19(1):31. doi: 10.1016/j.ejmhg.2017.06.005.

Nold MF, Nold-Petry CA, Zepp JA, Palmer BE, Bufler P, Dinarello CA. (2010). IL-37 is a fundamental inhibitor of innate immunity. *Nature Immunology.* 11(11):1014.

Nsengiyumva V, Fernando ME, Moxon JV, Krishna SM, Pinchbeck J, Omer SM et al. (2015) The association of circulating 25-hydroxyvitamin D concentration with peripheral arterial disease – a meta-analysis of observational studies. *Atherosclerosis.* 243(2):645. doi: 10.1016/j.atherosclerosis.2015.10.011.

Oda K, Tanaka N, Arai T, Araki J, Song Y, Zhang L et al. (2007) Polymorphisms in pro-and anti-inflammatory cytokine genes and susceptibility to atherosclerosis – a pathological study of 1503 consecutive autopsy cases. *Human Molecular Genetics.* 16(6):592. doi: 10.1093/hmg/ddl483.

Oh SH, Kweon SS, Choi JS, Rhee J, Lee YH, Nam HS et al. (2016) Association between vitamin D status and risk of peripheral arterial disease – the Dong-gu study. *Chonnam Medical Journal.* 52(3):212.

Oram JF, Heinecke JW. (2005) ATP-binding cassette transporter A1 – a cell cholesterol exporter that protects against cardiovascular disease. *Physiological Reviews.* 85(4):343. doi: 10.1152/physrev.00005.2005.

Pai JK, Mukamal KJ, Rexrode KM, Rimm EB. (2008) C-reactive protein (CRP) gene polymorphisms, CRP levels, and risk of incident coronary heart disease in two nested case-control studies. *PLoS One.* 3(1):e1395.

Parichatikanond W, Luangmonkong T, Mangmool S, Kurose H. (2020) Therapeutic targets for the treatment of cardiac fibrosis and cancer – focusing on TGF-β signaling. *Frontiers in Cardiovascular Medicine.* 7. doi: 10.3389/fcvm.2020.00034.

Parks JS, Chung S, Shelness GS. (2012) Hepatic ABC transporters and triglyceride metabolism. *Current Opinion in Lipidology.* 23(3):196. doi: 10.1097/MOL.0b013e328352dd1a.

Peters MJ, Van Halm VP, Voskuyl AE, Smulders YM, Boers M, Lems WF et al. (2009) Does rheumatoid arthritis equal diabetes mellitus as an independent risk factor for cardiovascular disease? A prospective study. *Arthritis Care & Research – Official Journal of the American College of Rheumatology.* 61(11):1571. doi: 10.1002/art.24836.

Pirillo A, Norata GD, Catapano AL. (2013) High-density lipoprotein subfractions-what the clinicians need to know. *Cardiology.* 124(2):116. doi: 10.1159/000346463.

Pohlers D, Brenmoehl J, Löffler I, Müller CK, Leipner C, Schultze-Mosgau S et al. (2009) TGF-β and fibrosis in different organs—molecular pathway imprints. *Biochimica et Biophysica Acta (BBA)-Molecular Basis of Disease.* 1792(-8):746. doi: 10.1016/j.bbadis.2009.06.004.

Posadas-Sánchez R, Pérez-Hernández N, Angeles-Martínez J, López-Bautista F, Villarreal-Molina T, Rodríguez-Pérez JM et al. (2017) Interleukin 35 polymorphisms are associated with decreased risk of premature coronary artery disease, metabolic parameters, and IL-35 levels – the genetics of atherosclerotic disease (GEA) study. *Mediators of Inflammation*. doi: 10.1155/2017/6012795.

Rafieian-Kopaei M, Setorki M, Doudi M, Baradaran A, Nasri H. (2014) Atherosclerosis – process, indicators, risk factors and new hopes. *International Journal of Preventive Medicine*. 5(8):927.

Rai H, Sinha N, Kumar S, Sharma AK, Agrawal S. (2016) Interleukin-1 gene cluster polymorphisms and their association with coronary artery disease – separate evidences from the largest case-control study amongst North Indians and an updated meta-analysis. *PLoS One*. 11(4):e0153480. doi: 10.1371/journal.pone.0153480.

Rajasingh J, Thangavel J, Siddiqui MR, Gomes I, Gao XP, Kishore R, et al. (2011) Improvement of cardiac function in mouse myocardial infarction after transplantation of epigenetically-modified bone marrow progenitor cells. *PLoS One*. 6(7):e22550. doi: 10.1371/journal.pone.0022550.

Rao M, Guo D, Jaber BL, Tighiouart H, Pereira BJ, Balakrishnan VS et al. (2004) Transforming growth factor-beta 1 gene polymorphisms and cardiovascular disease in hemodialysis patients. *Kidney International*. 66(1):419.

Rao M, Wong C, Kanetsky P, Girndt M, Stenvinkel P, Reilly M et al. (2007) Cytokine gene polymorphism and progression of renal and cardiovascular diseases. *Kidney International*. 72(5):549. doi: 10.1038/sj.ki.5002391.

Reiner AP, Wurfel MM, Lange LA, Carlson CS, Nord AS, Carty CL et al. (2008) Polymorphisms of the IL1-receptor antagonist gene (IL1RN) are associated with multiple markers of systemic inflammation. *Arteriosclerosis, Thrombosis, and Vascular Biology*. 28(7):1407. doi: 10.1161/ATVBAHA.108.167437.

Rippe JM. (2019) Lifestyle strategies for risk factor reduction, prevention, and treatment of cardiovascular disease. *American Journal of Lifestyle Medicine*. 13(2):204. doi: 10.1177%2F1559827618812395.

Rogus J, Beck JD, Offenbacher S, Huttner K, Iacoviello L, Latella MC et al. (2008) IL1B gene promoter haplotype pairs predict clinical levels of interleukin-1β and C-reactive protein. *Human Genetics*. 123(4):387. doi: 10.1007%2Fs00439-008-0488–6.

Rosen CJ, Adams JS, Bikle DD, Black DM, Demay MB, Manson JE et al. (2012) The nonskeletal effects of vitamin D – an Endocrine Society scientific statement. *Endocrine Reviews*. 33(3):456. doi: 10.1210/er.2012–1000.

Sabat R, Grütz G, Warszawska K, Kirsch S, Witte E, Wolk K et al. (2010) Biology of interleukin-10. *Cytokine & Growth Factor Reviews*. 21(5):331. doi: 10.1016/j.cytogfr.2010.09.002.

Salazar J, Martínez MS, Chávez-Castillo M, Núñez V, Añez R, Torres Y et al. (2014) C-reactive protein – an in-depth look into structure, function, and regulation. *International Scholarly Research Notices*. 2014. doi: 10.1155/2014/653045.

Scheller J, Chalaris A, Schmidt-Arras D, Rose-John S. (2011) The pro-and anti-inflammatory properties of the cytokine interleukin-6. *Biochimica et Biophysica Acta (BBA)-Molecular Cell Research*. 1813(5):878. doi: 10.1016/j.bbamcr.2011.01.034.

Schulze MB, Shai I, Rimm EB, Li T, Rifai N, Hu FB. (2005) Adiponectin and future coronary heart disease events among men with type 2 diabetes. *Diabetes*. 54(2):534. doi: 10.2337/diabetes.54.2.534.

Shanker J, Kakkar VV. (2010) Implications of genetic polymorphisms in inflammation-induced atherosclerosis. *The Open Cardiovascular Medicine Journal*. 4:30. doi: 10.2174/1874192401004020030.

Sharma S, Kulk N, Nold MF, Gräf R, Kim SH, Reinhardt D et al. (2008) The IL-1 family member 7b translocates to the nucleus and down-regulates proinflammatory cytokines. *The Journal of Immunology*. 180(8):5477. doi: 10.4049/jimmunol.180.8.5477.

Sharma S, Sharma A, Kumar S, Sharma SK, Ghosh B. (2006) Association of TNF haplotypes with asthma, serum IgE levels, and correlation with serum TNF-α levels. *American Journal of Respiratory Cell and Molecular Biology*. 35(-4):488. doi: 10.1165/rcmb.2006–0084OC.

Shibata R, Ouchi N, Murohara T. (2009) Adiponectin and cardiovascular disease. *Circulation Journal*. 0902260284. doi: 10.1253/circj.CJ–09–0057.

Shih HH, Zhang S, Cao W, Hahn A, Wang J, Paulsen JE et al. (2009) CRP is a novel ligand for the oxidized LDL receptor LOX-1. *American Journal of Physiology-Heart and Circulatory Physiology*. 296(5):H1643. doi: 10.1152/ajpheart.00938.2008.

Shrivastava AK, Singh HV, Raizada A, Singh SK. (2015) C-reactive protein, inflammation and coronary heart disease. *The Egyptian Heart Journal*. 67(2):89. doi: 10.1016/j.ehj.2014.11.005.

Sie MP, Uitterlinden AG, Bos MJ, Arp PP, Breteler MM, Koudstaal PJ et al. (2006) TGF-β1 polymorphisms and risk of myocardial infarction and stroke – the Rotterdam study. *Stroke*. 37(11):2667. doi: 10.1161/01.STR.0000244779.30070.1a.

Song HJ, Zhou XH, Guo L, Tian FL, Guo XF, Sun YX. (2015) Association of phosphodiesterase 4D gene and interleukin-6 receptor gene polymorphisms with ischemic stroke in a Chinese hypertensive population. *Genetics and Molecular Research*. 14(4):19396. doi: 10.4238/2015.

Srirangan S, Choy EH. (2010). The role of interleukin 6 in the pathophysiology of rheumatoid arthritis. *Therapeutic Advances in Musculoskeletal Disease*. 2(5):247. doi: 10.1177%2F1759720X10378372.

Steen EH, Wang X, Balaji S, Butte MJ, Bollyky PL, Keswani SG. (2020) The role of the anti-inflammatory cytokine interleukin-10 in tissue fibrosis. *Advances in Wound Care*. 9(4):184. doi: 10.1089/wound.2019.1032.

Stenvinkel P, Ketteler M, Johnson RJ, Lindholm B, Pecoits-Filho R, Riella M et al. (2005) IL-10, IL-6, and TNF-α – central factors in the altered cytokine network of uremia—the good, the bad, and the ugly. *Kidney International*. 67(-4):1216. doi: 10.1111/j.1523–1755.2005.00200.x.

Stuart SDF, De Jesus NM, Lindsey ML, Ripplinger CM. (2016) The crossroads of inflammation, fibrosis, and arrhythmia following myocardial infarction. *Journal of Molecular and Cellular Cardiology*. 91:114. doi: 10.1016/j.yjmcc.2015.12.024.

Szekely Y, Arbel Y. (2018) A review of Interleukin-1 in heart disease – where do we stand today? *Cardiology and Therapy*. 7(1):25–44.doi: 10.1007/s40119-018-0104–3.

Tanaka T, Narazaki M, Kishimoto T. (2014) IL-6 in inflammation, immunity, and disease. *Cold Spring Harbor Perspectives in Biology*. 6(10):a016295. doi: 10.1101/cshperspect.a016295.

Taube A, Schlich R, Sell H, Eckardt K, Eckel J. (2012) Inflammation and metabolic dysfunction – links to cardiovascular diseases. *American Journal of Physiology-Heart and Circulatory Physiology.* 302(11):H2148. doi: 10.1152/ajpheart.00907.2011.

Terry CF, Loukaci V, Green FR. (2000) Cooperative influence of genetic polymorphisms on interleukin 6 transcriptional regulation. *Journal of Biological Chemistry.* 275(24):18138. doi: 10.1074/jbc.M000379200.

Tiret L, Godefroy T, Lubos E, Nicaud V, Tregouet DA, Barbaux S et al. (2005). Genetic analysis of the interleukin-18 system highlights the role of the interleukin-18 gene in cardiovascular disease. *Circulation.* 112(5):643. doi: 10.1161/CIRCU LATIONAHA.104.519702.

Tousoulis D, Oikonomou E, Economou EK, Crea F, Kaski JC. (2016). Inflammatory cytokines in atherosclerosis – current therapeutic approaches. *European Heart Journal.* 37(-22):1723. doi: 10.1093/eurheartj/ehv759.

Towne JE, Renshaw BR, Douangpanya J, Lipsky BP, Shen M, Gabel CA et al. (2011) Interleukin-36 (IL-36) ligands require processing for full agonist (IL-36α, IL-36β, and IL-36γ) or antagonist (IL-36Ra) activity. *Journal of Biological Chemistry.* 286(49):42594. doi: 10.1074/jbc.M111.267922.

Travers JG, Kamal FA, Robbins J, Yutzey KE, Blaxall BC. (2016) Cardiac fibrosis – the fibroblast awakens. *Circulation Research.* 118(6):1021. doi: 10.1161/CIRCRESAHA.115. 306565.

Trégouët DA, König IR, Erdmann J, Munteanu A, Braund PS, Hall AS et al. (2009) Genome-wide haplotype association study identifies the SLC22A3-LPAL2-LPA gene cluster as a risk locus for coronary artery disease. *Nature Genetics.* 41(3):283.

Trigueros-Motos L, van Capelleveen JC, Torta F, Castaño D, Zhang LH, Chai EC et al. (2017) ABCA8 regulates cholesterol efflux and high-density lipoprotein cholesterol levels. *Arteriosclerosis, Thrombosis, and Vascular Biology.* 37(-11):2147. doi: 10.1161/ATVBAHA.117.309574.

Tsimikas S, Duff GW, Berger PB, Rogus J, Huttner K, Clopton P et al. (2014) Pro-inflammatory interleukin-1 genotypes potentiate the risk of coronary artery disease and cardiovascular events mediated by oxidized phospholipids and lipoprotein (a). *Journal of the American College of Cardiology.* 63(17):1724. doi: 10.1016/j.jacc.2013.12.030.

Turiano NA, Mroczek DK, Moynihan J, Chapman BP. (2013) Big 5 personality traits and interleukin-6 – evidence for "healthy Neuroticism" in a US population sample. *Brain, Behavior, and Immunity.* 28:83. doi: 10.1016/j.bbi.2012.10.020.

Tyagi N, Mohanty A, Kaur C, Kabi A, Kumari S, Soni S, et al. (2018) Interleukin-6 gene Polymorphisms And Coronary Artery Disease-A. *European Journal of Biomedical.* 5(12):526.

Uciechowski P, Dempke WC. (2020) Interleukin-6 – a masterplayer in the cytokine network. *Oncology.* 98(3):131. doi: 10.1159/000505099.

Urschel K, Cicha I. (2015) TNF-α in the cardiovascular system – from physiology to therapy. *International Journal of Interferon, Cytokine and Mediator Research.* 7:9–25.

Van Putten S, Shafieyan Y, Hinz B. (2016) Mechanical control of cardiac myofibroblasts. *Journal of Molecular and Cellular Cardiology.* 93:133. doi: 10.1016/j.yjmcc.2015.11.025.

Van Tassell BW, Toldo S, Mezzaroma E, Abbate A. (2013) Targeting interleukin-1 in heart disease. *Circulation.* 128(-17):1910. doi: 10.1161/CIRCULATIONAHA.113.003199.

Vasan RS, Sullivan LM, Roubenoff R, Dinarello CA, Harris T, Benjamin EJ et al. (2003) Inflammatory markers and risk of heart failure in elderly subjects without prior myocardial infarction – the Framingham Heart Study. *Circulation.* 107(11):1486. doi: 10.1161/01.CIR.0000057810.48709.F6.

Veerdonk FVD, Netea M. (2013). New insights in the immunobiology of IL-1 family members. *Frontiers in Immunology.* 4:167. doi: 10.3389/fimmu.2013.00167.

Vergeer M, Holleboom AG, Kastelein JJ, Kuivenhoven JA. (2010) The HDL hypothesis – does high-density lipoprotein protect from atherosclerosis? *Journal of Lipid Research.* 51(-8):2058. doi: 10.1194/jlr.R001610.

Vignali DA, Kuchroo VK. (2012) IL-12 family cytokines – immunological playmakers. *Nature Immunology.* 13(8): 722.

Voskoboinik A, Costello BT, La Gerche A, Prabhu S, Wong G, Flannery MD et al. (2019) Relation of alcohol consumption to left ventricular fibrosis using cardiac magnetic resonance imaging. *The American Journal of Cardiology.* 123(3):460. doi: 10.1016/j.amjcard.2018.10.026.

Wang J, Wei L, Yang X, Zhong J. (2019) Roles of growth differentiation factor 15 in atherosclerosis and coronary artery disease. *Journal of the American Heart Association.* 8(17):e012826. doi: 10.1161/JAHA.119.012826.

Weber C, Noels H. (2011) Atherosclerosis – current pathogenesis and therapeutic options. *Nature Medicine.* 17(11):1410.

Weng KP, Ho TY, Chiao YH, Cheng JT, Hsieh KS, Huang SH et al. (2010) Cytokine genetic polymorphisms and susceptibility to Kawasaki disease in Taiwanese children. *Circulation Journal.* 1010210952. doi: 10.1253/circj.CJ–10–0542.

Willerson JT, Ridker PM. (2004) Inflammation as a cardiovascular risk factor. *Circulation.* 109(21_suppl_1):II–2. doi: 10.1161/01.CIR.0000129535.04194.38.

World Health Organization. (2017) *Cardiovascular Diseases (CVDs) Fact Sheet.* Geneva: World Health Organization.

World Health Organization. (2014) *Global Status Report on Noncommunicable Diseases.* Geneva: World Health Organization Press.

Xuan Y, Wang L, Zhi H, Li X, Wei, P. (2016). Association between 3 IL-10 gene polymorphisms and cardiovascular disease risk – systematic review with meta-analysis and trial sequential analysis. *Medicine.* 95(6). doi: 10.1097/MD.0000000000002846.

Yadav U, Mahemuti A, Hu X, Abudureheman K, Xia Y, Tang B, Upur H. (2015) Single nucleotide polymorphisms in interleukin-6 and their association with venous thromboembolism. *Molecular Medicine Reports.* 11(6):4664. doi: 10.3892/mmr.2015.3248.

Yan L, Hu R, Tu S, Cheng WJ, Zheng Q, Wang JW et al. (2015). Meta-analysis of association between IL-6–634C/G polymorphism and osteoporosis. *Genetics and Molecular Research.* 14(4), 19225. doi: 10.4238/2015.December.29.32.

Yilmaz MI, Solak Y, Saglam M, Cayci T, Acikel C, Unal HU et al. (2014) The relationship between IL-10 levels and cardiovascular events in patients with CKD. *Clinical Journal of the American Society of Nephrology.* 9(7):1207. doi: 10.2215/CJN.08660813.

Yin D, Naji DH, Xia Y, Li S, Bai Y, Jiang G et al. (2017). Genomic variant in IL-37 confers a significant risk of coronary artery disease. *Scientific Reports.* 7:42175.

Yue Y, Meng K, Pu Y, Zhang X. (2017) Transforming growth factor beta (TGF-β) mediates cardiac fibrosis and induces diabetic cardiomyopathy. *Diabetes Research and Clinical Practice*. 133:124. doi: 10.1016/j.diabres.2017. 08.018.

Zhang RJ, Li XD, Zhang SW, Li XH, Wu L. (2017) IL-8–251A/T polymorphism contributes to coronary artery disease susceptibility in a Chinese population. *Genetics and Molecular Research*. 16(1). doi: 10.4238/gmr16018224.

Zhang X, Zhang B, Zhang M, Han Y, Zhao Y, Meng Z et al. (2011). Interleukin-8 gene polymorphism is associated with acute coronary syndrome in the Chinese Han population. *Cytokine*. 56(2):188. doi: 10.1016/j.cyto.2011.06.010.

Zhang YM, Zhong LJ, He BX, Li WC, Nie J, Wang X et al. (2006) The correlation between polymorphism at position-511C/T in the promoter region of interleukin 1B and the severity of coronary heart disease. *Zhonghua yi xue yi Chuan xue za zhi Zhonghua Yixue Yichuanxue Zazhi Chinese Journal of Medical Genetics*. 23(1):86.

Zhou L, Cai J, Liu G, Wei Y, Tang H. (2012) Associations between interleukin-1 gene polymorphisms and coronary heart disease risk – a meta-analysis. *PLoS One*. 7(9):e45641. doi: 10.1371/journal.pone.0045641.

Zhu Y, Xian X, Wang Z, Bi Y, Chen Q, Han X et al. (2018) Research progress on the relationship between atherosclerosis and inflammation. *Biomolecules*. 8(3):80. doi: 10.3390/biom8030080.

Zuo HP, Che L, Wu XZ. (2014) Correlation between the polymorphism of C-592A in interleukin-10 gene and genetic susceptibility to coronary heart disease. *Experimental & Clinical Cardiology*. 20:3440.

20

Role of Genetic Polymorphisms in the Etiopathogenesis of Esophageal Cancer

Ishrat Parveiz Bhat, Kulsum Akhtar, and Syed Mudassar
Department of Clinical Biochemistry, Sher-I-Kashmir Institute of Medical Sciences, Soura, Srinagar, India

CONTENTS

ABSTRACT Esophageal cancer (EC) has a high death rate and a poor prognosis. EC risk has been linked to a multitude of interactions between hereditary and environmental variables. The risk of EC has been associated with various environmental and genetic factors, as well as their interactions. The two most common histologic subtypes of EC – esophageal squamous cell carcinoma and esophageal adenocarcinoma – have highly diverse etiologies and substantial regional heterogeneity in EC prevalence. There are substantial differences in the risk factors that influence the occurrence and distribution of both the histological subtypes over the world. The carcinogen-producing materials taken through the food or acquired from the environment are mediated by a diverse group of xenobiotic-metabolizing enzymes. Carcinogens in the food or the environment can attach to DNA and create mutations, which can lead to cancer. Genetic polymorphisms in alleles involved in the production of such enzymes can result in differential expression or functionality of the encoding genes. Xenobiotic-metabolizing enzymes play important functions in detoxifying the carcinogens both in the surroundings and in the food. Potential genes involved in cell cycle regulation, DNA repair, and apoptosis have aroused great scientific interest in determining vulnerability to oncogenic exposure, immune responses to these hazards, and eventual cellular repair and programmed cell senescence capacity after DNA injury. Genetic polymorphisms could serve as molecular biomarkers, providing valuable insights into cancer development. The purpose of this chapter is to explain our current understanding of the mutations caused by risk factors related to the progression of EC in various populations; it focuses on genome variations, gene–environment interactions, and tumorigenesis.

KEYWORDS: *Esophageal Cancer; Esophageal Squamous Cell Carcinoma; Esophageal Adenocarcinoma; Genetic Polymorphism; Risk Factors; Xenobiotics.*

20.1 Introduction

Worldwide, esophageal cancer (EC) ranks seventh in terms of incidence (604,000 new cases) and sixth in terms of mortality (544,000 fatalities), accounting for the 1 in 18 cancer-related deaths in 2020 (Hyuna Sung, Siegel et al. 2021). Men account for over 70% of cases, with a two- to threefold variance in

DOI: 10.1201/9781003246244-20

prevalence and fatality rates between men and women. In transitioned countries, men's rates are higher than in transitional countries, while women's rates are comparable (Arnold, Abnet et al. 2020, Ferlay, Ervik et al. 2020). Interactions between various environmental and hereditary factors have been linked to the risk of EC. According to epidemiological studies, tobacco smoking, alcohol consumption, malnourishment, insufficient consumption of fresh fruits and vegetables, and frequent eating of pickled vegetables are the major environmental triggers for EC development (Van Blankenstein, Looman et al. 2005, Steevens, Schouten et al. 2010). However, not everyone who is exposed to these risks develops EC. Recent research studies have revealed that genetic polymorphisms have a role in the tumorigenesis and progression of EC. EC is divided into two histological subtypes: esophageal squamous cell carcinoma (ESCC) and esophageal adenocarcinoma (EAC), both of which are histologically and etiologically distinct. The two most frequent histologic subtypes ESCC and EAC, which have very different etiologies, show significant geographic variation in EC incidence. ESCC is the most common subtype in poor economies, whereas EAC is predominantly a condition of industrialized nations and is largely linked to weight gain, gastroesophageal reflux disease (GERD), and Barrett's esophagus (BE) (Mudan and Kang 2008). EC is the main cause of mortality and morbidity in both males and females of Bangladesh, as well as in males of Malawi and South Africa (Hyuna Sung, Siegel et al. 2021). In some areas of Asia where the risk is high (e.g., China), the incidence of ESCC is on the decline, which might be aided by monetary prosperity and nutritional advancements, while the decline in the EC prevalence in various growing economies (e.g., the USA, Australia, France, and the UK) is mainly attributed to cigarette smoking cessations (Thun, Linet et al. 2017). However, in lower-income nations, such as portions of Asian and African countries, the key risk variables for ESCC, which account for more than 90% of all EC cases, have yet to be identified, despite food factors (such as nutrient deficiencies and nitrosamines) being hypothesized (McCormack, Menya et al. 2017). Chewing of betel quid on the Indian subcontinent, intake of pickles (e.g., China), and drinking extremely hot meals and drinks are all potential risk factors for ESCC (e.g., in Uruguay, Iran, and Tanzania) (Thun, Linet et al. 2017). Excess body weight, GERD, and BE are among the primary risk factors for EAC in high-income nations, accounting for nearly two-thirds of EC cases (Thun, Linet et al. 2017). Incidence rates of EAC are therefore quickly increasing in high-income nations, due to obesity and increased gastric reflux illness, and partly due to low levels of chronic *H. pylori* infection, which has been inversely related to EAC (Arnold, Laversanne et al. 2017). Obesity is projected to become an increasingly major burden to the upcoming expense of EC in many high-income nations, with EAC overtaking ESCC in several (Arnold, Laversanne et al. 2017).

20.2 Pathogenesis of Esophageal Cancer

ESCC and EAC are the two most frequent histological forms of EC. Sarcomas and small cell carcinomas account for less than 1%–2% of all esophageal malignancies (Young, Percy et al. 1981).

ESCC, which is the most common form of EC in the world, shows an increased rate of incidence related to aging, culminating in the age group of 70s. ESCC affects both the middle and lower esophagus, with blacks having a threefold greater incidence than whites. Consumption of alcoholic beverages and consumption of tobacco-containing products are two primary risk variables. Although most studies demonstrate that alcohol use is the predominant risk factor, smoking in conjunction with alcohol intake may have a synergic impact, raising the risk level. After the comparison between males of same ethnicity and area and who don't smoke and drink and men who were both smokers and drinkers, the relative risk of black men was higher than of white men (Brown, Hoover et al. 1994). The process by which combining nicotine and alcohol increases the risk of EC has been well investigated. Alcohol harms DNA molecules by lowering metabolism rate within the cell, which reduces detoxifying property and promotes degradation and oxidation (Muwonge, Ramadas et al. 2008). Alcohol acts as a solvent for fat-soluble substances. As a result, tobacco's dangerous carcinogens have an easier time penetrating the esophagus epithelium (Mueller, Grufferman et al. 2006). Aromatic amines, nitrosamines, polycyclic aromatic hydrocarbons (PAHs), aldehydes, and phenols are some of the carcinogens found in tobacco. ESCC of the esophagus has also been linked to other carcinogens, such as nitrosamines present in salted vegetables and preserved salmon. The pathophysiology seems to be connected to squamous epithelial inflammation, which progresses to dysplasia and in situ malignancy (Mao, Zheng et al. 2011). EAC develops in the distal parts of esophagus about most of the time (Zhang 2013), and it has a strong relationship with GERD. If GERD is not treated, the stratified squamous epithelium that typically covers the esophagus is supplanted by a columnar epithelium, resulting in BE. Barrett metaplasia is thought to be caused by continuous production of stomach acid and bile at the gastroesophageal junction, causing esophageal injury (Spechler 2013). The precise nature of the metaplasia has yet to be discovered. BE is more prevalent in adults and less common in youngsters. It affects more men than women, and white people are more affected than Asian or African American people. The degree of esophageal metaplasia-lined esophagus may impact the risk of EAC, according to several studies. Patients with BE have a 50- to 100-fold increased risk of acquiring EC (Mao, Zheng et al. 2011). However, most of the individuals having BE would not acquire EAC; the yearly incidence of EAC for BE patients was found to be 0.12% (Hvid-Jensen, Pedersen et al. 2011). EAC is linked to high-grade dysplasia (HGD), which is also a major risk factor. Obesity is also associated with EAC, especially for people who have a predominant abdominal fat distribution. Through the secretion of adipokines and cytokines, hypertrophied adipocytes and inflammatory cells within fat deposits produce an environment of reduced inflammation and assist tumor formation (Nieman, Romero et al. 2013). In the tumor microenvironment, adipocytes provide energy and aid tumor growth and progression (Wang 2008). In comparison with ESCC, EAC has a better long-term prognosis after resection. The cumulative 5-year life expectancy for EAC was 47%, compared to 37% for ESCC in a study of 1,059 patients who had gone through resection (Siewert, Stein et al. 2001) (Figure 20.1).

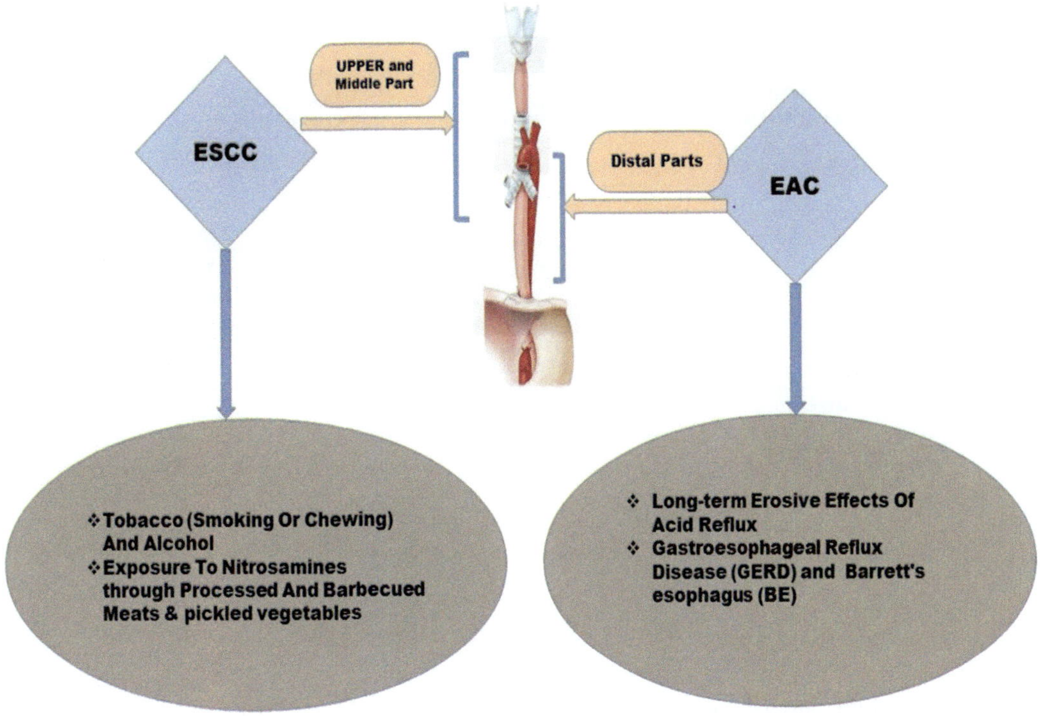

FIGURE 20.1 Pathophysiology and histological subtype location of esophageal cancer: Esophageal squamous cell carcinoma (ESCC) is most commonly seen in the top and central parts of the esophagus. Esophageal adenocarcinoma (EAC) is most common in the distal part of the esophagus and near the gastroesophageal junction. Both forms, however, may arise in any segment of esophagus.

20.3 Risk Factors for Esophageal Cancer

The two primary subtypes of EC have slightly distinct risk factors. Both the subtypes have some common risk factors such as age, gender, ethnicity, smoking, alcohol consumption, dietary factors, and hereditary factors. The two major risk variables for ESCC are tobacco smoking and alcoholic drinks, which have a strong synergistic effect (Prabhu, Obi et al. 2014). According to some estimates, roughly half of all instances are caused by smoking, one-third by alcohol, and more than three-quarters of all instances in males by a combined effects of smoking and excessive drinking (Montgomery, Basman et al. 2014). Alcohol-related risks seem to be connected to its aldehyde metabolite and variations in particular associated enzymes (Pennathur, Gibson et al. 2013). In Asia, such metabolic variations are quite prevalent. Other risk factors include drinking of hot beverages (above 65°C or 149°F) on a frequent basis, which can cause recurrent thermal damage to the esophagus mucosa, ingesting caustic chemicals (Montgomery, Basman et al. 2014), and excessive nitrosamine exposure (chemical substances present in both tobacco smoke and certain foods) (Pennathur, Gibson et al. 2013). Unhealthy eating habits appear to include nitrosamine exposure from processed and barbequed meat, pickled vegetables, and so on, as well as a limited consumption of fresh foods (Montgomery, Basman et al. 2014). Nutritional inadequacies, low socioeconomic position, and poor oral hygiene are other linked variables (Pennathur, Gibson et al. 2013). In Asia, chewing betel nut (areca) is a significant risk factor for ESCC (Akhtar 2013). The risk factors for ESCC are depicted in Figure 20.2.

GERD and BE are the two main variables that increase the risk of EAC. Long-term GERD may cause a variation in type of cell in the distal part of esophagus as a result of squamous lining erosion (de Jonge Pj, Grady et al. 2014). The process known as BE appears to occur roughly 20 years later in females than in males, probably owing to hormonal considerations (de Jonge Pj, Grady et al. 2014). Other risk factors for EAC including obesity (Kubo and Corley 2006, Renechan, Tyson et al. 2008), the absence of *Helicobacter pylori* infection (Rokkas, Pistiolas et al. 2007), and some drugs such as NSAIDs, PPIs, and statins, reduce the risk of EAC (Wheeler and Reed 2012). *H. pylori* infection (which is believed to impact more than 50% of the global population) is not associated with the risk of EAC and may be a preventive factor for it. Regardless of the fact that *H. pylori* is a source of GERD and a risk factor for gastric cancer, the infection relates to 50% lower incidence of EAC (Falk 2009, Lagergren and Lagergren 2013). The biological rationale for a protective effect is a little hazy (Falk 2009). One theory is that certain strains of *H. pylori* lower gastric acid, hence minimizing GERD injury (Harris 2019). Declining incidence of *H. pylori* infection in western countries in the past few years, which have since been related to decreasing home congestion, might play a role in the contemporaneous rise in EAC. Female hormones may help protect against EAC, which affects women less frequently than men and arises later in life, with an onset about 20 years later in women. The main risk factors for EAC are depicted in Figures 20.3 and 20.4.

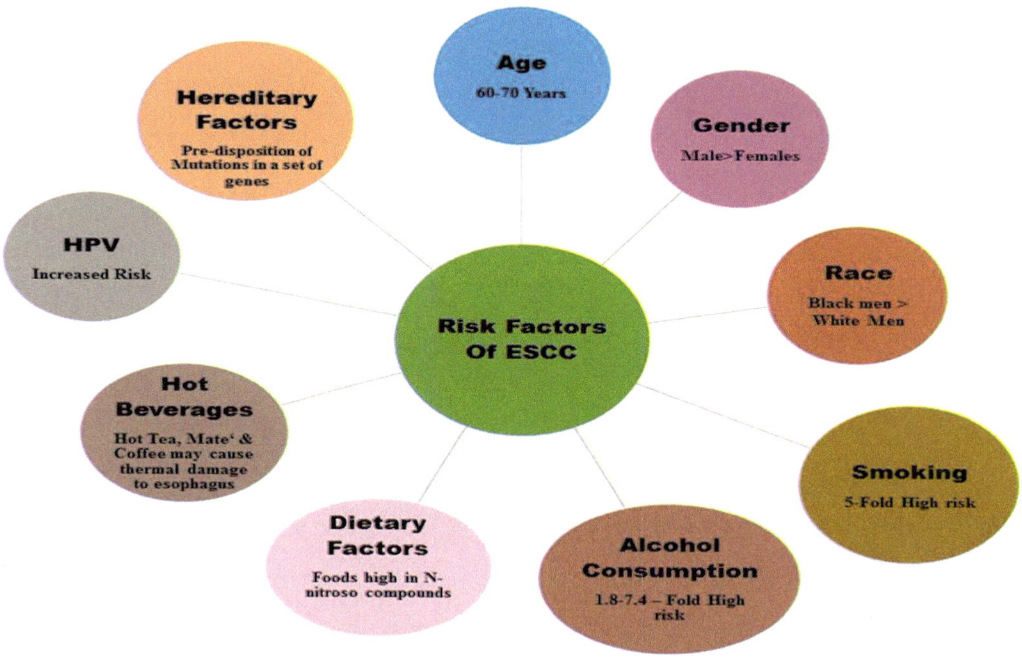

FIGURE 20.2 Risk factors associated with esophageal squamous cell carcinoma.

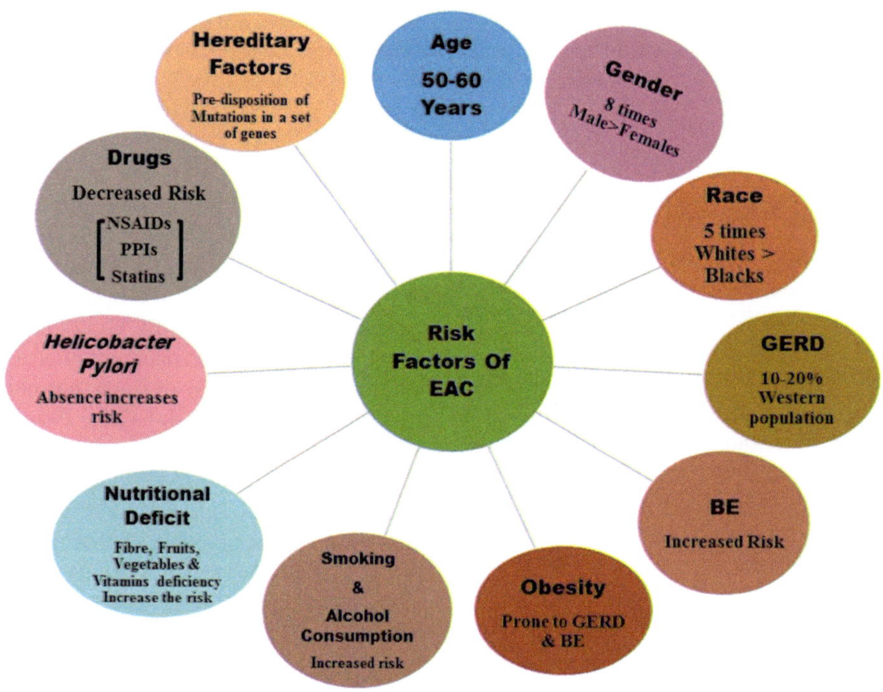

FIGURE 20.3 Risk factors associated with esophageal adenocarcinoma.

20.4 Genetic Background of Esophageal Cancer

20.4.1 Genes That Control Cell Division or Differentiation

The use of large sequencing technologies to assess complete mutational catalogs revealed significant genomic alterations in EC (Song, Li et al. 2014). Over 83% ESCCs had a somatic mutation in TP53, according to the first large-scale systematic investigation, which included whole-genome sequencing, whole-exome sequencing, and array comparative genomic hybridization (Song, Li et al. 2014). In 2%–10% of ESCCs, variations in genes that govern the cell cycle (Rb1, NFE2L2, CHEK1/2, and CDK2A) or that undergo differentiation (NOTCH1/3) have been identified (Gao, Chen et al. 2014, Lin, Hao et al. 2014, Song, Li et al. 2014). Furthermore, several genes which regulate the cell cycle show overexpression in ESCCs (e.g., CCND1 is overexpressed in 46.4%, CDK-1/4 is overexpressed in 23.6%, and MDM2 is overexpressed in 5.7%),

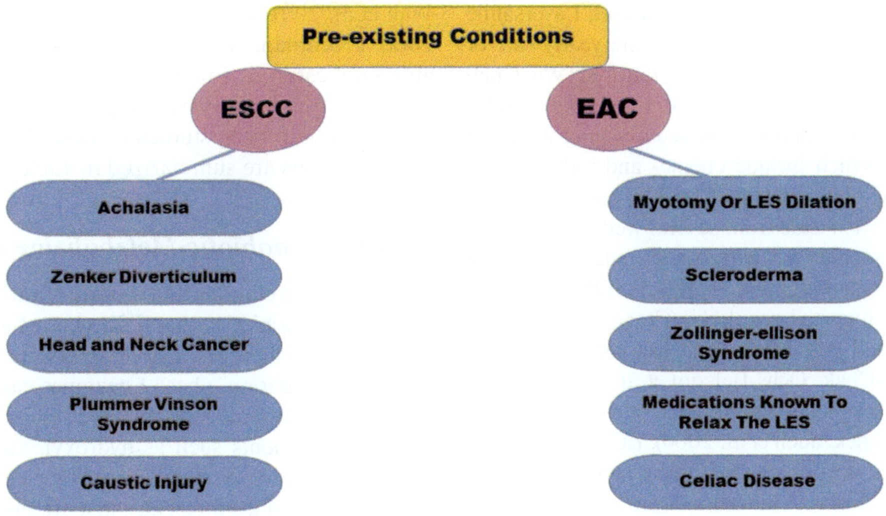

FIGURE 20.4 Pre-existing clinical conditions related to the risk of esophageal cancer.

indicating that they are involved in the ESCC tumorigenesis (Song, Li et al. 2014).

20.4.2 RAS Signaling or Receptor Tyrosine Kinase (RTK) and Epidermal Growth Factor Receptor (EGFR)

EGFR shows overexpression in 59%–76% of ESCC patients and is linked to a poor prognosis (Gao, Meng et al. 2014, Zhang, Zhu et al. 2014). Furthermore, genetic variations in downstream components of EGFR such as RTK, Ras, and Akt pathways are found in 78.6% of ESCC patients (Song, Li et al. 2014). One more study enrolled 193 ESCC patients to assess their EGFR levels and gene amplification status. The findings revealed that 49.2% of the study population showed increased levels of EGFR, linked to clinical staging and lymph node metastasis in a substantial way.

20.4.3 Genetic Variations in VEGF Signaling Pathway

Various cells generate vascular endothelial growth factor (VEGF), formerly named as vascular permeability factor (VPF), which encourages the blood vessel formation. VEGF is a growth factor that belongs to the platelet-derived growth factor (PDGF) family of cystine-knot growth factors. VEGF is a key signaling protein involved in both vasculogenesis and angiogenesis.

Angiogenesis is a precondition to tumorigenesis and might even improve the nutrition supplies supporting tumor development, increasing the chances for malignant transformation. Immunohistochemical analyses of VEGF in 128 GE junction adenocarcinoma patients showed a strong association with tumor staging and lymph node metastatic condition, with overexpression of VEGF-C correlated with poor overall survival (Xie, Zhai et al. 2013). FLT1 gene polymorphism (rs3794405) in EAC contributed to 45%–60% high death risk (Eng, Azad et al. 2015). In a study, a polymorphic form of VEGF 936 C>T has been found in EAC in a significant portion of patient samples (Lorenzen, Panzram et al. 2011).

20.4.4 Epigenetic Factors

The growth of ESCC and many other malignancies is linked to epigenetic alterations such as DNA methylation, chromatin remodeling, and genome imprint ablation. In ESCC, hypermethylation was observed in the promoter areas of APC, RB1, and CDKN2A. Overexpression of p53 is connected to methylation of CDKN2A, which encodes p16 and may influence RB1 gene expression (Taghavi, Biramijamal et al. 2010). These cell cycle regulatory mechanisms may aid in the progression of EC. Pyrosequencing analysis of the glutathione peroxidase 7 (GPX7) promoter region (from −162 to +138) from adenocarcinoma of esophagus displayed position-specific hypermethylation between CpG sites +13 and +64 (Peng, Hu et al. 2014). Cyclin E, which is encoded by the cyclin E1 gene, served very essential function on the development of EAC in its initial stages.

20.5 Genetic Polymorphisms and Esophageal Cancer

Genetic polymorphisms include the transmission of traits having more than one allele regulated by only one gene locus, with the less frequent allele having a prevalence of 1% or more, an inherited variation in DNA sequence. The presence of two or more genotypes simultaneously in a population is called genetic polymorphism (Nagasubramanian, Innocenti et al. 2003, Efferth and Volm 2005, Lee, Lockhart et al. 2005). SNPs are the basic and most prevalent variations in the human genome, which occur in any part of the gene, i.e., both the sense and antisense sections, and can cause changes in gene expression or protein function, affecting cell homeostasis and raising the chances of malignancy (Dutt and Beroukhim 2007, Mao, Young et al. 2007). Microsatellite instability and chromosomal insertions, deletions, or duplications are examples of genetic polymorphisms that can result in decreased or enhanced enzyme activity. Currently, ESCC and EAC are the most common tumor forms, accounting for the majority of EC occurrences. Surprisingly, although the frequency of ESCC is decreasing, the incidence of EAC is progressively rising. The most prevalent malignant histology affecting the esophagus is currently EAC (Arnold, Laversanne et al. 2017). In the last three decades alone, the yearly rate of EAC in North America has climbed three to fourfold (Wu, Chen et al. 2006, 2007). In spite of the adoption of localized endoscopic therapies, innovative surgical procedures, and intensive multimedia regimens which include chemo- and radiotherapy, the prognosis for EAC remains dismal, with 5-year life expectancy of about 10%–15%. Aside from several well-known clinical risk factors for esophageal malignancies, such as alcohol consumption, tobacco smoking, and GERD, familial EC accumulations have been reported regularly (Crane, Locke et al. 2008). BE, a recognized predictor of EC, has also been found to cluster in families (Chak, Ochs-Balcom et al. 2006). People who have a familial background have greater chances of getting this illness (Akbari, Malekzadeh et al. 2006). EC is particularly distinctive in that the two primary histologic subtypes, ESCC and EAC, are notably different in terms of prevalence, clinical characteristics, and etiology. As a result, previous studies have tended to look at these histologies independently, with the bulk of them focusing on ESCC. In a recent meta-analysis, to evaluate the risk of EC, it was reported that SNPs found in five genes showed a strong correlation, which include *CYP1A1* rs1048943, *EGF* rs444903, *MMP2* rs243865, *PLCE1* rs2274223 (for EC), and *HOTAIR* rs920778 (for ESCC) (Tian, Liu et al. 2019).

20.5.1 Genetic Polymorphisms in ESCC

Many of the molecular epidemiologic studies that served to uncover potential genes implicated in EC were shaped by prior understanding of extrinsic risk variables associated with ESCC, such as tobacco use, alcohol consumption, nitrosamine exposure, diet deficient in vital micronutrients, and HPV (Wu, Wan et al. 2001, Engel, Chow et al. 2003, Bahmanyar and Ye 2006, Freedman, Abnet et al. 2007). For example, gene variants implicated in enzyme-mediated activity or intracellular pathways that modify the impact of certain environmental exposures have been studied extensively in EC susceptibility. These findings are summarized in Table 20.2.

20.5.1.1 Xenobiotic-Metabolizing Enzymes (XMEs)

Our body gets a continuous exposure of various dietary and environmental xenobiotics that need phase I and phase II enzyme activations to show the cytotoxic effects (Singh and Michael 2009). By oxidation, phase I enzymes catalyze the bioconversion of parental molecules into hydrophilic intermediates, generating substituents such as hydroxyl, amine, and sulfhydryl groups, which are then conjugated by phase II enzymes. Phase II enzymes produce molecules that are more hydrophilic, making them easier for phase III transporters to eliminate in the urine. Although phase I and phase II enzymes are involved in detoxifying processes, they often produce volatile and highly reactive intermediary chemicals, which, if not eliminated rapidly, can attach irreversibly to DNA and form mutagenic and/or carcinogenic DNA adducts (Liska 1998). The cytochrome P450 (CYP) family of enzymes, which catalyze mostly oxidative processes and consequently operate as monooxygenases, oxidases, and peroxidases, account for around 75% of phase I enzymes (Ingelman-Sundberg, Sim et al. 2007). Glutathione S-transferases (GSTs), N-acetyltransferases (NATs), sulfotransferases (SULTs), and UDP-glucuronosyltransferases (UGTs) are among the phase II enzymes (Singh and Michael 2009). Table 20.1 lists the major XMEs (phase I, phase II, and phase III enzymes) implicated in xenobiotics in humans.

TABLE 20.1
Human Xenobiotic-Metabolizing Enzymes and Their Members

Phase I	**Oxidation:**
	Cytochrome P450 monooxygenases
	Flavin-containing monooxygenases
	Alcohol dehydrogenases and aldehyde dehydrogenases
	Monoamine oxidase
	Peroxidase co-oxidation
	Reduction:
	NADPH cytochrome P50 reductase
	Hydrolysis:
	Epoxide hydrolase
	Esterases
	Amidases
Phase II	**Methylation:**
	Methyltransferases
	Sulfation:
	Sulfotransferases
	Acetylation:
	N-acetyltransferases
	Glucuronidation:
	UDP-glucuronosyltransferases
	Glutathione conjugation:
	Glutathione S-transferases
	Glycine conjugation:
	XM-ligase
	Glycine N-acyltransferases
Phase III transporter	Multidrug resistance protein family (MRP-Family)

20.5.1.1.1 Cytochrome P450 Gene Family

The cytochrome P450 (CYP) gene family makes up the majority of phase I enzymes, which catalyze mostly oxidative processes and consequently operate as monooxygenases, oxidases, and peroxidases. These CYP family enzymes catalyze the initial stage in the metabolism of carcinogens found in cigarette smoke, exhaust gas, and the food, such as PAHs, N-nitrosamines, and heterocyclic amines (HCAs) (Gonzalez and Gelboin 1994). CYPs catalyze oxidative processes by introducing an oxygen atom into an organic substrate, resulting in highly reactive compounds that could readily attach to DNA and trigger mutations. These might, however, participate in reductive processes and produce less hazardous chemicals. The CYPs are found in a variety of tissues and organs, including the mucosa of the esophagus (Lechevrel, Casson et al. 1999). The majority of CYPs engaged in xenobiotic biotransformation are stimulated by chemical molecules (e.g., dioxin and benzopyrene) or nutritional components (e.g., ethanol and cruciferous vegetables), and cellular oxidants can inactivate them (Wong, Eiserich et al. 2001, Murray 2006). Three important receptors in cytosol, the pregnane X receptor (PXR), the constitutive androstane receptor (CAR), and the aryl hydrocarbon receptor (AhR), regulate CYP expression at the transcriptional level by detecting the potency of carcinogenic substances inside the cytosol and stimulating the expression of CYPs to improve xenobiotic eradication (Pascussi, Gerbal-Chaloin et al. 2008). A variety of genetic variations in the AhR gene have been shown to alter CYP function and perhaps risk of cancer development (Lamba 2008).

The CYP superfamily members are categorized based on their substrate specificity and amino acid sequence similarity. CYP1A1, CYP1A2, CYP1B1, CYP2A6, CYP2E1, and CYP3A4 are the key CYPs implicated in the metabolism of carcinogenic agents in human body (Ding and Kaminsky 2003) (Table 20.2).

EC susceptibility has been studied using genetic variants in the CYP1A1, CYP2E1, and CYP3A enzymes. In a study carried out on Taiwanese people, after the adjustment for relevant confounding factors, patients with the CYP1A1*2C variant (462V/V) had greater than twofold increased risk of getting ESCC than those having the wild-type allele (462 I/I) (Wu, Lee et al. 2002). Likewise, in Kashmir Valley, the mutant allele 462Val/Val was linked to increased ESCC risk, where smokers and people with CYP2E1 c1/c1 mutation showed even higher chances of getting this cancer (Bhat, Shah et al. 2014). CYP1A1*2A polymorphism, on the other hand, is not substantially linked with ESCC risk in such groups, despite controlling for various factors. In South Africans, the CYP2E1*6 heterozygous allele is linked to a higher risk of ESCC (Li, Dandara et al. 2005). CYP3A activity is mostly influenced by genetic variations in the CYP3A4 and CYP3A5 alleles, which are located within a 231-kb area on chromosome 7q21.1 and produce enzymes with considerable substrate specificity overlap (Gellner, Eiselt et al. 2001). In the human intestine and liver, CYP3A4 is the most prevalent CYP enzyme (Westlind, Löfberg et al. 1999, Canaparo, Finnström et al. 2007) although no function is seen in the esophagus (Lechevrel, Casson et al. 1999). The relationship between CYP3A5 variants and the chances of having ESCC in two demographic groups of South Africa revealed that people of mixed lineage who possess the CYP3A5*3 variant had much lower chances of getting ESCC than those who carry the wild-type gene; however, in the Black Xhosa community, there was no substantial connection was found. The CYP3A5 homozygous alleles (CYP3A5*3/*3, *3/*6, *3/7*, *6/*6, and *6/*7) found in the mixed ancestry populations were all linked to ESCC (Dandara, Ballo et al. 2005).

20.5.1.1.2 Epoxide Hydrolases

Epoxides are three-membered cyclic ethers, which are formed in living organisms or injected into them. XMEs, most notably CYPs, produce epoxides as the principal metabolite of aromatic and alkenic substances (Snyder, Chepiga et al. 1993). Epoxide hydrolases (EHs) comprise phase I enzymes that convert epoxides produced by aromatic chemical degradation to trans-dihydrodiols, which are eliminated from the body by phase II enzymes. There are four types of EHs in humans, which include epoxide hydrolase 1 (EPHX1) present in microsome and the three others present in cytosol, which are epoxide hydrolase 2 (EPHX2), epoxide hydrolase 3 (EPHX3), and epoxide hydrolase 4 (EPHX4) (Morisseau and Hammock 2005).

Two functional SNPs in the EPHX1 gene, 337T4C in exon 3 rs1051740 and 415A4G in exon 4 rs2234922, cause Tyr113His and His139Arg amino acid changes in the EPHX1 protein, resulting in a 40% drop and a 25% rise in enzymatic activity, respectively, related to the wild-type allele (Hassett, Alcher et al. 1994). The impact of EPHX1 variations in EC has been evaluated in case–control studies, with mixed findings. The carriers of the His139Arg polymorphic allele (139His/Arg and 139Arg/Arg) and the 139Arg allele have more chances of getting ESCC in a high-risk Indian community in Assam, compared to the wild-type genotype. Tobacco chewers and betel nut chewers had significant associations with the 139His/Arg genotype (Ihsan, Chattopadhyay et al. 2010). The 113Tyr/His heterozygous genotype, on the other hand, is linked to a lower incidence of ESCC, whereas the 113His/His homozygous allele is preventive in smokers, betel nut chewers, and alcohol drinkers (Jain, Tilak et al. 2008).

20.5.1.1.3 Glutathione S-Transferases

Glutathione S-transferases (GSTs) are phase II detoxifying enzymes that catalyze the conjugation of glutathione (GSH) to a broad range of electrophilic molecules, such as oxidative stress products and toxins from the environment, to neutralize their electrophilic sites and produce substances which are less reactive and easy to remove from the body (Hayes, Flanagan et al. 2005). GST could boost the bioactivation of some substances, resulting in the production of more cytotoxic and genotoxic metabolites. The pattern of biotransformation by GSTs into mutagenic or non-mutagenic species is determined by the substrate (Landi 2000). Although the liver has the highest concentration of GSTs in the human body, some isozymes are abundant in non-hepatic organs, such as the esophagus (Rowe, Nieves et al. 1997). The three primary subfamilies of human GSTs belong to cytosol, mitochondria, and membrane-bound microsomes. GSTs in the cytosol and

TABLE 20.2

Association of Genetic Polymorphisms with the Risk of ESCC

Genes Involved	Increased Risk	Reduced Risk	No Association
CYP Family			
CYP1A1	CYP1A1 *1/*2C CYP1A1 *1/*2A CYP1B1	-	CYP1A1 *1/*2A
CYP2E1	CYP2E1*1D CYP2E1*5B CYP2E1*6	-	CYP2E1 c1/c1
CYP3A	CYP3A5*3: (CYP3A5*1/*1,*1/*3, *1/*6, and *1/*7)	CYP3A5*3: (CYP3A5*3/*3,*3/*6, *3/*7, *6/*6, and *6/*7)	-
EHs			
EPHX1 EPHX2	His139Arg (139His/Arg, 139Arg/Arg) Tyr113His	113Tyr/His 113His/His 139Arg/Arg	-
GSTs			
GSTM1	(Null) GSTM1*0 (Non-null) Wt/Wt and Wt/Del	GSTM1 Del/Del GSTM1 A/B	GSTM1*0
GSTM3	GSTM3 (*A/*B+*B/*B) GSTM3*A/*A	GSTM3 (*A/*B) GSTM3*B	-
GSTT1	GSTT1*0 GSTT1*0 Del	GSTT1 Ile/Val	GSTT1*0
GSTT2	-	-	GSTT2B*0
GSTP1	GSTP1 Ile105Val GSTP1 105Ile/Ile	GSTP1 Ile/Val	GSTP1 Ile105Val
NATs			
NAT1	NAT1*4/*4, NAT1*4/*3 NAT1*3/*3, NAT1*10/*4, and NAT1*10/*3	-	NAT1*10/*10
NAT2	-	NAT2 haplotype C(481)A(590)G(857) NAT2*5 (341T>C)	NAT2 slow acetylator
SULTs			
	SULT1A1*2 SULT1A1*1/*2 SULT1A1*2/*2	SULTIA1 G213A	-
UGTs			
UGT1A	UGT1A1	UGT1A8	-
UGT2 (UGT2A and UGT2B)	-	UGT2B4	-
Enzymes of Repair Mechanism, Cell Cycle Regulation, or Apoptosis			
MMP genes	MMP-1 (rs1799750) MMP-2 (rs243865) MMP-9 (P574R GG)	-	-
EGF	EGF +61A>G (rs4444903)	-	-
PLCE1	PLCE1 A>G (rs2274223)	-	-
HOTAIR	HOTAIR (rs920778)	-	-
Fas	Fas A670 G Fas-L T844C	-	-
MDM	MDM2 T309 G	-	-
ECRG1	ECRG1 Arg290Gln ECRG1 TCA4/TCA3	-	-
p21	p21 Arg31Ser	-	-
TP53	p53 Arg72Pro (rs1042522)	p53 Arg72Pro	-
XRCC	XRCC1 Arg194Trp (rs1799782)	XRCC1 Arg280His XRCC1 Arg194Trp	-
XPD	-	XPD Asp312Asn	-

(Continued)

TABLE 20.2 (*Continued*)

Association of Genetic Polymorphisms with the Risk of ESCC

Genes Involved	Increased Risk	Reduced Risk	No Association
Others			
ADH1B	ADH1B Arg47His (rs1229984)	-	-
ALDH2	ALDH2 *1/*2 (rs671)	-	-
MTHFR	MTHFR A1298C (rs1801131)	MTHFR C677 T	-
CCND1	CCND1 (rs603965)	CCND1 G870A	-
COX-2	COX-2 (rs20417)	-	-
ERCC1	ERCC1C8092A & 118C/T	-	-
ERCC2	ERCC2 Asp312Asn (rs1799793)	-	-
NQO1 607C>T	NQO1 607C>T (rs1800566)	-	-

mitochondria are essential for the biotransformation of reactive exogenous and endogenous compounds, as well as for the chemical and oxidative stress defense. GSTs in microsomes play a key role in the synthesis of leukotrienes and prostaglandins. Many microsomal GSTs could also facilitate the conjugation of glutathione to a variety of electrophilic chemicals, although less is reported about the genetic polymorphisms in this GST subgroup. Based on amino acid sequence similarity, human cytosolic GSTs have been classified into six classes: alpha (α), mu (μ), omega (Ω), pi (π), theta (θ), and zeta (ζ), with each class having one or more isozymes encoded by a group of genes located on distinct chromosomes. Humans have only one type of mitochondrial GST, known as kappa. Different classes of enzymes metabolize a wide range of xenobiotics with a high degree of overlap in substrate specificity, although other substrates are extremely selective for specific GSTs (Hayes, Flanagan et al. 2005).

GSTM1/3 (μ-class), GSTT1 (θ-class), and GSTP1 (π-class) are the most polymorphic cytosolic GSTs in humans and have received the greatest attention in terms of susceptibility to EC. A variety of case–control studies in high-risk regions of China have looked at the link between the GSTM1 deletion polymorphism and the risk of ESCC. In Jiangsu Province, the GSTM1 null genotype (GSTM1*0 allele) was linked to an elevated incidence of ESCC (Gao, Takezaki et al. 2002), while no such link was seen in Huaian (Wang, Tang et al. 2006). In comparison, the GSTM1 non-null genotypes (Wt/Wt and Wt/Del) are considerably overrepresented in ESCC patients in Henan Province relative to controls and it was showed that the GSTM1 null genotype has a protective effect (Tan, Song et al. 2000). The GSTM1 homozygous deletion (Del/Del) is not linked to ESCC risk in Japanese males, even when stratified by smoking (Yokoyama, Kato et al. 2002), but it is linked to a considerable increase in risk in the Kazakhstan population (Djansugurova, Perfilyeva et al. 2013).

Studies conducted in India reported that the chances of ESCC linked to the GSTM1 homozygous deletion are substantially high in Assam, but not in New Delhi. GSTM1*0 has been found to have substantial gene–environment interactions with alcohol intake in New Delhi and betel nut chewing in Assam (Sharma, Das et al. 2013). A meta-analysis of 20 case–control studies on EC (ESCC+EAC) in which 1,947 cases and 3,508 controls were included showed a link with GSTM1*0, but an additional subgroup analysis by ethnicity and histological subtypes did not show any meaningful connection of GSTM1 deletion with EC risk (Zhuo, Zhang et al. 2009). In northern India, GSTM3 variants when combined (*A/*B+*B/*B) were related to an increased risk of EC (ESCC+ EAC) in comparison with the wild-type GSTM3*A/*A genotype, with a somewhat greater risk among men (Jain, Kumar et al. 2007). On the other hand, a case–control study conducted in Kashmir Valley found a lower risk of EC (ESCC+EAC) linked with either GSTM3 heterozygous genotype *A/*B or the GSTM3*B allele (Malik, Upadhyay et al. 2010). In diverse ethnicities, the impact of the GSTT1 deletion polymorphism in EC has been studied thoroughly. Although a strong relationship of EAC with smoking has been identified in individuals having the GSTT1 non-null genotype, the GSTT1*0 genotype appears to have no impact on the risk of ESCC or EAC in Caucasians (Wideroff, Vaughan et al. 2007, Zendehdel, Bahmanyar et al. 2009, Dura, Salomon et al. 2013).

20.5.1.1.4 N-Acetyltransferases

N-acetyltransferase (NAT) belongs to the category of enzymes that facilitate the transfer of acetyl groups from acetyl-CoA to arylamines, arylhydroxylamines, and arylhydrazines. NATs are the major phase II metabolizing enzymes involved in arylamine carcinogen metabolism found in overcooked meat and tobacco smoke. NATs catalyze either N-acetylation (detoxification) or O-acetylation (bioactivation) of aromatic amines, depending on the type of the substrate (Hein 2002). There are two types of NAT genes in humans, NAT-1 and NAT-2, 170 kb distanced on chromosome 8p21.3–23.1 (Hickman, Risch et al. 1994). NAT1 is found in nearly every tissue, including the esophagus, whereas NAT2 is found mostly in the liver, small intestine, and colon (Husain, Zhang et al. 2007). A considerable number of

SNPs have been found in the coding and noncoding regions of NAT1 and NAT2 genes. The wild-type NAT1*4 genotype was found to be the most common type throughout the communities investigated so far, whereas the NAT1*10 genotype, characterized by the existence of two SNPs (1088T4A [rs1057126] and 1095C4A [rs15561]) within the 3′-UTR, causes a change in the consensus polyadenylation signal and has substantially been linked to increased acetylation activity when compared to the wild-type allele (Wang, Para et al. 2011). SNPs of the NAT1 gene (1088T4A and 1095C4A) investigated in the communities of South African Black Xhosa and mixed ancestry were linked to ESCC risk. The NAT1 slow/intermediate acetylator status including the mixture of variants NAT1*4/*3, NAT1*4/*4, NAT1*3/*3, NAT1*10/*4, and NAT1*10/*3 was linked to a considerably higher chances of getting ESCC in Black Xhosa population (Matejcic, Vogelsang et al. 2015).

Some case–control studies on the polymorphism of NAT2 in EC predisposition that have been undertaken to date have shown inconclusive results. As per a case–control study conducted in Japan, the NAT2 slow acetylator was found to be associated with ESCC risk when compared to the rapid acetylator status. Slow acetylator variants of NAT2 such as NAT2*5, NAT2*6, and NAT2*7 showed an elevated risk of ESCC in Kashmir Valley, but the haplotype C variant of NAT2 C(481)A(590)G(857) has been linked to a lower risk (Malik, Upadhyay et al. 2009). No correlation between NAT2 slow acetylator and acetylation has been identified in the presence or absence of tobacco, alcohol, and occupational exposure in the Taiwanese (Lee, Lee et al. 2000) and North Indian populations (Jain, Kumar et al. 2007). In a study conducted in South Africa, the NAT2*5 (341T4C) showed a preventive effect against ESCC among mixed ancestry people, but no significant correlation was seen in people of black ancestry (Matejcic, Vogelsang et al. 2015).

20.5.1.1.5 Sulfotransferases (SULTs)

Sulfotransferase enzyme group is a class of phase II enzymes that catalyze the transfer of sulfates from 3′-phosphoaden osine-5′-phosphosulfate (PAPS) to nucleophilic molecules (Glatt, Boeing et al. 2001). SULT enzymes are divided into membrane-bound SULT group, which is found in Golgi apparatus and has a role in macromolecule sulfonation, and cytosolic SULT group, which catalyzes the sulfonation of small endogenous substances such as bile acids, neurotransmitters and steroids and also xenobiotic molecules. In humans, four cytosolic SULT enzymes – SULT1, SULT2, SULT4, and SULT6 – have been found and all of these have 13 isoenzymes that vary in substrate specificity and tissue distribution (Li, Lindsay et al. 2008). SULT1A1, 1A2, 1A3, 1B1, 1C2, 1C4, and 2A1 are the main SULT enzymes that help in metabolizing the ambient toxins.

Although the significance of SULT1A1*2 in EC susceptibility is unknown, a case–control study conducted in a South African community found that homozygous SULT1A1*2/*2 variant is associated with ESCC among people of mixed ancestry, particularly among men. SULT1A1*2/*2 variant, when combined with CYP3A5 heterozygous variants such as CYP3A5*1/*2, *1/*3, *1/*4, *1/*5, *1/*6, and *1/*7, was linked to a high chance of ESCC, which is considerably higher in tobacco smokers than in non-smokers (Dandara, Li et al. 2006).

20.5.1.1.6 UDP-Glucuronosyltransferases

The glucuronidation process, a significant step in phase II metabolism, is catalyzed by uridine diphosphate glucuronosyltransferases (UGTs). UGTs catalyze the conversion of glucuronic acid from 5-diphosphoglucuronic acid to glucuronide, which are eventually excreted. Human UGTs have been divided into two major subclasses based on amino acid sequence and genetic organization resemblances: UGT1A and UGT2 (UGT2A and UGT2B) (Mackenzie, Bock et al. 2005). Each UGT subgroup has two or more isoforms, and the most effective UGTs engaged in the detoxification process are UGT1A1, 1A3, 1A4, 1A6, 1A9, 2B4, and 2B7 (Burchell, Lockley et al. 2005).

In a case–control study, the high activity genotypes of UGT1A1 and UGT1A8 were strongly related to a higher and decreased risk of ESCC, respectively. When compared to low-activity genotypes, high- and medium-activity UGT2B4 genotypes were linked with a considerably lower incidence of ESCC. Polymorphisms in UGT genes that result in changed enzyme activity did not appear to change the risk of EAC in Caucasians (Dura, Salomon et al. 2012).

20.5.1.2 Enzymes Involved in Repair Mechanisms, Cell Cycle Regulation, and Apoptosis

The other set of potential genes that have sparked the science community's curiosity include those involved in cell cycle regulation, repair mechanisms, and apoptotic processes, such as TP53, cyclin D1, and CCND1 (Vos, Adams et al. 2003), as well as nucleotide excision repair (NER) and base excision repair (BER) genes (Lee, Lee et al. 2001, Xing, Qi et al. 2002, Hao, Wang et al. 2004, Ratnasinghe, Abnet et al. 2004). All of these genes have been found to be important in determining the sensitivity to carcinogen exposure, physiological reactivity to these hazards, or eventual capabilities for regenerative repairs and induction of apoptosis in the aftermath of DNA damage. None, however, were shown to have a consistent effect on ESCC incidence. Matrix metalloproteinase (MMP) genes are prime illustrations of a pathway that degrades extracellular matrix constituents and so plays an essential role in malignant transformation. Mutations in the MMP gene sequences have a role in altering the MMP production or activity. ESCC cases and controls have significantly different alleles for P574R polymorphic form of MMP-9; according to a research in the Chinese population, P574R GG genotype, for example, was persistently correlated with a considerably higher ESCC incidence as compared to the P574R CC variants (Yu, Zhou et al. 2004, Wu, Zhang et al. 2008). A SNP (rs243865) found in the promoter region of the MMP-2 gene was graded as high evidence of relation with EC risk in a meta-analysis (Tian, Liu et al. 2019).

Similarly, epidermal growth factor (EGF) plays a role in cell differentiation and proliferation, and mutations in EGF and their high expression have been linked to an increased risk of thoracic and gastrointestinal cancers (Hamai, Matsumura et al. 2005, Lim, Kim et al. 2005, Tanabe, Lemoine et al. 2008). Polymorphisms in EGFR 497Arg>Lys and EGF +61A>G (rs4444903), in particular, have been demonstrated to affect cell cycle progression, apoptosis, angiogenesis, and metastasis (Upadhyay, Jain et al. 2008). Interestingly, whereas the EGF +61A/A genotype has been linked to an increased risk of ESCC, there is no link between the EGFR 497Arg/Arg genotype and ESCC risk. The G allele of EGF +61A>G (rs4444903) polymorphism was rated as strong evidence of correlation with 1.38-fold increased risk of EC based on 1,713 sample size (Tian, Liu et al. 2019).

Phospholipase C epsilon 1 (PLCE1), which is found on chromosome 10q23, is involved in cell proliferation, differentiation, gene expression, and oncogenesis (Wing, Bourdon et al. 2003, Thatcher 2010). Based on a sample size of over 20,000, this SNP (rs2274223) was evaluated as strong evidence of relation with higher risk of EC for a common A-to-G transition of PLCE1 that may enhance the expression of PLCE1 protein (Wing, Bourdon et al. 2003, Bunney, Harris et al. 2006). In a meta-analysis study, this SNP was shown to have a substantial connection with EC risk in Asians, but not in Caucasians (Tian, Liu et al. 2019).

The HOX transcript antisense RNA (HOTAIR) gene, which is found on chromosome 12q13.13, has the potential to turn normal cells into cancerous ones (Ge, Jiang et al. 2017). Compared to the rs920778 CC genotype, the TT genotype was marginally significantly related to ESCC susceptibility in the Chinese population (Zhang, Zhou et al. 2014, Ge, Jiang et al. 2017). Based on 4,221 samples, the SNP (rs920778) was rated as strong evidence of true association with ESCC risk in the T allele of HOTAIR, which may induce genome-wide retargeting of polycomb repressive complex 2 and trimethylation of histone H3 at lysine 27 (H3K27me3), and deregulation of multiple downstream genes involved in the development and progression of ESCC.

20.5.1.3 Others

20.5.1.3.1 Alcohol Dehydrogenase and Aldehyde Dehydrogenase 2

Alcohol intake is a key risk factor for EC, and ESCC is one of the most prevalent malignant tumors (Yokoyama, Muramatsu et al. 1996, Chitra, Ashok et al. 2004, Garavello, Negri et al. 2005). Alcohol dehydrogenase (ADH) converts alcohol to acetaldehyde, which is then converted to acetate by acetaldehyde dehydrogenase (ALDH) (Yang, Matsuo et al. 2005, Ishikawa, Miyatsu et al. 2006). The ADH gene on chromosome 4 is a major cytosolic enzyme for ethanol (Zuo, Wang et al. 2013). It encodes at least seven ADH isoforms (ADH1–ADH7), each with somewhat distinct characteristics (Luo, Kranzler et al. 2008), and may be implicated in numerous neurotransmitter metabolic pathways (Svensson, Some et al. 1999). On human chromosome 4q21-q23, the ADH1B (commonly known as ADH2) gene is found. It is the gene that controls the majority of ADH activity in the liver when it comes to ethanol (Strommer 2011). SNPs in this gene may affect ethanol metabolism, resulting in differing alcohol clearance rates among people who exhibit the ADH1B variations (Yoshida, Hsu et al. 1991). Several investigations have found a link between the ADH1B polymorphism and EC susceptibility, although the results have not been consistent. For example, Ito et al. found that the ADH1B Arg47His variant was linked to EC in Japanese people and could be used in personalized prevention programs (Ito, Oze et al. 2015), whereas Ma et al. found no link between ADH1B gene variations and the risk of ESCC in Chinese people (Ma, Lv et al. 2010). Although meta-analyses to investigate this connection were conducted (Yang, Yokoyama et al. 2010, Zhang, Mai et al. 2010, Zhang, Jiang et al. 2014), only a few publications were chosen, only the Asian population was analyzed, and some repeat participants were included. Furthermore, due to diverse ethnic origins, Asians and Caucasians may have varied EC incidence, distribution, and susceptibilities (Zhang, Jin et al. 2012).

The ALDH2 gene, which is involved in the removal of acetaldehyde, which is formed as an intermediate product in alcohol metabolism in the body, is one of the most consistent bits of evidence supporting the impact of SNPs on ESCC risk. The ALDH2 gene, which codes for ALDH2, is made up of 13 exons on chromosome 12. A single substitution of Glu-to-Lys amino acid at position 487, which can be linked to a G-to-A transition (db SNP: rs671) in exon 12 of the gene, causes ALDH2 deficiency, which is a dominant characteristic. Most East Asian communities carry this defective allele (ALDH2*487Lys) (Wall 2005, Nishimura, Kimura et al. 2006), and around 45% of Chinese and Japanese people have the inactive ALDH2 phenotype (Isse, Oyama et al. 2005, Yokoyama, Kato et al. 2006). People who have the ALDH *1/*2 heterozygous genotype have a significantly higher risk of ESCC, with an estimated OR of 3.2 (Lewis and Smith 2005). ALDH *1/*2 polymorphism, which is common in East Asian population, but uncommon in other people, encodes for an inactive enzyme that causes high blood acetaldehyde levels following alcohol use.

In clinical studies, it's also linked to a flushing reaction after consuming alcohol. More notably, this polymorphism demonstrates a gene–environment interaction, since the elevated ESCC risk associated with the ALDH *1/*2 heterozygous genotype is substantially correlated with the amount of alcohol consumed (Lewis and Smith 2005). The homozygous *2/*2 mutation, on the other hand, is linked to a decreased incidence of ESCC, which has been linked to the total intolerance to alcohol that occurs in people with this genotype (Lewis and Smith 2005).

20.5.1.3.2 Methylenetetrahydrofolate Reductase

The C677T polymorphism of the methylenetetrahydrofolate reductase (MTHFR) gene is another example of the impact of gene–environment interactions (Wang, Guo et al. 2005, Wang, Miao et al. 2005, Li, Diao et al. 2008, Wang, Sasco et al. 2008). When compared to the CC genotype, the MTHFR C677T TT and TC genotypes significantly increased the incidence of ESCC (OR 1.58) (Li, Diao et al. 2008). Moreover, there was a substantial association between the TT and TC genotypes and recognized ESCC risk factors such as alcohol use, tobacco, and family background (Li, Diao et al. 2008). As a result, people with these genes, when combined with other risk factors, had a higher chance of dysplasia and ESCC. Regardless of the robustness of certain findings, only a few genes and interactions have been consistently linked to ESCC vulnerability in many investigations. Unfortunately, such a wide range of results make it impossible to draw any definitive conclusions from the data.

20.5.2 Genetic Polymorphisms in EAC

In contrast to ESCC, there is little evidence available on the molecular epidemiologic patterns of EAC. Table 20.3

TABLE 20.3

Association of Genetic Polymorphisms with the Risk of EAC

Gene Involved	SNPs Associated
GSTT1	Deletion (*1/*2)
GSTP1	Ile104Val
GSTM3	*A/*B
XPD	Lys751Gln
Cyclin D1	G870A
P73	5′-UTR G4A+C14 T
NQO1	C609T
ADH3	*1/*2
VEGF	460C/T, +405C/G, and +936C/T
ERCC1	C8092A & 118C/T
MGMT	rs12268840 & I143V
EGF	A61G

summarizes the results of some SNPs associated with EAC as well as particular polymorphism information. There were 24 genetic polymorphism associations with EAC risk identified so far. The majority of molecular epidemiology studies have focused on patient populations in North America and Europe, which is consistent with previous studies on the usual epidemiologic characteristics of EAC (e.g., incidence proportion and prevalence rates). A modest number of instances, ranging from 50 to 300, were included in the majority of these research studies (Casson, Zheng et al. 2005, Jain, Kumar et al. 2006, 2007, Murphy, Hughes et al. 2007). Similar to ESCC, polymorphisms in repair mechanism genes (such as XPD, ERCC1, and XPC), cell cycle regulation and p53 pathway genes (such as cyclin D1 and p73), and phase I and II enzymes involved in carcinogen activation or detoxification have been studied (e.g., GSTT1, GSTP1, and GSTM3) (Casson, Zheng et al. 2005, Jain, Kumar et al. 2006, 2007, Murphy, Hughes et al. 2007, Tse, Zhai et al. 2008).

The VEGF has also piqued the curiosity of scientists. In the process of tumor growth and metastasis, VEGF is a key regulator of angiogenesis. VEGF gene polymorphisms have been linked to changes in the expression of VEGF. Preliminary analysis has found that functional polymorphisms in the VEGF gene (e.g., −460 C/T and +936 C/T) are linked to the incidence of EAC, as expected (Zhai, Liu et al. 2008). The +936CT and TT genotypes, for example, were strongly related to the high risk of EAC as compared to the +936CC genotype (OR 1.49) (Zhai, Liu et al. 2008). Moreover, −460CT+CC was correlated with elevated EAC risk in smokers and −460T+TT was related to lower EAC risk in non-smokers (Zhai, Liu et al. 2008). Likewise, research on the EGF A61G polymorphisms has found that the G/G genotype increases the incidence of EAC. This situation could worsen in EAC patient who also have GERD or BE, indicating the existence of a gene–environment connection (Lanuti, Liu et al. 2008).

The principal biological defense against alkylating DNA damage is the DNA repair protein O6-methylguanine-DNA methyltransferase (MGMT). Each of the five MGMT SNPs (rs12269324, rs12268840, L84F, I143V, and K178R) was linked to an elevated risk of EAC (Doecke, Zhao et al. 2008). Two mutant MGMT genotypes rs12268840 and I143V were shown to have strong correlations. Homozygous carriers of MGMT rs12268840 who also experienced recurrent acid reflux had considerably increased risk of EAC, demonstrating the importance of gene–environment interactions (Doecke, Zhao et al. 2008). There were some parallels for smoking. Other DNA repair genes have been linked to this discovery. The homozygous variants ERCC1 genotype have been linked to lower EAC risk, whereas the homozygous variant XRCC1 genotype has been linked to a greater EAC risk.

Investigations of EAC have produced mixed results, similar to those of ESCC. The dearth of uniformity in numerous studies that have frequently looked at the identical gene variants cannot be overlooked, and it has made assimilating new genetic knowledge challenging. While some speculate that the disparities in results are attributable to variances in clinical groups between studies, they might also be aberrations of small sample sizes, which usually result in high percentages of false positives. In contrast, a number of genetic variants have been found to have no connection with EAC susceptibility, but given the small sample sizes of these investigations, these might be false-negative results.

20.6 Role of Genetic Polymorphisms in EC Prognosis

A strategy for accurately evaluating the outcomes of EC patients and identifying who could benefit from more aggressive treatments would be a useful tool. The ultimate objective is to use the genes to build prognostic models (e.g., nomograms) that can give helpful prognostic information to patients and clinicians when paired with clinical data from patients, allowing them to make better management and supportive care decisions. Despite vigorous utilization of local treatment, surgery, chemo- or radiotherapy, or a combination of modalities, overall survival rates for EC remain in the 15%–25% range (Crane, Locke et al. 2008). Furthermore, patients undergoing these therapies commonly have significant morbidity and problems, such as local toxicity (e.g., dysphagia, esophagitis, and deformity) and general side effects (e.g., weight loss and stress), all of which can reduce the standard of living (Kleinberg and Forastiere 2007). It will be difficult to create predictive tools dependent on genetic polymorphisms. Although some overlap is to be anticipated, some variations may have a good influence on survival, but a negative effect on toxicity.

Currently, the literature related to the role of polymorphisms in EC prognosis is restricted to a few studies. Studies of EC outcomes, like studies of EC risk, have typically been classified into sections based on histopathology and ethnic origin: (1) studies of gene polymorphisms and EAC results in whites and (2) analysis of polymorphic variations and ESCC outcomes in Asians. The majority of researches to date have been constrained by limited sample sizes in the form of a case–control study series and retrospective cohort studies. Overall survival and disease-free survival were chosen as the primary outcome measures in majority of the studies. The selection for polymorphism studies in such situations was made based on previously investigated data from general cancer risk-associated studies. In other situations, the choice was made based on prior understanding of probable intracellular pathways and targets thought to be significant in tumorigenesis.

Polymorphic variants of importance are found in a variety of pathways, including DNA repair (e.g., XRCC1) (Wu, Gu et al. 2006) and xenobiotic metabolism (e.g., GSTT1 and GSTM1). Due to the widespread usage of 5-fluorouracil in treating esophageal malignancies, much attention has been centered primarily on pharmacogenetics, notably the folate route (e.g., genetic polymorphism in MTHFR and thymidylate synthase) (Liao, Liu et al. 2006). Furthermore, because cisplatin chemotherapy is commonly used to treat ECs, the GSTs and DNA repair mechanisms have received a considerable attention (Harpole, Moore et al. 2001, Wu, Gu et al. 2006). Unfortunately, the investigation of these routes has yielded no clear pattern of outcomes. A few polymorphic variations of other routes (for example, IL-1b, L-Myc, and IL-6) have also been studied, although only in small studies (Shibuta, Inoue et al. 2000). Similarly, another study discovered a predictive link between a polymorphic version of the gene encoding the multidrug-resistant protein MDR1 (Wu, Gu et al. 2006). Again, no independent validation of these findings has been published or is accessible for evaluation in the literature.

20.7 Future Perspectives

Several reports have demonstrated the functional roles of SNPs in the development of EC. To fully leverage SNPs' prognostic and therapeutic potential, genetic variables and their connection with EC must be thoroughly explored. The effective and precise recognition of SNPs could be achieved only if research groups conduct large sample studies to provide appropriate statistical power to find real relationships, assuming they exist. It might demand multidisciplinary collaboration, implying the necessity to form specialist molecular epidemiological working groups with this goal in mind. Studies showed that the ADH1B Arg47His polymorphism may be a risk factor for EC in Asians. The Arg/Arg genotype, when paired with alcohol use, cigarette smoking, and men, may significantly increase the risk of EC. However, other studies with more ethnicities should be included in future. ALDH2 might be used as a therapeutic target in EC treatment.

Some of the western nations have a declining prevalence of ESCC, most likely owing to reduced cigarette smoking and some reduction in strong alcohol intake. ESCC, on the other hand, remains a major cause of cancer-related deaths globally. The prevalence of ESCC in underprivileged populations with inadequate research infrastructure has been a major hurdle to advancement, although investigations in Iran (Pourshams, Khademi et al. 2010), East Africa (McCormack, Menya et al. 2017), India (Dar, Bhat et al. 2012), and elsewhere have supplemented findings from decades of study in China. These ongoing investigations should lead to a better understanding of the very lethal and understudied malignancy that mostly affects the poorest members of our society. Despite the poor prognosis of individuals with EAC with the existing therapy, the preventive measures of this emergent ailment is critical. Minimizing the exposure to the main pre-identified modifiable risk factors may have a considerable influence on the incidence of adenocarcinoma. Clinicians' prime objective should be risk factor reduction and early identification. The acquisition of new biomarkers that can predict the success of neoadjuvant chemo- and radiotherapy remains important since it can help with therapy selection for individual patients at the time of diagnosis. The intriguing endeavor is the use of DNA microarrays and "SNP chips," which could allow researchers to evaluate several genetic variations at the same time. This sophisticated technique, which can efficiently analyze several hundred to even a million SNPs, allows for the study of complicated gene–gene interactions. While such innovative strategies are definitely advantageous, adequate research design and statistical considerations remain critical in assuring the validity of these high-throughput procedures.

ABBREVIATIONS

ADH: Alcohol dehydrogenase
AhR: Aryl hydrocarbon receptor
ALDH: Acetaldehyde dehydrogenase
BE: Barrett's esophagus
BER: Base excision repair
CAR: Constitutive androstane receptor
CYP: Cytochrome P450
EAC: Esophageal adenocarcinoma
EC: Esophageal cancer
EGF: Epidermal growth factor
EHs: Epoxide hydrolases
ESCC: Esophageal squamous cell carcinoma
GERD: Gastroesophageal reflux disease
GSH: Glutathione
GSTs: Glutathione S-transferases
GWAS: Genome-wide association studies
HGD: High-grade dysplasia
HOTAIR: HOX transcript antisense RNA
HPV: Human papilloma virus
LES: Lower esophageal sphincter
MGMT: O6-Methylguanine-DNA methyltransferase
MMP: Matrix metalloproteinase
MTHFR: Methylenetetrahydrofolate reductase
NAT: N-acetyltransferase
NER: Nucleotide excision repair
PAPS: 3′-Phosphoadenosine-5′-phosphosulfate
PLCE1: Phospholipase C epsilon 1
PXR: Pregnane X receptor
SNPs: Single nucleotide polymorphisms
SULTs: Sulfotransferases
UGTs: Uridine diphosphate glucuronosyltransferases
VEGF: Vascular endothelial growth factor
XMEs: Xenobiotic-metabolizing enzymes

REFERENCES

Akbari, M. R., R. Malekzadeh, D. Nasrollahzadeh, D. Amanian, P. Sun, F. Islami, M. Sotoudeh, S. Semnani, P. Boffeta and S. M. J. Dawsey (2006). Familial risks of esophageal cancer among the Turkmen population of the Caspian littoral of Iran. **119**(5): 1047–1051.

Akhtar, S. J. C. C. (2013). Areca nut chewing and esophageal squamous-cell carcinoma risk in Asians: a meta-analysis of case–control studies. **24**(2): 257–265.

Arnold, M., C. C. Abnet, R. E. Neale, J. Vignat, E. L. Giovannucci, K. A. McGlynn and F. J. G. Bray (2020). Global burden of 5 major types of gastrointestinal cancer. **159**(1): 335–349.e315.

Arnold, M., M. Laversanne, L. M. Brown, S. S. Devesa and F. Bray (2017). Predicting the future burden of Esophageal cancer by histological subtype: international trends in incidence up to 2030. **112**(8): 1247–1255.

Bahmanyar, S. and W. J. N. Ye (2006). Dietary patterns and risk of squamous-cell carcinoma and adenocarcinoma of the esophagus and adenocarcinoma of the gastric cardia: a population-based case-control study in Sweden. **54**(2): 171–178.

Bhat, G. A., I. A. Shah, M. A. Makhdoomi, B. Iqbal, R. Rafiq, S. Nabi, A. Masood, M. M. Lone and N. A. Dar (2014). CYP1A1 and CYP2E1 genotypes and risk of esophageal squamous cell carcinoma in a high-incidence region, Kashmir. **35**(6): 5323–5330.

Brown, L. M., R. N. Hoover, R. S. Greenberg, J. B. Schoenberg, A. G. Schwartz, G. M. Swanson, J. M. Liff, D. T. Silverman, R. B. Hayes and L. M. J. Pottern (1994). Are racial differences in squamous cell esophageal cancer explained by alcohol and tobacco use? **86**(17): 1340–1345.

Bunney, T. D., R. Harris, N. L. Gandarillas, M. B. Josephs, S. M. Roe, S. C. Sorli, H. F. Paterson, F. Rodrigues-Lima, D. Esposito and C. P. Ponting (2006). Structural and mechanistic insights into RAS association domains of phospholipase C epsilon. **21**(4): 495–507.

Burchell, B., D. J. Lockley, A. Staines, Y. Uesawa and M. W. Coughtrie (2005). Substrate specificity of human hepatic UDP-glucuronosyltransferases. **400**: 46–57.

Canaparo, R., N. Finnström, L. Serpe, A. Nordmark, E. Muntoni, M. Eandi, A. Rane, G. P. J. and C. Zara (2007). Expression of CYP3A isoforms and P-glycoprotein in human stomach, jejunum and ileum. **34**(11): 1138–1144.

Casson, A. G., Z. Zheng, S. C. Evans, L. Geldenhuys, S. V. van Zanten, P. J. Veugelers, G. A. Porter and D. L. Guernsey (2005). Cyclin D1 polymorphism (G870A) and risk for esophageal adenocarcinoma. *Cancer* **104**(4): 730–739.

Chak, A., H. Ochs-Balcom, G. Falk, W. M. Grady, M. Kinnard, J. E. Willis, R. Elston, C. J. C. E. Eng and P. Biomarkers (2006). Familiality in Barrett's esophagus, adenocarcinoma of the esophagus, and adenocarcinoma of the gastroesophageal junction. **15**(9): 1668–1673.

Chitra, S., L. Ashok, L. Anand, V. Srinivasan and V. J. Jayanthi (2004). Risk factors for esophageal cancer in Coimbatore, southern India: a hospital-based case-control study. **23**(1): 19–21.

Crane, S. J., G. R. Locke, 3rd, W. S. Harmsen, A. R. Zinsmeister, Y. Romero and N. J. Talley (2008). Survival trends in patients with gastric and esophageal adenocarcinomas: a population-based study. *Mayo Clin Proc* **83**(10): 1087–1094.

Dandara, C., R. Ballo and M. I. Parker (2005). CYP3A5 genotypes and risk of oesophageal cancer in two South African populations. **225**(2): 275–282.

Dandara, C., D.-P. Li, G. Walther and M. I. J. C. Parker (2006). Gene–environment interaction: the role of SULT1A1 and CYP3A5 polymorphisms as risk modifiers for squamous cell carcinoma of the oesophagus. **27**(4): 791–797.

Dar, N., G. Bhat, I. Shah, B. Iqbal, M. Kakhdoomi, I. Nisar, R. Rafiq, S. Iqbal, A. Bhat and S. J. Nabi (2012). Hookah smoking, nass chewing, and oesophageal squamous cell carcinoma in Kashmir, India. **107**(9): 1618–1623.

de Jonge Pj, v. B. M., W. J. B. s. o. e. Grady, cancer risk and i. f. m. G. (2014) Kuipers Ej. **63**: 191–202.

Ding, X. and L. S. Kaminsky (2003). Human extrahepatic cytochromes P450: function in xenobiotic metabolism and tissue-selective chemical toxicity in the respiratory and gastrointestinal tracts. **43**(1): 149–173.

Djansugurova, L. B., A. V. Perfilyeva, G. S. Zhunusova, K. B. Djantaeva, O. A. Iksan and E. M. Khussainova (2013). The determination of genetic markers of age-related cancer pathologies in populations from Kazakhstan. **4**: 70.

Doecke, J., Z. Z. Zhao, N. Pandeya, S. Sadeghi, M. Stark, A. C. Green, N. K. Hayward, P. M. Webb and D. C. Whiteman (2008). Polymorphisms in MGMT and DNA repair genes and the risk of esophageal adenocarcinoma. *Int J Cancer* **123**(1): 174–180.

Dura, P., J. Salomon, R. H. Te Morsche, H. M. Roelofs, J. O. Kristinsson, T. Wobbes, B. J. Witteman, A. C. Tan, J. P. Drenth and W. H. Peters (2012). High enzyme activity UGT1A1 or low activity UGT1A8 and UGT2B4 genotypes increase esophageal cancer risk. **40**(6): 1789–1796.

Dura, P., J. Salomon, R. H. Te Morsche, H. M. Roelofs, J. O. Kristinsson, T. Wobbes, B. J. Witteman, A. C. Tan, J. P. Drenth and W. H. Peters (2013). No role for glutathione S-transferase genotypes in Caucasian esophageal squamous cell or adenocarcinoma etiology: an European case–control study. **13**(1): 1–7.

Dutt, A. and R. J. Beroukhim (2007). Single nucleotide polymorphism array analysis of cancer. **19**(1): 43–49.

Efferth, T. and M. J. P. Volm (2005). Pharmacogenetics for individualized cancer chemotherapy. **107**(2): 155–176.

Eng, L., A. K. Azad, X. Qiu, Q. Q. Kong, D. Cheng, N. Ying, A. Tse, Q. Kuang, L. Dodbiba, D. J. Renouf, S. Marsh, S. Savas, H. J. Mackay, J. J. Knox, G. E. Darling, R. K. Wong, W. Xu, G. Liu and O. O. Faluyi (2015). Discovery and validation of vascular endothelial growth factor (VEGF) pathway polymorphisms in esophageal adenocarcinoma outcome. *Carcinogenesis* **36**(9): 956–962.

Engel, L. S., W. H. Chow, T. L. Vaughan, M. D. Gammon, H. A. Risch, J. L. Stanford, J. B. Schoenberg, S. T. Mayne, R. Dubrow and H. J. Rotterdam (2003). Population attributable risks of esophageal and gastric cancers. **95**(18): 1404–1413.

Falk, G. W. J. S. O. C. (2009). Risk factors for esophageal cancer development. **18**(3): 469–485.

Ferlay, J., M. Ervik, F. Lam, M. Colombet, L. Mery, M. Piñeros, A. Znaor, I. Soerjomataram and F. Bray (2020). *Global Cancer Observatory: Cancer Today*. International Agency for Research on Cancer, Lyon.

Freedman, N. D., C. C. Abnet, M. F. Leitzmann, T. Mouw, A. F. Subar, A. R. Hollenbeck and A. J. Schatzkin (2007). A prospective study of tobacco, alcohol, and the risk of esophageal and gastric cancer subtypes. **165**(12): 1424–1433.

Gao, Y.-B., Z.-L. Chen, J.-G. Li, X.-D. Hu, X.-J. Shi, Z.-M. Sun, F. Zhang, Z.-R. Zhao, Z.-T. Li and Z.-Y. Liu (2014). Genetic landscape of esophageal squamous cell carcinoma. **46**(10): 1097–1102.

Gao, Z., X. Meng, D. Mu, X. Sun and J. J. Yu (2014). Prognostic significance of epidermal growth factor receptor in locally advanced esophageal squamous cell carcinoma for patients receiving chemoradiotherapy. **7**(4): 1118–1122.

Gao, C.-M., T. Takezaki, J.-Z. Wu, Z.-Y. Li, Y.-T. Liu, S.-P. Li, J.-H. Ding, P. Su, X. Hu and T.-L. Xu (2002). Glutathione-S-transferases M1 (GSTM1) and GSTT1 genotype, smoking, consumption of alcohol and tea and risk of esophageal and stomach cancers: a case-control study of a high-incidence area in Jiangsu Province, China. **188**(1–2): 95–102.

Garavello, W., E. Negri, R. Talamini, F. Levi, P. Zambon, L. Dal Maso, C. Bosetti, S. Franceschi, C. J. C. E. La Vecchia and P. Biomarkers (2005). Family history of cancer, its combination with smoking and drinking, and risk of squamous cell carcinoma of the esophagus. **14**(6): 1390–1393.

Ge, Y., R. Jiang, M. Zhang, H. Wang, L. Zhang, J. Tang and C. J. Liang (2017). Analyzing 37,900 samples shows significant association between HOTAIR polymorphisms and cancer susceptibility: a meta-analysis. **32**(2): 231–242.

Gellner, K., R. Eiselt, E. Hustert, H. Arnold, I. Koch, M. Haberl, C. J. Deglmann, O. Burk, D. Buntefuss and S. J. P. Escher (2001). Genomic organization of the human CYP3A locus: identification of a new, inducible CYP3A gene. **11**(2): 111–121.

Glatt, H., H. Boeing, C. E. Engelke, L. Ma, A. Kuhlow, U. Pabel, D. Pomplun, W. Teubner and W. J. Meinl (2001). Human cytosolic sulphotransferases: genetics, characteristics, toxicological aspects. **482**(1–2): 27–40.

Gonzalez, F. J. and H. V. Gelboin (1994). Role of human cytochromes P450 in the metabolic activation of chemical carcinogens and toxins. **26**(1–2): 165–183.

Hamai, Y., S. Matsumura, K. Matsusaki, Y. Kitadai, K. Yoshida, Y. Yamaguchi, K. Imai, K. Nakachi, T. Toge and W. J. P. Yasui (2005). A single nucleotide polymorphism in the 5′ untranslated region of the EGF gene is associated with occurrence and malignant progression of gastric cancer. **72**(3): 133–138.

Hao, B., H. Wang, K. Zhou, Y. Li, X. Chen, G. Zhou, Y. Zhu, X. Miao, W. Tan and Q. J. C. Wei (2004). Identification of genetic variants in base excision repair pathway and their associations with risk of esophageal squamous cell carcinoma. **64**(12): 4378–4384.

Harpole, D. H., Jr., M. B. Moore, J. E. Herndon, 2nd, T. Aloia, T. A. D'Amico, T. Sporn, A. Parr, I. Linoila and C. Allegra (2001). The prognostic value of molecular marker analysis in patients treated with trimodality therapy for esophageal cancer. *Clin Cancer Res* **7**(3): 562–569.

Harris, R. E. (2019). *Epidemiology of Chronic Disease: Global Perspectives*. Jones & Bartlett Learning.

Hassett, C., L. Alcher, J. S. Sidhu and C. J. J. Omieclnskl (1994). Human microsomal epoxide hydrolase: genetic poloymorphism and functional expression in vitro of amino acid variants. **3**(3): 421–428.

Hayes, J. D., J. U. Flanagan and I. R. J. Jowsey (2005). Glutathione transferases. **45**: 51–88.

Hein, D. W. (2002). Molecular genetics and function of NAT1 and NAT2: role in aromatic amine metabolism and carcinogenesis. **506**: 65–77.

Hickman, D., A. Risch, V. Buckle, N. Spurr, S. Jeremiah, A. McCarthy and E. J. B. J. Sim (1994). Chromosomal localization of human genes for arylamine N-acetyltransferase. **297**(3): 441–445.

Husain, A., X. Zhang, M. A. Doll, D. F. Barker and D. W. Hein (2007). Identification of N-acetyltransferase 2 (NAT2) transcription start sites and quantitation of NAT2-specific mRNA in human tissues. **35**(5): 721–727.

Hvid-Jensen F, L. Pedersen, A. M. Drewes, H. T. Sørensen and P. Funch-Jensen (2011). Incidence of adenocarcinoma among patients with Barrett's esophagus. *N Engl J Med* **365**: 1375–1383.

Hyuna Sung, J. F., R. L. Siegel, M. Laversanne, I. Soerjomataram, A. Jemal and F. Bray (2021). Global cancer statistics 2020: GLOBOCAN estimates of incidence and mortality worldwide for 36 cancers in 185 countries. *CA Cancer J Clin* **71**: 209–249.

Ihsan, R., I. Chattopadhyay, R. Phukan, A. K. Mishra, J. Purkayastha, J. Sharma, E. Zomawia, Y. Verma, J. Mahanta and S. J. J. Saxena (2010). Role of epoxide hydrolase 1 gene polymorphisms in esophageal cancer in a high-risk area in India. **25**(8): 1456–1462.

Ingelman-Sundberg, M., S. C. Sim, A. Gomez and C. J. P. Rodriguez-Antona (2007). Influence of cytochrome P450 polymorphisms on drug therapies: pharmacogenetic, pharmacoepigenetic and clinical aspects. **116**(3): 496–526.

Ishikawa, H., Y. Miyatsu, K. Kurihara and K. J. M. R. F. Yokoyama (2006). Gene–environmental interactions between alcohol-drinking behavior and ALDH2 and CYP2E1 polymorphisms and their impact on micronuclei frequency in human lymphocytes. **594**(1–2): 1–9.

Isse, T., T. Oyama, K. Matsuno, I. Uchiyama and T. J. Kawamoto (2005). Aldehyde dehydrogenase 2 activity affects symptoms produced by an intraperitoneal acetaldehyde injection, but not acetaldehyde lethality. **30**(4): 315–328.

Ito, H., I. Oze, S. Hosono, M. Watanabe, H. Tanaka and K. Matsuo (2015). The risk prediction for esophageal cancer by drinking, smoking, and the polymorphisms of ALDH2 and ADH1B, AACR.

Jain, M., S. Kumar, P. Lal, A. Tiwari, U. Ghoshal and B. J. Mittal (2007a). Association of genetic polymorphisms of N-acetyltransferase 2 and susceptibility to esophageal cancer in north Indian population. **25**(5): 340–346.

Jain, M., S. Kumar, P. Lal, A. Tiwari, U. C. Ghoshal and B. Mittal (2007b). Role of GSTM3 polymorphism in the risk of developing esophageal cancer. *Cancer Epidemiol Biomarkers Prev* **16**(1): 178–181.

Jain, M., S. Kumar, N. Rastogi, P. Lal, U. C. Ghoshal, A. Tiwari, M. C. Pant, M. Q. Baiq and B. J. Mittal (2006). GSTT1, GSTM1 and GSTP1 genetic polymorphisms and interaction with tobacco, alcohol and occupational exposure in esophageal cancer patients from North India. **242**(1): 60–67.

Jain, M., A. R. Tilak, R. Upadhyay, A. Kumar and B. J. T. Mittal (2008). Microsomal epoxide hydrolase (EPHX1), slow (exon 3, 113His) and fast (exon 4, 139Arg) alleles confer susceptibility to squamous cell esophageal cancer. **230**(2): 247–251.

Kleinberg, L. and A. A. Forastiere (2007). Chemoradiation in the management of esophageal cancer. *J Clin Oncol* **25**(26): 4110–4117.

Kubo, A. and D. A. J. C. E. Corley (2006). Body mass index and adenocarcinomas of the esophagus or gastric cardia: a systematic review and meta-analysis. **15**(5): 872–878.

Lagergren, J. and P. J. Lagergren (2013). Recent developments in esophageal adenocarcinoma. **63**(4): 232–248.

Lamba, J. K. (2008). Pharmacogenetics of the constitutive androstane receptor.

Landi, S. J. (2000). Mammalian class theta GST and differential susceptibility to carcinogens: a review. **463**(3): 247–283.

Lanuti, M., G. Liu, J. M. Goodwin, R. Zhai, B. C. Fuchs, K. Asomaning, L. Su, N. S. Nishioka, K. K. Tanabe and D. C. Christiani (2008). A functional epidermal growth factor (EGF) polymorphism, EGF serum levels, and esophageal adenocarcinoma risk and outcome. *Clin Cancer Res* **14**(10): 3216–3222.

Lechevrel, M., A. G. Casson, C. R. Wolf, L. J. Hardie, M. B. Flinterman, R. Montesano and C. P. J. C. Wild (1999). Characterization of cytochrome P450 expression in human oesophageal mucosa. **20**(2): 243–248.

Lee, J. M., Y. C. Lee, S. Y. Yang, W. L. Shi, C. J. Lee, S. P. Luh, C. J. Chen, C. Y. Hsieh and M. T. Wu (2000). Genetic polymorphisms of p53 and GSTP1, but not NAT2, are associated with susceptibility to squamous-cell carcinoma of the esophagus. **89**(5): 458–464.

Lee, J. M., Y. C. Lee, S. Y. Yang, P. W. Yang, S. P. Luh, C. J. Lee, C. J. Chen and M. T. Wu (2001). Genetic polymorphisms of XRCC1 and risk of the esophageal cancer. **95**(4): 240–246.

Lee, W., A. C. Lockhart, R. B. Kim and M. L. J. T. O. Rothenberg (2005). Cancer pharmacogenomics: powerful tools in cancer chemotherapy and drug development. **10**(2): 104–111.

Lewis, S. J. and G. D. Smith (2005). Alcohol, ALDH2, and esophageal cancer: a meta-analysis which illustrates the potentials and limitations of a Mendelian randomization approach. **14**(8): 1967–1971.

Li, D., C. Dandara and M. I. Parker (2005). Association of cytochrome P450 2E1 genetic polymorphisms with squamous cell carcinoma of the oesophagus. **43**(4): 370–375.

Li, D., Y. Diao, H. Li, X. Fang and H. J. Li (2008). Association of the polymorphisms of MTHFR C677T, VDR C352T, and MPO G463A with risk for esophageal squamous cell dysplasia and carcinoma. **39**(6): 594–600.

Li, Y., J. Lindsay, L.-L. Wang and S.-F. Zhou (2008). Structure, function and polymorphism of human cytosolic sulfotransferases. **9**(2): 99–105.

Liao, Z., H. Liu, S. G. Swisher, L. Wang, T. T. Wu, A. M. Correa, J. A. Roth, J. D. Cox, R. Komaki, J. A. Ajani and Q. Wei (2006). Polymorphism at the 3'-UTR of the thymidylate synthase gene: a potential predictor for outcomes in Caucasian patients with esophageal adenocarcinoma treated with preoperative chemoradiation. *Int J Radiat Oncol Biol Phys* **64**(3): 700–708.

Lim, Y. J., J.-W. Kim, J. Y. Song, M.-S. Hong, S.-Y. Jin, S. H. Yoon, H. J. Park, B.-K. Choe, J. J. Lee and S.-V. Yim (2005). Epidermal growth factor gene polymorphism is different between schizophrenia and lung cancer patients in Korean population. **374**(3): 157–160.

Lin, D.-C., J.-J. Hao, Y. Nagata, L. Xu, L. Shang, X. Meng, Y. Sato, Y. Okuno, A. M. Varela and L.-W. Ding (2014). Genomic and molecular characterization of esophageal squamous cell carcinoma. **46**(5): 467–473.

Liska, D. J. J. (1998). The detoxification enzyme systems. **3**(3): 187–198.

Lorenzen, S., B. Panzram, G. Keller, F. Lordick, K. Herrmann, K. Becker, R. Langer, M. Schwaiger, J. R. Siewert and K. Ott (2011). Association of the VEGF 936C>T polymorphism with FDG uptake, clinical, histopathological, and metabolic response in patients with adenocarcinomas of the esophagogastric junction. *Mol Imaging Biol* **13**(1): 178–186.

Luo, X., H. R. Kranzler, L. Zuo, H. Zhang, S. Wang and J. J. Gelernter (2008). ADH7 variation modulates extraversion and conscientiousness in substance-dependent subjects. **147**(2): 179–186.

Ma, W.-J., G.-D. Lv, S.-T. Zheng, C.-G. Huang, Q. Liu, X. Wang, R.-Y. Lin, I. Sheyhidin and X.-M. Lu (2010). DNA polymorphism and risk of esophageal squamous cell carcinoma in a population of North Xinjiang, China. **16**(5): 641.

Mackenzie, P. I., K. W. Bock, B. Burchell, C. Guillemette, S.-i. Ikushiro, T. Iyanagi, J. O. Miners, I. S. Owens and D. W. Nebert (2005). Nomenclature update for the mammalian UDP glycosyltransferase (UGT) gene superfamily. **15**(10): 677–685.

Malik, M. A., R. Upadhyay, R. D. Mittal, S. A. Zargar and B. J. N. Mittal (2010). Association of xenobiotic metabolizing enzymes genetic polymorphisms with esophageal cancer in Kashmir Valley and influence of environmental factors. **62**(6): 734–742.

Malik, M. A., R. Upadhyay, D. R. Modi, S. A. Zargar and B. J. Mittal (2009). Association of NAT2 gene polymorphisms with susceptibility to esophageal and gastric cancers in the Kashmir Valley. **40**(5): 416–423.

Mao, X., B. D. Young and Y.-J. Lu (2007). The application of single nucleotide polymorphism microarrays in cancer research. **8**(4): 219–228.

Mao, W.-M., W.-H. Zheng and Z.-Q. Ling (2011). Epidemiologic risk factors for esophageal cancer development. **12**(10): 2461–2466.

Matejcic, M., M. Vogelsang, Y. Wang and I. M. (2015). NAT1 and NAT2 genetic polymorphisms and environmental exposure as risk factors for oesophageal squamous cell carcinoma: a case-control study. **15**(1): 1–14.

McCormack, V., D. Menya, M. Munishi, C. Dzamalala, N. Gasmelseed, M. Leon Roux, M. Assefa, O. Osano, M. Watts and A. J. I. Mwasamwaja (2017). Informing etiologic research priorities for squamous cell esophageal cancer in Africa: a review of setting-specific exposures to known and putative risk factors. **140**(2): 259–271.

Montgomery, E., F. Basman, P. Brennan and R. J. Malekzadeh (2014). Oesophageal cancer. **15**: 528–543.

Morisseau, C. and B. D. Hammock (2005). Epoxide hydrolases: mechanisms, inhibitor designs, and biological roles. **45**: 311–333.

Mudan, S. and J.-Y. Kang (2008). Epidemiology and clinical presentation in esophageal cancer. 1–10. S. Rankin (Ed.).

Mueller, N., S. Grufferman, D. Schottenfeld and J. Fraumeni (2006). *Cancer Epidemiology and Prevention*. Oxford University Press, New York.

Murphy, S. J., A. E. Hughes, C. C. Patterson, L. A. Anderson, R. G. Watson, B. T. Johnston, H. Comber, J. McGuigan, J. V. Reynolds and L. J. Murray (2007). A population-based association study of SNPs of GSTP1, MnSOD, GPX2 and Barrett's esophagus and esophageal adenocarcinoma. *Carcinogenesis* **28**(6): 1323–1328.

Murray, M. J. C (2006). Altered CYP expression and function in response to dietary factors: potential roles in disease pathogenesis. **7**(1): 67–81.

Muwonge, R., K. Ramadas, R. Sankila, S. Thara, G. Thomas, J. Vinoda and R. J. Sankaranarayanan (2008). Role of tobacco smoking, chewing and alcohol drinking in the risk of oral cancer in Trivandrum, India: a nested case-control design using incident cancer cases. **44**(5): 446–454.

Nagasubramanian, R., F. Innocenti and M. J. Ratain (2003). Pharmacogenetics in cancer treatment. **54**(1): 437–452.

Nieman, K. M., I. L. Romero, B. Van Houten and E. J. Lengyel (2013). Adipose tissue and adipocytes support tumorigenesis and metastasis. **1831**(10): 1533–1541.

Nishimura, F. T., Y. Kimura, S. Abe, T. Fukunaga and K. J. Saijoh (2006). Effect of-361 G/A polymorphism of aldehyde dehydrogenase-2 gene on alcohol metabolism and its expression in human peripheral blood leukocytes. **41**(2): 108–119.

Pascussi, J.-M., S. Gerbal-Chaloin, C. Duret, M. Daujat-Chavanieu, M.-J. Vilarem and P. J. A. Maurel (2008). The tangle of nuclear receptors that controls xenobiotic metabolism and transport: crosstalk and consequences. **48**: 1–32.

Peng, D., T. Hu, M. Soutto, A. Belkhiri, A. Zaika and W. J. G. El-Rifai (2014). Glutathione peroxidase 7 has potential tumour suppressor functions that are silenced by location-specific methylation in oesophageal adenocarcinoma. **63**(-4): 540–551.

Pennathur, A., M. K. Gibson, B. A. Jobe and J. D. J. T. L. Luketich (2013). Oesophageal carcinoma. **381**(9864): 400–412.

Pourshams, A., H. Khademi, A. F. Malekshah, F. Islami, M. Nouraei, A. R. Sadjadi, E. Jafari, N. Rakhshani, R. Salahi and S. J. Semnani (2010). Cohort profile: the Golestan Cohort Study—a prospective study of oesophageal cancer in northern Iran. **39**(1): 52–59.

Prabhu, A., K. O. Obi and J. H. Rubenstein (2014). The synergistic effects of alcohol and tobacco consumption on the risk of esophageal squamous cell carcinoma: a meta-analysis. *Am J Gastroenterol* **109**(6): 822–827.

Ratnasinghe, L. D., C. Abnet, Y.-L. Qiao, R. Modali, R. Stolzenberg-Solomon, Z.-W. Dong, S. M. Dawsey, S. D. Mark and P. R. Taylor (2004). Polymorphisms of XRCC1 and risk of esophageal and gastric cardia cancer. **216**(2): 157–164.

Renechan, A., M. Tyson, M. Egger, R. Heller and M. J. L. Zwalhen (2008). Body-mass index and incidence of cancer: a systematic review and meta-analysis of prospective observation studies. **371**: 569–578.

Rokkas, T., D. Pistiolas, P. Sechopoulos, I. Robotis and G. J. C. G. Margantinis (2007). Relationship between Helicobacter pylori infection and esophageal neoplasia: a meta-analysis. **5**(12): 1413–1417. e1412.

Rowe, J. D., E. Nieves and I. J. Listowsky (1997). Subunit diversity and tissue distribution of human glutathione S-transferases: interpretations based on electrospray ionization-MS and peptide sequence-specific antisera. **325**(2): 481–486.

Sharma, A., B. C. Das, A. Sehgal, R. Mehrotra, P. Kar, S. Sardana, R. Phukan, J. Mahanta, J. Purkayastha and S. J. Saxena (2013). GSTM1 and GSTT1 polymorphism and susceptibility to esophageal cancer in high-and low-risk regions of India. **34**(5): 3249–3257.

Shibuta, K., H. Inoue, K. Sato, A. Matsuyama, H. Ueo, Y. Tanaka, K. Mafune, G. F. Barnard and M. Mori (2000). L-myc restriction fragment length polymorphism in Japanese patients with esophageal cancer. *Jpn J Cancer Res* **91**(2): 199–203.

Siewert, J. R., H. J. Stein, M. Feith, B. L. Bruecher, H. Bartels and U. J. Fink (2001). Histologic tumor type is an independent prognostic parameter in esophageal cancer: lessons from more than 1,000 consecutive resections at a single center in the Western world. **234**(3): 360.

Singh, M. S. and M. J. Michael (2009). Role of xenobiotic metabolic enzymes in cancer epidemiology. 243–264.

Snyder, R., T. Chepiga, C. S. Yang, H. Thomas, K. Platt and F. J. T. Oesch (1993). Benzene metabolism by reconstituted cytochromes P450 2B1 and 2E1 and its modulation by cytochrome b5, microsomal epoxide hydrolase, and glutathione transferases: evidence for an important role of microsomal epoxide hydrolase in the formation of hydroquinone. **122** (2): 172–181.

Song, Y., L. Li, Y. Ou, Z. Gao, E. Li, X. Li, W. Zhang, J. Wang, L. Xu and Y. J. N. Zhou (2014). Identification of genomic alterations in oesophageal squamous cell cancer. **509** (7498): 91–95.

Spechler, S. J. J. J. (2013). Barrett esophagus and risk of esophageal cancer: a clinical review. **310**(6): 627–636.

Steevens, J., L. J. Schouten, R. A. Goldbohm and P. A. J. G. van den Brandt (2010). Alcohol consumption, cigarette smoking and risk of subtypes of oesophageal and gastric cancer: a prospective cohort study. **59**(01): 39–48.

Strommer, J. J. (2011). The plant ADH gene family. **66**(1): 128–142.

Svensson, S., M. Some, A. Lundsjö, A. Helander, T. Cronholm and J. O. Höög (1999). Activities of human alcohol dehydrogenases in the metabolic pathways of ethanol and serotonin. **262**(2): 324–329.

Taghavi, N., F. Biramijamal, M. Sotoudeh, H. Khademi, R. Malekzadeh, O. Moaven, B. Memar, A. A'rabi and M. R. Abbaszadegan (2010). p16 INK4a hypermethylation and p53, p16 and MDM2 protein expression in esophageal squamous cell carcinoma. **10**(1): 1–9.

Tan, W., N. Song, G.-Q. Wang, Q. Liu, H.-J. Tang, F. F. Kadlubar and D.-X. Lin (2000). Impact of genetic polymorphisms in cytochrome P450 2E1 and glutathione S-transferases M1, T1, and P1 on susceptibility to esophageal cancer among high-risk individuals in China. **9**(6): 551–556.

Tanabe, K. K., A. Lemoine, D. M. Finkelstein, H. Kawasaki, T. Fujii, R. T. Chung, G. Y. Lauwers, Y. Kulu, A. Muzikansky and D. J. J. Kuruppu (2008). Epidermal growth factor gene functional polymorphism and the risk of hepatocellular carcinoma in patients with cirrhosis. **299**(1): 53–60.

Thatcher, J. D. (2010). The inositol trisphosphate (IP3) signal transduction pathway. **3**(119): tr3–tr3.

Thun, M., M. S. Linet, J. R. Cerhan, C. A. Haiman and D. Schottenfeld (2017). *Cancer Epidemiology and Prevention*. Oxford University Press.

Tian, J., C. Liu, G. Liu, C. Zuo and H. J. Chen (2019). Cumulative evidence for association between genetic polymorphisms and esophageal cancer susceptibility: a review with evidence from meta-analysis and genome-wide association studies. **8**(3): 1289–1305.

Tse, D., R. Zhai, W. Zhou, R. S. Heist, K. Asomaning, L. Su, T. J. Lynch, J. C. Wain, D. C. Christiani and G. Liu (2008). Polymorphisms of the NER pathway genes, ERCC1 and XPD are associated with esophageal adenocarcinoma risk. *Cancer Causes Control* **19**(10): 1077–1083.

Upadhyay, R., M. Jain, S. Kumar, U. C. Ghoshal and B. J. Mittal (2008). Interaction of EGFR 497Arg> Lys with EGF+ 61A> G polymorphism: modulation of risk in esophageal cancer. **17**(4): 167–174.

Van Blankenstein, M., C. W. Looman, W. C. Hop and P. J. Bytzer (2005). The incidence of adenocarcinoma and squamous cell carcinoma of the esophagus: Barrett's esophagus makes a difference. **100**(4): 766–774.

Vos, M., C. H. Adams, T. C. Victor and P. D. van Helden (2003). Polymorphisms and mutations found in the regions flanking exons 5 to 8 of the TP53 gene in a population at high risk for esophageal cancer in South Africa. **140**(1): 23–30.

Wall, T. L. (2005). Genetic associations of alcohol and aldehyde dehydrogenase with alcohol dependence and their mechanisms of action. **27**(6): 700–703.

Wang, K. J. (2008). Practice Parameters Committee of the American College of Gastroenterology. Updated guidelines 2008 for the diagnosis, surveillance and therapy of Barrett's esophagus. **103**: 788–797.

Wang, L.-D., R.-F. Guo, Z.-M. Fan, X. He, S.-S. Gao, H.-Q. Guo, K. Matsuo, L.-M. Yin and J.-L. Li (2005). Association of methylenetetrahydrofolate reductase and thymidylate synthase promoter polymorphisms with genetic susceptibility to esophageal and cardia cancer in a Chinese high-risk population. **18**(3): 177–184.

Wang, L., X. Miao, W. Tan, X. Lu, P. Zhao, X. Zhao, Y. Shan, H. Li and D. J. Lin (2005). Genetic polymorphisms in methylenetetrahydrofolate reductase and thymidylate synthase and risk of pancreatic cancer. **3**(8): 743–751.

Wang, D., M. F. Para, S. L. Koletar and W. J. P. Sadee (2011). Human N-acetyltransferase 1 (NAT1)* 10 and* 11 alleles increase protein expression via distinct mechanisms and associate with sulfamethoxazole-induced hypersensitivity. **21**(10): 652.

Wang, J., A. J. Sasco, C. Fu, H. Xue, G. Guo, Z. Hua, Q. Zhou, Q. Jiang and B. J. Xu (2008). Aberrant DNA methylation of P16, MGMT, and hMLH1 genes in combination with MTHFR C677T genetic polymorphism in esophageal squamous cell carcinoma. **17**(1): 118–125.

Wang, Z., L. Tang, G. Sun, Y. Tang, Y. Xie, S. Wang, X. Hu, W. Gao, S. B. Cox and J.-S. Wang (2006). Etiological study of esophageal squamous cell carcinoma in an endemic region: a population-based case control study in Huaian, China. **6**(1): 1–9.

Westlind, A., L. Löfberg, N. Tindberg, T. B. Andersson and M. J. B. Ingelman-Sundberg (1999). Interindividual differences in hepatic expression of CYP3A4: relationship to genetic polymorphism in the 5′-upstream regulatory region. **259**(1): 201–205.

Wheeler, J. B. and C. E. Reed (2012). Epidemiology of esophageal cancer. **92**(5): 1077–1087.

Wideroff, L., T. L. Vaughan, F. M. Farin, M. D. Gammon, H. Risch, J. L. Stanford and W.-H. Chow (2007). GST, NAT1, CYP1A1 polymorphisms and risk of esophageal and gastric adenocarcinomas. **31**(3): 233–236.

Wing, M. R., D. M. Bourdon and T. K. Harden (2003). PLC-ε: a shared effector protein in Ras-, Rho-, and Gαβγ-mediated signaling. **3**(5): 273.

Wong, P. S.-Y., J. P. Eiserich, S. Reddy, C. L. Lopez, C. E. Cross and A. J. van der Vliet (2001). Inactivation of glutathione S-transferases by nitric oxide-derived oxidants: exploring a role for tyrosine nitration. **394**(2): 216–228.

Wu, X., V. W. Chen, P. A. Andrews, B. Ruiz and P. J. Correa (2007). Incidence of esophageal and gastric cancers among Hispanics, non-Hispanic whites and non-Hispanic blacks in the United States: subsite and histology differences. **18**(6): 585–593.

Wu, X., V. W. Chen, B. Ruiz, P. Andrews, L. J. Su and P. J. Correa (2006). Incidence of esophageal and gastric carcinomas among American Asians/Pacific Islanders, whites, and blacks: subsite and histology differences. **106**(3): 683–692.

Wu, X., J. Gu, T. T. Wu, S. G. Swisher, Z. Liao, A. M. Correa, J. Liu, C. J. Etzel, C. I. Amos, M. Huang, S. S. Chiang, L. Milas, W. N. Hittelman and J. A. Ajani (2006). Genetic variations in radiation and chemotherapy drug action pathways predict clinical outcomes in esophageal cancer. *J Clin Oncol* **24**(23): 3789–3798.

Wu, M., J. Lee, D. Wu, C. Ho, Y. Wang, Y. Lee, H. Hsu and E. J. Kao (2002). Genetic polymorphisms of cytochrome P4501A1 and oesophageal squamous-cell carcinoma in Taiwan. **87**(5): 529–532.

Wu, A. H., P. Wan and L. J. Bernstein (2001). A multiethnic population-based study of smoking, alcohol and body size and risk of adenocarcinomas of the stomach and esophagus (United States). **12**(8): 721–732.

Wu, J., L. Zhang, H. Luo, Z. Zhu, C. Zhang and Y. J. D. Hou (2008). Association of matrix metalloproteinases-9 gene polymorphisms with genetic susceptibility to esophageal squamous cell carcinoma. **27**(10): 553–557.

Xie, L. X., T. T. Zhai, L. P. Yang, E. Yang, X. H. Zhang, J. Y. Chen and H. J. Zhang (2013). Lymphangiogenesis and prognostic significance of vascular endothelial growth factor C in gastro-oesophageal junction adenocarcinoma. **94**(1): 39–46.

Xing, D., J. Qi, X. Miao, W. Lu, W. Tan and D. J. Lin (2002). Polymorphisms of DNA repair genes XRCC1 and XPD and their associations with risk of esophageal squamous cell carcinoma in a Chinese population. **100**(5): 600–605.

Yang, C., K. Matsuo, H. Ito, K. Hirose, K. Wakai, T. Saito, M. Shinoda, S. Hatooka, K. Mizutani and K. J. Tajima (2005). Esophageal cancer risk by ALDH2 and ADH2 polymorphisms and alcohol consumption: exploration of gene-environment and gene-gene interactions. **6**(3): 256.

Yang, S.-J., A. Yokoyama, T. Yokoyama, Y.-C. Huang, S.-Y. Wu, Y. Shao, J. Niu, J. Wang, Y. Liu and X.-Q. Zhou (2010). Relationship between genetic polymorphisms of ALDH2 and ADH1B and esophageal cancer risk: a meta-analysis. **16**(33): 4210.

Yokoyama, A., H. Kato, T. Yokoyama, H. Igaki, T. Tsujinaka, M. Muto, T. Omori, Y. Kumagai, M. Yokoyama and H. J. A. C. Watanabe (2006). Esophageal squamous cell carcinoma and aldehyde dehydrogenase-2 genotypes in Japanese females. **30**(3): 491–500.

Yokoyama, A., H. Kato, T. Yokoyama, T. Tsujinaka, M. Muto, T. Omori, T. Haneda, Y. Kumagai, H. Igaki and M. J. C. Yokoyama (2002). Genetic polymorphisms of alcohol and aldehyde dehydrogenases and glutathione S-transferase M1 and drinking, smoking, and diet in Japanese men with esophageal squamous cell carcinoma. **23**(11): 1851–1859.

Yokoyama, A., T. Muramatsu, T. Ohmori, S. Higuchi, M. Hayashida and H. J. C. E. Ishii (1996). Esophageal cancer and aldehyde dehydrogenase-2 genotypes in Japanese males. **5**(2): 99–102.

Yoshida, A., L. C. Hsu and M. J. Yasunami (1991). Genetics of human alcohol-metabolizing enzymes. **40**: 255–287.

Young Jr, J., C. Percy, A. Asire, J. Berg, M. Cusano, L. Gloeckler, J. Horm, W. Lourie Jr, E. Pollack and E. J. Shambaugh (1981). Cancer incidence and mortality in the United States, 1973–77. (57): 1–187.

Yu, C., Y. Zhou, X. Miao, P. Xiong, W. Tan and D. J. Lin (2004). Functional haplotypes in the promoter of matrix metalloproteinase-2 predict risk of the occurrence and metastasis of esophageal cancer. **64**(20): 7622–7628.

Zendehdel, K., S. Bahmanyar, S. McCarthy, O. Nyren, B. Andersson and W. J. Ye (2009). Genetic polymorphisms of glutathione S-transferase genes GSTP1, GSTM1, and GSTT1 and risk of esophageal and gastric cardia cancers. **20**(10): 2031–2038.

Zhai, R., G. Liu, K. Asomaning, L. Su, M. H. Kulke, R. S. Heist, N. S. Nishioka, T. J. Lynch, J. C. Wain, X. Lin and D. C. Christiani (2008). Genetic polymorphisms of VEGF, interactions with cigarette smoking exposure and esophageal adenocarcinoma risk. *Carcinogenesis* **29**(12): 2330–2334.

Zhang, Y. J. (2013). Epidemiology of esophageal cancer. **19**(34): 5598.

Zhang, L., Y. Jiang, Q. Wu, Q. Li, D. Chen, L. Xu, C. Zhang, M. Zhang and L. J. Ye (2014). Gene–environment interactions on the risk of esophageal cancer among Asian populations with the G48A polymorphism in the alcohol dehydrogenase-2 gene: a meta-analysis. **35**(5): 4705–4717.

Zhang, H.-Z., G.-F. Jin and H.-B. Shen (2012). Epidemiologic differences in esophageal cancer between Asian and Western populations. **31**(6): 281.

Zhang, G., R. Mai and B. J. Huang (2010). ADH1B Arg47His polymorphism is associated with esophageal cancer risk in high-incidence Asian population: evidence from a meta-analysis. **5**(10): e13679.

Zhang, X., L. Zhou, G. Fu, F. Sun, J. Shi, J. Wei, C. Lu, C. Zhou, Q. Yuan and M. J. C. Yang (2014). The identification of an ESCC susceptibility SNP rs920778 that regulates the expression of lncRNA HOTAIR via a novel intronic enhancer. **35**(9): 2062–2067.

Zhang, W., H. Zhu, X. Liu, Q. Wang, X. Zhang, J. He, K. Sun, X. Liu, Z. Zhou and N. J. Xu (2014). Epidermal growth factor receptor is a prognosis predictor in patients with esophageal squamous cell carcinoma. **98**(2): 513–519.

Zhuo, W.-L., Y.-S. Zhang, Y. Wang, X.-L. Zhuo, B. Zhu, L. Cai and Z.-T. Chen (2009). Association studies of CYP1A1 and GSTM1 polymorphisms with esophageal cancer risk: evidence-based meta-analyses. **40**(3): 169–179.

Zuo, L., K. Wang, X.-Y. Zhang, X. Pan, G. Wang, Y. Tan, C. Zhong, J. H. Krystal, H. Zhang and X. J. Luo (2013). Association between common alcohol dehydrogenase gene (ADH) variants and schizophrenia and autism. **132**(7): 735–743.

21

Cytokine Polymorphisms and Their Role in Modulating Cancer Risk

Mosin S. Khan
Government Medical College Srinagar & Associated Hospitals, Karan Nagar, Srinagar, Jammu and Kashmir, India
Government Medical College Baramulla & Associated Hospitals, Kanth Bagh, Baramulla, Jammu and Kashmir, India

Iqra Farooq, Sunia Faiz and Suhail S. Lone
Department of Biochemistry, Government Medical College
Srinagar & Associated Hospitals, Karan Nagar, Srinagar, Jammu and Kashmir, India

Sabhiya Majid
Department of Biochemistry & Multidisciplinary Research Unit (MRU-ICMR),
Government Medical College Srinagar & Associated Hospitals, Karan Nagar, Srinagar, Jammu and Kashmir, India

Waseem Qureshi
Registrar Academics, Government Medical College Srinagar & Associated Hospitals, Karan Nagar, Srinagar, Jammu and Kashmir, India

CONTENTS

DOI: 10.1201/9781003246244-21

ABSTRACT Cytokines have been known by several markers determined by the cell types that synthesize them or their functional properties. For example, cytokines that are obtained fundamentally from mononuclear cells such as macrophages have been stated as monokines, while the cytokines synthesized by activated T lymphocytes are called lymphokines. Cancer in humans is unpredictable, as is its response to treatment. Intrinsic genomic heterogeneity and instability in cancer cells could account for such behaviors. Individual variances in human genetic makeup, on the other hand, have an impact on the host and cancer cells relationship and may contribute to a wide range of outcomes. Polymorphisms in humans have been shown to influence immunological responses, and among immune-related genes, cytokines are polymorphic. Some cytokine and cytokine receptor polymorphisms may have direct functional significance by influencing gene expression, whereas others may demarcate a genetic relationship to a specific haplotype associated with specific cytokines and cytokine receptors. Some cytokine and cytokine receptor polymorphisms may have direct functional importance by affecting the gene expression directly or indirectly, while others may demarcate a genetic association with a specific haplotype linked with a specific clinical condition. Because they participate in immune and inflammatory responses, cytokines hold promise as therapeutic targets. Functional polymorphisms in various cytokines such as *IL-1, IL-4, IL-13, TNF-α,* and *interferons* have been investigated concerning many types of cancers. The basic elements of the various types of cytokines and their functions will be examined in this chapter. The widespread and basic role of cytokines in modulating cancer risk will be thoroughly investigated. Overall, the importance of cytokine polymorphisms in disease, specifically cancer, will be summarized in this chapter.

KEYWORDS: *Interleukin 1; Interleukin 6; Interferons; Cytokines; Polymorphism; Mutation.*

21.1 Introduction

Cytokines are vast and small proteins of the unchained category (~5–20 kDa), which play an essential role in cell signaling. Cytokines are extracellular glycoprotein/polypeptide signaling chemicals formed by a variety of immune cells and are important for assisting and managing immune responses. Figure 21.1 depicts the mechanism of action of cytokines. More than one type of cell results in the synthesis of a given cytokine (John, 2010). These polypeptides interact with signaling chemicals and cells, causing them to migrate toward regions of infection, inflammation, and trauma, as well as with key lymphocyte growth factors and other biological processes. Cytokines are unable to penetrate the cell's lipid bilayer and enter the cytoplasm. Cytokines are implicated in paracrine, autocrine, and endocrine signaling as immunomodulating agents. Cytokines can react at their source (autocrine action), in neighboring cells (paracrine action), or in distant cells (paracrine action) (endocrine action). They play a role in the maturation and regulation of immune system cells in this way. Figure 21.2 shows a schematic representation of cytokine acting in three different ways.

21.2 Classification of Cytokines

There are four different types of cytokines: family of four-helix bundles (InterPro: IPR009079): A group of four helices makes up the three-dimensional structure of member cytokines. This family is broken further into three subfamilies: (1) The IL-2 subfamily is a large family. Erythropoietin (EPO) and thrombopoietin (TPO) are two non-immunological cytokines that make up this complex (Leonard, 2001). Cytokines are further divided into short- and long-chain according to their topology. Some members of the interferon subfamily (2) as part of their receptors share the same gamma chain (Reche, 2019). (3) IL-10 is a member of the interleukin 10 (IL-10) family. The proteins IL-1 and IL-18 are included in the IL-1 family. The IL-17 family has yet to be fully described, although component cytokines have a unique role in stimulating T cell proliferation that results in cytotoxicity. Cell surface receptors maintain a balance between humoral and cell-based mediated immune responses in cytokines. Cytokines play an essential role in both health and disease, especially in the immunological responses of the host to sepsis, trauma, cancer, infection, inflammation, and reproduction.

21.3 History of Cytokines

Cytokines are a family of proteins released by immune system cells. These signaling molecules help in the regulation of inflammation, immunity, and hematopoiesis. Chemokines, interferons, and TNF (tumor necrosis factor) are examples of these. These are produced by many cells, including macrophages, B cells, and T cells, as well as other cells such as endothelial cells. They work by interacting with cell surface receptors and have a high affinity for them.

Hoskins' study in the 1930s found that rabbits, who had previously been infected with the herpes simplex virus, were protected from future infections with the same virus. Findlay and MacCallum reported in 1937 that infections caused by antigenically distinct viruses were resistant in animals infected with the virus, supporting and confirming the current understanding of IFN's role at the time. Their result was finally established in 1957 when Isaacs and Lindenmann discovered that when cells are infected by a virus, they generate a protein that can render other cells resistant to other viruses, based on their cell culture research. Glasgow discovered in 1966 that cytokine plays an important role in reinfection. As a result, the idea of "immune induction" of interferon had gained widespread acceptance by the end of the 1960s.

Two important studies emerged in the early 1970s: classical virus-induced interferon (now known as type I IFN) and immune-induced interferon (currently known as type II IFN). The IFN-α and IFN-β were coined in 1980 to designate the "classical interferons." IFN-γ was known as "immune-induced interferon" and was initially classified as part of the IFN family because of its ability to "hinder" viral infections. The IFN-lambda has been reported as the third form of IFN (type III IFN) in the recent decade. Interleukins IL-28A and IL-28B (IFN-2 and IFN-3, respectively) and IL-29 (IFN-1) are other names for this kind (Platanias, 2005; Billiau and Matthys, 2009; De Andrea et al., 2002; Fitzgerald-Bocarsly and Feng, 2007).

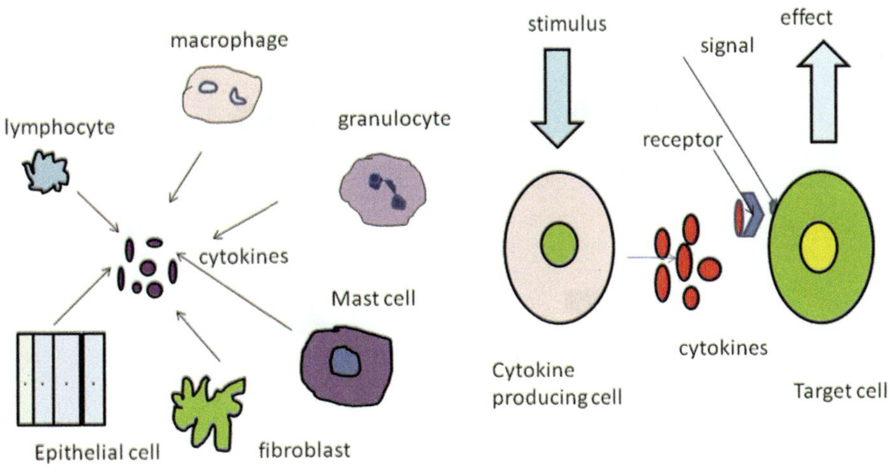

FIGURE 21.1 Mechanism of action of cytokines.

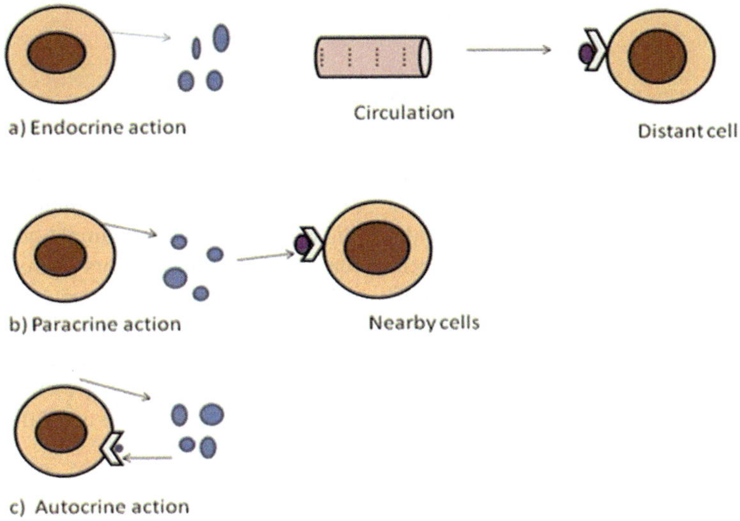

FIGURE 21.2 Schematic representation of cytokine acting in three different ways.

Dudley Dumonde coined the word "lymphokine" to characterize proteins released by lymphocytes in 1969, and later "monokines" (Dumonde et al., 1969). Stanley Cohen published an article in 1974 documenting MIF synthesis in virus-infected allantoic membrane and kidney cells. As a result, he coined the word "cytokine" (Cohen et al., 1974).

21.4 Types of Cytokines

Cytokines come in a wide range of kinds and include the following:

- Chemokines that mediate chemoattraction (chemotaxis) between cells.
- Interferons (IFNs), involved in antiviral responses.
- Interleukins (ILs), first used by scientists to describe cytokines whose primary targets are white blood cells (leukocytes). T-helper cells are liable for the vast majority of them.
- Monokines, synthesized exclusively by monocytes.
- Lymphokines, synthesized by lymphocytes and TNF (Zhang and An, 2007; Dinarello, 2007; Turner et al., 2014; Shea et al., 2002).

IFN (IFN-α, IFN-β, IFN-γ, and others), IL (IL-1, IL-2, and others), TNF (TNF-α and TNF-β), and others will be classified according to their kind, followed by a description of their function and a brief overview.

21.5 Cytokine Signaling

The human body's control relies heavily on cytokine signaling. The vast majority of cytokines are proteins synthesized by glial cells in the brain system and are necessary for intracellular signaling. Most cytokines function as local regulators, alerting and activating lymphocytes. Several cytokine signaling pathways are influenced by hormones such as growth hormones and leptin, the hormone that governs fat storage. To

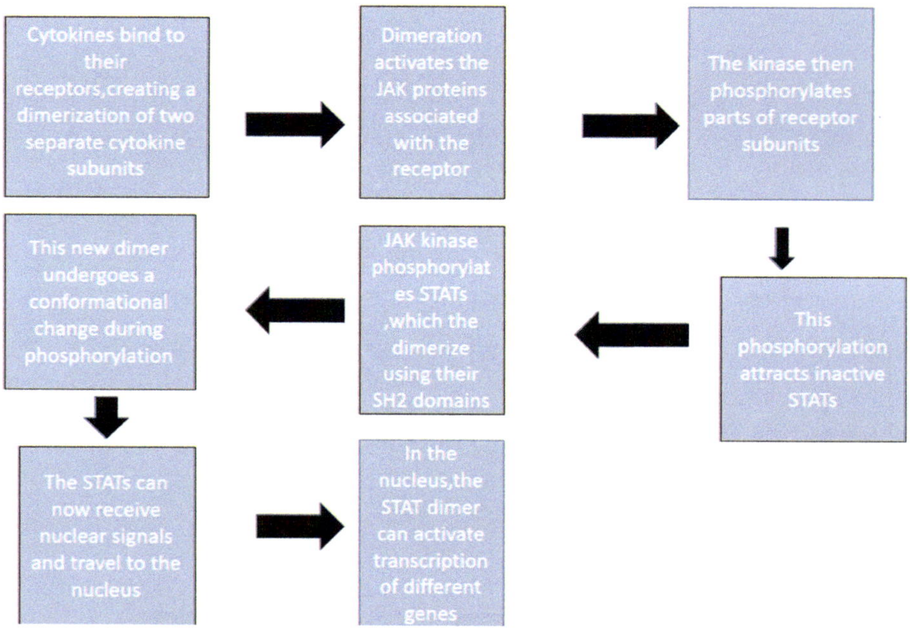

FLOWCHART 21.1 Flowchart describing the main steps involved in JAK/STAT signaling.

keep the human body healthy, the immune system relies on cytokine signaling. Foreign particles are engulfed by macrophages and dendritic cells, which then deliver cytokine signals to surrounding dormant lymphocytes. These cells have been programmed to identify specific antigens. The combination of macrophages and lymphocyte activation via cytokine signaling helps the body maintain homeostasis. Cytokines bind to receptors on target cells, triggering an intracellular signaling cascade. The protein kinase transduction cascade is the most prevalent of these pathways. Once the cytokine binds to the receptor located in the cell membrane, the process of phosphorylation activates inactive protein kinases.

21.6 Role of Cytokines in Cancer

Cytokines are implicated in balancing the spread and growth of cancers (Negus and Balkwill, 1996). Cancer cells can produce cytokines constitutively. These cytokines may create an environment that promotes cancer growth, and they may work in an autocrine or paracrine manner on supportive tissues, including blood vessels and fibroblasts. Normal cells, such as tumor-associated macrophages, may be triggered by the cytokines, which synthesize additional cytokines that assist the malignant process (TAM) (Mantovani, 1994) and endothelial cells (Ethier, 1995). There is a need for elucidating data from cultivated cell lines studies to assess the involvement of growth factors in breast cancer, such as TGF-α. However, several studies have shown that pro-inflammatory cytokines are lethal to solid tumors and that they grow in response to IL-1 (Ito et al., 1993), IL-2 (Han et al., 1996), and IL-6 (Naylor et al., 1993). These studies reported the expression of TNF gene in 71% of ovarian cancer biopsies (Naylor et al., 1993). TNF receptors were also found to be linked to tumor cells, suggesting that TNF may function in an autocrine manner in some cases. Many tumors synthesize granulocyte colony-stimulating factor, which is another example of autocrine growth. Cells are convicted by some tumors to make growth factors using the cytokines employed by some cancers. Negus et al. concluded that TAM is accumulated due to epithelial ovarian tumor cells synthesizing monocyte chemoattractant protein-1 (MCP-1), which stimulates cancer growth by synthesizing IL-1, IL-6, TNF, TGF-β, epidermal growth factor (EGF), and PDGF (Negus et al., 1995).

21.7 Cytokine Polymorphisms and Cancer

As far as treatment response goes, human cancer is an unpredictable disease. Cancer cells' inherent genetic heterogeneity and instability may explain some of their behavior. Individual diversity in human genetic makeup, on the other hand, could impact the connection between host and cancer cells. Human gene polymorphism has an essential role in immunological responses. Many cytokines are polymorphic. Some cytokine and cytokine receptor polymorphisms may have direct functional significance by influencing gene expression and/or function directly or indirectly, while others may simply classify a genetic association with a specific haplotype associated with a specific clinical disease. The impacts of cytokine gene polymorphism on the expression of the gene have been investigated at the two levels in vitro gene expression and in vivo disease connections. Only a small number of studies have integrated the two approaches. The size of the genomic domains that need to be studied, which include both coding and noncoding regions, is a fundamental constraint to the study of cytokine polymorphism in general. Several technologies are available for the genetic typing of known polymorphism(s) that could be applied to cytokine and cytokine receptor studies (Wang et al.,

2003). However, in clinical research, many people may need to be screened, and methods that can quickly detect known and unknown SNPs could greatly improve screening efficiency (Kwok, 2001).

Several studies accompanied so far reported the function of cytokine polymorphisms in modifying susceptibility to various cancers, including prostate, melanoma, breast, stomach, and esophageal cancers (McCarron et al., 2002; Howell et al., 2001, 2002; Giordani et al., 2003; Savage et al., 2004). Among these studies, many have reported a decrease in the rate of cancer linked with high synthesizing IL-10 phenotypes. This explains the role of IL-10 as an immunosuppressive cytokine. On the other hand, limited human and extensive experimental observations suggest that IL-10 may play a dual role in the oncogenic process by promoting cancer growth, on the one hand, and preconditioning the tumor microenvironment to immune rejection, on the other (Giordani et al., 2003; Savage et al., 2004).

The set of cytokines involved in cancer susceptibility are discussed in detail below.

21.7.1 Interleukin 1

Interleukin 1 (IL-1) family in humans is coded by the IL-1A gene and is known as hematopoietic 1 (Nicklin et al., 1994). The IL-1 family includes 11 cytokines that induce a compound grid of pro-inflammatory cytokines and, via integrin expression on endothelial cells and leukocytes, start and govern inflammatory responses (Dinarello, 2011).

21.7.1.1 Interleukin 1 Structure and Function

The pro-inflammatory effects of IL-1α and IL-1β are the strongest and most investigated. They have IL-1Ra as a natural antagonist (IL-1 receptor antagonist). They all have beta-trefoil folds in their structure and bind to the IL-1 receptor (IL-1R), which stimulates signaling through the MyD88 adaptor. The pro-inflammatory activity of IL-1α and IL-1β is controlled by IL-1Ra (Dinarello, 2011; Murzin et al., 1992; Gosavi et al., 2008). On human chromosome two, nine members of the IL-1 superfamily are grouped. These are produced through a sequence of gene duplications of a proto-IL-1β ligand, according to chromosomal and sequence analyses (Rivers-Auty et al., 2018). Although IL-18 and IL-33 are found on distinct chromosomes, there is little evidence from sequencing or chromosomal research to suggest that they are related to other members of the IL-1 superfamily. Because they have a similar structure, differ in their assignment, and signal through the same receptors, IL-33 and IL-18 are classed as members of the IL-1 superfamily (Rivers-Auty et al., 2018; Schmitz et al., 2005; Dao et al., 1996; Schmitz et al., 2005; Dao et al., 1996; Schmitz et al., 2005; Dao et al., 1996). Figure 21.3 contains a schematic representation of receptors for IL-1.

B lymphocytes, NK cells, microglia, and epithelial cells all produce IL-1, which starts in tissue monocytes, macrophages, dendritic cells, and fibroblasts. Their powerful inflammatory reaction protects the body from infection. These cytokines enable immunological competent cells on endothelial cells to migrate to infection sites by increasing the expression of adhesion factors expression (also known as diapedesis). By affecting the activity of the thermoregulatory center, they raise body temperature (fever) (hypothalamus). As a result, it acts as a natural pyrogen. IL-1 produces hyperalgesia, vasodilation, and hypotension in addition to causing fever (Sorrentino et al., 2015).

21.7.1.2 Reported SNPs in Interleukin 1

IL-1 gene group contains about 1,500 SNPs. Furthermore, this region contains several allele markers, such as 222/223 and gz5/gz6 in IL-1α as well as a variable number tandem repeat (VNTR) for IL-1 and IL-1RN. There are 148 SNPs in the

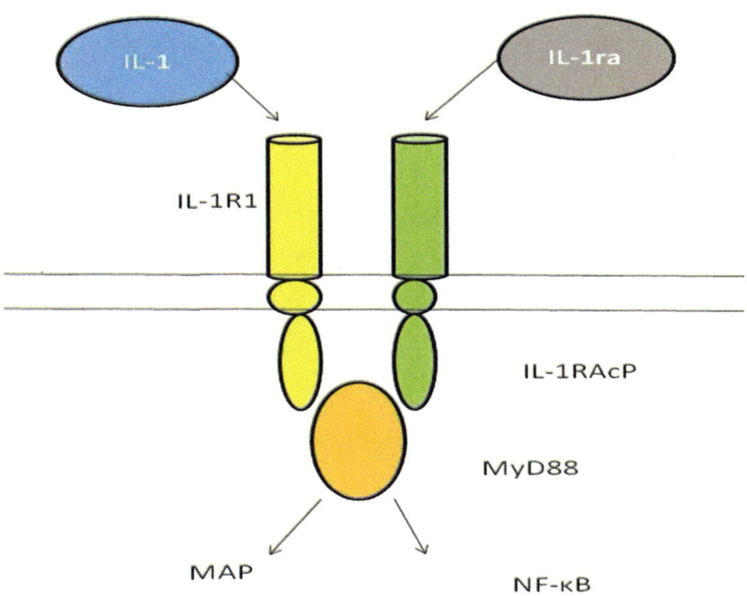

FIGURE 21.3 Schematic representation of receptors for IL-1.

IL-1 gene, according to reports. At location single+*G4845T (rs17561)* base polymorphism, alanine-to-serine is the most investigated SNP in the IL-1α gene. The following polymorphisms have been found in the IL-1β gene: (1) *rs1143634* at position *+3954*, as well as two promoter SNPs: (2) *T31C (rs1143627)* and (3) *–C511T (rs16944)*. The IL-1RN gene has so far been found to have 140 SNPs. IL-1RN gene polymorphisms that have received the most attention are the penta-allelic 86-bp tandem repeats – VNTR in intron 2 and intron 3 and *rs419598* at location *+2018* in exon 2 of the IL-1RN gene. The first of all interleukins, interleukin 1 (IL-1), was discovered. Interleukin 1 alpha (IL-1α), IL-1 beta (IL-1β), and IL-1 receptor antagonist (IL-1Ra), which are produced by multiple genes and work together to control innate immunity and acquired inflammation, are the most important cytokines in our bodies (Dinarello, 2018). IL-1 is primarily synthesized by monocytes and macrophages, which act as pro-proteins in inflammation and may play an essential role in cancer invasion, tumor growth, and distant metastasis (Malik and Kanneganti, 2018).

21.7.1.3 Interleukin 1 Polymorphisms in Cancer

IL-1α has been linked to lung cancer, prostate cancer, cervical cancer, colon cancer, and breast cancer in previous studies (Malik and Kanneganti, 2018; Kuan and Ziegler, 2018; Matsuo et al., 2009; Seddighzadeh et al., 2003; Song et al., 2016; Sorrentino et al., 2015; Xu et al., 2010; Schmidt et al., 2015; Ma et al., 2017; Cheng et al., 2012). These findings suggest that IL-1α may have a role in the progression and therapy of cancer. IL-1A is located on chromosome 2q and has six introns and seven exons, encoding the IL-1α protein. Three SNPs on the IL-1A gene have varied functional roles: *rs17561*, *rs1800587*, and *rs3783553*. A mutation in the fifth exon of IL-1A, together with an amino acid change, causes the *+G4845 T (rs17561)* variation (Kawaguchi et al., 2007). The IL-1A promoter region *–C889T (rs1800587)* has been shown to impact IL-1 expression (McDowell et al., 1995; Dominici et al., 2002). In the Thai population, the *IL-1 rs16944*C allele is a genetic predictor of HCC development in chronic hepatitis B patients, with a possible link to elevated IL-1 production in the liver (Hirankarn et al., 2006). In the Japanese population, the presence of HCC with chronic HCV infection is linked to the *rs1143627* T allele or *rs16944/rs1143627* haplotype CT. The *IL-1 rs1143627* T allele and the MMP-3 5A allele are both associated with a poor prognosis in HCC patients, suggesting that these gene polymorphisms could be used to predict prognosis (Wang et al., 2003). Furthermore, SNPs found in the IL-1A 3′-untranslated regions (UTRs) affect IL-1α expression by affecting miR-122 binding. It's called *rs3783553*, and it's an insertion/deletion variation. Many studies have found a link between IL-1A polymorphisms and various types of cancer; however, the conclusions are still unreliable. As a result, an updated meta-analysis of all malignancies is needed to investigate the impact of IL-1A polymorphisms in tumorigenesis.

21.7.2 Interleukin 6

IL-6 is both a pro-inflammatory and anti-inflammatory cytokine. The IL-6 gene in humans codes for it (Ferguson et al., 1988). Osteoblasts also produce IL-6, which helps osteoclasts form faster. Smooth muscle cells release IL-6 in the tunica media of many blood vessels. IL-6 works as an anti-inflammatory myokine by inhibiting TNF-alpha and IL-1 and activating IL-1ra and IL-10. Previous research has shown that IL-6 can function as an inflammatory signal for severe COVID-19 infection with a poor prognosis (Cardiac Rhythm News, 2020). Figure 21.4 shows the IL-6 receptor system.

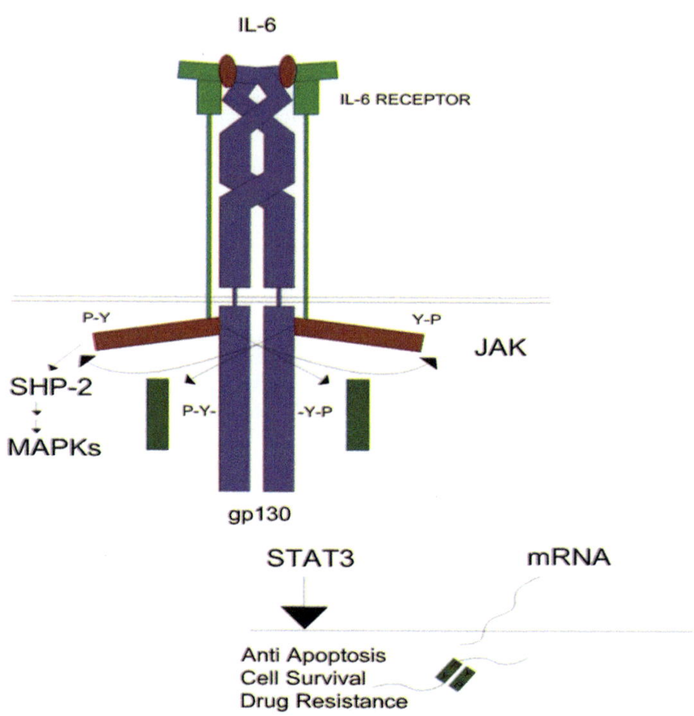

FIGURE 21.4 IL-6 receptor system.

21.7.2.1 Interleukin 6 Structure and Function

In the human and mouse genomes, the IL-6 gene is found on chromosomes 7p21 (Sehgal et al., 1986) and 5 (Mock et al., 1989). Inside exons are all conserved across species within the gene's protein-coding domain (Northemann et al., 1989). Outside the coding area, however, variation exists at the 5′ boundary of exon 1 and the 3′ border of exon 5. This preservation of genomic organization can also be seen in the genome. In exchange for certain microbial chemicals known as pathogen-associated molecular patterns (PAMPs), macrophages produce IL-6. Pattern recognition receptors, which include Toll-like receptors, are a group of innate immune system sensing molecules that bind to these PAMPs (TLRs). These are found in the intracellular compartments and cell surface, and they contribute to the production of inflammatory cytokines through intracellular signaling cascades. The main mediator of the acute phase response and fever is IL-6. In the bone marrow, IL-6 stimulates the production of acute phase proteins and neutrophils. It promotes the proliferation of B cells while inhibiting the growth of regulatory T cells (Banks et al., 1994) and initiate PGE2 production in the hypothalamus, causing the body's temperature set point to change. Because IL-6 is also produced by adipocytes, it could be one of the reasons why fat people have higher endogenous CRP levels (Bastard et al., 1999).

Given that IL-6 gene deletion produces mature-onset obesity in mice, IL-6 may act as a tonic suppressor of body fat (Wallenius et al., 2002; Matthews et al., 2010). Furthermore, IL-6 has effects on the CNS that can reduce body fat mass (Wallenius et al., 2002). In rodents, the hypothalamus and hindbrain provide the anti-obesity effect of IL-6 (Wallenius et al., 2002; Sadagurski et al., 2010; Campbell et al., 2014). In obese patients, IL-6 trans-signaling improves energy and glucose homeostasis. IL-6 in the CNS was found to somewhat mitigate the suppression of food intake and body weight induced by glucagon-like peptide 1 (GLP-1) receptor stimulation in several animal experiments (Shirazi et al., 2013). Outside of the CNS, IL-6 appears to boost GLP-1 production in the endocrine pancreas and the gut (Ellingsgaard et al., 2011). Amylin is another hormone that helps people lose weight. Amylin-induced IL-6 synthesis in the ventromedial hypothalamus (VMH) may be a method used to promote weight loss (Le Foll et al., 2015).

21.7.2.2 Reported SNPs in Interleukin 6

Several SNPs have been discovered in the region upstream of the IL-6 promoter. The most studied is *rs1800795* (formerly known as *–174G/C*), which is situated in the proximal promoter. *rs1800795* G/C alleles have been linked to IL-6 levels and disease risk in a variety of conditions, according to numerous studies (Cole, 2010). The *rs1800795* genotype has been found to influence synovial fibroblast IL-6 production. The most well-known polymorphism in the IL-6 gene is the transition at position *G174C (rs1800795)*. Several studies have linked higher IL-6 production to the GG genotype (wild type). The rs1800795 major allele had previously been found to have a higher IL-6-expressing variation (G). Two SNPs in the promoter region were investigated: *G174C (rs1800795)* and *G572C (rs1800796)*. IL-6 SNPs affect IL-6 gene transcription as well as serum, plasma, and tissue IL-6 concentrations, according to one study. They may also play a significant role in the development of certain bronchial asthma phenotypes.

21.7.2.3 Interleukin 6 Polymorphisms and Cancer

Breast cancer development is linked with elevated levels of interleukin 6 (IL-6) (Lithgow and Covington, 2005). A study reported that C carriers had a greater risk of early events than homozygous GG carriers in breast cancer patients (Markkula et al., 2014). IL-6 promoter polymorphisms and cancer risk have been the subject of several recent meta-analyses. The *rs1800795* polymorphism was found to increase the risk of prostate cancer and cervical cancer in two meta-analyses (Liu et al., 2017). Several IL-6 SNPs have been linked to the risk of CRC (Landi et al., 2003). The most researched SNP is *rs1800795*. *rs1800795* is found in the gene's promoter region (–174 n) and is involved in the binding of transcriptional factors, regulating transcription, RNA elongation, splicing, and maturation (Hugo et al., 2006). According to studies, the C allele increases the risk of CRC in both smokers and non-smokers. There was a major linkage between the distribution of genes of the *rs1800795* polymorphism and CRC risk in non-smokers, drinkers, and individuals consuming a low-fat diet. These investigations revealed that screening of *rs1800795* polymorphism in specific groups based on allele and genotype distributions in various groups could be a promising strategy for predicting CRC risk. Additional studies demonstrated substantial positive relationships between the *rs1800795* polymorphism and CRC risk in TNM stage I and TNM stage II. These grades indicated that *rs1800795* polymorphism could be a positive marker for CRC prognosis. In addition, IL-6 is known to promote tumor growth and metastasis. IL-6 plays a role in the advancement of uterine cancers (Ferdeghini et al., 1994). In IL-6, the investigations found SNPs *rs1800796*, *rs2069837*, *rs1524107*, *rs2066992*, and *rs2069840*, as well as six SNPs in IL-4: *rs2243250*, *rs2227284*, *rs2243267*, *rs2243270*, *rs2243283*, and *rs2243289*. *rs1524107* and *rs2066992* were shown to be associated with EC risk among these SNPs (Ferdeghini et al., 1994). Also, IL-6 has a role in the promotion of neoplastic transformation.

21.7.3 Interleukin 4

IL-4 is a pleiotropic cytokine that regulates multiple immune systems. It is involved in IgM-to-IgE isotype shift and IgG4 regulation in IgE. IL-4 is the primary molecule implicated in a variety of pathogenic pathways in allergy illness.

21.7.3.1 Interleukin 4 Structure and Function

On chromosome 5q31, the IL-4 gene is found. It's near to several other immunoregulatory genes, including IL-13, IL-5, IL-3, and IL-9. A linkage signal between chromosome 5q31 and total and specific serum IgE has been revealed in several genome-wide screenings (Meyers et al., 1994), bronchial hyperresponsiveness (Doull et al., 1996), eosinophilia (Martinez et al., 1998), and other asthma- and allergy-related

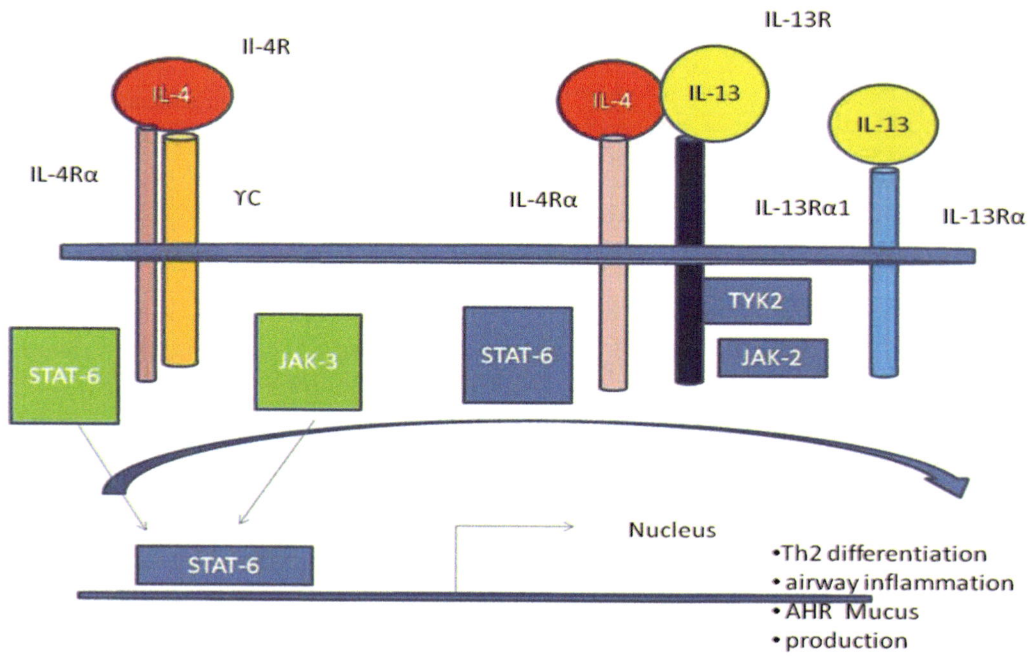

FIGURE 21.5 The interleukin 4 (IL-4)/IL-13/signal transducer and activator of transcription 6 (STAT6) signaling pathways are depicted in this diagram.

traits (Palmer et al., 1998). IL-4 is a cytokine that enables naive helper T cells (Th0 cells) to differentiate into Th2 cells, making it a candidate gene for atopic illnesses. It's a part of allergy-related pathways (Arai et al., 1989). Mast cells, Th2 cells, eosinophils, and basophils all produce IL-4 (Gadani et al., 2012). Interleukin 4 increases the development of B cells into plasma cells as well as the multiplication of activated B and T cells. Both humoral and adaptive immunity relies on this molecule. IL-4, which also boosts MHC class II production, aids B cell class switching to IgE. IL-4 suppresses Th1 cells, macrophages, IFN-gamma, and dendritic cells. Increased production of IL-4 has been linked to allergies (Hershey et al., 1997). A schematic representation of the interleukin 4 (IL-4)/IL-13/signal transducer and activator of transcription 6 (STAT6) signaling pathways is shown in Figure 21.5.

21.7.3.2 Reported SNPs in Interleukin 4

Many SNPs in IL-4 have been discovered, which may impact the role of mononuclear cells in the synthesis of IL-4 and other cytokines (Bartova et al., 2014). According to a study, TT for *SNP rs2243250* and GG for *SNP rs2227284* were the most prevalent IL-4 genotypes in asthma and allergic rhinitis patients (Micheal et al., 2013). Shang et al. discovered that the *IL-4 rs2243250 SNP* is related to elevated blood IL-4 levels, which may enhance the incidence of atopic dermatitis in children (Shang et al., 2016). In Chinese men, the SNP in *IL-4 rs2070874* has been identified as a risk factor for hepatocellular carcinoma (Lu et al., 2014).

Previous research has found that the TC genotype in the *IL-4 rs2243250 SNP*, as well as the TG and TT genotypes in the *rs2227284 SNP*, protect against migraine. These polymorphisms may influence TH2 release of IL-4 affinity to cell targets, resulting in a Th1/Th2 cytokine imbalance that may affect migraine pain spread. As a result, IL-4 could be used as a target for migraine-specific preventive medicines in the future. In certain studies, *−C589T* was linked to total serum IgE levels, asthma, and atopic dermatitis (Noguchi et al., 1998).

21.7.3.3 Interleukin 4 Polymorphisms and Cancer

The IL-4 gene has been fully sequenced, and a polymorphism in the IL-4 promoter region (*C-589T*) has been recognized (Rosenwasser et al., 1995). In the gene's 5′-UTR, a second single nucleotide polymorphism (SNP *−C33T*) was discovered (Suzuki et al., 1999). Interleukin 4 (IL-4) is an immunosuppressive and anti-angiogenic cytokine that plays an important role in antitumor immune responses. Certain IL-4 gene polymorphisms are thought to be good candidates for cancer susceptibility prognosis. Several researchers have looked into the potential association between IL-4 SNPs *−C590 T (rs2243250)* in the 5′-UTR, *−C34T (rs2070874)* in the 5′-UTR, and *rs79071878* (intron 3, 70-bp VNTR) and cancer risk; however, the findings have been mixed (Baniyash, 2006; Aggarwal et al., 2009; Bidwell et al., 2001; Wang and Joyce, 2010).

The IL-4R gene variant rs1801275 was linked to an increased risk of glioblastoma (Schwartzbaum et al., 2005). Because IL-4 increases the proliferation of Th2 cells in bladder cancer patients and suppresses Th1 responses, the *IL-4 rs2243250* polymorphism demonstrated a substantial connection with early multiple carcinomas in bladder cancer patients (Agarwal et al., 2006). To manage tumors by lysing cancer cells and worsening immune surveillance against tumors, the generation of effector cytotoxic T lymphocytes is required (Agarwal et al., 2006). It was found that patients with the *IL-4–590T* allele produce more IL-4 (Rosenwasser et al., 1995), which could affect immunosuppression against malignancies (Th1 inhibition). As a result, IL-4 could play a role in cancer cells escaping lymphocyte destruction and the growth of numerous tumors in the bladder urothelium (Tan and Coussens, 2007). Four IL-4 polymorphisms have recently been discovered to be linked to prostate cancer. Three polymorphisms *(rs2243228, rs2243250,* and *rs2227284)* were shown to be connected to cancer risk, whereas *rs2070874* was found to be linked to cancer-related fatalities. The IL-4 polymorphism may have a role in the development of androgen-independent tumors (Lee et al., 2008).

21.7.4 Interleukin 13

Interleukin 13 (IL-13) is a subtype of the interleukin family. It is involved in a variety of allergic reactions. These interleukins have an effect on B cells and monocytes, where they induce IgE class switching and upregulate class II expression.

21.7.4.1 Interleukin 13 Structure and Function

Interleukin 13 (IL-13) is a protein encoded by the IL-13 gene in humans (Minty et al., 1993). It is positioned on chromosome 5q31 with a length of 1.4 kb. It folds into four alpha-helical bundles and has a molecular weight of 13 kDa (Rael and Lockey, 2011). In several tissues, IL-13 is recognized to have a role in the physiologic changes generated by allergic inflammation. IL-13 is predominantly linked to the growth of airway illness. It contains anti-inflammatory properties. In the airways, IL-13 belongs to the matrix metalloproteinases (MMPs) family of protein-degrading enzymes. IL-13 persuades these MMPs as part of a protective mechanism against asphyxiation due to severe allergic inflammation (Minty et al., 1993).

Interleukin 13 plays an important role in goblet cell metaplasia (Fahy, 2002). In disorders characterized by high type 2 cytokines, such as chronic schistosomiasis and atopic dermatitis, IL-13 has been reported to be a potent fibrogenic program. This fibrogenic program is thought to be heavily reliant on PDGFR+ fibroblasts receiving direct IL-13 signaling via IL-4R (Gieseck et al., 2016).

21.7.4.2 Reported SNPs in Interleukin 13

Several SNPs in the IL-13 gene have been linked to asthma and allergic rhinitis (Bottema et al., 2010; Heinzmann et al., 2000). Graves et al. found seven polymorphisms in the IL-13 gene's different sections, two of which were in the promoter region (*−C1055T* and *–A1512C*). In the Dutch population, the TT genotype *−C1055T* was the most common. In a recent study of the *C1055T* SNP in asthma, a strong linkage was discovered among Caucasians (Graves et al., 2000). tOne SNP was discovered in the IL-13 gene's third intron, while three others were discovered in the 3′-UTR (*+G2525A*, *+C2580A*, and *+T2749C*). Another key SNP, Gln144Arg, commonly known as *+G2044A*, has been linked with asthma as well as elevated total serum IgE levels in the coding region of exon 4 of IL-13 (Graves et al., 2000). Individuals with the GA genotype in the *+G 2044A* SNP have a 40% higher risk of asthma than those with the homozygous GG. Furthermore, Asians were thought to be at a higher risk than Caucasians (Graves et al., 2000). In a meta-analysis of IL-13 SNPs in allergic rhinitis, the *+2044A* variant was revealed to play a significant role (Bunyavanich et al., 2011). The genetic association of IL-13 polymorphisms with asthma and associated features has been established in most previous investigations across varied ethnic communities. According to the research, IL-13 is a key possibility for its link to asthma and atopy.

21.7.4.3 Interleukin 13 Polymorphisms and Cancer

IL-13 is an anti-inflammatory and immunomodulatory factor that can generate cancer-specific immune responses (Formentini et al., 2012). Most glioma cell lines and glioblastoma tumor tissues have elevated levels of IL-13 (Zhu et al., 2012). By suppressing IL-13, the IL-4R/IL-13 signaling pathway has a functional role in tumor immune-surveillance and the development of cancer immunotherapy (Terabe et al., 2000). Furthermore, Liu's group discovered that IL-13 inhibits the proliferation of low-grade glioma cell lines in human glioma cell lines (Liu et al., 2000). According to a study, IL-13 stimulates the formation of the Reed-Sternberg cell, which can delay tumor immunosurveillance in Hodgkin's disease (Skinnider et al., 2001).

Although the *IL-13 rs20541* SNP has been revealed to be a shielding factor for HCC in studies, the practical applicability of the *IL-13 rs20541* SNP has yet to be discovered. Many researchers have found a link between *IL-13 rs20541* and cancers such as renal cell carcinoma (Chu et al., 2012), bladder cancer (Chu et al., 2012), and glioblastoma (Jin et al., 2013); however, none has found a link between *rs20541* and the existence of ULM. Chu et al. reported that in the Chinese population, the protective impact of the TT genotype of the rs20541 SNP has a protective role in the spread of renal cell carcinoma. The *rs20541* is a functional polymorphism in exon 4 that causes a change in the amino acid sequence from arginine (Arg) to glutamine (Gln) at codon 130, which may affect IL-13 transcription. The A allele (Gln) of *rs20541* negotiated IgE production and raised blood IgE levels in 1,399 children from three distinct populations, according to a study (Graves et al., 2000). Several epidemiological studies show a link between IgE levels and the risk of glioma (Schoemaker et al., 2007; Biggar et al., 2009), non-Hodgkin's lymphoma (Biggar et al., 2009), and pancreatic cancer (Biggar et al., 2009; van Toorenenbergen et al., 2010).

21.7.5 Tumor Necrosis Factor Alpha

TNF-α was discovered in 1975 as a chemical that causes tumor necrosis in vitro, hence the name. TNF expression is reportedly stimulated by immune system cells. These findings explain the TNF and the TNF receptor superfamily, which includes more than 40 members and includes TNF-α (also known as TNF) and TNF-β (also known as lymphotoxin). The binding of this family of cytokines to their specific receptors mediates inflammatory events (Abbas et al., 2007; Abbas and Lichtman 2012; Brenner et al., 2015; Ma et al., 2016). Figure 21.6 represents the signaling pathway of TNFR1.

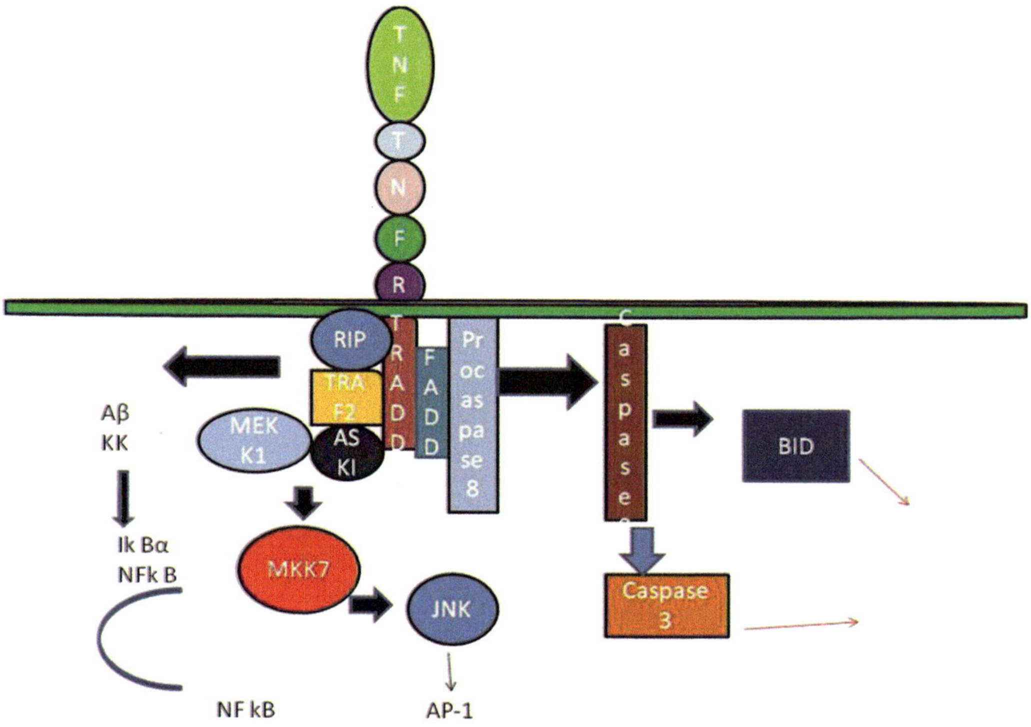

FIGURE 21.6 Signaling pathway of TNFR1. Red arrows represent multiple steps.

21.7.5.1 TNF Structure and Function

The TNF gene superfamily is found on chromosome 6p of the human major histocompatibility complex. In multiple studies, it has been associated with atopic asthma (Randolph et al., 2005; Migita et al., 2005). In its promoter region, several polymorphisms have been discovered (Smith and Humphries, 2009), with TNF-*G308A* being the variation with the most impact on this condition. Aside from inflammatory events, regulatory T cell failure has been connected to the progression of complex hereditary diseases such as atopy and asthma. The inactivation of cells near the antigen is a sign of peripheral T cell tolerance, which limits both proliferative response and cytokine production (Deniz and Erten, 2008; Palomares et al., 2010). Regulatory T cells are a group of T cell subtypes examined in humans (Jutel et al., 2003).

It maintains a state of tolerance to harmless chemicals while also controlling overactive immune responses that can lead to detrimental immunological disease (Gabry et al., 2011). TNF-β, a type II transmembrane protein, is essential for lymph node maturation and Peyer's patches, as well as the survival of secondary lymphoid organs. TNF-β expression is aided by lymphocytes (Drutskaya et al., 2010; Kuprash et al., 2002). TNF plays numerous physiologic roles in the immune system, inflammation, and the survival and death of various cells. Cytokines are primarily responsible for activating and transporting cells to infection sites as well as destroying pathogens such as bacteria and viruses. TNF also penetrates vascular endothelial cells, causing them to express adhesion molecules which allow immune system cells to attach to the blood vessel wall. TNF increases the affinity of leukocytes for their ligands, the expression of IL-1, and the activation of immune system cells' microbicidal capacities by inducing the synthesis of chemokines. This infection can recur if low levels of this cytokine are present (Abbas et al., 2007; Ma et al, 2016; Kalliolias and Ivashkiv, 2016; Guadagni et al., 2007; Sedger and McDermott, 2014).

TNF has a role in various autoimmune diseases, such as rheumatoid arthritis and inflammatory bowel disease. A large number of autoimmune, neoplastic, and other diseases occur due to a decrease in its production. Therefore, biological molecules that target TNF are used in the management of these diseases (Abbas et al., 2007; Brenner et al., 2015; Kuprash et al., 2002). TNF also promotes tumor cell necrosis by inducing programmed cell death, which has cytolytic properties. The intracellular proteins activate caspase-3, caspase-6, caspase-7, and other cytosolic substrates by causing the release of additional proteins such as pro-caspase-8. By interacting with latent DNase, these proteins encourage genomic DNA degradation and cell death. TNF may also promote carcinogenesis and tumor stabilization by activating NF-κB in tumor cells and encouraging the synthesis of IL-6 to evade immune system defenses, according to reports (Brenner et al., 2015; Kuprash et al., 2002; Kalliolias and Ivashkiv, 2016; Guadagni et al., 2007). In high quantities, TNF can lower blood glucose levels and cause intravascular thrombosis. TNF causes fever by promoting the synthesis of prostaglandins by hypothalamic cells (Abbas et al., 2007; Kuprash et al., 2002).

21.7.5.2 Reported SNPs in TNF

Various common polymorphisms have been found in the promoter region of cytokines, including *G1082A* and *C592A* for IL-10 and *C509T* for TGF-β genes, the majority of which have been linked to allergic diseases (Hobbs et al., 1998; Raedler et al., 2013; Zhang et al., 2010; Nie et al., 2012). Inflammation is frequent in the etiology of breast cancer, and it leads to metastasis. The majority of SNPs are found in the intronic region of genes, but only some SNPs are found in exons and have been associated with chronic disorders such as cancer (Lodish et al., 2002). There are around 16 polymorphisms in and around the TNF gene. The promoter polymorphisms *308 G/A* and *238 G/A* have been investigated the most. Studies on non-Hodgkin's lymphoma (Warzocha et al., 1998) and myeloma (Warzocha et al., 1998) show that these polymorphisms influence the promoter area, increasing TNF protein (Davies et al., 2000).

21.7.5.3 TNF Polymorphisms and Cancer

TNF-*α 308G>A* promoter polymorphism has a significant impact on transcriptional activity and cytokine production levels. Increased TNF synthesis is connected to the functional polymorphism with the *308A* allele. Transitioning from the G to the A allele could improve transcriptional factor binding to the polymorphism location. In various malignancies, such as pancreatic cancer (Talar-Wojnarowska et al., 2009), colorectal cancer (Kayacan et al., 2006), and non-small cell lung carcinoma (Kayacan et al., 2006), researchers have identified a higher serum and plasma TNF level in patients than in control group. A study found a relationship between the *LTA+252* genotype *AG/GG* and an increased risk of breast cancer (Park et al., 2002). TNF-α synthesis is controlled by genetic polymorphisms at the −238 *(rs361525)*, *−308 (rs1800629)*, *−857 (rs1799724)*, and *−1031 (rs1799964)* locations, according to the research (Lindholm et al., 2008). According to several meta-analyses, the TNF-*α-308* polymorphism is a risk factor for a variety of malignancies, including breast, gastric, and hepatic cancers (Yang et al., 2010). TNF-*α-238* polymorphism was found to have a protective impact against cancer in a study (Jang et al., 2001).

21.7.6 Interferons

Interferons are a type of cytokine that belongs to a larger category. Type I IFNs, type II IFNs, and type III IFNs are the three major classes. IFN-α (further divided into 13 distinct subtypes such as IFN-1, 2, 4, 5, 6, 7, 8, 10, 13, 14, 16, 17, and 21) and IFN-ASβ are the two most common type I IFNs. Since they interfere with viral replication, they are termed interferons. Figure 21.7 represents the schematic representation of the interferon effect on neighboring cells.

21.7.6.1 Interferon Structure and Function

Interferons (IFNs) are classified into 13 subtypes of IFN-alpha and a single species of IFN-beta and IFN-omega. Polypeptide secondary and tertiary folding is governed by their 20% overall sequence homology. According to three-dimensional models, type I IFNs have a globular structure made up of a bundle of five alpha-helices that may produce two polypeptide domains. The disulfide bond stabilizes both domains in a bioactive state. Type I IFNs improve natural killer cell lysis capacity, resulting in an antiviral effect that reduces viral replication and MHC class I molecule expression on cells infected with the virus

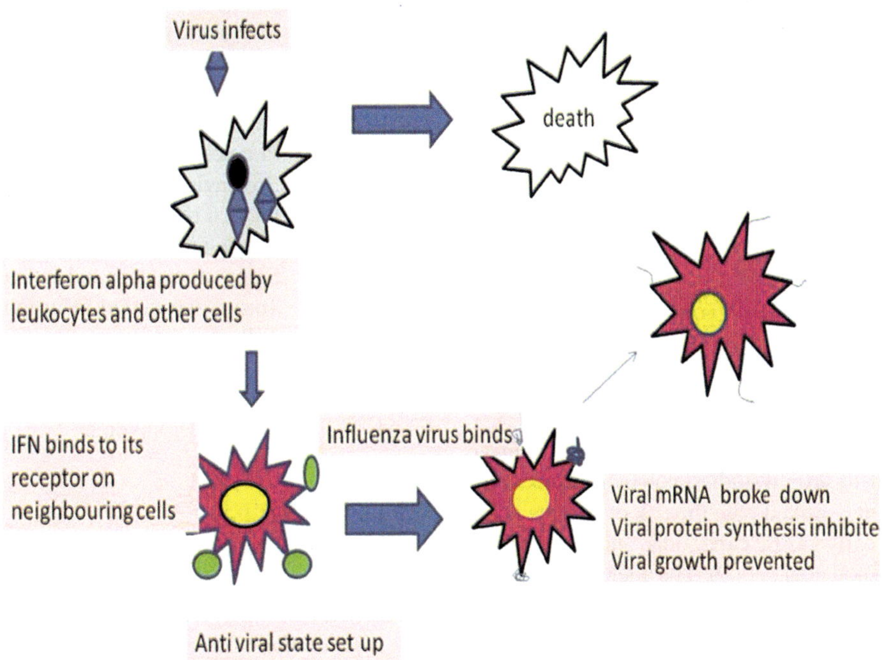

FIGURE 21.7 Schematic representation showing the action of interferons.

while simultaneously restoring Th1 cell development. This type of interferon is increased and easily measurable in the blood during infection. In contrast, type II IFN is represented by only one gene, IFN-γ.

In both innate and adaptive immunological responses, this cytokine serves a functional role in macrophage activation. Type III IFNs, also known as IL-28/IL-29, have biological actions similar to type I IFNs and have an important function in viral infection defense (Abbas et al., 2015; Ank et al., 2006; Parham and Janeway, 2009). IFNs serve a role in immune monitoring for malignant cells in addition to being the initial line of defense against viral infections. Type I interferons have a strong antiviral effect that is linked to several physiological alterations. The function of type I interferons, of which IFN-α and IFN-β are the most important, can be grouped into three categories. *First*, these cytokines increase viral replication defense in all cells by activating cellular genes, resulting in the degradation of viral mRNA and suppression of viral protein translation. *Second*, in virus-infected cells, they enhance the expression of ligands for NK cell receptors. *Third*, they cause NK cells to destroy virus-infected cells (Gonzalez-Navajas et al., 2012; Schroder et al., 2004). IFN-γ has a role in macrophage activation, causing these cells to activate microbicide effector activities. IFN-γ-stimulated macrophages had increased pinocytosis and phagocytosis, as well as better germ-killing abilities. IFN-γ also inhibits cell development and can initiate apoptosis (Schroder et al., 2004; Green et al., 2017).

21.7.6.2 Reported SNPs in Interferon

The *IFNGR1–56C/T SNP* and ocular AD were discovered to have a strong genetic relationship. *IFNGR1* SNPs have previously been linked to serum IgE concentration in individuals with atopic asthma, according to previous studies (Gao et al., 1999). The *IFNGR1* gene acts as an organ-specific susceptibility gene for ocular AD, in addition to its effect on blood IgE levels. Infection with *Helicobacter pylori* is linked to the *IFNGR1–56C/T SNP* (Thye et al., 2003) in African populations, as well as cerebral malaria (Koch et al., 2002). Higher blood H. pylori antibody concentrations were linked to the 56T genotype, and *56C/T* heterozygosity offered protection against cerebral malaria infection.

21.7.6.3 Interferon Polymorphisms in Cancer

In IFNG signaling, the *IFNGR1* protein plays a function (Chen et al., 2012). The reduced expression of IFNGR1 protein has been linked with esophageal cancer clinicopathological features (Wang et al., 2008) and ovarian cancer diagnosis (Duncan et al., 2007). According to several studies, polymorphisms in the *IFNGR1* gene are strongly linked with the risk of chronic hepatitis B virus infection, early gastric carcinoma, and rectal cancer (Canedo et al., 2008; Zhou et al., 2009). The allele of *rs2234711* in the promoter of *IFNGR*1 was shown to be linked to the risk of CRC in the study. In addition to early gastric cancer, *rs2234711* plays a function in chronic hepatitis B virus infection. Previously, it was discovered that *rs2234711* has a functional effect on stimulating B cell lines and that the C allele is linked to lower *IFNGR1* gene activity. Also, *rs2234711* is present in the binding site of POLB, which has been linked with CRC (Donigan et al., 2012; Watsuki et al., 2009). The study suggested that the functional variant *rs2234711* might influence CRC causation by balancing the expression or function of *IFNGR1*.

21.8 Conclusions

In this chapter, the most important characteristics of the various types of cytokines, as well as their roles, were discussed. As a result, the broad and novel role of cytokines in cancer risk modulation could be comprehensively explored. The chapter also covered cytokine polymorphisms in cancer, as well as SNPs linked to cytokines.

ABBREVIATIONS

CNS:	Central nervous system
COVID-19:	Coronavirus disease 19
CRP:	C-reactive protein
GCSF:	Granulocyte colony-stimulating factor
GLP-1:	Glucagon-like peptide 1
Gp-130:	Glycoprotein 130
IFN:	Interferon
IFNGR1:	Interferon gamma receptor 1
IgE:	Immunoglobulin E
IL-1:	Interleukin 1
IL-1Rα:	Interleukin 1 receptor alpha
IL-1Rβ:	Interleukin 1 receptor beta
IL-1RN:	Interleukin 1 receptor antagonist
IL-6:	Interleukin 6
IL-13:	Interleukin 13
INF:	Insulin-like growth factor
MAP:	Mitogen-activated protein
MCP-1:	Monocyte chemoattractant protein 1
MHC:	Major histocompatibility complex
NF-κb:	Nuclear factor kappa B
PAMPs:	Pathogen-associated molecular patterns
PDGF:	Platelet-derived growth factor
PGE2:	Prostaglandin E2
PRRs:	Pattern recognition receptors
SNPs:	Single nucleotide polymorphisms
STAT3:	Signal transducer and activator of transcription 3
TAM:	Tumor-associated macrophages
TLRs:	Toll-like receptors
TNF-α:	Tumor necrosis factor alpha
TRAP-1:	TFNR-associated factor 1
ULM:	Uterine leiomyoma

REFERENCES

Abbas AK, Lichtman AH (2012) *Imunologia Celular e Molecular.* 7th ed. Rio de Janeiro: Elsevier.

Abbas AK, Lichtman AH, Pillai S (2007) *Cellular and Molecular Immunology.* 6th ed. Philadelphia: Saunders Elsevier, 566.

Abbas AK, Lichtman AH, Pillai S (2015) *Cellular and Molecular Immunology.* 8th ed. Philadelphia: Elsevier Saunders, 535.

Agarwal A, Verma S, Burra U, Murthy NS, Mohanty NK, Saxena S (2006) Flow cytometric analysis of Th1 and Th2 cytokines in PBMCs as a parameter of immunological dysfunction in patients of superficial transitional cell carcinoma of the bladder. *Cancer Immunol Immunother* 55:734–43.

Aggarwal BB, Vijayalekshmi RV, Sung B (2009) Targeting inflammatory pathways for prevention and therapy of cancer. *Clin Cancer Res* 15:425–30.

Ank N, West H, Bartholdy C, Eriksson K, Thomsen AR, Paludan SR (2006) Lambda interferon (IFN-lambda), a type III IFN, is induced by viruses and IFNs and displays potent antiviral activity against select virus infections in vivo. *J Virol* 80:4501–9.

Arai N, Nomura D, Villaret D, DeWaal Malefijt R, Seiki M, Yoshida M (1989) Complete nucleotide sequence of the chromosomal gene for human-4 and its expression. *J Immunol* 142:274–82.

Baniyash M (2006) Chronic inflammation, immunosuppression, and cancer. *Semin Cancer Biol* 16:80–8.

Banks WA, Kastin AJ, Gutierrez EG (1994) Penetration of interleukin-6 across the murine blood-brain barrier. *Neurosci Lett* 179(1–2):53–6.

Bartova J, Linhartova PB, Podzimek S, Janatova T, Svobodova K, Fassmann A (2014) The effect of IL-4 gene polymorphisms on cytokine production in patients with chronic periodontitis and healthy controls. *Mediators Inflamm*. ID 185757.

Bastard JP, Jardel C, Delattre J, Hainque B, Bruckert E, Oberlin F (1999) Evidence for a link between adipose tissue interleukin-6 content and serum C-reactive protein concentrations in obese subjects. *Circulation* 99(16):2221–2.

Bidwell J, Keen L, Gallagher G, Kimberly R, Huizinga T, McDermott MF, Oksenberg J, McNicholl J, Pociot F, Hardt C (2001) Cytokine gene polymorphism in human disease. *Genes Immun* 2:61–70.

Biggar PJ, Christiansen M, Rostgaard K, Smedby KE, Adami HO, Glimeliuss B (2009) Immunoglobulin subclass levels in patients with non-Hodgkin lymphoma. *Int J Cancer* 124(11):2616–20.

Billiau A, Matthys P (2009) Interferon-gamma: a historical perspective. *Cytokine & Growth Fact Rev* 20:97–113.

Bottema RB, Nolte IM, Howard TD, Koppelman GH, Dubois AJ, de Meer G (2010) Interleukin 13 and Interleukin 4 Receptor Polymorphisms in Rhinitis and Asthma. *Int Arch Allergy Immunol* 153(3):259–67.

Brenner D, Blaser H, Mak TW (2015) Regulation of tumor necrosis factor signaling. *Nat Rev Immunol* 15:362–74.

Bunyavanich S, Shargorodsky J, Celedon JC (2011) A meta-analysis of Th2 pathway genetic variants and risk for allergic rhinitis. *Pediatr Allergy Immunol* 22(4):378–87.

Campbell IL, Erta M, Lim SL, Frausto R, May U, Rose-John S, Scheller J, Hidalgo J (2014) Trans-signaling is a dominant mechanism for the pathogenic actions of interleukin-6 in the brain. *J Neurosci* 34(7):2503–13.

Canedo P, Corso G, Pereira F, Lunet N, Suriano G (2008) The interferon-gamma receptor 1 (IFNGR1) –56C/T gene polymorphism is associated with increased risk of early gastric carcinoma. *Gut* 57:1504–8.

Chen C, Guo L, Shi M, Hu M, Yu M (2012) Modulation of IFN-gamma receptor 1 expression by AP-2alpha influences IFN-gamma sensitivity of cancer cells. *Am J Pathol* 180:661–71.

Cheng J, Li L, Liu Y (2012) Interleukin-1alpha induces immunosuppression by mesenchymal stem cells promoting the growth of prostate cancer cells. *Mol Med Rep* 6:955–60.

Chu H, Ma L, Wang M, Shi D, Qin C, Yuan L (2012) The polymorphisms of IL-4, IL-4R and IL-13 genes and bladder cancer risk in a Chinese population: a case-control study. *Mol Biol Rep* 39(5):5349–57.

Cohen S, Bigazzi PE, Yoshida T (1974) Similarities of T cell function in cell-mediated immunity and antibody production. *Cell Immunol* 12(1):150–9.

Cole SW (2010) Computational identification of gene-social environment interaction at the human IL6 locus. *Proc Natl Acad Sci USA* 107(12):5681–6.

Cytokine in John Lackie (2010) *A Dictionary of Biomedicine*. Oxford University Press. ISBN 978-0199549351.

Dao T, Ohashi K, Kayano T, Kurimoto M, Okamura H (1996) Interferon-gamma-inducing factor, a novel cytokine, enhances Fas ligand-mediated cytotoxicity of murine T helper 1 cell. *Cell Immunol* 173:230.

Davies FE, Rollinson SJ, Rawstron AC, Roman E, Richards S, Drayson M (2000) High-producer haplotypes of tumor necrosis factor-alpha and lymphotoxin alpha are associated with an increased risk of myeloma and have improved progression-free survival after treatment. *J Clin Oncol* 18:2843–51.

De Andrea M, Ravera R, Gioia D, Gariglio M, Landolfo S (2002) The interferon system: an overview. *Eur J Pediatr Neurol* 6:A41–6.

Deniz G, Erten G (2008) Regulatory NK cells suppress antigen-specific T cell response. *J Immunol* 180(2):850–7.

Dinarello CA (2007) Historical review of cytokines. *Eur J Immunol* 37:S34–45.

Dinarello CA(2011) Interleukin-1 in the pathogenesis and treatment of inflammatory diseases. *Blood* 117(14):3720–32.

Dinarello CA (2018) Overview of the IL-1 family in innate inflammation and acquired immunity. *Immunol Rev* 281:8–27.

Dominici R, Cattaneo M, Malferrari G (2002) Cloning and functional analysis of the allelic polymorphism in the transcription regulatory region of interleukin-1 alpha. *Immunogenetics* 54:82–6.

Donigan KA, Sun KW, Nemec AA, Murphy DL, Cong X (2012) Human POLB gene is mutated in high percentage of colorectal tumors. *J Biol Chem* 287:23830–9.

Doull IJ, Lawrence S, Watson M, Begishvili T, Beasley RW, Lampe F (1996) Allelic association of gene markers on chromosomes 5q and 11q with atopy and bronchial hyperresponsiveness. *Am J Respir Crit Care Med* 153:1280–4.

Drutskaya MS, Efimov GA, Kruglov AA, Kuprash DV, Nedospasov SA (2010) Tumor necrosis factor, lymphotoxin and cancer. *IUBMB Life* 62:283–9.

Dumonde DC, Wolstencroft RA, Panayi GS, Matthew M, Morley J, Howson WT (1969) Lymphokines non-antibody mediators of cellular immunity generated by lymphocyte activation. *Nature* 224(5214):38–50.

Duncan TJ, Rolland P, Deen S, Scott IV, Liu DT (2007) Loss of IFN gamma receptor is an independent prognostic factor in ovarian cancer. *Clin Cancer Res* 13:4139–45.

Ellingsgaard H, Hauselmann I, Schuler B, Habib AM, Baggio LL, Meier DT (2011) Interleukin-6 enhances insulin secretion by increasing glucagon-like peptide-1 secretion from L cells and alpha cells. *Nat Med* 17(11):1481–9.

Ethier SP (1995) Growth factor synthesis and human breast cancer progression. *J Natl Cancer Inst* 87:964–73.

Fahy JV (2002) Goblet cell and mucin gene abnormalities in asthma. *Chest* 122(6):320–6.

Ferdeghini M, Gadducci A, Prontera C, Bonuccelli A, Annicchiarico C, Fanucchi A, Bianchi R (1994) Serum interleukin-6 levels in uterine malignancies. *Anticancer Res* 14(2B):735.

Ferguson Smith AC, Chen YF, Newman MS, May LT, Sehgal PB, Ruddle FH (1988) Regional localization of the interferon-beta 2/B-cell stimulatory factor 2/hepatocyte stimulating factor gene to human chromosome 7p15-p21. *Genomics* 2(3):203–8.

Fitzgerald-Bocarsly P, Feng D (2007) The role of type I interferon production by dendritic cells in host defense. *Biochimie* 89:843–55.

Formentini A, Braun P, Fricke H, Link KH, Henne-Bruns D (2012) Expression of interleukin-4 and interleukin-13 and their receptors in colorectal cancer. *Int J Colorect Dis* 27(10):1369–76.

Gabry L, Christensen JR, Wu X, Kissenpfennig A, Malissen B (2011) Integrated T-cell receptor and costimulatory signals determine TGF–dependent differentiation and maintenance of Foxp3+ regulatory T cell. *Eur J Immunol* 41(5):1242–8.

Gadani, Sachin P, Cronk, James C, Norris, Geoffrey T, Kipnis, Jonathan (2012) Interleukin-4: a cytokine to remember. *J Immunol* 189(9):4213–9.

Gao PS, Mao XQ, Jouanguy E (1999) Nonpathogenic common variants of IFNGR1 and IFNGR2 in association with total serum IgE levels. *Biochem Biophys Res Commun* 263:425–9.

Gieseck RL, Ramalingam TR, Hart KM, Vannella KM, Cantu DA, Lu WY, Ferreira-González S, Forbes SJ, Vallier L, Wynn TA (2016) Interleukin-13 activates distinct cellular pathways leading to ductular reaction, steatosis, and fibrosis. *Immunity* 45(1):145–58.

Giordani L, Bruzzi P, Lasalandra C, Quaranta M, Schittulli F, Della Ragione F (2003) Association of breast cancer and polymorphisms of interleukin-10 and tumor necrosis factor-alpha genes. *Clin Chem* 49:1664–7.

Gonzalez-Navajas JM, Lee J, David M, Raz E (2012) Immunomodulatory functions of type I interferons. *Nat Rev Immunol* 12:125–35.

Gosavi S, Whitford PC, Jennings PA, Onuchic JN (2008) Extracting function from a beta-trefoil folding motif. *Proc Nat Acad Sci USA* 105(30):10384–9.

Graves PE, Kabesch M, Halonen M, Holberg CJ, Baldini M, Fritzsch C (2000) A cluster of seven tightly linked polymorphisms in the IL-13 gene is associated with total serum IgE levels in three populations of white children. *J Allergy Clin Immunol* 105(3):506–13.

Green DS, Young HA, Valencia JC (2017) Current prospects of type II interferon gamma signaling and autoimmunity. *J Biol Chem* 292:13925–334.

Guadagni F, Ferroni P, Palmirotta R, Portarena I, Formica V, Roselli M (2007) Review TNF/VEGF cross-talk in chronic inflammation-related cancer initiation and progression: an early target in anticancer therapeutic strategy. *In Vivo (Athens, Greece)* 21:147–61.

Han D, Pottin-Clemenceau, CImro MA (1996) IL2 triggers a tumor progression process in a melanoma cell line MELP derived from a patient whose metastasis increased in size during IL2/INFalpha biotherapy. *Oncogene* 12:1015–23.

Heinzmann A, Mao XQ, Akaiwa M, Kreomer RT, Gao PS, Ohshima K (2000) Genetic variants of IL-13 signalling and human asthma and atopy. *Hum Mol Genet* 9(4):549–9.

Hershey GK Friedrich MF, Esswein LA, Thomas ML, Chatila TA (1997) The association of atopy with a gain-of-function mutation in the alpha subunit of the interleukin-4 receptor. *N Engl J Med* 337(24):1720.

Hirankarn N, Kimkong I, Kummee P (2006) Interleukin 1beta gene polymorphism associated with hepatocellular carcinoma in hepatitis B virus infection. *World J Gastroenterol* 12:776–9.

Hobbs K, Negri J, Klinnert M, Rosenwasser L J, Borish L (1998) Interleukin-10 and transforming growth factor-β promoter polymorphisms in allergies and asthma. *Am J Respirat Crit Care Med* 158(6) :1958–62.

Howell WM, Bateman AC, Turner SJ, Collins A, Theaker JM (2002) Influence of vascular endothelial growth factor single nucleotide polymorphisms on tumour development in cutaneous malignant melanoma. *Genes Immun* 3:229–32.

Howell WM, Turner SJ, Bateman AC, Theaker JM (2001) IL10 promoter polymorphisms influence tumour development in cutaneous malignant melanoma. *Genes Immun* 2:25–31.

Hugo H, Cures A, Suraweera N (2006) Mutations in the MYB intron I regulatory sequence increase transcription in colon cancers. *Genes Chromosomes Cancer* 45:1143–54.

Ito R, Kitadai Y, Kyo E (1993) Interleukin 1α acts as an autocrine growth stimulator for human gastric carcinoma cells. *Cancer Res* 55:4102–6.

Jang WH, Yang YI, Yea SS, Lee YJ, Chun JH (2001) The –238 tumor necrosis factor-alpha promoter polymorphism is associated with decreased susceptibility to cancers. *Cancer Lett* 166:41–6.

Jin T, Li X, Zhang J, Wang H, Geng T, Li G (2013) Genetic association between selected cytokine genes and glioblastoma in the Han Chinese population. *BMC Cancer* 13:236.

Jutel M, Akdis M, Budak F (2003) IL-10 and TGF cooperatein the regulatory T cell response tomucosal allergens in normal immunity and specific immunotherapy. *Eur J Immunol* 33(5):1205–14.

Kalliolias GD, Ivashkiv LB (2016) TNF biology, pathogenic mechanisms and emerging therapeutic strategies. *Nat Rev Rheumatol* 12:49–62.

Kawaguchi Y, Tochimoto A, Hara M (2007) Contribution of single nucleotide polymorphisms of the IL1A gene to the cleavage of precursor IL-1alpha and its transcription activity. *Immunogenetics* 59:441–8.

Kayacan O, Karnak D, Beder S, Gullu E, Tutkak H, Senler FC (2006) Impact of TNF-alpha and IL-6 levels on development of cachexia in newly diagnosed NSCLC patients. *Am J Clin Oncol* 29:328–35.

Koch O, Awomoyi A, Usen S, Jallow M, Richardson A (2002) IFNGR1 gene promoter polymorphisms and susceptibility to cerebral malaria. *J Infect Dis* 185:1684–7.

Kuan EL, Ziegler SF (2018) A tumor-myeloid cell axis, mediated via the cytokines IL-1alpha and TSLP, promotes the progression of breast cancer. *Nat Immunol* 19:366–74.

Kuprash DV, Alimzhanov MB, Tumanov AV, Grivennikov SI, Shakhov AN, Drutskaya LN (2002) Redundancy in tumor necrosis factor (TNF) and lymphotoxin (LT) signaling in vivo: mice with inactivation of the entire TNF/LT locus versus single-knockout mice. *Mol Cell Biol* 22:8626–34.

Kwok PY (2001) Genetic association by whole-genome analysis. *Science* 294:2669–70.

Landi S, Moreno V, Gioia-Patricola L (2003) Association of common polymorphisms in inflammatory genes interleukin (IL)6, IL8, tumor necrosis factor alpha, NFKB1, and peroxisome proliferator-activated receptor gamma with colorectal cancer. *Cancer Res* 63:3560–6.

Lee SO, Pinder E, Chun JY, Lou W, Sun M, Gao AC (2008) Interleukin-4 stimulates androgen-independent growth in LNCaP human prostate cancer cells. *Prostate* 68: 85–91.

Le Foll C, Johnson MD, Dunn-Meynell AA, Boyle CN, Lutz TA, Levin BE (2015) Amylin-induced central IL-6 production enhances ventromedial hypothalamic leptin signaling. *Diabetes* 64(5):1621–31.

Leonard WJ (2001) Cytokines and immunodeficiency diseases. *Nat Rev Immunol* 3:200–8.

Lindholm E, Bakhtadze E, Cilio C, Agardh E, Groop L (2008) Association between LTA, TNF and AGER polymorphisms and late diabetic complications. *PLoS One* 3:e2546.

Lithgow D, Covington C (2005) Chronic inflammation and breast pathology. *Biol Res Nurs* 7:118–29.

Liu TZ, Guo ZQ, Wang T, Cao Y, Huang D, Wang XH (2017) Meta-analysis of the role of IL-6 rs1800795 polymorphism in the susceptibility to prostate cancer: evidence based on 17 studies. *Medicine (Baltimore)* 96:e6126.

Liu H, Jacobs BS, Liu J, Prayson RA, Estes ML (2000) Interleukin-13 sensitivity and receptor phenotypes of human glial cell lines: non-neoplastic glia and low-grade astrocytoma differ from malignant glioma. *Cancer Immunol Immunother* 49(6):319–24.

Lodish H, Berk A, Zipursky SL, Matsudaira P, Baltimore D, Damell J (2002) Genetic analysis in molecular biology. *Cal Molecular Biology*:255–93.

Lu Y, Wu Z, Qiliu P, Liping M (2014) Role of IL-4 gene polymorphisms in HBV-related hepatocellular carcinoma in a Chinese population. *PLoS One* 9(10):e110061

Ma J, Sun X, Guo T (2017) Interleukin-1 receptor antagonist inhibits angiogenesis via blockage IL-1alpha/PI3K/NF-kappabeta pathway in human colon cancer cell. *CMAR* 9:481–93.

Ma K, Zhang H, Baloch Z (2016) Pathogenetic and therapeutic applications of tumor necrosis factor-alpha (TNF-alpha) in major depressive disorder: a systematic review. *Int J Mol Sci* 17:733.

Malik A, Kanneganti TD (2018) Function and regulation of IL-1α in inflammatory diseases and cancer. *Immunol Rev* 281:124–37.

Mantovani A (1994) Tumor-associated macrophages in neoplastic progression. *Lab Invest* 71:5–16.

Markkula A, Simonsson M, Ingvar C, Rose C, Jernstrom H (2014) IL6 genotype, tumour ER-status, and treatment predicted disease-free survival in a prospective breast cancer cohort. *BMC Cancer* 14:759.

Martinez FD, Solomon S, Holberg CJ, Graves PE, Baldini M, EricksonRP (1998) Linkage of circulating eosinophils to markers on chromosome 5q. *Am J Respir Crit Care Med* 158:1739–44.

Matsuo Y, Sawai H, Ma J (2009) IL-1alpha secreted by colon cancer cells enhances angiogenesis: the relationship between IL-1alpha release and tumor cells' potential for liver metastasis. *J Surg Oncol* 99:361–7.

Matthews VB, Allen TL, Risis S, Chan MH, Henstridge DC, Watson N (2010) Interleukin-6-deficient mice develop hepatic inflammation and systemic insulin resistance. *Diabetologia* 53(11):2431–41

McCarron SL, Edwards S, Evans PR, Gibbs R, Dearnaley DP, Dowe A (2002) Influence of cytokine gene polymorphism on the development of prostate cancer. *Cancer Res* 62:3369–72.

McDowell TL, Symons JA, Ploski R (1995) A genetic association between juvenile rheumatoid arthritis and a novel interleukin-1 alpha polymorphism. *Arthritis Rheumat* 38: 221–8.

Meyers DA, Postma DS, Panhuysen CI, Xu J, Amelung PJ, Levitt RC (1994) Evidence for a locus regulating total serum IgE levels mapping tochromosome 5. *Genomics* 23:464–70.

Micheal S, Minhas K, Ishaque M, Ahmed F, Ahmed A (2013) IL-4 gene polymorphisms and their association with atopic asthma and allergic rhinitis in Pakistani patients. *J Investig Allergol Clin Immunol* 23(2):107–11.

Migita O, Noguchi E, Koga M (2005) Haplotype analysis of a100kb region spanning TNF-LTA identifies a polymorphism in the LTA promoter region that is associated with atopic asthma susceptibility in Japan. *Clin Exp Allergy* 35(6):790–6.

Minty A, Chalon P, Derocq JM, Dumont X, Guillemot JC, Kaghad M, Labit C, Leplatois P, Liauzun P, Miloux B (1993) Interleukin-13 is a new human lymphokine regulating inflammatory and immune responses. *Nature* 362 (6417):248–50.

Mock BA, Nordan RP, Justice MJ, Kozak C, Jenkins NA, Copeland NG, Clark SC, Wong G, Rudikoff S (1989) The murine IL-6 gene maps to the proximal region of chromosome 5. *J Immunol* 142:1372–6.

Murzin AG, Lesk AM, Chothia C (1992) beta-Trefoil fold. Patterns of structure and sequence in the Kunitz inhibitors interleukins-1 beta and 1 alpha and fibroblast growth factors. *J Mol Biol* 223(2):531–43.

Naylor MS, Stamp GW, Foulkes WD, Eccles D, Balkwill FR (1993) Tumor necrosis factor and its receptors in human ovarian cancer. *J Clin Invest* 91:2194–206.

Negus RP, Balkwill FR (1996) Cytokines in tumour growth, migration and metastasis. *World J Urol* 14:157–65.

Negus RP, Stamp GW, Relf MG, Burke F, Malik ST, Bernasconi S, Allavena P, Sozzani S, Mantovani A, Balkwill FR (1995) The detection and localization of monocyte chemoattractant protein-1 (MCP-1) in human ovarian cancer. *J Clin Invest* 95:2391–6.

Nicklin MJ, Weith A, Duff GW (1994) A physical map of the region encompassing the human interleukin-1 alpha, interleukin-1 beta, and interleukin-1 receptor antagonist genes. *Genomics* 19(2):382–4.

Nie W, Fang Z, Li B, Xiu QY (2012) Interleukin-10 promoter polymorphisms and asthma risk a meta-analysis. *Cytokine* 60(3):849–55.

Noguchi E, Shibasaki M, Arinami T, Takeda K, Yokouchi Y, KawashimaT (1998) Association of asthma and the interleukin-4 promoter gene in Japanese. *Clin Exp Allergy* 28:449–53

Northemann W, Braciak TA, Hattori M, Lee F, Fey GH (1989) Structure of the rat interleukin-6 gene and its expression in macrophage-derived cells. *J Biol Chem* 264: 16072–82.

Palmer LJ, Daniels SE, Rye PJ, Gibson NA, Tay GK, Cookson WO (1998) Linkage of chromosome 5q and 11q gene markers to asthma-associated quantitative traits in Australian children. *Am J Respir Crit Care Med* 158:1825–30.

Palomares O, Yaman G, Azkur AK, Akkoc T, Akdis M (2010) Role of Treg in immune regulation of allergic diseases. *Eur J Immunol* 40(5):1232–40.

Parham P, Janeway C (2009). *The Immune System*. 3rd ed. London, New York: Garland Science, 506.

Park KS, Mok JW, Ko HE, TokunagaK, LeeMH (2002) Polymorphisms of tumor necrosis factor A and B in brast cancer. *Eur J Immunogenet* 29:7–10.

Platanias LC (2005) Mechanisms of type-I- and type-II-interferon-mediated signaling. *Nat Rev Immunol* 5:375–86.

Raedler D, Illi S, Pinto LA (2013) IL10 polymorphisms influence-neonatalimmune responses, atopic dermatitis, and wheeze at age 3 year. *J Allergy Clin Immunol* 131(3):789–96.

Rael EL, Lockey RF (2011) Interleukin-13 signaling and its role in asthma. *World Allergy Organ J* 4(3):54–64.

Randolph AC, Lange C, Silverman EK, Lazarus R (2005) Extended haplotype in the tumor necrosis factor gene cluster is associated with asthma and asthma-related phenotypes. *Am J Respir Crit Care Med* 172(6):687–92.

Reche PA (2019) The tertiary structure of γc cytokines dictates receptor sharing. *Cytokine* 116:161–8.

Rivers-Auty J, Daniels MJ, Colliver I, Robertson DL, Brough D (2018) Redefining the ancestral origins of the interleukin-1 superfamily. *Nat Commun* 9(1):1156.

Rosenwasser LJ, Klemm DJ, Dresback JK, Inamura H, Mascali JJ, Klinnert M (1995) Promoter polymorphisms in the chromosome 5 gene cluster in asthma and atopy. *Clin Exp Allergy* 125(2):74–8.

Sadagurski M, Norquay L, Farhang J, D'Aquino K, Copps K, White MF (2010) Human IL6 enhances leptin action in mice. *Diabetologia* 53(3):525–35.

Savage SA, Abnet CC, Mark SD, Qiao YL, Dong ZW, Dawsey SM (2004) Genetic variants of the interleukin 8 and interleukin-8 receptor B genes and risk for gastric and esophageal cancer in high-risk Chinese population. *Cancer Epidemiol Biomarkers Prev* 13:2251–7.

Schmidt SV, Seibert S, Walch-Ruckheim B (2015) RIPK3 expression in cervical cancer cells is required for PolyIC-induced necroptosis, IL-1alpha release, and efficient paracrine dendritic cell activation. *Oncotarget* 6:8635–47.

Schmitz J, Owyang A, Oldham E, Song Y, Murphy E, McClanahan TK (2005) IL-33, an interleukin-1-like cytokine that signals via the IL-1 receptor-related protein ST2 and induces T helper type 2-associated cytokines. *Immunity* 23(5):479–90.

Schoemaker MJ, Swerdlow AJ, Hepworth SJ, van Tongeren M, Muir KR, McKinney PA (2007) History of allergic disease and risk of meningioma. *Am J Epidemiol* 165(5):477–85.

Schroder K, Hertzog PJ, Ravasi T, Hume DA (2004) Interferon-gamma an overview of signals, mechanisms and functions. *J Leukoc Biol* (75):163–89.

Schwartzbaum J, Ahlbom A, Malmer B, Lönn S, Brookes AJ, Doss H (2005) Polymorphisms associated with asthma are inversely related to glioblastoma multiforme. *Cancer Res* 65:6459–65.

Seddighzadeh M, Larsson P, Ulfgren AC (2003) Low IL-1alpha expression in bladder cancer tissue and survival. *Eur Urol* 43:362–8.

Sedger LM, McDermott MF (2014) TNF and TNF-receptors: from mediators of cell death and inflammation to therapeutic giants – past, present and future. *Cytokine Growth Factor Rev* 25:453–72.

Sehgal PB, Zilberstein A, Ruggieri R, May LT, Ferguson-Smith A, Slate DL, Revel M, Ruddel FH (1986) Human chromosome 7 carries the beta 2 interferon gene. *Proc Natl Acad Sci USA* 83:5219–22.

Shang H, Cao XL, Wan XJ, Meng J, Guo LH (2016) IL-4 gene polymorphism may contribute to an increased risk of atopic dermatitis in children. *Dis Markers*. 1021942.

Shea JJ, Ma A, Lipsky P (2002) Cytokines and autoimmunity. *Nat Rev Immunol* 22:37–45.

Shirazi R, Palsdottir V, Collander J, Anesten F, Vogel H, Langlet F, Jaschke A, Schürmann A, Prévot V, Shao R, Jansson JO, Skibicka KP (2013) Glucagon-like peptide 1 receptor induced suppression of food intake, and body weight is mediated by central IL-1 and IL-6. *Proc Nat Acad Sci USA* 110(40):16199–204.

Skinnider BF, Kapp U, Mak TW (2001) Interleukin 13 a growth factor in Hodgkin lymphoma. *Int Arch Allergy Immunol* 126(4):267–76.

Smith AJP, Humphries SE (2009) Cytokine and cytokine receptor gene polymorphisms and their functionality. *Cytokine Growth Factor Rev* 20(1):43–59.

Song Z, Lin Y, Ye X (2016) Expression of IL-1alpha and IL-6 is associated with progression and prognosis of human cervical cancer. *Med Sci Monit* 22:4475–81.

Sorrentino R, Terlizzi M, Di Crescenzo VG (2015) Human lung cancer-derived immunosuppressive plasmacytoid dendritic cells release IL-1alpha in an AIM2 inflammasome-dependent manner. *Am J Pathol* 185:3115–24.

Suzuki I, Yamaguchi E, Hizawa N, Itoh A, Kawakami Y (1999) A new polymorphism in the 5' flanking region of the human interleukin IL-4 gene. *Immunogenetics* 49:738–9.

Talar-Wojnarowska R, Gasiorowska A, Smolarz B, Romanowicz-Makowska H, Kulig A, Malecka-Panas E (2009) Tumor necrosis factor alpha and interferon gamma genes polymorphisms and serum levels in pancreatic adenocarcinoma. *Neoplasma* 56:56–62.

Tan TT, Coussens LM (2007) Humoral immunity, inflammation and cancer. *Curr Opin Immunol* 19:209–16.

Terabe M, Matsui S, Noben-Trauth N, Chen H, Watson C (2000) NKT cell-mediated repression of tumor immunosurveillance by IL-13 and the IL-4R-STAT6 pathway. *Nat Immunol* 1(6): 515–20.

Thye T, Burchard GD, Nilius M, Muller-Myhsok B, Horstmann RD (2003) Genomewide linkage analysis identifies polymorphism in the human interferon-gamma receptor affecting Helicobacter pylori infection. *Am J Hum Genet* 72:448–53.

Turner MD, Nedjai B, Hurst T, Pennington DJ (2014) Cytokines and chemokines at the crossroads of cell signalling and inflammatory disease. *Biochimica et Biophysica Acta (BBA)* 1843:2563–82.

Van Toorenenbergen AW, van Heerde MJ, van Buuren HR (2010) Potential value of serum total IgE for differentiation between autoimmune pancreatitis and pancreatic cancer. *Scand J Immunol* 72(5):444–8.

Wallenius K, Wallenius V, Sunter D, Dickson SL, Jansson JO (2002) Intracerebroventricular interleukin-6 treatment decreases body fat in rats. *Biochem Biophys Res Commun* 293(1):560–5.

Wang HW, Joyce JA (2010) Alternative activation of tumor-associated macrophages by IL-4 priming for protumoral functions. *Cell Cycle* 9:4824–35.

Wang Y, Kato N, Hoshida Y (2003) Interleukin-1beta gene polymorphisms associated with hepatocellular carcinoma in hepatitis C virus infection. *Hepatology* 37:65–71.

Wang Y, Liu D, Chen P, Koeffler HP, Tong X (2008) Negative feedback regulation of IFN-gamma pathway by IFN regulatory factor 2 in esophageal cancers. *Cancer Res* 68:1136–43.

Warzocha K, Ribeiro P, Bienvenu J, Roy P, Charlot C, Rigal D (1998) Genetic polymorphisms in the tumor necrosis factor locus influence non-Hodgkin's lymphoma outcome. *Blood* 91:3574–81.

Watsuki M, Mimori K, Yokobori T, Tanaka F, Tahara K (2009) A platinum agent resistance gene, POLB, is a prognostic indicator in colorectal cancer. *J Surg Oncol* 100:261–6.

Xu D, Matsuo Y, Ma J (2010) Cancer cell-derived IL-1alpha promotes HGF secretion by stromal cells and enhances metastatic potential in pancreatic cancer cells. *J Surg Oncol* 102:469–77.

Yang Y, Luo C, Feng R, Bi S (2010) The TNF-alpha, IL-1B and IL-10 polymorphisms and risk for hepatocellular carcinoma: a meta-analysis. *J Cancer Res Clin Oncol*.

Zhang JM, An J (2007) Cytokines, inflammation, and pain. *Inte Anesthesiol Clin* 245:27–37.

Zhang Y, Zhang J, Huang J (2010) Polymorphisms in the transforming growth factor-$\beta 1$ gene and the risk of asthma: a meta-analysis. *Respirology* 15(4):643–50.

Zhou J, Chen DQ, Poon VK, Zeng Y, Ng F (2009) A regulatory polymorphism in interferon-gamma receptor 1 promoter is associated with the susceptibility to chronic hepatitis B virus infection. *Immunogenetics* 61:423–30.

Zhu VF, Yang J, Lebrun DG, Li M (2012) Understanding the role of cytokines in glioblastoma multiforme pathogenesis. *Cancer Lett* 316(2):139–50.

22

Gastric Cancer and Genetic Polymorphisms in Modulating the Susceptibility

Shagufta Sheikh
Department of Biochemistry, Government Degree College for Women, Anantnag, Jammu and Kashmir, India

Sanjeev Rana
School of Biotechnology, Shri Mata Vaishno Devi University (SMVDU), Katra, J&K, India

Hilal Ahmad Wani
Department of Biochemistry, Government Degree College, Sumbal, J&K, India

Aarif Ali
Department of Clinical Biochemistry, School of Biological Sciences, University of Kashmir, Hazratbal, Jammu and Kashmir, India

Nouroz Sehar
Centre for Translational and Clinical Research, School of Chemical & Life Sciences, Jamia Hamdard, Hamdard Nagar, New Delhi, India

Sabhiya Majid and Nisar A. Naikoo
Department of Biochemistry & Multidisciplinary Research Unit (MRU-ICMR), Government Medical College Srinagar & Associated Hospitals, Karan Nagar, Srinagar, Jammu and Kashmir, India

Bashayr M. Alsuwayni, Azher Arafah, and Muneeb U. Rehman
Department of Clinical Pharmacy, College of Pharmacy, King Saud University, Riyadh, Saudi Arabia

CONTENTS

DOI: 10.1201/9781003246244-22

ABSTRACT Cancer still remains a leading cause of death, with incredible capability to defy the defenses of the body as well as medical interventions. One of the types of cancers that have resulted in the death of people around the globe is gastric cancer. Cancer disease is the culmination of a protracted complex course involving numerous environmental and genetic components. Genetic variation among individuals and cancer menace has been associated with specific allelomorphs of various genes, i.e., genetic polymorphisms which are reported to modulate the processes and progression of carcinogenesis. Recent findings have suggested that carcinogen metabolism, inflammatory response, cell proliferation regulation, mucosal protection, and antioxidant protection have all been suggested as probable biomarkers to forecast gastric cancer risk in recent years. Numerous studies done on various populations have been attributed to the single nucleotide polymorphisms in the candidate genes. In this chapter, we attempt to summarize the association of the most extensively studied polymorphisms in some of the candidate genes involved in immunogenic, DNA repair, cell signaling, and detoxification pathways, which could possibly lead to progression and development of gastric cancer.

KEYWORDS: *Gastric Cancer; Genetic Polymorphism; Allelomorphs; Candidate Genes; Immunogenic; DNA Repair; Cell Signaling; Detoxification.*

22.1 Introduction

> Tumors are a re-emergence of our inner Metazoan 1.0, a throwback to an ancient world when multicellular life was simpler....... In that sense, cancer is an accident waiting to happen.
>
> *Paul Davies*

Humans have known about cancer from the dawn of time. It is a root cause of death and a significant impediment to longer expectancy of life in almost every country of the world (Sung et al. 2021). With more than 1 million new cases projected in 2020 and a million estimated fatalities (approximately 1 in every 13 deaths), gastric cancer (GC) is a prominent malignancy worldwide ranking fifth in incidence and fourth in death (Hyuna et al. 2021). Cancer can develop in any area of the body and is named by the area where it initially manifested itself. Gastric cancer (GC), commonly described as stomach cancer, is a kind of cancer that attacks the stomach and its surrounding structures, mainly the esophagus and small intestine (Rugge et al. 2013). GC develops when stomach cells grow out of control and form tumors, invading normal tissues and moving to other parts of the body (Li et al. 2018). Cancer cells arise as a result of many genetic changes. These changes are caused by a multitude of factors. Environmental exposure to cancer-causing chemicals, lifestyle choices, and genes acquired from parents can all play a significant role (Ruddon et al. 2007). Most of the time, there is no evident cause. GCs usually grow over a long period of time. Precancerous changes in the stomach's inner lining (mucosa) frequently occur before a genuine cancer develops. Because these early changes rarely create symptoms, they go undiscovered for a long time (Gujar et al. 2016). GCs are categorized based on the kind of cells that give rise to them. Adenocarcinoma (AC), or stomach lining cancer, accounts for 90%–95% of GCs. These malignancies arise from gland cells in the stomach's inner lining (the mucosa) (Khanna et al. 2015). GIST, lymphoma, and carcinoid tumors are some of the other cancers that can occur in the stomach (Gullo et al. 2020). There are two main types of stomach ACs: The intestinal type that affects the intestines has a slightly better prognosis (outlook) (Dulak et al. 2012). Certain gene alterations are more common in cancer cells, which might contribute to the development of personalized drug therapy to treat them. The dispersed form has a greater chance of spreading fast. It's less common than intestinal cancer; however, it is difficult to treat.

TABLE 22.1

New Cases and Deaths for Some Cancers in 2020

Site of Cancer	No. of New Cases	No. of New Deaths
Breast of female	2,261,419	684,996
Lung	2,206,771	1,796,144
Prostate	1,414,259	375,304
Skin non-melanoma	1,198,073	63,731
Colon	1,148,515	576,858
Stomach	1,089,103	768,793

Source: GLOBOCAN (2020).

22.1.1 Worldwide Scenario of Gastric Cancer Incidence and Mortality

Gastric cancer (GC) still remains a major malignancy worldwide, despite a global drop in the number of cases, with over 1 million new cases reported in the year 2020 and an estimated 769,000 fatalities (Gullo et al. 2021). One in every thirteen people dies as a result of it. GC is the fifth most common cancer in incidence and fourth in mortality globally (Table 22.1). The incidence of stomach cancer varies greatly depending on geography and culture. Men had twofold greater rates than women in developed countries, but this ratio is 1.83 in developing countries (Buchholz et al. 2021). GC is the most frequently detected cancer in men and the foremost reason of cancer death in a number of South Central Asian nations (Gupta and Shaukat 2021). Eastern Asia and Eastern Europe have the highest incidence rates, while rates in Northern Europe and Northern America are normally low and similar to African regions. The countries such as Japan and Mongolia reported the highest prevalence in men and women, respectively. In industrialized nations, stomach cancer is more typically diagnosed. In high-middle HDI (Human Development Index) nations, the average incidence for males is twenty per lakh, whereas it is 6.6 per 100,000 in low-middle HDI countries. GC is the third most frequent cancer in men in India (Rahman et al. 2014).

The typical incidence of stomach cancer varies a lot depending on where you live (region) and who you are (gender). It is 32.1/100,000 among males and 13.2 among females in East Asia. In North America, the occurrence is 5.6/100,000 (Rawla and Barsouk 2019). Only 4.7 yearly diagnoses per 100,000

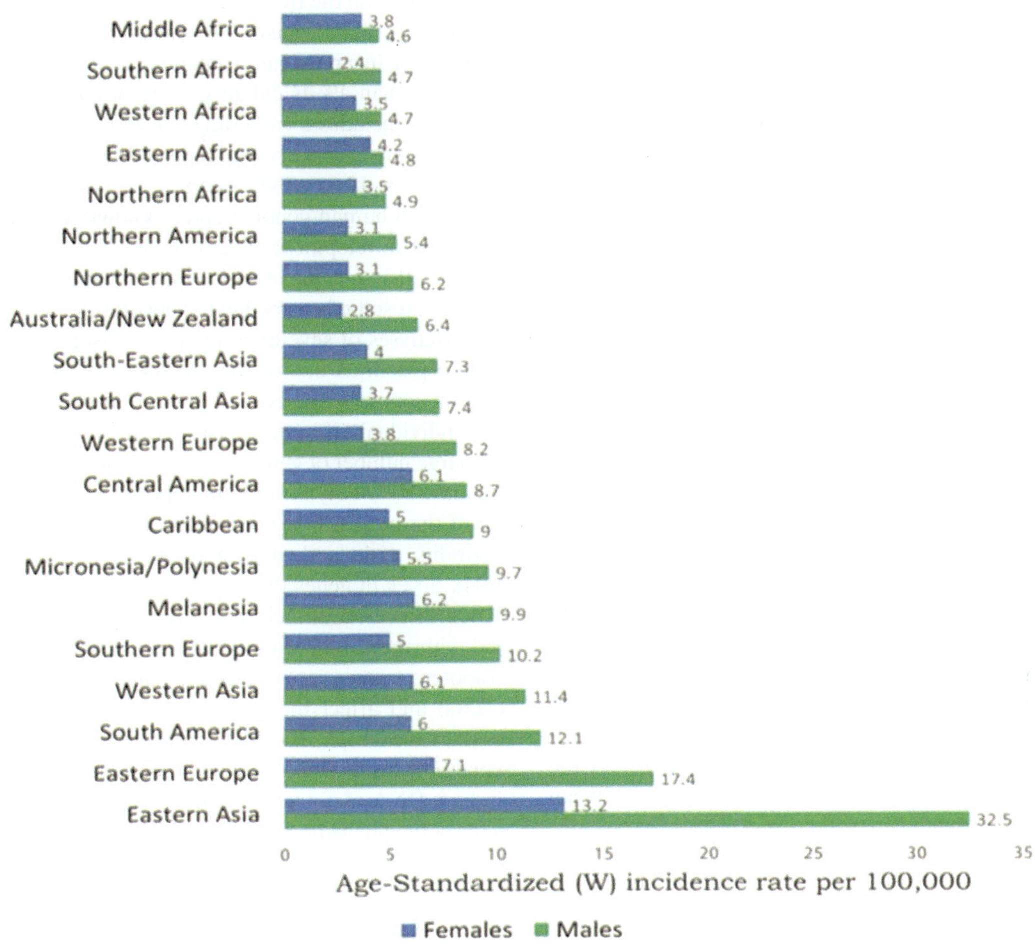

FIGURE 22.1 Region- and sex-specific incidence and age-standardized rates of gastric cancer in 2020. (Source: GLOBOCAN 2020.)

males are recorded in North and East Africa. With about 60 per 100,000 new male cases annually, the Korean Republic has the highest incidence (Antoni et al. 2017). Despite the fact that female prevalence rates are lesser with only 25 per 100,000, they follow the same geographical pattern.

Till mid-1990s, GC was the major reason of cancer-related deaths worldwide. Each year, 783,000 men die from GC, ranking it as the third most fatal cancer in men worldwide. GC is responsible for 8.3% of all cancer fatalities. From birth up to the age of 74, males have a 1.36% chance of dying from stomach cancer, whereas females have a 0.57% chance (Rawla and Barsouk 2019).

Males have a greater mortality rate from stomach cancer (Yang et al. 2014). Eastern and Central Asia, as well as Latin America, have high mortality rates, which coincide with high incidence, because of low survival rates and limited options of treatment, particularly in underdeveloped countries. It appears that lowering incidence is the key to decreasing death rate. GC is the main reason of death in men mainly in Central and Eastern Asian countries such as Kyrgyzstan, Turkmenistan, and Iran (Rawla and Barsouk 2019). GC is the foremost reason of death in females in four countries, as shown in Figure 22.1.

22.1.2 Etiology of Gastric Cancer

Since ACs account for 95% of stomach cancers, followed by primary gastric lymphoma (Parsonnet et al. 1994), based on their anatomical location, these adenocarcinomas are grouped into cardia (upper stomach) and non-cardia (lower stomach) cancers, despite the fact that GC is frequently described as a single entity (Hayakawa et al. 2016). Risk factors, carcinogenesis, and epidemiologic patterns differ among these entities. Chronic gastritis, stomach lining inflammation triggered by various environmental factors, is known to be the main cause of non-cardia malignancies (Khatoon et al. 2016). Despite the fact that the pathophysiology of cardia cancer is unknown, research suggests that it has a dual etiology, with some tumors related to infection by *Helicobacter pylori* and others related to extra body weight and gastroesophageal reflux disease (GERD) injury, with features similar to esophageal AC (Rawla and Barsouk 2019), so obesity and GERD are risk factors exclusive to cardia gastric cancer, whereas infection by *Helicobacter pylori*, poor socioeconomic status, and dietetic variables are risk factors unique to non-cardia GC.

The most prevalent reason for non-cardia stomach cancer is *Helicobacter pylori* infection, which is considered to account for virtually all instances (Fock 2014). *Helicobacter pylori* infection is extremely common, infecting half of the population all over the globe, and its geographic variation relates reasonably to stomach cancer risk. Despite the fact that malignancies of the gastric cardia are connected with stomach atrophy in the presence of *H. pylori* infection, cancer of cardia is not normally related to *Helicobacter pylori* infection and may even be inversely linked in some populations (Malfertheiner et al. 2005). It is reported that only 5% of infected hosts are documented to develop cancer, owing to variances in bacterial genomes, infection age, and climatic influences (Go 2002). In addition to *Helicobacter pylori*, alcohol consumption, smoking of cigarettes, and salt-preserved foods are also established risk factors for non-cardia stomach cancer. Low fruit consumption, along with a highly processed meat, barbecued or grilled meat, and seafood diet may upsurge the risk.

In addition to the above risk factors, genetic factors have a crucial role in gastric carcinogenesis (Bornschein et al. 2011). Even though the majority of identified cases are sporadic, inheritance is vitally important in the formation and progression of GC (Conteduca et al. 2013). In a wide range of genes, distinct allele variants (polymorphisms) are linked to individual genetic susceptibility. They have the ability to alter the impact of environmental exposure and the interaction between genes and the environment.

22.2 Genetic Polymorphisms in Relation to Gastric Cancer Susceptibility

The DNA in human genome consists of 3.1 billion nucleotides in a haploid set of 23 chromosomes. Since the human genome is quite enormous, there is high possibility of occurrence of polymorphic DNA. Suppose we examine DNA from a large number of individuals; there is high chance that base pair sequence would be the same for all of the DNA; hence, such portion of DNA is referred to as monomorphic. However, there would be some sections of DNA where the sequence of base pairs differs. Such portion of DNA is called polymorphic DNA or simply polymorphisms, sometimes also referred to as "spelling variations." It is this spelling order of nucleotides in DNA that in turn determines the sequence of proteins in a polypeptide chain determining the life functions. Therefore, variation in the sequence of four types of bases in DNA sequences in the genome is responsible for the differentiation and biological diversity in living organisms on earth. Due to the degeneracy in genetic code, some variation or spelling changes in the coding parts (exons) of sequences have evolutionary role and contribute to biological diversity in populations. However, other changes are translated in the polypeptide, resulting in the formation of protein with new functions and characteristics. The altered protein can stop or start the gene function or result in the partial protein sequence that's not sufficient to carry out functional activity. These variations alter the function of the cell and generate the difference within an organism and population.

Depending on the frequency, alleles are called either mutants, or common polymorphism. The mutant allele has a frequency of less than 1% in a population, whereas alleles having a frequency of greater than 1% is called genetic polymorphism. Genomic variability includes a wide range of variations from single base change, many base pairs, and repeated sequence. Single nucleotide polymorphism (SNP) is the commonest type of genetic variation in human genome and is known to play a significant role in the development of various diseases and thus can serve as an excellent genetic marker for the mapping of disease. SNPs arise as a result of errors due to substitution insertion and deletion. Analyses of several comparative diseases suggest genetic polymorphism as one of the factors associated with susceptibility to common diseases, difference in phenotypic traits, and inter-individual differences (Harro 2010).

A large number of genes associated with various cancer types contain SNPs. Polymorphic mutations or changes in DNA are one of the most important genetic determinants in illness susceptibility, risk, and genetic predisposition (Lamberts et al. 2004). SNPs in genes that regulate DNA repair, cell cycle regulation, metabolism, cell signaling, and immunity are associated with the susceptibility to cancer. SNPs are found in gene promoters, exons, introns, and 5′- and 3′-untranslated regions (UTRs) and affect the gene expression by a variety of processes; for example, the SNPs in promoter area influence gene expression via modifying promoter activity, transcription factor binding, DNA methylation, and histone changes. Furthermore, changes in epigenetic regulation caused by gene polymorphism contribute to intricacy behind cancer risk posed by these SNPs. Distinct SNPs situated in different genomic regions, involved in many genetic and epigenetic pathways, have been investigated and explored. The mechanisms underlying the effects of SNPs that result in cancer susceptibility are critical to understanding the molecular pathogenesis of various cancers.

Genome-wide association studies (GWAS) are used to find sequence changes in human genome, screen SNPs connected to human illnesses, and learn more about genetic variants and cancer risk (Witte 2010). Millions of SNPs have been identified, and two-stage GWAS have shown relationships between candidate gene SNPs and disease susceptibility (Nishizawa et al. 2015). Thus, genetic linkage analysis and association studies, which need control and case samples to identify contender genes for investigation based on the information of the pathophysiology of the ailment in question, are two approaches to investigate genes and genetic polymorphisms involved in disease susceptibility.

SNPs or single nucleotide polymorphisms are a frequent sort of genetic alteration that may hasten the onset of GC. Genes associated with crucial pathways of gastric carcinogenesis are studied in the research of genetic vulnerability to GC (González and Agudo 2012). This comprises genes involved in gastric mucosal protection, metabolization, detoxification of carcinogenic substances, DNA repair, cell proliferation and adhesion, oxidative damage repair (Palli et al. 2010), *H. pylori* identification in gastric epithelial cells, and genes involved in the activation of signaling pathways that cause pro-inflammatory and apoptotic responses (Alzahrani et al. 2014). This chapter aims to provide comprehensive insights into the assessment of the strength of significant relations between some important genetic polymorphisms and the onset of GC.

22.2.1 Genetic Variability in Metabolic Enzymes Involved in the Detoxification of Carcinogens

The majority of enzymes engaged in the metabolism of xenobiotics are polymorphic and have been linked to cancer susceptibility (Hatagima 2002).

Polymorphisms may affect the impact of environmental exposures, and these gene-environment interactions may explain some of the global diversity in stomach cancer incidence. Individual differences in stomach cancer risk have been linked in the recent decades to distinct variant alleles of different genes found in a considerable proportion of the population. The most important genes that are of critical importance are the ones involved in xenobiotic metabolism whose various variants seem very important in understanding the development of GC.

22.2.1.1 Cytochrome P450 Gene Polymorphism

Cytochrome P450 (CYP450) refers to a family of enzymes found in the smooth endoplasmic reticulum of hepatocytes and small intestinal epithelial cells. The CYP450 monooxygenases are the most prevalent isoenzymes in phase I metabolism (CYP1A1 and CYP2E1). They fight cancer by detoxifying a wide range of potentially cytotoxic/genotoxic compounds and participating in the biotransformation of xenobiotics such as PAHs (polycyclic aromatic hydrocarbons), dioxins, and nitrosamines such as those found in the smoke of tobacco (Hatagima 2002). CYP450 enzymes represent a large family of highly polymorphic isozymes. They are also known as heme thiolate proteins, monooxygenases that catalyze the incorporation of one oxygen atom into the substrate and convert it to a more water-soluble product. CYP450 catalyzes the reaction shown in Figure 22.2.

However, 57 genes and more than 59 pseudogenes have been identified in CYP450 superfamily and have been divided into 18 families and 43 subfamilies based on amino acid sequence. The significant enzymes engaged in the metabolic processes of xenobiotics and linked to GC vulnerability are discussed below (Hatagima 2002).

***CYP1A1* (CYP450 Family 1, Subfamily A, Polypeptide 1)**

CYP1A1 (Chr. 15q24) plays a significant role in the activation of tobacco carcinogens such as PAHs, aromatic amines (AAs), and benzo[a]pyrene. The *CYP1A1* gene (Chr. 15q24) has two non-synonymous polymorphisms that have been linked to the growth of cancer: a transition from T to C produced by the Msp1 enzyme cutting the gene to generate the *CYP1A1*2A* mutant allele, and an A instead of G substitution at 462 codon in seventh exon resulting in *CYP1A1**2C (Ile462Val, rs1048943) (Preissner et al. 2013). These changes are considered to have an impact on *CYP1A1* expression and activity, as well as the balance between metabolic activation and toxic chemical elimination, resulting in infection susceptibility (Gattás et al. 2006). A study carried out on Colombian patients indicated no association of the polymorphism with GC. Another study identified 90 GC patients with *2A allele to be at 50% lower risk in China (Roth et al. 2004). A meta-analysis of 11 studies on CYP1A1 MspI and 8 studies on CYP1A1 Ile462Val polymorphisms found no association with GC (Xue et al. 2014).

***CYP2E1* (CYP450 Family 2, Subfamily E, Polypeptide 1)**

The *CYP2E1* gene, located on Chr.10q26.3, encodes an enzyme involved in the metabolism of drugs, hormones, and xenobiotic toxins. It accounts for 7% of total CYP450 enzymes in the liver. SNPs in the *CYP2E1* genes can lead to alteration of the function of these enzymes, resulting in the activation of carcinogens, which involve in the tumor initiation. Three polymorphisms in the CYP2E1 gene have received the greatest attention. CYP2E*5 is located in the 5 0-regulatory region. It comprises two variations: G1293C (PstI) and C1053T (RsaI), with DraI detecting CYP2E*6 (T7632A). *CYP2E1*5A* and *CYP2E1*5B* (c2 allele) mutant allele cause polymorphic overexpression of the gene associated with the formation of tumor (Soya et al. 2005).

Many investigations have been conducted to establish the relationship between CYP2E1 polymorphism and GC. However, the findings were inconclusive, and racial variances were found. Few studies identified a substantial link between CYP2E1 RsaI polymorphism and stomach cancer risk (Boccia et al. 2007), whereas others found no link (Park et al. 2003).

22.2.1.2 Glutathione S-Transferase Gene Polymorphism

GSTs are phase II enzymes that detoxify potentially carcinogenic substances (Chow et al. 2007). Cytosolic GSTs are dimeric proteins, and at present, eight distinct classes of the soluble cytoplasmic mammalian glutathione S-transferases have been identified: alpha, kappa, mu, omega, pi, sigma, theta, and zeta. The active site of a GST molecule is composed of two subunits, the glutathione-binding site and the hydrophobic substrate-binding site. The residues forming the glutathione-binding site are conserved in the different classes, while those forming the substrate-binding site vary considerably, leading to a wide substrate specificity. The incorporation of glutathione upsurges the molecules' water excretability and solubility (Ketterer et al. 1988). The most deliberate substrates for GSTs are cancer-causing PAH diolepoxides, but these also perform critical defensive part contrary to oxidative stress (Dasari et al. 2018). Certain carcinogens, such as halogenated hydrocarbons, are activated by GSTs. GSTs might be involved also in the instigation of some carcinogenic

$$\text{RH (Substrate)} + O_2 + \text{NADPH} + H^+ \rightarrow \text{ROH (product)} + \text{NADP}^+ + H_2O$$

FIGURE 22.2 CYP-dependent oxygenation reaction. It involves the addition of a single oxygen atom to the substrate molecule in the presence of CYP450 monooxygenase and NADPH.

compounds such as hydrocarbons that are halogenated. In the gastrointestinal tract, the genes that encode three major GST isoenzymes for metabolizing enzymes of phase II (GSTM1, GSTP1, and GSTT1) are broadly expressed and polymorphic (García-Gonzalez et al. 2012). Allelic variation has been found in genes encoding for these GSTs.

GSTM1 (Glutathione S-transferase Mu1)

This gene encodes a glutathione S-transferase that belongs to the mu class. The mu class of enzymes functions in the detoxification of electrophilic compounds, including carcinogens, therapeutic drugs, environmental toxins, and products of oxidative stress, by conjugation with glutathione. The genes encoding the mu class of enzymes are organized in a gene cluster on chromosome 1p13.3 and are known to be highly polymorphic. These genetic variations can change an individual's susceptibility to carcinogens and toxins as well as affect the toxicity and efficacy of certain drugs.

The homozygous deletion of *GSTM1* (null genotype) gene has been found to lead to the entire non-appearance of the respective activity of the enzyme (Bhattacharjee et al. 2013) involved in the detoxification of chemical compounds that are supposed to be present in food, agrochemicals, cigarette smoke, and antineoplastic medications such as nitrosamines and PAHs. The gene removal has been projected to have arisen by unequal crossover. Allelic variants of GSTM, *i.e.,* A and B, have also been found to exist, but based on the current knowledge, they have no consequences in the catalytic activity of the enzyme (Widersten et al. 1991). In addition, a gene duplication has been found to exist in Saudi-Arabians (McLellan et al. 1997). The frequency of the null genotype is around 50% in Caucasians and Asians, but only 27% in Africans.

GSTPI (Glutathione S-transferase P)

The gene *GSTP1* has been found to be located on chromosome 11q at a very small region (Dong 2006). In addition to the *GSTP1*A* wild-type allele, two other *GSTP1*B* and **C* variant alleles have been noticed (Gsur et al. 2001). In both the variants at nucleotide 313, a point mutation leads to a change of a single amino acid at codon 105 (isoleucine to valine). Additionally, *GSTP1*C* also bears alternative point mutation leading to alanine instead of valine change corresponding to 114 codons. Both of these affected codons exist in nearby proximity of the binding site, which is hydrophobic in nature. The Val105 variant has been proved with either low or high affinity and specific activity based on the substrate. In comparison, the Ala114Val polymorphism has not been designated to impact the enzyme activity. Almost 5%–10% of Caucasian population have been revealed to be homozygous for allele corresponding to Val. Women with the Val/Val *GSTP1* genotype have been reported to have a better prognosis, which is likely owing to the decreased chemotherapeutic metabolic activity.

22.2.2 Genetic Variability of Proteins Involved in Mechanisms of DNA Repair

DNA repair pathways are answerable for preserving genomic integrity and protecting against carcinogenesis (Bray and West 2005; Chaudhuri and Nussenzweig 2017). On the other hand, inherited mutations or functional polymorphisms in DNA repair genes may affect the host's DNA repair ability and hence regulate the risk of carcinogenesis (Spitz et al. 2001). Multiple DNA repair genes have been connected to coding polymorphisms, which have been associated with hereditary vulnerability to a variety of human malignancies, including stomach cancer (Goode et al. 2002). Numerous studies suggest that frequent DNA repair gene SNPs are linked to a variety of sporadic malignancies, including GC (D'Errico et al. 2017). Understanding DNA repair gene polymorphisms may aid researchers in determining not just the risk of individuals being exposed to carcinogens in the environment, but also their receptivity to various DNA repair pathway-targeted therapies.

22.2.2.1 XRCC1 (X-ray repair cross-complementing group 1)

The protein-encoding gene maps on chromosome 19 at 19q13.2 and is an abundant nuclear zinc finger protein that is part of a DNA-binding protein complex. It is involved in the efficient sister chromatin exchange and repair of DNA single-strand breaks formed by exposure to DNA radiations and alkylating agents. Although the XRCC1 protein itself has no enzymatic activity, it interacts with the key enzymes involved in the base excision repair (BER), including poly (ADP-ribose) polymerase (Caldecott 2019), DNA polymerase β, DNA ligase III, and AP endonuclease (APE1), a rate-limiting enzyme in base excision repair. Therefore, polymorphisms causing amino acid substitutions may impair the interactions of XRCC1 with other enzymatic proteins and hence alter base excision repair activity.

In spite of the fact that the *XRCC1* gene has over 300 SNPs, three SNPs are particularly prevalent, leading to amino acid substitutions in codons 194 (rs1799782), 280 (rs25489), and 399 (rs25489) (rs25487) (Annamaneni et al. 2013). These non-conservative amino acid alterations could affect the function of XRCC1 and raise the risk of DNA damage. The Arg399Gln polymorphism, which results in a G-to-A nucleotide alteration in the PARP-binding region of the XRCC1 protein, is the only one that affects complex assembly and/or repair effectiveness.

XRCC3 (X-ray repair cross-complementing group 3)

The *XRCC3* gene is another essential gene in the DNA repair process, with chromosomal location 14q32.33 (Yuan et al. 2016). It acts on double-stranded DNA breaks by interacting with and stabilizing Rad51, a component of the process of homologous repair. It avoids chromosomal aberration by repairing lost information at a breakage site (Li et al. 2006). Several SNPs have been discovered in the XRCC3 gene. The non-synonymous SNP C722T (rs861539, Thr241Met) has received the greatest attention. Several papers have established a connection between this polymorphism and a higher risk of breast, skin, and lung cancer, despite the fact that nothing is clear about the functional significances of this polymorphism (Andreassen et al. 2002).

XPD (Xeroderma pigmentosum complementary D) gene

XPD is one of the most critical proteins in NER pathway, located on chromosome 19q13.32. It's also called excision repair cross-complementing group 2 (ERCC2) (Fleming et al. 2012). *XPD* is an ATP-dependent DNA-unwinding helicase that is involved in NER pathway and basal transcription (Egly and Coin 2011). Amino acid variations in distinct XPD domains, as part of the TFIIH transcription factor/complex, may modify the diverse interactions of protein and result in different phenotypic expressions. In the *XPD* gene, several SNPs have been reported, with Lys751Gln (rs13181) being one of the most common SNPs that result in an amino acid change. The *XPD751* polymorphism alters the electrical configuration of amino acids.

The existence of a link between *XRCC1* and *XPD* and an elevated risk of GC has been reported in a number of studies. Many studies have shown the existence of connection of *XRCC1* and *XPD* with an augmented risk of GC. The data from these studies suggested that polymorphisms in DNA repair genes may influence cancer susceptibility by altering repair capability. *XRCC1* and *XPD* have been examined the most often among the known DNA repair gene polymorphisms (Matullo et al. 2001). Previous research on the linkage between these polymorphisms and the danger of stomach cancer has yielded mixed results. The existence of a link between *XRCC1* and *XPD* and an elevated risk of GC has been testified in a number of studies. In other population-based researches, however, analyses of the same variations revealed no connection; for example, a previous research has found no significant link between *XRCC1 Arg399Gln* genotypes and GC risk in various populations. The same was discovered to be true for the *XPD Lys751Gln* polymorphism in prior investigations in Polish, Turkish, Italian, and Chinese populations. Some studies suggested that having all three polymorphic alleles OF XRCC1 (Arg194Trp, Arg399Gln, and Thr241Met alleles) in one patient increased the risk of developing chronic gastritis as a result of the process of inflammation led by *Helicobacter pylori* or the effect of environmental variables, or mutations buildup in the DNA of stomach epithelial cells. The existence of polymorphisms leads to a reduction in repair capability (Marin et al. 2016). The findings of these researches suggest that polymorphisms in the *XRCC1*, *XRCC3*, and *XPD* repair genes are linked to the growth of GC and that their combined effect may upsurge vulnerability to this dreadful disease and other malignancies also (Palli et al. 2010). To monitor and validate gene-to-gene interactions, which may be crucial in the disease's pathophysiology, many additional gene-based polymorphisms in DNA repair pathway should be included (Monroy et al. 2011).

22.2.2.2 APE1 (Apurinic/apyrimidinic endonuclease 1)

The AP endonuclease 1 (APE1) protein, also known as APE, APEX, HAP1, and REF-1, is located on chromosome 14q11.2 (Luo et al. 2014). It is a multifunctional protein that plays a key role in the BER pathway by hydrolyzing the phosphodiester backbone immediately 5′ to the AP site. APE1 can function as a 3′-phosphodiesterase to induce repair of DNA single-strand breaks caused either directly by reactive oxygen species, or indirectly by the enzymatic removal of damaged bases. APE1 is also recognized as a transcriptional coactivator for several transcription factors implicated in cancer formation, and it is thought to be a prospective approach for anticancer treatment (Tell et al. 2010).

Till date, many epidemiological studies have suggested genetic variations in *APE1* may confer individuals' susceptibility to cancer. A total of 18 SNPs in *APE1* have been identified, of which two functional SNPs (−656 T>G in the promoter region, rs1760944, and 1349 T>G in the fifth exon, rs1130409) have been wildly investigated. It was reported that (−656 T>G) polymorphism influenced the transcriptional activity of *APE1* and contributed to cancer susceptibility in lungs. As to (1349 T>G) polymorphism, functional studies suggested that the G allele may have altered endonuclease- and DNA-binding activity, reduced ability to communicate with other BER proteins, and decreased capacity to repair DNA oxidative damage (Au et al. 2003). Numerous studies investigated the influence of (1349 T>G) polymorphism on cancer risk; however, the results of these studies remain inconclusive.

Using an additive model, a study reveals substantial evidence for a link amid the G allele of *APE1* polymorphism and stomach cancer risk, with a 1.77 times amplified risk of stomach cancer in a diverse population with sample size of 2113 (Tian et al. 2019). In an analysis of subgroup, the mutant G allele amplified the risk of stomach cancer in comparison with wild-type T allele. This mutation enhances the growth of cancer through reducing DNA repair activity.

22.2.2.3 ERCC5 (Excision repair cross-complementing group 5)

The *ERCC5* gene (also known as *XPG*, *UVDR*, *XPGC*, *COFS3*, *ERCM2*, and *ERCC5–201*) is located on the human chromosome 13q32–33, which encodes a single-strand specific DNA endonuclease that makes the 3′-incision in DNA excision repair following UV-induced damage (Wang et al. 2013). ERCC5 (XPG DNA repair endonuclease) gene is an indispensable component of the nucleotide excision repair (NER) pathway, which belongs to the flap structure-specific endonuclease 1 (FEN1) family (Wang et al. 2013). ERCC5 is one of the main genes activated by p53 and is involved in DNA repair (Kannan et al. 2000; Neamatzadeh et al. 2015). Mutations in the ERCC5 gene cause either the cancer-prone disorders such as Xeroderma pigmentosum (XP), a skin disorder characterized by hypersensitivity to UV light and increased susceptibility to skin cancer following UV exposure, or the severe neurodevelopmental disorder such as Cockayne syndrome (CS) (Ma et al. 2012; Wang et al. 2013).

To date, several SNPs in *ERCC5* gene have been identified and have been studied for their association with the risk of different cancers. It has been discovered that five polymorphisms, rs751402 C>T, rs2094258 C>T, rs1047768 T>C, rs2296147 T>C, and rs873601 G>A, were the most commonly explored for their connections with the risk of GC. *ERCC* polymorphism (rs751402), which has a C-to-T transition, may change transcription domain-related *ERCC5* repair capacity, which could explain its link to stomach cancer. There is considerable evidence that some SNPs of ERCC5 gene are correlated to GC occurrence. However, the associations between ERCC5 gene polymorphisms and GC risk were still conflicting.

A meta-analysis included 33 case-control studies from 15 publications, with a total of 16,783 cancer cases and 23,063 controls. The combined findings revealed that the ERCC5 gene polymorphisms rs751402 C>T, rs2296147 T>C, and rs873601 G>A are related to a higher risk of GC; however, this risk was not found for the other two SNPs (rs2094258 C>T and rs1047768 T>C) (Namazi et al. 2017). On the contrary, a recent meta-analysis based on 47 case-control studies discovered that the ERCC5 gene rs873601 G>A polymorphism was strongly related to an elevated risk of various malignancies (Huang et al. 2017).

In another study, a polymorphism (rs751402) in the promoter region of *ERCC 5*, which regulates its expression and activity during transcription, has been discovered in human cells that were healthy (Tian et al. 2019). In a dominant model, this SNP was found to be strongly linked to an augmented menace of stomach cancer in Asians (Abdi et al. 2020).

22.2.3 Genetic Variability of Immunogenic Factors

Genetic polymorphisms (SNPs) in genes that influence the immune system have been linked to GC by a number of researchers (Zabaleta 2012). The important cytokines that undergo polymorphic change and thereby influence the expression of cytokines are *interleukin (IL)* gene *(IL-10) TNF-α (tumor necrosis factor alpha)* and *interleukin-1* (IL-1) (Sugimoto et al. 2007). The *IL-1* family of cytokines is made up of three genes: *IL-1A*, *IL-1B*, and *IL-1RN* (Banerjee and Saxena 2012). These genes encode for proteins interleukin-1α, interleukin-1β, and endogenous interleukin-1 (IL-1ra) correspondingly.

Interleukins

Interleukin-10 (IL-10) gene is located on chromosome 1q31–32, with five exons and four introns and its receptor to chromosome 11. It is a pleiotropic cytokine with prominent anti-inflammatory properties regulating B cell proliferation and differentiation and exhibiting immunoregulatory activities. In the case of *IL-10*, it shows SNPs at three positions: the 1082 position (G/A) (rs1800896), the 819 position (C/ T) (rs1800871), and the 592 position (C/A) (Hsu et al. 2015). It is reported that 592 (rs1800872) polymorphism affects the activity of transcriptional factors and their expression and hence can modulate the cancer susceptibility.

IL-10–819 C/T (rs1800871) promoter polymorphism has been shown to have a decreased incidence of stomach cancer in Mexican participant samples (Vargas-Alarcón et al. 2016). On the other hand, *IL-10–1082 (interleukin-10–1082)* and *IL-10–592 (interleukin-10–592)* polymorphisms have been seen to be strongly linked to GC in Asian cultures.

The development of GC is also linked to polymorphisms in the *IL-1* and *IL-1RN* genes (Shin et al. 2008). Those found in the *IL-1B* gene are linked to cytosine's inhibitory effect on stomach acid output. This suppression makes it easier for bacteria such as *H. pylori* to colonize and infect the stomach. In individuals having *H. pylori* infection and being carriers of the *IL-1RNA2-A2* or *IL-1B511TT* genotypes, IL-1 levels in the stomach mucosa are greater than normal, according to one research (Koosirirat et al. 2010). The *IL-1* and *IL-1RN* genotypes have been linked to the growth of cancer at various levels (Ahmed et al. 2017). Polymorphisms at various sites such as –31 (C/T), –511 (T/C), and +3954 (C/T) are widely investigated. Findings of a study reported that there is a substantial variance in the incidence of the genotype CC at –31 position among *H. pylori*-infested patients (28.57%) and uninfested individuals (8.33%). This shows that having the C genotype is linked to a higher risk of *H. pylori* infection. According to other demographic studies, harboring the allele *IL-1B-31*C* increases the likelihood of developing precancerous lesions. In Colombia, according to a meta-analysis conducted in Caucasian populations, patients harboring the genotype of TT for *IL-1B 511* gene exhibited a favorable correlation with the cancer (Ma et al. 2017), in spite of the fact that some publications reveal considerably higher frequencies of the *IL-1B+3954T* allele or the *IL-1B+3954C/T* heterozygous genotype in GC cases. The pro-inflammatory activity and linkage with the chance of stomach cancer of the *IL-1B+3954* polymorphic variant has not been clearly shown. A meta-analysis revealed that the *IL-1B+3954T* genotypic variant is not linked to an augmented risk of GC growth (Wang et al. 2007). In distinction to what has been stated in Brazil, where genotypic studies of various polymorphisms in case-control research setups revealed that the *IL-1RN*2* allele is the most common in patients having ulcers in stomach, the polymorphisms detected for the gene IL-1RN were not connected with stomach cancer.

In addition to the above, polymorphisms in the *interleukin-17 (G-197A)* promoter region is associated with GC in Iranian population (Rafiei et al. 2013). The presence of a single or multiple (G-197A) polymorphisms was found to have a substantial impact on the probability of developing GC in the patient population (Zeyaullah et al. 2021).

TNF (Tumor necrosis factor)

The TNF family is another group of pro-inflammatory defensive cytokines. It has two types TNF-α (tumor necrosis factor alpha) and TNF-β (tumor necrosis factor beta). Chronic inflammation is likely to have an important role in the development of GC and other cancers (Hasselbalch 2012). According to Epplein et al., increased circulating levels of cytokines related to inflammations such as TNF-α may increase the chance of stomach cancer (Epplein et al. 2013). The development of GC is linked to TNF-α-induced protein release by *H. pylori* (Suganuma et al. 2012). TNF-α/TNFR1 signaling activation in tumor microenvironment is thought to promote GC development by activating Noxol and GNA14 (Kong et al. 2021). TNF-α, which is found on chromosome 2, has four exons. There are numerous polymorphism sites in the gene, including the well-studied *TNF-238* (rs361525) and 308 (rs361525) (rs1800629).

A meta-analysis of 24 independent researches examining the connections between *TNF-A* polymorphisms and stomach cancer risk were found in multiple databases. *TNF-A –308G/A*, *TNF-A –238G/A*, and *TNF-A –857C/T* are the most widely studied polymorphisms. Only western populations showed a statistically meaningful link between *TNF-A –308GG* and *GC* (Gorouhi et al. 2008).

Another meta-analysis including 4,084 cases and 7,010 controls revealed that there was no substantial link between the *TNF-A –238G/A* (rs361525) polymorphism and augmented GC chance. Under five models, the *TNFA* (rs361525) polymorphism enhanced the risk of stomach cancer in Asians, but lessened the chance of stomach cancer in Caucasians in allelic and dominant models. Methods of genotyping demonstrated an elevated risk with other approaches in subgroup analysis. Finally, one of the meta-analyses revealed that the *TNF-A* (rs361525) polymorphic variant is linked to the incidence of stomach cancer, particularly in Asians (Tianshu Xu et al. 2018).

Toll-like receptors (TLRs)

TLRs are type I transmembrane glycomembrane and intracellular signaling domains. The extracellular domains have repeated leucine-rich repeat (LRR) modules and are responsible for binding the so-called pathogen-associated molecular patterns (PAMPs). For example, lipoproteins or lipopeptides are recognized by TLR2 in complex with TLR1 or TLR6; viral double-stranded RNAs are recognized by TLR3; and lipopolysaccharides are recognized by TLR4. Their instigation leads to the production of chemokines, antigen-presenting molecules, costimulatory molecules, and inflammatory cytokines (Kaisho and Akira 2006). Different *TLR* gene variants have been linked to GC, with ethnicity being the most important factor. Among the TLRs, it has been reported that the TLR2, lipoproteins bacterial receptor, and the TLR4, the lipopolysaccharide (LPS) receptor, are involved in the response to infection by *H. pylori* on gastric epithelial cells. Both the TLR2 and the TLR4 promote transcription of genes involved in immune activation, including nuclear factor kappa B (*NF-κB)* and also mintage-activated protein (*MAP*) kinase pathways. TLR4 is upregulated in *H. pylori*-infected gastric epithelial cell lines and macrophages, and TLR4 protein expression has been shown in chronic active gastritis, precancerous lesions, and gastric carcinoma cells. In response to *H. pylori* infection, TLR2 activates NF-κB in epithelial cells, triggering the production of interleukin-8 (IL-8), macrophage inflammatory protein-3, and growth-regulated oncogene alpha. Thus, functionally relevant mutations in TLR genes may change the host immune response to pathogens such as *H. pylori* infection.

Different *TLR* gene variants have been linked to GC, with ethnicity being the most important factor (Meliț et al. 2019). *TLR2* has been shown to be the most commonly expressed gene in stomach cancers of all *TLRs*. The specific polymorphism *TLR2*–196 to –174 del/del genotype has been reported to show decreased transcriptional activity of the *TLR2* gene. Likewise, the polymorphisms *in TLR4* are implicated in increased susceptibility to various diseases and also infection with the *H. pylori* associated with GC and its precursors.

TLR4 levels were shown to be associated with an augmented risk of GC (Castano-Rodriguez et al. 2014). Two SNPs in *TLR4*+896A/G (rs4986790) and +1196C/T (rs4986791) have received special attention in some studies, although the results are still controversial. In a study, the participants who had a *TLR-4* polymorphism of +1196C/T (rs4986791) Thr399Ile had a higher susceptibility to GC (De Oliveira et al. 2013), but insignificant linkage for *TLR4*+896A/G (rs4986790) *(Asp299Gly)* gene polymorphism was noticed (Zeyaullah et al. 2021).

HSP70 (Heat shock protein 70)

Heat shock proteins (HSPs) are a set of evolutionarily conserved proteins. HSPs have been classified into various subfamilies based on their molecular weight or systematic gene symbols. Broadly, in mammals, four major families of HSPs are recognized: HSP90 (HSP90α, HSP90β, and GRP94), HSP70 (HSP70, HSC70, mHSP70, and GRP78), HSP60, and small HSPs (HSP27, αB-crystallin, and HSP22) (Kampinga et al. 2009). Accumulation of misfolded protein in stressed cells triggers expressions of HSPs, which prevent protein aggregation and facilitate refolding or elimination of misfolded proteins in their capacities as chaperones (Kim et al. 2007). Recently, it has been reported that HSPs are significantly associated with cancers (Calderwood et al. 2006).

One of the most conservative and well-known HSP families is the HSP70 family. In the human HSP70 family, there are three major genes (HSPA1A, HSPA1B, and HSPA1L) and the coding proteins are designated as HSP70–1, HSP70–2, and HSP70-hom, respectively. All three genes were found on chromosome 6 in the class III area of the human major histocompatibility complex (MHC). The HSPA1A and HSPA1B genes both encode the heat-inducible HSP70 protein, but the HSPA1L gene produces a non-heat-inducible protein with 90% sequence similarity to the HSP70 protein. It has been hypothesized that HSP70 plays a significant role in tumor formation, therapy, and prognosis, and that it has different immunologic pathways that influence tumor cells and promote cell proliferation. HSP70 expression is excessively high in cancer cells, and the protein may play a role in oncogenesis and chemotherapy resistance (Kumar et al. 2016). In these genes, many SNPs have been identified. The most investigated areas are at HSP70–2 locations +1267*A/G* (rs1061581), HSP70-hom places +2437 *T/C* (rs2227956), and HSP70–1 positions +190*G/C* (rs1043618). These SNPs may influence HSP70 expression or function, contributing to disease risk and stress tolerance (Favatier et al. 1997).

The HSP70–2 A+1267G polymorphism is a synonymous mutation located in the coding region and is likely to affect the secondary structure of mRNA, thus affecting the stability of mRNA and protein expression, whereas the HSP70-hom T>C polymorphism has non-synonymous mutations that lead to a Met-to-Thr substitution at position 493 in the peptide-binding domain and may affect substrate-binding specificity and chaperone activity.

To date, several studies have investigated the association between the three HSP70 polymorphisms and risk of cancer. However, the results remain controversial. In genes *HSP70–1 (–p190G/C)*, *HSP70–2 (–p1267A/G)*, and *HSP70-HOM (–p2437T/C)* polymorphisms, only Cp190 allelic variant of the gene *HSP70–1* showed a substantial connection with stomach cancer. Despite the fact that polymorphisms in the genes of *HSP70–2* and *HSP70-HOM* were found to have no link to GC, a collective investigation of the frequencies of various *SP70* genotypic combinations revealed that *HSP70–1 G/G*, *HSP70–2 A/G*, and *HSP70 HOMT/T* were not found in all patients with stomach cancer, indicating a protective role against disease development (Ferrer-Ferrer et al. 2013). However, results from another meta-analysis suggested that the HSP70–2 polymorphism rather than HSP70-hom and HSP70–1 polymorphisms was associated with the risk of cancer (He et al. 2014).

22.2.4 Genetic Variations in Tumor Suppressor Genes and Apoptotic Pathway

GC formation and progression are influenced by genetic variables such as changes in oncogenes and tumor suppressor genes (McLean and El-Omar 2014). Also, apoptosis is a regulatory process that inhibits over-proliferation in normal cells, and abnormalities in the apoptosis pathway have been related to tumorigenesis (Matés and Sánchez-Jiménez 2000). In this regard, some important genes are described below.

22.2.4.1 TP53 (Tumor Protein)

The *P53* gene, also called "the guardian of the genome," is mapped to chromosome 17 (17q21–q22). It regulates the cell cycle, apoptosis, DNA repair, transcription, genome stability, and senescence acting as a tumor suppressor gene (Rufini et al. 2013). In the cell, p53 protein binds to DNA, if the DNA is damaged by toxic chemicals, UV radiations, etc., and the protein plays a critical role in repairing the DNA and preventing the inheritance of this damaged DNA to next generation of cells. If the DNA cannot be repaired, this protein prevents the cell from dividing and signals it to undergo apoptosis. This p53 protein stimulates another gene to produce a protein called p21 that interacts with a cell division-stimulating protein (cdk2). When p21 is complexed with cdk2, the cell cannot pass through to the next stage of cell division. However, mutant p53 can no longer bind DNA in an effective way, and as a consequence, the p21 protein is not made available to act as the "stop signal" for cell division. Thus, cells divide uncontrollably and form tumors.

Germline polymorphisms are also important in the genesis of cancer, in addition to somatic *P53* mutations (Assumpção et al. 2008). *P53* polymorphism at codon 72 (Arg72Pro) –342 C/G polymorphism has been identified as a frequent variant that enhances cancer susceptibility (Liu et al. 2013). The substitution of a guanine base for cytosine in this codon leads to the replacement of an arginine amino acid to proline, which influences the activity of the resulting protein. The proline variant is effective in the repair of DNA damage, while the arginine variant leads to a strong induction of apoptosis. The outcomes of studies on the link between this polymorphism and cancer susceptibility and progression differ depending on ethnicity (O'Donnell and Dolan 2009). There was no detectable genotype of the *P53* polymorphism in a stomach cancer cohort in a Korean population when compared to various groups having or not having *H. pylori*-linked chronic gastritis (Kim et al. 2010). Similar findings were reported from a study on Iranian population. The genotypic frequencies of *P53* polymorphism in the patients of Iran with stomach cancer did not differ substantially compared to healthy controls (Hedayatizadeh Omran et al. 2018). The association between the polymorphism in codon 72 of *P53* gene and stomach cancer was reported in a Chinese population, demonstrating that this polymorphism was linked to the stage and location of stomach cancer (Zhou et al. 2010). Moreover, a meta-analysis involving 1,665 stomach cancer patients and 2,358 healthy controls found that advanced GC patients had high frequencies of the Arg/Arg allele, while patients with cardia stomach cancer had substantially higher frequencies of the Pro/Pro allele than healthy controls. Researchers came to the supposition that a polymorphism in the codon 72 of *P53* gene is linked to stomach cancer in Asian people (Malakar et al. 2014). Similar findings were conveyed in a meta-analysis carried out by Liu et al. (2012).

CASP8 (Cysteinyl aspartate-specific protease)

Caspases are a family of cysteinyl aspartate-specific proteases, which are synthesized as zymogens. They are primary regulating enzymes in the apoptosis cascade. CASP8 is located at chromosome 2q33–34 and was discovered as a component of the CD95 (Fas/APO-1) death-inducing signaling complex (DISC). Caspase 8 has a number of isoforms, but only two forms are predominantly expressed in different tissues and cell lines, i.e., procaspase 8a and procaspase 8b. They act as a main initiator caspase in death receptor-induced apoptosis. Procaspase 8 mainly localizes to the cytosol, close to membrane, mitochondria, and centrosomes. Caspase 8 plays a central role in death receptor-induced cell apoptosis (Jin et al. 2009). On stimulation, procaspase 8 is recruited to CD95 or DISC. Procaspase 8 activation involves several steps leading to the generation of the active caspase 8 heterotetramer $p10_2$–$p18_2$. Active caspase 8 can either directly cleave effector caspase 3 and 7, or cleaves bid, which eventually leads to the release of cytochrome c and apoptosome formation followed by cleavage of effector caspase 3, 6, and 7 by procaspase 9. In addition to apoptosis, CAS8 has a role in programmed necrosis as well. Various somatic mutations of CAS8 are identified in different carcinomas, and all lead to catalytically inactive form. Inactivating CAS8 mutations are detected at different sites in about 10% of advanced gastric carcinoma (Soung et al. 2005).

The *CASP8–652 6N* ins/del polymorphism (rs3834129) is a six-nucleotide insertion/deletion variant in the *CASP8* promoter region that causes *CASP8* expression to be reduced (Mc Fie et al. 2018). T lymphocyte "activation-induced cell death" (AICD) activity is reduced when caspase 8 function is impaired, which is critical in cancer cell immune surveillance. This SNP was highly related to GC in the additive model, with a 1.14-fold lower risk of GC in the study with 1,701 population size. By inactivating the binding site of transcription factor stimulatory protein 1105, the variation reduces *CASP8* transcription, potentially modifying immune surveillance and lowering the risk of GC (Vendramini-Costa and Carvalho 2012). Although this SNP was only studied in case-control studies, its link to GC risk was clearly established in the meta-analysis by Jie Tian et al. (2019). Another meta-analysis with the highest statistical power published to date looked at the connection between the *CASP8–652 6N* ins/del polymorphism and cancer risk in 49 studies with 33,494 cases and 36,397 controls (Jiarong et al. 2017).

22.2.5 Genetic Variations in Cell Signaling Pathways and Gastric Cancer Susceptibility

Cell signaling is a complex communication mechanism that governs and coordinates fundamental cellular activities (Wei et al. 2004). Development, tissue repair, immunology, and good tissue homeostasis are all dependent on a cell's ability to sense and respond appropriately to its surroundings.

Cell surface and transmembrane receptors, which are key components of signaling pathways, aid signal transmission. The tyrosine kinase receptor family encompasses a diverse range of cell surface receptors (RTKs). RTKs regulate a variety of essential cellular functions, including cell cycle, cell migration, cell metabolism, and cell survival, as well as cell proliferation (Schlessinger 2000). The *RAS/RAF/MEK/ERK* pathway, commonly known as the *mitogen-activated protein kinase (MAPK)* pathway, regulates essential biological processes such as cell proliferation, differentiation, survival, and death in physiological conditions (Kolch 2000). The occurrence and advancement of malignancies is caused by MAPK signaling dysfunction, which is mostly caused by somatic mutations (Rodriguez-Viciana et al. 2006).

22.2.5.1 RAS (rat sarcoma virus)

The Ras proto-oncogenes that encode small guanine nucleotide-binding proteins with intrinsic GTPase activity have been closely related to GC. Ras-regulated signal pathways govern actin cytoskeletal integrity, cell proliferation, differentiation, cell adhesion, apoptosis, and cell migration. Ras and Ras-related proteins are frequently unregulated in malignancies, resulting in enhanced invasion and metastasis as well as reduced apoptosis. The three Ras genes in humans (*HRAS*, *KRAS*, and *NRAS*) are the most common oncogenes in human cancer; mutations that permanently activate Ras are found in 20%–25% of all malignancies. *KRAS* is the most often mutant gene (20%), followed by *NRAS* (8%) and *HRAS* (7%) (Liu et al. 2018). H-ras is a member of the RAS gene family. This family of genes has a key role in the development and progression of many human cancers. An important SNP has been observed in H-ras gene, which is caused by point mutations in the coding nucleotide position 81 (Gonzalez-Hormazabal et al. 2019). This SNP may be a risk factor for some cancers, especially GA as previous described.

In a Chinese population, Zhang et al. discovered that *HRAS* rs12628 (or T81C) was strongly linked with GC under the dominant model (Zhang et al. 2008; Khan and Mudassar 2021). However, not much data are available to support the association between polymorphisms in the (*RTK/RAS/RAF/MEK/ERK*) signaling pathway and GC.

22.2.5.2 ERBB2 (V-erb-b2 erythroblastic leukemia viral oncogene homolog 2)

C-ERBB2, also known as HER2, is located on a chromosome 17q12. It encodes p185, a tyrosine kinase cell surface receptor structurally similar to human growth factor receptor (EGFR). CERBB2 is overexpressed in 8%–30% of GCs. C-ERBB2 expression is more common in intestinal type tumors and is a marker of invasion and metastasis. Somatic mutations within kinase domain have been reported in gastric tumors and in other solid tumors as well (Wang 2017).

The polymorphism *HER2* (I655V), found in the coding region of the transmembrane domain, is the most studied variation of this gene (Mutluhan et al. 2008). It causes a shift from A to G and is linked to an increased risk of cancer because it can disrupt the development of the receptor's active dimers. It has been observed that the overexpression of the HER2 membrane receptor occurs in between 13% and 25% of GC patients.

Studies of the *EGFR* gene have discovered a polymorphic variant (*EGFR-R521K*) resulting from a G for A substitution at codon 521 in the extracellular subdomain of the gene, which results in a change in amino acid arginine to a lysine (Kallel et al. 2009). The mutant allele (*EGFR-R521K*) is thought to inhibit tyrosine kinase activity, ligand binding, growth stimulation, and the production of several proto-oncogenes (Ruhe et al. 2007).

A total of 17 SNPs have been reported, and the polymorphism *HER2* (I655V), found in the coding region of the transmembrane domain, is the most studied variation of this gene (Mutluhan et al. 2008). It causes a shift from A to G and is linked to an increased risk of cancer because it can disrupt the development of the receptor's active dimers. It has been observed that the overexpression of the HER2 membrane receptor occurs in between 13% and 25% of GC patients. Another (2326G/A) transition that resulted in gly776ser substitution was found to be associated with gastric carcinoma by cancer genome project and collaborative group.

22.2.5.3 CDH1 (Cadherin 1, type1, E-cadherin epithelial)

Cell-cell adhesions play a key role in cancer invasion and metastasis, as well as controlling the morphogenesis of normal and malignant tissues (Martin et al. 2013). E-cadherin, also known as CDH1, having chromosomal locus 16q22.1, is a trans-membrane glycoprotein found in epithelial cells that play a role in calcium-dependent cell-cell adhesion (Canel et al. 2013). Continued expression and functional activity of E-cadherin are required for the cells to remain tightly associated in the epithelium. In several human carcinomas, *CDH1* has been identified as a crucial invasion suppressor gene, and its downregulation has been associated with tumor aggressiveness and metastatic potential. Approximately 30%–40% of HDGC kindred harbor germline mutations in CDHI.

The most widely studied polymorphic variant of CDH1 gene is the (C/A rs16260) SNP and is located at −160 from the *CDH1* gene promoter's transcriptional start point, and the −160A allele lowers the transcriptional efficiency by 68% in vitro (Nourbehesht 2016). The possibility of this mutation influencing stomach cancer risk has attracted people's interest. Despite the fact that a single included study found a substantial link between specific genotypes and stomach cancer susceptibility (Pharoah et al. 2004), the findings of these investigations were inconsistent, if not outright conflicting. The results are still debatable and confusing. The AA genotype was shown to be related to a considerably higher incidence of stomach cancer in a Mexican population, and a similar outcome was found in an Omani population in 2010. In a Taiwanese population, Wu et al. discovered that the prevalence of the variation AA genotype in GC patients was much lower than that in controls, conferring a fivefold decrease in the risk of GC relative to the C genotype. In a meta-analysis that included 4,218 cases and 5,461 controls, Jiang et al. studied the link between the *CDH1 C-160A* polymorphism and the risk of stomach cancer. The entire data analysis from 22 case-control studies revealed no link between the *CDH1 C-160A* polymorphism and the risk of stomach cancer (Jiang et al. 2015).

22.2.5.4 DNTM1 (DNA (cytosine-5)-methyltransferase 1) gene

DNMT1, found on human chromosome 19p13.2, encodes a 1,632-amino acid protein, which catalyzes the transfer of methyl groups to specific CpG structures in DNA and has been linked to the development of carcinoma (Klein et al. 2013). DNMT1 overexpression has been associated with an increased risk of GC (Tan and Yeoh 2015). The *AKT-NFB* and *STAT3* signaling pathways have been connected to DNMT1 overexpression, which has been associated with aberrant DNA methylation on tumor suppressor genes and linked to an increase in GC formation.

There was strong evidence for an association between SNP rs16999593 and GC risk. This SNP causes a histidine-to-arginine substitution at position 97 (His97Arg) in the translated sequence, which might impair DNMT1 function and thereby increase GC susceptibility. In a population of 2,647 Asians, there was substantial evidence for a connection between SNP rs16999593 and GC risk; however, this research sample was restricted to Asian people, including a substantial proportion of Chinese patients (Tian et al. 2019).

22.3 Conclusions

Human malignancies are influenced by genetic polymorphisms, and a close link between cancer and genetic polymorphism for tumor has been revealed in a number of studies. The findings from many studies suggest that SNPs in low penetrance genes could be key modulators of GC. The list of gene mutations in GC is continually growing. The development of stomach cancer is thought to necessitate an average of 4.18 genetic alterations. The SNPs could not only help in understanding the etiology better, but can lead to the identification of population groups who are susceptible to the disease. In this review, we have attempted to include results from meta-analyses, rather than individual studies reporting measurements that are expected to have some degree of error. It can also contradict results from different studies and identify pattern among study results. In one such meta-analysis limited to Asian population, nine variants of genes such as *APE*, *DNMT1*, *ERCC5*, *GSTT1* null/presence, *MDM2*, *PPARG*, *TLR4*, *IL-17F*, and *CASP8* were significantly associated with GC. The risk was 1.77 times more pronounced for the gene *APE*, with respect to G allele, and this variant is thought to promote GC by decreasing the DNA's ability to repair the damage. However, there is no evidence that polymorphisms in the *RTK* pathway are linked to GC. Inherited gene mutations, as in the *GSTM1 null phenotype* or the *CDH1* gene, have been linked to an increased risk of GC. Hereditary diffuse gastric cancer (HDGC) is caused by the loss of one of the *CDH1* gene copies, which is an autosomal dominantly inherited disorder in which malignant cells proliferate beneath the stomach lining and eventually undergo metastasis.

Polymorphisms in the interleukin genes *IL-17* and *IL-10*, which are particularly common in Asian people, are also linked to an increased risk of the neoplasm. Similarly, familial adenomatous polyposis (FAP), the most frequent form of familial intestine GC, is caused by germline mutations in the *APC* gene and is an autosomal dominant genetic susceptibility to adenomatous polyps. Nonetheless, the lifetime chance of getting stomach cancer in FAP patients is less than 1%, showing that the polyps must be transformed into ACs by an environmental factor. Another inherited condition that raises the risk of stomach cancer is Lynch syndrome. All of these genetic variations were found to play a role in 1%–3% of cases. The most important reason in regional variations in stomach cancer incidence is thought to have a strong environmental component. In most regions of the world, stomach cancer incidence and mortality rates have dropped in the previous four decades, according to a comprehensive assessment of 74 studies, while rates have increased in some indigenous population groups. Indigenous Siberians, Mapuche in South America, Inuit in Greenland, Canada and Alaska, and Maoris in New Zealand all have high incidences of stomach cancer. Most meta-analyses for the same polymorphism have recently produced conflicting results, raising the likelihood of false-positive relationships. Although large-scale meta-analyses may indicate actual relationships between genetic variations and GC risk, the validity and strength of these associations must be further evaluated.

ABBREVIATIONS

A:	Adenine
AC:	Adenocarcinoma
AICD:	Activation-induced cell death
AKT:	Serine-/threonine-specific protein kinase
Ala:	Alanine
APC:	Adenomatous polyposis coli
APE1:	Apurinic/apyrimidinic endonuclease
Arg:	Arginine
C:	Cytosine
CASP:	Caspase 8
CDH:	Cadherin
c-MET:	Tyrosine-protein kinase Met
CYP450:	Cytochrome P450
del:	Deletion
DNA:	Deoxyribonucleic acid
DNMT:	DNA methyltransferase
EGFR:	Epidermal growth factor receptor
ErbB:	Erythroblastic oncogene B
ERK:	Extracellular signal-regulated kinase
ErrB:	Erythroblastic leukemia viral oncogene
FAP:	Familial adenomatous polyposis
FGFR2:	Fibroblast growth factor receptor 2
G:	Guanine
GC:	Gastric cancer
GERD:	Gastroesophageal reflux disease
GIST:	Gastrointestinal stroma tumor
Gln:	Glutamine
GLOBOCAN:	Global Cancer Observatory
GNA14:	Guanine nucleotide-binding protein 14

GST: Glutathione S-transferase
GSTM: Glutathione S-transferase mu
GSTP: Glutathione S-transferase pi
GSTT: Glutathione S-transferase theta
GWAS: Genome-wide association studies
HDGC: Hereditary diffuse gastric cancer
HDI: Human Development Index
Her: Human epidermal growth factor receptor
His: Histidine
H. pylori: *Helicobacter pylori*
HRAS: Harvey rat sarcoma virus
HSP70: Heat shock protein 70
IL: Interleukin
ins: Insertion
KRAS: Kirsten rat sarcoma virus
Lys: Lysine
MAPK: Mitogen-activated protein kinase
MDM2: Mouse double minute 2
MEK: Mitogen-activated protein kinase
Met: Methionine
Mspl: Merozoite surface protein
NER: Nucleotide excision repair
neu: Neural
Noxo1: NADPH oxidase organizer 1
NRAS: Neuroblastoma rat sarcoma virus
PAHs: Polycyclic aromatic hydrocarbons
PDGFR: Platelet-derived growth factor receptor
PPARG: Peroxisome proliferator-activated receptor gamma
PRR: Pseudo-response regulator
Rad: Restriction site-associated DNA
RAF: Rapidly accelerated fibrosarcoma
RAS: Rat sarcoma virus
rs: Reference SNP cluster ID
RTK: Receptor tyrosine kinase
SNP: Single nucleotide polymorphism
STAT3: Signal transducer and activator of transcription
T: Thymine
Thr: Threonine
TLR: Toll-like receptor
TNF: Tumor necrosis factor
TP53: Tumor protein P53
Val: Valine
XPD: Xeroderma pigmentosum
XRCC: X-ray repair cross-complementing group

REFERENCES

Abdi E, Latifi-Navid S, Zahri S, Kholghi-Oskooei V, Mostafaiy B, Yazdanbod A and Pourfarzi F. SNP-SNP interactions of oncogenic long non-coding RNAs HOTAIR and HOTTIP on gastric cancer susceptibility. *Scientific Reports*. 2020; 10(1), pp. 1–12.

Ahmed AB, Zidi S, Sghaier I, Ghazouani E, Mezlini A, Almawi W and Loueslati BY. Common variants in IL-1RN, IL-1β and TNF-α and the risk of ovarian cancer: a case control study. *Central-European Journal of Immunology*. 2017; 42(2), p. 150.

Alzahrani S, Lina TT, Gonzalez J, Pinchuk IV, Beswick EJ and Reyes VE. Effect of Helicobacter pylori on gastric epithelial cells. *World Journal of Gastroenterology: WJG*. 2014; 20(36), p. 12767.

Andreassen CN, Alsner J and Overgaard J. Does variability in normal tissue reactions after radiotherapy have a genetic basis–where and how to look for it? *Radiotherapy and Oncology*. 2002; 64(2), pp. 131–40.

Annamaneni S, Gorre M, Kagita S, Addepalli K, Digumarti RR, Satti V and Battini MR. Association of XRCC1 gene polymorphisms with chronic myeloid leukemia in the population of Andhra Pradesh, India. *Hematology*. 2013; 18(3), pp. 163–8.

Antoni S, Ferlay J, Soerjomataram I, Znaor A, Jemal A and Bray F. Bladder cancer incidence and mortality: a global overview and recent trends. *European Urology*. 2017; 71(1), pp. 96–108.

Assumpção JG, Seidinger AL, Mastellaro MJ, Ribeiro RC, Zambetti GP, Ganti R, Srivastava K, Shurtleff S, Pei D, Zeferino LC and Dufloth RM. Association of the germline TP53 R337H mutation with breast cancer in southern Brazil. *BMC Cancer*. 2008; 8(1), pp. 1–6.

Au WW, Salama SA and Sierra-Torres CH. Functional characterization of polymorphisms in DNA repair genes using cytogenetic challenge assays. *Environmental Health Perspectives*. 2003; 111(15), pp. 1843–50.

Banerjee M and Saxena M. Interleukin-1 (IL-1) family of cytokines: role in type 2 diabetes. *Clinica Chimica Acta*. 2012; 413(15–16), pp. 1163–70.

Bhattacharjee P, Paul S, Banerjee M, Patra D, Banerjee P, Ghoshal N, Bandyopadhyay A and Giri AK. Functional compensation of glutathione S-transferase M1 (GSTM1) null by another GST superfamily member, GSTM2. *Scientific Reports*. 2013; 3(1), pp. 1–6.

Boccia S, Sayed-Tabatabaei FA, Persiani R, Gianfagna F, Rausei S, Arzani D, La Greca A, D'Ugo D, La Torre G, van Duijn CM and Ricciardi G. Polymorphisms in metabolic genes, their combination and interaction with tobacco smoke and alcohol consumption and risk of gastric cancer: a case-control study in an Italian population. *BMC Cancer*. 2007; 7(1), pp. 1–8.

Bornschein J and Malfertheiner P. Gastric carcinogenesis. *Langenbeck's Archives of Surgery*. 2011; 396(6), pp. 729–42.

Bray CM and West CE. DNA repair mechanisms in plants: crucial sensors and effectors for the maintenance of genome integrity. *New Phytologist*. 2005; 168(3), pp. 511–28.

Buchholz K, Antosik P, Grzanka D, Gagat M, Smolińska M, Grzanka A, Gzil A, Kasperska A and Klimaszewska-Wiśniewska A. Expression of the body-weight signaling players: GDF15, GFRAL and RET and their clinical relevance in gastric cancer. *Journal of Cancer*. 2021; 12(15), p. 4698.

Cai J., Ye Q., Luo S., Zhuang Z., He K., Zhuo Z., Wan X., Cheng J. *CASP8* -652 6N insertion/deletion polymorphism and overall cancer risk: evidence from 49 studies. *Oncotarget*. 2017; 8, pp. 56780–56790.

Caldecott KW. XRCC1 protein; form and function. *DNA Repair*. 2019; 81, p. 102664.

Calderwood SK, Khaleque MA, Sawyer DB and Ciocca DR. Heat shock proteins in cancer: chaperones of tumorigenesis. *Trends in Biochemical Sciences*. 2006; 31(3), pp. 164–72.

Canel M, Serrels A, Frame MC and Brunton VG. E-cadherin–integrin crosstalk in cancer invasion and metastasis. *Journal of Cell Science*. 2013; 126(2), pp. 393–401.

Castano-Rodriguez N, Kaakoush NO, Pardo AL, Goh KL, Fock KM and Mitchell HM. Genetic polymorphisms in the toll-like receptor signalling pathway in Helicobacter pylori infection and related gastric cancer. *Human Immunology.* 2014; 75(8), pp. 808–15.

Chaudhuri AR and Nussenzweig A. The multifaceted roles of PARP1 in DNA repair and chromatin remodelling. *Nature Reviews Molecular Cell Biology.* 2017; 18(10), pp. 610–21.

Chen B, Cao L, Zhou Y, Yang P, Wan HW, Jia GQ, Liu L and Wu XT. Glutathione S-transferase T1 (GSTT1) gene polymorphism and gastric cancer susceptibility: a meta-analysis of epidemiologic studies. *Digestive Diseases and Sciences.* 2010; 55(7), pp. 1831–8.

Chow HHS, Hakim IA, Vining DR, Crowell JA, Tome ME, Ranger-Moore J, Cordova CA, Mikhael DM, Briehl MM and Alberts DS. Modulation of human glutathione s-transferases by polyphenon e intervention. *Cancer Epidemiology and Prevention Biomarkers.* 2007; 16(8), pp. 1662–6.

Compe E and Egly JM. TFIIH: when transcription met DNA repair. *Nature Reviews Molecular Cell Biology.* 2012; 13(6), pp. 343–54.

Conteduca V, Sansonno D, Lauletta G, Russi S, Ingravallo G and Dammacco F. H. Pylori infection and gastric cancer: state of the art. *International Journal of Oncology.* 2013; 42(1), pp. 5–18.

Dasari S, Ganjayi MS, Yellanurkonda P, Basha S and Meriga B. Role of glutathione S-transferases in detoxification of a polycyclic aromatic hydrocarbon, methylcholanthrene. *Chemico-Biological Interactions.* 2018; 294, pp. 81–90.

D'Errico M, Parlanti E, Pascucci B, Fortini P, Baccarini S, Simonelli V and Dogliotti E. Single nucleotide polymorphisms in DNA glycosylases: from function to disease. *Free Radical Biology and Medicine.* 2017; 107, pp. 278–91.

De Oliveira JG, Rossi AFT, Nizato DM, Miyasaki K and Silva AE. Profiles of gene polymorphisms in cytokines and toll-like receptors with higher risk for gastric cancer. *Digestive Diseases and Sciences.* 2013; 58(4), pp. 978–88.

Dong JT. Prevalent mutations in prostate cancer. *Journal of Cellular Biochemistry.* 2006; 97(3), pp. 433–47.

Dulak AM, Schumacher SE, Van Lieshout J, Imamura Y, Fox C, Shim B, Ramos AH, Saksena G, Baca SC, Baselga J and Tabernero J. Gastrointestinal adenocarcinomas of the esophagus, stomach, and colon exhibit distinct patterns of genome instability and oncogenesis. *Cancer Research.* 2012; 72(17), pp. 4383–93.

Egly JM and Coin F. A history of TFIIH: two decades of molecular biology on a pivotal transcription/repair factor. *DNA Repair.* 2011; 10(7), pp. 714–21.

Epplein M, Xiang YB, Cai Q, Peek RM Jr, Li H, Correa P, Gao J, Wu J, Michel A, Pawlita M, Zheng W, Shu XO. Circulating cytokines and gastric cancer risk. Cancer Causes Control. 2013 Dec;24(12), pp. 2245–50. doi: 10.1007/s10552-013-0284-z. Epub 2013 Sep 20. PMID: 24052422; PMCID: PMC3828746.

Favatier F, Bornman L, Hightower LE, Günther E, Polla BS. Variation in hsp gene expression and Hsp polymorphism: do they contribute to differential disease susceptibility and stress tolerance? *Cell Stress & Chaperones.* 1997; 2(3), p. 141.

Ferrer-Ferrer M, Malespin-Bendana W, Ramirez V, González MI, Carvajal A and Une C. Polymorphisms in genes coding for HSP-70 are associated with gastric cancer and duodenal ulcer in a population at high risk of gastric cancer in Costa Rica. *Archives of Medical Research.* 2013; 44(6), pp. 467–74.

Fleming ND, Agadjanian H, Nassanian H, Miller CW, Orsulic S, Karlan BY and Walsh CS. Xeroderma pigmentosum complementation group C single-nucleotide polymorphisms in the nucleotide excision repair pathway correlate with prolonged progression-free survival in advanced ovarian cancer. *Cancer.* 2012; 118(3), pp. 689–97.

Fock KM. The epidemiology and prevention of gastric cancer. *Alimentary Pharmacology & Therapeutics.* 2014; 40(3), pp. 250–60.

García-González MA, Quintero E, Bujanda L, Nicolás D, Benito R, Strunk M, Santolaria S, Sopeña F, Badía M, Hijona E and Pérez-Aísa MA. Relevance of GSTM1, GSTT1, and GSTP1 gene polymorphisms to gastric cancer susceptibility and phenotype. *Mutagenesis.* 2012; 27(6), pp. 771–7.

Gattás GJF, de Carvalho MB, Siraque MS, Curioni OA, Kohler P, Eluf-Neto J and Wünsch-Filho V. Genetic polymorphisms of CYP1A1, CYP2E1, GSTM1, and GSTT1 associated with head and neck cancer. *Head & Neck: Journal for the Sciences and Specialties of the Head and Neck.* 2006; 28(-9), pp. 819–26.

Go MF. Natural history and epidemiology of Helicobacter pylori infection. *Alimentary Pharmacology & Therapeutics.* 2002; 16, pp. 3–15.

González CA and Agudo A. Carcinogenesis, prevention and early detection of gastric cancer: where we are and where we should go. *International Journal of Cancer.* 2012; 130(4), pp. 745–53.

Gonzalez-Hormazabal P, Musleh M, Bustamante M, Stambuk J, Pisano R, Valladares H, Lanzarini E, Chiong H, Rojas J, Suazo J and Castro VG. Polymorphisms in RAS/RAF/MEK/ERK pathway are associated with gastric cancer. *Genes.* 2019; 10(1), p. 20.

Goode EL, Ulrich CM and Potter JD. Polymorphisms in DNA repair genes and associations with cancer risk. *Cancer Epidemiology and Prevention Biomarkers.* 2002; 11(12), pp. 1513–30.

Gorouhi F., Islami F., Bahrami H., Kamangar F. Tumour-necrosis factor-A polymorphisms and gastric cancer risk: a meta-analysis. *British Journal of Cancer.* 2008;98(8):1443–1451.

Gsur A, Haidinger G, Hinteregger S, Bernhofer G, Schatzl G, Madersbacher S, Marberger M, Vutuc C and Micksche M. Polymorphisms of glutathione-S-transferase genes (GSTP1, GSTM1 and GSTT1) and prostate-cancer risk. *International Journal of Cancer.* 2001; 95(3), pp. 152–5.

Gujar AA, Gujar GA, Veeramachaneni R and Indurkar PS. Endoscopic early detection of premalignant changes in stomach with mucosal biopsy in patients with symptomatology of acid peptic disease. *International Journal of Research in Medical Sciences.* 2016; 4(2), p. 533.

Gullo I, Grillo F, Mastracci L, Vanoli A, Carneiro F, Saragoni L, Limarzi F, Ferro J, Parente P and Fassan M. Precancerous lesions of the stomach, gastric cancer and hereditary gastric cancer syndromes. *Pathologica-Journal of the Italian Society of Anatomic Pathology and Diagnostic Cytopathology.* 2020; 112(3), pp. 166–85.

Gullo, I., Oliveira, C., van der Post, R. S., van Dieren, J. M., Davis, J. L., Bougen-Zhukov, N., Guilford, P., & Carneiro,

F. (2021). Updated perspective and directions on hereditary diffuse gastric cancer. In B. Jenkins (Ed.), *Research and clinical applications of targeting gastric neoplasms*. (pp. 217–258). London, UK: Elsevier. doi: 10.1016/B978-0-323-85563-1.00011-3

Gupta S and Shaukat A. Management of intestinal metaplasia and gastric cancer. *Practical Gastroenterology*. 2021, p. 29.

Harro J. Inter-individual differences in neurobiology as vulnerability factors for affective disorders: implications for psychopharmacology. *Pharmacology & Therapeutics*. 2010; 125(3), pp. 402–22.

Hasselbalch HC. Perspectives on chronic inflammation in essential thrombocythemia, polycythemia vera, and myelofibrosis: is chronic inflammation a trigger and driver of clonal evolution and development of accelerated atherosclerosis and second cancer? *Blood, The Journal of the American Society of Hematology*. 2012; 119(14), pp. 3219–25.

Hatagima A. Genetic polymorphisms and metabolism of endocrine disruptors in cancer susceptibility. *Cadernos de saude publica*. 2002; 18(2), pp. 357–77.

Hayakawa Y, Sethi N, Sepulveda AR, Bass AJ and Wang TC. Oesophageal adenocarcinoma and gastric cancer: should we mind the gap? *Nature Reviews Cancer*. 2016; 16(5), pp. 305–18.

He L, Deng T, Luo HS. Heat shock protein 70 gene polymorphisms and cancer risk: a meta-analysis. *The Scientific World Journal*. 2014; 2014. Article ID 540309, 9 pages, 2014. https://doi.org/10.1155/2014/540309

Hedayatizadeh Omran A, Alizadeh Navaei R, Janbabaei G, Omrani Nava V, Hasheminasab Y, Amjadi O and Tehrani M. Association of P53 gene polymorphism with gastric cancer in Northern Iran as a high risk region. *Biomedical Reports*. 2018; 8(5), pp. 433–8.

Hsu HJ, Yang YH, Shieh TY, Chen CH, Kao YH, Yang CF and Ko ECC. TGF-β1 and IL-10 single nucleotide polymorphisms as risk factors for oral cancer in Taiwanese. *The Kaohsiung Journal of Medical Sciences*. 2015; 31(3), pp. 123–9.

Huang J, Liu X, Tang LL, Long JT, Zhu J, Hua RX, Li J. XPG gene polymorphisms and cancer susceptibility: evidence from 47 studies. *Oncotarget*. 2017; 8(23), p. 37263.

Jiang B, Zhu K, Shao H, Bao C, Ou J and Sun W. Lack of association between the CDH1 polymorphism and gastric cancer susceptibility: a meta-analysis. *Scientific Reports*. 2015; 5(1), pp. 1–6.

Jin Z, Li Y, Pitti R, Lawrence D, Pham VC, Lill JR and Ashkenazi A. Cullin3-based polyubiquitination and p62-dependent aggregation of caspase-8 mediate extrinsic apoptosis signaling. *Cell*. 2009; 137(4), pp. 721–35.

Kaisho T and Akira S. Toll-like receptor function and signaling. *Journal of Allergy and Clinical Immunology*. 2006; 117(5), pp. 979–87.

Kallel I, Rebai M, Khabir A, Farid NR, Rebaï A. Genetic polymorphisms in the EGFR (R521K) and estrogen receptor (T594T) genes, EGFR and ErbB-2 protein expression, and breast cancer risk in Tunisia. J Biomed Biotechnol. 2009;2009:753683.

Kampinga HH, Hageman J, Vos MJ, Kubota H, Tanguay RM, Bruford EA, Cheetham ME, Chen B, Hightower LE. Guidelines for the nomenclature of the human heat shock proteins. *Cell Stress and Chaperones*. 2009; 14(1), pp. 105–11.

Ketterer, B. (1988). Protective role of glutathione and glutathione transferases in mutagenesis and carcinogenesis. Mutat. *Res.* **202**, 343–361.

Khan, M.S., Mudassar, S. (2021). Thyroid Cancer and SNPs. In: Sameer, A.S., Banday, M.Z., Nissar, S. (eds) Genetic Polymorphism and cancer susceptibility. Springer, Singapore. https://doi.org/10.1007/978-981-33-6699-2_8

Khanna P, Chua PJ, Bay BH and Baeg GH. The JAK/STAT signaling cascade in gastric carcinoma. *International Journal of Oncology*. 2015; 47(5), pp. 1617–26.

Kannan K, Amariglio N, Rechavi G, et al (2000). Pro le of gene expression regulated by induced p53: connection to the TGF-beta family. *FEBS Lett*, **470**, 77–82.

Khatoon J, Rai RP and Prasad KN. Role of Helicobacter pylori in gastric cancer: updates. *World Journal of Gastrointestinal Oncology*. 2016; 8(2), p. 147.

Kim N, Cho SI, Lee HS, Park JH, Kim JH, Kim JS, Jung HC and Song IS. The discrepancy between genetic polymorphism of p53 codon 72 and the expression of p53 protein in Helicobacter pylori-associated gastric cancer in Korea. *Digestive Diseases and Sciences*. 2010; 55(1), p. 101.

Kim HJ, Hwang NR, Lee KJ. Heat shock responses for understanding diseases of protein denaturation. Mol Cells. 2007 Apr 30;23(2):123-31. PMID: 17464187.

Klein CJ, Bird T, Ertekin-Taner N, Lincoln S, Hjorth R, Wu Y, Kwok J, Mer G, Dyck PJ and Nicholson GA. DNMT1 mutation hot spot causes varied phenotypes of HSAN1 with dementia and hearing loss. *Neurology*. 2013; 80(9), pp. 824–8.

Kolch W. Meaningful relationships: the regulation of the Ras/Raf/MEK/ERK pathway by protein interactions. *Biochemical Journal*. 2000; 351(2), pp. 289–305.

Kong W, Yin G, Zheng S, Liu X, Zhu A, Yu P, Zhang J, Shan Y, Ying R, Jin H. Long noncoding RNA (lncRNA) HOTAIR: Pathogenic roles and therapeutic opportunities in gastric cancer. Genes Dis. 2021 Aug 21;9(5):1269–1280.

Koosirirat C, Linpisarn S, Changsom D, Chawansuntati K and Wipasa J. Investigation of the anti-inflammatory effect of Curcuma longa in Helicobacter pylori-infected patients. *International Immunopharmacology*. 2010; 10(7), pp. 815–8.

Kumar S, Stokes III J, Singh UP, Gunn KS, Acharya A, Manne U and Mishra M. Targeting Hsp70: a possible therapy for cancer. *Cancer Letters*. 2016; 374(1), pp. 156–66.

van Rossum EF, Lamberts SW. Polymorphisms in the glucocorticoid receptor gene and their associations with metabolic parameters and body composition. *Recent Progress in Hormone Research*. 2004;59:333-57. doi: 10.1210/rp.59.1.333. PMID: 14749509.

Li W and Ma H. Double-stranded DNA breaks and gene functions in recombination and meiosis. *Cell Research*. 2006; 16(5), pp. 402–12.

Li W, Ng JMK, Wong CC, Ng EKW and Yu J. Molecular alterations of cancer cell and tumour microenvironment in metastatic gastric cancer. *Oncogene*. 2018; 37(36), pp. 4903–20.

Liu KJ, Qi HZ, Yao HL, Lei SL, Lei ZD, Li TG, Zhao H. An updated meta-analysis of the p53 codon 72 polymorphism and gastric cancer risk. *Molecular Biology Reports*. 2012; 39. pp. 8265–75. doi: 10.1007/s11033-012-1674-0.

Liu J, Tang X, Li M, Lu C, Shi J, Zhou L, Yuan Q and Yang M. Functional MDM4 rs4245739 genetic variant, alone and in combination with P53 Arg72Pro polymorphism, contributes to breast cancer susceptibility. *Breast Cancer Research and Treatment*. 2013; 140(1), pp. 151–7.

Liu F, Yang X, Geng M and Huang M. Targeting ERK, an Achilles' Heel of the MAPK pathway, in cancer therapy. *Acta Pharmaceutica Sinica B*. 2018; 8(4), pp. 552–62.

Luo H, Li Z, Qing Y, Zhang SH, Peng Y, Li Q and Wang D. Single nucleotide polymorphisms of DNA base-excision repair genes (APE1, OGG1 and XRCC1) associated with breast cancer risk in a Chinese population. *Asian Pacific Journal of Cancer Prevention*. 2014; 15(3), pp. 1133–40.

Ma H, Yu H, Liu Z, et al. (2012). Polymorphisms of XPG/ERCC5 and risk of squamous cell carcinoma of the head and neck. *Pharmacogenet Genomics*, **22**, 50–7.

Ma J, Wu D, Hu X, Li J, Cao M and Dong W. Associations between cytokine gene polymorphisms and susceptibility to Helicobacter pylori infection and Helicobacter pylori related gastric cancer, peptic ulcer disease: a meta-analysis. *PLoS One*. 2017; 12(4), p. e0176463.

Malakar M, Devi KR, Phukan RK, Kaur T, Deka M, Puia L, Sailo L, Lalhmangaihi T, Barua D, Rajguru SK and Mahanta J. p53 codon 72 polymorphism interactions with dietary and tobacco related habits and risk of stomach cancer in Mizoram, India. *Asian Pacific Journal of Cancer Prevention*. 2014; 15(2), pp. 717–23.

Malfertheiner P, Sipponen P, Naumann M, Moayyedi P, Mégraud F, Xiao SD, Sugano K and Nyrén O. Helicobacter pylori eradication has the potential to prevent gastric cancer: a state-of-the-art critique. *American College of Gastroenterology| ACG*. 2005; 100(9), pp. 2100–15.

Marin JJG, Al-Abdulla R, Lozano E, Briz O, Bujanda L, M Banales J and Macias RIR. Mechanisms of resistance to chemotherapy in gastric cancer. *Anti-Cancer Agents in Medicinal Chemistry (Formerly Current Medicinal Chemistry-Anti-Cancer Agents)*. 2016; 16(3), pp. 318–34.

Martin TA, Ye L, Sanders AJ, Lane J and Jiang WG. Cancer invasion and metastasis: molecular and cellular perspective. *Madame Curie Bioscience Database* [Internet]. Landes Bioscience. 2013.

Matés JM and Sánchez-Jiménez FM. Role of reactive oxygen species in apoptosis: implications for cancer therapy. *The International Journal of Biochemistry & Cell Biology*. 2000; 32(2), pp. 157–70.

Matullo G, Guarrera S, Carturan S, Peluso M, Malaveille C, Davico L, Piazza A and Vineis P. DNA repair gene polymorphisms, bulky DNA adducts in white blood cells and bladder cancer in a case-control study. *International Journal of Cancer*. 2001; 92(4), pp. 562–7.

Mc Fie S, Abrahams S, Patricios J, Suter J, Posthumus M and September AV. Inflammatory and apoptotic signalling pathways and concussion severity: a genetic association study. *Journal of Sports Sciences*. 2018; 36(19), pp. 2226–34.

McLean MH and El-Omar EM. Genetics of gastric cancer. *Nature Reviews Gastroenterology & hepatology*. 2014; 11(11), pp. 664–74.

McLellan RA, Oscarson M, Seidegård J, Evans DA, Ingelman-Sundberg M. Frequent occurrence of CYP2D6 gene duplication in Saudi Arabians. Pharmacogenetics. 1997 Jun;7(3):187–91. doi: 10.1097/00008571-199706000-00003. PMID: 9241658.

Meliţ LE, Mărginean CO, Mărginean CD and Mărginean MO. "The relationship between toll-like receptors and Helicobacter pylori-related gastropathies: still a controversial topic." "*Journal of Immunology Research*. 2019. Article ID 8197048, 10pages, 2019. https://doi.org/10.1155/2019/8197048

Monroy CM, Cortes AC, Lopez M, Rourke E, Etzel CJ, Younes A, Strom SS and El-Zein R. Hodgkin lymphoma risk: role of genetic polymorphisms and gene–gene interactions in DNA repair pathways. *Molecular Carcinogenesis*. 2011; 50(11), pp. 825–34.

Mutluhan H, Akbas E, Erdogan NE, Soylemez F, Senli MS, Polat A, Helvacı I and Seyrek E. The influence of HER2 genotypes as molecular markers on breast cancer outcome. *DNA and Cell Biology*. 2008; 27(10), pp. 575–9.

Namazi A, Forat-Yazdi M, Jafari MA, Foroughi E, Farahnak S, Nasiri R, Zare-Shehneh M, Neamatzadeh H. Association between polymorphisms of ERCC5 gene and susceptibility to gastric cancer: a systematic review and meta-analysis. *Asian Pacific Journal of Cancer Prevention: APJCP*. 2017; 18(10), p. 2611.

Neamatzadeh H, Soleimanizad R, Atefi A, et al (2015). Association between p53 codon 72 (Arg72Pro) polymorphism and primary open-angle glaucoma in Iranian patients. *Iran Biomed J*, **19**, 51–6.

Nishizawa D, Kasai S, Hasegawa J, Sato N, Yamada H, Tanioka F, Nagashima M, Katoh R, Satoh Y, Tagami M and Ujike H. Associations between the orexin (hypocretin) receptor 2 gene polymorphism Val308Ile and nicotine dependence in genome-wide and subsequent association studies. *Molecular Brain*. 2015; 8(1), pp. 1–17.

Nourbehesht N. Investigating genetic aberrations in metastatic gastric cancer and in primary renal cell carcinoma. McGill University (Canada). 2016.

O'Donnell PH and Dolan ME. Cancer pharmacoethnicity: ethnic differences in susceptibility to the effects of chemotherapy. *Clinical Cancer Research*. 2009; 15(15), pp. 4806–14.

Palli D, Polidoro S, D'Errico M, Saieva C, Guarrera S, Calcagnile AS, Sera F, Allione A, Gemma S, Zanna I and Filomena A. Polymorphic DNA repair and metabolic genes: a multigenic study on gastric cancer. *Mutagenesis*. 2010; 25(6), pp. 569–75.

Park GT, Lee OY, Kwon SJ, Lee CG, Yoon BC, Hahm JS, Lee MH, Hoo Lee D, Kee CS, Sun HS. Analysis of CYP2E1 polymorphism for the determination of genetic susceptibility to gastric cancer in Koreans. *Journal of Gastroenterology and Hepatology*. 2003; 18(11), pp. 1257–63.

Parsonnet J, Hansen S, Rodriguez L, Gelb AB, Warnke RA, Jellum E, Orentreich N, Vogelman JH and Friedman GD. Helicobacter pylori infection and gastric lymphoma. *New England Journal of Medicine*. 1994; 330(18), pp. 1267–71.

Pharoah PD, Dunning AM, Ponder BA and Easton DF. Association studies for finding cancer-susceptibility genetic variants. *Nature Reviews Cancer*. 2004; 4(11), pp. 850–60.

Piazuelo MB and Correa P. Gastric cancer: overview. *Colombia Medica*. 2013; 44(3), pp. 192–201.

Preissner SC, Hoffmann MF, Preissner R, Dunkel M, Gewiess A and Preissner S. Polymorphic cytochrome P450 enzymes (CYPs) and their role in personalized therapy. *PLoS One*. 2013; 8(12), p. e82562.

Qidwai T and Khan F. Tumour necrosis factor gene polymorphism and disease prevalence. *Scandinavian Journal of Immunology*. 2011; 74(6), pp. 522–47.

Rafiei A, Hosseini V, Janbabai G, Ghorbani A, Ajami A, Farzmandfar T, Azizi MD, Gilbreath JJ and Merrell DS.

Polymorphism in the interleukin-17A promoter contributes to gastric cancer. *World Journal of Gastroenterology: WJG.* 2013; 19(34), p. 5693.

Rahman R, Asombang AW and Ibdah JA. Characteristics of gastric cancer in Asia. *World Journal of Gastroenterology: WJG.* 2014; 20(16), p. 4483.

Rawla P and Barsouk A. Epidemiology of gastric cancer: global trends, risk factors and prevention. *Przeglad gastroenterologiczny.* 2019; 14(1), p. 26.

Rodriguez-Viciana P, Tetsu O, Tidyman WE, Estep AL, Conger BA, Santa Cruz M, McCormick F and Rauen KA. Germline mutations in genes within the MAPK pathway cause cardio-facio-cutaneous syndrome. *Science.* 2006; 311(5765), pp. 1287–90.

Rong G, Zhu Y, Tang W, Qiu H and Zhang S. The correlation of microRNA-499 rs3746444 T> C locus with the susceptibility of gastric cancer: from a case–control study to a meta-analysis. *Bioscience Reports.* 2021; 41(1), p. BSR20203461.

Roth MJ, Abnet CC, Johnson LL, Mark SD, Dong ZW, Taylor PR, Dawsey SM, Qiao YL. Polymorphic variation of Cyp1A1 is associated with the risk of gastric cardia cancer: a prospective case–cohort study of cytochrome P-450 1A1 and GST enzymes. *Cancer Causes & Control.* 2004;15(10), pp. 1077–83.

Ruddon RW. *Cancer Biology.* Oxford University Press. 2007.

Rufini A, Tucci P, Celardo I and Melino G. Senescence and aging: the critical roles of p53. *Oncogene.* 2013; 32(43), pp. 5129–43.

Rugge M, Capelle LG, Cappellesso R, Nitti D and Kuipers EJ. Precancerous lesions in the stomach: from biology to clinical patient management. *Best Practice & Research Clinical Gastroenterology.* 2013; 27(2), pp. 205–23.

Ruhe JE, Streit S, Hart S, Wong CH, Specht K, Knyazev P, Knyazeva T, Tay LS, Loo HL, Foo P and Wong W. Genetic alterations in the tyrosine kinase transcriptome of human cancer cell lines. *Cancer Research.* 2007; 67(23), pp. 11368–76.

Schlessinger J. Cell signaling by receptor tyrosine kinases. *Cell.* 2000; 103(2), pp. 211–25.

Shin WG, Jang JS, Kim HS, Kim SJ, Kim KH, Jang MK, Lee JH, Kim HJ and Kim HY. Polymorphisms of interleukin-1 and interleukin-2 genes in patients with gastric cancer in Korea. *Journal of Gastroenterology and Hepatology.* 2008; 23(10), pp. 1567–73.

Soung YH, Lee JW, Kim SY, Sung YJ, Park WS, Nam SW, Kim SH, Lee JY, Yoo NJ, Lee SH. Caspase-8 gene is frequently inactivated by the frameshift somatic mutation 1225_1226delTG in hepatocellular carcinomas. *Oncogene.* 2005; 24(1), pp. 141–7.

Soya SS, Padmaja N and Adithan C. Genetic polymorphisms of CYP2E1and GSTP1in a South Indian population-comparison with North Indians, Caucasians and Chinese. *Asian Pacific Journal of Cancer Prevention.* 2005; 6, pp. 315–9.

Spitz MR, Wu X, Wang Y, Wang LE, Shete S, Amos CI, Guo Z, Lei L, Mohrenweiser H and Wei Q. Modulation of nucleotide excision repair capacity by XPD polymorphisms in lung cancer patients. *Cancer Research.* 2001; 61(4), pp. 1354–7.

Suganuma, M., Watanabe, T., Yamaguchi, K., Takahashi, A., and Fujiki, H. (2012). Human gastric cancer development with TNF-alpha-inducing protein secreted from *Helicobacter pylori. Cancer Lett.* 322, 133–138. doi: 10.1016/j.canlet.2012.03.027.

Sugasawa K. Xeroderma pigmentosum genes: functions inside and outside DNA repair. *Carcinogenesis.* 2008; 29(3), pp. 455–65.

Sugimoto M, Furuta T, Shirai N, Nakamura A, Xiao F, Kajimura M, Sugimura H and Hishida A. Different effects of polymorphisms of tumor necrosis factor-alpha and interleukin-1 beta on development of peptic ulcer and gastric cancer. *Journal of Gastroenterology and Hepatology.* 2007; 22(1), pp. 51–9.

Sung H, Ferlay J, Siegel RL, Laversanne M, Soerjomataram I, Jemal A and Bray F. Global cancer statistics 2020: GLOBOCAN estimates of incidesnce and mortality worldwide for 36 cancers in 185 countries. *CA: A Cancer Journal for Clinicians.* 2021; 71(3), pp. 209–49.

Tan P and Yeoh KG. Genetics and molecular pathogenesis of gastric adenocarcinoma. *Gastroenterology.* 2015; 149(5), pp. 1153–62.

Tell G, Fantini D, Quadrifoglio F. Understanding different functions of mammalian AP endonuclease (APE1) as a promising tool for cancer treatment. *Cellular and Molecular Life Sciences.* 2010; 67(21), pp. 3589–608.

Tian J, Liu G, Zuo C, Liu C, He W and Chen H. Genetic polymorphisms and gastric cancer risk: a comprehensive review synopsis from meta-analysis and genome-wide association studies. *Cancer Biology & Medicine.* 2019; 16(2), p. 361–389.

Vargas-Alarcón G, Juárez-Cedillo E, Martínez-Rodríguez N, Fragoso JM, García-Hernández N and Juárez-Cedillo T. Association of interleukin-10 polymorphisms with risk factors of Alzheimer's disease and other dementias (SADEM study). *Immunology Letters.* 2016; 177, pp. 47–52.

Vendramini-Costa BD and Carvalho EJ. Molecular link mechanisms between inflammation and cancer. *Current Pharmaceutical Design.* 2012; 18(26), pp. 3831–52.

Wang Z. ErbB receptors and cancer. *ErbB Receptor Signaling.* 2017, pp. 3–35.

Wang M, Chu H, Zhang Z, et al. (2013). Molecular epidemiology of DNA repair gene polymorphisms and head and neck cancer. *J Biomed Res*, **27**, 179–92.

Wang P, Xia HHX, Zhang JY, Dai LP, Xu XQ and Wang KJ. Association of interleukin-1 gene polymorphisms with gastric cancer: a meta-analysis. *International Journal of Cancer.* 2007; 120(3), pp. 552–62.

Wei CJ, Xu X and Lo CW. Connexins and cell signaling in development and disease. *Annual Review of Cell and Developmental Biology.* 2004; 20, pp. 811–38.

Widersten M., Holmström E., Mannervik B., Cysteine residues are not essential for the catalytic activity of human class Mu glutathione transferase M1a-1a, *FEBS Letters.* 1991, Volume 293, Issues 1–2, 1991, pp. 156–159, ISSN 0014-5793, https://doi.org/10.1016/0014-5793(91)81175-8.

Widersten M., Holmström E., Mannervik B., Cysteine residues are not essential for the catalytic activity of human class Mu glutathione transferase M1a-1a, FEBS Letters. 1991, Volume 293, Issues 1–2, 1991, pp. 156–159, ISSN 0014-5793, https://doi.org/10.1016/0014-5793(91)81175-8.

Witte JS. Genome-wide association studies and beyond. *Annual Review of Public Health.* 2010; 31, pp. 9–20.

Xu T, Kong Z, Zhao H. Relationship Between Tumor Necrosis Factor-α rs361525 Polymorphism and Gastric Cancer Risk: A Meta-Analysis. Front Physiol. 2018 May 15;9:469. doi: 10.3389/fphys.2018.00469. PMID: 29867530; PMCID: PMC5962813.

Xue H, Lu Y, Xue Z, Lin B, Chen J, Tang F, Huang G. The effect of CYP1A1 and CYP1A2 polymorphisms on gastric cancer risk among different ethnicities: a systematic review and meta-analysis. *Tumor Biology*. 2014;35(5), pp. 4741–56.

Yang L, Zheng R, Wang N, Yuan Y, Liu S, Li H, Zhang S, Zeng H and Chen W. Incidence and mortality of stomach cancer in China, 2014. *Chinese Journal of Cancer Research*. 2018; 30(3), p. 291.

Yuan K, Huo M, Sun Y, Wu H, Chen H, Wang Y and Fu R. Association between x-ray repair cross-complementing group 3 (XRCC3) genetic polymorphisms and papillary thyroid cancer susceptibility in a Chinese Han population. *Tumor Biology*. 2016; 37(1), pp. 979–87.

Zabaleta J. Multifactorial etiology of gastric cancer. *Cancer Epigenetics*. 2012, pp. 411–35.

Zeyaullah, M & AlShahrani, Abdullah & Ahmad, Irfan. (2021). Association of Helicobacter pylori Infection and Host Cytokine Gene Polymorphism with Gastric Cancer. *Canadian Journal of Gastroenterology and Hepatology*. 2021; 1–12. 10.1155/2021/8810620.

Zhang, Yongjing & Jin, Mingjuan & Liu, Bing & Ma, Xinyuan & Yao, Kaiyan & Li, Qilong. (2008). Association between H-RAS T81C genetic polymorphism and gastrointestinal cancer risk: A population based case-control study in China. *BMC Cancer*. 8. 256. 10.1186/1471-2407-8-256.

Zhou Y, Li N, Zhuang W and Wu X. p53 Codon 72 polymorphism and gastric cancer risk in a Chinese Han population. *Genetic Testing and Molecular Biomarkers*. 2010; 14(6), pp. 829–83.

23

Cytokine Gene Polymorphisms and Their Role in Autoimmune Diseases

Kulsum Akhtar and Ishrat Parveiz Bhat
Department of Clinical Biochemistry, Sher-I-Kashmir Institute of Medical Sciences, Soura, Srinagar, India

Syed Sameer Aga
Basic Medical Sciences, College of Medicine, King Saud bin Abdulaziz University for Health Sciences (KSAU-HS), King Abdullah International Medical Research Center (KAIMRC), Ministry of National Guard Health Affairs (NGHA), Jeddah, Kingdom of Saudi Arabia
Infinity Biochemistry Pvt. Ltd. Srinagar, Kashmir, India

Syed Douhath Yousuf and Syed Mudassar
Department of Clinical Biochemistry, Sher-I-Kashmir Institute of Medical Sciences, Soura, Srinagar, India

CONTENTS

ABSTRACT Cytokines are small non-structural proteins with low molecular weights that regulate inflammation and immunity in a complex manner. It has long been assumed that haematopoietic cells, lymphoid cells, and different pro-inflammatory and anti-inflammatory cells are involved in the formation of the immunological and inflammatory response and that cytokines mediate the intricate interactions of these cells. Cytokines act as the intercellular messengers of the immune system. Cytokines have evolved over the years and now include the interferons, the interleukins, the chemokines, the tumour necrosis factor family, and adipokines. Cytokines, formerly known as "soluble factors," are produced by and elicit a response from every cell except the red blood cell. Pleiotropism is the hallmark of a cytokine. Beyond innate and adaptive immunity, cytokines have a significant role in various functions, including immune cell differentiation, inflammation, tumourigenesis, and viral pathogenesis. In addition to inflammation, immunity, and infections, cytokines have now extended their scope to atherosclerosis and cancer. As a result, cytokines may serve as diagnostic, prognostic, and therapeutic biomarkers for health and disease. Cytokines play a significant role in the pathogenesis of autoimmune diseases. For most naturally occurring autoimmune diseases, the precise triggers for the breakdown of self-tolerance and the subsequent steps leading to the occurrence of pathogenic autoimmune responses are still unknown.

KEYWORDS: *Cytokines; Tumour Necrosis Factor; Adipokines; Pleiotropism; Autoimmune Diseases.*

23.1 Introduction

Cytokines are low-molecular-weight (≈6–70 kDa), soluble proteins secreted by a range of cells such as lymphocytes, macrophages, natural killer (NK) cells, and mast cells. They play an essential role in the immune response and act as indispensable mediators associated with the communication network of the

DOI: 10.1201/9781003246244-23

immune system (O'Shea and Murray, 2008; Kulbe et al., 2012). A single cytokine can be secreted by a variety of cell types and operate on several cell types, resulting in multiple biological actions (Sprague and Khalil, 2009). In the last 25 years, cytokines have secured a vital place as diagnostic, prognostic, and therapeutic agents in human disease. Even though cytokines are being studied in almost every biological discipline, the domains of inflammation, immunology, atherosclerosis, and cancer are dominated by cytokine-mediated effects. Chemokines and their receptors, for example, have influenced inflammation, HIV-1 pathogenesis, lymphocyte trafficking, and autoimmunity. Some chronic diseases appear to be fuelled by "autoinflammatory" pathways, while others display traditional "autoimmune" characteristics (Jin et al., 2007). Cytokines play a role in stem cell differentiation, vaccine efficacy, and allograft rejection. Pleiotropism, or the ability to have multiple biological properties, is the hallmark of cytokines.

The word "cytokine" includes interferons, interleukins, the chemokine family, mesenchymal growth factors, the tumour necrosis factor family, and adipokines (Dinarello, 2007). Cytokines can be classified according to their functions. Some cytokines, for example, are lymphocyte growth factors, while others serve as pro-inflammatory or anti-inflammatory molecules, while some others polarise the immune response to antigen (Dinarello, 2007). Cytokine and its receptor genes have sparked great interest as potential genetic risk factors for autoimmune diseases. Autoimmune diseases occur by an abnormal immune response against innocent substances present in our body. As multifactorial diseases, they are distinguished by modest genetic risk heritability and complex interactions with a variety of environmental influences. Human autoimmune disorders are a set of pathologies in which the immune system responds to its antigens, causing clinical damage or destruction of body tissue (Kunz and Ibrahim, 2009; Wang et al., 2015). SSC (systemic sclerosis), JIA (juvenile idiopathic arthritis), BD (Behcet's disease), RA (rheumatoid arthritis), MS (multiple sclerosis), GD (Graves' disease), SLE (systemic lupus erythematosus), and TID (type 1 diabetes) are examples of autoimmune disorders (Kunz and Ibrahim, 2009; Wang et al., 2015).

23.2 History of Cytokines

Before the emergence of receptors and signalling cascades, cytokines emerged from their earliest forms as intracellular molecules. Invertebrates such as starfish and Drosophila have been shown to exhibit cytokine-like activities, and they play an important role in host defence and repair. Poikilothermic lizards have been shown to have a cytokine-mediated rise in body temperature as a survival mechanism (Bernheim and Kluger, 1976). Some cytokines operate as transcription factors as well as extracellular ligands for certain receptors. The infection has wreaked havoc on societies throughout history, and one could argue that cytokine-mediated events are at the root of the pathological processes of these epidemics. The generation of "pus" by the host is the source of cytokine biology. Exudates containing pus, as well as the presence of fever, local swelling, and pain, were well described by ancient writers even before the microscope. Pus was visible and could be analysed later in the lab. The field began with an interest in insoluble "factors," as they were called, which were examined as products of white blood cells (pus) in the mid-1940s. Most cytokines are soluble substances produced by one cell that acts on another cell. Today, we know that cytokines can also serve as integral membrane proteins and that some of them are never released from the cell. Initially, cytokines were referred to as "lymphokines" to separate them from "monokines" in an attempt to categorise these soluble factors by their primary sources, but this nomenclature was short-lived, and "cytokines" was adopted. Every cell, except red blood cells, can produce as well as respond to cytokines.

The initial relationship between the disease (in most instances, infection) and the host response (fever) was established using soluble components produced by neutrophils from the peritoneal cavity of rabbits. Fever, viral infection resistance, an increased white blood cell count, the creation of acute phase proteins, and cancer cell death were among the biological features of these soluble substances. Soluble factors as lymphocyte function regulators were not examined until the mid-1970s when Igal Gery and Byron Waksman described the "lymphocyte activating factor" for the first time (Gery and Waksman, 1972) and later as a T cell growth factor (Mier and Gallo, 1980). A paradigm emerged that a disease process causes cells to produce these soluble factors (cytokines), and it is the property of these "factors" that account for the manifestations of the disease. The concept is best understood in the context of the host's non-specific physiological and immunological changes that occur as a result of infection. The "innate immune response" is nothing more than a cytokine storm to infection or cell damage triggered by Toll-like receptors (TLRs). The endocrine system of hormones was used to define a cytokine as a soluble factor produced by one cell and acting on another cell to cause a change in the function of the target cell. In some ways, cytokines can be thought of as "hormones" of immune and inflammatory responses. However, there are a few characteristics of cytokines that cannot be included in this definition. Hormones, for example, are the primary product of a specific tissue or cell, whereas cytokines are produced by nearly all cells (Dinarello, 2007).

23.3 Classification of Cytokines and Their Clinical Significance

Cytokines are secreted proteins that have growth, differentiation, and activation functions. Cytokines regulate and determine the nature of immune responses. Generally, cytokines can be classified into the following categories: tumour necrosis factors (TNFs); interleukins (ILs); lymphokines; monokines; interferons (IFNs)—involved in antiviral responses and include IFN-α, IFN-β, IFN-γ; colony-stimulating factors (CSFs)—support cell growth in semi-solid media and include granulocyte G-CSF, monocyte M-CSF, and granulocyte–monocyte GM-CSF; transforming growth factors (TGFs)—include TGF-b1, TGF-b2, and TGF-b3; bone morphogenetic proteins

(BMPs); activins and inhibins; and chemokines involved in chemotaxis. The chemokine superfamily is divided into four subfamilies (XC, CC, CXC, and CX3C) based on the presence of a conserved cysteine residue at the NH2 terminus, and the variable region "X" (Sprague and Khalil, 2009).

Based on their cellular source, cytokines are categorised as type 1 cytokines, which are produced by a cluster of differentiation 4 (CD4)+ T-helper 1 (Thl) cells and include IL-2, IL-12, IFN-γ, and TNF-γ; type 2 cytokines are produced by CD4+ Th2 cells and include IL-4, IL-5, IL-6, IL-10, and IL-13 (Sprague and Khalil, 2009) (Tables 23.1 and 23.2; Figures 23.1 and 23.2). Depending on their role, cytokines may also be categorised as pro-inflammatory or anti-inflammatory cytokines (Sprague and Khalil, 2009). Pro-inflammatory cytokines, which include IL-1β, IL-6, IL-8, IL-12, TNF-α, and interferons, among others, enhance inflammatory reactions and tend to stimulate immunocompetent cells. In contrast, they inhibit inflammation and suppress immune cells (Boshtam et al., 2017). Some cytokines (such as IL-6) possess both pro- and anti-inflammatory properties.

23.3.1 Pro-inflammatory Cytokines

Cytokines, which induce an acute phase response to protect the host from irritation, injury, and infection, are the primary regulators of the inflammatory response. The production of pro-inflammatory cytokines such as IL-1β, IL-6, IL-8, IL-12, IFN-γ, and TNF-α from the same cell or distinct cells initiates this reaction. The primary function of these cytokines is to communicate to surrounding tissues the presence of infection or damage. Furthermore, these cytokines can enter the systemic circulation, causing immune cell activation as well as substantial physiologic changes in the host such as fever and the acute phase reaction (Boshtam et al., 2017). Pro-inflammatory cytokines have immune properties that can help protect the host against invasion by bacteria and other microbes in the immediate environment (Dinarello, 2000). Pro-inflammatory cytokines, which are released from macrophages, have a significant role in defence against infection (Cheng et al., 2014). Macrophages are the first line of defence in the host's fight against bacterial infection, and they play a key role in the genesis of adaptive immune responses. It is worth considering that an excessive pro-inflammatory response may lead to chronic inflammation and disrupt pathways responsible for biological homeostasis, leading to detrimental health problems such as cancer, diabetes, cardiovascular diseases, Parkinson's disease, and ageing and age-related diseases. Autoimmune diseases (Miossec, 1997; Raphael et al., 2015), T1D, RA, inflammatory bowel disease (IBD), and MS are examples of diseases in which the immune system wrongly attacks one's own body. Pro-inflammatory interferons play a crucial role in the genesis of autoimmune diseases. Pro-inflammatory cytokines such as IL-1, IFN-γ, and TNF-α have been found to trigger chemokines that attract neutrophils, which are critical contributors to

TABLE 23.1

Classification of Cytokines Based on Cellular Source

Type	Cellular Source	Cytokines
Type 1 cytokines	Produced by the cluster of differentiation 4 (CD4)+ T-helper 1 (Th1) cells	IL-2, IL-12, IFN-γ, and TNF-β
Type 2 cytokines	Produced by CD4+ Th2 cells	IL-4, IL-5, IL-6, IL-10, and IL-13

TABLE 23.2

Classification of Cytokines Based on Role

Type	Cytokines
Pro-inflammatory cytokines	IL-1β, IL-6, IL-8, IL-12, TNF-α, and interferons, among others
Anti-inflammatory cytokines	IL-4, IL-6, IL-10, IL-11, IL-13, IL-1 receptor antagonist (IL-1RA), and TGF-β

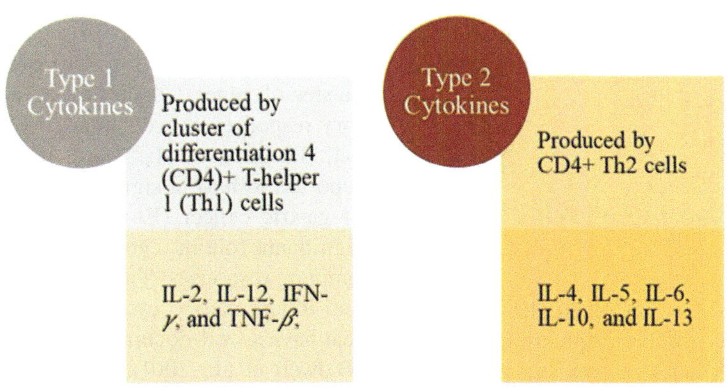

FIGURE 23.1 Classification of cytokines based on cellular source.

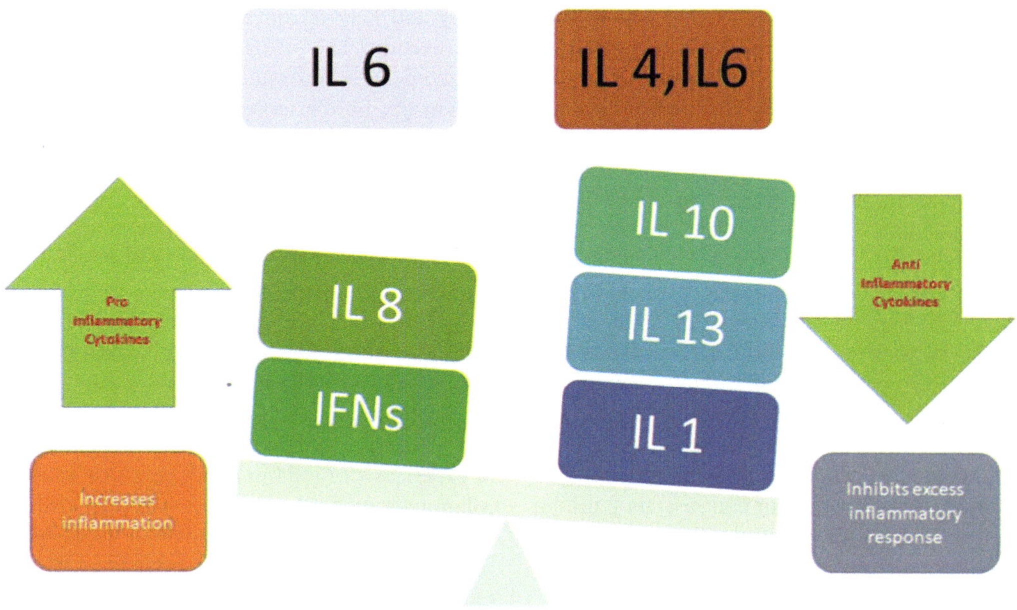

FIGURE 23.2 Pro-inflammatory and anti-inflammatory cytokines.

the formation of reactive oxygen species (ROS) and carcinogenesis, during the growth and spread of tumours (Dinarello, 2000).

23.3.2 Anti-inflammatory Cytokines

The anti-inflammatory cytokines such as IL-1 receptor antagonist, IL-4, IL-6, IL-10, IL-11, IL-13, and TGF-β are a group of immunoregulatory molecules that prevent pro-inflammatory cytokines from causing excess inflammatory response (Opal and DePalo, 2000). IL-10, for example, is a potent anti-inflammatory cytokine with immunoregulatory roles that suppress the production of various pro-inflammatory cytokines. IL-10 also has an anti-inflammatory effect on eosinophils, basophils, and mast cells and, hence, plays a significant role in allergy and asthma control and regulation (Al-Dabbagh et al., 2013). Under physiologic settings, these cytokines control the potentially harmful effects of sustained or excess expression of pro-inflammatory reactions. These anti-inflammatory cytokines have already been shown to be useful in a variety of clinical settings associated with excess inflammation. Anti-inflammatory cytokines, for example, can be used as drugs to treat inflammation-related disorders (Table 23.3).

23.4 Gene Polymorphisms in Cytokine Genes

23.4.1 Effects

Single nucleotide polymorphisms (SNPs) in the regulatory regions of cytokine genes influence their expression levels and can act as disease modifiers. The polymorphisms in cytokine genes lead to inter-individual differences in their production, resulting in variations in immunological responses and thereby leading to various diseases (D'Alfonso et al., 2000). Many cytokines and immune cells accurately regulate the dynamic balance of the body's immune function under normal physiological settings. Once the body's immune balance becomes abnormal, various autoimmune diseases will occur (Klecha et al., 2008). The abnormal expression of a variety of cytokines such as HLA (human leucocyte antigen), CD152, LYP, FcRL3, CD40, and their genes will inhibit autoimmune tolerance (McLachlan and Rapoport, 2014). TNF-α gene polymorphisms have been reported to be strongly associated with autoimmune and inflammatory disorders (Ruddle, 1992; Spies et al., 1986). Cytokines play a role in the inflammatory response associated with innate immunity and the acquired immune response. During *H. pylori* infection, cytokines such as INF-γ, interleukin 1h (IL-1h), and tumour necrosis factor-α (TNF-α) are found to be upregulated. Genes encoding cytokines harbour polymorphic regions, which affect gene transcription and thereby influence inflammatory processes in response to infectious disease (Hurme et al., 1998).

Polymorphisms in human IL-10, IL-1B, TNF-α, IFN-γ, and IL-1 receptor antagonist (IL-1RN) genes have been shown to affect cytokine expression. Several cytokine polymorphisms, such as IL-1h, IL-10, TNF-α, and others, have been evaluated as risk factors for gastric cancer. IL-1 encompasses a cluster of genes, which have a significant role in inflammatory responses as well as controlling the acidity of the stomach (El-Omar et al., 2000). IL-1 polymorphisms have been reported to have a significant association with increased risk of gastric cancer (El-Omar et al., 2000). Cytokines have a significant role in regulating both humoral and cell-mediated immune responses. The promoter regions of several pro- and anti-inflammatory cytokines genes harbour polymorphisms that have a well-documented influence in cytokine production (Bidwell et al., 2001). These promoter polymorphisms may lead to either high- or low-level production of a given cytokine, which may have a profound effect on the anti-tumour immune responses in pancreatic cancer. Moreover, particular cytokines

TABLE 23.3

Summary of Characteristics and SNPs of Different Cytokines

Cytokines	Cytokine Type	Biological Functions	SNPs
IL-1β	Pro-inflammatory	Principal mediator of the systemic effects of IL-1; it affects IL-6-induced gene expression	–511 C>T +3953 C>T
IL-6	Pro-inflammatory and anti-inflammatory	Inducer of the acute phase response as well as specific cellular and humoral immune responses. Inhibition of TNF and IL-1 production by macrophages	–174 C>G
IL-8	Pro-inflammatory	Pro-inflammatory mediators that orchestrate the recruitment of leucocytes to sites of inflammation	–251 T>A
IL-12	Pro-inflammatory	Th cell differentiation TNF-α, IFN-γ synthesis	–1188 A>C +1159 C>A
TNF-α	Pro-inflammatory	Pro-inflammatory, neutrophil activation, bone resorption, anticoagulant, tumour necrosis, activate and increase permeability, stimulate adhesion molecules	–308 G>A –237 G>A
IFN-γ	Pro-inflammatory	Pro-inflammatory promotes Th1 immune responses/secretion of Th1-associated cytokines, inhibits ECM synthesis by SMC MHC I expression	+874 A>T –5644 G>A
IL-1RA	Anti-inflammatory	Inhibition of IL-1α- and IL-1β-mediated cellular activation at the IL-1 cellular receptor level	+4845 G>T –889 C>T
IL-4	Anti-inflammatory	Promotes Th2 lymphocyte development; inhibition of LPS-induced pro-inflammatory cytokine synthesis	–590 C>T +148 A>G
IL-10	Anti-inflammatory	Inhibition of monocyte/macrophage and neutrophil cytokine production and inhibition of Th1-type lymphocyte responses	–1117 T>C –627 C>A –854 T>C
IL-11	Anti-inflammatory	Inhibits pro-inflammatory cytokine response by monocytes/macrophages and promotes Th2 lymphocyte response	A1>2 dinucleotide rep.
IL-13	Anti-inflammatory	Shares homology with IL-4 and shares the IL-4 receptor; attenuation of monocyte/macrophage function	+2044 –1111 C>T

(*e.g.* IL-8 and IL-10) may have a significant role in the development of tumours via their action on pathways of tumour angiogenesis (McCarron et al., 2002). SNPs in certain genes with significant amounts of variation among ethnic groups all over the world have been linked to the pathogenesis of diabetes. Therefore, finding polymorphic variants of cytokine genes in different populations, as well as genotypic connections between SNPs and gene-gene interactions, will be useful as indicators of T2DM (type 2 diabetes mellitus) susceptibility. Cytokines are mediators of inflammation, namely interleukin (IL)-1β, 1Ra, 18, 4, 6, 10, tumour necrosis factor-α, and adiponectin, which are responsible for immune responses in disease pathogenesis, including T2DM (Banerjee and Saxena, 2014) (Figure 23.3).

The pathogenesis of COVID-19 is linked to a significant inflammatory response, which activates a complex group of mediators including interleukins (Chen et al., 2020). With the progression of the disease, excessive production of pro-inflammatory cytokines results in a cytokine storm, which is responsible for the severe progression of the disease and the acute organ injuries which culminate into organ failures (Öztürk et al., 2020). SNPs are found to be effective in the pathways that play a significant role in the host's resistance to the diseases, and in the susceptibility to disease and the severity of disease (Öztürk et al., 2020). SNPs in IL-6, IL-6R, IL-10, IL-17A, and IL-17F genes may have an important role in the clinical course of COVID-19 infection, and these polymorphisms have a profound influence on survival/mortality rates due to COVID-19 infection (Mao et al., 2017; Pacha et al., 2020). The humoral immune response is mediated by IL-4, and polymorphisms in the IL-4 (position 2590) and its receptor gene (nucleotide 1902) have been linked to atopy (Hershey et al., 1997; Rosenwasser et al., 1995). Promotor region variation of the IL-6 gene has been correlated with juvenile rheumatoid arthritis (Fishman et al., 1998). Polymorphisms in IL-10 (position 21082) has been associated with SLE (Lazarus et al., 1997) and RA (Hajeer et al., 1998). IL-6 is a major inflammatory cytokine, and increased systemic levels are seen in many diseases, including autoimmune thyroiditis disease (ATD) (Ajjan et al., 1996).

23.5 Cytokine Polymorphisms and Autoimmune Diseases

Over the second half of the twentieth century, the incidence, as well as the prevalence of autoimmune diseases, has seen an escalating trend, posing a major health concern. These diseases are mostly chronic and can be fatal. The causes of autoimmune diseases remain largely unknown. The recent advancement in knowledge has substantially expanded our understanding of the aetiology and pathogenesis of autoimmune diseases, and it is well documented that the environment, genes, and immunity all play a role in the development of autoimmunity. Cytokines, which include both pro-inflammatory and ant-inflammatory cytokines, have a role in the pathogenesis of various diseases in many ways, including controlling inflammation and angiogenesis (Guan et al., 2017) (Table 23.4).

Cytokine and cytokine receptor genes have allured great interest as potential genes for autoimmune diseases (Vandenbroeck, 2006). Since 2007, the use of genome-wide association studies (GWAS) has aided in the accurate identification of allelic variants in over 35 cytokine loci as susceptibility factors for various autoimmune diseases (Vandenbroeck, 2012).

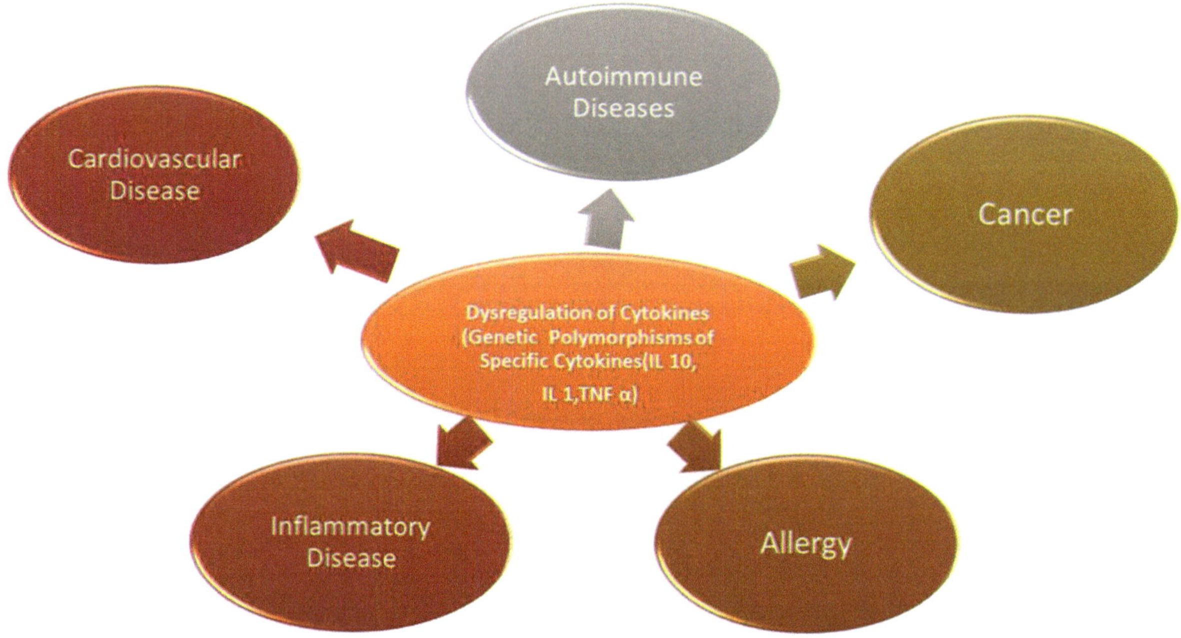

FIGURE 23.3 Cytokine regulation in health and disease.

TABLE 23.4
Multiple Cytokines Related to Different Biological Conditions

Disease	Related Cytokines
Autoimmune diseases	IL-1, IL-2, IL-6, IL-12, IL-15, IL-16, IL-17, IL-18, IL-23, TNF-α, IFN-α, IFN-γ
Allergy	IL-1, IL-4, IL-5, IL-9, IL-10, IL-13
Alzheimer's disease	TNF-α, TGF-β, IL-1, IL-4, IL-6, IL-10
Atherosclerosis	TNF-α, IFN-γ, TGF-β, IL-1, IL-2, IL-4, IL-5, IL-6, IL-8, IL-10, IL-12, IL-17, IL-18, IL-20, IL-33, IL-37
Cardiovascular diseases	TNF-α, TGF-β, IL-1, IL-6, IL-10, IL-17, IL-18
Cancer	TNF-α, IL-6, IL-10, IL-12, 1L-17, IL-23
Depression	TNF-α, IFN-γ, IL-1, IL-2, IL-6
Gastrointestinal diseases	TNF-α, IFN-γ, TGF-β, IL-1, IL-4, IL-6, IL-8, IL-10
Sepsis	TNF-α, IFN-γ, TGF-β, MIF, IL-1, IL-6, IL-4, IL-10, IL-12
Ageing	IL-6, IL-8, IL-10, IL-13, TNF-α, IFN-γ

23.5.1 Interleukins

The term "interleukin" (IL) has been used to describe a group of cytokines with complex immunomodulatory effects, such as cell proliferation, maturation, migration, and adhesion. These cytokines are also critical in the development and activation of immune cells (Brocker et al., 2010). ILs constitute a large family of immunomodulatory proteins that cause a wide range of reactions in cells and tissues. Within mammals, these cytokines make up a vast number of known immunological "second messenger" molecules (Commins et al., 2010). Instead of acting as an endocrine signal, ILs have a paracrine or autocrine effect. ILs have both pro-inflammatory and anti-inflammatory properties. ILs are critical mediators of the physiological response to infection and also contribute substantially to the pathophysiology of a variety of diseases. As a result, they constitute a class of proteins that could be useful as therapeutic targets. The term "interleukin" was first coined in 1979 to designate a group of secreted molecules produced by leucocytes in a letter to the editor of the Journal of Immunology (Brocker et al., 2010). Functional features, sequence homology, and receptor chain similarity are the key criteria that are used to classify interleukins. Furthermore, the structural features of cytokines 6 (for example, ILs) are important factors for classifying them into distinct cytokine groups (Akdis et al., 2016).

Cytokines are categorised into two types: type/class I and type/class II cytokines.

Type I ILs possess a four-helix bundle joint by one short and two long loops. These four anti-parallel compact blocks of α-helices are oriented with up-up/down-down directions (Brocker et al., 2010). Additionally, the type I cytokines (ILs) are divided into two groups based on the length of the four-helix bundle, short or long: those with more than 165 amino acids (AAs) and those with less than 165 AAs. IL-2, IL-3, IL-4, IL-5, IL-7, IL-9, IL-13, IL-15, IL-21, colony-stimulating factor 2 (CSF-2), and thymic stromal lymphopoietin (TSLP) are examples of short-chain type I cytokines, while long-chain type I cytokines include IL-6, IL-11, IL-12A, IL-23A, IL-27A,

IL-31, oncostatin M (OSM), leukaemia inhibitory factor (LIF), cardiotrophin 1 (CTF-1), ciliary neurotrophic factor (CNTF), and cardiotrophin-like cytokine factor 1 (CLCF-1).

Type II ILs have a six- to seven-helix bundle with six to seven compact blocks of α-helices with an anti-parallel spatial arrangement. IL-10, IL-19, IL-20, IL-22, IL-24, IL-26, IL-28A, IL-28B, and IL-29 are examples of type II ILs (Brocker et al., 2010). By interacting with a variety of receptors, such as TLRs, ILs play a role in both pro- and anti-inflammatory responses. ILs and TLRs have a role in homeostasis, cancers, autoimmune disorders, and infectious diseases (Behzadi et al., 2021; Gajdács and Behzadi, 2020; Ryff et al., 2008). Because of the importance of ILs and their role in different parts of the human body, the occurrence of any mutation or polymorphism in their genes may result in a wide range of diseases and disorders, including autoimmune diseases. Among these abnormalities, SNPs can present as important disruptive factors. Because of the significance of ILs and their role in various parts of the human body, any mutation or polymorphism in their genes can cause a variety of diseases and disorders, including autoimmune diseases. SNPs can represent as significant disruptive factors among these disorders.

The IL-2RA locus was one of the first to be identified as an autoimmune risk factor in early GWAS. Thus far, this gene locus has been identified as a risk factor for a variety of autoimmune disorders including MS (International Multiple Sclerosis Genetics Consortium et al., 2007) and T1D (Barrett et al., 2009). The alpha-chain of the IL-7 receptor is involved in V(D)J recombination during lymphocyte development and regulates the access of the TCRγ locus by STAT5. A GWAS has identified rs6897932, a non-synonymous missense SNP in the sixth exon of the IL-7R gene that changes T244 as a risk factor for MS (International Multiple Sclerosis Genetics Consortium et al., 2007). The disulphide-bonded p35 (coded for by IL-12A) and p40 (coded for by IL-12B) subunits of IL-12, a heterodimeric cytokine, are critical for cell-mediated immunity and Th1 differentiation (Collison et al., 2007). IL-12 has been linked to the development and pathogenesis of a variety of autoimmune diseases (Kastelein et al., 2007; Paunović et al., 2008). IL-12A, IL-12B, and IL-12RB2, three genes that code for proteins involved in the early stages of the IL-12 binding and signalling cascade, have appeared as risk factors in GWAS on autoimmune diseases.

The bulk of observed risk SNPs are found 5′ from the IL-12B gene at different locations, and IL-12B has been identified in GWAS on six autoimmune diseases. The risk of psoriasis was found to be linked to SNP located in an intronic sequence of the STAT2 gene, which is located 3.7 kb downstream from IL-23A (Nair et al., 2009). SNPs, which are located in the IL-27 gene region that codes for the p28 subunit, which heterodimerises with the EBI3 subunit to generate the highly pleiotropic regulatory cytokine IL-27 (Collison and Vignali, 2008), were reported to be associated with the risk of T1D (Barrett et al., 2009). Through GWAS, polymorphisms within or near numerous other cytokine and chemokine gene loci have been linked to a variety of autoimmune disorders. In two GWAS on BD, promoter and intronic variants in IL-10 were found to have genome-wide significance (Mizuki et al., 2010; Remmers et al., 2010). The associated rs1518111 variant was reported to be associated with altered mRNA and protein expression (Remmers et al., 2010). The risk factor for ulcerative colitis (UC) has been linked to the T allele of the rs3024505 SNP, which is located close to the 3′-untranslated region of IL-10 (Franke et al., 2010) and SLE (Gateva et al., 2009).

23.5.2 Tumour Necrosis Factor (TNF)

TNF is a multifunctional pro-inflammatory cytokine that plays a significant role in autoimmune disorders such as RA, pernicious anaemia, and diabetes mellitus. The TNF gene is located on chromosome 6 (6p21.31) in the class III region of the human major histocompatibility complex (MHC), spanning about 3 kb and containing four exons. TNF-α production is regulated at both the transcriptional and post-transcriptional levels, with regulatory sequences within the 59 ends of the gene influencing the transcription rate (Spriggs et al., 1992). TNF represents two homologous proteins that are derived primarily by mononuclear phagocytes (TNF-α) and lymphocytes (TNF-β) (Beutler and Cerami, 1989). Neutrophils, lymphocytes, natural killer (NK) cells, endothelium, and mast cells all produce TNF-α. TNF-α is processed as a membrane-bound protein from which the soluble active factor is cleaved by using the enzyme TNF-α-converting enzyme (Perez et al., 1990).

Several SNPs in the human TNF-α gene promoter region have been discovered as having the potential to create structural alterations inside regulatory regions, which could affect the function or regulation of TNF-α production. Because of its placement within MHC and biological activities, polymorphisms within this locus may play a role in the development and pathogenesis of a variety of autoimmune and infectious disorders. The combination of these polymorphisms may play a role in autoimmune disorders. Increased TNF-α transcript levels with 2238 AA and GA genotypes compared to GG genotypes were observed in a genotype-phenotypic investigation of SLE patients (Suarez et al., 2005). TNF-α metabolism disturbances have been well documented and reported to be linked to a variety of autoimmune and infectious diseases, including RA (Elliott et al., 1994), SLE (Jacob et al., 1990), and Crohn's disease (CD) (van Dullemen et al., 1995). TNF-α 2238G/A and 2308G/A polymorphisms were found to have a substantial correlation with GD in a study comprising 22 SNPs (Simmonds et al., 2004). According to several studies, sequence variation in the regulatory region of the TNF-*α* gene has been associated with a variety of autoimmune disorders (van Heel et al., 2002). A number of SNPs have been observed in the *TNF-α* gene (Richardson et al., 2001). Among these SNPs, one G (guanine) > A (adenine) polymorphism is found upstream of the gene at position -308 and is reported to affect TNF-α levels. The A allele has a stronger transcriptional activity than the TNF-α -308G variant and is commonly linked to autoimmune disorders (Fan et al., 2011; Wilson et al., 1997). The location of the gene within the MHC and the possible role of -308 G > A polymorphism in the TNF-*α* gene promoter activity have raised the notion that this polymorphism could affect immunologic homeostasis and have a role in the aetiology of coeliac disease (CeD).

23.5.3 Interferons

IFNs were originally described in 1957 as molecules that interfere with viral replication (Isaacs and Lindenmann, 1957). Since then, a vast body of evidence suggests that IFN plays a role in viral infection responses by direct actions on infected cells and by strongly affecting the behaviour of innate and adaptive immune responses (Biron, 2001). IFNs are engaged in a variety of biological processes having crucial roles in infectious diseases, cancer, inflammation, and autoimmunity in addition to antiviral activities. There are three types of IFNs in the IFN family: type I, type II, and type III. Multiple biological processes are mediated by these IFNs (Hall and Rosen, 2010; Stark et al., 1998). Antiviral actions are known to be regulated by IFNs. IFNs from all three classes appear to have distinct roles in the development of autoimmune disorders. Various gene polymorphisms linked with an increased risk of disease development are shared between autoimmune disorders (Teruel and Alarcón-Riquelme, 2016). Interestingly, most of them are found in the IFN pathway (Jiang et al., 2020). When signals are propagated, these polymorphisms may affect IFN production. Polymorphisms of IRF5, IRF7, and IRF8 have been linked to SLE risk in large-scale association studies (Cui et al., 2013). IFN levels were also shown to be higher in patients with RA, scleroderma, and primary Sjögren's disease (pSS) (Hooks et al., 1979).

Another key aspect of active interferon-mediated signalling in SLE patients is the overexpression of type I interferon-regulated genes, known as the interferon signature, which is especially prevalent in individuals with severe disease (Baechler et al., 2003). pSS, dermatomycosis, psoriasis, and some RA patients have all been shown to have an interferon signature (Baechler et al., 2006). MS has recently been linked to the SNP rs17445836, which is situated 61 kb downstream of the IRF8 gene (De Jager et al., 2009). Polymorphisms in the IRF5 gene have been linked to SLE and various other systemic autoimmune diseases (Matta and Barnes, 2020; Miceli-Richard et al., 2009; Sigurdsson et al., 2005). STAT4 (signal transducer and activator of transcription 4) is a transcription factor that is activated by a variety of cytokines, including type I IFNs, and is involved in T cell development and IFN-γ production. STAT4 polymorphisms have been associated with an increased risk of RA, SLE, and other diseases (Remmers et al., 2007).

23.6 Conclusions

To summarise, cytokines are ubiquitous molecules that serve as vital messengers for and between immune cells, assisting in the immune system's delicate and intricate equilibrium. Embryonic development, disease aetiology, non-specific response to infection, specific response to antigen, changes in cognitive functioning, autoimmunity, and the advancement of degenerative processes of ageing are all affected by cytokines. Such regulation of immune homeostasis is critical for both health and disease, and disruption of this balance leads to a variety of chronic pathophysiological conditions. A myriad of SNPs in cytokine genes have shown a well-documented role in autoimmune disorders. Autoimmune disorders are the consequence of complex interactions between different immune cell types, including both T and B lymphocytes, as well as professional antigen-presenting cells such as macrophages and dendritic cells. In a wide range of autoimmune diseases, these cellular interactions culminate in auto-aggressive responses that target a number of cell types in various tissues and organs. Although the aetiology of most autoimmune disorders is unknown, studies conducted in recent years have made significant progress in understanding how the many immune cell types implicated in autoimmunity communicate with one another, how they initiate autoimmune inflammation, and how they cause tissue damage. In autoimmunity, cytokines and chemokines orchestrate the recruitment, survival, expansion, effector function, and contraction of autoreactive lymphocytes as important parts of this communication network (Santamaria, 2003). As a result, it is important that therapeutic targeting of cytokine pathways holds enormous potential for patients suffering from a variety of chronic conditions that are difficult to treat. Differential cytokine production is a recent advancement in the area that can be used to define human disease as "autoimmune" or "autoinflammatory," affecting therapeutic approaches (Dinarello, 2007). To explain the underlying mechanism of diverse immune responses, including the cytokine storm in COVID-19 patients, more research on genetic variants in COVID-19-infected patients with varying degrees of severity is required.

ABBREVIATIONS

ATD:	Autoimmune thyroid disorder
BD:	Behcet's disease
BMPs:	Bone morphogenetic proteins
CD:	Crohn's disease
CeD:	Coeliac disease
CLCF-1:	Cardiotrophin-like cytokine factor 1
CNTF:	Ciliary neurotrophic factor
COVID-19:	Coronavirus disease 19
CSFs:	Colony-stimulating factors
CSF-2:	Colony-stimulating factor 2
CTF-1:	Cardiotrophin 1
GD:	Graves' disease
GWAS:	Genome-wide association studies
IBD:	Inflammatory bowel disease
IFNs:	Interferons
ILs:	Interleukins
JIA:	Juvenile idiopathic arthritis
LIF:	Leukaemia inhibitory factor
MHC:	Major histocompatibility complex
MS:	Multiple sclerosis
NK cells:	Natural killer cells
OSM:	Oncostatin M
pSS:	Primary Sjögren's syndrome
RA:	Rheumatoid arthritis
ROS:	Reactive oxygen species
SLE:	Systemic lupus erythematosus
SNPs:	Single nucleotide polymorphisms
SSC:	Systemic sclerosis

STAT: Signal transducer and activator of transcription
T2DM: Type 2 diabetes mellitus
TGF: Transforming growth factor
TID: Type 1 diabetes
TLRs: Toll-like receptors
TNF: Tumour necrosis factor
TSLP: Thymic stromal lymphopoietin

REFERENCES

Ajjan RA, Watson PF, Weetman AP. Cytokines and thyroid function. *Adv Neuroimmunol* 1996;6:359–86. https://doi.org/10.1016/s0960-5428(97)00027–7.

Akdis M, Aab A, Altunbulakli C, Azkur K, Costa RA, Crameri R, et al. Interleukins (from IL-1 to IL-38), interferons, transforming growth factor β, and TNF-α: receptors, functions, and roles in diseases. *J Allergy Clin Immunol* 2016;138:984–1010. https://doi.org/10.1016/j.jaci.2016.06.033.

Al-Dabbagh M, Lapphra K, Scheifele DW, Halperin SA, Langley JM, Cho P, et al. Elevated inflammatory mediators in adults with oculorespiratory syndrome following influenza immunization: a public health agency of Canada/Canadian Institutes of Health Research Influenza Research Network Study. *Clin Vaccine Immunol* 2013;20:1108–14. https://doi.org/10.1128/CVI.00659–12.

Baechler EC, Batliwalla FM, Karypis G, Gaffney PM, Ortmann WA, Espe KJ, et al. Interferon-inducible gene expression signature in peripheral blood cells of patients with severe lupus. *Proc Natl Acad Sci U S A* 2003;100:2610–5. https://doi.org/10.1073/pnas.0337679100.

Baechler EC, Batliwalla FM, Reed AM, Peterson EJ, Gaffney PM, Moser KL, et al. Gene expression profiling in human autoimmunity. *Immunol Rev* 2006;210:120–37. https://doi.org/10.1111/j.0105-2896.2006.00367.x.

Banerjee M, Saxena M. Genetic polymorphisms of cytokine genes in type 2 diabetes mellitus. *World J Diabetes* 2014;5:493–504. https://doi.org/10.4239/wjd.v5.i4.493.

Barrett JC, Clayton DG, Concannon P, Akolkar B, Cooper JD, Erlich HA, et al. Genome-wide association study and meta-analysis find that over 40 loci affect risk of type 1 diabetes. *Nat Genet* 2009;41:703–7. https://doi.org/10.1038/ng.381.

Behzadi P, García-Perdomo HA, Karpiński TM. Toll-like receptors: general molecular and structural biology. *J Immunol Res* 2021;2021:e9914854. https://doi.org/10.1155/2021/9914854.

Bernheim HA, Kluger MJ. Fever and antipyresis in the lizard Dipsosaurus dorsalis. *Am J Physiol* 1976;231:198–203. https://doi.org/10.1152/ajplegacy.1976.231.1.198.

Beutler B, Cerami A. The biology of cachectin/TNF–a primary mediator of the host response. *Annu Rev Immunol* 1989;7:625–55. https://doi.org/10.1146/annurev.iy.07.040189.003205.

Bidwell J, Keen L, Gallagher G, Kimberly R, Huizinga T, McDermott MF, et al. Cytokine gene polymorphism in human disease: on-line databases, supplement 1. *Genes Immun* 2001;2. https://doi.org/10.1038/sj.gene.6363733.

Biron CA. Interferons alpha and beta as immune regulators–a new look. *Immunity* 2001;14:661–4. https://doi.org/10.1016/s1074-7613(01)00154–6.

Boshtam M, Asgary S, Kouhpayeh S, Shariati L, Khanahmad H. Aptamers against pro- and anti-inflammatory cytokines: a review. *Inflammation* 2017;40. https://doi.org/10.1007/s10753-016-0477–1.

Brocker C, Thompson D, Matsumoto A, Nebert DW, Vasiliou V. Evolutionary divergence and functions of the human interleukin (IL) gene family. *Hum Genomics* 2010;5:30–55. https://doi.org/10.1186/1479-7364-5-1-30.

Chen G, Wu D, Guo W, Cao Y, Huang D, Wang H, et al. Clinical and immunological features of severe and moderate coronavirus disease 2019. *J Clin Invest* 2020;130:2620–9. https://doi.org/10.1172/JCI137244.

Cheng A, Yan H, Han C, Wang W, Tian Y, Chen X. Polyphenols from blueberries modulate inflammation cytokines in LPS-induced RAW264.7 macrophages. *Int J Biol Macromol* 2014;69:382–7. https://doi.org/10.1016/j.ijbiomac.2014.05.071.

Collison LW, Vignali DAA. Interleukin-35: odd one out or part of the family? *Immunol Rev* 2008;226:248–62. https://doi.org/10.1111/j.1600-065X.2008.00704.x.

Collison LW, Workman CJ, Kuo TT, Boyd K, Wang Y, Vignali KM, et al. The inhibitory cytokine IL-35 contributes to regulatory T-cell function. *Nature* 2007;450:566–9. https://doi.org/10.1038/nature06306.

Commins SP, Borish L, Steinke JW. Immunologic messenger molecules: cytokines, interferons, and chemokines. *J Allergy Clin Immunol* 2010;125:S53–72. https://doi.org/10.1016/j.jaci.2009.07.008.

Cui Y, Sheng Y, Zhang X. Genetic susceptibility to SLE: recent progress from GWAS. *J Autoimmun* 2013;41:25–33. https://doi.org/10.1016/j.jaut.2013.01.008.

D'Alfonso S, Rampi M, Bocchio D, Colombo G, Scorza-Smeraldi R, Momigliano-Richardi P. Systemic lupus erythematosus candidate genes in the Italian population: evidence for a significant association with interleukin-10. *Arthritis Rheum* 2000;43:120–8. https://doi.org/10.1002/1529-0131(-200001)43:1<120::AID-ANR15>3.0.CO;2–3.

De Jager PL, Jia X, Wang J, de Bakker PIW, Ottoboni L, Aggarwal NT, et al. Meta-analysis of genome scans and replication identify CD6, IRF8 and TNFRSF1A as new multiple sclerosis susceptibility loci. *Nat Genet* 2009;41:776–82. https://doi.org/10.1038/ng.401.

Dinarello CA. Proinflammatory cytokines. *Chest* 2000;118:503–8. https://doi.org/10.1378/chest.118.2.503.

Dinarello CA. Historical Review of Cytokines. *Eur J Immunol* 2007;37:S34–45. https://doi.org/10.1002/eji.200737772.

van Dullemen HM, van Deventer SJ, Hommes DW, Bijl HA, Jansen J, Tytgat GN, et al. Treatment of Crohn's disease with anti-tumor necrosis factor chimeric monoclonal antibody (cA2). *Gastroenterology* 1995;109:129–35. https://doi.org/10.1016/0016–5085(95)90277–5.

Elliott MJ, Maini RN, Feldmann M, Long-Fox A, Charles P, Bijl H, et al. Repeated therapy with monoclonal antibody to tumour necrosis factor alpha (cA2) in patients with rheumatoid arthritis. *Lancet* 1994;344:1125–7. https://doi.org/10.1016/s0140-6736(94)90632–7.

El-Omar EM, Carrington M, Chow WH, McColl KE, Bream JH, Young HA, et al. Interleukin-1 polymorphisms associated with increased risk of gastric cancer. *Nature* 2000;404:398–402. https://doi.org/10.1038/35006081.

Fan W, Maoqing W, Wangyang C, Fulan H, Dandan L, Jiaojiao R, et al. Relationship between the polymorphism of tumor necrosis factor-α-308 G>A and susceptibility to inflammatory bowel diseases and colorectal cancer: a meta-analysis. *Eur J Hum Genet* 2011;19:432–7. https://doi.org/10.1038/ejhg.2010.159.

Fishman D, Faulds G, Jeffery R, Mohamed-Ali V, Yudkin JS, Humphries S, et al. The effect of novel polymorphisms in the interleukin-6 (IL-6) gene on IL-6 transcription and plasma IL-6 levels, and an association with systemic-onset juvenile chronic arthritis. *J Clin Invest* 1998;102:1369–76.

Franke A, McGovern DPB, Barrett JC, Wang K, Radford-Smith GL, Ahmad T, et al. Genome-wide meta-analysis increases to 71 the number of confirmed Crohn's disease susceptibility loci. *Nat Genet* 2010;42:1118–25. https://doi.org/10.1038/ng.717.

Gajdács M, Behzadi P. Introductory chapter: cytokines–the diamonds and pearls of biological systems, 2020, pp. 1–7. https://doi.org/10.5772/intechopen.93197.

Gateva V, Sandling JK, Hom G, Taylor KE, Chung SA, Sun X, et al. A large-scale replication study identifies TNIP1, PRDM1, JAZF1, UHRF1BP1 and IL10 as risk loci for systemic lupus erythematosus. *Nat Genet* 2009;41:1228–33. https://doi.org/10.1038/ng.468.

Gery I, Waksman BH. Potentiation of the T-lymphocyte response to mitogens. II. The cellular source of potentiating mediator(s). *J Exp Med* 1972;136:143–55. https://doi.org/10.1084/jem.136.1.143.

Guan Q, Gao X, Wang J, Sun Y, Shekhar S. Cytokines in autoimmune disease. *Mediat Inflam* 2017;2017:e5089815. https://doi.org/10.1155/2017/5089815.

Hajeer AH, Lazarus M, Turner D, Mageed RA, Vencovsky J, Sinnott P, et al. IL-10 gene promoter polymorphisms in rheumatoid arthritis. *Scand J Rheumatol* 1998;27:142–5. https://doi.org/10.1080/030097498441029.

Hall JC, Rosen A. Type I interferons: crucial participants in disease amplification in autoimmunity. *Nat Rev Rheumatol* 2010;6:40–9. https://doi.org/10.1038/nrrheum.2009.237.

van Heel DA, Udalova IA, De Silva AP, McGovern DP, Kinouchi Y, Hull J, et al. Inflammatory bowel disease is associated with a TNF polymorphism that affects an interaction between the OCT1 and NF(-kappa)B transcription factors. *Hum Mol Genet* 2002;11:1281–9. https://doi.org/10.1093/hmg/11.11.1281.

Hershey GK, Friedrich MF, Esswein LA, Thomas ML, Chatila TA. The association of atopy with a gain-of-function mutation in the alpha subunit of the interleukin-4 receptor. *N Engl J Med* 1997;337:1720–5. https://doi.org/10.1056/NEJM199712113372403.

Hooks JJ, Moutsopoulos HM, Geis SA, Stahl NI, Decker JL, Notkins AL. Immune interferon in the circulation of patients with autoimmune disease. *N Engl J Med* 1979;301:5–8. https://doi.org/10.1056/NEJM197907053010102.

Hurme M, Lahdenpohja N, Santtila S. Gene polymorphisms of interleukins 1 and 10 in infectious and autoimmune diseases. *Ann Med* 1998;30:469–73. https://doi.org/10.3109/07853899809002488.

International Multiple Sclerosis Genetics Consortium, Hafler DA, Compston A, Sawcer S, Lander ES, Daly MJ, et al. Risk alleles for multiple sclerosis identified by a genomewide study. *N Engl J Med* 2007;357:851–62. https://doi.org/10.1056/NEJMoa073493.

Isaacs A, Lindenmann J. Virus interference. I. The interferon. *Proc R Soc Lond B Biol Sci* 1957;147:258–67. https://doi.org/10.1098/rspb.1957.0048.

Jacob CO, Fronek Z, Lewis GD, Koo M, Hansen JA, McDevitt HO. Heritable major histocompatibility complex class II-associated differences in production of tumor necrosis factor alpha: relevance to genetic predisposition to systemic lupus erythematosus. *Proc Natl Acad Sci U S A* 1990;87:1233–7. https://doi.org/10.1073/pnas.87.3.1233.

Jiang J, Zhao M, Chang C, Wu H, Lu Q. Type I Interferons in the pathogenesis and treatment of autoimmune diseases. *Clin Rev Allergy Immunol* 2020;59:248–72. https://doi.org/10.1007/s12016-020-08798-2.

Jin Y, Mailloux CM, Gowan K, Riccardi SL, LaBerge G, Bennett DC, et al. NALP1 in vitiligo-associated multiple autoimmune disease. *N Engl J Med* 2007;356:1216–25. https://doi.org/10.1056/NEJMoa061592.

Kastelein RA, Hunter CA, Cua DJ. Discovery and biology of IL-23 and IL-27: related but functionally distinct regulators of inflammation. *Annu Rev Immunol* 2007;25:221–42. https://doi.org/10.1146/annurev.immunol.22.012703.104758.

Klecha AJ, Barreiro Arcos ML, Frick L, Genaro AM, Cremaschi G. Immune-endocrine interactions in autoimmune thyroid diseases. *Neuroimmunomodulation* 2008;15:68–75. https://doi.org/10.1159/000135626.

Kulbe H, Chakravarty P, Leinster DA, Charles KA, Kwong J, Thompson RG, et al. A dynamic inflammatory cytokine network in the human ovarian cancer microenvironment. *Cancer Res* 2012;72:66–75. https://doi.org/10.1158/0008-5472.CAN-11-2178.

Kunz M, Ibrahim SM. Cytokines and cytokine profiles in human autoimmune diseases and animal models of autoimmunity. *Mediators Inflamm* 2009;2009:979258. https://doi.org/10.1155/2009/979258.

Lazarus M, Hajeer AH, Turner D, Sinnott P, Worthington J, Ollier WE, et al. Genetic variation in the interleukin 10 gene promoter and systemic lupus erythematosus. *J Rheumatol* 1997;24:2314–7.

Mao Z-R, Zhang S-L, Feng B. Association of IL-10 (-819T/C, -592A/C and -1082A/G) and IL-6–174G/C gene polymorphism and the risk of pneumonia-induced sepsis. *Biomarkers* 2017;22:106–12. https://doi.org/10.1080/1354750X.2016.1210677.

Matta B, Barnes BJ. Coordination between innate immune cells, type I IFNs and IRF5 drives SLE pathogenesis. *Cytokine* 2020;132:154731. https://doi.org/10.1016/j.cyto.2019.05.018.

McCarron SL, Edwards S, Evans PR, Gibbs R, Dearnaley DP, Dowe A, et al. Influence of cytokine gene polymorphisms on the development of prostate cancer. *Cancer Res* 2002;62:3369–72.

McLachlan SM, Rapoport B. Breaking tolerance to thyroid antigens: changing concepts in thyroid autoimmunity. *Endocr Rev* 2014;35:59–105. https://doi.org/10.1210/er.2013–1055.

Miceli-Richard C, Gestermann N, Ittah M, Comets E, Loiseau P, Puechal X, et al. The CGGGG insertion/deletion polymorphism of the IRF5 promoter is a strong risk factor for primary Sjögren's syndrome. *Arthritis Rheum* 2009;60:1991–7. https://doi.org/10.1002/art.24662.

Mier JW, Gallo RC. Purification and some characteristics of human T-cell growth factor from phytohemagglutinin-stimulated lymphocyte-conditioned media. *Proc Natl Acad Sci U S A* 1980;77:6134–8. https://doi.org/10.1073/pnas.77.10.6134.

Miossec P. Cytokine-induced autoimmune disorders. *Drug Saf* 1997;17:93–104.https://doi.org/10.2165/00002018–199717020–00002.

Mizuki N, Meguro A, Ota M, Ohno S, Shiota T, Kawagoe T, et al. Genome-wide association studies identify IL23R-IL12RB2 and IL10 as Behçet's disease susceptibility loci. *Nat Genet* 2010;42:703–6. https://doi.org/10.1038/ng.624.

Nair RP, Duffin KC, Helms C, Ding J, Stuart PE, Goldgar D, et al. Genome-wide scan reveals association of psoriasis with IL-23 and NF-κB pathways. *Nat Genet* 2009;41:199–204. https://doi.org/10.1038/ng.311.

Opal SM, DePalo VA. Anti-inflammatory cytokines. *Chest* 2000;117:1162–72. https://doi.org/10.1378/chest.117.4.1162.

O'Shea JJ, Murray PJ. Cytokine signaling modules in inflammatory responses. Immunity 2008;28. https://doi.org/10.1016/j.immuni.2008.03.002.

Öztürk R, Taşova Y, Ayaz A. COVID-19: pathogenesis, genetic polymorphism, clinical features and laboratory findings. *Turk J Med Sci* 2020;50:638–57. https://doi.org/10.3906/sag–2005–287.

Pacha O, Sallman MA, Evans SE. COVID-19: a case for inhibiting IL-17? *Nat Rev Immunol* 2020;20:345–6. https://doi.org/10.1038/s41577-020-0328–z.

Paunović V, Carroll HP, Vandenbroeck K, Gadina M. Signalling, inflammation and arthritis: crossed signals: the role of interleukin (IL)-12, -17, -23 and -27 in autoimmunity. *Rheumatology (Oxford, England)* 2008;47. https://doi.org/10.1093/rheumatology/kem352.

Perez C, Albert I, DeFay K, Zachariades N, Gooding L, Kriegler M. A nonsecretable cell surface mutant of tumor necrosis factor (TNF) kills by cell-to-cell contact. *Cell* 1990;63:251–8. https://doi.org/10.1016/0092-8674(90)90158–b.

Raphael I, Nalawade S, Eagar TN, Forsthuber TG. T cell subsets and their signature cytokines in autoimmune and inflammatory diseases. *Cytokine* 2015;74:5–17. https://doi.org/10.1016/j.cyto.2014.09.011.

Remmers EF, Cosan F, Kirino Y, Ombrello MJ, Abaci N, Satorius C, et al. Genome-wide association study identifies variants in the MHC class I, IL10, and IL23R-IL12RB2 regions associated with Behçet's disease. *Nat Genet* 2010;42:698–702. https://doi.org/10.1038/ng.625.

Remmers EF, Plenge RM, Lee AT, Graham RR, Hom G, Behrens TW, et al. STAT4 and the risk of rheumatoid arthritis and systemic lupus erythematosus. *N Engl J Med* 2007;357:977–86. https://doi.org/10.1056/NEJMoa073003.

Richardson A, Sisay-Joof F, Ackerman H, Usen S, Katundu P, Taylor T, et al. Nucleotide diversity of the TNF gene region in an African village. *Genes Immun* 2001;2:343–8. https://doi.org/10.1038/sj.gene.6363789.

Rosenwasser LJ, Klemm DJ, Dresback JK, Inamura H, Mascali JJ, Klinnert M, et al. Promoter polymorphisms in the chromosome 5 gene cluster in asthma and atopy. *Clin Exp Allergy* 1995;25 Suppl 2:74–8; discussion 95–6. https://doi.org/10.1111/j.1365-2222.1995.tb00428.x.

Ruddle NH. Tumor necrosis factor (TNF-alpha) and lymphotoxin (TNF-beta). *Curr Opin Immunol* 1992;4:327–32. https://doi.org/10.1016/0952-7915(92)90084–r.

Ryff J-C, Bordens RW, Pestka S. *Interferons and Interleukins. Pharmaceutical Biotechnology.* 3rd ed., CRC Press; 2008.

Santamaria P. Cytokines and chemokines in autoimmune disease: an overview. *Adv Exp Med Biol* 2003;520:1–7. https://doi.org/10.1007/978-1-4615-0171-8_1.

Sigurdsson S, Nordmark G, Göring HHH, Lindroos K, Wiman A-C, Sturfelt G, et al. Polymorphisms in the tyrosine kinase 2 and interferon regulatory factor 5 genes are associated with systemic lupus erythematosus. *Am J Hum Genet* 2005;76:528–37. https://doi.org/10.1086/428480.

Simmonds MJ, Heward JM, Howson JMM, Foxall H, Nithiyananthan R, Franklyn JA, et al. A systematic approach to the assessment of known TNF-α polymorphisms in Graves' disease. *Genes Immun* 2004;5:267–73. https://doi.org/10.1038/sj.gene.6364066.

Spies T, Morton CC, Nedospasov SA, Fiers W, Pious D, Strominger JL. Genes for the tumor necrosis factors alpha and beta are linked to the human major histocompatibility complex. *Proc Natl Acad Sci U S A* 1986;83:8699–702.

Sprague AH, Khalil RA. Inflammatory cytokines in vascular dysfunction and vascular disease. *Biochem Pharmacol* 2009;78:539–52. https://doi.org/10.1016/j.bcp.2009.04.029.

Spriggs DR, Deutsch S, Kufe DW. Genomic structure, induction, and production of TNF-alpha. *Immunol Ser* 1992;56:3–34.

Stark GR, Kerr IM, Williams BR, Silverman RH, Schreiber RD. How cells respond to interferons. *Annu Rev Biochem* 1998;67:227–64. https://doi.org/10.1146/annurev.biochem.67.1.227.

Suarez A, Lopez P, Mozo L, Gutierrez C. Differential effect of IL10 and TNFα genotypes on determining susceptibility to discoid and systemic lupus erythematosus. *Ann Rheum Dis* 2005;64:1605–10. https://doi.org/10.1136/ard.2004.035048.

Teruel M, Alarcón-Riquelme ME. The genetic basis of systemic lupus erythematosus: what are the risk factors and what have we learned. *J Autoimmun* 2016;74:161–75. https://doi.org/10.1016/j.jaut.2016.08.001.

Vandenbroeck K, editor. *Cytokine Gene Polymorphisms in Multifactorial Conditions.* Boca Raton: CRC Press; 2006. https://doi.org/10.1201/9781420005325.

Vandenbroeck K. Cytokine gene polymorphisms and human autoimmune disease in the era of genome-wide association studies. *J Interferon Cytokine Res* 2012;32:139–51. https://doi.org/10.1089/jir.2011.0103.

Wang L, Wang F-S, Gershwin ME. Human autoimmune diseases: a comprehensive update. *J Intern Med* 2015;278:369–95. https://doi.org/10.1111/joim.12395.

Wilson AG, Symons JA, McDowell TL, McDevitt HO, Duff GW. Effects of a polymorphism in the human tumor necrosis factor alpha promoter on transcriptional activation. *Proc Natl Acad Sci U S A* 1997;94:3195–9. https://doi.org/10.1073/pnas.94.7.3195.

24

Genomic Disturbances Associated with Penile Cancer and Novel Therapies

María Juliana Chaves-Medina
UROGIV Research Group. School of Medicine, Universidad del Valle,y Cali, CO

Herney Andrés García-Perdomo
Division of Urology/Urooncology, Department of Surgery, School of Medicine, Universidad del Valle, Cali, CO

CONTENTS

ABSTRACT The prevalence of penile cancer (PeCa) is more common in patients aged between 50 and 70 years; this disease is rare and aggressive. It is more common in developing countries, and it is also associated with a high prevalence of human papillomavirus (HPV) infection. Nowadays, molecular pathways are beginning to be known for different investigators to use novel therapies for future diagnosis and treatment. The knowledge of single nucleotide polymorphism is a new proposal in scientific studies that contributes to the better survival of the patients. This brief review has the most updated knowledge on PeCa for understanding the molecular markers and their application in monitoring and predicting disease.

24.1 Introduction

Penile cancer (PeCa) is considered a rare and aggressive neoplasm. It has an incidence of 0,1–0,9 men/100,000 diagnosed in Europe and the USA per year. It is approximately up to 19 men per 100,000 in Asia, Africa, and South America (Kayes et al., 2007; Montes-Cardona & García-Perdomo, 2017). High incidence, morbidity, and mortality are commonly seen in young patients in developing countries (Adashek et al., 2019; Marchioni et al., 2018).

The papillary variant of squamous cell carcinoma (SCC) is the most predominant tumor type in PeCa, accounting for 95% of all cases (Kayes et al., 2007). Melanomas, sarcomas, lymphomas, and basal cell carcinomas account for the remaining 5%. Some are considered pre-malignant, such as erythroplasia of Queyrat, Bowen's disease, leukoplakia, balanitis xerotica obliterans (BXO), and Buschke-Lowenstein's tumor (Kayes et al., 2007). Furthermore, factors impacting prognosis (Table 24.1) include older age, tumor grade, and regional lymph node invasion at diagnosis. Perhaps, acknowledging the molecular structure of PeCa could help understand prognosis and develop novel therapies for treatment. This brief review has the most updated knowledge on PeCa for understanding the molecular markers and their application in monitoring and predicting disease.

24.2 Pathology and Pathogenesis of PeCa

PeCa diagnosis is based on physical examination of primary lesions and regional lymph nodes evaluation (Marchioni et al., 2018). The first changes begin at the glans and involve the foreskin; they can frequently be detected in the form of an exophytic, superficial, or ulcerated painless lesion. This disease progresses slowly, and people generally die from local complications such as growing infection, necrosis, and finally, erosion of the femoral vessel (Adashek et al., 2019; Albersen & Spiess, 2019).

Significant risk factors for penile cancer include phimosis, smoking, inflammatory conditions, and human papillomavirus (HPV) infection. The combination of these factors suggests that this malignant disease involves multiple molecular pathways—moreover, chronic inflammation, oncogenic viral products, and carcinogenic cofactors (Kayes et al., 2007).

SCC accounts for 95% of all malignancies; half of these come from the glans non-keratinized epithelium or prepuce inner layer. On the contrary, other malignancies (melanoma

DOI: 10.1201/9781003246244-24

TABLE 24.1

Risk Factors and Protective Factors for PeCa

Risk Factors	Protective Factors
Smoking	Circumcision
Inflammatory conditions	High educational level
HPV infection	Conservative sexual behavior
Phimosis	
Obesity	
Multiple sexual partners	
HIV	

and sarcoma) or metastases are sporadic. In penile SCC, various histologic subtypes are distinguished. The new Union for International Cancer Control (UICC) classification distinguishes between HPV-associated and non-HPV-associated penile carcinomas (Hakenberg et al., 2018).

Chronic inflammations such as lichen sclerosus, phimosis, and BXO, are strongly associated with a high prevalence. In epidemiological studies, chronic cigarette smoking, obesity, and poor personal hygiene are other risk factors (Hakenberg et al., 2018) (6); in addition, phimosis and chronic inflammations such as balanoposthitis are the most commonly found factors. According to Hakenberg et al., circumcision in childhood significantly reduces the prevalence of this disease. Also, removing the foreskin decreases the non-keratinized penile skin's exposed surface by 50%, which could be a protective factor against this illness.

24.3 Single Nucleotide Polymorphism Molecular Pathways

Nowadays, it is clear that PeCa disease is associated with HPV infection, the one that is predominantly linked as high-risk subtypes 16 and 18, which strongly correlates with the carcinogenic processes in generating in situ and invasive cancers of epithelial tumors. Stoehr et al. (2018) reviewed a cohort of healthy Caucasian males and a cohort with the disease. They showed the association of oncogenic proteins E6 and E7 in the pathways controlling cellular differentiation, proliferation, and apoptosis (Chahoud et al., 2019; Stoehr et al., 2018; Van Der Kwast et al., 2020). The E7 protein is bound with the inactivation of the tumor suppressor retinoblastoma (Rb), and the E6 protein is associated with the cellular p53 protein followed by degradation at the ubiquitin-proteasome pathway (Stoehr et al., 2018).

Stoehr et al. (2018) studied the degradation of the E6 protein, which showed the influence and control of single nucleotide polymorphism (SNP) in exon 4 of the *p53* gene (Stoehr et al., 2018). At position 215 of the p53 cDNA, either guanine (5′-*CGC*-3′, encoding for the amino acid arginine) or cytosine (5′-*CCC*-3′, encoding for the amino acid proline) base is located, resulting in a polymorphic amino acid site at codon 72 of P53 (p.Arg72Pro) (Stoehr et al., 2018).

In this investigation, it is clear that the cells with p.Arg72 consequently are more resistant to apoptotic stimuli, but the p53 pathway not only happens in PeCa. It could be found in other types of cancer; however, there is no association with HPV infection (Stoehr et al., 2018). Generally, verrucous tumors do not associate with HPV infection, leading to the hypothesis that they depend on other genetic and molecular changes (Kayes et al., 2007; Stoehr et al., 2018; Zhou et al., 2018).

Besides, Castren et al. analyzed the expression profile of pre-malignant conditions and evaluated the profiles of invasive tumors. The authors concluded no correlation between P53 expression and HPV infection because they did not overexpress the p53 pathway. It means that MDM2 protein is the one that controls p53 overexpression and not the other way around, saying that the overexpression of MDM2 is essential to control the progression of PeCa (Kayes et al., 2007).

Studies exploring the expression of p.Arg72Pro in invasive penile cancer have not shown a relationship with the grade or stage of the disease. However, substantial evidence from other studies refutes these data (Stoehr et al., 2018). Humbey et al. reviewed this hypothesis in men from France with penile cancer (Humbey et al., 2003; Kayes et al., 2007). A link between HPV infection and penile cancer was confirmed, but this study concluded that the Arg/Arg genotype at codon 72 was not a risk factor in this series. Considerable research has been done, but only limited evidence exists (Gentile et al., 2018; Humbey et al., 2003).

Nevertheless, molecular biological changes may be relevant to prognosis. In penile cancer, loss of heterozygosity adjacent to tumor suppressor genes (2q, 6p, 8q, 9p, 12q, 17p13) is commonly observed and occurs even more frequently in lymph node metastases (3p, 6p, 6q, 8q, 9p, 11q, 12q, 15q, 17p, 18q) (Chahoud et al., 2019; Kayes et al., 2007). Also, regarding prognosis and copy number alterations (CNAs), Yu et al. found them present in 54% of MYCN. Additional targets were TP53 (54%), Bmi1 (62.85), FAK (54%), and MDM2 (51.4%). Consequently, they found a worse overall survival for those with MYCN, and FAK amplifications, considering them independent prognostic factors for PeCa (Yu et al., 2020).

A p16 damage related to the ROS/RNS pathway may play a crucial role in penile carcinogenesis, specifically in the context of chronic inflammation (Chipollini et al., 2017) (p16 promoter hypermethylation and loss of heterozygosity) (13). Other tumor suppressor genes, such as KAI1 and nm23H1, are associated with metastatic spread (Hakenberg et al., 2018).

Other clinically relevant gene alterations include CDKN2A, NOTCH1, PIK3CA, CCND1, SOX2, EGFR amplification, BRCA2 insertions/deletions, and FBXW7 point mutations (Chahoud et al., 2019). Chahoud et al. reported the whole-exome sequencing (WES) analysis of 34 patients, where the most common mutations were found in TP53 (35%), NOTCH1 (35%), CDKN2A (23%), PIK3CA (21%), and CASP8 (21%) (10,12), this pathway is an independent predictor of lymph node metastases. Also, patients with negative p53 had better survival. However, the overall outcome was considerably worse if tumors were positive for p53 and HPV DNA (1) (Table 24.2) (Chahoud et al., 2019; Stecca et al., 2021).

The relation between p53 expression and HPV infection was also analyzed. Studies showed that p53 staining was an independent predictor of lymph node metastases. Also, patients with negative p53 had significantly better 5-year and 10-year overall survival. However, the overall outcome was

TABLE 24.2

Molecular Disturbances Reported for Penile Carcinomas

Inflammation	Tumor Suppressor Genes	Oncogenes	Growth Factors/Receptors
COX-2	P53	PI3K/PTEN/AKT	HPV E5/E7
PGE-2	P16	HER-3/HER-4	EGFR, VEGF
	PTEN, PTCH1	NOTCH1, NOTCH2, CCND1, SOX2,	TSP-1
	Rb/E2F	FBXW7	RASSF1A
	KAI1, nm23H1	MYC, ROS1	CAFs

considerably worse if tumors were positive for p53 and HPV DNA (Kayes et al., 2007) (Table 24.2). Accordingly, whether p53 alterations are of prognostic relevance has not yet been conclusively established (Hakenberg et al., 2018).

Feber et al. (2016) performed the WES of PeCa, and they found genes containing novel mutations. As a result, CSN1/GPS1 was mutated in 11% of cases, with a single case retaining two CSN1 alterations. They found that CSN1 suppresses G-proteins and mitogen-activated signal transduction, and it might be an essential part of the COP9 signalosome complex, suggesting gene silencing and changes in epigenetic regulation (Feber et al., 2016).

On the other hand, Nastazia et al. (2018) studied polymorphism rs34536443 in a Pro-to-Ala change at amino acid position 1,104 of the encoded tyrosine kinase 2 variant (Tyk2). This protein affects its enzymatic activity and confers protection against autoimmune and cancer diseases (Lesgidou et al., 2018). However, the details of the underlined mechanism are unknown; they suggested that the reduction in the enzymatic activity of the resulting protein variant due to stabilization of inactive conformations dysregulation of these pathways through somatic mutations of the PIK3CA, HRAS, and KRAS genes increases, but these are not established at squamous cell carcinoma cases (Andersson et al., 2008; Lesgidou et al., 2018).

Andersson et al. (2008) found somatic mutations in 11 of 28 penile cancer samples (39%). There were mutations in the *PIK3CA* (29%), *HRAS* (7%), and *KRAS* (3%) genes (Friedenreich et al., 2016). Besides, the growth of receptor mutations is among the most common events in various cancers and tyrosine kinase activity. Activating the tyrosine kinase receptor stimulates RAF signal, which increases mitogen-activated (MEK) and kinase pathway (ERK). This activation regulates transcription factors and controls the cell cycle determining survival, growth, differentiation, and transformation (Friedenreich et al., 2016).

PIK3CA mutations located in exon 9 were G1624A or G1633A with the first nucleotide of the coding mRNA sequence as the reference, corresponding to amino acid changes of E542K and E545K (Adashek et al., 2019; Friedenreich et al., 2016). The *HRAS* gene 1 mutation was found in exon 2 (G34A). Another one was found in exon 3 (A182T) with the first nucleotide of the coding sequence as the reference, corresponding to amino acid changes of G12S and Q61L (Adashek et al., 2019). The single mutation identified in the *KRAS* gene was a nucleotide change of G34A, corresponding to a G12S amino acid change.

Somatic mutations in the *PIK3CA* gene did not coincide with any particular tumor grade, pT stage, or N stage. In addition, *HRAS* and *KRAS* gene mutations were found in pathologically characterized advanced tumors, which play a crucial role in cell migration and invasion. On the other hand, the number of HPV-infected samples did not significantly differ between mutated and non-mutated samples (Friedenreich et al., 2016).

Cytogenetic analysis, genomic hybridization techniques, and the use of serine protease inhibitors such as SERPINB and SERPINB4 as serological markers have been assessed in several studies involving penile cancer (Kayes et al., 2007). Guerrero et al. suggested that the hypermethylation of thrombospondin 1 (TSP-1), RASSF1A, and p16 genes likely contributes to disease progression (Kuasne et al., 2021).

24.4 Chronic Inflammation and Chromosomal Imbalance

Other essential mediators in inflammation-induced penile carcinogenesis are cyclooxygenase-2 (COX-2) and prostaglandin E2 (PGE2). COX-2 is significantly expressed in PeCa; overproduction of prostaglandins and thromboxanes play a critical role in the proliferation, angiogenesis, and activation of EGFR (Andersson et al., 2008).

Kuasne et al. (2021) established a cell line derived from fibroblast overexpressed markers related to the amplification of cancer-associated fibroblasts (CAFs) signature (*MMP2*, *REAB3B*, *COL6A1*, *COL6A2*, *CTSK*, *THY1*, *PDGFRA*, *DCN*, and fibroblast activation *ACTA2*). Primary tumors presented fewer chromosomal imbalances, with the highest number of altered regions. In addition, primary tumor marked similar chromosomal imbalances, including a homozygous loss of 9p21.3 (*CDKN2A* and *CDKN2B* genes), a gain of 4q12 (*PDGFRA*), an amplification of 11p15.5 (*H19* and *IGF2* genes), and a loss of 6p25.3 (*DUSP22*), which were linked to morphological features, migration, and amplification (10,17).

MicroRNAs (miRNAs) frequently regulate gene expression and post-transcriptional response involved in cell growth, axonogenesis, and angiogenesis. Integrative analysis revealed *MMP1* and *MMP12* might be regulated by *hsa-miR-145-5p*, which was underexpressed in PeCa tissues (Andersson et al., 2008). The hsa-miR-145 (miRNAs) underexpression did not predict a poorer prognosis.

The target *MMP1* showed increased expression in patients with lymph node metastasis. A recent study reported loss of *miR-1*, *miR-101*, and *miR-204* associated with lymph node metastasis and unfavorable prognosis (Andersson et al., 2008), as well as alterations, binding with the transformation of normal cells to malignant lesions.

Zhou and coworkers (2018) analyzed the biological and molecular characteristics of penile squamous carcinoma cell

TABLE 24.3

SNVs and Indels of Critical Genes in Penile Cancer Cell Lines (Zhou et al., 2018)

Cells	Penl1	Penl2	149RCa	149RM	LM156
TP53	R244P*, P33R	R234H	R209Q*, R71C*, P33R	R209Q*, R71C*, P33R	P33R
PTCH1	T1195S	T1195S	P1315L	P1315L	P1315L
CDKN2A	R80X*	R80X*	H83Y	H83Y	-
BRCA1		S1566G			
NOTCH2	N46S*	P6fs	N46S, S220R, P227H, T235S	N46SΔ	N46S
NOTCH1	G427D$	C467F$	-	-	D1200N*$
FBXW7	-	-	S25X	-	-
ROS1	-	P224S*	-	-	-
MSH6	-	R922Q	-	-	-
Migration Potential	↑	↑↑	↑	↑	↑↑
Metastatic LNs	↑	↑↑	↑↑	↓	↑
Chemotherapy	Sensitivity	Sensitivity	Resistance	Resistance	Resistance

X—stop codon, *—validated by Sanger sequencing, Δ—not validated by Sanger sequencing, other variations (without labels * or Δ)—not tested by Sanger sequencing, $—variants that were not indexed in the COSMIC70 database.

line models. Five cell lines, Penl1, Penl2, 149RCa, 149RM, and LM156, were sequenced. Those analyses showed that these cell lines exhibited a similar mutational spectrum, containing 1229–1290 SNPs (Table 24.3). All cell lines consist of pure, immortalized, and epithelial-like cells without HPV infection (Zhou et al., 2018).

Cell line models were compared to the National Center for Biotechnology Information (NCBI) human reference genome, and transition mutations (C:G>T:A and T:A>C:G) were the most frequent substitution types (Chipollini et al., 2017). All cell lines had CDKN2A deletion/mutation, MYC amplification, and TP53 mutations (Chipollini et al., 2017) (Table 24.3).

24.5 Novel Therapies

Recently, different authors have reported molecular markers and new pathways that are potentially helpful for developing novel therapies; these epigenetic mutations are pointed out as potential therapeutic targets, including checkpoint inhibitors in cell cycle control (Friedenreich et al., 2016; Vanthoor et al., 2021).

Until now, we know that circumcision in childhood reduces the risk of invasive penile cancer. The effect of circumcision could be primarily mediated by eliminating phimosis rather than decreasing the HPV risk of transmission. In addition, prophylactic HPV vaccination is another primary prevention strategy that has led to a significant decrease in HPV-associated infections and cancers; however, it is still a matter of debate in PeCa.

Novel therapies involve adoptive T cell therapy by harvesting patient-specific T cells derived from primary or metastatic focus. These HPV-targeted tumor-infiltrating lymphocytes (TILs) could then be transferred back to donor patients to induce an anti-tumor immune response (Andersson et al., 2008; Vanthoor et al., 2021).

Cell tumors present at least 80% SNPs, and the overexpression of CAF markers is the new target in a clinical trial, focusing on downstream proteins (Chipollini et al., 2017; Vanthoor et al., 2021). Also, alterations in the DNA repair pathway (BRCA2 and ATM GA) and tyrosine kinase constitute the potential for personalized, targeted therapy to be evaluated in this rare and aggressive cancer.

The mutations in NOTCH1 lead to loss of function disturbances, rendering the cancer cells vulnerable to PI3K/mTOR inhibition. Alterations in the mTOR pathway (NF1 GA and PTEN) suggest possible targeting with inhibitors rapamycin (mTOR)/phosphatidylinositol 3-kinase pathway (Stoehr et al., 2018). Positive cells for PD-L1 expression on infiltrating immune cells constitute a valid rationale to test the activity of programmed cell death protein 1 (PD-1) immune checkpoint blockade. Udager et al. first reported the frequent *PD-L1* expression in penile cancer (Andersson et al., 2008; Stecca et al., 2021); approximately 62% of tumors were tested positive for *PD-L1* and associated with poor disease-specific survival. Trials are being planned using anti-*PD-1* (NCT02837042) or *PD-L1* antibodies (NCT02721732) that may clarify the role of checkpoint inhibition in PeCa (Andersson et al., 2008; Kuasne et al., 2021; Stecca et al., 2021).

Immunotherapy using checkpoint inhibitors, such as PD-1, anti-PD-L1, and anti-cytotoxic T lymphocyte antigen 4 (CTLA4), has emerged as promising treatment options for various cancers, including other HPV-associated ones (Kuasne et al., 2021).

The HER/PTEN/Akt pathway was also disrupted. This is an exciting opportunity for clinical evaluation with multiple available targeted treatment options such as dacomitinib, which is one of these targeted treatment options that have been approved for the treatment of patients with epidermal growth factor receptor exon 19 deletion or exon 21 L858R substitution mutations (Friedenreich et al., 2016; Vanthoor et al., 2021). Another trial of irreversible kinase inhibitors of humans using afatinib (NCT02541903) was performed. Trials provide insights into targeting the *HER* pathway with preliminary data showing anti-tumor activity linking molecular alterations with clinical response (Chipollini et al., 2017).

Zhou et al. (2018) thought that HRAS G13V/G13R mutations and PI3KCA amplification might relate to resistance to anti-EGFR. They detected overexpression in most of these cell lines and EGFR amplification in two cell lines, but they were all resistant to EGFR inhibitors. Activating HRAS and PI3KCA mutations correlates with anti-EGFR therapy resistance, but this needs more investigation (Zhou et al., 2018).

Zhu et al. evaluated the efficacy of kinase inhibitors (sorafenib and sunitinib) in six patients who had previously received at least two chemotherapy regimens. One partial response and four stable disease responses were observed; they concluded that molecularly defined groups might benefit from *VEGF*-targeted therapy or, conversely, may be specifically sensitive to taxanes (Chipollini et al., 2017). However, more research is needed for this treatment.

24.6 Conclusions

PeCa is a rare disease with a poor prognosis and high impact on quality of life. A few years ago, the research incorporated molecular panels to understand oncogenesis that can identify therapeutic targets, but was limited by a few pieces of research. Incorporating molecular therapies represents a potential revolution in diagnosis and prognosis of disease. Furthermore, additional prospective studies are necessary to investigate factors that determine the presence of metastatic disease and prognosis.

ABBREVIATIONS

A:	Adenine
BXO:	Balanitis xerotica obliterans
C:	Cytosine
COX-2:	Cyclooxygenase-2
CTLA4:	Cytotoxic T lymphocyte antigen 4
EGFR:	Epidermal growth factor receptor
ERK:	Kinase pathway
DNA:	Deoxyribonucleic acid
G:	Guanine
HPV:	Human papillomavirus
MEK:	Mitogen-activated kinase
miRNA:	MicroRNA
mRNA:	Messenger RNA
mTOR:	Mammalian target of rapamycin
NCBI:	National Center for Biotechnology Information
PD-1:	Programmed cell death protein 1
PD-L1:	Programmed death ligand 1
PeCa:	Penile cancer
PGE2:	Prostaglandin E2
Rb:	Retinoblastoma
RNA:	Ribonucleic acid
SCC:	Squamous cell carcinoma
SNP:	Single nucleotide polymorphism
T:	Thiamine
TSP-1:	Thrombospondin 1
Tyk2:	Tyrosine kinase 2 variant
UICC:	Union for International Cancer Control
WES:	Whole-exome sequencing

REFERENCES

Adashek, J. J., Necchi, A., & Spiess, P. E. (2019). Updates in the molecular epidemiology and systemic approaches to penile cancer. In *Urologic Oncology: Seminars and Original Investigations* (Vol. 37, Issue 7, pp. 403–408). Elsevier Inc. https://doi.org/10.1016/j.urolonc.2019.04.012

Albersen, M., & Spiess, P. E. (2019). Editorial: Penile cancer. In *Current Opinion in Urology* (Vol. 29, Issue 2, pp. 143–144). Lippincott Williams and Wilkins. https://doi.org/10.1097/MOU.0000000000000586

Andersson, P., Kolaric, A., Windahl, T., Kirrander, P., Söderkvist, P., & Karlsson, M. G. (2008). PIK3CA, HRAS and KRAS Gene mutations in human penile cancer. *Journal of Urology, 179*(5), 2030–2034. https://doi.org/10.1016/j.juro.2007.12.040

Chahoud, J., Pickering, C. R., & Pettaway, C. A. (2019). Genetics and penile cancer: Recent developments and implications. *Current Opinion in Urology, 29*(4), 364–370. https://doi.org/10.1097/MOU.0000000000000640

Chipollini, J., Chaing, S., Azizi, M., Kidd, L. C., Kim, P., & Spiess, P. E. (2017). Advances in understanding of penile carcinogenesis: The search for actionable targets. In *International Journal of Molecular Sciences* (Vol. 18, Issue 8). MDPI AG. https://doi.org/10.3390/ijms18081777

Feber, A., Worth, D. C., Chakravarthy, A., Winter, P. de, Shah, K., Arya, M., Saqib, M., Nigam, R., Malone, P. R., Tan, W. S., Rodney, S., Freeman, A., Jameson, C., Wilson, G. A., Powles, T., Beck, S., Fenton, T., Sharp, T. V., Muneer, A., & Kelly, J. D. (2016). CSN1 Somatic mutations in penile squamous cell carcinoma. *Cancer Research, 76*(16), 4720–4727. https://doi.org/10.1158/0008-5472.CAN-15-3134

Friedenreich, C. M., Neilson, H. K., Farris, M. S., & Courneya, K. S. (2016). Physical activity and cancer outcomes: A precision medicine approach. *Clinical Cancer Research, 22*(19), 4766–4775. https://doi.org/10.1158/1078-0432.CCR-16-0067

Gentile, V., Vicini, P., Giacomelli, L., Cardillo, M. R., Pierangeli, A., & Degener, A. M. (2018). Detection of human papillomavirus DNA, P53 and KI67 expression in penile carcinomas. *International Journal of Immunopathology and Pharmacology, 19*(1), 209–215. https://doi.org/10.1177/205873920601900121

Hakenberg, O. W., Dräger, D. L., Erbersdobler, A., Naumann, C. M., Jünemann, K. P., & Protzel, C. (2018). übersichtsarbeit: Diagnostik und therapie des Peniskarzinoms. In *Deutsches Arzteblatt International* (Vol. 115, Issue 39, pp. 646–652). Deutscher Arzte-Verlag GmbH. https://doi.org/10.3238/arztebl.2018.0646

Humbey, O., Cairey-Remonnay, S., Guérrini, J. S., Algros, M. P., Mougin, C., Bittard, H., & Aubin, F. (2003). Detection of the human papillomavirus and analysis of the TP53 polymorphism of exon 4 at codon 72 in penile squamous cell carcinomas. *European Journal of Cancer, 39*(5), 684–690. https://doi.org/10.1016/S0959–8049(02)00835-3

Kayes, O., Ahmed, H. U., Arya, M., & Minhas, S. (2007). Molecular and genetic pathways in penile cancer. *Lancet Oncology, 8*(5), 420–429. https://doi.org/10.1016/S1470-2045(07)70137-7

Kuasne, H., Canto, L. M. do, Aagaard, M. M., Muñoz, J. J. M., Jamblinne, C. De, Marchi, F. A., Scapulatempo-Neto, C., Faria, E. F., Lopes, A., Carréno, S., & Rogatto, S. R. (2021). Penile cancer-derived cells molecularly characterized as models to guide targeted therapies. *Cells, 10*(4), 814. https://doi.org/10.3390/cells10040814

Lesgidou, N., Eliopoulos, E., Goulielmos, G. N., & Vlassi, M. (2018). Insights on the alteration of functionality of a tyrosine kinase 2 variant: A molecular dynamics study. *Bioinformatics (Oxford, England), 34*(17), i781–i786. https://doi.org/10.1093/BIOINFORMATICS/BTY556

Marchioni, M., Berardinelli, F., De Nunzio, C., Spiess, P., Porpiglia, F., Schips, L., & Cindolo, L. (2018). New insight in penile cancer. *Minerva Urologica e Nefrologica*, *70*(6), 559–569. https://doi.org/10.23736/S0393-2249.18.03215-0

Montes-Cardona, C., & García-Perdomo, H. A. (2017). Incidence of penile cancer worldwide : Systematic review and meta-analysis. *Rev Panam Salud Publica*, *41*, e117. https://doi.org/10.26633/RPSP.2017.117

Stecca, C. E., Alt, M., Jiang, D. M., Chung, P., Crook, J. M., Kulkarni, G. S., & Sridhar, S. S. (2021). Recent advances in the management of penile cancer: A contemporary review of the literature. *Oncology and Therapy*, *9*(1), 21–39. https://doi.org/10.1007/s40487-020-00135-z

Stoehr, R., Weisser, R., Wendler, O., Giedl, J., Daifalla, K., Gaisa, N. T., Richter, G., Campean, V., Burger, M., Wullich, B., & Hartmann, A. (2018). P53 codon 72 polymorphism and risk for squamous cell carcinoma of the penis: A Caucasian case-control study. *Journal of Cancer*, *9*(22), 4234–4241. https://doi.org/10.7150/jca.26050

Van Der Kwast, T. H., Egevad, L., Kristiansen, G., & Grignon, D. J. (2020). The International Society of Urological Pathology consultation on molecular pathology of urogenital cancer. *American Journal of Surgical Pathology*, *44*(7), 859–861. https://doi.org/10.1097/PAS.0000000000001496

Vanthoor, J., Vos, G., & Albersen, M. (2021). Penile cancer: Potential target for immunotherapy? *World Journal of Urology*, *39*(5), 1405–1411. https://doi.org/10.1007/s00345-020-03510-7

Yu, Y., Gao, C., Chen, Y., Wang, M., Zhang, J., Ma, X., Liu, S., Yuan, H., Li, Z., & Niu, H. (2020). Copy number analysis reveal genetic risks of penile cancer. *Frontiers in Oncology*, *10*, 2788. https://doi.org/10.3389/FONC.2020.596261/BIBTEX

Zhou, Q. H., Deng, C. Z., Li, Z. S., Chen, J. P., Yao, K., Huang, K. B., Liu, T. Y., Liu, Z. W., Qin, Z. K., Zhou, F. J., Huang, W., Han, H., & Liu, R. Y. (2018). Molecular characterization and integrative genomic analysis of a panel of newly established penile cancer cell lines article. *Cell Death and Disease*, *9*(6). https://doi.org/10.1038/s41419-018-0736-1

25

Role of Genetic Polymorphisms in Gastric Cancer Susceptibility

Syed Douhath Yousuf and Syed Mudassar
Department of Clinical Biochemistry, Sher-I-Kashmir Institute of Medical Sciences, Soura, Srinagar, India

CONTENTS

DOI: 10.1201/9781003246244-25

ABSTRACT Gastric cancer is among the most prevalent and lethal malignancies, having a grim prognosis all over the world. It is the third most common cause of cancer-related mortality. Gastric cancer is a complex condition that is influenced by both environmental and hereditary factors. It is best described as an infection-induced, inflammation-driven disease that strikes people with genetic predisposition after several decades. Different alleles of different genes (polymorphisms) that regulate carcinogenesis pathways and disease progression have been related to individual propensity to cancer risk. Gastric cancer is connected to hereditary cancer syndromes in 3% of cases. The majority of occurrences of stomach cancer are sporadic, resulting from a complex interaction of acquired genetic defects and environmental risk factors. When paired with exposure to environmental risk factors, single nucleotide polymorphisms are one of the acquired genetic defects that might accelerate the development of malignant disease. Several classes of genes involved in the development of stomach cancer display polymorphism. In the past, the genes involved were discovered using a candidate gene method, and more recently, genome-wide association studies have been used. This chapter discusses the genetic profile of gastric cancer, with an emphasis on how single nucleotide polymorphisms influence gastric cancer propensity and outcomes.

KEYWORDS: *Gastric Cancer; Single Nucleotide Polymorphism; Mutations; Genetic Profile; Risk Factors; Epigenetic; GWAS.*

25.1 Introduction

Gastric Cancer (GC) is a heterogeneous, aggressive solid tumour malignancy that is a global health hazard. In terms of total mortality, GC is the world's third greatest cause of cancer-related deaths, trailing only lung and colorectal cancers (Bray et al., 2018). Geographical location and ethnicity have an impact on the prevalence of GC. The locations with the highest occurrence are Central and South America, Eastern Europe, and East Asia (China and Japan), while Australia and New Zealand, Southern Asia, North and East Africa, and North America are considered low incidence zones. Eastern Asia has the highest prevalence worldwide (Ang and Fock, 2014). In the USA, the incidence varies by ethnicity, with Asians having the highest prevalence and non-Hispanic whites having the lowest. It also reveals a sex predisposition, with men two to three times more likely than women to be impacted. The prognosis for GC is bleak, with 5-year survival rates of just 5.3% in patients with advanced cancer and 68.8% in individuals with localised disease (Chandra et al., 2021). Nonetheless, since World War II, the global incidence of stomach cancer has been dropping (Bertuccio et al., 2009). Improved living standards and early detection measures have been credited with this. As a risk factor, *Helicobacter pylori* infection is critical for tactical eradication in high-risk locations as a preventive intervention (Lee et al., 2006).

The aetiology of GC is complex, with both genetic and environmental variables having a part in its aetiopathogenesis (Yusefi et al., 2018). Over the last several decades, tremendous progress has been made in our understanding of the genetic, epigenetic, transcriptomic, and metabolic abnormalities that underpin this ailment. Individual diversity in chances of cancer has been linked to distinct alleles of different genes (polymorphisms) that regulate carcinogenesis pathways and the likelihood of development as a genetic illness. This has allowed for a better knowledge of the disease's neoplastic course by describing and proposing important molecular indicators as prognostic markers. We want to examine the genetic aspects of gastric cancer and gain new and developing information about this illness in this chapter with particular emphasis on how single mutations can affect or predict the risk of gastric cancer.

25.1.1 Risk Factors

There are a number of risk variables at play. There are two types of risk factors: non-modifiable risk factors and modifiable risk factors. Non-modifiable risk factors include those that a person is born with and has no grip over it such as advanced age, male sex, ethnic background, personal or family history of stomach polyps, and the presence of predisposing syndromes, and modifiable risk factors include those that a person can handle, for example voluptuous habits such as tobacco smoking and alcohol abuse; exposure to radiations; and *Helicobacter pylori* infection (Cavatorta et al., 2018) (Figure 25.1).

25.1.2 Gastric Cancer Classification

Lauren's GC classification, which was first published in 1965, is still the most often used among GC classifications. Intestinal and diffuse GCs were identified as two main histological subgroups of GC. Visible glands and cohesiveness between tumour cells distinguish intestinal cancer. The diffuse subtype includes cells that are poorly cohesive and infiltrate the stomach wall diffusely, with little or no gland development. The cells are normally small and spherical, and they form a signet ring. The signet ring cell carcinoma is classified as a diffuse subtype in this classification. Later, to characterise rare histology, a separate group of undetermined type was added. The intestinal subtype has been linked to intestinal metaplasia of the gastric mucosa and the presence of *H. pylori* infection, whereas the diffuse GC subtype is derived from normal gastric mucosa.

World Health Organization (WHO) categorisation (Flejou, 2011) was published in 2010. Multiple subgroups of gastric adenocarcinoma (GA), such as tubular, mucinous, papillary, and mixed carcinoma, are analogous to the indeterminate type in the Lauren categorisation system. The signet ring cell carcinoma is a form of poorly cohesive cancer. The remaining adenocarcinomas of the stomach are categorised as rare. Tubular adenocarcinoma is the most prevalent GC subtype, followed by papillary and mucinous kinds, according to WHO classification.

The majority of stomach cancers (90%–95%) are adenocarcinomas. Intestinal gastric malignancies account for around 53% of all stomach malignancies. Around 18% of all stomach malignancies are diffuse gastric tumours. Gastro-oesophageal junction adenocarcinoma accounts for roughly 18% of all adenocarcinomas. Gastrointestinal stromal tumours (GISTs) are rare stomach tumours that begin in cells termed interstitial

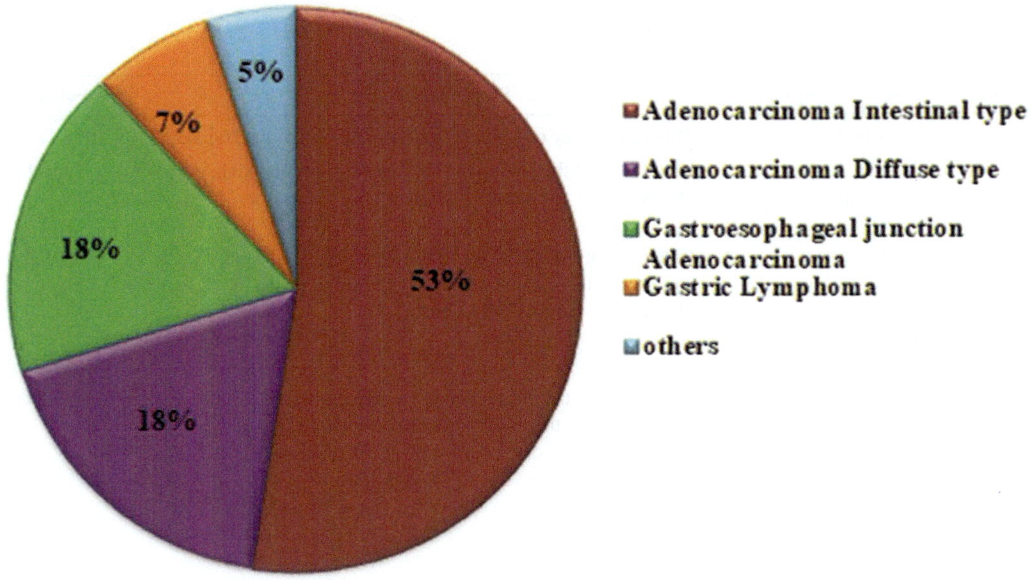

FIGURE 25.1 Various risk factors associated with gastric cancer.

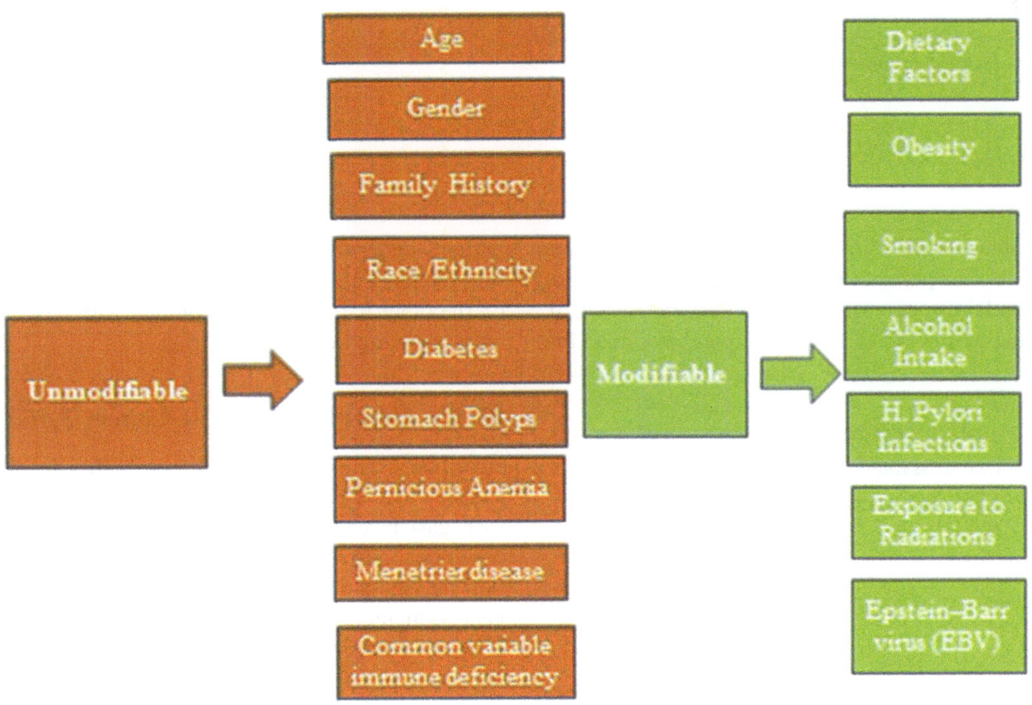

FIGURE 25.2 Gastric cancer types by relative incidence.

cells of Cajal in the stomach wall. Neuroendocrine tumours/carcinoids are a rare, slow-growing stomach tumour that begins in the cells of the neurological and endocrine systems. Lymphomas account for roughly 7% of all stomach cancers (Qu et al., 2015).

Figure 25.2 shows the different kinds of stomach cancer and their relative occurrence.

Molecular Classification:

Attempts to characterise GC at the molecular level have been made in recent years in the goal of informing specific therapeutic approaches. Three types, viz. proliferative, metabolic, and mesenchymal cells, were identified based on the mRNA analysis. These have a variety of molecular and genetic features. Proliferative tumours have high levels of genomic instability, TP53 mutations, DNA hypomethylation, and high activity of the E2F, MYC, and RAS oncogenic pathways. The metabolic subtype demonstrates the expression of genes found in normal gastric mucosa; therefore, cancer cell expression is identical to that of normal cells. The mesenchymal subtype is defined by low CDH1 expression and abnormal PI3K-AKT-mTOR pathway activation (Lei et al., 2013).

TABLE 25.1
Staging and Survival of Gastric Cancers

TNM Staging	Stage Grouping	Stage Description	Survival Rate (%)
Stage 0	T0, N0, M0	This malignancy only affects the inside lining of the stomach.	
Stage IA	T1, N0, M0	Penetration of the second or third layers of the stomach, or the second layer and nearby lymph nodes.	90–100
Stage IB	T1, N1, M0 T2, N0, M0		
Stage IIA	T1, N2, M0 T2, N1, M0 T3, N0, M0	If the cancer has spread to the second layer and distant lymph nodes, the third layer and only close lymph nodes, or all four layers but not the lymph nodes.	75–85
Stage IIB	T1, N3a, M0 T2, N2, M0		
Stage IIIA	T2, N3a, M0 T3, N2, M0 T4a, N1, M0 T4b, N0, M0	Penetration to the third layer and further far lymph nodes, or penetration to the fourth layer and surrounding tissues, as well as adjacent or farther away lymph nodes.	30–40
Stage IIIB	T1 or T2, N3b, M0 T3, N3a, M0 T4a, N3a, M0 T4b, N1 or N2, M0		
Stage IIIC	T3 or T4a, N3b, M0 T4b, N3a or N3b, M0		
Stage IV	any T, any N, M1	Cancer has spread to nearby tissues and lymph nodes, or it has advanced to other organs.	<5

T, tumour; N, node; M, metastasis.

A rigorous molecular investigation of 295 primary GAs was undertaken in 2014 as part of the Cancer Genome Atlas project, and a molecular category of GA was produced (Bass et al., 2014). According to this classification, GC can be divided into four subtypes:

1. Tumours positive for Epstein–Barr virus (EBV)
2. Microsatellite unstable tumours (MSI)
3. Genomically stable tumours (GS)
4. Tumours with chromosomal instability (CIN).

This classification paved the way for trials of targeted therapies in GC.

25.1.3 Staging

After discovering the cancer cells in the tissue sample, the next step is to stage the disease or determine its extent (Wagner et al., 2017) (see Table 25.1).

25.2 Genetic Profile of Gastric Cancer

The aetiopathogenesis of GC is complex, with environmental and genetic factors influencing the disease's course (Machlowska et al., 2020). The onset of sickness at a young age and the presence of familial clustering are both evidence of inherited susceptibility. According to studies, 30% of patients under the age of 30 have a cancer-related family history (Rona et al., 2017). The vast majority of stomach cancers, on the other hand, strike at random, with only around 15% of cases showing familial clustering.

25.2.1 Hereditary Gastric Cancer Syndromes

Hereditary cancer syndromes are linked to 3% of stomach cancer cases (McLean and El-Omar, 2014). Hereditary diffuse gastric cancer (HDGC) is the most common form of hereditary gastric cancer. It's an autosomal dominant condition caused by a heterozygous germline mutation in the CDH1 gene, which codes for E-cadherin. Cadherin 1 plays a role in cell–cell adhesion (Oliveira et al., 2009). Gastric adenocarcinoma and proximal polyposis of the stomach (GAPPS), a new hereditary gastric cancer disease, has been reported (Worthley et al., 2012). A malignant growth of fundic gland polyps in numerous families characterises this syndrome. GAPPS is caused by a genetic abnormality that is still unclear. Gastric cancer is also a common extra-colonic symptom in hereditary non-polyposis colon cancer (HNPCC). HNPCC is caused by faulty DNA repair caused by mutations in DNA mismatch repair genes such as MSH2 or MLH1 (Caldas et al., 1999). Other inheritable syndromes associated with an increased risk of gastric cancer include Li–Fraumeni syndrome (caused by a germline mutation in TP53) and Peutz–Jeghers syndrome (caused by a mutation in STK11, resulting in dysregulated signal transduction and blocking of the mTOR pathway) (Hearle et al., 2006; Van Lier et al., 2010).

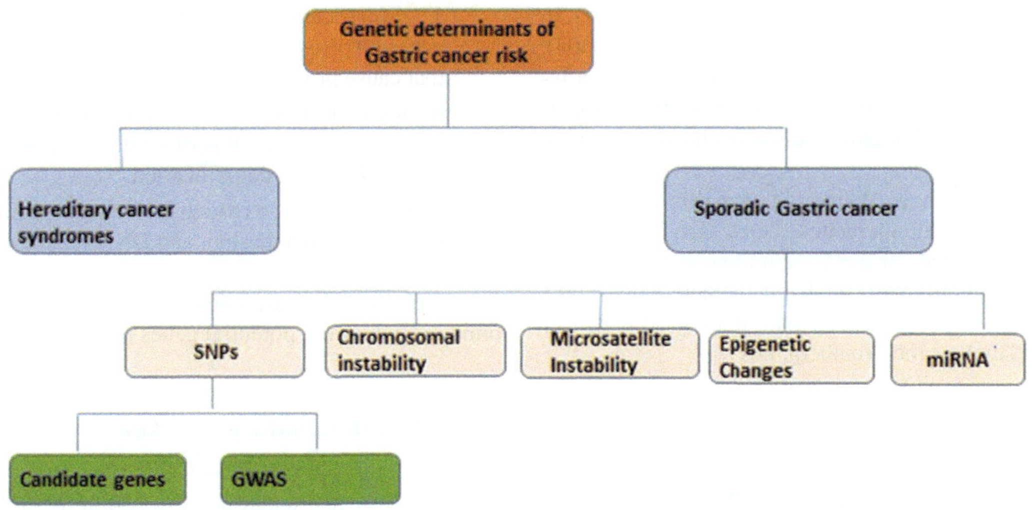

FIGURE 25.3 Genetic determinants of gastric cancer.

25.2.2 Acquired Genetic Abnormalities in Gastric Cancer

As previously noted, stomach cancer is mostly sporadic and results from a complex combination between acquired genetic abnormalities and environmental risk factors. The processes that lead to acquired genetic defects (Figure 25.3) are as follows.

25.2.2.1 Chromosomal Insufficiency

Cell replication involves a dynamic process called chromosomal instability, which can result in transitory or long-lasting alterations. This process can result in the loss or gain of entire chromosomes, as well as a change in the amount of copies of DNA (aneuploidy). Alternatively, the instability could simply affect a portion of a chromosome, causing essential gene families to lose or gain function. Oncogenes, tumour suppressor genes, DNA repair genes, and cell cycle checkpoint genes may all play a role in the genesis of malignant transformation. This process can result in translocation, amplification, deletion, or allelic loss (loss of heterozygosity). There are numerous chromosomal abnormalities in gastric cancer (Hudler, 2012). Intestinal cancer copies number gains at 8q, 17q, and 20q, and diffuse gastric cancer copy number gains at 12q and 13q are two examples. Similar changes have been linked to patient survival and other clinicopathological variables (Buffart et al., 2011; Kim et al., 2003; Wu et al., 2001).

25.2.2.2 Microsatellite Instability

In HNPCC, microsatellite instability is crucial for the establishment of a clinical phenotype. Aside from that, microsatellite instability is found in up to 50% of sporadic stomach cancers. It's mostly caused by epigenetic changes in mismatch repair genes, particularly MSH1 (Ottini et al., 2006). The most prevalent epigenetic change in this condition is hypermethylation of the promoter region, which hinders DNA mismatch repair and results in many mutations inside simple nucleotide repeats. The degree of microsatellite instability, which is assessed by the frequency of mutations in microsatellite markers, is also used to classify gastric tumours (Falchetti et al., 2008). Tumours with a high level of microsatellite instability are more likely to be of the intestinal subtype, positioned antrally, and have a better prognosis than those with a stable or low level of microsatellite instability (Choi et al., 2014).

25.2.2.3 Epigenetic Changes

This modification induces changes in the expression of target genes in addition to changes in DNA sequence. Hypomethylation or hypermethylation of DNA CpG islands, chromatin remodelling, histone acetylation or methylation, and changes in miRNA expression or binding can all affect downstream oncogenes and tumour suppressor genes, favouring malignant cell transformation. As a result, such miRNAs can operate as "oncomiRs," preventing tumour suppressor genes from being expressed (Song and Ajani, 2013). The significance of these epigenetic changes in the aetiology of stomach cancer has long been recognised (Gigek et al., 2012). In a gastric cancer cell line, the overexpression of miR-181a, which affects the target gene KLF6in and induces preproliferative and antiapoptotic effects, is an excellent example (Zhang X et al., 2012).

25.2.2.4 Single Nucleotide Polymorphisms (SNPs)

SNPs (pronounced "snips") are changes in a single DNA building block called a nucleotide (for example, in a certain length of DNA, an SNP might replace the nucleotide cytosine (C) with the nucleotide thymine (T)). SNPs are the most common sort of genetic variation. A SNP is defined as a single nucleotide polymorphism with a minor allele frequency of greater than 1% in at least one population (Risch, 2000). The majority of SNPs are "silent," meaning they have no impact on the function or expression of a gene. SNPs are the result of point mutations that are preferentially preserved in populations. Haplotype blocks, which are divided into distinct blocks of linkage disequilibrium (LD), are inherited from SNPs in the same chromosomal region (Erichsen and Chanock, 2004). Specific genetic variants, as well as the most common polymorphisms, differ by ethnicity and, when paired with exposure to environmental risk factors, can hasten the development of cancer.

By mechanisms such as disrupting signalling pathways, altering host-environment interactions (such as in the response to intraluminal pathogens or dietary components), or disrupting cell cycle, proliferation, or cell death characteristics, SNPs can promote a malignant transformation in gastric cancer pathogenesis (Hanahan and Weinberg, 2011).

25.3 Methodologies for Studying Genetic Polymorphism in Gastric Cancer

The research of genetic variants (i.e., SNPs) in human disease has been done using two separate methodologies. SNPs are chosen from genes that "make sense" in the first technique, known as the candidate gene strategy. In other words, they are consistent with a reasonable understanding of biology. In this technique, the relationship is investigated within selected candidate genes in a hypothesis-driven manner based on disease pathophysiology. Despite being the most apparent technique, studying candidate genes has two drawbacks: a small sample size and insufficient statistical power. Real-world prospective validation, which necessitates large, population-based research, has been missing due to its labour- and resource-intensive nature (Wadhwa et al., 2013). A meta-analysis using previously known results is a practical means of getting more precise and genuine association.

Hypothesis-free genome-wide association studies (GWAS) are the second and more modern technique for high-throughput genetic analysis. This technique has revealed additional susceptibility loci, providing fresh molecular insights into the aetiology of stomach cancer (McLean and El-Omar, 2014). GWAS systems are meant to account for tagged SNPs with substantial LD among many variations, rather than focusing on operational polymorphisms with known mechanistic importance (Yu and Chen, 2015). This technique has a high probability of false negatives due to multiple testing and a lack of information about the functional value of novel susceptibility loci. Both approaches appear to be complementary in the knowledge of genetic predispositions to cancer. We must concentrate on determining the genuine significance of candidate gene discoveries as well as the mechanism of properly conducted GWAS.

We'll go over the research on potential gene SNPs in gastric cancer first and then look at the most current GWAS results.

25.3.1 SNPs in Candidate Genes

Genes implicated in important processes of gastric carcinogenesis were studied using SNPs in candidate genes in GC (Rosero et al., 2016). Genetic polymorphisms in GC can be categorised into three broad groups:

1. Genes involved in inflammatory and immunological responses to infection, specifically genes involved in *H. pylori* identification in gastric epithelial cells and genes involved in the activation of signalling pathways that cause pro-inflammatory and apoptotic responses.
2. Genes responsible for the protection of the stomach mucosa, as well as those engaged in carcinogen metabolism and detoxification.
3. Genes involved in oxidative damage repair, cell proliferation and adhesion, and DNA repair.

The following is a summary of the genetic polymorphisms found in each of the potential genes listed above.

25.3.1.1 Inflammatory Cytokine Gene Polymorphisms

H. pylori (HP) is a widespread infection that can cause peptic ulcers and stomach cancer. The host's reaction to *H. pylori* determines the clinical symptoms of the infection. Hypochlorhydria due to gastritis predominating in the stomach body precedes the premalignant phenotype, gastric atrophy. HP is one of the probable microorganisms that might cause inflammasome dysregulation, which can lead to tumour growth. Soluble effector molecules produced from HP that are carcinogenic or oncogenic have been linked to an abnormal modulation of inflammatory cytokines. The diagnosis and screening of precancerous suspected patients may be aided by HP-induced changes in cytokine gene polymorphism (El-Omar et al., 1997) (see Table 25.2).

25.3.1.2 IL-1B

IL-1B, a pro-inflammatory cytokine that is upregulated in response to *H. pylori* infection, is a potent regulator of stomach acid secretion and hence important in the development of hypochlorhydria (El-Omar, 2001). The inhibitory impact of this cytokine on stomach acid output has been linked to polymorphisms in the IL-1β gene. This suppression makes it easier for bacteria such as *H. pylori* to colonise and infect the stomach, as well as the emergence of pre-neoplastic conditions that can progress to cancer. As a result, it's an important candidate gene for research into the effects of SNPs. Several SNPs in the IL-1B gene, which is found on chromosome 2q13, have been investigated. Premalignant stomach abnormalities have been associated with IL-1B-31*C or IL-1B-511*T genotypes, as well as IL-1RN*2/*2 genotypes.

In a range of ethnic cultures, this inherited predisposition to the risk of stomach cancer has been shown (Zhao et al., 2012). Several meta-analyses have confirmed this link, notably in white populations (Camargo et al., 2006; Vincenzi et al., 2008; Xue et al., 2010). In a meta-analysis that included 35 studies and a total of 5,503 cases and 7,865 controls, Kamangar et al. found no overall link between IL-1B *511 T, IL-1B *31 C, or IL-1RN *2 alleles and the risk of gastric cancer; however, there was substantial heterogeneity among research results. There was no consistent association in the subgroup analysis by anatomic subsite (cardia vs. non-cardia), histology (intestinal vs. diffuse histology), ethnicity (western vs. East Asian studies), or studies that satisfied certain quality criteria (Kamangar et al., 2006).

TABLE 25.2

SNPs in Inflammatory Cytokine Genes and Risk of GC

S. No.	Genes	SNPs	Effects on Gastric Cancer	References
1	IL-1B	IL-1B-31*C IL-1B-511*T IL-1RN*2/*2	Genotypes have all been linked to premalignant stomach alterations.	Zhao et al. (2012), Kamangar et al. (2006)
2	IL-17A	187G>A (rs22759133)	This SNP has been linked to an increased risk of gastric cancer in a number of studies.	Qinghai et al. (2014), Rafiei et al. (2013), Shibata et al. (2009)
3	IL-2	rs2069762	IL-2 gene polymorphism has no link to gastric ulcers, gastric cancer, or duodenal ulcers. More commonly found genotype 330 TT was closely associated with the development of gastric mucosal atrophy and could confer vulnerability to gastric cancer.	Shin et al. (2008), Togawa et al. (2005)
4	IL-8	rs4073	A link was found between the common A allele and a higher incidence of GC and stomach ulcers. It was found that the AA genotype is linked to an increased risk of developing stomach cancer and may confer a higher susceptibility in Asian populations, notably the Chinese Han population.	Ohyauchi et al. (2005), Xue et al. (2012)
5	TNF-α	308 G>A	In Caucasians, a substantial link between the A allele and the A/A genotype and stomach cancer has been shown.	Gorouhi et al. (2008)
6	TNF-β	TNF-β p252 (G/A)	Individuals with non-atrophic gastritis, intestinal metaplasia, gastric cancer, duodenal ulcers, and asymptomatic patients have been linked to the TNF-β p 252 (G/A) polymorphism in the Mexican population.	Partida-Rodrguez et al. (2010), Yoon et al. (2010)
7	TLR-4	rs4986790 Asp299Gly	Both SNPs have been linked to stomach cancer caused by *Helicobacter pylori* and its antecedents.	Arbour et al. (2000), Hold et al. (2007)
8	Foxp3	rs3761548	This polymorphism plays a role in GA development by affecting Treg function and immunomodulatory cytokine production.	Ezzeddini et al. (2021)

25.3.1.3 IL-17

Another cytokine gene that has been examined is IL-17, which is found on chromosome 6p12. It is predominantly produced by type 1 T helper cells in the inflammatory milieu, aside from natural killer T cells, neutrophils, and CD8+ cytotoxic T cells. Several studies have connected the 187G>A polymorphism in the IL-17A gene (rs22759133) to an increased risk of gastric cancer (Qinghai et al., 2014; Rafiei et al., 2013; Shibata et al., 2009). Polymorphisms in the IL-17A gene were found to have a positive association with *H. pylori* infection and an increased risk of gastric cancer in a Chinese population ($P=0.036$) (Zhang X et al., 2014).

25.3.1.4 IL-2

IL-2, a 15.5–16 kDa protein that regulates the activities of white blood cells (leucocytes, often lymphocytes) that are responsible for immunity, is another possibility involved in the development of gastric cancer. It is encoded by a gene on chromosome 4. IL-2 is involved in the body's natural reaction to microbial infection as well as the distinction between "non-self" and "self." The –330 (rs2069762) polymorphism in the IL-2 gene promoter region regulates IL-2 production and plays a key role in IL-2 inducible expression, despite mixed findings from multiple studies (John et al., 1998). The IL-2 gene polymorphism shows no connection to stomach ulcers, gastric cancer, or duodenal ulcers, according to Shin et al. (2008). According to Togawa et al., the more prevalent genotype 330 TT is linked to the development of gastric mucosal atrophy and may increase the risk of gastric cancer. Both investigations, on the other hand, came to the same conclusion: these polymorphisms have nothing to do with the presence of *H. pylori* (Togawa et al., 2005).

25.3.1.5 IL-8

Interleukin 8 (IL-8) is a cytokine found on chromosome 4 that is implicated in the inflammatory and immunological response to *H. pylori*. The role of IL-8, also known as neutrophil chemotactic factor, is dual. It causes target cells, mostly neutrophils, but also other granulocytes, to migrate towards the infection site by inducing chemotaxis. Once they've arrived, IL-8 induces phagocytosis. IL-8 is also recognised to be a powerful angiogenesis promoter. Individuals with gastrointestinal disease who are infected with *H. pylori* have higher levels of IL-8 expression. The 251A>T (rs4073) polymorphism in the IL-8 gene has been shown to influence susceptibility to *H. pylori*-related gastrointestinal disorders in the Japanese population. In addition, Ohyauchi et al. (2005) discovered a relationship between the common A allele and an increased risk of GC and stomach ulcers. Xue et al. (2012) discovered that the AA genotype is associated with an elevated risk of stomach cancer and may confer a higher susceptibility in Asian populations, particularly the Chinese Han population, in a meta-analysis.

25.3.1.6 TNF Family

Another group of pro-inflammatory defensive cytokines is the TNF family. The gene is found in the major histocompatibility complex class III region on the short arm of chromosome 6. The two types are tumour necrosis factor alpha (TNF-α) and tumour necrosis factor beta (TNF-β). TNFR1 is expressed ubiquitously on most cell types, whereas TNFR2 is found largely in endothelial, epithelial, and immune cell subsets. TNFR1 signalling increases apoptosis and is pro-inflammatory, whereas TNFR2 signalling stimulates cell proliferation and is anti-inflammatory. Overexpression of TNF-α as a result of chronic inflammatory and autoimmune illnesses has been linked to the presence of certain polymorphisms in the TNF gene.

One of the most studied allele variants of this gene is the transition from G to A identified at the –308 position of the gene promoter. A significant relationship between the A allele and the A/A genotype and stomach cancer has been discovered in Caucasians (Gorouhi et al., 2008). SNPs have been discovered in the gene encoding TNF-β, which has cytotoxic properties against tumour cells. One of these is a G-for-A alteration in the first intron at position p252, which results in TNF-β overexpression. The TNF-β p252 (G/A) polymorphism has been associated with non-atrophic gastritis, intestinal metaplasia, gastric cancer, duodenal ulcers, and asymptomatic patients in the Mexican population (Partida-Rodrguez et al., 2010; Yoon et al., 2010).

25.3.1.7 HuGE Study

In 2011, Persson and colleagues published the systematic review and meta-analysis Human Genome Epidemiology (HuGE) (Persson et al., 2011). They intended to see whether SNPs in inflammatory genes such as IL-1B, IL-1RN, IL-8, IL-10, and TNF were associated. They were based on studies that took place between 1990 and 2006. The studies comprised cross-sectional, case-control, cohort, or case-case studies with specific definitions of cases and controls, as well as reporting of outcomes and risk estimates. A total of 180 references were chosen for the review. Subgroup analyses were conducted using anatomic site, histologic type, geographic location (Asian vs. non-Asian population), *H. pylori* infection status, and study quality criteria. The main conclusions of these meta-analyses are as follows:

- A number of studies have connected the presence of IL-1RN*2 to an increased risk of gastric cancer.
- This risk was only seen in non-Asian ethnicities and was found in intestinal and diffuse tumours, particularly those with a distant location.
- In Asian populations, IL-1B-31C has been associated with a decreased risk of stomach cancer.
- IL-1B-511T carriers have also been linked to a greater incidence of cardia gastric cancer, albeit this association appears to be limited to non-Asians.
- IL-10–1082G had radically different effects in Asians (increased risk) and non-Asians (decreased risk), despite the fact that high-quality research did not support this conclusion (decreased risk).
- Finally, there was strong evidence that non-Asians who were TNFA308A carriers had a higher incidence of diffuse tumours.

25.3.1.8 Toll-Like Receptors (TLRs)

TLRs are a protein family that plays an important function in the innate immune system. They are single-pass membrane-spanning receptors that recognise structurally conserved microorganism-produced chemicals and are commonly expressed on sentinel cells such as macrophages and dendritic cells. The gene is found on chromosome 9 of the human genome. TLR4, also known as the lipopolysaccharide (LPS) receptor, was first discovered on gastric epithelial cells as a putative *H. pylori* signalling receptor (Su et al., 2003). The development of gastrointestinal malignancies has been linked to TLRs and innate immune responses (Fukata and Abreu, 2008). Non-cardia malignancy has been associated with a functional variant in the TLR4 gene (Hold et al., 2007). There are a lot of polymorphisms in TLRs (El-Omar et al., 2008). Two proposed pathways linking TLRs to gastrointestinal malignancies are TLR activation of cyclooxygenase-2 and nuclear factor kappa B (NF-κB), a crucial inflammatory transcription factor. Arbour et al. discovered a functional polymorphism at position 896 in exon 4 of the TLR4 gene (dbSNP ID: rs4986790) (Arbour et al., 2000). This missense mutation causes the amino acid substitution Asp299Gly in the third exon of the TLR4 gene. Later, it was discovered that SNP Asp299Gly co-segregates with SNP Thr399Ile, which is also located in TLR4's third exon. In Caucasians, these SNPs are found in approximately 10% of the population. Both SNPs have been associated with stomach cancer induced by *Helicobacter pylori* and its antecedents (Hold et al., 2007).

25.3.1.9 Foxp3 Polymorphisms

T regulatory cells (Tregs) and related cytokines play a key function in suppressing the immune response to cancer. On genotyping, Foxp3 gene patients with the AC+AA genotype are more likely than those with the CC genotype to develop GA. In addition, GA patients had significantly greater levels of IL-10, IL-35, and TGF-β than controls, with diffuse-type GA patients having significantly higher levels than intestinal-type GA patients. The levels of IL-35 and IL-10 in GA patients with the CC, AC, and AA genotypes differed considerably. Serum IL-35, IL-10, and TGF-β levels were revealed to have a prognostic effect in GA patients. The polymorphism (rs3761548) in the Foxp3 gene (on chromosome X) may influence GA propensity as well as serum levels of IL-10, IL-35, and TGF-β. This polymorphism is important for GA development (Ezzeddini et al., 2021).

25.3.2 Genes Involved in DNA Repair Mechanisms

Cells engage in a dynamic process of DNA damage and repair in response to intrinsic and extrinsic stresses. The genome is kept in good form by a series of DNA repair genes, and mutations in these genes can lead to the accumulation of defective DNA, which can lead to cancer. Basic single-strand break and cleavage repair methods require the repair gene XRCC1 (X-ray repair cross-complementing group 1), which is found on chromosome 19. A link between two allele variants in the XRCC1 gene (an A/G substitution at position 280 of exon 9 (Arg/His) and a G/A substitution at position 399 of exon 10 (Arg/Gln)) and susceptibility to gastric cancer was discovered in a Chinese population (Qiao et al., 2013) (see Table 25.3).

The MTHFR gene, found on chromosome 1p36, produces 5,10-methylenetetrahydrofolate reductase, a rate-limiting enzyme

TABLE 25.3

SNPs in Various Genes Involved in DNA Repair, Metabolisation and Detoxification, and Risk of GC

S. No.				
1.	XRCC1	A/G substitution at position 280 of exon 9 (Arg/His) G/A substitution at position 399 of exon 10 (Arg/Gln)	In a Chinese population, a correlation was found between two allele variations in the XRCC1 gene and susceptibility to gastric cancer.	Qiao et al. (2013)
2.	MTHFR	C677T	The MTHFR 677 T allele was found to be related to a considerably higher risk of developing stomach cancer.	Chen et al. (2015)
3.	Cytochrome P450	CYP2E1	The c2 allele, whether heterozygous or homozygous, was linked to an increased risk of GC.	Gao et al. (2002)
4.	Glutathione S-transferases (GSTs)	GSTP1 GSTT1 GSTM1	The active protein is absent in two common deletion polymorphisms in the GSTT1 and GSTM1 genes. These SNPs have been shown to be linked to a high risk of GC because of the reduced ability to detoxify carcinogenic chemicals.	Garca-González et al. (2012), Hayes and Pulford (2008)

for folic acid metabolism and DNA methylation. Reduced thymidylate synthase (TS) activity has been linked to MTHFR gene mutations in cancer cells, which is a crucial and rate-limiting enzyme for guaranteeing an adequate supply of DNA for accurate DNA synthesis and repair (Fodingeer et al., 2001). As a result, SNPs in the MTHFR gene may be linked to the risk of cancer. Alanine is substituted with valine in a common MTHFR C677T mutation, resulting in a thermolabile variant with only 30% of the activity of the wild-type MTHFR enzyme. The link between this gene mutation and the risk of stomach cancer has been studied. A meta-analysis that comprised 28 eligible trials with 5,757 cases and 8,501 controls looked at the strength of the connection between the MTHFR C677T polymorphism and stomach cancer risk. The MTHFR 677 T allele has been linked to a significantly increased risk of developing stomach cancer. Subgroup studies found significant associations among Asians, but not in Caucasians, suggesting that ethnicity may play a role in stomach cancer risk (Chen et al., 2015).

25.3.2.1 Genes Involved in Metabolisation and Detoxification of Carcinogenic Compounds

Xenobiotics such as polycyclic aromatic hydrocarbons (PAHs), nitrosamines, and dioxins like those present in tobacco smoke are biotransformed by enzymes such as cytochrome P450 and glutathione S-transferases (GSTs). As a result, they protect against cancer. Polymorphisms in both sets of genes, as well as their link to GC, have been discovered. Castano Molina et al. (2009) discovered polymorphisms in the CYP2E1 gene on chromosome 10 (alleles c1 and c2). The c2 allele has been related to a higher risk of GC, whether heterozygous or homozygous. In the presence of this polymorphism, smoking increased susceptibility to GC, indicating an environmental modifying effect (see Table 25.3).

Similar findings were published by Gao, who found that homozygous and heterozygous individuals with this polymorphism experience elevated risks when they smoke and consume a lot of meat (Gao et al., 2002). GSTP1, GSTT1, and GSTM1 are the three principal GST isoenzymes, and their genes have been revealed to be highly polymorphic. In two common deletion polymorphisms in the GSTT1 and GSTM1 genes, located on chromosomes 22 and 1p13, respectively, the active protein is missing. The most recent studies (Garcia-Gonzalez et al., 2012; Hayes and Pulford, 2008) have looked at these and have shown that they are linked to a high risk of GC due to a lower ability to metabolise carcinogenic substances.

25.3.2.2 Genes Involved in Tumour Suppression/Apoptosis

The human nuclear phosphoprotein p53 is a tetrameric transcription factor. It turns on the transcription of multiple genes in a mechanism that destroys damaged cells that have become hazardous to the organism swiftly. As a result, changes in this gene, which is found on chromosome 17, may raise the risk of cancer. The p53 polymorphism in exon 4, codon 72, has received the most attention. It is the arginine (Arg) of a proline (Pro). According to ethnicity, studies examining the link between this polymorphism and cancer susceptibility and progression have yielded varying results. There was no link between the Pro/Pro polymorphism and the risk of developing GC in Colombian and Costa Rican groups; however, the Pro/Pro genotype was associated with the risk of developing diffuse GC in Japanese populations (Alpizar-Alpizar et al., 2005; Cardona Rivas et al., 2007; Hiyama et al., 2002) (see Table 25.4).

25.3.2.3 Genes Involved in Proliferation, Migration, Metabolism, Cell Differentiation, and Survival

Cells have receptors that bind signalling chemicals such as growth factors for diverse cellular processes such as proliferation, development, and differentiation. HERs are tyrosine kinase-activated cellular receptors that regulate a range of cellular processes downstream. The HER family includes HER-1, also known as epidermal growth factor receptor

TABLE 25.4

SNPs in Genes Involved in Tumour Suppression/Apoptosis and Risk of GC

S. No.	Gene	SNP	Effects on GC	Reference
1.	P53	P53 polymorphism in exon 4, codon 72. It represents a proline's arginine (Arg) (Pro).	In Colombian and Costa Rican populations, there was no link between the Pro/Pro polymorphism and the risk of developing GC, but the Pro/Pro genotype was linked to the risk of developing diffuse GC in Japanese populations.	Alpizar-Alpizar et al. (2005), Cardona Rivas et al. (2007), Hiyama et al. (2002)

TABLE 25.5

SNPs in Genes Involved in Proliferation, Migration, Metabolism, Cell Differentiation, Survival, DNA Methylation, and Risk of GC

S. No.	Gene	SNP	Effect on GC	Reference
1.	ERBB2 gene	rs2643194 rs2934971 rs1058808	SNPs were linked to a higher risk of GC in the Mexican population.	Vázquez Ibarra et al. (2019)
2.	DNMTs	rs16999593 rs11254413 rs7560488 rs36012910	All four SNPs were linked to GC. The DNMT2 rs11254413 G allele, on the other hand, had a protective effect against GC. DNMT1 rs16999593 exhibited a statistically significant connection with GC development.	Neves et al. (2016)

(EGFR/ERbB1), HER-2 (ERbB2), HER-3, and HER-4. ERbB2 (also known as HER-2) is a proto-oncogene that encodes a 185-kDa tyrosine kinase receptor whose phosphorylation initiates signalling pathways that lead to cell division, proliferation, differentiation, and anti-apoptosis signalling. It is found on chromosome 17 (see Table 25.5).

According to the findings of an open-label, international, phase III randomised controlled trial (trastuzumab for gastric cancer, ToGA) published in 2010, the humanised anti-HER-2 monoclonal antibody trastuzumab statistically significantly prolonged the overall survival in patients with HER-2-positive advanced GEA. As a result, it is now the only biomarker used to determine the most effective treatment for patients with advanced gastro-oesophageal adenocarcinoma (GEA). The HER-2 membrane receptor has been reported to be overexpressed in approximately 13%–25% of GC patients, particularly in intestinal tumours and tumours near the oesophagogastric junction. Polymorphisms in the HER gene have been associated with this overexpression.

The polymorphism HER-2-I655V, which is present in the transmembrane domain's coding region, is the most well studied. Because it can alter the development of the receptor's active dimers, it induces a shift from A to G and is associated with an increased risk of cancer (Ramos-de la Medina et al., 2006). The SNPs rs2643194, rs2934971, and rs1058808 in the ERBB2 gene have recently been associated with a higher risk of GC in the Mexican population (Vázquez Ibarra et al., 2019).

25.3.2.4 *DNA Methyltransferases (DNMTs)*

Epigenetic alterations, such as aberrant DNA methylation, have been associated with gastric cancer. Promoter methylation is catalysed by DNA methyltransferases (DNMTs). In the context of GC and epigenetic changes, SNPs in DNMTs may influence protein expression and consequently DNA regulation, as well as stomach cancer (GC) risk. In a comprehensive review and meta-analysis involving 11 studies and a total of 24 SNPs in DNMTs, only 4 SNPs in DNMTs, DNMT1 rs16999593, DNMT2 rs11254413, DNMT3A rs7560488, and DNMT3A rs36012910, were associated with GC. On the other hand, the DNMT2 rs11254413 G allele demonstrated a protective effect against GC. When SNPs from several studies were combined, only DNMT1 rs16999593, situated on chromosome 19, showed a statistically significant link with GC development (OR = 1.31; 95% CI = 1.08–1.60; p = 0.006 for TC+CC genotypes) (Neves et al., 2016) (see Table 25.5).

25.3.2.5 *Single Nucleotide Polymorphisms in microRNAs*

Among the reported SNPs, correlations between SNPs in genes encoding microRNAs (miRNAs) or their binding sites are of special interest. MiRNAs are noncoding RNA molecules that are single-stranded and have about 22 nucleotides. MiRNAs bind to complementary sequences in messenger RNA's 3′-untranslated regions, inhibiting their stability and translational efficiency and so lowering gene post-transcriptional activity. Because a single miRNA can have several targets, even minor changes in its expression can have a significant impact on a wide range of genes, including many oncogenes and tumour suppressor genes. SNPs in miRNAs may also have a role in gastric cancer susceptibility by altering miRNA expression or function, leading to aberrant gene expression. The SNPs rs2910164 in miR-146a and rs11614913 in miR-196a2 have been discovered to be risk markers for stomach cancer (Hua et al., 2014; Saunders et al., 2007; Lovat et al., 2011) (see Table 25.6).

TABLE 25.6

SNPs in miRNA and Risk of GC

S. No.	miRNA	SNP	Effect on GC	Reference
1.	miR-146a	rs2910164	These SNPs have been identified as risk factors for stomach cancer. They cause aberrant gene expression by affecting the expression or function of miRNAs.	Hua et al. (2014), Saunders et al. (2007)
2.	miR-196a2	rs11614913		Lovat et al. (2011)

TABLE 25.7

GWAS for Studying SNP of Various Genes and Risk of GC

S. No.	Gene	SNP	Effect on GC	Reference
1.	PSCA	rs2976392	This polymorphism was strongly linked with diffuse GC.	Sakamoto et al. (2008), Wang et al. (2012c)
		rs2294008	C>T polymorphism is a risk factor for GC.	Wang et al. (2015)
2.	PLCE1	rs2274223	SNP was found to be related to gastric cancer risk. However, it was stronger for Asians and tumours originating in the cardia region.	Abnet et al. (2010), Mocellin et al. (2015)
		rs1304295	SNP was found to be related to gastric cancer risk.	
		rs13361707	The SNP was associated with a higher risk of non-cardia GC.	
		rs9841504	Higher risk of non-cardia GC among the 17 SNPs examined.	
3.	PRKAA1	rs13361707	This SNP elevates the GC risk.	Cai et al. (2017), Zhang et al. (2011)
4.	LRFN-2	rs2494938	This SNP has been linked to non-cardia stomach cancer.	Jin et al. (2012)
		rs10484761	The SNP has been linked to GC prognosis, with allele C being linked to poor survival.	Kang et al. (2014)
5.	MUC1	rs4072037	G>A polymorphism in exon 2 appears to confer an increased risk of gastric cancer, whereas the G allele appears to be protective.	Jia et al. (2010), Silva et al. (2001), Zheng et al. (2013)
		rs80142782	The SNP is linked to increased vulnerability to non-cardia GC.	Wang et al. (2017)

25.3.2.6 GWAS

A GWAS was used to look for SNPs connected to the disease throughout the entire genome. As a result of these studies, previously undiscovered genes linked to stomach cancer have been uncovered.

In 2008, the Study Group of the Millennium Genome Project for Cancer was the first to scan more than 80,000 SNPs in 188 GC patients and 752 controls from Japan. This led to the finding of –rs2976392, an intronic A>G SNP in the prostate stem cell antigen (PSCA) gene that was found to be strongly associated with widespread GC (1.62-fold) (Sakamoto et al., 2008; Wang et al., 2012c). It is located on chromosome 8. A glycosylphosphatidylinositol-anchored cell membrane glycoprotein is encoded by this gene. PSCA's role in gastric tissue is uncertain, while it is thought to have a tumour-suppressing impact on the gastric epithelium (Saeki et al., 2010). Following this screening, a number of PSCA association validation studies were done in diverse populations. Furthermore, rs2294008 has been linked to the risk of stomach cancer in various meta-analyses. According to a meta-analysis by Wang et al., which comprised 26 studies with 30,050 cases and 51,671 controls (Asian and Caucasian ethnicities), the C>T polymorphism is a risk factor for GC (Wang et al., 2015) (see Table 25.7).

Another meta-analysis by Mocellin et al. discovered a significant connection between PSCA and tumour susceptibility, but the results were limited by heterogeneity. Only in a subgroup analysis of the same data was the T allele connected to an increased risk of non-cardia GC (Mocellin et al., 2015). Three more studies in Hispanic and Eastern European groups (Garcia-Gonzalez et al., 2015; Sun et al., 2014) have corroborated the association between the T allele and increased cancer risk (Kupcinskas et al., 2014).

In a Chinese population genome-wide association study of ESCC was performed that included 1,077 oesophageal cancer cases and 1,733 controls to identified a previously unknown susceptibility locus for ESCC: PLCE1 at 10q23. Further association was confirmed in 2,766 cases of GA and the 11,013 control subjects. PLCE1 had important biological implications for both ESCC and GA (Wang et al., 2010).

GC risk has been associated with polymorphisms in the Phospholipase C epsilon 1 (PLCE1) gene (which codes for pancreas-enriched phospholipase C). PLCE1 has 39 exons and is found on the q arm of chromosome 10 at location 23.33. PLCE1 catalyses the hydrolysis of polyphosphoinositides such as phosphatidylinositol 4,5-bisphosphate (PtdIns(4,5)P2) to produce the second messengers Ins(1,4,5)P3 and diacylglycerol. These substances set off a series of intracellular reactions that lead to cell proliferation, differentiation, and gene expression. As an oncogene, PLCE1 has previously been associated with skin and intestinal cancers. Eighteen "hits" were validated in 2,766 cases of gastric cancer, and the SNPs PLCE1 rs2274223 and C20orf54 rs1304295 were found to be linked to gastric cancer risk. The PLCE1 rs2274223 SNP was identified as a risk allele for stomach cancer in another investigation of 2,240 cases (Abnet et al., 2010). Although PLCE1 SNPs have been related to a higher incidence of gastric cancer and a worse prognosis in Chinese patients (Wang et al., 2012a; Zhang et al., 2011), white individuals (Luo et al., 2011) have not been affected (Palmer et al., 2012). So far, nine meta-analyses of PCLE gene polymorphisms have been published. The largest meta-analysis, which included 11 studies and included both Asians and Caucasians, discovered a relationship between the PLCE1 rs2274223 SNP and GC risk; however, the link was higher for Asians and tumours originating in the cardia region (Mocellin et al., 2015).

Susceptibility to non-cardia stomach cancer was studied in a GWAS of over 1 million SNPs in a thousand Chinese patients. The T>C rs13361707 at 5p13.1 and the C>G rs9841504 at 3q13.31 were found to have a greater risk of non-cardia GC among the 18 "hits," and the SNPs PLCE1 rs2274223 and C20orf54 rs1304295 were shown to be connected to gastric cancer risk.

There have been nine meta-analyses of PCLE gene polymorphisms published so far. The largest meta-analysis, which included 11 papers with both Asians and Caucasians, found a correlation between the PLCE1 rs2274223 SNP and GC risk; however, the link was stronger for Asians and tumours originating in the cardia region (Mocellin et al., 2015). Seventeen SNPs were investigated. The first SNP codes for the AMP-activated protein kinase alpha 1 catalytic subunit and is situated in PRKAA1's first intron. In the same region, PTGER4, which codes for a prostaglandin E receptor 4, has also been linked to GC. In the 3q pointed region, the zinc finger and BTB domain-containing 20 (ZBTB20) gene is associated with GC susceptibility (Shi et al., 2011).

The association between two genes on the 5p13.1 chromosome and an increased risk of gastric cancer was confirmed in two studies in a Korean population, but the link between the ZBTB20 polymorphism and an increased risk of gastric cancer was not replicated (Kim et al., 2014; Song et al., 2013). Kim et al. (2014) revealed four new PRKAA1 SNPs associated with an increased risk of GC, two upstream and two downstream. On the other hand, the functional processes that cause the higher risk are unknown. However, in the case of SNP PRKAA1 (rs13361707), a substantial interaction has been described, leading to an increased GC risk (Cai et al., 2017). Finally, GC has been associated with the loss of chromosomes 1q, 8q, and 20q (Zhang et al., 2011).

Non-cardia stomach cancer has also been associated with SNP rs2494938 in the starting region of LRFN2, which encodes for a member of synaptic adhesion-like molecules and is located at 6p21.1. This association was revealed in a study that integrated data from three GWAS to look for pleiotropic loci in three types of cancer (Jin et al., 2012). T>C (rs10484761), another polymorphism in the 6p21.1 region, has been connected to GC prognosis, with allele C being associated with poor survival (Kang et al., 2014). Several GWAS in various geographical and ethnic populations have discovered a link between gastric cancer and SNPs in the Mucin 1 (MUC1) gene, which codes for the cell surface glycoprotein mucin 1, using both a candidate gene approach and detection within susceptibility loci (Jia et al., 2010; Silva et al., 2001).

Exon 2's rs4072037 (G>A) polymorphism appears to increase the risk of gastric cancer, while the G allele appears to protect against it. The protective effect of the G allele was confirmed in a meta-analysis, with a 28% decreased risk of gastric cancer overall (Zheng et al., 2013). This meta-analysis included ten studies with over 6,000 cases of gastric cancer and 10,000 healthy controls from various ethnic and geographical groups. In a meta-analysis involving 4,220 cases and 6,384 controls, these findings were confirmed (Duan et al., 2014).

A meta-analysis combining data from two GWAS conducted on Chinese with non-cardia GC confirmed MUC1 locus vulnerability. It also discovered two new loci: a 5q14.3 region containing the intron of a long noncoding RNA with unknown function and an rs80142782 gene associated with ASHL1, an epigenetic regulator of histone methylation protein gene on 1q22. Both loci were associated with greater sensitivity to non-cardia GC in the group studied (Wang et al., 2017).

GWAS can also be used to investigate bacteria. A GWAS on 173 *H. pylori* isolates from the European population (hpEurope) with clinical illness (e.g. gastritis) has recently been completed, with 49 of them having stomach cancer. SNPs and genes revealed differed in frequency between those with stomach cancer and those with gastritis. Non-synonymous alterations in several less well-studied genes, as well as the presence of babA and genes in the cag pathogenicity island, one of the primary virulence factors of *H. pylori*, were associated with the gastric cancer phenotype. The authors devised a simple risk score to identify strains that are likely to cause cancer. This procedure appears to be exciting and has a lot of therapeutic potential, but it has to be developed and verified first.

25.4 Clinical Implications of Genetic Polymorphisms

In order to have any clinical importance, genetic association studies employing SNPs should educate us about two fundamental aspects of cancer, namely susceptibility and outcomes. The findings include prognostic information on survival, complications, and the effectiveness of pharmaceutical interventions. Although technology and bioinformatics advances have increased our ability to analyse a huge number of genes in a single study, it is still simply a preliminary look at a complicated process. This is because SNP findings must be replicated before they can be accepted, and certainly before they can be used in therapeutic trials. We need to know everything there is to know about the scope and structure of genetic variations in different populations, including SNPs, haplotypes, and LD for genes, as well as cancer-related pathways. This must be followed by translational research and, eventually, validation in population-wide clinical studies if it is to be employed in early cancer diagnosis and treatment (Erichsen and Chanock, 2004).

25.4.1 Association of Variants with Treatment Strategy

Although the majority of genetic research so far have focused on discovering propensity variations, some have also looked for variants that could predict treatment outcomes or toxicities. Individual reactions to cancer treatment techniques may differ, and genetic variations have been proposed as a possible explanation. Thus, the presence of the thymidylate synthase (TS) 3′-UTR 6 bp nucleotide fragment can be associated with GC susceptibility to 5-FU-based chemotherapy (CTx) (Lu et al., 2006), allowing this polymorphism profile to be used to guide the selection of 5-FU-based CTx for advanced GC (Lu et al., 2006). Ott and colleagues studied the relationship between DNA polymorphisms in the TS and 5,10-methylenetetrahydrofolate reductase (MTHFR) genes and response and survival in patients with locally advanced GC treated with 5-FU-based preoperative CTx (Ott et al., 2006). In CTx patients, the 2rpt/2rpt (p=0.002) and 2rpt/3rpt (p=0.004) genotypes were found to have a significant survival benefit. After CTx, the TS tandem repeat polymorphism was connected to patient survival. In a cytotoxicity investigation published in 2011, Ha et al. used MKN1 and SNU638 clones transfected with the G allele of deoxyribonuclease II beta rs3738573. Docetaxel sensitivity was shown to be higher in the G allele than in the C allele. The T allele of the 5-hydroxytryptamine receptor IE rs3828741 was also more paclitaxel sensitive than the C allele in AGS and SNU638 clones transfected with it (Ha et al., 2011).

25.4.2 Association of Variants with Prognosis

The prognosis of GC patients has been linked to some genetic differences. One of these is the rs2294008T variant, which predicts a worse overall prognosis in diffuse-type GC patients (hazard ratio = 1.85) (Garcia-Gonzalez et al., 2015). Furthermore, among Caucasians, the PSCA rs2294008 polymorphism is connected to the prognosis of diffuse GC. SNP rs4135385 AG/AA genotypes were significantly connected to improved GC survival compared to the GG genotype (adjusted hazard ratio = 0.80), according to Wang et al. (2012c), with the association being higher among patients with non-cardia GC. Polymorphisms in the MTHFR gene, MAP3K1 gene, and TLR-9 gene have all been found to be useful prognostic indicators for GC patients (Blank et al., 2014; Wang et al., 2013; Yu and Chen, 2015).

25.4.3 Functional Investigation

Only a small number of SNPs in GC have been functionally interpreted. However, functional characterisation of risk loci after GWAS is becoming more relevant. The intricate interplay between genetics and the environment, however, makes it a difficult task in GC and other malignancies (Freedman et al., 2011). The functional analysis of the PSCA protein is a nice example. It's present in glandular crypts with stem cell progenitors, which are assumed to be the GC's initiator cell. PSCA is known to be downregulated at both the gene and protein levels in GC, and this has been verified in a number of case-control studies in Asian ethnic groups, however with varied degrees of frequency and the risk of both intestinal and diffuse histological subtypes. In vitro transfection investigations later established PSCA's significance in epithelial cell growth prevention (Sakamoto et al., 2008).

25.5 Future Perspectives

Although much current molecular knowledge about GC aetiology hasn't had a direct impact on patient treatment in everyday clinical practise, it has enhanced our understanding of the disease's pathogenesis and discovered possible therapeutic targets or biomarkers to aid diagnosis and prognosis. As technology progresses, large-scale GWAS will continue to find genetic risk markers, which might be utilised in prospective studies to stratify risk in a suitable population. Following the GWAS, research should concentrate on a detailed genetic epidemiology evaluation of the associated locus, its functional significance, and in vitro and in vivo experimental explanations of the molecular mechanisms of causality (Chakravarti et al., 2013). Another source of unexplained risk is rare, low-penetrance genetic variants. The next goal is to employ genetic findings to establish consistent and useful implications for the prevention and treatment of GC.

25.6 Conclusions

In this review, we focused on genetic polymorphisms and covered GC epidemiology, risk factors, classification, and the current knowledge of the disease's genetic landscape. The research of gastric cancer's genetic base has led to a better understanding of the disease's pathophysiology, as well as the role of infection and chronic inflammation in the disease's development. GC appears to be an "infection-induced, inflammation-driven malignancy that develops over decades in persons with a hereditary vulnerability." The key to prevention is to interrupt this process early on, either by preventing infection, or by removing it before it causes irreversible damage. The wider genetic landscape in gastric cancer was explained in this chapter, as well as the most relevant candidate genes for gastric cancer susceptibility that have been investigated thus far. We looked at the most recent findings on GWAS in stomach cancer. A brief description of how genetic studies can inform outcomes was provided in addition to susceptibility. Individualised medicine combined with targeted therapy appears to be a viable treatment option for this illness. Molecular profiling could help identify which populations in a community should be evaluated and treated with eradication therapy before permanent effects such as gastric atrophy and intestinal metaplasia occur. If it were able to select groups at increased risk of stomach cancer, as established by a robust genetic profile, mass eradication therapy would be unnecessary. The ultimate goal of studying the molecular basis of any disease is to develop individualised treatments.

ABBREVIATIONS

CDH1:	Cadherin 1 gene
CIN:	Chromosomal instability
CTx:	Chemotherapy
DM:	Diabetes mellitus
DNMTs:	DNA methyltransferases
EBV:	Epstein–Barr virus
GAPPS:	Gastric adenocarcinoma and proximal polyposis of the stomach
GC:	Gastric cancer
GS:	Genomically stable tumours

GSTs:	Glutathione S-transferases
GWAS:	Genome-wide association studies
HDGC:	Hereditary diffuse gastric cancer
HER:	Human epidermal growth factor receptor
HNPCC:	Hereditary non-polyposis colon cancer
H. pylori:	*Helicobacter pylori*
HuGE:	Human Genome Epidemiology
IL:	Interleukin
MAP3K1:	Mitogen-activated protein kinase kinase kinase 1
miRNAs:	MicroRNAs
MSI:	Microsatellite unstable tumours
MTHFR:	5,10-Methylenetetrahydrofolate reductase
PLCE1:	Phospholipase C epsilon 1
PSCA:	Prostate stem cell antigen
SNPs:	Single nucleotide polymorphisms
TLRs:	Toll-like receptors
TNF:	Tumour necrosis factor
WCRF/AICR:	World Cancer Research Fund/American Institute for Cancer Research
WHO:	World Health Organization
XRCC1:	X-ray repair cross-complementing protein 1

REFERENCES

Abnet CC, Freedman ND, Hu N, Wang Z, Yu K, Shu XO, Yuan JM, Zheng W, Dawsey SM, Dong LM, Lee MP, Ding T, Qiao YL, Gao YT, Koh WP, Xiang YB, Tang ZZ, Fan JH, Wang C, Wheeler W, Gail MH, Yeager M, Yuenger J, Hutchinson A, Jacobs KB, Giffen CA, Burdett L, Fraumeni JF Jr, Tucker MA, Chow WH, Goldstein AM, Chanock SJ, Taylor PR. A shared susceptibility locus in PLCE1 at 10q23 for gastric adenocarcinoma and esophageal squamous cell carcinoma. *Nat Genet.* 2010;42(9):764–767. doi: 10.1038/ng.649.

Alpizar-Alpizar W, Sierra R, Cuenca P, Une C, Mena F, Ignacio Perez-Perez G. Asociacion del polimorfismo del codon 72 del gen p53 con el riesgo de cáncer gástrico en una población de alto riesgo de Costa Rica. *Revista de biología tropical.* 2005;53(3–4):317–324. doi: 10.15517/rbt.v53i3-4.14406.

Ang TL, Fock KM. Clinical epidemiology of gastric cancer. *Singapore Med J.* 2014;55(12):621–628. doi: 10.11622/smedj.2014174.

Arbour NC, Lorenz E, Schutte BC, Zabner J, Kline JN, Jones M, Frees K, Watt JL, Schwartz DA. TLR4 mutations are associated with endotoxin hyporesponsiveness in humans. *Nat Genet.* 2000;25(2):187–191. doi: 10.1038/76048.

Bass AJ, Thorsson V, Shmulevich I, Reynolds SM, Miller M, Bernard B, Shen H. Comprehensive molecular characterization of gastric adenocarcinoma. *Nature.* 2014;513(-7517):202–209. doi: 10.1038/nature13480.

Bertuccio P, Chatenoud L, Levi F, Praud D, Ferlay J, Negri E, Malvezzi M, La Vecchia, C. Recent patterns in gastric cancer: a global overview. *Int J Cancer.* 2009;125(3):666–673. doi: 10.1002/ijc.24290.

Blank S, Rachakonda S, Keller G, Weichert W, Lordick F, Langer R, Kumar R, Ott K. A retrospective comparative exploratory study on two methylentetrahydrofolate reductase (MTHFR) polymorphisms in esophagogastric cancer: the A1298C MTHFR polymorphism is an independent prognostic factor only in neoadjuvantly treated gastric cancer patients. *BMC Cancer.* 2014;14(1):1–15. doi: 10.1186/1471-2407-14-58.

Bray F, Ferlay J, Soerjomataram I, Siegel RL, Torre LA, Jemal A. Global cancer statistics 2018: GLOBOCAN estimates of incidence and mortality worldwide for 36 cancers in 185 countries. *CA: Cancer J Clin.* 2018;68(6):394–424. doi: 10.3322/caac.21492.

Buffart TE, Louw M, Van Grieken NC, Tijssen M, Carvalho B, Ylstra B, van der Merwe SW, Meijer GA. Gastric cancers of Western European and African patients show different patterns of genomic instability. *BMC Med Genom.* 2011;4:7. doi: 10.1186/1755-8794-4-7.

Cai M, Dai S, Chen W, Xia C, Lu L, Dai, S, Zhou L. Environmental factors, seven GWAS-identified susceptibility loci, and risk of gastric cancer and its precursors in a Chinese population. *Cancer Med.* 2017;6(3):708–720. doi: 10.1002/cam4.1038.

Caldas C, Carneiro F, Lynch HT, Yokota J, Wiesner GL, Powell SM, Jankowski JA. Familial gastric cancer: overview and guidelines for management. *J Med Genet.* 1999;36(12): 873–880.

Camargo MC, Mera R, Correa P, Peek RM, Fontham ET, Goodman KJ, Schneider BG. Interleukin-1β and interleukin-1 receptor antagonist gene polymorphisms and gastric cancer: a meta-analysis. *Cancer Epidemiol Prev Biomarkers.* 2006;15(-9):1674–1687. doi: 10.1158/1055-9965.

Cardona Rivas DE, Molina C, Marin Marmolejo JC. Cancer gastrico, tabaquismo, consumo de licor, estrato socioeconomico y polimorfismo en el codon 72 del gen P53 en una poblacion de Manizales. *Biosalud.* 2007; 6: 33–44.

Castano Molina E, Santacoloma M, Arango L, Camargo M. CYP1A1, CYP2E1 y riesgo a cancer gastrico en una poblacion colombiana de alta incidencia. *Acta Biologica Colombiana.* 2009;14(3):203–215.

Cavatorta O, Scida S, Miraglia C, Barchi A, Nouvenne A, Leandro G, Meschi T, De Angelis GL and Di Mario F. Epidemiology of gastric cancer and risk factors. *Acta Biomed.* 2018; 89(8-s): 82–87. doi: 10.23750/abm.v89i8-S.7966.

Chakravarti A, Clark AG, Mootha VK. Distilling pathophysiology from complex disease genetics. *Cell.* 2013;155(1):21–26. doi: 10.1016/j.cell.2013.09.001.

Chandra R, Balachandar N, Wang S, Reznik S, Zeh H, Porembka M. The changing face of gastric cancer: epidemiologic trends and advances in novel therapies. *Cancer Gene Therapy.* 2021;28(5):390–399. doi: 10.1038/s41417-020-00234-z

Chen L, Lu N, Zhang BH, Weng LI, Lu J. Association between the MTHFR C677T polymorphism and gastric cancer susceptibility: a meta-analysis of 5,757 cases and 8,501 controls. *Oncol Lett.* 2015;10(2):1159–1165. doi: 10.3892/ol.2015.3356.

Choi YY, Bae JM, An JY, Kwon IG, Cho I, Shin HB, Cheong JH. Is microsatellite instability a prognostic marker in gastric cancer?: a systematic review with metaanalysis. *J Surg Oncol.* 2014;110(2):129–135. doi: 10.1002/jso.23618.

Duan F, Song C, Dai L, Cui S, Zhang X, Zhao X. Novel functional variants locus in PLCE1 and susceptibility to digestive tract cancer in the Chinese population: a meta-analysis. *Int J Biol Mark.* 2014;29(4):301–309. doi: 10.5301/jbm.5000090.

El-Omar EM. The importance of interleukin 1β in Helicobacter pylori associated disease. *Gut.* 2001;48(6):743–747. doi: 10.1136/gut.48.6.743.

El-Omar E, Ng MT, Hold GL. Polymorphisms in Toll-like receptor genes and risk of cancer. *Oncogene.* 2008;27(2):244–252. doi: 10.1038/sj.onc.1210912.

El-Omar EM, Oien K, El-Nujumi A, Gillen D, Wirz A, Dahill S, McColl K. Helicobacter pylori infection and chronic gastric acid hyposecretion. *Gastroenterology.* 1997;113(1):15–24. doi: 10.1016/s0016-5085(97)70075-1.

Erichsen H, Chanock S. SNPs in cancer research and treatment. *Br J Cancer.* 2004;90(4):747–751. doi: 10.1038/sj.bjc.6601574.

Ezzeddini R, Somi MH, Taghikhani M, Moaddab SY, Shirazi KM, Shirmohammadi M, Farrokhi AS. Association of Foxp3 rs3761548 polymorphism with cytokines concentration in gastric adenocarcinoma patients. *Cytokine.* 2021;138:155351. doi: 10.1016/j.cyto.2020.155351.

Falchetti M, Saieva C, Lupi R, Masala G, Rizzolo P, Zanna I, Nesi G. Gastric cancer with high-level microsatellite instability: target gene mutations, clinicopathologic features, and long-term survival. *Human Pathol.* 2008;39(6):925–932. doi: 10.1016/j.humpath.2007.10.024.

Flejou JF. WHO Classification of digestive tumors: the fourth edition. *Ann Pathol.* 2011;31(5):27–31. doi: 10.1016/j.annpat.2011.08.001.

Fodingeer M, Wagner OF, Horl WH, Sunder-Plassmann G. Recent insights into the molecular genetics of the homocysteine metabolism. *Kidney Int Suppl.* 2001;78:238–242. doi: 10.1046/j.1523-1755.2001.59780238.x.

Freedman ML, Monteiro AN, Gayther SA, Coetzee GA, Risch A, Plass C, Duggan D. Principles for the post-GWAS functional characterization of cancer risk loci. *Nat Genet.* 2011;43(6):513–518. doi: 10.1038/ng.840.

Fukata M, Abreu MT. Role of Toll-like receptors in gastrointestinal malignancies. *Oncogene.* 2008;27(2):234–243. doi: 10.1038/sj.onc.1210908.

Gao C, Takezaki T, Wu J, Li Z, Wang J, Ding J, Tajima K. Interaction between cytochrome P-450 2E1 polymorphisms and environmental factors with risk of esophageal and stomach cancers in Chinese. *Cancer Epidemiol Prev Biomarkers.* 2002;11(1):29–34.

Garcia-Gonzalez MA, Bujanda L, Quintero E, Santolaria S, Benito R, Strunk M, Nicolas-Perez D. Association of PSCA rs2294008 gene variants with poor prognosis and increased susceptibility to gastric cancer and decreased risk of duodenal ulcer disease. *Int J Cancer.* 2015;137(6):1362–1373. doi: 10.1002/ijc.29500.

Garcia-Gonzalez MA, Quintero E, Bujanda L, Nicolas D, Benito R, Strunk M, Hijona E. Relevance of GSTM1, GSTT1, and GSTP1 gene polymorphisms to gastric cancer susceptibility and phenotype. *Mutagenesis.* 2012;27(6):771–777. doi: 10.1093/mutage/ges049.

Gigek CO, Chen ES, Calcagno DQ, Wisnieski F, Burbano RR, Smith MAC. Epigenetic mechanisms in gastric cancer. *Epigenomics.* 2012;4(3):279–294. doi: 10.2217/epi.12.22.

Gorouhi F, Islami F, Bahrami H, Kamangar F. Tumour-necrosis factor – a polymorphisms and gastric cancer risk: a meta-analysis. *Br J Cancer.* 2008;98(8):1443–1451. doi 10.1038/sj.bjc.6604277.

Ha YJ, Yoon SN, Jeon YJ, Cho DH, Roh SA, Kim BS, Kim JC. Genome-wide identification of chemosensitive single nucleotide polymorphism markers in gastric cancer. *Anticancer Res.* 2011;31(12):4329–4338.

Hanahan D, Weinberg RA. Hallmarks of cancer: the next generation. *Cell.* 2011;144(5):646–674. doi: 10.1016/j.cell.2011.02.013.

Hayes JD, Pulford DJ. The glutathione S-transferase supergene family: regulation of GST and the contribution of the Isoenzymes to cancer chemoprotection and drug resistance part I. *Crit Rev Biochem Mol Biol.* 2008;30(6):445–520. doi: 10.3109/10409239509083491.

Hearle N, Schumacher V, Menko FH, Olschwang S, Boardman LA, Gille JJ, Lim W. Frequency and spectrum of cancers in the Peutz-Jeghers syndrome. *Clin Cancer Res.* 2006;12(-10):3209–3215. doi: 10.1158/1078-0432.CCR-06-0083.

Hiyama T, Tanaka S, Kitadai Y, Ito M, Sumii M, Yoshihara M, Chayama K. p53 Codon 72 polymorphism in gastric cancer susceptibility in patients with Helicobacter pylori-associated chronic gastritis. *Int J Cancer.* 2002;100(3):304–308. doi: 10.1002/ijc.10483.

Hold GL, Rabkin CS, Chow WH, Smith MG, Gammon MD, Risch HA, Zatonski W. A functional polymorphism of toll-like receptor 4 gene increases risk of gastric carcinoma and its precursors. *Gastroenterology.* 2007;132(3):905–912. doi: 10.1053/j.gastro.2006.12.026.

Hua HB, Yan TT, Sun QM. miRNA polymorphisms and risk of gastric cancer in Asian population. *World J Gastroenterol.* 2014;20(19):5700. doi: 10.3748/wjg.v20.i19.5700.

Hudler P. Genetic aspects of gastric cancer instability. *Sci World J.* 2012;2012:761909. doi: 10.1100/2012/761909.

Jia Y, Persson C, Hou L, Zheng Z, Yeager M, Lissowska J, Ye W. A comprehensive analysis of common genetic variation in MUC1, MUC5AC, MUC6 genes and risk of stomach cancer. *Cancer Causes & Control.* 2010;21(2):313–321. doi: 10.1007/s10552-009-9463-3.

Jin G, Ma H, Wu C, Dai J, Zhang R, Shi Y, Zhou Y. Genetic variants at 6p21. 1 and 7p15. 3 are associated with risk of multiple cancers in Han Chinese. *Am J Human Genet.* 2012;91(5):928–934. doi: 10.1016/j.ajhg.2012.09.009.

John S, Turner D, Donn R, Sinnott P, Worthington J, Ollier W, Hajeer A. Two novel biallelic polymorphisms in the IL-2 gene. *Eur J Immunogenet.* 1998;25(6):419–420. doi: 10.1046/j.1365-2370.1998.00139.x.

Kamangar F, Cheng C, Abnet CC, Rabkin CS. Interleukin-1B polymorphisms and gastric cancer risk a meta-analysis. *Cancer Epidemiol Prev Biomarkers.* 2006;15(10):1920–1928. doi: 10.1158/1055-9965.EPI-06-0267.

Kang M, Ding X, Xu M, Zhu H, Liu S, Wang M, Zhou J. Genetic variation rs10484761 on 6p21. 1 derived from a genome-wide association study is associated with gastric cancer survival in a Chinese population. *Gene.* 2014;536(1):59–64. doi: 10.1016/j.gene.2013.11.087.

Kim KM, Kwon MS, Hong SJ, Min KO, Seo EJ, Lee KY, Rhyu, MG. Genetic classification of intestinal-type and diffuse-type gastric cancers based on chromosomal loss and microsatellite instability. *Virchows Archiv.* 2003;443(4):491–500. doi: 10.1007/s00428-003-0840-0.

Kim YD, Yim DH, Eom SY, Moon SI, Yun HY, Song YJ, Kim BS. Risk of gastric cancer is associated with PRKAA1 gene polymorphisms in Koreans. *World J Gastroenterol.* 2014;20(26):8592–8598. doi: 10.3748/wjg.v20.i26.8592.

Kupcinskas J, Wex T, Link A, Bartuseviciute R, Dedelaite M, Kevalaite G, Jonaitis L. PSCA and MUC1 gene polymorphisms are linked with gastric cancer and pre-malignant gastric conditions. *Anticancer Res.* 2014;34(12):7167–7175.

Lee KJ, Inoue M, Otani T, Iwasaki M, Sasazuki S, Tsugane S. Gastric cancer screening and subsequent risk of gastric cancer: a large-scale population-based cohort study, with a 13-year follow-up in Japan. *Int J Cancer.* 2006;118(9):2315–2321. doi: 10.1002/ijc.21664.

Lei Z, Tan IB, Das K, Deng N, Zouridis H, Pattison S, Ooi CH. Identification of molecular subtypes of gastric cancer with different responses to PI3-kinase inhibitors and 5-fluorouracil. *Gastroenterology.* 2013;145(3):554–565. doi: 10.1053/j.gastro.2013.05.010.

Lovat F, Valeri N, Croce CM. MicroRNAs in the pathogenesis of cancer. *Semin Oncol.* 2011;38(6):724–733. doi: 10.1053/j.seminoncol.2011.08.006.

Lu JW, Gao CM, Wu JZ, Cao HX, Tajima K, Feng JF. Polymorphism in the 3′-untranslated region of the thymidylate synthase gene and sensitivity of stomach cancer to fluoropyrimidine-based chemotherapy. *J Human Genet.* 2006;51(3):155–160. doi: 10.1007/s10038-005-0339-4.

Luo D, Gao Y, Wang S, Wang M, Wu D, Wang W, Tan Y. Genetic variation in PLCE1 is associated with gastric cancer survival in a Chinese population. *J Gastroenterol.* 2011;46(-11):1260–1266. doi: 10.1007/s00535-011-0445-3.

Machlowska J, Baj J, Sitarz M, Maciejewski R, Sitarz R. Gastric cancer: epidemiology, risk factors, classification, genomic characteristics and treatment strategies. *Int J Mol Sci.* 2020;21(11):4012. doi: 10.3390/ijms21114012.

McLean MH, El-Omar EM. Genetics of Gastric cancer. *Nat Rev Gastroenterol Hepatol.* 2014.11(11):664–674. doi: 10.1038/nrgastro.2014.143.

Mocellin S, Verdi D, Pooley KA, Nitti D. Genetic variation and gastric cancer risk: a field synopsis and meta-analysis. *Gut.* 2015;64(8):1209–1219. doi: 10.1136/gutjnl-2015-309168.

Neves M, Ribeiro J, Medeiros R, Sousa H. Genetic polymorphism in DNMTs and gastric cancer: a systematic review and meta-analysis. *Porto Biomed J.* 2016;1(5):164–172. doi: 10.1016/j.pbj.2016.10.005.

Ohyauchi M., Imatani A, Yonechi M, Asano N, Miura A, Iijima K, Shimosegawa T. The polymorphism interleukin 8 251 A/T influences the susceptibility of Helicobacter pylori related gastric diseases in the Japanese population. *Gut.* 2005;54(3):330–335 doi: 10.1136/gut.2003.033050.

Oliveira C, Senz J, Kaurah P, Pinheiro H, Sanges R, Haegert A, Corso G, Schouten J, R. Fitzgerald R, Vogelsang H, Keller G, Dwerryhouse S, Grimmer D, Chin SF, Yang HF, Jackson CE, Seruca R, Roviello F, Stupka E, Caldas C and Huntsman D. Germline CDH1 deletions in hereditary diffuse gastric cancer families. *Hum Mol Genet.*2009;18(9):1545–55. doi: 10.1093/hmg/ddp046.

Ott K, Vogelsang H, Marton N, Becker K, Lordick F, Kobl M, Fink U. The thymidylate synthase tandem repeat promoter polymorphism: a predictor for tumor-related survival in neoadjuvant treated locally advanced gastric cancer. *Int J Cancer.* 2006;119(12):2885–2894. doi: 10.1002/ijc.22235.

Ottini L, Falchetti M, Lupi R, Rizzolo P, Agnese V, Colucci G, Russo A. Patterns of genomic instability in gastric cancer: clinical implications and perspectives. *Ann Oncol.* 2006;17(7):97–102. doi: 10.1093/annonc/mdl960.

Palmer AJ, Lochhead P, Hold GL, Rabkin CS, Vaughan TL, Lissowska J, El-Omar EM. Genetic variation in C20 or f54, PLCE1 and MUC1 and risk of upper gastrointestinal cancers in Caucasian populations. *Eur J Cancer Prev.* 2012;21(-6):541. doi: 10.1097/CEJ.0b013e3283529b79.

Partida-Rodriguez O, Torres J, Flores-Luna L, Camorlinga M, Nieves-Ramirez M, Lazcano E, Perez-Rodriguez M. Polymorphisms in TNF and HSP-70 show a significant association with gastric cancer and duodenal ulcer. *Int J Cancer.* 2010;126(8):1861–1868. doi: 10.1002/ijc.24773.

Persson C, Canedo P, Machado JC, El-Omar EM, Forman D. Polymorphisms in inflammatory response genes and their association with gastric cancer: a HuGE systematic review and meta-analyses. *Am J Epidemiol.* 2011;173(3) 259–270. doi: 10.1093/aje/kwq370.

Qiao W, Wang T, Zhang L, Tang Q, Wang D, Sun H. Association study of single nucleotide polymorphisms in XRCC1 gene with the risk of gastric cancer in Chinese population. *Int J Biol Sci.* 2013;9(7):753. doi: 10.7150/ijbs.6783.

Qinghai Z, Yanying W, Yunfang C, Xukui Z, Xiaoqiao Z. Effect of interleukin-17A and interleukin-17F gene polymorphisms on the risk of gastric cancer in a Chinese population. *Gene.* 2014;537(2):328–332. doi: 10.1016/j.gene.2013.11.007.

Qu Q, Xuan W, Fan GH. Roles of resolvins in the resolution of acute inflammation. *Cell Biol Int.* 2015;39(1):3–22. doi: 10.1016/j.gene.2013.11.007.

Rafiei A, Hosseini V, Janbabai G, Ghorbani A, Ajami A, Farzmandfar T, Merrell DS. Polymorphism in the interleukin-17A promoter contributes to gastric cancer. *World J Gastroenterol.* 2013;19(34):5693. doi: 10.3748/wjg.v19.i34.5693.

Ramos-de la Medina A, More H, Medina-Franco H, Humar B, Gamboa A, Ortiz L, Guilford, P. Single nucleotide polymorphisms (SNPs) at CDH1 promoter region in familial gastric cancer. *Rev Esp Enferm Dig.* 2006;98(1):36–41. doi: 10.4321/s1130-01082006000100005.

Risch NJ. Searching for genetic determinants in the new millennium. *Nature.* 2000;405(6788):847–856. doi: 10.1038/35015718.

Rona KA, Schwameis K, Zehetner J, Samakar K, Green K, Samaan J, Lipham JC. Gastric cancer in the young: an advanced disease with poor prognostic features. *J Surg Oncol.* 2017;115(-4):371–375. doi: 10.1002/jso.24533.

Rosero GCY, Corredor M, Mejia OL. Polimorfismos en genes implicados en el desarrollo de cancer gastrico: revision. *Revista colombiana de Gastroenterologia.* 2016;31(4):391–402.

Saeki N, Gu J, Yoshida T, Wu X. Prostate stem cell antigen: a Jekyll and Hyde molecule? *Clin Cancer Res.* 2010;16(-14):3533–3538. doi: 10.1158/1078-0432.CCR-09-3169.

Sakamoto H, Yoshimura K, Saeki N, Katai H, Shimoda T, Matsuno Y, Kato S. Genetic variation in PSCA is associated with susceptibility to diffuse-type gastric cancer. *Nat Genet.* 2008;40(6):730–740. doi: 10.1038/ng.

Saunders MA, Liang H, Li WH. Human polymorphism at microRNAs and microRNA target sites. *Proc Nat Acad Sci.* 2007;104(9):3300–3305. doi: 10.1073/pnas.0611347104.

Shi Y, Hu Z, Wu C, Dai J, Li H, Dong J, Lu F. A genome-wide association study identifies new susceptibility loci for non-cardia gastric cancer at 3q13.31 and 5p13.1. *Nat Genet.* 2011;43(12):1215–1218. doi: 10.1038/ng.978.

Shibata T, Tahara T, Hirata I, Arisawa T. Genetic polymorphism of interleukin-17A and-17F genes in gastric carcinogenesis. *Human Immunol.* 2009;70(7):547–551. doi: 10.1016/j.humimm.2009.04.030.

Shin WG, Jang JS, Kim HS, Kim SJ, Kim KH, Jang MK, Kim HY. Polymorphisms of interleukin-1 and interleukin-2 genes in patients with gastric cancer in Korea. *J Gastroenterol Hepatol.* 2008;23(10):1567–1573. doi: 10.1111/j.1440-1746.2008.05479.x

Silva F, Carvalho F, Peixoto A, Seixas M, Almeida R, Carneiro F, Swallow DM. MUC1 gene polymorphism in the gastric carcinogenesis pathway. *Eur J Human Genet.* 2001;9(7):-548–552. doi: 10.1038/sj.ejhg.5200677.

Song S, Ajani JA. The role of microRNAs in cancers of the upper gastrointestinal tract. *Nat Rev Gastroenterol Hepatol.* 2013;10(2):109–118. doi: 10.1038/nrgastro.2012.210.

Song HR, Kim HN, Kweon SS, Choi JS, Shim HJ, Cho SH, Choi YD. Genetic variations in the PRKAA1 and ZBTB20 genes and gastric cancer susceptibility in a Korean population. *Molecularcarcinogenesis.* 2013;52(S1):155–160. doi: 10.1002/mc.22063.

Su B, Ceponis PJ, Lebel S, Huynh H, Sherman PM. Helicobacter pylori activates Toll-like receptor 4 expression in gastrointestinal epithelial cells. *Infect Immun.* 2003;71(6):3496–3502. doi: 10.1128/IAI.71.6.3496-3502.2003.

Sun Y, Gu J, Ajani JA, Chang DW, Wu X, Stroehlein JR. Genetic and intermediate phenotypic susceptibility markers of gastric cancer in Hispanic Americans: a case-control study. *Cancer.* 2014;120(19):3040–3048. doi: 10.1002/cncr.28792.

Togawa S, Joh T, Itoh M, Katsuda N, Ito H, Matsuo K, Hamajima N. Interleukin-2 gene polymorphisms associated with increased risk of gastric atrophy from Helicobacter pylori infection. *Helicobacter.* 2005;10(3):172–178. doi: 10.1111/j.1523-5378.2005.00308.x

Van Lier M, Wagner A, Mathus Vliegen E, Kuipers E, Steyerberg E, Van Leerdam M. High cancer risk in Peutz–Jeghers syndrome: a systematic review and surveillance recommendations. *Am J Gastroenterol.* 2010;105(6):1258–1264. doi: 10.1038/ajg.2009.725.

Vázquez Ibarra KC, Bustos Carpinteyro AR, Garcia-Ruvalcaba A, Magaaa-Torres MT, Gutierrez-Aguilar R, Marin-Contreras ME, Sanchez-Lopez JY. The ERBB2 gene polymorphisms rs2643194, rs2934971, and rs1058808 are associated with increased risk of gastric cancer. *Braz J Med Biol Res.* 2019;52(5):8379–8379. doi: 10.1590/1414-431X20198379.

Vincenzi B, Patti G, Galluzzo S, Pantano F, Venditti O, Santini D, Marra M. Interleukin 1β-511T gene (IL1β) polymorphism is correlated with gastric cancer in the Caucasian population: results from a meta-analysis. *Oncol Rep.* 2008;20(5):1213–1220.

Wadhwa R, Song S, Lee JS, Yao Y, Wei Q, Ajani JA. Gastric cancer—molecular and clinical dimensions. *Nat Rev Clin Oncol.* 2013;10(11):643–655. doi: 10.1038/nrclinonc.2013.170.

Wagner AD, Syn NL, Moehler M, Grothe W, Yong WP, Tai BC, Unverzagt S. Chemotherapy for advanced gastric cancer. *Cochr Database System Rev.* 2017;29(8). doi: 10.1002/14651858.CD004064.pub4.

Wang Z, Dai J, Hu N, Miao X, Abnet CC, Yang M, Zhu X. Identification of new susceptibility loci for gastric noncardia adenocarcinoma: pooled results from two Chinese genome-wide association studies. *Gut.* 2017;66(4):581–587. doi: 10.1136/gutjnl-2015-310612.

Wang S, Tian Y, Wu D, Zhu H, Luo D, Gong W, Zhang Z. Genetic variation of CTNNB1 gene is associated with susceptibility and prognosis of gastric cancer in a Chinese population. *Mutagenesis.* 2012b;27(6):623–630. doi: 10.1093/mutage/ges027.

Wang M, Wang XJ, Ma YF, Ma XB, Dai ZM, Lv Y, Dai ZJ. PSCA rs2294008 C> T polymorphism contributes to gastric and bladder cancer risk. *Ther Clin Risk Manag.* 2015;11:237. doi: 10.2147/TCRM.S77089.

Wang X, Xue L, Yang Y, Xu L, Zhang G. TLR9 promoter polymorphism is associated with both an increased susceptibility to gastric carcinoma and poor prognosis. *PLoS One.* 2013;8(6):65731. doi: 10.1371/journal.pone.0065731.

Wang M, Zhang R, He J, Qiu L, Li J, Wang Y, Yang J. Potentially functional variants of PLCE1 identified by GWASs contribute to gastric adenocarcinoma susceptibility in an eastern Chinese population. *PLoS One.* 2012a;7(3):e31932. doi: 10.1371/journal.pone.0031932.

Wang T, Zhang L, Li H, Wang B, Chen K. Prostate stem cell antigen polymorphisms and susceptibility to gastric cancer: a systematic review and meta-analysis. *Cancer Epidemiol Prev Biomarkers.* 2012c;21(5):843–850. doi: 10.1158/1055-9965.EPI-11-1176.

Wang LD, Zhou FY, Li XM, Sun LD, Song X, Jin Y, Cui J. Genome-wide association study of esophageal squamous cell carcinoma in Chinese subjects identifies a susceptibility locus at PLCE1. *Nat Genet.* 2010;42(9):759–763. doi: 10.1038/ng.648.

Worthley D, Phillips K, Wayte N, Schrader K, Healey S, Kaurah P, Moore D. Gastric adenocarcinoma and proximal polyposis of the stomach (GAPPS): a new autosomal dominant syndrome. *Gut.* 2012;61(5):774–779. doi: 10.1136/gutjnl-2011-300348.

Wu MS, Chang MC, Huang SP, Tseng CC, Sheu JC, Lin YW, Lin JT. Correlation of histologic subtypes and replication error phenotype with comparative genomic hybridization in gastric cancer. *Genes, Chromosomes Cancer.* 2001;30(1):80–86. doi: 10.1002/1098-2264(2000)9999:9999.

Xue H, Lin B, Ni P, Xu H, Huang G. Interleukin-1B and interleukin-1 RN polymorphisms and gastric carcinoma risk: a meta-analysis. *J Gastroenterol Hepatol.* 2010;25(10):1604–1617. doi: 10.1111/j.1440-1746.2010.06428.x.

Xue H, Liu J, Lin B, Wang Z, Sun J, Huang G. A meta-analysis of interleukin-8-251 promoter polymorphism associated with gastric cancer risk. *PLoS One.* 2012;7(1):e28083. doi: 10.1371/journal.pone.0028083.

Yoon JH, Song JH, Kang YH, Park YK, Zhang C, Nam SW, Park WS. TNF-α and TNF-β polymorphisms with susceptibility to gastric cancer in a Korean population. *Mol Cell Toxicol.* 2010;6(2):161–167. doi 10.1007/s13273-010-0023-4.

Yu CY, Chen HY. Genetic variations and gastric cancer. *Gastrointest Tumor.* 2015;2(3):90-97. doi: 10.1159/000431265.

Yusefi AR, Lankarani KB, Bastani P, Radinmanesh M, Kavosi Z. Risk factors for gastric cancer: a systematic review. *Asian Pacific J Cancer Prev.* 2018;19(3):591–603. doi: 10.22034/APJCP.2018.19.3.591.

Zhang H, Jin G, Li H, Ren C, Ding Y, Zhang Q, Xu Y. Genetic variants at 1q22 and 10q23 reproducibly associated with gastric cancer susceptibility in a Chinese population. *Carcinogenesis.* 2011;32(6):848–852. doi: 10.1093/carcin/bgr051.

Zhang X, Nie Y, Du Y, Cao J, Shen B, Li Y. MicroRNA-181a promotes gastric cancer by negatively regulating tumor suppressor KLF6. *Tumor Biol.* 2012;33(5):1589–1597. doi: 10.1007/s13277-012-0414-3.

Zhang X, Zheng L, Sun Y, Zhang X. Analysis of the association of interleukin-17 gene polymorphisms with gastric cancer risk and interaction with Helicobacter pylori infection in a Chinese population. *Tumor Biol.* 2014;35(2):1575–1580. doi: 10.1007/s13277-013-1217-x.

Zhao JD, Geng PL, Li ZQ, Cui S, Zhao JH, Wang LJ, Shen GS. Associations between interleukin-1 polymorphisms and gastric cancers among three ethnicities. *World J Gastroenterol.* 2012:18(47):7093–7099. doi: 10.3748/wjg.v18.i47.7093.

Zheng L, Zhu C, Gu J, Xi P, Du J, Jin G. Functional polymorphism rs4072037 in MUC1 gene contributes to the susceptibility to gastric cancer: evidence from pooled 6,580 cases and 10,324 controls. *Mol Biol Rep*. 2013;40(10):5791–5796. doi: 10.1007/s11033-013-2682-4.

Index

Note: Bold page numbers refer to tables; *italic* page numbers refer to figures.

9781032153162